SEARS AND ZEMANSKY'S

UNIVERSITY PHYSICS

VOLUME 1
13TH EDITION

HUGH D. YOUNG
CARNEGIE MELLON
UNIVERSITY

ROGER A. FREEDMAN
UNIVERSITY
OF CALIFORNIA,
SANTA BARBARA

CONTRIBUTING AUTHOR
A. LEWIS FORD
TEXAS A&M UNIVERSITY

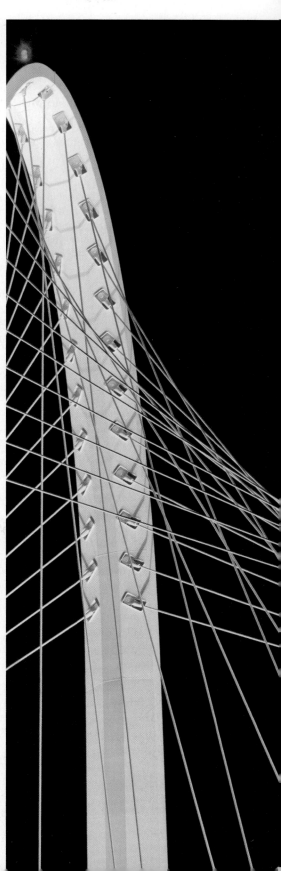

Addison-Wesley
Boston Columbus Indianapolis
New York San Francisco Upper Saddle River
Amsterdam Cape Town Dubai London
Madrid Milan Munich Paris Montréal Toronto
Delhi Mexico City São Paulo Sydney
Hong Kong Seoul Singapore Taipei Tokyo

Publisher: Jim Smith
Executive Editor: Nancy Whilton
Project Editor: Chandrika Madhavan
Director of Development: Michael Gillespie
Editorial Manager: Laura Kenney
Senior Development Editor: Margot Otway
Editorial Assistant: Steven Le
Associate Media Producer: Kelly Reed
Managing Editor: Corinne Benson
Production Project Manager: Beth Collins

Production Management and Composition: Nesbitt Graphics
Copyeditor: Carol Reitz
Interior Designer: Elm Street Publishing Services
Cover Designer: Derek Bacchus
Illustrators: Rolin Graphics
Senior Art Editor: Donna Kalal
Photo Researcher: Eric Shrader
Manufacturing Buyer: Jeff Sargent
Senior Marketing Manager: Kerry Chapman

Cover Photo Credits: Getty Images/Mirko Cassanelli; Mirko Cassanelli

Library of Congress Cataloging-in-Publication Data

Young, Hugh D.
 Sears and Zemansky's university physics : with modern physics. -- 13th ed.
/ Hugh D. Young, Roger A. Freedman ; contributing author, A. Lewis Ford.
    p. cm.
 Includes bibliographical references and index.
 ISBN-13: 978-0-321-69686-1 (student ed. : alk. paper)
 ISBN-10: 0-321-69686-7 (student ed. : alk. paper)
 ISBN-13: 978-0-321-69685-4 (exam copy)
 ISBN-10: 0-321-69685-9 (exam copy)
 1. Physics--Textbooks. I. Freedman, Roger A. II. Ford, A. Lewis (Albert
Lewis) III. Sears, Francis Weston, 1898-1975. University physics. IV. Title.
V. Title: University physics.
 QC21.3.Y68 2012
 530--dc22

                                                2010044896

ISBN 13: 978-0-321-73338-2; ISBN 10: 0-321-73338-X (Student edition)
ISBN 13: 978-0-321-69685-4; ISBN 10: 0-321-69685-9 (Exam copy)

Addison-Wesley
is an imprint of

BRIEF CONTENTS

VOLUME 1: Chapters 1–20 • VOLUME 2: Chapters 21–37 • VOLUME 3: Chapters 37–44

Build **Skills**

Learn basic and advanced skills that help solve a broad range of physics problems.

Problem-Solving Strategies coach students in how to approach specific types of problems.

▼

This text's **uniquely extensive set** of **Examples** enables students to explore problem-solving challenges in exceptional detail.

Consistent

The **Identify / Set Up / Execute / Evaluate** format, used in all Examples, encourages students to tackle problems thoughtfully rather than skipping to the math.

Focused

All Examples and Problem-Solving Strategies are revised to be more concise and focused.

Visual

Most Examples employ a diagram—often a **pencil sketch** that shows what a student should draw.

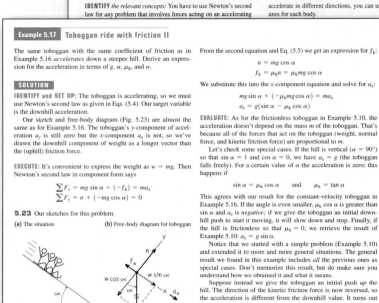

Problem-Solving Strategy 5.2 Newton's Second Law: Dynamics of Particles

IDENTIFY *the relevant concepts:* You have to use Newton's second law for *any* problem that involves forces acting on an accelerating

Example 5.17 Toboggan ride with friction II

The same toboggan with the same coefficient of friction as in Example 5.16 *accelerates* down a steeper hill. Derive an expression for the acceleration in terms of g, α, μ_k, and w.

SOLUTION

IDENTIFY and SET UP: The toboggan is accelerating, so we must use Newton's second law as given in Eqs. (5.4). Our target variable is the downhill acceleration.

Our sketch and free-body diagram (Fig. 5.23) are almost the same as for Example 5.16. The toboggan's y-component of acceleration a_y is still zero but the x-component a_x is not, so we've drawn the downhill component of weight as a longer vector than the (uphill) friction force.

EXECUTE: It's convenient to express the weight as $w = mg$. Then Newton's second law in component form says

$$\sum F_x = mg \sin \alpha + (-f_k) = ma_x$$
$$\sum F_y = n + (-mg \cos \alpha) = 0$$

5.23 Our sketches for this problem.

(a) The situation   (b) Free-body diagram for toboggan

From the second equation and Eq. (5.5) we get an expression for f_k:

$$n = mg \cos \alpha$$
$$f_k = \mu_k n = \mu_k mg \cos \alpha$$

We substitute this into the x-component equation and solve for a_x:

$$mg \sin \alpha + (-\mu_k mg \cos \alpha) = ma_x$$
$$a_x = g(\sin \alpha - \mu_k \cos \alpha)$$

EVALUATE: As for the frictionless toboggan in Example 5.10, the acceleration doesn't depend on the mass m of the toboggan. That's because all of the forces that act on the toboggan (weight, normal force, and kinetic friction force) are proportional to m.

Let's check some special cases. If the hill is vertical ($\alpha = 90°$) so that $\sin \alpha = 1$ and $\cos \alpha = 0$, we have $a_x = g$ (the toboggan falls freely). For a certain value of α the acceleration is zero; this happens if

$$\sin \alpha = \mu_k \cos \alpha \qquad \text{and} \qquad \mu_k = \tan \alpha$$

This agrees with our result for the constant-velocity toboggan in Example 5.16. If the angle is even smaller, $\mu_k \cos \alpha$ is greater than $\sin \alpha$ and a_x is *negative*; if we give the toboggan an initial downhill push to start it moving, it will slow down and stop. Finally, if the hill is frictionless so that $\mu_k = 0$, we retrieve the result of Example 5.10: $a_x = g \sin \alpha$.

Notice that we started with a simple problem (Example 5.10) and extended it to more and more general situations. The general result we found in this example includes *all* the previous ones as special cases. Don't memorize this result, but do make sure you understand how we obtained it and what it means.

Suppose instead we give the toboggan an initial push *up* the hill. The direction of the kinetic friction force is now reversed, so the acceleration is different from the downhill value. It turns out that the expression for a_x is the same as for downhill motion except that the minus sign becomes plus. Can you show this?

NEW! Video Tutor Solution for Every Example
Each Example is explained and solved by an instructor in a Video Tutor solution provided in the Study Area of MasteringPhysics® and in the Pearson eText.

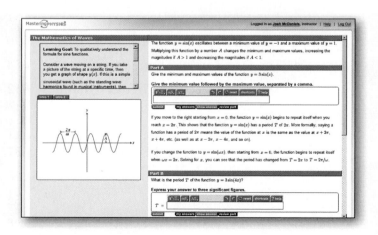

NEW! Mathematics Review Tutorials
MasteringPhysics offers an extensive set of assignable mathematics review tutorials—covering differential and integral calculus as well as algebra and trigonometry.

Build **Confidence**

Develop problem-solving confidence through a range of practice options—from guided to unguided.

NEW! Bridging Problems
At the start of each problem set, a Bridging Problem helps students make the leap from routine exercises to challenging problems with confidence and ease.

Each Bridging Problem poses a moderately difficult, multi-concept problem, which often draws on earlier chapters. In place of a full solution, it provides a skeleton solution guide consisting of questions and hints.

A full solution is explained in a **Video Tutor**, provided in the Study Area of MasteringPhysics® and in the Pearson eText.

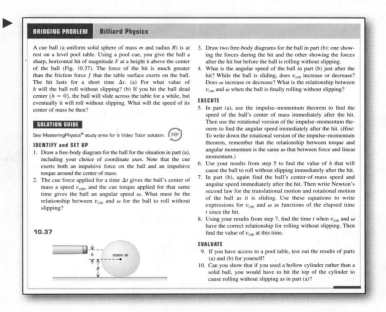

BRIDGING PROBLEM   Billiard Physics

A cue ball (a uniform solid sphere of mass m and radius R) is at rest on a level pool table. Using a pool cue, you give the ball a sharp, horizontal hit of magnitude F at a height h above the center of the ball (Fig. 10.37). The force of the hit is much greater than the friction force f that the table surface exerts on the ball. The hit lasts for a short time Δt. (a) For what value of h will the ball roll without slipping? (b) If you hit the ball dead center ($h = 0$), the ball will slide across the table for a while, but eventually it will roll without slipping. What will the speed of its center of mass be then?

SOLUTION GUIDE

See MasteringPhysics® study area for a Video Tutor solution.

IDENTIFY and SET UP
1. Draw a free-body diagram for the ball for the situation in part (a), including your choice of coordinate axes. Note that the cue exerts both an impulsive force on the ball and an impulsive torque around the center of mass.
2. The cue force applied for a time Δt gives the ball's center of mass a speed v_{cm}, and the cue torque applied for that same time gives the ball an angular speed ω. What must be the relationship between v_{cm} and ω for the ball to roll without slipping?

10.37

3. Draw two free-body diagrams for the ball in part (b): one showing the forces during the hit and the other showing the forces after the hit but before the ball is rolling without slipping.
4. What is the angular speed of the ball in part (b) just after the hit? While the ball is sliding, does v_{cm} increase or decrease? Does ω increase or decrease? What is the relationship between v_{cm} and ω when the ball is finally rolling without slipping?

EXECUTE
5. In part (a), use the impulse–momentum theorem to find the speed of the ball's center of mass immediately after the hit. Then use the rotational version of the impulse–momentum theorem to find the angular speed immediately after the hit. (*Hint:* To write down the rotational version of the impulse–momentum theorem, remember that the relationship between torque and angular momentum is the same as that between force and linear momentum.)
6. Use your results from step 5 to find the value of h that will cause the ball to roll without slipping immediately after the hit.
7. In part (b), again find the ball's center-of-mass speed and angular speed immediately after the hit. Then write Newton's second law for the translational motion and rotational motion of the ball as it is sliding. Use these equations to write expressions for v_{cm} and ω as functions of the elapsed time t since the hit.
8. Using your results from step 7, find the time t when v_{cm} and ω have the correct relationship for rolling without slipping. Then find the value of v_{cm} at this time.

EVALUATE
9. If you have access to a pool table, test out the results of parts (a) and (b) for yourself!
10. Can you show that if you used a hollow cylinder rather than a solid ball, you would have to hit the top of the cylinder to cause rolling without slipping as in part (a)?

14.95 • **CP** In Fig. P14.95 the upper ball is released from rest, collides with the stationary lower ball, and sticks to it. The strings are both 50.0 cm long. The upper ball has mass 2.00 kg, and it is initially 10.0 cm higher than the lower ball, which has mass 3.00 kg. Find the frequency and maximum angular displacement of the motion after the collision.

Figure P14.95

10.0 cm

14.96 •• **CP BIO** *T. rex.* Model the leg of the *T. rex* in Example 14.10 (Section 14.6) as two uniform rods, each 1.55 m long, joined rigidly end to end. Let the lower rod have mass M and the upper rod mass $2M$. The composite object is pivoted about the top of the upper rod. Compute the oscillation period of this object for small-amplitude oscillations. Compare your result to that of Example 14.10.

14.97 •• **CALC** A slender, uniform, metal rod with mass M is pivoted without friction about an axis through its midpoint and perpendicular to the rod. A horizontal spring with force constant k is attached to the lower end of the rod, with the other end of the spring attached to a rigid support. If the rod is displaced by a small angle Θ from the vertical (Fig. P14.97) and released, show that it moves in angular SHM and calculate the period. (*Hint:* Assume that the angle Θ is small enough for the approximations $\sin \Theta \approx \Theta$ and $\cos \Theta \approx 1$ to be valid. The motion is simple harmonic if $d^2\theta/dt^2 = -\omega^2\theta$, and the period is then $T = 2\pi/\omega$.)

Figure P14.97

θ

In response to professors, the **Problem Sets** now include more biomedically oriented problems (BIO), more difficult problems requiring calculus (CALC), and more cumulative problems that draw on earlier chapters (CP).

About 20% of problems are new or revised. These revisions are driven by detailed student-performance data gathered nationally through MasteringPhysics.

Problem difficulty is now indicated by a three-dot ranking system based on data from MasteringPhysics.

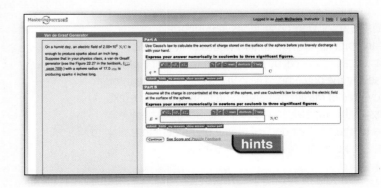

NEW! Enhanced End-of-Chapter Problems in MasteringPhysics
Select end-of-chapter problems will now offer additional support such as problem-solving strategy hints, relevant math review and practice, and links to the eText. These new enhanced problems bridge the gap between guided tutorials and traditional homework problems.

Bring Physics **to Life**

Deepen knowledge of physics by building connections to the real world.

NEW! Applications of Physics
Throughout the text, free-standing captioned photos apply physics to real situations, with particular emphasis on applications of biomedical and general interest.

Application Tendons Are Nonideal Springs
Muscles exert forces via the tendons that attach them to bones. A tendon consists of long, stiffly elastic collagen fibers. The graph shows how the tendon from the hind leg of a wallaby (a small kangaroo) stretches in response to an applied force. The tendon does not exhibit the simple, straight-line behavior of an ideal spring, so the work it does has to be found by integration [Eq. (6.7)]. Note that the tendon exerts less force while relaxing than while stretching. As a result, the relaxing tendon does only about 93% of the work that was done to stretch it.

Application Moment of Inertia of a Bird's Wing
When a bird flaps its wings, it rotates the wings up and down around the shoulder. A hummingbird has small wings with a small moment of inertia, so the bird can make its wings move rapidly (up to 70 beats per second). By contrast, the Andean condor (*Vultur gryphus*) has immense wings that are hard to move due to their large moment of inertia. Condors flap their wings at about one beat per second on takeoff, but at most times prefer to soar while holding their wings steady.

Application Listening for Turbulent Flow
Normal blood flow in the human aorta is laminar, but a small disturbance such as a heart pathology can cause the flow to become turbulent. Turbulence makes noise, which is why listening to blood flow with a stethoscope is a useful diagnostic technique.

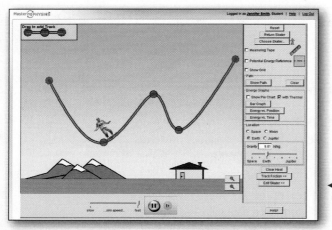

NEW! PhET Simulations and Tutorials
Sixteen assignable PhET Tutorials enable students to make connections between real-life phenomena and the underlying physics. 76 PhET simulations are provided in the Study Area of MasteringPhysics® and in the Pearson eText.

The comprehensive library of ActivPhysics applets and applet-based tutorials is also available.

NEW! Video Tutor Demonstrations and Tutorials
"Pause and predict" demonstration videos of key physics concepts engage students by asking them to submit a prediction before seeing the outcome. These videos are available through the Study Area of MasteringPhysics and in the Pearson eText. A set of assignable tutorials based on these videos challenge students to transfer their understanding of the demonstration to a related problem situation.

Biomedically Based End-of-Chapter Problems
To serve biosciences students, the text adds a substantial number of problems based on biological and biomedical situations.

Figure **E21.23**

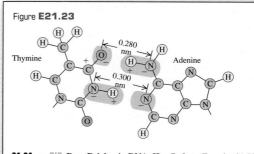

21.24 •• **BIO Base Pairing in DNA, II.** Refer to Exercise 21.23. Figure E21.24 shows the bonding of the cytosine and guanine molecules. The O—H and H—N distances are each 0.110 nm. In this case, assume that the bonding is due only to the forces along the O—H—O, N—H—N, and O—H—N combinations, and assume also that these three combinations are parallel to each other. Calculate the *net* force that cytosine exerts on guanine due to the preceding three combinations. Is this force attractive or repulsive?

Make a Difference with
MasteringPhysics®

www.masteringphysics.com

MasteringPhysics is the most effective and widely used online science tutorial, homework, and assessment system available.

NEW! Pre-Built Assignments ▶
For every chapter in the book, MasteringPhysics now provides pre-built assignments that cover the material with a tested mix of tutorials and end-of-chapter problems of graded difficulty. Professors may use these assignments as-is or take them as a starting point for modification.

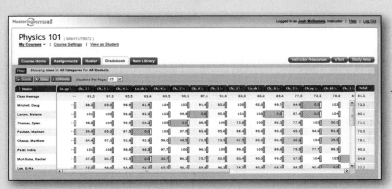

◀ **Gradebook**
- Every assignment is graded automatically.
- Shades of red highlight vulnerable students and challenging assignments.

Class Performance on Assignment ▶
Click on a problem to see which step your students struggled with most, and even their most common wrong answers. Compare results at every stage with the national average or with your previous class.

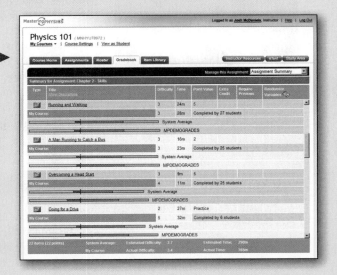

◀ **Gradebook Diagnostics**
This screen provides your favorite weekly diagnostics. With a single click, charts summarize the most difficult problems, vulnerable students, grade distribution, and even improvement in scores over the course.

ABOUT THE AUTHORS

Hugh D. Young is Emeritus Professor of Physics at Carnegie Mellon University. He earned both his undergraduate and graduate degrees from that university. He earned his Ph.D. in fundamental particle theory under the direction of the late Richard Cutkosky. He joined the faculty of Carnegie Mellon in 1956 and retired in 2004. He also had two visiting professorships at the University of California, Berkeley.

Dr. Young's career has centered entirely on undergraduate education. He has written several undergraduate-level textbooks, and in 1973 he became a coauthor with Francis Sears and Mark Zemansky for their well-known introductory texts. In addition to his role on Sears and Zemansky's *University Physics,* he is also author of Sears and Zemansky's *College Physics.*

Dr. Young earned a bachelor's degree in organ performance from Carnegie Mellon in 1972 and spent several years as Associate Organist at St. Paul's Cathedral in Pittsburgh. He has played numerous organ recitals in the Pittsburgh area. Dr. Young and his wife, Alice, usually travel extensively in the summer, especially overseas and in the desert canyon country of southern Utah.

Roger A. Freedman is a Lecturer in Physics at the University of California, Santa Barbara. Dr. Freedman was an undergraduate at the University of California campuses in San Diego and Los Angeles, and did his doctoral research in nuclear theory at Stanford University under the direction of Professor J. Dirk Walecka. He came to UCSB in 1981 after three years teaching and doing research at the University of Washington.

At UCSB, Dr. Freedman has taught in both the Department of Physics and the College of Creative Studies, a branch of the university intended for highly gifted and motivated undergraduates. He has published research in nuclear physics, elementary particle physics, and laser physics. In recent years, he has worked to make physics lectures a more interactive experience through the use of classroom response systems.

In the 1970s Dr. Freedman worked as a comic book letterer and helped organize the San Diego Comic-Con (now the world's largest popular culture convention) during its first few years. Today, when not in the classroom or slaving over a computer, Dr. Freedman can be found either flying (he holds a commercial pilot's license) or driving with his wife, Caroline, in their 1960 Nash Metropolitan convertible.

A. Lewis Ford is Professor of Physics at Texas A&M University. He received a B.A. from Rice University in 1968 and a Ph.D. in chemical physics from the University of Texas at Austin in 1972. After a one-year postdoc at Harvard University, he joined the Texas A&M physics faculty in 1973 and has been there ever since. Professor Ford's research area is theoretical atomic physics, with a specialization in atomic collisions. At Texas A&M he has taught a variety of undergraduate and graduate courses, but primarily introductory physics.

HOW TO SUCCEED IN PHYSICS BY REALLY TRYING

Mark Hollabaugh *Normandale Community College*

Physics encompasses the large and the small, the old and the new. From the atom to galaxies, from electrical circuitry to aerodynamics, physics is very much a part of the world around us. You probably are taking this introductory course in calculus-based physics because it is required for subsequent courses you plan to take in preparation for a career in science or engineering. Your professor wants you to learn physics and to enjoy the experience. He or she is very interested in helping you learn this fascinating subject. That is part of the reason your professor chose this textbook for your course. That is also the reason Drs. Young and Freedman asked me to write this introductory section. We want you to succeed!

The purpose of this section of *University Physics* is to give you some ideas that will assist your learning. Specific suggestions on how to use the textbook will follow a brief discussion of general study habits and strategies.

Preparation for This Course

If you had high school physics, you will probably learn concepts faster than those who have not because you will be familiar with the language of physics. If English is a second language for you, keep a glossary of new terms that you encounter and make sure you understand how they are used in physics. Likewise, if you are farther along in your mathematics courses, you will pick up the mathematical aspects of physics faster. Even if your mathematics is adequate, you may find a book such as Arnold D. Pickar's *Preparing for General Physics: Math Skill Drills and Other Useful Help (Calculus Version)* to be useful. Your professor may actually assign sections of this math review to assist your learning.

Learning to Learn

Each of us has a different learning style and a preferred means of learning. Understanding your own learning style will help you to focus on aspects of physics that may give you difficulty and to use those components of your course that will help you overcome the difficulty. Obviously you will want to spend more time on those aspects that give you the most trouble. If you learn by hearing, lectures will be very important. If you learn by explaining, then working with other students will be useful to you. If solving problems is difficult for you, spend more time learning how to solve problems. Also, it is important to understand and develop good study habits. Perhaps the most important thing you can do for yourself is to set aside adequate, regularly scheduled study time in a distraction-free environment.

Answer the following questions for yourself:
- Am I able to use fundamental mathematical concepts from algebra, geometry and trigonometry? (If not, plan a program of review with help from your professor.)
- In similar courses, what activity has given me the most trouble? (Spend more time on this.) What has been the easiest for me? (Do this first; it will help to build your confidence.)

- Do I understand the material better if I read the book before or after the lecture? (You may learn best by skimming the material, going to lecture, and then undertaking an in-depth reading.)
- Do I spend adequate time in studying physics? (A rule of thumb for a class like this is to devote, on the average, 2.5 hours out of class for each hour in class. For a course meeting 5 hours each week, that means you should spend about 10 to 15 hours per week studying physics.)
- Do I study physics every day? (Spread that 10 to 15 hours out over an entire week!) At what time of the day am I at my best for studying physics? (Pick a specific time of the day and stick to it.)
- Do I work in a quiet place where I can maintain my focus? (Distractions will break your routine and cause you to miss important points.)

Working with Others

Scientists or engineers seldom work in isolation from one another but rather work cooperatively. You will learn more physics and have more fun doing it if you work with other students. Some professors may formalize the use of cooperative learning or facilitate the formation of study groups. You may wish to form your own informal study group with members of your class who live in your neighborhood or dorm. If you have access to e-mail, use it to keep in touch with one another. Your study group is an excellent resource when reviewing for exams.

Lectures and Taking Notes

An important component of any college course is the lecture. In physics this is especially important because your professor will frequently do demonstrations of physical principles, run computer simulations, or show video clips. All of these are learning activities that will help you to understand the basic principles of physics. Don't miss lectures, and if for some reason you do, ask a friend or member of your study group to provide you with notes and let you know what happened.

Take your class notes in outline form, and fill in the details later. It can be very difficult to take word for word notes, so just write down key ideas. Your professor may use a diagram from the textbook. Leave a space in your notes and just add the diagram later. After class, edit your notes, filling in any gaps or omissions and noting things you need to study further. Make references to the textbook by page, equation number, or section number.

Make sure you ask questions in class, or see your professor during office hours. Remember the only "dumb" question is the one that is not asked. Your college may also have teaching assistants or peer tutors who are available to help you with difficulties you may have.

Examinations

Taking an examination is stressful. But if you feel adequately prepared and are well-rested, your stress will be lessened. Preparing for an exam is a continual process; it begins the moment the last exam is over. You should immediately go over the exam and understand any mistakes you made. If you worked a problem and made substantial errors, try this: Take a piece of paper and divide it down the middle with a line from top to bottom. In one column, write the proper solution to the problem. In the other column, write what you did and why, if you know, and why your solution was incorrect. If you are uncertain why you made your mistake, or how to avoid making it again, talk with your professor. Physics continually builds on fundamental ideas and it is important to correct any misunderstandings immediately. *Warning:* While cramming at the last minute may get you through the present exam, you will not adequately retain the concepts for use on the next exam.

PREFACE

This book is the product of more than six decades of leadership and innovation in physics education. When the first edition of *University Physics* by Francis W. Sears and Mark W. Zemansky was published in 1949, it was revolutionary among calculus-based physics textbooks in its emphasis on the fundamental principles of physics and how to apply them. The success of *University Physics* with generations of several million students and educators around the world is a testament to the merits of this approach, and to the many innovations it has introduced subsequently.

In preparing this new Thirteenth Edition, we have further enhanced and developed *University Physics* to assimilate the best ideas from education research with enhanced problem-solving instruction, pioneering visual and conceptual pedagogy, the first systematically enhanced problems, and the most pedagogically proven and widely used online homework and tutorial system in the world.

New to This Edition

- Included in each chapter, **Bridging Problems** provide a transition between the single-concept Examples and the more challenging end-of-chapter problems. Each Bridging Problem poses a difficult, multiconcept problem, which often incorporates physics from earlier chapters. In place of a full solution, it provides a skeleton **Solution Guide** consisting of questions and hints, which helps train students to approach and solve challenging problems with confidence.
- **All Examples, Conceptual Examples, and Problem-Solving Strategies are revised** to enhance conciseness and clarity for today's students.
- The **core modern physics chapters** (Chapters 38–41) are revised extensively to provide a more idea-centered, less historical approach to the material. Chapters 42–44 are also revised significantly.
- **The fluid mechanics chapter now precedes the chapters on gravitation and periodic motion,** so that the latter immediately precedes the chapter on mechanical waves.
- **Additional bioscience applications** appear throughout the text, mostly in the form of marginal photos with explanatory captions, to help students see how physics is connected to many breakthroughs and discoveries in the biosciences.
- The **text has been streamlined** for tighter and more focused language.
- **Using data from MasteringPhysics, changes to the end-of-chapter content** include the following:
  - **15%–20% of problems are new.**
  - The number and level of **calculus-requiring problems** has been increased.
  - Most chapters include **five to seven biosciences-related problems.**
  - The number of **cumulative problems** (those incorporating physics from earlier chapters) has been increased.
- **Over 70 PhET simulations** are linked to the Pearson eText and provided in the Study Area of the MasteringPhysics website (with icons in the print text). These powerful simulations allow students to interact productively with the physics concepts they are learning. PhET clicker questions are also included on the Instructor Resource DVD.
- **Video Tutors bring key content to life throughout the text:**
  - **Dozens of Video Tutors feature "pause-and-predict" demonstrations of key physics concepts** and incorporate assessment as the student progresses to actively engage the student in understanding the key conceptual ideas underlying the physics principles.

*Standard, Extended,
and Three-Volume Editions*

With MasteringPhysics:
- **Standard Edition:** Chapters 1–37 (ISBN 978-0-321-69688-5)
- **Extended Edition:** Chapters 1–44 (ISBN 978-0-321-67546-0)

Without MasteringPhysics:
- **Standard Edition:** Chapters 1–37 (ISBN 978-0-321-69689-2)
- **Extended Edition:** Chapters 1–44 (ISBN 978-0-321-69686-1)
- **Volume 1:** Chapters 1–20 (ISBN 978-0-321-73338-2)
- **Volume 2:** Chapters 21–37 (ISBN 978-0-321-75121-8)
- **Volume 3:** Chapters 37–44 (ISBN 978-0-321-75120-1)

- **Every Worked Example in the book is accompanied by a Video Tutor Solution** that walks students through the problem-solving process, providing a virtual teaching assistant on a round-the-clock basis.
- **All of these Video Tutors play directly through links within the Pearson eText.** Many also appear in the Study Area within MasteringPhysics.

Key Features of *University Physics*

- Deep and extensive **problem sets** cover a wide range of difficulty and exercise both physical understanding and problem-solving expertise. Many problems are based on complex real-life situations.
- This text offers a larger number of **Examples** and **Conceptual Examples** than any other leading calculus-based text, allowing it to explore problem-solving challenges not addressed in other texts.
- A research-based **problem-solving approach (Identify, Set Up, Execute, Evaluate)** is used not just in every Example but also in the Problem-Solving Strategies and throughout the Student and Instructor Solutions Manuals and the Study Guide. This consistent approach teaches students to tackle problems thoughtfully rather than cutting straight to the math.
- **Problem-Solving Strategies** coach students in how to approach specific types of problems.
- The **Figures** use a simplified graphical style to focus on the physics of a situation, and they incorporate **explanatory annotation.** Both techniques have been demonstrated to have a strong positive effect on learning.
- Figures that illustrate Example solutions often take the form of black-and-white **pencil sketches,** which directly represent what a student should draw in solving such a problem.
- The popular **Caution paragraphs** focus on typical misconceptions and student problem areas.
- End-of-section **Test Your Understanding** questions let students check their grasp of the material and use a multiple-choice or ranking-task format to probe for common misconceptions.
- **Visual Summaries** at the end of each chapter present the key ideas in words, equations, and thumbnail pictures, helping students to review more effectively.

Instructor Supplements

Note: For convenience, all of the following instructor supplements (except for the Instructor Resource DVD) can be downloaded from the Instructor Area, accessed via the left-hand navigation bar of MasteringPhysics (www.masteringphysics.com).

Instructor Solutions, prepared by A. Lewis Ford (Texas A&M University) and Wayne Anderson, contain complete and detailed solutions to all end-of-chapter problems. All solutions follow consistently the same Identify/Set Up/Execute/Evaluate problem-solving framework used in the textbook. Download only from the MasteringPhysics Instructor Area or from the Instructor Resource Center (www.pearsonhighered.com/irc).

The cross-platform **Instructor Resource DVD** (ISBN 978-0-321-69661-8) provides a comprehensive library of more than 420 applets from ActivPhysics OnLine as well as all line figures from the textbook in JPEG format. In addition, all the key equations, problem-solving strategies, tables, and chapter summaries are provided in editable Word format. In-class weekly multiple-choice questions for use with various Classroom Response Systems (CRS) are also provided, based on the Test Your Understanding questions in the text. Lecture outlines in PowerPoint are also included along with over 70 PhET simulations.

MasteringPhysics® (www.masteringphysics.com) is the most advanced, educationally effective, and widely used physics homework and tutorial system in the world. Eight years in development, it provides instructors with a library of extensively pre-tested end-of-chapter problems and rich, multipart, multistep tutorials that incorporate a wide variety of answer types, wrong answer feedback, individualized help (comprising hints or simpler sub-problems upon request), all driven by the largest metadatabase of student problem-solving in the world. NSF-sponsored published research (and subsequent studies) show that MasteringPhysics has dramatic educational results. MasteringPhysics allows instructors to build wide-ranging homework assignments of just the right difficulty and length and provides them with efficient tools to analyze both class trends, and the work of any student in unprecedented detail.

MasteringPhysics routinely provides instant and individualized feedback and guidance to more than 100,000 students every day. A wide range of tools and support make MasteringPhysics fast and easy for instructors and students to learn to use. Extensive class tests show that by the end of their course, an unprecedented eight of nine students recommend MasteringPhysics as their preferred way to study physics and do homework.

MasteringPhysics enables instructors to:
- Quickly build homework assignments that combine regular end-of-chapter problems and tutoring (through additional multi-step tutorial problems that offer wrong-answer feedback and simpler problems upon request).
- Expand homework to include the widest range of automatically graded activities available—from numerical problems with randomized values, through algebraic answers, to free-hand drawing.
- Choose from a wide range of nationally pre-tested problems that provide accurate estimates of time to complete and difficulty.
- After an assignment is completed, quickly identify not only the problems that were the trickiest for students but the individual problem types where students had trouble.
- Compare class results against the system's worldwide average for each problem assigned, to identify issues to be addressed with just-in-time teaching.
- Check the work of an individual student in detail, including time spent on each problem, what wrong answers they submitted at each step, how much help they asked for, and how many practice problems they worked.

ActivPhysics OnLine™ (which is accessed through the Study Area within www.masteringphysics.com) provides a comprehensive library of more than 420 tried and tested ActivPhysics applets updated for web delivery using the latest online technologies. In addition, it provides a suite of highly regarded applet-based tutorials developed by education pioneers Alan Van Heuvelen and Paul D'Alessandris. Margin icons throughout the text direct students to specific exercises that complement the textbook discussion.

   The online exercises are designed to encourage students to confront misconceptions, reason qualitatively about physical processes, experiment quantitatively, and learn to think critically. The highly acclaimed ActivPhysics OnLine companion workbooks help students work through complex concepts and understand them more clearly. More than 420 applets from the ActivPhysics OnLine library are also available on the Instructor Resource DVD for this text.

The **Test Bank** contains more than 2,000 high-quality problems, with a range of multiple-choice, true/false, short-answer, and regular homework-type questions. Test files are provided both in TestGen (an easy-to-use, fully networkable program for creating and editing quizzes and exams) and Word format. Download only from the MasteringPhysics Instructor Area or from the Instructor Resource Center (www.pearsonhighered.com/irc).

Five Easy Lessons: Strategies for Successful Physics Teaching (ISBN 978-0-805-38702-5) by Randall D. Knight (California Polytechnic State University, San Luis Obispo) is packed with creative ideas on how to enhance any physics course. It is an invaluable companion for both novice and veteran physics instructors.

Student Supplements

The **Study Guide** by Laird Kramer reinforces the text's emphasis on problem-solving strategies and student misconceptions. The *Study Guide for Volume 1* (ISBN 978-0-321-69665-6) covers Chapters 1–20, and the *Study Guide for Volumes 2 and 3* (ISBN 978-0-321-69669-4) covers Chapters 21–44.

The **Student Solutions Manual** by Lewis Ford (Texas A&M University) and Wayne Anderson contains detailed, step-by-step solutions to more than half of the odd-numbered end-of-chapter problems from the textbook. All solutions follow consistently the same Identify/Set Up/Execute/Evaluate problem-solving framework used in the textbook. The *Student Solutions Manual for Volume 1* (ISBN 978-0-321-69668-7) covers Chapters 1–20, and the *Student Solutions Manual for Volumes 2 and 3* (ISBN 978-0-321-69667-0) covers Chapters 21–44.

 MasteringPhysics® (www.masteringphysics.com) is a homework, tutorial, and assessment system based on years of research into how students work physics problems and precisely where they need help. Studies show that students who use MasteringPhysics significantly increase their scores compared to hand-written homework. MasteringPhysics achieves this improvement by providing students with instantaneous feedback specific to their wrong answers, simpler sub-problems upon request when they get stuck, and partial credit for their method(s). This individualized, 24/7 Socratic tutoring is recommended by nine out of ten students to their peers as the most effective and time-efficient way to study.

Pearson eText is available through MasteringPhysics, either automatically when MasteringPhysics is packaged with new books, or available as a purchased upgrade online. Allowing students access to the text wherever they have access to the Internet, Pearson eText comprises the full text, including figures that can be enlarged for better viewing. With eText, students are also able to pop up definitions and terms to help with vocabulary and the reading of the material. Students can also take notes in eText using the annotation feature at the top of each page.

Pearson Tutor Services (www.pearsontutorservices.com). Each student's subscription to MasteringPhysics also contains complimentary access to Pearson Tutor Services, powered by Smarthinking, Inc. By logging in with their MasteringPhysics ID and password, students will be connected to highly qualified e-instructors who provide additional interactive online tutoring on the major concepts of physics. Some restrictions apply; offer subject to change.

 ActivPhysics OnLine™ (which is accessed through the Study Area within www.masteringphysics.com) provides students with a suite of highly regarded applet-based tutorials (see above). The following workbooks help students work through complex concepts and understand them more clearly.

ActivPhysics OnLine Workbook, Volume 1: Mechanics * Thermal Physics * Oscillations & Waves (978-0-805-39060-5)

ActivPhysics OnLine Workbook, Volume 2: Electricity & Magnetism * Optics * Modern Physics (978-0-805-39061-2)

Acknowledgments

We would like to thank the hundreds of reviewers and colleagues who have offered valuable comments and suggestions over the life of this textbook. The continuing success of *University Physics* is due in large measure to their contributions.

Edward Adelson (Ohio State University), Ralph Alexander (University of Missouri at Rolla), J. G. Anderson, R. S. Anderson, Wayne Anderson (Sacramento City College), Alex Azima (Lansing Community College), Dilip Balamore (Nassau Community College), Harold Bale (University of North Dakota), Arun Bansil (Northeastern University), John Barach (Vanderbilt University), J. D. Barnett, H. H. Barschall, Albert Bartlett (University of Colorado), Marshall Bartlett (Hollins University), Paul Baum (CUNY, Queens College), Frederick Becchetti (University of Michigan), B. Bederson, David Bennum (University of Nevada, Reno), Lev I. Berger (San Diego State University), Robert Boeke (William Rainey Harper College), S. Borowitz, A. C. Braden, James Brooks (Boston University), Nicholas E. Brown (California Polytechnic State University, San Luis Obispo), Tony Buffa (California Polytechnic State University, San Luis Obispo), A. Capecelatro, Michael Cardamone (Pennsylvania State University), Duane Carmony (Purdue University), Troy Carter (UCLA), P. Catranides, John Cerne (SUNY at Buffalo), Tim Chupp (University of Michigan), Shinil Cho (La Roche College), Roger Clapp (University of South Florida), William M. Cloud (Eastern Illinois University), Leonard Cohen (Drexel University), W. R. Coker (University of Texas, Austin), Malcolm D. Cole (University of Missouri at Rolla), H. Conrad, David Cook (Lawrence University), Gayl Cook (University of Colorado), Hans Courant (University of Minnesota), Bruce A. Craver (University of Dayton), Larry Curtis (University of Toledo), Jai Dahiya (Southeast Missouri State University), Steve Detweiler (University of Florida), George Dixon (Oklahoma State University), Donald S. Duncan, Boyd Edwards (West Virginia University), Robert Eisenstein (Carnegie Mellon University), Amy Emerson Missourn (Virginia Institute of Technology), William Faissler (Northeastern University), William Fasnacht (U.S. Naval Academy), Paul Feldker (St. Louis Community College), Carlos Figueroa (Cabrillo College), L. H. Fisher, Neil Fletcher (Florida State University), Robert Folk, Peter Fong (Emory University), A. Lewis Ford (Texas A&M University), D. Frantszog, James R. Gaines (Ohio State University), Solomon Gartenhaus (Purdue University), Ron Gautreau (New Jersey Institute of Technology), J. David Gavenda (University of Texas, Austin), Dennis Gay (University of North Florida), James Gerhart (University of Washington), N. S. Gingrich, J. L. Glathart, S. Goodwin, Rich Gottfried (Frederick Community College), Walter S. Gray (University of Michigan), Paul Gresser (University of Maryland), Benjamin Grinstein (UC San Diego), Howard Grotch (Pennsylvania State University), John Gruber (San Jose State University), Graham D. Gutsche (U.S. Naval Academy), Michael J. Harrison (Michigan State University), Harold Hart (Western Illinois University), Howard Hayden (University of Connecticut), Carl Helrich (Goshen College), Laurent Hodges (Iowa State University), C. D. Hodgman, Michael Hones (Villanova University), Keith Honey (West Virginia Institute of Technology), Gregory Hood (Tidewater Community College), John Hubisz (North Carolina State University), M. Iona, John Jaszczak (Michigan Technical University), Alvin Jenkins (North Carolina State University), Robert P. Johnson (UC Santa Cruz), Lorella Jones (University of Illinois), John Karchek (GMI Engineering & Management Institute), Thomas Keil (Worcester Polytechnic Institute), Robert Kraemer (Carnegie Mellon University), Jean P. Krisch (University of Michigan), Robert A. Kromhout, Andrew Kunz (Marquette University), Charles Lane (Berry College), Thomas N. Lawrence (Texas State University), Robert J. Lee, Alfred Leitner (Rensselaer Polytechnic University), Gerald P. Lietz (De Paul University), Gordon Lind (Utah State University), S. Livingston, Elihu Lubkin (University of Wisconsin, Milwaukee), Robert Luke (Boise State University), David Lynch (Iowa State University), Michael Lysak (San Bernardino Valley College), Jeffrey Mallow (Loyola University), Robert Mania (Kentucky State University), Robert Marchina (University of Memphis), David Markowitz (University of Connecticut), R. J. Maurer, Oren Maxwell (Florida International University), Joseph L. McCauley (University of Houston), T. K. McCubbin, Jr. (Pennsylvania State University), Charles McFarland (University of Missouri at Rolla), James Mcguire (Tulane University), Lawrence McIntyre (University of Arizona), Fredric Messing (Carnegie-Mellon University), Thomas Meyer (Texas A&M University), Andre Mirabelli (St. Peter's College, New Jersey), Herbert Muether (S.U.N.Y., Stony Brook), Jack Munsee (California State University, Long Beach), Lorenzo Narducci (Drexel University), Van E. Neie (Purdue University), David A. Nordling (U. S. Naval Academy), Benedict Oh (Pennsylvania State University), L. O. Olsen, Jim Pannell (DeVry Institute of Technology), W. F. Parks (University of Missouri), Robert Paulson (California State University, Chico), Jerry Peacher (University of Missouri at Rolla), Arnold Perlmutter (University of Miami), Lennart Peterson (University of Florida), R. J. Peterson (University of Colorado, Boulder), R. Pinkston, Ronald Poling (University of Minnesota), J. G. Potter, C. W. Price (Millersville University), Francis Prosser (University of Kansas), Shelden H. Radin, Roberto Ramos (Drexel University), Michael Rapport (Anne Arundel Community College), R. Resnick, James A. Richards, Jr., John S. Risley (North Carolina State University), Francesc Roig (University of California, Santa Barbara), T. L. Rokoske, Richard Roth (Eastern Michigan University), Carl Rotter (University of West Virginia), S. Clark Rowland (Andrews University), Rajarshi Roy (Georgia Institute of Technology), Russell A. Roy (Santa Fe Community College), Dhiraj Sardar (University of Texas, San Antonio), Bruce Schumm (UC Santa Cruz), Melvin Schwartz (St. John's University), F. A. Scott, L. W. Seagondollar, Paul Shand (University of Northern Iowa), Stan Shepherd (Pennsylvania State University), Douglas Sherman (San Jose State), Bruce Sherwood (Carnegie Mellon University), Hugh Siefkin (Greenville College), Tomasz Skwarnicki (Syracuse University), C. P. Slichter, Charles W. Smith (University of Maine, Orono), Malcolm Smith (University of Lowell), Ross Spencer (Brigham Young University), Julien Sprott (University of Wisconsin), Victor Stanionis (Iona College), James Stith (American Institute of Physics), Chuck Stone (North Carolina A&T State

University), Edward Strother (Florida Institute of Technology), Conley Stutz (Bradley University), Albert Stwertka (U.S. Merchant Marine Academy), Kenneth Szpara-DeNisco (Harrisburg Area Community College), Martin Tiersten (CUNY, City College), David Toot (Alfred University), Somdev Tyagi (Drexel University), F. Verbrugge, Helmut Vogel (Carnegie Mellon University), Robert Webb (Texas A & M), Thomas Weber (Iowa State University), M. Russell Wehr, (Pennsylvania State University), Robert Weidman (Michigan Technical University), Dan Whalen (UC San Diego), Lester V. Whitney, Thomas Wiggins (Pennsylvania State University), David Willey (University of Pittsburgh, Johnstown), George Williams (University of Utah), John Williams (Auburn University), Stanley Williams (Iowa State University), Jack Willis, Suzanne Willis (Northern Illinois University), Robert Wilson (San Bernardino Valley College), L. Wolfenstein, James Wood (Palm Beach Junior College), Lowell Wood (University of Houston), R. E. Worley, D. H. Ziebell (Manatee Community College), George O. Zimmerman (Boston University)

In addition, we both have individual acknowledgments we would like to make.

I want to extend my heartfelt thanks to my colleagues at Carnegie Mellon, especially Professors Robert Kraemer, Bruce Sherwood, Ruth Chabay, Helmut Vogel, and Brian Quinn, for many stimulating discussions about physics pedagogy and for their support and encouragement during the writing of several successive editions of this book. I am equally indebted to the many generations of Carnegie Mellon students who have helped me learn what good teaching and good writing are, by showing me what works and what doesn't. It is always a joy and a privilege to express my gratitude to my wife Alice and our children Gretchen and Rebecca for their love, support, and emotional sustenance during the writing of several successive editions of this book. May all men and women be blessed with love such as theirs. — H. D. Y.

I would like to thank my past and present colleagues at UCSB, including Rob Geller, Carl Gwinn, Al Nash, Elisabeth Nicol, and Francesc Roig, for their whole-hearted support and for many helpful discussions. I owe a special debt of gratitude to my early teachers Willa Ramsay, Peter Zimmerman, William Little, Alan Schwettman, and Dirk Walecka for showing me what clear and engaging physics teaching is all about, and to Stuart Johnson for inviting me to become a co-author of *University Physics* beginning with the 9th edition. I want to express special thanks to the editorial staff at Addison-Wesley and their partners: to Nancy Whilton for her editorial vision; to Margot Otway for her superb graphic sense and careful development of this edition; to Peter Murphy for his contributions to the worked examples; to Jason J. B. Harlow for his careful reading of the page proofs; and to Chandrika Madhavan, Steven Le, and Cindy Johnson for keeping the editorial and production pipeline flowing. Most of all, I want to express my gratitude and love to my wife Caroline, to whom I dedicate my contribution to this book. Hey, Caroline, the new edition's done at last — let's go flying! — R. A. F.

Please Tell Us What You Think!

We welcome communications from students and professors, especially concerning errors or deficiencies that you find in this edition. We have devoted a lot of time and effort to writing the best book we know how to write, and we hope it will help you to teach and learn physics. In turn, you can help us by letting us know what still needs to be improved! Please feel free to contact us either electronically or by ordinary mail. Your comments will be greatly appreciated.

December 2010

Hugh D. Young
Department of Physics
Carnegie Mellon University
Pittsburgh, PA 15213
hdy@andrew.cmu.edu

Roger A. Freedman
Department of Physics
University of California, Santa Barbara
Santa Barbara, CA 93106-9530
airboy@physics.ucsb.edu
http://www.physics.ucsb.edu/~airboy/

DETAILED CONTENTS

MECHANICS

UNITS, PHYSICAL QUANTITIES, AND VECTORS

1

? Being able to predict the path of a thunderstorm is essential for minimizing the damage it does to lives and property. If a thunderstorm is moving at 20 km/h in a direction 53° north of east, how far north does the thunderstorm move in 1 h?

LEARNING GOALS

By studying this chapter, you will learn:

- Three fundamental quantities of physics and the units physicists use to measure them.

- How to keep track of significant figures in your calculations.

- The difference between scalars and vectors, and how to add and subtract vectors graphically.

- What the components of a vector are, and how to use them in calculations.

- What unit vectors are, and how to use them with components to describe vectors.

- Two ways of multiplying vectors.

Physics is one of the most fundamental of the sciences. Scientists of all disciplines use the ideas of physics, including chemists who study the structure of molecules, paleontologists who try to reconstruct how dinosaurs walked, and climatologists who study how human activities affect the atmosphere and oceans. Physics is also the foundation of all engineering and technology. No engineer could design a flat-screen TV, an interplanetary spacecraft, or even a better mousetrap without first understanding the basic laws of physics.

The study of physics is also an adventure. You will find it challenging, sometimes frustrating, occasionally painful, and often richly rewarding. If you've ever wondered why the sky is blue, how radio waves can travel through empty space, or how a satellite stays in orbit, you can find the answers by using fundamental physics. You will come to see physics as a towering achievement of the human intellect in its quest to understand our world and ourselves.

In this opening chapter, we'll go over some important preliminaries that we'll need throughout our study. We'll discuss the nature of physical theory and the use of idealized models to represent physical systems. We'll introduce the systems of units used to describe physical quantities and discuss ways to describe the accuracy of a number. We'll look at examples of problems for which we can't (or don't want to) find a precise answer, but for which rough estimates can be useful and interesting. Finally, we'll study several aspects of vectors and vector algebra. Vectors will be needed throughout our study of physics to describe and analyze physical quantities, such as velocity and force, that have direction as well as magnitude.

1.1 The Nature of Physics

Physics is an *experimental* science. Physicists observe the phenomena of nature and try to find patterns that relate these phenomena. These patterns are called physical theories or, when they are very well established and widely used, physical laws or principles.

CAUTION **The meaning of the word "theory"** Calling an idea a theory does *not* mean that it's just a random thought or an unproven concept. Rather, a theory is an explanation of natural phenomena based on observation and accepted fundamental principles. An example is the well-established theory of biological evolution, which is the result of extensive research and observation by generations of biologists. ▌

To develop a physical theory, a physicist has to learn to ask appropriate questions, design experiments to try to answer the questions, and draw appropriate conclusions from the results. Figure 1.1 shows two famous facilities used for physics experiments.

Legend has it that Galileo Galilei (1564–1642) dropped light and heavy objects from the top of the Leaning Tower of Pisa (Fig. 1.1a) to find out whether their rates of fall were the same or different. From examining the results of his experiments (which were actually much more sophisticated than in the legend), he made the inductive leap to the principle, or theory, that the acceleration of a falling body is independent of its weight.

The development of physical theories such as Galileo's often takes an indirect path, with blind alleys, wrong guesses, and the discarding of unsuccessful theories in favor of more promising ones. Physics is not simply a collection of facts and principles; it is also the *process* by which we arrive at general principles that describe how the physical universe behaves.

No theory is ever regarded as the final or ultimate truth. The possibility always exists that new observations will require that a theory be revised or discarded. It is in the nature of physical theory that we can disprove a theory by finding behavior that is inconsistent with it, but we can never prove that a theory is always correct.

Getting back to Galileo, suppose we drop a feather and a cannonball. They certainly do *not* fall at the same rate. This does not mean that Galileo was wrong; it means that his theory was incomplete. If we drop the feather and the cannonball *in a vacuum* to eliminate the effects of the air, then they do fall at the same rate. Galileo's theory has a **range of validity:** It applies only to objects for which the force exerted by the air (due to air resistance and buoyancy) is much less than the weight. Objects like feathers or parachutes are clearly outside this range.

Often a new development in physics extends a principle's range of validity. Galileo's analysis of falling bodies was greatly extended half a century later by Newton's laws of motion and law of gravitation.

1.1 Two research laboratories. **(a)** According to legend, Galileo investigated falling bodies by dropping them from the Leaning Tower in Pisa, Italy, and he studied pendulum motion by observing the swinging of the chandelier in the adjacent cathedral. **(b)** The Large Hadron Collider (LHC) in Geneva, Switzerland, the world's largest particle accelerator, is used to explore the smallest and most fundamental constituents of matter. This photo shows a portion of one of the LHC's detectors (note the worker on the yellow platform).

(a)

(b)

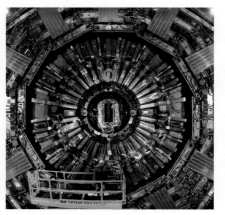

1.2 Solving Physics Problems

At some point in their studies, almost all physics students find themselves thinking, "I understand the concepts, but I just can't solve the problems." But in physics, truly understanding a concept *means* being able to apply it to a variety of problems. Learning how to solve problems is absolutely essential; you don't *know* physics unless you can *do* physics.

How do you learn to solve physics problems? In every chapter of this book you will find *Problem-Solving Strategies* that offer techniques for setting up and solving problems efficiently and accurately. Following each *Problem-Solving Strategy* are one or more worked *Examples* that show these techniques in action. (The *Problem-Solving Strategies* will also steer you away from some *incorrect* techniques that you may be tempted to use.) You'll also find additional examples that aren't associated with a particular *Problem-Solving Strategy.* In addition, at the end of each chapter you'll find a *Bridging Problem* that uses more than one of the key ideas from the chapter. Study these strategies and problems carefully, and work through each example for yourself on a piece of paper.

Different techniques are useful for solving different kinds of physics problems, which is why this book offers dozens of *Problem-Solving Strategies.* No matter what kind of problem you're dealing with, however, there are certain key steps that you'll always follow. (These same steps are equally useful for problems in math, engineering, chemistry, and many other fields.) In this book we've organized these steps into four stages of solving a problem.

All of the *Problem-Solving Strategies* and *Examples* in this book will follow these four steps. (In some cases we will combine the first two or three steps.) We encourage you to follow these same steps when you solve problems yourself. You may find it useful to remember the acronym ***I SEE***—short for *Identify, Set up, Execute,* and *Evaluate.*

Problem-Solving Strategy 1.1   Solving Physics Problems

IDENTIFY *the relevant concepts:* Use the physical conditions stated in the problem to help you decide which physics concepts are relevant. Identify the **target variables** of the problem—that is, the quantities whose values you're trying to find, such as the speed at which a projectile hits the ground, the intensity of a sound made by a siren, or the size of an image made by a lens. Identify the known quantities, as stated or implied in the problem. This step is essential whether the problem asks for an algebraic expression or a numerical answer.

SET UP *the problem:* Given the concepts you have identified and the known and target quantities, choose the equations that you'll use to solve the problem and decide how you'll use them. Make sure that the variables you have identified correlate exactly with those in the equations. If appropriate, draw a sketch of the situation described in the problem. (Graph paper, ruler, protractor, and compass will help you make clear, useful sketches.) As best you can,

estimate what your results will be and, as appropriate, predict what the physical behavior of a system will be. The worked examples in this book include tips on how to make these kinds of estimates and predictions. If this seems challenging, don't worry—you'll get better with practice!

EXECUTE *the solution:* This is where you "do the math." Study the worked examples to see what's involved in this step.

EVALUATE *your answer:* Compare your answer with your estimates, and reconsider things if there's a discrepancy. If your answer includes an algebraic expression, assure yourself that it represents what would happen if the variables in it were taken to extremes. For future reference, make note of any answer that represents a quantity of particular significance. Ask yourself how you might answer a more general or more difficult version of the problem you have just solved.

Idealized Models

In everyday conversation we use the word "model" to mean either a small-scale replica, such as a model railroad, or a person who displays articles of clothing (or the absence thereof). In physics a **model** is a simplified version of a physical system that would be too complicated to analyze in full detail.

For example, suppose we want to analyze the motion of a thrown baseball (Fig. 1.2a). How complicated is this problem? The ball is not a perfect sphere (it has raised seams), and it spins as it moves through the air. Wind and air resistance influence its motion, the ball's weight varies a little as its distance from the center of the earth changes, and so on. If we try to include all these things, the analysis gets hopelessly complicated. Instead, we invent a simplified version of the problem. We neglect the size and shape of the ball by representing it as a point object, or **particle.** We neglect air resistance by making the ball move in a vacuum, and we make the weight constant. Now we have a problem that is simple enough to deal with (Fig. 1.2b). We will analyze this model in detail in Chapter 3.

We have to overlook quite a few minor effects to make an idealized model, but we must be careful not to neglect too much. If we ignore the effects of gravity completely, then our model predicts that when we throw the ball up, it will go in a straight line and disappear into space. A useful model is one that simplifies a problem enough to make it manageable, yet keeps its essential features.

1.2 To simplify the analysis of (a) a baseball in flight, we use (b) an idealized model.

(a) A real baseball in flight

Baseball spins and has a complex shape.

Air resistance and wind exert forces on the ball.

Direction of motion

Gravitational force on ball depends on altitude.

(b) An idealized model of the baseball

Baseball is treated as a point object (particle).

No air resistance.

Direction of motion

Gravitational force on ball is constant.

The validity of the predictions we make using a model is limited by the validity of the model. For example, Galileo's prediction about falling bodies (see Section 1.1) corresponds to an idealized model that does not include the effects of air resistance. This model works fairly well for a dropped cannonball, but not so well for a feather.

Idealized models play a crucial role throughout this book. Watch for them in discussions of physical theories and their applications to specific problems.

1.3 Standards and Units

As we learned in Section 1.1, physics is an experimental science. Experiments require measurements, and we generally use numbers to describe the results of measurements. Any number that is used to describe a physical phenomenon quantitatively is called a **physical quantity.** For example, two physical quantities that describe you are your weight and your height. Some physical quantities are so fundamental that we can define them only by describing how to measure them. Such a definition is called an **operational definition.** Two examples are measuring a distance by using a ruler and measuring a time interval by using a stopwatch. In other cases we define a physical quantity by describing how to calculate it from other quantities that we *can* measure. Thus we might define the average speed of a moving object as the distance traveled (measured with a ruler) divided by the time of travel (measured with a stopwatch).

When we measure a quantity, we always compare it with some reference standard. When we say that a Ferrari 458 Italia is 4.53 meters long, we mean that it is 4.53 times as long as a meter stick, which we define to be 1 meter long. Such a standard defines a **unit** of the quantity. The meter is a unit of distance, and the second is a unit of time. When we use a number to describe a physical quantity, we must always specify the unit that we are using; to describe a distance as simply "4.53" wouldn't mean anything.

To make accurate, reliable measurements, we need units of measurement that do not change and that can be duplicated by observers in various locations. The system of units used by scientists and engineers around the world is commonly called "the metric system," but since 1960 it has been known officially as the **International System,** or **SI** (the abbreviation for its French name, *Système International*). Appendix A gives a list of all SI units as well as definitions of the most fundamental units.

1.3 The measurements used to determine (a) the duration of a second and (b) the length of a meter. These measurements are useful for setting standards because they give the same results no matter where they are made.

(a) Measuring the second

Microwave radiation with a frequency of exactly 9,192,631,770 cycles per second ...

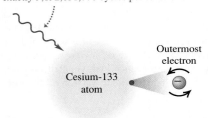

Outermost electron

Cesium-133 atom

... causes the outermost electron of a cesium-133 atom to reverse its spin direction.

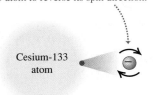

Cesium-133 atom

An atomic clock uses this phenomenon to tune microwaves to this exact frequency. It then counts 1 second for each 9,192,631,770 cycles.

(b) Measuring the meter

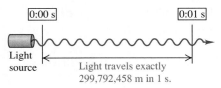

0:00 s   0:01 s

Light source

Light travels exactly 299,792,458 m in 1 s.

Time

From 1889 until 1967, the unit of time was defined as a certain fraction of the mean solar day, the average time between successive arrivals of the sun at its highest point in the sky. The present standard, adopted in 1967, is much more precise. It is based on an atomic clock, which uses the energy difference between the two lowest energy states of the cesium atom. When bombarded by microwaves of precisely the proper frequency, cesium atoms undergo a transition from one of these states to the other. One **second** (abbreviated s) is defined as the time required for 9,192,631,770 cycles of this microwave radiation (Fig. 1.3a).

Length

In 1960 an atomic standard for the meter was also established, using the wavelength of the orange-red light emitted by atoms of krypton (^{86}Kr) in a glow discharge tube. Using this length standard, the speed of light in vacuum was measured to be 299,792,458 m/s. In November 1983, the length standard was changed again so that the speed of light in vacuum was *defined* to be precisely

299,792,458 m/s. Hence the new definition of the **meter** (abbreviated m) is the distance that light travels in vacuum in 1/299,792,458 second (Fig. 1.3b). This provides a much more precise standard of length than the one based on a wavelength of light.

Mass

The standard of mass, the **kilogram** (abbreviated kg), is defined to be the mass of a particular cylinder of platinum–iridium alloy kept at the International Bureau of Weights and Measures at Sèvres, near Paris (Fig. 1.4). An atomic standard of mass would be more fundamental, but at present we cannot measure masses on an atomic scale with as much accuracy as on a macroscopic scale. The *gram* (which is not a fundamental unit) is 0.001 kilogram.

Unit Prefixes

Once we have defined the fundamental units, it is easy to introduce larger and smaller units for the same physical quantities. In the metric system these other units are related to the fundamental units (or, in the case of mass, to the gram) by multiples of 10 or $\frac{1}{10}$. Thus one kilometer (1 km) is 1000 meters, and one centimeter (1 cm) is $\frac{1}{100}$ meter. We usually express multiples of 10 or $\frac{1}{10}$ in exponential notation: $1000 = 10^3$, $\frac{1}{1000} = 10^{-3}$, and so on. With this notation, $1\,km = 10^3\,m$ and $1\,cm = 10^{-2}\,m$.

The names of the additional units are derived by adding a **prefix** to the name of the fundamental unit. For example, the prefix "kilo-," abbreviated k, always means a unit larger by a factor of 1000; thus

$$1 \text{ kilometer} = 1 \text{ km} = 10^3 \text{ meters} = 10^3 \text{ m}$$
$$1 \text{ kilogram} = 1 \text{ kg} = 10^3 \text{ grams} = 10^3 \text{ g}$$
$$1 \text{ kilowatt} = 1 \text{ kW} = 10^3 \text{ watts} = 10^3 \text{ W}$$

A table on the inside back cover of this book lists the standard SI prefixes, with their meanings and abbreviations.

Table 1.1 gives some examples of the use of multiples of 10 and their prefixes with the units of length, mass, and time. Figure 1.5 shows how these prefixes are used to describe both large and small distances.

The British System

Finally, we mention the British system of units. These units are used only in the United States and a few other countries, and in most of these they are being replaced by SI units. British units are now officially defined in terms of SI units, as follows:

Length: 1 inch = 2.54 cm (exactly)

Force: 1 pound = 4.448221615260 newtons (exactly)

1.4 The international standard kilogram is the metal object carefully enclosed within these nested glass containers.

Table 1.1 Some Units of Length, Mass, and Time

Length	Mass	Time
1 nanometer $= 1$ nm $= 10^{-9}$ m (a few times the size of the largest atom)	1 microgram $= 1\,\mu$g $= 10^{-6}$ g $= 10^{-9}$ kg (mass of a very small dust particle)	1 nanosecond $= 1$ ns $= 10^{-9}$ s (time for light to travel 0.3 m)
1 micrometer $= 1\,\mu$m $= 10^{-6}$ m (size of some bacteria and living cells)	1 milligram $= 1$ mg $= 10^{-3}$ g $= 10^{-6}$ kg (mass of a grain of salt)	1 microsecond $= 1\,\mu$s $= 10^{-6}$ s (time for space station to move 8 mm)
1 millimeter $= 1$ mm $= 10^{-3}$ m (diameter of the point of a ballpoint pen)	1 gram $= 1$ g $= 10^{-3}$ kg (mass of a paper clip)	1 millisecond $= 1$ ms $= 10^{-3}$ s (time for sound to travel 0.35 m)
1 centimeter $= 1$ cm $= 10^{-2}$ m (diameter of your little finger)		
1 kilometer $= 1$ km $= 10^3$ m (a 10-minute walk)		

1.5 Some typical lengths in the universe. (f) is a scanning tunneling microscope image of atoms on a crystal surface; (g) is an artist's impression.

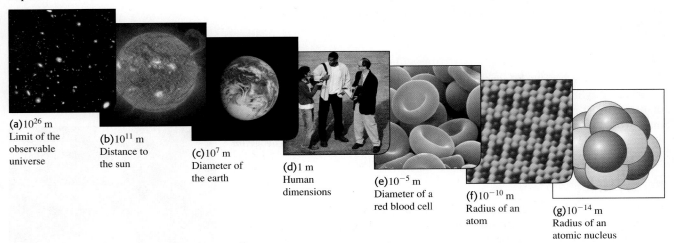

(a)10^{26} m
Limit of the observable universe

(b)10^{11} m
Distance to the sun

(c)10^7 m
Diameter of the earth

(d)1 m
Human dimensions

(e)10^{-5} m
Diameter of a red blood cell

(f)10^{-10} m
Radius of an atom

(g)10^{-14} m
Radius of an atomic nucleus

1.6 Many everyday items make use of both SI and British units. An example is this speedometer from a U.S.-built automobile, which shows the speed in both kilometers per hour (inner scale) and miles per hour (outer scale).

The newton, abbreviated N, is the SI unit of force. The British unit of time is the second, defined the same way as in SI. In physics, British units are used only in mechanics and thermodynamics; there is no British system of electrical units.

In this book we use SI units for all examples and problems, but we occasionally give approximate equivalents in British units. As you do problems using SI units, you may also wish to convert to the approximate British equivalents if they are more familiar to you (Fig. 1.6). But you should try to *think* in SI units as much as you can.

1.4 Unit Consistency and Conversions

We use equations to express relationships among physical quantities, represented by algebraic symbols. Each algebraic symbol always denotes both a number and a unit. For example, d might represent a distance of 10 m, t a time of 5 s, and v a speed of 2 m/s.

An equation must always be **dimensionally consistent.** You can't add apples and automobiles; two terms may be added or equated only if they have the same units. For example, if a body moving with constant speed v travels a distance d in a time t, these quantities are related by the equation

$$d = vt$$

If d is measured in meters, then the product vt must also be expressed in meters. Using the above numbers as an example, we may write

$$10 \text{ m} = \left(2 \frac{\text{m}}{\cancel{s}}\right)(5 \, \cancel{s})$$

Because the unit 1/s on the right side of the equation cancels the unit s, the product has units of meters, as it must. In calculations, units are treated just like algebraic symbols with respect to multiplication and division.

> **CAUTION** **Always use units in calculations** When a problem requires calculations using numbers with units, *always* write the numbers with the correct units and carry the units through the calculation as in the example above. This provides a very useful check. If at some stage in a calculation you find that an equation or an expression has inconsistent units, you know you have made an error somewhere. In this book we will *always* carry units through all calculations, and we strongly urge you to follow this practice when you solve problems.

Problem-Solving Strategy 1.2   Solving Physics Problems

IDENTIFY *the relevant concepts:* In most cases, it's best to use the fundamental SI units (lengths in meters, masses in kilograms, and times in seconds) in every problem. If you need the answer to be in a different set of units (such as kilometers, grams, or hours), wait until the end of the problem to make the conversion.

SET UP *the problem* and **EXECUTE** *the solution:* Units are multiplied and divided just like ordinary algebraic symbols. This gives us an easy way to convert a quantity from one set of units to another: Express the same physical quantity in two different units and form an equality.

For example, when we say that 1 min = 60 s, we don't mean that the number 1 is equal to the number 60; rather, we mean that 1 min represents the same physical time interval as 60 s. For this reason, the ratio $(1\ \text{min})/(60\ \text{s})$ equals 1, as does its reciprocal $(60\ \text{s})/(1\ \text{min})$. We may multiply a quantity by either of these

factors (which we call *unit multipliers*) without changing that quantity's physical meaning. For example, to find the number of seconds in 3 min, we write

$$3\ \text{min} = (3\ \cancel{\text{min}})\left(\frac{60\ \text{s}}{1\ \cancel{\text{min}}}\right) = 180\ \text{s}$$

EVALUATE *your answer:* If you do your unit conversions correctly, unwanted units will cancel, as in the example above. If, instead, you had multiplied 3 min by $(1\ \text{min})/(60\ \text{s})$, your result would have been the nonsensical $\frac{1}{20}\ \text{min}^2/\text{s}$. To be sure you convert units properly, you must write down the units at *all* stages of the calculation.

Finally, check whether your answer is reasonable. For example, the result 3 min = 180 s is reasonable because the second is a smaller unit than the minute, so there are more seconds than minutes in the same time interval.

Example 1.1   Converting speed units

The world land speed record is 763.0 mi/h, set on October 15, 1997, by Andy Green in the jet-engine car *Thrust SSC*. Express this speed in meters per second.

SOLUTION

IDENTIFY, SET UP, and EXECUTE: We need to convert the units of a speed from mi/h to m/s. We must therefore find unit multipliers that relate (i) miles to meters and (ii) hours to seconds. In Appendix E (or inside the front cover of this book) we find the equalities 1 mi = 1.609 km, 1 km = 1000 m, and 1 h = 3600 s. We set up the conversion as follows, which ensures that all the desired cancellations by division take place:

$$763.0\ \text{mi/h} = \left(763.0\ \frac{\cancel{\text{mi}}}{\cancel{\text{h}}}\right)\left(\frac{1.609\ \cancel{\text{km}}}{1\ \cancel{\text{mi}}}\right)\left(\frac{1000\ \text{m}}{1\ \cancel{\text{km}}}\right)\left(\frac{1\ \cancel{\text{h}}}{3600\ \text{s}}\right)$$

$$= 341.0\ \text{m/s}$$

EVALUATE: Green's was the first supersonic land speed record (the speed of sound in air is about 340 m/s). This example shows a useful rule of thumb: A speed expressed in m/s is a bit less than half the value expressed in mi/h, and a bit less than one-third the value expressed in km/h. For example, a normal freeway speed is about 30 m/s = 67 mi/h = 108 km/h, and a typical walking speed is about 1.4 m/s = 3.1 mi/h = 5.0 km/h.

Example 1.2   Converting volume units

The world's largest cut diamond is the First Star of Africa (mounted in the British Royal Sceptre and kept in the Tower of London). Its volume is 1.84 cubic inches. What is its volume in cubic centimeters? In cubic meters?

SOLUTION

IDENTIFY, SET UP, and EXECUTE: Here we are to convert the units of a volume from cubic inches (in.^{3}) to both cubic centimeters (cm^3) and cubic meters (m^3). Appendix E gives us the equality 1 in. = 2.540 cm, from which we obtain 1 in.^{3} = $(2.54\ \text{cm})^3$. We then have

$$1.84\ \text{in.}^3 = (1.84\ \text{in.}^3)\left(\frac{2.54\ \text{cm}}{1\ \text{in.}}\right)^3$$

$$= (1.84)(2.54)^3\ \frac{\cancel{\text{in.}^3}\ \text{cm}^3}{\cancel{\text{in.}^3}} = 30.2\ \text{cm}^3$$

Appendix F also gives us 1 m = 100 cm, so

$$30.2\ \text{cm}^3 = (30.2\ \text{cm}^3)\left(\frac{1\ \text{m}}{100\ \text{cm}}\right)^3$$

$$= (30.2)\left(\frac{1}{100}\right)^3\frac{\cancel{\text{cm}^3}\ \text{m}^3}{\cancel{\text{cm}^3}} = 30.2 \times 10^{-6}\ \text{m}^3$$

$$= 3.02 \times 10^{-5}\ \text{m}^3$$

EVALUATE: Following the pattern of these conversions, you can show that 1 in.$^3 \approx 16\ \text{cm}^3$ and that 1 m$^3 \approx 60{,}000\ \text{in.}^3$. These approximate unit conversions may be useful for future reference.

1.5 Uncertainty and Significant Figures

1.7 This spectacular mishap was the result of a very small percent error—traveling a few meters too far at the end of a journey of hundreds of thousands of meters.

Table 1.2 Using Significant Figures

Multiplication or division:
Result may have no more significant figures than the starting number with the fewest significant figures:

$$\frac{0.745 \times 2.2}{3.885} = 0.42$$

$$1.32578 \times 10^{7} \times 4.11 \times 10^{-3} = 5.45 \times 10^{4}$$

Addition or subtraction:
Number of significant figures is determined by the starting number with the largest uncertainty (i.e., fewest digits to the right of the decimal point):

$$27.153 + 138.2 - 11.74 = 153.6$$

1.8 Determining the value of π from the circumference and diameter of a circle.

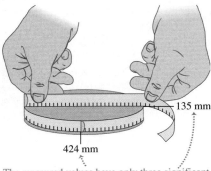

135 mm

424 mm

The measured values have only three significant figures, so their calculated ratio (π) also has only three significant figures.

Measurements always have uncertainties. If you measure the thickness of the cover of a hardbound version of this book using an ordinary ruler, your measurement is reliable only to the nearest millimeter, and your result will be 3 mm. It would be *wrong* to state this result as 3.00 mm; given the limitations of the measuring device, you can't tell whether the actual thickness is 3.00 mm, 2.85 mm, or 3.11 mm. But if you use a micrometer caliper, a device that measures distances reliably to the nearest 0.01 mm, the result will be 2.91 mm. The distinction between these two measurements is in their **uncertainty.** The measurement using the micrometer caliper has a smaller uncertainty; it's a more accurate measurement. The uncertainty is also called the **error** because it indicates the maximum difference there is likely to be between the measured value and the true value. The uncertainty or error of a measured value depends on the measurement technique used.

We often indicate the **accuracy** of a measured value—that is, how close it is likely to be to the true value—by writing the number, the symbol $\pm$, and a second number indicating the uncertainty of the measurement. If the diameter of a steel rod is given as 56.47 $\pm$ 0.02 mm, this means that the true value is unlikely to be less than 56.45 mm or greater than 56.49 mm. In a commonly used shorthand notation, the number 1.6454(21) means 1.6454 $\pm$ 0.0021. The numbers in parentheses show the uncertainty in the final digits of the main number.

We can also express accuracy in terms of the maximum likely **fractional error** or **percent error** (also called *fractional uncertainty* and *percent uncertainty*). A resistor labeled "47 ohms $\pm$ 10%" probably has a true resistance that differs from 47 ohms by no more than 10% of 47 ohms—that is, by about 5 ohms. The resistance is probably between 42 and 52 ohms. For the diameter of the steel rod given above, the fractional error is (0.02 mm)/(56.47 mm), or about 0.0004; the percent error is (0.0004)(100%), or about 0.04%. Even small percent errors can sometimes be very significant (Fig. 1.7).

In many cases the uncertainty of a number is not stated explicitly. Instead, the uncertainty is indicated by the number of meaningful digits, or **significant figures,** in the measured value. We gave the thickness of the cover of this book as 2.91 mm, which has three significant figures. By this we mean that the first two digits are known to be correct, while the third digit is uncertain. The last digit is in the hundredths place, so the uncertainty is about 0.01 mm. Two values with the *same* number of significant figures may have *different* uncertainties; a distance given as 137 km also has three significant figures, but the uncertainty is about 1 km.

When you use numbers that have uncertainties to compute other numbers, the computed numbers are also uncertain. When numbers are multiplied or divided, the number of significant figures in the result can be no greater than in the factor with the fewest significant figures. For example, 3.1416 $\times$ 2.34 $\times$ 0.58 = 4.3. When we add and subtract numbers, it's the location of the decimal point that matters, not the number of significant figures. For example, 123.62 + 8.9 = 132.5. Although 123.62 has an uncertainty of about 0.01, 8.9 has an uncertainty of about 0.1. So their sum has an uncertainty of about 0.1 and should be written as 132.5, not 132.52. Table 1.2 summarizes these rules for significant figures.

As an application of these ideas, suppose you want to verify the value of π, the ratio of the circumference of a circle to its diameter. The true value of this ratio to ten digits is 3.141592654. To test this, you draw a large circle and measure its circumference and diameter to the nearest millimeter, obtaining the values 424 mm and 135 mm (Fig. 1.8). You punch these into your calculator and obtain the quotient (424 mm)/(135 mm) = 3.140740741. This may seem to disagree with the true value of π, but keep in mind that each of your measurements has three significant figures, so your measured value of π can have only three significant figures. It should be stated simply as 3.14. Within the limit of three significant figures, your value does agree with the true value.

In the examples and problems in this book we usually give numerical values with three significant figures, so your answers should usually have no more than three significant figures. (Many numbers in the real world have even less accuracy. An automobile speedometer, for example, usually gives only two significant figures.) Even if you do the arithmetic with a calculator that displays ten digits, it would be wrong to give a ten-digit answer because it misrepresents the accuracy of the results. Always round your final answer to keep only the correct number of significant figures or, in doubtful cases, one more at most. In Example 1.1 it would have been wrong to state the answer as 341.01861 m/s. Note that when you reduce such an answer to the appropriate number of significant figures, you must *round,* not *truncate.* Your calculator will tell you that the ratio of 525 m to 311 m is 1.688102894; to three significant figures, this is 1.69, not 1.68.

When we calculate with very large or very small numbers, we can show significant figures much more easily by using **scientific notation,** sometimes called **powers-of-10 notation.** The distance from the earth to the moon is about 384,000,000 m, but writing the number in this form doesn't indicate the number of significant figures. Instead, we move the decimal point eight places to the left (corresponding to dividing by 10^8) and multiply by 10^8; that is,

$$384{,}000{,}000 \text{ m} = 3.84 \times 10^8 \text{ m}$$

In this form, it is clear that we have three significant figures. The number 4.00×10^{-7} also has three significant figures, even though two of them are zeros. Note that in scientific notation the usual practice is to express the quantity as a number between 1 and 10 multiplied by the appropriate power of 10.

When an integer or a fraction occurs in a general equation, we treat that number as having no uncertainty at all. For example, in the equation $v_x^2 = v_{0x}^2 + 2a_x(x - x_0)$, which is Eq. (2.13) in Chapter 2, the coefficient 2 is *exactly* 2. We can consider this coefficient as having an infinite number of significant figures ($2.000000\ldots$). The same is true of the exponent 2 in v_x^2 and v_{0x}^2.

Finally, let's note that **precision** is not the same as *accuracy.* A cheap digital watch that gives the time as 10:35:17 A.M. is very *precise* (the time is given to the second), but if the watch runs several minutes slow, then this value isn't very *accurate.* On the other hand, a grandfather clock might be very accurate (that is, display the correct time), but if the clock has no second hand, it isn't very precise. A high-quality measurement is both precise *and* accurate.

Example 1.3 Significant figures in multiplication

The rest energy E of an object with rest mass m is given by Einstein's famous equation $E = mc^2$, where c is the speed of light in vacuum. Find E for an electron for which (to three significant figures) $m = 9.11 \times 10^{-31}$ kg. The SI unit for E is the joule (J); $1 \text{ J} = 1 \text{ kg} \cdot \text{m}^2/\text{s}^2$.

SOLUTION

IDENTIFY and SET UP: Our target variable is the energy E. We are given the value of the mass m; from Section 1.3 (or Appendix F) the speed of light is $c = 2.99792458 \times 10^8$ m/s.

EXECUTE: Substituting the values of m and c into Einstein's equation, we find

$$
\begin{aligned}
E &= (9.11 \times 10^{-31} \text{ kg})(2.99792458 \times 10^8 \text{ m/s})^2 \\
&= (9.11)(2.99792458)^2(10^{-31})(10^8)^2 \text{ kg} \cdot \text{m}^2/\text{s}^2 \\
&= (81.87659678)(10^{[-31+(2\times 8)]}) \text{ kg} \cdot \text{m}^2/\text{s}^2 \\
&= 8.187659678 \times 10^{-14} \text{ kg} \cdot \text{m}^2/\text{s}^2
\end{aligned}
$$

Since the value of m was given to only three significant figures, we must round this to

$$E = 8.19 \times 10^{-14} \text{ kg} \cdot \text{m}^2/\text{s}^2 = 8.19 \times 10^{-14} \text{ J}$$

EVALUATE: While the rest energy contained in an electron may seem ridiculously small, on the atomic scale it is tremendous. Compare our answer to 10^{-19} J, the energy gained or lost by a single atom during a typical chemical reaction. The rest energy of an electron is about 1,000,000 times larger! (We'll discuss the significance of rest energy in Chapter 37.)

Test Your Understanding of Section 1.5   The density of a material is equal to its mass divided by its volume. What is the density (in kg/m^3) of a rock of mass 1.80 kg and volume 6.0×10^{-4} m^3? (i) 3×10^3 kg/m^3; (ii) 3.0×10^3 kg/m^3; (iii) 3.00×10^3 kg/m^3; (iv) 3.000×10^3 kg/m^3; (v) any of these—all of these answers are mathematically equivalent.

1.6 Estimates and Orders of Magnitude

We have stressed the importance of knowing the accuracy of numbers that represent physical quantities. But even a very crude estimate of a quantity often gives us useful information. Sometimes we know how to calculate a certain quantity, but we have to guess at the data we need for the calculation. Or the calculation might be too complicated to carry out exactly, so we make some rough approximations. In either case our result is also a guess, but such a guess can be useful even if it is uncertain by a factor of two, ten, or more. Such calculations are often called **order-of-magnitude estimates.** The great Italian-American nuclear physicist Enrico Fermi (1901–1954) called them "back-of-the-envelope calculations."

Exercises 1.16 through 1.25 at the end of this chapter are of the estimating, or order-of-magnitude, variety. Most require guesswork for the needed input data. Don't try to look up a lot of data; make the best guesses you can. Even when they are off by a factor of ten, the results can be useful and interesting.

Example 1.4   **An order-of-magnitude estimate**

You are writing an adventure novel in which the hero escapes across the border with a billion dollars' worth of gold in his suitcase. Could anyone carry that much gold? Would it fit in a suitcase?

SOLUTION

IDENTIFY, SET UP, and EXECUTE: Gold sells for around $400 an ounce. (The price has varied between $200 and $1000 over the past decade or so.) An ounce is about 30 grams; that's worth remembering. So ten dollars' worth of gold has a mass of $\frac{1}{40}$ ounce, or around one gram. A billion (10^9) dollars' worth of gold

is a hundred million (10^8) grams, or a hundred thousand (10^5) kilograms. This corresponds to a weight in British units of around 200,000 lb, or 100 tons. No human hero could lift that weight!

Roughly what is the *volume* of this gold? The density of gold is much greater than that of water (1 g/cm^3), or 1000 kg/m^3; if its density is 10 times that of water, this much gold will have a volume of 10 m^3, many times the volume of a suitcase.

EVALUATE: Clearly your novel needs rewriting. Try the calculation again with a suitcase full of five-carat (1-gram) diamonds, each worth $100,000. Would this work?

Application **Scalar Temperature, Vector Wind**
This weather station measures temperature, a scalar quantity that can be positive or negative (say, +20°C or −5°C) but has no direction. It also measures wind velocity, which is a vector quantity with both magnitude and direction (for example, 15 km/h from the west).

Test Your Understanding of Section 1.6   Can you estimate the total number of teeth in all the mouths of everyone (students, staff, and faculty) on your campus? (*Hint:* How many teeth are in your mouth? Count them!)

1.7 Vectors and Vector Addition

Some physical quantities, such as time, temperature, mass, and density, can be described completely by a single number with a unit. But many other important quantities in physics have a *direction* associated with them and cannot be described by a single number. A simple example is describing the motion of an airplane: We must say not only how fast the plane is moving but also in what direction. The speed of the airplane combined with its direction of motion together constitute a quantity called *velocity*. Another example is *force*, which in physics means a push or pull exerted on a body. Giving a complete description of a force means describing both how hard the force pushes or pulls on the body and the direction of the push or pull.

When a physical quantity is described by a single number, we call it a **scalar quantity.** In contrast, a **vector quantity** has both a **magnitude** (the "how much" or "how big" part) and a direction in space. Calculations that combine scalar quantities use the operations of ordinary arithmetic. For example, 6 kg + 3 kg = 9 kg, or 4 × 2 s = 8 s. However, combining vectors requires a different set of operations.

To understand more about vectors and how they combine, we start with the simplest vector quantity, **displacement.** Displacement is simply a change in the position of an object. Displacement is a vector quantity because we must state not only how far the object moves but also in what direction. Walking 3 km north from your front door doesn't get you to the same place as walking 3 km southeast; these two displacements have the same magnitude but different directions.

We usually represent a vector quantity such as displacement by a single letter, such as $\vec{A}$ in Fig. 1.9a. In this book we always print vector symbols in ***boldface italic type with an arrow above them.*** We do this to remind you that vector quantities have different properties from scalar quantities; the arrow is a reminder that vectors have direction. When you handwrite a symbol for a vector, *always* write it with an arrow on top. If you don't distinguish between scalar and vector quantities in your notation, you probably won't make the distinction in your thinking either, and hopeless confusion will result.

We always *draw* a vector as a line with an arrowhead at its tip. The length of the line shows the vector's magnitude, and the direction of the line shows the vector's direction. Displacement is always a straight-line segment directed from the starting point to the ending point, even though the object's actual path may be curved (Fig. 1.9b). Note that displacement is not related directly to the total *distance* traveled. If the object were to continue on past P_2 and then return to P_1, the displacement for the entire trip would be *zero* (Fig. 1.9c).

If two vectors have the same direction, they are **parallel.** If they have the same magnitude *and* the same direction, they are *equal,* no matter where they are located in space. The vector $\vec{A}'$ from point P_3 to point P_4 in Fig. 1.10 has the same length and direction as the vector $\vec{A}$ from P_1 to P_2. These two displacements are equal, even though they start at different points. We write this as $\vec{A}' = \vec{A}$ in Fig. 1.10; the boldface equals sign emphasizes that equality of two vector quantities is not the same relationship as equality of two scalar quantities. Two vector quantities are equal only when they have the same magnitude *and* the same direction.

The vector $\vec{B}$ in Fig. 1.10, however, is not equal to $\vec{A}$ because its direction is *opposite* to that of $\vec{A}$. We define the **negative of a vector** as a vector having the same magnitude as the original vector but the *opposite* direction. The negative of vector quantity $\vec{A}$ is denoted as $-\vec{A}$, and we use a boldface minus sign to emphasize the vector nature of the quantities. If $\vec{A}$ is 87 m south, then $-\vec{A}$ is 87 m north. Thus we can write the relationship between $\vec{A}$ and $\vec{B}$ in Fig. 1.10 as $\vec{A} = -\vec{B}$ or $\vec{B} = -\vec{A}$. When two vectors $\vec{A}$ and $\vec{B}$ have opposite directions, whether their magnitudes are the same or not, we say that they are **antiparallel.**

We usually represent the *magnitude* of a vector quantity (in the case of a displacement vector, its length) by the same letter used for the vector, but in *light italic type* with *no* arrow on top. An alternative notation is the vector symbol with vertical bars on both sides:

$$\text{(Magnitude of } \vec{A}) = A = |\vec{A}| \tag{1.1}$$

The magnitude of a vector quantity is a scalar quantity (a number) and is *always positive.* Note that a vector can never be equal to a scalar because they are different kinds of quantities. The expression "$\vec{A} = 6\text{ m}$" is just as wrong as "2 oranges = 3 apples"!

When drawing diagrams with vectors, it's best to use a scale similar to those used for maps. For example, a displacement of 5 km might be represented in a diagram by a vector 1 cm long, and a displacement of 10 km by a vector 2 cm long. In a diagram for velocity vectors, a vector that is 1 cm long might represent

1.9 Displacement as a vector quantity. A displacement is always a straight-line segment directed from the starting point to the ending point, even if the path is curved.

(a) We represent a displacement by an arrow pointing in the direction of displacement.

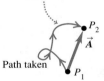

Ending position: P_2

Displacement $\vec{A}$

Starting position: P_1

Handwritten notation: $\vec{A}$

(b) Displacement depends only on the starting and ending positions—not on the path taken.

P_2

$\vec{A}$

Path taken

P_1

(c) Total displacement for a round trip is 0, regardless of the distance traveled.

P_1

1.10 The meaning of vectors that have the same magnitude and the same or opposite direction.

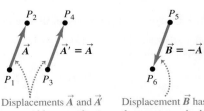

P_2 P_4 P_5

$\vec{A}$ $\vec{A}' = \vec{A}$ $\vec{B} = -\vec{A}$

P_1 P_3 P_6

Displacements $\vec{A}$ and $\vec{A}'$ are equal because they have the same length and direction.

Displacement $\vec{B}$ has the same magnitude as $\vec{A}$ but opposite direction; $\vec{B}$ is the negative of $\vec{A}$.

1.11 Three ways to add two vectors. As shown in (b), the order in vector addition doesn't matter; vector addition is commutative.

(a) We can add two vectors by placing them head to tail.

(b) Adding them in reverse order gives the same result.

(c) We can also add them by constructing a parallelogram.

a velocity of magnitude 5 m/s. A velocity of 20 m/s would then be represented by a vector 4 cm long.

Vector Addition and Subtraction

Suppose a particle undergoes a displacement $\vec{A}$ followed by a second displacement $\vec{B}$. The final result is the same as if the particle had started at the same initial point and undergone a single displacement $\vec{C}$ (Fig. 1.11a). We call displacement $\vec{C}$ the **vector sum,** or **resultant,** of displacements $\vec{A}$ and $\vec{B}$. We express this relationship symbolically as

$$\vec{C} = \vec{A} + \vec{B} \tag{1.2}$$

The boldface plus sign emphasizes that adding two vector quantities requires a geometrical process and is not the same operation as adding two scalar quantities such as $2 + 3 = 5$. In vector addition we usually place the *tail* of the *second* vector at the *head,* or tip, of the *first* vector (Fig. 1.11a).

If we make the displacements $\vec{A}$ and $\vec{B}$ in reverse order, with $\vec{B}$ first and $\vec{A}$ second, the result is the same (Fig. 1.11b). Thus

$$\vec{C} = \vec{B} + \vec{A} \quad \text{and} \quad \vec{A} + \vec{B} = \vec{B} + \vec{A} \tag{1.3}$$

This shows that the order of terms in a vector sum doesn't matter. In other words, vector addition obeys the commutative law.

Figure 1.11c shows another way to represent the vector sum: If vectors $\vec{A}$ and $\vec{B}$ are both drawn with their tails at the same point, vector $\vec{C}$ is the diagonal of a parallelogram constructed with $\vec{A}$ and $\vec{B}$ as two adjacent sides.

> **CAUTION** **Magnitudes in vector addition** It's a common error to conclude that if $\vec{C} = \vec{A} + \vec{B}$, then the magnitude C should equal the magnitude A plus the magnitude B. In general, this conclusion is *wrong;* for the vectors shown in Fig. 1.11, you can see that $C < A + B$. The magnitude of $\vec{A} + \vec{B}$ depends on the magnitudes of $\vec{A}$ and $\vec{B}$ *and* on the angle between $\vec{A}$ and $\vec{B}$ (see Problem 1.90). Only in the special case in which $\vec{A}$ and $\vec{B}$ are *parallel* is the magnitude of $\vec{C} = \vec{A} + \vec{B}$ equal to the sum of the magnitudes of $\vec{A}$ and $\vec{B}$ (Fig. 1.12a). When the vectors are *antiparallel* (Fig. 1.12b), the magnitude of $\vec{C}$ equals the *difference* of the magnitudes of $\vec{A}$ and $\vec{B}$. Be careful about distinguishing between scalar and vector quantities, and you'll avoid making errors about the magnitude of a vector sum.

1.12 (a) Only when two vectors $\vec{A}$ and $\vec{B}$ are parallel does the magnitude of their sum equal the sum of their magnitudes: $C = A + B$. (b) When $\vec{A}$ and $\vec{B}$ are antiparallel, the magnitude of their sum equals the *difference* of their magnitudes: $C = |A - B|$.

(a) The sum of two parallel vectors

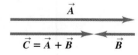

(b) The sum of two antiparallel vectors

When we need to add more than two vectors, we may first find the vector sum of any two, add this vectorially to the third, and so on. Figure 1.13a shows three vectors $\vec{A}$, $\vec{B}$, and $\vec{C}$. In Fig. 1.13b we first add $\vec{A}$ and $\vec{B}$ to give a vector sum $\vec{D}$; we then add vectors $\vec{C}$ and $\vec{D}$ by the same process to obtain the vector sum $\vec{R}$:

$$\vec{R} = (\vec{A} + \vec{B}) + \vec{C} = \vec{D} + \vec{C}$$

1.13 Several constructions for finding the vector sum $\vec{A} + \vec{B} + \vec{C}$.

(a) To find the sum of these three vectors ...

(b) we could add $\vec{A}$ and $\vec{B}$ to get $\vec{D}$ and then add $\vec{C}$ to $\vec{D}$ to get the final sum (resultant) $\vec{R}$, ...

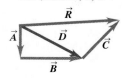

(c) or we could add $\vec{B}$ and $\vec{C}$ to get $\vec{E}$ and then add $\vec{A}$ to $\vec{E}$ to get $\vec{R}$, ...

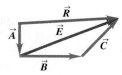

(d) or we could add $\vec{A}$, $\vec{B}$, and $\vec{C}$ to get $\vec{R}$ directly, ...

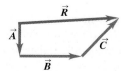

(e) or we could add $\vec{A}$, $\vec{B}$, and $\vec{C}$ in any other order and still get $\vec{R}$.

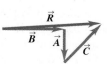

1.14 To construct the vector difference $\vec{A} - \vec{B}$, you can either place the tail of $-\vec{B}$ at the head of $\vec{A}$ or place the two vectors $\vec{A}$ and $\vec{B}$ head to head.

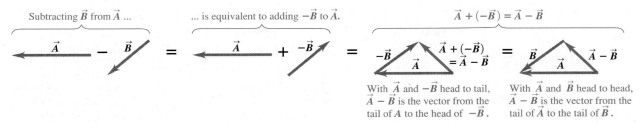

Subtracting $\vec{B}$ from $\vec{A}$ ... ... is equivalent to adding $-\vec{B}$ to $\vec{A}$. $\vec{A} + (-\vec{B}) = \vec{A} - \vec{B}$

With $\vec{A}$ and $-\vec{B}$ head to tail, $\vec{A} - \vec{B}$ is the vector from the tail of $\vec{A}$ to the head of $-\vec{B}$.

With $\vec{A}$ and $\vec{B}$ head to head, $\vec{A} - \vec{B}$ is the vector from the tail of $\vec{A}$ to the tail of $\vec{B}$.

Alternatively, we can first add $\vec{B}$ and $\vec{C}$ to obtain vector $\vec{E}$ (Fig. 1.13c), and then add $\vec{A}$ and $\vec{E}$ to obtain $\vec{R}$:

$$\vec{R} = \vec{A} + (\vec{B} + \vec{C}) = \vec{A} + \vec{E}$$

We don't even need to draw vectors $\vec{D}$ and $\vec{E}$; all we need to do is draw $\vec{A}$, $\vec{B}$, and $\vec{C}$ in succession, with the tail of each at the head of the one preceding it. The sum vector $\vec{R}$ extends from the tail of the first vector to the head of the last vector (Fig. 1.13d). The order makes no difference; Fig. 1.13e shows a different order, and we invite you to try others. We see that vector addition obeys the associative law.

We can *subtract* vectors as well as add them. To see how, recall that vector $-\vec{A}$ has the same magnitude as $\vec{A}$ but the opposite direction. We define the difference $\vec{A} - \vec{B}$ of two vectors $\vec{A}$ and $\vec{B}$ to be the vector sum of $\vec{A}$ and $-\vec{B}$:

$$\vec{A} - \vec{B} = \vec{A} + (-\vec{B}) \tag{1.4}$$

Figure 1.14 shows an example of vector subtraction.

A vector quantity such as a displacement can be multiplied by a scalar quantity (an ordinary number). The displacement $2\vec{A}$ is a displacement (vector quantity) in the same direction as the vector $\vec{A}$ but twice as long; this is the same as adding $\vec{A}$ to itself (Fig. 1.15a). In general, when a vector $\vec{A}$ is multiplied by a scalar c, the result $c\vec{A}$ has magnitude $|c|A$ (the absolute value of c multiplied by the magnitude of the vector $\vec{A}$). If c is positive, $c\vec{A}$ is in the same direction as $\vec{A}$; if c is negative, $c\vec{A}$ is in the direction opposite to $\vec{A}$. Thus $3\vec{A}$ is parallel to $\vec{A}$, while $-3\vec{A}$ is antiparallel to $\vec{A}$ (Fig. 1.15b).

A scalar used to multiply a vector may also be a physical quantity. For example, you may be familiar with the relationship $\vec{F} = m\vec{a}$; the net force $\vec{F}$ (a vector quantity) that acts on a body is equal to the product of the body's mass m (a scalar quantity) and its acceleration $\vec{a}$ (a vector quantity). The direction of $\vec{F}$ is the same as that of $\vec{a}$ because m is positive, and the magnitude of $\vec{F}$ is equal to the mass m (which is positive) multiplied by the magnitude of $\vec{a}$. The unit of force is the unit of mass multiplied by the unit of acceleration.

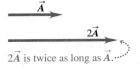

MasteringPHYSICS

PhET: Vector Addition

1.15 Multiplying a vector (a) by a positive scalar and (b) by a negative scalar.

(a) Multiplying a vector by a positive scalar changes the magnitude (length) of the vector, but not its direction.

$2\vec{A}$ is twice as long as $\vec{A}$.

(b) Multiplying a vector by a negative scalar changes its magnitude and reverses its direction.

$-3\vec{A}$ is three times as long as $\vec{A}$ and points in the opposite direction.

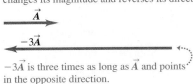

Example 1.5 **Addition of two vectors at right angles**

A cross-country skier skis 1.00 km north and then 2.00 km east on a horizontal snowfield. How far and in what direction is she from the starting point?

SOLUTION

IDENTIFY and SET UP: The problem involves combining two displacements at right angles to each other. In this case, vector addition amounts to solving a right triangle, which we can do using the Pythagorean theorem and simple trigonometry. The target variables are the skier's straight-line distance and direction from her starting point. Figure 1.16 is a scale diagram of the two displacements and the resultant net displacement. We denote the direction from the starting point by the angle ϕ (the Greek letter phi). The displacement appears to be about 2.4 km. Measurement with a protractor indicates that ϕ is about 63°.

EXECUTE: The distance from the starting point to the ending point is equal to the length of the hypotenuse:

$$\sqrt{(1.00 \text{ km})^2 + (2.00 \text{ km})^2} = 2.24 \text{ km}$$

Continued

1.16 The vector diagram, drawn to scale, for a ski trip.

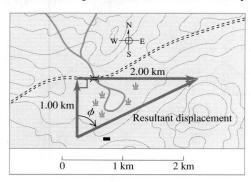

A little trigonometry (from Appendix B) allows us to find angle ϕ:

$$\tan \phi = \frac{\text{Opposite side}}{\text{Adjacent side}} = \frac{2.00 \text{ km}}{1.00 \text{ km}}$$

$$\phi = 63.4°$$

We can describe the direction as 63.4° east of north or $90° - 63.4° = 26.6°$ north of east.

EVALUATE: Our answers (2.24 km and $\phi = 63.4°$) are close to our predictions. In the more general case in which you have to add two vectors *not* at right angles to each other, you can use the law of cosines in place of the Pythagorean theorem and use the law of sines to find an angle corresponding to ϕ in this example. (You'll find these trigonometric rules in Appendix B.) We'll see more techniques for vector addition in Section 1.8.

Test Your Understanding of Section 1.7 Two displacement vectors, $\vec{S}$ and $\vec{T}$, have magnitudes $S = 3$ m and $T = 4$ m. Which of the following could be the magnitude of the difference vector $\vec{S} - \vec{T}$? (There may be more than one correct answer.) (i) 9 m; (ii) 7 m; (iii) 5 m; (iv) 1 m; (v) 0 m; (vi) −1 m.

1.8 Components of Vectors

In Section 1.7 we added vectors by using a scale diagram and by using properties of right triangles. Measuring a diagram offers only very limited accuracy, and calculations with right triangles work only when the two vectors are perpendicular. So we need a simple but general method for adding vectors. This is called the method of *components*.

To define what we mean by the components of a vector $\vec{A}$, we begin with a rectangular (Cartesian) coordinate system of axes (Fig. 1.17a). We then draw the vector with its tail at O, the origin of the coordinate system. We can represent any vector lying in the xy-plane as the sum of a vector parallel to the x-axis and a vector parallel to the y-axis. These two vectors are labeled $\vec{A}_x$ and $\vec{A}_y$ in Fig. 1.17a; they are called the **component vectors** of vector $\vec{A}$, and their vector sum is equal to $\vec{A}$. In symbols,

$$\vec{A} = \vec{A}_x + \vec{A}_y \tag{1.5}$$

1.17 Representing a vector $\vec{A}$ in terms of (a) component vectors $\vec{A}_x$ and $\vec{A}_y$ and (b) components A_x and A_y (which in this case are both positive).

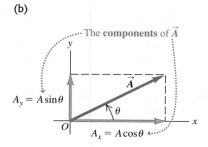

Since each component vector lies along a coordinate-axis direction, we need only a single number to describe each one. When $\vec{A}_x$ points in the positive x-direction, we define the number A_x to be equal to the magnitude of $\vec{A}_x$. When $\vec{A}_x$ points in the negative x-direction, we define the number A_x to be equal to the negative of that magnitude (the magnitude of a vector quantity is itself never negative). We define the number A_y in the same way. The two numbers A_x and A_y are called the **components** of $\vec{A}$ (Fig. 1.17b).

CAUTION **Components are not vectors** The components A_x and A_y of a vector $\vec{A}$ are just numbers; they are *not* vectors themselves. This is why we print the symbols for components in light italic type with *no* arrow on top instead of in boldface italic with an arrow, which is reserved for vectors.

We can calculate the components of the vector $\vec{A}$ if we know its magnitude A and its direction. We'll describe the direction of a vector by its angle relative to some reference direction. In Fig. 1.17b this reference direction is the positive x-axis, and the angle between vector $\vec{A}$ and the positive x-axis is θ (the Greek letter theta). Imagine that the vector $\vec{A}$ originally lies along the $+x$-axis and that you then rotate it to its correct direction, as indicated by the arrow in Fig. 1.17b on the angle θ. If this rotation is from the $+x$-axis toward the $+y$-axis, as shown in Fig. 1.17b, then θ is *positive;* if the rotation is from the $+x$-axis toward the $-y$-axis, θ is *negative.* Thus the $+y$-axis is at an angle of 90°, the $-x$-axis at 180°, and the $-y$-axis at 270° (or $-90°$). If θ is measured in this way, then from the definition of the trigonometric functions,

$$\frac{A_x}{A} = \cos\theta \qquad \text{and} \qquad \frac{A_y}{A} = \sin\theta$$

$$A_x = A\cos\theta \qquad \text{and} \qquad A_y = A\sin\theta \tag{1.6}$$

(θ measured from the $+x$-axis, rotating toward the $+y$-axis)

In Fig. 1.17b A_x and A_y are positive. This is consistent with Eqs. (1.6); θ is in the first quadrant (between 0° and 90°), and both the cosine and the sine of an angle in this quadrant are positive. But in Fig. 1.18a the component B_x is negative. Again, this agrees with Eqs. (1.6); the cosine of an angle in the second quadrant is negative. The component B_y is positive ($\sin\theta$ is positive in the second quadrant). In Fig. 1.18b both C_x and C_y are negative (both $\cos\theta$ and $\sin\theta$ are negative in the third quadrant).

CAUTION **Relating a vector's magnitude and direction to its components** Equations (1.6) are correct *only* when the angle θ is measured from the positive x-axis as described above. If the angle of the vector is given from a different reference direction or using a different sense of rotation, the relationships are different. Be careful! Example 1.6 illustrates this point. ▌

1.18 The components of a vector may be positive or negative numbers.

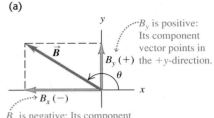

(a)

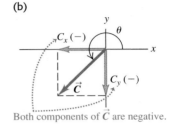

(b)

Both components of $\vec{C}$ are negative.

Example 1.6 **Finding components**

(a) What are the x- and y-components of vector $\vec{D}$ in Fig. 1.19a? The magnitude of the vector is $D = 3.00$ m, and the angle $\alpha = 45°$. (b) What are the x- and y-components of vector $\vec{E}$ in Fig. 1.19b? The magnitude of the vector is $E = 4.50$ m, and the angle $\beta = 37.0°$.

SOLUTION

IDENTIFY and SET UP: We can use Eqs. (1.6) to find the components of these vectors, but we have to be careful: Neither of the angles α or β in Fig. 1.19 is measured from the $+x$-axis toward the $+y$-axis. We estimate from the figure that the lengths of the com-

1.19 Calculating the x- and y-components of vectors.

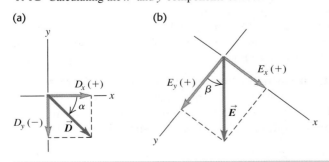

ponents in part (a) are both roughly 2 m, and that those in part (b) are 3 m and 4 m. We've indicated the signs of the components in the figure.

EXECUTE: (a) The angle α (the Greek letter alpha) between the positive x-axis and $\vec{D}$ is measured toward the *negative* y-axis. The angle we must use in Eqs. (1.6) is $\theta = -\alpha = -45°$. We then find

$$D_x = D\cos\theta = (3.00 \text{ m})(\cos(-45°)) = +2.1 \text{ m}$$

$$D_y = D\sin\theta = (3.00 \text{ m})(\sin(-45°)) = -2.1 \text{ m}$$

Had you been careless and substituted $+45°$ for θ in Eqs. (1.6), your result for D_y would have had the wrong sign.

(b) The x- and y-axes in Fig. 1.19b are at right angles, so it doesn't matter that they aren't horizontal and vertical, respectively. But to use Eqs. (1.6), we must use the angle $\theta = 90.0° - \beta = 90.0° - 37.0° = 53.0°$. Then we find

$$E_x = E\cos 53.0° = (4.50 \text{ m})(\cos 53.0°) = +2.71 \text{ m}$$

$$E_y = E\sin 53.0° = (4.50 \text{ m})(\sin 53.0°) = +3.59 \text{ m}$$

EVALUATE: Our answers to both parts are close to our predictions. But ask yourself this: Why do the answers in part (a) correctly have only two significant figures?

Doing Vector Calculations Using Components

Using components makes it relatively easy to do various calculations involving vectors. Let's look at three important examples.

1. Finding a vector's magnitude and direction from its components. We can describe a vector completely by giving either its magnitude and direction or its x- and y-components. Equations (1.6) show how to find the components if we know the magnitude and direction. We can also reverse the process: We can find the magnitude and direction if we know the components. By applying the Pythagorean theorem to Fig. 1.17b, we find that the magnitude of vector $\vec{A}$ is

$$A = \sqrt{A_x^2 + A_y^2} \qquad (1.7)$$

(We always take the positive root.) Equation (1.7) is valid for any choice of x-axis and y-axis, as long as they are mutually perpendicular. The expression for the vector direction comes from the definition of the tangent of an angle. If θ is measured from the positive x-axis, and a positive angle is measured toward the positive y-axis (as in Fig. 1.17b), then

$$\tan \theta = \frac{A_y}{A_x} \qquad \text{and} \qquad \theta = \arctan \frac{A_y}{A_x} \qquad (1.8)$$

We will always use the notation arctan for the inverse tangent function. The notation $\tan^{-1}$ is also commonly used, and your calculator may have an INV or 2ND button to be used with the TAN button.

CAUTION **Finding the direction of a vector from its components** There's one slight complication in using Eqs. (1.8) to find θ: Any two angles that differ by 180° have the same tangent. Suppose $A_x = 2$ m and $A_y = -2$ m as in Fig. 1.20; then $\tan \theta = -1$. But both 135° and 315° (or $-45°$) have tangents of -1. To decide which is correct, we have to look at the individual components. Because A_x is positive and A_y is negative, the angle must be in the fourth quadrant; thus $\theta = 315°$ (or $-45°$) is the correct value. Most pocket calculators give $\arctan(-1) = -45°$. In this case that is correct; but if instead we have $A_x = -2$ m and $A_y = 2$ m, then the correct angle is 135°. Similarly, when A_x and A_y are both negative, the tangent is positive, but the angle is in the third quadrant. You should *always* draw a sketch like Fig. 1.20 to check which of the two possibilities is the correct one.

1.20 Drawing a sketch of a vector reveals the signs of its x- and y-components.

Suppose that $\tan \theta = \dfrac{A_y}{A_x} = -1$. What is θ?

Two angles have tangents of -1: 135° and 315°. Inspection of the diagram shows that θ must be 315°.

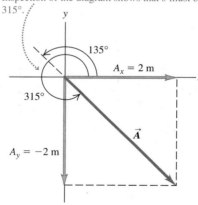

2. Multiplying a vector by a scalar. If we multiply a vector $\vec{A}$ by a scalar c, each component of the product $\vec{D} = c\vec{A}$ is the product of c and the corresponding component of $\vec{A}$:

$$D_x = cA_x \qquad D_y = cA_y \qquad \text{(components of } \vec{D} = c\vec{A}) \qquad (1.9)$$

For example, Eq. (1.9) says that each component of the vector $2\vec{A}$ is twice as great as the corresponding component of the vector $\vec{A}$, so $2\vec{A}$ is in the same direction as $\vec{A}$ but has twice the magnitude. Each component of the vector $-3\vec{A}$ is three times as great as the corresponding component of the vector $\vec{A}$ but has the opposite sign, so $-3\vec{A}$ is in the opposite direction from $\vec{A}$ and has three times the magnitude. Hence Eqs. (1.9) are consistent with our discussion in Section 1.7 of multiplying a vector by a scalar (see Fig. 1.15).

3. Using components to calculate the vector sum (resultant) of two or more vectors. Figure 1.21 shows two vectors $\vec{A}$ and $\vec{B}$ and their vector sum $\vec{R}$, along with the x- and y-components of all three vectors. You can see from the diagram that the x-component R_x of the vector sum is simply the sum $(A_x + B_x)$

of the x-components of the vectors being added. The same is true for the y-components. In symbols,

$$R_x = A_x + B_x \qquad R_y = A_y + B_y \qquad \text{(components of } \vec{R} = \vec{A} + \vec{B}) \qquad \text{(1.10)}$$

Figure 1.21 shows this result for the case in which the components A_x, A_y, B_x, and B_y are all positive. You should draw additional diagrams to verify for yourself that Eqs. (1.10) are valid for *any* signs of the components of $\vec{A}$ and $\vec{B}$.

If we know the components of any two vectors $\vec{A}$ and $\vec{B}$, perhaps by using Eqs. (1.6), we can compute the components of the vector sum $\vec{R}$. Then if we need the magnitude and direction of $\vec{R}$, we can obtain them from Eqs. (1.7) and (1.8) with the A's replaced by R's.

We can extend this procedure to find the sum of any number of vectors. If $\vec{R}$ is the vector sum of $\vec{A}, \vec{B}, \vec{C}, \vec{D}, \vec{E}, \ldots$, the components of $\vec{R}$ are

$$R_x = A_x + B_x + C_x + D_x + E_x + \cdots$$
$$R_y = A_y + B_y + C_y + D_y + E_y + \cdots \qquad \text{(1.11)}$$

We have talked only about vectors that lie in the xy-plane, but the component method works just as well for vectors having any direction in space. We can introduce a z-axis perpendicular to the xy-plane; then in general a vector $\vec{A}$ has components A_x, A_y, and A_z in the three coordinate directions. Its magnitude A is

$$A = \sqrt{A_x^2 + A_y^2 + A_z^2} \qquad \text{(1.12)}$$

Again, we always take the positive root. Also, Eqs. (1.11) for the components of the vector sum $\vec{R}$ have an additional member:

$$R_z = A_z + B_z + C_z + D_z + E_z + \cdots$$

We've focused on adding *displacement* vectors, but the method is applicable to all vector quantities. When we study the concept of force in Chapter 4, we'll find that forces are vectors that obey the same rules of vector addition that we've used with displacement.

1.21 Finding the vector sum (resultant) of $\vec{A}$ and $\vec{B}$ using components.

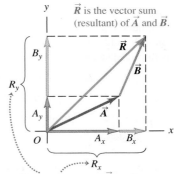

The components of $\vec{R}$ are the sums of the components of $\vec{A}$ and $\vec{B}$:

$$R_y = A_y + B_y \qquad R_x = A_x + B_x$$

Problem-Solving Strategy 1.3 | Vector Addition

IDENTIFY *the relevant concepts:* Decide what the target variable is. It may be the magnitude of the vector sum, the direction, or both.

SET UP *the problem:* Sketch the vectors being added, along with suitable coordinate axes. Place the tail of the first vector at the origin of the coordinates, place the tail of the second vector at the head of the first vector, and so on. Draw the vector sum $\vec{R}$ from the tail of the first vector (at the origin) to the head of the last vector. Use your sketch to estimate the magnitude and direction of $\vec{R}$. Select the mathematical tools you'll use for the full calculation: Eqs. (1.6) to obtain the components of the vectors given, if necessary, Eqs. (1.11) to obtain the components of the vector sum, Eq. (1.12) to obtain its magnitude, and Eqs. (1.8) to obtain its direction.

EXECUTE *the solution* as follows:
1. Find the x- and y-components of each individual vector and record your results in a table, as in Example 1.7 below. If a vector is described by a magnitude A and an angle θ, measured

from the $+x$-axis toward the $+y$-axis, then its components are given by Eqs. 1.6:

$$A_x = A \cos \theta \qquad A_y = A \sin \theta$$

If the angles of the vectors are given in some other way, perhaps using a different reference direction, convert them to angles measured from the $+x$-axis as in Example 1.6 above.
2. Add the individual x-components algebraically (including signs) to find R_x, the x-component of the vector sum. Do the same for the y-components to find R_y. See Example 1.7 below.
3. Calculate the magnitude R and direction θ of the vector sum using Eqs. (1.7) and (1.8):

$$R = \sqrt{R_x^2 + R_y^2} \qquad \theta = \arctan \frac{R_y}{R_x}$$

EVALUATE *your answer:* Confirm that your results for the magnitude and direction of the vector sum agree with the estimates you made from your sketch. The value of θ that you find with a calculator may be off by 180°; your drawing will indicate the correct value.

Example 1.7 Adding vectors using their components

Three players on a reality TV show are brought to the center of a large, flat field. Each is given a meter stick, a compass, a calculator, a shovel, and (in a different order for each contestant) the following three displacements:

$$\vec{A}: 72.4 \text{ m}, 32.0° \text{ east of north}$$

$$\vec{B}: 57.3 \text{ m}, 36.0° \text{ south of west}$$

$$\vec{C}: 17.8 \text{ m due south}$$

The three displacements lead to the point in the field where the keys to a new Porsche are buried. Two players start measuring immediately, but the winner first *calculates* where to go. What does she calculate?

SOLUTION

IDENTIFY and SET UP: The goal is to find the sum (resultant) of the three displacements, so this is a problem in vector addition. Figure 1.22 shows the situation. We have chosen the +x-axis as

1.22 Three successive displacements $\vec{A}$, $\vec{B}$, and $\vec{C}$ and the resultant (vector sum) displacement $\vec{R} = \vec{A} + \vec{B} + \vec{C}$.

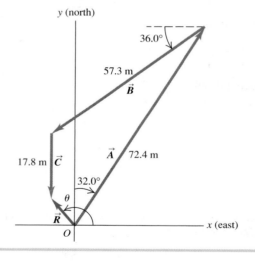

east and the +y-axis as north. We estimate from the diagram that the vector sum $\vec{R}$ is about 10 m, 40° west of north (which corresponds to $\theta \approx 130°$).

EXECUTE: The angles of the vectors, measured from the +x-axis toward the +y-axis, are $(90.0° - 32.0°) = 58.0°$, $(180.0° + 36.0°) = 216.0°$, and $270.0°$, respectively. We may now use Eqs. (1.6) to find the components of $\vec{A}$:

$$A_x = A \cos \theta_A = (72.4 \text{ m})(\cos 58.0°) = 38.37 \text{ m}$$

$$A_y = A \sin \theta_A = (72.4 \text{ m})(\sin 58.0°) = 61.40 \text{ m}$$

We've kept an extra significant figure in the components; we'll round to the correct number of significant figures at the end of our calculation. The table below shows the components of all the displacements, the addition of the components, and the other calculations.

Distance	Angle	x-component	y-component
$A = 72.4$ m	58.0°	38.37 m	61.40 m
$B = 57.3$ m	216.0°	−46.36 m	−33.68 m
$C = 17.8$ m	270.0°	0.00 m	−17.80 m
		$R_x = -7.99$ m	$R_y = 9.92$ m

$$R = \sqrt{(-7.99 \text{ m})^2 + (9.92 \text{ m})^2} = 12.7 \text{ m}$$

$$\theta = \arctan \frac{9.92 \text{ m}}{-7.99 \text{ m}} = -51°$$

Comparing to Fig. 1.22 shows that the calculated angle is clearly off by 180°. The correct value is $\theta = 180° - 51° = 129°$, or 39° west of north.

EVALUATE: Our calculated answers for R and θ agree with our estimates. Notice how drawing the diagram in Fig. 1.22 made it easy to avoid a 180° error in the direction of the vector sum.

Example 1.8 A simple vector addition in three dimensions

After an airplane takes off, it travels 10.4 km west, 8.7 km north, and 2.1 km up. How far is it from the takeoff point?

SOLUTION

Let the +x-axis be east, the +y-axis north, and the +z-axis up. Then the components of the airplane's displacement are $A_x = -10.4$ km, $A_y = 8.7$ km, and $A_z = 2.1$ km. From Eq. (1.12), the magnitude of the displacement is

$$A = \sqrt{(-10.4 \text{ km})^2 + (8.7 \text{ km})^2 + (2.1 \text{ km})^2} = 13.7 \text{ km}$$

Test Your Understanding of Section 1.8 Two vectors $\vec{A}$ and $\vec{B}$ both lie in the xy-plane. (a) Is it possible for $\vec{A}$ to have the same magnitude as $\vec{B}$ but different components? (b) Is it possible for $\vec{A}$ to have the same components as $\vec{B}$ but a different magnitude? ❙

1.9 Unit Vectors

A **unit vector** is a vector that has a magnitude of 1, with no units. Its only purpose is to *point*—that is, to describe a direction in space. Unit vectors provide a convenient notation for many expressions involving components of vectors. We will always include a caret or "hat" (^) in the symbol for a unit vector to distinguish it from ordinary vectors whose magnitude may or may not be equal to 1.

In an x-y coordinate system we can define a unit vector $\hat{\imath}$ that points in the direction of the positive x-axis and a unit vector $\hat{\jmath}$ that points in the direction of the positive y-axis (Fig. 1.23a). Then we can express the relationship between component vectors and components, described at the beginning of Section 1.8, as follows:

$$\vec{A}_x = A_x\hat{\imath}$$
$$\vec{A}_y = A_y\hat{\jmath} \qquad (1.13)$$

Similarly, we can write a vector $\vec{A}$ in terms of its components as

$$\vec{A} = A_x\hat{\imath} + A_y\hat{\jmath} \qquad (1.14)$$

Equations (1.13) and (1.14) are vector equations; each term, such as $A_x\hat{\imath}$, is a vector quantity (Fig. 1.23b).

Using unit vectors, we can express the vector sum $\vec{R}$ of two vectors $\vec{A}$ and $\vec{B}$ as follows:

$$\vec{A} = A_x\hat{\imath} + A_y\hat{\jmath}$$
$$\vec{B} = B_x\hat{\imath} + B_y\hat{\jmath}$$
$$\vec{R} = \vec{A} + \vec{B}$$
$$= (A_x\hat{\imath} + A_y\hat{\jmath}) + (B_x\hat{\imath} + B_y\hat{\jmath}) \qquad (1.15)$$
$$= (A_x + B_x)\hat{\imath} + (A_y + B_y)\hat{\jmath}$$
$$= R_x\hat{\imath} + R_y\hat{\jmath}$$

Equation (1.15) restates the content of Eqs. (1.10) in the form of a single vector equation rather than two component equations.

If the vectors do not all lie in the xy-plane, then we need a third component. We introduce a third unit vector $\hat{k}$ that points in the direction of the positive z-axis (Fig. 1.24). Then Eqs. (1.14) and (1.15) become

$$\vec{A} = A_x\hat{\imath} + A_y\hat{\jmath} + A_z\hat{k}$$
$$\vec{B} = B_x\hat{\imath} + B_y\hat{\jmath} + B_z\hat{k} \qquad (1.16)$$

$$\vec{R} = (A_x + B_x)\hat{\imath} + (A_y + B_y)\hat{\jmath} + (A_z + B_z)\hat{k}$$
$$= R_x\hat{\imath} + R_y\hat{\jmath} + R_z\hat{k} \qquad (1.17)$$

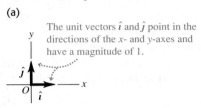

1.23 (a) The unit vectors $\hat{\imath}$ and $\hat{\jmath}$. (b) Expressing a vector $\vec{A}$ in terms of its components.

(a)

The unit vectors $\hat{\imath}$ and $\hat{\jmath}$ point in the directions of the x- and y-axes and have a magnitude of 1.

(b)

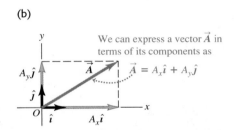

We can express a vector $\vec{A}$ in terms of its components as
$$\vec{A} = A_x\hat{\imath} + A_y\hat{\jmath}$$

1.24 The unit vectors $\hat{\imath}, \hat{\jmath}$, and $\hat{k}$.

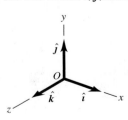

Example 1.9 | **Using unit vectors**

Given the two displacements

$$\vec{D} = (6.00\hat{\imath} + 3.00\hat{\jmath} - 1.00\hat{k})\,\text{m} \quad \text{and}$$

$$\vec{E} = (4.00\hat{\imath} - 5.00\hat{\jmath} + 8.00\hat{k})\,\text{m}$$

find the magnitude of the displacement $2\vec{D} - \vec{E}$.

SOLUTION

IDENTIFY and SET UP: We are to multiply the vector $\vec{D}$ by 2 (a scalar) and subtract the vector $\vec{E}$ from the result, so as to obtain the vector $\vec{F} = 2\vec{D} - \vec{E}$. Equation (1.9) says that to multiply $\vec{D}$ by 2, we multiply each of its components by 2. We can use Eq. (1.17) to do the subtraction; recall from Section 1.7 that subtracting a vector is the same as adding the negative of that vector.

EXECUTE: We have

$$\vec{F} = 2(6.00\hat{\imath} + 3.00\hat{\jmath} - 1.00\hat{k})\,\text{m} - (4.00\hat{\imath} - 5.00\hat{\jmath} + 8.00\hat{k})\,\text{m}$$

$$= [(12.00 - 4.00)\hat{\imath} + (6.00 + 5.00)\hat{\jmath} + (-2.00 - 8.00)\hat{k}]\,\text{m}$$

$$= (8.00\hat{\imath} + 11.00\hat{\jmath} - 10.00\hat{k})\,\text{m}$$

From Eq. (1.12) the magnitude of $\vec{F}$ is

$$F = \sqrt{F_x^2 + F_y^2 + F_z^2}$$

$$= \sqrt{(8.00\,\text{m})^2 + (11.00\,\text{m})^2 + (-10.00\,\text{m})^2}$$

$$= 16.9\,\text{m}$$

EVALUATE: Our answer is of the same order of magnitude as the larger components that appear in the sum. We wouldn't expect our answer to be much larger than this, but it could be much smaller.

Test Your Understanding of Section 1.9   Arrange the following vectors in order of their magnitude, with the vector of largest magnitude first. (i) $\vec{A} = (3\hat{\imath} + 5\hat{\jmath} - 2\hat{k})$ m; (ii) $\vec{B} = (-3\hat{\imath} + 5\hat{\jmath} - 2\hat{k})$ m; (iii) $\vec{C} = (3\hat{\imath} - 5\hat{\jmath} - 2\hat{k})$ m; (iv) $\vec{D} = (3\hat{\imath} + 5\hat{\jmath} + 2\hat{k})$ m.

1.10 Products of Vectors

Vector addition develops naturally from the problem of combining displacements and will prove useful for calculating many other vector quantities. We can also express many physical relationships by using *products* of vectors. Vectors are not ordinary numbers, so ordinary multiplication is not directly applicable to vectors. We will define two different kinds of products of vectors. The first, called the *scalar product,* yields a result that is a scalar quantity. The second, the *vector product,* yields another vector.

1.25 Calculating the scalar product of two vectors, $\vec{A} \cdot \vec{B} = AB\cos\phi$.

(a)

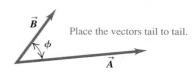

Place the vectors tail to tail.

(b) $\vec{A} \cdot \vec{B}$ equals $A(B\cos\phi)$.

(Magnitude of $\vec{A}$) times (Component of $\vec{B}$ in direction of $\vec{A}$)

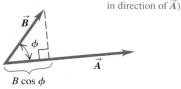

$B\cos\phi$

(c) $\vec{A} \cdot \vec{B}$ also equals $B(A\cos\phi)$

(Magnitude of $\vec{B}$) times (Component of $\vec{A}$ in direction of $\vec{B}$)

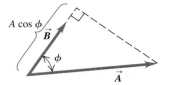

$A\cos\phi$

Scalar Product

The **scalar product** of two vectors $\vec{A}$ and $\vec{B}$ is denoted by $\vec{A} \cdot \vec{B}$. Because of this notation, the scalar product is also called the **dot product.** Although $\vec{A}$ and $\vec{B}$ are vectors, the quantity $\vec{A} \cdot \vec{B}$ is a scalar.

To define the scalar product $\vec{A} \cdot \vec{B}$ we draw the two vectors $\vec{A}$ and $\vec{B}$ with their tails at the same point (Fig. 1.25a). The angle ϕ (the Greek letter phi) between their directions ranges from $0°$ to $180°$. Figure 1.25b shows the projection of the vector $\vec{B}$ onto the direction of $\vec{A}$; this projection is the component of $\vec{B}$ in the direction of $\vec{A}$ and is equal to $B\cos\phi$. (We can take components along any direction that's convenient, not just the x- and y-axes.) We define $\vec{A} \cdot \vec{B}$ to be the magnitude of $\vec{A}$ multiplied by the component of $\vec{B}$ in the direction of $\vec{A}$. Expressed as an equation,

$$\vec{A} \cdot \vec{B} = AB\cos\phi = |\vec{A}||\vec{B}|\cos\phi \quad \begin{array}{l}\text{(definition of the scalar} \\ \text{(dot) product)}\end{array} \quad (1.18)$$

Alternatively, we can define $\vec{A} \cdot \vec{B}$ to be the magnitude of $\vec{B}$ multiplied by the component of $\vec{A}$ in the direction of $\vec{B}$, as in Fig. 1.25c. Hence $\vec{A} \cdot \vec{B} = B(A\cos\phi) = AB\cos\phi$, which is the same as Eq. (1.18).

The scalar product is a scalar quantity, not a vector, and it may be positive, negative, or zero. When ϕ is between $0°$ and $90°$, $\cos\phi > 0$ and the scalar product is positive (Fig. 1.26a). When ϕ is between 90° and 180° so that $\cos\phi < 0$, the component of $\vec{B}$ in the direction of $\vec{A}$ is negative, and $\vec{A} \cdot \vec{B}$ is negative (Fig. 1.26b). Finally, when $\phi = 90°$, $\vec{A} \cdot \vec{B} = 0$ (Fig. 1.26c). *The scalar product of two perpendicular vectors is always zero.*

For any two vectors $\vec{A}$ and $\vec{B}$, $AB\cos\phi = BA\cos\phi$. This means that $\vec{A} \cdot \vec{B} = \vec{B} \cdot \vec{A}$. The scalar product obeys the commutative law of multiplication; the order of the two vectors does not matter.

We will use the scalar product in Chapter 6 to describe work done by a force. When a constant force $\vec{F}$ is applied to a body that undergoes a displacement $\vec{s}$, the work W (a scalar quantity) done by the force is given by

$$W = \vec{F} \cdot \vec{s}$$

The work done by the force is positive if the angle between $\vec{F}$ and $\vec{s}$ is between 0° and 90°, negative if this angle is between 90° and 180°, and zero if $\vec{F}$ and $\vec{s}$ are perpendicular. (This is another example of a term that has a special meaning in physics; in everyday language, "work" isn't something that can be positive or negative.) In later chapters we'll use the scalar product for a variety of purposes, from calculating electric potential to determining the effects that varying magnetic fields have on electric circuits.

Calculating the Scalar Product Using Components

We can calculate the scalar product $\vec{A} \cdot \vec{B}$ directly if we know the x-, y-, and z-components of $\vec{A}$ and $\vec{B}$. To see how this is done, let's first work out the scalar products of the unit vectors. This is easy, since $\hat{\imath}$, $\hat{\jmath}$, and $\hat{k}$ all have magnitude 1 and are perpendicular to each other. Using Eq. (1.18), we find

$$\hat{\imath} \cdot \hat{\imath} = \hat{\jmath} \cdot \hat{\jmath} = \hat{k} \cdot \hat{k} = (1)(1)\cos 0° = 1$$
$$\hat{\imath} \cdot \hat{\jmath} = \hat{\imath} \cdot \hat{k} = \hat{\jmath} \cdot \hat{k} = (1)(1)\cos 90° = 0 \tag{1.19}$$

Now we express $\vec{A}$ and $\vec{B}$ in terms of their components, expand the product, and use these products of unit vectors:

$$\begin{aligned} \vec{A} \cdot \vec{B} = {} & (A_x\hat{\imath} + A_y\hat{\jmath} + A_z\hat{k}) \cdot (B_x\hat{\imath} + B_y\hat{\jmath} + B_z\hat{k}) \\ = {} & A_x\hat{\imath} \cdot B_x\hat{\imath} + A_x\hat{\imath} \cdot B_y\hat{\jmath} + A_x\hat{\imath} \cdot B_z\hat{k} \\ & + A_y\hat{\jmath} \cdot B_x\hat{\imath} + A_y\hat{\jmath} \cdot B_y\hat{\jmath} + A_y\hat{\jmath} \cdot B_z\hat{k} \\ & + A_z\hat{k} \cdot B_x\hat{\imath} + A_z\hat{k} \cdot B_y\hat{\jmath} + A_z\hat{k} \cdot B_z\hat{k} \\ = {} & A_xB_x\hat{\imath} \cdot \hat{\imath} + A_xB_y\hat{\imath} \cdot \hat{\jmath} + A_xB_z\hat{\imath} \cdot \hat{k} \\ & + A_yB_x\hat{\jmath} \cdot \hat{\imath} + A_yB_y\hat{\jmath} \cdot \hat{\jmath} + A_yB_z\hat{\jmath} \cdot \hat{k} \\ & + A_zB_x\hat{k} \cdot \hat{\imath} + A_zB_y\hat{k} \cdot \hat{\jmath} + A_zB_z\hat{k} \cdot \hat{k} \end{aligned} \tag{1.20}$$

From Eqs. (1.19) we see that six of these nine terms are zero, and the three that survive give simply

$$\vec{A} \cdot \vec{B} = A_xB_x + A_yB_y + A_zB_z \qquad \begin{array}{l}\text{(scalar (dot) product in} \\ \text{terms of components)}\end{array} \tag{1.21}$$

Thus *the scalar product of two vectors is the sum of the products of their respective components.*

The scalar product gives a straightforward way to find the angle ϕ between any two vectors $\vec{A}$ and $\vec{B}$ whose components are known. In this case we can use Eq. (1.21) to find the scalar product of $\vec{A}$ and $\vec{B}$. Example 1.11 on the next page shows how to do this.

1.26 The scalar product $\vec{A} \cdot \vec{B} = AB\cos\phi$ can be positive, negative, or zero, depending on the angle between $\vec{A}$ and $\vec{B}$.

(a)

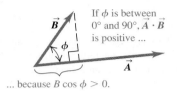

If ϕ is between 0° and 90°, $\vec{A} \cdot \vec{B}$ is positive ...

... because $B\cos\phi > 0$.

(b)

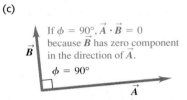

If ϕ is between 90° and 180°, $\vec{A} \cdot \vec{B}$ is negative ...

... because $B\cos\phi < 0$.

(c)

If $\phi = 90°$, $\vec{A} \cdot \vec{B} = 0$ because $\vec{B}$ has zero component in the direction of $\vec{A}$.

$\phi = 90°$

Example 1.10 Calculating a scalar product

Find the scalar product $\vec{A} \cdot \vec{B}$ of the two vectors in Fig. 1.27. The magnitudes of the vectors are $A = 4.00$ and $B = 5.00$.

SOLUTION

IDENTIFY and SET UP: We can calculate the scalar product in two ways: using the magnitudes of the vectors and the angle between them (Eq. 1.18), and using the components of the vectors (Eq. 1.21). We'll do it both ways, and the results will check each other.

1.27 Two vectors in two dimensions.

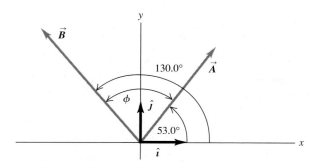

EXECUTE: The angle between the two vectors is $\phi = 130.0° - 53.0° = 77.0°$, so Eq. (1.18) gives us

$$\vec{A} \cdot \vec{B} = AB \cos \phi = (4.00)(5.00) \cos 77.0° = 4.50$$

To use Eq. (1.21), we must first find the components of the vectors. The angles of $\vec{A}$ and $\vec{B}$ are given with respect to the $+x$-axis and are measured in the sense from the $+x$-axis to the $+y$-axis, so we can use Eqs. (1.6):

$$A_x = (4.00) \cos 53.0° = 2.407$$
$$A_y = (4.00) \sin 53.0° = 3.195$$
$$B_x = (5.00) \cos 130.0° = -3.214$$
$$B_y = (5.00) \sin 130.0° = 3.830$$

As in Example 1.7, we keep an extra significant figure in the components and round at the end. Equation (1.21) now gives us

$$\vec{A} \cdot \vec{B} = A_x B_x + A_y B_y + A_z B_z$$
$$= (2.407)(-3.214) + (3.195)(3.830) + (0)(0) = 4.50$$

EVALUATE: Both methods give the same result, as they should.

Example 1.11 Finding an angle with the scalar product

Find the angle between the vectors

$$\vec{A} = 2.00\hat{i} + 3.00\hat{j} + 1.00\hat{k} \quad \text{and}$$
$$\vec{B} = -4.00\hat{i} + 2.00\hat{j} - 1.00\hat{k}$$

SOLUTION

IDENTIFY and SET UP: We're given the x-, y-, and z-components of two vectors. Our target variable is the angle ϕ between them (Fig. 1.28). To find this, we'll solve Eq. (1.18), $\vec{A} \cdot \vec{B} = AB \cos \phi$, for ϕ in terms of the scalar product $\vec{A} \cdot \vec{B}$ and the magnitudes A and B. We can evaluate the scalar product using Eq. (1.21),

1.28 Two vectors in three dimensions.

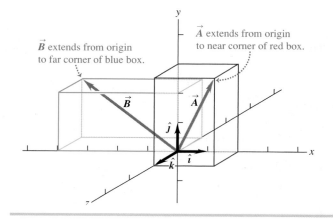

$\vec{B}$ extends from origin to far corner of blue box.

$\vec{A}$ extends from origin to near corner of red box.

$\vec{A} \cdot \vec{B} = A_x B_x + A_y B_y + A_z B_z$, and we can find A and B using Eq. (1.7).

EXECUTE: We solve Eq. (1.18) for $\cos \phi$ and write $\vec{A} \cdot \vec{B}$ using Eq. (1.21). Our result is

$$\cos \phi = \frac{\vec{A} \cdot \vec{B}}{AB} = \frac{A_x B_x + A_y B_y + A_z B_z}{AB}$$

We can use this formula to find the angle between *any* two vectors $\vec{A}$ and $\vec{B}$. Here we have $A_x = 2.00$, $A_y = 3.00$, and $A_z = 1.00$, and $B_x = -4.00$, $B_y = 2.00$, and $B_z = -1.00$. Thus

$$\vec{A} \cdot \vec{B} = A_x B_x + A_y B_y + A_z B_z$$
$$= (2.00)(-4.00) + (3.00)(2.00) + (1.00)(-1.00)$$
$$= -3.00$$

$$A = \sqrt{A_x^2 + A_y^2 + A_z^2} = \sqrt{(2.00)^2 + (3.00)^2 + (1.00)^2}$$
$$= \sqrt{14.00}$$

$$B = \sqrt{B_x^2 + B_y^2 + B_z^2} = \sqrt{(-4.00)^2 + (2.00)^2 + (-1.00)^2}$$
$$= \sqrt{21.00}$$

$$\cos \phi = \frac{A_x B_x + A_y B_y + A_z B_z}{AB} = \frac{-3.00}{\sqrt{14.00}\sqrt{21.00}} = -0.175$$

$$\phi = 100°$$

EVALUATE: As a check on this result, note that the scalar product $\vec{A} \cdot \vec{B}$ is negative. This means that ϕ is between 90° and 180° (see Fig. 1.26), which agrees with our answer.

Vector Product

The **vector product** of two vectors $\vec{A}$ and $\vec{B}$, also called the **cross product,** is denoted by $\vec{A} \times \vec{B}$. As the name suggests, the vector product is itself a vector. We'll use this product in Chapter 10 to describe torque and angular momentum; in Chapters 27 and 28 we'll use it to describe magnetic fields and forces.

To define the vector product $\vec{A} \times \vec{B}$, we again draw the two vectors $\vec{A}$ and $\vec{B}$ with their tails at the same point (Fig. 1.29a). The two vectors then lie in a plane. We define the vector product to be a vector quantity with a direction perpendicular to this plane (that is, perpendicular to both $\vec{A}$ and $\vec{B}$) and a magnitude equal to $AB \sin \phi$. That is, if $\vec{C} = \vec{A} \times \vec{B}$, then

$$C = AB \sin \phi \quad \text{(magnitude of the vector (cross) product of } \vec{A} \text{ and } \vec{B}) \qquad (1.22)$$

We measure the angle ϕ from $\vec{A}$ toward $\vec{B}$ and take it to be the smaller of the two possible angles, so ϕ ranges from 0° to 180°. Then $\sin \phi \geq 0$ and C in Eq. (1.22) is never negative, as must be the case for a vector magnitude. Note also that when $\vec{A}$ and $\vec{B}$ are parallel or antiparallel, $\phi = 0$ or 180° and $C = 0$. That is, *the vector product of two parallel or antiparallel vectors is always zero.* In particular, *the vector product of any vector with itself is zero.*

CAUTION **Vector product vs. scalar product** Be careful not to confuse the expression $AB \sin \phi$ for the magnitude of the vector product $\vec{A} \times \vec{B}$ with the similar expression $AB \cos \phi$ for the scalar product $\vec{A} \cdot \vec{B}$. To see the difference between these two expressions, imagine that we vary the angle between $\vec{A}$ and $\vec{B}$ while keeping their magnitudes constant. When $\vec{A}$ and $\vec{B}$ are parallel, the magnitude of the vector product will be zero and the scalar product will be maximum. When $\vec{A}$ and $\vec{B}$ are perpendicular, the magnitude of the vector product will be maximum and the scalar product will be zero. ▮

There are always *two* directions perpendicular to a given plane, one on each side of the plane. We choose which of these is the direction of $\vec{A} \times \vec{B}$ as follows. Imagine rotating vector $\vec{A}$ about the perpendicular line until it is aligned with $\vec{B}$, choosing the smaller of the two possible angles between $\vec{A}$ and $\vec{B}$. Curl the fingers of your right hand around the perpendicular line so that the fingertips point in the direction of rotation; your thumb will then point in the direction of $\vec{A} \times \vec{B}$. Figure 1.29a shows this **right-hand rule** and describes a second way to think about this rule.

Similarly, we determine the direction of $\vec{B} \times \vec{A}$ by rotating $\vec{B}$ into $\vec{A}$ as in Fig. 1.29b. The result is a vector that is *opposite* to the vector $\vec{A} \times \vec{B}$. The vector product is *not* commutative! In fact, for any two vectors $\vec{A}$ and $\vec{B}$,

$$\vec{A} \times \vec{B} = -\vec{B} \times \vec{A} \qquad (1.23)$$

Just as we did for the scalar product, we can give a geometrical interpretation of the magnitude of the vector product. In Fig. 1.30a, $B \sin \phi$ is the component of vector $\vec{B}$ that is *perpendicular* to the direction of vector $\vec{A}$. From Eq. (1.22) the magnitude of $\vec{A} \times \vec{B}$ equals the magnitude of $\vec{A}$ multiplied by the component of $\vec{B}$ perpendicular to $\vec{A}$. Figure 1.30b shows that the magnitude of $\vec{A} \times \vec{B}$ also equals the magnitude of $\vec{B}$ multiplied by the component of $\vec{A}$ perpendicular to $\vec{B}$. Note that Fig. 1.30 shows the case in which ϕ is between 0° and 90°; you should draw a similar diagram for ϕ between 90° and 180° to show that the same geometrical interpretation of the magnitude of $\vec{A} \times \vec{B}$ still applies.

Calculating the Vector Product Using Components

If we know the components of $\vec{A}$ and $\vec{B}$, we can calculate the components of the vector product using a procedure similar to that for the scalar product. First we work out the multiplication table for the unit vectors $\hat{\imath}, \hat{\jmath}$, and $\hat{k}$, all three of which

1.29 (a) The vector product $\vec{A} \times \vec{B}$ determined by the right-hand rule. (b) $\vec{B} \times \vec{A} = -\vec{A} \times \vec{B}$; the vector product is anticommutative.

(a) Using the right-hand rule to find the direction of $\vec{A} \times \vec{B}$

① Place $\vec{A}$ and $\vec{B}$ tail to tail.

② Point fingers of right hand along $\vec{A}$, with palm facing $\vec{B}$.

③ Curl fingers toward $\vec{B}$.

④ Thumb points in direction of $\vec{A} \times \vec{B}$.

(b) $\vec{B} \times \vec{A} = -\vec{A} \times \vec{B}$ (the vector product is anticommutative)

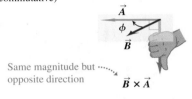

Same magnitude but ⋯⋯⋯ opposite direction

1.30 Calculating the magnitude $AB \sin \phi$ of the vector product of two vectors, $\vec{A} \times \vec{B}$.

(a)

(Magnitude of $\vec{A} \times \vec{B}$) equals $A(B \sin \phi)$.

(Magnitude of $\vec{A}$) times (Component of $\vec{B}$ perpendicular to $\vec{A}$)

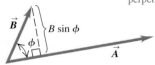

(b)

(Magnitude of $\vec{A} \times \vec{B}$) also equals $B(A \sin \phi)$.

(Magnitude of $\vec{B}$) times (Component of $\vec{A}$ perpendicular to $\vec{B}$)

1.31 (a) We will always use a right-handed coordinate system, like this one. (b) We will never use a left-handed coordinate system (in which $\hat{\imath} \times \hat{\jmath} = -\hat{k}$, and so on).

(a) A right-handed coordinate system

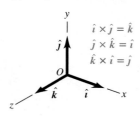

$$\hat{\imath} \times \hat{\jmath} = \hat{k}$$
$$\hat{\jmath} \times \hat{k} = \hat{\imath}$$
$$\hat{k} \times \hat{\imath} = \hat{\jmath}$$

(b) A left-handed coordinate system; we will not use these.

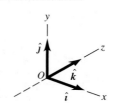

are perpendicular to each other (Fig. 1.31a). The vector product of any vector with itself is zero, so

$$\hat{\imath} \times \hat{\imath} = \hat{\jmath} \times \hat{\jmath} = \hat{k} \times \hat{k} = \mathbf{0}$$

The boldface zero is a reminder that each product is a zero *vector*—that is, one with all components equal to zero and an undefined direction. Using Eqs. (1.22) and (1.23) and the right-hand rule, we find

$$\hat{\imath} \times \hat{\jmath} = -\hat{\jmath} \times \hat{\imath} = \hat{k}$$
$$\hat{\jmath} \times \hat{k} = -\hat{k} \times \hat{\jmath} = \hat{\imath} \qquad (1.24)$$
$$\hat{k} \times \hat{\imath} = -\hat{\imath} \times \hat{k} = \hat{\jmath}$$

You can verify these equations by referring to Fig. 1.31a.

Next we express $\vec{A}$ and $\vec{B}$ in terms of their components and the corresponding unit vectors, and we expand the expression for the vector product:

$$\begin{aligned}
\vec{A} \times \vec{B} &= (A_x\hat{\imath} + A_y\hat{\jmath} + A_z\hat{k}) \times (B_x\hat{\imath} + B_y\hat{\jmath} + B_z\hat{k}) \\
&= A_x\hat{\imath} \times B_x\hat{\imath} + A_x\hat{\imath} \times B_y\hat{\jmath} + A_x\hat{\imath} \times B_z\hat{k} \\
&\quad + A_y\hat{\jmath} \times B_x\hat{\imath} + A_y\hat{\jmath} \times B_y\hat{\jmath} + A_y\hat{\jmath} \times B_z\hat{k} \\
&\quad + A_z\hat{k} \times B_x\hat{\imath} + A_z\hat{k} \times B_y\hat{\jmath} + A_z\hat{k} \times B_z\hat{k}
\end{aligned} \qquad (1.25)$$

We can also rewrite the individual terms in Eq. (1.25) as $A_x\hat{\imath} \times B_y\hat{\jmath} = (A_xB_y)\hat{\imath} \times \hat{\jmath}$, and so on. Evaluating these by using the multiplication table for the unit vectors in Eqs. (1.24) and then grouping the terms, we get

$$\vec{A} \times \vec{B} = (A_yB_z - A_zB_y)\hat{\imath} + (A_zB_x - A_xB_z)\hat{\jmath} + (A_xB_y - A_yB_x)\hat{k} \quad (1.26)$$

Thus the components of $\vec{C} = \vec{A} \times \vec{B}$ are given by

$$C_x = A_yB_z - A_zB_y \qquad C_y = A_zB_x - A_xB_z \qquad C_z = A_xB_y - A_yB_x$$
$$\text{(components of } \vec{C} = \vec{A} \times \vec{B}) \qquad (1.27)$$

The vector product can also be expressed in determinant form as

$$\vec{A} \times \vec{B} = \begin{vmatrix} \hat{\imath} & \hat{\jmath} & \hat{k} \\ A_x & A_y & A_z \\ B_x & B_y & B_z \end{vmatrix}$$

If you aren't familiar with determinants, don't worry about this form.

With the axis system of Fig. 1.31a, if we reverse the direction of the z-axis, we get the system shown in Fig. 1.31b. Then, as you may verify, the definition of the vector product gives $\hat{\imath} \times \hat{\jmath} = -\hat{k}$ instead of $\hat{\imath} \times \hat{\jmath} = \hat{k}$. In fact, all vector products of the unit vectors $\hat{\imath}, \hat{\jmath},$ and $\hat{k}$ would have signs opposite to those in Eqs. (1.24). We see that there are two kinds of coordinate systems, differing in the signs of the vector products of unit vectors. An axis system in which $\hat{\imath} \times \hat{\jmath} = \hat{k}$, as in Fig. 1.31a, is called a **right-handed system.** The usual practice is to use *only* right-handed systems, and we will follow that practice throughout this book.

Example 1.12   **Calculating a vector product**

Vector $\vec{A}$ has magnitude 6 units and is in the direction of the $+x$-axis. Vector $\vec{B}$ has magnitude 4 units and lies in the xy-plane, making an angle of $30°$ with the $+x$-axis (Fig. 1.32). Find the vector product $\vec{C} = \vec{A} \times \vec{B}$.

SOLUTION

IDENTIFY and SET UP: We'll find the vector product in two ways, which will provide a check of our calculations. First we'll use Eq. (1.22) and the right-hand rule; then we'll use Eqs. (1.27) to find the vector product using components.

1.32 Vectors $\vec{A}$ and $\vec{B}$ and their vector product $\vec{C} = \vec{A} \times \vec{B}$. The vector $\vec{B}$ lies in the xy-plane.

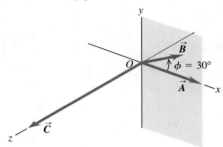

EXECUTE: From Eq. (1.22) the magnitude of the vector product is

$$AB \sin \phi = (6)(4)(\sin 30°) = 12$$

By the right-hand rule, the direction of $\vec{A} \times \vec{B}$ is along the $+z$-axis (the direction of the unit vector $\hat{k}$), so we have $\vec{C} = \vec{A} \times \vec{B} = 12\hat{k}$.

To use Eqs. (1.27), we first determine the components of $\vec{A}$ and $\vec{B}$:

$A_x = 6$	$A_y = 0$	$A_z = 0$
$B_x = 4 \cos 30° = 2\sqrt{3}$	$B_y = 4 \sin 30° = 2$	$B_z = 0$

Then Eqs. (1.27) yield

$$C_x = (0)(0) - (0)(2) = 0$$
$$C_y = (0)(2\sqrt{3}) - (6)(0) = 0$$
$$C_z = (6)(2) - (0)(2\sqrt{3}) = 12$$

Thus again we have $\vec{C} = 12\hat{k}$.

EVALUATE: Both methods give the same result. Depending on the situation, one or the other of the two approaches may be the more convenient one to use.

Test Your Understanding of Section 1.10 Vector $\vec{A}$ has magnitude 2 and vector $\vec{B}$ has magnitude 3. The angle ϕ between $\vec{A}$ and $\vec{B}$ is known to be $0°$, $90°$, or $180°$. For each of the following situations, state what the value of ϕ must be. (In each situation there may be more than one correct answer.) (a) $\vec{A} \cdot \vec{B} = 0$; (b) $\vec{A} \times \vec{B} = 0$; (c) $\vec{A} \cdot \vec{B} = 6$; (d) $\vec{A} \cdot \vec{B} = -6$; (e) (Magnitude of $\vec{A} \times \vec{B}$) = 6. ❙

Physical quantities and units: Three fundamental physical quantities are mass, length, and time. The corresponding basic SI units are the kilogram, the meter, and the second. Derived units for other physical quantities are products or quotients of the basic units. Equations must be dimensionally consistent; two terms can be added only when they have the same units. (See Examples 1.1 and 1.2.)

Significant figures: The accuracy of a measurement can be indicated by the number of significant figures or by a stated uncertainty. The result of a calculation usually has no more significant figures than the input data. When only crude estimates are available for input data, we can often make useful order-of-magnitude estimates. (See Examples 1.3 and 1.4.)

Significant figures in magenta

$$\pi = \frac{C}{2r} = \frac{0.424 \text{ m}}{2(0.06750 \text{ m})} = 3.14$$

$$123.62 + 8.9 = 132.5$$

Scalars, vectors, and vector addition: Scalar quantities are numbers and combine with the usual rules of arithmetic. Vector quantities have direction as well as magnitude and combine according to the rules of vector addition. The negative of a vector has the same magnitude but points in the opposite direction. (See Example 1.5.)

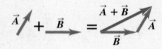

Vector components and vector addition: Vector addition can be carried out using components of vectors. The x-component of $\vec{R} = \vec{A} + \vec{B}$ is the sum of the x-components of $\vec{A}$ and $\vec{B}$, and likewise for the y- and z-components. (See Examples 1.6–1.8.)

$$R_x = A_x + B_x$$
$$R_y = A_y + B_y \qquad (1.10)$$
$$R_z = A_z + B_z$$

Unit vectors: Unit vectors describe directions in space. A unit vector has a magnitude of 1, with no units. The unit vectors $\hat{\imath}$, $\hat{\jmath}$, and $\hat{k}$, aligned with the x-, y-, and z-axes of a rectangular coordinate system, are especially useful. (See Example 1.9.)

$$\vec{A} = A_x\hat{\imath} + A_y\hat{\jmath} + A_z\hat{k} \qquad (1.16)$$

Scalar product: The scalar product $C = \vec{A} \cdot \vec{B}$ of two vectors $\vec{A}$ and $\vec{B}$ is a scalar quantity. It can be expressed in terms of the magnitudes of $\vec{A}$ and $\vec{B}$ and the angle ϕ between the two vectors, or in terms of the components of $\vec{A}$ and $\vec{B}$. The scalar product is commutative; $\vec{A} \cdot \vec{B} = \vec{B} \cdot \vec{A}$. The scalar product of two perpendicular vectors is zero. (See Examples 1.10 and 1.11.)

$$\vec{A} \cdot \vec{B} = AB\cos\phi = |\vec{A}||\vec{B}|\cos\phi \qquad (1.18)$$
$$\vec{A} \cdot \vec{B} = A_xB_x + A_yB_y + A_zB_z \qquad (1.21)$$

Scalar product $\vec{A} \cdot \vec{B} = AB \cos\phi$

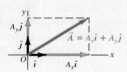

Vector product: The vector product $\vec{C} = \vec{A} \times \vec{B}$ of two vectors $\vec{A}$ and $\vec{B}$ is another vector $\vec{C}$. The magnitude of $\vec{A} \times \vec{B}$ depends on the magnitudes of $\vec{A}$ and $\vec{B}$ and the angle ϕ between the two vectors. The direction of $\vec{A} \times \vec{B}$ is perpendicular to the plane of the two vectors being multiplied, as given by the right-hand rule. The components of $\vec{C} = \vec{A} \times \vec{B}$ can be expressed in terms of the components of $\vec{A}$ and $\vec{B}$. The vector product is not commutative; $\vec{A} \times \vec{B} = -\vec{B} \times \vec{A}$. The vector product of two parallel or antiparallel vectors is zero. (See Example 1.12.)

$$C = AB\sin\phi \qquad (1.22)$$
$$C_x = A_yB_z - A_zB_y$$
$$C_y = A_zB_x - A_xB_z \qquad (1.27)$$
$$C_z = A_xB_y - A_yB_x$$

$\vec{A} \times \vec{B}$ is perpendicular to the plane of $\vec{A}$ and $\vec{B}$.

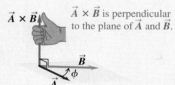

(Magnitude of $\vec{A} \times \vec{B}$) $= AB \sin\phi$

BRIDGING PROBLEM   **Vectors on the Roof**

An air-conditioning unit is fastened to a roof that slopes at an angle of 35° above the horizontal (Fig. 1.33). Its weight is a force on the air conditioner that is directed vertically downward. In order that the unit not crush the roof tiles, the component of the unit's weight perpendicular to the roof cannot exceed 425 N. (One newton, or 1 N, is the SI unit of force. It is equal to 0.2248 lb.) (a) What is the maximum allowed weight of the unit? (b) If the fasteners fail, the unit slides 1.50 m along the roof before it comes to a halt against a ledge. How much work does the weight force do on the unit during its slide if the unit has the weight calculated in part (a)? As we described in Section 1.10, the work done by a force $\vec{F}$ on an object that undergoes a displacement $\vec{s}$ is $W = \vec{F} \cdot \vec{s}$.

1.33 An air-conditioning unit on a slanted roof.

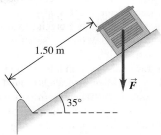

SOLUTION GUIDE

See MasteringPhysics® study area for a Video Tutor solution.

IDENTIFY and SET UP

1. This problem involves vectors and components. What are the known quantities? Which aspect(s) of the weight vector (magnitude, direction, and/or particular components) represent the target variable for part (a)? Which aspect(s) must you know to solve part (b)?

2. Make a sketch based on Fig. 1.33. Add x- and y-axes, choosing the positive direction for each. Your axes don't have to be horizontal and vertical, but they do have to be mutually perpendicular. Make the most convenient choice.

3. Choose the equations you'll use to determine the target variables.

EXECUTE

4. Use the relationship between the magnitude and direction of a vector and its components to solve for the target variable in

part (a). Be careful: Is 35° the correct angle to use in the equation? (*Hint:* Check your sketch.)

5. Make sure your answer has the correct number of significant figures.

6. Use the definition of the scalar product to solve for the target variable in part (b). Again, make sure to use the correct number of significant figures.

EVALUATE

7. Did your answer to part (a) include a vector component whose absolute value is greater than the magnitude of the vector? Is that possible?

8. There are two ways to find the scalar product of two vectors, one of which you used to solve part (b). Check your answer by repeating the calculation using the other way. Do you get the same answer?

Problems   For instructor-assigned homework, go to www.masteringphysics.com

•, ••, •••: Problems of increasing difficulty. **CP**: Cumulative problems incorporating material from earlier chapters. **CALC**: Problems requiring calculus. **BIO**: Biosciences problems.

DISCUSSION QUESTIONS

Q1.1 How many correct experiments do we need to disprove a theory? How many do we need to prove a theory? Explain.

Q1.2 A guidebook describes the rate of climb of a mountain trail as 120 meters per kilometer. How can you express this as a number with no units?

Q1.3 Suppose you are asked to compute the tangent of 5.00 meters. Is this possible? Why or why not?

Q1.4 A highway contractor stated that in building a bridge deck he poured 250 yards of concrete. What do you think he meant?

Q1.5 What is your height in centimeters? What is your weight in newtons?

Q1.6 The U.S. National Institute of Standards and Technology (NIST) maintains several accurate copies of the international standard kilogram. Even after careful cleaning, these national standard

kilograms are gaining mass at an average rate of about 1 μg/y (y = year) when compared every 10 years or so to the standard international kilogram. Does this apparent change have any importance? Explain.

Q1.7 What physical phenomena (other than a pendulum or cesium clock) could you use to define a time standard?

Q1.8 Describe how you could measure the thickness of a sheet of paper with an ordinary ruler.

Q1.9 The quantity $\pi = 3.14159\ldots$ is a number with no dimensions, since it is a ratio of two lengths. Describe two or three other geometrical or physical quantities that are dimensionless.

Q1.10 What are the units of volume? Suppose another student tells you that a cylinder of radius r and height h has volume given by $\pi r^3 h$. Explain why this cannot be right.

Q1.11 Three archers each fire four arrows at a target. Joe's four arrows hit at points 10 cm above, 10 cm below, 10 cm to the left, and 10 cm to the right of the center of the target. All four of Moe's arrows hit within 1 cm of a point 20 cm from the center, and Flo's four arrows all hit within 1 cm of the center. The contest judge says that one of the archers is precise but not accurate, another archer is accurate but not precise, and the third archer is both accurate and precise. Which description goes with which archer? Explain your reasoning.

Q1.12 A circular racetrack has a radius of 500 m. What is the displacement of a bicyclist when she travels around the track from the north side to the south side? When she makes one complete circle around the track? Explain your reasoning.

Q1.13 Can you find two vectors with different lengths that have a vector sum of zero? What length restrictions are required for three vectors to have a vector sum of zero? Explain your reasoning.

Q1.14 One sometimes speaks of the "direction of time," evolving from past to future. Does this mean that time is a vector quantity? Explain your reasoning.

Q1.15 Air traffic controllers give instructions to airline pilots telling them in which direction they are to fly. These instructions are called "vectors." If these are the only instructions given, is the name "vector" used correctly? Why or why not?

Q1.16 Can you find a vector quantity that has a magnitude of zero but components that are different from zero? Explain. Can the magnitude of a vector be less than the magnitude of any of its components? Explain.

Q1.17 (a) Does it make sense to say that a vector is *negative*? Why? (b) Does it make sense to say that one vector is the negative of another? Why? Does your answer here contradict what you said in part (a)?

Q1.18 If $\vec{C}$ is the vector sum of $\vec{A}$ and $\vec{B}$, $\vec{C} = \vec{A} + \vec{B}$, what must be true about the directions and magnitudes of $\vec{A}$ and $\vec{B}$ if $C = A + B$? What must be true about the directions and magnitudes of $\vec{A}$ and $\vec{B}$ if $C = 0$?

Q1.19 If $\vec{A}$ and $\vec{B}$ are nonzero vectors, is it possible for $\vec{A} \cdot \vec{B}$ and $\vec{A} \times \vec{B}$ *both* to be zero? Explain.

Q1.20 What does $\vec{A} \cdot \vec{A}$, the scalar product of a vector with itself, give? What about $\vec{A} \times \vec{A}$, the vector product of a vector with itself?

Q1.21 Let $\vec{A}$ represent any nonzero vector. Why is $\vec{A}/A$ a unit vector, and what is its direction? If θ is the angle that $\vec{A}$ makes with the +x-axis, explain why $(\vec{A}/A) \cdot \hat{\imath}$ is called the *direction cosine* for that axis.

Q1.22 Which of the following are legitimate mathematical operations: (a) $\vec{A} \cdot (\vec{B} - \vec{C})$; (b) $(\vec{A} - \vec{B}) \times \vec{C}$; (c) $\vec{A} \cdot (\vec{B} \times \vec{C})$; (d) $\vec{A} \times (\vec{B} \times \vec{C})$; (e) $\vec{A} \times (\vec{B} \cdot \vec{C})$? In each case, give the reason for your answer.

Q1.23 Consider the two repeated vector products $\vec{A} \times (\vec{B} \times \vec{C})$ and $(\vec{A} \times \vec{B}) \times \vec{C}$. Give an example that illustrates the general rule that these two vector products do not have the same magnitude or direction. Can you choose the vectors $\vec{A}$, $\vec{B}$, and $\vec{C}$ such that these two vector products *are* equal? If so, give an example.

Q1.24 Show that, no matter what $\vec{A}$ and $\vec{B}$ are, $\vec{A} \cdot (\vec{A} \times \vec{B}) = 0$. (*Hint:* Do not look for an elaborate mathematical proof. Rather look at the definition of the direction of the cross product.)

Q1.25 (a) If $\vec{A} \cdot \vec{B} = 0$, does it necessarily follow that $A = 0$ or $B = 0$? Explain. (b) If $\vec{A} \times \vec{B} = \mathbf{0}$, does it necessarily follow that $A = 0$ or $B = 0$? Explain.

Q1.26 If $\vec{A} = \mathbf{0}$ for a vector in the xy-plane, does it follow that $A_x = -A_y$? What *can* you say about A_x and A_y?

EXERCISES

Section 1.3 Standards and Units
Section 1.4 Unit Consistency and Conversions

1.1 • Starting with the definition 1 in. = 2.54 cm, find the number of (a) kilometers in 1.00 mile and (b) feet in 1.00 km.

1.2 •• According to the label on a bottle of salad dressing, the volume of the contents is 0.473 liter (L). Using only the conversions 1 L = 1000 cm^3 and 1 in. = 2.54 cm, express this volume in cubic inches.

1.3 •• How many nanoseconds does it take light to travel 1.00 ft in vacuum? (This result is a useful quantity to remember.)

1.4 •• The density of gold is 19.3 g/cm^3. What is this value in kilograms per cubic meter?

1.5 • The most powerful engine available for the classic 1963 Chevrolet Corvette Sting Ray developed 360 horsepower and had a displacement of 327 cubic inches. Express this displacement in liters (L) by using only the conversions 1 L = 1000 cm^3 and 1 in. = 2.54 cm.

1.6 •• A square field measuring 100.0 m by 100.0 m has an area of 1.00 hectare. An acre has an area of 43,600 ft^2. If a country lot has an area of 12.0 acres, what is the area in hectares?

1.7 • How many years older will you be 1.00 gigasecond from now? (Assume a 365-day year.)

1.8 • While driving in an exotic foreign land you see a speed limit sign on a highway that reads 180,000 furlongs per fortnight. How many miles per hour is this? (One furlong is $\frac{1}{8}$ mile, and a fortnight is 14 days. A furlong originally referred to the length of a plowed furrow.)

1.9 • A certain fuel-efficient hybrid car gets gasoline mileage of 55.0 mpg (miles per gallon). (a) If you are driving this car in Europe and want to compare its mileage with that of other European cars, express this mileage in km/L (L = liter). Use the conversion factors in Appendix E. (b) If this car's gas tank holds 45 L, how many tanks of gas will you use to drive 1500 km?

1.10 • The following conversions occur frequently in physics and are very useful. (a) Use 1 mi = 5280 ft and 1 h = 3600 s to convert 60 mph to units of ft/s. (b) The acceleration of a freely falling object is 32 ft/s^2. Use 1 ft = 30.48 cm to express this acceleration in units of m/s^2. (c) The density of water is 1.0 g/cm^3. Convert this density to units of kg/m^3.

1.11 •• **Neptunium.** In the fall of 2002, a group of scientists at Los Alamos National Laboratory determined that the critical mass of neptunium-237 is about 60 kg. The critical mass of a fissionable material is the minimum amount that must be brought together to start a chain reaction. This element has a density of 19.5 g/cm^3. What would be the radius of a sphere of this material that has a critical mass?

1.12 • **BIO** (a) The recommended daily allowance (RDA) of the trace metal magnesium is 410 mg/day for males. Express this quantity in μg/day. (b) For adults, the RDA of the amino acid lysine is 12 mg per kg of body weight. How many grams per day should a 75-kg adult receive? (c) A typical multivitamin tablet can contain 2.0 mg of vitamin B_2 (riboflavin), and the RDA is 0.0030 g/day. How many such tablets should a person take each day to get the proper amount of this vitamin, assuming that he gets none from any other sources? (d) The RDA for the trace element selenium is 0.000070 g/day. Express this dose in mg/day.

Section 1.5 Uncertainty and Significant Figures

1.13 •• Figure 1.7 shows the result of unacceptable error in the stopping position of a train. (a) If a train travels 890 km from Berlin to Paris and then overshoots the end of the track by 10 m, what is the percent error in the total distance covered? (b) Is it correct to write the total distance covered by the train as 890,010 m? Explain.

1.14 • With a wooden ruler you measure the length of a rectangular piece of sheet metal to be 12 mm. You use micrometer calipers to measure the width of the rectangle and obtain the value 5.98 mm. Give your answers to the following questions to the correct number of significant figures. (a) What is the area of the rectangle? (b) What is the ratio of the rectangle's width to its length? (c) What is the perimeter of the rectangle? (d) What is the difference between the length and width? (e) What is the ratio of the length to the width?

1.15 •• A useful and easy-to-remember approximate value for the number of seconds in a year is $\pi \times 10^7$. Determine the percent error in this approximate value. (There are 365.24 days in one year.)

Section 1.6 Estimates and Orders of Magnitude

1.16 • How many gallons of gasoline are used in the United States in one day? Assume that there are two cars for every three people, that each car is driven an average of 10,000 mi per year, and that the average car gets 20 miles per gallon.

1.17 •• BIO A rather ordinary middle-aged man is in the hospital for a routine check-up. The nurse writes the quantity 200 on his medical chart but forgets to include the units. Which of the following quantities could the 200 plausibly represent? (a) his mass in kilograms; (b) his height in meters; (c) his height in centimeters; (d) his height in millimeters; (e) his age in months.

1.18 • How many kernels of corn does it take to fill a 2-L soft drink bottle?

1.19 • How many words are there in this book?

1.20 • BIO Four astronauts are in a spherical space station. (a) If, as is typical, each of them breathes about 500 cm^3 of air with each breath, approximately what volume of air (in cubic meters) do these astronauts breathe in a year? (b) What would the diameter (in meters) of the space station have to be to contain all this air?

1.21 • BIO How many times does a typical person blink her eyes in a lifetime?

1.22 • BIO How many times does a human heart beat during a lifetime? How many gallons of blood does it pump? (Estimate that the heart pumps 50 cm^3 of blood with each beat.)

1.23 • In Wagner's opera *Das Rheingold,* the goddess Freia is ransomed for a pile of gold just tall enough and wide enough to hide her from sight. Estimate the monetary value of this pile. The density of gold is 19.3 g/cm^3, and its value is about $10 per gram (although this varies).

1.24 • You are using water to dilute small amounts of chemicals in the laboratory, drop by drop. How many drops of water are in a 1.0-L bottle? (*Hint:* Start by estimating the diameter of a drop of water.)

1.25 • How many pizzas are consumed each academic year by students at your school?

Section 1.7 Vectors and Vector Addition

1.26 •• Hearing rattles from a snake, you make two rapid displacements of magnitude 1.8 m and 2.4 m. In sketches (roughly to scale), show how your two displacements might add up to give a resultant of magnitude (a) 4.2 m; (b) 0.6 m; (c) 3.0 m.

1.27 •• A postal employee drives a delivery truck along the route shown in Fig. E1.27. Determine the magnitude and direction of the resultant displacement by drawing a scale diagram. (See also Exercise 1.34 for a different approach to this same problem.)

Figure **E1.27**

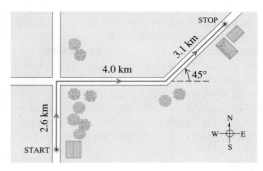

1.28 •• For the vectors $\vec{A}$ and $\vec{B}$ in Fig. E1.28, use a scale drawing to find the magnitude and direction of (a) the vector sum $\vec{A} + \vec{B}$ and (b) the vector difference $\vec{A} - \vec{B}$. Use your answers to find the magnitude and direction of (c) $-\vec{A} - \vec{B}$ and (d) $\vec{B} - \vec{A}$. (See also Exercise 1.35 for a different approach to this problem.)

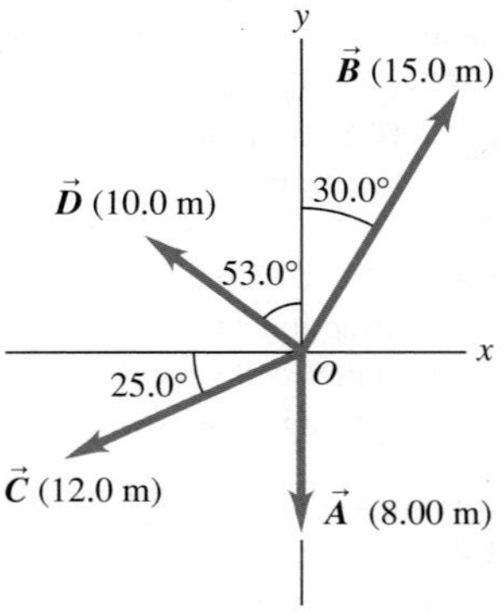

1.29 •• A spelunker is surveying a cave. She follows a passage 180 m straight west, then 210 m in a direction 45° east of south, and then 280 m at 30° east of north. After a fourth unmeasured displacement, she finds herself back where she started. Use a scale drawing to determine the magnitude and direction of the fourth displacement. (See also Problem 1.69 for a different approach to this problem.)

Section 1.8 Components of Vectors

1.30 •• Let the angle θ be the angle that the vector $\vec{A}$ makes with the $+x$-axis, measured counterclockwise from that axis. Find the angle θ for a vector that has the following components: (a) $A_x = 2.00$ m, $A_y = -1.00$ m; (b) $A_x = 2.00$ m, $A_y = 1.00$ m; (c) $A_x = -2.00$ m, $A_y = 1.00$ m; (d) $A_x = -2.00$ m, $A_y = -1.00$ m.

1.31 • Compute the x- and y-components of the vectors $\vec{A}, \vec{B}, \vec{C},$ and $\vec{D}$ in Fig. E1.28.

1.32 • Vector $\vec{A}$ is in the direction 34.0° clockwise from the $-y$-axis. The x-component of $\vec{A}$ is $A_x = -16.0$ m. (a) What is the y-component of $\vec{A}$? (b) What is the magnitude of $\vec{A}$?

1.33 • Vector $\vec{A}$ has y-component $A_y = +13.0$ m. $\vec{A}$ makes an angle of 32.0° counterclockwise from the $+y$-axis. (a) What is the x-component of $\vec{A}$? (b) What is the magnitude of $\vec{A}$?

1.34 •• A postal employee drives a delivery truck over the route shown in Fig. E1.27. Use the method of components to determine the magnitude and direction of her resultant displacement. In a vector-addition diagram (roughly to scale), show that the resultant displacement found from your diagram is in qualitative agreement with the result you obtained using the method of components.

1.35 • For the vectors $\vec{A}$ and $\vec{B}$ in Fig. E1.28, use the method of components to find the magnitude and direction of (a) the vector sum $\vec{A} + \vec{B}$; (b) the vector sum $\vec{B} + \vec{A}$; (c) the vector difference $\vec{A} - \vec{B}$; (d) the vector difference $\vec{B} - \vec{A}$.

1.36 • Find the magnitude and direction of the vector represented by the following pairs of components: (a) $A_x = -8.60$ cm,

Figure **E1.28**

$A_y = 5.20$ cm; (b) $A_x = -9.70$ m, $A_y = -2.45$ m; (c) $A_x = 7.75$ km, $A_y = -2.70$ km.

1.37 •• A disoriented physics professor drives 3.25 km north, then 2.90 km west, and then 1.50 km south. Find the magnitude and direction of the resultant displacement, using the method of components. In a vector-addition diagram (roughly to scale), show that the resultant displacement found from your diagram is in qualitative agreement with the result you obtained using the method of components.

1.38 •• Two ropes in a vertical plane exert equal-magnitude forces on a hanging weight but pull with an angle of 86.0° between them. What pull does each one exert if their resultant pull is 372 N directly upward?

1.39 •• Vector $\vec{A}$ is 2.80 cm long and is 60.0° above the x-axis in the first quadrant. Vector $\vec{B}$ is 1.90 cm long and is 60.0° below the x-axis in the fourth quadrant (Fig. E1.39). Use components to find the magnitude and direction of (a) $\vec{A} + \vec{B}$; (b) $\vec{A} - \vec{B}$; (c) $\vec{B} - \vec{A}$. In each case, sketch the vector addition or subtraction and show that your numerical answers are in qualitative agreement with your sketch.

Figure **E1.39**

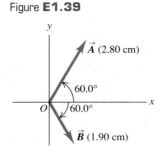

Section 1.9 Unit Vectors

1.40 • In each case, find the x- and y-components of vector $\vec{A}$: (a) $\vec{A} = 5.0\hat{\imath} - 6.3\hat{\jmath}$; (b) $\vec{A} = 11.2\hat{\jmath} - 9.91\hat{\imath}$; (c) $\vec{A} = -15.0\hat{\imath} + 22.4\hat{\jmath}$; (d) $\vec{A} = 5.0\vec{B}$, where $\vec{B} = 4\hat{\imath} - 6\hat{\jmath}$.

1.41 •• Write each vector in Fig. E1.28 in terms of the unit vectors $\hat{\imath}$ and $\hat{\jmath}$.

1.42 •• Given two vectors $\vec{A} = 4.00\hat{\imath} + 7.00\hat{\jmath}$ and $\vec{B} = 5.00\hat{\imath} - 2.00\hat{\jmath}$, (a) find the magnitude of each vector; (b) write an expression for the vector difference $\vec{A} - \vec{B}$ using unit vectors; (c) find the magnitude and direction of the vector difference $\vec{A} - \vec{B}$. (d) In a vector diagram show $\vec{A}$, $\vec{B}$, and $\vec{A} - \vec{B}$, and also show that your diagram agrees qualitatively with your answer in part (c).

1.43 •• (a) Write each vector in Fig. E1.43 in terms of the unit vectors $\hat{\imath}$ and $\hat{\jmath}$. (b) Use unit vectors to express the vector $\vec{C}$, where $\vec{C} = 3.00\vec{A} - 4.00\vec{B}$. (c) Find the magnitude and direction of $\vec{C}$.

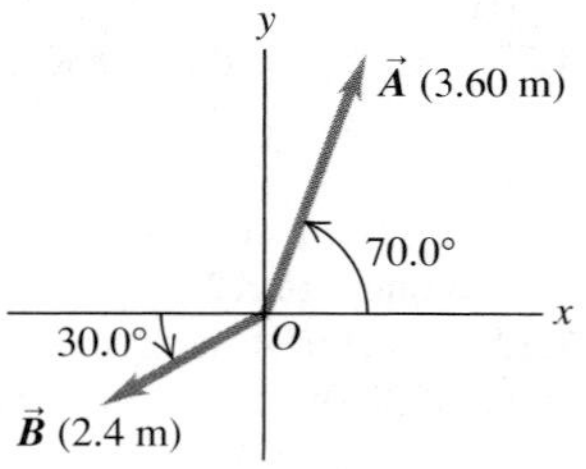

1.44 •• (a) Is the vector $(\hat{\imath} + \hat{\jmath} + \hat{k})$ a unit vector? Justify your answer. (b) Can a unit vector have any components with magnitude greater than unity? Can it have any negative components? In each case justify your answer. (c) If $\vec{A} = a(3.0\hat{\imath} + 4.0\hat{\jmath})$, where a is a constant, determine the value of a that makes $\vec{A}$ a unit vector.

Figure **E1.43**

Section 1.10 Products of Vectors

1.45 • For the vectors $\vec{A}$, $\vec{B}$, and $\vec{C}$ in Fig. E1.28, find the scalar products (a) $\vec{A} \cdot \vec{B}$; (b) $\vec{B} \cdot \vec{C}$; (c) $\vec{A} \cdot \vec{C}$.

1.46 •• (a) Find the scalar product of the two vectors $\vec{A}$ and $\vec{B}$ given in Exercise 1.42. (b) Find the angle between these two vectors.

1.47 •• Find the angle between each of the following pairs of vectors:

(a) $\vec{A} = -2.00\hat{\imath} + 6.00\hat{\jmath}$ and $\vec{B} = 2.00\hat{\imath} - 3.00\hat{\jmath}$
(b) $\vec{A} = 3.00\hat{\imath} + 5.00\hat{\jmath}$ and $\vec{B} = 10.00\hat{\imath} + 6.00\hat{\jmath}$
(c) $\vec{A} = -4.00\hat{\imath} + 2.00\hat{\jmath}$ and $\vec{B} = 7.00\hat{\imath} + 14.00\hat{\jmath}$

1.48 •• Find the vector product $\vec{A} \times \vec{B}$ (expressed in unit vectors) of the two vectors given in Exercise 1.42. What is the magnitude of the vector product?

1.49 • For the vectors $\vec{A}$ and $\vec{D}$ in Fig. E1.28, (a) find the magnitude and direction of the vector product $\vec{A} \times \vec{D}$; (b) find the magnitude and direction of $\vec{D} \times \vec{A}$.

1.50 • For the two vectors in Fig. E1.39, (a) find the magnitude and direction of the vector product $\vec{A} \times \vec{B}$; (b) find the magnitude and direction of $\vec{B} \times \vec{A}$.

1.51 • For the two vectors $\vec{A}$ and $\vec{B}$ in Fig. E1.43, (a) find the scalar product $\vec{A} \cdot \vec{B}$; (b) find the magnitude and direction of the vector product $\vec{A} \times \vec{B}$.

1.52 • The vector $\vec{A}$ is 3.50 cm long and is directed into this page. Vector $\vec{B}$ points from the lower right corner of this page to the upper left corner of this page. Define an appropriate right-handed coordinate system, and find the three components of the vector product $\vec{A} \times \vec{B}$, measured in cm². In a diagram, show your coordinate system and the vectors $\vec{A}$, $\vec{B}$, and $\vec{A} \times \vec{B}$.

1.53 • Given two vectors $\vec{A} = -2.00\hat{\imath} + 3.00\hat{\jmath} + 4.00\hat{k}$ and $\vec{B} = 3.00\hat{\imath} + 1.00\hat{\jmath} - 3.00\hat{k}$, do the following. (a) Find the magnitude of each vector. (b) Write an expression for the vector difference $\vec{A} - \vec{B}$ using unit vectors. (c) Find the magnitude of the vector difference $\vec{A} - \vec{B}$. Is this the same as the magnitude of $\vec{B} - \vec{A}$? Explain.

PROBLEMS

1.54 • An acre, a unit of land measurement still in wide use, has a length of one furlong ($\frac{1}{8}$ mi) and a width one-tenth of its length. (a) How many acres are in a square mile? (b) How many square feet are in an acre? See Appendix E. (c) An acre-foot is the volume of water that would cover 1 acre of flat land to a depth of 1 foot. How many gallons are in 1 acre-foot?

1.55 •• **An Earthlike Planet.** In January 2006 astronomers reported the discovery of a planet comparable in size to the earth orbiting another star and having a mass about 5.5 times the earth's mass. It is believed to consist of a mixture of rock and ice, similar to Neptune. If this planet has the same density as Neptune (1.76 g/cm³), what is its radius expressed (a) in kilometers and (b) as a multiple of earth's radius? Consult Appendix F for astronomical data.

1.56 •• **The Hydrogen Maser.** You can use the radio waves generated by a hydrogen maser as a standard of frequency. The frequency of these waves is 1,420,405,751.786 hertz. (A hertz is another name for one cycle per second.) A clock controlled by a hydrogen maser is off by only 1 s in 100,000 years. For the following questions, use only three significant figures. (The large number of significant figures given for the frequency simply illustrates the remarkable accuracy to which it has been measured.) (a) What is the time for one cycle of the radio wave? (b) How many cycles occur in 1 h? (c) How many cycles would have occurred during the age of the earth, which is estimated to be 4.6×10^9 years? (d) By how many seconds would a hydrogen maser clock be off after a time interval equal to the age of the earth?

1.57 • **BIO** **Breathing Oxygen.** The density of air under standard laboratory conditions is 1.29 kg/m³, and about 20% of that air consists of oxygen. Typically, people breathe about $\frac{1}{2}$ L of air per breath. (a) How many grams of oxygen does a person breathe in a day? (b) If this air is stored uncompressed in a cubical tank, how long is each side of the tank?

1.58 ••• A rectangular piece of aluminum is 7.60 ± 0.01 cm long and 1.90 ± 0.01 cm wide. (a) Find the area of the rectangle and the uncertainty in the area. (b) Verify that the fractional uncertainty in the area is equal to the sum of the fractional uncertainties in the length and in the width. (This is a general result; see Challenge Problem 1.98.)

1.59 ••• As you eat your way through a bag of chocolate chip cookies, you observe that each cookie is a circular disk with a diameter of 8.50 ± 0.02 cm and a thickness of 0.050 ± 0.005 cm. (a) Find the average volume of a cookie and the uncertainty in the volume. (b) Find the ratio of the diameter to the thickness and the uncertainty in this ratio.

1.60 • **BIO** Biological tissues are typically made up of 98% water. Given that the density of water is $1.0 \times 10^3 \text{ kg/m}^3$, estimate the mass of (a) the heart of an adult human; (b) a cell with a diameter of 0.5 μm; (c) a honey bee.

1.61 • **BIO** Estimate the number of atoms in your body. (*Hint:* Based on what you know about biology and chemistry, what are the most common types of atom in your body? What is the mass of each type of atom? Appendix D gives the atomic masses for different elements, measured in atomic mass units; you can find the value of an atomic mass unit, or 1 u, in Appendix E.)

1.62 ••• How many dollar bills would you have to stack to reach the moon? Would that be cheaper than building and launching a spacecraft? (*Hint:* Start by folding a dollar bill to see how many thicknesses make 1.0 mm.)

1.63 ••• How much would it cost to paper the entire United States (including Alaska and Hawaii) with dollar bills? What would be the cost to each person in the United States?

1.64 • **Stars in the Universe.** Astronomers frequently say that there are more stars in the universe than there are grains of sand on all the beaches on the earth. (a) Given that a typical grain of sand is about 0.2 mm in diameter, estimate the number of grains of sand on all the earth's beaches, and hence the approximate number of stars in the universe. It would be helpful to consult an atlas and do some measuring. (b) Given that a typical galaxy contains about 100 billion stars and there are more than 100 billion galaxies in the known universe, estimate the number of stars in the universe and compare this number with your result from part (a).

1.65 ••• Two workers pull horizontally on a heavy box, but one pulls twice as hard as the other. The larger pull is directed at 25.0° west of north, and the resultant of these two pulls is 460.0 N directly northward. Use vector components to find the magnitude of each of these pulls and the direction of the smaller pull.

1.66 •• Three horizontal ropes pull on a large stone stuck in the ground, producing the vector forces $\vec{A}$, $\vec{B}$, and $\vec{C}$ shown in Fig. P1.66. Find the magnitude and direction of a fourth force on the stone that will make the vector sum of the four forces zero.

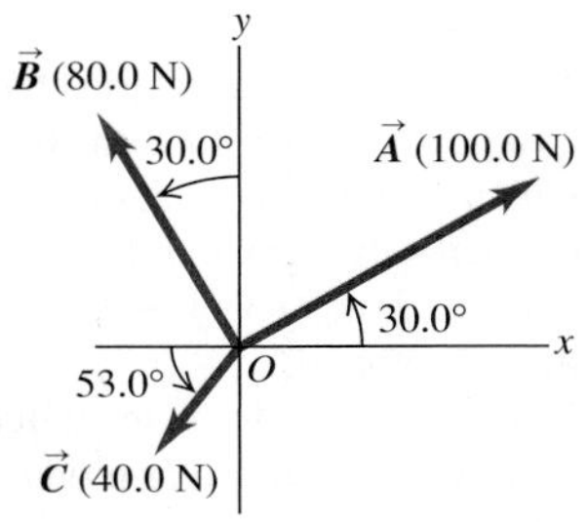

1.67 •• You are to program a robotic arm on an assembly line to move in the xy-plane. Its first displacement is $\vec{A}$; its second displacement is $\vec{B}$, of magnitude 6.40 cm and direction 63.0° measured in the sense from the +x-axis toward the −y-axis. The resultant $\vec{C} = \vec{A} + \vec{B}$ of the two displacements should also have a magnitude of 6.40 cm, but a direction 22.0° measured in the sense

Figure **P1.66**

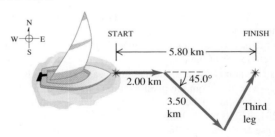

from the +x-axis toward the +y-axis. (a) Draw the vector-addition diagram for these vectors, roughly to scale. (b) Find the components of $\vec{A}$. (c) Find the magnitude and direction of $\vec{A}$.

1.68 ••• **Emergency Landing.** A plane leaves the airport in Galisteo and flies 170 km at 68° east of north and then changes direction to fly 230 km at 48° south of east, after which it makes an immediate emergency landing in a pasture. When the airport sends out a rescue crew, in which direction and how far should this crew fly to go directly to this plane?

1.69 ••• As noted in Exercise 1.29, a spelunker is surveying a cave. She follows a passage 180 m straight west, then 210 m in a direction 45° east of south, and then 280 m at 30° east of north. After a fourth unmeasured displacement she finds herself back where she started. Use the method of components to determine the magnitude and direction of the fourth displacement. Draw the vector-addition diagram and show that it is in qualitative agreement with your numerical solution.

1.70 •• (a) Find the magnitude and direction of the vector $\vec{R}$ that is the sum of the three vectors $\vec{A}$, $\vec{B}$, and $\vec{C}$ in Fig. E1.28. In a diagram, show how $\vec{R}$ is formed from these three vectors. (b) Find the magnitude and direction of the vector $\vec{S} = \vec{C} - \vec{A} - \vec{B}$. In a diagram, show how $\vec{S}$ is formed from these three vectors.

1.71 •• A rocket fires two engines simultaneously. One produces a thrust of 480 N directly forward, while the other gives a 513-N thrust at 32.4° above the forward direction. Find the magnitude and direction (relative to the forward direction) of the resultant force that these engines exert on the rocket.

1.72 •• A sailor in a small sailboat encounters shifting winds. She sails 2.00 km east, then 3.50 km southeast, and then an additional distance in an unknown direction. Her final position is 5.80 km directly east of the starting point (Fig. P1.72). Find the magnitude and direction of the third leg of the journey. Draw the vector-addition diagram and show that it is in qualitative agreement with your numerical solution.

Figure **P1.72**

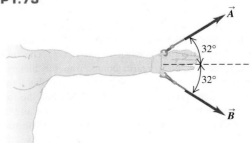

1.73 ••• **BIO** **Dislocated Shoulder.** A patient with a dislocated shoulder is put into a traction apparatus as shown in Fig. P1.73. The pulls $\vec{A}$ and $\vec{B}$ have equal magnitudes and must combine to produce an outward traction force of 5.60 N on the patient's arm. How large should these pulls be?

Figure **P1.73**

1.74 ••• On a training flight, a student pilot flies from Lincoln, Nebraska, to Clarinda, Iowa, then to St. Joseph, Missouri, and then to Manhattan, Kansas (Fig. P1.74). The directions are shown relative to north: 0° is north, 90° is east, 180° is south, and 270° is west. Use the method of components to find (a) the distance she has to fly from Manhattan to get back to Lincoln, and (b) the direction (relative to north) she must fly to get there. Illustrate your solutions with a vector diagram.

Figure **P1.74**

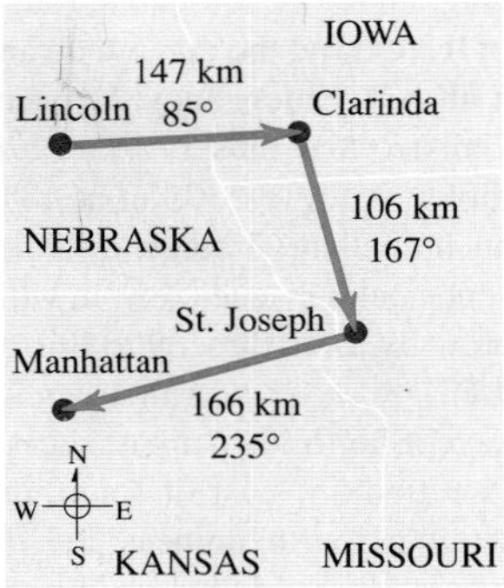

1.75 •• **Equilibrium.** We say an object is in *equilibrium* if all the forces on it balance (add up to zero). Figure P1.75 shows a beam weighing 124 N that is supported in equilibrium by a 100.0-N pull and a force $\vec{F}$ at the floor. The third force on the beam is the 124-N weight that acts vertically downward. (a) Use vector components to find the magnitude and direction of $\vec{F}$. (b) Check the reasonableness of your answer in part (a) by doing a graphical solution approximately to scale.

Figure **P1.75**

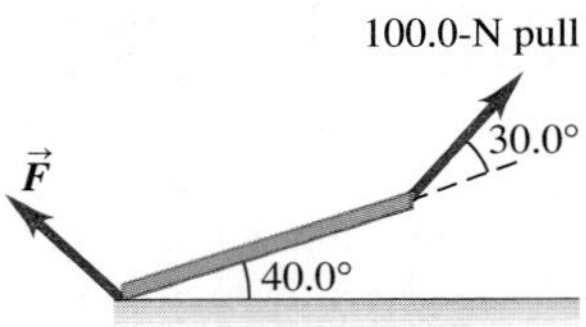

1.76 ••• **Getting Back.** An explorer in the dense jungles of equatorial Africa leaves his hut. He takes 40 steps northeast, then 80 steps 60° north of west, then 50 steps due south. Assume his steps all have equal length. (a) Sketch, roughly to scale, the three vectors and their resultant. (b) Save the explorer from becoming hopelessly lost in the jungle by giving him the displacement, calculated using the method of components, that will return him to his hut.

1.77 ••• A graphic artist is creating a new logo for her company's website. In the graphics program she is using, each pixel in an image file has coordinates (x, y), where the origin $(0, 0)$ is at the upper left corner of the image, the $+x$-axis points to the right, and the $+y$-axis points down. Distances are measured in pixels. (a) The artist draws a line from the pixel location $(10, 20)$ to the location $(210, 200)$. She wishes to draw a second line that starts at $(10, 20)$, is 250 pixels long, and is at an angle of 30° measured clockwise from the first line. At which pixel location should this second line end? Give your answer to the nearest pixel. (b) The artist now draws an arrow that connects the lower right end of the first line to the lower right end of the second line. Find the length and direction of this arrow. Draw a diagram showing all three lines.

1.78 ••• A ship leaves the island of Guam and sails 285 km at 40.0° north of west. In which direction must it now head and how far must it sail so that its resultant displacement will be 115 km directly east of Guam?

1.79 •• **BIO** **Bones and Muscles.** A patient in therapy has a forearm that weighs 20.5 N and that lifts a 112.0-N weight. These two forces have direction vertically downward. The only other significant forces on his forearm come from the biceps muscle (which acts perpendicularly to the forearm) and the force at the elbow. If the biceps produces a pull of 232 N when the forearm is raised 43° above the horizontal, find the magnitude and direction of the force that the elbow exerts on the forearm. (The sum of the elbow force and the biceps force must balance the weight of the arm and the weight it is carrying, so their vector sum must be 132.5 N, upward.)

1.80 ••• You are hungry and decide to go to your favorite neighborhood fast-food restaurant. You leave your apartment and take the elevator 10 flights down (each flight is 3.0 m) and then go 15 m south to the apartment exit. You then proceed 0.2 km east, turn north, and go 0.1 km to the entrance of the restaurant. (a) Determine the displacement from your apartment to the restaurant. Use unit vector notation for your answer, being sure to make clear your choice of coordinates. (b) How far did you travel along the path you took from your apartment to the restaurant, and what is the magnitude of the displacement you calculated in part (a)?

1.81 •• While following a treasure map, you start at an old oak tree. You first walk 825 m directly south, then turn and walk 1.25 km at 30.0° west of north, and finally walk 1.00 km at 40.0° north of east, where you find the treasure: a biography of Isaac Newton! (a) To return to the old oak tree, in what direction should you head and how far will you walk? Use components to solve this problem. (b) To see whether your calculation in part (a) is reasonable, check it with a graphical solution drawn roughly to scale.

1.82 •• A fence post is 52.0 m from where you are standing, in a direction 37.0° north of east. A second fence post is due south from you. What is the distance of the second post from you, if the distance between the two posts is 80.0 m?

1.83 •• A dog in an open field runs 12.0 m east and then 28.0 m in a direction 50.0° west of north. In what direction and how far must the dog then run to end up 10.0 m south of her original starting point?

1.84 ••• Ricardo and Jane are standing under a tree in the middle of a pasture. An argument ensues, and they walk away in different directions. Ricardo walks 26.0 m in a direction 60.0° west of north. Jane walks 16.0 m in a direction 30.0° south of west. They then stop and turn to face each other. (a) What is the distance between them? (b) In what direction should Ricardo walk to go directly toward Jane?

1.85 ••• John, Paul, and George are standing in a strawberry field. Paul is 14.0 m due west of John. George is 36.0 m from Paul, in a direction 37.0° south of east from Paul's location. How far is George from John? What is the direction of George's location from that of John?

1.86 ••• You are camping with two friends, Joe and Karl. Since all three of you like your privacy, you don't pitch your tents close together. Joe's tent is 21.0 m from yours, in the direction 23.0° south of east. Karl's tent is 32.0 m from yours, in the direction 37.0° north of east. What is the distance between Karl's tent and Joe's tent?

1.87 •• Vectors $\vec{A}$ and $\vec{B}$ have scalar product −6.00 and their vector product has magnitude +9.00. What is the angle between these two vectors?

1.88 •• **Bond Angle in Methane.** In the methane molecule, CH_4, each hydrogen atom is at a corner of a regular tetrahedron with the carbon atom at the center. In coordinates where one of the C–H bonds is in the direction of $\hat{\imath} + \hat{\jmath} + \hat{k}$, an adjacent C–H bond is in the $\hat{\imath} - \hat{\jmath} - \hat{k}$ direction. Calculate the angle between these two bonds.

1.89 •• Vector $\vec{A}$ has magnitude 12.0 m and vector $\vec{B}$ has magnitude 16.0 m. The scalar product $\vec{A} \cdot \vec{B}$ is 90.0 m². What is the magnitude of the vector product between these two vectors?

1.90 •• When two vectors $\vec{A}$ and $\vec{B}$ are drawn from a common point, the angle between them is ϕ. (a) Using vector techniques, show that the magnitude of their vector sum is given by

$$\sqrt{A^2 + B^2 + 2AB \cos \phi}$$

(b) If $\vec{A}$ and $\vec{B}$ have the same magnitude, for which value of ϕ will their vector sum have the same magnitude as $\vec{A}$ or $\vec{B}$?

1.91 •• A cube is placed so that one corner is at the origin and three edges are along the x-, y-, and z-axes of a coordinate system (Fig. P1.91). Use vectors to compute (a) the angle between the edge along the z-axis (line ab) and the diagonal from the origin to the opposite corner (line ad), and (b) the angle between line ac (the diagonal of a face) and line ad.

Figure **P1.91**

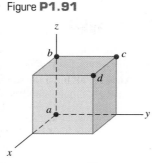

1.92 •• Vector $\vec{A}$ has magnitude 6.00 m and vector $\vec{B}$ has magnitude 3.00 m. The vector product between these two vectors has magnitude 12.0 m^2. What are the two possible values for the scalar product of these two vectors? For each value of $\vec{A} \cdot \vec{B}$, draw a sketch that shows $\vec{A}$ and $\vec{B}$ and explain why the vector products in the two sketches are the same but the scalar products differ.

1.93 •• The scalar product of vectors $\vec{A}$ and $\vec{B}$ is $+48.0$ m^2. Vector $\vec{A}$ has magnitude 9.00 m and direction 28.0° west of south. If vector $\vec{B}$ has direction 39.0° south of east, what is the magnitude of $\vec{B}$?

1.94 ••• Obtain a *unit vector* perpendicular to the two vectors given in Exercise 1.53.

1.95 •• You are given vectors $\vec{A} = 5.0\hat{\imath} - 6.5\hat{\jmath}$ and $\vec{B} = -3.5\hat{\imath} + 7.0\hat{\jmath}$. A third vector $\vec{C}$ lies in the xy-plane. Vector $\vec{C}$ is perpendicular to vector $\vec{A}$, and the scalar product of $\vec{C}$ with $\vec{B}$ is 15.0. From this information, find the components of vector $\vec{C}$.

1.96 •• Two vectors $\vec{A}$ and $\vec{B}$ have magnitudes $A = 3.00$ and $B = 3.00$. Their vector product is $\vec{A} \times \vec{B} = -5.00\hat{k} + 2.00\hat{\imath}$. What is the angle between $\vec{A}$ and $\vec{B}$?

1.97 •• Later in our study of physics we will encounter quantities represented by $(\vec{A} \times \vec{B}) \cdot \vec{C}$. (a) Prove that for any three vectors $\vec{A}$, $\vec{B}$, and $\vec{C}$, $\vec{A} \cdot (\vec{B} \times \vec{C}) = (\vec{A} \times \vec{B}) \cdot \vec{C}$. (b) Calculate $(\vec{A} \times \vec{B}) \cdot \vec{C}$ for the three vectors $\vec{A}$ with magnitude $A = 5.00$ and angle $\theta_A = 26.0°$ measured in the sense from the $+x$-axis toward the $+y$-axis, $\vec{B}$ with $B = 4.00$ and $\theta_B = 63.0°$, and $\vec{C}$ with magnitude 6.00 and in the $+z$-direction. Vectors $\vec{A}$ and $\vec{B}$ are in the xy-plane.

CHALLENGE PROBLEMS

1.98 ••• The length of a rectangle is given as $L \pm l$ and its width as $W \pm w$. (a) Show that the uncertainty in its area A is $a = Lw + lW$. Assume that the uncertainties l and w are small, so that the product lw is very small and you can ignore it. (b) Show that the fractional uncertainty in the area is equal to the sum of the fractional uncertainty in length and the fractional uncertainty in width. (c) A rectangular solid has dimensions $L \pm l$, $W \pm w$, and $H \pm h$. Find the fractional uncertainty in the volume, and show that it equals the sum of the fractional uncertainties in the length, width, and height.

1.99 ••• **Completed Pass.** At Enormous State University (ESU), the football team records its plays using vector displacements, with the origin taken to be the position of the ball before the play starts. In a certain pass play, the receiver starts at $+1.0\hat{\imath} - 5.0\hat{\jmath}$, where the units are yards, $\hat{\imath}$ is to the right, and

$\hat{\jmath}$ is downfield. Subsequent displacements of the receiver are $+9.0\hat{\imath}$ (in motion before the snap), $+11.0\hat{\jmath}$ (breaks downfield), $-6.0\hat{\imath} + 4.0\hat{\jmath}$ (zigs), and $+12.0\hat{\imath} + 18.0\hat{\jmath}$ (zags). Meanwhile, the quarterback has dropped straight back to a position $-7.0\hat{\jmath}$. How far and in which direction must the quarterback throw the ball? (Like the coach, you will be well advised to diagram the situation before solving it numerically.)

1.100 ••• **Navigating in the Solar System.** The *Mars Polar Lander* spacecraft was launched on January 3, 1999. On December 3, 1999, the day *Mars Polar Lander* touched down on the Martian surface, the positions of the earth and Mars were given by these coordinates:

	x	y	z
Earth	0.3182 AU	0.9329 AU	0.0000 AU
Mars	1.3087 AU	-0.4423 AU	-0.0414 AU

In these coordinates, the sun is at the origin and the plane of the earth's orbit is the xy-plane. The earth passes through the $+x$-axis once a year on the autumnal equinox, the first day of autumn in the northern hemisphere (on or about September 22). One AU, or *astronomical unit,* is equal to 1.496×10^8 km, the average distance from the earth to the sun. (a) In a diagram, show the positions of the sun, the earth, and Mars on December 3, 1999. (b) Find the following distances in AU on December 3, 1999: (i) from the sun to the earth; (ii) from the sun to Mars; (iii) from the earth to Mars. (c) As seen from the earth, what was the angle between the direction to the sun and the direction to Mars on December 3, 1999? (d) Explain whether Mars was visible from your location at midnight on December 3, 1999. (When it is midnight at your location, the sun is on the opposite side of the earth from you.)

1.101 ••• **Navigating in the Big Dipper.** All the stars of the Big Dipper (part of the constellation Ursa Major) may appear to be the same distance from the earth, but in fact they are very far from each other. Figure P1.101 shows the distances from the earth to each of these stars. The distances are given in light-years (ly), the distance that light travels in one year. One light-year equals 9.461×10^{15} m. (a) Alkaid and Merak are 25.6° apart in the earth's sky. In a diagram, show the relative positions of Alkaid, Merak, and our sun. Find the distance in light-years from Alkaid to Merak. (b) To an inhabitant of a planet orbiting Merak, how many degrees apart in the sky would Alkaid and our sun be?

Figure **P1.101**

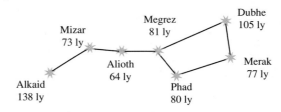

1.102 ••• The vector $\vec{r} = x\hat{\imath} + y\hat{\jmath} + z\hat{k}$, called the *position vector,* points from the origin $(0, 0, 0)$ to an arbitrary point in space with coordinates (x, y, z). Use what you know about vectors to prove the following: All points (x, y, z) that satisfy the equation $Ax + By + Cz = 0$, where A, B, and C are constants, lie in a plane that passes through the origin and that is perpendicular to the vector $A\hat{\imath} + B\hat{\jmath} + C\hat{k}$. Sketch this vector and the plane.

Answers

Chapter Opening Question ?

Take the $+x$-axis to point east and the $+y$-axis to point north. Then what we are trying to find is the y-component of the velocity vector, which has magnitude $v = 20$ km/h and is at an angle $\theta = 53°$ measured from the $+x$-axis toward the $+y$-axis. From Eqs. (1.6) we have $v_y = v\sin\theta = (20 \text{ km/h})\sin 53° = 16$ km/h. So the thunderstorm moves 16 km north in 1 h.

Test Your Understanding Questions

1.5 Answer: (ii) Density $= (1.80 \text{ kg})/(6.0 \times 10^{-4} \text{ m}^3) = 3.0 \times 10^3$ kg/m^3. When we multiply or divide, the number with the fewest significant figures controls the number of significant figures in the result.

1.6 The answer depends on how many students are enrolled at your campus.

1.7 Answers: (ii), (iii), and (iv) The vector $-\vec{T}$ has the same magnitude as the vector $\vec{T}$, so $\vec{S} - \vec{T} = \vec{S} + (-\vec{T})$ is the *sum* of one vector of magnitude 3 m and one of magnitude 4 m. This sum has magnitude 7 m if $\vec{S}$ and $-\vec{T}$ are parallel and magnitude 1 m if $\vec{S}$ and $-\vec{T}$ are antiparallel. The magnitude of $\vec{S} - \vec{T}$ is 5 m if $\vec{S}$ and $-\vec{T}$ are perpendicular, so that the vectors $\vec{S}$, $\vec{T}$, and $\vec{S} - \vec{T}$ form a 3–4–5 right triangle. Answer (i) is impossible because the magnitude of the sum of two vectors cannot be greater than the sum of the magnitudes; answer (v) is impossible because the sum of two vectors can be zero only if the two vectors are antiparallel and have the same magnitude; and answer (vi) is impossible because the magnitude of a vector cannot be negative.

1.8 Answers: (a) yes, (b) no Vectors $\vec{A}$ and $\vec{B}$ can have the same magnitude but different components if they point in different directions. If they have the same components, however, they are the same vector ($\vec{A} = \vec{B}$) and so must have the same magnitude.

1.9 Answer: all have the same magnitude The four vectors $\vec{A}$, $\vec{B}$, $\vec{C}$, and $\vec{D}$ all point in different directions, but all have the same magnitude:

$$A = B = C = D = \sqrt{(\pm 3 \text{ m})^2 + (\pm 5 \text{ m})^2 + (\pm 2 \text{ m})^2}$$
$$= \sqrt{9 \text{ m}^2 + 25 \text{ m}^2 + 4 \text{ m}^2} = \sqrt{38 \text{ m}^2} = 6.2 \text{ m}$$

1.10 Answers: (a) $\phi = 90°$, **(b)** $\phi = 0°$ or $\phi = 180°$, **(c)** $\phi = 0°$, **(d)** $\phi = 180°$, **(e)** $\phi = 90°$ (a) The scalar product is zero only if $\vec{A}$ and $\vec{B}$ are perpendicular. (b) The vector product is zero only if $\vec{A}$ and $\vec{B}$ are either parallel or antiparallel. (c) The scalar product is equal to the product of the magnitudes ($\vec{A} \cdot \vec{B} = AB$) only if $\vec{A}$ and $\vec{B}$ are parallel. (d) The scalar product is equal to the negative of the product of the magnitudes ($\vec{A} \cdot \vec{B} = -AB$) only if $\vec{A}$ and $\vec{B}$ are antiparallel. (e) The magnitude of the vector product is equal to the product of the magnitudes [(magnitude of $\vec{A} \times \vec{B}$) $= AB$] only if $\vec{A}$ and $\vec{B}$ are perpendicular.

Bridging Problem

Answers: (a) 5.2×10^2 N
(b) 4.5×10^2 N $\cdot$ m

MOTION ALONG A STRAIGHT LINE

2

? A bungee jumper speeds up during the first part of his fall, then slows to a halt as the bungee cord stretches and becomes taut. Is it accurate to say that the jumper is *accelerating* as he slows during the final part of his fall?

LEARNING GOALS

By studying this chapter, you will learn:

- How to describe straight-line motion in terms of average velocity, instantaneous velocity, average acceleration, and instantaneous acceleration.

- How to interpret graphs of position versus time, velocity versus time, and acceleration versus time for straight-line motion.

- How to solve problems involving straight-line motion with constant acceleration, including free-fall problems.

- How to analyze straight-line motion when the acceleration is not constant.

What distance must an airliner travel down a runway before reaching takeoff speed? When you throw a baseball straight up in the air, how high does it go? When a glass slips from your hand, how much time do you have to catch it before it hits the floor? These are the kinds of questions you will learn to answer in this chapter. We are beginning our study of physics with *mechanics,* the study of the relationships among force, matter, and motion. In this chapter and the next we will study *kinematics,* the part of mechanics that enables us to describe motion. Later we will study *dynamics,* which relates motion to its causes.

In this chapter we concentrate on the simplest kind of motion: a body moving along a straight line. To describe this motion, we introduce the physical quantities *velocity* and *acceleration.* In physics these quantities have definitions that are more precise and slightly different from the ones used in everyday language. Both velocity and acceleration are *vectors:* As you learned in Chapter 1, this means that they have both magnitude and direction. Our concern in this chapter is with motion along a straight line only, so we won't need the full mathematics of vectors just yet. But using vectors will be essential in Chapter 3 when we consider motion in two or three dimensions.

We'll develop simple equations to describe straight-line motion in the important special case when the acceleration is constant. An example is the motion of a freely falling body. We'll also consider situations in which the acceleration varies during the motion; in this case, it's necessary to use integration to describe the motion. (If you haven't studied integration yet, Section 2.6 is optional.)

2.1 Displacement, Time, and Average Velocity

Suppose a drag racer drives her AA-fuel dragster along a straight track (Fig. 2.1). To study the dragster's motion, we need a coordinate system. We choose the *x*-axis to lie along the dragster's straight-line path, with the origin *O* at the starting line. We also choose a point on the dragster, such as its front end, and represent the entire dragster by that point. Hence we treat the dragster as a **particle.**

A useful way to describe the motion of the particle that represents the dragster is in terms of the change in the particle's coordinate *x* over a time interval. Suppose that 1.0 s after the start the front of the dragster is at point P_1, 19 m from the origin, and 4.0 s after the start it is at point P_2, 277 m from the origin. The *displacement* of the particle is a vector that points from P_1 to P_2 (see Section 1.7). Figure 2.1 shows that this vector points along the *x*-axis. The *x*-component of the displacement is the change in the value of *x*, $(277\text{ m} - 19\text{ m}) = 258\text{ m}$, that took place during the time interval of $(4.0\text{ s} - 1.0\text{ s}) = 3.0\text{ s}$. We define the dragster's **average velocity** during this time interval as a *vector* quantity whose *x*-component is the change in *x* divided by the time interval: $(258\text{ m})/(3.0\text{ s}) = 86\text{ m/s}$.

In general, the average velocity depends on the particular time interval chosen. For a 3.0-s time interval *before* the start of the race, the average velocity would be zero because the dragster would be at rest at the starting line and would have zero displacement.

Let's generalize the concept of average velocity. At time t_1 the dragster is at point P_1, with coordinate x_1, and at time t_2 it is at point P_2, with coordinate x_2. The displacement of the dragster during the time interval from t_1 to t_2 is the vector from P_1 to P_2. The *x*-component of the displacement, denoted Δx, is the change in the coordinate *x*:

$$\Delta x = x_2 - x_1 \qquad (2.1)$$

The dragster moves along the *x*-axis only, so the *y*- and *z*-components of the displacement are equal to zero.

> **CAUTION** **The meaning of Δx** Note that Δx is *not* the product of Δ and *x*; it is a single symbol that means "the change in the quantity *x*." We always use the Greek capital letter Δ (delta) to represent a *change* in a quantity, equal to the *final* value of the quantity minus the *initial* value—never the reverse. Likewise, the time interval from t_1 to t_2 is Δt, the change in the quantity t: $\Delta t = t_2 - t_1$ (final time minus initial time). ▮

The *x*-component of average velocity, or **average *x*-velocity,** is the *x*-component of displacement, Δx, divided by the time interval Δt during which

2.1 Positions of a dragster at two times during its run.

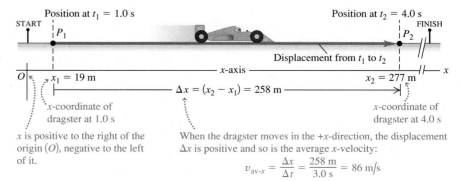

Position at $t_1 = 1.0$ s

START

P_1

Position at $t_2 = 4.0$ s

FINISH

P_2

Displacement from t_1 to t_2

x-axis

O $x_1 = 19$ m

$x_2 = 277$ m

$\Delta x = (x_2 - x_1) = 258$ m

x-coordinate of dragster at 1.0 s

x-coordinate of dragster at 4.0 s

x is positive to the right of the origin (O), negative to the left of it.

When the dragster moves in the +*x*-direction, the displacement Δx is positive and so is the average *x*-velocity:

$$v_{\text{av-}x} = \frac{\Delta x}{\Delta t} = \frac{258\text{ m}}{3.0\text{ s}} = 86\text{ m/s}$$

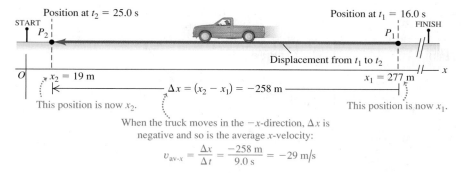

2.2 Positions of an official's truck at two times during its motion. The points P_1 and P_2 now indicate the positions of the truck, and so are the reverse of Fig. 2.1.

the displacement occurs. We use the symbol $v_{\text{av-}x}$ for average x-velocity (the subscript "av" signifies average value and the subscript x indicates that this is the x-component):

$$v_{\text{av-}x} = \frac{x_2 - x_1}{t_2 - t_1} = \frac{\Delta x}{\Delta t} \qquad \text{(average } x\text{-velocity, straight-line motion)} \quad \text{(2.2)}$$

As an example, for the dragster $x_1 = 19$ m, $x_2 = 277$ m, $t_1 = 1.0$ s, and $t_2 = 4.0$ s, so Eq. (2.2) gives

$$v_{\text{av-}x} = \frac{277 \text{ m} - 19 \text{ m}}{4.0 \text{ s} - 1.0 \text{ s}} = \frac{258 \text{ m}}{3.0 \text{ s}} = 86 \text{ m/s}$$

The average x-velocity of the dragster is positive. This means that during the time interval, the coordinate x increased and the dragster moved in the positive x-direction (to the right in Fig. 2.1).

If a particle moves in the *negative* x-direction during a time interval, its average velocity for that time interval is negative. For example, suppose an official's truck moves to the left along the track (Fig. 2.2). The truck is at $x_1 = 277$ m at $t_1 = 16.0$ s and is at $x_2 = 19$ m at $t_2 = 25.0$ s. Then $\Delta x = (19 \text{ m} - 277 \text{ m}) = -258$ m and $\Delta t = (25.0 \text{ s} - 16.0 \text{ s}) = 9.0$ s. The x-component of average velocity is $v_{\text{av-}x} = \Delta x / \Delta t = (-258 \text{ m})/(9.0 \text{ s}) = -29$ m/s. Table 2.1 lists some simple rules for deciding whether the x-velocity is positive or negative.

Table 2.1 Rules for the Sign of x-Velocity

If the x-coordinate is:	... the x-velocity is:
Positive & increasing (getting more positive)	Positive: Particle is moving in $+x$-direction
Positive & decreasing (getting less positive)	Negative: Particle is moving in $-x$-direction
Negative & increasing (getting less negative)	Positive: Particle is moving in $+x$-direction
Negative & decreasing (getting more negative)	Negative: Particle is moving in $-x$-direction

Note: These rules apply to both the average x-velocity $v_{\text{av-}x}$ and the instantaneous x-velocity v_x (to be discussed in Section 2.2).

CAUTION **Choice of the positive x-direction** You might be tempted to conclude that positive average x-velocity must mean motion to the right, as in Fig. 2.1, and that negative average x-velocity must mean motion to the left, as in Fig. 2.2. But that's correct *only* if the positive x-direction is to the right, as we chose it to be in Figs. 2.1 and 2.2. Had we chosen the positive x-direction to be to the left, with the origin at the finish line, the dragster would have negative average x-velocity and the official's truck would have positive average x-velocity. In most problems the direction of the coordinate axis will be yours to choose. Once you've made your choice, you *must* take it into account when interpreting the signs of $v_{\text{av-}x}$ and other quantities that describe motion! ▮

With straight-line motion we sometimes call Δx simply the displacement and $v_{\text{av-}x}$ simply the average velocity. But be sure to remember that these are really the x-components of vector quantities that, in this special case, have *only* x-components. In Chapter 3, displacement, velocity, and acceleration vectors will have two or three nonzero components.

Figure 2.3 is a graph of the dragster's position as a function of time—that is, an **x-t graph.** The curve in the figure *does not* represent the dragster's path in space; as Fig. 2.1 shows, the path is a straight line. Rather, the graph is a pictorial way to represent how the dragster's position changes with time. The points p_1 and p_2 on the graph correspond to the points P_1 and P_2 along the dragster's path. Line p_1p_2 is the hypotenuse of a right triangle with vertical side $\Delta x = x_2 - x_1$

2.3 The position of a dragster as a function of time.

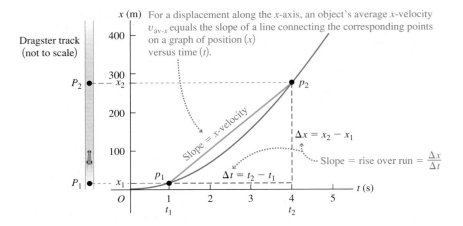

Dragster track (not to scale)

x (m) For a displacement along the x-axis, an object's average x-velocity v_{av-x} equals the slope of a line connecting the corresponding points on a graph of position (x) versus time (t).

Slope = x-velocity

$\Delta x = x_2 - x_1$

Slope = rise over run = $\dfrac{\Delta x}{\Delta t}$

$\Delta t = t_2 - t_1$

Table 2.2 Typical Velocity Magnitudes

A snail's pace	10^{-3} m/s
A brisk walk	2 m/s
Fastest human	11 m/s
Freeway speeds	30 m/s
Fastest car	341 m/s
Random motion of air molecules	500 m/s
Fastest airplane	1000 m/s
Orbiting communications satellite	3000 m/s
Electron orbiting in a hydrogen atom	2×10^6 m/s
Light traveling in a vacuum	3×10^8 m/s

and horizontal side $\Delta t = t_2 - t_1$. The average x-velocity $v_{av-x} = \Delta x / \Delta t$ of the dragster equals the *slope* of the line $p_1 p_2$—that is, the ratio of the triangle's vertical side Δx to its horizontal side Δt.

The average x-velocity depends only on the total displacement $\Delta x = x_2 - x_1$ that occurs during the time interval $\Delta t = t_2 - t_1$, not on the details of what happens during the time interval. At time t_1 a motorcycle might have raced past the dragster at point P_1 in Fig. 2.1, then blown its engine and slowed down to pass through point P_2 at the same time t_2 as the dragster. Both vehicles have the same displacement during the same time interval and so have the same average x-velocity.

If distance is given in meters and time in seconds, average velocity is measured in meters per second (m/s). Other common units of velocity are kilometers per hour (km/h), feet per second (ft/s), miles per hour (mi/h), and knots (1 knot = 1 nautical mile/h = 6080 ft/h). Table 2.2 lists some typical velocity magnitudes.

Test Your Understanding of Section 2.1 Each of the following automobile trips takes one hour. The positive x-direction is to the east. (i) Automobile A travels 50 km due east. (ii) Automobile B travels 50 km due west. (iii) Automobile C travels 60 km due east, then turns around and travels 10 km due west. (iv) Automobile D travels 70 km due east. (v) Automobile E travels 20 km due west, then turns around and travels 20 km due east. (a) Rank the five trips in order of average x-velocity from most positive to most negative. (b) Which trips, if any, have the same average x-velocity? (c) For which trip, if any, is the average x-velocity equal to zero?

2.4 The winner of a 50-m swimming race is the swimmer whose average velocity has the greatest magnitude—that is, the swimmer who traverses a displacement Δx of 50 m in the shortest elapsed time Δt.

2.2 Instantaneous Velocity

Sometimes the average velocity is all you need to know about a particle's motion. For example, a race along a straight line is really a competition to see whose average velocity, v_{av-x}, has the greatest magnitude. The prize goes to the competitor who can travel the displacement Δx from the start to the finish line in the shortest time interval, Δt (Fig. 2.4).

But the average velocity of a particle during a time interval can't tell us how fast, or in what direction, the particle was moving at any given time during the interval. To do this we need to know the **instantaneous velocity,** or the velocity at a specific instant of time or specific point along the path.

CAUTION **How long is an instant?** Note that the word "instant" has a somewhat different definition in physics than in everyday language. You might use the phrase "It lasted just an instant" to refer to something that lasted for a very short time interval. But in physics an instant has no duration at all; it refers to a single value of time.

To find the instantaneous velocity of the dragster in Fig. 2.1 at the point P_1, we move the second point P_2 closer and closer to the first point P_1 and compute the average velocity $v_{\text{av-}x} = \Delta x/\Delta t$ over the ever-shorter displacement and time interval. Both Δx and Δt become very small, but their ratio does not necessarily become small. In the language of calculus, the limit of $\Delta x/\Delta t$ as Δt approaches zero is called the **derivative** of x with respect to t and is written dx/dt. *The instantaneous velocity is the limit of the average velocity as the time interval approaches zero; it equals the instantaneous rate of change of position with time.* We use the symbol v_x, with no "av" subscript, for the instantaneous velocity along the x-axis, or the **instantaneous x-velocity:**

$$v_x = \lim_{\Delta t \to 0} \frac{\Delta x}{\Delta t} = \frac{dx}{dt} \quad \text{(instantaneous x-velocity, straight-line motion)} \quad (2.3)$$

The time interval Δt is always positive, so v_x has the same algebraic sign as Δx. A positive value of v_x means that x is increasing and the motion is in the positive x-direction; a negative value of v_x means that x is decreasing and the motion is in the negative x-direction. A body can have positive x and negative v_x, or the reverse; x tells us where the body is, while v_x tells us how it's moving (Fig. 2.5). The rules that we presented in Table 2.1 (Section 2.1) for the sign of average x-velocity $v_{\text{av-}x}$ also apply to the sign of instantaneous x-velocity v_x.

Instantaneous velocity, like average velocity, is a vector quantity; Eq. (2.3) defines its x-component. In straight-line motion, all other components of instantaneous velocity are zero. In this case we often call v_x simply the instantaneous velocity. (In Chapter 3 we'll deal with the general case in which the instantaneous velocity can have nonzero x-, y-, and z-components.) When we use the term "velocity," we will always mean instantaneous rather than average velocity.

The terms "velocity" and "speed" are used interchangeably in everyday language, but they have distinct definitions in physics. We use the term **speed** to denote distance traveled divided by time, on either an average or an instantaneous basis. Instantaneous *speed,* for which we use the symbol v with *no* subscripts, measures how fast a particle is moving; instantaneous *velocity* measures how fast *and* in what direction it's moving. Instantaneous speed is the magnitude of instantaneous velocity and so can never be negative. For example, a particle with instantaneous velocity $v_x = 25$ m/s and a second particle with $v_x = -25$ m/s are moving in opposite directions at the same instantaneous speed 25 m/s.

2.5 Even when he's moving forward, this cyclist's instantaneous x-velocity can be negative—if he's traveling in the negative x-direction. In any problem, the choice of which direction is positive and which is negative is entirely up to you.

CAUTION **Average speed and average velocity** Average speed is *not* the magnitude of average velocity. When César Cielo set a world record in 2009 by swimming 100.0 m in 46.91 s, his average speed was $(100.0\text{ m})/(46.91\text{ s}) = 2.132$ m/s. But because he swam two lengths in a 50-m pool, he started and ended at the same point and so had zero total displacement and zero average *velocity!* Both average speed and instantaneous speed are scalars, not vectors, because these quantities contain no information about direction. ∎

Example 2.1 **Average and instantaneous velocities**

A cheetah is crouched 20 m to the east of an observer (Fig. 2.6a). At time $t = 0$ the cheetah begins to run due east toward an antelope that is 50 m to the east of the observer. During the first 2.0 s of the attack, the cheetah's coordinate x varies with time according to the equation $x = 20$ m $+ (5.0$ m/s$^2)t^2$. (a) Find the cheetah's displacement between $t_1 = 1.0$ s and $t_2 = 2.0$ s. (b) Find its average velocity during that interval. (c) Find its instantaneous velocity at $t_1 = 1.0$ s by taking $\Delta t = 0.1$ s, then 0.01 s, then 0.001 s. (d) Derive an expression for the cheetah's instantaneous velocity as a function of time, and use it to find v_x at $t = 1.0$ s and $t = 2.0$ s.

SOLUTION

IDENTIFY and SET UP: Figure 2.6b shows our sketch of the cheetah's motion. We use Eq. (2.1) for displacement, Eq. (2.2) for average velocity, and Eq. (2.3) for instantaneous velocity.

Continued

2.6 A cheetah attacking an antelope from ambush. The animals are not drawn to the same scale as the axis.

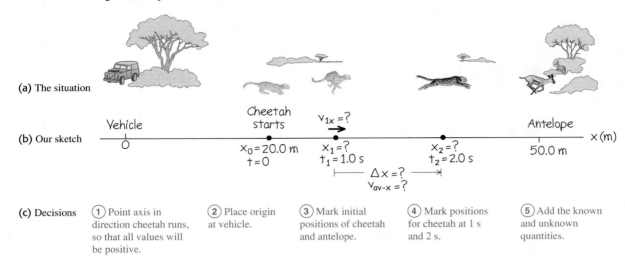

(a) The situation

(b) Our sketch

(c) Decisions

    ① Point axis in direction cheetah runs, so that all values will be positive.

    ② Place origin at vehicle.

    ③ Mark initial positions of cheetah and antelope.

    ④ Mark positions for cheetah at 1 s and 2 s.

    ⑤ Add the known and unknown quantities.

EXECUTE: (a) At $t_1 = 1.0$ s and $t_2 = 2.0$ s the cheetah's positions x_1 and x_2 are

$$x_1 = 20 \text{ m} + (5.0 \text{ m/s}^2)(1.0 \text{ s})^2 = 25 \text{ m}$$
$$x_2 = 20 \text{ m} + (5.0 \text{ m/s}^2)(2.0 \text{ s})^2 = 40 \text{ m}$$

The displacement during this 1.0-s interval is

$$\Delta x = x_2 - x_1 = 40 \text{ m} - 25 \text{ m} = 15 \text{ m}$$

(b) The average x-velocity during this interval is

$$v_{\text{av-}x} = \frac{x_2 - x_1}{t_2 - t_1} = \frac{40 \text{ m} - 25 \text{ m}}{2.0 \text{ s} - 1.0 \text{ s}} = \frac{15 \text{ m}}{1.0 \text{ s}} = 15 \text{ m/s}$$

(c) With $\Delta t = 0.1$ s the time interval is from $t_1 = 1.0$ s to a new $t_2 = 1.1$ s. At t_2 the position is

$$x_2 = 20 \text{ m} + (5.0 \text{ m/s}^2)(1.1 \text{ s})^2 = 26.05 \text{ m}$$

The average x-velocity during this 0.1-s interval is

$$v_{\text{av-}x} = \frac{26.05 \text{ m} - 25 \text{ m}}{1.1 \text{ s} - 1.0 \text{ s}} = 10.5 \text{ m/s}$$

Following this pattern, you can calculate the average x-velocities for 0.01-s and 0.001-s intervals: The results are 10.05 m/s and 10.005 m/s. As Δt gets smaller, the average x-velocity gets closer to 10.0 m/s, so we conclude that the instantaneous x-velocity at $t = 1.0$ s is 10.0 m/s. (We suspended the rules for significant-figure counting in these calculations.)

(d) To find the instantaneous x-velocity as a function of time, we take the derivative of the expression for x with respect to t. The derivative of a constant is zero, and for any n the derivative of t^n is nt^{n-1}, so the derivative of t^2 is $2t$. We therefore have

$$v_x = \frac{dx}{dt} = (5.0 \text{ m/s}^2)(2t) = (10 \text{ m/s}^2)t$$

At $t = 1.0$ s, this yields $v_x = 10$ m/s, as we found in part (c); at $t = 2.0$ s, $v_x = 20$ m/s.

EVALUATE: Our results show that the cheetah picked up speed from $t = 0$ (when it was at rest) to $t = 1.0$ s $(v_x = 10 \text{ m/s})$ to $t = 2.0$ s $(v_x = 20 \text{ m/s})$. This makes sense; the cheetah covered only 5 m during the interval $t = 0$ to $t = 1.0$ s, but it covered 15 m during the interval $t = 1.0$ s to $t = 2.0$ s.

MasteringPHYSICS

ActivPhysics 1.1: Analyzing Motion Using Diagrams

Finding Velocity on an x-t Graph

We can also find the x-velocity of a particle from the graph of its position as a function of time. Suppose we want to find the x-velocity of the dragster in Fig. 2.1 at point P_1. As point P_2 in Fig. 2.1 approaches point P_1, point p_2 in the x-t graphs of Figs. 2.7a and 2.7b approaches point p_1 and the average x-velocity is calculated over shorter time intervals Δt. In the limit that $\Delta t \rightarrow 0$, shown in Fig. 2.7c, the slope of the line p_1p_2 equals the slope of the line tangent to the curve at point p_1. Thus, *on a graph of position as a function of time for straight-line motion, the instantaneous x-velocity at any point is equal to the slope of the tangent to the curve at that point.*

If the tangent to the x-t curve slopes upward to the right, as in Fig. 2.7c, then its slope is positive, the x-velocity is positive, and the motion is in the positive x-direction. If the tangent slopes downward to the right, the slope of the x-t graph

2.7 Using an *x-t* graph to go from (a), (b) average *x*-velocity to (c) instantaneous *x*-velocity v_x. In (c) we find the slope of the tangent to the *x-t* curve by dividing any vertical interval (with distance units) along the tangent by the corresponding horizontal interval (with time units).

(a)

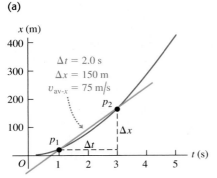

$\Delta t = 2.0$ s
$\Delta x = 150$ m
$v_{av\text{-}x} = 75$ m/s

As the average *x*-velocity $v_{av\text{-}x}$ is calculated over shorter and shorter time intervals ...

(b)

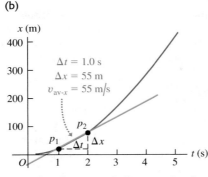

$\Delta t = 1.0$ s
$\Delta x = 55$ m
$v_{av\text{-}x} = 55$ m/s

... its value $v_{av\text{-}x} = \Delta x/\Delta t$ approaches the instantaneous *x*-velocity.

(c)

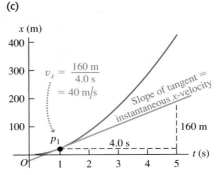

$v_x = \dfrac{160 \text{ m}}{4.0 \text{ s}} = 40$ m/s

Slope of tangent = instantaneous *x*-velocity

160 m

4.0 s

The instantaneous *x*-velocity v_x at any given point equals the slope of the tangent to the *x-t* curve at that point.

2.8 (a) The *x-t* graph of the motion of a particular particle. The slope of the tangent at any point equals the velocity at that point.
(b) A motion diagram showing the position and velocity of the particle at each of the times labeled on the *x-t* graph.

(a) *x-t* graph

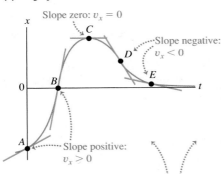

Slope zero: $v_x = 0$

Slope negative: $v_x < 0$

Slope positive: $v_x > 0$

(b) Particle's motion

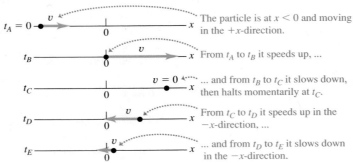

$t_A = 0$ — The particle is at $x < 0$ and moving in the +*x*-direction.

t_B — From t_A to t_B it speeds up, ...

t_C — $v = 0$ ... and from t_B to t_C it slows down, then halts momentarily at t_C.

t_D — From t_C to t_D it speeds up in the −*x*-direction, ...

t_E — ... and from t_D to t_E it slows down in the −*x*-direction.

The steeper the slope (positive or negative) of an object's *x-t* graph, the greater is the object's speed in the positive or negative *x*-direction.

and the *x*-velocity are negative, and the motion is in the negative *x*-direction. When the tangent is horizontal, the slope and the *x*-velocity are zero. Figure 2.8 illustrates these three possibilities.

Figure 2.8 actually depicts the motion of a particle in two ways: as (a) an *x-t* graph and (b) a **motion diagram** that shows the particle's position at various instants (like frames from a video of the particle's motion) as well as arrows to represent the particle's velocity at each instant. We will use both *x-t* graphs and motion diagrams in this chapter to help you understand motion. You will find it worth your while to draw *both* an *x-t* graph and a motion diagram as part of solving any problem involving motion.

2.9 An *x-t* graph for a particle.

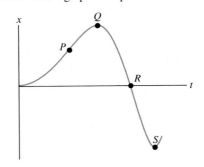

Test Your Understanding of Section 2.2 Figure 2.9 is an *x-t* graph of the motion of a particle. (a) Rank the values of the particle's *x*-velocity v_x at the points *P, Q, R,* and *S* from most positive to most negative. (b) At which points is v_x positive? (c) At which points is v_x negative? (d) At which points is v_x zero? (e) Rank the values of the particle's *speed* at the points *P, Q, R,* and *S* from fastest to slowest. ❙

2.3 Average and Instantaneous Acceleration

Just as velocity describes the rate of change of position with time, *acceleration* describes the rate of change of velocity with time. Like velocity, acceleration is a vector quantity. When the motion is along a straight line, its only nonzero component is along that line. As we'll see, acceleration in straight-line motion can refer to either speeding up or slowing down.

Average Acceleration

Let's consider again a particle moving along the x-axis. Suppose that at time t_1 the particle is at point P_1 and has x-component of (instantaneous) velocity v_{1x}, and at a later time t_2 it is at point P_2 and has x-component of velocity v_{2x}. So the x-component of velocity changes by an amount $\Delta v_x = v_{2x} - v_{1x}$ during the time interval $\Delta t = t_2 - t_1$.

We define the **average acceleration** of the particle as it moves from P_1 to P_2 to be a vector quantity whose x-component $a_{\text{av-}x}$ (called the **average x-acceleration**) equals Δv_x, the change in the x-component of velocity, divided by the time interval Δt:

$$a_{\text{av-}x} = \frac{v_{2x} - v_{1x}}{t_2 - t_1} = \frac{\Delta v_x}{\Delta t} \qquad \text{(average x-acceleration, straight-line motion)} \qquad (2.4)$$

For straight-line motion along the x-axis we will often call $a_{\text{av-}x}$ simply the average acceleration. (We'll encounter the other components of the average acceleration vector in Chapter 3.)

If we express velocity in meters per second and time in seconds, then average acceleration is in meters per second per second, or $(\text{m/s})/\text{s}$. This is usually written as m/s^2 and is read "meters per second squared."

CAUTION Acceleration vs. velocity Be very careful not to confuse acceleration with velocity! Velocity describes how a body's position changes with time; it tells us how fast and in what direction the body moves. Acceleration describes how the velocity changes with time; it tells us how the speed and direction of motion are changing. It may help to remember the phrase "acceleration is to velocity as velocity is to position." It can also help to imagine yourself riding along with the moving body. If the body accelerates forward and gains speed, you feel pushed backward in your seat; if it accelerates backward and loses speed, you feel pushed forward. If the velocity is constant and there's no acceleration, you feel neither sensation. (We'll see the reason for these sensations in Chapter 4.)

Example 2.2 | **Average acceleration**

An astronaut has left an orbiting spacecraft to test a new personal maneuvering unit. As she moves along a straight line, her partner on the spacecraft measures her velocity every 2.0 s, starting at time $t = 1.0$ s:

t	v_x	t	v_x
1.0 s	0.8 m/s	9.0 s	−0.4 m/s
3.0 s	1.2 m/s	11.0 s	−1.0 m/s
5.0 s	1.6 m/s	13.0 s	−1.6 m/s
7.0 s	1.2 m/s	15.0 s	−0.8 m/s

Find the average x-acceleration, and state whether the speed of the astronaut increases or decreases over each of these 2.0-s time intervals: (a) $t_1 = 1.0$ s to $t_2 = 3.0$ s; (b) $t_1 = 5.0$ s to $t_2 = 7.0$ s; (c) $t_1 = 9.0$ s to $t_2 = 11.0$ s; (d) $t_1 = 13.0$ s to $t_2 = 15.0$ s.

SOLUTION

IDENTIFY and SET UP: We'll use Eq. (2.4) to determine the average acceleration $a_{\text{av-}x}$ from the change in velocity over each time interval. To find the changes in speed, we'll use the idea that speed v is the magnitude of the instantaneous velocity v_x.

The upper part of Fig. 2.10 is our graph of the x-velocity as a function of time. On this v_x-t graph, the slope of the line connecting the endpoints of each interval is the average x-acceleration $a_{av-x} = \Delta v_x / \Delta t$ for that interval. The four slopes (and thus the *signs* of the average accelerations) are, respectively, positive, negative, negative, and positive. The third and fourth slopes (and thus the average accelerations themselves) have greater magnitude than the first and second.

2.10 Our graphs of x-velocity versus time (top) and average x-acceleration versus time (bottom) for the astronaut.

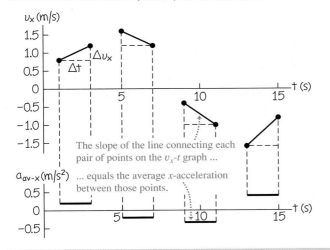

EXECUTE: Using Eq. (2.4), we find:

(a) $a_{av-x} = (1.2 \text{ m/s} - 0.8 \text{ m/s})/(3.0 \text{ s} - 1.0 \text{ s}) = 0.2 \text{ m/s}^2$. The speed (magnitude of instantaneous x-velocity) increases from 0.8 m/s to 1.2 m/s.

(b) $a_{av-x} = (1.2 \text{ m/s} - 1.6 \text{ m/s})/(7.0 \text{ s} - 5.0 \text{ s}) = -0.2 \text{ m/s}^2$. The speed decreases from 1.6 m/s to 1.2 m/s.

(c) $a_{av-x} = [-1.0 \text{ m/s} - (-0.4 \text{ m/s})]/(11.0 \text{ s} - 9.0 \text{ s}) = -0.3 \text{ m/s}^2$. The speed increases from 0.4 m/s to 1.0 m/s.

(d) $a_{av-x} = [-0.8 \text{ m/s} - (-1.6 \text{ m/s})]/(15.0 \text{ s} - 13.0 \text{ s}) = 0.4 \text{ m/s}^2$. The speed decreases from 1.6 m/s to 0.8 m/s.

In the lower part of Fig. 2.10, we graph the values of a_{av-x}.

EVALUATE: The signs and relative magnitudes of the average accelerations agree with our qualitative predictions. For future reference, note this connection among speed, velocity, and acceleration: Our results show that when the average x-acceleration has the *same* direction (same algebraic sign) as the initial velocity, as in intervals (a) and (c), the astronaut goes faster. When a_{av-x} has the *opposite* direction (opposite algebraic sign) from the initial velocity, as in intervals (b) and (d), she slows down. Thus positive x-acceleration means speeding up if the x-velocity is positive [interval (a)] but slowing down if the x-velocity is negative [interval (d)]. Similarly, negative x-acceleration means speeding up if the x-velocity is negative [interval (c)] but slowing down if the x-velocity is positive [interval (b)].

Instantaneous Acceleration

We can now define **instantaneous acceleration** following the same procedure that we used to define instantaneous velocity. As an example, suppose a race car driver is driving along a straightaway as shown in Fig. 2.11. To define the instantaneous acceleration at point P_1, we take the second point P_2 in Fig. 2.11 to be closer and closer to P_1 so that the average acceleration is computed over shorter and shorter time intervals. *The instantaneous acceleration is the limit of the average acceleration as the time interval approaches zero.* In the language of calculus, *instantaneous acceleration equals the derivative of velocity with time.* Thus

$$a_x = \lim_{\Delta t \to 0} \frac{\Delta v_x}{\Delta t} = \frac{dv_x}{dt} \quad \begin{array}{l}\text{(instantaneous } x\text{-acceleration,} \\ \text{straight-line motion)}\end{array} \quad (2.5)$$

Note that a_x in Eq. (2.5) is really the x-component of the acceleration vector, or the **instantaneous x-acceleration;** in straight-line motion, all other components of this vector are zero. From now on, when we use the term "acceleration," we will always mean instantaneous acceleration, not average acceleration.

2.11 A Grand Prix car at two points on the straightaway.

Example 2.3 **Average and instantaneous accelerations**

Suppose the x-velocity v_x of the car in Fig. 2.11 at any time t is given by the equation

$$v_x = 60 \text{ m/s} + (0.50 \text{ m/s}^3)t^2$$

(a) Find the change in x-velocity of the car in the time interval $t_1 = 1.0$ s to $t_2 = 3.0$ s. (b) Find the average x-acceleration in this time interval. (c) Find the instantaneous x-acceleration at time $t_1 = 1.0$ s by taking Δt to be first 0.1 s, then 0.01 s, then 0.001 s. (d) Derive an expression for the instantaneous x-acceleration as a function of time, and use it to find a_x at $t = 1.0$ s and $t = 3.0$ s.

SOLUTION

IDENTIFY and SET UP: This example is analogous to Example 2.1 in Section 2.2. (Now is a good time to review that example.) In Example 2.1 we found the average x-velocity from the change in position over shorter and shorter time intervals, and we obtained an expression for the instantaneous x-velocity by differentiating the position as a function of time. In this example we have an exact parallel. Using Eq. (2.4), we'll find the average x-*acceleration* from the change in x-*velocity* over a time interval. Likewise, using Eq. (2.5), we'll obtain an expression for the instantaneous x-*acceleration* by differentiating the x-*velocity* as a function of time.

EXECUTE: (a) Before we can apply Eq. (2.4), we must find the x-velocity at each time from the given equation. At $t_1 = 1.0$ s and $t_2 = 3.0$ s, the velocities are

$$v_{1x} = 60 \text{ m/s} + (0.50 \text{ m/s}^3)(1.0 \text{ s})^2 = 60.5 \text{ m/s}$$
$$v_{2x} = 60 \text{ m/s} + (0.50 \text{ m/s}^3)(3.0 \text{ s})^2 = 64.5 \text{ m/s}$$

The change in x-velocity Δv_x between $t_1 = 1.0$ s and $t_2 = 3.0$ s is

$$\Delta v_x = v_{2x} - v_{1x} = 64.5 \text{ m/s} - 60.5 \text{ m/s} = 4.0 \text{ m/s}$$

(b) The average x-acceleration during this time interval of duration $t_2 - t_1 = 2.0$ s is

$$a_{\text{av-}x} = \frac{v_{2x} - v_{1x}}{t_2 - t_1} = \frac{4.0 \text{ m/s}}{2.0 \text{ s}} = 2.0 \text{ m/s}^2$$

During this time interval the x-velocity and average x-acceleration have the same algebraic sign (in this case, positive), and the car speeds up.

(c) When $\Delta t = 0.1$ s, we have $t_2 = 1.1$ s. Proceeding as before, we find

$$v_{2x} = 60 \text{ m/s} + (0.50 \text{ m/s}^3)(1.1 \text{ s})^2 = 60.605 \text{ m/s}$$
$$\Delta v_x = 0.105 \text{ m/s}$$
$$a_{\text{av-}x} = \frac{\Delta v_x}{\Delta t} = \frac{0.105 \text{ m/s}}{0.1 \text{ s}} = 1.05 \text{ m/s}^2$$

You should follow this pattern to calculate $a_{\text{av-}x}$ for $\Delta t = 0.01$ s and $\Delta t = 0.001$ s; the results are $a_{\text{av-}x} = 1.005 \text{ m/s}^2$ and $a_{\text{av-}x} = 1.0005 \text{ m/s}^2$, respectively. As Δt gets smaller, the average x-acceleration gets closer to 1.0 m/s^2, so the instantaneous x-acceleration at $t = 1.0$ s is 1.0 m/s^2.

(d) By Eq. (2.5) the instantaneous x-acceleration is $a_x = dv_x/dt$. The derivative of a constant is zero and the derivative of t^2 is $2t$, so

$$a_x = \frac{dv_x}{dt} = \frac{d}{dt}[60 \text{ m/s} + (0.50 \text{ m/s}^3)t^2]$$
$$= (0.50 \text{ m/s}^3)(2t) = (1.0 \text{ m/s}^3)t$$

When $t = 1.0$ s,

$$a_x = (1.0 \text{ m/s}^3)(1.0 \text{ s}) = 1.0 \text{ m/s}^2$$

When $t = 3.0$ s,

$$a_x = (1.0 \text{ m/s}^3)(3.0 \text{ s}) = 3.0 \text{ m/s}^2$$

EVALUATE: Neither of the values we found in part (d) is equal to the average x-acceleration found in part (b). That's because the car's instantaneous x-acceleration varies with time. The rate of change of acceleration with time is sometimes called the "jerk."

Finding Acceleration on a v_x-t Graph or an x-t Graph

In Section 2.2 we interpreted average and instantaneous x-velocity in terms of the slope of a graph of position versus time. In the same way, we can interpret average and instantaneous x-acceleration by using a graph with instantaneous velocity v_x on the vertical axis and time t on the horizontal axis—that is, a v_x-t **graph** (Fig. 2.12). The points on the graph labeled p_1 and p_2 correspond to points P_1 and P_2 in Fig. 2.11. The average x-acceleration $a_{\text{av-}x} = \Delta v_x/\Delta t$ during this interval is the slope of the line p_1p_2. As point P_2 in Fig. 2.11 approaches point P_1, point p_2 in the v_x-t graph of Fig. 2.12 approaches point p_1, and the slope of the line p_1p_2 approaches the slope of the line tangent to the curve at point p_1. Thus, *on a graph of x-velocity as a function of time, the instantaneous x-acceleration at any point is equal to the slope of the tangent to the curve at that point.* Tangents drawn at different points along the curve in Fig. 2.12 have different slopes, so the instantaneous x-acceleration varies with time.

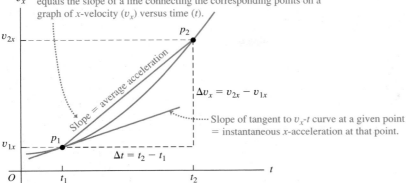

For a displacement along the x-axis, an object's average x-acceleration equals the slope of a line connecting the corresponding points on a graph of x-velocity (v_x) versus time (t).

$\Delta v_x = v_{2x} - v_{1x}$

Slope of tangent to v_x-t curve at a given point = instantaneous x-acceleration at that point.

$\Delta t = t_2 - t_1$

2.12 A v_x-t graph of the motion in Fig. 2.11.

CAUTION **The signs of x-acceleration and x-velocity** By itself, the algebraic sign of the x-acceleration does *not* tell you whether a body is speeding up or slowing down. You must compare the signs of the x-velocity and the x-acceleration. When v_x and a_x have the *same* sign, the body is speeding up. If both are positive, the body is moving in the positive direction with increasing speed. If both are negative, the body is moving in the negative direction with an x-velocity that is becoming more and more negative, and again the speed is increasing. When v_x and a_x have *opposite* signs, the body is slowing down. If v_x is positive and a_x is negative, the body is moving in the positive direction with decreasing speed; if v_x is negative and a_x is positive, the body is moving in the negative direction with an x-velocity that is becoming less negative, and again the body is slowing down. Table 2.3 summarizes these ideas, and Fig. 2.13 illustrates some of these possibilities. ▮

The term "deceleration" is sometimes used for a decrease in speed. Because it may mean positive or negative a_x, depending on the sign of v_x, we avoid this term.

We can also learn about the acceleration of a body from a graph of its *position* versus time. Because $a_x = dv_x/dt$ and $v_x = dx/dt$, we can write

$$a_x = \frac{dv_x}{dt} = \frac{d}{dt}\left(\frac{dx}{dt}\right) = \frac{d^2x}{dt^2} \qquad (2.6)$$

Table 2.3 Rules for the Sign of x-Acceleration

If x-velocity is:	. . . x-acceleration is:
Positive & increasing (getting more positive)	Positive: Particle is moving in $+x$-direction & speeding up
Positive & decreasing (getting less positive)	Negative: Particle is moving in $+x$-direction & slowing down
Negative & increasing (getting less negative)	Positive: Particle is moving in $-x$-direction & slowing down
Negative & decreasing (getting more negative)	Negative: Particle is moving in $-x$-direction & speeding up

Note: These rules apply to both the average x-acceleration $a_{\text{av-}x}$ and the instantaneous x-acceleration a_x.

2.13 (a) A v_x-t graph of the motion of a different particle from that shown in Fig. 2.8. The slope of the tangent at any point equals the x-acceleration at that point. (b) A motion diagram showing the position, velocity, and acceleration of the particle at each of the times labeled on the v_x-t graph. The positions are consistent with the v_x-t graph; for instance, from t_A to t_B the velocity is negative, so at t_B the particle is at a more negative value of x than at t_A.

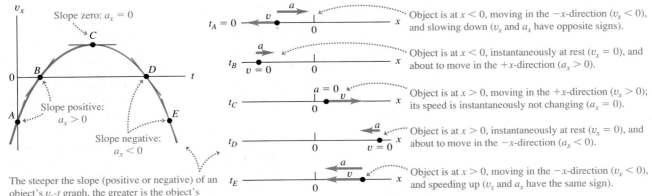

2.14 (a) The same *x-t* graph as shown in Fig. 2.8a. The *x*-velocity is equal to the *slope* of the graph, and the acceleration is given by the *concavity* or *curvature* of the graph. (b) A motion diagram showing the position, velocity, and acceleration of the particle at each of the times labeled on the *x-t* graph.

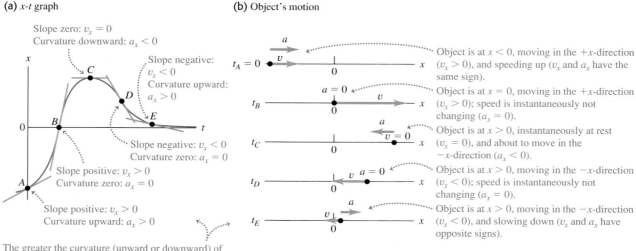

(a) *x-t* graph

Slope zero: $v_x = 0$
Curvature downward: $a_x < 0$

Slope negative:
$v_x < 0$
Curvature upward:
$a_x > 0$

Slope negative: $v_x < 0$
Curvature zero: $a_x = 0$

Slope positive: $v_x > 0$
Curvature zero: $a_x = 0$

Slope positive: $v_x > 0$
Curvature upward: $a_x > 0$

The greater the curvature (upward or downward) of an object's *x-t* graph, the greater is the object's acceleration in the positive or negative *x*-direction.

(b) Object's motion

$t_A = 0$ — Object is at $x < 0$, moving in the $+x$-direction ($v_x > 0$), and speeding up (v_x and a_x have the same sign).

t_B — Object is at $x = 0$, moving in the $+x$-direction ($v_x > 0$); speed is instantaneously not changing ($a_x = 0$).

t_C — Object is at $x > 0$, instantaneously at rest ($v_x = 0$), and about to move in the $-x$-direction ($a_x < 0$).

t_D — Object is at $x > 0$, moving in the $-x$-direction ($v_x < 0$); speed is instantaneously not changing ($a_x = 0$).

t_E — Object is at $x > 0$, moving in the $-x$-direction ($v_x < 0$), and slowing down (v_x and a_x have opposite signs).

That is, a_x is the second derivative of x with respect to t. The second derivative of any function is directly related to the *concavity* or *curvature* of the graph of that function (Fig. 2.14). Where the *x-t* graph is concave up (curved upward), the *x*-acceleration is positive and v_x is increasing; at a point where the *x-t* graph is concave down (curved downward), the *x*-acceleration is negative and v_x is decreasing. At a point where the *x-t* graph has no curvature, such as an inflection point, the *x*-acceleration is zero and the velocity is not changing. Figure 2.14 shows all three of these possibilities.

Examining the curvature of an *x-t* graph is an easy way to decide what the *sign* of acceleration is. This technique is less helpful for determining numerical values of acceleration because the curvature of a graph is hard to measure accurately.

2.15 A motion diagram for a particle moving in a straight line in the positive *x*-direction with constant positive *x*-acceleration a_x. The position, velocity, and acceleration are shown at five equally spaced times.

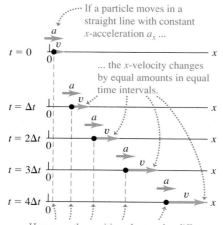

If a particle moves in a straight line with constant *x*-acceleration a_x ...

$t = 0$

... the *x*-velocity changes by equal amounts in equal time intervals.

$t = \Delta t$

$t = 2\Delta t$

$t = 3\Delta t$

$t = 4\Delta t$

However, the position changes by *different* amounts in equal time intervals because the velocity is changing.

Test Your Understanding of Section 2.3 Look again at the *x-t* graph in Fig. 2.9 at the end of Section 2.2. (a) At which of the points P, Q, R, and S is the *x*-acceleration a_x positive? (b) At which points is the *x*-acceleration negative? (c) At which points does the *x*-acceleration appear to be zero? (d) At each point state whether the velocity is increasing, decreasing, or not changing. ❙

2.4 Motion with Constant Acceleration

The simplest kind of accelerated motion is straight-line motion with *constant* acceleration. In this case the velocity changes at the same rate throughout the motion. As an example, a falling body has a constant acceleration if the effects of the air are not important. The same is true for a body sliding on an incline or along a rough horizontal surface, or for an airplane being catapulted from the deck of an aircraft carrier.

Figure 2.15 is a motion diagram showing the position, velocity, and acceleration for a particle moving with constant acceleration. Figures 2.16 and 2.17 depict this same motion in the form of graphs. Since the *x*-acceleration is constant, the a_x-*t* **graph** (graph of *x*-acceleration versus time) in Fig. 2.16 is a horizontal line. The graph of *x*-velocity versus time, or v_x-*t* graph, has a constant *slope* because the acceleration is constant, so this graph is a straight line (Fig. 2.17).

When the x-acceleration a_x is constant, the average x-acceleration $a_{\text{av-}x}$ for any time interval is the same as a_x. This makes it easy to derive equations for the position x and the x-velocity v_x as functions of time. To find an expression for v_x, we first replace $a_{\text{av-}x}$ in Eq. (2.4) by a_x:

$$a_x = \frac{v_{2x} - v_{1x}}{t_2 - t_1} \qquad (2.7)$$

Now we let $t_1 = 0$ and let t_2 be any later time t. We use the symbol v_{0x} for the x-velocity at the initial time $t = 0$; the x-velocity at the later time t is v_x. Then Eq. (2.7) becomes

$$a_x = \frac{v_x - v_{0x}}{t - 0} \qquad \text{or}$$

$$v_x = v_{0x} + a_x t \qquad \text{(constant x-acceleration only)} \qquad (2.8)$$

In Eq. (2.8) the term $a_x t$ is the product of the constant rate of change of x-velocity, a_x, and the time interval t. Therefore it equals the *total* change in x-velocity from the initial time $t = 0$ to the later time t. The x-velocity v_x at any time t then equals the initial x-velocity v_{0x} (at $t = 0$) plus the change in x-velocity $a_x t$ (Fig. 2.17).

Equation (2.8) also says that the change in x-velocity $v_x - v_{0x}$ of the particle between $t = 0$ and any later time t equals the *area* under the a_x-t graph between those two times. You can verify this from Fig. 2.16: Under this graph is a rectangle of vertical side a_x, horizontal side t, and area $a_x t$. From Eq. (2.8) this is indeed equal to the change in velocity $v_x - v_{0x}$. In Section 2.6 we'll show that even if the x-acceleration is not constant, the change in x-velocity during a time interval is still equal to the area under the a_x-t curve, although in that case Eq. (2.8) does not apply.

Next we'll derive an equation for the position x as a function of time when the x-acceleration is constant. To do this, we use two different expressions for the average x-velocity $v_{\text{av-}x}$ during the interval from $t = 0$ to any later time t. The first expression comes from the definition of $v_{\text{av-}x}$, Eq. (2.2), which is true whether or not the acceleration is constant. We call the position at time $t = 0$ the *initial position*, denoted by x_0. The position at the later time t is simply x. Thus for the time interval $\Delta t = t - 0$ the displacement is $\Delta x = x - x_0$, and Eq. (2.2) gives

$$v_{\text{av-}x} = \frac{x - x_0}{t} \qquad (2.9)$$

We can also get a second expression for $v_{\text{av-}x}$ that is valid only when the x-acceleration is constant, so that the x-velocity changes at a constant rate. In this case the average x-velocity for the time interval from 0 to t is simply the average of the x-velocities at the beginning and end of the interval:

$$v_{\text{av-}x} = \frac{v_{0x} + v_x}{2} \qquad \text{(constant x-acceleration only)} \qquad (2.10)$$

(This equation is *not* true if the x-acceleration varies during the time interval.) We also know that with constant x-acceleration, the x-velocity v_x at any time t is given by Eq. (2.8). Substituting that expression for v_x into Eq. (2.10), we find

$$v_{\text{av-}x} = \tfrac{1}{2}(v_{0x} + v_{0x} + a_x t)$$

$$= v_{0x} + \tfrac{1}{2}a_x t \qquad \text{(constant x-acceleration only)} \qquad (2.11)$$

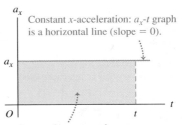

2.16 An acceleration-time $(a_x\text{-}t)$ graph for straight-line motion with constant positive x-acceleration a_x.

Area under a_x-t graph $= v_x - v_{0x}$ = change in x-velocity from time 0 to time t.

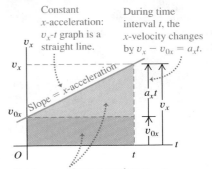

2.17 A velocity-time $(v_x\text{-}t)$ graph for straight-line motion with constant positive x-acceleration a_x. The initial x-velocity v_{0x} is also positive in this case.

Total area under v_x-t graph $= x - x_0$ = change in x-coordinate from time 0 to time t.

MasteringPHYSICS

PhET: Forces in 1 Dimension
ActivPhysics 1.1: Analyzing Motion Using Diagrams
ActivPhysics 1.2: Analyzing Motion Using Graphs
ActivPhysics 1.3: Predicting Motion from Graphs
ActivPhysics 1.4: Predicting Motion from Equations
ActivPhysics 1.5: Problem-Solving Strategies for Kinematics
ActivPhysics 1.6: Skier Races Downhill

Finally, we set Eqs. (2.9) and (2.11) equal to each other and simplify:

$$v_{0x} + \frac{1}{2}a_x t = \frac{x - x_0}{t} \qquad \text{or}$$

$$x = x_0 + v_{0x}t + \frac{1}{2}a_x t^2 \qquad \text{(constant } x\text{-acceleration only)} \qquad (2.12)$$

Here's what Eq. (2.12) tells us: If at time $t = 0$ a particle is at position x_0 and has x-velocity v_{0x}, its new position x at any later time t is the sum of three terms—its initial position x_0, plus the distance $v_{0x}t$ that it would move if its x-velocity were constant, plus an additional distance $\frac{1}{2}a_x t^2$ caused by the change in x-velocity.

A graph of Eq. (2.12)—that is, an x-t graph for motion with constant x-acceleration (Fig. 2.18a)—is always a *parabola*. Figure 2.18b shows such a graph. The curve intercepts the vertical axis (x-axis) at x_0, the position at $t = 0$. The slope of the tangent at $t = 0$ equals v_{0x}, the initial x-velocity, and the slope of the tangent at any time t equals the x-velocity v_x at that time. The slope and x-velocity are continuously increasing, so the x-acceleration a_x is positive; you can also see this because the graph in Fig. 2.18b is concave up (it curves upward). If a_x is negative, the x-t graph is a parabola that is concave down (has a downward curvature).

If there is zero x-acceleration, the x-t graph is a straight line; if there is a constant x-acceleration, the additional $\frac{1}{2}a_x t^2$ term in Eq. (2.12) for x as a function of t curves the graph into a parabola (Fig. 2.19a). We can analyze the v_x-t graph in the same way. If there is zero x-acceleration this graph is a horizontal line (the x-velocity is constant); adding a constant x-acceleration gives a slope to the v_x-t graph (Fig. 2.19b).

2.18 (a) Straight-line motion with constant acceleration. (b) A position-time (x-t) graph for this motion (the same motion as is shown in Figs. 2.15, 2.16, and 2.17). For this motion the initial position x_0, the initial velocity v_{0x}, and the acceleration a_x are all positive.

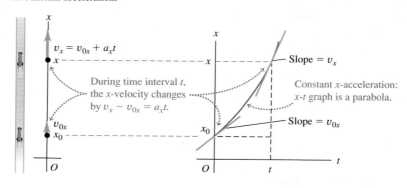

(a) A race car moves in the x-direction with constant acceleration.

(b) The x-t graph

2.19 (a) How a constant x-acceleration affects a body's (a) x-t graph and (b) v_x-t graph.

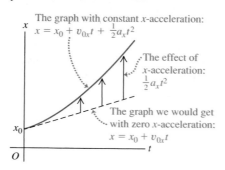

(a) An x-t graph for an object moving with positive constant x-acceleration

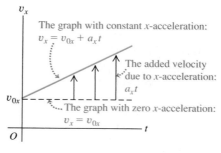

(b) The v_x-t graph for the same object

Just as the change in x-velocity of the particle equals the area under the a_x-t graph, the displacement—that is, the change in position—equals the area under the v_x-t graph. To be specific, the displacement $x - x_0$ of the particle between $t = 0$ and any later time t equals the area under the v_x-t graph between those two times. In Fig. 2.17 we divide the area under the graph into a dark-colored rectangle (vertical side v_{0x}, horizontal side t, and area $v_{0x}t$) and a light-colored right triangle (vertical side $a_x t$, horizontal side t, and area $\frac{1}{2}(a_x t)(t) = \frac{1}{2}a_x t^2$). The total area under the v_x-t graph is

$$x - x_0 = v_{0x}t + \tfrac{1}{2}a_x t^2$$

in agreement with Eq. (2.12).

The displacement during a time interval is always equal to the area under the v_x-t curve. This is true even if the acceleration is *not* constant, although in that case Eq. (2.12) does not apply. (We'll show this in Section 2.6.)

It's often useful to have a relationship for position, x-velocity, and (constant) x-acceleration that does not involve the time. To obtain this, we first solve Eq. (2.8) for t and then substitute the resulting expression into Eq. (2.12):

$$t = \frac{v_x - v_{0x}}{a_x}$$

$$x = x_0 + v_{0x}\left(\frac{v_x - v_{0x}}{a_x}\right) + \tfrac{1}{2}a_x\left(\frac{v_x - v_{0x}}{a_x}\right)^2$$

We transfer the term x_0 to the left side and multiply through by $2a_x$:

$$2a_x(x - x_0) = 2v_{0x}v_x - 2v_{0x}^2 + v_x^2 - 2v_{0x}v_x + v_{0x}^2$$

Finally, simplifying gives us

$$v_x^2 = v_{0x}^2 + 2a_x(x - x_0) \qquad \text{(constant } x\text{-acceleration only)} \qquad \text{(2.13)}$$

We can get one more useful relationship by equating the two expressions for $v_{\text{av-}x}$, Eqs. (2.9) and (2.10), and multiplying through by t. Doing this, we obtain

$$x - x_0 = \left(\frac{v_{0x} + v_x}{2}\right)t \qquad \text{(constant } x\text{-acceleration only)} \qquad \text{(2.14)}$$

Note that Eq. (2.14) does not contain the x-acceleration a_x. This equation can be handy when a_x is constant but its value is unknown.

Equations (2.8), (2.12), (2.13), and (2.14) are the *equations of motion with constant acceleration* (Table 2.4). By using these equations, we can solve *any* problem involving straight-line motion of a particle with constant acceleration.

For the particular case of motion with constant x-acceleration depicted in Fig. 2.15 and graphed in Figs. 2.16, 2.17, and 2.18, the values of x_0, v_{0x}, and a_x are all positive. We invite you to redraw these figures for cases in which one, two, or all three of these quantities are negative.

MasteringPHYSICS

PhET: The Moving Man
ActivPhysics 1.8: Seat Belts Save Lives
ActivPhysics 1.9: Screeching to a Halt
ActivPhysics 1.11: Car Starts, Then Stops
ActivPhysics 1.12: Solving Two-Vehicle Problems
ActivPhysics 1.13: Car Catches Truck
ActivPhysics 1.14: Avoiding a Rear-End Collision

Table 2.4 Equations of Motion with Constant Acceleration

Equation		Includes Quantities		
$v_x = v_{0x} + a_x t$ (2.8)	t		v_x	a_x
$x = x_0 + v_{0x}t + \tfrac{1}{2}a_x t^2$ (2.12)	t	x		a_x
$v_x^2 = v_{0x}^2 + 2a_x(x - x_0)$ (2.13)		x	v_x	a_x
$x - x_0 = \left(\dfrac{v_{0x} + v_x}{2}\right)t$ (2.14)	t	x	v_x	

Problem-Solving Strategy 2.1 | Motion with Constant Acceleration

IDENTIFY *the relevant concepts:* In most straight-line motion problems, you can use the constant-acceleration equations (2.8), (2.12), (2.13), and (2.14). If you encounter a situation in which the acceleration *isn't* constant, you'll need a different approach (see Section 2.6).

SET UP *the problem* using the following steps:
1. Read the problem carefully. Make a motion diagram showing the location of the particle at the times of interest. Decide where to place the origin of coordinates and which axis direction is positive. It's often helpful to place the particle at the origin at time $t = 0$; then $x_0 = 0$. Remember that your choice of the positive axis direction automatically determines the positive directions for *x*-velocity and *x*-acceleration. If *x* is positive to the right of the origin, then v_x and a_x are also positive toward the right.
2. Identify the physical quantities (times, positions, velocities, and accelerations) that appear in Eqs. (2.8), (2.12), (2.13), and (2.14) and assign them appropriate symbols — x, x_0, v_x, v_{0x}, and a_x, or symbols related to those. Translate the prose into physics: "*When* does the particle arrive at its highest point" means "What is the value of *t* when *x* has its maximum value?" In Example 2.4 below, "*Where* is the motorcyclist when his velocity is 25 m/s?" means "What is the value of *x* when $v_x = 25$ m/s?" Be alert for implicit information. For example, "A car sits at a stop light" usually means $v_{0x} = 0$.
3. Make a list of the quantities such as x, x_0, v_x, v_{0x}, a_x, and t. Some of them will be known and some will be unknown.

Write down the values of the known quantities, and decide which of the unknowns are the target variables. Make note of the *absence* of any of the quantities that appear in the four constant-acceleration equations.
4. Use Table 2.4 to identify the applicable equations. (These are often the equations that don't include any of the absent quantities that you identified in step 3.) Usually you'll find a single equation that contains only one of the target variables. Sometimes you must find two equations, each containing the same two unknowns.
5. Sketch graphs corresponding to the applicable equations. The v_x-t graph of Eq. (2.8) is a straight line with slope a_x. The x-t graph of Eq. (2.12) is a parabola that's concave up if a_x is positive and concave down if a_x is negative.
6. On the basis of your accumulated experience with such problems, and taking account of what your sketched graphs tell you, make any qualitative and quantitative predictions you can about the solution.

EXECUTE *the solution:* If a single equation applies, solve it for the target variable, *using symbols only*; then substitute the known values and calculate the value of the target variable. If you have two equations in two unknowns, solve them simultaneously for the target variables.

EVALUATE *your answer:* Take a hard look at your results to see whether they make sense. Are they within the general range of values that you expected?

Example 2.4 | Constant-acceleration calculations

A motorcyclist heading east through a small town accelerates at a constant 4.0 m/s^2 after he leaves the city limits (Fig. 2.20). At time $t = 0$ he is 5.0 m east of the city-limits signpost, moving east at 15 m/s. (a) Find his position and velocity at $t = 2.0$ s. (b) Where is he when his velocity is 25 m/s?

SOLUTION

IDENTIFY and SET UP: The *x*-acceleration is constant, so we can use the constant-acceleration equations. We take the signpost as the origin of coordinates ($x = 0$) and choose the positive *x*-axis to point east (see Fig. 2.20, which is also a motion diagram). The known variables are the initial position and velocity, $x_0 = 5.0$ m and $v_{0x} = 15$ m/s, and the acceleration, $a_x = 4.0$ m/s^2. The unknown target variables in part (a) are the values of the position x and the *x*-velocity v_x at $t = 2.0$ s; the target variable in part (b) is the value of x when $v_x = 25$ m/s.

EXECUTE: (a) Since we know the values of x_0, v_{0x}, and a_x, Table 2.4 tells us that we can find the position x at $t = 2.0$ s by using

2.20 A motorcyclist traveling with constant acceleration.

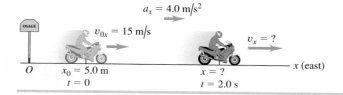

Eq. (2.12) and the *x*-velocity v_x at this time by using Eq. (2.8):

$$x = x_0 + v_{0x}t + \tfrac{1}{2}a_x t^2$$
$$= 5.0 \text{ m} + (15 \text{ m/s})(2.0 \text{ s}) + \tfrac{1}{2}(4.0 \text{ m/s}^2)(2.0 \text{ s})^2$$
$$= 43 \text{ m}$$

$$v_x = v_{0x} + a_x t$$
$$= 15 \text{ m/s} + (4.0 \text{ m/s}^2)(2.0 \text{ s}) = 23 \text{ m/s}$$

(b) We want to find the value of x when $v_x = 25$ m/s, but we don't know the time when the motorcycle has this velocity. Table 2.4 tells us that we should use Eq. (2.13), which involves x, v_x, and a_x but does *not* involve *t*:

$$v_x^2 = v_{0x}^2 + 2a_x(x - x_0)$$

Solving for x and substituting the known values, we find

$$x = x_0 + \frac{v_x^2 - v_{0x}^2}{2a_x}$$
$$= 5.0 \text{ m} + \frac{(25 \text{ m/s})^2 - (15 \text{ m/s})^2}{2(4.0 \text{ m/s}^2)} = 55 \text{ m}$$

EVALUATE: You can check the result in part (b) by first using Eq. (2.8), $v_x = v_{0x} + a_x t$, to find the time at which $v_x = 25$ m/s, which turns out to be $t = 2.5$ s. You can then use Eq. (2.12), $x = x_0 + v_{0x}t + \tfrac{1}{2}a_x t^2$, to solve for x. You should find $x = 55$ m, the same answer as above. That's the long way to solve the problem, though. The method we used in part (b) is much more efficient.

Example 2.5 Two bodies with different accelerations

A motorist traveling with a constant speed of 15 m/s (about 34 mi/h) passes a school-crossing corner, where the speed limit is 10 m/s (about 22 mi/h). Just as the motorist passes the school-crossing sign, a police officer on a motorcycle stopped there starts in pursuit with a constant acceleration of 3.0 m/s² (Fig. 2.21a). (a) How much time elapses before the officer passes the motorist? (b) What is the officer's speed at that time? (c) At that time, what distance has each vehicle traveled?

SOLUTION

IDENTIFY and SET UP: The officer and the motorist both move with constant acceleration (equal to zero for the motorist), so we can use the constant-acceleration formulas. We take the origin at the sign, so $x_0 = 0$ for both, and we take the positive direction to the right. Let x_P and x_M represent the positions of the officer and the motorist at any time; their initial velocities are $v_{P0x} = 0$ and $v_{M0x} = 15$ m/s, and their accelerations are $a_{Px} = 3.0$ m/s² and $a_{Mx} = 0$. Our target variable in part (a) is the time when the officer passes the motorist—that is, when the two vehicles are at the same position x; Table 2.4 tells us that Eq. (2.12) is useful for this part. In part (b) we're looking for the officer's speed v (the magnitude of his velocity) at the time found in part (a). We'll use Eq. (2.8) for this part. In part (c) we'll use Eq. (2.12) again to find the position of either vehicle at this same time.

Figure 2.21b shows an x-t graph for both vehicles. The straight line represents the motorist's motion, $x_M = x_{M0} + v_{M0x}t = v_{M0x}t$. The graph for the officer's motion is the right half of a concave–up parabola:

$$x_P = x_{P0} + v_{P0x}t + \tfrac{1}{2}a_{Px}t^2 = \tfrac{1}{2}a_{Px}t^2$$

A good sketch will show that the officer and motorist are at the same position ($x_P = x_M$) at about $t = 10$ s, at which time both have traveled about 150 m from the sign.

EXECUTE: (a) To find the value of the time t at which the motorist and police officer are at the same position, we set $x_P = x_M$ by equating the expressions above and solving that equation for t:

$$v_{M0x}t = \tfrac{1}{2}a_{Px}t^2$$

$$t = 0 \quad \text{or} \quad t = \frac{2v_{M0x}}{a_{Px}} = \frac{2(15 \text{ m/s})}{3.0 \text{ m/s}^2} = 10 \text{ s}$$

Both vehicles have the same x-coordinate at *two* times, as Fig. 2.21b indicates. At $t = 0$ the motorist passes the officer; at $t = 10$ s the officer passes the motorist.

(b) We want the magnitude of the officer's x-velocity v_{Px} at the time t found in part (a). Substituting the values of v_{P0x} and a_{Px} into Eq. (2.8) along with $t = 10$ s from part (a), we find

$$v_{Px} = v_{P0x} + a_{Px}t = 0 + (3.0 \text{ m/s}^2)(10 \text{ s}) = 30 \text{ m/s}$$

The officer's speed is the absolute value of this, which is also 30 m/s.

(c) In 10 s the motorist travels a distance

$$x_M = v_{M0x}t = (15 \text{ m/s})(10 \text{ s}) = 150 \text{ m}$$

and the officer travels

$$x_P = \tfrac{1}{2}a_{Px}t^2 = \tfrac{1}{2}(3.0 \text{ m/s}^2)(10 \text{ s})^2 = 150 \text{ m}$$

This verifies that they have gone equal distances when the officer passes the motorist.

EVALUATE: Our results in parts (a) and (c) agree with our estimates from our sketch. Note that at the time when the officer passes the motorist, they do *not* have the same velocity. At this time the motorist is moving at 15 m/s and the officer is moving at 30 m/s. You can also see this from Fig. 2.21b. Where the two x-t curves cross, their slopes (equal to the values of v_x for the two vehicles) are different.

Is it just coincidence that when the two vehicles are at the same position, the officer is going twice the speed of the motorist? Equation (2.14), $x - x_0 = [(v_{0x} + v_x)/2]t$, gives the answer. The motorist has constant velocity, so $v_{M0x} = v_{Mx}$, and the distance $x - x_0$ that the motorist travels in time t is $v_{M0x}t$. The officer has zero initial velocity, so in the same time t the officer travels a distance $\tfrac{1}{2}v_{Px}t$. If the two vehicles cover the same distance in the same amount of time, the two values of $x - x_0$ must be the same. Hence when the officer passes the motorist $v_{M0x}t = \tfrac{1}{2}v_{Px}t$ and $v_{Px} = 2v_{M0x}$—that is, the officer has exactly twice the motorist's velocity. Note that this is true no matter what the value of the officer's acceleration.

2.21 (a) Motion with constant acceleration overtaking motion with constant velocity. (b) A graph of x versus t for each vehicle.

(b)

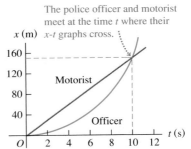

(a)

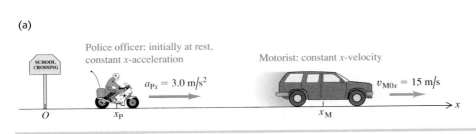

Test Your Understanding of Section 2.4 Four possible v_x-t graphs are shown for the two vehicles in Example 2.5. Which graph is correct?

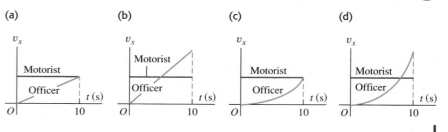

(a)   (b)   (c)   (d)

2.5 Freely Falling Bodies

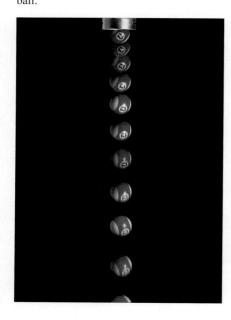

2.22 Multiflash photo of a freely falling ball.

MasteringPHYSICS

PhET: Lunar Lander
ActivPhysics 1.7: Balloonist Drops Lemonade
ActivPhysics 1.10: Pole-Vaulter Lands

The most familiar example of motion with (nearly) constant acceleration is a body falling under the influence of the earth's gravitational attraction. Such motion has held the attention of philosophers and scientists since ancient times. In the fourth century B.C., Aristotle thought (erroneously) that heavy bodies fall faster than light bodies, in proportion to their weight. Nineteen centuries later, Galileo (see Section 1.1) argued that a body should fall with a downward acceleration that is constant and independent of its weight.

Experiment shows that if the effects of the air can be neglected, Galileo is right; all bodies at a particular location fall with the same downward acceleration, regardless of their size or weight. If in addition the distance of the fall is small compared with the radius of the earth, and if we ignore small effects due to the earth's rotation, the acceleration is constant. The idealized motion that results under all of these assumptions is called **free fall,** although it includes rising as well as falling motion. (In Chapter 3 we will extend the discussion of free fall to include the motion of projectiles, which move both vertically and horizontally.)

Figure 2.22 is a photograph of a falling ball made with a stroboscopic light source that produces a series of short, intense flashes. As each flash occurs, an image of the ball at that instant is recorded on the photograph. There are equal time intervals between flashes, so the average velocity of the ball between successive flashes is proportional to the distance between corresponding images. The increasing distances between images show that the velocity is continuously changing; the ball is accelerating downward. Careful measurement shows that the velocity change is the same in each time interval, so the acceleration of the freely falling ball is constant.

The constant acceleration of a freely falling body is called the **acceleration due to gravity,** and we denote its magnitude with the letter g. We will frequently use the approximate value of g at or near the earth's surface:

$$g = 9.8 \text{ m/s}^2 = 980 \text{ cm/s}^2 = 32 \text{ ft/s}^2 \qquad \text{(approximate value near the earth's surface)}$$

The exact value varies with location, so we will often give the value of g at the earth's surface to only two significant figures. On the surface of the moon, the acceleration due to gravity is caused by the attractive force of the moon rather than the earth, and $g = 1.6 \text{ m/s}^2$. Near the surface of the sun, $g = 270 \text{ m/s}^2$.

CAUTION *g is always a positive number* Because g is the *magnitude* of a vector quantity, it is always a *positive* number. If you take the positive direction to be upward, as we do in Example 2.6 and in most situations involving free fall, the acceleration is negative (downward) and equal to $-g$. Be careful with the sign of g, or else you'll have no end of trouble with free-fall problems. ▮

In the following examples we use the constant-acceleration equations developed in Section 2.4. You should review Problem-Solving Strategy 2.1 in that section before you study the next examples.

Example 2.6   A freely falling coin

A one-euro coin is dropped from the Leaning Tower of Pisa and falls freely from rest. What are its position and velocity after 1.0 s, 2.0 s, and 3.0 s?

SOLUTION

IDENTIFY and SET UP: "Falls freely" means "falls with constant acceleration due to gravity," so we can use the constant-acceleration equations. The right side of Fig. 2.23 shows our motion diagram for the coin. The motion is vertical, so we use a vertical

2.23 A coin freely falling from rest.

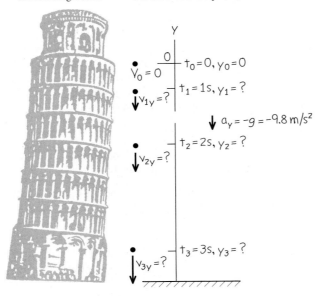

coordinate axis and call the coordinate y instead of x. We take the origin O at the starting point and the *upward* direction as positive. The initial coordinate y_0 and initial y-velocity v_{0y} are both zero. The y-acceleration is downward (in the negative y-direction), so $a_y = -g = -9.8 \text{ m/s}^2$. (Remember that, by definition, g itself is a positive quantity.) Our target variables are the values of y and v_y at the three given times. To find these, we use Eqs. (2.12) and (2.8) with x replaced by y. Our choice of the upward direction as positive means that all positions and velocities we calculate will be negative.

EXECUTE: At a time t after the coin is dropped, its position and y-velocity are

$$y = y_0 + v_{0y}t + \tfrac{1}{2}a_yt^2 = 0 + 0 + \tfrac{1}{2}(-g)t^2 = (-4.9 \text{ m/s}^2)t^2$$

$$v_y = v_{0y} + a_yt = 0 + (-g)t = (-9.8 \text{ m/s}^2)t$$

When $t = 1.0$ s, $y = (-4.9 \text{ m/s}^2)(1.0 \text{ s})^2 = -4.9$ m and $v_y = (-9.8 \text{ m/s}^2)(1.0 \text{ s}) = -9.8$ m/s; after 1 s, the coin is 4.9 m below the origin (y is negative) and has a downward velocity (v_y is negative) with magnitude 9.8 m/s.

We can find the positions and y-velocities at 2.0 s and 3.0 s in the same way. The results are $y = -20$ m and $v_y = -20$ m/s at $t = 2.0$ s, and $y = -44$ m and $v_y = -29$ m/s at $t = 3.0$ s.

EVALUATE: All our answers are negative, as we expected. If we had chosen the positive y-axis to point downward, the acceleration would have been $a_y = +g$ and all our answers would have been positive.

Example 2.7   Up-and-down motion in free fall

You throw a ball vertically upward from the roof of a tall building. The ball leaves your hand at a point even with the roof railing with an upward speed of 15.0 m/s; the ball is then in free fall. On its way back down, it just misses the railing. Find (a) the ball's position and velocity 1.00 s and 4.00 s after leaving your hand; (b) the ball's velocity when it is 5.00 m above the railing; (c) the maximum height reached; (d) the ball's acceleration when it is at its maximum height.

SOLUTION

IDENTIFY and SET UP: The words "in free fall" mean that the acceleration is due to gravity, which is constant. Our target variables are position [in parts (a) and (c)], velocity [in parts (a) and (b)], and acceleration [in part (d)]. We take the origin at the point where the ball leaves your hand, and take the positive direction to be upward (Fig. 2.24). The initial position y_0 is zero, the initial y-velocity v_{0y} is +15.0 m/s, and the y-acceleration is $a_y = -g = -9.80 \text{ m/s}^2$.

In part (a), as in Example 2.6, we'll use Eqs. (2.12) and (2.8) to find the position and velocity as functions of time. In part (b) we must find the velocity at a given *position* (no time is given), so we'll use Eq. (2.13).

Figure 2.25 shows the y-t and v_y-t graphs for the ball. The y-t graph is a concave-down parabola that rises and then falls, and the v_y-t graph is a downward-sloping straight line. Note that the ball's velocity is zero when it is at its highest point.

EXECUTE: (a) The position and y-velocity at time t are given by Eqs. (2.12) and (2.8) with x's replaced by y's:

$$y = y_0 + v_{0y}t + \tfrac{1}{2}a_yt^2 = y_0 + v_{0y}t + \tfrac{1}{2}(-g)t^2$$
$$= (0) + (15.0 \text{ m/s})t + \tfrac{1}{2}(-9.80 \text{ m/s}^2)t^2$$
$$v_y = v_{0y} + a_yt = v_{0y} + (-g)t$$
$$= 15.0 \text{ m/s} + (-9.80 \text{ m/s}^2)t$$

Continued

When $t = 1.00$ s, these equations give $y = +10.1$ m and $v_y = +5.2$ m/s. That is, the ball is 10.1 m above the origin (y is positive) and moving upward (v_y is positive) with a speed of 5.2 m/s. This is less than the initial speed because the ball slows as it ascends. When $t = 4.00$ s, those equations give $y = -18.4$ m and $v_y = -24.2$ m/s. The ball has passed its highest point and is 18.4 m *below* the origin (y is negative). It is moving *downward* (v_y is negative) with a speed of 24.2 m/s. The ball gains speed as it descends; Eq. (2.13) tells us that it is moving at the initial 15.0-m/s speed as it moves downward past the ball's launching point, and continues to gain speed as it descends further.

(b) The y-velocity at any position y is given by Eq. (2.13) with x's replaced by y's:

$$v_y{}^2 = v_{0y}{}^2 + 2a_y(y - y_0) = v_{0y}{}^2 + 2(-g)(y - 0)$$
$$= (15.0 \text{ m/s})^2 + 2(-9.80 \text{ m/s}^2)y$$

When the ball is 5.00 m above the origin we have $y = +5.00$ m, so

$$v_y{}^2 = (15.0 \text{ m/s})^2 + 2(-9.80 \text{ m/s}^2)(5.00 \text{ m}) = 127 \text{ m}^2/\text{s}^2$$
$$v_y = \pm 11.3 \text{ m/s}$$

We get *two* values of v_y because the ball passes through the point $y = +5.00$ m twice, once on the way up (so v_y is positive) and once on the way down (so v_y is negative) (see Figs. 2.24 and 2.25a).

(c) At the instant at which the ball reaches its maximum height y_1, its y-velocity is momentarily zero: $v_y = 0$. We use Eq. (2.13) to find y_1. With $v_y = 0$, $y_0 = 0$, and $a_y = -g$, we get

$$0 = v_{0y}{}^2 + 2(-g)(y_1 - 0)$$
$$y_1 = \frac{v_{0y}{}^2}{2g} = \frac{(15.0 \text{ m/s})^2}{2(9.80 \text{ m/s}^2)} = +11.5 \text{ m}$$

(d) **CAUTION** **A free-fall misconception** It's a common misconception that at the highest point of free-fall motion, where the velocity is zero, the acceleration is also zero. If this were so, once the ball reached the highest point it would hang there suspended in midair! Remember that acceleration is the rate of change of velocity, and the ball's velocity is continuously changing. At every point, including the highest point, and at any velocity, including zero, the acceleration in free fall is always $a_y = -g = -9.80$ m/s^2.

EVALUATE: A useful way to check any free-fall problem is to draw the y-t and v_y-t graphs as we did in Fig. 2.25. Note that these are graphs of Eqs. (2.12) and (2.8), respectively. Given the numerical values of the initial position, initial velocity, and acceleration, you can easily create these graphs using a graphing calculator or an online mathematics program.

2.24 Position and velocity of a ball thrown vertically upward.

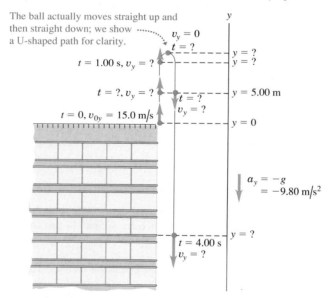

The ball actually moves straight up and then straight down; we show a U-shaped path for clarity.

$v_y = 0$
$t = ?$
$t = 1.00$ s, $v_y = ?$
$y = ?$
$y = ?$
$t = ?, v_y = ?$
$t = ?$
$y = 5.00$ m
$v_y = ?$
$t = 0, v_{0y} = 15.0$ m/s
$y = 0$
$a_y = -g = -9.80$ m/s^2
$y = ?$
$t = 4.00$ s
$v_y = ?$

2.25 (a) Position and (b) velocity as functions of time for a ball thrown upward with an initial speed of 15 m/s.

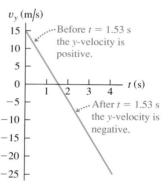

(a) y-t graph (curvature is downward because $a_y = -g$ is negative)

y (m)
Before $t = 1.53$ s the ball moves upward.
After $t = 1.53$ s the ball moves downward.

(b) v_y-t graph (straight line with negative slope because $a_y = -g$ is constant and negative)

v_y (m/s)
Before $t = 1.53$ s the y-velocity is positive.
After $t = 1.53$ s the y-velocity is negative.

Example 2.8 **Two solutions or one?**

At what time after being released has the ball in Example 2.7 fallen 5.00 m below the roof railing?

SOLUTION

IDENTIFY and SET UP: We treat this as in Example 2.7, so y_0, v_{0y}, and $a_y = -g$ have the same values as there. In this example, however, the target variable is the time at which the ball is at $y = -5.00$ m.

The best equation to use is Eq. (2.12), which gives the position y as a function of time t:

$$y = y_0 + v_{0y}t + \tfrac{1}{2}a_y t^2 = y_0 + v_{0y}t + \tfrac{1}{2}(-g)t^2$$

This is a *quadratic* equation for t, which we want to solve for the value of t when $y = -5.00$ m.

EXECUTE: We rearrange the equation so that is has the standard form of a quadratic equation for an unknown x, $Ax^2 + Bx + C = 0$:

$$(\tfrac{1}{2}g)t^2 + (-v_{0y})t + (y - y_0) = At^2 + Bt + C = 0$$

By comparison, we identify $A = \tfrac{1}{2}g$, $B = -v_{0y}$, and $C = y - y_0$. The quadratic formula (see Appendix B) tells us that this equation has *two* solutions:

$$t = \frac{-B \pm \sqrt{B^2 - 4AC}}{2A}$$
$$= \frac{-(-v_{0y}) \pm \sqrt{(-v_{0y})^2 - 4(\tfrac{1}{2}g)(y - y_0)}}{2(\tfrac{1}{2}g)}$$
$$= \frac{v_{0y} \pm \sqrt{v_{0y}{}^2 - 2g(y - y_0)}}{g}$$

Substituting the values $y_0 = 0$, $v_{0y} = +15.0$ m/s, $g = 9.80$ m/s^2, and $y = -5.00$ m, we find

$$t = \frac{(15.0 \text{ m/s}) \pm \sqrt{(15.0 \text{ m/s})^2 - 2(9.80 \text{ m/s}^2)(-5.00 \text{ m} - 0)}}{9.80 \text{ m/s}^2}$$

You can confirm that the numerical answers are $t = +3.36$ s and $t = -0.30$ s. The answer $t = -0.30$ s doesn't make physical sense, since it refers to a time *before* the ball left your hand at $t = 0$. So the correct answer is $t = +3.36$ s.

EVALUATE: Why did we get a second, fictitious solution? The explanation is that constant-acceleration equations like Eq. (2.12) are based on the assumption that the acceleration is constant for *all* values of time, whether positive, negative, or zero. Hence the solution $t = -0.30$ s refers to an imaginary moment when a freely falling ball was 5.00 m below the roof railing and rising to meet your hand. Since the ball didn't leave your hand and go into free fall until $t = 0$, this result is pure fiction.

You should repeat these calculations to find the times when the ball is 5.00 m *above* the origin ($y = +5.00$ m). The two answers are $t = +0.38$ s and $t = +2.68$ s. These are both positive values of t, and both refer to the real motion of the ball after leaving your hand. At the earlier time the ball passes through $y = +5.00$ m moving upward; at the later time it passes through this point moving downward. [Compare this with part (b) of Example 2.7, and again refer to Fig. 2.25a.]

You should also solve for the times when $y = +15.0$ m. In this case, both solutions involve the square root of a negative number, so there are *no* real solutions. Again Fig. 2.25a shows why; we found in part (c) of Example 2.7 that the ball's maximum height is $y = +11.5$ m, so it *never* reaches $y = +15.0$ m. While a quadratic equation such as Eq. (2.12) always has two solutions, in some situations one or both of the solutions will not be physically reasonable.

Test Your Understanding of Section 2.5 If you toss a ball upward with a certain initial speed, it falls freely and reaches a maximum height h a time t after it leaves your hand. (a) If you throw the ball upward with double the initial speed, what new maximum height does the ball reach? (i) $h\sqrt{2}$; (ii) $2h$; (iii) $4h$; (iv) $8h$; (v) $16h$. (b) If you throw the ball upward with double the initial speed, how long does it take to reach its new maximum height? (i) $t/2$; (ii) $t/\sqrt{2}$; (iii) t; (iv) $t\sqrt{2}$; (v) $2t$.

2.6 Velocity and Position by Integration

This section is intended for students who have already learned a little integral calculus. In Section 2.4 we analyzed the special case of straight-line motion with constant acceleration. When a_x is not constant, as is frequently the case, the equations that we derived in that section are no longer valid (Fig. 2.26). But even when a_x varies with time, we can still use the relationship $v_x = dx/dt$ to find the x-velocity v_x as a function of time if the position x is a known function of time. And we can still use $a_x = dv_x/dt$ to find the x-acceleration a_x as a function of time if the x-velocity v_x is a known function of time.

In many situations, however, position and velocity are not known functions of time, while acceleration is (Fig. 2.27). How can we find the position and velocity in straight-line motion from the acceleration function $a_x(t)$?

We first consider a graphical approach. Figure 2.28 is a graph of x-acceleration versus time for a body whose acceleration is not constant. We can divide the time interval between times t_1 and t_2 into many smaller intervals, calling a typical one Δt. Let the average x-acceleration during Δt be $a_{\text{av-}x}$. From Eq. (2.4) the change in x-velocity Δv_x during Δt is

$$\Delta v_x = a_{\text{av-}x} \, \Delta t$$

Graphically, Δv_x equals the area of the shaded strip with height $a_{\text{av-}x}$ and width Δt—that is, the area under the curve between the left and right sides of Δt. The total change in x-velocity during any interval (say, t_1 to t_2) is the sum of the x-velocity changes Δv_x in the small subintervals. So the total x-velocity change is represented graphically by the *total* area under the a_x-t curve between the vertical

2.26 When you push your car's accelerator pedal to the floorboard, the resulting acceleration is *not* constant: The greater the car's speed, the more slowly it gains additional speed. A typical car takes twice as long to accelerate from 50 km/h to 100 km/h as it does to accelerate from 0 to 50 km/h.

2.27 The inertial navigation system (INS) on board a long-range airliner keeps track of the airliner's acceleration. The pilots input the airliner's initial position and velocity before takeoff, and the INS uses the acceleration data to calculate the airliner's position and velocity throughout the flight.

lines t_1 and t_2. (In Section 2.4 we showed this for the special case in which the acceleration is constant.)

In the limit that all the Δt's become very small and their number very large, the value of $a_{\text{av-}x}$ for the interval from any time t to $t + \Delta t$ approaches the instantaneous x-acceleration a_x at time t. In this limit, the area under the a_x-t curve is the *integral* of a_x (which is in general a function of t) from t_1 to t_2. If v_{1x} is the x-velocity of the body at time t_1 and v_{2x} is the velocity at time t_2, then

$$v_{2x} - v_{1x} = \int_{v_{1x}}^{v_{2x}} dv_x = \int_{t_1}^{t_2} a_x\, dt \tag{2.15}$$

The change in the x-velocity v_x is the time integral of the x-acceleration a_x.

We can carry out exactly the same procedure with the curve of x-velocity versus time. If x_1 is a body's position at time t_1 and x_2 is its position at time t_2, from Eq. (2.2) the displacement Δx during a small time interval Δt is equal to $v_{\text{av-}x}\,\Delta t$, where $v_{\text{av-}x}$ is the average x-velocity during Δt. The total displacement $x_2 - x_1$ during the interval $t_2 - t_1$ is given by

$$x_2 - x_1 = \int_{x_1}^{x_2} dx = \int_{t_1}^{t_2} v_x\, dt \tag{2.16}$$

The change in position x—that is, the displacement—is the time integral of x-velocity v_x. Graphically, the displacement between times t_1 and t_2 is the area under the v_x-t curve between those two times. [This is the same result that we obtained in Section 2.4 for the special case in which v_x is given by Eq. (2.8).]

If $t_1 = 0$ and t_2 is any later time t, and if x_0 and v_{0x} are the position and velocity, respectively, at time $t = 0$, then we can rewrite Eqs. (2.15) and (2.16) as follows:

2.28 An a_x-t graph for a body whose x-acceleration is not constant.

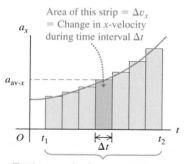

Area of this strip = Δv_x
= Change in x-velocity during time interval Δt

Total area under the x-t graph from t_1 to t_2
= Net change in x-velocity from t_1 to t_2

$$v_x = v_{0x} + \int_0^t a_x\, dt \tag{2.17}$$

$$x = x_0 + \int_0^t v_x\, dt \tag{2.18}$$

Here x and v_x are the position and x-velocity at time t. If we know the x-acceleration a_x as a function of time and we know the initial velocity v_{0x}, we can use Eq. (2.17) to find the x-velocity v_x at any time; in other words, we can find v_x as a function of time. Once we know this function, and given the initial position x_0, we can use Eq. (2.18) to find the position x at any time.

Example 2.9 | **Motion with changing acceleration**

Sally is driving along a straight highway in her 1965 Mustang. At $t = 0$, when she is moving at 10 m/s in the positive x-direction, she passes a signpost at $x = 50$ m. Her x-acceleration as a function of time is

$$a_x = 2.0 \text{ m/s}^2 - (0.10 \text{ m/s}^3)t$$

(a) Find her x-velocity v_x and position x as functions of time. (b) When is her x-velocity greatest? (c) What is that maximum x-velocity? (d) Where is the car when it reaches that maximum x-velocity?

SOLUTION

IDENTIFY and SET UP: The x-acceleration is a function of time, so we *cannot* use the constant-acceleration formulas of Section 2.4. Instead, we use Eq. (2.17) to obtain an expression for v_x as a function of time, and then use that result in Eq. (2.18) to find an expression for x as a function of t. We'll then be able to answer a variety of questions about the motion.

EXECUTE: (a) At $t = 0$, Sally's position is $x_0 = 50$ m and her x-velocity is $v_{0x} = 10$ m/s. To use Eq. (2.17), we note that the integral of t^n (except for $n = -1$) is $\int t^n dt = \frac{1}{n+1} t^{n+1}$. Hence we find

$$v_x = 10 \text{ m/s} + \int_0^t [2.0 \text{ m/s}^2 - (0.10 \text{ m/s}^3)t] dt$$

$$= 10 \text{ m/s} + (2.0 \text{ m/s}^2)t - \tfrac{1}{2}(0.10 \text{ m/s}^3)t^2$$

Now we use Eq. (2.18) to find x as a function of t:

$$x = 50 \text{ m} + \int_0^t \left[10 \text{ m/s} + (2.0 \text{ m/s}^2)t - \tfrac{1}{2}(0.10 \text{ m/s}^3)t^2 \right] dt$$

$$= 50 \text{ m} + (10 \text{ m/s})t + \tfrac{1}{2}(2.0 \text{ m/s}^2)t^2 - \tfrac{1}{6}(0.10 \text{ m/s}^3)t^3$$

Figure 2.29 shows graphs of a_x, v_x, and x as functions of time as given by the equations above. Note that for any time t, the slope of the v_x-t graph equals the value of a_x and the slope of the x-t graph equals the value of v_x.

(b) The maximum value of v_x occurs when the x-velocity stops increasing and begins to decrease. At that instant, $dv_x/dt = a_x = 0$. So we set the expression for a_x equal to zero and solve for t:

$$0 = 2.0 \text{ m/s}^2 - (0.10 \text{ m/s}^3)t$$

$$t = \frac{2.0 \text{ m/s}^2}{0.10 \text{ m/s}^3} = 20 \text{ s}$$

(c) We find the maximum x-velocity by substituting $t = 20$ s, the time from part (b) when velocity is maximum, into the equation for v_x from part (a):

$$v_{\text{max-}x} = 10 \text{ m/s} + (2.0 \text{ m/s}^2)(20 \text{ s}) - \tfrac{1}{2}(0.10 \text{ m/s}^3)(20 \text{ s})^2$$

$$= 30 \text{ m/s}$$

(d) To find the car's position at the time that we found in part (b), we substitute $t = 20$ s into the expression for x from part (a):

$$x = 50 \text{ m} + (10 \text{ m/s})(20 \text{ s}) + \tfrac{1}{2}(2.0 \text{ m/s}^2)(20 \text{ s})^2$$

$$- \tfrac{1}{6}(0.10 \text{ m/s}^3)(20 \text{ s})^3$$

$$= 517 \text{ m}$$

EVALUATE: Figure 2.29 helps us interpret our results. The top graph shows that a_x is positive between $t = 0$ and $t = 20$ s and negative after that. It is zero at $t = 20$ s, the time at which v_x is maximum (the high point in the middle graph). The car speeds up until $t = 20$ s (because v_x and a_x have the same sign) and slows down after $t = 20$ s (because v_x and a_x have opposite signs).

Since v_x is maximum at $t = 20$ s, the x-t graph (the bottom graph in Fig. 2.29) has its maximum positive slope at this time. Note that the x-t graph is concave up (curved upward) from $t = 0$ to $t = 20$ s, when a_x is positive. The graph is concave down (curved downward) after $t = 20$ s, when a_x is negative.

2.29 The position, velocity, and acceleration of the car in Example 2.9 as functions of time. Can you show that if this motion continues, the car will stop at $t = 44.5$ s?

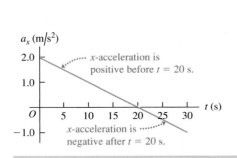

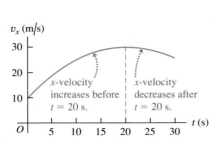

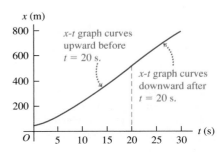

Test Your Understanding of Section 2.6 If the x-acceleration a_x is increasing with time, will the v_x-t graph be (i) a straight line, (ii) concave up (i.e., with an upward curvature), or (iii) concave down (i.e., with a downward curvature)?

Straight-line motion, average and instantaneous x-velocity: When a particle moves along a straight line, we describe its position with respect to an origin O by means of a coordinate such as x. The particle's average x-velocity $v_{\text{av-}x}$ during a time interval $\Delta t = t_2 - t_1$ is equal to its displacement $\Delta x = x_2 - x_1$ divided by Δt. The instantaneous x-velocity v_x at any time t is equal to the average x-velocity for the time interval from t to $t + \Delta t$ in the limit that Δt goes to zero. Equivalently, v_x is the derivative of the position function with respect to time. (See Example 2.1.)

$$v_{\text{av-}x} = \frac{x_2 - x_1}{t_2 - t_1} = \frac{\Delta x}{\Delta t} \quad (2.2)$$

$$v_x = \lim_{\Delta t \to 0} \frac{\Delta x}{\Delta t} = \frac{dx}{dt} \quad (2.3)$$

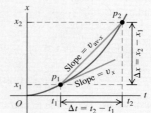

Average and instantaneous x-acceleration: The average x-acceleration $a_{\text{av-}x}$ during a time interval Δt is equal to the change in velocity $\Delta v_x = v_{2x} - v_{1x}$ during that time interval divided by Δt. The instantaneous x-acceleration a_x is the limit of $a_{\text{av-}x}$ as Δt goes to zero, or the derivative of v_x with respect to t. (See Examples 2.2 and 2.3.)

$$a_{\text{av-}x} = \frac{v_{2x} - v_{1x}}{t_2 - t_1} = \frac{\Delta v_x}{\Delta t} \quad (2.4)$$

$$a_x = \lim_{\Delta t \to 0} \frac{\Delta v_x}{\Delta t} = \frac{dv_x}{dt} \quad (2.5)$$

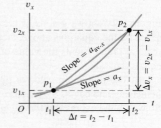

Straight-line motion with constant acceleration: When the x-acceleration is constant, four equations relate the position x and the x-velocity v_x at any time t to the initial position x_0, the initial x-velocity v_{0x} (both measured at time $t = 0$), and the x-acceleration a_x. (See Examples 2.4 and 2.5.)

Constant x-acceleration only:

$$v_x = v_{0x} + a_x t \quad (2.8)$$

$$x = x_0 + v_{0x}t + \tfrac{1}{2}a_x t^2 \quad (2.12)$$

$$v_x^2 = v_{0x}^2 + 2a_x(x - x_0) \quad (2.13)$$

$$x - x_0 = \left(\frac{v_{0x} + v_x}{2}\right)t \quad (2.14)$$

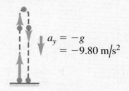

Freely falling bodies: Free fall is a case of motion with constant acceleration. The magnitude of the acceleration due to gravity is a positive quantity, g. The acceleration of a body in free fall is always downward. (See Examples 2.6–2.8.)

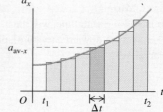

$$a_y = -g$$
$$= -9.80 \text{ m/s}^2$$

Straight-line motion with varying acceleration: When the acceleration is not constant but is a known function of time, we can find the velocity and position as functions of time by integrating the acceleration function. (See Example 2.9.)

$$v_x = v_{0x} + \int_0^t a_x \, dt \quad (2.17)$$

$$x = x_0 + \int_0^t v_x \, dt \quad (2.18)$$

The superhero Green Lantern steps from the top of a tall building. He falls freely from rest to the ground, falling half the total distance to the ground during the last 1.00 s of his fall. What is the height h of the building?

SOLUTION GUIDE

See MasteringPhysics® study area for a Video Tutor solution.

IDENTIFY and SET UP

1. You're told that Green Lantern falls freely from rest. What does this imply about his acceleration? About his initial velocity?
2. Choose the direction of the positive y-axis. It's easiest to make the same choice we used for freely falling objects in Section 2.5.
3. You can divide Green Lantern's fall into two parts: from the top of the building to the halfway point and from the halfway point to the ground. You know that the second part of the fall lasts 1.00 s. Decide what you would need to know about Green

Lantern's motion at the halfway point in order to solve for the target variable h. Then choose two equations, one for the first part of the fall and one for the second part, that you'll use together to find an expression for h. (There are several pairs of equations that you could choose.)

EXECUTE

4. Use your two equations to solve for the height h. Note that heights are always positive numbers, so your answer should be positive.

EVALUATE

5. To check your answer for h, use one of the free-fall equations to find how long it takes Green Lantern to fall (i) from the top of the building to half the height and (ii) from the top of the building to the ground. If your answer for h is correct, time (ii) should be 1.00 s greater than time (i). If it isn't, you'll need to go back and look for errors in how you found h.

Problems   For instructor-assigned homework, go to www.masteringphysics.com

•, ••, •••: Problems of increasing difficulty. **CP**: Cumulative problems incorporating material from earlier chapters. **CALC**: Problems requiring calculus. **BIO**: Biosciences problems.

DISCUSSION QUESTIONS

Q2.1 Does the speedometer of a car measure speed or velocity? Explain.

Q2.2 The top diagram in Fig. Q2.2 represents a series of high-speed photographs of an insect flying in a straight line from left to right (in the positive x-direction). Which of the graphs in Fig. Q2.2 most plausibly depicts this insect's motion?

Figure **Q2.2**

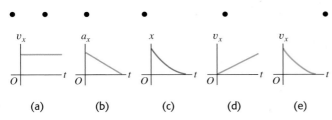

| (a) | (b) | (c) | (d) | (e) |

Q2.3 Can an object with constant acceleration reverse its direction of travel? Can it reverse its direction *twice*? In each case, explain your reasoning.

Q2.4 Under what conditions is average velocity equal to instantaneous velocity?

Q2.5 Is it possible for an object (a) to be slowing down while its acceleration is increasing in magnitude; (b) to be speeding up while its acceleration is decreasing? In each case, explain your reasoning.

Q2.6 Under what conditions does the magnitude of the average velocity equal the average speed?

Q2.7 When a Dodge Viper is at Elwood's Car Wash, a BMW Z3 is at Elm and Main. Later, when the Dodge reaches Elm and Main,

the BMW reaches Elwood's Car Wash. How are the cars' average velocities between these two times related?

Q2.8 A driver in Massachusetts was sent to traffic court for speeding. The evidence against the driver was that a policewoman observed the driver's car alongside a second car at a certain moment, and the policewoman had already clocked the second car as going faster than the speed limit. The driver argued, "The second car was passing me. I was not speeding." The judge ruled against the driver because, in the judge's words, "If two cars were side by side, you were both speeding." If you were a lawyer representing the accused driver, how would you argue this case?

Q2.9 Can you have a zero displacement and a nonzero average velocity? A nonzero velocity? Illustrate your answers on an x-t graph.

Q2.10 Can you have zero acceleration and nonzero velocity? Explain using a v_x-t graph.

Q2.11 Can you have zero velocity and nonzero average acceleration? Zero velocity and nonzero acceleration? Explain using a v_x-t graph, and give an example of such motion.

Q2.12 An automobile is traveling west. Can it have a velocity toward the west and at the same time have an acceleration toward the east? Under what circumstances?

Q2.13 The official's truck in Fig. 2.2 is at $x_1 = 277$ m at $t_1 = 16.0$ s and is at $x_2 = 19$ m at $t_2 = 25.0$ s. (a) Sketch *two* different possible x-t graphs for the motion of the truck. (b) Does the average velocity v_{av-x} during the time interval from t_1 to t_2 have the same value for both of your graphs? Why or why not?

Q2.14 Under constant acceleration the average velocity of a particle is half the sum of its initial and final velocities. Is this still true if the acceleration is *not* constant? Explain.

Q2.15 You throw a baseball straight up in the air so that it rises to a maximum height much greater than your height. Is the magnitude of the acceleration greater while it is being thrown or after it leaves your hand? Explain.

Q2.16 Prove these statements: (a) As long as you can neglect the effects of the air, if you throw anything vertically upward, it will have the same speed when it returns to the release point as when it was released. (b) The time of flight will be twice the time it takes to get to its highest point.

Q2.17 A dripping water faucet steadily releases drops 1.0 s apart. As these drops fall, will the distance between them increase, decrease, or remain the same? Prove your answer.

Q2.18 If the initial position and initial velocity of a vehicle are known and a record is kept of the acceleration at each instant, can you compute the vehicle's position after a certain time from these data? If so, explain how this might be done.

Q2.19 From the top of a tall building you throw one ball straight up with speed v_0 and one ball straight down with speed v_0. (a) Which ball has the greater speed when it reaches the ground? (b) Which ball gets to the ground first? (c) Which ball has a greater displacement when it reaches the ground? (d) Which ball has traveled the greater distance when it hits the ground?

Q2.20 A ball is dropped from rest from the top of a building of height h. At the same instant, a second ball is projected vertically upward from ground level, such that it has zero speed when it reaches the top of the building. When the two balls pass each other, which ball has the greater speed, or do they have the same speed? Explain. Where will the two balls be when they are alongside each other: at height $h/2$ above the ground, below this height, or above this height? Explain.

Q2.21 An object is thrown straight up into the air and feels no air resistance. How is it possible for the object to have an acceleration when it has stopped moving at its highest point?

Q2.22 When you drop an object from a certain height, it takes time T to reach the ground with no air resistance. If you dropped it from three times that height, how long (in terms of T) would it take to reach the ground?

EXERCISES

Section 2.1 Displacement, Time, and Average Velocity

2.1 • A car travels in the $+x$-direction on a straight and level road. For the first 4.00 s of its motion, the average velocity of the car is $v_{\text{av-}x} = 6.25$ m/s. How far does the car travel in 4.00 s?

2.2 •• In an experiment, a shearwater (a seabird) was taken from its nest, flown 5150 km away, and released. The bird found its way back to its nest 13.5 days after release. If we place the origin in the nest and extend the $+x$-axis to the release point, what was the bird's average velocity in m/s (a) for the return flight, and (b) for the whole episode, from leaving the nest to returning?

2.3 •• **Trip Home.** You normally drive on the freeway between San Diego and Los Angeles at an average speed of 105 km/h (65 mi/h), and the trip takes 2 h and 20 min. On a Friday afternoon, however, heavy traffic slows you down and you drive the same distance at an average speed of only 70 km/h (43 mi/h). How much longer does the trip take?

2.4 •• **From Pillar to Post.** Starting from a pillar, you run 200 m east (the $+x$-direction) at an average speed of 5.0 m/s, and then run 280 m west at an average speed of 4.0 m/s to a post. Calculate (a) your average speed from pillar to post and (b) your average velocity from pillar to post.

2.5 • Starting from the front door of your ranch house, you walk 60.0 m due east to your windmill, and then you turn around and slowly walk 40.0 m west to a bench where you sit and watch the sunrise. It takes you 28.0 s to walk from your house to the windmill and then 36.0 s to walk from the windmill to the bench. For the entire trip from your front door to the bench, what are (a) your average velocity and (b) your average speed?

2.6 •• A Honda Civic travels in a straight line along a road. Its distance x from a stop sign is given as a function of time t by the equation $x(t) = \alpha t^2 - \beta t^3$, where $\alpha = 1.50$ m/s^2 and $\beta = 0.0500$ m/s^3. Calculate the average velocity of the car for each time interval: (a) $t = 0$ to $t = 2.00$ s; (b) $t = 0$ to $t = 4.00$ s; (c) $t = 2.00$ s to $t = 4.00$ s.

Section 2.2 Instantaneous Velocity

2.7 • **CALC** A car is stopped at a traffic light. It then travels along a straight road so that its distance from the light is given by $x(t) = bt^2 - ct^3$, where $b = 2.40$ m/s^2 and $c = 0.120$ m/s^3. (a) Calculate the average velocity of the car for the time interval $t = 0$ to $t = 10.0$ s. (b) Calculate the instantaneous velocity of the car at $t = 0$, $t = 5.0$ s, and $t = 10.0$ s. (c) How long after starting from rest is the car again at rest?

2.8 • **CALC** A bird is flying due east. Its distance from a tall building is given by $x(t) = 28.0$ m $+ (12.4$ m/s$)t - (0.0450$ m/s$^3)t^3$. What is the instantaneous velocity of the bird when $t = 8.00$ s?

2.9 •• A ball moves in a straight line (the x-axis). The graph in Fig. E2.9 shows this ball's velocity as a function of time. (a) What are the ball's average speed and average velocity during the first 3.0 s? (b) Suppose that the ball moved in such a way that the graph segment after 2.0 s was -3.0 m/s instead of $+3.0$ m/s. Find the ball's average speed and average velocity in this case.

Figure **E2.9**

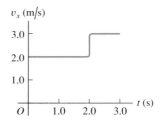

2.10 • A physics professor leaves her house and walks along the sidewalk toward campus. After 5 min it starts to rain and she returns home. Her distance from her house as a function of time is shown in Fig. E2.10. At which of the labeled points is her velocity (a) zero? (b) constant and positive? (c) constant and negative? (d) increasing in magnitude? (e) decreasing in magnitude?

Figure **E2.10**

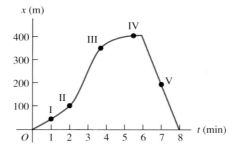

Figure **E2.11**

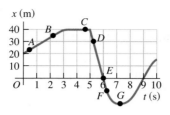

2.11 •• A test car travels in a straight line along the x-axis. The graph in Fig. E2.11 shows the car's position x as a function of time. Find its instantaneous velocity at points A through G.

Section 2.3 Average and Instantaneous Acceleration

2.12 • Figure E2.12 shows the velocity of a solar-powered car as a function of time. The driver accelerates from a stop sign, cruises for 20 s at a constant speed of 60 km/h, and then brakes to come to a stop 40 s after leaving the stop sign. (a) Compute the average acceleration during the following time intervals: (i) $t = 0$ to $t = 10$ s; (ii) $t = 30$ s to $t = 40$ s; (iii) $t = 10$ s to $t = 30$ s; (iv) $t = 0$ to $t = 40$ s. (b) What is the instantaneous acceleration at $t = 20$ s and at $t = 35$ s?

Figure **E2.12**

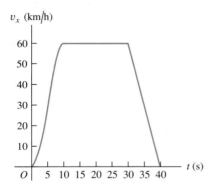

2.13 • **The Fastest (and Most Expensive) Car!** The table shows test data for the Bugatti Veyron, the fastest car made. The car is moving in a straight line (the x-axis).

Time (s)	0	2.1	20.0	53
Speed (mi/h)	0	60	200	253

(a) Make a v_x-t graph of this car's velocity (in mi/h) as a function of time. Is its acceleration constant? (b) Calculate the car's average acceleration (in m/s^2) between (i) 0 and 2.1 s; (ii) 2.1 s and 20.0 s; (iii) 20.0 s and 53 s. Are these results consistent with your graph in part (a)? (Before you decide to buy this car, it might be helpful to know that only 300 will be built, it runs out of gas in 12 minutes at top speed, and it costs $1.25 million!)

2.14 •• **CALC** A race car starts from rest and travels east along a straight and level track. For the first 5.0 s of the car's motion, the eastward component of the car's velocity is given by $v_x(t) = (0.860 \text{ m/s}^3)t^2$. What is the acceleration of the car when $v_x = 16.0$ m/s?

2.15 • **CALC** A turtle crawls along a straight line, which we will call the x-axis with the positive direction to the right. The equation for the turtle's position as a function of time is $x(t) = 50.0 \text{ cm} + (2.00 \text{ cm/s})t - (0.0625 \text{ cm/s}^2)t^2$. (a) Find the turtle's initial velocity, initial position, and initial acceleration. (b) At what time t

is the velocity of the turtle zero? (c) How long after starting does it take the turtle to return to its starting point? (d) At what times t is the turtle a distance of 10.0 cm from its starting point? What is the velocity (magnitude and direction) of the turtle at each of these times? (e) Sketch graphs of x versus t, v_x versus t, and a_x versus t, for the time interval $t = 0$ to $t = 40$ s.

2.16 • An astronaut has left the International Space Station to test a new space scooter. Her partner measures the following velocity changes, each taking place in a 10-s interval. What are the magnitude, the algebraic sign, and the direction of the average acceleration in each interval? Assume that the positive direction is to the right. (a) At the beginning of the interval the astronaut is moving toward the right along the x-axis at 15.0 m/s, and at the end of the interval she is moving toward the right at 5.0 m/s. (b) At the beginning she is moving toward the left at 5.0 m/s, and at the end she is moving toward the left at 15.0 m/s. (c) At the beginning she is moving toward the right at 15.0 m/s, and at the end she is moving toward the left at 15.0 m/s.

2.17 • **CALC** A car's velocity as a function of time is given by $v_x(t) = \alpha + \beta t^2$, where $\alpha = 3.00$ m/s and $\beta = 0.100$ m/s^3. (a) Calculate the average acceleration for the time interval $t = 0$ to $t = 5.00$ s. (b) Calculate the instantaneous acceleration for $t = 0$ and $t = 5.00$ s. (c) Draw v_x-t and a_x-t graphs for the car's motion between $t = 0$ and $t = 5.00$ s.

2.18 •• **CALC** The position of the front bumper of a test car under microprocessor control is given by $x(t) = 2.17 \text{ m} + (4.80 \text{ m/s}^2)t^2 - (0.100 \text{ m/s}^6)t^6$. (a) Find its position and acceleration at the instants when the car has zero velocity. (b) Draw x-t, v_x-t, and a_x-t graphs for the motion of the bumper between $t = 0$ and $t = 2.00$ s.

Section 2.4 Motion with Constant Acceleration

2.19 •• An antelope moving with constant acceleration covers the distance between two points 70.0 m apart in 7.00 s. Its speed as it passes the second point is 15.0 m/s. (a) What is its speed at the first point? (b) What is its acceleration?

2.20 •• **BIO Blackout?** A jet fighter pilot wishes to accelerate from rest at a constant acceleration of 5g to reach Mach 3 (three times the speed of sound) as quickly as possible. Experimental tests reveal that he will black out if this acceleration lasts for more than 5.0 s. Use 331 m/s for the speed of sound. (a) Will the period of acceleration last long enough to cause him to black out? (b) What is the greatest speed he can reach with an acceleration of 5g before blacking out?

2.21 • **A Fast Pitch.** The fastest measured pitched baseball left the pitcher's hand at a speed of 45.0 m/s. If the pitcher was in contact with the ball over a distance of 1.50 m and produced constant acceleration, (a) what acceleration did he give the ball, and (b) how much time did it take him to pitch it?

2.22 •• **A Tennis Serve.** In the fastest measured tennis serve, the ball left the racquet at 73.14 m/s. A served tennis ball is typically in contact with the racquet for 30.0 ms and starts from rest. Assume constant acceleration. (a) What was the ball's acceleration during this serve? (b) How far did the ball travel during the serve?

2.23 •• **BIO Automobile Airbags.** The human body can survive an acceleration trauma incident (sudden stop) if the magnitude of the acceleration is less than 250 m/s^2. If you are in an automobile accident with an initial speed of 105 km/h (65 mi/h) and you are stopped by an airbag that inflates from the dashboard, over what distance must the airbag stop you for you to survive the crash?

2.24 • BIO If a pilot accelerates at more than 4g, he begins to "gray out" but doesn't completely lose consciousness. (a) Assuming constant acceleration, what is the shortest time that a jet pilot starting from rest can take to reach Mach 4 (four times the speed of sound) without graying out? (b) How far would the plane travel during this period of acceleration? (Use 331 m/s for the speed of sound in cold air.)

2.25 • BIO **Air-Bag Injuries.** During an auto accident, the vehicle's air bags deploy and slow down the passengers more gently than if they had hit the windshield or steering wheel. According to safety standards, the bags produce a maximum acceleration of 60g that lasts for only 36 ms (or less). How far (in meters) does a person travel in coming to a complete stop in 36 ms at a constant acceleration of 60g?

2.26 • BIO **Prevention of Hip Fractures.** Falls resulting in hip fractures are a major cause of injury and even death to the elderly. Typically, the hip's speed at impact is about 2.0 m/s. If this can be reduced to 1.3 m/s or less, the hip will usually not fracture. One way to do this is by wearing elastic hip pads. (a) If a typical pad is 5.0 cm thick and compresses by 2.0 cm during the impact of a fall, what constant acceleration (in m/s^2 and in g's) does the hip undergo to reduce its speed from 2.0 m/s to 1.3 m/s? (b) The acceleration you found in part (a) may seem rather large, but to fully assess its effects on the hip, calculate how long it lasts.

2.27 • BIO **Are We Martians?** It has been suggested, and not facetiously, that life might have originated on Mars and been carried to the earth when a meteor hit Mars and blasted pieces of rock (perhaps containing primitive life) free of the surface. Astronomers know that many Martian rocks have come to the earth this way. (For information on one of these, search the Internet for "ALH 84001.") One objection to this idea is that microbes would have to undergo an enormous lethal acceleration during the impact. Let us investigate how large such an acceleration might be. To escape Mars, rock fragments would have to reach its escape velocity of 5.0 km/s, and this would most likely happen over a distance of about 4.0 m during the meteor impact. (a) What would be the acceleration (in m/s^2 and g's) of such a rock fragment, if the acceleration is constant? (b) How long would this acceleration last? (c) In tests, scientists have found that over 40% of *Bacillius subtilis* bacteria survived after an acceleration of 450,000g. In light of your answer to part (a), can we rule out the hypothesis that life might have been blasted from Mars to the earth?

2.28 • **Entering the Freeway.** A car sits in an entrance ramp to a freeway, waiting for a break in the traffic. The driver accelerates with constant acceleration along the ramp and onto the freeway. The car starts from rest, moves in a straight line, and has a speed of 20 m/s (45 mi/h) when it reaches the end of the 120-m-long ramp. (a) What is the acceleration of the car? (b) How much time does it take the car to travel the length of the ramp? (c) The traffic on the freeway is moving at a constant speed of 20 m/s. What distance does the traffic travel while the car is moving the length of the ramp?

2.29 •• **Launch of the Space Shuttle.** At launch the space shuttle weighs 4.5 million pounds. When it is launched from rest, it takes 8.00 s to reach 161 km/h, and at the end of the first 1.00 min its speed is 1610 km/h. (a) What is the average acceleration (in m/s^2) of the shuttle (i) during the first 8.00 s, and (ii) between 8.00 s and the end of the first 1.00 min? (b) Assuming the acceleration is constant during each time interval (but not necessarily the same in both intervals), what distance does the shuttle travel (i) during the first 8.00 s, and (ii) during the interval from 8.00 s to 1.00 min?

2.30 •• A cat walks in a straight line, which we shall call the x-axis with the positive direction to the right. As an observant physicist, you make measurements of this cat's motion and construct a graph of the feline's velocity as a function of time (Fig. E2.30). (a) Find the cat's velocity at $t = 4.0$ s and at $t = 7.0$ s. (b) What is the cat's acceleration at $t = 3.0$ s? At $t = 6.0$ s? At $t = 7.0$ s? (c) What distance does the cat move during the first 4.5 s? From $t = 0$ to $t = 7.5$ s? (d) Sketch clear graphs of the cat's acceleration and position as functions of time, assuming that the cat started at the origin.

Figure **E2.30**

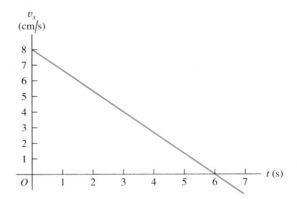

2.31 •• The graph in Fig. E2.31 shows the velocity of a motorcycle police officer plotted as a function of time. (a) Find the instantaneous acceleration at $t = 3$ s, at $t = 7$ s, and at $t = 11$ s. (b) How far does the officer go in the first 5 s? The first 9 s? The first 13 s?

Figure **E2.31**

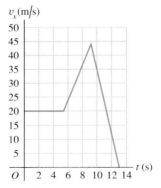

2.32 • Two cars, A and B, move along the x-axis. Figure E2.32 is a graph of the positions of A and B versus time. (a) In motion diagrams (like Figs. 2.13b and 2.14b), show the position, velocity, and acceleration of each of the two cars at $t = 0$, $t = 1$ s, and $t = 3$ s. (b) At what time(s), if any, do A and B have the same position? (c) Graph velocity versus time for both A and B. (d) At what time(s), if any, do A and B have the same velocity? (e) At what time(s), if any, does car A pass car B? (f) At what time(s), if any, does car B pass car A?

Figure **E2.32**

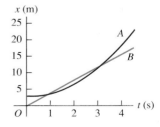

2.33 •• **Mars Landing.** In January 2004, NASA landed exploration vehicles on Mars. Part of the descent consisted of the following stages:

Stage A: Friction with the atmosphere reduced the speed from 19,300 km/h to 1600 km/h in 4.0 min.
Stage B: A parachute then opened to slow it down to 321 km/h in 94 s.
Stage C: Retro rockets then fired to reduce its speed to zero over a distance of 75 m.

Assume that each stage followed immediately after the preceding one and that the acceleration during each stage was constant. (a) Find the rocket's acceleration (in m/s^2) during each stage. (b) What total distance (in km) did the rocket travel during stages A, B, and C?

2.34 • At the instant the traffic light turns green, a car that has been waiting at an intersection starts ahead with a constant acceleration of $3.20 \ m/s^2$. At the same instant a truck, traveling with a constant speed of 20.0 m/s, overtakes and passes the car. (a) How far beyond its starting point does the car overtake the truck? (b) How fast is the car traveling when it overtakes the truck? (c) Sketch an x-t graph of the motion of both vehicles. Take $x = 0$ at the intersection. (d) Sketch a v_x-t graph of the motion of both vehicles.

Section 2.5 Freely Falling Bodies

2.35 •• (a) If a flea can jump straight up to a height of 0.440 m, what is its initial speed as it leaves the ground? (b) How long is it in the air?

2.36 •• A small rock is thrown vertically upward with a speed of 18.0 m/s from the edge of the roof of a 30.0-m-tall building. The rock doesn't hit the building on its way back down and lands in the street below. Air resistance can be neglected. (a) What is the speed of the rock just before it hits the street? (b) How much time elapses from when the rock is thrown until it hits the street?

2.37 • A juggler throws a bowling pin straight up with an initial speed of 8.20 m/s. How much time elapses until the bowling pin returns to the juggler's hand?

2.38 •• You throw a glob of putty straight up toward the ceiling, which is 3.60 m above the point where the putty leaves your hand. The initial speed of the putty as it leaves your hand is 9.50 m/s. (a) What is the speed of the putty just before it strikes the ceiling? (b) How much time from when it leaves your hand does it take the putty to reach the ceiling?

2.39 •• A tennis ball on Mars, where the acceleration due to gravity is $0.379g$ and air resistance is negligible, is hit directly upward and returns to the same level 8.5 s later. (a) How high above its original point did the ball go? (b) How fast was it moving just after being hit? (c) Sketch graphs of the ball's vertical position, vertical velocity, and vertical acceleration as functions of time while it's in the Martian air.

2.40 •• **Touchdown on the Moon.** A lunar lander is making its descent to Moon Base I (Fig. E2.40). The lander descends slowly under the retrothrust of its descent engine. The engine is cut off when the lander is 5.0 m above the surface and has a downward speed of 0.8 m/s. With the engine off,

Figure **E2.40**

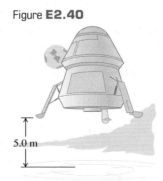

5.0 m

the lander is in free fall. What is the speed of the lander just before it touches the surface? The acceleration due to gravity on the moon is $1.6 \ m/s^2$.

2.41 •• **A Simple Reaction-Time Test.** A meter stick is held vertically above your hand, with the lower end between your thumb and first finger. On seeing the meter stick released, you grab it with these two fingers. You can calculate your reaction time from the distance the meter stick falls, read directly from the point where your fingers grabbed it. (a) Derive a relationship for your reaction time in terms of this measured distance, d. (b) If the measured distance is 17.6 cm, what is the reaction time?

2.42 •• A brick is dropped (zero initial speed) from the roof of a building. The brick strikes the ground in 2.50 s. You may ignore air resistance, so the brick is in free fall. (a) How tall, in meters, is the building? (b) What is the magnitude of the brick's velocity just before it reaches the ground? (c) Sketch a_y-t, v_y-t, and y-t graphs for the motion of the brick.

2.43 •• **Launch Failure.** A 7500-kg rocket blasts off vertically from the launch pad with a constant upward acceleration of $2.25 \ m/s^2$ and feels no appreciable air resistance. When it has reached a height of 525 m, its engines suddenly fail so that the only force acting on it is now gravity. (a) What is the maximum height this rocket will reach above the launch pad? (b) How much time after engine failure will elapse before the rocket comes crashing down to the launch pad, and how fast will it be moving just before it crashes? (c) Sketch a_y-t, v_y-t, and y-t graphs of the rocket's motion from the instant of blast-off to the instant just before it strikes the launch pad.

2.44 •• A hot-air balloonist, rising vertically with a constant velocity of magnitude 5.00 m/s, releases a sandbag at an instant when the balloon is 40.0 m above the ground (Fig. E2.44). After it is released, the sandbag is in free fall. (a) Compute the position and velocity of the sandbag at 0.250 s and 1.00 s after its release. (b) How many seconds after its release will the bag strike the ground? (c) With what magnitude of velocity does it strike the ground? (d) What is the greatest height above the ground that the sandbag reaches? (e) Sketch a_y-t, v_y-t, and y-t graphs for the motion.

Figure **E2.44**

$v = 5.00 \ m/s$

40.0 m to ground

2.45 • **BIO** The rocket-driven sled *Sonic Wind No. 2,* used for investigating the physiological effects of large accelerations, runs on a straight, level track 1070 m (3500 ft) long. Starting from rest, it can reach a speed of 224 m/s (500 mi/h) in 0.900 s. (a) Compute the acceleration in m/s^2, assuming that it is constant. (b) What is the ratio of this acceleration to that of a freely falling body (g)? (c) What distance is covered in 0.900 s? (d) A magazine article states that at the end of a certain run, the speed of the sled decreased from 283 m/s (632 mi/h) to zero in 1.40 s and that during this time the magnitude of the acceleration was greater than $40g$. Are these figures consistent?

2.46 • An egg is thrown nearly vertically upward from a point near the cornice of a tall building. It just misses the cornice on the way down and passes a point 30.0 m below its starting point 5.00 s after it leaves the thrower's hand. Air resistance may be ignored.

(a) What is the initial speed of the egg? (b) How high does it rise above its starting point? (c) What is the magnitude of its velocity at the highest point? (d) What are the magnitude and direction of its acceleration at the highest point? (e) Sketch a_y-t, v_y-t, and y-t graphs for the motion of the egg.

2.47 •• A 15-kg rock is dropped from rest on the earth and reaches the ground in 1.75 s. When it is dropped from the same height on Saturn's satellite Enceladus, it reaches the ground in 18.6 s. What is the acceleration due to gravity on Enceladus?

2.48 • A large boulder is ejected vertically upward from a volcano with an initial speed of 40.0 m/s. Air resistance may be ignored. (a) At what time after being ejected is the boulder moving at 20.0 m/s upward? (b) At what time is it moving at 20.0 m/s downward? (c) When is the displacement of the boulder from its initial position zero? (d) When is the velocity of the boulder zero? (e) What are the magnitude and direction of the acceleration while the boulder is (i) moving upward? (ii) Moving downward? (iii) At the highest point? (f) Sketch a_y-t, v_y-t, and y-t graphs for the motion.

2.49 •• Two stones are thrown vertically upward from the ground, one with three times the initial speed of the other. (a) If the faster stone takes 10 s to return to the ground, how long will it take the slower stone to return? (b) If the slower stone reaches a maximum height of H, how high (in terms of H) will the faster stone go? Assume free fall.

Section 2.6 Velocity and Position by Integration

2.50 • CALC For constant a_x, use Eqs. (2.17) and (2.18) to find v_x and x as functions of time. Compare your results to Eqs. (2.8) and (2.12).

2.51 • CALC A rocket starts from rest and moves upward from the surface of the earth. For the first 10.0 s of its motion, the vertical acceleration of the rocket is given by $a_y = (2.80 \text{ m/s}^3)t$, where the +y-direction is upward. (a) What is the height of the rocket above the surface of the earth at $t = 10.0$ s? (b) What is the speed of the rocket when it is 325 m above the surface of the earth?

2.52 •• CALC The acceleration of a bus is given by $a_x(t) = \alpha t$, where $\alpha = 1.2 \text{ m/s}^3$. (a) If the bus's velocity at time $t = 1.0$ s is 5.0 m/s, what is its velocity at time $t = 2.0$ s? (b) If the bus's position at time $t = 1.0$ s is 6.0 m, what is its position at time $t = 2.0$ s? (c) Sketch a_x-t, v_x-t, and x-t graphs for the motion.

2.53 •• CALC The acceleration of a motorcycle is given by $a_x(t) = At - Bt^2$, where $A = 1.50 \text{ m/s}^3$ and $B = 0.120 \text{ m/s}^4$. The motorcycle is at rest at the origin at time $t = 0$. (a) Find its position and velocity as functions of time. (b) Calculate the maximum velocity it attains.

2.54 •• BIO **Flying Leap of the Flea.** High-speed motion pictures (3500 frames/second) of a jumping, 210-μg flea yielded the data used to plot the graph given in Fig. E2.54. (See "The Flying Leap of the Flea" by M. Rothschild, Y. Schlein, K. Parker, C. Neville, and S. Sternberg in the November 1973 *Scientific American*.) This flea was about 2 mm long and jumped at a nearly vertical takeoff angle. Use the graph to answer the questions. (a) Is the acceleration of the flea ever zero? If so, when? Justify your answer. (b) Find the maximum height the flea reached in the first 2.5 ms. (c) Find the flea's acceleration at 0.5 ms, 1.0 ms, and 1.5 ms. (d) Find the flea's height at 0.5 ms, 1.0 ms, and 1.5 ms.

Figure **E2.54**

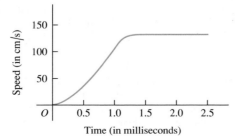

Time (in milliseconds)

PROBLEMS

2.55 • BIO A typical male sprinter can maintain his maximum acceleration for 2.0 s and his maximum speed is 10 m/s. After reaching this maximum speed, his acceleration becomes zero and then he runs at constant speed. Assume that his acceleration is constant during the first 2.0 s of the race, that he starts from rest, and that he runs in a straight line. (a) How far has the sprinter run when he reaches his maximum speed? (b) What is the magnitude of his average velocity for a race of the following lengths: (i) 50.0 m, (ii) 100.0 m, (iii) 200.0 m?

2.56 •• On a 20-mile bike ride, you ride the first 10 miles at an average speed of 8 mi/h. What must your average speed over the next 10 miles be to have your average speed for the total 20 miles be (a) 4 mi/h? (b) 12 mi/h? (c) Given this average speed for the first 10 miles, can you possibly attain an average speed of 16 mi/h for the total 20-mile ride? Explain.

2.57 •• CALC The position of a particle between $t = 0$ and $t = 2.00$ s is given by $x(t) = (3.00 \text{ m/s}^3)t^3 - (10.0 \text{ m/s}^2)t^2 + (9.00 \text{ m/s})t$. (a) Draw the x-t, v_x-t, and a_x-t graphs of this particle. (b) At what time(s) between $t = 0$ and $t = 2.00$ s is the particle instantaneously at rest? Does your numerical result agree with the v_x-t graph in part (a)? (c) At each time calculated in part (b), is the acceleration of the particle positive or negative? Show that in each case the same answer is deduced from $a_x(t)$ and from the v_x-t graph. (d) At what time(s) between $t = 0$ and $t = 2.00$ s is the velocity of the particle instantaneously not changing? Locate this point on the v_x-t and a_x-t graphs of part (a). (e) What is the particle's greatest distance from the origin ($x = 0$) between $t = 0$ and $t = 2.00$ s? (f) At what time(s) between $t = 0$ and $t = 2.00$ s is the particle *speeding up* at the greatest rate? At what time(s) between $t = 0$ and $t = 2.00$ s is the particle *slowing down* at the greatest rate? Locate these points on the v_x-t and a_x-t graphs of part (a).

2.58 •• CALC A lunar lander is descending toward the moon's surface. Until the lander reaches the surface, its height above the surface of the moon is given by $y(t) = b - ct + dt^2$, where $b = 800$ m is the initial height of the lander above the surface, $c = 60.0$ m/s, and $d = 1.05 \text{ m/s}^2$. (a) What is the initial velocity of the lander, at $t = 0$? (b) What is the velocity of the lander just before it reaches the lunar surface?

2.59 ••• **Earthquake Analysis.** Earthquakes produce several types of shock waves. The most well known are the P-waves (P for *primary* or *pressure*) and the S-waves (S for *secondary* or *shear*). In the earth's crust, the P-waves travel at around 6.5 km/s, while the S-waves move at about 3.5 km/s. The actual speeds vary depending on the type of material they are going through. The time delay between the arrival of these two waves at a seismic recording station tells geologists how far away the earthquake occurred. If the time delay is 33 s, how far from the seismic station did the earthquake occur?

2.60 •• **Relay Race.** In a relay race, each contestant runs 25.0 m while carrying an egg balanced on a spoon, turns around, and comes back to the starting point. Edith runs the first 25.0 m in 20.0 s. On the return trip she is more confident and takes only 15.0 s. What is the magnitude of her average velocity for (a) the first 25.0 m? (b) The return trip? (c) What is her average velocity for the entire round trip? (d) What is her average speed for the round trip?

2.61 ••• A rocket carrying a satellite is accelerating straight up from the earth's surface. At 1.15 s after liftoff, the rocket clears the top of its launch platform, 63 m above the ground. After an additional 4.75 s, it is 1.00 km above the ground. Calculate the magnitude of the average velocity of the rocket for (a) the 4.75-s part of its flight and (b) the first 5.90 s of its flight.

2.62 ••• The graph in Fig. P2.62 describes the acceleration as a function of time for a stone rolling down a hill starting from rest. (a) Find the stone's velocity at $t = 2.5$ s and at $t = 7.5$ s. (b) Sketch a graph of the stone's velocity as a function of time.

Figure **P2.62**

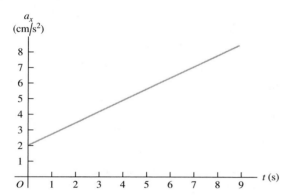

2.63 •• Dan gets on Interstate Highway I–80 at Seward, Nebraska, and drives due west in a straight line and at an average velocity of magnitude 88 km/h. After traveling 76 km, he reaches the Aurora exit (Fig. P2.63). Realizing he has gone too far, he turns around and drives due east 34 km back to the York exit at an average velocity of magnitude 72 km/h. For his whole trip from Seward to the York exit, what are (a) his average speed and (b) the magnitude of his average velocity?

Figure **P2.63**

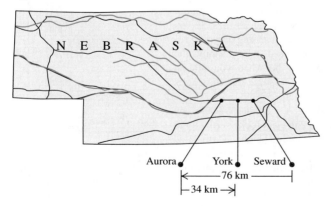

2.64 ••• A subway train starts from rest at a station and accelerates at a rate of 1.60 m/s² for 14.0 s. It runs at constant speed for 70.0 s and slows down at a rate of 3.50 m/s² until it stops at the next station. Find the *total* distance covered.

2.65 •• A world-class sprinter accelerates to his maximum speed in 4.0 s. He then maintains this speed for the remainder of a 100-m race, finishing with a total time of 9.1 s. (a) What is the runner's average acceleration during the first 4.0 s? (b) What is his average

acceleration during the last 5.1 s? (c) What is his average acceleration for the entire race? (d) Explain why your answer to part (c) is not the average of the answers to parts (a) and (b).

2.66 •• A sled starts from rest at the top of a hill and slides down with a constant acceleration. At some later time the sled is 14.4 m from the top, 2.00 s after that it is 25.6 m from the top, 2.00 s later 40.0 m from the top, and 2.00 s later it is 57.6 m from the top. (a) What is the magnitude of the average velocity of the sled during each of the 2.00-s intervals after passing the 14.4-m point? (b) What is the acceleration of the sled? (c) What is the speed of the sled when it passes the 14.4-m point? (d) How much time did it take to go from the top to the 14.4-m point? (e) How far did the sled go during the first second after passing the 14.4-m point?

2.67 • A gazelle is running in a straight line (the x-axis). The graph in Fig. P2.67 shows this animal's velocity as a function of time. During the first 12.0 s, find (a) the total distance moved and (b) the displacement of the gazelle. (c) Sketch an a_x-t graph showing this gazelle's acceleration as a function of time for the first 12.0 s.

Figure **P2.67**

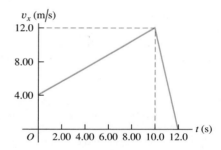

2.68 • A rigid ball traveling in a straight line (the x-axis) hits a solid wall and suddenly rebounds during a brief instant. The v_x-t graph in Fig. P2.68 shows this ball's velocity as a function of time. During the first 20.0 s of its motion, find (a) the total distance the ball moves and (b) its displacement. (c) Sketch a graph of a_x-t for this ball's motion. (d) Is the graph shown really vertical at 5.00 s? Explain.

Figure **P2.68**

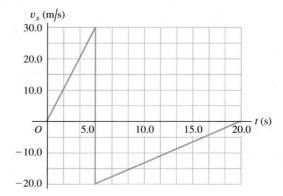

2.69 ••• A ball starts from rest and rolls down a hill with uniform acceleration, traveling 150 m during the second 5.0 s of its motion. How far did it roll during the first 5.0 s of motion?

2.70 •• **Collision.** The engineer of a passenger train traveling at 25.0 m/s sights a freight train whose caboose is 200 m ahead on

Figure **P2.70**

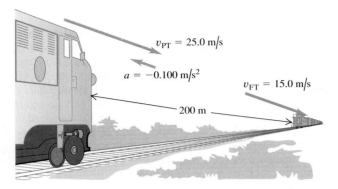

$v_{PT} = 25.0$ m/s

$a = -0.100$ m/s^2

$v_{FT} = 15.0$ m/s

200 m

the same track (Fig. P2.70). The freight train is traveling at 15.0 m/s in the same direction as the passenger train. The engineer of the passenger train immediately applies the brakes, causing a constant acceleration of 0.100 m/s^2 in a direction opposite to the train's velocity, while the freight train continues with constant speed. Take $x = 0$ at the location of the front of the passenger train when the engineer applies the brakes. (a) Will the cows nearby witness a collision? (b) If so, where will it take place? (c) On a single graph, sketch the positions of the front of the passenger train and the back of the freight train.

2.71 ••• Large cockroaches can run as fast as 1.50 m/s in short bursts. Suppose you turn on the light in a cheap motel and see one scurrying directly away from you at a constant 1.50 m/s. If you start 0.90 m behind the cockroach with an initial speed of 0.80 m/s toward it, what minimum constant acceleration would you need to catch up with it when it has traveled 1.20 m, just short of safety under a counter?

2.72 •• Two cars start 200 m apart and drive toward each other at a steady 10 m/s. On the front of one of them, an energetic grasshopper jumps back and forth between the cars (he has strong legs!) with a constant horizontal velocity of 15 m/s relative to the ground. The insect jumps the instant he lands, so he spends no time resting on either car. What total distance does the grasshopper travel before the cars hit?

2.73 • An automobile and a truck start from rest at the same instant, with the automobile initially at some distance behind the truck. The truck has a constant acceleration of 2.10 m/s^2, and the automobile an acceleration of 3.40 m/s^2. The automobile overtakes the truck after the truck has moved 40.0 m. (a) How much time does it take the automobile to overtake the truck? (b) How far was the automobile behind the truck initially? (c) What is the speed of each when they are abreast? (d) On a single graph, sketch the position of each vehicle as a function of time. Take $x = 0$ at the initial location of the truck.

2.74 ••• Two stunt drivers drive directly toward each other. At time $t = 0$ the two cars are a distance D apart, car 1 is at rest, and car 2 is moving to the left with speed v_0. Car 1 begins to move at $t = 0$, speeding up with a constant acceleration a_x. Car 2 continues to move with a constant velocity. (a) At what time do the two cars collide? (b) Find the speed of car 1 just before it collides with car 2. (c) Sketch x-t and v_x-t graphs for car 1 and car 2. For each of the two graphs, draw the curves for both cars on the same set of axes.

2.75 •• A marble is released from one rim of a hemispherical bowl of diameter 50.0 cm and rolls down and up to the opposite rim in 10.0 s. Find (a) the average speed and (b) the average velocity of the marble.

2.76 •• **CALC** An object's velocity is measured to be $v_x(t) = \alpha - \beta t^2$, where $\alpha = 4.00$ m/s and $\beta = 2.00$ m/s^3. At $t = 0$ the object is at $x = 0$. (a) Calculate the object's position and acceleration as functions of time. (b) What is the object's maximum *positive* displacement from the origin?

2.77 •• **Passing.** The driver of a car wishes to pass a truck that is traveling at a constant speed of 20.0 m/s (about 45 mi/h). Initially, the car is also traveling at 20.0 m/s and its front bumper is 24.0 m behind the truck's rear bumper. The car accelerates at a constant 0.600 m/s^2, then pulls back into the truck's lane when the rear of the car is 26.0 m ahead of the front of the truck. The car is 4.5 m long and the truck is 21.0 m long. (a) How much time is required for the car to pass the truck? (b) What distance does the car travel during this time? (c) What is the final speed of the car?

2.78 •• On Planet X, you drop a 25-kg stone from rest and measure its speed at various times. Then you use the data you obtained to construct a graph of its speed v as a function of time t (Fig. P2.78). From the information in the graph, answer the following questions: (a) What is g on Planet X? (b) An astronaut drops a piece of equipment from rest out of the landing module, 3.5 m above the surface of Planet X. How long will it take this equipment to reach the ground, and how fast will it be moving when it gets there? (c) How fast would an astronaut have to project an object straight upward to reach a height of 18.0 m above the release point, and how long would it take to reach that height?

Figure **P2.78**

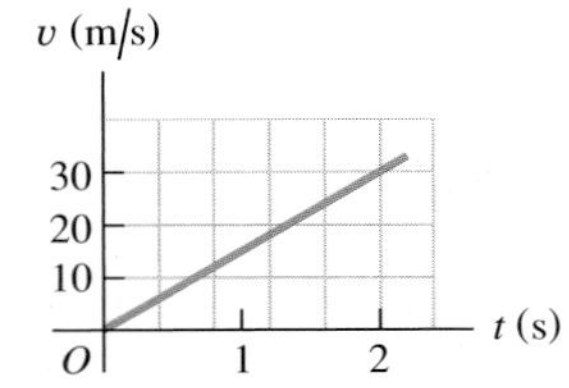

v (m/s)

30
20
10

O 1   2 t (s)

2.79 ••• **CALC** The acceleration of a particle is given by $a_x(t) = -2.00$ m/s^2 + (3.00 m/s^3)t. (a) Find the initial velocity v_{0x} such that the particle will have the same x-coordinate at $t = 4.00$ s as it had at $t = 0$. (b) What will be the velocity at $t = 4.00$ s?

2.80 • **Egg Drop.** You are on the roof of the physics building, 46.0 m above the ground (Fig. P2.80). Your physics professor, who is 1.80 m tall, is walking alongside the building at a constant speed of 1.20 m/s. If you wish to drop an egg on your professor's head, where should the professor be when you release the egg? Assume that the egg is in free fall.

Figure **P2.80**

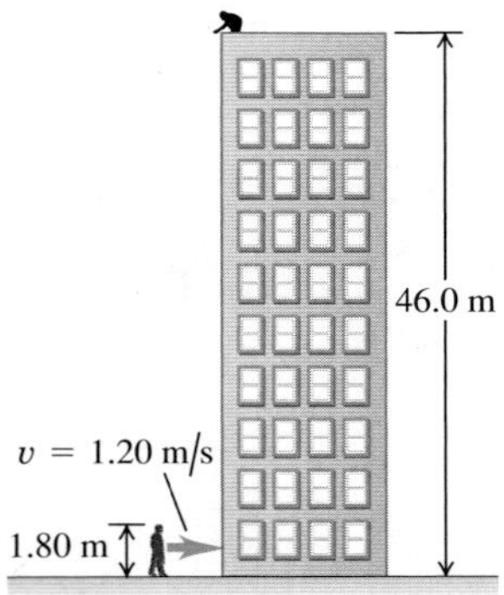

46.0 m

$v = 1.20$ m/s

1.80 m

2.81 • A certain volcano on earth can eject rocks vertically to a maximum height H. (a) How high (in terms of H) would these rocks go if a volcano on Mars ejected them with the same initial velocity? The acceleration due to gravity on Mars is 3.71 m/s^2, and you can neglect air resistance on both planets. (b) If the rocks are in the air for a time T on earth, for how long (in terms of T) will they be in the air on Mars?

2.82 •• An entertainer juggles balls while doing other activities. In one act, she throws a ball vertically upward, and while it is in the air, she runs to and from a table 5.50 m away at a constant speed of 2.50 m/s, returning just in time to catch the falling ball. (a) With what minimum initial speed must she throw the ball upward to accomplish this feat? (b) How high above its initial position is the ball just as she reaches the table?

2.83 • Visitors at an amusement park watch divers step off a platform 21.3 m (70 ft) above a pool of water. According to the announcer, the divers enter the water at a speed of 56 mi/h (25 m/s). Air resistance may be ignored. (a) Is the announcer correct in this claim? (b) Is it possible for a diver to leap directly upward off the board so that, missing the board on the way down, she enters the water at 25.0 m/s? If so, what initial upward speed is required? Is the required initial speed physically attainable?

2.84 ••• A flowerpot falls off a windowsill and falls past the window below. You may ignore air resistance. It takes the pot 0.420 s to pass from the top to the bottom of this window, which is 1.90 m high. How far is the top of the window below the windowsill from which the flowerpot fell?

2.85 ••• **Look Out Below.** Sam heaves a 16-lb shot straight upward, giving it a constant upward acceleration from rest of 35.0 m/s^2 for 64.0 cm. He releases it 2.20 m above the ground. You may ignore air resistance. (a) What is the speed of the shot when Sam releases it? (b) How high above the ground does it go? (c) How much time does he have to get out of its way before it returns to the height of the top of his head, 1.83 m above the ground?

2.86 ••• **A Multistage Rocket.** In the first stage of a two-stage rocket, the rocket is fired from the launch pad starting from rest but with a constant acceleration of 3.50 m/s^2 upward. At 25.0 s after launch, the second stage fires for 10.0 s, which boosts the rocket's velocity to 132.5 m/s upward at 35.0 s after launch. This firing uses up all the fuel, however, so after the second stage has finished firing, the only force acting on the rocket is gravity. Air resistance can be neglected. (a) Find the maximum height that the stage-two rocket reaches above the launch pad. (b) How much time after the end of the stage-two firing will it take for the rocket to fall back to the launch pad? (c) How fast will the stage-two rocket be moving just as it reaches the launch pad?

2.87 •• **Juggling Act.** A juggler performs in a room whose ceiling is 3.0 m above the level of his hands. He throws a ball upward so that it just reaches the ceiling. (a) What is the initial velocity of the ball? (b) What is the time required for the ball to reach the ceiling? At the instant when the first ball is at the ceiling, the juggler throws a second ball upward with two-thirds the initial velocity of the first. (c) How long after the second ball is thrown do the two balls pass each other? (d) At what distance above the juggler's hand do they pass each other?

2.88 •• A physics teacher performing an outdoor demonstration suddenly falls from rest off a high cliff and simultaneously shouts "Help." When she has fallen for 3.0 s, she hears the echo of her shout from the valley floor below. The speed of sound is 340 m/s. (a) How tall is the cliff? (b) If air resistance is neglected, how fast will she be moving just before she hits the ground? (Her actual speed will be less than this, due to air resistance.)

2.89 ••• A helicopter carrying Dr. Evil takes off with a constant upward acceleration of 5.0 m/s^2. Secret agent Austin Powers jumps on just as the helicopter lifts off the ground. After the two men struggle for 10.0 s, Powers shuts off the engine and steps out of the helicopter. Assume that the helicopter is in free fall after its engine is shut off, and ignore the effects of air resistance. (a) What is the maximum height above ground reached by the helicopter? (b) Powers deploys a jet pack strapped on his back 7.0 s after leaving the helicopter, and then he has a constant downward acceleration with magnitude 2.0 m/s^2. How far is Powers above the ground when the helicopter crashes into the ground?

2.90 •• **Cliff Height.** You are climbing in the High Sierra where you suddenly find yourself at the edge of a fog-shrouded cliff. To find the height of this cliff, you drop a rock from the top and 10.0 s later hear the sound of it hitting the ground at the foot of the cliff. (a) Ignoring air resistance, how high is the cliff if the speed of sound is 330 m/s? (b) Suppose you had ignored the time it takes the sound to reach you. In that case, would you have overestimated or underestimated the height of the cliff? Explain your reasoning.

2.91 ••• **Falling Can.** A painter is standing on scaffolding that is raised at constant speed. As he travels upward, he accidentally nudges a paint can off the scaffolding and it falls 15.0 m to the ground. You are watching, and measure with your stopwatch that it takes 3.25 s for the can to reach the ground. Ignore air resistance. (a) What is the speed of the can just before it hits the ground? (b) Another painter is standing on a ledge, with his hands 4.00 m above the can when it falls off. He has lightning-fast reflexes and if the can passes in front of him, he can catch it. Does he get the chance?

2.92 •• Determined to test the law of gravity for himself, a student walks off a skyscraper 180 m high, stopwatch in hand, and starts his free fall (zero initial velocity). Five seconds later, Superman arrives at the scene and dives off the roof to save the student. Superman leaves the roof with an initial speed v_0 that he produces by pushing himself downward from the edge of the roof with his legs of steel. He then falls with the same acceleration as any freely falling body. (a) What must the value of v_0 be so that Superman catches the student just before they reach the ground? (b) On the same graph, sketch the positions of the student and of Superman as functions of time. Take Superman's initial speed to have the value calculated in part (a). (c) If the height of the skyscraper is less than some minimum value, even Superman can't reach the student before he hits the ground. What is this minimum height?

2.93 ••• During launches, rockets often discard unneeded parts. A certain rocket starts from rest on the launch pad and accelerates upward at a steady 3.30 m/s^2. When it is 235 m above the launch pad, it discards a used fuel canister by simply disconnecting it. Once it is disconnected, the only force acting on the canister is gravity (air resistance can be ignored). (a) How high is the rocket when the canister hits the launch pad, assuming that the rocket does not change its acceleration? (b) What total distance did the canister travel between its release and its crash onto the launch pad?

2.94 •• A ball is thrown straight up from the ground with speed v_0. At the same instant, a second ball is dropped from rest from a height H, directly above the point where the first ball was thrown upward. There is no air resistance. (a) Find the time at which the two balls collide. (b) Find the value of H in terms of v_0 and g so that at the instant when the balls collide, the first ball is at the highest point of its motion.

2.95 • **CALC** Two cars, A and B, travel in a straight line. The distance of A from the starting point is given as a function of time by $x_A(t) = \alpha t + \beta t^2$, with $\alpha = 2.60$ m/s and $\beta = 1.20$ m/s^2. The distance of B from the starting point is $x_B(t) = \gamma t^2 - \delta t^3$, with $\gamma = 2.80$ m/s^2 and $\delta = 0.20$ m/s^3. (a) Which car is ahead just after they leave the starting point? (b) At what time(s) are the cars at the same point? (c) At what time(s) is the distance from A to B neither increasing nor decreasing? (d) At what time(s) do A and B have the same acceleration?

CHALLENGE PROBLEMS

2.96 ••• In the vertical jump, an athlete starts from a crouch and jumps upward to reach as high as possible. Even the best athletes spend little more than 1.00 s in the air (their "hang time"). Treat the athlete as a particle and let y_{max} be his maximum height above the floor. To explain why he seems to hang in the air, calculate the ratio of the time he is above $y_{max}/2$ to the time it takes him to go from the floor to that height. You may ignore air resistance.

2.97 ••• **Catching the Bus.** A student is running at her top speed of 5.0 m/s to catch a bus, which is stopped at the bus stop. When the student is still 40.0 m from the bus, it starts to pull away, moving with a constant acceleration of 0.170 m/s². (a) For how much time and what distance does the student have to run at 5.0 m/s before she overtakes the bus? (b) When she reaches the bus, how fast is the bus traveling? (c) Sketch an x-t graph for both the student and the bus. Take $x = 0$ at the initial position of the student. (d) The equations you used in part (a) to find the time have a second solution, corresponding to a later time for which the student and bus are again at the same place if they continue their specified motions. Explain the significance of this second solution. How fast is the bus traveling at this point? (e) If the student's top speed is 3.5 m/s, will she catch the bus? (f) What is the *minimum* speed the student must have to just catch up with the bus? For what time and what distance does she have to run in that case?

2.98 ••• An alert hiker sees a boulder fall from the top of a distant cliff and notes that it takes 1.30 s for the boulder to fall the last third of the way to the ground. You may ignore air resistance.

(a) What is the height of the cliff in meters? (b) If in part (a) you get two solutions of a quadratic equation and you use one for your answer, what does the other solution represent?

2.99 ••• A ball is thrown straight up from the edge of the roof of a building. A second ball is dropped from the roof 1.00 s later. You may ignore air resistance. (a) If the height of the building is 20.0 m, what must the initial speed of the first ball be if both are to hit the ground at the same time? On the same graph, sketch the position of each ball as a function of time, measured from when the first ball is thrown. Consider the same situation, but now let the initial speed v_0 of the first ball be given and treat the height h of the building as an unknown. (b) What must the height of the building be for both balls to reach the ground at the same time (i) if v_0 is 6.0 m/s and (ii) if v_0 is 9.5 m/s? (c) If v_0 is greater than some value v_{max}, a value of h does not exist that allows both balls to hit the ground at the same time. Solve for v_{max}. The value v_{max} has a simple physical interpretation. What is it? (d) If v_0 is less than some value v_{min}, a value of h does not exist that allows both balls to hit the ground at the same time. Solve for v_{min}. The value v_{min} also has a simple physical interpretation. What is it?

Answers

Chapter Opening Question

Yes. Acceleration refers to *any* change in velocity, including both speeding up and slowing down.

Test Your Understanding Questions

2.1 Answer to (a): (iv), (i) and (iii) (tie), (v), (ii); answer to (b): (i) and (iii); answer to (c): (v) In (a) the average x-velocity is $v_{av-x} = \Delta x/\Delta t$. For all five trips, $\Delta t = 1$ h. For the individual trips, we have (i) $\Delta x = +50$ km, $v_{av-x} = +50$ km/h; (ii) $\Delta x = -50$ km, $v_{av-x} = -50$ km/h; (iii) $\Delta x = 60$ km $- 10$ km $= +50$ km, $v_{av-x} = +50$ km/h; (iv) $\Delta x = +70$ km, $v_{av-x} = +70$ km/h; (v) $\Delta x = -20$ km $+ 20$ km $= 0$, $v_{av-x} = 0$. In (b) both have $v_{av-x} = +50$ km/h.

2.2 Answers: (a) P, Q and S (tie), R The x-velocity is **(b)** positive when the slope of the x-t graph is positive (P), **(c)** negative when the slope is negative (R), and **(d)** zero when the slope is zero (Q and S). **(e)** R, P, Q and S (tie) The speed is greatest when the slope of the x-t graph is steepest (either positive or negative) and zero when the slope is zero.

2.3 Answers: (a) S, where the x-t graph is curved upward (concave up). **(b)** Q, where the x-t graph is curved downward (concave down). **(c)** P and R, where the x-t graph is not curved either up or down. **(d)** At P, $a_x = 0$ (velocity is **not changing**); at Q, $a_x < 0$

(velocity is **decreasing,** i.e., changing from positive to zero to negative); at R, $a_x = 0$ (velocity is **not changing**); and at S, $a_x > 0$ (velocity is **increasing,** i.e., changing from negative to zero to positive).

2.4 Answer: (b) The officer's x-acceleration is constant, so her v_x-t graph is a straight line, and the officer's motorcycle is moving faster than the motorist's car when the two vehicles meet at $t = 10$ s.

2.5 Answers: (a) (iii) Use Eq. (2.13) with x replaced by y and $a_y = g$; $v_y^2 = v_{0y}^2 - 2g(y - y_0)$. The starting height is $y_0 = 0$ and the y-velocity at the maximum height $y = h$ is $v_y = 0$, so $0 = v_{0y}^2 - 2gh$ and $h = v_{0y}^2/2g$. If the initial y-velocity is increased by a factor of 2, the maximum height increases by a factor of $2^2 = 4$ and the ball goes to height $4h$. **(b) (v)** Use Eq. (2.8) with x replaced by y and $a_y = g$; $v_y = v_{0y} - gt$. The y-velocity at the maximum height is $v_y = 0$, so $0 = v_{0y} - gt$ and $t = v_{0y}/g$. If the initial y-velocity is increased by a factor of 2, the time to reach the maximum height increases by a factor of 2 and becomes $2t$.

2.6 Answer: (ii) The acceleration a_x is equal to the slope of the v_x-t graph. If a_x is increasing, the slope of the v_x-t graph is also increasing and the graph is concave up.

Bridging Problem

Answer: $h = 57.1$ m

MOTION IN TWO OR THREE DIMENSIONS

? If a cyclist is going around a curve at constant speed, is he accelerating? If so, in which direction is he accelerating?

What determines where a batted baseball lands? How do you describe the motion of a roller coaster car along a curved track or the flight of a circling hawk? Which hits the ground first: a baseball that you simply drop or one that you throw horizontally?

We can't answer these kinds of questions using the techniques of Chapter 2, in which particles moved only along a straight line. Instead, we need to extend our descriptions of motion to two- and three-dimensional situations. We'll still use the vector quantities displacement, velocity, and acceleration, but now these quantities will no longer lie along a single line. We'll find that several important kinds of motion take place in two dimensions only—that is, in a *plane*. We can describe these motions with two components of position, velocity, and acceleration.

We also need to consider how the motion of a particle is described by different observers who are moving relative to each other. The concept of *relative velocity* will play an important role later in the book when we study collisions, when we explore electromagnetic phenomena, and when we introduce Einstein's special theory of relativity.

This chapter merges the vector mathematics of Chapter 1 with the kinematic language of Chapter 2. As before, we are concerned with describing motion, not with analyzing its causes. But the language you learn here will be an essential tool in later chapters when we study the relationship between force and motion.

3.1 Position and Velocity Vectors

3.1 The position vector $\vec{r}$ from the origin to point P has components x, y, and z. The path that the particle follows through space is in general a curve (Fig. 3.2).

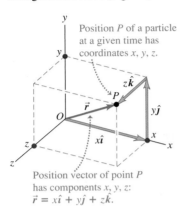

Position P of a particle at a given time has coordinates x, y, z.

Position vector of point P has components x, y, z:
$\vec{r} = x\hat{\imath} + y\hat{\jmath} + z\hat{k}$.

To describe the *motion* of a particle in space, we must first be able to describe the particle's *position*. Consider a particle that is at a point P at a certain instant. The **position vector** $\vec{r}$ of the particle at this instant is a vector that goes from the origin of the coordinate system to the point P (Fig. 3.1). The Cartesian coordinates x, y, and z of point P are the x-, y-, and z-components of vector $\vec{r}$. Using the unit vectors we introduced in Section 1.9, we can write

$$\vec{r} = x\hat{\imath} + y\hat{\jmath} + z\hat{k} \quad \text{(position vector)} \tag{3.1}$$

During a time interval Δt the particle moves from P_1, where its position vector is $\vec{r}_1$, to P_2, where its position vector is $\vec{r}_2$. The change in position (the displacement) during this interval is $\Delta\vec{r} = \vec{r}_2 - \vec{r}_1 = (x_2 - x_1)\hat{\imath} + (y_2 - y_1)\hat{\jmath} + (z_2 - z_1)\hat{k}$. We define the **average velocity** $\vec{v}_{\text{av}}$ during this interval in the same way we did in Chapter 2 for straight-line motion, as the displacement divided by the time interval:

$$\vec{v}_{\text{av}} = \frac{\vec{r}_2 - \vec{r}_1}{t_2 - t_1} = \frac{\Delta\vec{r}}{\Delta t} \quad \text{(average velocity vector)} \tag{3.2}$$

Dividing a vector by a scalar is really a special case of *multiplying* a vector by a scalar, described in Section 1.7; the average velocity $\vec{v}_{\text{av}}$ is equal to the displacement vector $\Delta\vec{r}$ multiplied by $1/\Delta t$, the reciprocal of the time interval. Note that the x-component of Eq. (3.2) is $v_{\text{av-}x} = (x_2 - x_1)/(t_2 - t_1) = \Delta x/\Delta t$. This is just Eq. (2.2), the expression for average x-velocity that we found in Section 2.1 for one-dimensional motion.

We now define **instantaneous velocity** just as we did in Chapter 2: It is the limit of the average velocity as the time interval approaches zero, and it equals the instantaneous rate of change of position with time. The key difference is that position $\vec{r}$ and instantaneous velocity $\vec{v}$ are now both vectors:

$$\vec{v} = \lim_{\Delta t \to 0} \frac{\Delta\vec{r}}{\Delta t} = \frac{d\vec{r}}{dt} \quad \text{(instantaneous velocity vector)} \tag{3.3}$$

3.2 The average velocity $\vec{v}_{\text{av}}$ between points P_1 and P_2 has the same direction as the displacement $\Delta\vec{r}$.

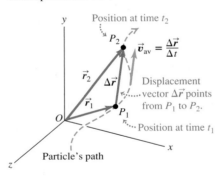

Position at time t_2

$\vec{v}_{\text{av}} = \dfrac{\Delta\vec{r}}{\Delta t}$

Displacement vector $\Delta\vec{r}$ points from P_1 to P_2.

Position at time t_1

Particle's path

The *magnitude* of the vector $\vec{v}$ at any instant is the *speed* v of the particle at that instant. The *direction* of $\vec{v}$ at any instant is the same as the direction in which the particle is moving at that instant.

Note that as $\Delta t \to 0$, points P_1 and P_2 in Fig. 3.2 move closer and closer together. In this limit, the vector $\Delta\vec{r}$ becomes tangent to the path. The direction of $\Delta\vec{r}$ in this limit is also the direction of the instantaneous velocity $\vec{v}$. This leads to an important conclusion: *At every point along the path, the instantaneous velocity vector is tangent to the path at that point* (Fig. 3.3).

It's often easiest to calculate the instantaneous velocity vector using components. During any displacement $\Delta\vec{r}$, the changes Δx, Δy, and Δz in the three coordinates of the particle are the *components* of $\Delta\vec{r}$. It follows that the components v_x, v_y, and v_z of the instantaneous velocity $\vec{v}$ are simply the time derivatives of the coordinates x, y, and z. That is,

3.3 The vectors $\vec{v}_1$ and $\vec{v}_2$ are the instantaneous velocities at the points P_1 and P_2 shown in Fig. 3.2.

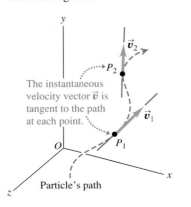

The instantaneous velocity vector $\vec{v}$ is tangent to the path at each point.

Particle's path

$$v_x = \frac{dx}{dt} \qquad v_y = \frac{dy}{dt} \qquad v_z = \frac{dz}{dt} \quad \begin{array}{l}\text{(components of}\\ \text{instantaneous velocity)}\end{array} \tag{3.4}$$

The x-component of $\vec{v}$ is $v_x = dx/dt$, which is the same as Eq. (2.3)—the expression for instantaneous velocity for straight-line motion that we obtained in Section 2.2. Hence Eq. (3.4) is a direct extension of the idea of instantaneous velocity to motion in three dimensions.

We can also get Eq. (3.4) by taking the derivative of Eq. (3.1). The unit vectors $\hat{i}$, $\hat{j}$, and $\hat{k}$ are constant in magnitude and direction, so their derivatives are zero, and we find

$$\vec{v} = \frac{d\vec{r}}{dt} = \frac{dx}{dt}\hat{i} + \frac{dy}{dt}\hat{j} + \frac{dz}{dt}\hat{k} \qquad (3.5)$$

This shows again that the components of $\vec{v}$ are dx/dt, dy/dt, and dz/dt.

The magnitude of the instantaneous velocity vector $\vec{v}$—that is, the speed—is given in terms of the components v_x, v_y, and v_z by the Pythagorean relation:

$$|\vec{v}| = v = \sqrt{v_x^2 + v_y^2 + v_z^2} \qquad (3.6)$$

Figure 3.4 shows the situation when the particle moves in the xy-plane. In this case, z and v_z are zero. Then the speed (the magnitude of $\vec{v}$) is

$$v = \sqrt{v_x^2 + v_y^2}$$

and the direction of the instantaneous velocity $\vec{v}$ is given by the angle α (the Greek letter alpha) in the figure. We see that

$$\tan\alpha = \frac{v_y}{v_x} \qquad (3.7)$$

(We always use Greek letters for angles. We use α for the direction of the instantaneous velocity vector to avoid confusion with the direction θ of the *position* vector of the particle.)

The instantaneous velocity vector is usually more interesting and useful than the average velocity vector. From now on, when we use the word "velocity," we will always mean the instantaneous velocity vector $\vec{v}$ (rather than the average velocity vector). Usually, we won't even bother to call $\vec{v}$ a vector; it's up to you to remember that velocity is a vector quantity with both magnitude and direction.

3.4 The two velocity components for motion in the xy-plane.

The instantaneous velocity vector $\vec{v}$ is always tangent to the path.

v_x and v_y are the x- and y-components of $\vec{v}$.

Example 3.1   Calculating average and instantaneous velocity

A robotic vehicle, or rover, is exploring the surface of Mars. The stationary Mars lander is the origin of coordinates, and the surrounding Martian surface lies in the xy-plane. The rover, which we represent as a point, has x- and y-coordinates that vary with time:

$$x = 2.0 \text{ m} - (0.25 \text{ m/s}^2)t^2$$
$$y = (1.0 \text{ m/s})t + (0.025 \text{ m/s}^3)t^3$$

(a) Find the rover's coordinates and distance from the lander at $t = 2.0$ s. (b) Find the rover's displacement and average velocity vectors for the interval $t = 0.0$ s to $t = 2.0$ s. (c) Find a general expression for the rover's instantaneous velocity vector $\vec{v}$. Express $\vec{v}$ at $t = 2.0$ s in component form and in terms of magnitude and direction.

SOLUTION

IDENTIFY and SET UP: This problem involves motion in two dimensions, so we must use the vector equations obtained in this section. Figure 3.5 shows the rover's path (dashed line). We'll use Eq. (3.1) for position $\vec{r}$, the expression $\Delta\vec{r} = \vec{r}_2 - \vec{r}_1$ for displacement, Eq. (3.2) for average velocity, and Eqs. (3.5), (3.6), and (3.7)

3.5 At $t = 0.0$ s the rover has position vector $\vec{r}_0$ and instantaneous velocity vector $\vec{v}_0$. Likewise, $\vec{r}_1$ and $\vec{v}_1$ are the vectors at $t = 1.0$ s; $\vec{r}_2$ and $\vec{v}_2$ are the vectors at $t = 2.0$ s.

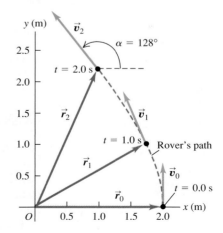

Continued

for instantaneous velocity and its magnitude and direction. The target variables are stated in the problem.

EXECUTE: (a) At $t = 2.0$ s the rover's coordinates are

$$x = 2.0 \text{ m} - (0.25 \text{ m/s}^2)(2.0 \text{ s})^2 = 1.0 \text{ m}$$
$$y = (1.0 \text{ m/s})(2.0 \text{ s}) + (0.025 \text{ m/s}^3)(2.0 \text{ s})^3 = 2.2 \text{ m}$$

The rover's distance from the origin at this time is

$$r = \sqrt{x^2 + y^2} = \sqrt{(1.0 \text{ m})^2 + (2.2 \text{ m})^2} = 2.4 \text{ m}$$

(b) To find the displacement and average velocity over the given time interval, we first express the position vector $\vec{r}$ as a function of time t. From Eq. (3.1) this is

$$\vec{r} = x\hat{\imath} + y\hat{\jmath}$$
$$= [2.0 \text{ m} - (0.25 \text{ m/s}^2)t^2]\hat{\imath}$$
$$+ [(1.0 \text{ m/s})t + (0.025 \text{ m/s}^3)t^3]\hat{\jmath}$$

At $t = 0.0$ s the position vector $\vec{r}_0$ is

$$\vec{r}_0 = (2.0 \text{ m})\hat{\imath} + (0.0 \text{ m})\hat{\jmath}$$

From part (a), the position vector $\vec{r}_2$ at $t = 2.0$ s is

$$\vec{r}_2 = (1.0 \text{ m})\hat{\imath} + (2.2 \text{ m})\hat{\jmath}$$

The displacement from $t = 0.0$ s to $t = 2.0$ s is therefore

$$\Delta \vec{r} = \vec{r}_2 - \vec{r}_0 = (1.0 \text{ m})\hat{\imath} + (2.2 \text{ m})\hat{\jmath} - (2.0 \text{ m})\hat{\imath}$$
$$= (-1.0 \text{ m})\hat{\imath} + (2.2 \text{ m})\hat{\jmath}$$

During this interval the rover moves 1.0 m in the negative x-direction and 2.2 m in the positive y-direction. From Eq. (3.2), the average velocity over this interval is the displacement divided by the elapsed time:

$$\vec{v}_{av} = \frac{\Delta \vec{r}}{\Delta t} = \frac{(-1.0 \text{ m})\hat{\imath} + (2.2 \text{ m})\hat{\jmath}}{2.0 \text{ s} - 0.0 \text{ s}}$$
$$= (-0.50 \text{ m/s})\hat{\imath} + (1.1 \text{ m/s})\hat{\jmath}$$

The components of this average velocity are $v_{av\text{-}x} = -0.50$ m/s and $v_{av\text{-}y} = 1.1$ m/s.

(c) From Eq. (3.4) the components of *instantaneous* velocity are the time derivatives of the coordinates:

$$v_x = \frac{dx}{dt} = (-0.25 \text{ m/s}^2)(2t)$$
$$v_y = \frac{dy}{dt} = 1.0 \text{ m/s} + (0.025 \text{ m/s}^3)(3t^2)$$

Hence the instantaneous velocity vector is

$$\vec{v} = v_x\hat{\imath} + v_y\hat{\jmath} = (-0.50 \text{ m/s}^2)t\hat{\imath}$$
$$+ [1.0 \text{ m/s} + (0.075 \text{ m/s}^3)t^2]\hat{\jmath}$$

At $t = 2.0$ s the velocity vector $\vec{v}_2$ has components

$$v_{2x} = (-0.50 \text{ m/s}^2)(2.0 \text{ s}) = -1.0 \text{ m/s}$$
$$v_{2y} = 1.0 \text{ m/s} + (0.075 \text{ m/s}^3)(2.0 \text{ s})^2 = 1.3 \text{ m/s}$$

The magnitude of the instantaneous velocity (that is, the speed) at $t = 2.0$ s is

$$v_2 = \sqrt{v_{2x}^2 + v_{2y}^2} = \sqrt{(-1.0 \text{ m/s})^2 + (1.3 \text{ m/s})^2}$$
$$= 1.6 \text{ m/s}$$

Figure 3.5 shows the direction of the velocity vector $\vec{v}_2$, which is at an angle α between 90° and 180° with respect to the positive x-axis. From Eq. (3.7) we have

$$\arctan \frac{v_y}{v_x} = \arctan \frac{1.3 \text{ m/s}}{-1.0 \text{ m/s}} = -52°$$

This is off by 180°; the correct value of the angle is $\alpha = 180° - 52° = 128°$, or 38° west of north.

EVALUATE: Compare the components of *average* velocity that we found in part (b) for the interval from $t = 0.0$ s to $t = 2.0$ s ($v_{av\text{-}x} = -0.50$ m/s, $v_{av\text{-}y} = 1.1$ m/s) with the components of *instantaneous* velocity at $t = 2.0$ s that we found in part (c) ($v_{2x} = -1.0$ m/s, $v_{2y} = 1.3$ m/s). The comparison shows that, just as in one dimension, the average velocity vector $\vec{v}_{av}$ over an interval is in general *not* equal to the instantaneous velocity $\vec{v}$ at the end of the interval (see Example 2.1).

Figure 3.5 shows the position vectors $\vec{r}$ and instantaneous velocity vectors $\vec{v}$ at $t = 0.0$ s, 1.0 s, and 2.0 s. (You should calculate these quantities for $t = 0.0$ s and $t = 1.0$ s.) Notice that $\vec{v}$ is tangent to the path at every point. The magnitude of $\vec{v}$ increases as the rover moves, which means that its speed is increasing.

Test Your Understanding of Section 3.1 In which of these situations *would* the average velocity vector $\vec{v}_{av}$ over an interval be equal to the instantaneous velocity $\vec{v}$ at the end of the interval? (i) a body moving along a curved path at constant speed; (ii) a body moving along a curved path and speeding up; (iii) a body moving along a straight line at constant speed; (iv) a body moving along a straight line and speeding up.

3.2 The Acceleration Vector

Now let's consider the *acceleration* of a particle moving in space. Just as for motion in a straight line, acceleration describes how the velocity of the particle changes. But since we now treat velocity as a vector, acceleration will describe changes in the velocity magnitude (that is, the speed) *and* changes in the direction of velocity (that is, the direction in which the particle is moving).

In Fig. 3.6a, a car (treated as a particle) is moving along a curved road. The vectors $\vec{v}_1$ and $\vec{v}_2$ represent the car's instantaneous velocities at time t_1, when the car

3.6 (a) A car moving along a curved road from P_1 to P_2. (b) How to obtain the change in velocity $\Delta\vec{v} = \vec{v}_2 - \vec{v}_1$ by vector subtraction. (c) The vector $\vec{a}_{av} = \Delta\vec{v}/\Delta t$ represents the average acceleration between P_1 and P_2.

(a)

(b)

(c)

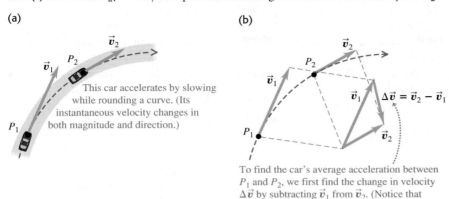

This car accelerates by slowing while rounding a curve. (Its instantaneous velocity changes in both magnitude and direction.)

To find the car's average acceleration between P_1 and P_2, we first find the change in velocity $\Delta\vec{v}$ by subtracting $\vec{v}_1$ from $\vec{v}_2$. (Notice that $\vec{v}_1 + \Delta\vec{v} = \vec{v}_2$.)

The average acceleration has the same direction as the change in velocity, $\Delta\vec{v}$.

is at point P_1, and at time t_2, when the car is at point P_2. The two velocities may differ in both magnitude and direction. During the time interval from t_1 to t_2, the *vector change in velocity* is $\vec{v}_2 - \vec{v}_1 = \Delta\vec{v}$, so $\vec{v}_2 = \vec{v}_1 + \Delta\vec{v}$ (Fig. 3.6b). We define the **average acceleration** $\vec{a}_{av}$ of the car during this time interval as the velocity change divided by the time interval $t_2 - t_1 = \Delta t$:

$$\vec{a}_{av} = \frac{\vec{v}_2 - \vec{v}_1}{t_2 - t_1} = \frac{\Delta\vec{v}}{\Delta t} \qquad \text{(average acceleration vector)} \qquad (3.8)$$

Average acceleration is a *vector* quantity in the same direction as the vector $\Delta\vec{v}$ (Fig. 3.6c). The x-component of Eq. (3.8) is $a_{av\text{-}x} = (v_{2x} - v_{1x})/(t_2 - t_1) = \Delta v_x/\Delta t$, which is just Eq. (2.4) for the average acceleration in straight-line motion.

As in Chapter 2, we define the **instantaneous acceleration** $\vec{a}$ (a *vector* quantity) at point P_1 as the limit of the average acceleration vector when point P_2 approaches point P_1, so $\Delta\vec{v}$ and Δt both approach zero (Fig. 3.7). The instantaneous acceleration is also equal to the instantaneous rate of change of velocity with time:

$$\vec{a} = \lim_{\Delta t \to 0} \frac{\Delta\vec{v}}{\Delta t} = \frac{d\vec{v}}{dt} \qquad \text{(instantaneous acceleration vector)} \qquad (3.9)$$

The velocity vector $\vec{v}$, as we have seen, is tangent to the path of the particle. The instantaneous acceleration vector $\vec{a}$, however, does *not* have to be tangent to the path. Figure 3.7a shows that if the path is curved, $\vec{a}$ points toward the concave side of the path—that is, toward the inside of any turn that the particle is making. The acceleration is tangent to the path only if the particle moves in a straight line (Fig. 3.7b).

CAUTION **Any particle following a curved path is accelerating** When a particle is moving in a curved path, it always has nonzero acceleration, even when it moves with constant speed. This conclusion may seem contrary to your intuition, but it's really just contrary to the everyday use of the word "acceleration" to mean that speed is increasing. The more precise definition given in Eq. (3.9) shows that there is a nonzero acceleration whenever the velocity vector changes in any way, whether there is a change of speed, direction, or both. ▮

To convince yourself that a particle has a nonzero acceleration when moving on a curved path with constant speed, think of your sensations when you ride in a car. When the car accelerates, you tend to move inside the car in a

3.7 (a) Instantaneous acceleration $\vec{a}$ at point P_1 in Fig. 3.6. (b) Instantaneous acceleration for motion along a straight line.

(a) Acceleration: curved trajectory

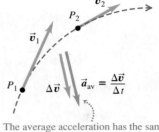

To find the instantaneous acceleration $\vec{a}$ at P_1 ...

... we take the limit of $\vec{a}_{av}$ as P_2 approaches P_1 ...

... meaning that $\Delta\vec{v}$ and Δt approach 0.

$\vec{a} = \lim\limits_{\Delta t \to 0} \dfrac{\Delta\vec{v}}{\Delta t}$

Acceleration points to concave side of path.

(b) Acceleration: straight-line trajectory

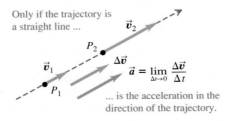

Only if the trajectory is a straight line ...

$\vec{a} = \lim\limits_{\Delta t \to 0} \dfrac{\Delta\vec{v}}{\Delta t}$

... is the acceleration in the direction of the trajectory.

direction *opposite* to the car's acceleration. (We'll discover the reason for this behavior in Chapter 4.) Thus you tend to slide toward the back of the car when it accelerates forward (speeds up) and toward the front of the car when it accelerates backward (slows down). If the car makes a turn on a level road, you tend to slide toward the outside of the turn; hence the car has an acceleration toward the inside of the turn.

We will usually be interested in the instantaneous acceleration, not the average acceleration. From now on, we will use the term "acceleration" to mean the instantaneous acceleration vector $\vec{a}$.

Each component of the acceleration vector is the derivative of the corresponding component of velocity:

$$a_x = \frac{dv_x}{dt} \qquad a_y = \frac{dv_y}{dt} \qquad a_z = \frac{dv_z}{dt} \qquad \text{(components of instantaneous acceleration)} \qquad (3.10)$$

In terms of unit vectors,

$$\vec{a} = \frac{dv_x}{dt}\hat{i} + \frac{dv_y}{dt}\hat{j} + \frac{dv_z}{dt}\hat{k} \qquad (3.11)$$

The *x*-component of Eqs. (3.10) and (3.11), $a_x = dv_x/dt$, is the expression from Section 2.3 for instantaneous acceleration in one dimension, Eq. (2.5). Figure 3.8 shows an example of an acceleration vector that has both *x*- and *y*-components.

Since each component of velocity is the derivative of the corresponding coordinate, we can express the components a_x, a_y, and a_z of the acceleration vector $\vec{a}$ as

$$a_x = \frac{d^2x}{dt^2} \qquad a_y = \frac{d^2y}{dt^2} \qquad a_z = \frac{d^2z}{dt^2} \qquad (3.12)$$

The acceleration vector $\vec{a}$ itself is

$$\vec{a} = \frac{d^2x}{dt^2}\hat{i} + \frac{d^2y}{dt^2}\hat{j} + \frac{d^2z}{dt^2}\hat{k} \qquad (3.13)$$

3.8 When the arrow is released, its acceleration vector has both a horizontal component (a_x) and a vertical component (a_y).

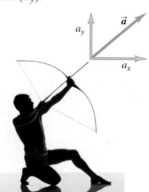

Example 3.2 **Calculating average and instantaneous acceleration**

Let's return to the motions of the Mars rover in Example 3.1. (a) Find the components of the average acceleration for the interval $t = 0.0$ s to $t = 2.0$ s. (b) Find the instantaneous acceleration at $t = 2.0$ s.

SOLUTION

IDENTIFY and SET UP: In Example 3.1 we found the components of the rover's instantaneous velocity at any time *t*:

$$v_x = \frac{dx}{dt} = (-0.25 \text{ m/s}^2)(2t) = (-0.50 \text{ m/s}^2)t$$

$$v_y = \frac{dy}{dt} = 1.0 \text{ m/s} + (0.025 \text{ m/s}^3)(3t^2)$$

$$= 1.0 \text{ m/s} + (0.075 \text{ m/s}^3)t^2$$

We'll use the vector relationships among velocity, average acceleration, and instantaneous acceleration. In part (a) we determine the values of v_x and v_y at the beginning and end of the interval and

then use Eq. (3.8) to calculate the components of the average acceleration. In part (b) we obtain expressions for the instantaneous acceleration components at any time *t* by taking the time derivatives of the velocity components as in Eqs. (3.10).

EXECUTE: (a) In Example 3.1 we found that at $t = 0.0$ s the velocity components are

$$v_x = 0.0 \text{ m/s} \qquad v_y = 1.0 \text{ m/s}$$

and that at $t = 2.00$ s the components are

$$v_x = -1.0 \text{ m/s} \qquad v_y = 1.3 \text{ m/s}$$

Thus the components of average acceleration in the interval $t = 0.0$ s to $t = 2.0$ s are

$$a_{\text{av-}x} = \frac{\Delta v_x}{\Delta t} = \frac{-1.0 \text{ m/s} - 0.0 \text{ m/s}}{2.0 \text{ s} - 0.0 \text{ s}} = -0.50 \text{ m/s}^2$$

$$a_{\text{av-}y} = \frac{\Delta v_y}{\Delta t} = \frac{1.3 \text{ m/s} - 1.0 \text{ m/s}}{2.0 \text{ s} - 0.0 \text{ s}} = 0.15 \text{ m/s}^2$$

(b) Using Eqs. (3.10), we find

$$a_x = \frac{dv_x}{dt} = -0.50 \text{ m/s}^2 \qquad a_y = \frac{dv_y}{dt} = (0.075 \text{ m/s}^3)(2t)$$

Hence the instantaneous acceleration vector $\vec{a}$ at time t is

$$\vec{a} = a_x\hat{\imath} + a_y\hat{\jmath} = (-0.50 \text{ m/s}^2)\hat{\imath} + (0.15 \text{ m/s}^3)t\hat{\jmath}$$

At $t = 2.0$ s the components of acceleration and the acceleration vector are

$$a_x = -0.50 \text{ m/s}^2 \qquad a_y = (0.15 \text{ m/s}^3)(2.0 \text{ s}) = 0.30 \text{ m/s}^2$$

$$\vec{a} = (-0.50 \text{ m/s}^2)\hat{\imath} + (0.30 \text{ m/s}^2)\hat{\jmath}$$

The magnitude of acceleration at this time is

$$a = \sqrt{a_x^2 + a_y^2}$$
$$= \sqrt{(-0.50 \text{ m/s}^2)^2 + (0.30 \text{ m/s}^2)^2} = 0.58 \text{ m/s}^2$$

A sketch of this vector (Fig. 3.9) shows that the direction angle β of $\vec{a}$ with respect to the positive x-axis is between $90°$ and $180°$. From Eq. (3.7) we have

$$\arctan\frac{a_y}{a_x} = \arctan\frac{0.30 \text{ m/s}^2}{-0.50 \text{ m/s}^2} = -31°$$

Hence $\beta = 180° + (-31°) = 149°$.

EVALUATE: Figure 3.9 shows the rover's path and the velocity and acceleration vectors at $t = 0.0$ s, 1.0 s, and 2.0 s. (You should use

the results of part (b) to calculate the instantaneous acceleration at $t = 0.0$ s and $t = 1.0$ s for yourself.) Note that $\vec{v}$ and $\vec{a}$ are *not* in the same direction at any of these times. The velocity vector $\vec{v}$ is tangent to the path at each point (as is always the case), and the acceleration vector $\vec{a}$ points toward the concave side of the path.

3.9 The path of the robotic rover, showing the velocity and acceleration at $t = 0.0$ s ($\vec{v}_0$ and $\vec{a}_0$), $t = 1.0$ s ($\vec{v}_1$ and $\vec{a}_1$), and $t = 2.0$ s ($\vec{v}_2$ and $\vec{a}_2$).

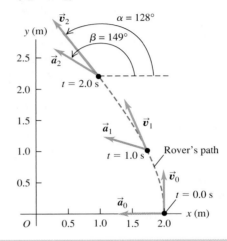

Parallel and Perpendicular Components of Acceleration

Equations (3.10) tell us about the components of a particle's instantaneous acceleration vector $\vec{a}$ along the x-, y-, and z-axes. Another useful way to think about $\vec{a}$ is in terms of its component *parallel* to the particle's path—that is, parallel to the velocity—and its component *perpendicular* to the path—and hence perpendicular to the velocity (Fig. 3.10). That's because the parallel component $a_\parallel$ tells us about changes in the particle's *speed,* while the perpendicular component $a_\perp$ tells us about changes in the particle's *direction of motion.* To see why the parallel and perpendicular components of $\vec{a}$ have these properties, let's consider two special cases.

In Fig. 3.11a the acceleration vector is in the same direction as the velocity $\vec{v}_1$, so $\vec{a}$ has only a parallel component $a_\parallel$ (that is, $a_\perp = 0$). The velocity change $\Delta\vec{v}$ during a small time interval Δt is in the same direction as $\vec{a}$ and hence in the same direction as $\vec{v}_1$. The velocity $\vec{v}_2$ at the end of Δt is in the same direction as $\vec{v}_1$ but has greater magnitude. Hence during the time interval Δt the particle in Fig. 3.11a moved in a straight line with increasing speed (compare Fig. 3.7b).

In Fig. 3.11b the acceleration is *perpendicular* to the velocity, so $\vec{a}$ has only a perpendicular component $a_\perp$ (that is, $a_\parallel = 0$). In a small time interval Δt, the

3.10 The acceleration can be resolved into a component $a_\parallel$ parallel to the path (that is, along the tangent to the path) and a component $a_\perp$ perpendicular to the path (that is, along the normal to the path).

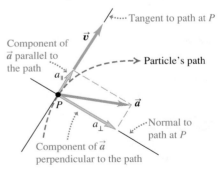

3.11 The effect of acceleration directed **(a)** parallel to and **(b)** perpendicular to a particle's velocity.

(a) Acceleration parallel to velocity

Changes only *magnitude* of velocity: speed changes; direction doesn't.

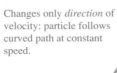

(b) Acceleration perpendicular to velocity

Changes only *direction* of velocity: particle follows curved path at constant speed.

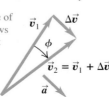

velocity change $\Delta\vec{v}$ is very nearly perpendicular to $\vec{v}_1$, and so $\vec{v}_1$ and $\vec{v}_2$ have different directions. As the time interval Δt approaches zero, the angle ϕ in the figure also approaches zero, $\Delta\vec{v}$ becomes perpendicular to *both* $\vec{v}_1$ and $\vec{v}_2$, and $\vec{v}_1$ and $\vec{v}_2$ have the same magnitude. In other words, the speed of the particle stays the same, but the direction of motion changes and the path of the particle curves.

In the most general case, the acceleration $\vec{a}$ has components *both* parallel and perpendicular to the velocity $\vec{v}$, as in Fig. 3.10. Then the particle's speed will change (described by the parallel component $a_\parallel$) *and* its direction of motion will change (described by the perpendicular component $a_\perp$) so that it follows a curved path.

Figure 3.12 shows a particle moving along a curved path for three different situations: constant speed, increasing speed, and decreasing speed. If the speed is constant, $\vec{a}$ is perpendicular, or *normal*, to the path and to $\vec{v}$ and points toward the concave side of the path (Fig. 3.12a). If the speed is increasing, there is still a perpendicular component of $\vec{a}$, but there is also a parallel component having the same direction as $\vec{v}$ (Fig. 3.12b). Then $\vec{a}$ points ahead of the normal to the path. (This was the case in Example 3.2.) If the speed is decreasing, the parallel component has the direction opposite to $\vec{v}$, and $\vec{a}$ points behind the normal to the path (Fig. 3.12c; compare Fig. 3.7a). We will use these ideas again in Section 3.4 when we study the special case of motion in a circle.

PhET: Maze Game

3.12 Velocity and acceleration vectors for a particle moving through a point P on a curved path with (a) constant speed, (b) increasing speed, and (c) decreasing speed.

(a) When speed is constant along a curved path ...

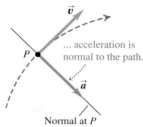

(b) When speed is increasing along a curved path ...

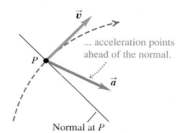

(c) When speed is decreasing along a curved path ...

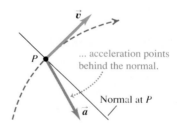

Example 3.3   Calculating parallel and perpendicular components of acceleration

For the rover of Examples 3.1 and 3.2, find the parallel and perpendicular components of the acceleration at $t = 2.0$ s.

SOLUTION

IDENTIFY and SET UP: We want to find the components of the acceleration vector $\vec{a}$ that are parallel and perpendicular to the velocity vector $\vec{v}$. We found the directions of $\vec{v}$ and $\vec{a}$ in Examples 3.1 and 3.2, respectively; Fig. 3.9 shows the results. From these directions we can find the angle between the two vectors and the components of $\vec{a}$ with respect to the direction of $\vec{v}$.

EXECUTE: From Example 3.2, at $t = 2.0$ s the particle has an acceleration of magnitude 0.58 m/s^2 at an angle of $149°$ with respect to the positive x-axis. In Example 3.1 we found that at this time the velocity vector is at an angle of $128°$ with respect to the positive x-axis. The angle between $\vec{a}$ and $\vec{v}$ is therefore $149° - 128° = 21°$ (Fig. 3.13). Hence the components of acceleration parallel and perpendicular to $\vec{v}$ are

$$a_\parallel = a\cos 21° = (0.58 \text{ m/s}^2)\cos 21° = 0.54 \text{ m/s}^2$$
$$a_\perp = a\sin 21° = (0.58 \text{ m/s}^2)\sin 21° = 0.21 \text{ m/s}^2$$

3.13 The parallel and perpendicular components of the acceleration of the rover at $t = 2.0$ s.

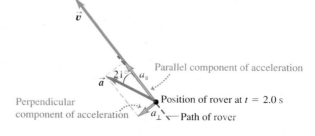

EVALUATE: The parallel component $a_\parallel$ is positive (in the same direction as $\vec{v}$), which means that the speed is increasing at this instant. The value $a_\parallel = +0.54$ m/s^2 tells us that the speed is increasing at this instant at a rate of 0.54 m/s per second. The perpendicular component $a_\perp$ is not zero, which means that at this instant the rover is turning—that is, it is changing direction and following a curved path.

Conceptual Example 3.4   **Acceleration of a skier**

A skier moves along a ski-jump ramp (Fig. 3.14a). The ramp is straight from point A to point C and curved from point C onward. The skier speeds up as she moves downhill from point A to point E, where her speed is maximum. She slows down after passing point E. Draw the direction of the acceleration vector at each of the points B, D, E, and F.

SOLUTION

Figure 3.14b shows our solution. At point B the skier is moving in a straight line with increasing speed, so her acceleration points downhill, in the same direction as her velocity. At points D, E, and F the skier is moving along a curved path, so her acceleration has a component perpendicular to the path (toward the concave side of the path) at each of these points. At point D there is also an acceleration component in the direction of her motion because she is speeding up. So the acceleration vector points *ahead* of the normal to her path at point D, as Fig. 3.14b shows. At point E, the skier's speed is instantaneously not changing; her speed is maximum at this point, so its derivative is zero. There is therefore no parallel component of $\vec{a}$, and the acceleration is perpendicular to her motion. At point F there is an acceleration component *opposite to* the direction of her motion because now she's slowing down. The acceleration vector therefore points *behind* the normal to her path.

In the next section we'll consider the skier's acceleration after she flies off the ramp.

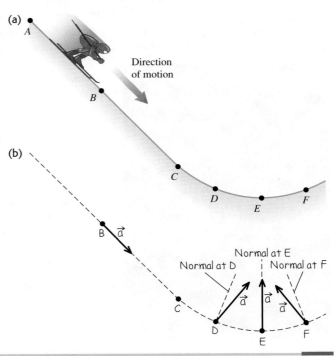

3.14 (a) The skier's path. (b) Our solution.

Test Your Understanding of Section 3.2 A sled travels over the crest of a snow-covered hill. The sled slows down as it climbs up one side of the hill and gains speed as it descends on the other side. Which of the vectors (1 through 9) in the figure correctly shows the direction of the sled's acceleration at the crest? (Choice 9 is that the acceleration is zero.)

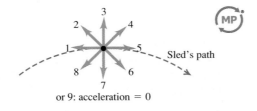

or 9: acceleration = 0

3.3 Projectile Motion

A **projectile** is any body that is given an initial velocity and then follows a path determined entirely by the effects of gravitational acceleration and air resistance. A batted baseball, a thrown football, a package dropped from an airplane, and a bullet shot from a rifle are all projectiles. The path followed by a projectile is called its **trajectory.**

To analyze this common type of motion, we'll start with an idealized model, representing the projectile as a particle with an acceleration (due to gravity) that is constant in both magnitude and direction. We'll neglect the effects of air resistance and the curvature and rotation of the earth. Like all models, this one has limitations. Curvature of the earth has to be considered in the flight of long-range missiles, and air resistance is of crucial importance to a sky diver. Nevertheless, we can learn a lot from analysis of this simple model. For the remainder of this chapter the phrase "projectile motion" will imply that we're ignoring air resistance. In Chapter 5 we will see what happens when air resistance cannot be ignored.

Projectile motion is always confined to a vertical plane determined by the direction of the initial velocity (Fig. 3.15). This is because the acceleration due to

3.15 The trajectory of an idealized projectile.

- A projectile moves in a vertical plane that contains the initial velocity vector $\vec{v}_0$.
- Its trajectory depends only on $\vec{v}_0$ and on the downward acceleration due to gravity.

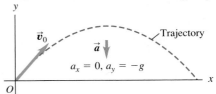

3.16 The red ball is dropped from rest, and the yellow ball is simultaneously projected horizontally; successive images in this stroboscopic photograph are separated by equal time intervals. At any given time, both balls have the same y-position, y-velocity, and y-acceleration, despite having different x-positions and x-velocities.

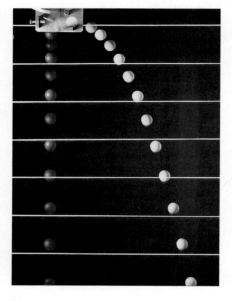

ActivPhysics 3.1: Solving Projectile Motion Problems
ActivPhysics 3.2: Two Balls Falling
ActivPhysics 3.3: Changing the x-velocity
ActivPhysics 3.4: Projecting x-y-Accelerations

gravity is purely vertical; gravity can't accelerate the projectile sideways. Thus projectile motion is *two-dimensional*. We will call the plane of motion the xy-coordinate plane, with the x-axis horizontal and the y-axis vertically upward.

The key to analyzing projectile motion is that we can treat the x- and y-coordinates separately. The x-component of acceleration is zero, and the y-component is constant and equal to $-g$. (By definition, g is always positive; with our choice of coordinate directions, a_y is negative.) So *we can analyze projectile motion as a combination of horizontal motion with constant velocity and vertical motion with constant acceleration*. Figure 3.16 shows two projectiles with different x-motion but identical y-motion; one is dropped from rest and the other is projected horizontally, but both projectiles fall the same distance in the same time.

We can then express all the vector relationships for the projectile's position, velocity, and acceleration by separate equations for the horizontal and vertical components. The components of $\vec{a}$ are

$$a_x = 0 \qquad a_y = -g \qquad \text{(projectile motion, no air resistance)} \quad (3.14)$$

Since the x-acceleration and y-acceleration are both constant, we can use Eqs. (2.8), (2.12), (2.13), and (2.14) directly. For example, suppose that at time $t = 0$ our particle is at the point (x_0, y_0) and that at this time its velocity components have the initial values v_{0x} and v_{0y}. The components of acceleration are $a_x = 0$, $a_y = -g$. Considering the x-motion first, we substitute 0 for a_x in Eqs. (2.8) and (2.12). We find

$$v_x = v_{0x} \quad (3.15)$$

$$x = x_0 + v_{0x}t \quad (3.16)$$

For the y-motion we substitute y for x, v_y for v_x, v_{0y} for v_{0x}, and $a_y = -g$ for a_x:

$$v_y = v_{0y} - gt \quad (3.17)$$

$$y = y_0 + v_{0y}t - \tfrac{1}{2}gt^2 \quad (3.18)$$

It's usually simplest to take the initial position (at $t = 0$) as the origin; then $x_0 = y_0 = 0$. This might be the position of a ball at the instant it leaves the thrower's hand or the position of a bullet at the instant it leaves the gun barrel.

Figure 3.17 shows the trajectory of a projectile that starts at (or passes through) the origin at time $t = 0$, along with its position, velocity, and velocity

3.17 If air resistance is negligible, the trajectory of a projectile is a combination of horizontal motion with constant velocity and vertical motion with constant acceleration.

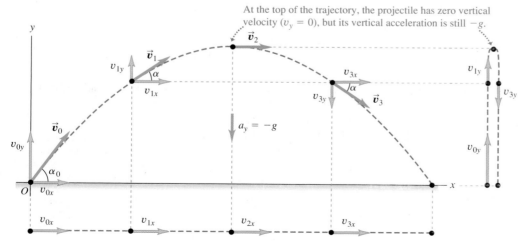

At the top of the trajectory, the projectile has zero vertical velocity ($v_y = 0$), but its vertical acceleration is still $-g$.

Vertically, the projectile is in constant-acceleration motion in response to the earth's gravitational pull. Thus its vertical velocity *changes* by equal amounts during equal time intervals.

Horizontally, the projectile is in constant-velocity motion: Its horizontal acceleration is zero, so it moves equal x-distances in equal time intervals.

components at equal time intervals. The x-component of acceleration is zero, so v_x is constant. The y-component of acceleration is constant and not zero, so v_y changes by equal amounts in equal times, just the same as if the projectile were launched vertically with the same initial y-velocity.

We can also represent the initial velocity $\vec{v}_0$ by its magnitude v_0 (the initial speed) and its angle α_0 with the positive x-axis (Fig. 3.18). In terms of these quantities, the components v_{0x} and v_{0y} of the initial velocity are

$$v_{0x} = v_0 \cos \alpha_0 \qquad v_{0y} = v_0 \sin \alpha_0 \qquad \text{(3.19)}$$

If we substitute these relationships in Eqs. (3.15) through (3.18) and set $x_0 = y_0 = 0$, we find

$$x = (v_0 \cos \alpha_0)t \qquad \text{(projectile motion)} \qquad \text{(3.20)}$$

$$y = (v_0 \sin \alpha_0)t - \tfrac{1}{2}gt^2 \qquad \text{(projectile motion)} \qquad \text{(3.21)}$$

$$v_x = v_0 \cos \alpha_0 \qquad \text{(projectile motion)} \qquad \text{(3.22)}$$

$$v_y = v_0 \sin \alpha_0 - gt \qquad \text{(projectile motion)} \qquad \text{(3.23)}$$

These equations describe the position and velocity of the projectile in Fig. 3.17 at any time t.

We can get a lot of information from Eqs. (3.20) through (3.23). For example, at any time the distance r of the projectile from the origin (the magnitude of the position vector $\vec{r}$) is given by

$$r = \sqrt{x^2 + y^2} \qquad \text{(3.24)}$$

The projectile's speed (the magnitude of its velocity) at any time is

$$v = \sqrt{v_x^2 + v_y^2} \qquad \text{(3.25)}$$

The *direction* of the velocity, in terms of the angle α it makes with the positive x-direction (see Fig. 3.17), is given by

$$\tan \alpha = \frac{v_y}{v_x} \qquad \text{(3.26)}$$

The velocity vector $\vec{v}$ is tangent to the trajectory at each point.

We can derive an equation for the trajectory's shape in terms of x and y by eliminating t. From Eqs. (3.20) and (3.21), which assume $x_0 = y_0 = 0$, we find $t = x/(v_0 \cos \alpha_0)$ and

$$y = (\tan \alpha_0)x - \frac{g}{2v_0^2 \cos^2 \alpha_0}x^2 \qquad \text{(3.27)}$$

Don't worry about the details of this equation; the important point is its general form. Since v_0, $\tan \alpha_0$, $\cos \alpha_0$, and g are constants, Eq. (3.27) has the form

$$y = bx - cx^2$$

where b and c are constants. This is the equation of a *parabola*. In our simple model of projectile motion, the trajectory is always a parabola (Fig. 3.19).

When air resistance *isn't* always negligible and has to be included, calculating the trajectory becomes a lot more complicated; the effects of air resistance depend on velocity, so the acceleration is no longer constant. Figure 3.20 shows a

3.18 The initial velocity components v_{0x} and v_{0y} of a projectile (such as a kicked soccer ball) are related to the initial speed v_0 and initial angle α_0.

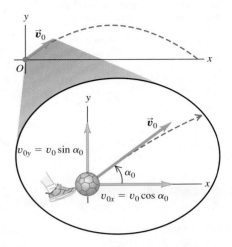

MasteringPHYSICS

PhET: Projectile Motion
ActivPhysics 3.5: Initial Velocity Components
ActivPhysics 3.6: Target Practice I
ActivPhysics 3.7: Target Practice II

3.19 The nearly parabolic trajectories of **(a)** a bouncing ball and **(b)** blobs of molten rock ejected from a volcano.

(a) Successive images of ball are separated by equal time intervals.

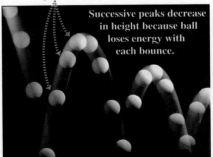

Successive peaks decrease in height because ball loses energy with each bounce.

(b)

Trajectories are nearly parabolic.

3.20 Air resistance has a large cumulative effect on the motion of a baseball. In this simulation we allow the baseball to fall below the height from which it was thrown (for example, the baseball could have been thrown from a cliff).

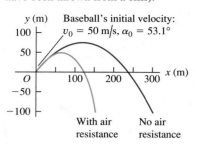

computer simulation of the trajectory of a baseball both without air resistance and with air resistance proportional to the square of the baseball's speed. We see that air resistance has a very large effect; the maximum height and range both decrease, and the trajectory is no longer a parabola. (If you look closely at Fig. 3.19b, you'll see that the trajectories of the volcanic blobs deviate in a similar way from a parabolic shape.)

Conceptual Example 3.5 | Acceleration of a skier, continued

Let's consider again the skier in Conceptual Example 3.4. What is her acceleration at each of the points G, H, and I in Fig. 3.21a *after* she flies off the ramp? Neglect air resistance.

SOLUTION

Figure 3.21b shows our answer. The skier's acceleration changed from point to point while she was on the ramp. But as soon as she

leaves the ramp, she becomes a projectile. So at points G, H, and I, and indeed at *all* points after she leaves the ramp, the skier's acceleration points vertically downward and has magnitude g. No matter how complicated the acceleration of a particle before it becomes a projectile, its acceleration as a projectile is given by $a_x = 0$, $a_y = -g$.

3.21 (a) The skier's path during the jump. (b) Our solution.

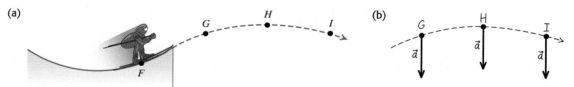

Problem-Solving Strategy 3.1 | Projectile Motion

NOTE: The strategies we used in Sections 2.4 and 2.5 for straight-line, constant-acceleration problems are also useful here.

IDENTIFY *the relevant concepts:* The key concept to remember is that throughout projectile motion, the acceleration is downward and has a constant magnitude g. Note that the projectile-motion equations don't apply to *throwing* a ball, because during the throw the ball is acted on by both the thrower's hand and gravity. These equations apply only *after* the ball leaves the thrower's hand.

SET UP *the problem* using the following steps:
1. Define your coordinate system and make a sketch showing your axes. Usually it's easiest to make the x-axis horizontal and the y-axis upward, and to place the origin at the initial ($t = 0$) position where the body first becomes a projectile (such as where a ball leaves the thrower's hand). Then the components of the (constant) acceleration are $a_x = 0$, $a_y = -g$, and the initial position is $x_0 = 0$, $y_0 = 0$.
2. List the unknown and known quantities, and decide which unknowns are your target variables. For example, you might be given the initial velocity (either the components or the magnitude and direction) and asked to find the coordinates and velocity components at some later time. In any case, you'll be using

Eqs. (3.20) through (3.23). (Equations (3.24) through (3.27) may be useful as well.) Make sure that you have as many equations as there are target variables to be found.
3. State the problem in words and then translate those words into symbols. For example, *when* does the particle arrive at a certain point? (That is, at what value of t?) *Where* is the particle when its velocity has a certain value? (That is, what are the values of x and y when v_x or v_y has the specified value?) Since $v_y = 0$ at the highest point in a trajectory, the question "When does the projectile reach its highest point?" translates into "What is the value of t when $v_y = 0$?" Similarly, "When does the projectile return to its initial elevation?" translates into "What is the value of t when $y = y_0$?"

EXECUTE *the solution:* Find the target variables using the equations you chose. Resist the temptation to break the trajectory into segments and analyze each segment separately. You don't have to start all over when the projectile reaches its highest point! It's almost always easier to use the same axes and time scale throughout the problem. If you need numerical values, use $g = 9.80 \text{ m/s}^2$.

EVALUATE *your answer:* As always, look at your results to see whether they make sense and whether the numerical values seem reasonable.

Example 3.6 A body projected horizontally

A motorcycle stunt rider rides off the edge of a cliff. Just at the edge his velocity is horizontal, with magnitude 9.0 m/s. Find the motorcycle's position, distance from the edge of the cliff, and velocity 0.50 s after it leaves the edge of the cliff.

SOLUTION

IDENTIFY and SET UP: Figure 3.22 shows our sketch of the motorcycle's trajectory. He is in projectile motion as soon as he leaves the edge of the cliff, which we choose to be the origin of coordinates so $x_0 = 0$ and $y_0 = 0$. His initial velocity $\vec{v}_0$ at the edge of the cliff is horizontal (that is, $\alpha_0 = 0$), so its components are $v_{0x} = v_0 \cos \alpha_0 = 9.0$ m/s and $v_{0y} = v_0 \sin \alpha_0 = 0$. To find the motorcycle's position at $t = 0.50$ s, we use Eqs. (3.20) and (3.21); we then find the distance from the origin using Eq. (3.24). Finally, we use Eqs. (3.22) and (3.23) to find the velocity components at $t = 0.50$ s.

EXECUTE: From Eqs. (3.20) and (3.21), the motorcycle's x- and y-coordinates at $t = 0.50$ s are

$$x = v_{0x}t = (9.0 \text{ m/s})(0.50 \text{ s}) = 4.5 \text{ m}$$

$$y = -\tfrac{1}{2}gt^2 = -\tfrac{1}{2}(9.80 \text{ m/s}^2)(0.50 \text{ s})^2 = -1.2 \text{ m}$$

The negative value of y shows that the motorcycle is below its starting point.

From Eq. (3.24), the motorcycle's distance from the origin at $t = 0.50$ s is

$$r = \sqrt{x^2 + y^2} = \sqrt{(4.5 \text{ m})^2 + (-1.2 \text{ m})^2} = 4.7 \text{ m}$$

From Eqs. (3.22) and (3.23), the velocity components at $t = 0.50$ s are

$$v_x = v_{0x} = 9.0 \text{ m/s}$$
$$v_y = -gt = (-9.80 \text{ m/s}^2)(0.50 \text{ s}) = -4.9 \text{ m/s}$$

3.22 Our sketch for this problem.

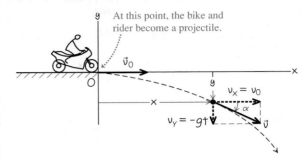

At this point, the bike and rider become a projectile.

The motorcycle has the same horizontal velocity v_x as when it left the cliff at $t = 0$, but in addition there is a downward (negative) vertical velocity v_y. The velocity vector at $t = 0.50$ s is

$$\vec{v} = v_x\hat{\imath} + v_y\hat{\jmath} = (9.0 \text{ m/s})\hat{\imath} + (-4.9 \text{ m/s})\hat{\jmath}$$

From Eq. (3.25), the speed (magnitude of the velocity) at $t = 0.50$ s is

$$v = \sqrt{v_x^2 + v_y^2}$$
$$= \sqrt{(9.0 \text{ m/s})^2 + (-4.9 \text{ m/s})^2} = 10.2 \text{ m/s}$$

From Eq. (3.26), the angle α of the velocity vector is

$$\alpha = \arctan \frac{v_y}{v_x} = \arctan\left(\frac{-4.9 \text{ m/s}}{9.0 \text{ m/s}}\right) = -29°$$

The velocity is 29° below the horizontal.

EVALUATE: Just as in Fig. 3.17, the motorcycle's horizontal motion is unchanged by gravity; the motorcycle continues to move horizontally at 9.0 m/s, covering 4.5 m in 0.50 s. The motorcycle initially has zero vertical velocity, so it falls vertically just like a body released from rest and descends a distance $\tfrac{1}{2}gt^2 = 1.2$ m in 0.50 s.

Example 3.7 Height and range of a projectile I: A batted baseball

A batter hits a baseball so that it leaves the bat at speed $v_0 = 37.0$ m/s at an angle $\alpha_0 = 53.1°$. (a) Find the position of the ball and its velocity (magnitude and direction) at $t = 2.00$ s. (b) Find the time when the ball reaches the highest point of its flight, and its height h at this time. (c) Find the *horizontal range R*—that is, the horizontal distance from the starting point to where the ball hits the ground.

SOLUTION

IDENTIFY and SET UP: As Fig. 3.20 shows, air resistance strongly affects the motion of a baseball. For simplicity, however, we'll ignore air resistance here and use the projectile-motion equations to describe the motion. The ball leaves the bat at $t = 0$ a meter or so above ground level, but we'll neglect this distance and assume that it starts at ground level ($y_0 = 0$). Figure 3.23 shows our

3.23 Our sketch for this problem.

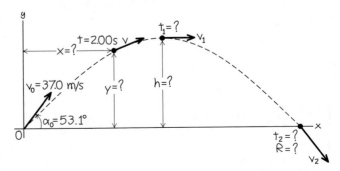

sketch of the ball's trajectory. We'll use the same coordinate system as in Figs. 3.17 and 3.18, so we can use Eqs. (3.20) through

Continued

(3.23). Our target variables are (a) the position and velocity of the ball 2.00 s after it leaves the bat, (b) the time t when the ball is at its maximum height (that is, when $v_y = 0$) and the y-coordinate at this time, and (c) the x-coordinate when the ball returns to ground level ($y = 0$).

EXECUTE: (a) We want to find x, y, v_x, and v_y at $t = 2.00$ s. The initial velocity of the ball has components

$$v_{0x} = v_0 \cos \alpha_0 = (37.0 \text{ m/s})\cos 53.1° = 22.2 \text{ m/s}$$
$$v_{0y} = v_0 \sin \alpha_0 = (37.0 \text{ m/s})\sin 53.1° = 29.6 \text{ m/s}$$

From Eqs. (3.20) through (3.23),

$$x = v_{0x}t = (22.2 \text{ m/s})(2.00 \text{ s}) = 44.4 \text{ m}$$

$$y = v_{0y}t - \tfrac{1}{2}gt^2$$
$$= (29.6 \text{ m/s})(2.00 \text{ s}) - \tfrac{1}{2}(9.80 \text{ m/s}^2)(2.00 \text{ s})^2$$
$$= 39.6 \text{ m}$$

$$v_x = v_{0x} = 22.2 \text{ m/s}$$
$$v_y = v_{0y} - gt = 29.6 \text{ m/s} - (9.80 \text{ m/s}^2)(2.00 \text{ s})$$
$$= 10.0 \text{ m/s}$$

The y-component of velocity is positive at $t = 2.00$ s, so the ball is still moving upward (Fig. 3.23). From Eqs. (3.25) and (3.26), the magnitude and direction of the velocity are

$$v = \sqrt{v_x^2 + v_y^2} = \sqrt{(22.2 \text{ m/s})^2 + (10.0 \text{ m/s})^2}$$
$$= 24.4 \text{ m/s}$$

$$\alpha = \arctan\left(\frac{10.0 \text{ m/s}}{22.2 \text{ m/s}}\right) = \arctan 0.450 = 24.2°$$

The direction of the velocity (the direction of the ball's motion) is 24.2° above the horizontal.

(b) At the highest point, the vertical velocity v_y is zero. Call the time when this happens t_1; then

$$v_y = v_{0y} - gt_1 = 0$$
$$t_1 = \frac{v_{0y}}{g} = \frac{29.6 \text{ m/s}}{9.80 \text{ m/s}^2} = 3.02 \text{ s}$$

The height h at the highest point is the value of y at time t_1:

$$h = v_{0y}t_1 - \tfrac{1}{2}gt_1^2$$
$$= (29.6 \text{ m/s})(3.02 \text{ s}) - \tfrac{1}{2}(9.80 \text{ m/s}^2)(3.02 \text{ s})^2$$
$$= 44.7 \text{ m}$$

(c) We'll find the horizontal range in two steps. First, we find the time t_2 when $y = 0$ (the ball is at ground level):

$$y = 0 = v_{0y}t_2 - \tfrac{1}{2}gt_2^2 = t_2\left(v_{0y} - \tfrac{1}{2}gt_2\right)$$

This is a quadratic equation for t_2. It has two roots:

$$t_2 = 0 \quad \text{and} \quad t_2 = \frac{2v_{0y}}{g} = \frac{2(29.6 \text{ m/s})}{9.80 \text{ m/s}^2} = 6.04 \text{ s}$$

The ball is at $y = 0$ at both times. The ball *leaves* the ground at $t_2 = 0$, and it hits the ground at $t_2 = 2v_{0y}/g = 6.04$ s.

The horizontal range R is the value of x when the ball returns to the ground at $t_2 = 6.04$ s:

$$R = v_{0x}t_2 = (22.2 \text{ m/s})(6.04 \text{ s}) = 134 \text{ m}$$

The vertical component of velocity when the ball hits the ground is

$$v_y = v_{0y} - gt_2 = 29.6 \text{ m/s} - (9.80 \text{ m/s}^2)(6.04 \text{ s})$$
$$= -29.6 \text{ m/s}$$

That is, v_y has the same magnitude as the initial vertical velocity v_{0y} but the opposite direction (down). Since v_x is constant, the angle $\alpha = -53.1°$ (below the horizontal) at this point is the negative of the initial angle $\alpha_0 = 53.1°$.

EVALUATE: It's often useful to check results by getting them in a different way. For example, we can also find the maximum height in part (b) by applying the constant-acceleration formula Eq. (2.13) to the y-motion:

$$v_y^2 = v_{0y}^2 + 2a_y(y - y_0) = v_{0y}^2 - 2g(y - y_0)$$

At the highest point, $v_y = 0$ and $y = h$. You should solve this equation for h; you should get the same answer that we obtained in part (b). (Do you?)

Note that the time to hit the ground, $t_2 = 6.04$ s, is exactly twice the time to reach the highest point, $t_1 = 3.02$ s. Hence the time of descent equals the time of ascent. This is *always* true if the starting and end points are at the same elevation and if air resistance can be neglected.

Note also that $h = 44.7$ m in part (b) is comparable to the 52.4-m height above the playing field of the roof of the Hubert H. Humphrey Metrodome in Minneapolis, and the horizontal range $R = 134$ m in part (c) is greater than the 99.7-m distance from home plate to the right-field fence at Safeco Field in Seattle. In reality, due to air resistance (which we have neglected) a batted ball with the initial speed and angle we've used here won't go as high or as far as we've calculated (see Fig. 3.20).

Example 3.8 **Height and range of a projectile II: Maximum height, maximum range**

Find the maximum height h and horizontal range R (see Fig. 3.23) of a projectile launched with speed v_0 at an initial angle α_0 between 0° and 90°. For a given v_0, what value of α_0 gives maximum height? What value gives maximum horizontal range?

SOLUTION

IDENTIFY and SET UP: This is almost the same as parts (b) and (c) of Example 3.7, except that now we want general expressions for h and R. We also want the values of α_0 that give the maximum values of h and R. In part (b) of Example 3.7 we found that the projectile reaches the high point of its trajectory (so that $v_y = 0$) at time $t_1 = v_{0y}/g$, and in part (c) we found that the projectile returns to its starting height (so that $y = y_0$) at time $t_2 = 2v_{0y}/g = 2t_1$. We'll use Eq. (3.21) to find the y-coordinate h at t_1 and Eq. (3.20) to find the x-coordinate R at time t_2. We'll express our answers in terms of the launch speed v_0 and launch angle α_0 using Eqs. (3.19).

EXECUTE: From Eqs. (3.19), $v_{0x} = v_0 \cos \alpha_0$ and $v_{0y} = v_0 \sin \alpha_0$. Hence we can write the time t_1 when $v_y = 0$ as

$$t_1 = \frac{v_{0y}}{g} = \frac{v_0 \sin \alpha_0}{g}$$

Equation (3.21) gives the height $y = h$ at this time:

$$h = (v_0 \sin \alpha_0)\left(\frac{v_0 \sin \alpha_0}{g}\right) - \frac{1}{2}g\left(\frac{v_0 \sin \alpha_0}{g}\right)^2$$

$$= \frac{v_0^2 \sin^2 \alpha_0}{2g}$$

For a given launch speed v_0, the maximum value of h occurs for $\sin \alpha_0 = 1$ and $\alpha_0 = 90°$—that is, when the projectile is launched straight up. (If it is launched horizontally, as in Example 3.6, $\alpha_0 = 0$ and the maximum height is zero!)

The time t_2 when the projectile hits the ground is

$$t_2 = \frac{2v_{0y}}{g} = \frac{2v_0 \sin \alpha_0}{g}$$

The horizontal range R is the value of x at this time. From Eq. (3.20), this is

$$R = (v_0 \cos \alpha_0)t_2 = (v_0 \cos \alpha_0)\frac{2v_0 \sin \alpha_0}{g}$$

$$= \frac{v_0^2 \sin 2\alpha_0}{g}$$

(We used the trigonometric identity $2 \sin \alpha_0 \cos \alpha_0 = \sin 2\alpha_0$, found in Appendix B.) The maximum value of $\sin 2\alpha_0$ is 1; this occurs when $2\alpha_0 = 90°$ or $\alpha_0 = 45°$. This angle gives the maximum range for a given initial speed if air resistance can be neglected.

EVALUATE: Figure 3.24 is based on a composite photograph of three trajectories of a ball projected from a small spring gun at angles of 30°, 45°, and 60°. The initial speed v_0 is approximately the same in all three cases. The horizontal range is greatest for the 45° angle. The ranges are nearly the same for the 30° and 60° angles: Can you prove that for a given value of v_0 the range is the same for both an initial angle α_0 and an initial angle $90° - \alpha_0$? (This is not the case in Fig. 3.24 due to air resistance.)

CAUTION **Height and range of a projectile** We don't recommend memorizing the above expressions for h, R, and $R_{\max}$. They are applicable only in the special circumstances we have described. In particular, the expressions for the range R and maximum range $R_{\max}$ can be used *only* when launch and landing heights are equal. There are many end-of-chapter problems to which these equations do *not* apply.

3.24 A launch angle of 45° gives the maximum horizontal range. The range is shorter with launch angles of 30° and 60°.

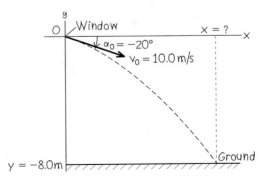

A 45° launch angle gives the greatest range; other angles fall shorter.

Launch angle:
$\alpha_0 = 30°$
$\alpha_0 = 45°$
$\alpha_0 = 60°$

Example 3.9 Different initial and final heights

You throw a ball from your window 8.0 m above the ground. When the ball leaves your hand, it is moving at 10.0 m/s at an angle of 20° below the horizontal. How far horizontally from your window will the ball hit the ground? Ignore air resistance.

SOLUTION

IDENTIFY and SET UP: As in Examples 3.7 and 3.8, we want to find the horizontal coordinate of a projectile when it is at a given y-value. The difference here is that this value of y is *not* the same as the initial value. We again choose the x-axis to be horizontal and the y-axis to be upward, and place the origin of coordinates at the point where the ball leaves your hand (Fig. 3.25). We have $v_0 = 10.0$ m/s and $\alpha_0 = -20°$ (the angle is negative because the initial velocity is below the horizontal). Our target variable is the value of x when the ball reaches the ground at $y = -8.0$ m. We'll use Eq. (3.21) to find the time t when this happens, then use Eq. (3.20) to find the value of x at this time.

3.25 Our sketch for this problem.

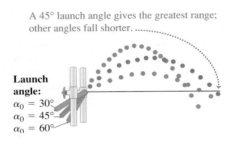

EXECUTE: To determine t, we rewrite Eq. (3.21) in the standard form for a quadratic equation for t:

$$\tfrac{1}{2}gt^2 - (v_0 \sin \alpha_0)t + y = 0$$

Continued

The roots of this equation are

$$t = \frac{v_0 \sin \alpha_0 \pm \sqrt{(-v_0 \sin \alpha_0)^2 - 4(\frac{1}{2}g)y}}{2(\frac{1}{2}g)}$$

$$= \frac{v_0 \sin \alpha_0 \pm \sqrt{v_0^2 \sin^2 \alpha_0 - 2gy}}{g}$$

$$= \frac{\left[\begin{array}{c}(10.0 \text{ m/s}) \sin(-20°) \\ \pm \sqrt{(10.0 \text{ m/s})^2 \sin^2(-20°) - 2(9.80 \text{ m/s}^2)(-8.0 \text{ m})}\end{array}\right]}{9.80 \text{ m/s}^2}$$

$$= -1.7 \text{ s} \quad \text{or} \quad 0.98 \text{ s}$$

We discard the negative root, since it refers to a time before the ball left your hand. The positive root tells us that the ball reaches the ground at $t = 0.98$ s. From Eq. (3.20), the ball's x-coordinate at that time is

$$x = (v_0 \cos \alpha_0)t = (10.0 \text{ m/s})[\cos(-20°)](0.98 \text{ s})$$
$$= 9.2 \text{ m}$$

The ball hits the ground a horizontal distance of 9.2 m from your window.

EVALUATE: The root $t = -1.7$ s is an example of a "fictional" solution to a quadratic equation. We discussed these in Example 2.8 in Section 2.5; you should review that discussion.

Example 3.10 **The zookeeper and the monkey**

A monkey escapes from the zoo and climbs a tree. After failing to entice the monkey down, the zookeeper fires a tranquilizer dart directly at the monkey (Fig. 3.26). The monkey lets go at the instant the dart leaves the gun. Show that the dart will *always* hit the monkey, provided that the dart reaches the monkey before he hits the ground and runs away.

SOLUTION

IDENTIFY and SET UP: We have *two* bodies in projectile motion: the dart and the monkey. They have different initial positions and initial velocities, but they go into projectile motion at the same time $t = 0$. We'll first use Eq. (3.20) to find an expression for the time t when the x-coordinates x_{monkey} and x_{dart} are equal. Then we'll use that expression in Eq. (3.21) to see whether y_{monkey} and y_{dart} are also equal at this time; if they are, the dart hits the monkey. We

make the usual choice for the x- and y-directions, and place the origin of coordinates at the muzzle of the tranquilizer gun (Fig. 3.26).

EXECUTE: The monkey drops straight down, so $x_{\text{monkey}} = d$ at all times. From Eq. (3.20), $x_{\text{dart}} = (v_0 \cos \alpha_0)t$. We solve for the time t when these x-coordinates are equal:

$$d = (v_0 \cos \alpha_0)t \quad \text{so} \quad t = \frac{d}{v_0 \cos \alpha_0}$$

We must now show that $y_{\text{monkey}} = y_{\text{dart}}$ at this time. The monkey is in one-dimensional free fall; its position at any time is given by Eq. (2.12), with appropriate symbol changes. Figure 3.26 shows that the monkey's initial height above the dart-gun's muzzle is $y_{\text{monkey}-0} = d \tan \alpha_0$, so

$$y_{\text{monkey}} = d \tan \alpha_0 - \frac{1}{2}gt^2$$

3.26 The tranquilizer dart hits the falling monkey.

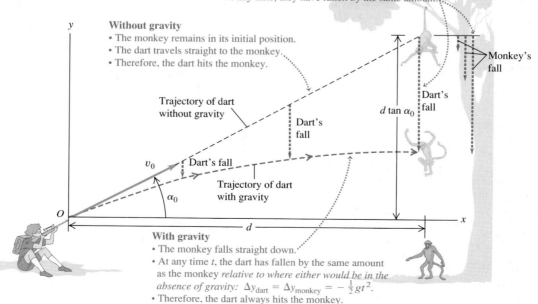

Dashed arrows show how far the dart and monkey have fallen at specific times relative to where they would be without gravity. At any time, they have fallen by the same amount.

Without gravity
• The monkey remains in its initial position.
• The dart travels straight to the monkey.
• Therefore, the dart hits the monkey.

Trajectory of dart without gravity

Dart's fall

$d \tan \alpha_0$

Monkey's fall

Dart's fall

v_0

Dart's fall

α_0

Trajectory of dart with gravity

d

With gravity
• The monkey falls straight down.
• At any time t, the dart has fallen by the same amount as the monkey *relative to where either would be in the absence of gravity:* $\Delta y_{\text{dart}} = \Delta y_{\text{monkey}} = -\frac{1}{2}gt^2$.
• Therefore, the dart always hits the monkey.

From Eq. (3.21),

$$y_{\text{dart}} = (v_0 \sin \alpha_0)t - \tfrac{1}{2}gt^2$$

Comparing these two equations, we see that we'll have $y_{\text{monkey}} = y_{\text{dart}}$ (and a hit) if $d \tan \alpha_0 = (v_0 \sin \alpha_0)t$ at the time when the two x-coordinates are equal. To show that this happens, we replace t with $d/(v_0 \cos \alpha_0)$, the time when $x_{\text{monkey}} = x_{\text{dart}}$. Sure enough, we find that

$$(v_0 \sin \alpha_0)t = (v_0 \sin \alpha_0)\,\frac{d}{v_0 \cos \alpha_0} = d \tan \alpha_0$$

EVALUATE: We've proved that the y-coordinates of the dart and the monkey are equal at the same time that their x-coordinates are equal; a dart aimed at the monkey *always* hits it, no matter what v_0 is (provided the monkey doesn't hit the ground first). This result is independent of the value of g, the acceleration due to gravity. With no gravity ($g = 0$), the monkey would remain motionless, and the dart would travel in a straight line to hit him. With gravity, both fall the same distance $gt^2/2$ below their $t = 0$ positions, and the dart still hits the monkey (Fig. 3.26).

Test Your Understanding of Section 3.3

In Example 3.10, suppose the tranquilizer dart has a relatively low muzzle velocity so that the dart reaches a maximum height at a point P before striking the monkey, as shown in the figure. When the dart is at point P, will the monkey be (i) at point A (higher than P), (ii) at point B (at the same height as P), or (iii) at point C (lower than P)? Ignore air resistance.

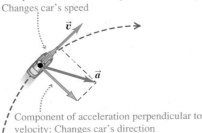

3.4 Motion in a Circle

When a particle moves along a curved path, the direction of its velocity changes. As we saw in Section 3.2, this means that the particle *must* have a component of acceleration perpendicular to the path, even if its speed is constant (see Fig. 3.11b). In this section we'll calculate the acceleration for the important special case of motion in a circle.

Uniform Circular Motion

When a particle moves in a circle with *constant speed*, the motion is called **uniform circular motion.** A car rounding a curve with constant radius at constant speed, a satellite moving in a circular orbit, and an ice skater skating in a circle with constant speed are all examples of uniform circular motion (Fig. 3.27c; compare Fig. 3.12a). There is no component of acceleration parallel (tangent) to the path; otherwise, the speed would change. The acceleration vector is perpendicular (normal) to the path and hence directed inward (never outward!) toward the center of the circular path. This causes the direction of the velocity to change without changing the speed.

3.27 A car moving along a circular path. If the car is in uniform circular motion as in (c), the speed is constant and the acceleration is directed toward the center of the circular path (compare Fig. 3.12).

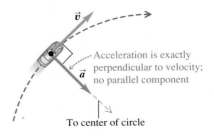

(a) Car speeding up along a circular path

Component of acceleration parallel to velocity: Changes car's speed

Component of acceleration perpendicular to velocity: Changes car's direction

(b) Car slowing down along a circular path

Component of acceleration perpendicular to velocity: Changes car's direction

Component of acceleration parallel to velocity: Changes car's speed

(c) Uniform circular motion: Constant speed along a circular path

Acceleration is exactly perpendicular to velocity; no parallel component

To center of circle

3.28 Finding the velocity change $\Delta\vec{v}$, average acceleration $\vec{a}_{av}$, and instantaneous acceleration $\vec{a}_{rad}$ for a particle moving in a circle with constant speed.

(a) A particle moves a distance Δs at constant speed along a circular path.

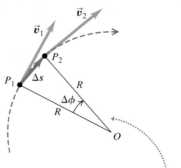

(b) The corresponding change in velocity and average acceleration

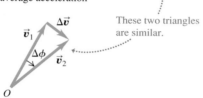

These two triangles are similar.

(c) The instantaneous acceleration

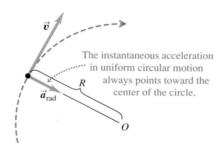

The instantaneous acceleration in uniform circular motion always points toward the center of the circle.

We can find a simple expression for the magnitude of the acceleration in uniform circular motion. We begin with Fig. 3.28a, which shows a particle moving with constant speed in a circular path of radius R with center at O. The particle moves from P_1 to P_2 in a time Δt. The vector change in velocity $\Delta\vec{v}$ during this time is shown in Fig. 3.28b.

The angles labeled $\Delta\phi$ in Figs. 3.28a and 3.28b are the same because $\vec{v}_1$ is perpendicular to the line OP_1 and $\vec{v}_2$ is perpendicular to the line OP_2. Hence the triangles in Figs. 3.28a and 3.28b are *similar*. The ratios of corresponding sides of similar triangles are equal, so

$$\frac{|\Delta\vec{v}|}{v_1} = \frac{\Delta s}{R} \quad \text{or} \quad |\Delta\vec{v}| = \frac{v_1}{R}\Delta s$$

The magnitude a_{av} of the average acceleration during Δt is therefore

$$a_{av} = \frac{|\Delta\vec{v}|}{\Delta t} = \frac{v_1}{R}\frac{\Delta s}{\Delta t}$$

The magnitude a of the *instantaneous* acceleration $\vec{a}$ at point P_1 is the limit of this expression as we take point P_2 closer and closer to point P_1:

$$a = \lim_{\Delta t\to 0}\frac{v_1}{R}\frac{\Delta s}{\Delta t} = \frac{v_1}{R}\lim_{\Delta t\to 0}\frac{\Delta s}{\Delta t}$$

If the time interval Δt is short, Δs is the distance the particle moves along its curved path. So the limit of $\Delta s/\Delta t$ is the speed v_1 at point P_1. Also, P_1 can be any point on the path, so we can drop the subscript and let v represent the speed at any point. Then

$$a_{rad} = \frac{v^2}{R} \quad \text{(uniform circular motion)} \quad (3.28)$$

We have added the subscript "rad" as a reminder that the direction of the instantaneous acceleration at each point is always along a radius of the circle (toward the center of the circle; see Figs. 3.27c and 3.28c). So we have found that *in uniform circular motion, the magnitude a_{rad} of the instantaneous acceleration is equal to the square of the speed v divided by the radius R of the circle. Its direction is perpendicular to $\vec{v}$ and inward along the radius.*

Because the acceleration in uniform circular motion is always directed toward the center of the circle, it is sometimes called **centripetal acceleration.** The word "centripetal" is derived from two Greek words meaning "seeking the center." Figure 3.29a shows the directions of the velocity and acceleration vectors at several points for a particle moving with uniform circular motion.

3.29 Acceleration and velocity **(a)** for a particle in uniform circular motion and **(b)** for a projectile with no air resistance.

(a) Uniform circular motion

(b) Projectile motion

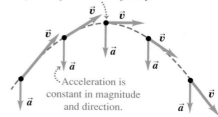

Acceleration has constant magnitude but varying direction.

Velocity and acceleration are always perpendicular.

Velocity and acceleration are perpendicular only at the peak of the trajectory.

Acceleration is constant in magnitude and direction.

CAUTION **Uniform circular motion vs. projectile motion** The acceleration in uniform circular motion (Fig. 3.29a) has some similarities to the acceleration in projectile motion without air resistance (Fig. 3.29b), but there are also some important differences. In both kinds of motion the *magnitude* of acceleration is the same at all times. However, in uniform circular motion the *direction* of $\vec{a}$ changes continuously so that it always points toward the center of the circle. (At the top of the circle the acceleration points down; at the bottom of the circle the acceleration points up.) In projectile motion, by contrast, the direction of $\vec{a}$ remains the same at all times. ▮

We can also express the magnitude of the acceleration in uniform circular motion in terms of the **period** T of the motion, the time for one revolution (one complete trip around the circle). In a time T the particle travels a distance equal to the circumference $2\pi R$ of the circle, so its speed is

$$v = \frac{2\pi R}{T} \tag{3.29}$$

When we substitute this into Eq. (3.28), we obtain the alternative expression

$$a_{rad} = \frac{4\pi^2 R}{T^2} \quad \text{(uniform circular motion)} \tag{3.30}$$

Mastering PHYSICS

PhET: Ladybug Revolution
PhET: Motion in 2D

Example 3.11 **Centripetal acceleration on a curved road**

An Aston Martin V8 Vantage sports car has a "lateral acceleration" of $0.96g = (0.96)(9.8 \text{ m/s}^2) = 9.4 \text{ m/s}^2$. This is the maximum centripetal acceleration the car can sustain without skidding out of a curved path. If the car is traveling at a constant 40 m/s (about 89 mi/h, or 144 km/h) on level ground, what is the radius R of the tightest unbanked curve it can negotiate?

SOLUTION

IDENTIFY, SET UP, and EXECUTE: The car is in uniform circular motion because it's moving at a constant speed along a curve that is a segment of a circle. Hence we can use Eq. (3.28) to solve for the target variable R in terms of the given centripetal acceleration

a_{rad} and speed v:

$$R = \frac{v^2}{a_{rad}} = \frac{(40 \text{ m/s})^2}{9.4 \text{ m/s}^2} = 170 \text{ m (about 560 ft)}$$

This is the *minimum* radius because a_{rad} is the *maximum* centripetal acceleration.

EVALUATE: The minimum turning radius R is proportional to the *square* of the speed, so even a small reduction in speed can make R substantially smaller. For example, reducing v by 20% (from 40 m/s to 32 m/s) would decrease R by 36% (from 170 m to 109 m).

Another way to make the minimum turning radius smaller is to *bank* the curve. We'll investigate this option in Chapter 5.

Example 3.12 **Centripetal acceleration on a carnival ride**

Passengers on a carnival ride move at constant speed in a horizontal circle of radius 5.0 m, making a complete circle in 4.0 s. What is their acceleration?

SOLUTION

IDENTIFY and SET UP: The speed is constant, so this is uniform circular motion. We are given the radius $R = 5.0$ m and the period $T = 4.0$ s, so we can use Eq. (3.30) to calculate the acceleration directly, or we can calculate the speed v using Eq. (3.29) and then find the acceleration using Eq. (3.28).

EXECUTE: From Eq. (3.30),

$$a_{rad} = \frac{4\pi^2(5.0 \text{ m})}{(4.0 \text{ s})^2} = 12 \text{ m/s}^2 = 1.3g$$

We can check this answer by using the second, roundabout approach. From Eq. (3.29), the speed is

$$v = \frac{2\pi R}{T} = \frac{2\pi(5.0 \text{ m})}{4.0 \text{ s}} = 7.9 \text{ m/s}$$

The centripetal acceleration is then

$$a_{rad} = \frac{v^2}{R} = \frac{(7.9 \text{ m/s})^2}{5.0 \text{ m}} = 12 \text{ m/s}^2$$

EVALUATE: As in Example 3.11, the direction of $\vec{a}$ is always toward the center of the circle. The magnitude of $\vec{a}$ is relatively mild as carnival rides go; some roller coasters subject their passengers to accelerations as great as $4g$.

3.30 A particle moving in a vertical loop with a varying speed, like a roller coaster car.

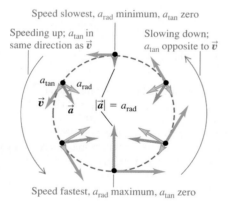

Speed slowest, a_{rad} minimum, a_{tan} zero

Speeding up; a_{tan} in same direction as $\vec{v}$

Slowing down; a_{tan} opposite to $\vec{v}$

a_{tan} a_{rad}

$\vec{v}$ $\vec{a}$ $|\vec{a}| = a_{rad}$

Speed fastest, a_{rad} maximum, a_{tan} zero

Nonuniform Circular Motion

We have assumed throughout this section that the particle's speed is constant as it goes around the circle. If the speed varies, we call the motion **nonuniform circular motion.** In nonuniform circular motion, Eq. (3.28) still gives the *radial* component of acceleration $a_{rad} = v^2/R$, which is always *perpendicular* to the instantaneous velocity and directed toward the center of the circle. But since the speed v has different values at different points in the motion, the value of a_{rad} is not constant. The radial (centripetal) acceleration is greatest at the point in the circle where the speed is greatest.

In nonuniform circular motion there is also a component of acceleration that is *parallel* to the instantaneous velocity (see Figs. 3.27a and 3.27b). This is the component $a_{\parallel}$ that we discussed in Section 3.2; here we call this component a_{tan} to emphasize that it is *tangent* to the circle. The tangential component of acceleration a_{tan} is equal to the rate of change of *speed.* Thus

$$a_{rad} = \frac{v^2}{R} \quad \text{and} \quad a_{tan} = \frac{d|\vec{v}|}{dt} \quad \text{(nonuniform circular motion)} \quad (3.31)$$

The tangential component is in the same direction as the velocity if the particle is speeding up, and in the opposite direction if the particle is slowing down (Fig. 3.30). If the particle's speed is constant, $a_{tan} = 0$.

CAUTION **Uniform vs. nonuniform circular motion** Note that the two quantities

$$\frac{d|\vec{v}|}{dt} \quad \text{and} \quad \left|\frac{d\vec{v}}{dt}\right|$$

are *not* the same. The first, equal to the tangential acceleration, is the rate of change of speed; it is zero whenever a particle moves with constant speed, even when its direction of motion changes (such as in *uniform* circular motion). The second is the magnitude of the vector acceleration; it is zero only when the particle's acceleration *vector* is zero—that is, when the particle moves in a straight line with constant speed. In *uniform* circular motion $|d\vec{v}/dt| = a_{rad} = v^2/r$; in *nonuniform* circular motion there is also a tangential component of acceleration, so $|d\vec{v}/dt| = \sqrt{a_{rad}^2 + a_{tan}^2}$. |

Test Your Understanding of Section 3.4 Suppose that the particle in Fig. 3.30 experiences four times the acceleration at the bottom of the loop as it does at the top of the loop. Compared to its speed at the top of the loop, is its speed at the bottom of the loop (i) $\sqrt{2}$ times as great; (ii) 2 times as great; (iii) $2\sqrt{2}$ times as great; (iv) 4 times as great; or (v) 16 times as great?

3.5 Relative Velocity

You've no doubt observed how a car that is moving slowly forward appears to be moving backward when you pass it. In general, when two observers measure the velocity of a moving body, they get different results if one observer is moving relative to the other. The velocity seen by a particular observer is called the velocity *relative* to that observer, or simply **relative velocity.** Figure 3.31 shows a situation in which understanding relative velocity is extremely important.

We'll first consider relative velocity along a straight line, then generalize to relative velocity in a plane.

Relative Velocity in One Dimension

A passenger walks with a velocity of 1.0 m/s along the aisle of a train that is moving with a velocity of 3.0 m/s (Fig. 3.32a). What is the passenger's velocity?

It's a simple enough question, but it has no single answer. As seen by a second passenger sitting in the train, she is moving at 1.0 m/s. A person on a bicycle standing beside the train sees the walking passenger moving at 1.0 m/s + 3.0 m/s = 4.0 m/s. An observer in another train going in the opposite direction would give still another answer. We have to specify which observer we mean, and we speak of the velocity *relative* to a particular observer. The walking passenger's velocity relative to the train is 1.0 m/s, her velocity relative to the cyclist is 4.0 m/s, and so on. Each observer, equipped in principle with a meter stick and a stopwatch, forms what we call a **frame of reference.** Thus a frame of reference is a coordinate system plus a time scale.

Let's use the symbol A for the cyclist's frame of reference (at rest with respect to the ground) and the symbol B for the frame of reference of the moving train. In straight-line motion the position of a point P relative to frame A is given by $x_{P/A}$ (the position of P with respect to A), and the position of P relative to frame B is given by $x_{P/B}$ (Fig. 3.32b). The position of the origin of B with respect to the origin of A is $x_{B/A}$. Figure 3.32b shows that

$$x_{P/A} = x_{P/B} + x_{B/A} \tag{3.32}$$

In words, the coordinate of P relative to A equals the coordinate of P relative to B plus the coordinate of B relative to A.

The x-velocity of P relative to frame A, denoted by $v_{P/A\text{-}x}$, is the derivative of $x_{P/A}$ with respect to time. The other velocities are similarly obtained. So the time derivative of Eq. (3.32) gives us a relationship among the various velocities:

$$\frac{dx_{P/A}}{dt} = \frac{dx_{P/B}}{dt} + \frac{dx_{B/A}}{dt} \quad \text{or}$$

$$v_{P/A\text{-}x} = v_{P/B\text{-}x} + v_{B/A\text{-}x} \quad \text{(relative velocity along a line)} \tag{3.33}$$

Getting back to the passenger on the train in Fig. 3.32, we see that A is the cyclist's frame of reference, B is the frame of reference of the train, and point P represents the passenger. Using the above notation, we have

$$v_{P/B\text{-}x} = +1.0 \text{ m/s} \quad v_{B/A\text{-}x} = +3.0 \text{ m/s}$$

From Eq. (3.33) the passenger's velocity $v_{P/A}$ relative to the cyclist is

$$v_{P/A\text{-}x} = +1.0 \text{ m/s} + 3.0 \text{ m/s} = +4.0 \text{ m/s}$$

as we already knew.

In this example, both velocities are toward the right, and we have taken this as the positive x-direction. If the passenger walks toward the *left* relative to the train, then $v_{P/B\text{-}x} = -1.0$ m/s, and her x-velocity relative to the cyclist is $v_{P/A\text{-}x} = -1.0$ m/s + 3.0 m/s = +2.0 m/s. The sum in Eq. (3.33) is always an algebraic sum, and any or all of the x-velocities may be negative.

When the passenger looks out the window, the stationary cyclist on the ground appears to her to be moving backward; we can call the cyclist's velocity relative to her $v_{A/P\text{-}x}$. Clearly, this is just the negative of the *passenger's* velocity relative to the *cyclist*, $v_{P/A\text{-}x}$. In general, if A and B are any two points or frames of reference,

$$v_{A/B\text{-}x} = -v_{B/A\text{-}x} \tag{3.34}$$

3.31 Airshow pilots face a complicated problem involving relative velocities. They must keep track of their motion relative to the air (to maintain enough airflow over the wings to sustain lift), relative to each other (to keep a tight formation without colliding), and relative to their audience (to remain in sight of the spectators).

3.32 (a) A passenger walking in a train. (b) The position of the passenger relative to the cyclist's frame of reference and the train's frame of reference.

(a)

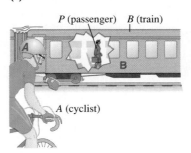

(b)

Problem-Solving Strategy 3.2 Relative Velocity

IDENTIFY *the relevant concepts:* Whenever you see the phrase "velocity relative to" or "velocity with respect to," it's likely that the concepts of relative velocity will be helpful.

SET UP *the problem:* Sketch and label each frame of reference in the problem. Each moving body has its own frame of reference; in addition, you'll almost always have to include the frame of reference of the earth's surface. (Statements such as "The car is traveling north at 90 km/h" implicitly refer to the car's velocity relative to the surface of the earth.) Use the labels to help identify the target variable. For example, if you want to find the *x*-velocity of a car (*C*) with respect to a bus (*B*), your target variable is $v_{C/B\text{-}x}$.

EXECUTE *the solution:* Solve for the target variable using Eq. (3.33). (If the velocities aren't along the same direction, you'll need to use the vector form of this equation, derived later in this section.) It's

important to note the order of the double subscripts in Eq. (3.33): $v_{B/A\text{-}x}$ means "*x*-velocity of *B* relative to *A*." These subscripts obey a kind of algebra, as Eq. (3.33) shows. If we regard each one as a fraction, then the fraction on the left side is the *product* of the fractions on the right side: $P/A = (P/B)(B/A)$. You can apply this rule to any number of frames of reference. For example, if there are three different frames of reference *A*, *B*, and *C*, Eq. (3.33) becomes

$$v_{P/A\text{-}x} = v_{P/C\text{-}x} + v_{C/B\text{-}x} + v_{B/A\text{-}x}$$

EVALUATE *your answer:* Be on the lookout for stray minus signs in your answer. If the target variable is the *x*-velocity of a car relative to a bus ($v_{C/B\text{-}x}$), make sure that you haven't accidentally calculated the *x*-velocity of the *bus* relative to the *car* ($v_{B/C\text{-}x}$). If you've made this mistake, you can recover using Eq. (3.34).

Example 3.13 Relative velocity on a straight road

You drive north on a straight two-lane road at a constant 88 km/h. A truck in the other lane approaches you at a constant 104 km/h (Fig. 3.33). Find (a) the truck's velocity relative to you and (b) your velocity relative to the truck. (c) How do the relative velocities change after you and the truck pass each other? Treat this as a one-dimensional problem.

SOLUTION

IDENTIFY and SET UP: In this problem about relative velocities along a line, there are three reference frames: you (Y), the truck (T), and the earth's surface (E). Let the positive *x*-direction be north (Fig. 3.33). Then your *x*-velocity relative to the earth is $v_{Y/E\text{-}x} = +88$ km/h. The truck is initially approaching you, so it is moving south and its *x*-velocity with respect to the earth is $v_{T/E\text{-}x} = -104$ km/h. The target variables in parts (a) and (b) are $v_{T/Y\text{-}x}$ and $v_{Y/T\text{-}x}$, respectively. We'll use Eq. (3.33) to find the first target variable and Eq. (3.34) to find the second.

EXECUTE: (a) To find $v_{T/Y\text{-}x}$, we write Eq. (3.33) for the known $v_{T/E\text{-}x}$ and rearrange:

$$v_{T/E\text{-}x} = v_{T/Y\text{-}x} + v_{Y/E\text{-}x}$$
$$v_{T/Y\text{-}x} = v_{T/E\text{-}x} - v_{Y/E\text{-}x}$$
$$= -104 \text{ km/h} - 88 \text{ km/h} = -192 \text{ km/h}$$

The truck is moving at 192 km/h in the negative *x*-direction (south) relative to you.

(b) From Eq. (3.34),

$$v_{Y/T\text{-}x} = -v_{T/Y\text{-}x} = -(-192 \text{ km/h}) = +192 \text{ km/h}$$

3.33 Reference frames for you and the truck.

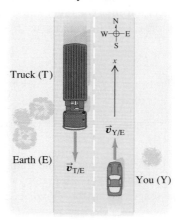

You are moving at 192 km/h in the positive *x*-direction (north) relative to the truck.

(c) The relative velocities do *not* change after you and the truck pass each other. The relative *positions* of the bodies don't matter. After it passes you the truck is still moving at 192 km/h toward the south relative to you, even though it is now moving away from you instead of toward you.

EVALUATE: To check your answer in part (b), use Eq. (3.33) directly in the form $v_{Y/T\text{-}x} = v_{Y/E\text{-}x} + v_{E/T\text{-}x}$. (The *x*-velocity of the earth with respect to the truck is the opposite of the *x*-velocity of the truck with respect to the earth: $v_{E/T\text{-}x} = -v_{T/E\text{-}x}$.) Do you get the same result?

Relative Velocity in Two or Three Dimensions

We can extend the concept of relative velocity to include motion in a plane or in space by using vector addition to combine velocities. Suppose that the passenger in Fig. 3.32a is walking not down the aisle of the railroad car but from one side of the car to the other, with a speed of 1.0 m/s (Fig. 3.34a). We can again describe the passenger's position *P* in two different frames of reference: *A* for the stationary ground observer and B for the moving train. But instead of coordinates x, we use position vectors $\vec{r}$ because the problem is now two-dimensional. Then, as Fig. 3.34b shows,

$$\vec{r}_{P/A} = \vec{r}_{P/B} + \vec{r}_{B/A} \qquad (3.35)$$

Just as we did before, we take the time derivative of this equation to get a relationship among the various velocities; the velocity of P relative to A is $\vec{v}_{P/A} = d\vec{r}_{P/A}/dt$ and so on for the other velocities. We get

$$\vec{v}_{P/A} = \vec{v}_{P/B} + \vec{v}_{B/A} \qquad \text{(relative velocity in space)} \qquad (3.36)$$

Equation (3.36) is known as the *Galilean velocity transformation*. It relates the velocity of a body P with respect to frame A and its velocity with respect to frame B ($\vec{v}_{P/A}$ and $\vec{v}_{P/B}$, respectively) to the velocity of frame B with respect to frame A ($\vec{v}_{B/A}$). If all three of these velocities lie along the same line, then Eq. (3.36) reduces to Eq. (3.33) for the components of the velocities along that line.

If the train is moving at $v_{B/A} = 3.0$ m/s relative to the ground and the passenger is moving at $v_{P/B} = 1.0$ m/s relative to the train, then the passenger's velocity vector $\vec{v}_{P/A}$ relative to the ground is as shown in Fig. 3.34c. The Pythagorean theorem then gives us

$$v_{P/A} = \sqrt{(3.0 \text{ m/s})^2 + (1.0 \text{ m/s})^2} = \sqrt{10 \text{ m}^2/\text{s}^2} = 3.2 \text{ m/s}$$

Figure 3.34c also shows that the *direction* of the passenger's velocity vector relative to the ground makes an angle ϕ with the train's velocity vector $\vec{v}_{B/A}$, where

$$\tan \phi = \frac{v_{P/B}}{v_{B/A}} = \frac{1.0 \text{ m/s}}{3.0 \text{ m/s}} \qquad \text{and} \qquad \phi = 18°$$

As in the case of motion along a straight line, we have the general rule that if A and B are *any* two points or frames of reference,

$$\vec{v}_{A/B} = -\vec{v}_{B/A} \qquad (3.37)$$

The velocity of the passenger relative to the train is the negative of the velocity of the train relative to the passenger, and so on.

In the early 20th century Albert Einstein showed in his special theory of relativity that the velocity-addition relationship given in Eq. (3.36) has to be modified when speeds approach the speed of light, denoted by c. It turns out that if the passenger in Fig. 3.32a could walk down the aisle at $0.30c$ and the train could move at $0.90c$, then her speed relative to the ground would be not $1.20c$ but $0.94c$; nothing can travel faster than light! We'll return to the special theory of relativity in Chapter 37.

3.34 (a) A passenger walking across a railroad car. (b) Position of the passenger relative to the cyclist's frame and the train's frame. (c) Vector diagram for the velocity of the passenger relative to the ground (the cyclist's frame), $\vec{v}_{P/A}$.

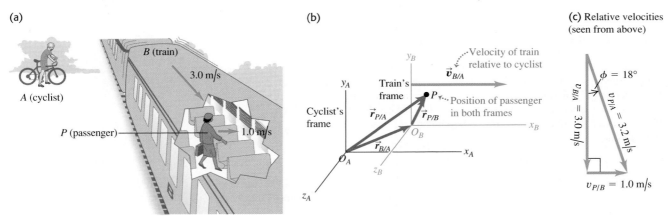

Example 3.14 Flying in a crosswind

An airplane's compass indicates that it is headed due north, and its airspeed indicator shows that it is moving through the air at 240 km/h. If there is a 100-km/h wind from west to east, what is the velocity of the airplane relative to the earth?

SOLUTION

IDENTIFY and SET UP: This problem involves velocities in two dimensions (northward and eastward), so it is a relative velocity problem using vectors. We are given the magnitude and direction of the velocity of the plane (P) relative to the air (A). We are also given the magnitude and direction of the wind velocity, which is the velocity of the air A with respect to the earth (E):

$$\vec{v}_{P/A} = 240 \text{ km/h} \quad \text{due north}$$
$$\vec{v}_{A/E} = 100 \text{ km/h} \quad \text{due east}$$

We'll use Eq. (3.36) to find our target variables: the magnitude and direction of the velocity $\vec{v}_{P/E}$ of the plane relative to the earth.

EXECUTE: From Eq. (3.36) we have

$$\vec{v}_{P/E} = \vec{v}_{P/A} + \vec{v}_{A/E}$$

Figure 3.35 shows that the three relative velocities constitute a right-triangle vector addition; the unknowns are the speed $v_{P/E}$ and the angle α. We find

$$v_{P/E} = \sqrt{(240 \text{ km/h})^2 + (100 \text{ km/h})^2} = 260 \text{ km/h}$$
$$\alpha = \arctan\left(\frac{100 \text{ km/h}}{240 \text{ km/h}}\right) = 23° \text{ E of N}$$

EVALUATE: You can check the results by taking measurements on the scale drawing in Fig. 3.35. The crosswind increases the speed of the airplane relative to the earth, but pushes the airplane off course.

3.35 The plane is pointed north, but the wind blows east, giving the resultant velocity $\vec{v}_{P/E}$ relative to the earth.

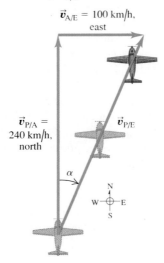

Example 3.15 Correcting for a crosswind

With wind and airspeed as in Example 3.14, in what direction should the pilot head to travel due north? What will be her velocity relative to the earth?

SOLUTION

IDENTIFY and SET UP: Like Example 3.14, this is a relative velocity problem with vectors. Figure 3.36 is a scale drawing of the situation. Again the vectors add in accordance with Eq. (3.36) and form a right triangle:

$$\vec{v}_{P/E} = \vec{v}_{P/A} + \vec{v}_{A/E}$$

As Fig. 3.36 shows, the pilot points the nose of the airplane at an angle β into the wind to compensate for the crosswind. This angle, which tells us the direction of the vector $\vec{v}_{P/A}$ (the velocity of the airplane relative to the air), is one of our target variables. The other target variable is the speed of the airplane over the ground, which is the magnitude of the vector $\vec{v}_{P/E}$ (the velocity of the airplane relative to the earth). The known and unknown quantities are

$$\vec{v}_{P/E} = \text{magnitude unknown} \quad \text{due north}$$
$$\vec{v}_{P/A} = 240 \text{ km/h} \quad \text{direction unknown}$$
$$\vec{v}_{A/E} = 100 \text{ km/h} \quad \text{due east}$$

3.36 The pilot must point the plane in the direction of the vector $\vec{v}_{P/A}$ to travel due north relative to the earth.

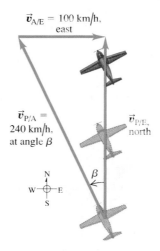

We'll solve for the target variables using Fig. 3.36 and trigonometry.

EXECUTE: From Fig. 3.36 the speed $v_{P/E}$ and the angle β are

$$v_{P/E} = \sqrt{(240 \text{ km/h})^2 - (100 \text{ km/h})^2} = 218 \text{ km/h}$$

$$\beta = \arcsin\left(\frac{100 \text{ km/h}}{240 \text{ km/h}}\right) = 25°$$

The pilot should point the airplane 25° west of north, and her ground speed is then 218 km/h.

EVALUATE: There were two target variables—the magnitude of a vector and the direction of a vector—in both this example and Example 3.14. In Example 3.14 the magnitude and direction referred to the *same* vector ($\vec{v}_{P/E}$); here they refer to *different* vectors ($\vec{v}_{P/E}$ and $\vec{v}_{P/A}$).

While we expect a *headwind* to reduce an airplane's speed relative to the ground, this example shows that a *crosswind* does, too. That's an unfortunate fact of aeronautical life.

Test Your Understanding of Section 3.5 Suppose the nose of an airplane is pointed due east and the airplane has an airspeed of 150 km/h. Due to the wind, the airplane is moving due *north* relative to the ground and its speed relative to the ground is 150 km/h. What is the velocity of the air relative to the earth? (i) 150 km/h from east to west; (ii) 150 km/h from south to north; (iii) 150 km/h from southeast to northwest; (iv) 212 km/h from east to west; (v) 212 km/h from south to north; (vi) 212 km/h from southeast to northwest; (vii) there is no possible wind velocity that could cause this.

Position, velocity, and acceleration vectors: The position vector $\vec{r}$ of a point P in space is the vector from the origin to P. Its components are the coordinates x, y, and z.

The average velocity vector $\vec{v}_{av}$ during the time interval Δt is the displacement $\Delta \vec{r}$ (the change in the position vector $\vec{r}$) divided by Δt. The instantaneous velocity vector $\vec{v}$ is the time derivative of $\vec{r}$, and its components are the time derivatives of x, y, and z. The instantaneous speed is the magnitude of $\vec{v}$. The velocity $\vec{v}$ of a particle is always tangent to the particle's path. (See Example 3.1.)

The average acceleration vector $\vec{a}_{av}$ during the time interval Δt equals $\Delta \vec{v}$ (the change in the velocity vector $\vec{v}$) divided by Δt. The instantaneous acceleration vector $\vec{a}$ is the time derivative of $\vec{v}$, and its components are the time derivatives of v_x, v_y, and v_z. (See Example 3.2.)

The component of acceleration parallel to the direction of the instantaneous velocity affects the speed, while the component of $\vec{a}$ perpendicular to $\vec{v}$ affects the direction of motion. (See Examples 3.3 and 3.4.)

$$\vec{r} = x\hat{\imath} + y\hat{\jmath} + z\hat{k} \tag{3.1}$$

$$\vec{v}_{av} = \frac{\vec{r}_2 - \vec{r}_1}{t_2 - t_1} = \frac{\Delta \vec{r}}{\Delta t} \tag{3.2}$$

$$\vec{v} = \lim_{\Delta t \to 0} \frac{\Delta \vec{r}}{\Delta t} = \frac{d\vec{r}}{dt} \tag{3.3}$$

$$v_x = \frac{dx}{dt} \quad v_y = \frac{dy}{dt} \quad v_z = \frac{dz}{dt} \tag{3.4}$$

$$\vec{a}_{av} = \frac{\vec{v}_2 - \vec{v}_1}{t_2 - t_1} = \frac{\Delta \vec{v}}{\Delta t} \tag{3.8}$$

$$\vec{a} = \lim_{\Delta t \to 0} \frac{\Delta \vec{v}}{\Delta t} = \frac{d\vec{v}}{dt} \tag{3.9}$$

$$a_x = \frac{dv_x}{dt}$$

$$a_y = \frac{dv_y}{dt} \tag{3.10}$$

$$a_z = \frac{dv_z}{dt}$$

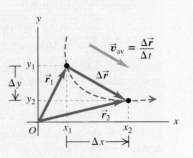

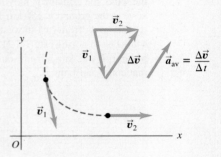

Projectile motion: In projectile motion with no air resistance, $a_x = 0$ and $a_y = -g$. The coordinates and velocity components are simple functions of time, and the shape of the path is always a parabola. We usually choose the origin to be at the initial position of the projectile. (See Examples 3.5–3.10.)

$$x = (v_0 \cos \alpha_0) t \tag{3.20}$$

$$y = (v_0 \sin \alpha_0) t - \tfrac{1}{2} g t^2 \tag{3.21}$$

$$v_x = v_0 \cos \alpha_0 \tag{3.22}$$

$$v_y = v_0 \sin \alpha_0 - gt \tag{3.23}$$

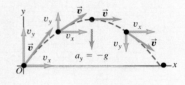

Uniform and nonuniform circular motion: When a particle moves in a circular path of radius R with constant speed v (uniform circular motion), its acceleration $\vec{a}$ is directed toward the center of the circle and perpendicular to $\vec{v}$. The magnitude a_{rad} of the acceleration can be expressed in terms of v and R or in terms of R and the period T (the time for one revolution), where $v = 2\pi R/T$. (See Examples 3.11 and 3.12.)

If the speed is not constant in circular motion (nonuniform circular motion), there is still a radial component of $\vec{a}$ given by Eq. (3.28) or (3.30), but there is also a component of $\vec{a}$ parallel (tangential) to the path. This tangential component is equal to the rate of change of speed, dv/dt.

$$a_{rad} = \frac{v^2}{R} \tag{3.28}$$

$$a_{rad} = \frac{4\pi^2 R}{T^2} \tag{3.30}$$

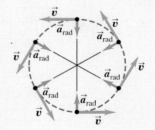

Relative velocity: When a body P moves relative to a body (or reference frame) B, and B moves relative to A, we denote the velocity of P relative to B by $\vec{v}_{P/B}$, the velocity of P relative to A by $\vec{v}_{P/A}$, and the velocity of B relative to A by $\vec{v}_{B/A}$. If these velocities are all along the same line, their components along that line are related by Eq. (3.33). More generally, these velocities are related by Eq. (3.36). (See Examples 3.13–3.15.)

$$v_{P/A\text{-}x} = v_{P/B\text{-}x} + v_{B/A\text{-}x} \tag{3.33}$$
(relative velocity along a line)

$$\vec{v}_{P/A} = \vec{v}_{P/B} + \vec{v}_{B/A} \tag{3.36}$$
(relative velocity in space)

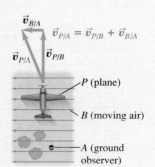

BRIDGING PROBLEM | **Launching Up an Incline**

You fire a ball with an initial speed v_0 at an angle ϕ above the surface of an incline, which is itself inclined at an angle θ above the horizontal (Fig. 3.37). (a) Find the distance, measured along the incline, from the launch point to the point when the ball strikes the incline. (b) What angle ϕ gives the maximum range, measured along the incline? Ignore air resistance.

SOLUTION GUIDE

See MasteringPhysics® study area for a Video Tutor solution.

IDENTIFY and SET UP

1. Since there's no air resistance, this is a problem in projectile motion. The goal is to find the point where the ball's parabolic trajectory intersects the incline.
2. Choose the x- and y-axes and the position of the origin. When in doubt, use the suggestions given in Problem-Solving Strategy 3.1 in Section 3.3.
3. In the projectile equations from Section 3.3, the launch angle α_0 is measured from the horizontal. What is this angle in terms of θ and ϕ? What are the initial x- and y-components of the ball's initial velocity?
4. You'll need to write an equation that relates x and y for points along the incline. What is this equation? (This takes just geometry and trigonometry, not physics.)

3.37 Launching a ball from an inclined ramp.

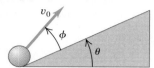

EXECUTE

5. Write the equations for the x-coordinate and y-coordinate of the ball as functions of time t.
6. When the ball hits the incline, x and y are related by the equation that you found in step 4. Based on this, at what time t does the ball hit the incline?
7. Based on your answer from step 6, at what coordinates x and y does the ball land on the incline? How far is this point from the launch point?
8. What value of ϕ gives the *maximum* distance from the launch point to the landing point? (Use your knowledge of calculus.)

EVALUATE

9. Check your answers for the case $\theta = 0$, which corresponds to the incline being horizontal rather than tilted. (You already know the answers for this case. Do you know why?)

Problems   For instructor-assigned homework, go to www.masteringphysics.com

•, ••, •••: Problems of increasing difficulty. **CP**: Cumulative problems incorporating material from earlier chapters. **CALC**: Problems requiring calculus. **BIO**: Biosciences problems.

DISCUSSION QUESTIONS

Q3.1 A simple pendulum (a mass swinging at the end of a string) swings back and forth in a circular arc. What is the direction of the acceleration of the mass when it is at the ends of the swing? At the midpoint? In each case, explain how you obtain your answer.

Q3.2 Redraw Fig. 3.11a if $\vec{a}$ is antiparallel to $\vec{v}_1$. Does the particle move in a straight line? What happens to its speed?

Q3.3 A projectile moves in a parabolic path without air resistance. Is there any point at which $\vec{a}$ is parallel to $\vec{v}$? Perpendicular to $\vec{v}$? Explain.

Q3.4 When a rifle is fired at a distant target, the barrel is not lined up exactly on the target. Why not? Does the angle of correction depend on the distance to the target?

Q3.5 At the same instant that you fire a bullet horizontally from a rifle, you drop a bullet from the height of the barrel. If there is no air resistance, which bullet hits the ground first? Explain.

Q3.6 A package falls out of an airplane that is flying in a straight line at a constant altitude and speed. If you could ignore air resistance, what would be the path of the package as observed by the pilot? As observed by a person on the ground?

Q3.7 Sketch the six graphs of the x- and y-components of position, velocity, and acceleration versus time for projectile motion with $x_0 = y_0 = 0$ and $0 < \alpha_0 < 90°$.

Q3.8 If a jumping frog can give itself the same initial speed regardless of the direction in which it jumps (forward or straight up), how is the maximum vertical height to which it can jump related to its maximum horizontal range $R_{max} = v_0^2/g$?

Q3.9 A projectile is fired upward at an angle θ above the horizontal with an initial speed v_0. At its maximum height, what are its velocity vector, its speed, and its acceleration vector?

Q3.10 In uniform circular motion, what are the *average* velocity and *average* acceleration for one revolution? Explain.

Q3.11 In uniform circular motion, how does the acceleration change when the speed is increased by a factor of 3? When the radius is decreased by a factor of 2?

Q3.12 In uniform circular motion, the acceleration is perpendicular to the velocity at every instant. Is this still true when the motion is not uniform—that is, when the speed is not constant?

Q3.13 Raindrops hitting the side windows of a car in motion often leave diagonal streaks even if there is no wind. Why? Is the explanation the same or different for diagonal streaks on the windshield?

Q3.14 In a rainstorm with a strong wind, what determines the best position in which to hold an umbrella?

Q3.15 You are on the west bank of a river that is flowing north with a speed of 1.2 m/s. Your swimming speed relative to the water is 1.5 m/s, and the river is 60 m wide. What is your path relative to the earth that allows you to cross the river in the shortest time? Explain your reasoning.

Q3.16 A stone is thrown into the air at an angle above the horizontal and feels negligible air resistance. Which graph in Fig. Q3.16 best depicts the stone's *speed v* as a function of time *t* while it is in the air?

Figure **Q3.16**

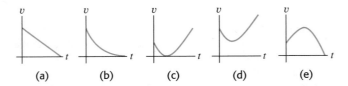

EXERCISES

Section 3.1 Position and Velocity Vectors

3.1 • A squirrel has *x*- and *y*-coordinates (1.1 m, 3.4 m) at time $t_1 = 0$ and coordinates (5.3 m, −0.5 m) at time $t_2 = 3.0$ s. For this time interval, find (a) the components of the average velocity, and (b) the magnitude and direction of the average velocity.

3.2 • A rhinoceros is at the origin of coordinates at time $t_1 = 0$. For the time interval from $t_1 = 0$ to $t_2 = 12.0$ s, the rhino's average velocity has *x*-component −3.8 m/s and *y*-component 4.9 m/s. At time $t_2 = 12.0$ s, (a) what are the *x*- and *y*-coordinates of the rhino? (b) How far is the rhino from the origin?

3.3 •• **CALC** A web page designer creates an animation in which a dot on a computer screen has a position of $\vec{r} = [4.0 \text{ cm} + (2.5 \text{ cm/s}^2)t^2]\hat{\imath} + (5.0 \text{ cm/s})t\hat{\jmath}$. (a) Find the magnitude and direction of the dot's average velocity between $t = 0$ and $t = 2.0$ s. (b) Find the magnitude and direction of the instantaneous velocity at $t = 0$, $t = 1.0$ s, and $t = 2.0$ s. (c) Sketch the dot's trajectory from $t = 0$ to $t = 2.0$ s, and show the velocities calculated in part (b).

3.4 • **CALC** The position of a squirrel running in a park is given by $\vec{r} = [(0.280 \text{ m/s})t + (0.0360 \text{ m/s}^2)t^2]\hat{\imath} + (0.0190 \text{ m/s}^3)t^3\hat{\jmath}$. (a) What are $v_x(t)$ and $v_y(t)$, the *x*- and *y*-components of the velocity of the squirrel, as functions of time? (b) At $t = 5.00$ s, how far is the squirrel from its initial position? (c) At $t = 5.00$ s, what are the magnitude and direction of the squirrel's velocity?

Section 3.2 The Acceleration Vector

3.5 • A jet plane is flying at a constant altitude. At time $t_1 = 0$ it has components of velocity $v_x = 90$ m/s, $v_y = 110$ m/s. At time $t_2 = 30.0$ s the components are $v_x = -170$ m/s, $v_y = 40$ m/s. (a) Sketch the velocity vectors at t_1 and t_2. How do these two vectors differ? For this time interval calculate (b) the components of the average acceleration, and (c) the magnitude and direction of the average acceleration.

3.6 •• A dog running in an open field has components of velocity $v_x = 2.6$ m/s and $v_y = -1.8$ m/s at $t_1 = 10.0$ s. For the time interval from $t_1 = 10.0$ s to $t_2 = 20.0$ s, the average acceleration of the dog has magnitude 0.45 m/s² and direction 31.0° measured from the +*x*-axis toward the +*y*-axis. At $t_2 = 20.0$ s, (a) what are the *x*- and *y*-components of the dog's velocity? (b) What are the magnitude and direction of the dog's velocity? (c) Sketch the velocity vectors at t_1 and t_2. How do these two vectors differ?

3.7 •• **CALC** The coordinates of a bird flying in the *xy*-plane are given by $x(t) = \alpha t$ and $y(t) = 3.0 \text{ m} - \beta t^2$, where $\alpha = 2.4$ m/s and $\beta = 1.2$ m/s². (a) Sketch the path of the bird between $t = 0$ and $t = 2.0$ s. (b) Calculate the velocity and acceleration vectors of the bird as functions of time. (c) Calculate the magnitude and direction of the bird's velocity and acceleration at $t = 2.0$ s. (d) Sketch the velocity and acceleration vectors at $t = 2.0$ s. At this instant, is the bird speeding up, is it slowing down, or is its speed instantaneously not changing? Is the bird turning? If so, in what direction?

Section 3.3 Projectile Motion

3.8 • **CALC** A remote-controlled car is moving in a vacant parking lot. The velocity of the car as a function of time is given by $\vec{v} = [5.00 \text{ m/s} - (0.0180 \text{ m/s}^3)t^2]\hat{\imath} + [2.00 \text{ m/s} + (0.550 \text{ m/s}^2)t]\hat{\jmath}$. (a) What are $a_x(t)$ and $a_y(t)$, the *x*- and *y*-components of the velocity of the car as functions of time? (b) What are the magnitude and direction of the velocity of the car at $t = 8.00$ s? (b) What are the magnitude and direction of the acceleration of the car at $t = 8.00$ s?

3.9 • A physics book slides off a horizontal tabletop with a speed of 1.10 m/s. It strikes the floor in 0.350 s. Ignore air resistance. Find (a) the height of the tabletop above the floor; (b) the horizontal distance from the edge of the table to the point where the book strikes the floor; (c) the horizontal and vertical components of the book's velocity, and the magnitude and direction of its velocity, just before the book reaches the floor. (d) Draw *x-t*, *y-t*, v_x-*t*, and v_y-*t* graphs for the motion.

3.10 •• A daring 510-N swimmer dives off a cliff with a running horizontal leap, as shown in Fig. E3.10. What must her minimum speed be just as she leaves the top of the cliff so that she will miss the ledge at the bottom, which is 1.75 m wide and 9.00 m below the top of the cliff?

Figure **E3.10**

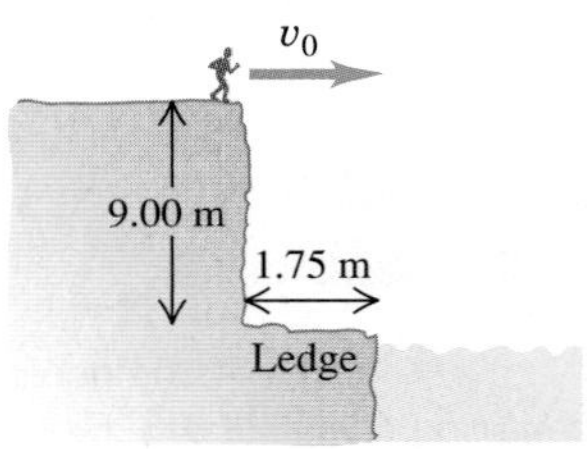

v_0

9.00 m

1.75 m

Ledge

3.11 • Two crickets, Chirpy and Milada, jump from the top of a vertical cliff. Chirpy just drops and reaches the ground in 3.50 s, while Milada jumps horizontally with an initial speed of 95.0 cm/s. How far from the base of the cliff will Milada hit the ground?

3.12 • A rookie quarterback throws a football with an initial upward velocity component of 12.0 m/s and a horizontal velocity component of 20.0 m/s. Ignore air resistance. (a) How much time is required for the football to reach the highest point of the trajectory? (b) How high is this point? (c) How much time (after it is thrown) is required for the football to return to its original level? How does this compare with the time calculated in part (a)? (d) How far has the football traveled horizontally during this time? (e) Draw *x-t*, *y-t*, v_x-*t*, and v_y-*t* graphs for the motion.

3.13 •• **Leaping the River I.** A car traveling on a level horizontal road comes to a bridge during a storm and finds the bridge washed out. The driver must get to the other side, so he decides to try leaping it with his car. The side of the road the car is on is 21.3 m above the river, while the opposite side is a mere 1.8 m above the river. The river itself is a raging torrent 61.0 m wide. (a) How fast should the car be traveling at the time it leaves the road in order just to clear the river and land safely on the opposite side? (b) What is the speed of the car just before it lands on the other side?

3.14 • **BIO** **The Champion Jumper of the Insect World.** The froghopper, *Philaenus spumarius,* holds the world record for insect jumps. When leaping at an angle of 58.0° above the horizontal, some of the tiny critters have reached a maximum height of 58.7 cm above the level ground. (See *Nature,* Vol. 424, July 31, 2003, p. 509.) (a) What was the takeoff speed for such a leap? (b) What horizontal distance did the froghopper cover for this world-record leap?

3.15 •• Inside a starship at rest on the earth, a ball rolls off the top of a horizontal table and lands a distance D from the foot of the table. This starship now lands on the unexplored Planet X. The commander, Captain Curious, rolls the same ball off the same table with the same initial speed as on earth and finds that it lands a distance $2.76D$ from the foot of the table. What is the acceleration due to gravity on Planet X?

3.16 • On level ground a shell is fired with an initial velocity of 50.0 m/s at 60.0° above the horizontal and feels no appreciable air resistance. (a) Find the horizontal and vertical components of the shell's initial velocity. (b) How long does it take the shell to reach its highest point? (c) Find its maximum height above the ground. (d) How far from its firing point does the shell land? (e) At its highest point, find the horizontal and vertical components of its acceleration and velocity.

3.17 • A major leaguer hits a baseball so that it leaves the bat at a speed of 30.0 m/s and at an angle of 36.9° above the horizontal. You can ignore air resistance. (a) At what *two* times is the baseball at a height of 10.0 m above the point at which it left the bat? (b) Calculate the horizontal and vertical components of the baseball's velocity at each of the two times calculated in part (a). (c) What are the magnitude and direction of the baseball's velocity when it returns to the level at which it left the bat?

3.18 • A shot putter releases the shot some distance above the level ground with a velocity of 12.0 m/s, 51.0° above the horizontal. The shot hits the ground 2.08 s later. You can ignore air resistance. (a) What are the components of the shot's acceleration while in flight? (b) What are the components of the shot's velocity at the beginning and at the end of its trajectory? (c) How far did she throw the shot horizontally? (d) Why does the expression for R in Example 3.8 *not* give the correct answer for part (c)? (e) How high was the shot above the ground when she released it? (f) Draw x-t, y-t, v_x-t, and v_y-t graphs for the motion.

3.19 •• **Win the Prize.** In a carnival booth, you win a stuffed giraffe if you toss a quarter into a small dish. The dish is on a shelf above the point where the quarter leaves your hand and is a horizontal distance of 2.1 m from this point (Fig. E3.19). If you toss the coin with a velocity of 6.4 m/s at an angle of 60° above the horizontal, the coin lands in the dish. You can ignore air resistance. (a) What is the height of the shelf above the point where the

Figure **E3.19**

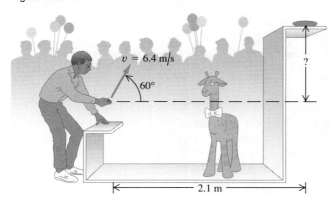

quarter leaves your hand? (b) What is the vertical component of the velocity of the quarter just before it lands in the dish?

3.20 •• Suppose the departure angle α_0 in Fig. 3.26 is 42.0° and the distance d is 3.00 m. Where will the dart and monkey meet if the initial speed of the dart is (a) 12.0 m/s? (b) 8.0 m/s? (c) What will happen if the initial speed of the dart is 4.0 m/s? Sketch the trajectory in each case.

3.21 •• A man stands on the roof of a 15.0-m-tall building and throws a rock with a velocity of magnitude 30.0 m/s at an angle of 33.0° above the horizontal. You can ignore air resistance. Calculate (a) the maximum height above the roof reached by the rock; (b) the magnitude of the velocity of the rock just before it strikes the ground; and (c) the horizontal range from the base of the building to the point where the rock strikes the ground. (d) Draw x-t, y-t, v_x-t, and v_y-t graphs for the motion.

3.22 • Firemen are shooting a stream of water at a burning building using a high-pressure hose that shoots out the water with a speed of 25.0 m/s as it leaves the end of the hose. Once it leaves the hose, the water moves in projectile motion. The firemen adjust the angle of elevation α of the hose until the water takes 3.00 s to reach a building 45.0 m away. You can ignore air resistance; assume that the end of the hose is at ground level. (a) Find the angle of elevation α. (b) Find the speed and acceleration of the water at the highest point in its trajectory. (c) How high above the ground does the water strike the building, and how fast is it moving just before it hits the building?

3.23 •• A 124-kg balloon carrying a 22-kg basket is descending with a constant downward velocity of 20.0 m/s. A 1.0-kg stone is thrown from the basket with an initial velocity of 15.0 m/s perpendicular to the path of the descending balloon, as measured relative to a person at rest in the basket. The person in the basket sees the stone hit the ground 6.00 s after being thrown. Assume that the balloon continues its downward descent with the same constant speed of 20.0 m/s. (a) How high was the balloon when the rock was thrown out? (b) How high is the balloon when the rock hits the ground? (c) At the instant the rock hits the ground, how far is it from the basket? (d) Just before the rock hits the ground, find its horizontal and vertical velocity components as measured by an observer (i) at rest in the basket and (ii) at rest on the ground.

Section 3.4 Motion in a Circle

3.24 •• BIO **Dizziness.** Our balance is maintained, at least in part, by the endolymph fluid in the inner ear. Spinning displaces this fluid, causing dizziness. Suppose a dancer (or skater) is spinning at a very fast 3.0 revolutions per second about a vertical axis through the center of his head. Although the distance varies from person to person, the inner ear is approximately 7.0 cm from the axis of spin. What is the radial acceleration (in m/s² and in g's) of the endolymph fluid?

3.25 •• The earth has a radius of 6380 km and turns around once on its axis in 24 h. (a) What is the radial acceleration of an object at the earth's equator? Give your answer in m/s² and as a fraction of g. (b) If a_{rad} at the equator is greater than g, objects will fly off the earth's surface and into space. (We will see the reason for this in Chapter 5.) What would the period of the earth's rotation have to be for this to occur?

3.26 •• A model of a helicopter rotor has four blades, each 3.40 m long from the central shaft to the blade tip. The model is rotated in a wind tunnel at 550 rev/min. (a) What is the linear speed of the blade tip, in m/s? (b) What is the radial acceleration of the blade tip expressed as a multiple of the acceleration of gravity, g?

3.27 • **BIO Pilot Blackout in a Power Dive.** A jet plane comes in for a downward dive as shown in Fig. E3.27. The bottom part of the path is a quarter circle with a radius of curvature of 350 m. According to medical tests, pilots lose consciousness at an acceleration of 5.5g. At what speed (in m/s and in mph) will the pilot black out for this dive?

Figure **E3.27**

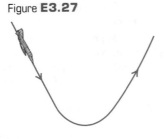

3.28 • The radius of the earth's orbit around the sun (assumed to be circular) is 1.50×10^8 km, and the earth travels around this orbit in 365 days. (a) What is the magnitude of the orbital velocity of the earth, in m/s? (b) What is the radial acceleration of the earth toward the sun, in m/s²? (c) Repeat parts (a) and (b) for the motion of the planet Mercury (orbit radius $= 5.79 \times 10^7$ km, orbital period $= 88.0$ days).

3.29 • A Ferris wheel with radius 14.0 m is turning about a horizontal axis through its center (Fig. E3.29). The linear speed of a passenger on the rim is constant and equal to 7.00 m/s. What are the magnitude and direction of the passenger's acceleration as she passes through (a) the lowest point in her circular motion? (b) The highest point in her circular motion? (c) How much time does it take the Ferris wheel to make one revolution?

Figure **E3.29**

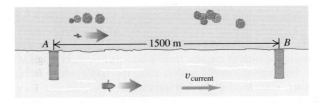

3.30 •• **BIO Hypergravity.** At its Ames Research Center, NASA uses its large "20-G" centrifuge to test the effects of very large accelerations ("hypergravity") on test pilots and astronauts. In this device, an arm 8.84 m long rotates about one end in a horizontal plane, and the astronaut is strapped in at the other end. Suppose that he is aligned along the arm with his head at the outermost end. The maximum sustained acceleration to which humans are subjected in this machine is typically 12.5g. (a) How fast must the astronaut's head be moving to experience this maximum acceleration? (b) What is the *difference* between the acceleration of his head and feet if the astronaut is 2.00 m tall? (c) How fast in rpm (rev/min) is the arm turning to produce the maximum sustained acceleration?

Section 3.5 Relative Velocity

3.31 • A "moving sidewalk" in an airport terminal building moves at 1.0 m/s and is 35.0 m long. If a woman steps on at one end and walks at 1.5 m/s relative to the moving sidewalk, how much time does she require to reach the opposite end if she walks (a) in the same direction the sidewalk is moving? (b) In the opposite direction?

3.32 • A railroad flatcar is traveling to the right at a speed of 13.0 m/s relative to an observer standing on the ground. Someone is riding a motor scooter on the flatcar (Fig. E3.32). What is the velocity (magnitude and direction) of the motor scooter relative to the flatcar if its velocity relative to the observer on the ground is (a) 18.0 m/s to the right? (b) 3.0 m/s to the left? (c) zero?

Figure **E3.32**

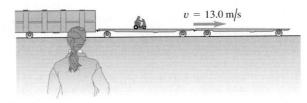

3.33 •• A canoe has a velocity of 0.40 m/s southeast relative to the earth. The canoe is on a river that is flowing 0.50 m/s east relative to the earth. Find the velocity (magnitude and direction) of the canoe relative to the river.

3.34 • Two piers, A and B, are located on a river: B is 1500 m downstream from A (Fig. E3.34). Two friends must make round trips from pier A to pier B and return. One rows a boat at a constant speed of 4.00 km/h relative to the water; the other walks on the shore at a constant speed of 4.00 km/h. The velocity of the river is 2.80 km/h in the direction from A to B. How much time does it take each person to make the round trip?

Figure **E3.34**

3.35 • **Crossing the River I.** A river flows due south with a speed of 2.0 m/s. A man steers a motorboat across the river; his velocity relative to the water is 4.2 m/s due east. The river is 800 m wide. (a) What is his velocity (magnitude and direction) relative to the earth? (b) How much time is required to cross the river? (c) How far south of his starting point will he reach the opposite bank?

3.36 • **Crossing the River II.** (a) In which direction should the motorboat in Exercise 3.35 head in order to reach a point on the opposite bank directly east from the starting point? (The boat's speed relative to the water remains 4.2 m/s.) (b) What is the velocity of the boat relative to the earth? (c) How much time is required to cross the river?

3.37 •• The nose of an ultralight plane is pointed south, and its airspeed indicator shows 35 m/s. The plane is in a 10-m/s wind blowing toward the southwest relative to the earth. (a) In a vector-addition diagram, show the relationship of $\vec{v}_{P/E}$ (the velocity of the plane relative to the earth) to the two given vectors. (b) Letting x be east and y be north, find the components of $\vec{v}_{P/E}$. (c) Find the magnitude and direction of $\vec{v}_{P/E}$.

3.38 •• An airplane pilot wishes to fly due west. A wind of 80.0 km/h (about 50 mi/h) is blowing toward the south. (a) If the airspeed of the plane (its speed in still air) is 320.0 km/h (about 200 mi/h), in which direction should the pilot head? (b) What is the speed of the plane over the ground? Illustrate with a vector diagram.

3.39 •• **BIO Bird Migration.** Canadian geese migrate essentially along a north–south direction for well over a thousand kilometers in some cases, traveling at speeds up to about 100 km/h. If one such bird is flying at 100 km/h relative to the air, but there is a

40 km/h wind blowing from west to east, (a) at what angle relative to the north–south direction should this bird head so that it will be traveling directly southward relative to the ground? (b) How long will it take the bird to cover a ground distance of 500 km from north to south? (*Note:* Even on cloudy nights, many birds can navigate using the earth's magnetic field to fix the north–south direction.)

PROBLEMS

3.40 •• An athlete starts at point *A* and runs at a constant speed of 6.0 m/s around a circular track 100 m in diameter, as shown in Fig. P3.40. Find the *x*- and *y*-components of this runner's average velocity and average acceleration between points (a) *A* and *B*, (b) *A* and *C*, (c) *C* and *D*, and (d) *A* and *A* (a full lap). (e) Calculate the magnitude of the runner's average velocity

Figure **P3.40**

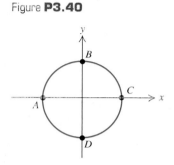

between *A* and *B*. Is his average speed equal to the magnitude of his average velocity? Why or why not? (f) How can his velocity be changing if he is running at constant speed?

3.41 • **CALC** A rocket is fired at an angle from the top of a tower of height $h_0 = 50.0$ m. Because of the design of the engines, its position coordinates are of the form $x(t) = A + Bt^2$ and $y(t) = C + Dt^3$, where A, B, C, and D are constants. Furthermore, the acceleration of the rocket 1.00 s after firing is $\vec{a} = (4.00\hat{i} + 3.00\hat{j})$ m/s^2. Take the origin of coordinates to be at the base of the tower. (a) Find the constants A, B, C, and D, including their SI units. (b) At the instant after the rocket is fired, what are its acceleration vector and its velocity? (c) What are the *x*- and *y*-components of the rocket's velocity 10.0 s after it is fired, and how fast is it moving? (d) What is the position vector of the rocket 10.0 s after it is fired?

3.42 ••• **CALC** A faulty model rocket moves in the *xy*-plane (the positive *y*-direction is vertically upward). The rocket's acceleration has components $a_x(t) = \alpha t^2$ and $a_y(t) = \beta - \gamma t$, where $\alpha = 2.50$ m/s^4, $\beta = 9.00$ m/s^2, and $\gamma = 1.40$ m/s^3. At $t = 0$ the rocket is at the origin and has velocity $\vec{v}_0 = v_{0x}\hat{i} + v_{0y}\hat{j}$ with $v_{0x} = 1.00$ m/s and $v_{0y} = 7.00$ m/s. (a) Calculate the velocity and position vectors as functions of time. (b) What is the maximum height reached by the rocket? (c) Sketch the path of the rocket. (d) What is the horizontal displacement of the rocket when it returns to $y = 0$?

3.43 •• **CALC** If $\vec{r} = bt^2\hat{i} + ct^3\hat{j}$, where b and c are positive constants, when does the velocity vector make an angle of 45.0° with the *x*- and *y*-axes?

3.44 •• **CALC** The position of a dragonfly that is flying parallel to the ground is given as a function of time by $\vec{r} = [2.90 \text{ m} + (0.0900 \text{ m/s}^2)t^2]\hat{i} - (0.0150 \text{ m/s}^3)t^3\hat{j}$. (a) At what value of *t* does the velocity vector of the insect make an angle of 30.0° clockwise from the +*x*-axis? (b) At the time calculated in part (a), what are the magnitude and direction of the acceleration vector of the insect?

3.45 •• **CP CALC** A small toy airplane is flying in the *xy*-plane parallel to the ground. In the time interval $t = 0$ to $t = 1.00$ s, its velocity as a function of time is given by $\vec{v} = (1.20 \text{ m/s}^2)t\hat{i} + [12.0 \text{ m/s} - (2.00 \text{ m/s}^2)t]\hat{j}$. At what

value of *t* is the velocity of the plane perpendicular to its acceleration?

3.46 •• **CALC** A bird flies in the *xy*-plane with a velocity vector given by $\vec{v} = (\alpha - \beta t^2)\hat{i} + \gamma t\hat{j}$, with $\alpha = 2.4$ m/s, $\beta = 1.6$ m/s^3, and $\gamma = 4.0$ m/s^2. The positive *y*-direction is vertically upward. At $t = 0$ the bird is at the origin. (a) Calculate the position and acceleration vectors of the bird as functions of time. (b) What is the bird's altitude (*y*-coordinate) as it flies over $x = 0$ for the first time after $t = 0$?

3.47 ••• **CP** A test rocket is launched by accelerating it along a 200.0-m incline at 1.25 m/s^2 starting from rest at point *A* (Fig. P3.47). The incline rises at 35.0° above the horizontal, and at the instant the rocket leaves it, its engines turn off and it is subject only to gravity (air resistance can be ignored). Find (a) the maximum height above the ground that the rocket reaches, and (b) the greatest horizontal range of the rocket beyond point *A*.

Figure **P3.47**

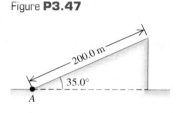

3.48 • **Martian Athletics.** In the long jump, an athlete launches herself at an angle above the ground and lands at the same height, trying to travel the greatest horizontal distance. Suppose that on earth she is in the air for time *T*, reaches a maximum height *h*, and achieves a horizontal distance *D*. If she jumped in *exactly* the same way during a competition on Mars, where g_{Mars} is 0.379 of its earth value, find her time in the air, maximum height, and horizontal distance. Express each of these three quantities in terms of its earth value. Air resistance can be neglected on both planets.

3.49 •• **Dynamite!** A demolition crew uses dynamite to blow an old building apart. Debris from the explosion flies off in all directions and is later found at distances as far as 50 m from the explosion. Estimate the maximum speed at which debris was blown outward by the explosion. Describe any assumptions that you make.

3.50 ••• **BIO Spiraling Up.** It is common to see birds of prey rising upward on thermals. The paths they take may be spiral-like. You can model the spiral motion as uniform circular motion combined with a constant upward velocity. Assume a bird completes a circle of radius 6.00 m every 5.00 s and rises vertically at a constant rate of 3.00 m/s. Determine: (a) the speed of the bird relative to the ground; (b) the bird's acceleration (magnitude and direction); and (c) the angle between the bird's velocity vector and the horizontal.

3.51 •• A jungle veterinarian with a blow-gun loaded with a tranquilizer dart and a sly 1.5-kg monkey are each 25 m above the ground in trees 70 m apart. Just as the hunter shoots horizontally at the monkey, the monkey drops from the tree in a vain attempt to escape being hit. What must the minimum muzzle velocity of the dart have been for the hunter to have hit the monkey before it reached the ground?

3.52 ••• A movie stuntwoman drops from a helicopter that is 30.0 m above the ground and moving with a constant velocity whose components are 10.0 m/s upward and 15.0 m/s horizontal and toward the south. You can ignore air resistance. (a) Where on the ground (relative to the position of the helicopter when she drops) should the stuntwoman have placed the foam mats that break her fall? (b) Draw *x-t*, *y-t*, v_x-*t*, and v_y-*t* graphs of her motion.

3.53 •• In fighting forest fires, airplanes work in support of ground crews by dropping water on the fires. A pilot is practicing by dropping a canister of red dye, hoping to hit a target on the ground below. If the plane is flying in a horizontal path 90.0 m above the ground and with a speed of 64.0 m/s (143 mi/h), at what horizontal distance from the target should the pilot release the canister? Ignore air resistance.

3.54 •• A cannon, located 60.0 m from the base of a vertical 25.0-m-tall cliff, shoots a 15-kg shell at 43.0° above the horizontal toward the cliff. (a) What must the minimum muzzle velocity be for the shell to clear the top of the cliff? (b) The ground at the top of the cliff is level, with a constant elevation of 25.0 m above the cannon. Under the conditions of part (a), how far does the shell land past the edge of the cliff?

3.55 •• An airplane is flying with a velocity of 90.0 m/s at an angle of 23.0° above the horizontal. When the plane is 114 m directly above a dog that is standing on level ground, a suitcase drops out of the luggage compartment. How far from the dog will the suitcase land? You can ignore air resistance.

3.56 ••• As a ship is approaching the dock at 45.0 cm/s, an important piece of landing equipment needs to be thrown to it before it can dock. This equipment is thrown at 15.0 m/s at 60.0° above the horizontal from the top of a tower at the edge of the water, 8.75 m above the ship's deck (Fig. P3.56). For this equipment to land at the front of the ship, at what distance D from the dock should the ship be when the equipment is thrown? Air resistance can be neglected.

Figure **P3.56**

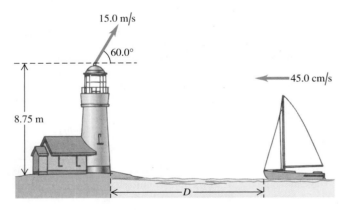

3.57 • **CP CALC** A toy rocket is launched with an initial velocity of 12.0 m/s in the horizontal direction from the roof of a 30.0-m-tall building. The rocket's engine produces a horizontal acceleration of $(1.60 \text{ m/s}^3)t$, in the same direction as the initial velocity, but in the vertical direction the acceleration is g, downward. Air resistance can be neglected. What horizontal distance does the rocket travel before reaching the ground?

3.58 •• **An Errand of Mercy.** An airplane is dropping bales of hay to cattle stranded in a blizzard on the Great Plains. The pilot releases the bales at 150 m above the level ground when the plane is flying at 75 m/s in a direction 55° above the horizontal. How far in front of the cattle should the pilot release the hay so that the bales land at the point where the cattle are stranded?

3.59 ••• **The Longest Home Run.** According to the *Guinness Book of World Records,* the longest home run ever measured was hit by Roy "Dizzy" Carlyle in a minor league game. The ball traveled 188 m (618 ft) before landing on the ground outside the ballpark. (a) Assuming the ball's initial velocity was in a direction 45° above the horizontal and ignoring air resistance, what did the initial speed of the ball need to be to produce such a home run if the ball was hit at a point 0.9 m (3.0 ft) above ground level? Assume that the ground was perfectly flat. (b) How far

would the ball be above a fence 3.0 m (10 ft) high if the fence was 116 m (380 ft) from home plate?

3.60 ••• A water hose is used to fill a large cylindrical storage tank of diameter D and height $2D$. The hose shoots the water at 45° above the horizontal from the same level as the base of the tank and is a distance 6D away (Fig. P3.60). For what *range* of launch speeds (v_0) will the water enter the tank? Ignore air resistance, and express your answer in terms of D and g.

Figure **P3.60**

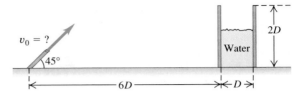

3.61 •• A projectile is being launched from ground level with no air resistance. You want to avoid having it enter a temperature inversion layer in the atmosphere a height h above the ground. (a) What is the maximum launch speed you could give this projectile if you shot it straight up? Express your answer in terms of h and g. (b) Suppose the launcher available shoots projectiles at twice the maximum launch speed you found in part (a). At what maximum angle above the horizontal should you launch the projectile? (c) How far (in terms of h) from the launcher does the projectile in part (b) land?

3.62 •• **Kicking a Field Goal.** In U.S. football, after a touchdown the team has the opportunity to earn one more point by kicking the ball over the bar between the goal posts. The bar is 10.0 ft above the ground, and the ball is kicked from ground level, 36.0 ft horizontally from the bar (Fig. P3.62). Football regulations are stated in English units, but convert them to SI units for this problem. (a) There is a minimum angle above the ground such that if the ball is launched below this angle, it can never clear the bar, no matter how fast it is kicked. What is this angle? (b) If the ball is kicked at 45.0° above the horizontal, what must its initial speed be if it is to just clear the bar? Express your answer in m/s and in km/h.

Figure **P3.62**

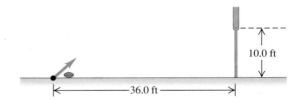

3.63 •• A grasshopper leaps into the air from the edge of a vertical cliff, as shown in Fig. P3.63. Use information from the figure to find (a) the initial speed of the grasshopper and (b) the height of the cliff.

3.64 •• **A World Record.** In the shot put, a standard track-and-field event, a 7.3-kg object (the shot) is thrown by releasing it at approximately 40° over a straight left leg. The world record for distance, set by Randy Barnes in 1990, is 23.11 m. Assuming that Barnes released the shot put at 40.0° from a height of 2.00 m above the ground, with what speed, in m/s and in mph, did he release it?

3.65 ••• **Look Out!** A snowball rolls off a barn roof that
slopes downward at an angle of
40° (Fig. P3.65). The edge of the
roof is 14.0 m above the ground,
and the snowball has a speed of
7.00 m/s as it rolls off the roof.
Ignore air resistance. (a) How far
from the edge of the barn does
the snowball strike the ground if
it doesn't strike anything else
while falling? (b) Draw x-t, y-t,
v_x-t, and v_y-t graphs for the
motion in part (a). (c) A man 1.9 m
tall is standing 4.0 m from the
edge of the barn. Will he be hit by the snowball?

Figure **P3.65**

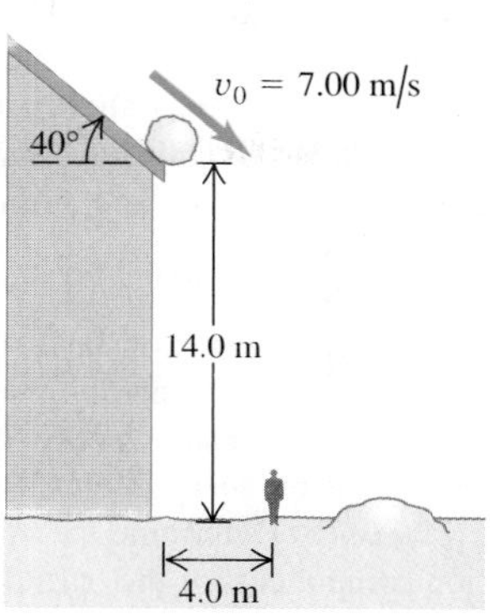

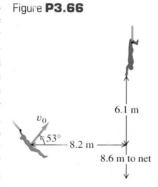

3.66 ••• **On the Flying Trapeze.**
A new circus act is called the
Texas Tumblers. Lovely Mary
Belle swings from a trapeze, projects herself at an angle of 53°, and
is supposed to be caught by Joe
Bob, whose hands are 6.1 m above
and 8.2 m horizontally from her
launch point (Fig. P3.66). You
can ignore air resistance. (a)
What initial speed v_0 must Mary
Belle have just to reach Joe Bob?
(b) For the initial speed calculated
in part (a), what are the magnitude

Figure **P3.66**

and direction of her velocity when Mary Belle reaches Joe Bob?
(c) Assuming that Mary Belle has the initial speed calculated in
part (a), draw x-t, y-t, v_x-t, and v_y-t graphs showing the motion of
both tumblers. Your graphs should show the motion up until the point
where Mary Belle reaches Joe Bob. (d) The night of their debut performance, Joe Bob misses her completely as she flies past. How far
horizontally does Mary Belle travel, from her initial launch point,
before landing in the safety net 8.6 m below her starting point?

3.67 •• **Leaping the River II.** A physics professor did daredevil
stunts in his spare time. His last stunt was an attempt to jump
across a river on a motorcycle (Fig. P3.67). The takeoff ramp was
inclined at 53.0°, the river was 40.0 m wide, and the far bank was
15.0 m lower than the top of the ramp. The river itself was 100 m
below the ramp. You can ignore air resistance. (a) What should his
speed have been at the top of the ramp to have just made it to the
edge of the far bank? (b) If his speed was only half the value found
in part (a), where did he land?

Figure **P3.67**

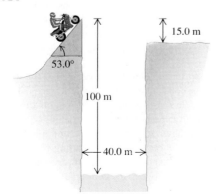

3.68 •• A rock is thrown from the roof of a building with a velocity v_0 at an angle of α_0 from the horizontal. The building has
height h. You can ignore air resistance. Calculate the magnitude of
the velocity of the rock just before it strikes the ground, and show
that this speed is independent of α_0.

3.69 • A 5500-kg cart carrying a vertical rocket launcher moves
to the right at a constant speed of 30.0 m/s along a horizontal
track. It launches a 45.0-kg rocket vertically upward with an initial
speed of 40.0 m/s relative to the cart. (a) How high will the rocket
go? (b) Where, relative to the cart, will the rocket land? (c) How
far does the cart move while the rocket is in the air? (d) At what
angle, relative to the horizontal, is the rocket traveling just as it
leaves the cart, as measured by an observer at rest on the ground?
(e) Sketch the rocket's trajectory as seen by an observer (i) stationary on the cart and (ii) stationary on the ground.

3.70 • A 2.7-kg ball is thrown upward with an initial speed of
20.0 m/s from the edge of a 45.0-m-high cliff. At the instant the
ball is thrown, a woman starts running away from the base of the
cliff with a constant speed of 6.00 m/s. The woman runs in a
straight line on level ground, and air resistance acting on the ball
can be ignored. (a) At what angle above the horizontal should the
ball be thrown so that the runner will catch it just before it hits
the ground, and how far does the woman run before she catches
the ball? (b) Carefully sketch the ball's trajectory as viewed by
(i) a person at rest on the ground and (ii) the runner.

3.71 • A 76.0-kg boulder is rolling horizontally at the top of a
vertical cliff that is 20 m above the surface of a lake, as shown in
Fig. P3.71. The top of the vertical face of a dam is located 100 m
from the foot of the cliff, with the top of the dam level with the surface of the water in the lake. A level plain is 25 m below the top of
the dam. (a) What must be the minimum speed of the rock just as it
leaves the cliff so it will travel to the plain without striking the
dam? (b) How far from the foot of the dam does the rock hit the
plain?

Figure **P3.71**

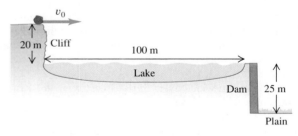

3.72 •• **Tossing Your Lunch.** Henrietta is going off to her
physics class, jogging down the sidewalk at 3.05 m/s. Her hus-
band Bruce suddenly realizes that she left in such a hurry that she
forgot her lunch of bagels, so he runs to the window of their apartment, which is 38.0 m above the street level and directly above the
sidewalk, to throw them to her. Bruce throws them horizontally
9.00 s after Henrietta has passed below the window, and she
catches them on the run. You can ignore air resistance. (a) With
what initial speed must Bruce throw the bagels so Henrietta can
catch them just before they hit the ground? (b) Where is Henrietta
when she catches the bagels?

3.73 ••• Two tanks are engaged in a training exercise on level
ground. The first tank fires a paint-filled training round with a
muzzle speed of 250 m/s at 10.0° above the horizontal while
advancing toward the second tank with a speed of 15.0 m/s relative to the ground. The second tank is retreating at 35.0 m/s relative to the ground, but is hit by the shell. You can ignore air

resistance and assume the shell hits at the same height above ground from which it was fired. Find the distance between the tanks (a) when the round was first fired and (b) at the time of impact.

3.74 ••• **CP Bang!** A student sits atop a platform a distance h above the ground. He throws a large firecracker horizontally with a speed v. However, a wind blowing parallel to the ground gives the firecracker a constant horizontal acceleration with magnitude a. This results in the firecracker reaching the ground directly under the student. Determine the height h in terms of v, a, and g. You can ignore the effect of air resistance on the vertical motion.

3.75 •• In a Fourth of July celebration, a firework is launched from ground level with an initial velocity of 25.0 m/s at 30.0° from the *vertical*. At its maximum height it explodes in a starburst into many fragments, two of which travel forward initially at 20.0 m/s at $\pm 53.0°$ with respect to the horizontal, both quantities measured *relative to the original firework just before it exploded*. With what angles with respect to the horizontal do the two fragments initially move right after the explosion, as measured by a spectator standing on the ground?

3.76 • When it is 145 m above the ground, a rocket traveling vertically upward at a constant 8.50 m/s relative to the ground launches a secondary rocket at a speed of 12.0 m/s at an angle of 53.0° above the horizontal, both quantities being measured by an astronaut sitting in the rocket. After it is launched the secondary rocket is in free-fall. (a) Just as the secondary rocket is launched, what are the horizontal and vertical components of its velocity relative to (i) the astronaut sitting in the rocket and (ii) Mission Control on the ground? (b) Find the initial speed and launch angle of the secondary rocket as measured by Mission Control. (c) What maximum height above the ground does the secondary rocket reach?

3.77 ••• In an action-adventure film, the hero is supposed to throw a grenade from his car, which is going 90.0 km/h, to his enemy's car, which is going 110 km/h. The enemy's car is 15.8 m in front of the hero's when he lets go of the grenade. If the hero throws the grenade so its initial velocity relative to him is at an angle of 45° above the horizontal, what should the magnitude of the initial velocity be? The cars are both traveling in the same direction on a level road. You can ignore air resistance. Find the magnitude of the velocity both relative to the hero and relative to the earth.

3.78 • A 400.0-m-wide river flows from west to east at 30.0 m/min. Your boat moves at 100.0 m/min relative to the water no matter which direction you point it. To cross this river, you start from a dock at point A on the south bank. There is a boat landing directly opposite at point B on the north bank, and also one at point C, 75.0 m downstream from B (Fig. P3.78). (a) Where on the north shore will you land if you point your boat perpendicular to the water current, and what distance will you have traveled? (b) If you initially aim your boat directly toward point C and do not change that bearing relative to the shore, where on the north shore will you

Figure **P3.78**

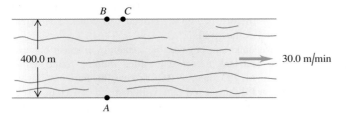

land? (c) To reach point C: (i) at what bearing must you aim your boat, (ii) how long will it take to cross the river, (iii) what distance do you travel, and (iv) and what is the speed of your boat as measured by an observer standing on the river bank?

3.79 • **CALC Cycloid.** A particle moves in the xy-plane. Its coordinates are given as functions of time by

$$x(t) = R(\omega t - \sin \omega t) \qquad y(t) = R(1 - \cos \omega t)$$

where R and ω are constants. (a) Sketch the trajectory of the particle. (This is the trajectory of a point on the rim of a wheel that is rolling at a constant speed on a horizontal surface. The curve traced out by such a point as it moves through space is called a *cycloid*.) (b) Determine the velocity components and the acceleration components of the particle at any time t. (c) At which times is the particle momentarily at rest? What are the coordinates of the particle at these times? What are the magnitude and direction of the acceleration at these times? (d) Does the magnitude of the acceleration depend on time? Compare to uniform circular motion.

3.80 •• A projectile is fired from point A at an angle above the horizontal. At its highest point, after having traveled a horizontal distance D from its launch point, it suddenly explodes into two identical fragments that travel horizontally with equal but opposite velocities as measured *relative to the projectile just before it exploded*. If one fragment lands back at point A, how far from A (in terms of D) does the other fragment land?

3.81 •• An airplane pilot sets a compass course due west and maintains an airspeed of 220 km/h. After flying for 0.500 h, she finds herself over a town 120 km west and 20 km south of her starting point. (a) Find the wind velocity (magnitude and direction). (b) If the wind velocity is 40 km/h due south, in what direction should the pilot set her course to travel due west? Use the same airspeed of 220 km/h.

3.82 •• **Raindrops.** When a train's velocity is 12.0 m/s eastward, raindrops that are falling vertically with respect to the earth make traces that are inclined 30.0° to the vertical on the windows of the train. (a) What is the horizontal component of a drop's velocity with respect to the earth? With respect to the train? (b) What is the magnitude of the velocity of the raindrop with respect to the earth? With respect to the train?

3.83 ••• In a World Cup soccer match, Juan is running due north toward the goal with a speed of 8.00 m/s relative to the ground. A teammate passes the ball to him. The ball has a speed of 12.0 m/s and is moving in a direction 37.0° east of north, relative to the ground. What are the magnitude and direction of the ball's velocity relative to Juan?

3.84 •• An elevator is moving upward at a constant speed of 2.50 m/s. A bolt in the elevator ceiling 3.00 m above the elevator floor works loose and falls. (a) How long does it take for the bolt to fall to the elevator floor? What is the speed of the bolt just as it hits the elevator floor (b) according to an observer in the elevator? (c) According to an observer standing on one of the floor landings of the building? (d) According to the observer in part (c), what distance did the bolt travel between the ceiling and the floor of the elevator?

3.85 • **CP** Suppose the elevator in Problem 3.84 starts from rest and maintains a constant upward acceleration of 4.00 m/s², and the bolt falls out the instant the elevator begins to move. (a) How long does it take for the bolt to reach the floor of the elevator? (b) Just as it reaches the floor, how fast is the bolt moving according to an observer (i) in the elevator? (ii) Standing on the floor landings of the building? (c) According to each observer in part (b), how far has the bolt traveled between the ceiling and floor of the elevator?

3.86 •• Two soccer players, Mia and Alice, are running as Alice passes the ball to Mia. Mia is running due north with a speed of 6.00 m/s. The velocity of the ball relative to Mia is 5.00 m/s in a direction 30.0° east of south. What are the magnitude and direction of the velocity of the ball relative to the ground?

3.87 ••• **Projectile Motion on an Incline.** Refer to the Bridging Problem in Chapter 3. (a) An archer on ground that has a constant upward slope of 30.0° aims at a target 60.0 m farther up the incline. The arrow in the bow and the bull's-eye at the center of the target are each 1.50 m above the ground. The initial velocity of the arrow just after it leaves the bow has magnitude 32.0 m/s. At what angle above the *horizontal* should the archer aim to hit the bull's-eye? If there are two such angles, calculate the smaller of the two. You might have to solve the equation for the angle by iteration—that is, by trial and error. How does the angle compare to that required when the ground is level, with 0 slope? (b) Repeat the problem for ground that has a constant *downward* slope of 30.0°.

CHALLENGE PROBLEMS

3.88 ••• CALC A projectile is thrown from a point P. It moves in such a way that its distance from P is always increasing. Find the maximum angle above the horizontal with which the projectile could have been thrown. You can ignore air resistance.

3.89 ••• Two students are canoeing on a river. While heading upstream, they accidentally drop an empty bottle overboard. They then continue paddling for 60 minutes, reaching a point 2.0 km farther upstream. At this point they realize that the bottle is missing and, driven by ecological awareness, they turn around and head downstream. They catch up with and retrieve the bottle (which has been moving along with the current) 5.0 km downstream from the turn-around point. (a) Assuming a constant paddling effort throughout, how fast is the river flowing? (b) What would the canoe speed in a still lake be for the same paddling effort?

3.90 ••• CP A rocket designed to place small payloads into orbit is carried to an altitude of 12.0 km above sea level by a converted airliner. When the airliner is flying in a straight line at a constant speed of 850 km/h, the rocket is dropped. After the drop, the airliner maintains the same altitude and speed and continues to fly in a straight line. The rocket falls for a brief time, after which its rocket motor turns on. Once its rocket motor is on, the combined effects of thrust and gravity give the rocket a constant acceleration of magnitude 3.00g directed at an angle of 30.0° above the horizontal. For reasons of safety, the rocket should be at least 1.00 km in front of the airliner when it climbs through the airliner's altitude. Your job is to determine the minimum time that the rocket must fall before its engine starts. You can ignore air resistance. Your answer should include (i) a diagram showing the flight paths of both the rocket and the airliner, labeled at several points with vectors for their velocities and accelerations; (ii) an *x-t* graph showing the motions of both the rocket and the airliner; and (iii) a *y-t* graph showing the motions of both the rocket and the airliner. In the diagram and the graphs, indicate when the rocket is dropped, when the rocket motor turns on, and when the rocket climbs through the altitude of the airliner.

Answers

Chapter Opening Question

A cyclist going around a curve at constant speed has an acceleration directed toward the inside of the curve (see Section 3.2, especially Fig. 3.12a).

Test Your Understanding Questions

3.1 Answer: (iii) If the instantaneous velocity $\vec{v}$ is constant over an interval, its value at any point (including the end of the interval) is the same as the average velocity $\vec{v}_{av}$ over the interval. In (i) and (ii) the direction of $\vec{v}$ at the end of the interval is tangent to the path at that point, while the direction of $\vec{v}_{av}$ points from the beginning of the path to its end (in the direction of the net displacement). In (iv) $\vec{v}$ and $\vec{v}_{av}$ are both directed along the straight line, but $\vec{v}$ has a greater magnitude because the speed has been increasing.

3.2 Answer: vector 7 At the high point of the sled's path, the speed is minimum. At that point the speed is neither increasing nor decreasing, and the parallel component of the acceleration (that is, the horizontal component) is zero. The acceleration has only a perpendicular component toward the inside of the sled's curved path. In other words, the acceleration is downward.

3.3 Answer: (i) If there were no gravity ($g = 0$), the monkey would not fall and the dart would follow a straight-line path (shown as a dashed line). The effect of gravity is to make the monkey and the dart both fall the same distance $\frac{1}{2}gt^2$ below their $g = 0$ positions. Point A is the same distance below the monkey's initial position as point P is below the dashed straight line, so point A is where we would find the monkey at the time in question.

3.4 Answer: (ii) At both the top and bottom of the loop, the acceleration is purely radial and is given by Eq. (3.28). The radius R is the same at both points, so the difference in acceleration is due purely to differences in speed. Since a_{rad} is proportional to the square of v, the speed must be twice as great at the bottom of the loop as at the top.

3.5 Answer: (vi) The effect of the wind is to cancel the airplane's eastward motion and give it a northward motion. So the velocity of the air relative to the ground (the wind velocity) must have one 150-km/h component to the west and one 150-km/h component to the north. The combination of these is a vector of magnitude $\sqrt{(150 \text{ km/h})^2 + (150 \text{ km/h})^2} = 212$ km/h that points to the northwest.

Bridging Problem

Answers: (a) $R = \dfrac{2v_0^2}{g} \dfrac{\cos(\theta + \phi)\sin\phi}{\cos^2\theta}$ (b) $\phi = 45° - \dfrac{\theta}{2}$

4 NEWTON'S LAWS OF MOTION

LEARNING GOALS

By studying this chapter, you will learn:

- What the concept of force means in physics, and why forces are vectors.

- The significance of the net force on an object, and what happens when the net force is zero.

- The relationship among the net force on an object, the object's mass, and its acceleration.

- How the forces that two bodies exert on each other are related.

? This pit crew member is pushing a race car forward. Is the race car pushing back on him? If so, does it push back with the same magnitude of force or a different amount?

We've seen in the last two chapters how to use the language and mathematics of *kinematics* to describe motion in one, two, or three dimensions. But what *causes* bodies to move the way that they do? For example, how can a tugboat push a cruise ship that's much heavier than the tug? Why is it harder to control a car on wet ice than on dry concrete? The answers to these and similar questions take us into the subject of **dynamics,** the relationship of motion to the forces that cause it.

In this chapter we will use two new concepts, *force* and *mass,* to analyze the principles of dynamics. These principles were clearly stated for the first time by Sir Isaac Newton (1642–1727); today we call them **Newton's laws of motion.** The first law states that when the net force on a body is zero, its motion doesn't change. The second law relates force to acceleration when the net force is *not* zero. The third law is a relationship between the forces that two interacting bodies exert on each other.

Newton did not *derive* the three laws of motion, but rather *deduced* them from a multitude of experiments performed by other scientists, especially Galileo Galilei (who died the same year Newton was born). These laws are truly fundamental, for they cannot be deduced or proved from other principles. Newton's laws are the foundation of **classical mechanics** (also called **Newtonian mechanics**); using them, we can understand most familiar kinds of motion. Newton's laws need modification only for situations involving extremely high speeds (near the speed of light) or very small sizes (such as within the atom).

Newton's laws are very simple to state, yet many students find these laws difficult to grasp and to work with. The reason is that before studying physics, you've spent years walking, throwing balls, pushing boxes, and doing dozens of things that involve motion. Along the way, you've developed a set of "common sense" ideas about motion and its causes. But many of these "common sense" ideas don't stand up to logical analysis. A big part of the job of this chapter—and of the rest of our study of physics—is helping you to recognize how "common sense" ideas can sometimes lead you astray, and how to adjust your understanding of the physical world to make it consistent with what experiments tell us.

4.1 Force and Interactions

In everyday language, a **force** is a push or a pull. A better definition is that a force is an *interaction* between two bodies or between a body and its environment (Fig. 4.1). That's why we always refer to the force that one body *exerts* on a second body. When you push on a car that is stuck in the snow, you exert a force on the car; a steel cable exerts a force on the beam it is hoisting at a construction site; and so on. As Fig. 4.1 shows, force is a *vector* quantity; you can push or pull a body in different directions.

When a force involves direct contact between two bodies, such as a push or pull that you exert on an object with your hand, we call it a **contact force.** Figures 4.2a, 4.2b, and 4.2c show three common types of contact forces. The **normal force** (Fig. 4.2a) is exerted on an object by any surface with which it is in contact. The adjective *normal* means that the force always acts perpendicular to the surface of contact, no matter what the angle of that surface. By contrast, the **friction force** (Fig. 4.2b) exerted on an object by a surface acts *parallel* to the surface, in the direction that opposes sliding. The pulling force exerted by a stretched rope or cord on an object to which it's attached is called a **tension force** (Fig. 4.2c). When you tug on your dog's leash, the force that pulls on her collar is a tension force.

In addition to contact forces, there are **long-range forces** that act even when the bodies are separated by empty space. The force between two magnets is an example of a long-range force, as is the force of gravity (Fig. 4.2d); the earth pulls a dropped object toward it even though there is no direct contact between the object and the earth. The gravitational force that the earth exerts on your body is called your **weight.**

To describe a force vector $\vec{F}$, we need to describe the *direction* in which it acts as well as its *magnitude,* the quantity that describes "how much" or "how hard" the force pushes or pulls. The SI unit of the magnitude of force is the *newton,* abbreviated N. (We'll give a precise definition of the newton in Section 4.3.) Table 4.1 lists some typical force magnitudes.

Table 4.1 Typical Force Magnitudes

Sun's gravitational force on the earth	3.5×10^{22} N
Thrust of a space shuttle during launch	3.1×10^{7} N
Weight of a large blue whale	1.9×10^{6} N
Maximum pulling force of a locomotive	8.9×10^{5} N
Weight of a 250-lb linebacker	1.1×10^{3} N
Weight of a medium apple	1 N
Weight of smallest insect eggs	2×10^{-6} N
Electric attraction between the proton and the electron in a hydrogen atom	8.2×10^{-8} N
Weight of a very small bacterium	1×10^{-18} N
Weight of a hydrogen atom	1.6×10^{-26} N
Weight of an electron	8.9×10^{-30} N
Gravitational attraction between the proton and the electron in a hydrogen atom	3.6×10^{-47} N

4.1 Some properties of forces.

- A force is a push or a pull.
- A force is an interaction between two objects or between an object and its environment.
- A force is a vector quantity, with magnitude and direction.

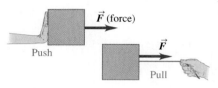

$\vec{F}$ (force)

Push

$\vec{F}$

Pull

4.2 Four common types of forces.

(a) **Normal force** $\vec{n}$: When an object rests or pushes on a surface, the surface exerts a push on it that is directed perpendicular to the surface.

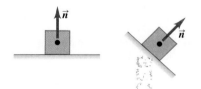

$\vec{n}$

$\vec{n}$

(b) **Friction force** $\vec{f}$: In addition to the normal force, a surface may exert a frictional force on an object, directed parallel to the surface.

$\vec{n}$

$\vec{f}$

(c) **Tension force** $\vec{T}$: A pulling force exerted on an object by a rope, cord, etc.

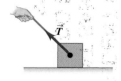

$\vec{T}$

(d) **Weight** $\vec{w}$: The pull of gravity on an object is a long-range force (a force that acts over a distance).

$\vec{w}$

4.3 Using a vector arrow to denote the force that we exert when **(a)** pulling a block with a string or **(b)** pushing a block with a stick.

(a) A 10-N pull directed 30° above the horizontal

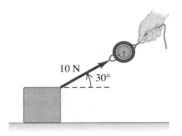

10 N
30°

(b) A 10-N push directed 45° below the horizontal

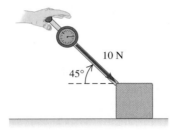

10 N
45°

4.4 Superposition of forces.

Two forces $\vec{F}_1$ and $\vec{F}_2$ acting on a body at point O have the same effect as a single force $\vec{R}$ equal to their vector sum.

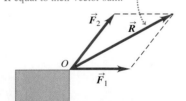

$\vec{F}_2$
$\vec{R}$
O
$\vec{F}_1$

A common instrument for measuring force magnitudes is the *spring balance.* It consists of a coil spring enclosed in a case with a pointer attached to one end. When forces are applied to the ends of the spring, it stretches by an amount that depends on the force. We can make a scale for the pointer by using a number of identical bodies with weights of exactly 1 N each. When one, two, or more of these are suspended simultaneously from the balance, the total force stretching the spring is 1 N, 2 N, and so on, and we can label the corresponding positions of the pointer 1 N, 2 N, and so on. Then we can use this instrument to measure the magnitude of an unknown force. We can also make a similar instrument that measures pushes instead of pulls.

Figure 4.3 shows a spring balance being used to measure a pull or push that we apply to a box. In each case we draw a vector to represent the applied force. The length of the vector shows the magnitude; the longer the vector, the greater the force magnitude.

Superposition of Forces

When you throw a ball, there are at least two forces acting on it: the push of your hand and the downward pull of gravity. Experiment shows that when two forces $\vec{F}_1$ and $\vec{F}_2$ act at the same time at the same point on a body (Fig. 4.4), the effect on the body's motion is the same as if a single force $\vec{R}$ were acting equal to the vector sum of the original forces: $\vec{R} = \vec{F}_1 + \vec{F}_2$. More generally, *any number of forces applied at a point on a body have the same effect as a single force equal to the vector sum of the forces.* This important principle is called **superposition of forces.**

The principle of superposition of forces is of the utmost importance, and we will use it throughout our study of physics. For example, in Fig. 4.5a, force $\vec{F}$ acts on a body at point O. The component vectors of $\vec{F}$ in the directions Ox and Oy are $\vec{F}_x$ and $\vec{F}_y$. When $\vec{F}_x$ and $\vec{F}_y$ are applied simultaneously, as in Fig. 4.5b, the effect is exactly the same as the effect of the original force $\vec{F}$. Hence *any force can be replaced by its component vectors, acting at the same point.*

It's frequently more convenient to describe a force $\vec{F}$ in terms of its x- and y-components F_x and F_y rather than by its component vectors (recall from Section 1.8 that *component vectors* are vectors, but *components* are just numbers). For the case shown in Fig. 4.5, both F_x and F_y are positive; for other orientations of the force $\vec{F}$, either F_x or F_y may be negative or zero.

Our coordinate axes don't have to be vertical and horizontal. Figure 4.6 shows a crate being pulled up a ramp by a force $\vec{F}$, represented by its components F_x and F_y parallel and perpendicular to the sloping surface of the ramp.

4.5 The force $\vec{F}$, which acts at an angle θ from the x-axis, may be replaced by its rectangular component vectors $\vec{F}_x$ and $\vec{F}_y$.

(a) Component vectors: $\vec{F}_x$ and $\vec{F}_y$
Components: $F_x = F \cos\theta$ and $F_y = F \sin\theta$

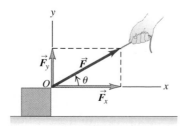

y
$\vec{F}_y$
$\vec{F}$
θ
O
x
$\vec{F}_x$

(b) Component vectors $\vec{F}_x$ and $\vec{F}_y$ together have the same effect as original force $\vec{F}$.

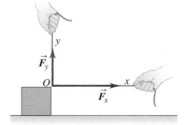

y
$\vec{F}_y$
O
x
$\vec{F}_x$

CAUTION **Using a wiggly line in force diagrams** In Fig. 4.6 we draw a wiggly line through the force vector $\vec{F}$ to show that we have replaced it by its x- and y-components. Otherwise, the diagram would include the same force twice. We will draw such a wiggly line in any force diagram where a force is replaced by its components. Look for this wiggly line in other figures in this and subsequent chapters.

We will often need to find the vector sum (resultant) of *all* the forces acting on a body. We call this the **net force** acting on the body. We will use the Greek letter Σ (capital sigma, equivalent to the Roman S) as a shorthand notation for a sum. If the forces are labeled $\vec{F}_1$, $\vec{F}_2$, $\vec{F}_3$, and so on, we abbreviate the sum as

$$\vec{R} = \vec{F}_1 + \vec{F}_2 + \vec{F}_3 + \cdots = \sum \vec{F} \tag{4.1}$$

We read $\sum \vec{F}$ as "the vector sum of the forces" or "the net force." The component version of Eq. (4.1) is the pair of component equations

$$R_x = \sum F_x \qquad R_y = \sum F_y \tag{4.2}$$

Here $\sum F_x$ is the sum of the x-components and $\sum F_y$ is the sum of the y-components (Fig. 4.7). Each component may be positive or negative, so be careful with signs when you evaluate these sums. (You may want to review Section 1.8.)

Once we have R_x and R_y we can find the magnitude and direction of the net force $\vec{R} = \sum \vec{F}$ acting on the body. The magnitude is

$$R = \sqrt{R_x^2 + R_y^2}$$

and the angle θ between $\vec{R}$ and the $+x$-axis can be found from the relationship $\tan \theta = R_y/R_x$. The components R_x and R_y may be positive, negative, or zero, and the angle θ may be in any of the four quadrants.

In three-dimensional problems, forces may also have z-components; then we add the equation $R_z = \sum F_z$ to Eq. (4.2). The magnitude of the net force is then

$$R = \sqrt{R_x^2 + R_y^2 + R_z^2}$$

4.6 F_x and F_y are the components of $\vec{F}$ parallel and perpendicular to the sloping surface of the inclined plane.

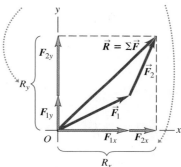

4.7 Finding the components of the vector sum (resultant) $\vec{R}$ of two forces $\vec{F}_1$ and $\vec{F}_2$.

Example 4.1 **Superposition of forces**

Three professional wrestlers are fighting over a champion's belt. Figure 4.8a shows the horizontal force each wrestler applies to the belt, as viewed from above. The forces have magnitudes $F_1 = 250$ N, $F_2 = 50$ N, and $F_3 = 120$ N. Find the x- and y-components of the net force on the belt, and find its magnitude and direction.

4.8 (a) Three forces acting on a belt. (b) The net force $\vec{R} = \sum \vec{F}$ and its components.

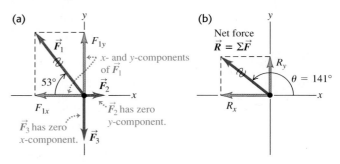

SOLUTION

IDENTIFY and SET UP: This is a problem in vector addition in which the vectors happen to represent forces. We want to find the x- and y-components of the net force $\vec{R}$, so we'll use the component method of vector addition expressed by Eqs. (4.2). Once we know the components of $\vec{R}$, we can find its magnitude and direction.

EXECUTE: From Fig. 4.8a the angles between the three forces $\vec{F}_1$, $\vec{F}_2$, and $\vec{F}_3$ and the $+x$-axis are $\theta_1 = 180° - 53° = 127°$, $\theta_2 = 0°$, and $\theta_3 = 270°$. The x- and y-components of the three forces are

$$F_{1x} = (250 \text{ N}) \cos 127° = -150 \text{ N}$$
$$F_{1y} = (250 \text{ N}) \sin 127° = 200 \text{ N}$$
$$F_{2x} = (50 \text{ N}) \cos 0° = 50 \text{ N}$$

$$F_{2y} = (50 \text{ N}) \sin 0° = 0 \text{ N}$$
$$F_{3x} = (120 \text{ N}) \cos 270° = 0 \text{ N}$$
$$F_{3y} = (120 \text{ N}) \sin 270° = -120 \text{ N}$$

From Eqs. (4.2) the net force $\vec{R} = \sum \vec{F}$ has components

$$R_x = F_{1x} + F_{2x} + F_{3x} = (-150 \text{ N}) + 50 \text{ N} + 0 \text{ N} = -100 \text{ N}$$
$$R_y = F_{1y} + F_{2y} + F_{3y} = 200 \text{ N} + 0 \text{ N} + (-120 \text{ N}) = 80 \text{ N}$$

Continued

The net force has a negative x-component and a positive y-component, as shown in Fig. 4.8b.

The magnitude of $\vec{R}$ is

$$R = \sqrt{R_x^2 + R_y^2} = \sqrt{(-100 \text{ N})^2 + (80 \text{ N})^2} = 128 \text{ N}$$

To find the angle between the net force and the $+x$-axis, we use Eq. (1.8):

$$\theta = \arctan \frac{R_y}{R_x} = \arctan\left(\frac{80 \text{ N}}{-100 \text{ N}}\right) = \arctan(-0.80)$$

The arctangent of -0.80 is $-39°$, but Fig. 4.8b shows that the net force lies in the second quadrant. Hence the correct solution is $\theta = -39° + 180° = 141°$.

EVALUATE: The net force is *not* zero. Your intuition should suggest that wrestler 1 (who exerts the largest force on the belt, $F_1 = 250$ N) will walk away with it when the struggle ends.

You should check the direction of $\vec{R}$ by adding the vectors $\vec{F}_1$, $\vec{F}_2$, and $\vec{F}_3$ graphically. Does your drawing show that $\vec{R} = \vec{F}_1 + \vec{F}_2 + \vec{F}_3$ points in the second quadrant as we found above?

4.9 The slicker the surface, the farther a puck slides after being given an initial velocity. On an air-hockey table **(c)** the friction force is practically zero, so the puck continues with almost constant velocity.

(a) Table: puck stops short.

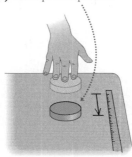

(b) Ice: puck slides farther.

(c) Air-hockey table: puck slides even farther.

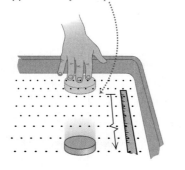

Test Your Understanding of Section 4.1 Figure 4.6 shows a force $\vec{F}$ acting on a crate. With the x- and y-axes shown in the figure, which statement about the components of the *gravitational* force that the earth exerts on the crate (the crate's weight) is *correct*? (i) The x- and y-components are both positive. (ii) The x-component is zero and the y-component is positive. (iii) The x-component is negative and the y-component is positive. (iv) The x- and y-components are both negative. (v) The x-component is zero and the y-component is negative. (vi) The x-component is positive and the y-component is negative.

4.2 Newton's First Law

How do the forces that act on a body affect its motion? To begin to answer this question, let's first consider what happens when the net force on a body is *zero*. You would almost certainly agree that if a body is at rest, and if no net force acts on it (that is, no net push or pull), that body will remain at rest. But what if there is zero net force acting on a body in *motion*?

To see what happens in this case, suppose you slide a hockey puck along a horizontal tabletop, applying a horizontal force to it with your hand (Fig. 4.9a). After you stop pushing, the puck *does not* continue to move indefinitely; it slows down and stops. To keep it moving, you have to keep pushing (that is, applying a force). You might come to the "common sense" conclusion that bodies in motion naturally come to rest and that a force is required to sustain motion.

But now imagine pushing the puck across a smooth surface of ice (Fig. 4.9b). After you quit pushing, the puck will slide a lot farther before it stops. Put it on an air-hockey table, where it floats on a thin cushion of air, and it moves still farther (Fig. 4.9c). In each case, what slows the puck down is *friction,* an interaction between the lower surface of the puck and the surface on which it slides. Each surface exerts a frictional force on the puck that resists the puck's motion; the difference in the three cases is the magnitude of the frictional force. The ice exerts less friction than the tabletop, so the puck travels farther. The gas molecules of the air-hockey table exert the least friction of all. If we could eliminate friction completely, the puck would never slow down, and we would need no force at all to keep the puck moving once it had been started. Thus the "common sense" idea that a force is required to sustain motion is *incorrect*.

Experiments like the ones we've just described show that when *no* net force acts on a body, the body either remains at rest *or* moves with constant velocity in a straight line. Once a body has been set in motion, no net force is needed to keep it moving. We call this observation *Newton's first law of motion:*

> **Newton's first law of motion:** A body acted on by no net force moves with constant velocity (which may be zero) and zero acceleration.

The tendency of a body to keep moving once it is set in motion results from a property called **inertia.** You use inertia when you try to get ketchup out of a bottle by shaking it. First you start the bottle (and the ketchup inside) moving forward; when you jerk the bottle back, the ketchup tends to keep moving forward and, you hope, ends up on your burger. The tendency of a body at rest to remain at rest is also due to inertia. You may have seen a tablecloth yanked out from under the china without breaking anything. The force on the china isn't great enough to make it move appreciably during the short time it takes to pull the tablecloth away.

It's important to note that the *net* force is what matters in Newton's first law. For example, a physics book at rest on a horizontal tabletop has two forces acting on it: an upward supporting force, or normal force, exerted by the tabletop (see Fig. 4.2a) and the downward force of the earth's gravitational attraction (a long-range force that acts even if the tabletop is elevated above the ground; see Fig. 4.2d). The upward push of the surface is just as great as the downward pull of gravity, so the *net* force acting on the book (that is, the vector sum of the two forces) is zero. In agreement with Newton's first law, if the book is at rest on the tabletop, it remains at rest. The same principle applies to a hockey puck sliding on a horizontal, frictionless surface: The vector sum of the upward push of the surface and the downward pull of gravity is zero. Once the puck is in motion, it continues to move with constant velocity because the *net* force acting on it is zero.

Here's another example. Suppose a hockey puck rests on a horizontal surface with negligible friction, such as an air-hockey table or a slab of wet ice. If the puck is initially at rest and a single horizontal force $\vec{F}_1$ acts on it (Fig. 4.10a), the puck starts to move. If the puck is in motion to begin with, the force changes its speed, its direction, or both, depending on the direction of the force. In this case the net force is equal to $\vec{F}_1$, which is *not* zero. (There are also two vertical forces: the earth's gravitational attraction and the upward normal force exerted by the surface. But as we mentioned earlier, these two forces cancel.)

Now suppose we apply a second force $\vec{F}_2$ (Fig. 4.10b), equal in magnitude to $\vec{F}_1$ but opposite in direction. The two forces are negatives of each other, $\vec{F}_2 = -\vec{F}_1$, and their vector sum is zero:

$$\sum \vec{F} = \vec{F}_1 + \vec{F}_2 = \vec{F}_1 + (-\vec{F}_1) = 0$$

Again, we find that if the body is at rest at the start, it remains at rest; if it is initially moving, it continues to move in the same direction with constant speed. These results show that in Newton's first law, *zero net force is equivalent to no force at all.* This is just the principle of superposition of forces that we saw in Section 4.1.

When a body is either at rest or moving with constant velocity (in a straight line with constant speed), we say that the body is in **equilibrium.** For a body to be in equilibrium, it must be acted on by no forces, or by several forces such that their vector sum—that is, the net force—is zero:

$$\sum \vec{F} = 0 \qquad \text{(body in equilibrium)} \qquad [4.3]$$

For this to be true, each component of the net force must be zero, so

$$\sum F_x = 0 \qquad \sum F_y = 0 \qquad \text{(body in equilibrium)} \qquad [4.4]$$

We are assuming that the body can be represented adequately as a point particle. When the body has finite size, we also have to consider *where* on the body the forces are applied. We will return to this point in Chapter 11.

4.10 (a) A hockey puck accelerates in the direction of a net applied force $\vec{F}_1$. (b) When the net force is zero, the acceleration is zero, and the puck is in equilibrium.

(a) A puck on a frictionless surface accelerates when acted on by a single horizontal force.

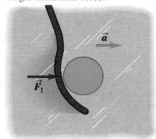

(b) An object acted on by forces whose vector sum is zero behaves as though no forces act on it.

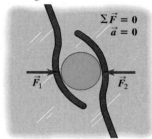

Application Sledding with Newton's First Law
The downward force of gravity acting on the child and sled is balanced by an upward normal force exerted by the ground. The adult's foot exerts a forward force that balances the backward force of friction on the sled. Hence there is no net force on the child and sled, and they slide with a constant velocity.

Conceptual Example 4.2 | **Zero net force means constant velocity**

In the classic 1950 science fiction film *Rocketship X-M*, a spaceship is moving in the vacuum of outer space, far from any star or planet, when its engine dies. As a result, the spaceship slows down and stops. What does Newton's first law say about this scene?

SOLUTION

After the engine dies there are no forces acting on the spaceship, so according to Newton's first law it will *not* stop but will continue to move in a straight line with constant speed. Some science fiction movies are based on accurate science; this is not one of them.

Conceptual Example 4.3 | **Constant velocity means zero net force**

You are driving a Maserati GranTurismo S on a straight testing track at a constant speed of 250 km/h. You pass a 1971 Volkswagen Beetle doing a constant 75 km/h. On which car is the net force greater?

SOLUTION

The key word in this question is "net." Both cars are in equilibrium because their velocities are constant; Newton's first law therefore says that the *net* force on each car is *zero*.

This seems to contradict the "common sense" idea that the faster car must have a greater force pushing it. Thanks to your

Maserati's high-power engine, it's true that the track exerts a greater forward force on your Maserati than it does on the Volkswagen. But a *backward* force also acts on each car due to road friction and air resistance. When the car is traveling with constant velocity, the vector sum of the forward and backward forces is zero. There is more air resistance on the fast-moving Maserati than on the slow-moving Volkswagen, which is why the Maserati's engine must be more powerful than that of the Volkswagen.

Inertial Frames of Reference

In discussing relative velocity in Section 3.5, we introduced the concept of *frame of reference*. This concept is central to Newton's laws of motion. Suppose you are in a bus that is traveling on a straight road and speeding up. If you could stand in the aisle on roller skates, you would start moving *backward* relative to the bus as the bus gains speed. If instead the bus was slowing to a stop, you would start moving forward down the aisle. In either case, it looks as though Newton's first law is not obeyed; there is no net force acting on you, yet your velocity changes. What's wrong?

The point is that the bus is accelerating with respect to the earth and is *not* a suitable frame of reference for Newton's first law. This law is valid in some frames of reference and not valid in others. A frame of reference in which Newton's first law *is* valid is called an **inertial frame of reference.** The earth is at least approximately an inertial frame of reference, but the bus is not. (The earth is not a completely inertial frame, owing to the acceleration associated with its rotation and its motion around the sun. These effects are quite small, however; see Exercises 3.25 and 3.28.) Because Newton's first law is used to define what we mean by an inertial frame of reference, it is sometimes called the *law of inertia.*

Figure 4.11 helps us understand what you experience when riding in a vehicle that's accelerating. In Fig. 4.11a, a vehicle is initially at rest and then begins to accelerate to the right. A passenger on roller skates (which nearly eliminate the effects of friction) has virtually no net force acting on her, so she tends to remain at rest relative to the inertial frame of the earth. As the vehicle accelerates around her, she moves backward relative to the vehicle. In the same way, a passenger in a vehicle that is slowing down tends to continue moving with constant velocity relative to the earth, and so moves forward relative to the vehicle (Fig. 4.11b). A vehicle is also accelerating if it moves at a constant speed but is turning (Fig. 4.11c). In this case a passenger tends to continue moving relative to

4.11 Riding in an accelerating vehicle.

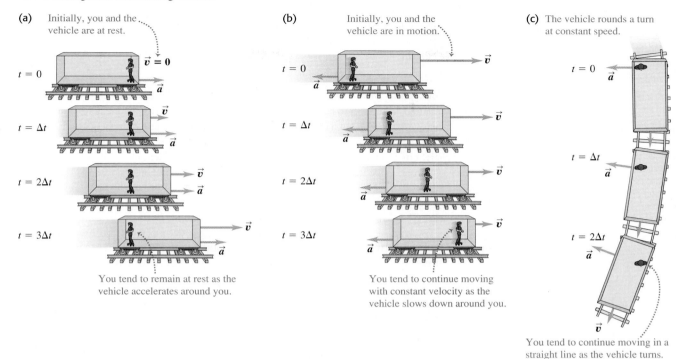

(a) Initially, you and the vehicle are at rest.

$t = 0$ $\vec{v} = 0$ $\vec{a}$

$t = \Delta t$ $\vec{v}$ $\vec{a}$

$t = 2\Delta t$ $\vec{v}$ $\vec{a}$

$t = 3\Delta t$ $\vec{v}$ $\vec{a}$

You tend to remain at rest as the vehicle accelerates around you.

(b) Initially, you and the vehicle are in motion.

$t = 0$ $\vec{v}$ $\vec{a}$

$t = \Delta t$ $\vec{v}$ $\vec{a}$

$t = 2\Delta t$ $\vec{v}$ $\vec{a}$

$t = 3\Delta t$ $\vec{v}$ $\vec{a}$

You tend to continue moving with constant velocity as the vehicle slows down around you.

(c) The vehicle rounds a turn at constant speed.

$t = 0$ $\vec{a}$

$t = \Delta t$ $\vec{a}$

$t = 2\Delta t$ $\vec{a}$

$\vec{v}$

You tend to continue moving in a straight line as the vehicle turns.

the earth at constant speed in a straight line; relative to the vehicle, the passenger moves to the side of the vehicle on the outside of the turn.

In each case shown in Fig. 4.11, an observer in the vehicle's frame of reference might be tempted to conclude that there *is* a net force acting on the passenger, since the passenger's velocity *relative to the vehicle* changes in each case. This conclusion is simply wrong; the net force on the passenger is indeed zero. The vehicle observer's mistake is in trying to apply Newton's first law in the vehicle's frame of reference, which is *not* an inertial frame and in which Newton's first law isn't valid (Fig. 4.12). In this book we will use *only* inertial frames of reference.

We've mentioned only one (approximately) inertial frame of reference: the earth's surface. But there are many inertial frames. If we have an inertial frame of reference A, in which Newton's first law is obeyed, then *any* second frame of reference B will also be inertial if it moves relative to A with constant velocity $\vec{v}_{B/A}$. We can prove this using the relative-velocity relationship Eq. (3.36) from Section 3.5:

$$\vec{v}_{P/A} = \vec{v}_{P/B} + \vec{v}_{B/A}$$

Suppose that P is a body that moves with constant velocity $\vec{v}_{P/A}$ with respect to an inertial frame A. By Newton's first law the net force on this body is zero. The velocity of P relative to another frame B has a different value, $\vec{v}_{P/B} = \vec{v}_{P/A} - \vec{v}_{B/A}$. But if the relative velocity $\vec{v}_{B/A}$ of the two frames is constant, then $\vec{v}_{P/B}$ is constant as well. Thus B is also an inertial frame; the velocity of P in this frame is constant, and the net force on P is zero, so Newton's first law is obeyed in B. Observers in frames A and B will disagree about the velocity of P, but they will agree that P has a constant velocity (zero acceleration) and has zero net force acting on it.

4.12 From the frame of reference of the car, it seems as though a force is pushing the crash test dummies forward as the car comes to a sudden stop. But there is really no such force: As the car stops, the dummies keep moving forward as a consequence of Newton's first law.

There is no single inertial frame of reference that is preferred over all others for formulating Newton's laws. If one frame is inertial, then every other frame moving relative to it with constant velocity is also inertial. Viewed in this light, the state of rest and the state of motion with constant velocity are not very different; both occur when the vector sum of forces acting on the body is zero.

Test Your Understanding of Section 4.2 In which of the following situations is there zero net force on the body? (i) an airplane flying due north at a steady 120 m/s and at a constant altitude; (ii) a car driving straight up a hill with a 3° slope at a constant 90 km/h; (iii) a hawk circling at a constant 20 km/h at a constant height of 15 m above an open field; (iv) a box with slick, frictionless surfaces in the back of a truck as the truck accelerates forward on a level road at 5 m/s². ∎

4.3 Newton's Second Law

Newton's first law tells us that when a body is acted on by zero net force, it moves with constant velocity and zero acceleration. In Fig. 4.13a, a hockey puck is sliding to the right on wet ice. There is negligible friction, so there are no horizontal forces acting on the puck; the downward force of gravity and the upward normal force exerted by the ice surface sum to zero. So the net force $\Sigma \vec{F}$ acting on the puck is zero, the puck has zero acceleration, and its velocity is constant.

But what happens when the net force is *not* zero? In Fig. 4.13b we apply a constant horizontal force to a sliding puck in the same direction that the puck is moving. Then $\Sigma \vec{F}$ is constant and in the same horizontal direction as $\vec{v}$. We find that during the time the force is acting, the velocity of the puck changes at a constant rate;

4.13 Exploring the relationship between the acceleration of a body and the net force acting on the body (in this case, a hockey puck on a frictionless surface).

(a) A puck moving with constant velocity (in equilibrium): $\Sigma \vec{F} = 0$, $\vec{a} = 0$

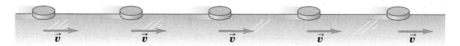

(b) A constant net force in the direction of motion causes a constant acceleration in the same direction as the net force.

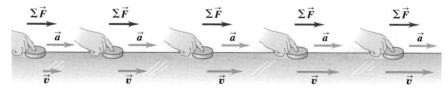

(c) A constant net force opposite the direction of motion causes a constant acceleration in the same direction as the net force.

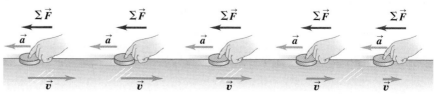

that is, the puck moves with constant acceleration. The speed of the puck increases, so the acceleration $\vec{a}$ is in the same direction as $\vec{v}$ and $\Sigma\vec{F}$.

In Fig. 4.13c we reverse the direction of the force on the puck so that $\Sigma\vec{F}$ acts opposite to $\vec{v}$. In this case as well, the puck has an acceleration; the puck moves more and more slowly to the right. The acceleration $\vec{a}$ in this case is to the left, in the same direction as $\Sigma\vec{F}$. As in the previous case, experiment shows that the acceleration is constant if $\Sigma\vec{F}$ is constant.

We conclude that *a net force acting on a body causes the body to accelerate in the same direction as the net force.* If the magnitude of the net force is constant, as in Figs. 4.13b and 4.13c, then so is the magnitude of the acceleration.

These conclusions about net force and acceleration also apply to a body moving along a curved path. For example, Fig. 4.14 shows a hockey puck moving in a horizontal circle on an ice surface of negligible friction. A rope is attached to the puck and to a stick in the ice, and this rope exerts an inward tension force of constant magnitude on the puck. The net force and acceleration are both constant in magnitude and directed toward the center of the circle. The speed of the puck is constant, so this is uniform circular motion, as discussed in Section 3.4.

Figure 4.15a shows another experiment to explore the relationship between acceleration and net force. We apply a constant horizontal force to a puck on a frictionless horizontal surface, using the spring balance described in Section 4.1 with the spring stretched a constant amount. As in Figs. 4.13b and 4.13c, this horizontal force equals the net force on the puck. If we change the magnitude of the net force, the acceleration changes in the same proportion. Doubling the net force doubles the acceleration (Fig. 4.15b), halving the net force halves the acceleration (Fig. 4.15c), and so on. Many such experiments show that *for any given body, the magnitude of the acceleration is directly proportional to the magnitude of the net force acting on the body.*

Mass and Force

Our results mean that for a given body, the *ratio* of the magnitude $|\Sigma\vec{F}|$ of the net force to the magnitude $a = |\vec{a}|$ of the acceleration is constant, regardless of the magnitude of the net force. We call this ratio the *inertial mass,* or simply the **mass,** of the body and denote it by m. That is,

$$m = \frac{|\Sigma\vec{F}|}{a} \quad \text{or} \quad |\Sigma\vec{F}| = ma \quad \text{or} \quad a = \frac{|\Sigma\vec{F}|}{m} \quad (4.5)$$

Mass is a quantitative measure of inertia, which we discussed in Section 4.2. The last of the equations in Eqs. (4.5) says that the greater its mass, the more a body "resists" being accelerated. When you hold a piece of fruit in your hand at the supermarket and move it slightly up and down to estimate its heft, you're applying a force and seeing how much the fruit accelerates up and down in response. If a force causes a large acceleration, the fruit has a small mass; if the same force causes only a small acceleration, the fruit has a large mass. In the same way, if you hit a table-tennis ball and then a basketball with the same force, the basketball has much smaller acceleration because it has much greater mass.

The SI unit of mass is the **kilogram.** We mentioned in Section 1.3 that the kilogram is officially defined to be the mass of a cylinder of platinum–iridium alloy kept in a vault near Paris. We can use this standard kilogram, along with Eqs. (4.5), to define the **newton:**

One newton is the amount of net force that gives an acceleration of 1 meter per second squared to a body with a mass of 1 kilogram.

4.14 A top view of a hockey puck in uniform circular motion on a frictionless horizontal surface.

Puck moves at constant speed around circle.

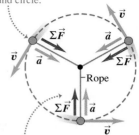

At all points, the acceleration $\vec{a}$ and the net force $\Sigma\vec{F}$ point in the same direction—always toward the center of the circle.

4.15 For a body of a given mass m, the magnitude of the body's acceleration is directly proportional to the magnitude of the net force acting on the body.

(a) A constant net force $\Sigma\vec{F}$ causes a constant acceleration $\vec{a}$.

(b) Doubling the net force doubles the acceleration.

(c) Halving the force halves the acceleration.

This definition allows us to calibrate the spring balances and other instruments used to measure forces. Because of the way we have defined the newton, it is related to the units of mass, length, and time. For Eqs. (4.5) to be dimensionally consistent, it must be true that

$$1 \text{ newton} = (1 \text{ kilogram})(1 \text{ meter per second squared})$$

or

$$1 \text{ N} = 1 \text{ kg} \cdot \text{m/s}^2$$

We will use this relationship many times in the next few chapters, so keep it in mind.

We can also use Eqs. (4.5) to compare a mass with the standard mass and thus to *measure* masses. Suppose we apply a constant net force $\Sigma \vec{F}$ to a body having a known mass m_1 and we find an acceleration of magnitude a_1 (Fig. 4.16a). We then apply the same force to another body having an unknown mass m_2, and we find an acceleration of magnitude a_2 (Fig. 4.16b). Then, according to Eqs. (4.5),

$$m_1 a_1 = m_2 a_2$$

$$\frac{m_2}{m_1} = \frac{a_1}{a_2} \quad \text{(same net force)} \quad (4.6)$$

For the same net force, the ratio of the masses of two bodies is the inverse of the ratio of their accelerations. In principle we could use Eq. (4.6) to measure an unknown mass m_2, but it is usually easier to determine mass indirectly by measuring the body's *weight*. We'll return to this point in Section 4.4.

When two bodies with masses m_1 and m_2 are fastened together, we find that the mass of the composite body is always $m_1 + m_2$ (Fig. 4.16c). This additive property of mass may seem obvious, but it has to be verified experimentally. Ultimately, the mass of a body is related to the number of protons, electrons, and neutrons it contains. This wouldn't be a good way to *define* mass because there is no practical way to count these particles. But the concept of mass is the most fundamental way to characterize the quantity of matter in a body.

4.16 For a given net force $\Sigma \vec{F}$ acting on a body, the acceleration is inversely proportional to the mass of the body. Masses add like ordinary scalars.

(a) A known force $\Sigma \vec{F}$ causes an object with mass m_1 to have an acceleration $\vec{a}_1$.

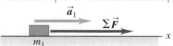

(b) Applying the same force $\Sigma \vec{F}$ to a second object and noting the acceleration allow us to measure the mass.

(c) When the two objects are fastened together, the same method shows that their composite mass is the sum of their individual masses.

Stating Newton's Second Law

We've been careful to state that the *net* force on a body is what causes that body to accelerate. Experiment shows that if a combination of forces $\vec{F}_1$, $\vec{F}_2$, $\vec{F}_3$, and so on is applied to a body, the body will have the same acceleration (magnitude and direction) as when only a single force is applied, if that single force is equal to the vector sum $\vec{F}_1 + \vec{F}_2 + \vec{F}_3 + \cdots$. In other words, the principle of superposition of forces (see Fig. 4.4) also holds true when the net force is not zero and the body is accelerating.

Equations (4.5) relate the magnitude of the net force on a body to the magnitude of the acceleration that it produces. We have also seen that the direction of the net force is the same as the direction of the acceleration, whether the body's path is straight or curved. Newton wrapped up all these relationships and experimental results in a single concise statement that we now call *Newton's second law of motion:*

Newton's second law of motion: **If a net external force acts on a body, the body accelerates. The direction of acceleration is the same as the direction of the net force. The mass of the body times the acceleration of the body equals the net force vector.**

In symbols,

$$\sum \vec{F} = m\vec{a} \quad \text{(Newton's second law of motion)} \quad \text{(4.7)}$$

An alternative statement is that the acceleration of a body is in the same direction as the net force acting on the body, and is equal to the net force divided by the body's mass:

$$\vec{a} = \frac{\sum \vec{F}}{m}$$

Newton's second law is a fundamental law of nature, the basic relationship between force and motion. Most of the remainder of this chapter and all of the next are devoted to learning how to apply this principle in various situations.

Equation (4.7) has many practical applications (Fig. 4.17). You've actually been using it all your life to measure your body's acceleration. In your inner ear, microscopic hair cells sense the magnitude and direction of the force that they must exert to cause small membranes to accelerate along with the rest of your body. By Newton's second law, the acceleration of the membranes—and hence that of your body as a whole—is proportional to this force and has the same direction. In this way, you can sense the magnitude and direction of your acceleration even with your eyes closed!

Using Newton's Second Law

There are at least four aspects of Newton's second law that deserve special attention. First, Eq. (4.7) is a *vector* equation. Usually we will use it in component form, with a separate equation for each component of force and the corresponding component of acceleration:

$$\sum F_x = ma_x \qquad \sum F_y = ma_y \qquad \sum F_z = ma_z \qquad \begin{array}{l}\text{(Newton's second}\\ \text{law of motion)}\end{array} \quad \text{(4.8)}$$

This set of component equations is equivalent to the single vector equation (4.7). Each component of the net force equals the mass times the corresponding component of acceleration.

Second, the statement of Newton's second law refers to *external* forces. By this we mean forces exerted on the body by other bodies in its environment. It's impossible for a body to affect its own motion by exerting a force on itself; if it were possible, you could lift yourself to the ceiling by pulling up on your belt! That's why only external forces are included in the sum $\sum \vec{F}$ in Eqs. (4.7) and (4.8).

Third, Eqs. (4.7) and (4.8) are valid only when the mass m is *constant*. It's easy to think of systems whose masses change, such as a leaking tank truck, a rocket ship, or a moving railroad car being loaded with coal. But such systems are better handled by using the concept of momentum; we'll get to that in Chapter 8.

Finally, Newton's second law is valid only in inertial frames of reference, just like the first law. Thus it is not valid in the reference frame of any of the accelerating vehicles in Fig. 4.11; relative to any of these frames, the passenger accelerates even though the net force on the passenger is zero. We will usually assume that the earth is an adequate approximation to an inertial frame, although because of its rotation and orbital motion it is not precisely inertial.

CAUTION $m\vec{a}$ **is not a force** You must keep in mind that even though the vector $m\vec{a}$ is equal to the vector sum $\sum \vec{F}$ of all the forces acting on the body, the vector $m\vec{a}$ is *not* a force. Acceleration is a *result* of a nonzero net force; it is not a force itself. It's "common sense" to think that there is a "force of acceleration" that pushes you back into your seat

4.17 The design of high-performance motorcycles depends fundamentally on Newton's second law. To maximize the forward acceleration, the designer makes the motorcycle as light as possible (that is, minimizes the mass) and uses the most powerful engine possible (thus maximizing the forward force).

Lightweight body (small m)

Powerful engine (large F)

Application Blame Newton's Second Law
This car stopped because of Newton's second law: The tree exerted an external force on the car, giving the car an acceleration that changed its velocity to zero.

Mastering**PHYSICS**

ActivPhysics 2.1.3: Tension Change
ActivPhysics 2.1.4: Sliding on an Incline

when your car accelerates forward from rest. But *there is no such force;* instead, your inertia causes you to tend to stay at rest relative to the earth, and the car accelerates around you (see Fig. 4.11a). The "common sense" confusion arises from trying to apply Newton's second law where it isn't valid, in the noninertial reference frame of an accelerating car. We will always examine motion relative to *inertial* frames of reference only.

In learning how to use Newton's second law, we will begin in this chapter with examples of straight-line motion. Then in Chapter 5 we will consider more general cases and develop more detailed problem-solving strategies.

Example 4.4 **Determining acceleration from force**

A worker applies a constant horizontal force with magnitude 20 N to a box with mass 40 kg resting on a level floor with negligible friction. What is the acceleration of the box?

SOLUTION

IDENTIFY and SET UP: This problem involves force and acceleration, so we'll use Newton's second law. In *any* problem involving forces, the first steps are to choose a coordinate system and to identify all of the forces acting on the body in question. It's usually convenient to take one axis either along or opposite the direction of the body's acceleration, which in this case is horizontal. Hence we take the $+x$-axis to be in the direction of the applied horizontal force (that is, the direction in which the box accelerates) and the $+y$-axis to be upward (Fig. 4.18). In most force problems that you'll encounter (including this one), the force vectors all lie in a plane, so the z-axis isn't used.

The forces acting on the box are (i) the horizontal force $\vec{F}$ exerted by the worker, of magnitude 20 N; (ii) the weight $\vec{w}$ of the box—that is, the downward gravitational force exerted by the earth; and (iii) the upward supporting force $\vec{n}$ exerted by the floor. As in Section 4.2, we call $\vec{n}$ a *normal* force because it is normal (perpendicular) to the surface of contact. (We use an italic letter n to avoid confusion with the abbreviation N for newton.) Friction is negligible, so no friction force is present.

The box doesn't move vertically, so the y-acceleration is zero: $a_y = 0$. Our target variable is the x-acceleration, a_x. We'll find it using Newton's second law in component form, Eqs. (4.8).

EXECUTE: From Fig. 4.18 only the 20-N force exerted by the worker has a nonzero x-component. Hence the first of Eqs. (4.8) tells us that

$$\sum F_x = F = 20 \text{ N} = ma_x$$

4.18 Our sketch for this problem. The tiles under the box are freshly waxed, so we assume that friction is negligible.

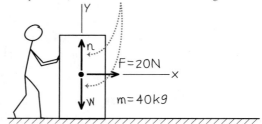

The box has no vertical acceleration, so the vertical components of the net force sum to zero. Nevertheless, for completeness, we show the vertical forces acting on the box.

The x-component of acceleration is therefore

$$a_x = \frac{\sum F_x}{m} = \frac{20 \text{ N}}{40 \text{ kg}} = \frac{20 \text{ kg} \cdot \text{m/s}^2}{40 \text{ kg}} = 0.50 \text{ m/s}^2$$

EVALUATE: The acceleration is in the $+x$-direction, the same direction as the net force. The net force is constant, so the acceleration is also constant. If we know the initial position and velocity of the box, we can find its position and velocity at any later time from the constant-acceleration equations of Chapter 2.

To determine a_x, we didn't need the y-component of Newton's second law from Eqs. (4.8), $\sum F_y = ma_y$. Can you use this equation to show that the magnitude n of the normal force in this situation is equal to the weight of the box?

Example 4.5 **Determining force from acceleration**

A waitress shoves a ketchup bottle with mass 0.45 kg to her right along a smooth, level lunch counter. The bottle leaves her hand moving at 2.8 m/s, then slows down as it slides because of a constant horizontal friction force exerted on it by the countertop. It slides for 1.0 m before coming to rest. What are the magnitude and direction of the friction force acting on the bottle?

SOLUTION

IDENTIFY and SET UP: This problem involves forces and acceleration (the slowing of the ketchup bottle), so we'll use Newton's second law to solve it. As in Example 4.4, we choose a coordinate system and identify the forces acting on the bottle (Fig. 4.19). We choose the $+x$-axis to be in the direction that the bottle slides, and

4.19 Our sketch for this problem.

We draw one diagram for the bottle's motion and one showing the forces on the bottle.

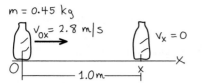

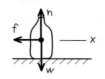

take the origin to be where the bottle leaves the waitress's hand. The friction force $\vec{f}$ slows the bottle down, so its direction must be opposite the direction of the bottle's velocity (see Fig. 4.13c).

Our target variable is the magnitude f of the friction force. We'll find it using the x-component of Newton's second law from Eqs. (4.8). We aren't told the x-component of the bottle's acceleration, a_x, but we know that it's constant because the friction force that causes the acceleration is constant. Hence we can calculate a_x using a constant-acceleration formula from Section 2.4. We know the bottle's initial and final x-coordinates ($x_0 = 0$ and $x = 1.0$ m) and its initial and final x-velocity ($v_{0x} = 2.8$ m/s and $v_x = 0$), so the easiest equation to use is Eq. (2.13), $v_x^2 = v_{0x}^2 + 2a_x(x - x_0)$.

EXECUTE: We solve Eq. (2.13) for a_x:

$$a_x = \frac{v_x^2 - v_{0x}^2}{2(x - x_0)} = \frac{(0 \text{ m/s})^2 - (2.8 \text{ m/s})^2}{2(1.0 \text{ m} - 0 \text{ m})} = -3.9 \text{ m/s}^2$$

The negative sign means that the bottle's acceleration is toward the *left* in Fig. 4.19, opposite to its velocity; this is as it must be, because the bottle is slowing down. The net force in the x-direction is the x-component $-f$ of the friction force, so

$$\sum F_x = -f = ma_x = (0.45 \text{ kg})(-3.9 \text{ m/s}^2)$$
$$= -1.8 \text{ kg} \cdot \text{m/s}^2 = -1.8 \text{ N}$$

The negative sign shows that the net force on the bottle is toward the left. The *magnitude* of the friction force is $f = 1.8$ N.

EVALUATE: As a check on the result, try repeating the calculation with the $+x$-axis to the *left* in Fig. 4.19. You'll find that $\sum F_x$ is equal to $+f = +1.8$ N (because the friction force is now in the $+x$-direction), and again you'll find $f = 1.8$ N. The answers for the *magnitudes* of forces don't depend on the choice of coordinate axes!

Some Notes on Units

A few words about units are in order. In the cgs metric system (not used in this book), the unit of mass is the gram, equal to 10^{-3} kg, and the unit of distance is the centimeter, equal to 10^{-2} m. The cgs unit of force is called the *dyne:*

$$1 \text{ dyne} = 1 \text{ g} \cdot \text{cm/s}^2 = 10^{-5} \text{ N}$$

In the British system, the unit of force is the *pound* (or pound-force) and the unit of mass is the *slug* (Fig. 4.20). The unit of acceleration is 1 foot per second squared, so

$$1 \text{ pound} = 1 \text{ slug} \cdot \text{ft/s}^2$$

The official definition of the pound is

$$1 \text{ pound} = 4.448221615260 \text{ newtons}$$

It is handy to remember that a pound is about 4.4 N and a newton is about 0.22 pound. Another useful fact: A body with a mass of 1 kg has a weight of about 2.2 lb at the earth's surface.

Table 4.2 lists the units of force, mass, and acceleration in the three systems.

Test Your Understanding of Section 4.3 Rank the following situations in order of the magnitude of the object's acceleration, from lowest to highest. Are there any cases that have the same magnitude of acceleration? (i) a 2.0-kg object acted on by a 2.0-N net force; (ii) a 2.0-kg object acted on by an 8.0-N net force; (iii) an 8.0-kg object acted on by a 2.0-N net force; (iv) an 8.0-kg object acted on by a 8.0-N net force. ∎

4.4 Mass and Weight

One of the most familiar forces is the *weight* of a body, which is the gravitational force that the earth exerts on the body. (If you are on another planet, your weight is the gravitational force that planet exerts on you.) Unfortunately, the terms *mass* and *weight* are often misused and interchanged in everyday conversation. It is absolutely essential for you to understand clearly the distinctions between these two physical quantities.

4.20 Despite its name, the English unit of mass has nothing to do with the type of slug shown here. A common garden slug has a mass of about 15 grams, or about 10^{-3} slug.

Table 4.2 Units of Force, Mass, and Acceleration

System of Units	Force	Mass	Acceleration
SI	newton (N)	kilogram (kg)	m/s²
cgs	dyne (dyn)	gram (g)	cm/s²
British	pound (lb)	slug	ft/s²

Mass characterizes the *inertial* properties of a body. Mass is what keeps the china on the table when you yank the tablecloth out from under it. The greater the mass, the greater the force needed to cause a given acceleration; this is reflected in Newton's second law, $\sum \vec{F} = m\vec{a}$.

Weight, on the other hand, is a *force* exerted on a body by the pull of the earth. Mass and weight are related: Bodies having large mass also have large weight. A large stone is hard to throw because of its large *mass,* and hard to lift off the ground because of its large *weight.*

To understand the relationship between mass and weight, note that a freely falling body has an acceleration of magnitude g. Newton's second law tells us that a force must act to produce this acceleration. If a 1-kg body falls with an acceleration of 9.8 m/s² the required force has magnitude

$$F = ma = (1 \text{ kg})(9.8 \text{ m/s}^2) = 9.8 \text{ kg} \cdot \text{m/s}^2 = 9.8 \text{ N}$$

The force that makes the body accelerate downward is its weight. Any body near the surface of the earth that has a mass of 1 kg *must* have a weight of 9.8 N to give it the acceleration we observe when it is in free fall. More generally, a body with mass m must have weight with magnitude w given by

$$w = mg \qquad \text{(magnitude of the weight of a body of mass } m) \qquad (4.9)$$

4.21 The relationship of mass and weight.

Falling body, mass m Hanging body, mass m

$\vec{T}$

$\vec{a} = \vec{g}$ $\vec{a} = 0$

Weight
$\vec{w} = m\vec{g}$ $\sum\vec{F} = \vec{w}$ Weight
$\vec{w} = m\vec{g}$ $\sum\vec{F} = 0$

• The relationship of mass to weight: $\vec{w} = m\vec{g}$.
• This relationship is the same whether a body is falling or stationary.

Hence the magnitude w of a body's weight is directly proportional to its mass m. The weight of a body is a force, a vector quantity, and we can write Eq. (4.9) as a vector equation (Fig. 4.21):

$$\vec{w} = m\vec{g} \qquad (4.10)$$

Remember that g is the *magnitude* of $\vec{g}$, the acceleration due to gravity, so g is always a positive number, by definition. Thus w, given by Eq. (4.9), is the *magnitude* of the weight and is also always positive.

CAUTION **A body's weight acts at all times** It is important to understand that the weight of a body acts on the body *all the time,* whether it is in free fall or not. If we suspend an object from a rope, it is in equilibrium, and its acceleration is zero. But its weight, given by Eq. (4.10), is still pulling down on it (Fig. 4.21). In this case the rope pulls up on the object, applying an upward force. The *vector sum* of the forces is zero, but the weight still acts. ▮

Conceptual Example 4.6   **Net force and acceleration in free fall**

In Example 2.6, a one-euro coin was dropped from rest from the Leaning Tower of Pisa. If the coin falls freely, so that the effects of the air are negligible, how does the net force on the coin vary as it falls?

SOLUTION

In free fall, the acceleration $\vec{a}$ of the coin is constant and equal to $\vec{g}$. Hence by Newton's second law the net force $\sum\vec{F} = m\vec{a}$ is also constant and equal to $m\vec{g}$, which is the coin's weight $\vec{w}$ (Fig. 4.22). The coin's velocity changes as it falls, but the net force acting on it is constant. (If this surprises you, reread Conceptual Example 4.3.)

The net force on a freely falling coin is constant even if you initially toss it upward. The force that your hand exerts on the coin to toss it is a contact force, and it disappears the instant the coin

leaves your hand. From then on, the only force acting on the coin is its weight $\vec{w}$.

4.22 The acceleration of a freely falling object is constant, and so is the net force acting on the object.

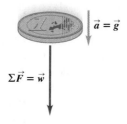

$\vec{a} = \vec{g}$

$\sum\vec{F} = \vec{w}$

Variation of g with Location

We will use $g = 9.80 \text{ m/s}^2$ for problems set on the earth (or, if the other data in the problem are given to only two significant figures, $g = 9.8 \text{ m/s}^2$). In fact, the value of g varies somewhat from point to point on the earth's surface—from about 9.78 to 9.82 m/s^2—because the earth is not perfectly spherical and because of effects due to its rotation and orbital motion. At a point where $g = 9.80 \text{ m/s}^2$, the weight of a standard kilogram is $w = 9.80 \text{ N}$. At a different point, where $g = 9.78 \text{ m/s}^2$, the weight is $w = 9.78 \text{ N}$ but the mass is still 1 kg. The weight of a body varies from one location to another; the mass does not.

If we take a standard kilogram to the surface of the moon, where the acceleration of free fall (equal to the value of g at the moon's surface) is 1.62 m/s^2, its weight is 1.62 N, but its mass is still 1 kg (Fig. 4.23). An 80.0-kg astronaut has a weight on earth of $(80.0 \text{ kg})(9.80 \text{ m/s}^2) = 784 \text{ N}$, but on the moon the astronaut's weight would be only $(80.0 \text{ kg})(1.62 \text{ m/s}^2) = 130 \text{ N}$. In Chapter 13 we'll see how to calculate the value of g at the surface of the moon or on other worlds.

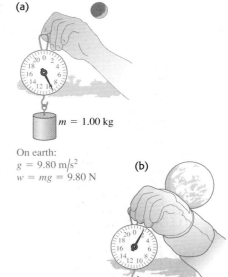

4.23 The weight of a 1-kilogram mass (a) on earth and (b) on the moon.

(a)

$m = 1.00 \text{ kg}$

On earth:
$g = 9.80 \text{ m/s}^2$
$w = mg = 9.80 \text{ N}$

(b)

On the moon:
$g = 1.62 \text{ m/s}^2$
$w = mg = 1.62 \text{ N}$
$m = 1.00 \text{ kg}$

Measuring Mass and Weight

In Section 4.3 we described a way to compare masses by comparing their accelerations when they are subjected to the same net force. Usually, however, the easiest way to measure the mass of a body is to measure its weight, often by comparing with a standard. Equation (4.9) says that two bodies that have the same weight at a particular location also have the same mass. We can compare weights very precisely; the familiar equal-arm balance (Fig. 4.24) can determine with great precision (up to 1 part in 10^6) when the weights of two bodies are equal and hence when their masses are equal.

The concept of mass plays two rather different roles in mechanics. The weight of a body (the gravitational force acting on it) is proportional to its mass; we call the property related to gravitational interactions *gravitational mass*. On the other hand, we call the inertial property that appears in Newton's second law the *inertial mass*. If these two quantities were different, the acceleration due to gravity might well be different for different bodies. However, extraordinarily precise experiments have established that in fact the two *are* the same to a precision of better than one part in 10^{12}.

CAUTION **Don't confuse mass and weight** The SI units for mass and weight are often misused in everyday life. Incorrect expressions such as "This box weighs 6 kg" are nearly universal. What is meant is that the *mass* of the box, probably determined indirectly by *weighing,* is 6 kg. Be careful to avoid this sloppy usage in your own work! In SI units, weight (a force) is measured in newtons, while mass is measured in kilograms.

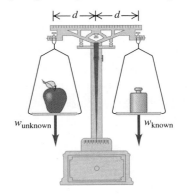

4.24 An equal-arm balance determines the mass of a body (such as an apple) by comparing its weight to a known weight.

w_{unknown}

w_{known}

Example 4.7 **Mass and weight**

A 2.49×10^4 N Rolls-Royce Phantom traveling in the $+x$-direction makes an emergency stop; the x-component of the net force acting on it is -1.83×10^4 N. What is its acceleration?

SOLUTION

IDENTIFY and SET UP: Our target variable is the x-component of the car's acceleration, a_x. We use the x-component portion of Newton's second law, Eqs. (4.8), to relate force and acceleration. To do this, we need to know the car's mass. The newton is a unit for

force, however, so 2.49×10^4 N is the car's *weight,* not its mass. Hence we'll first use Eq. (4.9) to determine the car's mass from its weight. The car has a positive x-velocity and is slowing down, so its x-acceleration will be negative.

EXECUTE: The mass of the car is

$$m = \frac{w}{g} = \frac{2.49 \times 10^4 \text{ N}}{9.80 \text{ m/s}^2} = \frac{2.49 \times 10^4 \text{ kg} \cdot \text{m/s}^2}{9.80 \text{ m/s}^2}$$
$$= 2540 \text{ kg}$$

Continued

Then $\sum F_x = ma_x$ gives

$$a_x = \frac{\sum F_x}{m} = \frac{-1.83 \times 10^4 \text{ N}}{2540 \text{ kg}} = \frac{-1.83 \times 10^4 \text{ kg} \cdot \text{m/s}^2}{2540 \text{ kg}}$$

$$= -7.20 \text{ m/s}^2$$

EVALUATE: The negative sign means that the acceleration vector points in the negative x-direction, as we expected. The magnitude of this acceleration is pretty high; passengers in this car will experience a lot of rearward force from their shoulder belts.

The acceleration is also equal to $-0.735g$. The number -0.735 is also the ratio of -1.83×10^4 N (the x-component of the net force) to 2.49×10^4 N (the weight). In fact, the acceleration of a body, expressed as a multiple of g, is *always* equal to the ratio of the net force on the body to its weight. Can you see why?

Test Your Understanding of Section 4.4 Suppose an astronaut landed on a planet where $g = 19.6 \text{ m/s}^2$. Compared to earth, would it be easier, harder, or just as easy for her to walk around? Would it be easier, harder, or just as easy for her to catch a ball that is moving horizontally at 12 m/s? (Assume that the astronaut's spacesuit is a lightweight model that doesn't impede her movements in any way.)

4.5 Newton's Third Law

A force acting on a body is always the result of its interaction with another body, so forces always come in pairs. You can't pull on a doorknob without the doorknob pulling back on you. When you kick a football, the forward force that your foot exerts on the ball launches it into its trajectory, but you also feel the force the ball exerts back on your foot. If you kick a boulder, the pain you feel is due to the force that the boulder exerts on your foot.

In each of these cases, the force that you exert on the other body is in the opposite direction to the force that body exerts on you. Experiments show that whenever two bodies interact, the two forces that they exert on each other are always *equal in magnitude* and *opposite in direction.* This fact is called *Newton's third law of motion:*

> **Newton's third law of motion:** If body A exerts a force $\vec{F}_{A \text{ on } B}$ on body B (an "action"), then body B exerts a force $\vec{F}_{B \text{ on } A}$ on body A (a "reaction"). These two forces have the same magnitude but are opposite in direction. These two forces act on *different* bodies.

4.25 If body A exerts a force $\vec{F}_{A \text{ on } B}$ on body B, then body B exerts a force $\vec{F}_{B \text{ on } A}$ on body A that is equal in magnitude and opposite in direction: $\vec{F}_{A \text{ on } B} = -\vec{F}_{B \text{ on } A}$.

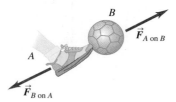

For example, in Fig. 4.25 $\vec{F}_{A \text{ on } B}$ is the force applied *by* body A (first subscript) *on* body B (second subscript), and $\vec{F}_{B \text{ on } A}$ is the force applied *by* body B (first subscript) *on* body A (second subscript). The mathematical statement of Newton's third law is

$$\vec{F}_{A \text{ on } B} = -\vec{F}_{B \text{ on } A} \qquad \text{(Newton's third law of motion)} \qquad [4.11]$$

It doesn't matter whether one body is inanimate (like the soccer ball in Fig. 4.25) and the other is not (like the kicker): They necessarily exert forces on each other that obey Eq. (4.11).

In the statement of Newton's third law, "action" and "reaction" are the two opposite forces (in Fig. 4.25, $\vec{F}_{A \text{ on } B}$ and $\vec{F}_{B \text{ on } A}$); we sometimes refer to them as an **action–reaction pair.** This is *not* meant to imply any cause-and-effect relationship; we can consider either force as the "action" and the other as the "reaction." We often say simply that the forces are "equal and opposite," meaning that they have equal magnitudes and opposite directions.

CAUTION **The two forces in an action–reaction pair act on different bodies** We stress that the two forces described in Newton's third law act on *different* bodies. This is important in problems involving Newton's first or second law, which involve the forces that act on a single body. For instance, the net force on the soccer ball in Fig. 4.25 is the vector sum of the weight of the ball and the force $\vec{F}_{A \text{ on } B}$ exerted by the kicker. You wouldn't include the force $\vec{F}_{B \text{ on } A}$ because this force acts on the kicker, not on the ball.

In Fig. 4.25 the action and reaction forces are *contact* forces that are present only when the two bodies are touching. But Newton's third law also applies to *long-range* forces that do not require physical contact, such as the force of gravitational attraction. A table-tennis ball exerts an upward gravitational force on the earth that's equal in magnitude to the downward gravitational force the earth exerts on the ball. When you drop the ball, both the ball and the earth accelerate toward each other. The net force on each body has the same magnitude, but the earth's acceleration is microscopically small because its mass is so great. Nevertheless, it does move!

Conceptual Example 4.8 **Which force is greater?**

After your sports car breaks down, you start to push it to the nearest repair shop. While the car is starting to move, how does the force you exert on the car compare to the force the car exerts on you? How do these forces compare when you are pushing the car along at a constant speed?

SOLUTION

Newton's third law says that in *both* cases, the force you exert on the car is equal in magnitude and opposite in direction to the force the car exerts on you. It's true that you have to push harder to get the car going than to keep it going. But no matter how hard you push on the car, the car pushes just as hard back on you. Newton's third law gives the same result whether the two bodies are at rest, moving with constant velocity, or accelerating.

You may wonder how the car "knows" to push back on you with the same magnitude of force that you exert on it. It may help to visualize the forces you and the car exert on each other as interactions between the atoms at the surface of your hand and the atoms at the surface of the car. These interactions are analogous to miniature springs between adjacent atoms, and a compressed spring exerts equally strong forces on both of its ends.

Fundamentally, though, the reason we know that objects of different masses exert equally strong forces on each other is that experiment tells us so. Physics isn't merely a collection of rules and equations; rather, it's a systematic description of the natural world based on experiment and observation.

Conceptual Example 4.9 **Applying Newton's third law: Objects at rest**

An apple sits at rest on a table, in equilibrium. What forces act on the apple? What is the reaction force to each of the forces acting on the apple? What are the action–reaction pairs?

SOLUTION

Figure 4.26a shows the forces acting on the apple. $\vec{F}_{\text{earth on apple}}$ is the weight of the apple—that is, the downward gravitational force exerted *by* the earth *on* the apple. Similarly, $\vec{F}_{\text{table on apple}}$ is the upward force exerted *by* the table *on* the apple.

Figure 4.26b shows one of the action–reaction pairs involving the apple. As the earth pulls down on the apple, with force $\vec{F}_{\text{earth on apple}}$, the apple exerts an equally strong upward pull on the earth $\vec{F}_{\text{apple on earth}}$. By Newton's third law (Eq. 4.11) we have

$$\vec{F}_{\text{apple on earth}} = -\vec{F}_{\text{earth on apple}}$$

Also, as the table pushes up on the apple with force $\vec{F}_{\text{table on apple}}$, the corresponding reaction is the downward force $\vec{F}_{\text{apple on table}}$

4.26 The two forces in an action–reaction pair always act on different bodies.

(a) The forces acting on the apple

(b) The action–reaction pair for the interaction between the apple and the earth

(c) The action–reaction pair for the interaction between the apple and the table

(d) We eliminate one of the forces acting on the apple

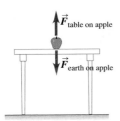

$\vec{F}_{\text{apple on earth}} = -\vec{F}_{\text{earth on apple}}$

Action–reaction pairs always represent a mutual interaction of two different objects.

$\vec{F}_{\text{apple on table}} = -\vec{F}_{\text{table on apple}}$

The two forces on the apple CANNOT be an action–reaction pair because they act on the same object. We see that if we eliminate one, the other remains.

Continued

exerted by the apple on the table (Fig. 4.26c). For this action–reaction pair we have

$$\vec{F}_{\text{apple on table}} = -\vec{F}_{\text{table on apple}}$$

The two forces acting on the apple, $\vec{F}_{\text{table on apple}}$ and $\vec{F}_{\text{earth on apple}}$, are *not* an action–reaction pair, despite being equal in magnitude and opposite in direction. They do not represent the mutual interaction of two bodies; they are two different forces acting on the *same* body. Figure 4.26d shows another way to see this. If we suddenly yank the table out from under the apple, the forces $\vec{F}_{\text{apple on table}}$ and $\vec{F}_{\text{table on apple}}$ suddenly become zero, but $\vec{F}_{\text{apple on earth}}$ and $\vec{F}_{\text{earth on apple}}$ are unchanged (the gravitational interaction is still present). Because $\vec{F}_{\text{table on apple}}$ is now zero, it can't be the negative of the nonzero $\vec{F}_{\text{earth on apple}}$, and these two forces can't be an action–reaction pair. *The two forces in an action–reaction pair **never** act on the same body.*

Conceptual Example 4.10 Applying Newton's third law: Objects in motion

A stonemason drags a marble block across a floor by pulling on a rope attached to the block (Fig. 4.27a). The block is not necessarily in equilibrium. How are the various forces related? What are the action–reaction pairs?

SOLUTION

We'll use the subscripts B for the block, R for the rope, and M for the mason. In Fig. 4.27b the vector $\vec{F}_{\text{M on R}}$ represents the force exerted by the *mason* on the *rope*. The corresponding reaction is the equal and opposite force $\vec{F}_{\text{R on M}}$ exerted by the *rope* on the *mason*. Similarly, $\vec{F}_{\text{R on B}}$ represents the force exerted by the *rope* on the *block,* and the corresponding reaction is the equal and opposite force $\vec{F}_{\text{B on R}}$ exerted by the *block* on the *rope.* For these two action–reaction pairs, we have

$$\vec{F}_{\text{R on M}} = -\vec{F}_{\text{M on R}} \quad \text{and} \quad \vec{F}_{\text{B on R}} = -\vec{F}_{\text{R on B}}$$

Be sure you understand that the forces $\vec{F}_{\text{M on R}}$ and $\vec{F}_{\text{B on R}}$ (Fig. 4.27c) are *not* an action–reaction pair, because both of these forces act on the *same* body (the rope); an action and its reaction *must* always act on *different* bodies. Furthermore, the forces $\vec{F}_{\text{M on R}}$ and $\vec{F}_{\text{B on R}}$ are not necessarily equal in magnitude. Applying Newton's second law to the rope, we get

$$\sum \vec{F} = \vec{F}_{\text{M on R}} + \vec{F}_{\text{B on R}} = m_{\text{rope}}\vec{a}_{\text{rope}}$$

If the block and rope are accelerating (speeding up or slowing down), the rope is not in equilibrium, and $\vec{F}_{\text{M on R}}$ must have a different magnitude than $\vec{F}_{\text{B on R}}$. By contrast, the action–reaction forces $\vec{F}_{\text{M on R}}$ and $\vec{F}_{\text{R on M}}$ are always equal in magnitude, as are $\vec{F}_{\text{R on B}}$ and $\vec{F}_{\text{B on R}}$. Newton's third law holds whether or not the bodies are accelerating.

In the special case in which the rope is in equilibrium, the forces $\vec{F}_{\text{M on R}}$ and $\vec{F}_{\text{B on R}}$ are equal in magnitude, and they are opposite in direction. But this is an example of Newton's *first* law, not his third; these are two forces on the same body, not forces of two bodies on each other. Another way to look at this is that in equilibrium, $\vec{a}_{\text{rope}} = 0$ in the preceding equation. Then $\vec{F}_{\text{B on R}} = -\vec{F}_{\text{M on R}}$ because of Newton's first or second law.

Another special case is if the rope is accelerating but has negligibly small mass compared to that of the block or the mason. In this case, $m_{\text{rope}} = 0$ in the above equation, so again $\vec{F}_{\text{B on R}} = -\vec{F}_{\text{M on R}}$. Since Newton's third law says that $\vec{F}_{\text{B on R}}$ *always* equals $-\vec{F}_{\text{R on B}}$ (they are an action–reaction pair), in this "massless-rope" case $\vec{F}_{\text{R on B}}$ also equals $\vec{F}_{\text{M on R}}$.

For both the "massless-rope" case and the case of the rope in equilibrium, the force of the rope on the block is equal in magnitude and direction to the force of the mason on the rope (Fig. 4.27d). Hence we can think of the rope as "transmitting" to the block the force the mason exerts on the rope. This is a useful point of view, but remember that it is valid *only* when the rope has negligibly small mass or is in equilibrium.

4.27 Identifying the forces that act when a mason pulls on a rope attached to a block.

(a) The block, the rope, and the mason

(b) The action–reaction pairs

$\vec{F}_{\text{R on M}}$ $\vec{F}_{\text{M on R}}$

$\vec{F}_{\text{B on R}}$ $\vec{F}_{\text{R on B}}$

(c) *Not* an action–reaction pair

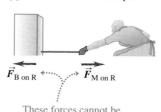

$\vec{F}_{\text{B on R}}$ $\vec{F}_{\text{M on R}}$

These forces cannot be an action–reaction pair because they act on the same object (the rope).

(d) Not necessarily equal

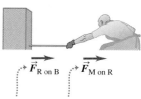

$\vec{F}_{\text{R on B}}$ $\vec{F}_{\text{M on R}}$

These forces are equal only if the rope is in equilibrium (or can be treated as massless).

Conceptual Example 4.11 A Newton's third law paradox?

We saw in Conceptual Example 4.10 that the stonemason pulls as hard on the rope–block combination as that combination pulls back on him. Why, then, does the block move while the stonemason remains stationary?

SOLUTION

To resolve this seeming paradox, keep in mind the difference between Newton's *second* and *third* laws. The only forces involved in Newton's second law are those that act *on* a given body. The vector sum of these forces determines the body's acceleration, if any. By contrast, Newton's third law relates the forces that two *different* bodies exert on *each other*. The third law alone tells you nothing about the motion of either body.

If the rope–block combination is initially at rest, it begins to slide if the stonemason exerts a force $\vec{F}_{M \text{ on } R}$ that is *greater* in magnitude than the friction force that the floor exerts on the block (Fig. 4.28). (The block has a smooth underside, which helps to minimize friction.) Then there is a net force to the right on the rope–block combination, and it accelerates to the right. By contrast, the stonemason *doesn't* move because the net force acting on him is *zero*. His shoes have nonskid soles that don't slip on the floor, so the friction force that the floor exerts on him is strong enough to balance the pull of the rope on him, $\vec{F}_{R \text{ on } M}$. (Both the block and the stonemason also experience a downward force of gravity and an upward normal force exerted by the floor. These forces balance each other and cancel out, so we haven't included them in Fig. 4.28.)

Once the block is moving at the desired speed, the stonemason doesn't need to pull as hard; he must exert only enough force to balance the friction force on the block. Then the net force on the

4.28 The horizontal forces acting on the block–rope combination (left) and the mason (right). (The vertical forces are not shown.)

These forces are an action–reaction pair. They have the same magnitude but act on different objects.

Friction force of floor on block

$\vec{F}_{M \text{ on } R}$ $\vec{F}_{R \text{ on } M}$

Friction force of floor on mason

Block + rope

The block begins sliding if $\vec{F}_{M \text{ on } R}$ overcomes the friction force on the block.

Mason

The mason remains at rest if $\vec{F}_{R \text{ on } M}$ is balanced by the friction force on the mason.

moving block is zero, and the block continues to move toward the mason at a constant velocity, in accordance with Newton's first law.

So the block accelerates but the stonemason doesn't because different amounts of friction act on them. If the floor were freshly waxed, so that there was little friction between the floor and the stonemason's shoes, pulling on the rope might start the block sliding to the right *and* start him sliding to the left.

The moral of this example is that when analyzing the motion of a body, you must remember that only the forces acting *on* a body determine its motion. From this perspective, Newton's third law is merely a tool that can help you determine what those forces are.

A body that has pulling forces applied at its ends, such as the rope in Fig. 4.27, is said to be in *tension*. The **tension** at any point is the magnitude of force acting at that point (see Fig. 4.2c). In Fig. 4.27b the tension at the right end of the rope is the magnitude of $\vec{F}_{M \text{ on } R}$ (or of $\vec{F}_{R \text{ on } M}$), and the tension at the left end equals the magnitude of $\vec{F}_{B \text{ on } R}$ (or of $\vec{F}_{R \text{ on } B}$). If the rope is in equilibrium and if no forces act except at its ends, the tension is the *same* at both ends and throughout the rope. Thus, if the magnitudes of $\vec{F}_{B \text{ on } R}$ and $\vec{F}_{M \text{ on } R}$ are 50 N each, the tension in the rope is 50 N (*not* 100 N). The *total* force vector $\vec{F}_{B \text{ on } R} + \vec{F}_{M \text{ on } R}$ acting on the rope in this case is zero!

We emphasize once more a fundamental truth: The two forces in an action–reaction pair *never* act on the same body. Remembering this simple fact can often help you avoid confusion about action–reaction pairs and Newton's third law.

Test Your Understanding of Section 4.5 You are driving your car on a country road when a mosquito splatters on the windshield. Which has the greater magnitude: the force that the car exerted on the mosquito or the force that the mosquito exerted on the car? Or are the magnitudes the same? If they are different, how can you reconcile this fact with Newton's third law? If they are equal, why is the mosquito splattered while the car is undamaged?

4.6 Free-Body Diagrams

Newton's three laws of motion contain all the basic principles we need to solve a wide variety of problems in mechanics. These laws are very simple in form, but the process of applying them to specific situations can pose real challenges. In this brief section we'll point out three key ideas and techniques to use in any problems involving Newton's laws. You'll learn others in Chapter 5, which also extends the use of Newton's laws to cover more complex situations.

1. *Newton's first and second laws apply to a specific body.* Whenever you use Newton's first law, $\sum \vec{F} = 0$, for an equilibrium situation or Newton's second law, $\sum \vec{F} = m\vec{a}$, for a nonequilibrium situation, you must decide at the beginning to which body you are referring. This decision may sound trivial, but it isn't.

2. *Only forces acting on the body matter.* The sum $\sum \vec{F}$ includes all the forces that act *on* the body in question. Hence, once you've chosen the body to analyze, you have to identify all the forces acting on it. Don't get confused between the forces acting on a body and the forces exerted by that body on some other body. For example, to analyze a person walking, you would include in $\sum \vec{F}$ the force that the ground exerts on the person as he walks, but *not* the force that the person exerts on the ground (Fig. 4.29). These forces form an action–reaction pair and are related by Newton's third law, but only the member of the pair that acts on the body you're working with goes into $\sum \vec{F}$.

3. *Free-body diagrams are essential to help identify the relevant forces.* A **free-body diagram** is a diagram showing the chosen body by itself, "free" of its surroundings, with vectors drawn to show the magnitudes and directions of all the forces applied to the body by the various other bodies that interact with it. We have already shown some free-body diagrams in Figs. 4.18, 4.19, 4.21, and 4.26a. Be careful to include all the forces acting *on* the body, but be equally careful *not* to include any forces that the body exerts on any other body. In particular, the two forces in an action–reaction pair must *never* appear in the same free-body diagram because they never act on the same body. Furthermore, forces that a body exerts on itself are never included, since these can't affect the body's motion.

4.29 The simple act of walking depends crucially on Newton's third law. To start moving forward, you push backward on the ground with your foot. As a reaction, the ground pushes forward on your foot (and hence on your body as a whole) with a force of the same magnitude. This *external* force provided by the ground is what accelerates your body forward.

CAUTION **Forces in free-body diagrams** When you have a complete free-body diagram, you *must* be able to answer this question for each force: What other body is applying this force? If you can't answer that question, you may be dealing with a nonexistent force. Be especially on your guard to avoid nonexistent forces such as "the force of acceleration" or "the $m\vec{a}$ force," discussed in Section 4.3. ▌

When a problem involves more than one body, you have to take the problem apart and draw a separate free-body diagram for each body. For example, Fig. 4.27c shows a separate free-body diagram for the rope in the case in which the rope is considered massless (so that no gravitational force acts on it). Figure 4.28 also shows diagrams for the block and the mason, but these are *not* complete free-body diagrams because they don't show all the forces acting on each body. (We left out the vertical forces—the weight force exerted by the earth and the upward normal force exerted by the floor.)

Figure 4.30 presents three real-life situations and the corresponding complete free-body diagrams. Note that in each situation a person exerts a force on something in his or her surroundings, but the force that shows up in the person's free-body diagram is the surroundings pushing back *on* the person.

Test Your Understanding of Section 4.6 The buoyancy force shown in Fig. 4.30c is one half of an action–reaction pair. What force is the other half of this pair? (i) the weight of the swimmer; (ii) the forward thrust force; (iii) the backward drag force; (iv) the downward force that the swimmer exerts on the water; (v) the backward force that the swimmer exerts on the water by kicking.

4.30 Examples of free-body diagrams. Each free-body diagram shows all of the external forces that act on the object in question.

(a)

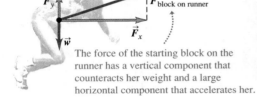

The force of the starting block on the runner has a vertical component that counteracts her weight and a large horizontal component that accelerates her.

(b)

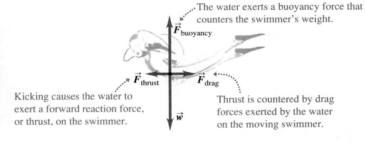

To jump up, this player will push down against the floor, increasing the upward reaction force $\vec{n}$ of the floor on him.

This player is a freely falling object.

(c)

The water exerts a buoyancy force that counters the swimmer's weight.

$\vec{F}_{\text{buoyancy}}$

$\vec{F}_{\text{thrust}}$ $\vec{F}_{\text{drag}}$

Kicking causes the water to exert a forward reaction force, or thrust, on the swimmer.

Thrust is countered by drag forces exerted by the water on the moving swimmer.

$\vec{w}$

Force as a vector: Force is a quantitative measure of the interaction between two bodies. It is a vector quantity. When several forces act on a body, the effect on its motion is the same as when a single force, equal to the vector sum (resultant) of the forces, acts on the body. (See Example 4.1.)

$$\vec{R} = \vec{F}_1 + \vec{F}_2 + \vec{F}_3 + \cdots = \sum \vec{F} \quad (4.1)$$

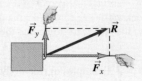

The net force on a body and Newton's first law: Newton's first law states that when the vector sum of all forces acting on a body (the *net force*) is zero, the body is in equilibrium and has zero acceleration. If the body is initially at rest, it remains at rest; if it is initially in motion, it continues to move with constant velocity. This law is valid only in inertial frames of reference. (See Examples 4.2 and 4.3.)

$$\sum \vec{F} = 0 \quad (4.3)$$

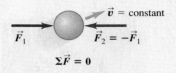

Mass, acceleration, and Newton's second law: The inertial properties of a body are characterized by its *mass*. The acceleration of a body under the action of a given set of forces is directly proportional to the vector sum of the forces (the *net force*) and inversely proportional to the mass of the body. This relationship is Newton's second law. Like Newton's first law, this law is valid only in inertial frames of reference. The unit of force is defined in terms of the units of mass and acceleration. In SI units, the unit of force is the newton (N), equal to $1 \text{ kg} \cdot \text{m/s}^2$. (See Examples 4.4 and 4.5.)

$$\sum \vec{F} = m\vec{a} \quad (4.7)$$

$$\sum F_x = ma_x$$

$$\sum F_y = ma_y \quad (4.8)$$

$$\sum F_z = ma_z$$

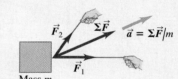

Weight: The weight $\vec{w}$ of a body is the gravitational force exerted on it by the earth. Weight is a vector quantity. The magnitude of the weight of a body at any specific location is equal to the product of its mass m and the magnitude of the acceleration due to gravity g at that location. While the weight of a body depends on its location, the mass is independent of location. (See Examples 4.6 and 4.7.)

$$w = mg \quad (4.9)$$

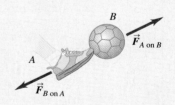

Newton's third law and action–reaction pairs: Newton's third law states that when two bodies interact, they exert forces on each other that at each instant are equal in magnitude and opposite in direction. These forces are called action and reaction forces. Each of these two forces acts on only one of the two bodies; they never act on the same body. (See Examples 4.8–4.11.)

$$\vec{F}_{A \text{ on } B} = -\vec{F}_{B \text{ on } A} \quad (4.11)$$

BRIDGING PROBLEM Links in a Chain

A student suspends a chain consisting of three links, each of mass $m = 0.250$ kg, from a light rope. She pulls upward on the rope, so that the rope applies an upward force of 9.00 N to the chain. (a) Draw a free-body diagram for the entire chain, considered as a body, and one for each of the three links. (b) Use the diagrams of part (a) and Newton's laws to find (i) the acceleration of the chain, (ii) the force exerted by the top link on the middle link, and (iii) the force exerted by the middle link on the bottom link. Treat the rope as massless.

SOLUTION GUIDE

See MasteringPhysics® study area for a Video Tutor solution.

IDENTIFY and SET UP

1. There are four objects of interest in this problem: the chain as a whole and the three individual links. For each of these four objects, make a list of the external forces that act on it. Besides the force of gravity, your list should include only forces exerted by other objects that *touch* the object in question.
2. Some of the forces in your lists form action–reaction pairs (one pair is the force of the top link on the middle link and the force of the middle link on the top link). Identify all such pairs.
3. Use your lists to help you draw a free-body diagram for each of the four objects. Choose the coordinate axes.

4. Use your lists to decide how many unknowns there are in this problem. Which of these are target variables?

EXECUTE

5. Write a Newton's second law equation for each of the four objects, and write a Newton's third law equation for each action–reaction pair. You should have at least as many equations as there are unknowns (see step 4). Do you?
6. Solve the equations for the target variables.

EVALUATE

7. You can check your results by substituting them back into the equations from step 6. This is especially important to do if you ended up with more equations in step 5 than you used in step 6.
8. Rank the force of the rope on the chain, the force of the top link on the middle link, and the force of the middle link on the bottom link in order from smallest to largest magnitude. Does this ranking make sense? Explain.
9. Repeat the problem for the case where the upward force that the rope exerts on the chain is only 7.35 N. Is the ranking in step 8 the same? Does this make sense?

Problems

For instructor-assigned homework, go to www.masteringphysics.com

•, ••, •••: Problems of increasing difficulty. **CP**: Cumulative problems incorporating material from earlier chapters. **CALC**: Problems requiring calculus. **BIO**: Biosciences problems.

DISCUSSION QUESTIONS

Q4.1 Can a body be in equilibrium when only one force acts on it? Explain.

Q4.2 A ball thrown straight up has zero velocity at its highest point. Is the ball in equilibrium at this point? Why or why not?

Q4.3 A helium balloon hovers in midair, neither ascending nor descending. Is it in equilibrium? What forces act on it?

Q4.4 When you fly in an airplane at night in smooth air, there is no sensation of motion, even though the plane may be moving at 800 km/h (500 mi/h). Why is this?

Q4.5 If the two ends of a rope in equilibrium are pulled with forces of equal magnitude and opposite direction, why is the total tension in the rope not zero?

Q4.6 You tie a brick to the end of a rope and whirl the brick around you in a horizontal circle. Describe the path of the brick after you suddenly let go of the rope.

Q4.7 When a car stops suddenly, the passengers tend to move forward relative to their seats. Why? When a car makes a sharp turn, the passengers tend to slide to one side of the car. Why?

Q4.8 Some people say that the "force of inertia" (or "force of momentum") throws the passengers forward when a car brakes sharply. What is wrong with this explanation?

Q4.9 A passenger in a moving bus with no windows notices that a ball that has been at rest in the aisle suddenly starts to move toward the rear of the bus. Think of two different possible explanations, and devise a way to decide which is correct.

Q4.10 Suppose you chose the fundamental SI units to be force, length, and time instead of mass, length, and time. What would be the units of mass in terms of those fundamental units?

Q4.11 Some of the ancient Greeks thought that the "natural state" of an object was to be at rest, so objects would seek their natural state by coming to rest if left alone. Explain why this incorrect view can actually seem quite plausible in the everyday world.

Q4.12 Why is the earth only approximately an inertial reference frame?

Q4.13 Does Newton's second law hold true for an observer in a van as it speeds up, slows down, or rounds a corner? Explain.

Q4.14 Some students refer to the quantity $m\vec{a}$ as "the force of acceleration." Is it correct to refer to this quantity as a force? If so, what exerts this force? If not, what is a better description of this quantity?

Q4.15 The acceleration of a falling body is measured in an elevator traveling upward at a constant speed of 9.8 m/s. What result is obtained?

Q4.16 You can play catch with a softball in a bus moving with constant speed on a straight road, just as though the bus were at rest. Is this still possible when the bus is making a turn at constant speed on a level road? Why or why not?

Q4.17 Students sometimes say that the force of gravity on an object is 9.8 m/s². What is wrong with this view?

Q4.18 The head of a hammer begins to come loose from its wooden handle. How should you strike the handle on a concrete sidewalk to reset the head? Why does this work?

Q4.19 Why can it hurt your foot more to kick a big rock than a small pebble? *Must* the big rock hurt more? Explain.

Q4.20 "It's not the fall that hurts you; it's the sudden stop at the bottom." Translate this saying into the language of Newton's laws of motion.

Q4.21 A person can dive into water from a height of 10 m without injury, but a person who jumps off the roof of a 10-m-tall building and lands on a concrete street is likely to be seriously injured. Why is there a difference?

Q4.22 Why are cars designed to crumple up in front and back for safety? Why not for side collisions and rollovers?

Q4.23 When a bullet is fired from a rifle, what is the origin of the force that accelerates the bullet?

Q4.24 When a string barely strong enough lifts a heavy weight, it can lift the weight by a steady pull; but if you jerk the string, it will break. Explain in terms of Newton's laws of motion.

Q4.25 A large crate is suspended from the end of a vertical rope. Is the tension in the rope greater when the crate is at rest or when it is moving upward at constant speed? If the crate is traveling upward, is the tension in the rope greater when it is speeding up or when it is slowing down? In each case explain in terms of Newton's laws of motion.

Q4.26 Which feels a greater pull due to the earth's gravity, a 10-kg stone or a 20-kg stone? If you drop them, why does the 20-kg stone not fall with twice the acceleration of the 10-kg stone? Explain your reasoning.

Q4.27 Why is it incorrect to say that 1.0 kg *equals* 2.2 lb?

Q4.28 A horse is hitched to a wagon. Since the wagon pulls back on the horse just as hard as the horse pulls on the wagon, why doesn't the wagon remain in equilibrium, no matter how hard the horse pulls?

Q4.29 True or false? You exert a push P on an object and it pushes back on you with a force F. If the object is moving at constant velocity, then F is equal to P, but if the object is being accelerated, then P must be greater than F.

Q4.30 A large truck and a small compact car have a head-on collision. During the collision, the truck exerts a force $\vec{F}_{\text{T on C}}$ on the car, and the car exerts a force $\vec{F}_{\text{C on T}}$ on the truck. Which force has the larger magnitude, or are they the same? Does your answer depend on how fast each vehicle was moving before the collision? Why or why not?

Q4.31 When a car comes to a stop on a level highway, what force causes it to slow down? When the car increases its speed on the same highway, what force causes it to speed up? Explain.

Q4.32 A small compact car is pushing a large van that has broken down, and they travel along the road with equal velocities and accelerations. While the car is speeding up, is the force it exerts on the van larger than, smaller than, or the same magnitude as the force the van exerts on it? Which object, the car or the van, has the larger net force on it, or are the net forces the same? Explain.

Q4.33 Consider a tug-of-war between two people who pull in opposite directions on the ends of a rope. By Newton's third law, the force that A exerts on B is just as great as the force that B exerts on A. So what determines who wins? (*Hint:* Draw a free-body diagram showing all the forces that act on each person.)

Q4.34 On the moon, $g = 1.62$ m/s². If a 2-kg brick drops on your foot from a height of 2 m, will this hurt more, or less, or the same if it happens on the moon instead of on the earth? Explain. If a 2-kg brick is thrown and hits you when it is moving horizontally at 6 m/s, will this hurt more, less, or the same if it happens on the moon instead of

on the earth? Explain. (On the moon, assume that you are inside a pressurized structure, so you are not wearing a spacesuit.)

Q4.35 A manual for student pilots contains the following passage: "When an airplane flies at a steady altitude, neither climbing nor descending, the upward lift force from the wings equals the airplane's weight. When the airplane is climbing at a steady rate, the upward lift is greater than the weight; when the airplane is descending at a steady rate, the upward lift is less than the weight." Are these statements correct? Explain.

Q4.36 If your hands are wet and no towel is handy, you can remove some of the excess water by shaking them. Why does this get rid of the water?

Q4.37 If you are squatting down (such as when you are examining the books on the bottom shelf in a library or bookstore) and suddenly get up, you can temporarily feel light-headed. What do Newton's laws of motion have to say about why this happens?

Q4.38 When a car is hit from behind, the passengers can receive a whiplash. Use Newton's laws of motion to explain what causes this to occur.

Q4.39 In a head-on auto collision, passengers not wearing seat belts can be thrown through the windshield. Use Newton's laws of motion to explain why this happens.

Q4.40 In a head-on collision between a compact 1000-kg car and a large 2500-kg car, which one experiences the greater force? Explain. Which one experiences the greater acceleration? Explain why. Now explain why passengers in the small car are more likely to be injured than those in the large car, even if the bodies of both cars are equally strong.

Q4.41 Suppose you are in a rocket with no windows, traveling in deep space far from any other objects. Without looking outside the rocket or making any contact with the outside world, explain how you could determine if the rocket is (a) moving forward at a constant 80% of the speed of light and (b) accelerating in the forward direction.

EXERCISES

Section 4.1 Force and Interactions

4.1 • Two forces have the same magnitude F. What is the angle between the two vectors if their sum has a magnitude of (a) $2F$? (b) $\sqrt{2}F$? (c) zero? Sketch the three vectors in each case.

4.2 • Workmen are trying to free an SUV stuck in the mud. To extricate the vehicle, they use three horizontal ropes, producing the force vectors shown in Fig. E4.2. (a) Find the x- and y-components of each of the three pulls. (b) Use the components to find the magnitude and direction of the resultant of the three pulls.

Figure E4.2

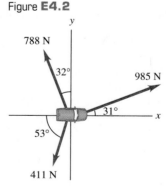

788 N
985 N
32°
31°
53°
411 N

4.3 • **BIO Jaw Injury.** Due to a jaw injury, a patient must wear a strap (Fig. E4.3) that produces a net upward force of 5.00 N on his chin. The tension is the same throughout the strap. To what tension must the strap be adjusted to provide the necessary upward force?

Figure **E4.3**

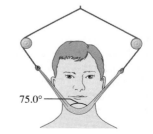

75.0°

4.4 • A man is dragging a trunk up the loading ramp of a mover's truck. The ramp has a slope angle of 20.0°, and the man pulls upward with a force $\vec{F}$ whose direction makes an angle of 30.0° with the ramp (Fig. E4.4). (a) How large a force $\vec{F}$ is necessary for the component F_x parallel to the ramp to be 60.0 N? (b) How large will the component F_y perpendicular to the ramp then be?

Figure **E4.4**

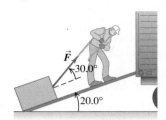

4.5 •• Two dogs pull horizontally on ropes attached to a post; the angle between the ropes is 60.0°. If dog A exerts a force of 270 N and dog B exerts a force of 300 N, find the magnitude of the resultant force and the angle it makes with dog A's rope.

4.6 • Two forces, $\vec{F}_1$ and $\vec{F}_2$, act at a point. The magnitude of $\vec{F}_1$ is 9.00 N, and its direction is 60.0° above the x-axis in the second quadrant. The magnitude of $\vec{F}_2$ is 6.00 N, and its direction is 53.1° below the x-axis in the third quadrant. (a) What are the x- and y-components of the resultant force? (b) What is the magnitude of the resultant force?

Section 4.3 Newton's Second Law

4.7 •• A 68.5-kg skater moving initially at 2.40 m/s on rough horizontal ice comes to rest uniformly in 3.52 s due to friction from the ice. What force does friction exert on the skater?

4.8 •• You walk into an elevator, step onto a scale, and push the "up" button. You also recall that your normal weight is 625 N. Start answering each of the following questions by drawing a free-body diagram. (a) If the elevator has an acceleration of magnitude 2.50 m/s², what does the scale read? (b) If you start holding a 3.85-kg package by a light vertical string, what will be the tension in this string once the elevator begins accelerating?

4.9 • A box rests on a frozen pond, which serves as a frictionless horizontal surface. If a fisherman applies a horizontal force with magnitude 48.0 N to the box and produces an acceleration of magnitude 3.00 m/s², what is the mass of the box?

4.10 •• A dockworker applies a constant horizontal force of 80.0 N to a block of ice on a smooth horizontal floor. The frictional force is negligible. The block starts from rest and moves 11.0 m in 5.00 s. (a) What is the mass of the block of ice? (b) If the worker stops pushing at the end of 5.00 s, how far does the block move in the next 5.00 s?

4.11 • A hockey puck with mass 0.160 kg is at rest at the origin ($x = 0$) on the horizontal, frictionless surface of the rink. At time $t = 0$ a player applies a force of 0.250 N to the puck, parallel to the x-axis; he continues to apply this force until $t = 2.00$ s. (a) What are the position and speed of the puck at $t = 2.00$ s? (b) If the same force is again applied at $t = 5.00$ s, what are the position and speed of the puck at $t = 7.00$ s?

4.12 • A crate with mass 32.5 kg initially at rest on a warehouse floor is acted on by a net horizontal force of 140 N. (a) What acceleration is produced? (b) How far does the crate travel in 10.0 s? (c) What is its speed at the end of 10.0 s?

4.13 • A 4.50-kg toy cart undergoes an acceleration in a straight line (the x-axis). The graph in Fig. E4.13 shows this acceleration as a function of time. (a) Find the

Figure **E4.13**

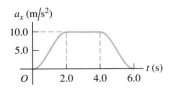

maximum net force on this cart. When does this maximum force occur? (b) During what times is the net force on the cart a constant? (c) When is the net force equal to zero?

4.14 • A 2.75-kg cat moves in a straight line (the x-axis). Figure E4.14 shows a graph of the x-component of this cat's velocity as a function of time. (a) Find the maximum net force on this cat. When does this force occur? (b) When is the net force on the cat equal to zero? (c) What is the net force at time 8.5 s?

Figure **E4.14**

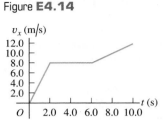

4.15 • A small 8.00-kg rocket burns fuel that exerts a time-varying upward force on the rocket as the rocket moves upward from the launch pad. This force obeys the equation $F = A + Bt^2$. Measurements show that at $t = 0$, the force is 100.0 N, and at the end of the first 2.00 s, it is 150.0 N. (a) Find the constants A and B, including their SI units. (b) Find the *net* force on this rocket and its acceleration (i) the instant after the fuel ignites and (ii) 3.00 s after fuel ignition. (c) Suppose you were using this rocket in outer space, far from all gravity. What would its acceleration be 3.00 s after fuel ignition?

4.16 • An electron (mass = 9.11×10^{-31} kg) leaves one end of a TV picture tube with zero initial speed and travels in a straight line to the accelerating grid, which is 1.80 cm away. It reaches the grid with a speed of 3.00×10^6 m/s. If the accelerating force is constant, compute (a) the acceleration; (b) the time to reach the grid; (c) the net force, in newtons. (You can ignore the gravitational force on the electron.)

Section 4.4 Mass and Weight

4.17 • Superman throws a 2400-N boulder at an adversary. What horizontal force must Superman apply to the boulder to give it a horizontal acceleration of 12.0 m/s²?

4.18 • BIO (a) An ordinary flea has a mass of 210 μg. How many newtons does it weigh? (b) The mass of a typical froghopper is 12.3 mg. How many newtons does it weigh? (c) A house cat typically weighs 45 N. How many pounds does it weigh, and what is its mass in kilograms?

4.19 • At the surface of Jupiter's moon Io, the acceleration due to gravity is $g = 1.81$ m/s². A watermelon weighs 44.0 N at the surface of the earth. (a) What is the watermelon's mass on the earth's surface? (b) What are its mass and weight on the surface of Io?

4.20 • An astronaut's pack weighs 17.5 N when she is on earth but only 3.24 N when she is at the surface of an asteroid. (a) What is the acceleration due to gravity on this asteroid? (b) What is the mass of the pack on the asteroid?

Section 4.5 Newton's Third Law

4.21 • BIO World-class sprinters can accelerate out of the starting blocks with an acceleration that is nearly horizontal and has magnitude 15 m/s². How much horizontal force must a 55-kg sprinter exert on the starting blocks during a start to produce this acceleration? Which body exerts the force that propels the sprinter: the blocks or the sprinter herself?

4.22 A small car (mass 380 kg) is pushing a large truck (mass 900 kg) due east on a level road. The car exerts a horizontal force of 1200 N on the truck. What is the magnitude of the force that the truck exerts on the car?

4.23 Boxes A and B are in contact on a horizontal, frictionless surface, as shown in Fig. E4.23. Box A has mass 20.0 kg and box B has mass 5.0 kg. A horizontal force of 100 N is exerted on box A. What is the magnitude of the force that box A exerts on box B?

Figure **E4.23**

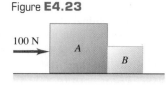

100 N

4.24 •• The upward normal force exerted by the floor is 620 N on an elevator passenger who weighs 650 N. What are the reaction forces to these two forces? Is the passenger accelerating? If so, what are the magnitude and direction of the acceleration?

4.25 •• A student with mass 45 kg jumps off a high diving board. Using 6.0×10^{24} kg for the mass of the earth, what is the acceleration of the earth toward her as she accelerates toward the earth with an acceleration of 9.8 m/s^2? Assume that the net force on the earth is the force of gravity she exerts on it.

Section 4.6 Free-Body Diagrams

4.26 • An athlete throws a ball of mass m directly upward, and it feels no appreciable air resistance. Draw a free-body diagram of this ball while it is free of the athlete's hand and (a) moving upward; (b) at its highest point; (c) moving downward. (d) Repeat parts (a), (b), and (c) if the athlete throws the ball at a 60° angle above the horizontal instead of directly upward.

4.27 •• Two crates, A and B, sit at rest side by side on a frictionless horizontal surface. The crates have masses m_A and m_B. A horizontal force $\vec{F}$ is applied to crate A and the two crates move off to the right. (a) Draw clearly labeled free-body diagrams for crate A and for crate B. Indicate which pairs of forces, if any, are third-law action–reaction pairs. (b) If the magnitude of force $\vec{F}$ is less than the total weight of the two crates, will it cause the crates to move? Explain.

4.28 •• A person pulls horizontally on block B in Fig. E4.28, causing both blocks to move together as a unit. While this system is moving, make a carefully labeled free-body diagram of block A if (a) the table is frictionless and (b) there is friction between block B and the table and the pull is equal to the friction force on block B due to the table.

Figure **E4.28**

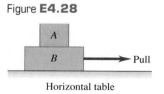

Horizontal table

4.29 • A ball is hanging from a long string that is tied to the ceiling of a train car traveling eastward on horizontal tracks. An observer inside the train car sees the ball hang motionless. Draw a clearly labeled free-body diagram for the ball if (a) the train has a uniform velocity, and (b) the train is speeding up uniformly. Is the net force on the ball zero in either case? Explain.

4.30 •• CP A .22 rifle bullet, traveling at 350 m/s, strikes a large tree, which it penetrates to a depth of 0.130 m. The mass of the bullet is 1.80 g. Assume a constant retarding force. (a) How much time is required for the bullet to stop? (b) What force, in newtons, does the tree exert on the bullet?

4.31 •• A chair of mass 12.0 kg is sitting on the horizontal floor; the floor is not frictionless. You push on the chair with a force $F = 40.0$ N that is directed at an angle of 37.0° below the horizontal and the chair slides along the floor. (a) Draw a clearly labeled free-body diagram for the chair. (b) Use your diagram and Newton's laws to calculate the normal force that the floor exerts on the chair.

4.32 •• A skier of mass 65.0 kg is pulled up a snow-covered slope at constant speed by a tow rope that is parallel to the ground. The ground slopes upward at a constant angle of 26.0° above the horizontal, and you can ignore friction. (a) Draw a clearly labeled free-body diagram for the skier. (b) Calculate the tension in the tow rope.

PROBLEMS

4.33 CP A 4.80-kg bucket of water is accelerated upward by a cord of negligible mass whose breaking strength is 75.0 N. If the bucket starts from rest, what is the minimum time required to raise the bucket a vertical distance of 12.0 m without breaking the cord?

4.34 ••• A large box containing your new computer sits on the bed of your pickup truck. You are stopped at a red light. The light turns green and you stomp on the gas and the truck accelerates. To your horror, the box starts to slide toward the back of the truck. Draw clearly labeled free-body diagrams for the truck and for the box. Indicate pairs of forces, if any, that are third-law action–reaction pairs. (The bed of the truck is *not* frictionless.)

4.35 • Two horses pull horizontally on ropes attached to a stump. The two forces $\vec{F}_1$ and $\vec{F}_2$ that they apply to the stump are such that the net (resultant) force $\vec{R}$ has a magnitude equal to that of $\vec{F}_1$ and makes an angle of 90° with $\vec{F}_1$. Let $F_1 = 1300$ N and $R = 1300$ N also. Find the magnitude of $\vec{F}_2$ and its direction (relative to $\vec{F}_1$).

4.36 •• CP You have just landed on Planet X. You take out a 100-g ball, release it from rest from a height of 10.0 m, and measure that it takes 2.2 s to reach the ground. You can ignore any force on the ball from the atmosphere of the planet. How much does the 100-g ball weigh on the surface of Planet X?

4.37 •• Two adults and a child want to push a wheeled cart in the direction marked x in Fig. P4.37. The two adults push with horizontal forces $\vec{F}_1$ and $\vec{F}_2$ as shown in the figure. (a) Find the magnitude and direction of the *smallest* force that the child should exert. You can ignore the effects of friction. (b) If the child exerts the minimum force found in part (a), the cart accelerates at 2.0 m/s^2 in the $+x$-direction. What is the weight of the cart?

Figure **P4.37**

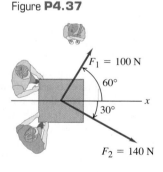

$F_1 = 100$ N

60°

30°

x

$F_2 = 140$ N

4.38 • CP An oil tanker's engines have broken down, and the wind is blowing the tanker straight toward a reef at a constant speed of 1.5 m/s (Fig. P4.38). When the tanker is 500 m from the reef, the wind dies down just as the engineer gets the engines going again. The rudder is stuck, so the only choice is to try to accelerate straight backward away from the reef. The mass of the tanker and cargo is 3.6×10^7 kg, and the engines produce a net horizontal force of 8.0×10^4 N on the tanker. Will the ship hit the reef? If it does, will the oil be safe? The hull can withstand an impact at a speed of 0.2 m/s or less. You can ignore the retarding force of the water on the tanker's hull.

Figure **P4.38**

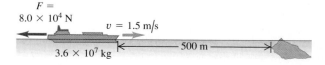

$F = 8.0 \times 10^4$ N

$v = 1.5$ m/s

3.6 × 10^7 kg

500 m

4.39 •• **CP BIO** **A Standing Vertical Jump.** Basketball player Darrell Griffith is on record as attaining a standing vertical jump of 1.2 m (4 ft). (This means that he moved upward by 1.2 m after his feet left the floor.) Griffith weighed 890 N (200 lb). (a) What is his speed as he leaves the floor? (b) If the time of the part of the jump before his feet left the floor was 0.300 s, what was his average acceleration (magnitude and direction) while he was pushing against the floor? (c) Draw his free-body diagram (see Section 4.6). In terms of the forces on the diagram, what is the net force on him? Use Newton's laws and the results of part (b) to calculate the average force he applied to the ground.

4.40 ••• **CP** An advertisement claims that a particular automobile can "stop on a dime." What net force would actually be necessary to stop a 850-kg automobile traveling initially at 45.0 km/h in a distance equal to the diameter of a dime, which is 1.8 cm?

4.41 •• **BIO** **Human Biomechanics.** The fastest pitched baseball was measured at 46 m/s. Typically, a baseball has a mass of 145 g. If the pitcher exerted his force (assumed to be horizontal and constant) over a distance of 1.0 m, (a) what force did he produce on the ball during this record-setting pitch? (b) Draw free-body diagrams of the ball during the pitch and just *after* it left the pitcher's hand.

4.42 •• **BIO** **Human Biomechanics.** The fastest served tennis ball, served by "Big Bill" Tilden in 1931, was measured at 73.14 m/s. The mass of a tennis ball is 57 g, and the ball is typically in contact with the tennis racquet for 30.0 ms, with the ball starting from rest. Assuming constant acceleration, (a) what force did Big Bill's tennis racquet exert on the tennis ball if he hit it essentially horizontally? (b) Draw free-body diagrams of the tennis ball during the serve and just after it moved free of the racquet.

4.43 • Two crates, one with mass 4.00 kg and the other with mass 6.00 kg, sit on the frictionless surface of a frozen pond, connected by a light rope (Fig. P4.43). A woman wearing golf shoes (so she can get traction on the ice) pulls horizontally on the 6.00-kg crate with a force F that gives the crate an acceleration of 2.50 m/s^2. (a) What is the acceleration of the 4.00-kg crate? (b) Draw a free-body diagram for the 4.00-kg crate. Use that diagram and Newton's second law to find the tension T in the rope that connects the two crates. (c) Draw a free-body diagram for the 6.00-kg crate. What is the direction of the net force on the 6.00-kg crate? Which is larger in magnitude, force T or force F? (d) Use part (c) and Newton's second law to calculate the magnitude of the force F.

Figure **P4.43**

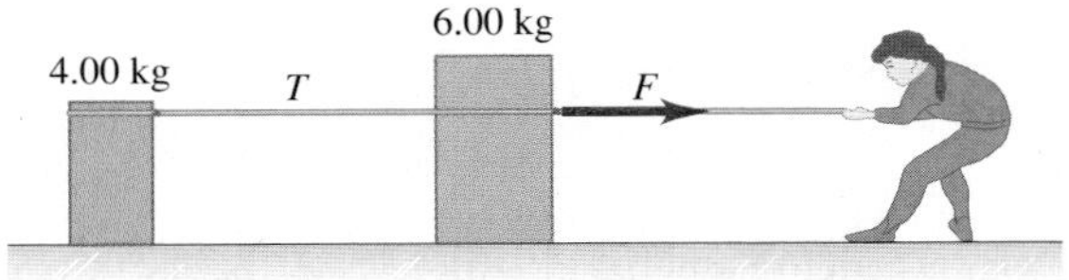

4.44 • An astronaut is tethered by a strong cable to a spacecraft. The astronaut and her spacesuit have a total mass of 105 kg, while the mass of the cable is negligible. The mass of the spacecraft is 9.05×10^4 kg. The spacecraft is far from any large astronomical bodies, so we can ignore the gravitational forces on it and the astronaut. We also assume that both the spacecraft and the astronaut are initially at rest in an inertial reference frame. The astronaut then pulls on the cable with a force of 80.0 N. (a) What force does the cable exert on the astronaut? (b) Since $\Sigma\vec{F} = m\vec{a}$, how can a "massless" ($m = 0$) cable exert a force? (c) What is the astronaut's acceleration? (d) What force does the cable exert on the spacecraft? (e) What is the acceleration of the spacecraft?

4.45 • **CALC** To study damage to aircraft that collide with large birds, you design a test gun that will accelerate chicken-sized objects so that their displacement along the gun barrel is given by $x = (9.0 \times 10^3 \text{ m/s}^2)t^2 - (8.0 \times 10^4 \text{ m/s}^3)t^3$. The object leaves the end of the barrel at $t = 0.025$ s. (a) How long must the gun barrel be? (b) What will be the speed of the objects as they leave the end of the barrel? (c) What net force must be exerted on a 1.50-kg object at (i) $t = 0$ and (ii) $t = 0.025$ s?

4.46 •• A spacecraft descends vertically near the surface of Planet X. An upward thrust of 25.0 kN from its engines slows it down at a rate of 1.20 m/s^2, but it speeds up at a rate of 0.80 m/s^2 with an upward thrust of 10.0 kN. (a) In each case, what is the direction of the acceleration of the spacecraft? (b) Draw a free-body diagram for the spacecraft. In each case, speeding up or slowing down, what is the direction of the net force on the spacecraft? (c) Apply Newton's second law to each case, slowing down or speeding up, and use this to find the spacecraft's weight near the surface of Planet X.

4.47 •• **CP** A 6.50-kg instrument is hanging by a vertical wire inside a space ship that is blasting off at the surface of the earth. This ship starts from rest and reaches an altitude of 276 m in 15.0 s with constant acceleration. (a) Draw a free-body diagram for the instrument during this time. Indicate which force is greater. (b) Find the force that the wire exerts on the instrument.

4.48 •• Suppose the rocket in Problem 4.47 is coming in for a vertical landing instead of blasting off. The captain adjusts the engine thrust so that the magnitude of the rocket's acceleration is the same as it was during blast-off. Repeat parts (a) and (b).

4.49 •• **BIO** **Insect Dynamics.** The froghopper (*Philaenus spumarius*), the champion leaper of the insect world, has a mass of 12.3 mg and leaves the ground (in the most energetic jumps) at 4.0 m/s from a vertical start. The jump itself lasts a mere 1.0 ms before the insect is clear of the ground. Assuming constant acceleration, (a) draw a free-body diagram of this mighty leaper while the jump is taking place; (b) find the force that the ground exerts on the froghopper during its jump; and (c) express the force in part (b) in terms of the froghopper's weight.

4.50 • A loaded elevator with very worn cables has a total mass of 2200 kg, and the cables can withstand a maximum tension of 28,000 N. (a) Draw the free-body force diagram for the elevator. In terms of the forces on your diagram, what is the net force on the elevator? Apply Newton's second law to the elevator and find the maximum upward acceleration for the elevator if the cables are not to break. (b) What would be the answer to part (a) if the elevator were on the moon, where $g = 1.62$ m/s^2?

4.51 •• **CP** **Jumping to the Ground.** A 75.0-kg man steps off a platform 3.10 m above the ground. He keeps his legs straight as he falls, but at the moment his feet touch the ground his knees begin to bend, and, treated as a particle, he moves an additional 0.60 m before coming to rest. (a) What is his speed at the instant his feet touch the ground? (b) Treating him as a particle, what is his acceleration (magnitude and direction) as he slows down, if the acceleration is assumed to be constant? (c) Draw his free-body diagram (see Section 4.6). In terms of the forces on the diagram, what is the net force on him? Use Newton's laws and the results of part (b) to calculate the average force his feet exert on the ground while he slows down. Express this force in newtons and also as a multiple of his weight.

4.52 ••• **CP** A 4.9-N hammer head is stopped from an initial downward velocity of 3.2 m/s in a distance of 0.45 cm by a nail in a pine board. In addition to its weight, there is a 15-N downward force on the hammer head applied by the person using the hammer. Assume that the acceleration of the hammer head is constant while it is in contact with the nail and moving downward. (a) Draw a free-body diagram for the hammer head. Identify the reaction force to each action force in the diagram. (b) Calculate the downward force $\vec{F}$ exerted by the hammer head on the nail while the hammer head is in contact with the nail and moving downward. (c) Suppose the nail is in hardwood and the distance the hammer head travels in coming to rest is only 0.12 cm. The downward forces on the hammer head are the same as in part (b). What then is the force $\vec{F}$ exerted by the hammer head on the nail while the hammer head is in contact with the nail and moving downward?

4.53 •• A uniform cable of weight w hangs vertically downward, supported by an upward force of magnitude w at its top end. What is the tension in the cable (a) at its top end; (b) at its bottom end; (c) at its middle? Your answer to each part must include a free-body diagram. (*Hint:* For each question choose the body to analyze to be a section of the cable or a point along the cable.) (d) Graph the tension in the rope versus the distance from its top end.

4.54 •• The two blocks in Fig. P4.54 are connected by a heavy uniform rope with a mass of 4.00 kg. An upward force of 200 N is applied as shown. (a) Draw three free-body diagrams: one for the 6.00-kg block, one for the 4.00-kg rope, and another one for the 5.00-kg block. For each force, indicate what body exerts that force. (b) What is the acceleration of the system? (c) What is the tension at the top of the heavy rope? (d) What is the tension at the midpoint of the rope?

Figure **P4.54**

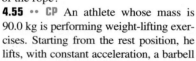
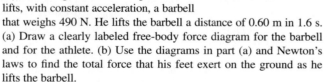

$F = 200$ N

6.00 kg

4.00 kg

5.00 kg

4.55 •• CP An athlete whose mass is 90.0 kg is performing weight-lifting exercises. Starting from the rest position, he lifts, with constant acceleration, a barbell that weighs 490 N. He lifts the barbell a distance of 0.60 m in 1.6 s. (a) Draw a clearly labeled free-body force diagram for the barbell and for the athlete. (b) Use the diagrams in part (a) and Newton's laws to find the total force that his feet exert on the ground as he lifts the barbell.

4.56 ••• A hot-air balloon consists of a basket, one passenger, and some cargo. Let the total mass be M. Even though there is an upward lift force on the balloon, the balloon is initially accelerating downward at a rate of $g/3$. (a) Draw a free-body diagram for the descending balloon. (b) Find the upward lift force in terms of the initial total weight Mg. (c) The passenger notices that he is heading straight for a waterfall and decides he needs to go up. What fraction of the total weight must he drop overboard so that the balloon accelerates *upward* at a rate of $g/2$? Assume that the upward lift force remains the same.

4.57 CP Two boxes, A and B, are connected to each end of a light vertical rope, as shown in Fig. P4.57. A constant upward force $F = 80.0$ N is applied to box A. Starting from rest, box B descends 12.0 m in 4.00 s. The tension in the rope connecting the two boxes is 36.0 N. (a) What is the mass of box B? (b) What is the mass of box A?

Figure **P4.57**

4.58 ••• CALC The position of a 2.75×10^5-N training helicopter under test is given by $\vec{r} = (0.020 \text{ m/s}^3)t^3\hat{\imath} + (2.2 \text{ m/s})t\hat{\jmath} - (0.060 \text{ m/s}^2)t^2\hat{k}$. Find the net force on the helicopter at $t = 5.0$ s.

4.59 • CALC An object with mass m moves along the x-axis. Its position as a function of time is given by $x(t) = At - Bt^3$, where A and B are constants. Calculate the net force on the object as a function of time.

4.60 • CALC An object with mass m initially at rest is acted on by a force $\vec{F} = k_1\hat{\imath} + k_2t^3\hat{\jmath}$, where k_1 and k_2 are constants. Calculate the velocity $\vec{v}(t)$ of the object as a function of time.

4.61 •• CP CALC A mysterious rocket-propelled object of mass 45.0 kg is initially at rest in the middle of the horizontal, frictionless surface of an ice-covered lake. Then a force directed east and with magnitude $F(t) = (16.8 \text{ N/s})t$ is applied. How far does the object travel in the first 5.00 s after the force is applied?

CHALLENGE PROBLEMS

4.62 ••• CALC An object of mass m is at rest in equilibrium at the origin. At $t = 0$ a new force $\vec{F}(t)$ is applied that has components

$$F_x(t) = k_1 + k_2y \qquad F_y(t) = k_3t$$

where k_1, k_2, and k_3 are constants. Calculate the position $\vec{r}(t)$ and velocity $\vec{v}(t)$ vectors as functions of time.

Answers

Chapter Opening Question ?

Newton's third law tells us that the car pushes on the crew member just as hard as the crew member pushes on the car, but in the opposite direction. This is true whether the car's engine is on and the car is moving forward partly under its own power, or the engine is off and being propelled by the crew member's push alone. The force magnitudes are different in the two situations, but in either case the push of the car on the crew member is just as strong as the push of the crew member on the car.

Test Your Understanding Questions

4.1 Answer: (iv) The gravitational force on the crate points straight downward. In Fig. 4.6 the x-axis points up and to the right, and the y-axis points up and to the left. Hence the gravitational force has both an x-component and a y-component, and both are negative.

4.2 Answer: (i), (ii), and (iv) In (i), (ii), and (iv) the body is not accelerating, so the net force on the body is zero. [In (iv), the box remains stationary as seen in the inertial reference frame of the ground as the truck accelerates forward, like the skater in Fig. 4.11a.] In (iii), the hawk is moving in a circle; hence it is accelerating and is *not* in equilibrium.

4.3 Answer: (iii), (i) and (iv) (tie), (ii) The acceleration is equal to the net force divided by the mass. Hence the magnitude of the acceleration in each situation is

(i) $a = (2.0 \text{ N})/(2.0 \text{ kg}) = 1.0 \text{ m/s}^2$;

(ii) $a = (8.0 \text{ N})/(2.0 \text{ N}) = 4.0 \text{ m/s}^2$;

(iii) $a = (2.0 \text{ N})/(8.0 \text{ kg}) = 0.25 \text{ m/s}^2$;

(iv) $a = (8.0 \text{ N})/(8.0 \text{ kg}) = 1.0 \text{ m/s}^2$.

4.4 It would take twice the effort for the astronaut to walk around because her weight on the planet would be twice as much as on the earth. But it would be just as easy to catch a ball moving horizontally. The ball's *mass* is the same as on earth, so the horizontal force the astronaut would have to exert to bring it to a stop (i.e., to give it the same acceleration) would also be the same as on earth.

4.5 By Newton's third law, the two forces have equal magnitudes. Because the car has much greater mass than the mosquito, it undergoes only a tiny, imperceptible acceleration in response to the force of the impact. By contrast, the mosquito, with its minuscule mass, undergoes a catastrophically large acceleration.

4.6 Answer: (iv) The buoyancy force is an *upward* force that the *water* exerts on the *swimmer*. By Newton's third law, the other half of the action–reaction pair is a *downward* force that the *swimmer* exerts on the *water* and has the same magnitude as the buoyancy force. It's true that the weight of the swimmer is also downward and has the same magnitude as the buoyancy force; however, the weight acts on the same body (the swimmer) as the buoyancy force, and so these forces aren't an action–reaction pair.

Bridging Problem

Answers: **(a)** *See a Video Tutor solution on MasteringPhysics®*
(b) (i) 2.20 m/s²; (ii) 6.00 N; (iii) 3.00 N

5 APPLYING NEWTON'S LAWS

LEARNING GOALS

By studying this chapter, you will learn:

- How to use Newton's first law to solve problems involving the forces that act on a body in equilibrium.

- How to use Newton's second law to solve problems involving the forces that act on an accelerating body.

- The nature of the different types of friction forces—static friction, kinetic friction, rolling friction, and fluid resistance—and how to solve problems that involve these forces.

- How to solve problems involving the forces that act on a body moving along a circular path.

- The key properties of the four fundamental forces of nature.

? This skydiver is descending under a parachute at a steady rate. In this situation, which has a greater magnitude: the force of gravity or the upward force of the air on the skydiver?

We saw in Chapter 4 that Newton's three laws of motion, the foundation of classical mechanics, can be stated very simply. But *applying* these laws to situations such as an iceboat skating across a frozen lake, a toboggan sliding down a hill, or an airplane making a steep turn requires analytical skills and problem-solving technique. In this chapter we'll help you extend the problem-solving skills you began to develop in Chapter 4.

We'll begin with equilibrium problems, in which we analyze the forces that act on a body at rest or moving with constant velocity. We'll then consider bodies that are not in equilibrium, for which we'll have to deal with the relationship between forces and motion. We'll learn how to describe and analyze the contact force that acts on a body when it rests on or slides over a surface. We'll also analyze the forces that act on a body that moves in a circle with constant speed. We close the chapter with a brief look at the fundamental nature of force and the classes of forces found in our physical universe.

5.1 Using Newton's First Law: Particles in Equilibrium

We learned in Chapter 4 that a body is in *equilibrium* when it is at rest or moving with constant velocity in an inertial frame of reference. A hanging lamp, a kitchen table, an airplane flying straight and level at a constant speed—all are examples of equilibrium situations. In this section we consider only equilibrium of a body that can be modeled as a particle. (In Chapter 11 we'll see how to analyze a body in equilibrium that can't be represented adequately as a particle, such as a bridge that's supported at various points along its span.) The essential physical principle is Newton's first law: When a particle is in equilibrium, the *net* force acting on it—that is, the vector sum of all the forces acting on it—must be zero:

$$\sum \vec{F} = 0 \qquad \text{(particle in equilibrium, vector form)} \qquad (5.1)$$

We most often use this equation in component form:

$$\sum F_x = 0 \qquad \sum F_y = 0 \qquad \text{(particle in equilibrium, component form)} \quad (5.2)$$

This section is about using Newton's first law to solve problems dealing with bodies in equilibrium. Some of these problems may seem complicated, but the important thing to remember is that *all* problems involving particles in equilibrium are done in the same way. Problem-Solving Strategy 5.1 details the steps you need to follow for any and all such problems. Study this strategy carefully, look at how it's applied in the worked-out examples, and try to apply it yourself when you solve assigned problems.

Problem-Solving Strategy 5.1 — Newton's First Law: Equilibrium of a Particle

IDENTIFY *the relevant concepts:* You must use Newton's *first* law for any problem that involves forces acting on a body in equilibrium—that is, either at rest or moving with constant velocity. For example, a car is in equilibrium when it's parked, but also when it's traveling down a straight road at a steady speed.

If the problem involves more than one body and the bodies interact with each other, you'll also need to use Newton's *third* law. This law allows you to relate the force that one body exerts on a second body to the force that the second body exerts on the first one.

Identify the target variable(s). Common target variables in equilibrium problems include the magnitude and direction (angle) of one of the forces, or the components of a force.

SET UP *the problem* using the following steps:
1. Draw a very simple sketch of the physical situation, showing dimensions and angles. You don't have to be an artist!
2. Draw a free-body diagram for each body that is in equilibrium. For the present, we consider the body as a particle, so you can represent it as a large dot. In your free-body diagram, *do not* include the other bodies that interact with it, such as a surface it may be resting on or a rope pulling on it.
3. Ask yourself what is interacting with the body by touching it or in any other way. On your free-body diagram, draw a force vector for each interaction. Label each force with a symbol for the *magnitude* of the force. If you know the angle at which a force is directed, draw the angle accurately and label it. Include the body's weight, unless the body has negligible mass. If the mass is given, use $w = mg$ to find the weight. A surface in contact with the body exerts a normal force perpendicular to the surface and possibly a friction force parallel to the surface. A rope or chain exerts a pull (never a push) in a direction along its length.
4. *Do not* show in the free-body diagram any forces exerted *by* the body on any other body. The sums in Eqs. (5.1) and (5.2)

include only forces that act *on* the body. For each force on the body, ask yourself "What other body causes that force?" If you can't answer that question, you may be imagining a force that isn't there.
5. Choose a set of coordinate axes and include them in your free-body diagram. (If there is more than one body in the problem, choose axes for each body separately.) Label the positive direction for each axis. If a body rests or slides on a plane surface, it usually simplifies things to choose axes that are parallel and perpendicular to this surface, even when the plane is tilted.

EXECUTE *the solution* as follows:
1. Find the components of each force along each of the body's coordinate axes. Draw a wiggly line through each force vector that has been replaced by its components, so you don't count it twice. The *magnitude* of a force is always positive, but its *components* may be positive or negative.
2. Set the sum of all x-components of force equal to zero. In a separate equation, set the sum of all y-components equal to zero. (*Never* add x- and y-components in a single equation.)
3. If there are two or more bodies, repeat all of the above steps for each body. If the bodies interact with each other, use Newton's third law to relate the forces they exert on each other.
4. Make sure that you have as many independent equations as the number of unknown quantities. Then solve these equations to obtain the target variables.

EVALUATE *your answer:* Look at your results and ask whether they make sense. When the result is a symbolic expression or formula, check to see that your formula works for any special cases (particular values or extreme cases for the various quantities) for which you can guess what the results ought to be.

Example 5.1 One-dimensional equilibrium: Tension in a massless rope

A gymnast with mass $m_G = 50.0$ kg suspends herself from the lower end of a hanging rope of negligible mass. The upper end of the rope is attached to the gymnasium ceiling. (a) What is the gymnast's weight? (b) What force (magnitude and direction) does the rope exert on her? (c) What is the tension at the top of the rope?

SOLUTION

IDENTIFY and SET UP: The gymnast and the rope are in equilibrium, so we can apply Newton's first law to both bodies. We'll use Newton's third law to relate the forces that they exert on each other. The target variables are the gymnast's weight, w_G; the force that the bottom of the rope exerts on the gymnast (call it $T_{R \text{ on } G}$); and the force that the ceiling exerts on the top of the rope (call it $T_{C \text{ on } R}$). Figure 5.1 shows our sketch of the situation and free-body diagrams for the gymnast and for the rope. We take the positive y-axis to be upward in each diagram. Each force acts in the vertical direction and so has only a y-component.

The forces $T_{R \text{ on } G}$ (the upward force of the rope on the gymnast, Fig. 5.1b) and $T_{G \text{ on } R}$ (the downward force of the gymnast on the rope, Fig. 5.1c) form an action–reaction pair. By Newton's third law, they must have the same magnitude.

5.1 Our sketches for this problem.

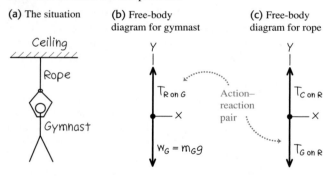

(a) The situation   **(b)** Free-body diagram for gymnast   **(c)** Free-body diagram for rope

Note that Fig. 5.1c includes only the forces that act *on* the rope. In particular, it doesn't include the force that the *rope* exerts on the *ceiling* (compare the discussion of the apple in Conceptual Example 4.9 in Section 4.5). Similarly, the force that the rope exerts on the ceiling doesn't appear in Fig. 5.1c.

EXECUTE: (a) The magnitude of the gymnast's weight is the product of her mass and the acceleration due to gravity, g:

$$w_G = m_G g = (50.0 \text{ kg})(9.80 \text{ m/s}^2) = 490 \text{ N}$$

(b) The gravitational force on the gymnast (her weight) points in the negative y-direction, so its y-component is $-w_G$. The upward force of the rope on the gymnast has unknown magnitude $T_{R \text{ on } G}$ and positive y-component $+T_{R \text{ on } G}$. We find this using Newton's first law:

$$\text{Gymnast:} \quad \sum F_y = T_{R \text{ on } G} + (-w_G) = 0 \quad \text{so}$$
$$T_{R \text{ on } G} = w_G = 490 \text{ N}$$

The rope pulls *up* on the gymnast with a force $T_{R \text{ on } G}$ of magnitude 490 N. (By Newton's third law, the gymnast pulls *down* on the rope with a force of the same magnitude, $T_{G \text{ on } R} = 490$ N.)

(c) We have assumed that the rope is weightless, so the only forces on it are those exerted by the ceiling (upward force of unknown magnitude $T_{C \text{ on } R}$) and by the gymnast (downward force of magnitude $T_{G \text{ on } R} = 490$ N). From Newton's first law, the *net* vertical force on the rope in equilibrium must be zero:

$$\text{Rope:} \quad \sum F_y = T_{C \text{ on } R} + (-T_{G \text{ on } R}) = 0 \quad \text{so}$$
$$T_{C \text{ on } R} = T_{G \text{ on } R} = 490 \text{ N}$$

EVALUATE: The *tension* at any point in the rope is the magnitude of the force that acts at that point. For this weightless rope, the tension $T_{G \text{ on } R}$ at the lower end has the same value as the tension $T_{C \text{ on } R}$ at the upper end. For such an ideal weightless rope, the tension has the same value at any point along the rope's length. (See the discussion in Conceptual Example 4.10 in Section 4.5.)

Example 5.2 One-dimensional equilibrium: Tension in a rope with mass

Find the tension at each end of the rope in Example 5.1 if the weight of the rope is 120 N.

SOLUTION

IDENTIFY and SET UP: As in Example 5.1, the target variables are the magnitudes $T_{G \text{ on } R}$ and $T_{C \text{ on } R}$ of the forces that act at the bottom and top of the rope, respectively. Once again, we'll apply Newton's first law to the gymnast and to the rope, and use Newton's third law to relate the forces that the gymnast and rope exert on each other. Again we draw separate free-body diagrams for the gymnast (Fig. 5.2a) and the rope (Fig. 5.2b). There is now a *third* force acting on the rope, however: the weight of the rope, of magnitude $w_R = 120$ N.

EXECUTE: The gymnast's free-body diagram is the same as in Example 5.1, so her equilibrium condition is also the same. From

Newton's third law, $T_{R \text{ on } G} = T_{G \text{ on } R}$, and we again have

$$\text{Gymnast:} \quad \sum F_y = T_{R \text{ on } G} + (-w_G) = 0 \quad \text{so}$$
$$T_{R \text{ on } G} = T_{G \text{ on } R} = w_G = 490 \text{ N}$$

The equilibrium condition $\sum F_y = 0$ for the rope is now

$$\text{Rope:} \quad \sum F_y = T_{C \text{ on } R} + (-T_{G \text{ on } R}) + (-w_R) = 0$$

Note that the y-component of $T_{C \text{ on } R}$ is positive because it points in the $+y$-direction, but the y-components of both $T_{G \text{ on } R}$ and w_R are negative. We solve for $T_{C \text{ on } R}$ and substitute the values $T_{G \text{ on } R} = T_{R \text{ on } G} = 490$ N and $w_R = 120$ N:

$$T_{C \text{ on } R} = T_{G \text{ on } R} + w_R = 490 \text{ N} + 120 \text{ N} = 610 \text{ N}$$

EVALUATE: When we include the weight of the rope, the tension is *different* at the rope's two ends: 610 N at the top and 490 N at the bottom. The force $T_{C\,on\,R} = 610$ N exerted by the ceiling has to hold up both the 490-N weight of the gymnast and the 120-N weight of the rope.

To see this more clearly, we draw a free-body diagram for a composite body consisting of the gymnast and rope together (Fig. 5.2c). Only two external forces act on this composite body: the force $T_{C\,on\,R}$ exerted by the ceiling and the total weight $w_G + w_R = 490$ N $+ 120$ N $= 610$ N. (The forces $T_{G\,on\,R}$ and $T_{R\,on\,G}$ are *internal* to the composite body. Newton's first law applies only to *external* forces, so these internal forces play no role.) Hence Newton's first law applied to this composite body is

Composite body: $\quad \sum F_y = T_{C\,on\,R} + [-(w_G + w_R)] = 0$

and so $T_{C\,on\,R} = w_G + w_R = 610$ N.

Treating the gymnast and rope as a composite body is simpler, but we can't find the tension $T_{G\,on\,R}$ at the bottom of the rope by this method. *Moral: Whenever you have more than one body in a problem involving Newton's laws, the safest approach is to treat each body separately.*

5.2 Our sketches for this problem, including the weight of the rope.

(a) Free-body diagram for gymnast   **(b)** Free-body diagram for rope   **(c)** Free-body diagram for gymnast and rope as a composite body

---

Example 5.3   Two-dimensional equilibrium

In Fig. 5.3a, a car engine with weight w hangs from a chain that is linked at ring O to two other chains, one fastened to the ceiling and the other to the wall. Find expressions for the tension in each of the three chains in terms of w. The weights of the ring and chains are negligible compared with the weight of the engine.

SOLUTION

IDENTIFY and SET UP: The target variables are the tension magnitudes T_1, T_2, and T_3 in the three chains (Fig. 5.3a). All the bodies are in equilibrium, so we'll use Newton's first law. We need three independent equations, one for each target variable. However, applying Newton's first law to just one body gives us only *two* equations, as in Eqs. (5.2). So we'll have to consider more than one body in equilibrium. We'll look at the engine (which is acted on by T_1) and the ring (which is acted on by all three chains and so is acted on by all three tensions).

Figures 5.3b and 5.3c show our free-body diagrams and choice of coordinate axes. There are two forces that act on the engine: its weight w and the upward force T_1 exerted by the vertical chain.

Three forces act on the ring: the tensions from the vertical chain (T_1), the horizontal chain (T_2), and the slanted chain (T_3). Because the vertical chain has negligible weight, it exerts forces of the same magnitude T_1 at both of its ends (see Example 5.1). (If the weight of this chain were not negligible, these two forces would have different magnitudes like the rope in Example 5.2.) The weight of the ring is also negligible, which is why it isn't included in Fig. 5.3c.

EXECUTE: The forces acting on the engine are along the y-axis only, so Newton's first law says

Engine: $\quad \sum F_y = T_1 + (-w) = 0 \quad$ and $\quad T_1 = w$

The horizontal and slanted chains don't exert forces on the engine itself because they are not attached to it. These forces do appear when we apply Newton's first law to the ring, however. In the free-body diagram for the ring (Fig. 5.3c), remember that T_1, T_2, and T_3 are the *magnitudes* of the forces. We resolve the force with magnitude T_3 into its x- and y-components. The ring is in equilibrium, so using Newton's first law we can write (separate)

5.3 (a) The situation. (b), (c) Our free-body diagrams.

(a) Engine, chains, and ring

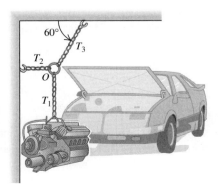

(b) Free-body diagram for engine   **(c)** Free-body diagram for ring O

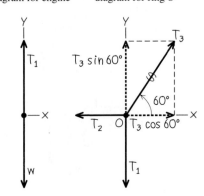

Continued

equations stating that the x- and y-components of the net force on the ring are zero:

Ring: $\sum F_x = T_3 \cos 60° + (-T_2) = 0$

Ring: $\sum F_y = T_3 \sin 60° + (-T_1) = 0$

Because $T_1 = w$ (from the engine equation), we can rewrite the second ring equation as

$$T_3 = \frac{T_1}{\sin 60°} = \frac{w}{\sin 60°} = 1.2w$$

We can now use this result in the first ring equation:

$$T_2 = T_3 \cos 60° = w \frac{\cos 60°}{\sin 60°} = 0.58w$$

EVALUATE: The chain attached to the ceiling exerts a force on the ring with a *vertical* component equal to T_1, which in turn is equal to w. But this force also has a horizontal component, so its magnitude T_3 is somewhat larger than w. This chain is under the greatest tension and is the one most susceptible to breaking.

To get enough equations to solve this problem, we had to consider not only the forces on the engine but also the forces acting on a second body (the ring connecting the chains). Situations like this are fairly common in equilibrium problems, so keep this technique in mind.

Example 5.4 An inclined plane

A car of weight w rests on a slanted ramp attached to a trailer (Fig. 5.4a). Only a cable running from the trailer to the car prevents the car from rolling off the ramp. (The car's brakes are off and its transmission is in neutral.) Find the tension in the cable and the force that the ramp exerts on the car's tires.

SOLUTION

IDENTIFY: The car is in equilibrium, so we use Newton's first law. The ramp exerts a separate force on each of the car's tires, but for simplicity we lump these forces into a single force. For a further simplification, we'll neglect any friction force the ramp exerts on the tires (see Fig. 4.2b). Hence the ramp only exerts a force on the car that is *perpendicular* to the ramp. As in Section 4.1, we call this force the *normal* force (see Fig. 4.2a). The two target variables are the magnitude n of the normal force and the magnitude T of the tension in the cable.

SET UP: Figure 5.4 shows the situation and a free-body diagram for the car. The three forces acting on the car are its weight (magnitude w), the tension in the cable (magnitude T), and the normal force (magnitude n). Note that the angle α between the ramp and the horizontal is equal to the angle α between the weight vector $\vec{w}$ and the downward normal to the plane of the ramp. Note also that we choose the x- and y-axes to be parallel and perpendicular to the ramp so that we only need to resolve one force (the weight) into x- and y-components. If we chose axes that were horizontal and vertical, we'd have to resolve both the normal force and the tension into components.

EXECUTE: To write down the x- and y-components of Newton's first law, we must first find the components of the weight. One complication is that the angle α in Fig. 5.4b is *not* measured from the $+x$-axis toward the $+y$-axis. Hence we *cannot* use Eqs. (1.6) directly to find the components. (You may want to review Section 1.8 to make sure that you understand this important point.)

One way to find the components of $\vec{w}$ is to consider the right triangles in Fig. 5.4b. The sine of α is the magnitude of the x-component of $\vec{w}$ (that is, the side of the triangle opposite α) divided by the magnitude w (the hypotenuse of the triangle). Similarly, the cosine of α is the magnitude of the y-component (the side of the triangle adjacent to α) divided by w. Both components are negative, so $w_x = -w \sin \alpha$ and $w_y = -w \cos \alpha$.

Another approach is to recognize that one component of $\vec{w}$ must involve $\sin \alpha$ while the other component involves $\cos \alpha$. To decide which is which, draw the free-body diagram so that the angle α is noticeably smaller or larger than 45°. (You'll have to fight the natural tendency to draw such angles as being close to 45°.) We've drawn Fig. 5.4b so that α is smaller than 45°, so $\sin \alpha$ is less than $\cos \alpha$. The figure shows that the x-component of $\vec{w}$ is smaller than the y-component, so the x-component must involve $\sin \alpha$ and the y-component must involve $\cos \alpha$. We again find $w_x = -w \sin \alpha$ and $w_y = -w \cos \alpha$.

In Fig. 5.4b we draw a wiggly line through the original vector representing the weight to remind us not to count it twice. Newton's first law gives us

$$\sum F_x = T + (-w \sin \alpha) = 0$$
$$\sum F_y = n + (-w \cos \alpha) = 0$$

(Remember that T, w, and n are all *magnitudes* of vectors and are therefore all positive.) Solving these equations for T and n, we find

$$T = w \sin \alpha$$
$$n = w \cos \alpha$$

EVALUATE: Our answers for T and n depend on the value of α. To check this dependence, let's look at some special cases. If the ramp is horizontal ($\alpha = 0$), we get $T = 0$ and $n = w$. As you might expect, no cable tension T is needed to hold the car, and the normal force n is equal in magnitude to the weight. If the ramp is vertical ($\alpha = 90°$), we get $T = w$ and $n = 0$. The cable tension T supports

5.4 A cable holds a car at rest on a ramp.

(a) Car on ramp

(b) Free-body diagram for car

We replace the weight by its components.

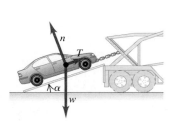

all of the car's weight, and there's nothing pushing the car against the ramp.

CAUTION **Normal force and weight may not be equal** It's a common error to automatically assume that the magnitude n of the normal force is equal to the weight w: Our result shows that this is *not* true in general. It's always best to treat n as a variable and solve for its value, as we have done here.

How would the answers for T and n be affected if the car were being pulled up the ramp at a constant speed? This, too, is an equilibrium situation, since the car's velocity is constant. So the calculation is the same, and T and n have the same values as when the car is at rest. (It's true that T must be greater than $w \sin \alpha$ to *start* the car moving up the ramp, but that's not what we asked.)

Example 5.5 **Equilibrium of bodies connected by cable and pulley**

Blocks of granite are to be hauled up a 15° slope out of a quarry, and dirt is to be dumped into the quarry to fill up old holes. To simplify the process, you design a system in which a granite block on a cart with steel wheels (weight w_1, including both block and cart) is pulled uphill on steel rails by a dirt-filled bucket (weight w_2, including both dirt and bucket) that descends vertically into the quarry (Fig. 5.5a). How must the weights w_1 and w_2 be related in order for the system to move with constant speed? Ignore friction in the pulley and wheels, and ignore the weight of the cable.

SOLUTION

IDENTIFY and SET UP: The cart and bucket each move with a constant velocity (in a straight line at constant speed). Hence each body is in equilibrium, and we can apply Newton's first law to each. Our target is an expression relating the weights w_1 and w_2.

Figure 5.5b shows our idealized model for the system, and Figs. 5.5c and 5.5d show our free-body diagrams. The two forces on the bucket are its weight w_2 and an upward tension exerted by the cable. As for the car on the ramp in Example 5.4, three forces act on the cart: its weight w_1, a normal force of magnitude n exerted by the rails, and a tension force from the cable. (We're ignoring friction, so we assume that the rails exert no force on the cart parallel to the incline.) Note that we orient the axes differently for each body; the choices shown are the most convenient ones.

We're assuming that the cable has negligible weight, so the tension forces that the cable exerts on the cart and on the bucket have the same magnitude T. As we did for the car in Example 5.4, we represent the weight of the cart in terms of its x- and y-components.

EXECUTE: Applying $\Sigma F_y = 0$ to the bucket in Fig. 5.5c, we find

$$\Sigma F_y = T + (-w_2) = 0 \quad \text{so} \quad T = w_2$$

Applying $\Sigma F_x = 0$ to the cart (and block) in Fig. 5.5d, we get

$$\Sigma F_x = T + (-w_1 \sin 15°) = 0 \quad \text{so} \quad T = w_1 \sin 15°$$

Equating the two expressions for T, we find

$$w_2 = w_1 \sin 15° = 0.26 w_1$$

EVALUATE: Our analysis doesn't depend at all on the direction in which the cart and bucket move. Hence the system can move with constant speed in *either* direction if the weight of the dirt and bucket is 26% of the weight of the granite block and cart. What would happen if w_2 were greater than $0.26 w_1$? If it were less than $0.26 w_1$?

Notice that we didn't need the equation $\Sigma F_y = 0$ for the cart and block. Can you use this to show that $n = w_1 \cos 15°$?

5.5 (a) The situation. (b) Our idealized model. (c), (d) Our free-body diagrams.

(a) Dirt-filled bucket pulls cart with granite block

Cart

Bucket

15°

(b) Idealized model of the system

w_1 Cart

15°

Bucket

w_2

(c) Free-body diagram for bucket

T

w_2

(d) Free-body diagram for cart

n

T

$w_1 \sin 15°$

15°

$w_1 \cos 15°$

w_1

Test Your Understanding of Section 5.1 A traffic light of weight w hangs from two lightweight cables, one on each side of the light. Each cable hangs at a 45° angle from the horizontal. What is the tension in each cable? (i) $w/2$; (ii) $w/\sqrt{2}$; (iii) w; (iv) $w\sqrt{2}$; (v) $2w$.

5.6 Correct and incorrect free-body diagrams for a falling body.

(a)

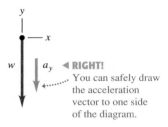

Only the force of gravity acts on this falling fruit.

(b) Correct free-body diagram

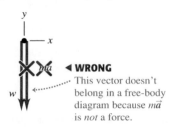

w a_y ◀ **RIGHT!**
You can safely draw the acceleration vector to one side of the diagram.

(c) Incorrect free-body diagram

$m\vec{a}$ ◀ **WRONG**
w This vector doesn't belong in a free-body diagram because $m\vec{a}$ is *not* a force.

5.2 Using Newton's Second Law: Dynamics of Particles

We are now ready to discuss *dynamics* problems. In these problems, we apply Newton's second law to bodies on which the net force is *not* zero. These bodies are *not* in equilibrium and hence are accelerating. The net force on the body is equal to the mass of the body times its acceleration:

$$\sum \vec{F} = m\vec{a} \qquad \text{(Newton's second law, vector form)} \qquad (5.3)$$

We most often use this relationship in component form:

$$\sum F_x = ma_x \qquad \sum F_y = ma_y \qquad \begin{array}{l}\text{(Newton's second law,}\\ \text{component form)}\end{array} \qquad (5.4)$$

The following problem-solving strategy is very similar to Problem-Solving Strategy 5.1 for equilibrium problems in Section 5.1. Study it carefully, watch how we apply it in our examples, and use it when you tackle the end-of-chapter problems. You can solve *any* dynamics problem using this strategy.

CAUTION $m\vec{a}$ **doesn't belong in free-body diagrams** Remember that the quantity $m\vec{a}$ is the *result* of forces acting on a body, *not* a force itself; it's not a push or a pull exerted by anything in the body's environment. When you draw the free-body diagram for an accelerating body (like the fruit in Fig. 5.6a), make sure you *never* include the "$m\vec{a}$ force" because *there is no such force* (Fig. 5.6c). You should review Section 4.3 if you're not clear on this point. Sometimes we draw the acceleration vector $\vec{a}$ *alongside* a free-body diagram, as in Fig. 5.6b. But we *never* draw the acceleration vector with its tail touching the body (a position reserved exclusively for the forces that act on the body).

Problem-Solving Strategy 5.2 **Newton's Second Law: Dynamics of Particles**

IDENTIFY *the relevant concepts:* You have to use Newton's second law for *any* problem that involves forces acting on an accelerating body.

Identify the target variable—usually an acceleration or a force. If the target variable is something else, you'll need to select another concept to use. For example, suppose the target variable is how fast a sled is moving when it reaches the bottom of a hill. Newton's second law will let you find the sled's acceleration; you'll then use the constant-acceleration relationships from Section 2.4 to find velocity from acceleration.

SET UP *the problem* using the following steps:
1. Draw a simple sketch of the situation that shows each moving body. For each body, draw a free-body diagram that shows all the forces acting *on* the body. (The acceleration of a body is determined by the forces that act on it, *not* by the forces that it exerts on anything else.) Make sure you can answer the question "What other body is applying this force?" for each force in your diagram. Never include the quantity $m\vec{a}$ in your free-body diagram; it's not a force!
2. Label each force with an algebraic symbol for the force's *magnitude*. Usually, one of the forces will be the body's weight; it's usually best to label this as $w = mg$.
3. Choose your *x*- and *y*-coordinate axes for each body, and show them in its free-body diagram. Be sure to indicate the positive direction for each axis. If you know the direction of the acceleration, it usually simplifies things to take one positive axis along that direction. If your problem involves two or more bodies that

accelerate in different directions, you can use a different set of axes for each body.
4. In addition to Newton's second law, $\sum \vec{F} = m\vec{a}$, identify any other equations you might need. For example, you might need one or more of the equations for motion with constant acceleration. If more than one body is involved, there may be relationships among their motions; for example, they may be connected by a rope. Express any such relationships as equations relating the accelerations of the various bodies.

EXECUTE *the solution* as follows:
1. For each body, determine the components of the forces along each of the body's coordinate axes. When you represent a force in terms of its components, draw a wiggly line through the original force vector to remind you not to include it twice.
2. Make a list of all the known and unknown quantities. In your list, identify the target variable or variables.
3. For each body, write a separate equation for each component of Newton's second law, as in Eqs. (5.4). In addition, write any additional equations that you identified in step 4 of "Set Up." (You need as many equations as there are target variables.)
4. Do the easy part—the math! Solve the equations to find the target variable(s).

EVALUATE *your answer:* Does your answer have the correct units? (When appropriate, use the conversion $1 \text{ N} = 1 \text{ kg} \cdot \text{m/s}^2$.) Does it have the correct algebraic sign? When possible, consider particular values or extreme cases of quantities and compare the results with your intuitive expectations. Ask, "Does this result make sense?"

Example 5.6 **Straight-line motion with a constant force**

An iceboat is at rest on a frictionless horizontal surface (Fig. 5.7a). A wind is blowing along the direction of the runners so that 4.0 s after the iceboat is released, it is moving at 6.0 m/s (about 22 km/h, or 13 mi/h). What constant horizontal force F_W does the wind exert on the iceboat? The combined mass of iceboat and rider is 200 kg.

SOLUTION

IDENTIFY and SET UP: Our target variable is one of the forces (F_W) acting on the accelerating iceboat, so we need to use Newton's second law. The forces acting on the iceboat and rider (considered as a unit) are the weight w, the normal force n exerted by the surface, and the horizontal force F_W. Figure 5.7b shows the free-body diagram. The net force and hence the acceleration are to the right, so we chose the positive x-axis in this direction. The acceleration isn't given; we'll need to find it. Since the wind is assumed to exert a constant force, the resulting acceleration is constant and we can use one of the constant-acceleration formulas from Section 2.4.

5.7 (a) The situation. (b) Our free-body diagram.

(a) Iceboat and rider on frictionless ice

(b) Free-body diagram for iceboat and rider

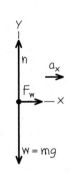

The iceboat starts at rest (its initial x-velocity is $v_{0x} = 0$) and it attains an x-velocity $v_x = 6.0$ m/s after an elapsed time $t = 4.0$ s. To relate the x-acceleration a_x to these quantities we use Eq. (2.8), $v_x = v_{0x} + a_x t$. There is no vertical acceleration, so we expect that the normal force on the iceboat is equal in magnitude to the iceboat's weight.

EXECUTE: The *known* quantities are the mass $m = 200$ kg, the initial and final x-velocities $v_{0x} = 0$ and $v_x = 6.0$ m/s, and the elapsed time $t = 4.0$ s. The three *unknown* quantities are the acceleration a_x, the normal force n, and the horizontal force F_W. Hence we need three equations.

The first two equations are the x- and y-equations for Newton's second law. The force F_W is in the positive x-direction, while the forces n and $w = mg$ are in the positive and negative y-directions, respectively. Hence we have

$$\sum F_x = F_W = ma_x$$
$$\sum F_y = n + (-mg) = 0 \quad \text{so} \quad n = mg$$

The third equation is the constant-acceleration relationship, Eq. (2.8):

$$v_x = v_{0x} + a_x t$$

To find F_W, we first solve this third equation for a_x and then substitute the result into the $\sum F_x$ equation:

$$a_x = \frac{v_x - v_{0x}}{t} = \frac{6.0 \text{ m/s} - 0 \text{ m/s}}{4.0 \text{ s}} = 1.5 \text{ m/s}^2$$
$$F_W = ma_x = (200 \text{ kg})(1.5 \text{ m/s}^2) = 300 \text{ kg} \cdot \text{m/s}^2$$

Since 1 kg $\cdot$ m/s^2 = 1 N, the final answer is

$$F_W = 300 \text{ N} \text{ (about 67 lb)}$$

EVALUATE: Our answers for F_W and n have the correct units for a force, and (as expected) the magnitude n of the normal force is equal to mg. Does it seem reasonable that the force F_W is substantially *less* than mg?

Example 5.7 **Straight-line motion with friction**

Suppose a constant horizontal friction force with magnitude 100 N opposes the motion of the iceboat in Example 5.6. In this case, what constant force F_W must the wind exert on the iceboat to cause the same constant x-acceleration $a_x = 1.5$ m/s^2?

SOLUTION

IDENTIFY and SET UP: Again the target variable is F_W. We are given the x-acceleration, so to find F_W all we need is Newton's second law. Figure 5.8 shows our new free-body diagram. The only difference from Fig. 5.7b is the addition of the friction force $\vec{f}$, which points opposite the motion. (Note that the *magnitude* $f = 100$ N is a positive quantity, but the *component* in the x-direction f_x is negative, equal to $-f$ or -100 N.) Because the wind must now overcome the friction force to yield the same acceleration as in Example 5.6, we expect our answer for F_W to be greater than the 300 N we found there.

5.8 Our free-body diagram for the iceboat and rider with a friction force $\vec{f}$ opposing the motion.

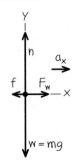

Continued

EXECUTE: Two forces now have x-components: the force of the wind and the friction force. The x-component of Newton's second law gives

$$\sum F_x = F_W + (-f) = ma_x$$
$$F_W = ma_x + f = (200 \text{ kg})(1.5 \text{ m/s}^2) + (100 \text{ N}) = 400 \text{ N}$$

EVALUATE: The required value of F_W is 100 N greater than in Example 5.6 because the wind must now push against an additional 100-N friction force.

Example 5.8 | Tension in an elevator cable

An elevator and its load have a combined mass of 800 kg (Fig. 5.9a). The elevator is initially moving downward at 10.0 m/s; it slows to a stop with constant acceleration in a distance of 25.0 m. What is the tension T in the supporting cable while the elevator is being brought to rest?

SOLUTION

IDENTIFY and SET UP: The target variable is the tension T, which we'll find using Newton's second law. As in Example 5.6, we'll determine the acceleration using a constant-acceleration formula. Our free-body diagram (Fig. 5.9b) shows two forces acting on the elevator: its weight w and the tension force T of the cable. The elevator is moving downward with decreasing speed, so its acceleration is upward; we chose the positive y-axis to be upward.

The elevator is moving in the negative y-direction, so its initial y-velocity v_{0y} and its y-displacement $y - y_0$ are both negative: $v_{0y} = -10.0$ m/s and $y - y_0 = -25.0$ m. The final y-velocity is $v_y = 0$. To find the y-acceleration a_y from this information, we'll use Eq. (2.13) in the form $v_y^2 = v_{0y}^2 + 2a_y(y - y_0)$. Once we have a_y, we'll substitute it into the y-component of Newton's second law from Eqs. (5.4) and solve for T. The net force must be upward to give an upward acceleration, so we expect T to be greater than the weight $w = mg = (800 \text{ kg})(9.80 \text{ m/s}^2) = 7840$ N.

EXECUTE: First let's write out Newton's second law. The tension force acts upward and the weight acts downward, so

$$\sum F_y = T + (-w) = ma_y$$

We solve for the target variable T:

$$T = w + ma_y = mg + ma_y = m(g + a_y)$$

5.9 (a) The situation. (b) Our free-body diagram.

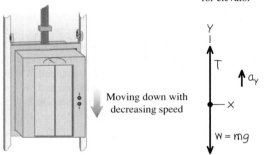

(a) Descending elevator

(b) Free-body diagram for elevator

Moving down with decreasing speed

$w = mg$

To determine a_y, we rewrite the constant-acceleration equation $v_y^2 = v_{0y}^2 + 2a_y(y - y_0)$:

$$a_y = \frac{v_y^2 - v_{0y}^2}{2(y - y_0)} = \frac{(0)^2 - (-10.0 \text{ m/s})^2}{2(-25.0 \text{ m})} = +2.00 \text{ m/s}^2$$

The acceleration is upward (positive), just as it should be.

Now we can substitute the acceleration into the equation for the tension:

$$T = m(g + a_y) = (800 \text{ kg})(9.80 \text{ m/s}^2 + 2.00 \text{ m/s}^2)$$
$$= 9440 \text{ N}$$

EVALUATE: The tension is greater than the weight, as expected. Can you see that we would get the same answers for a_y and T if the elevator were moving *upward* and *gaining* speed at a rate of 2.00 m/s²?

Example 5.9 | Apparent weight in an accelerating elevator

A 50.0-kg woman stands on a bathroom scale while riding in the elevator in Example 5.8. What is the reading on the scale?

SOLUTION

IDENTIFY and SET UP: The scale (Fig. 5.10a) reads the magnitude of the downward force exerted *by* the woman *on* the scale. By Newton's third law, this equals the magnitude of the upward normal force exerted *by* the scale *on* the woman. Hence our target variable is the magnitude n of the normal force. We'll find n by applying Newton's second law to the woman. We already know her acceleration; it's the same as the acceleration of the elevator, which we calculated in Example 5.8.

Figure 5.10b shows our free-body diagram for the woman. The forces acting on her are the normal force n exerted by the scale and her weight $w = mg = (50.0 \text{ kg})(9.80 \text{ m/s}^2) = 490$ N.

5.10 (a) The situation. (b) Our free-body diagram.

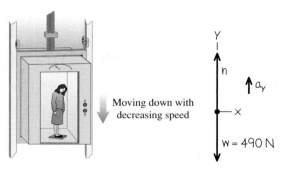

(a) Woman in a descending elevator

(b) Free-body diagram for woman

Moving down with decreasing speed

$w = 490$ N

(The tension force, which played a major role in Example 5.8, doesn't appear here because it doesn't act on the woman.) From Example 5.8, the y-acceleration of the elevator and of the woman is $a_y = +2.00 \text{ m/s}^2$. As in Example 5.8, the upward force on the body accelerating upward (in this case, the normal force on the woman) will have to be greater than the body's weight to produce the upward acceleration.

EXECUTE: Newton's second law gives

$$\sum F_y = n + (-mg) = ma_y$$
$$n = mg + ma_y = m(g + a_y)$$
$$= (50.0 \text{ kg})(9.80 \text{ m/s}^2 + 2.00 \text{ m/s}^2) = 590 \text{ N}$$

EVALUATE: Our answer for n means that while the elevator is stopping, the scale pushes up on the woman with a force of 590 N. By Newton's third law, she pushes down on the scale with the same force. So the scale reads 590 N, which is 100 N more than her actual weight. The scale reading is called the passenger's **apparent weight.** The woman *feels* the floor pushing up harder on her feet than when the elevator is stationary or moving with constant velocity.

What would the woman feel if the elevator were accelerating *downward,* so that $a_y = -2.00 \text{ m/s}^2$? This would be the case if the elevator were moving upward with decreasing speed or moving downward with increasing speed. To find the answer for this situation, we just insert the new value of a_y in our equation for n:

$$n = m(g + a_y) = (50.0 \text{ kg})[9.80 \text{ m/s}^2 + (-2.00 \text{ m/s}^2)]$$
$$= 390 \text{ N}$$

Now the woman feels as though she weighs only 390 N, or 100 N *less* than her actual weight w.

You can feel these effects yourself; try taking a few steps in an elevator that is coming to a stop after descending (when your apparent weight is greater than w) or coming to a stop after ascending (when your apparent weight is less than w).

Apparent Weight and Apparent Weightlessness

Let's generalize the result of Example 5.9. When a passenger with mass m rides in an elevator with y-acceleration a_y, a scale shows the passenger's apparent weight to be

$$n = m(g + a_y)$$

When the elevator is accelerating upward, a_y is positive and n is greater than the passenger's weight $w = mg$. When the elevator is accelerating downward, a_y is negative and n is less than the weight. If the passenger doesn't know the elevator is accelerating, she may feel as though her weight is changing; indeed, this is just what the scale shows.

The extreme case occurs when the elevator has a downward acceleration $a_y = -g$—that is, when it is in free fall. In that case $n = 0$ and the passenger *seems* to be weightless. Similarly, an astronaut orbiting the earth with a spacecraft experiences *apparent weightlessness* (Fig. 5.11). In each case, the person is not truly weightless because a gravitational force still acts. But the person's sensations in this free-fall condition are exactly the same as though the person were in outer space with no gravitational force at all. In both cases the person and the vehicle (elevator or spacecraft) fall together with the same acceleration g, so nothing pushes the person against the floor or walls of the vehicle.

5.11 Astronauts in orbit feel "weightless" because they have the same acceleration as their spacecraft—*not* because they are "outside the pull of the earth's gravity." (If no gravity acted on them, the astronauts and their spacecraft wouldn't remain in orbit, but would fly off into deep space.)

Example 5.10 **Acceleration down a hill**

A toboggan loaded with students (total weight w) slides down a snow-covered slope. The hill slopes at a constant angle α, and the toboggan is so well waxed that there is virtually no friction. What is its acceleration?

SOLUTION

IDENTIFY and SET UP: Our target variable is the acceleration, which we'll find using Newton's second law. There is no friction, so only two forces act on the toboggan: its weight w and the normal force n exerted by the hill.

Figure 5.12 shows our sketch and free-body diagram. As in Example 5.4, the surface is inclined, so the normal force is not vertical and is not equal in magnitude to the weight. Hence we must use both components of $\sum \vec{F} = m\vec{a}$ in Eqs. (5.4). We take axes parallel

5.12 Our sketches for this problem.

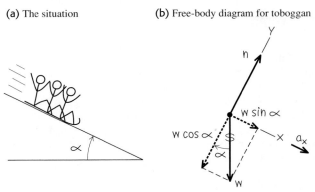

(a) The situation

(b) Free-body diagram for toboggan

Continued

and perpendicular to the surface of the hill, so that the acceleration (which is parallel to the hill) is along the positive x-direction.

EXECUTE: The normal force has only a y-component, but the weight has both x- and y-components: $w_x = w \sin \alpha$ and $w_y = -w \cos \alpha$. (In Example 5.4 we had $w_x = -w \sin \alpha$. The difference is that the positive x-axis was uphill in Example 5.4 but is downhill in Fig. 5.12b.) The wiggly line in Fig. 5.12b reminds us that we have resolved the weight into its components. The acceleration is purely in the $+x$-direction, so $a_y = 0$. Newton's second law in component form then tells us that

$$\sum F_x = w \sin \alpha = ma_x$$
$$\sum F_y = n - w \cos \alpha = ma_y = 0$$

Since $w = mg$, the x-component equation tells us that $mg \sin \alpha = ma_x$, or

$$a_x = g \sin \alpha$$

Note that we didn't need the y-component equation to find the acceleration. That's part of the beauty of choosing the x-axis to lie along the acceleration direction! The y-equation tells us the mag-

nitude of the normal force exerted by the hill on the toboggan:

$$n = w \cos \alpha = mg \cos \alpha$$

EVALUATE: Notice that the normal force n is not equal to the toboggan's weight (compare Example 5.4). Notice also that the mass m does not appear in our result for the acceleration. That's because the downhill force on the toboggan (a component of the weight) is proportional to m, so the mass cancels out when we use $\sum F_x = ma_x$ to calculate a_x. Hence *any* toboggan, regardless of its mass, slides down a frictionless hill with acceleration $g \sin \alpha$.

If the plane is horizontal, $\alpha = 0$ and $a_x = 0$ (the toboggan does not accelerate); if the plane is vertical, $\alpha = 90°$ and $a_x = g$ (the toboggan is in free fall).

CAUTION **Common free-body diagram errors** Figure 5.13 shows both the correct way (Fig. 5.13a) and a common *incorrect* way (Fig. 5.13b) to draw the free-body diagram for the toboggan. The diagram in Fig. 5.13b is wrong for two reasons: The normal force must be drawn perpendicular to the surface, and there's no such thing as the "$m\vec{a}$ force." If you remember that "normal" means "perpendicular" and that $m\vec{a}$ is not itself a force, you'll be well on your way to always drawing correct free-body diagrams.

5.13 Correct and incorrect free-body diagrams for a toboggan on a frictionless hill.

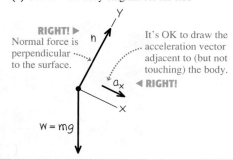

(a) Correct free-body diagram for the sled

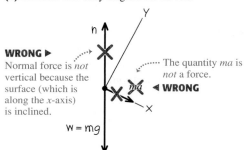

(b) Incorrect free-body diagram for the sled

Example 5.11 **Two bodies with the same acceleration**

You push a 1.00-kg food tray through the cafeteria line with a constant 9.0-N force. The tray pushes on a 0.50-kg carton of milk (Fig. 5.14a). The tray and carton slide on a horizontal surface so greasy that friction can be neglected. Find the acceleration of the tray and carton and the horizontal force that the tray exerts on the carton.

SOLUTION

IDENTIFY and SET UP: Our *two* target variables are the acceleration of the tray–carton system and the force of the tray on the carton. We'll use Newton's second law to get two equations, one for each target variable. We set up and solve the problem in two ways.

Method 1: We treat the milk carton (mass m_C) and tray (mass m_T) as separate bodies, each with its own free-body diagram (Figs. 5.14b and 5.14c). The force F that you exert on the tray doesn't appear in the free-body diagram for the carton, which is accelerated by the force (of magnitude $F_{\text{T on C}}$) exerted on it by the tray. By Newton's third law, the carton exerts a force of equal magnitude on the tray: $F_{\text{C on T}} = F_{\text{T on C}}$. We take the acceleration to

be in the positive x-direction; both the tray and milk carton move with the same x-acceleration a_x.

Method 2: We treat the tray and milk carton as a composite body of mass $m = m_T + m_C = 1.50 \text{ kg}$ (Fig. 5.14d). The only horizontal force acting on this body is the force F that you exert. The forces $F_{\text{T on C}}$ and $F_{\text{C on T}}$ don't come into play because they're *internal* to this composite body, and Newton's second law tells us that only *external* forces affect a body's acceleration (see Section 4.3). To find the magnitude $F_{\text{T on C}}$ we'll again apply Newton's second law to the carton, as in Method 1.

EXECUTE: *Method 1:* The x-component equations of Newton's second law are

Tray: $\quad \sum F_x = F - F_{\text{C on T}} = F - F_{\text{T on C}} = m_T a_x$
Carton: $\quad \sum F_x = F_{\text{T on C}} = m_C a_x$

These are two simultaneous equations for the two target variables a_x and $F_{\text{T on C}}$. (Two equations are all we need, which means that

5.14 Pushing a food tray and milk carton in the cafeteria line.

(a) A milk carton and a food tray   (b) Free-body diagram for milk carton   (c) Free-body diagram for food tray   (d) Free-body diagram for carton and tray as a composite body

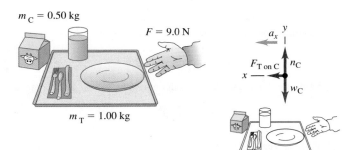

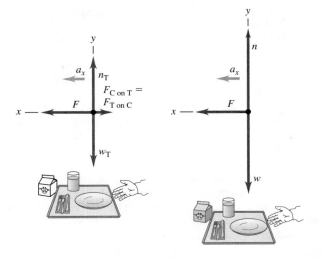

the y-components don't play a role in this example.) An easy way to solve the two equations for a_x is to add them; this eliminates $F_{\text{T on C}}$, giving

$$F = m_{\text{T}}a_x + m_{\text{C}}a_x = (m_{\text{T}} + m_{\text{C}})a_x$$

and

$$a_x = \frac{F}{m_{\text{T}} + m_{\text{C}}} = \frac{9.0\text{ N}}{1.00\text{ kg} + 0.50\text{ kg}} = 6.0\text{ m/s}^2 = 0.61g$$

Substituting this value into the carton equation gives

$$F_{\text{T on C}} = m_{\text{C}}a_x = (0.50\text{ kg})(6.0\text{ m/s}^2) = 3.0\text{ N}$$

Method 2: The x-component of Newton's second law for the composite body of mass m is

$$\sum F_x = F = ma_x$$

The acceleration of this composite body is

$$a_x = \frac{F}{m} = \frac{9.0\text{ N}}{1.50\text{ kg}} = 6.0\text{ m/s}^2$$

Then, looking at the milk carton by itself, we see that to give it an acceleration of 6.0 m/s² requires that the tray exert a force

$$F_{\text{T on C}} = m_{\text{C}}a_x = (0.50\text{ kg})(6.0\text{ m/s}^2) = 3.0\text{ N}$$

EVALUATE: The answers are the same with both methods. To check the answers, note that there are different forces on the two sides of the tray: $F = 9.0$ N on the right and $F_{\text{C on T}} = 3.0$ N on the left. The net horizontal force on the tray is $F - F_{\text{C on T}} = 6.0$ N, exactly enough to accelerate a 1.00-kg tray at 6.0 m/s².

Treating two bodies as a single, composite body works *only* if the two bodies have the same magnitude *and* direction of acceleration. If the accelerations are different we must treat the two bodies separately, as in the next example.

Example 5.12   **Two bodies with the same magnitude of acceleration**

Figure 5.15a shows an air-track glider with mass m_1 moving on a level, frictionless air track in the physics lab. The glider is connected to a lab weight with mass m_2 by a light, flexible, non-stretching string that passes over a stationary, frictionless pulley. Find the acceleration of each body and the tension in the string.

SOLUTION

IDENTIFY and SET UP: The glider and weight are accelerating, so again we must use Newton's second law. Our three target variables are the tension T in the string and the accelerations of the two bodies.

The two bodies move in different directions—one horizontal, one vertical—so we can't consider them together as we did the bodies in Example 5.11. Figures 5.15b and 5.15c show our free-body diagrams and coordinate systems. It's convenient to have both bodies accelerate in the positive axis directions,

5.15 (a) The situation. (b), (c) Our free-body diagrams.

(a) Apparatus   (b) Free-body diagram for glider   (c) Free-body diagram for weight

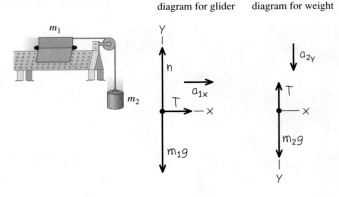

Continued

so we chose the positive *y*-direction for the lab weight to be downward.

We consider the string to be massless and to slide over the pulley without friction, so the tension *T* in the string is the same throughout and it applies a force of the same magnitude *T* to each body. (You may want to review Conceptual Example 4.10, in which we discussed the tension force exerted by a massless string.) The weights are m_1g and m_2g.

While the *directions* of the two accelerations are different, their *magnitudes* are the same. (That's because the string doesn't stretch, so the two bodies must move equal distances in equal times and their speeds at any instant must be equal. When the speeds change, they change at the same rate, so the accelerations of the two bodies must have the same magnitude *a*.) We can express this relationship as $a_{1x} = a_{2y} = a$, which means that we have only *two* target variables: *a* and the tension *T*.

What results do we expect? If $m_1 = 0$ (or, approximately, for m_1 much less than m_2) the lab weight will fall freely with acceleration *g*, and the tension in the string will be zero. For $m_2 = 0$ (or, approximately, for m_2 much less than m_1) we expect zero acceleration and zero tension.

EXECUTE: Newton's second law gives

Glider: $\qquad \sum F_x = T = m_1a_{1x} = m_1a$

Glider: $\qquad \sum F_y = n + (-m_1g) = m_1a_{1y} = 0$

Lab weight: $\quad \sum F_y = m_2g + (-T) = m_2a_{2y} = m_2a$

(There are no forces on the lab weight in the *x*-direction.) In these equations we've used $a_{1y} = 0$ (the glider doesn't accelerate vertically) and $a_{1x} = a_{2y} = a$.

The *x*-equation for the glider and the equation for the lab weight give us two simultaneous equations for *T* and *a*:

Glider: $\qquad T = m_1a$

Lab weight: $\qquad m_2g - T = m_2a$

We add the two equations to eliminate *T*, giving

$$m_2g = m_1a + m_2a = (m_1 + m_2)a$$

and so the magnitude of each body's acceleration is

$$a = \frac{m_2}{m_1 + m_2}g$$

Substituting this back into the glider equation $T = m_1a$, we get

$$T = \frac{m_1m_2}{m_1 + m_2}g$$

EVALUATE: The acceleration is in general less than *g*, as you might expect; the string tension keeps the lab weight from falling freely. The tension *T* is *not* equal to the weight m_2g of the lab weight, but is *less* by a factor of $m_1/(m_1 + m_2)$. If *T* were equal to m_2g, then the lab weight would be in equilibrium, and it isn't.

As predicted, the acceleration is equal to *g* for $m_1 = 0$ and equal to zero for $m_2 = 0$, and $T = 0$ for either $m_1 = 0$ or $m_2 = 0$.

CAUTION **Tension and weight may not be equal** It's a common mistake to assume that if an object is attached to a vertical string, the string tension must be equal to the object's weight. That was the case in Example 5.5, where the acceleration was zero, but it's not the case in this example! The only safe approach is *always* to treat the tension as a variable, as we did here. ∎

PhET: Lunar Lander
ActivPhysics 2.1.5: Car Race
ActivPhysics 2.2: Lifting a Crate
ActivPhysics 2.3: Lowering a Crate
ActivPhysics 2.4: Rocket Blasts Off
ActivPhysics 2.5: Modified Atwood Machine

Test Your Understanding of Section 5.2 Suppose you hold the glider in Example 5.12 so that it and the weight are initially at rest. You give the glider a push to the left in Fig. 5.15a and then release it. The string remains taut as the glider moves to the left, comes instantaneously to rest, then moves to the right. At the instant the glider has zero velocity, what is the tension in the string? (i) greater than in Example 5.12; (ii) the same as in Example 5.12; (iii) less than in Example 5.12, but greater than zero; (iv) zero. ∎

5.16 The sport of ice hockey depends on having the right amount of friction between a player's skates and the ice. If there were too much friction, the players would move too slowly; if there were too little friction, they would fall over.

5.3 Frictional Forces

We've seen several problems where a body rests or slides on a surface that exerts forces on the body. Whenever two bodies interact by direct contact (touching) of their surfaces, we describe the interaction in terms of *contact forces*. The normal force is one example of a contact force; in this section we'll look in detail at another contact force, the force of friction.

Friction is important in many aspects of everyday life. The oil in a car engine minimizes friction between moving parts, but without friction between the tires and the road we couldn't drive or turn the car. Air drag—the frictional force exerted by the air on a body moving through it—decreases automotive fuel economy but makes parachutes work. Without friction, nails would pull out, light bulbs would unscrew effortlessly, and ice hockey would be hopeless (Fig. 5.16).

Kinetic and Static Friction

When you try to slide a heavy box of books across the floor, the box doesn't move at all unless you push with a certain minimum force. Then the box starts moving, and you can usually keep it moving with less force than you needed to get it started. If you take some of the books out, you need less force than before to get it started or keep it moving. What general statements can we make about this behavior?

First, when a body rests or slides on a surface, we can think of the surface as exerting a single contact force on the body, with force components perpendicular and parallel to the surface (Fig. 5.17). The perpendicular component vector is the normal force, denoted by $\vec{n}$. The component vector parallel to the surface (and perpendicular to $\vec{n}$) is the **friction force,** denoted by $\vec{f}$. If the surface is friction-less, then $\vec{f}$ is zero but there is still a normal force. (Frictionless surfaces are an unattainable idealization, like a massless rope. But we can approximate a surface as frictionless if the effects of friction are negligibly small.) The direction of the friction force is always such as to oppose relative motion of the two surfaces.

The kind of friction that acts when a body slides over a surface is called a **kinetic friction force** $\vec{f}_k$. The adjective "kinetic" and the subscript "k" remind us that the two surfaces are moving relative to each other. The *magnitude* of the kinetic friction force usually increases when the normal force increases. This is why it takes more force to slide a box across the floor when it's full of books than when it's empty. Automotive brakes use the same principle: The harder the brake pads are squeezed against the rotating brake disks, the greater the braking effect. In many cases the magnitude of the kinetic friction force f_k is found experimen-tally to be approximately *proportional* to the magnitude n of the normal force. In such cases we represent the relationship by the equation

$$f_k = \mu_k n \qquad \text{(magnitude of kinetic friction force)} \qquad (5.5)$$

where μ_k (pronounced "mu-sub-k") is a constant called the **coefficient of kinetic friction.** The more slippery the surface, the smaller this coefficient. Because it is a quotient of two force magnitudes, μ_k is a pure number without units.

CAUTION Friction and normal forces are always perpendicular Remember that Eq. (5.5) is *not* a vector equation because $\vec{f}_k$ and $\vec{n}$ are always perpendicular. Rather, it is a scalar relationship between the magnitudes of the two forces. ∎

Equation (5.5) is only an approximate representation of a complex phenome-non. On a microscopic level, friction and normal forces result from the intermol-ecular forces (fundamentally electrical in nature) between two rough surfaces at points where they come into contact (Fig. 5.18). As a box slides over the floor, bonds between the two surfaces form and break, and the total number of such bonds varies; hence the kinetic friction force is not perfectly constant. Smoothing the surfaces can actually increase friction, since more molecules are able to inter-act and bond; bringing two smooth surfaces of the same metal together can cause a "cold weld." Lubricating oils work because an oil film between two surfaces (such as the pistons and cylinder walls in a car engine) prevents them from com-ing into actual contact.

Table 5.1 lists some representative values of μ_k. Although these values are given with two significant figures, they are only approximate, since friction forces can also depend on the speed of the body relative to the surface. For now we'll ignore this effect and assume that μ_k and f_k are independent of speed, in order to concentrate on the simplest cases. Table 5.1 also lists coefficients of static friction; we'll define these shortly.

Friction forces may also act when there is *no* relative motion. If you try to slide a box across the floor, the box may not move at all because the floor exerts an equal and opposite friction force on the box. This is called a **static friction force** $\vec{f}_s$. In Fig. 5.19a, the box is at rest, in equilibrium, under the action of its weight $\vec{w}$ and the upward normal force $\vec{n}$. The normal force is equal in magnitude to the weight ($n = w$) and is exerted on the box by the floor. Now we tie a rope

5.17 When a block is pushed or pulled over a surface, the surface exerts a contact force on it.

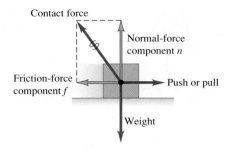

The friction and normal forces are really components of a single contact force.

5.18 The normal and friction forces arise from interactions between molecules at high points on the surfaces of the block and the floor.

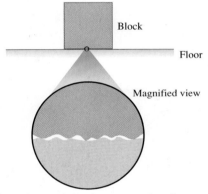

On a microscopic level, even smooth surfaces are rough; they tend to catch and cling.

Table 5.1 Approximate Coefficients of Friction

Materials	Coefficient of Static Friction, μ_s	Coefficient of Kinetic Friction, μ_k
Steel on steel	0.74	0.57
Aluminum on steel	0.61	0.47
Copper on steel	0.53	0.36
Brass on steel	0.51	0.44
Zinc on cast iron	0.85	0.21
Copper on cast iron	1.05	0.29
Glass on glass	0.94	0.40
Copper on glass	0.68	0.53
Teflon on Teflon	0.04	0.04
Teflon on steel	0.04	0.04
Rubber on concrete (dry)	1.0	0.8
Rubber on concrete (wet)	0.30	0.25

5.19 (a), (b), (c) When there is no relative motion, the magnitude of the static friction force f_s is less than or equal to $\mu_s n$. (d) When there is relative motion, the magnitude of the kinetic friction force f_k equals $\mu_k n$. (e) A graph of the friction force magnitude f as a function of the magnitude T of the applied force. The kinetic friction force varies somewhat as intermolecular bonds form and break.

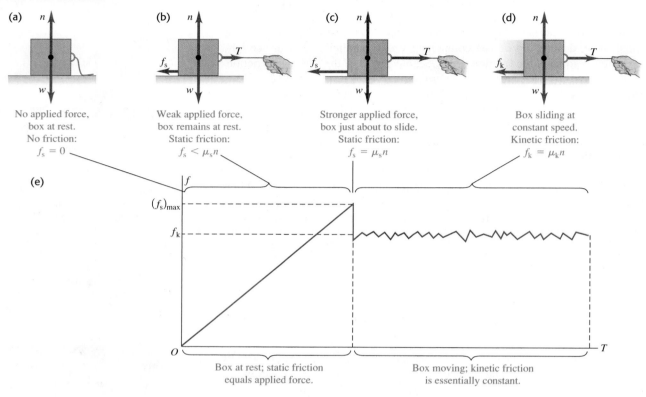

(a) No applied force, box at rest. No friction: $f_s = 0$

(b) Weak applied force, box remains at rest. Static friction: $f_s < \mu_s n$

(c) Stronger applied force, box just about to slide. Static friction: $f_s = \mu_s n$

(d) Box sliding at constant speed. Kinetic friction: $f_k = \mu_k n$

(e)

$(f_s)_{max}$

f_k

Box at rest; static friction equals applied force.

Box moving; kinetic friction is essentially constant.

to the box (Fig. 5.19b) and gradually increase the tension T in the rope. At first the box remains at rest because the force of static friction f_s also increases and stays equal in magnitude to T.

At some point T becomes greater than the maximum static friction force f_s the surface can exert. Then the box "breaks loose" (the tension T is able to break the bonds between molecules in the surfaces of the box and floor) and starts to slide. Figure 5.19c shows the forces when T is at this critical value. If T exceeds this value, the box is no longer in equilibrium. For a given pair of surfaces the maximum value of f_s depends on the normal force. Experiment shows that in many cases this maximum value, called $(f_s)_{max}$, is approximately *proportional* to n; we call the proportionality factor μ_s the **coefficient of static friction.** Table 5.1 lists some representative values of μ_s. In a particular situation, the actual force of static friction can have any magnitude between zero (when there is no other force parallel to the surface) and a maximum value given by $\mu_s n$. In symbols,

$$f_s \leq \mu_s n \qquad \text{(magnitude of static friction force)} \qquad (5.6)$$

Like Eq. (5.5), this is a relationship between magnitudes, *not* a vector relationship. The equality sign holds only when the applied force T has reached the critical value at which motion is about to start (Fig. 5.19c). When T is less than this value (Fig. 5.19b), the inequality sign holds. In that case we have to use the equilibrium conditions ($\sum \vec{F} = 0$) to find f_s. If there is no applied force ($T = 0$) as in Fig. 5.19a, then there is no static friction force either ($f_s = 0$).

As soon as the box starts to slide (Fig. 5.19d), the friction force usually *decreases* (Fig. 5.19e); it's easier to keep the box moving than to start it moving. Hence the coefficient of kinetic friction is usually *less* than the coefficient of static friction for any given pair of surfaces, as Table 5.1 shows.

Application Static Friction and Windshield Wipers

The squeak of windshield wipers on dry glass is a stick-slip phenomenon. The moving wiper blade sticks to the glass momentarily, then slides when the force applied to the blade by the wiper motor overcomes the maximum force of static friction. When the glass is wet from rain or windshield cleaning solution, friction is reduced and the wiper blade doesn't stick.

In some situations the surfaces will alternately stick (static friction) and slip (kinetic friction). This is what causes the horrible sound made by chalk held at the wrong angle while writing on the blackboard and the shriek of tires sliding on asphalt pavement. A more positive example is the motion of a violin bow against the string.

When a body slides on a layer of gas, friction can be made very small. In the linear air track used in physics laboratories, the gliders are supported on a layer of air. The frictional force is velocity dependent, but at typical speeds the effective coefficient of friction is of the order of 0.001.

Example 5.13 Friction in horizontal motion

You want to move a 500-N crate across a level floor. To start the crate moving, you have to pull with a 230-N horizontal force. Once the crate "breaks loose" and starts to move, you can keep it moving at constant velocity with only 200 N. What are the coefficients of static and kinetic friction?

SOLUTION

IDENTIFY and SET UP: The crate is in equilibrium both when it is at rest and when it is moving with constant velocity, so we use Newton's first law, as expressed by Eqs. (5.2). We use Eqs. (5.5) and (5.6) to find the target variables μ_s and μ_k.

Figures 5.20a and 5.20b show our sketch and free-body diagram for the instant just before the crate starts to move, when the static friction force has its maximum possible value

5.20 Our sketches for this problem.

(a) Pulling a crate

(b) Free-body diagram for crate just before it starts to move

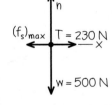

(c) Free-body diagram for crate moving at constant speed

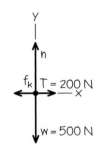

$(f_s)_{max} = \mu_s n$. Once the crate is moving, the friction force changes to its kinetic form (Fig. 5.20c). In both situations, four forces act on the crate: the downward weight (magnitude $w = 500$ N), the upward normal force (magnitude n) exerted by the floor, a tension force (magnitude T) to the right exerted by the rope, and a friction force to the left exerted by the ground. Because the rope in Fig. 5.20a is in equilibrium, the tension is the same at both ends. Hence the tension force that the rope exerts on the crate has the same magnitude as the force you exert on the rope. Since it's easier to keep the crate moving than to start it moving, we expect that $\mu_k < \mu_s$.

EXECUTE: Just before the crate starts to move (Fig. 5.20b), we have from Eqs. (5.2)

$$\sum F_x = T + (-(f_s)_{max}) = 0 \quad so \quad (f_s)_{max} = T = 230 \text{ N}$$
$$\sum F_y = n + (-w) = 0 \quad so \quad n = w = 500 \text{ N}$$

Now we solve Eq. (5.6), $(f_s)_{max} = \mu_s n$, for the value of μ_s:

$$\mu_s = \frac{(f_s)_{max}}{n} = \frac{230 \text{ N}}{500 \text{ N}} = 0.46$$

After the crate starts to move (Fig. 5.20c) we have

$$\sum F_x = T + (-f_k) = 0 \quad so \quad f_k = T = 200 \text{ N}$$
$$\sum F_y = n + (-w) = 0 \quad so \quad n = w = 500 \text{ N}$$

Using $f_k = \mu_k n$ from Eq. (5.5), we find

$$\mu_k = \frac{f_k}{n} = \frac{200 \text{ N}}{500 \text{ N}} = 0.40$$

EVALUATE: As expected, the coefficient of kinetic friction is less than the coefficient of static friction.

Example 5.14 Static friction can be less than the maximum

In Example 5.13, what is the friction force if the crate is at rest on the surface and a horizontal force of 50 N is applied to it?

SOLUTION

IDENTIFY and SET UP: The applied force is less than the maximum force of static friction, $(f_s)_{max} = 230$ N. Hence the crate remains at rest and the net force acting on it is zero. The target variable is the magnitude f_s of the friction force. The free-body diagram is the

same as in Fig. 5.20b, but with $(f_s)_{max}$ replaced by f_s and $T = 230$ N replaced by $T = 50$ N.

EXECUTE: From the equilibrium conditions, Eqs. (5.2), we have

$$\sum F_x = T + (-f_s) = 0 \quad so \quad f_s = T = 50 \text{ N}$$

EVALUATE: The friction force can prevent motion for any horizontal applied force up to $(f_s)_{max} = \mu_s n = 230$ N. Below that value, f_s has the same magnitude as the applied force.

Example 5.15 Minimizing kinetic friction

In Example 5.13, suppose you move the crate by pulling upward on the rope at an angle of 30° above the horizontal. How hard must you pull to keep it moving with constant velocity? Assume that $\mu_k = 0.40$.

SOLUTION

IDENTIFY and SET UP: The crate is in equilibrium because its velocity is constant, so we again apply Newton's first law. Since the crate is in motion, the floor exerts a *kinetic* friction force. The target variable is the magnitude T of the tension force.

Figure 5.21 shows our sketch and free-body diagram. The kinetic friction force f_k is still equal to $\mu_k n$, but now the normal

5.21 Our sketches for this problem.

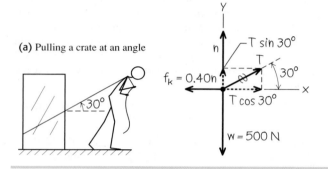

(b) Free-body diagram for moving crate

(a) Pulling a crate at an angle

force n is *not* equal in magnitude to the crate's weight. The force exerted by the rope has a vertical component that tends to lift the crate off the floor; this *reduces n* and so reduces f_k.

EXECUTE: From the equilibrium conditions and the equation $f_k = \mu_k n$, we have

$$\sum F_x = T\cos 30° + (-f_k) = 0 \quad \text{so} \quad T\cos 30° = \mu_k n$$
$$\sum F_y = T\sin 30° + n + (-w) = 0 \quad \text{so} \quad n = w - T\sin 30°$$

These are two equations for the two unknown quantities T and n. One way to find T is to substitute the expression for n in the second equation into the first equation and then solve the resulting equation for T:

$$T\cos 30° = \mu_k(w - T\sin 30°)$$
$$T = \frac{\mu_k w}{\cos 30° + \mu_k \sin 30°} = 188 \text{ N}$$

We can substitute this result into either of the original equations to obtain n. If we use the second equation, we get

$$n = w - T\sin 30° = (500 \text{ N}) - (188 \text{ N})\sin 30° = 406 \text{ N}$$

EVALUATE: As expected, the normal force is less than the 500-N weight of the box. It turns out that the tension required to keep the crate moving at constant speed is a little less than the 200-N force needed when you pulled horizontally in Example 5.13. Can you find an angle where the required pull is *minimum*? (See Challenge Problem 5.121.)

Example 5.16 Toboggan ride with friction I

Let's go back to the toboggan we studied in Example 5.10. The wax has worn off, so there is now a nonzero coefficient of kinetic friction μ_k. The slope has just the right angle to make the toboggan slide with constant velocity. Find this angle in terms of w and μ_k.

SOLUTION

IDENTIFY and SET UP: Our target variable is the slope angle α. The toboggan is in equilibrium because its velocity is constant, so we use Newton's first law in the form of Eqs. (5.2).

Three forces act on the toboggan: its weight, the normal force, and the kinetic friction force. The motion is downhill, so the friction force (which opposes the motion) is directed uphill. Figure 5.22 shows our sketch and free-body diagram (compare Fig. 5.12b in Example 5.10). The magnitude of the kinetic friction force is $f_k = \mu_k n$. We expect that the greater the value of μ_k, the steeper will be the required slope.

EXECUTE: The equilibrium conditions are

$$\sum F_x = w\sin \alpha + (-f_k) = w\sin \alpha - \mu_k n = 0$$
$$\sum F_y = n + (-w\cos \alpha) = 0$$

Rearranging these two equations, we get

$$\mu_k n = w\sin \alpha \quad \text{and} \quad n = w\cos \alpha$$

As in Example 5.10, the normal force is *not* equal to the weight. We eliminate n by dividing the first of these equations by the

5.22 Our sketches for this problem.

(a) The situation

(b) Free-body diagram for toboggan

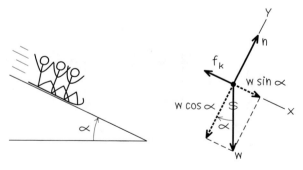

second, with the result

$$\mu_k = \frac{\sin \alpha}{\cos \alpha} = \tan \alpha \quad \text{so} \quad \alpha = \arctan \mu_k$$

EVALUATE: The weight w doesn't appear in this expression. *Any* toboggan, regardless of its weight, slides down an incline with constant speed if the coefficient of kinetic friction equals the tangent of the slope angle of the incline. The arctangent function increases as its argument increases, so it's indeed true that the slope angle α increases as μ_k increases.

Example 5.17   Toboggan ride with friction II

The same toboggan with the same coefficient of friction as in Example 5.16 *accelerates* down a steeper hill. Derive an expression for the acceleration in terms of g, α, μ_k, and w.

SOLUTION

IDENTIFY and SET UP: The toboggan is accelerating, so we must use Newton's second law as given in Eqs. (5.4). Our target variable is the downhill acceleration.

Our sketch and free-body diagram (Fig. 5.23) are almost the same as for Example 5.16. The toboggan's y-component of acceleration a_y is still zero but the x-component a_x is not, so we've drawn the downhill component of weight as a longer vector than the (uphill) friction force.

EXECUTE: It's convenient to express the weight as $w = mg$. Then Newton's second law in component form says

$$\sum F_x = mg \sin \alpha + (-f_k) = ma_x$$
$$\sum F_y = n + (-mg \cos \alpha) = 0$$

5.23 Our sketches for this problem.

(a) The situation

(b) Free-body diagram for toboggan

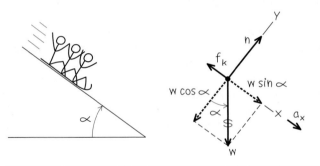

From the second equation and Eq. (5.5) we get an expression for f_k:

$$n = mg \cos \alpha$$
$$f_k = \mu_k n = \mu_k mg \cos \alpha$$

We substitute this into the x-component equation and solve for a_x:

$$mg \sin \alpha + (-\mu_k mg \cos \alpha) = ma_x$$
$$a_x = g(\sin \alpha - \mu_k \cos \alpha)$$

EVALUATE: As for the frictionless toboggan in Example 5.10, the acceleration doesn't depend on the mass m of the toboggan. That's because all of the forces that act on the toboggan (weight, normal force, and kinetic friction force) are proportional to m.

Let's check some special cases. If the hill is vertical ($\alpha = 90°$) so that $\sin \alpha = 1$ and $\cos \alpha = 0$, we have $a_x = g$ (the toboggan falls freely). For a certain value of α the acceleration is zero; this happens if

$$\sin \alpha = \mu_k \cos \alpha \qquad \text{and} \qquad \mu_k = \tan \alpha$$

This agrees with our result for the constant-velocity toboggan in Example 5.16. If the angle is even smaller, $\mu_k \cos \alpha$ is greater than $\sin \alpha$ and a_x is *negative;* if we give the toboggan an initial downhill push to start it moving, it will slow down and stop. Finally, if the hill is frictionless so that $\mu_k = 0$, we retrieve the result of Example 5.10: $a_x = g \sin \alpha$.

Notice that we started with a simple problem (Example 5.10) and extended it to more and more general situations. The general result we found in this example includes *all* the previous ones as special cases. Don't memorize this result, but do make sure you understand how we obtained it and what it means.

Suppose instead we give the toboggan an initial push *up* the hill. The direction of the kinetic friction force is now reversed, so the acceleration is different from the downhill value. It turns out that the expression for a_x is the same as for downhill motion except that the minus sign becomes plus. Can you show this?

Rolling Friction

It's a lot easier to move a loaded filing cabinet across a horizontal floor using a cart with wheels than to slide it. How much easier? We can define a **coefficient of rolling friction** μ_r, which is the horizontal force needed for constant speed on a flat surface divided by the upward normal force exerted by the surface. Transportation engineers call μ_r the *tractive resistance.* Typical values of μ_r are 0.002 to 0.003 for steel wheels on steel rails and 0.01 to 0.02 for rubber tires on concrete. These values show one reason railroad trains are generally much more fuel efficient than highway trucks.

Fluid Resistance and Terminal Speed

Sticking your hand out the window of a fast-moving car will convince you of the existence of **fluid resistance,** the force that a fluid (a gas or liquid) exerts on a body moving through it. The moving body exerts a force on the fluid to push it out of the way. By Newton's third law, the fluid pushes back on the body with an equal and opposite force.

The *direction* of the fluid resistance force acting on a body is always opposite the direction of the body's velocity relative to the fluid. The *magnitude* of the fluid resistance force usually increases with the speed of the body through the fluid.

5.24 A metal ball falling through a fluid (oil).

(a) Metal ball falling through oil

(b) Free-body diagram for ball in oil

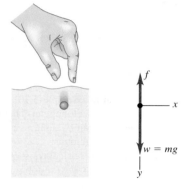

Application **Pollen and Fluid Resistance**

These spiky spheres are pollen grains from the ragweed flower (*Ambrosia psilostachya*) and a common cause of hay fever. Because of their small radius (about 10 μm = 0.01 mm), when they are released into the air the fluid resistance force on them is proportional to their speed. The terminal speed given by Eq. (5.9) is only about 1 cm/s. Hence even a moderate wind can keep pollen grains aloft and carry them substantial distances from their source.

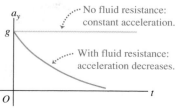

This is very different from the kinetic friction force between two surfaces in contact, which we can usually regard as independent of speed. For small objects moving at very low speeds, the magnitude f of the fluid resistance force is approximately proportional to the body's speed v:

$$f = kv \qquad \text{(fluid resistance at low speed)} \qquad (5.7)$$

where k is a proportionality constant that depends on the shape and size of the body and the properties of the fluid. Equation (5.7) is appropriate for dust particles falling in air or a ball bearing falling in oil. For larger objects moving through air at the speed of a tossed tennis ball or faster, the resisting force is approximately proportional to v^2 rather than to v. It is then called **air drag** or simply *drag*. Airplanes, falling raindrops, and bicyclists all experience air drag. In this case we replace Eq. (5.7) by

$$f = Dv^2 \qquad \text{(fluid resistance at high speed)} \qquad (5.8)$$

Because of the v^2 dependence, air drag increases rapidly with increasing speed. The air drag on a typical car is negligible at low speeds but comparable to or greater than rolling resistance at highway speeds. The value of D depends on the shape and size of the body and on the density of the air. You should verify that the units of the constant k in Eq. (5.7) are $N \cdot s/m$ or kg/s, and that the units of the constant D in Eq. (5.8) are $N \cdot s^2/m^2$ or kg/m.

Because of the effects of fluid resistance, an object falling in a fluid does *not* have a constant acceleration. To describe its motion, we can't use the constant-acceleration relationships from Chapter 2; instead, we have to start over using Newton's second law. As an example, suppose you drop a metal ball at the surface of a bucket of oil and let it fall to the bottom (Fig. 5.24a). The fluid resistance force in this situation is given by Eq. (5.7). What are the acceleration, velocity, and position of the metal ball as functions of time?

Figure 5.24b shows the free-body diagram. We take the positive y-direction to be downward and neglect any force associated with buoyancy in the oil. Since the ball is moving downward, its speed v is equal to its y-velocity v_y and the fluid resistance force is in the $-y$-direction. There are no x-components, so Newton's second law gives

$$\sum F_y = mg + (-kv_y) = ma_y$$

When the ball first starts to move, $v_y = 0$, the resisting force is zero, and the initial acceleration is $a_y = g$. As the speed increases, the resisting force also increases, until finally it is equal in magnitude to the weight. At this time $mg - kv_y = 0$, the acceleration becomes zero, and there is no further increase in speed. The final speed v_t, called the **terminal speed**, is given by $mg - kv_t = 0$, or

$$v_t = \frac{mg}{k} \qquad \text{(terminal speed, fluid resistance } f = kv) \qquad (5.9)$$

Figure 5.25 shows how the acceleration, velocity, and position vary with time. As time goes by, the acceleration approaches zero and the velocity approaches v_t

5.25 Graphs of the motion of a body falling without fluid resistance and with fluid resistance proportional to the speed.

Acceleration versus time

No fluid resistance: constant acceleration.

With fluid resistance: acceleration decreases.

Velocity versus time

No fluid resistance: velocity keeps increasing.

With fluid resistance: velocity has an upper limit.

Position versus time

No fluid resistance: parabolic curve.

With fluid resistance: curve straightens out.

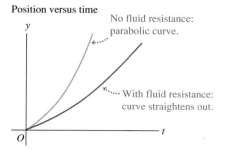

(remember that we chose the positive y-direction to be down). The slope of the graph of y versus t becomes constant as the velocity becomes constant.

To see how the graphs in Fig. 5.25 are derived, we must find the relationship between velocity and time during the interval before the terminal speed is reached. We go back to Newton's second law, which we rewrite using $a_y = dv_y/dt$:

$$m\frac{dv_y}{dt} = mg - kv_y$$

After rearranging terms and replacing mg/k by v_t, we integrate both sides, noting that $v_y = 0$ when $t = 0$:

$$\int_0^v \frac{dv_y}{v_y - v_t} = -\frac{k}{m}\int_0^t dt$$

which integrates to

$$\ln\frac{v_t - v_y}{v_t} = -\frac{k}{m}t \quad \text{or} \quad 1 - \frac{v_y}{v_t} = e^{-(k/m)t}$$

and finally

$$v_y = v_t[1 - e^{-(k/m)t}] \tag{5.10}$$

Note that v_y becomes equal to the terminal speed v_t only in the limit that $t \to \infty$; the ball cannot attain terminal speed in any finite length of time.

The derivative of v_y gives a_y as a function of time, and the integral of v_y gives y as a function of time. We leave the derivations for you to complete; the results are

$$a_y = ge^{-(k/m)t} \tag{5.11}$$

$$y = v_t\left[t - \frac{m}{k}(1 - e^{-(k/m)t})\right] \tag{5.12}$$

Now look again at Fig. 5.25, which shows graphs of these three relationships.

In deriving the terminal speed in Eq. (5.9), we assumed that the fluid resistance force is proportional to the speed. For an object falling through the air at high speeds, so that the fluid resistance is equal to Dv^2 as in Eq. (5.8), the terminal speed is reached when Dv^2 equals the weight mg (Fig. 5.26a). You can show that the terminal speed v_t is given by

$$v_t = \sqrt{\frac{mg}{D}} \quad \text{(terminal speed, fluid resistance } f = Dv^2) \tag{5.13}$$

This expression for terminal speed explains why heavy objects in air tend to fall faster than light objects. Two objects with the same physical size but different mass (say, a table-tennis ball and a lead ball with the same radius) have the same value of D but different values of m. The more massive object has a higher terminal speed and falls faster. The same idea explains why a sheet of paper falls faster if you first crumple it into a ball; the mass m is the same, but the smaller size makes D smaller (less air drag for a given speed) and v_t larger. Skydivers use the same principle to control their descent (Fig. 5.26b).

Figure 5.27 shows the trajectories of a baseball with and without air drag, assuming a coefficient $D = 1.3 \times 10^{-3}$ kg/m (appropriate for a batted ball at sea level). You can see that both the range of the baseball and the maximum height reached are substantially less than the zero-drag calculation would lead you to believe. Hence the baseball trajectory we calculated in Example 3.8 (Section 3.3) by ignoring air drag is unrealistic. Air drag is an important part of the game of baseball!

5.26 (a) Air drag and terminal speed. (b) By changing the positions of their arms and legs while falling, skydivers can change the value of the constant D in Eq. (5.8) and hence adjust the terminal speed of their fall [Eq. (5.13)].

(a) Free-body diagrams for falling with air drag

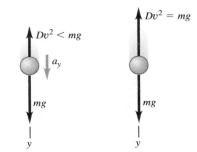

Before terminal speed: Object accelerating, drag force less than weight.

At terminal speed v_t: Object in equilibrium, drag force equals weight.

(b) A skydiver falling at terminal speed

5.27 Computer-generated trajectories of a baseball launched at 50 m/s at 35° above the horizontal. Note that the scales are different on the horizontal and vertical axes.

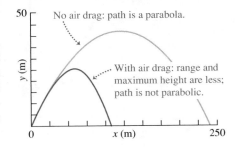

No air drag: path is a parabola.

With air drag: range and maximum height are less; path is not parabolic.

Example 5.18 Terminal speed of a skydiver

For a human body falling through air in a spread-eagle position (Fig. 5.26b), the numerical value of the constant D in Eq. (5.8) is about 0.25 kg/m. Find the terminal speed for a lightweight 50-kg skydiver.

SOLUTION

IDENTIFY and SET UP: This example uses the relationship among terminal speed, mass, and drag coefficient. We use Eq. (5.13) to find the target variable v_t.

EXECUTE: We find for $m = 50$ kg:

$$v_t = \sqrt{\frac{mg}{D}} = \sqrt{\frac{(50 \text{ kg})(9.8 \text{ m/s}^2)}{0.25 \text{ kg/m}}}$$

$$= 44 \text{ m/s (about 160 km/h, or 99 mi/h)}$$

EVALUATE: The terminal speed is proportional to the square root of the skydiver's mass. A skydiver with the same drag coefficient D but twice the mass would have a terminal speed $\sqrt{2} = 1.41$ times greater, or 63 m/s. (A more massive skydiver would also have more frontal area and hence a larger drag coefficient, so his terminal speed would be a bit less than 63 m/s.) Even the lightweight skydiver's terminal speed is quite high, so skydives don't last very long. A drop from 2800 m (9200 ft) to the surface at the terminal speed takes only $(2800 \text{ m})/(44 \text{ m/s}) = 64$ s.

When the skydiver deploys the parachute, the value of D increases greatly. Hence the terminal speed of the skydiver and parachute decreases dramatically to a much lower value.

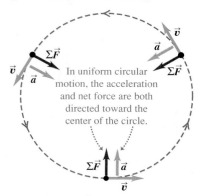

5.28 Net force, acceleration, and velocity in uniform circular motion.

In uniform circular motion, the acceleration and net force are both directed toward the center of the circle.

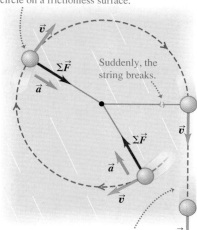

5.29 What happens if the inward radial force suddenly ceases to act on a body in circular motion?

A ball attached to a string whirls in a circle on a frictionless surface.

Suddenly, the string breaks.

No net force now acts on the ball, so it obeys Newton's first law—it moves in a straight line at constant velocity.

Test Your Understanding of Section 5.3 Consider a box that is placed on different surfaces. (a) In which situation(s) is there *no* friction force acting on the box? (b) In which situation(s) is there a *static* friction force acting on the box? (c) In which situation(s) is there a *kinetic* friction force on the box? (i) The box is at rest on a rough horizontal surface. (ii) The box is at rest on a rough tilted surface. (iii) The box is on the rough-surfaced flat bed of a truck—the truck is moving at a constant velocity on a straight, level road, and the box remains in the same place in the middle of the truck bed. (iv) The box is on the rough-surfaced flat bed of a truck—the truck is speeding up on a straight, level road, and the box remains in the same place in the middle of the truck bed. (v) The box is on the rough-surfaced flat bed of a truck—the truck is climbing a hill, and the box is sliding toward the back of the truck. ❙

5.4 Dynamics of Circular Motion

We talked about uniform circular motion in Section 3.4. We showed that when a particle moves in a circular path with constant speed, the particle's acceleration is always directed toward the center of the circle (perpendicular to the instantaneous velocity). The magnitude a_{rad} of the acceleration is constant and is given in terms of the speed v and the radius R of the circle by

$$a_{\text{rad}} = \frac{v^2}{R} \quad \text{(uniform circular motion)} \tag{5.14}$$

The subscript "rad" is a reminder that at each point the acceleration is radially inward toward the center of the circle, perpendicular to the instantaneous velocity. We explained in Section 3.4 why this acceleration is often called *centripetal acceleration*.

We can also express the centripetal acceleration a_{rad} in terms of the *period T*, the time for one revolution:

$$T = \frac{2\pi R}{v} \tag{5.15}$$

In terms of the period, a_{rad} is

$$a_{\text{rad}} = \frac{4\pi^2 R}{T^2} \quad \text{(uniform circular motion)} \tag{5.16}$$

Uniform circular motion, like all other motion of a particle, is governed by Newton's second law. To make the particle accelerate toward the center of the circle, the net force $\Sigma \vec{F}$ on the particle must always be directed toward the center (Fig. 5.28). The magnitude of the acceleration is constant, so the magnitude F_{net} of the net force must also be constant. If the inward net force stops acting, the particle flies off in a straight line tangent to the circle (Fig. 5.29).

The magnitude of the radial acceleration is given by $a_{rad} = v^2/R$, so the magnitude F_{net} of the net force on a particle with mass m in uniform circular motion must be

$$F_{net} = ma_{rad} = m\frac{v^2}{R} \quad \text{(uniform circular motion)} \quad (5.17)$$

Uniform circular motion can result from *any* combination of forces, just so the net force $\Sigma\vec{F}$ is always directed toward the center of the circle and has a constant magnitude. Note that the body need not move around a complete circle: Equation (5.17) is valid for *any* path that can be regarded as part of a circular arc.

CAUTION **Avoid using "centrifugal force"** Figure 5.30 shows both a correct free-body diagram for uniform circular motion (Fig. 5.30a) and a common *incorrect* diagram (Fig. 5.30b). Figure 5.30b is incorrect because it includes an extra outward force of magnitude $m(v^2/R)$ to "keep the body out there" or to "keep it in equilibrium." There are three reasons not to include such an outward force, usually called *centrifugal force* ("centrifugal" means "fleeing from the center"). First, the body does *not* "stay out there": It is in constant motion around its circular path. Because its velocity is constantly changing in direction, the body accelerates and is *not* in equilibrium. Second, if there *were* an additional outward force that balanced the inward force, the net force would be zero and the body would move in a straight line, not a circle (Fig. 5.29). And third, the quantity $m(v^2/R)$ is *not* a force; it corresponds to the $m\vec{a}$ side of $\Sigma\vec{F} = m\vec{a}$ and does not appear in $\Sigma\vec{F}$ (Fig. 5.30a). It's true that when you ride in a car that goes around a circular path, you tend to slide to the outside of the turn as though there was a "centrifugal force." But we saw in Section 4.2 that what really happens is that you tend to keep moving in a straight line, and the outer side of the car "runs into" you as the car turns (Fig. 4.11c). *In an inertial frame of reference there is no such thing as "centrifugal force."* We won't mention this term again, and we strongly advise you to avoid using it as well.

5.30 (a) Correct and (b) incorrect free-body diagrams for a body in uniform circular motion.

(a) Correct free-body diagram

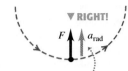

If you include the acceleration, draw it to one side of the body to show that it's not a force.

(b) Incorrect free-body diagram

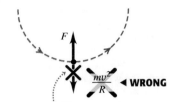

The quantity mv^2/R is *not* a force—it doesn't belong in a free-body diagram.

Example 5.19 **Force in uniform circular motion**

A sled with a mass of 25.0 kg rests on a horizontal sheet of essentially frictionless ice. It is attached by a 5.00-m rope to a post set in the ice. Once given a push, the sled revolves uniformly in a circle around the post (Fig. 5.31a). If the sled makes five complete revolutions every minute, find the force F exerted on it by the rope.

SOLUTION

IDENTIFY and SET UP: The sled is in uniform circular motion, so it has a constant radial acceleration. We'll apply Newton's second law to the sled to find the magnitude F of the force exerted by the rope (our target variable).

5.31 (a) The situation. (b) Our free-body diagram.

(a) A sled in uniform circular motion

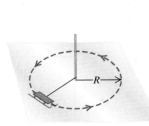

(b) Free-body diagram for sled

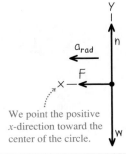

We point the positive x-direction toward the center of the circle.

Figure 5.31b shows our free-body diagram for the sled. The acceleration has only an x-component; this is toward the center of the circle, so we denote it as a_{rad}. The acceleration isn't given, so we'll need to determine its value using either Eq. (5.14) or Eq. (5.16).

EXECUTE: The force F appears in Newton's second law for the x-direction:

$$\Sigma F_x = F = ma_{rad}$$

We can find the centripetal acceleration a_{rad} using Eq. (5.16). The sled moves in a circle of radius $R = 5.00$ m with a period $T = (60.0\text{ s})/(5\text{ rev}) = 12.0$ s, so

$$a_{rad} = \frac{4\pi^2 R}{T^2} = \frac{4\pi^2(5.00\text{ m})}{(12.0\text{ s})^2} = 1.37\text{ m/s}^2$$

The magnitude F of the force exerted by the rope is then

$$F = ma_{rad} = (25.0\text{ kg})(1.37\text{ m/s}^2)$$
$$= 34.3\text{ kg}\cdot\text{m/s}^2 = 34.3\text{ N}$$

EVALUATE: You can check our value for a_{rad} by first finding the speed using Eq. (5.15), $v = 2\pi R/T$, and then using $a_{rad} = v^2/R$ from Eq. (5.14). Do you get the same result?

A greater force would be needed if the sled moved around the circle at a higher speed v. In fact, if v were doubled while R remained the same, F would be four times greater. Can you show this? How would F change if v remained the same but the radius R were doubled?

Example 5.20 **A conical pendulum**

An inventor designs a pendulum clock using a bob with mass m at the end of a thin wire of length L. Instead of swinging back and forth, the bob is to move in a horizontal circle with constant speed v, with the wire making a fixed angle β with the vertical direction (Fig. 5.32a). This is called a *conical pendulum* because the suspending wire traces out a cone. Find the tension F in the wire and the period T (the time for one revolution of the bob).

SOLUTION

IDENTIFY and SET UP: To find our target variables, the tension F and period T, we need two equations. These will be the horizontal and vertical components of Newton's second law applied to the bob. We'll find the radial acceleration of the bob using one of the circular motion equations.

Figure 5.32b shows our free-body diagram and coordinate system for the bob at a particular instant. There are just two forces on the bob: the weight mg and the tension F in the wire. Note that the

5.32 (a) The situation. (b) Our free-body diagram.

(a) The situation

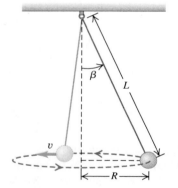

(b) Free-body diagram for pendulum bob

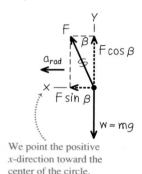

We point the positive x-direction toward the center of the circle.

center of the circular path is in the same horizontal plane as the bob, *not* at the top end of the wire. The horizontal component of tension is the force that produces the radial acceleration a_{rad}.

EXECUTE: The bob has zero vertical acceleration; the horizontal acceleration is toward the center of the circle, which is why we use the symbol a_{rad}. Newton's second law says

$$\sum F_x = F \sin \beta = ma_{rad}$$
$$\sum F_y = F \cos \beta + (-mg) = 0$$

These are two equations for the two unknowns F and β. The equation for $\sum F_y$ gives $F = mg/\cos \beta$; that's our target expression for F in terms of β. Substituting this result into the equation for $\sum F_x$ and using $\sin \beta/\cos \beta = \tan \beta$, we find

$$a_{rad} = g \tan \beta$$

To relate β to the period T, we use Eq. (5.16) for a_{rad}, solve for T, and insert $a_{rad} = g \tan \beta$:

$$a_{rad} = \frac{4\pi^2 R}{T^2} \quad \text{so} \quad T^2 = \frac{4\pi^2 R}{a_{rad}}$$

$$T = 2\pi \sqrt{\frac{R}{g \tan \beta}}$$

Figure 5.32a shows that $R = L \sin \beta$. We substitute this and use $\sin \beta/\tan \beta = \cos \beta$:

$$T = 2\pi \sqrt{\frac{L \cos \beta}{g}}$$

EVALUATE: For a given length L, as the angle β increases, $\cos \beta$ decreases, the period T becomes smaller, and the tension $F = mg/\cos \beta$ increases. The angle can never be 90°, however; this would require that $T = 0$, $F = \infty$, and $v = \infty$. A conical pendulum would not make a very good clock because the period depends on the angle β in such a direct way.

Example 5.21 **Rounding a flat curve**

The sports car in Example 3.11 (Section 3.4) is rounding a flat, unbanked curve with radius R (Fig. 5.33a). If the coefficient of static friction between tires and road is μ_s, what is the maximum speed v_{max} at which the driver can take the curve without sliding?

SOLUTION

IDENTIFY and SET UP: The car's acceleration as it rounds the curve has magnitude $a_{rad} = v^2/R$. Hence the maximum speed v_{max} (our target variable) corresponds to the maximum acceleration a_{rad} and to the maximum horizontal force on the car toward the center of its circular path. The only horizontal force acting on the car is the friction force exerted by the road. So to solve this problem we'll need Newton's second law, the equations of uniform circular motion, and our knowledge of the friction force from Section 5.3.

The free-body diagram in Fig. 5.33b includes the car's weight $w = mg$ and the two forces exerted by the road: the normal force n and the horizontal friction force f. The friction force must point toward the center of the circular path in order to cause the radial acceleration. The car doesn't slide toward or away from the center

of the circle, so the friction force is *static* friction, with a maximum magnitude $f_{max} = \mu_s n$ [see Eq. (5.6)].

5.33 (a) The situation. (b) Our free-body diagram.

(a) Car rounding flat curve

(b) Free-body diagram for car

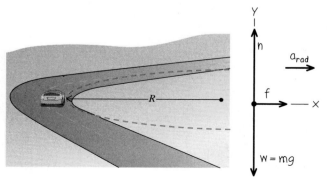

EXECUTE: The acceleration toward the center of the circular path is $a_{\text{rad}} = v^2/R$. There is no vertical acceleration. Thus we have

$$\sum F_x = f = ma_{\text{rad}} = m\frac{v^2}{R}$$

$$\sum F_y = n + (-mg) = 0$$

The second equation shows that $n = mg$. The first equation shows that the friction force *needed* to keep the car moving in its circular path increases with the car's speed. But the maximum friction force *available* is $f_{\text{max}} = \mu_s n = \mu_s mg$, and this determines the car's maximum speed. Substituting $\mu_s mg$ for f and v_{max} for v in the first equation, we find

$$\mu_s mg = m\frac{v^2_{\text{max}}}{R} \quad \text{so} \quad v_{\text{max}} = \sqrt{\mu_s gR}$$

As an example, if $\mu_s = 0.96$ and $R = 230$ m, we have

$$v_{\text{max}} = \sqrt{(0.96)(9.8 \text{ m/s}^2)(230 \text{ m})} = 47 \text{ m/s}$$

or about 170 km/h (100 mi/h). This is the maximum speed for this radius.

EVALUATE: If the car's speed is slower than $v_{\text{max}} = \sqrt{\mu_s gR}$, the required friction force is less than the maximum value $f_{\text{max}} = \mu_s mg$, and the car can easily make the curve. If we try to take the curve going *faster* than v_{max}, we will skid. We could still go in a circle without skidding at this higher speed, but the radius would have to be larger.

The maximum centripetal acceleration (called the "lateral acceleration" in Example 3.11) is equal to $\mu_s g$. That's why it's best to take curves at less than the posted speed limit if the road is wet or icy, either of which can reduce the value of μ_s and hence $\mu_s g$.

Example 5.22 · Rounding a banked curve

For a car traveling at a certain speed, it is possible to bank a curve at just the right angle so that no friction at all is needed to maintain the car's turning radius. Then a car can safely round the curve even on wet ice. (Bobsled racing depends on this same idea.) Your engineering firm plans to rebuild the curve in Example 5.21 so that a car moving at a chosen speed v can safely make the turn even with no friction (Fig. 5.34a). At what angle β should the curve be banked?

SOLUTION

IDENTIFY and SET UP: With no friction, the only forces acting on the car are its weight and the normal force. Because the road is banked, the normal force (which acts perpendicular to the road surface) has a horizontal component. This component causes the car's horizontal acceleration toward the center of the car's circular path. We'll use Newton's second law to find the target variable β.

Our free-body diagram (Fig. 5.34b) is very similar to the diagram for the conical pendulum in Example 5.20 (Fig. 5.32b). The normal force acting on the car plays the role of the tension force exerted by the wire on the pendulum bob.

EXECUTE: The normal force $\vec{n}$ is perpendicular to the roadway and is at an angle β with the vertical (Fig. 5.34b). Thus it has a vertical component $n \cos \beta$ and a horizontal component $n \sin \beta$.

The acceleration in the x-direction is the centripetal acceleration $a_{\text{rad}} = v^2/R$; there is no acceleration in the y-direction. Thus the equations of Newton's second law are

$$\sum F_x = n \sin \beta = ma_{\text{rad}}$$
$$\sum F_y = n \cos \beta + (-mg) = 0$$

From the $\sum F_y$ equation, $n = mg/\cos \beta$. Substituting this into the $\sum F_x$ equation and using $a_{\text{rad}} = v^2/R$, we get an expression for the bank angle:

$$\tan \beta = \frac{a_{\text{rad}}}{g} = \frac{v^2}{gR} \quad \text{so} \quad \beta = \arctan\frac{v^2}{gR}$$

EVALUATE: The bank angle depends on both the speed and the radius. For a given radius, no one angle is correct for all speeds. In the design of highways and railroads, curves are often banked for the average speed of the traffic over them. If $R = 230$ m and $v = 25$ m/s (equal to a highway speed of 88 km/h, or 55 mi/h), then

$$\beta = \arctan\frac{(25 \text{ m/s})^2}{(9.8 \text{ m/s}^2)(230 \text{ m})} = 15°$$

This is within the range of banking angles actually used in highways.

5.34 (a) The situation. (b) Our free-body diagram.

(a) Car rounding banked curve

(b) Free-body diagram for car

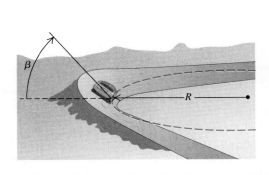
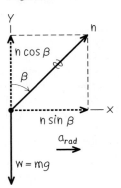

5.35 An airplane banks to one side in order to turn in that direction. The vertical component of the lift force $\vec{L}$ balances the force of gravity; the horizontal component of $\vec{L}$ causes the acceleration v^2/R.

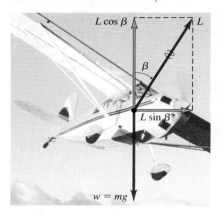

$L\cos\beta$ L

β

$L\sin\beta$

$w = mg$

ActivPhysics 4.2: Circular Motion Problem Solving
ActivPhysics 4.3: Cart Goes over Circular Path
ActivPhysics 4.4: Ball Swings on a String
ActivPhysics 4.5: Car Circles a Track

Banked Curves and the Flight of Airplanes

The results of Example 5.22 also apply to an airplane when it makes a turn in level flight (Fig. 5.35). When an airplane is flying in a straight line at a constant speed and at a steady altitude, the airplane's weight is exactly balanced by the lift force $\vec{L}$ exerted by the air. (The upward lift force that the air exerts on the wings is a reaction to the downward push the wings exert on the air as they move through it.) To make the airplane turn, the pilot banks the airplane to one side so that the lift force has a horizontal component as Fig. 5.35 shows. (The pilot also changes the angle at which the wings "bite" into the air so that the vertical component of lift continues to balance the weight.) The bank angle is related to the airplane's speed v and the radius R of the turn by the same expression as in Example 5.22: $\tan\beta = v^2/gR$. For an airplane to make a tight turn (small R) at high speed (large v), $\tan\beta$ must be large and the required bank angle β must approach 90°.

We can also apply the results of Example 5.22 to the *pilot* of an airplane. The free-body diagram for the pilot of the airplane is exactly as shown in Fig. 5.34b; the normal force $n = mg/\cos\beta$ is exerted on the pilot by the seat. As in Example 5.9, n is equal to the apparent weight of the pilot, which is greater than the pilot's true weight mg. In a tight turn with a large bank angle β, the pilot's apparent weight can be tremendous: $n = 5.8mg$ at $\beta = 80°$ and $n = 9.6mg$ at $\beta = 84°$. Pilots black out in such tight turns because the apparent weight of their blood increases by the same factor, and the human heart isn't strong enough to pump such apparently "heavy" blood to the brain.

Motion in a Vertical Circle

In Examples 5.19, 5.20, 5.21, and 5.22 the body moved in a horizontal circle. Motion in a *vertical* circle is no different in principle, but the weight of the body has to be treated carefully. The following example shows what we mean.

Example 5.23 | **Uniform circular motion in a vertical circle**

A passenger on a carnival Ferris wheel moves in a vertical circle of radius R with constant speed v. The seat remains upright during the motion. Find expressions for the force the seat exerts on the passenger at the top of the circle and at the bottom.

SOLUTION

IDENTIFY and SET UP: The target variables are n_T, the upward normal force the seat applies to the passenger at the top of the circle, and n_B, the normal force at the bottom. We'll find these using Newton's second law and the uniform circular motion equations.

Figure 5.36a shows the passenger's velocity and acceleration at the two positions. The acceleration always points toward the center of the circle—downward at the top of the circle and upward at the bottom of the circle. At each position the only forces acting are vertical: the upward normal force and the downward force of gravity. Hence we need only the vertical component of Newton's second law. Figures 5.36b and 5.36c show free-body diagrams for the two positions. We take the positive y-direction as upward in both cases (that is, *opposite* the direction of the acceleration at the top of the circle).

EXECUTE: At the top the acceleration has magnitude v^2/R, but its vertical component is negative because its direction is downward.

Hence $a_y = -v^2/R$ and Newton's second law tells us that

Top: $\quad \sum F_y = n_T + (-mg) = -m\dfrac{v^2}{R} \quad$ or

$$n_T = mg\left(1 - \frac{v^2}{gR}\right)$$

5.36 Our sketches for this problem.

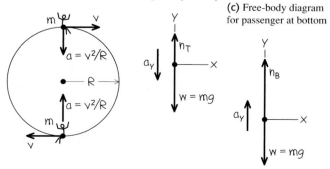

(a) Sketch of two positions

(b) Free-body diagram for passenger at top

(c) Free-body diagram for passenger at bottom

At the bottom the acceleration is upward, so $a_y = +v^2/R$ and Newton's second law says

$$\text{Bottom:} \qquad \sum F_y = n_B + (-mg) = +m\frac{v^2}{R} \qquad \text{or}$$

$$n_B = mg\left(1 + \frac{v^2}{gR}\right)$$

EVALUATE: Our result for n_T tells us that at the top of the Ferris wheel, the upward force the seat applies to the passenger is *smaller* in magnitude than the passenger's weight $w = mg$. If the ride goes fast enough that $g - v^2/R$ becomes zero, the seat applies *no* force, and the passenger is about to become airborne. If v becomes still larger, n_T becomes negative; this means that a *downward* force (such as from a seat belt) is needed to keep the passenger in the seat. By contrast, the normal force n_B at the bottom is always *greater* than the passenger's weight. You feel the seat pushing up on you more firmly than when you are at rest. You can see that n_T and n_B are the values of the passenger's *apparent weight* at the top and bottom of the circle (see Section 5.2).

When we tie a string to an object and whirl it in a vertical circle, the analysis in Example 5.23 isn't directly applicable. The reason is that v is *not* constant in this case; except at the top and bottom of the circle, the net force (and hence the acceleration) does *not* point toward the center of the circle (Fig. 5.37). So both $\sum \vec{F}$ and $\vec{a}$ have a component tangent to the circle, which means that the speed changes. Hence this is a case of *nonuniform* circular motion (see Section 3.4). Even worse, we can't use the constant-acceleration formulas to relate the speeds at various points because *neither* the magnitude nor the direction of the acceleration is constant. The speed relationships we need are best obtained by using the concept of energy. We'll consider such problems in Chapter 7.

Test Your Understanding of Section 5.4 Satellites are held in orbit by the force of our planet's gravitational attraction. A satellite in a small-radius orbit moves at a higher speed than a satellite in an orbit of large radius. Based on this information, what you can conclude about the earth's gravitational attraction for the satellite? (i) It increases with increasing distance from the earth. (ii) It is the same at all distances from the earth. (iii) It decreases with increasing distance from the earth. (iv) This information by itself isn't enough to answer the question. ❙

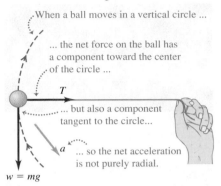

5.37 A ball moving in a vertical circle.

When a ball moves in a vertical circle ...

... the net force on the ball has a component toward the center of the circle ...

T

... but also a component tangent to the circle...

a ... so the net acceleration is not purely radial.

$w = mg$

5.5 The Fundamental Forces of Nature

We have discussed several kinds of forces—including weight, tension, friction, fluid resistance, and the normal force—and we will encounter others as we continue our study of physics. But just how many kinds of forces are there? Our current understanding is that all forces are expressions of just four distinct classes of *fundamental* forces, or interactions between particles (Fig. 5.38). Two are familiar in everyday experience. The other two involve interactions between subatomic particles that we cannot observe with the unaided senses.

Gravitational interactions include the familiar force of your *weight,* which results from the earth's gravitational attraction acting on you. The mutual gravitational attraction of various parts of the earth for each other holds our planet together (Fig. 5.38a). Newton recognized that the sun's gravitational attraction for the earth keeps the earth in its nearly circular orbit around the sun. In Chapter 13 we will study gravitational interactions in greater detail, and we will analyze their vital role in the motions of planets and satellites.

The second familiar class of forces, **electromagnetic interactions,** includes electric and magnetic forces. If you run a comb through your hair, the comb ends up with an electric charge; you can use the electric force exerted by this charge to pick up bits of paper. All atoms contain positive and negative electric charge, so atoms and molecules can exert electric forces on one another (Fig. 5.38b). Contact forces, including the normal force, friction, and fluid resistance, are the combination of all such forces exerted on the atoms of a body by atoms in its surroundings. *Magnetic* forces, such as those between magnets or between a magnet and a piece of iron, are actually the result of electric charges in motion. For example, an electromagnet causes magnetic interactions because electric

5.38 Examples of the fundamental interactions in nature. (a) The moon and the earth are held together and held in orbit by gravitational forces. (b) This molecule of bacterial plasmid DNA is held together by electromagnetic forces between its atoms. (c) The sun shines because in its core, strong forces between nuclear particles cause the release of energy. (d) When a massive star explodes into a supernova, a flood of energy is released by weak interactions between the star's nuclear particles.

(a) Gravitational forces hold planets together.

(b) Electromagnetic forces hold molecules together.

(c) Strong forces release energy to power the sun.

(d) Weak forces play a role in exploding stars.

charges move through its wires. We will study electromagnetic interactions in detail in the second half of this book.

On the atomic or molecular scale, gravitational forces play no role because electric forces are enormously stronger: The electrical repulsion between two protons is stronger than their gravitational attraction by a factor of about 10^{35}. But in bodies of astronomical size, positive and negative charges are usually present in nearly equal amounts, and the resulting electrical interactions nearly cancel out. Gravitational interactions are thus the dominant influence in the motion of planets and in the internal structure of stars.

The other two classes of interactions are less familiar. One, the **strong interaction,** is responsible for holding the nucleus of an atom together. Nuclei contain electrically neutral neutrons and positively charged protons. The electric force between charged protons tries to push them apart; the strong attractive force between nuclear particles counteracts this repulsion and makes the nucleus stable. In this context the strong interaction is also called the *strong nuclear force.* It has much shorter range than electrical interactions, but within its range it is much stronger. The strong interaction plays a key role in thermonuclear reactions that take place at the sun's core and generate the sun's heat and light (Fig. 5.38c).

Finally, there is the **weak interaction.** Its range is so short that it plays a role only on the scale of the nucleus or smaller. The weak interaction is responsible for a common form of radioactivity called beta decay, in which a neutron in a radioactive nucleus is transformed into a proton while ejecting an electron and a nearly massless particle called an antineutrino. The weak interaction between the antineutrino and ordinary matter is so feeble that an antineutrino could easily penetrate a wall of lead a million kilometers thick! Yet when a giant star undergoes a cataclysmic explosion called a supernova, most of the energy is released by way of the weak interaction (Fig. 5.38d).

In the 1960s physicists developed a theory that described the electromagnetic and weak interactions as aspects of a single *electroweak* interaction. This theory has passed every experimental test to which it has been put. Encouraged by this success, physicists have made similar attempts to describe the strong, electromagnetic, and weak interactions in terms of a single *grand unified theory* (GUT), and have taken steps toward a possible unification of all interactions into a *theory of everything* (TOE). Such theories are still speculative, and there are many unanswered questions in this very active field of current research.

Using Newton's first law: When a body is in equilibrium in an inertial frame of reference—that is, either at rest or moving with constant velocity—the vector sum of forces acting on it must be zero (Newton's first law). Free-body diagrams are essential in identifying the forces that act on the body being considered.

Newton's third law (action and reaction) is also frequently needed in equilibrium problems. The two forces in an action–reaction pair *never* act on the same body. (See Examples 5.1–5.5.)

The normal force exerted on a body by a surface is *not* always equal to the body's weight. (See Example 5.3.)

$$\sum \vec{F} = 0 \quad \text{(vector form)} \qquad (5.1)$$

$$\sum F_x = 0$$
$$\sum F_y = 0 \quad \text{(component form)} \qquad (5.2)$$

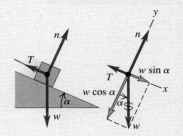

Using Newton's second law: If the vector sum of forces on a body is *not* zero, the body accelerates. The acceleration is related to the net force by Newton's second law.

Just as for equilibrium problems, free-body diagrams are essential for solving problems involving Newton's second law, and the normal force exerted on a body is not always equal to its weight. (See Examples 5.6–5.12.)

Vector form:

$$\sum \vec{F} = m\vec{a} \qquad (5.3)$$

Component form:

$$\sum F_x = ma_x \qquad \sum F_y = ma_y \qquad (5.4)$$

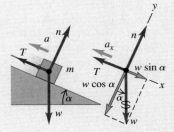

Friction and fluid resistance: The contact force between two bodies can always be represented in terms of a normal force $\vec{n}$ perpendicular to the surface of contact and a friction force $\vec{f}$ parallel to the surface.

When a body is sliding over the surface, the friction force is called *kinetic* friction. Its magnitude f_k is approximately equal to the normal force magnitude n multiplied by the coefficient of kinetic friction μ_k. When a body is *not* moving relative to a surface, the friction force is called *static* friction. The *maximum* possible static friction force is approximately equal to the magnitude n of the normal force multiplied by the coefficient of static friction μ_s. The *actual* static friction force may be anything from zero to this maximum value, depending on the situation. Usually μ_s is greater than μ_k for a given pair of surfaces in contact. (See Examples 5.13–5.17.)

Rolling friction is similar to kinetic friction, but the force of fluid resistance depends on the speed of an object through a fluid. (See Example 5.18.)

Magnitude of kinetic friction force:

$$f_k = \mu_k n \qquad (5.5)$$

Magnitude of static friction force:

$$f_s \le \mu_s n \qquad (5.6)$$

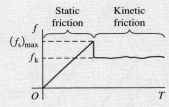

Forces in circular motion: In uniform circular motion, the acceleration vector is directed toward the center of the circle. The motion is governed by Newton's second law, $\sum \vec{F} = m\vec{a}$. (See Examples 5.19–5.23.)

Acceleration in uniform circular motion:

$$a_{\text{rad}} = \frac{v^2}{R} = \frac{4\pi^2 R}{T^2} \qquad (5.14), (5.16)$$

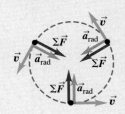

161

BRIDGING PROBLEM   In a Rotating Cone

A small block with mass m is placed inside an inverted cone that is rotating about a vertical axis such that the time for one revolution of the cone is T (Fig. 5.39). The walls of the cone make an angle β with the horizontal. The coefficient of static friction between the block and the cone is μ_s. If the block is to remain at a constant height h above the apex of the cone, what are (a) the maximum value of T and (b) the minimum value of T? (That is, find expressions for T_{max} and T_{min} in terms of β and h.)

SOLUTION GUIDE

See MasteringPhysics® Study Area for a Video Tutor solution.

IDENTIFY and SET UP

1. Although we want the block to not slide up or down on the inside of the cone, this is *not* an equilibrium problem. The block rotates with the cone and is in uniform circular motion, so it has an acceleration directed toward the center of its circular path.
2. Identify the forces on the block. What is the direction of the friction force when the cone is rotating as slowly as possible, so T has its maximum value T_{max}? What is the direction of the friction force when the cone is rotating as rapidly as possible, so T has its minimum value T_{min}? In these situations does the static friction force have its *maximum* magnitude? Why or why not?
3. Draw a free-body diagram for the block when the cone is rotating with $T = T_{max}$ and a free-body diagram when the cone is rotating with $T = T_{min}$. Choose coordinate axes, and remember that it's usually easiest to choose one of the axes to be in the direction of the acceleration.
4. What is the radius of the circular path that the block follows? Express this in terms of β and h.
5. Make a list of the unknown quantities, and decide which of these are the target variables.

5.39 A block inside a spinning cone.

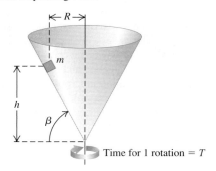

Time for 1 rotation = T

EXECUTE

6. Write Newton's second law in component form for the case in which the cone is rotating with $T = T_{max}$. Write the acceleration in terms of T_{max}, β, and h, and write the static friction force in terms of the normal force n.
7. Solve these equations for the target variable T_{max}.
8. Repeat steps 6 and 7 for the case in which the cone is rotating with $T = T_{min}$, and solve for the target variable T_{min}.

EVALUATE

9. You'll end up with some fairly complicated expressions for T_{max} and T_{min}, so check them over carefully. Do they have the correct units? Is the minimum time T_{min} less than the maximum time T_{max}, as it must be?
10. What do your expressions for T_{max} and T_{min} become if $\mu_s = 0$? Check your results by comparing with Example 5.22 in Section 5.4.

Problems

For instructor-assigned homework, go to www.masteringphysics.com

•, ••, •••: Problems of increasing difficulty. **CP**: Cumulative problems incorporating material from earlier chapters. **CALC**: Problems requiring calculus. **BIO**: Biosciences problems.

DISCUSSION QUESTIONS

Q5.1 A man sits in a seat that is suspended from a rope. The rope passes over a pulley suspended from the ceiling, and the man holds the other end of the rope in his hands. What is the tension in the rope, and what force does the seat exert on the man? Draw a free-body force diagram for the man.

Q5.2 "In general, the normal force is not equal to the weight." Give an example where these two forces are equal in magnitude, and at least two examples where they are not.

Q5.3 A clothesline hangs between two poles. No matter how tightly the line is stretched, it always sags a little at the center. Explain why.

Q5.4 A car is driven up a steep hill at constant speed. Discuss all the forces acting on the car. What pushes it up the hill?

Q5.5 For medical reasons it is important for astronauts in outer space to determine their body mass at regular intervals. Devise a scheme for measuring body mass in an apparently weightless environment.

Q5.6 To push a box up a ramp, is the force required smaller if you push horizontally or if you push parallel to the ramp? Why?

Q5.7 A woman in an elevator lets go of her briefcase but it does not fall to the floor. How is the elevator moving?

Q5.8 You can classify scales for weighing objects as those that use springs and those that use standard masses to balance unknown masses. Which group would be more accurate when used in an accelerating spaceship? When used on the moon?

Q5.9 When you tighten a nut on a bolt, how are you increasing the frictional force? How does a lock washer work?

Q5.10 A block rests on an inclined plane with enough friction to prevent it from sliding down. To start the block moving, is it easier to push it up the plane or down the plane? Why?

Q5.11 A crate of books rests on a level floor. To move it along the floor at a constant velocity, why do you exert a smaller force if you pull it at an angle θ above the horizontal than if you push it at the same angle below the horizontal?

Q5.12 In a world without friction, which of the following activities could you do (or not do)? Explain your reasoning. (a) drive around an unbanked highway curve; (b) jump into the air; (c) start walking on a horizontal sidewalk; (d) climb a vertical ladder; (e) change lanes on the freeway.

Q5.13 Walking on horizontal slippery ice can be much more tiring than walking on ordinary pavement. Why?

Q5.14 When you stand with bare feet in a wet bathtub, the grip feels fairly secure, and yet a catastrophic slip is quite possible. Explain this in terms of the two coefficients of friction.

Q5.15 You are pushing a large crate from the back of a freight elevator to the front as the elevator is moving to the next floor. In which situation is the force you must apply to move the crate the smallest and in which is it the largest: when the elevator is accelerating upward, when it is accelerating downward, or when it is traveling at constant speed? Explain.

Q5.16 The moon is accelerating toward the earth. Why isn't it getting closer to us?

Q5.17 An automotive magazine calls decreasing-radius curves "the bane of the Sunday driver." Explain.

Q5.18 You often hear people say that "friction always opposes motion." Give at least one example where (a) static friction *causes* motion, and (b) kinetic friction *causes* motion.

Q5.19 If there is a net force on a particle in uniform circular motion, why doesn't the particle's speed change?

Q5.20 A curve in a road has the banking angle calculated and posted for 80 km/h. However, the road is covered with ice so you cautiously plan to drive slower than this limit. What may happen to your car? Why?

Q5.21 You swing a ball on the end of a lightweight string in a horizontal circle at constant speed. Can the string ever be truly horizontal? If not, would it slope above the horizontal or below the horizontal? Why?

Q5.22 The centrifugal force is not included in the free-body diagrams of Figs. 5.34b and 5.35. Explain why not.

Q5.23 A professor swings a rubber stopper in a horizontal circle on the end of a string in front of his class. He tells Caroline, in the first row, that he is going to let the string go when the stopper is directly in front of her face. Should Caroline worry?

Q5.24 To keep the forces on the riders within allowable limits, loop-the-loop roller coaster rides are often designed so that the loop, rather than being a perfect circle, has a larger radius of curvature at the bottom than at the top. Explain.

Q5.25 A tennis ball drops from rest at the top of a tall glass cylinder, first with the air pumped out of the cylinder so there is no air resistance, and then a second time after the air has been readmitted to the cylinder. You examine multiflash photographs of the two drops. From these photos how can you tell which one is which, or can you?

Q5.26 If you throw a baseball straight upward with speed v_0, how does its speed, when it returns to the point from where you threw it, compare to v_0 (a) in the absence of air resistance and (b) in the presence of air resistance? Explain.

Q5.27 You throw a baseball straight upward. If air resistance is *not* ignored, how does the time required for the ball to go from the height at which it was thrown up to its maximum height compare to the time required for it to fall from its maximum height back down to the height from which it was thrown? Explain your answer.

Q5.28 You take two identical tennis balls and fill one with water. You release both balls simultaneously from the top of a tall building. If air resistance is negligible, which ball strikes the ground first? Explain. What is the answer if air resistance is *not* negligible?

Q5.29 A ball is dropped from rest and feels air resistance as it falls. Which of the graphs in Fig. Q5.29 best represents its acceleration as a function of time?

Figure **Q5.29**

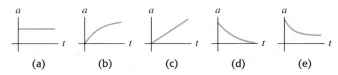

Q5.30 A ball is dropped from rest and feels air resistance as it falls. Which of the graphs in Fig. Q5.30 best represents its vertical velocity component as a function of time?

Figure **Q5.30**

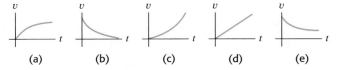

Q5.31 When does a baseball in flight have an acceleration with a positive upward component? Explain in terms of the forces on the ball and also in terms of the velocity components compared to the terminal speed. Do *not* ignore air resistance.

Q5.32 When a batted baseball moves with air drag, does it travel a greater horizontal distance while climbing to its maximum height or while descending from its maximum height back to the ground? Or is the horizontal distance traveled the same for both? Explain in terms of the forces acting on the ball.

Q5.33 "A ball is thrown from the edge of a high cliff. No matter what the angle at which it is thrown, due to air resistance, the ball will eventually end up moving vertically downward." Justify this statement.

EXERCISES

Section 5.1 Using Newton's First Law: Particles in Equilibrium

5.1 • Two 25.0-N weights are suspended at opposite ends of a rope that passes over a light, frictionless pulley. The pulley is attached to a chain that goes to the ceiling. (a) What is the tension in the rope? (b) What is the tension in the chain?

5.2 • In Fig. E5.2 each of the suspended blocks has weight w. The pulleys are frictionless and the ropes have negligible weight. Calculate, in each case, the tension T in the rope in terms of the weight w. In each case, include the free-body diagram or diagrams you used to determine the answer.

Figure **E5.2**

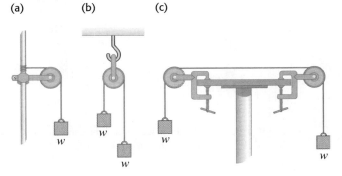

5.3 • A 75.0-kg wrecking ball hangs from a uniform heavy-duty chain having a mass of 26.0 kg. (a) Find the maximum and minimum tension in the chain. (b) What is the tension at a point three-fourths of the way up from the bottom of the chain?

5.4 •• **BIO** **Injuries to the Spinal Column.** In the treatment of spine injuries, it is often necessary to provide some tension along the spinal column to stretch the backbone. One device for doing this is the Stryker frame, illustrated in Fig. E5.4a. A weight W is attached to the patient (sometimes around a neck collar, as shown in Fig. E5.4b), and friction between the person's body and the bed prevents sliding. (a) If the coefficient of static friction between a 78.5-kg patient's body and the bed is 0.75, what is the maximum traction force along the spinal column that W can provide without causing the patient to slide? (b) Under the conditions of maximum traction, what is the tension in each cable attached to the neck collar?

Figure **E5.4**

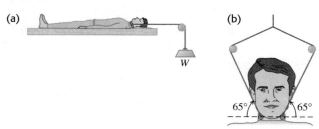

5.5 •• A picture frame hung against a wall is suspended by two wires attached to its upper corners. If the two wires make the same angle with the vertical, what must this angle be if the tension in each wire is equal to 0.75 of the weight of the frame? (Ignore any friction between the wall and the picture frame.)

5.6 •• A large wrecking ball is held in place by two light steel cables (Fig. E5.6). If the mass m of the wrecking ball is 4090 kg, what are (a) the tension T_B in the cable that makes an angle of 40° with the vertical and (b) the tension T_A in the horizontal cable?

Figure **E5.6**

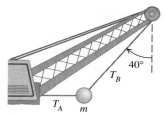

5.7 •• Find the tension in each cord in Fig. E5.7 if the weight of the suspended object is w.

Figure **E5.7**

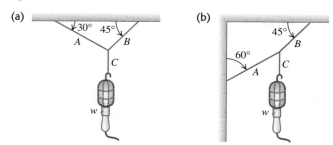

5.8 •• A 1130-kg car is held in place by a light cable on a very smooth (frictionless) ramp, as shown in Fig. E5.8. The cable makes an angle of 31.0° above the surface of the ramp, and the ramp itself rises at 25.0° above the horizontal. (a) Draw a free-body diagram for the car. (b) Find the tension in the cable. (c) How hard does the surface of the ramp push on the car?

Figure **E5.8**

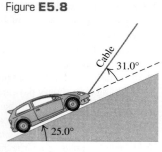

5.9 •• A man pushes on a piano with mass 180 kg so that it slides at constant velocity down a ramp that is inclined at 11.0° above the horizontal floor. Neglect any friction acting on the piano. Calculate the magnitude of the force applied by the man if he pushes (a) parallel to the incline and (b) parallel to the floor.

5.10 •• In Fig. E5.10 the weight w is 60.0 N. (a) What is the tension in the diagonal string? (b) Find the magnitudes of the horizontal forces $\vec{F}_1$ and $\vec{F}_2$ that must be applied to hold the system in the position shown.

Figure **E5.10**

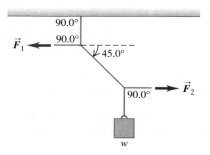

Section 5.2 Using Newton's Second Law: Dynamics of Particles

5.11 •• **BIO** **Stay Awake!** An astronaut is inside a 2.25×10^6 kg rocket that is blasting off vertically from the launch pad. You want this rocket to reach the speed of sound (331 m/s) as quickly as possible, but you also do not want the astronaut to black out. Medical tests have shown that astronauts are in danger of blacking out at an acceleration greater than $4g$. (a) What is the maximum thrust the engines of the rocket can have to just barely avoid blackout? Start with a free-body diagram of the rocket. (b) What force, in terms of her weight w, does the rocket exert on the astronaut? Start with a free-body diagram of the astronaut. (c) What is the shortest time it can take the rocket to reach the speed of sound?

5.12 •• A 125-kg (including all the contents) rocket has an engine that produces a constant vertical force (the *thrust*) of 1720 N. Inside this rocket, a 15.5-N electrical power supply rests on the floor. (a) Find the acceleration of the rocket. (b) When it has reached an altitude of 120 m, how hard does the floor push on the power supply? (*Hint:* Start with a free-body diagram for the power supply.)

5.13 •• **CP** *Genesis* **Crash.** On September 8, 2004, the *Genesis* spacecraft crashed in the Utah desert because its parachute did not open. The 210-kg capsule hit the ground at 311 km/h and penetrated the soil to a depth of 81.0 cm. (a) Assuming it to be constant, what was its acceleration (in m/s² and in g's) during the crash? (b) What force did the ground exert on the capsule during the crash? Express the force in newtons and as a multiple of the capsule's weight. (c) For how long did this force last?

5.14 • Three sleds are being pulled horizontally on frictionless horizontal ice using horizontal ropes (Fig. E5.14). The pull is of magnitude 125 N. Find (a) the acceleration of the system and (b) the tension in ropes A and B.

Figure **E5.14**

5.15 •• **Atwood's Machine.** A 15.0-kg load of bricks hangs from one end of a rope that passes over a small, frictionless pulley. A 28.0-kg counterweight is suspended from the other end of the rope, as shown in Fig. E5.15. The system is released from rest. (a) Draw two free-body diagrams, one for the load of bricks and one for the counterweight. (b) What is the magnitude of the upward acceleration of the load of bricks? (c) What is the tension in the rope while the load is moving? How does the tension compare to the weight of the load of bricks? To the weight of the counterweight?

Figure **E5.15**

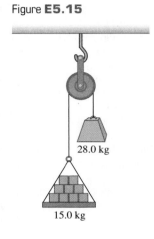

5.16 •• **CP** A 8.00-kg block of ice, released from rest at the top of a 1.50-m-long frictionless ramp, slides downhill, reaching a speed of 2.50 m/s at the bottom. (a) What is the angle between the ramp and the horizontal? (b) What would be the speed of the ice at the bottom if the motion were opposed by a constant friction force of 10.0 N parallel to the surface of the ramp?

5.17 •• A light rope is attached to a block with mass 4.00 kg that rests on a frictionless, horizontal surface. The horizontal rope passes over a frictionless, massless pulley, and a block with mass m is suspended from the other end. When the blocks are released, the tension in the rope is 10.0 N. (a) Draw two free-body diagrams, one for the 4.00-kg block and one for the block with mass m. (b) What is the acceleration of either block? (c) Find the mass m of the hanging block. (d) How does the tension compare to the weight of the hanging block?

5.18 •• **CP Runway Design.** A transport plane takes off from a level landing field with two gliders in tow, one behind the other. The mass of each glider is 700 kg, and the total resistance (air drag plus friction with the runway) on each may be assumed constant and equal to 2500 N. The tension in the towrope between the transport plane and the first glider is not to exceed 12,000 N. (a) If a speed of 40 m/s is required for takeoff, what minimum length of runway is needed? (b) What is the tension in the towrope between the two gliders while they are accelerating for the takeoff?

5.19 •• **CP** A 750.0-kg boulder is raised from a quarry 125 m deep by a long uniform chain having a mass of 575 kg. This chain is of uniform strength, but at any point it can support a maximum tension no greater than 2.50 times its weight without breaking. (a) What is the maximum acceleration the boulder can have and still get out of the quarry, and (b) how long does it take to be lifted out at maximum acceleration if it started from rest?

5.20 •• **Apparent Weight.** A 550-N physics student stands on a bathroom scale in an 850-kg (including the student) elevator that is supported by a cable. As the elevator starts moving, the scale reads 450 N. (a) Find the acceleration of the elevator (magnitude and direction). (b) What is the acceleration if the scale reads 670 N? (c) If the scale reads zero, should the student worry? Explain. (d) What is the tension in the cable in parts (a) and (c)?

5.21 •• **CP BIO Force During a Jump.** An average person can reach a maximum height of about 60 cm when jumping straight up from a crouched position. During the jump itself, the person's body from the knees up typically rises a distance of around 50 cm. To keep the calculations simple and yet get a reasonable result, assume that the *entire body* rises this much during the jump. (a) With what initial speed does the person leave the ground to reach a height of 60 cm? (b) Draw a free-body diagram of the person during the jump. (c) In terms of this jumper's weight w, what force does the ground exert on him or her during the jump?

5.22 •• **CP CALC** A 2540-kg test rocket is launched vertically from the launch pad. Its fuel (of negligible mass) provides a thrust force so that its vertical velocity as a function of time is given by $v(t) = At + Bt^2$, where A and B are constants and time is measured from the instant the fuel is ignited. At the instant of ignition, the rocket has an upward acceleration of 1.50 m/s^2 and 1.00 s later an upward velocity of 2.00 m/s. (a) Determine A and B, including their SI units. (b) At 4.00 s after fuel ignition, what is the acceleration of the rocket, and (c) what thrust force does the burning fuel exert on it, assuming no air resistance? Express the thrust in newtons and as a multiple of the rocket's weight. (d) What was the initial thrust due to the fuel?

5.23 •• **CP CALC** A 2.00-kg box is moving to the right with speed 9.00 m/s on a horizontal, frictionless surface. At $t = 0$ a horizontal force is applied to the box. The force is directed to the left and has magnitude $F(t) = (6.00 \text{ N/s}^2)t^2$. (a) What distance does the box move from its position at $t = 0$ before its speed is reduced to zero? (b) If the force continues to be applied, what is the speed of the box at $t = 3.00$ s?

5.24 •• **CP CALC** A 5.00-kg crate is suspended from the end of a short vertical rope of negligible mass. An upward force $F(t)$ is applied to the end of the rope, and the height of the crate above its initial position is given by $y(t) = (2.80 \text{ m/s})t + (0.610 \text{ m/s}^3)t^3$. What is the magnitude of the force F when $t = 4.00$ s?

Section 5.3 Frictional Forces

5.25 • **BIO The Trendelenburg Position.** In emergencies with major blood loss, the doctor will order the patient placed in the Trendelenburg position, in which the foot of the bed is raised to get maximum blood flow to the brain. If the coefficient of static friction between the typical patient and the bedsheets is 1.20, what is the maximum angle at which the bed can be tilted with respect to the floor before the patient begins to slide?

5.26 • In a laboratory experiment on friction, a 135-N block resting on a rough horizontal table is pulled by a horizontal wire. The pull gradually increases until the block begins to move and continues to increase thereafter. Figure E5.26 shows a graph of the friction force on this block as a function of the pull. (a) Identify the

Figure **E5.26**

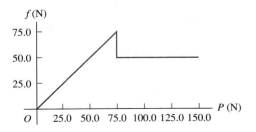

regions of the graph where static and kinetic friction occur. (b) Find the coefficients of static and kinetic friction between the block and the table. (c) Why does the graph slant upward in the first part but then level out? (d) What would the graph look like if a 135-N brick were placed on the box, and what would the coefficients of friction be in that case?

5.27 •• **CP** A stockroom worker pushes a box with mass 11.2 kg on a horizontal surface with a constant speed of 3.50 m/s. The coefficient of kinetic friction between the box and the surface is 0.20. (a) What horizontal force must the worker apply to maintain the motion? (b) If the force calculated in part (a) is removed, how far does the box slide before coming to rest?

5.28 •• A box of bananas weighing 40.0 N rests on a horizontal surface. The coefficient of static friction between the box and the surface is 0.40, and the coefficient of kinetic friction is 0.20. (a) If no horizontal force is applied to the box and the box is at rest, how large is the friction force exerted on the box? (b) What is the magnitude of the friction force if a monkey applies a horizontal force of 6.0 N to the box and the box is initially at rest? (c) What minimum horizontal force must the monkey apply to start the box in motion? (d) What minimum horizontal force must the monkey apply to keep the box moving at constant velocity once it has been started? (e) If the monkey applies a horizontal force of 18.0 N, what is the magnitude of the friction force and what is the box's acceleration?

5.29 •• A 45.0-kg crate of tools rests on a horizontal floor. You exert a gradually increasing horizontal push on it and observe that the crate just begins to move when your force exceeds 313 N. After that you must reduce your push to 208 N to keep it moving at a steady 25.0 cm/s. (a) What are the coefficients of static and kinetic friction between the crate and the floor? (b) What push must you exert to give it an acceleration of 1.10 m/s²? (c) Suppose you were performing the same experiment on this crate but were doing it on the moon instead, where the acceleration due to gravity is 1.62 m/s². (i) What magnitude push would cause it to move? (ii) What would its acceleration be if you maintained the push in part (b)?

5.30 •• Some sliding rocks approach the base of a hill with a speed of 12 m/s. The hill rises at 36° above the horizontal and has coefficients of kinetic and static friction of 0.45 and 0.65, respectively, with these rocks. (a) Find the acceleration of the rocks as they slide up the hill. (b) Once a rock reaches its highest point, will it stay there or slide down the hill? If it stays there, show why. If it slides down, find its acceleration on the way down.

5.31 •• You are lowering two boxes, one on top of the other, down the ramp shown in Fig. E5.31 by pulling on a rope parallel to the surface of the ramp. Both boxes move together at a constant speed of 15.0 cm/s. The coefficient of kinetic friction between the ramp and the lower box is 0.444, and the coefficient of static friction between the two boxes is 0.800. (a) What force do you need to exert to accomplish this? (b) What are the magnitude and direction of the friction force on the upper box?

Figure **E5.31**

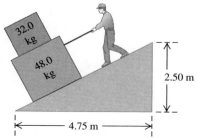

32.0 kg

48.0 kg

2.50 m

4.75 m

5.32 •• A pickup truck is carrying a toolbox, but the rear gate of the truck is missing, so the box will slide out if it is set moving. The coefficients of kinetic and static friction between the box and the bed of the truck are 0.355 and 0.650, respectively. Starting from rest, what is the shortest time this truck could accelerate uniformly to 30.0 m/s without causing the box to slide? Include a free-body diagram of the toolbox as part of your solution.

5.33 •• **CP** **Stopping Distance.** (a) If the coefficient of kinetic friction between tires and dry pavement is 0.80, what is the shortest distance in which you can stop an automobile by locking the brakes when traveling at 28.7 m/s (about 65 mi/h)? (b) On wet pavement the coefficient of kinetic friction may be only 0.25. How fast should you drive on wet pavement in order to be able to stop in the same distance as in part (a)? (*Note:* Locking the brakes is *not* the safest way to stop.)

5.34 •• Consider the system shown in Fig. E5.34. Block *A* weighs 45.0 N and block *B* weighs 25.0 N. Once block *B* is set into downward motion, it descends at a constant speed. (a) Calculate the coefficient of kinetic friction between block *A* and the tabletop. (b) A cat, also of weight 45.0 N, falls asleep on top of block *A*. If block *B* is now set into downward motion, what is its acceleration (magnitude and direction)?

Figure **E5.34**

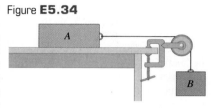

A

B

5.35 • Two crates connected by a rope lie on a horizontal surface (Fig. E5.35). Crate *A* has mass m_A and crate *B* has mass m_B. The coefficient of kinetic friction between each crate and the surface is μ_k. The crates are pulled to the right at constant velocity by a horizontal force $\vec{F}$. In terms of m_A, m_B, and μ_k, calculate (a) the magnitude of the force $\vec{F}$ and (b) the tension in the rope connecting the blocks. Include the free-body diagram or diagrams you used to determine each answer.

Figure **E5.35**

A

B

$\vec{F}$

5.36 •• **CP** A 25.0-kg box of textbooks rests on a loading ramp that makes an angle α with the horizontal. The coefficient of kinetic friction is 0.25, and the coefficient of static friction is 0.35. (a) As the angle α is increased, find the minimum angle at which the box starts to slip. (b) At this angle, find the acceleration once the box has begun to move. (c) At this angle, how fast will the box be moving after it has slid 5.0 m along the loading ramp?

5.37 •• **CP** As shown in Fig. E5.34, block *A* (mass 2.25 kg) rests on a tabletop. It is connected by a horizontal cord passing over a light, frictionless pulley to a hanging block *B* (mass 1.30 kg). The coefficient of kinetic friction between block *A* and the tabletop is 0.450. After the blocks are released from rest, find (a) the speed of each block after moving 3.00 cm and (b) the tension in the cord. Include the free-body diagram or diagrams you used to determine the answers.

5.38 •• A box with mass *m* is dragged across a level floor having a coefficient of kinetic friction μ_k by a rope that is pulled upward at an angle θ above the horizontal with a force of magnitude *F*. (a) In terms of *m*, μ_k, θ, and *g*, obtain an expression for the magnitude of the force required to move the box with constant speed. (b) Knowing that you are studying physics, a CPR instructor asks you how much force it would take to slide a 90-kg patient across a floor at constant speed by pulling on him at an angle of 25° above the horizontal. By dragging some weights wrapped in an old pair of pants down the hall with a spring balance, you find that $\mu_k = 0.35$. Use the result of part (a) to answer the instructor's question.

5.39 •• A large crate with mass m rests on a horizontal floor. The coefficients of friction between the crate and the floor are μ_s and μ_k. A woman pushes downward at an angle θ below the horizontal on the crate with a force $\vec{F}$. (a) What magnitude of force $\vec{F}$ is required to keep the crate moving at constant velocity? (b) If μ_s is greater than some critical value, the woman cannot start the crate moving no matter how hard she pushes. Calculate this critical value of μ_s.

5.40 •• You throw a baseball straight up. The drag force is proportional to v^2. In terms of g, what is the y-component of the ball's acceleration when its speed is half its terminal speed and (a) it is moving up? (b) It is moving back down?

5.41 • (a) In Example 5.18 (Section 5.3), what value of D is required to make $v_t = 42$ m/s for the skydiver? (b) If the skydiver's daughter, whose mass is 45 kg, is falling through the air and has the same D (0.25 kg/m) as her father, what is the daughter's terminal speed?

Section 5.4 Dynamics of Circular Motion

5.42 •• A small car with mass 0.800 kg travels at constant speed on the inside of a track that is a vertical circle with radius 5.00 m (Fig. E5.42). If the normal force exerted by the track on the car when it is at the top of the track (point B) is 6.00 N, what is the normal force on the car when it is at the bottom of the track (point A)?

Figure **E5.42**

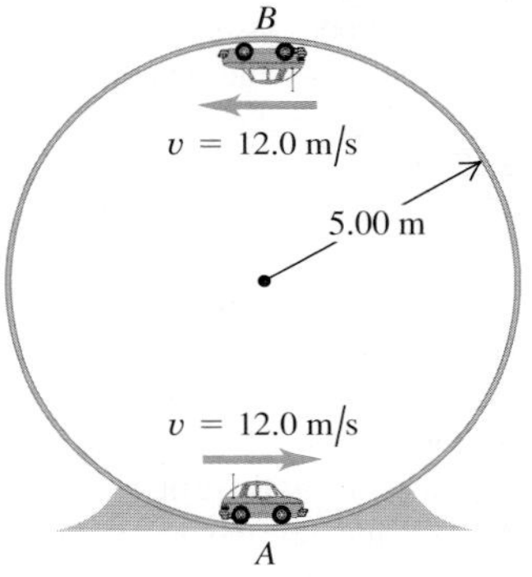

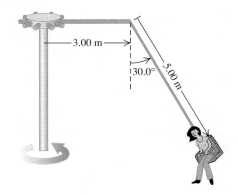

5.43 •• A machine part consists of a thin 40.0-cm-long bar with small 1.15-kg masses fastened by screws to its ends. The screws can support a maximum force of 75.0 N without pulling out. This bar rotates about an axis perpendicular to it at its center. (a) As the bar is turning at a constant rate on a horizontal, frictionless surface, what is the maximum speed the masses can have without pulling out the screws? (b) Suppose the machine is redesigned so that the bar turns at a constant rate in a vertical circle. Will one of the screws be more likely to pull out when the mass is at the top of the circle or at the bottom? Use a free-body diagram to see why. (c) Using the result of part (b), what is the greatest speed the masses can have without pulling a screw?

5.44 • A flat (unbanked) curve on a highway has a radius of 220.0 m. A car rounds the curve at a speed of 25.0 m/s. (a) What is the minimum coefficient of friction that will prevent sliding? (b) Suppose the highway is icy and the coefficient of friction between the tires and pavement is only one-third what you found in part (a). What should be the maximum speed of the car so it can round the curve safely?

5.45 •• A 1125-kg car and a 2250-kg pickup truck approach a curve on the expressway that has a radius of 225 m. (a) At what angle should the highway engineer bank this curve so that vehicles traveling at 65.0 mi/h can safely round it regardless of the condition of their tires? Should the heavy truck go slower than the

lighter car? (b) As the car and truck round the curve at find the normal force on each one due to the highway surface.

5.46 •• The "Giant Swing" at a county fair consists of a vertical central shaft with a number of horizontal arms attached at its upper end (Fig. E5.46). Each arm supports a seat suspended from a cable 5.00 m long, the upper end of the cable being fastened to the arm at a point 3.00 m from the central shaft. (a) Find the time of one revolution of the swing if the cable supporting a seat makes an angle of 30.0° with the vertical. (b) Does the angle depend on the weight of the passenger for a given rate of revolution?

Figure **E5.46**

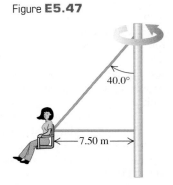

5.47 •• In another version of the "Giant Swing" (see Exercise 5.46), the seat is connected to two cables as shown in Fig. E5.47, one of which is horizontal. The seat swings in a horizontal circle at a rate of 32.0 rpm (rev/min). If the seat weighs 255 N and an 825-N person is sitting in it, find the tension in each cable.

Figure **E5.47**

5.48 •• A small button placed on a horizontal rotating platform with diameter 0.320 m will revolve with the platform when it is brought up to a speed of 40.0 rev/min, provided the button is no more than 0.150 m from the axis. (a) What is the coefficient of static friction between the button and the platform? (b) How far from the axis can the button be placed, without slipping, if the platform rotates at 60.0 rev/min?

5.49 •• **Rotating Space Stations.** One problem for humans living in outer space is that they are apparently weightless. One way around this problem is to design a space station that spins about its center at a constant rate. This creates "artificial gravity" at the outside rim of the station. (a) If the diameter of the space station is 800 m, how many revolutions per minute are needed for the "artificial gravity" acceleration to be 9.80 m/s²? (b) If the space station is a waiting area for travelers going to Mars, it might be desirable to simulate the acceleration due to gravity on the Martian surface (3.70 m/s²). How many revolutions per minute are needed in this case?

5.50 • The Cosmoclock 21 Ferris wheel in Yokohama City, Japan, has a diameter of 100 m. Its name comes from its 60 arms, each of which can function as a second hand (so that it makes one revolution every 60.0 s). (a) Find the speed of the passengers when the Ferris wheel is rotating at this rate. (b) A passenger weighs 882 N at the weight-guessing booth on the ground. What is his apparent weight at the highest and at the lowest point on the Ferris wheel? (c) What would be the time for one revolution if the passenger's apparent weight at the highest point were zero? (d) What then would be the passenger's apparent weight at the lowest point?

5.51 •• An airplane flies in a loop (a circular path in a vertical plane) of radius 150 m. The pilot's head always points toward the center of the loop. The speed of the airplane is not constant; the airplane goes slowest at the top of the loop and fastest at the bottom. (a) At the top of the loop, the pilot feels weightless. What is the speed of the airplane at this point? (b) At the bottom of the loop, the speed of the airplane is 280 km/h. What is the apparent weight of the pilot at this point? His true weight is 700 N.

5.52 •• A 50.0-kg stunt pilot who has been diving her airplane vertically pulls out of the dive by changing her course to a circle in a vertical plane. (a) If the plane's speed at the lowest point of the circle is 95.0 m/s, what is the minimum radius of the circle for the acceleration at this point not to exceed $4.00g$? (b) What is the apparent weight of the pilot at the lowest point of the pullout?

5.53 • **Stay Dry!** You tie a cord to a pail of water, and you swing the pail in a vertical circle of radius 0.600 m. What minimum speed must you give the pail at the highest point of the circle if no water is to spill from it?

5.54 •• A bowling ball weighing 71.2 N (16.0 lb) is attached to the ceiling by a 3.80-m rope. The ball is pulled to one side and released; it then swings back and forth as a pendulum. As the rope swings through the vertical, the speed of the bowling ball is 4.20 m/s. (a) What is the acceleration of the bowling ball, in magnitude and direction, at this instant? (b) What is the tension in the rope at this instant?

5.55 •• **BIO Effect on Blood of Walking.** While a person is walking, his arms swing through approximately a 45° angle in $\frac{1}{2}$ s. As a reasonable approximation, we can assume that the arm moves with constant speed during each swing. A typical arm is 70.0 cm long, measured from the shoulder joint. (a) What is the acceleration of a 1.0-g drop of blood in the fingertips at the bottom of the swing? (b) Draw a free-body diagram of the drop of blood in part (a). (c) Find the force that the blood vessel must exert on the drop of blood in part (a). Which way does this force point? (d) What force would the blood vessel exert if the arm were not swinging?

PROBLEMS

5.56 •• An adventurous archaeologist crosses between two rock cliffs by slowly going hand over hand along a rope stretched between the cliffs. He stops to rest at the middle of the rope (Fig. P5.56). The rope will break if the tension in it exceeds 2.50×10^4 N, and our hero's mass is 90.0 kg. (a) If the angle θ is 10.0°, find the tension in the rope. (b) What is the smallest value the angle θ can have if the rope is not to break?

Figure **P5.56**

5.57 ••• Two ropes are connected to a steel cable that supports a hanging weight as shown in Fig. P5.57. (a) Draw a free-body diagram showing all of the forces acting at the knot that connects the two ropes to the steel cable. Based on your force diagram, which of the two ropes will have the greater tension? (b) If the maximum tension either rope can sustain without breaking is 5000 N, determine the maximum value of the hanging weight that these ropes can safely support. You can ignore the weight of the ropes and the steel cable.

Figure **P5.57**

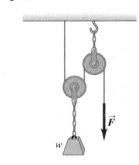

5.58 •• In Fig. P5.58 a worker lifts a weight w by pulling down on a rope with a force $\vec{F}$. The upper pulley is attached to the ceiling by a chain, and the lower pulley is attached to the weight by another chain. In terms of w, find the tension in each chain and the magnitude of the force $\vec{F}$ if the weight is lifted at constant speed. Include the free-body diagram or diagrams you used to determine your answers. Assume that the rope, pulleys, and chains all have negligible weights.

Figure **P5.58**

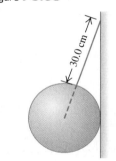

5.59 ••• A solid uniform 45.0-kg ball of diameter 32.0 cm is supported against a vertical, frictionless wall using a thin 30.0-cm wire of negligible mass, as shown in Fig. P5.59. (a) Draw a free-body diagram for the ball and use it to find the tension in the wire. (b) How hard does the ball push against the wall?

Figure **P5.59**

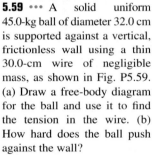

5.60 ••• A horizontal wire holds a solid uniform ball of mass m in place on a tilted ramp that rises 35.0° above the horizontal. The surface of this ramp is perfectly smooth, and the wire is directed away from the center of the ball (Fig. P5.60). (a) Draw a free-body diagram for the ball. (b) How hard does the surface of the ramp push on the ball? (c) What is the tension in the wire?

Figure **P5.60**

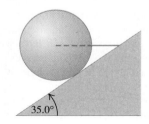

5.61 •• **CP BIO Forces During Chin-ups.** People who do chin-ups raise their chin just over a bar (the chinning bar), supporting themselves with only their arms. Typically, the body below the arms is raised by about 30 cm in a time of 1.0 s, starting from rest. Assume that the entire body of a 680-N person doing chin-ups is raised this distance and that half the 1.0 s is spent accelerating upward and the other half accelerating downward, uniformly in both cases. Draw a free-body diagram of the person's body, and then apply it to find the force his arms must exert on him during the accelerating part of the chin-up.

5.62 •• **CP BIO Prevention of Hip Injuries.** People (especially the elderly) who are prone to falling can wear hip pads to cushion the impact on their hip from a fall. Experiments have shown that if the speed at impact can be reduced to 1.3 m/s or less, the hip will usually not fracture. Let us investigate the worst-case scenario in which a 55-kg person completely loses her footing (such as on icy pavement) and falls a distance of 1.0 m, the distance from her hip to the ground. We shall assume that the person's entire body has the same acceleration, which, in reality, would not quite be true. (a) With what speed does her hip reach the ground? (b) A typical hip pad can reduce the person's speed to 1.3 m/s over a distance of 2.0 cm. Find the acceleration (assumed to be constant) of this person's hip while she is slowing down and the force the pad exerts on it. (c) The force in part (b) is very large. To see whether it is likely to cause injury, calculate how long it lasts.

5.63 ••• CALC A 3.00-kg box that is several hundred meters above the surface of the earth is suspended from the end of a short vertical rope of negligible mass. A time-dependent upward force is applied to the upper end of the rope, and this results in a tension in the rope of $T(t) = (36.0 \text{ N/s})t$. The box is at rest at $t = 0$. The only forces on the box are the tension in the rope and gravity. (a) What is the velocity of the box at (i) $t = 1.00$ s and (ii) $t = 3.00$ s? (b) What is the maximum distance that the box descends below its initial position? (c) At what value of t does the box return to its initial position?

5.64 •• CP A 5.00-kg box sits at rest at the bottom of a ramp that is 8.00 m long and that is inclined at $30.0°$ above the horizontal. The coefficient of kinetic friction is $\mu_k = 0.40$, and the coefficient of static friction is $\mu_s = 0.50$. What constant force F, applied parallel to the surface of the ramp, is required to push the box to the top of the ramp in a time of 4.00 s?

5.65 •• Two boxes connected by a light horizontal rope are on a horizontal surface, as shown in Fig. P5.35. The coefficient of kinetic friction between each box and the surface is $\mu_k = 0.30$. One box (box B) has mass 5.00 kg, and the other box (box A) has mass m. A force F with magnitude 40.0 N and direction $53.1°$ above the horizontal is applied to the 5.00-kg box, and both boxes move to the right with $a = 1.50 \text{ m/s}^2$. (a) What is the tension T in the rope that connects the boxes? (b) What is the mass m of the second box?

5.66 ••• A 6.00-kg box sits on a ramp that is inclined at $37.0°$ above the horizontal. The coefficient of kinetic friction between the box and the ramp is $\mu_k = 0.30$. What *horizontal* force is required to move the box up the incline with a constant acceleration of 4.20 m/s^2?

5.67 •• CP In Fig. P5.34 block A has mass m and block B has mass 6.00 kg. The coefficient of kinetic friction between block A and the tabletop is $\mu_k = 0.40$. The mass of the rope connecting the blocks can be neglected. The pulley is light and frictionless. When the system is released from rest, the hanging block descends 5.00 m in 3.00 s. What is the mass m of block A?

5.68 •• CP In Fig. P5.68 $m_1 = 20.0$ kg and $\alpha = 53.1°$. The coefficient of kinetic friction between the block and the incline is $\mu_k = 0.40$. What must be the mass m_2 of the hanging block if it is to descend 12.0 m in the first 3.00 s after the system is released from rest?

Figure **P5.68**

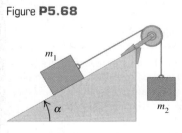

5.69 ••• CP Rolling Friction. Two bicycle tires are set rolling with the same initial speed of 3.50 m/s on a long, straight road, and the distance each travels before its speed is reduced by half is measured. One tire is inflated to a pressure of 40 psi and goes 18.1 m; the other is at 105 psi and goes 92.9 m. What is the coefficient of rolling friction μ_r for each? Assume that the net horizontal force is due to rolling friction only.

5.70 •• **A Rope with Mass.** A block with mass M is attached to the lower end of a vertical, uniform rope with mass m and length L. A constant upward force $\vec{F}$ is applied to the top of the rope, causing the rope and block to accelerate upward. Find the tension in the rope at a distance x from the top end of the rope, where x can have any value from 0 to L.

5.71 •• A block with mass m_1 is placed on an inclined plane with slope angle α and is connected to a second hanging block with mass m_2 by a cord passing over a small, frictionless pulley (Fig. P5.68). The coefficient of static friction is μ_s and the coefficient of kinetic friction is μ_k. (a) Find the mass m_2 for which block m_1 moves up the plane at constant speed once it is set in motion. (b) Find the mass m_2 for which block m_1 moves down the plane at constant speed once it is set in motion. (c) For what range of values of m_2 will the blocks remain at rest if they are released from rest?

5.72 •• Block A in Fig. P5.72 weighs 60.0 N. The coefficient of static friction between the block and the surface on which it rests is 0.25. The weight w is 12.0 N and the system is in equilibrium. (a) Find the friction force exerted on block A. (b) Find the maximum weight w for which the system will remain in equilibrium.

Figure **P5.72**

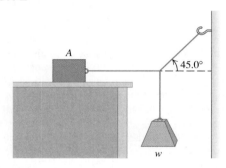

5.73 •• Block A in Fig. P5.73 weighs 2.40 N and block B weighs 3.60 N. The coefficient of kinetic friction between all surfaces is 0.300. Find the magnitude of the horizontal force $\vec{F}$ necessary to drag block B to the left at constant speed (a) if A rests on B and moves with it (Fig. P5.73a). (b) If A is held at rest (Fig. P5.73b).

Figure **P5.73**

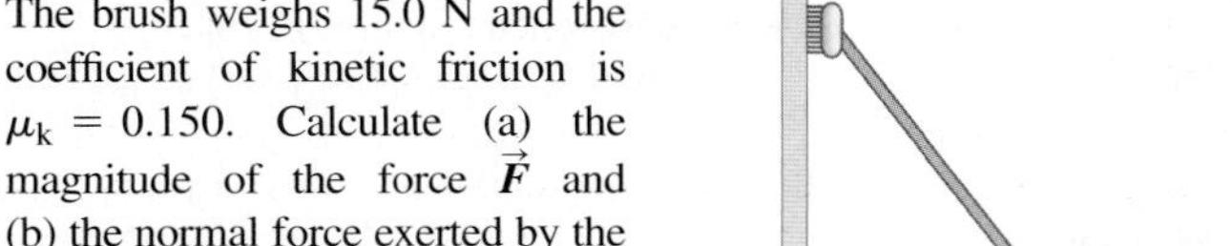

(a) 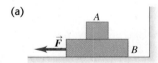   (b)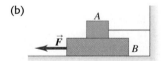

5.74 ••• A window washer pushes his scrub brush up a vertical window at constant speed by applying a force $\vec{F}$ as shown in Fig. P5.74. The brush weighs 15.0 N and the coefficient of kinetic friction is $\mu_k = 0.150$. Calculate (a) the magnitude of the force $\vec{F}$ and (b) the normal force exerted by the window on the brush.

5.75 •• BIO The Flying Leap of a Flea. High-speed motion pictures (3500 frames/second) of a jumping 210-μg flea yielded the data to plot the flea's acceleration as a function of time as

Figure **P5.74**

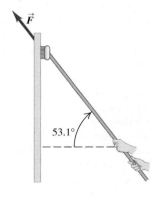

shown in Fig. P5.75. (See "The Flying Leap of the Flea," by M. Rothschild et al. in the November 1973 *Scientific American*.) This flea was about 2 mm long and jumped at a nearly vertical takeoff angle. Use the measurements shown on the graph to answer the questions. (a) Find the *initial* net external force on the flea. How does it compare to the flea's weight? (b) Find the *maximum* net external force on this jumping flea. When does this maximum force occur? (c) Use the graph to find the flea's maximum speed.

Figure **P5.75**

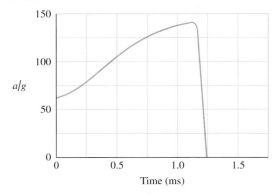

Time (ms)

5.76 •• **CP** A 25,000-kg rocket blasts off vertically from the earth's surface with a constant acceleration. During the motion considered in the problem, assume that g remains constant (see Chapter 13). Inside the rocket, a 15.0-N instrument hangs from a wire that can support a maximum tension of 45.0 N. (a) Find the minimum time for this rocket to reach the sound barrier (330 m/s) without breaking the inside wire and the maximum vertical thrust of the rocket engines under these conditions. (b) How far is the rocket above the earth's surface when it breaks the sound barrier?

5.77 ••• **CP CALC** You are standing on a bathroom scale in an elevator in a tall building. Your mass is 64 kg. The elevator starts from rest and travels upward with a speed that varies with time according to $v(t) = (3.0 \text{ m/s}^2)t + (0.20 \text{ m/s}^3)t^2$. When $t = 4.0$ s, what is the reading of the bathroom scale?

5.78 ••• **CP Elevator Design.** You are designing an elevator for a hospital. The force exerted on a passenger by the floor of the elevator is not to exceed 1.60 times the passenger's weight. The elevator accelerates upward with constant acceleration for a distance of 3.0 m and then starts to slow down. What is the maximum speed of the elevator?

5.79 •• **CP** You are working for a shipping company. Your job is to stand at the bottom of a 8.0-m-long ramp that is inclined at 37° above the horizontal. You grab packages off a conveyor belt and propel them up the ramp. The coefficient of kinetic friction between the packages and the ramp is $\mu_k = 0.30$. (a) What speed do you need to give a package at the bottom of the ramp so that it has zero speed at the top of the ramp? (b) Your coworker is supposed to grab the packages as they arrive at the top of the ramp, but she misses one and it slides back down. What is its speed when it returns to you?

5.80 •• A hammer is hanging by a light rope from the ceiling of a bus. The ceiling of the bus is parallel to the roadway. The bus is traveling in a straight line on a horizontal street. You observe that the hammer hangs at rest with respect to the bus when the angle between the rope and the ceiling of the bus is 67°. What is the acceleration of the bus?

5.81 ••• A steel washer is suspended inside an empty shipping crate from a light string attached to the top of the crate. The crate slides down a long ramp that is inclined at an angle of 37° above the horizontal. The crate has mass 180 kg. You are sitting inside the crate (with a flashlight); your mass is 55 kg. As the crate is sliding down the ramp, you find the washer is at rest with respect to the crate when the string makes an angle of 68° with the top of the crate. What is the coefficient of kinetic friction between the ramp and the crate?

5.82 • **CP Lunch Time!** You are riding your motorcycle one day down a wet street that slopes downward at an angle of 20° below the horizontal. As you start to ride down the hill, you notice a construction crew has dug a deep hole in the street at the bottom of the hill. A Siberian tiger, escaped from the City Zoo, has taken up residence in the hole. You apply the brakes and lock your wheels at the top of the hill, where you are moving with a speed of 20 m/s. The inclined street in front of you is 40 m long. (a) Will you plunge into the hole and become the tiger's lunch, or do you skid to a stop before you reach the hole? (The coefficients of friction between your motorcycle tires and the wet pavement are $\mu_s = 0.90$ and $\mu_k = 0.70$.) (b) What must your initial speed be if you are to stop just before reaching the hole?

5.83 ••• In the system shown in Fig. P5.34, block A has mass m_A, block B has mass m_B, and the rope connecting them has a *nonzero* mass m_{rope}. The rope has a total length L, and the pulley has a very small radius. You can ignore any sag in the horizontal part of the rope. (a) If there is no friction between block A and the tabletop, find the acceleration of the blocks at an instant when a length d of rope hangs vertically between the pulley and block B. As block B falls, will the magnitude of the acceleration of the system increase, decrease, or remain constant? Explain. (b) Let $m_A = 2.00$ kg, $m_B = 0.400$ kg, $m_{\text{rope}} = 0.160$ kg, and $L = 1.00$ m. If there is friction between block A and the tabletop, with $\mu_k = 0.200$ and $\mu_s = 0.250$, find the minimum value of the distance d such that the blocks will start to move if they are initially at rest. (c) Repeat part (b) for the case $m_{\text{rope}} = 0.040$ kg. Will the blocks move in this case?

5.84 ••• If the coefficient of static friction between a table and a uniform massive rope is μ_s, what fraction of the rope can hang over the edge of the table without the rope sliding?

5.85 •• A 40.0-kg packing case is initially at rest on the floor of a 1500-kg pickup truck. The coefficient of static friction between the case and the truck floor is 0.30, and the coefficient of kinetic friction is 0.20. Before each acceleration given below, the truck is traveling due north at constant speed. Find the magnitude and direction of the friction force acting on the case (a) when the truck accelerates at 2.20 m/s² northward and (b) when it accelerates at 3.40 m/s² southward.

5.86 • **CP Traffic Court.** You are called as an expert witness in the trial of a traffic violation. The facts are these: A driver slammed on his brakes and came to a stop with constant acceleration. Measurements of his tires and the skid marks on the pavement indicate that he locked his car's wheels, the car traveled 192 ft before stopping, and the coefficient of kinetic friction between the road and his tires was 0.750. The charge is that he was speeding in a 45-mi/h zone. He pleads innocent. What is your conclusion, guilty or innocent? How fast was he going when he hit his brakes?

5.87 ••• Two identical 15.0-kg balls, each 25.0 cm in diameter, are suspended by two 35.0-cm wires as shown in Fig. P5.87. The entire apparatus is supported by a single 18.0-cm wire, and the surfaces of the balls are perfectly smooth. (a) Find the tension in each of the three wires. (b) How hard does each ball push on the other one?

Figure **P5.87**

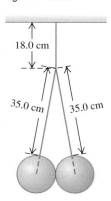

5.88 ⋅⋅ **CP Losing Cargo.** A 12.0-kg box rests on the flat floor of a truck. The coefficients of friction between the box and floor are $\mu_s = 0.19$ and $\mu_k = 0.15$. The truck stops at a stop sign and then starts to move with an acceleration of 2.20 m/s². If the box is 1.80 m from the rear of the truck when the truck starts, how much time elapses before the box falls off the truck? How far does the truck travel in this time?

5.89 ⋅⋅⋅ Block A in Fig. P5.89 weighs 1.90 N, and block B weighs 4.20 N. The coefficient of kinetic friction between all surfaces is 0.30. Find the magnitude of the horizontal force $\vec{F}$ necessary to drag block B to the left at constant speed if A and B are connected by a light, flexible cord passing around a fixed, frictionless pulley.

Figure **P5.89**

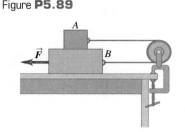

5.90 ⋅⋅⋅ **CP** You are part of a design team for future exploration of the planet Mars, where $g = 3.7$ m/s². An explorer is to step out of a survey vehicle traveling horizontally at 33 m/s when it is 1200 m above the surface and then fall freely for 20 s. At that time, a portable advanced propulsion system (PAPS) is to exert a constant force that will decrease the explorer's speed to zero at the instant she touches the surface. The total mass (explorer, suit, equipment, and PAPS) is 150 kg. Assume the change in mass of the PAPS to be negligible. Find the horizontal and vertical components of the force the PAPS must exert, and for what interval of time the PAPS must exert it. You can ignore air resistance.

5.91 ⋅⋅ Block A in Fig. P5.91 has a mass of 4.00 kg, and block B has mass 12.0 kg. The coefficient of kinetic friction between block B and the horizontal surface is 0.25. (a) What is the mass of block C if block B is moving to the right and speeding up with an acceleration of 2.00 m/s²? (b) What is the tension in each cord when block B has this acceleration?

Figure **P5.91**

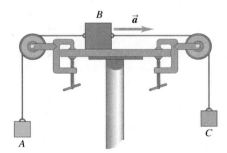

5.92 ⋅⋅ Two blocks connected by a cord passing over a small, frictionless pulley rest on frictionless planes (Fig. P5.92). (a) Which way will the system move when the blocks are released from rest? (b) What is the acceleration of the blocks? (c) What is the tension in the cord?

Figure **P5.92**

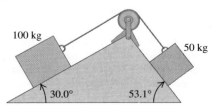

5.93 ⋅⋅ In terms of m_1, m_2, and g, find the acceleration of each block in Fig. P5.93. There is no friction anywhere in the system.

Figure **P5.93**

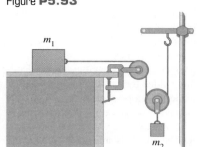

5.94 ⋅⋅⋅ Block B, with mass 5.00 kg, rests on block A, with mass 8.00 kg, which in turn is on a horizontal tabletop (Fig. P5.94). There is no friction between block A and the tabletop, but the coefficient of static friction between block A and block B is 0.750. A light string attached to block A passes over a frictionless, massless pulley, and block C is suspended from the other end of the string. What is the largest mass that block C can have so that blocks A and B still slide together when the system is released from rest?

Figure **P5.94**

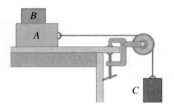

5.95 ⋅⋅⋅ Two objects with masses 5.00 kg and 2.00 kg hang 0.600 m above the floor from the ends of a cord 6.00 m long passing over a frictionless pulley. Both objects start from rest. Find the maximum height reached by the 2.00-kg object.

5.96 ⋅⋅ **Friction in an Elevator.** You are riding in an elevator on the way to the 18th floor of your dormitory. The elevator is accelerating upward with $a = 1.90$ m/s². Beside you is the box containing your new computer; the box and its contents have a total mass of 28.0 kg. While the elevator is accelerating upward, you push horizontally on the box to slide it at constant speed toward the elevator door. If the coefficient of kinetic friction between the box and the elevator floor is $\mu_k = 0.32$, what magnitude of force must you apply?

5.97 ⋅ A block is placed against the vertical front of a cart as shown in Fig. P5.97. What acceleration must the cart have so that block A does not fall? The coefficient of static friction between the block and the cart is μ_s. How would an observer on the cart describe the behavior of the block?

Figure **P5.97**

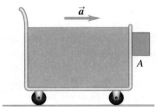

5.98 ⋅⋅⋅ Two blocks with masses 4.00 kg and 8.00 kg are connected by a string and slide down a 30.0° inclined plane (Fig. P5.98). The coefficient of kinetic friction between the

Figure **P5.98**

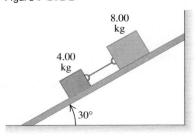

4.00-kg block and the plane is 0.25; that between the 8.00-kg block and the plane is 0.35. (a) Calculate the acceleration of each block. (b) Calculate the tension in the string. (c) What happens if the positions of the blocks are reversed, so the 4.00-kg block is above the 8.00-kg block?

5.99 ••• Block A, with weight $3w$, slides down an inclined plane S of slope angle $36.9°$ at a constant speed while plank B, with weight w, rests on top of A. The plank is attached by a cord to the wall (Fig. P5.99). (a) Draw a diagram of all the forces acting on block A. (b) If the coefficient of kinetic friction is the same between A and B and between S and A, determine its value.

Figure **P5.99**

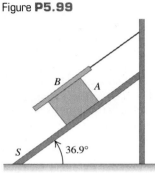

5.100 •• **Accelerometer.** The system shown in Fig. P5.100 can be used to measure the acceleration of the system. An observer riding on the platform measures the angle θ that the thread supporting the light ball makes with the vertical. There is no friction anywhere. (a) How is θ related to the acceleration of the system? (b) If $m_1 = 250$ kg and $m_2 = 1250$ kg, what is θ? (c) If you can vary m_1 and m_2, what is the largest angle θ you could achieve? Explain how you need to adjust m_1 and m_2 to do this.

Figure **P5.100**

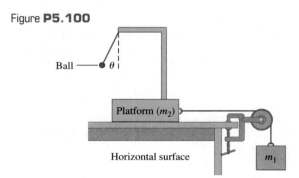

5.101 ••• **Banked Curve I.** A curve with a 120-m radius on a level road is banked at the correct angle for a speed of 20 m/s. If an automobile rounds this curve at 30 m/s, what is the minimum coefficient of static friction needed between tires and road to prevent skidding?

5.102 •• **Banked Curve II.** Consider a wet roadway banked as in Example 5.22 (Section 5.4), where there is a coefficient of static friction of 0.30 and a coefficient of kinetic friction of 0.25 between the tires and the roadway. The radius of the curve is $R = 50$ m. (a) If the banking angle is $\beta = 25°$, what is the *maximum* speed the automobile can have before sliding *up* the banking? (b) What is the *minimum* speed the automobile can have before sliding *down* the banking?

5.103 ••• Blocks A, B, and C are placed as in Fig. P5.103 and connected by ropes of negligible mass. Both A and B weigh 25.0 N each, and the coefficient of kinetic friction between each block and the surface is 0.35. Block C descends with constant velocity. (a) Draw two separate free-body diagrams showing the forces acting on A and on B. (b) Find the tension in the rope connecting blocks A and B. (c) What is the weight of block C? (d) If the rope connecting A and B were cut, what would be the acceleration of C?

Figure **P5.103**

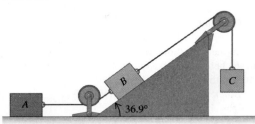

5.104 •• You are riding in a school bus. As the bus rounds a flat curve at constant speed, a lunch box with mass 0.500 kg, suspended from the ceiling of the bus by a string 1.80 m long, is found to hang at rest relative to the bus when the string makes an angle of $30.0°$ with the vertical. In this position the lunch box is 50.0 m from the center of curvature of the curve. What is the speed v of the bus?

5.105 • **The Monkey and Bananas Problem.** A 20-kg monkey has a firm hold on a light rope that passes over a frictionless pulley and is attached to a 20-kg bunch of bananas (Fig. P5.105). The monkey looks up, sees the bananas, and starts to climb the rope to get them. (a) As the monkey climbs, do the bananas move up, down, or remain at rest? (b) As the monkey climbs, does the distance between the monkey and the bananas decrease, increase, or remain constant? (c) The monkey releases her hold on the rope. What happens to the distance between the monkey and the bananas while she is falling? (d) Before reaching the ground, the monkey grabs the rope to stop her fall. What do the bananas do?

Figure **P5.105**

5.106 •• **CALC** You throw a rock downward into water with a speed of $3mg/k$, where k is the coefficient in Eq. (5.7). Assume that the relationship between fluid resistance and speed is as given in Eq. (5.7), and calculate the speed of the rock as a function of time.

5.107 •• A rock with mass $m = 3.00$ kg falls from rest in a viscous medium. The rock is acted on by a net constant downward force of 18.0 N (a combination of gravity and the buoyant force exerted by the medium) and by a fluid resistance force $f = kv$, where v is the speed in m/s and $k = 2.20$ N $\cdot$ s/m (see Section 5.3). (a) Find the initial acceleration a_0. (b) Find the acceleration when the speed is 3.00 m/s. (c) Find the speed when the acceleration equals $0.1a_0$. (d) Find the terminal speed v_t. (e) Find the coordinate, speed, and acceleration 2.00 s after the start of the motion. (f) Find the time required to reach a speed of $0.9v_t$.

5.108 •• **CALC** A rock with mass m slides with initial velocity v_0 on a horizontal surface. A retarding force F_R that the surface exerts on the rock is proportional to the square root of the instantaneous velocity of the rock ($F_R = -kv^{1/2}$). (a) Find expressions for the velocity and position of the rock as a function of time. (b) In terms of m, k, and v_0, at what time will the rock come to rest? (c) In terms of m, k, and v_0, what is the distance of the rock from its starting point when it comes to rest?

5.109 ••• You observe a 1350-kg sports car rolling along flat pavement in a straight line. The only horizontal forces acting on it are a constant rolling friction and air resistance (proportional to the square of its speed). You take the following data during a time interval of 25 s: When its speed is 32 m/s, the car slows down at a rate of -0.42 m/s^2, and when its speed is decreased to 24 m/s, it slows down at -0.30 m/s^2. (a) Find the coefficient of rolling friction and the air drag constant D. (b) At what constant speed will this car move down an incline that makes a $2.2°$ angle with the horizontal? (c) How is the constant speed for an incline of angle β related to the terminal speed of this sports car if the car drops off a high cliff? Assume that in both cases the air resistance force is proportional to the square of the speed, and the air drag constant is the same.

5.110 ••• The 4.00-kg block in Fig. P5.110 is attached to a vertical rod by means of two strings. When the system rotates about the axis of the rod, the strings are extended as shown in the diagram and the tension in the upper string is 80.0 N. (a) What is the tension in the lower cord? (b) How many revolutions per minute does the system make? (c) Find the number of revolutions per minute at which the lower cord just goes slack. (d) Explain what happens if the number of revolutions per minute is less than in part (c).

Figure **P5.110**

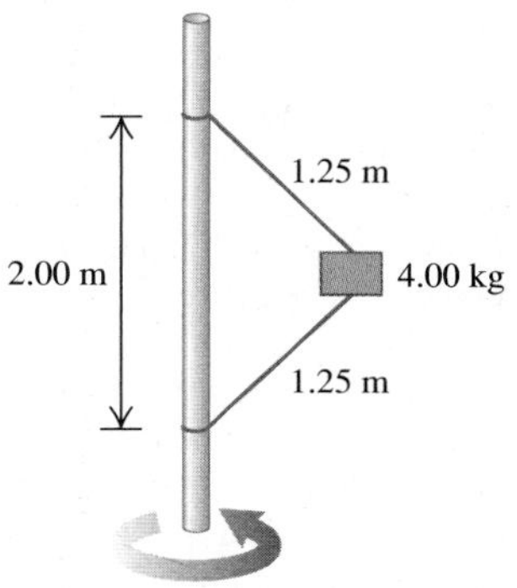

1.25 m

2.00 m

4.00 kg

1.25 m

5.111 ••• **CALC** Equation (5.10) applies to the case where the initial velocity is zero. (a) Derive the corresponding equation for $v_y(t)$ when the falling object has an initial downward velocity with magnitude v_0. (b) For the case where $v_0 < v_t$, sketch a graph of v_y as a function of t and label v_t on your graph. (c) Repeat part (b) for the case where $v_0 > v_t$. (d) Discuss what your result says about $v_y(t)$ when $v_0 = v_t$.

5.112 ••• **CALC** A small rock moves in water, and the force exerted on it by the water is given by Eq. (5.7). The terminal speed of the rock is measured and found to be 2.0 m/s. The rock is projected *upward* at an initial speed of 6.0 m/s. You can ignore the buoyancy force on the rock. (a) In the absence of fluid resistance, how high will the rock rise and how long will it take to reach this maximum height? (b) When the effects of fluid resistance are included, what are the answers to the questions in part (a)?

5.113 •• **Merry-Go-Round.** One December identical twins Jena and Jackie are playing on a large merry-go-round (a disk mounted parallel to the ground, on a vertical axle through its center) in their school playground in northern Minnesota. Each twin has mass 30.0 kg. The icy coating on the merry-go-round surface makes it frictionless. The merry-go-round revolves at a constant rate as the twins ride on it. Jena, sitting 1.80 m from the center of the merry-go-round, must hold on to one of the metal posts attached to the merry-go-round with a horizontal force of 60.0 N to keep from sliding off. Jackie is sitting at the edge, 3.60 m from the center. (a) With what horizontal force must Jackie hold on to keep from falling off? (b) If Jackie falls off, what will be her horizontal velocity when she becomes airborne?

5.114 •• A 70-kg person rides in a 30-kg cart moving at 12 m/s at the top of a hill that is in the shape of an arc of a circle with a radius of 40 m. (a) What is the apparent weight of the person as the cart passes over the top of the hill? (b) Determine the maximum speed that the cart may travel at the top of the hill without losing contact with the surface. Does your answer depend on the mass of the cart or the mass of the person? Explain.

5.115 •• On the ride "Spindletop" at the amusement park Six Flags Over Texas, people stood against the inner wall of a hollow vertical cylinder with radius 2.5 m. The cylinder started to rotate, and when it reached a constant rotation rate of 0.60 rev/s, the floor on which people were standing dropped about 0.5 m. The people remained pinned against the wall. (a) Draw a force diagram for a person on this ride, after the floor has dropped. (b) What minimum coefficient of static friction is required if the person on the ride is not to slide downward to the new position of the floor? (c) Does your answer in part (b) depend on the mass of the passenger? (*Note:* When the ride is over, the cylinder is slowly brought to rest. As it slows down, people slide down the walls to the floor.)

5.116 •• A passenger with mass 85 kg rides in a Ferris wheel like that in Example 5.23 (Section 5.4). The seats travel in a circle of radius 35 m. The Ferris wheel rotates at constant speed and makes one complete revolution every 25 s. Calculate the magnitude and direction of the net force exerted on the passenger by the seat when she is (a) one-quarter revolution past her lowest point and (b) one-quarter revolution past her highest point.

5.117 • **Ulterior Motives.** You are driving a classic 1954 Nash Ambassador with a friend who is sitting to your right on the passenger side of the front seat. The Ambassador has flat bench seats. You would like to be closer to your friend and decide to use physics to achieve your romantic goal by making a quick turn. (a) Which way (to the left or to the right) should you turn the car to get your friend to slide closer to you? (b) If the coefficient of static friction between your friend and the car seat is 0.35, and you keep driving at a constant speed of 20 m/s, what is the maximum radius you could make your turn and still have your friend slide your way?

5.118 •• A physics major is working to pay his college tuition by performing in a traveling carnival. He rides a motorcycle inside a hollow, transparent plastic sphere. After gaining sufficient speed, he travels in a vertical circle with a radius of 13.0 m. The physics major has mass 70.0 kg, and his motorcycle has mass 40.0 kg. (a) What minimum speed must he have at the top of the circle if the tires of the motorcycle are not to lose contact with the sphere? (b) At the bottom of the circle, his speed is twice the value calculated in part (a). What is the magnitude of the normal force exerted on the motorcycle by the sphere at this point?

5.119 •• A small bead can slide without friction on a circular hoop that is in a vertical plane and has a radius of 0.100 m. The hoop rotates at a constant rate of 4.00 rev/s about a vertical diameter (Fig. P5.119). (a) Find the angle β at which the bead is in vertical equilibrium. (Of course, it has a radial acceleration toward the axis.) (b) Is it possible for the bead to "ride" at the same elevation as the center of the hoop? (c) What will happen if the hoop rotates at 1.00 rev/s?

Figure **P5.119**

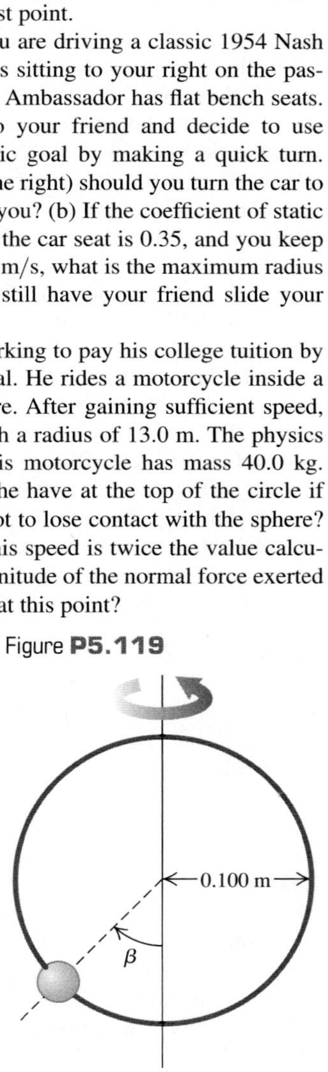

0.100 m

β

5.120 •• A small remote-controlled car with mass 1.60 kg moves at a constant speed of $v = 12.0$ m/s in a vertical circle inside a hollow metal cylinder that has a radius of 5.00 m (Fig. P5.120). What is the magnitude of the normal force exerted on the car by the walls of the cylinder at

(a) point *A* (at the bottom of the vertical circle) and (b) point *B* (at the top of the vertical circle)?

Figure **P5.120**

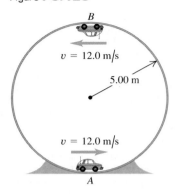

CHALLENGE PROBLEMS

5.121 ••• **CALC Angle for Minimum Force.** A box with weight *w* is pulled at constant speed along a level floor by a force $\vec{F}$ that is at an angle θ above the horizontal. The coefficient of kinetic friction between the floor and box is μ_k. (a) In terms of θ, μ_k, and *w*, calculate *F*. (b) For $w = 400$ N and $\mu_k = 0.25$, calculate *F* for θ ranging from $0°$ to $90°$ in increments of $10°$. Graph *F* versus θ. (c) From the general expression in part (a), calculate the value of θ for which the value of *F*, required to maintain constant speed, is a minimum. (*Hint:* At a point where a function is minimum, what are the first and second derivatives of the function? Here *F* is a function of θ.) For the special case of $w = 400$ N and $\mu_k = 0.25$, evaluate this optimal θ and compare your result to the graph you constructed in part (b).

5.122 ••• **Moving Wedge.** A wedge with mass *M* rests on a frictionless, horizontal tabletop. A block with mass *m* is placed on the wedge (Fig. P5.122a). There is no friction between the block and the wedge. The system is released from rest. (a) Calculate the acceleration of the wedge and the horizontal and vertical components of the acceleration of the block. (b) Do your answers to part (a) reduce to the correct results when *M* is very large? (c) As seen by a stationary observer, what is the shape of the trajectory of the block?

Figure **P5.122**

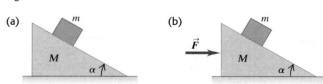

5.123 ••• A wedge with mass *M* rests on a frictionless horizontal tabletop. A block with mass *m* is placed on the wedge and a horizontal force $\vec{F}$ is applied to the wedge (Fig. P5.122b). What must the magnitude of $\vec{F}$ be if the block is to remain at a constant height above the tabletop?

5.124 ••• **CALC Falling Baseball.** You drop a baseball from the roof of a tall building. As the ball falls, the air exerts a drag force proportional to the square of the ball's speed ($f = Dv^2$). (a) In a diagram, show the direction of motion and indicate, with the aid of vectors, all the forces acting on the ball. (b) Apply Newton's second law and infer from the resulting equation the general properties of the motion. (c) Show that the ball acquires a terminal speed

that is as given in Eq. (5.13). (d) Derive the equation for the speed at any time. (*Note:*

$$\int \frac{dx}{a^2 - x^2} = \frac{1}{a}\operatorname{arctanh}\left(\frac{x}{a}\right)$$

where

$$\tanh(x) = \frac{e^x - e^{-x}}{e^x + e^{-x}} = \frac{e^{2x} - 1}{e^{2x} + 1}$$

defines the hyperbolic tangent.)

5.125 ••• **Double Atwood's Machine.** In Fig. P5.125 masses m_1 and m_2 are connected by a light string *A* over a light, frictionless pulley *B*. The axle of pulley *B* is connected by a second light string *C* over a second light, frictionless pulley *D* to a mass m_3. Pulley *D* is suspended from the ceiling by an attachment to its axle. The system is released from rest. In terms of m_1, m_2, m_3, and *g*, what are (a) the acceleration of block m_3; (b) the acceleration of pulley *B*; (c) the acceleration of block m_1; (d) the acceleration of block m_2; (e) the tension in string *A*; (f) the tension in string *C*? (g) What do your expressions give for the special case of $m_1 = m_2$ and $m_3 = m_1 + m_2$? Is this sensible?

Figure **P5.125**

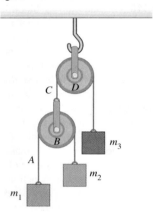

5.126 ••• The masses of blocks *A* and *B* in Fig. P5.126 are 20.0 kg and 10.0 kg, respectively. The blocks are initially at rest on the floor and are connected by a massless string passing over a massless and frictionless pulley. An upward force $\vec{F}$ is applied to the pulley. Find the accelerations $\vec{a}_A$ of block *A* and $\vec{a}_B$ of block *B* when *F* is (a) 124 N; (b) 294 N; (c) 424 N.

Figure **P5.126**

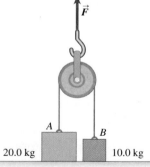

5.127 ••• A ball is held at rest at position *A* in Fig. P5.127 by two light strings. The horizontal string is cut and the ball starts swinging as a pendulum. Point *B* is the farthest to the right the ball goes as it swings back and forth. What is the ratio of the tension in the supporting string at position *B* to its value at *A* before the horizontal string was cut?

Figure **P5.127**

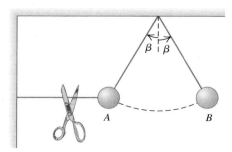

Answers

Chapter Opening Question

Neither; the upward force of the air has the *same* magnitude as the force of gravity. Although the skydiver and parachute are descending, their vertical velocity is constant and so their vertical acceleration is zero. Hence the net vertical force on the skydiver and parachute must also be zero, and the individual vertical forces must balance.

Test Your Understanding Questions

5.1 Answer: (ii) The two cables are arranged symmetrically, so the tension in either cable has the same magnitude T. The vertical component of the tension from each cable is $T\sin 45°$ (or, equivalently, $T\cos 45°$), so Newton's first law applied to the vertical forces tells us that $2T\sin 45° - w = 0$. Hence $T = w/(2\sin 45°) = w/\sqrt{2} = 0.71w$. Each cable supports half of the weight of the traffic light, but the tension is greater than $w/2$ because only the vertical component of the tension counteracts the weight.

5.2 Answer: (ii) No matter what the instantaneous velocity of the glider, its acceleration is constant and has the value found in Example 5.12. In the same way, the acceleration of a body in free fall is the same whether it is ascending, descending, or at the high point of its motion (see Section 2.5).

5.3 Answers to (a): (i), (iii); answers to (b): (ii), (iv); answer to (c): (v) In situations (i) and (iii) the box is not accelerating (so the net force on it must be zero) and there is no other force acting parallel to the horizontal surface; hence no friction force is needed to prevent sliding. In situations (ii) and (iv) the box would start to slide over the surface if no friction were present, so a static friction force must act to prevent this. In situation (v) the box is sliding over a rough surface, so a kinetic friction force acts on it.

5.4 Answer: (iii) A satellite of mass m orbiting the earth at speed v in an orbit of radius r has an acceleration of magnitude v^2/r, so the net force acting on it from the earth's gravity has magnitude $F = mv^2/r$. The farther the satellite is from earth, the greater the value of r, the smaller the value of v, and hence the smaller the values of v^2/r and of F. In other words, the earth's gravitational force decreases with increasing distance.

Bridging Problem

Answers: **(a)** $T_{\text{max}} = 2\pi\sqrt{\dfrac{h(\cos\beta + \mu_s\sin\beta)}{g\tan\beta(\sin\beta - \mu_s\cos\beta)}}$

(b) $T_{\text{min}} = 2\pi\sqrt{\dfrac{h(\cos\beta - \mu_s\sin\beta)}{g\tan\beta(\sin\beta + \mu_s\cos\beta)}}$

6 WORK AND KINETIC ENERGY

LEARNING GOALS

By studying this chapter, you will learn:

- What it means for a force to do work on a body, and how to calculate the amount of work done.

- The definition of the kinetic energy (energy of motion) of a body, and what it means physically.

- How the total work done on a body changes the body's kinetic energy, and how to use this principle to solve problems in mechanics.

- How to use the relationship between total work and change in kinetic energy when the forces are not constant, the body follows a curved path, or both.

- How to solve problems involving power (the rate of doing work).

? After finding a piece of breakfast cereal on the floor, this ant picked it up and carried it away. As the ant was lifting the piece of cereal, did the *cereal* do work on the *ant*?

Suppose you try to find the speed of an arrow that has been shot from a bow. You apply Newton's laws and all the problem-solving techniques that we've learned, but you run across a major stumbling block: After the archer releases the arrow, the bow string exerts a *varying* force that depends on the arrow's position. As a result, the simple methods that we've learned aren't enough to calculate the speed. Never fear; we aren't by any means finished with mechanics, and there are other methods for dealing with such problems.

The new method that we're about to introduce uses the ideas of *work* and *energy*. The importance of the energy idea stems from the *principle of conservation of energy*: Energy is a quantity that can be converted from one form to another but cannot be created or destroyed. In an automobile engine, chemical energy stored in the fuel is converted partially to the energy of the automobile's motion and partially to thermal energy. In a microwave oven, electromagnetic energy obtained from your power company is converted to thermal energy of the food being cooked. In these and all other processes, the *total* energy—the sum of all energy present in all different forms—remains the same. No exception has ever been found.

We'll use the energy idea throughout the rest of this book to study a tremendous range of physical phenomena. This idea will help you understand why a sweater keeps you warm, how a camera's flash unit can produce a short burst of light, and the meaning of Einstein's famous equation $E = mc^2$.

In this chapter, though, our concentration will be on mechanics. We'll learn about one important form of energy called *kinetic energy,* or energy of motion, and how it relates to the concept of *work*. We'll also consider *power,* which is the time rate of doing work. In Chapter 7 we'll expand the ideas of work and kinetic energy into a deeper understanding of the concepts of energy and the conservation of energy.

6.1 Work

You'd probably agree that it's hard work to pull a heavy sofa across the room, to lift a stack of encyclopedias from the floor to a high shelf, or to push a stalled car off the road. Indeed, all of these examples agree with the everyday meaning of *work*—any activity that requires muscular or mental effort.

In physics, work has a much more precise definition. By making use of this definition we'll find that in any motion, no matter how complicated, the total work done on a particle by all forces that act on it equals the change in its *kinetic energy*—a quantity that's related to the particle's speed. This relationship holds even when the forces acting on the particle aren't constant, a situation that can be difficult or impossible to handle with the techniques you learned in Chapters 4 and 5. The ideas of work and kinetic energy enable us to solve problems in mechanics that we could not have attempted before.

In this section we'll see how work is defined and how to calculate work in a variety of situations involving *constant* forces. Even though we already know how to solve problems in which the forces are constant, the idea of work is still useful in such problems. Later in this chapter we'll relate work and kinetic energy, and then apply these ideas to problems in which the forces are *not* constant.

The three examples of work described above—pulling a sofa, lifting encyclopedias, and pushing a car—have something in common. In each case you do work by exerting a *force* on a body while that body *moves* from one place to another—that is, undergoes a *displacement* (Fig. 6.1). You do more work if the force is greater (you push harder on the car) or if the displacement is greater (you push the car farther down the road).

The physicist's definition of work is based on these observations. Consider a body that undergoes a displacement of magnitude s along a straight line. (For now, we'll assume that any body we discuss can be treated as a particle so that we can ignore any rotation or changes in shape of the body.) While the body moves, a constant force $\vec{F}$ acts on it in the same direction as the displacement $\vec{s}$ (Fig. 6.2). We define the **work** W done by this constant force under these circumstances as the product of the force magnitude F and the displacement magnitude s:

$$W = Fs \quad \text{(constant force in direction of straight-line displacement)} \quad (6.1)$$

The work done on the body is greater if either the force F or the displacement s is greater, in agreement with our observations above.

CAUTION **Work** $= W$, **weight** $= w$ Don't confuse uppercase W (work) with lowercase w (weight). Though the symbols are similar, work and weight are different quantities. ▮

The SI unit of work is the **joule** (abbreviated J, pronounced "jool," and named in honor of the 19th-century English physicist James Prescott Joule). From Eq. (6.1) we see that in any system of units, the unit of work is the unit of force multiplied by the unit of distance. In SI units the unit of force is the newton and the unit of distance is the meter, so 1 joule is equivalent to 1 *newton-meter* (N·m):

$$1 \text{ joule} = (1 \text{ newton})(1 \text{ meter}) \quad \text{or} \quad 1 \text{ J} = 1 \text{ N} \cdot \text{m}$$

In the British system the unit of force is the pound (lb), the unit of distance is the foot (ft), and the unit of work is the *foot-pound* (ft·lb). The following conversions are useful:

$$1 \text{ J} = 0.7376 \text{ ft} \cdot \text{lb} \qquad 1 \text{ ft} \cdot \text{lb} = 1.356 \text{ J}$$

As an illustration of Eq. (6.1), think of a person pushing a stalled car. If he pushes the car through a displacement $\vec{s}$ with a constant force $\vec{F}$ in the direction

6.1 These people are doing work as they push on the stalled car because they exert a force on the car as it moves.

6.2 The work done by a constant force acting in the same direction as the displacement.

If a body moves through a displacement $\vec{s}$ while a constant force $\vec{F}$ acts on it in the same direction ...

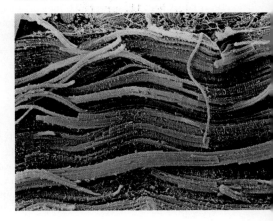

... the work done by the force on the body is $W = Fs$.

Application **Work and Muscle Fibers**
Our ability to do work with our bodies comes from our skeletal muscles. The fiberlike cells of skeletal muscle, shown in this micrograph, have the ability to shorten, causing the muscle as a whole to contract and to exert force on the tendons to which it attaches. Muscle can exert a force of about 0.3 N per square millimeter of cross-sectional area: The greater the cross-sectional area, the more fibers the muscle has and the more force it can exert when it contracts.

6.3 The work done by a constant force acting at an angle to the displacement.

Car moves through displacement $\vec{s}$ while a constant force $\vec{F}$ acts on it at an angle ϕ to the displacement.

$F_\perp$ does *no* work on car.
$F_\perp = F\sin\phi$
Only $F_\parallel$ does work on car:
$W = F_\parallel s = (F\cos\phi)s$
$= Fs\cos\phi$
$F_\parallel = F\cos\phi$

MasteringPHYSICS

ActivPhysics 5.1: Work Calculations

of motion, the amount of work he does on the car is given by Eq. (6.1): $W = Fs$. But what if the person pushes at an angle ϕ to the car's displacement (Fig. 6.3)? Then $\vec{F}$ has a component $F_\parallel = F\cos\phi$ in the direction of the displacement and a component $F_\perp = F\sin\phi$ that acts perpendicular to the displacement. (Other forces must act on the car so that it moves along $\vec{s}$, not in the direction of $\vec{F}$. We're interested only in the work that the person does, however, so we'll consider only the force he exerts.) In this case only the parallel component $F_\parallel$ is effective in moving the car, so we define the work as the product of this force component and the magnitude of the displacement. Hence $W = F_\parallel s = (F\cos\phi)s$, or

$$W = Fs\cos\phi \quad \text{(constant force, straight-line displacement)} \quad (6.2)$$

We are assuming that F and ϕ are constant during the displacement. If $\phi = 0$, so that $\vec{F}$ and $\vec{s}$ are in the same direction, then $\cos\phi = 1$ and we are back to Eq. (6.1).

Equation (6.2) has the form of the *scalar product* of two vectors, which we introduced in Section 1.10: $\vec{A} \cdot \vec{B} = AB\cos\phi$. You may want to review that definition. Hence we can write Eq. (6.2) more compactly as

$$W = \vec{F} \cdot \vec{s} \quad \text{(constant force, straight-line displacement)} \quad (6.3)$$

CAUTION **Work is a scalar** Here's an essential point: Work is a *scalar* quantity, even though it's calculated by using two vector quantities (force and displacement). A 5-N force toward the east acting on a body that moves 6 m to the east does exactly the same amount of work as a 5-N force toward the north acting on a body that moves 6 m to the north.

Example 6.1 | **Work done by a constant force**

(a) Steve exerts a steady force of magnitude 210 N (about 47 lb) on the stalled car in Fig. 6.3 as he pushes it a distance of 18 m. The car also has a flat tire, so to make the car track straight Steve must push at an angle of 30° to the direction of motion. How much work does Steve do? (b) In a helpful mood, Steve pushes a second stalled car with a steady force $\vec{F} = (160 \text{ N})\hat{\imath} - (40 \text{ N})\hat{\jmath}$. The displacement of the car is $\vec{s} = (14 \text{ m})\hat{\imath} + (11 \text{ m})\hat{\jmath}$. How much work does Steve do in this case?

SOLUTION

IDENTIFY and SET UP: In both parts (a) and (b), the target variable is the work W done by Steve. In each case the force is constant and the displacement is along a straight line, so we can use Eq. (6.2) or (6.3). The angle between $\vec{F}$ and $\vec{s}$ is given in part (a), so we can apply Eq. (6.2) directly. In part (b) both $\vec{F}$ and $\vec{s}$ are given in terms

of components, so it's best to calculate the scalar product using Eq. (1.21): $\vec{A} \cdot \vec{B} = A_xB_x + A_yB_y + A_zB_z$.

EXECUTE: (a) From Eq. (6.2),

$$W = Fs\cos\phi = (210 \text{ N})(18 \text{ m})\cos 30° = 3.3 \times 10^3 \text{ J}$$

(b) The components of $\vec{F}$ are $F_x = 160$ N and $F_y = -40$ N, and the components of $\vec{s}$ are $x = 14$ m and $y = 11$ m. (There are no z-components for either vector.) Hence, using Eqs. (1.21) and (6.3), we have

$$W = \vec{F} \cdot \vec{s} = F_x x + F_y y$$
$$= (160 \text{ N})(14 \text{ m}) + (-40 \text{ N})(11 \text{ m})$$
$$= 1.8 \times 10^3 \text{ J}$$

EVALUATE: In each case the work that Steve does is more than 1000 J. This shows that 1 joule is a rather small amount of work.

6.4 A constant force $\vec{F}$ can do positive, negative, or zero work depending on the angle between $\vec{F}$ and the displacement $\vec{s}$.

Direction of Force (or Force Component)	Situation	Force Diagram

(a) **Force $\vec{F}$ has a component in direction of displacement:**
$W = F_{\parallel}s = (F\cos\phi)s$
Work is *positive*.

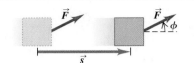

(b) **Force $\vec{F}$ has a component opposite to direction of displacement:**
$W = F_{\parallel}s = (F\cos\phi)s$
Work is *negative* (because $F\cos\phi$ is negative for $90° < \phi < 180°$).

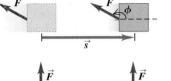

(c) **Force $\vec{F}$ (or force component $F_{\perp}$) is perpendicular to direction of displacement:** The force (or force component) does *no* work on the object.

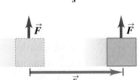

Work: Positive, Negative, or Zero

In Example 6.1 the work done in pushing the cars was positive. But it's important to understand that work can also be negative or zero. This is the essential way in which work as defined in physics differs from the "everyday" definition of work. When the force has a component in the *same direction* as the displacement (ϕ between zero and 90°), $\cos\phi$ in Eq. (6.2) is positive and the work W is *positive* (Fig. 6.4a). When the force has a component *opposite* to the displacement (ϕ between 90° and 180°), $\cos\phi$ is negative and the work is *negative* (Fig. 6.4b). When the force is *perpendicular* to the displacement, $\phi = 90°$ and the work done by the force is *zero* (Fig. 6.4c). The cases of zero work and negative work bear closer examination, so let's look at some examples.

There are many situations in which forces act but do zero work. You might think it's "hard work" to hold a barbell motionless in the air for 5 minutes (Fig. 6.5). But in fact, you aren't doing any work at all on the barbell because there is no displacement. You get tired because the components of muscle fibers in your arm do work as they continually contract and relax. This is work done by one part of the arm exerting force on another part, however, *not* on the barbell. (We'll say more in Section 6.2 about work done by one part of a body on another part.) Even when you walk with constant velocity on a level floor while carrying a book, you still do no work on it. The book has a displacement, but the (vertical) supporting force that you exert on the book has no component in the direction of the (horizontal) motion. Then $\phi = 90°$ in Eq. (6.2), and $\cos\phi = 0$. When a body slides along a surface, the work done on the body by the normal force is zero; and when a ball on a string moves in uniform circular motion, the work done on the ball by the tension in the string is also zero. In both cases the work is zero because the force has no component in the direction of motion.

What does it really mean to do *negative* work? The answer comes from Newton's third law of motion. When a weightlifter lowers a barbell as in Fig. 6.6a, his hands and the barbell move together with the same displacement $\vec{s}$. The barbell exerts a force $\vec{F}_{\text{barbell on hands}}$ on his hands in the same direction as the hands' displacement, so the work done by the *barbell* on his *hands* is positive (Fig. 6.6b). But by Newton's third law the weightlifter's hands exert an equal and opposite force $\vec{F}_{\text{hands on barbell}} = -\vec{F}_{\text{barbell on hands}}$ on the barbell (Fig. 6.6c). This force, which keeps the barbell from crashing to the floor, acts opposite to the barbell's displacement. Thus the work done by his *hands* on the *barbell* is negative.

6.5 A weightlifter does no work on a barbell as long as he holds it stationary.

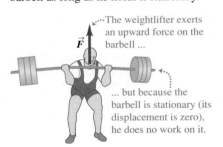

The weightlifter exerts an upward force on the barbell ...

$\vec{F}$

... but because the barbell is stationary (its displacement is zero), he does no work on it.

6.6 This weightlifter's hands do negative work on a barbell as the barbell does positive work on his hands.

(a) A weightlifter lowers a barbell to the floor.

(b) The barbell does *positive* work on the weightlifter's hands.

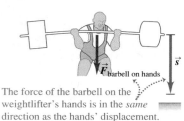

The force of the barbell on the weightlifter's hands is in the *same* direction as the hands' displacement.

(c) The weightlifter's hands do *negative* work on the barbell.

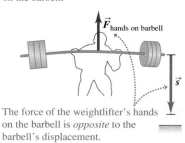

The force of the weightlifter's hands on the barbell is *opposite* to the barbell's displacement.

Because the weightlifter's hands and the barbell have the same displacement, the work that his hands do on the barbell is just the negative of the work that the barbell does on his hands. In general, when one body does negative work on a second body, the second body does an equal amount of *positive* work on the first body.

CAUTION **Keep track of who's doing the work** We always speak of work done *on* a particular body *by* a specific force. Always be sure to specify exactly what force is doing the work you are talking about. When you lift a book, you exert an upward force on the book and the book's displacement is upward, so the work done by the lifting force on the book is positive. But the work done by the *gravitational* force (weight) on a book being lifted is *negative* because the downward gravitational force is opposite to the upward displacement.

Total Work

How do we calculate work when *several* forces act on a body? One way is to use Eq. (6.2) or (6.3) to compute the work done by each separate force. Then, because work is a scalar quantity, the *total* work W_{tot} done on the body by all the forces is the algebraic sum of the quantities of work done by the individual forces. An alternative way to find the total work W_{tot} is to compute the vector sum of the forces (that is, the net force) and then use this vector sum as $\vec{F}$ in Eq. (6.2) or (6.3). The following example illustrates both of these techniques.

Example 6.2 **Work done by several forces**

A farmer hitches her tractor to a sled loaded with firewood and pulls it a distance of 20 m along level ground (Fig. 6.7a). The total weight of sled and load is 14,700 N. The tractor exerts a constant 5000-N force at an angle of 36.9° above the horizontal. A 3500-N friction force opposes the sled's motion. Find the work done by each force acting on the sled and the total work done by all the forces.

SOLUTION

IDENTIFY AND SET UP: Each force is constant and the sled's displacement is along a straight line, so we can calculate the work using the ideas of this section. We'll find the total work in two ways: (1) by adding the work done on the sled by each force and (2) by finding the work done by the net force on the sled. We first draw a free-body diagram showing all of the forces acting on the sled, and we choose a coordinate system (Fig. 6.7b). For each force—weight, normal force, force of the tractor, and friction force—we know the angle between the displacement (in the positive *x*-direction) and the force. Hence we can use Eq. (6.2) to calculate the work each force does.

As in Chapter 5, we'll find the net force by adding the components of the four forces. Newton's second law tells us that because the sled's motion is purely horizontal, the net force can have only a horizontal component.

EXECUTE: (1) The work W_w done by the weight is zero because its direction is perpendicular to the displacement (compare Fig. 6.4c). For the same reason, the work W_n done by the normal force is also zero. (Note that we don't need to calculate the magnitude *n* to conclude this.) So $W_w = W_n = 0$.

That leaves the work W_T done by the force F_T exerted by the tractor and the work W_f done by the friction force f. From Eq. (6.2),

$$W_T = F_T s \cos\phi = (5000\ \text{N})(20\ \text{m})(0.800) = 80{,}000\ \text{N}\cdot\text{m}$$
$$= 80\ \text{kJ}$$

The friction force $\vec{f}$ is opposite to the displacement, so for this force $\phi = 180°$ and $\cos\phi = -1$. Again from Eq. (6.2),

$$W_f = fs\cos 180° = (3500\ \text{N})(20\ \text{m})(-1) = -70{,}000\ \text{N}\cdot\text{m}$$
$$= -70\ \text{kJ}$$

6.7 Calculating the work done on a sled of firewood being pulled by a tractor.

(a)

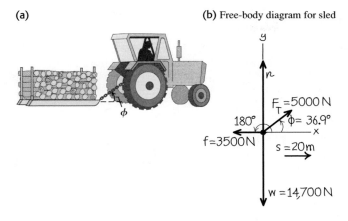

(b) Free-body diagram for sled

The total work W_{tot} done on the sled by all forces is the *algebraic* sum of the work done by the individual forces:

$$W_{\text{tot}} = W_w + W_n + W_T + W_f = 0 + 0 + 80 \text{ kJ} + (-70 \text{ kJ})$$
$$= 10 \text{ kJ}$$

(2) In the second approach, we first find the *vector* sum of all the forces (the net force) and then use it to compute the total work. The vector sum is best found by using components. From Fig. 6.7b,

$$\sum F_x = F_T \cos\phi + (-f) = (5000 \text{ N}) \cos 36.9° - 3500 \text{ N}$$
$$= 500 \text{ N}$$

$$\sum F_y = F_T \sin\phi + n + (-w)$$
$$= (5000 \text{ N}) \sin 36.9° + n - 14{,}700 \text{ N}$$

We don't need the second equation; we know that the y-component of force is perpendicular to the displacement, so it does no work. Besides, there is no y-component of acceleration, so $\sum F_y$ must be zero anyway. The total work is therefore the work done by the total x-component:

$$W_{\text{tot}} = (\sum \vec{F}) \cdot \vec{s} = (\sum F_x)s = (500 \text{ N})(20 \text{ m}) = 10{,}000 \text{ J}$$
$$= 10 \text{ kJ}$$

EVALUATE: We get the same result for W_{tot} with either method, as we should. Note also that the net force in the x-direction is *not* zero, and so the sled must accelerate as it moves. In Section 6.2 we'll return to this example and see how to use the concept of work to explore the sled's changes of speed.

Test Your Understanding of Section 6.1 An electron moves in a straight line toward the east with a constant speed of 8×10^7 m/s. It has electric, magnetic, and gravitational forces acting on it. During a 1-m displacement, the total work done on the electron is (i) positive; (ii) negative; (iii) zero; (iv) not enough information given to decide.

6.2 Kinetic Energy and the Work–Energy Theorem

PhET: The Ramp

The total work done on a body by external forces is related to the body's displacement—that is, to changes in its position. But the total work is also related to changes in the *speed* of the body. To see this, consider Fig. 6.8, which shows three examples of a block sliding on a frictionless table. The forces acting on the block are its weight $\vec{w}$, the normal force $\vec{n}$, and the force $\vec{F}$ exerted on it by the hand.

In Fig. 6.8a the net force on the block is in the direction of its motion. From Newton's second law, this means that the block speeds up; from Eq. (6.1), this also means that the total work W_{tot} done on the block is positive. The total work is *negative* in Fig. 6.8b because the net force opposes the displacement; in this case the block slows down. The net force is zero in Fig. 6.8c, so the speed of the block stays the same and the total work done on the block is zero. We can conclude that *when a particle undergoes a displacement, it speeds up if $W_{\text{tot}} > 0$, slows down if $W_{\text{tot}} < 0$, and maintains the same speed if $W_{\text{tot}} = 0$.*

Let's make these observations more quantitative. Consider a particle with mass m moving along the x-axis under the action of a constant net force with magnitude F directed along the positive x-axis (Fig. 6.9). The particle's acceleration is constant and given by Newton's second law, $F = ma_x$. Suppose the speed changes from v_1 to v_2 while the particle undergoes a displacement $s = x_2 - x_1$

6.8 The relationship between the total work done on a body and how the body's speed changes.

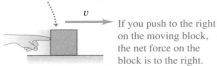

(a)

A block slides to the right on a frictionless surface.

If you push to the right on the moving block, the net force on the block is to the right.

• The total work done on the block during a displacement $\vec{s}$ is positive: $W_{\text{tot}} > 0$.
• The block speeds up.

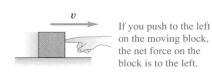

(b)

If you push to the left on the moving block, the net force on the block is to the left.

• The total work done on the block during a displacement $\vec{s}$ is negative: $W_{\text{tot}} < 0$.
• The block slows down.

(c)

If you push straight down on the moving block, the net force on the block is zero.

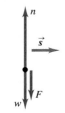

• The total work done on the block during a displacement $\vec{s}$ is zero: $W_{\text{tot}} = 0$.
• The block's speed stays the same.

6.9 A constant net force $\vec{F}$ does work on a moving body.

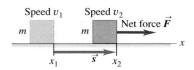

from point x_1 to x_2. Using a constant-acceleration equation, Eq. (2.13), and replacing v_{0x} by v_1, v_x by v_2, and $(x - x_0)$ by s, we have

$$v_2^2 = v_1^2 + 2a_x s$$

$$a_x = \frac{v_2^2 - v_1^2}{2s}$$

When we multiply this equation by m and equate ma_x to the net force F, we find

$$F = ma_x = m\frac{v_2^2 - v_1^2}{2s} \qquad \text{and}$$

$$Fs = \tfrac{1}{2}mv_2^2 - \tfrac{1}{2}mv_1^2 \tag{6.4}$$

The product Fs is the work done by the net force F and thus is equal to the total work W_{tot} done by all the forces acting on the particle. The quantity $\tfrac{1}{2}mv^2$ is called the **kinetic energy** K of the particle:

$$K = \tfrac{1}{2}mv^2 \qquad \text{(definition of kinetic energy)} \tag{6.5}$$

6.10 Comparing the kinetic energy $K = \tfrac{1}{2}mv^2$ of different bodies.

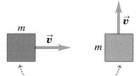

Same mass, same speed, different directions of motion: *same* kinetic energy

Twice the mass, same speed: *twice* the kinetic energy

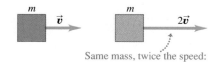

Same mass, twice the speed: *four times* the kinetic energy

Like work, the kinetic energy of a particle is a scalar quantity; it depends on only the particle's mass and speed, not its direction of motion (Fig. 6.10). A car (viewed as a particle) has the same kinetic energy when going north at 10 m/s as when going east at 10 m/s. Kinetic energy can never be negative, and it is zero only when the particle is at rest.

We can now interpret Eq. (6.4) in terms of work and kinetic energy. The first term on the right side of Eq. (6.4) is $K_2 = \tfrac{1}{2}mv_2^2$, the final kinetic energy of the particle (that is, after the displacement). The second term is the initial kinetic energy, $K_1 = \tfrac{1}{2}mv_1^2$, and the difference between these terms is the *change* in kinetic energy. So Eq. (6.4) says:

The work done by the net force on a particle equals the change in the particle's kinetic energy:

$$W_{\text{tot}} = K_2 - K_1 = \Delta K \qquad \text{(work–energy theorem)} \tag{6.6}$$

This result is the **work–energy theorem.**

The work–energy theorem agrees with our observations about the block in Fig. 6.8. When W_{tot} is *positive,* the kinetic energy *increases* (the final kinetic energy K_2 is greater than the initial kinetic energy K_1) and the particle is going faster at the end of the displacement than at the beginning. When W_{tot} is *negative,* the kinetic energy *decreases* (K_2 is less than K_1) and the speed is less after the displacement. When $W_{tot} = 0$, the kinetic energy stays the same ($K_1 = K_2$) and the speed is unchanged. Note that the work–energy theorem by itself tells us only about changes in *speed,* not velocity, since the kinetic energy doesn't depend on the direction of motion.

From Eq. (6.4) or Eq. (6.6), kinetic energy and work must have the same units. Hence the joule is the SI unit of both work and kinetic energy (and, as we will see later, of all kinds of energy). To verify this, note that in SI units the quantity $K = \frac{1}{2}mv^2$ has units $kg \cdot (m/s)^2$ or $kg \cdot m^2/s^2$; we recall that $1\ N = 1\ kg \cdot m/s^2$, so

$$1\ J = 1\ N \cdot m = 1\ (kg \cdot m/s^2) \cdot m = 1\ kg \cdot m^2/s^2$$

In the British system the unit of kinetic energy and of work is

$$1\ ft \cdot lb = 1\ ft \cdot slug \cdot ft/s^2 = 1\ slug \cdot ft^2/s^2$$

Because we used Newton's laws in deriving the work–energy theorem, we can use this theorem only in an inertial frame of reference. Note also that the work–energy theorem is valid in *any* inertial frame, but the values of W_{tot} and $K_2 - K_1$ may differ from one inertial frame to another (because the displacement and speed of a body may be different in different frames).

We've derived the work–energy theorem for the special case of straight-line motion with constant forces, and in the following examples we'll apply it to this special case only. We'll find in the next section that the theorem is valid in general, even when the forces are not constant and the particle's trajectory is curved.

Problem-Solving Strategy 6.1   Work and Kinetic Energy

IDENTIFY *the relevant concepts:* The work–energy theorem, $W_{tot} = K_2 - K_1$, is extremely useful when you want to relate a body's speed v_1 at one point in its motion to its speed v_2 at a different point. (It's less useful for problems that involve the *time* it takes a body to go from point 1 to point 2 because the work–energy theorem doesn't involve time at all. For such problems it's usually best to use the relationships among time, position, velocity, and acceleration described in Chapters 2 and 3.)

SET UP *the problem* using the following steps:
1. Identify the initial and final positions of the body, and draw a free-body diagram showing all the forces that act on the body.
2. Choose a coordinate system. (If the motion is along a straight line, it's usually easiest to have both the initial and final positions lie along one of the axes.)
3. List the unknown and known quantities, and decide which unknowns are your target variables. The target variable may be the body's initial or final speed, the magnitude of one of the forces acting on the body, or the body's displacement.

EXECUTE *the solution:* Calculate the work W done by each force. If the force is constant and the displacement is a straight line, you can use Eq. (6.2) or Eq. (6.3). (Later in this chapter we'll see how to handle varying forces and curved trajectories.) Be sure to check signs; W must be positive if the force has a component in the direction of the displacement, negative if the force has a component opposite to the displacement, and zero if the force and displacement are perpendicular.

Add the amounts of work done by each force to find the total work W_{tot}. Sometimes it's easier to calculate the vector sum of the forces (the net force) and then find the work done by the net force; this value is also equal to W_{tot}.

Write expressions for the initial and final kinetic energies, K_1 and K_2. Note that kinetic energy involves *mass,* not *weight*; if you are given the body's weight, use $w = mg$ to find the mass.

Finally, use Eq. (6.6), $W_{tot} = K_2 - K_1$, and Eq. (6.5), $K = \frac{1}{2}mv^2$, to solve for the target variable. Remember that the right-hand side of Eq. (6.6) represents the change of the body's kinetic energy between points 1 and 2; that is, it is the *final* kinetic energy minus the *initial* kinetic energy, never the other way around. (If you can predict the sign of W_{tot}, you can predict whether the body speeds up or slows down.)

EVALUATE *your answer:* Check whether your answer makes sense. Remember that kinetic energy $K = \frac{1}{2}mv^2$ can never be negative. If you come up with a negative value of K, perhaps you interchanged the initial and final kinetic energies in $W_{tot} = K_2 - K_1$ or made a sign error in one of the work calculations.

Example 6.3 Using work and energy to calculate speed

Let's look again at the sled in Fig. 6.7 and our results from Example 6.2. Suppose the sled's initial speed v_1 is 2.0 m/s. What is the speed of the sled after it moves 20 m?

SOLUTION

IDENTIFY and SET UP: We'll use the work–energy theorem, Eq. (6.6), $W_{tot} = K_2 - K_1$, since we are given the initial speed $v_1 = 2.0$ m/s and want to find the final speed v_2. Figure 6.11 shows our sketch of the situation. The motion is in the positive x-direction. In Example 6.2 we calculated the total work done by all the forces: $W_{tot} = 10$ kJ. Hence the kinetic energy of the sled and its load must increase by 10 kJ, and the speed of the sled must also increase.

EXECUTE: To write expressions for the initial and final kinetic energies, we need the mass of the sled and load. The combined *weight* is 14,700 N, so the mass is

$$m = \frac{w}{g} = \frac{14{,}700 \text{ N}}{9.8 \text{ m/s}^2} = 1500 \text{ kg}$$

Then the initial kinetic energy K_1 is

$$K_1 = \tfrac{1}{2}mv_1^2 = \tfrac{1}{2}(1500 \text{ kg})(2.0 \text{ m/s})^2 = 3000 \text{ kg} \cdot \text{m}^2/\text{s}^2$$

$$= 3000 \text{ J}$$

6.11 Our sketch for this problem.

The final kinetic energy K_2 is

$$K_2 = \tfrac{1}{2}mv_2^2 = \tfrac{1}{2}(1500 \text{ kg})v_2^2$$

The work–energy theorem, Eq. (6.6), gives

$$K_2 = K_1 + W_{tot} = 3000 \text{ J} + 10{,}000 \text{ J} = 13{,}000 \text{ J}$$

Setting these two expressions for K_2 equal, substituting 1 J = 1 kg $\cdot$ m^2/s^2, and solving for the final speed v_2, we find

$$v_2 = 4.2 \text{ m/s}$$

EVALUATE: The total work is positive, so the kinetic energy increases ($K_2 > K_1$) and the speed increases ($v_2 > v_1$).

This problem can also be solved without the work–energy theorem. We can find the acceleration from $\sum \vec{F} = m\vec{a}$ and then use the equations of motion for constant acceleration to find v_2. Since the acceleration is along the x-axis,

$$a = a_x = \frac{\sum F_x}{m} = \frac{500 \text{ N}}{1500 \text{ kg}} = 0.333 \text{ m/s}^2$$

Then, using Eq. (2.13),

$$v_2^2 = v_1^2 + 2as = (2.0 \text{ m/s})^2 + 2(0.333 \text{ m/s}^2)(20 \text{ m})$$

$$= 17.3 \text{ m}^2/\text{s}^2$$

$$v_2 = 4.2 \text{ m/s}$$

This is the same result we obtained with the work–energy approach, but there we avoided the intermediate step of finding the acceleration. You will find several other examples in this chapter and the next that *can* be done without using energy considerations but that are easier when energy methods are used. When a problem can be done by two methods, doing it by both methods (as we did here) is a good way to check your work.

Example 6.4 Forces on a hammerhead

The 200-kg steel hammerhead of a pile driver is lifted 3.00 m above the top of a vertical I-beam being driven into the ground (Fig. 6.12a). The hammerhead is then dropped, driving the I-beam 7.4 cm deeper into the ground. The vertical guide rails exert a constant 60-N friction force on the hammerhead. Use the work–energy theorem to find (a) the speed of the hammerhead just as it hits the I-beam and (b) the average force the hammerhead exerts on the I-beam. Ignore the effects of the air.

SOLUTION

IDENTIFY: We'll use the work–energy theorem to relate the hammerhead's speed at different locations and the forces acting on it. There are *three* locations of interest: point 1, where the hammerhead starts from rest; point 2, where it first contacts the I-beam; and point 3, where the hammerhead and I-beam come to a halt (Fig. 6.12a). The two target variables are the hammerhead's speed at point 2 and the average force the hammerhead exerts between points 2 and 3. Hence we'll apply the work–energy theorem

twice: once for the motion from 1 to 2, and once for the motion from 2 to 3.

SET UP: Figure 6.12b shows the vertical forces on the hammerhead as it falls from point 1 to point 2. (We can ignore any horizontal forces that may be present because they do no work as the hammerhead moves vertically.) For this part of the motion, our target variable is the hammerhead's final speed v_2.

Figure 6.12c shows the vertical forces on the hammerhead during the motion from point 2 to point 3. In addition to the forces shown in Fig. 6.12b, the I-beam exerts an upward normal force of magnitude n on the hammerhead. This force actually varies as the hammerhead comes to a halt, but for simplicity we'll treat n as a constant. Hence n represents the *average* value of this upward force during the motion. Our target variable for this part of the motion is the force that the *hammerhead* exerts on the I-beam; it is the reaction force to the normal force exerted by the I-beam, so by Newton's third law its magnitude is also n.

EXECUTE: (a) From point 1 to point 2, the vertical forces are the downward weight $w = mg = (200 \text{ kg})(9.8 \text{ m/s}^2) = 1960 \text{ N}$ and the upward friction force $f = 60 \text{ N}$. Thus the net downward force is $w - f = 1900 \text{ N}$. The displacement of the hammerhead from point 1 to point 2 is downward and equal to $s_{12} = 3.00 \text{ m}$. The total work done on the hammerhead between point 1 and point 2 is then

$$W_{tot} = (w - f)s_{12} = (1900 \text{ N})(3.00 \text{ m}) = 5700 \text{ J}$$

At point 1 the hammerhead is at rest, so its initial kinetic energy K_1 is zero. Hence the kinetic energy K_2 at point 2 equals the total work done on the hammerhead between points 1 and 2:

$$W_{tot} = K_2 - K_1 = K_2 - 0 = \tfrac{1}{2}mv_2^2 - 0$$

$$v_2 = \sqrt{\frac{2W_{tot}}{m}} = \sqrt{\frac{2(5700 \text{ J})}{200 \text{ kg}}} = 7.55 \text{ m/s}$$

This is the hammerhead's speed at point 2, just as it hits the I-beam.

(b) As the hammerhead moves downward from point 2 to point 3, its displacement is $s_{23} = 7.4 \text{ cm} = 0.074 \text{ m}$ and the net downward force acting on it is $w - f - n$ (Fig. 6.12c). The total work done on the hammerhead during this displacement is

$$W_{tot} = (w - f - n)s_{23}$$

The initial kinetic energy for this part of the motion is K_2, which from part (a) equals 5700 J. The final kinetic energy is $K_3 = 0$ (the hammerhead ends at rest). From the work–energy theorem,

$$W_{tot} = (w - f - n)s_{23} = K_3 - K_2$$

$$n = w - f - \frac{K_3 - K_2}{s_{23}}$$

$$= 1960 \text{ N} - 60 \text{ N} - \frac{0 \text{ J} - 5700 \text{ J}}{0.074 \text{ m}} = 79{,}000 \text{ N}$$

The downward force that the hammerhead exerts on the I-beam has this same magnitude, 79,000 N (about 9 tons)—more than 40 times the weight of the hammerhead.

EVALUATE: The net change in the hammerhead's kinetic energy from point 1 to point 3 is zero; a relatively small net force does positive work over a large distance, and then a much larger net force does negative work over a much smaller distance. The same thing happens if you speed up your car gradually and then drive it into a brick wall. The very large force needed to reduce the kinetic energy to zero over a short distance is what does the damage to your car—and possibly to you.

6.12 (a) A pile driver pounds an I-beam into the ground. (b), (c) Free-body diagrams. Vector lengths are not to scale.

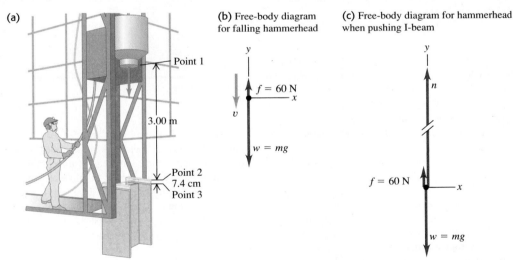

The Meaning of Kinetic Energy

Example 6.4 gives insight into the physical meaning of kinetic energy. The hammerhead is dropped from rest, and its kinetic energy when it hits the I-beam equals the total work done on it up to that point by the net force. This result is true in general: To accelerate a particle of mass m from rest (zero kinetic energy)

6.13 When a billiards player hits a cue ball at rest, the ball's kinetic energy after being hit is equal to the work that was done on it by the cue. The greater the force exerted by the cue and the greater the distance the ball moves while in contact with it, the greater the ball's kinetic energy.

up to a speed , the total work done on it must equal the change in kinetic energy from zero to $K = \frac{1}{2}mv^2$:

$$W_{\text{tot}} = K - 0 = K$$

So *the kinetic energy of a particle is equal to the total work that was done to accelerate it from rest to its present speed* (Fig. 6.13). The definition $K = \frac{1}{2}mv^2$, Eq. (6.5), wasn't chosen at random; it's the *only* definition that agrees with this interpretation of kinetic energy.

In the second part of Example 6.4 the kinetic energy of the hammerhead did work on the I-beam and drove it into the ground. This gives us another interpretation of kinetic energy: *The kinetic energy of a particle is equal to the total work that particle can do in the process of being brought to rest.* This is why you pull your hand and arm backward when you catch a ball. As the ball comes to rest, it does an amount of work (force times distance) on your hand equal to the ball's initial kinetic energy. By pulling your hand back, you maximize the distance over which the force acts and so minimize the force on your hand.

Conceptual Example 6.5 | **Comparing kinetic energies**

Two iceboats like the one in Example 5.6 (Section 5.2) hold a race on a frictionless horizontal lake (Fig. 6.14). The two iceboats have masses m and $2m$. The iceboats have identical sails, so the wind exerts the same constant force $\vec{F}$ on each iceboat. They start from rest and cross the finish line a distance s away. Which iceboat crosses the finish line with greater kinetic energy?

SOLUTION

If you use the definition of kinetic energy, $K = \frac{1}{2}mv^2$, Eq. (6.5), the answer to this problem isn't obvious. The iceboat of mass $2m$ has greater mass, so you might guess that it has greater kinetic energy at the finish line. But the lighter iceboat, of mass m, has greater acceleration and crosses the finish line with a greater speed, so you might guess that *this* iceboat has the greater kinetic energy. How can we decide?

The key is to remember that *the kinetic energy of a particle is equal to the total work done to accelerate it from rest.* Both iceboats travel the same distance s from rest, and only the horizontal force F in the direction of motion does work on either iceboat. Hence the total work done between the starting line and the finish line is the *same* for each iceboat, $W_{\text{tot}} = Fs$. At the finish line, each iceboat has a kinetic energy equal to the work W_{tot} done on it, because each iceboat started from rest. So both iceboats have the *same* kinetic energy at the finish line!

6.14 A race between iceboats.

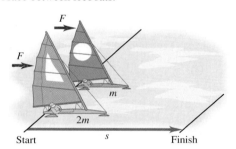

You might think this is a "trick" question, but it isn't. If you really understand the meanings of quantities such as kinetic energy, you can solve problems more easily and with better insight.

Notice that we didn't need to know anything about how much time each iceboat took to reach the finish line. This is because the work–energy theorem makes no direct reference to time, only to displacement. In fact the iceboat of mass m has greater acceleration and so takes less time to reach the finish line than does the iceboat of mass $2m$.

Work and Kinetic Energy in Composite Systems

In this section we've been careful to apply the work–energy theorem only to bodies that we can represent as *particles*—that is, as moving point masses. New subtleties appear for more complex systems that have to be represented as many particles with different motions. We can't go into these subtleties in detail in this chapter, but here's an example.

6.3 Work and Energy with Varying Forces **187**

Suppose a boy stands on frictionless roller skates on a level surface, facing a rigid wall (Fig. 6.15). He pushes against the wall, which makes him move to the right. The forces acting on him are his weight $\vec{w}$, the upward normal forces $\vec{n}_1$ and $\vec{n}_2$ exerted by the ground on his skates, and the horizontal force $\vec{F}$ exerted on him by the wall. There is no vertical displacement, so $\vec{w}$, $\vec{n}_1$, and $\vec{n}_2$ do no work. Force $\vec{F}$ accelerates him to the right, but the parts of his body where that force is applied (the boy's hands) do not move while the force acts. Thus the force $\vec{F}$ also does no work. Where, then, does the boy's kinetic energy come from?

The explanation is that it's not adequate to represent the boy as a single point mass. Different parts of the boy's body have different motions; his hands remain stationary against the wall while his torso is moving away from the wall. The various parts of his body interact with each other, and one part can exert forces and do work on another part. Therefore the *total* kinetic energy of this *composite* system of body parts can change, even though no work is done by forces applied by bodies (such as the wall) that are outside the system. In Chapter 8 we'll consider further the motion of a collection of particles that interact with each other. We'll discover that just as for the boy in this example, the total kinetic energy of such a system can change even when no work is done on any part of the system by anything outside it.

6.15 The external forces acting on a skater pushing off a wall. The work done by these forces is zero, but the skater's kinetic energy changes nonetheless.

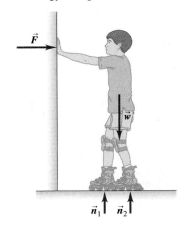

Test Your Understanding of Section 6.2 Rank the following bodies in order of their kinetic energy, from least to greatest. (i) a 2.0-kg body moving at 5.0 m/s; (ii) a 1.0-kg body that initially was at rest and then had 30 J of work done on it; (iii) a 1.0-kg body that initially was moving at 4.0 m/s and then had 20 J of work done on it; (iv) a 2.0-kg body that initially was moving at 10 m/s and then did 80 J of work on another body.

6.3 Work and Energy with Varying Forces

So far in this chapter we've considered work done by *constant forces* only. But what happens when you stretch a spring? The more you stretch it, the harder you have to pull, so the force you exert is *not* constant as the spring is stretched. We've also restricted our discussion to *straight-line* motion. There are many situations in which a body moves along a curved path and is acted on by a force that varies in magnitude, direction, or both. We need to be able to compute the work done by the force in these more general cases. Fortunately, we'll find that the work–energy theorem holds true even when varying forces are considered and when the body's path is not straight.

Work Done by a Varying Force, Straight-Line Motion

To add only one complication at a time, let's consider straight-line motion along the x-axis with a force whose x-component F_x may change as the body moves. (A real-life example is driving a car along a straight road with stop signs, so the driver has to alternately step on the gas and apply the brakes.) Suppose a particle moves along the x-axis from point x_1 to x_2 (Fig. 6.16a). Figure 6.16b is a graph of the x-component of force as a function of the particle's coordinate x. To find the work done by this force, we divide the total displacement into small segments Δx_a, Δx_b, and so on (Fig. 6.16c). We approximate the work done by the force during segment Δx_a as the average x-component of force F_{ax} in that segment multiplied by the x-displacement Δx_a. We do this for each segment and then add the results for all the segments. The work done by the force in the total displacement from x_1 to x_2 is approximately

$$W = F_{ax}\Delta x_a + F_{bx}\Delta x_b + \cdots$$

6.16 Calculating the work done by a varying force F_x in the x-direction as a particle moves from x_1 to x_2.

(a) Particle moving from x_1 to x_2 in response to a changing force in the x-direction

(b)

(c)

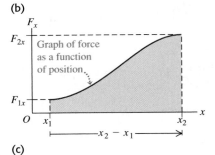

PhET: Molecular Motors
PhET: Stretching DNA

In the limit that the number of segments becomes very large and the width of each becomes very small, this sum becomes the *integral* of F_x from x_1 to x_2:

$$W = \int_{x_1}^{x_2} F_x \, dx \qquad \text{(varying x-component of force, straight-line displacement)} \qquad (6.7)$$

Note that $F_{ax} \Delta x_a$ represents the *area* of the first vertical strip in Fig. 6.16c and that the integral in Eq. (6.7) represents the area under the curve of Fig. 6.16b between x_1 and x_2. *On a graph of force as a function of position, the total work done by the force is represented by the area under the curve between the initial and final positions.* An alternative interpretation of Eq. (6.7) is that the work W equals the average force that acts over the entire displacement, multiplied by the displacement.

In the special case that F_x, the x-component of the force, is constant, it may be taken outside the integral in Eq. (6.7):

$$W = \int_{x_1}^{x_2} F_x \, dx = F_x \int_{x_1}^{x_2} dx = F_x(x_2 - x_1) \qquad \text{(constant force)}$$

But $x_2 - x_1 = s$, the total displacement of the particle. So in the case of a constant force F, Eq. (6.7) says that $W = Fs$, in agreement with Eq. (6.1). The interpretation of work as the area under the curve of F_x as a function of x also holds for a constant force; $W = Fs$ is the area of a rectangle of height F and width s (Fig. 6.17).

Now let's apply these ideas to the stretched spring. To keep a spring stretched beyond its unstretched length by an amount x, we have to apply a force of equal magnitude at each end (Fig. 6.18). If the elongation x is not too great, the force we apply to the right-hand end has an x-component directly proportional to x:

$$F_x = kx \qquad \text{(force required to stretch a spring)} \qquad (6.8)$$

6.17 The work done by a constant force F in the x-direction as a particle moves from x_1 to x_2.

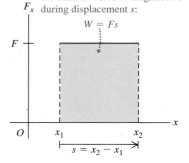

The rectangular area under the graph represents the work done by the constant force of magnitude F during displacement s:

where k is a constant called the **force constant** (or spring constant) of the spring. The units of k are force divided by distance: N/m in SI units and lb/ft in British units. A floppy toy spring such as a Slinky™ has a force constant of about 1 N/m; for the much stiffer springs in an automobile's suspension, k is about 10^5 N/m. The observation that force is directly proportional to elongation for elongations that are not too great was made by Robert Hooke in 1678 and is known as **Hooke's law.** It really shouldn't be called a "law," since it's a statement about a specific device and not a fundamental law of nature. Real springs don't always obey Eq. (6.8) precisely, but it's still a useful idealized model. We'll discuss Hooke's law more fully in Chapter 11.

To stretch a spring, we must do work. We apply equal and opposite forces to the ends of the spring and gradually increase the forces. We hold the left end stationary, so the force we apply at this end does no work. The force at the moving end *does* do work. Figure 6.19 is a graph of F_x as a function of x, the elongation of the spring. The work done by this force when the elongation goes from zero to a maximum value X is

$$W = \int_0^X F_x \, dx = \int_0^X kx \, dx = \tfrac{1}{2}kX^2 \qquad (6.9)$$

We can also obtain this result graphically. The area of the shaded triangle in Fig. 6.19, representing the total work done by the force, is equal to half the product of the base and altitude, or

$$W = \tfrac{1}{2}(X)(kX) = \tfrac{1}{2}kX^2$$

6.18 The force needed to stretch an ideal spring is proportional to the spring's elongation: $F_x = kx$.

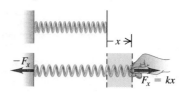

6.19 Calculating the work done to stretch a spring by a length X.

The area under the graph represents the work done on the spring as the spring is stretched from $x = 0$ to a maximum value X:

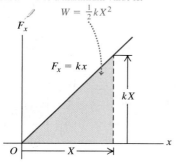

This equation also says that the work is the *average* force $kX/2$ multiplied by the total displacement X. We see that the total work is proportional to the *square* of the final elongation X. To stretch an ideal spring by 2 cm, you must do four times as much work as is needed to stretch it by 1 cm.

Equation (6.9) assumes that the spring was originally unstretched. If initially the spring is already stretched a distance x_1, the work we must do to stretch it to a greater elongation x_2 (Fig. 6.20a) is

$$W = \int_{x_1}^{x_2} F_x \, dx = \int_{x_1}^{x_2} kx \, dx = \tfrac{1}{2}kx_2^2 - \tfrac{1}{2}kx_1^2 \qquad (6.10)$$

You should use your knowledge of geometry to convince yourself that the trapezoidal area under the graph in Fig. 6.20b is given by the expression in Eq. (6.10).

If the spring has spaces between the coils when it is unstretched, then it can also be compressed, and Hooke's law holds for compression as well as stretching. In this case the force and displacement are in the opposite directions from those shown in Fig. 6.18, and so F_x and x in Eq. (6.8) are both negative. Since both F_x and x are reversed, the force again is in the same direction as the displacement, and the work done by F_x is again positive. So the total work is still given by Eq. (6.9) or (6.10), even when X is negative or either or both of x_1 and x_2 are negative.

CAUTION **Work done *on* a spring vs. work done *by* a spring** Note that Eq. (6.10) gives the work that *you* must do *on* a spring to change its length. For example, if you stretch a spring that's originally relaxed, then $x_1 = 0$, $x_2 > 0$, and $W > 0$: The force you apply to one end of the spring is in the same direction as the displacement, and the work you do is positive. By contrast, the work that the *spring* does on whatever it's attached to is given by the *negative* of Eq. (6.10). Thus, as you pull on the spring, the spring does negative work on you. Paying careful attention to the sign of work will eliminate confusion later on! ▮

6.20 Calculating the work done to stretch a spring from one extension to a greater one.

(a) Stretching a spring from elongation x_1 to elongation x_2

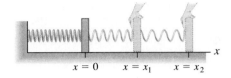

(b) Force-versus-distance graph

The trapezoidal area under the graph represents the work done on the spring to stretch it from $x = x_1$ to $x = x_2$: $W = \tfrac{1}{2}kx_2^2 - \tfrac{1}{2}kx_1^2$

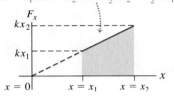

Example 6.6 | **Work done on a spring scale**

A woman weighing 600 N steps on a bathroom scale that contains a stiff spring (Fig. 6.21). In equilibrium, the spring is compressed 1.0 cm under her weight. Find the force constant of the spring and the total work done on it during the compression.

SOLUTION

IDENTIFY and SET UP: In equilibrium the upward force exerted by the spring balances the downward force of the woman's weight. We'll use this principle and Eq. (6.8) to determine the force constant k, and we'll use Eq. (6.10) to calculate the work W that the

6.21 Compressing a spring in a bathroom scale.

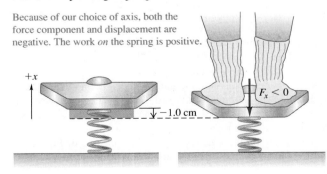

Because of our choice of axis, both the force component and displacement are negative. The work *on* the spring is positive.

$+x$

-1.0 cm

$F_x < 0$

woman does on the spring to compress it. We take positive values of x to correspond to elongation (upward in Fig. 6.21), so that the displacement of the end of the spring (x) and the x-component of the force that the woman exerts on it (F_x) are both negative. The applied force and the displacement are in the same direction, so the work done on the spring will be positive.

EXECUTE: The top of the spring is displaced by $x = -1.0$ cm $= -0.010$ m, and the woman exerts a force $F_x = -600$ N on the spring. From Eq. (6.8) the force constant is then

$$k = \frac{F_x}{x} = \frac{-600 \text{ N}}{-0.010 \text{ m}} = 6.0 \times 10^4 \, \text{N/m}$$

Then, using $x_1 = 0$ and $x_2 = -0.010$ m in Eq. (6.10), we have

$$W = \tfrac{1}{2}kx_2^2 - \tfrac{1}{2}kx_1^2$$

$$= \tfrac{1}{2}(6.0 \times 10^4 \, \text{N/m})(-0.010 \text{ m})^2 - 0 = 3.0 \text{ J}$$

EVALUATE: The work done is positive, as expected. Our arbitrary choice of the positive direction has no effect on the answer for W. You can test this by taking the positive x-direction to be downward, corresponding to compression. Do you get the same values for k and W as we found here?

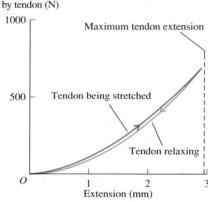

Force exerted
by tendon (N)

Work–Energy Theorem for Straight-Line Motion, Varying Forces

In Section 6.2 we derived the work–energy theorem, $W_{\text{tot}} = K_2 - K_1$, for the special case of straight-line motion with a constant net force. We can now prove that this theorem is true even when the force varies with position. As in Section 6.2, let's consider a particle that undergoes a displacement x while being acted on by a net force with x-component F_x, which we now allow to vary. Just as in Fig. 6.16, we divide the total displacement x into a large number of small segments Δx. We can apply the work–energy theorem, Eq. (6.6), to each segment because the value of F_x in each small segment is approximately constant. The change in kinetic energy in segment Δx_a is equal to the work $F_{ax}\Delta x_a$, and so on. The total change of kinetic energy is the sum of the changes in the individual segments, and thus is equal to the total work done on the particle during the entire displacement. So $W_{\text{tot}} = \Delta K$ holds for varying forces as well as for constant ones.

Here's an alternative derivation of the work–energy theorem for a force that may vary with position. It involves making a change of variable from x to v_x in the work integral. As a preliminary, we note that the acceleration a of the particle can be expressed in various ways, using $a_x = dv_x/dt$, $v_x = dx/dt$, and the chain rule for derivatives:

$$a_x = \frac{dv_x}{dt} = \frac{dv_x}{dx}\frac{dx}{dt} = v_x\frac{dv_x}{dx} \qquad (6.11)$$

From this result, Eq. (6.7) tells us that the total work done by the *net* force F_x is

$$W_{\text{tot}} = \int_{x_1}^{x_2} F_x \, dx = \int_{x_1}^{x_2} ma_x \, dx = \int_{x_1}^{x_2} mv_x\frac{dv_x}{dx} \, dx \qquad (6.12)$$

Now $(dv_x/dx)dx$ is the change in velocity dv_x during the displacement dx, so in Eq. (6.12) we can substitute dv_x for $(dv_x/dx)\,dx$. This changes the integration variable from x to v_x, so we change the limits from x_1 and x_2 to the corresponding x-velocities v_1 and v_2 at these points. This gives us

$$W_{\text{tot}} = \int_{v_1}^{v_2} mv_x \, dv_x$$

The integral of $v_x \, dv_x$ is just $v_x^2/2$. Substituting the upper and lower limits, we finally find

$$W_{\text{tot}} = \tfrac{1}{2}mv_2^2 - \tfrac{1}{2}mv_1^2 \qquad (6.13)$$

This is the same as Eq. (6.6), so the work–energy theorem is valid even without the assumption that the net force is constant.

Example 6.7 **Motion with a varying force**

An air-track glider of mass 0.100 kg is attached to the end of a horizontal air track by a spring with force constant 20.0 N/m (Fig. 6.22a). Initially the spring is unstretched and the glider is moving at 1.50 m/s to the right. Find the maximum distance d that the glider moves to the right (a) if the air track is turned on, so that there is no friction, and (b) if the air is turned off, so that there is kinetic friction with coefficient $\mu_k = 0.47$.

SOLUTION

IDENTIFY and SET UP: The force exerted by the spring is not constant, so we *cannot* use the constant-acceleration formulas of Chapter 2 to solve this problem. Instead, we'll use the

work–energy theorem, since the total work done involves the distance moved (our target variable). In Figs. 6.22b and 6.22c we choose the positive x-direction to be to the right (in the direction of the glider's motion). We take $x = 0$ at the glider's initial position (where the spring is unstretched) and $x = d$ (the target variable) at the position where the glider stops. The motion is purely horizontal, so only the horizontal forces do work. Note that Eq. (6.10) gives the work done by the *glider* on the *spring* as it stretches; to use the work–energy theorem we need the work done by the *spring* on the *glider*, which is the negative of Eq. (6.10). We expect the glider to move farther without friction than with friction.

6.22 (a) A glider attached to an air track by a spring. (b), (c) Our free-body diagrams.

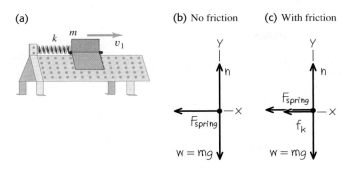

(a)

(b) No friction

(c) With friction

EXECUTE: (a) Equation (6.10) says that as the glider moves from $x_1 = 0$ to $x_2 = d$, it does an amount of work $W = \frac{1}{2}kd^2 - \frac{1}{2}k(0)^2 = \frac{1}{2}kd^2$ on the spring. The amount of work that the *spring* does on the *glider* is the negative of this, $-\frac{1}{2}kd^2$. The spring stretches until the glider comes instantaneously to rest, so the final kinetic energy K_2 is zero. The initial kinetic energy is $\frac{1}{2}mv_1^2$, where $v_1 = 1.50$ m/s is the glider's initial speed. From the work–energy theorem,

$$-\tfrac{1}{2}kd^2 = 0 - \tfrac{1}{2}mv_1^2$$

We solve for the distance d the glider moves:

$$d = v_1\sqrt{\frac{m}{k}} = (1.50 \text{ m/s})\sqrt{\frac{0.100 \text{ kg}}{20.0 \text{ N/m}}}$$
$$= 0.106 \text{ m} = 10.6 \text{ cm}$$

The stretched spring subsequently pulls the glider back to the left, so the glider is at rest only instantaneously.

(b) If the air is turned off, we must include the work done by the kinetic friction force. The normal force n is equal in magnitude to the weight of the glider, since the track is horizontal and there are

no other vertical forces. Hence the kinetic friction force has constant magnitude $f_k = \mu_k n = \mu_k mg$. The friction force is directed opposite to the displacement, so the work done by friction is

$$W_{\text{fric}} = f_k d \cos 180° = -f_k d = -\mu_k mgd$$

The total work is the sum of W_{fric} and the work done by the spring, $-\frac{1}{2}kd^2$. The work–energy theorem then says that

$$-\mu_k mgd - \tfrac{1}{2}kd^2 = 0 - \tfrac{1}{2}mv_1^2 \quad \text{or}$$
$$\tfrac{1}{2}kd^2 + \mu_k mgd - \tfrac{1}{2}mv_1^2 = 0$$

This is a quadratic equation for d. The solutions are

$$d = -\frac{\mu_k mg}{k} \pm \sqrt{\left(\frac{\mu_k mg}{k}\right)^2 + \frac{mv_1^2}{k}}$$

We have

$$\frac{\mu_k mg}{k} = \frac{(0.47)(0.100 \text{ kg})(9.80 \text{ m/s}^2)}{20.0 \text{ N/m}} = 0.02303 \text{ m}$$

$$\frac{mv_1^2}{k} = \frac{(0.100 \text{ kg})(1.50 \text{ m/s})^2}{20.0 \text{ N/m}} = 0.01125 \text{ m}^2$$

so

$$d = -(0.02303 \text{ m}) \pm \sqrt{(0.02303 \text{ m})^2 + 0.01125 \text{ m}^2}$$
$$= 0.086 \text{ m} \quad \text{or} \quad -0.132 \text{ m}$$

The quantity d is a positive displacement, so only the positive value of d makes sense. Thus with friction the glider moves a distance $d = 0.086$ m = 8.6 cm.

EVALUATE: Note that if we set $\mu_k = 0$, our algebraic solution for d in part (b) reduces to $d = v_1\sqrt{m/k}$, the zero-friction result from part (a). With friction, the glider goes a shorter distance. Again the glider stops instantaneously, and again the spring force pulls it toward the left; whether it moves or not depends on how great the *static* friction force is. How large would the coefficient of static friction μ_s have to be to keep the glider from springing back to the left?

Work–Energy Theorem for Motion Along a Curve

We can generalize our definition of work further to include a force that varies in direction as well as magnitude, and a displacement that lies along a curved path. Figure 6.23a shows a particle moving from P_1 to P_2 along a curve. We divide the curve between these points into many infinitesimal vector displacements, and we call a typical one of these $d\vec{l}$. Each $d\vec{l}$ is tangent to the path at its position. Let $\vec{F}$ be the force at a typical point along the path, and let ϕ be the angle between $\vec{F}$ and $d\vec{l}$ at this point. Then the small element of work dW done on the particle during the displacement $d\vec{l}$ may be written as

$$dW = F\cos\phi \, dl = F_\parallel \, dl = \vec{F} \cdot d\vec{l}$$

where $F_\parallel = F\cos\phi$ is the component of $\vec{F}$ in the direction parallel to $d\vec{l}$ (Fig. 6.23b). The total work done by $\vec{F}$ on the particle as it moves from P_1 to P_2 is then

$$W = \int_{P_1}^{P_2} F\cos\phi \, dl = \int_{P_1}^{P_2} F_\parallel \, dl = \int_{P_1}^{P_2} \vec{F} \cdot d\vec{l} \qquad \text{(work done on a curved path)} \qquad \text{(6.14)}$$

6.23 A particle moves along a curved path from point P_1 to P_2, acted on by a force $\vec{F}$ that varies in magnitude and direction.

(a)

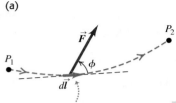

During an infinitesimal displacement $d\vec{l}$, the force $\vec{F}$ does work dW on the particle:
$$dW = \vec{F} \cdot d\vec{l} = F \cos \phi \, dl$$

(b)

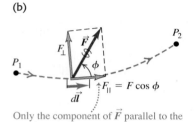

Only the component of $\vec{F}$ parallel to the displacement, $F_\parallel = F \cos \phi$, contributes to the work done by $\vec{F}$.

We can now show that the work–energy theorem, Eq. (6.6), holds true even with varying forces and a displacement along a curved path. The force $\vec{F}$ is essentially constant over any given infinitesimal segment $d\vec{l}$ of the path, so we can apply the work–energy theorem for straight-line motion to that segment. Thus the change in the particle's kinetic energy K over that segment equals the work $dW = F_\parallel \, dl = \vec{F} \cdot d\vec{l}$ done on the particle. Adding up these infinitesimal quantities of work from all the segments along the whole path gives the total work done, Eq. (6.14), which equals the total change in kinetic energy over the whole path. So $W_{\text{tot}} = \Delta K = K_2 - K_1$ is true *in general,* no matter what the path and no matter what the character of the forces. This can be proved more rigorously by using steps like those in Eqs. (6.11) through (6.13).

Note that only the component of the net force parallel to the path, $F_\parallel$, does work on the particle, so only this component can change the speed and kinetic energy of the particle. The component perpendicular to the path, $F_\perp = F \sin \phi$, has no effect on the particle's speed; it acts only to change the particle's direction.

The integral in Eq. (6.14) is called a *line integral*. To evaluate this integral in a specific problem, we need some sort of detailed description of the path and of the way in which $\vec{F}$ varies along the path. We usually express the line integral in terms of some scalar variable, as in the following example.

Example 6.8 | Motion on a curved path

At a family picnic you are appointed to push your obnoxious cousin Throckmorton in a swing (Fig. 6.24a). His weight is w, the length of the chains is R, and you push Throcky until the chains make an angle θ_0 with the vertical. To do this, you exert a varying horizontal force $\vec{F}$ that starts at zero and gradually increases just enough that Throcky and the swing move very slowly and remain very nearly in equilibrium throughout the process. What is the total work done on Throcky by all forces? What is the work done by the tension T in the chains? What is the work you do by exerting the force $\vec{F}$? (Neglect the weight of the chains and seat.)

SOLUTION

IDENTIFY and SET UP: The motion is along a curve, so we'll use Eq. (6.14) to calculate the work done by the net force, by the tension force, and by the force $\vec{F}$. Figure 6.24b shows our free-body diagram and coordinate system for some arbitrary point in Throcky's motion. We have replaced the sum of the tensions in the two chains with a single tension T.

EXECUTE: There are two ways to find the total work done during the motion: (1) by calculating the work done by each force and then adding those quantities, and (2) by calculating the work done by the net force. The second approach is far easier here because Throcky is in equilibrium at every point. Hence the net force on him is zero, the integral of the net force in Eq. (6.14) is zero, and the total work done on him is zero.

It's also easy to find the work done by the chain tension T because this force is perpendicular to the direction of motion at all points along the path. Hence at all points the angle between the chain tension and the displacement vector $d\vec{l}$ is 90° and the scalar product in Eq. (6.14) is zero. Thus the chain tension does zero work.

6.24 (a) Pushing cousin Throckmorton in a swing. (b) Our free-body diagram.

(a)

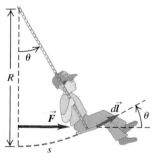

(b) Free-body diagram for Throckmorton (neglecting the weight of the chains and seat)

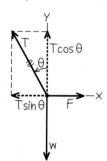

To compute the work done by $\vec{F}$, we need to know how this force varies with the angle θ. The net force on Throcky is zero, so $\Sigma F_x = 0$ and $\Sigma F_y = 0$. From Fig. 6.24b,

$$\Sigma F_x = F + (-T \sin \theta) = 0$$
$$\Sigma F_y = T \cos \theta + (-w) = 0$$

By eliminating T from these two equations, we obtain the magnitude $F = w \tan \theta$.

The point where $\vec{F}$ is applied moves through the arc s (Fig. 6.24a). The arc length s equals the radius R of the circular path multiplied by the length θ (in radians), so $s = R\theta$. Therefore the displacement $d\vec{l}$ corresponding to a small change of angle $d\theta$ has a magnitude $dl = ds = R\,d\theta$. The work done by $\vec{F}$ is then

$$W = \int \vec{F} \cdot d\vec{l} = \int F\cos\theta\,ds$$

Now we express F and ds in terms of the angle θ, whose value increases from 0 to θ_0:

$$W = \int_0^{\theta_0} (w\tan\theta)\cos\theta\,(R\,d\theta) = wR \int_0^{\theta_0} \sin\theta\,d\theta$$

$$= wR(1 - \cos\theta_0)$$

EVALUATE: If $\theta_0 = 0$, there is no displacement; then $\cos\theta_0 = 1$ and $W = 0$, as we should expect. If $\theta_0 = 90°$, then $\cos\theta_0 = 0$ and $W = wR$. In that case the work you do is the same as if you had lifted Throcky straight up a distance R with a force equal to his weight w. In fact (as you may wish to confirm), the quantity $R(1 - \cos\theta_0)$ is the increase in his height above the ground during the displacement, so for any value of θ_0 the work done by the force $\vec{F}$ is the change in height multiplied by the weight. This is an example of a more general result that we'll prove in Section 7.1.

We can check our results by writing the forces and the infinitesimal displacement $d\vec{l}$ in terms of their x- and y-components. Figure 6.24a shows that $d\vec{l}$ has a magnitude of ds, an x-component of $ds\cos\theta$, and a y-component of $ds\sin\theta$. Hence $d\vec{l} =$ $\hat{\imath}\,ds\cos\theta + \hat{\jmath}\,ds\sin\theta$. Similarly, we can write the three forces as

$$\vec{T} = \hat{\imath}(-T\sin\theta) + \hat{\jmath}T\cos\theta$$

$$\vec{w} = \hat{\jmath}(-w)$$

$$\vec{F} = \hat{\imath}F$$

We use Eq. (1.21) to calculate the scalar product of each of these forces with $d\vec{l}$:

$$\vec{T} \cdot d\vec{l} = (-T\sin\theta)(ds\cos\theta) + (T\cos\theta)(ds\sin\theta) = 0$$

$$\vec{w} \cdot d\vec{l} = (-w)(ds\sin\theta) = -w\sin\theta\,ds$$

$$\vec{F} \cdot d\vec{l} = F(ds\cos\theta) = F\cos\theta\,ds$$

Since $\vec{T} \cdot d\vec{l} = 0$, the integral of this quantity is zero and the work done by the chain tension is zero, just as we found above. Using $ds = R\,d\theta$, we find the work done by the force of gravity is

$$\int \vec{w} \cdot d\vec{l} = \int (-w\sin\theta)R\,d\theta = -wR \int_0^{\theta_0} \sin\theta\,d\theta$$

$$= -wR(1 - \cos\theta_0)$$

Gravity does negative work because this force pulls down while Throcky moves upward. Finally, the work done by the force $\vec{F}$ is the same integral $\int \vec{F} \cdot d\vec{l} = \int F\cos\theta\,ds$ that we calculated above. The method of components is often the most convenient way to calculate scalar products, so use it when it makes your life easier!

Test Your Understanding of Section 6.3 In Example 5.20 (Section 5.4) we examined a conical pendulum. The speed of the pendulum bob remains constant as it travels around the circle shown in Fig. 5.32a. (a) Over one complete circle, how much work does the tension force F do on the bob? (i) a positive amount; (ii) a negative amount; (iii) zero. (b) Over one complete circle, how much work does the weight do on the bob? (i) a positive amount; (ii) a negative amount; (iii) zero. ❙

6.4 Power

The definition of work makes no reference to the passage of time. If you lift a barbell weighing 100 N through a vertical distance of 1.0 m at constant velocity, you do $(100\text{ N})(1.0\text{ m}) = 100$ J of work whether it takes you 1 second, 1 hour, or 1 year to do it. But often we need to know how quickly work is done. We describe this in terms of *power*. In ordinary conversation the word "power" is often synonymous with "energy" or "force." In physics we use a much more precise definition: **Power** is the time *rate* at which work is done. Like work and energy, power is a scalar quantity.

When a quantity of work ΔW is done during a time interval Δt, the average work done per unit time or **average power** P_{av} is defined to be

$$P_{av} = \frac{\Delta W}{\Delta t} \qquad \text{(average power)} \qquad (6.15)$$

The rate at which work is done might not be constant. We can define **instantaneous power** P as the quotient in Eq. (6.15) as Δt approaches zero:

$$P = \lim_{\Delta t \to 0} \frac{\Delta W}{\Delta t} = \frac{dW}{dt} \qquad \text{(instantaneous power)} \qquad (6.16)$$

The SI unit of power is the **watt** (W), named for the English inventor James Watt. One watt equals 1 joule per second: $1\text{ W} = 1\text{ J/s}$ (Fig. 6.25). The kilowatt

6.25 The same amount of work is done in both of these situations, but the power (the rate at which work is done) is different.

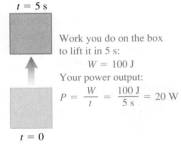

$t = 5\text{ s}$

Work you do on the box to lift it in 5 s:
$$W = 100\text{ J}$$
Your power output:
$$P = \frac{W}{t} = \frac{100\text{ J}}{5\text{ s}} = 20\text{ W}$$

$t = 0$

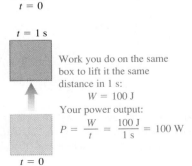

$t = 1\text{ s}$

Work you do on the same box to lift it the same distance in 1 s:
$$W = 100\text{ J}$$
Your power output:
$$P = \frac{W}{t} = \frac{100\text{ J}}{1\text{ s}} = 100\text{ W}$$

$t = 0$

6.26 The value of the horsepower derives from experiments by James Watt, who measured that a horse could do 33,000 foot-pounds of work per minute in lifting coal from a coal pit.

$(1 \text{ kW} = 10^3 \text{ W})$ and the megawatt $(1 \text{ MW} = 10^6 \text{ W})$ are also commonly used. In the British system, work is expressed in foot-pounds, and the unit of power is the foot-pound per second. A larger unit called the *horsepower* (hp) is also used (Fig. 6.26):

$$1 \text{ hp} = 550 \text{ ft} \cdot \text{lb/s} = 33,000 \text{ ft} \cdot \text{lb/min}$$

That is, a 1-hp motor running at full load does 33,000 ft · lb of work every minute. A useful conversion factor is

$$1 \text{ hp} = 746 \text{ W} = 0.746 \text{ kW}$$

The watt is a familiar unit of *electrical* power; a 100-W light bulb converts 100 J of electrical energy into light and heat each second. But there's nothing inherently electrical about a watt. A light bulb could be rated in horsepower, and an engine can be rated in kilowatts.

The *kilowatt-hour* (kW · h) is the usual commercial unit of electrical energy. One kilowatt-hour is the total work done in 1 hour (3600 s) when the power is 1 kilowatt (10^3 J/s), so

$$1 \text{ kW} \cdot \text{h} = (10^3 \text{ J/s})(3600 \text{ s}) = 3.6 \times 10^6 \text{ J} = 3.6 \text{ MJ}$$

The kilowatt-hour is a unit of *work* or *energy*, not power.

In mechanics we can also express power in terms of force and velocity. Suppose that a force $\vec{F}$ acts on a body while it undergoes a vector displacement $\Delta\vec{s}$. If $F_\parallel$ is the component of $\vec{F}$ tangent to the path (parallel to $\Delta\vec{s}$), then the work done by the force is $\Delta W = F_\parallel \Delta s$. The average power is

$$P_{av} = \frac{F_\parallel \Delta s}{\Delta t} = F_\parallel \frac{\Delta s}{\Delta t} = F_\parallel v_{av} \qquad (6.17)$$

Instantaneous power P is the limit of this expression as $\Delta t \to 0$:

$$P = F_\parallel v \qquad (6.18)$$

where v is the magnitude of the instantaneous velocity. We can also express Eq. (6.18) in terms of the scalar product:

$$P = \vec{F} \cdot \vec{v} \qquad \begin{array}{l}\text{(instantaneous rate at which}\\ \text{force } \vec{F} \text{ does work on a particle)}\end{array} \qquad (6.19)$$

Example 6.9 Force and power

Each of the four jet engines on an Airbus A380 airliner develops a thrust (a forward force on the airliner) of 322,000 N (72,000 lb). When the airplane is flying at 250 m/s (900 km/h, or roughly 560 mi/h), what horsepower does each engine develop?

SOLUTION

IDENTIFY, SET UP and EXECUTE: Our target variable is the instantaneous power P, which is the rate at which the thrust does work. We use Eq. (6.18). The thrust is in the direction of motion, so $F_\parallel$ is just equal to the thrust. At $v = 250$ m/s, the power developed by each engine is

$$P = F_\parallel v = (3.22 \times 10^5 \text{ N})(250 \text{ m/s}) = 8.05 \times 10^7 \text{ W}$$

$$= (8.05 \times 10^7 \text{ W})\frac{1 \text{ hp}}{746 \text{ W}} = 108,000 \text{ hp}$$

EVALUATE: The speed of modern airliners is directly related to the power of their engines (Fig. 6.27). The largest propeller-driven airliners of the 1950s had engines that developed about 3400 hp

6.27 (a) Propeller-driven and (b) jet airliners.

(a)

(b)

$(2.5 \times 10^6 \text{ W})$, giving them maximum speeds of about 600 km/h (370 mi/h). Each engine on an Airbus A380 develops more than 30 times more power, enabling it to fly at about 900 km/h (560 mi/h) and to carry a much heavier load.

If the engines are at maximum thrust while the airliner is at rest on the ground so that $v = 0$, the engines develop *zero* power. Force and power are not the same thing!

Example 6.10 A "power climb"

A 50.0-kg marathon runner runs up the stairs to the top of Chicago's 443-m-tall Willis Tower, the tallest building in the United States (Fig. 6.28). To lift herself to the top in 15.0 minutes, what must be her average power output? Express your answer in watts, in kilowatts, and in horsepower.

SOLUTION

IDENTIFY and SET UP: We'll treat the runner as a particle of mass m. Her average power output P_{av} must be enough to lift her at constant speed against gravity.

We can find P_{av} in two ways: (1) by determining how much work she must do and dividing that quantity by the elapsed time, as in Eq. (6.15), or (2) by calculating the average upward force she must exert (in the direction of the climb) and multiplying that quantity by her upward velocity, as in Eq. (6.17).

EXECUTE: (1) As in Example 6.8, lifting a mass m against gravity requires an amount of work equal to the weight mg multiplied by the height h it is lifted. Hence the work the runner must do is

$$W = mgh = (50.0 \text{ kg})(9.80 \text{ m/s}^2)(443 \text{ m})$$
$$= 2.17 \times 10^5 \text{ J}$$

She does this work in a time 15.0 min = 900 s, so from Eq. (6.15) the average power is

$$P_{av} = \frac{2.17 \times 10^5 \text{J}}{900 \text{ s}} = 241 \text{ W} = 0.241 \text{ kW} = 0.323 \text{ hp}$$

(2) The force exerted is vertical and the average vertical component of velocity is $(443 \text{ m})/(900 \text{ s}) = 0.492 \text{ m/s}$, so from Eq. (6.17) the average power is

6.28 How much power is required to run up the stairs of Chicago's Willis Tower in 15 minutes?

$$P_{av} = F_\| v_{av} = (mg)v_{av}$$
$$= (50.0 \text{ kg})(9.80 \text{ m/s}^2)(0.492 \text{ m/s}) = 241 \text{ W}$$

which is the same result as before.

EVALUATE: The runner's *total* power output will be several times greater than 241 W. The reason is that the runner isn't really a particle but a collection of parts that exert forces on each other and do work, such as the work done to inhale and exhale and to make her arms and legs swing. What we've calculated is only the part of her power output that lifts her to the top of the building.

Test Your Understanding of Section 6.4 The air surrounding an airplane in flight exerts a drag force that acts opposite to the airplane's motion. When the Airbus A380 in Example 6.9 is flying in a straight line at a constant altitude at a constant 250 m/s, what is the rate at which the drag force does work on it? (i) 432,000 hp; (ii) 108,000 hp; (iii) 0; (iv) −108,000 hp; (v) −432,000 hp.

Work done by a force: When a constant force $\vec{F}$ acts on a particle that undergoes a straight-line displacement $\vec{s}$, the work done by the force on the particle is defined to be the scalar product of $\vec{F}$ and $\vec{s}$. The unit of work in SI units is 1 joule = 1 newton-meter ($1\ J = 1\ N \cdot m$). Work is a scalar quantity; it can be positive or negative, but it has no direction in space. (See Examples 6.1 and 6.2.)

$$W = \vec{F} \cdot \vec{s} = Fs\cos\phi \quad \text{(6.2), (6.3)}$$

ϕ = angle between $\vec{F}$ and $\vec{s}$

$W = F_{\parallel}s$
$= (F\cos\phi)s$

$F_{\parallel} = F\cos\phi$

Kinetic energy: The kinetic energy K of a particle equals the amount of work required to accelerate the particle from rest to speed v. It is also equal to the amount of work the particle can do in the process of being brought to rest. Kinetic energy is a scalar that has no direction in space; it is always positive or zero. Its units are the same as the units of work: $1\ J = 1\ N \cdot m = 1\ kg \cdot m^2/s^2$.

$$K = \tfrac{1}{2}mv^2 \quad \text{(6.5)}$$

Doubling m doubles K.

Doubling v quadruples K.

The work–energy theorem: When forces act on a particle while it undergoes a displacement, the particle's kinetic energy changes by an amount equal to the total work done on the particle by all the forces. This relationship, called the work–energy theorem, is valid whether the forces are constant or varying and whether the particle moves along a straight or curved path. It is applicable only to bodies that can be treated as particles. (See Examples 6.3–6.5.)

$$W_{\text{tot}} = K_2 - K_1 = \Delta K \quad \text{(6.6)}$$

W_{tot} = Total work done on particle along path

$K_1 = \tfrac{1}{2}mv_1^2$

$K_2 = \tfrac{1}{2}mv_2^2 = K_1 + W_{\text{tot}}$

Work done by a varying force or on a curved path: When a force varies during a straight-line displacement, the work done by the force is given by an integral, Eq. (6.7). (See Examples 6.6 and 6.7.) When a particle follows a curved path, the work done on it by a force $\vec{F}$ is given by an integral that involves the angle ϕ between the force and the displacement. This expression is valid even if the force magnitude and the angle ϕ vary during the displacement. (See Example 6.8.)

$$W = \int_{x_1}^{x_2} F_x \, dx \quad \text{(6.7)}$$

$$W = \int_{P_1}^{P_2} F\cos\phi \, dl = \int_{P_1}^{P_2} F_{\parallel} \, dl$$

$$= \int_{P_1}^{P_2} \vec{F} \cdot d\vec{l} \quad \text{(6.14)}$$

Area = Work done by force during displacement

Power: Power is the time rate of doing work. The average power P_{av} is the amount of work ΔW done in time Δt divided by that time. The instantaneous power is the limit of the average power as Δt goes to zero. When a force $\vec{F}$ acts on a particle moving with velocity $\vec{v}$, the instantaneous power (the rate at which the force does work) is the scalar product of $\vec{F}$ and $\vec{v}$. Like work and kinetic energy, power is a scalar quantity. The SI unit of power is 1 watt = 1 joule/second ($1\ W = 1\ J/s$). (See Examples 6.9 and 6.10.)

$$P_{\text{av}} = \frac{\Delta W}{\Delta t} \quad \text{(6.15)}$$

$$P = \lim_{\Delta t \to 0} \frac{\Delta W}{\Delta t} = \frac{dW}{dt} \quad \text{(6.16)}$$

$$P = \vec{F} \cdot \vec{v} \quad \text{(6.19)}$$

$t = 5\ s$

$t = 0$

Work you do on the box to lift it in 5 s:
$W = 100\ J$
Your power output:
$P = \dfrac{W}{t} = \dfrac{100\ J}{5\ s}$
$= 20\ W$

BRIDGING PROBLEM   A Spring That Disobeys Hooke's Law

Consider a hanging spring of negligible mass that does *not* obey Hooke's law. When the spring is extended by a distance x, the force exerted by the spring has magnitude αx^2, where α is a positive constant. The spring is not extended when a block of mass m is attached to it. The block is then released, stretching the spring as it falls (Fig. 6.29). (a) How fast is the block moving when it has fallen a distance x_1? (b) At what rate does the spring do work on the block at this point? (c) Find the maximum distance x_2 that the spring stretches. (d) Will the block *remain* at the point found in part (c)?

6.29 The block is attached to a spring that does not obey Hooke's law.

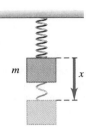

SOLUTION GUIDE

See MasteringPhysics® study area for a Video Tutor solution.

IDENTIFY and SET UP

1. The spring force in this problem isn't constant, so you have to use the work–energy theorem. You'll also need to use Eq. (6.7) to find the work done by the spring over a given displacement.
2. Draw a free-body diagram for the block, including your choice of coordinate axes. Note that x represents how far the spring is *stretched,* so choose the positive x-axis accordingly. On your coordinate axis, label the points $x = x_1$ and $x = x_2$.
3. Make a list of the unknown quantities, and decide which of these are the target variables.

EXECUTE

4. Calculate the work done on the block by the spring as the block falls an arbitrary distance x. (The integral isn't a difficult one. Use Appendix B if you need a reminder.) Is the work done by the spring positive, negative, or zero?

5. Calculate the work done on the block by any other forces as the block falls an arbitrary distance x. Is this work positive, negative, or zero?
6. Use the work–energy theorem to find the target variables. (You'll also need to use an equation for power.) *Hint:* When the spring is at its maximum stretch, what is the speed of the block?
7. To answer part (d), consider the *net* force that acts on the block when it is at the point found in part (c).

EVALUATE

8. We learned in Chapter 2 that after an object dropped from rest has fallen freely a distance x_1, its speed is $\sqrt{2gx_1}$. Use this to decide whether your answer in part (a) makes sense. In addition, ask yourself whether the algebraic sign of your answer in part (b) makes sense.
9. Find the value of x where the net force on the block would be zero. How does this compare to your result for x_2? Is this consistent with your answer in part (d)?

Problems

For instructor-assigned homework, go to www.masteringphysics.com

•, ••, •••: Problems of increasing difficulty. **CP:** Cumulative problems incorporating material from earlier chapters. **CALC:** Problems requiring calculus. **BIO:** Biosciences problems.

DISCUSSION QUESTIONS

Q6.1 The sign of many physical quantities depends on the choice of coordinates. For example, a_y for free-fall motion can be negative or positive, depending on whether we choose upward or downward as positive. Is the same thing true of work? In other words, can we make positive work negative by a different choice of coordinates? Explain.

Q6.2 An elevator is hoisted by its cables at constant speed. Is the total work done on the elevator positive, negative, or zero? Explain.

Q6.3 A rope tied to a body is pulled, causing the body to accelerate. But according to Newton's third law, the body pulls back on the rope with an equal and opposite force. Is the total work done then zero? If so, how can the body's kinetic energy change? Explain.

Q6.4 If it takes total work W to give an object a speed v and kinetic energy K, starting from rest, what will be the object's speed (in terms of v) and kinetic energy (in terms of K) if we do twice as much work on it, again starting from rest?

Q6.5 If there is a net nonzero force on a moving object, is it possible for the total work done on the object to be zero? Explain, with an example that illustrates your answer.

Q6.6 In Example 5.5 (Section 5.1), how does the work done on the bucket by the tension in the cable compare to the work done on the cart by the tension in the cable?

Q6.7 In the conical pendulum in Example 5.20 (Section 5.4), which of the forces do work on the bob while it is swinging?

Q6.8 For the cases shown in Fig. Q6.8, the object is released from rest at the top and feels no friction or air resistance. In

Figure **Q6.8**

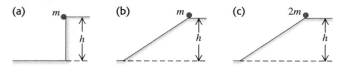

which (if any) cases will the mass have (i) the greatest speed at the bottom and (ii) the most work done on it by the time it reaches the bottom?

Q6.9 A force $\vec{F}$ is in the x-direction and has a magnitude that depends on x. Sketch a possible graph of F versus x such that the force does zero work on an object that moves from x_1 to x_2, even though the force magnitude is not zero at all x in this range.

Q6.10 Does the kinetic energy of a car change more when it speeds up from 10 to 15 m/s or from 15 to 20 m/s? Explain.

Q6.11 A falling brick has a mass of 1.5 kg and is moving straight downward with a speed of 5.0 m/s. A 1.5-kg physics book is sliding across the floor with a speed of 5.0 m/s. A 1.5-kg melon is traveling with a horizontal velocity component 3.0 m/s to the right and a vertical component 4.0 m/s upward. Do these objects all have the same velocity? Do these objects all have the same kinetic energy? For each question, give the reasoning behind your answer.

Q6.12 Can the *total* work done on an object during a displacement be negative? Explain. If the total work is negative, can its magnitude be larger than the initial kinetic energy of the object? Explain.

Q6.13 A net force acts on an object and accelerates it from rest to a speed v_1. In doing so, the force does an amount of work W_1. By what factor must the work done on the object be increased to produce three times the final speed, with the object again starting from rest?

Q6.14 A truck speeding down the highway has a lot of kinetic energy relative to a stopped state trooper, but no kinetic energy relative to the truck driver. In these two frames of reference, is the same amount of work required to stop the truck? Explain.

Q6.15 You are holding a briefcase by the handle, with your arm straight down by your side. Does the force your hand exerts do work on the briefcase when (a) you walk at a constant speed down a horizontal hallway and (b) you ride an escalator from the first to second floor of a building? In each case justify your answer.

Q6.16 When a book slides along a tabletop, the force of friction does negative work on it. Can friction ever do *positive* work? Explain. (*Hint:* Think of a box in the back of an accelerating truck.)

Q6.17 Time yourself while running up a flight of steps, and compute the average rate at which you do work against the force of gravity. Express your answer in watts and in horsepower.

Q6.18 Fractured Physics. Many terms from physics are badly misused in everyday language. In each case, explain the errors involved. (a) A *strong* person is called *powerful*. What is wrong with this use of *power*? (b) When a worker carries a bag of concrete along a level construction site, people say he did a lot of *work*. Did he?

Q6.19 An advertisement for a portable electrical generating unit claims that the unit's diesel engine produces 28,000 hp to drive an electrical generator that produces 30 MW of electrical power. Is this possible? Explain.

Q6.20 A car speeds up while the engine delivers constant power. Is the acceleration greater at the beginning of this process or at the end? Explain.

Q6.21 Consider a graph of instantaneous power versus time, with the vertical P-axis starting at $P = 0$. What is the physical significance of the area under the P-versus-t curve between vertical lines at t_1 and t_2? How could you find the average power from the graph? Draw a P-versus-t curve that consists of two straight-line sections and for which the peak power is equal to twice the average power.

Q6.22 A nonzero net force acts on an object. Is it possible for any of the following quantities to be constant: (a) the particle's speed; (b) the particle's velocity; (c) the particle's kinetic energy?

Q6.23 When a certain force is applied to an ideal spring, the spring stretches a distance x from its unstretched length and does work W. If instead twice the force is applied, what distance (in terms of x) does the spring stretch from its unstretched length, and how much work (in terms of W) is required to stretch it this distance?

Q6.24 If work W is required to stretch a spring a distance x from its unstretched length, what work (in terms of W) is required to stretch the spring an *additional* distance x?

EXERCISES

Section 6.1 Work

6.1 • You push your physics book 1.50 m along a horizontal tabletop with a horizontal push of 2.40 N while the opposing force of friction is 0.600 N. How much work does each of the following forces do on the book: (a) your 2.40-N push, (b) the friction force, (c) the normal force from the tabletop, and (d) gravity? (e) What is the net work done on the book?

6.2 • A tow truck pulls a car 5.00 km along a horizontal roadway using a cable having a tension of 850 N. (a) How much work does the cable do on the car if it pulls horizontally? If it pulls at 35.0° above the horizontal? (b) How much work does the cable do on the tow truck in both cases of part (a)? (c) How much work does gravity do on the car in part (a)?

6.3 • A factory worker pushes a 30.0-kg crate a distance of 4.5 m along a level floor at constant velocity by pushing horizontally on it. The coefficient of kinetic friction between the crate and the floor is 0.25. (a) What magnitude of force must the worker apply? (b) How much work is done on the crate by this force? (c) How much work is done on the crate by friction? (d) How much work is done on the crate by the normal force? By gravity? (e) What is the total work done on the crate?

6.4 •• Suppose the worker in Exercise 6.3 pushes downward at an angle of 30° below the horizontal. (a) What magnitude of force must the worker apply to move the crate at constant velocity? (b) How much work is done on the crate by this force when the crate is pushed a distance of 4.5 m? (c) How much work is done on the crate by friction during this displacement? (d) How much work is done on the crate by the normal force? By gravity? (e) What is the total work done on the crate?

6.5 •• A 75.0-kg painter climbs a ladder that is 2.75 m long leaning against a vertical wall. The ladder makes a 30.0° angle with the wall. (a) How much work does gravity do on the painter? (b) Does the answer to part (a) depend on whether the painter climbs at constant speed or accelerates up the ladder?

6.6 •• Two tugboats pull a disabled supertanker. Each tug exerts a constant force of 1.80×10^6 N, one 14° west of north and the other 14° east of north, as they pull the tanker 0.75 km toward the north. What is the total work they do on the supertanker?

6.7 • Two blocks are connected by a very light string passing over a massless and frictionless pulley (Fig. E6.7). Traveling at constant speed, the 20.0-N block moves 75.0 cm to the right and the 12.0-N block moves 75.0 cm downward. During this process, how much work is done (a) on the 12.0-N block by (i) gravity and (ii) the tension in the string? (b) On the 20.0-N block by (i) gravity,

(ii) the tension in the string, (iii) friction, and (iv) the normal force? (c) Find the total work done on each block.

Figure **E6.7**

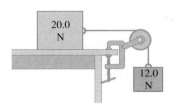

6.8 •• A loaded grocery cart is rolling across a parking lot in a strong wind. You apply a constant force $\vec{F} = (30 \text{ N})\hat{\imath} - (40 \text{ N})\hat{\jmath}$ to the cart as it undergoes a displacement $\vec{s} = (-9.0 \text{ m})\hat{\imath} - (3.0 \text{ m})\hat{\jmath}$. How much work does the force you apply do on the grocery cart?

6.9 • A 0.800-kg ball is tied to the end of a string 1.60 m long and swung in a vertical circle. (a) During one complete circle, starting anywhere, calculate the total work done on the ball by (i) the tension in the string and (ii) gravity. (b) Repeat part (a) for motion along the semicircle from the lowest to the highest point on the path.

6.10 •• An 8.00-kg package in a mail-sorting room slides 2.00 m down a chute that is inclined at 53.0° below the horizontal. The coefficient of kinetic friction between the package and the chute's surface is 0.40. Calculate the work done on the package by (a) friction, (b) gravity, and (c) the normal force. (d) What is the net work done on the package?

6.11 •• A boxed 10.0-kg computer monitor is dragged by friction 5.50 m up along the moving surface of a conveyor belt inclined at an angle of 36.9° above the horizontal. If the monitor's speed is a constant 2.10 cm/s, how much work is done on the monitor by (a) friction, (b) gravity, and (c) the normal force of the conveyor belt?

6.12 •• You apply a constant force $\vec{F} = (-68.0 \text{ N})\hat{\imath} + (36.0 \text{ N})\hat{\jmath}$ to a 380-kg car as the car travels 48.0 m in a direction that is 240.0° counterclockwise from the $+x$-axis. How much work does the force you apply do on the car?

Section 6.2 Kinetic Energy and the Work–Energy Theorem

6.13 •• **Animal Energy.** BIO Adult cheetahs, the fastest of the great cats, have a mass of about 70 kg and have been clocked running at up to 72 mph (32 m/s). (a) How many joules of kinetic energy does such a swift cheetah have? (b) By what factor would its kinetic energy change if its speed were doubled?

6.14 •• A 1.50-kg book is sliding along a rough horizontal surface. At point A it is moving at 3.21 m/s, and at point B it has slowed to 1.25 m/s. (a) How much work was done on the book between A and B? (b) If -0.750 J of work is done on the book from B to C, how fast is it moving at point C? (c) How fast would it be moving at C if $+0.750$ J of work were done on it from B to C?

6.15 • **Meteor Crater.** About 50,000 years ago, a meteor crashed into the earth near present-day Flagstaff, Arizona. Measurements from 2005 estimate that this meteor had a mass of about 1.4×10^8 kg (around 150,000 tons) and hit the ground at a speed of 12 km/s. (a) How much kinetic energy did this meteor deliver to the ground? (b) How does this energy compare to the energy released by a 1.0-megaton nuclear bomb? (A megaton bomb releases the same amount of energy as a million tons of TNT, and 1.0 ton of TNT releases 4.184×10^9 J of energy.)

6.16 • **Some Typical Kinetic Energies.** (a) In the Bohr model of the atom, the ground-state electron in hydrogen has an orbital speed of 2190 km/s. What is its kinetic energy? (Consult Appendix F.)

(b) If you drop a 1.0-kg weight (about 2 lb) from a height of 1.0 m, how many joules of kinetic energy will it have when it reaches the ground? (c) Is it reasonable that a 30-kg child could run fast enough to have 100 J of kinetic energy?

6.17 •• In Fig. E6.7 assume that there is no friction force on the 20.0-N block that sits on the tabletop. The pulley is light and frictionless. (a) Calculate the tension T in the light string that connects the blocks. (b) For a displacement in which the 12.0-N block descends 1.20 m, calculate the total work done on (i) the 20.0-N block and (ii) the 12.0-N block. (c) For the displacement in part (b), calculate the total work done on the system of the two blocks. How does your answer compare to the work done on the 12.0-N block by gravity? (d) If the system is released from rest, what is the speed of the 12.0-N block when it has descended 1.20 m?

6.18 • A 4.80-kg watermelon is dropped from rest from the roof of a 25.0-m-tall building and feels no appreciable air resistance. (a) Calculate the work done by gravity on the watermelon during its displacement from the roof to the ground. (b) Just before it strikes the ground, what is the watermelon's (i) kinetic energy and (ii) speed? (c) Which of the answers in parts (a) and (b) would be *different* if there were appreciable air resistance?

6.19 •• Use the work–energy theorem to solve each of these problems. You can use Newton's laws to check your answers. Neglect air resistance in all cases. (a) A branch falls from the top of a 95.0-m-tall redwood tree, starting from rest. How fast is it moving when it reaches the ground? (b) A volcano ejects a boulder directly upward 525 m into the air. How fast was the boulder moving just as it left the volcano? (c) A skier moving at 5.00 m/s encounters a long, rough horizontal patch of snow having coefficient of kinetic friction 0.220 with her skis. How far does she travel on this patch before stopping? (d) Suppose the rough patch in part (c) was only 2.90 m long? How fast would the skier be moving when she reached the end of the patch? (e) At the base of a frictionless icy hill that rises at 25.0° above the horizontal, a toboggan has a speed of 12.0 m/s toward the hill. How high vertically above the base will it go before stopping?

6.20 •• You throw a 20-N rock vertically into the air from ground level. You observe that when it is 15.0 m above the ground, it is traveling at 25.0 m/s upward. Use the work–energy theorem to find (a) the rock's speed just as it left the ground and (b) its maximum height.

6.21 •• You are a member of an Alpine Rescue Team. You must project a box of supplies up an incline of constant slope angle α so that it reaches a stranded skier who is a vertical distance h above the bottom of the incline. The incline is slippery, but there is some friction present, with kinetic friction coefficient μ_k. Use the work–energy theorem to calculate the minimum speed you must give the box at the bottom of the incline so that it will reach the skier. Express your answer in terms of g, h, μ_k, and α.

6.22 •• A mass m slides down a smooth inclined plane from an initial vertical height h, making an angle α with the horizontal. (a) The work done by a force is the sum of the work done by the components of the force. Consider the components of gravity parallel and perpendicular to the surface of the plane. Calculate the work done on the mass by each of the components, and use these results to show that the work done by gravity is exactly the same as if the mass had fallen straight down through the air from a height h. (b) Use the work–energy theorem to prove that the speed of the mass at the bottom of the incline is the same as if it had been dropped from height h, independent of the angle α of the incline. Explain how this speed can be independent of the slope angle. (c) Use the results of part (b) to find the speed of a rock that slides down an icy frictionless hill, starting from rest 15.0 m above the bottom.

6.23 · A sled with mass 8.00 kg moves in a straight line on a frictionless horizontal surface. At one point in its path, its speed is 4.00 m/s; after it has traveled 2.50 m beyond this point, its speed is 6.00 m/s. Use the work–energy theorem to find the force acting on the sled, assuming that this force is constant and that it acts in the direction of the sled's motion.

6.24 ·· A soccer ball with mass 0.420 kg is initially moving with speed 2.00 m/s. A soccer player kicks the ball, exerting a constant force of magnitude 40.0 N in the same direction as the ball's motion. Over what distance must the player's foot be in contact with the ball to increase the ball's speed to 6.00 m/s?

6.25 · A 12-pack of Omni-Cola (mass 4.30 kg) is initially at rest on a horizontal floor. It is then pushed in a straight line for 1.20 m by a trained dog that exerts a horizontal force with magnitude 36.0 N. Use the work–energy theorem to find the final speed of the 12-pack if (a) there is no friction between the 12-pack and the floor, and (b) the coefficient of kinetic friction between the 12-pack and the floor is 0.30.

6.26 · A batter hits a baseball with mass 0.145 kg straight upward with an initial speed of 25.0 m/s. (a) How much work has gravity done on the baseball when it reaches a height of 20.0 m above the bat? (b) Use the work–energy theorem to calculate the speed of the baseball at a height of 20.0 m above the bat. You can ignore air resistance. (c) Does the answer to part (b) depend on whether the baseball is moving upward or downward at a height of 20.0 m? Explain.

6.27 · A little red wagon with mass 7.00 kg moves in a straight line on a frictionless horizontal surface. It has an initial speed of 4.00 m/s and then is pushed 3.0 m in the direction of the initial velocity by a force with a magnitude of 10.0 N. (a) Use the work–energy theorem to calculate the wagon's final speed. (b) Calculate the acceleration produced by the force. Use this acceleration in the kinematic relationships of Chapter 2 to calculate the wagon's final speed. Compare this result to that calculated in part (a).

6.28 ·· A block of ice with mass 2.00 kg slides 0.750 m down an inclined plane that slopes downward at an angle of 36.9° below the horizontal. If the block of ice starts from rest, what is its final speed? You can ignore friction.

6.29 · **Stopping Distance.** A car is traveling on a level road with speed v_0 at the instant when the brakes lock, so that the tires slide rather than roll. (a) Use the work–energy theorem to calculate the minimum stopping distance of the car in terms of v_0, g, and the coefficient of kinetic friction μ_k between the tires and the road. (b) By what factor would the minimum stopping distance change if (i) the coefficient of kinetic friction were doubled, or (ii) the initial speed were doubled, or (iii) both the coefficient of kinetic friction and the initial speed were doubled?

6.30 ·· A 30.0-kg crate is initially moving with a velocity that has magnitude 3.90 m/s in a direction 37.0° west of north. How much work must be done on the crate to change its velocity to 5.62 m/s in a direction 63.0° south of east?

Section 6.3 Work and Energy with Varying Forces

6.31 · **BIO Heart Repair.** A surgeon is using material from a donated heart to repair a patient's damaged aorta and needs to know the elastic characteristics of this aortal material. Tests performed on a 16.0-cm strip of the donated aorta reveal that it stretches 3.75 cm when a 1.50-N pull is exerted on it. (a) What is the force constant of this strip of aortal material? (b) If the maximum distance it will be able to stretch when it replaces the aorta in the damaged heart is 1.14 cm, what is the greatest force it will be able to exert there?

6.32 ·· To stretch a spring 3.00 cm from its unstretched length, 12.0 J of work must be done. (a) What is the force constant of this spring? (b) What magnitude force is needed to stretch the spring 3.00 cm from its unstretched length? (c) How much work must be done to compress this spring 4.00 cm from its unstretched length, and what force is needed to compress it this distance?

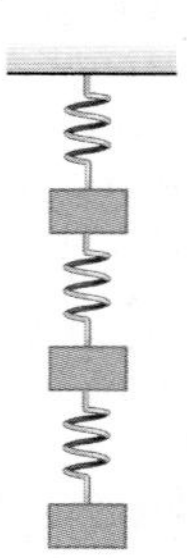

6.33 · Three identical 6.40-kg masses are hung by three identical springs, as shown in Fig. E6.33. Each spring has a force constant of 7.80 kN/m and was 12.0 cm long before any masses were attached to it. (a) Draw a free-body diagram of each mass. (b) How long is each spring when hanging as shown? (*Hint:* First isolate only the bottom mass. Then treat the bottom two masses as a system. Finally, treat all three masses as a system.)

Figure **E6.33**

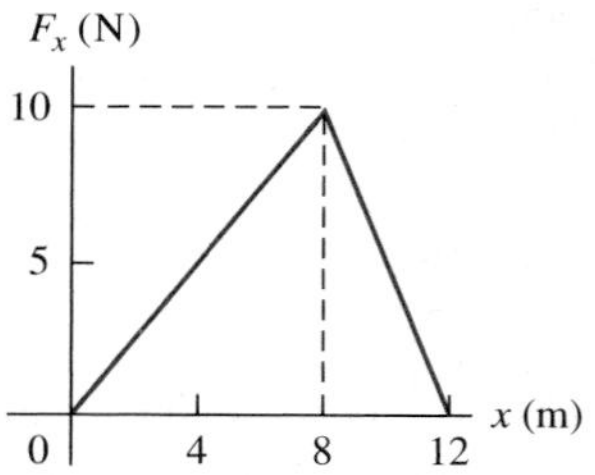

6.34 · A child applies a force $\vec{F}$ parallel to the x-axis to a 10.0-kg sled moving on the frozen surface of a small pond. As the child controls the speed of the sled, the x-component of the force she applies varies with the x-coordinate of the sled as shown in Fig. E6.34. Calculate the work done by the force $\vec{F}$ when the sled moves (a) from $x = 0$ to $x = 8.0$ m; (b) from $x = 8.0$ m to $x = 12.0$ m; (c) from $x = 0$ to 12.0 m.

Figure **E6.34**

6.35 ·· Suppose the sled in Exercise 6.34 is initially at rest at $x = 0$. Use the work–energy theorem to find the speed of the sled at (a) $x = 8.0$ m and (b) $x = 12.0$ m. You can ignore friction between the sled and the surface of the pond.

6.36 · A 2.0-kg box and a 3.0-kg box on a perfectly smooth horizontal floor have a spring of force constant 250 N/m compressed between them. If the initial compression of the spring is 6.0 cm, find the acceleration of each box the instant after they are released. Be sure to include free-body diagrams of each box as part of your solution.

6.37 ·· A 6.0-kg box moving at 3.0 m/s on a horizontal, frictionless surface runs into a light spring of force constant 75 N/cm. Use the work–energy theorem to find the maximum compression of the spring.

6.38 ·· **Leg Presses.** As part of your daily workout, you lie on your back and push with your feet against a platform attached to two stiff springs arranged side by side so that they are parallel to each other. When you push the platform, you compress the springs. You do 80.0 J of work when you compress the springs 0.200 m from their uncompressed length. (a) What magnitude of force must you apply to hold the platform in this position? (b) How much *additional* work must you do to move the platform 0.200 m *farther,* and what maximum force must you apply?

6.39 ·· (a) In Example 6.7 (Section 6.3) it was calculated that with the air track turned off, the glider travels 8.6 cm before it stops instantaneously. How large would the coefficient of static friction μ_s have to be to keep the glider from springing back to the left? (b) If the coefficient of static friction between the glider and the track is $\mu_s = 0.60$, what is the maximum initial speed v_1 that the glider can be given and still remain at rest after it stops instantaneously? With the air track turned off, the coefficient of kinetic friction is $\mu_k = 0.47$.

6.40 • A 4.00-kg block of ice is placed against a horizontal spring that has force constant $k = 200$ N/m and is compressed 0.025 m. The spring is released and accelerates the block along a horizontal surface. You can ignore friction and the mass of the spring. (a) Calculate the work done on the block by the spring during the motion of the block from its initial position to where the spring has returned to its uncompressed length. (b) What is the speed of the block after it leaves the spring?

6.41 • A force $\vec{F}$ is applied to a 2.0-kg radio-controlled model car parallel to the x-axis as it moves along a straight track. The x-component of the force varies with the x-coordinate of the car as shown in Fig. E6.41. Calculate the work done by the force $\vec{F}$ when the car moves from (a) $x = 0$ to $x = 3.0$ m; (b) $x = 3.0$ m to $x = 4.0$ m; (c) $x = 4.0$ m to $x = 7.0$ m; (d) $x = 0$ to $x = 7.0$ m; (e) $x = 7.0$ m to $x = 2.0$ m.

Figure **E6.41**

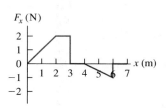

6.42 • Suppose the 2.0-kg model car in Exercise 6.41 is initially at rest at $x = 0$ and $\vec{F}$ is the net force acting on it. Use the work–energy theorem to find the speed of the car at (a) $x = 3.0$ m; (b) $x = 4.0$ m; (c) $x = 7.0$ m.

6.43 •• At a waterpark, sleds with riders are sent along a slippery, horizontal surface by the release of a large compressed spring. The spring with force constant $k = 40.0$ N/cm and negligible mass rests on the frictionless horizontal surface. One end is in contact with a stationary wall. A sled and rider with total mass 70.0 kg are pushed against the other end, compressing the spring 0.375 m. The sled is then released with zero initial velocity. What is the sled's speed when the spring (a) returns to its uncompressed length and (b) is still compressed 0.200 m?

6.44 • **Half of a Spring.** (a) Suppose you cut a massless ideal spring in half. If the full spring had a force constant k, what is the force constant of each half, in terms of k? (*Hint:* Think of the original spring as two equal halves, each producing the same force as the entire spring. Do you see why the forces must be equal?) (b) If you cut the spring into three equal segments instead, what is the force constant of each one, in terms of k?

6.45 •• A small glider is placed against a compressed spring at the bottom of an air track that slopes upward at an angle of 40.0° above the horizontal. The glider has mass 0.0900 kg. The spring has $k = 640$ N/m and negligible mass. When the spring is released, the glider travels a maximum distance of 1.80 m along the air track before sliding back down. Before reaching this maximum distance, the glider loses contact with the spring. (a) What distance was the spring originally compressed? (b) When the glider has traveled along the air track 0.80 m from its initial position against the compressed spring, is it still in contact with the spring? What is the kinetic energy of the glider at this point?

6.46 •• An ingenious bricklayer builds a device for shooting bricks up to the top of the wall where he is working. He places a brick on a vertical compressed spring with force constant $k = 450$ N/m and negligible mass. When the spring is released, the brick is propelled upward. If the brick has mass 1.80 kg and is to reach a maximum height of 3.6 m above its initial position on the compressed spring, what distance must the bricklayer compress the spring initially? (The brick loses contact with the spring when the spring returns to its uncompressed length. Why?)

6.47 •• **CALC** A force in the $+x$-direction with magnitude $F(x) = 18.0$ N $- (0.530$ N/m$)x$ is applied to a 6.00-kg box that is sitting on the horizontal, frictionless surface of a frozen lake. $F(x)$ is the only horizontal force on the box. If the box is initially at rest at $x = 0$, what is its speed after it has traveled 14.0 m?

Section 6.4 Power

6.48 •• A crate on a motorized cart starts from rest and moves with a constant eastward acceleration of $a = 2.80$ m/s². A worker assists the cart by pushing on the crate with a force that is eastward and has magnitude that depends on time according to $F(t) = (5.40$ N/s$)t$. What is the instantaneous power supplied by this force at $t = 5.00$ s?

6.49 • How many joules of energy does a 100-watt light bulb use per hour? How fast would a 70-kg person have to run to have that amount of kinetic energy?

6.50 •• **BIO** **Should You Walk or Run?** It is 5.0 km from your home to the physics lab. As part of your physical fitness program, you could run that distance at 10 km/h (which uses up energy at the rate of 700 W), or you could walk it leisurely at 3.0 km/h (which uses energy at 290 W). Which choice would burn up more energy, and how much energy (in joules) would it burn? Why is it that the more intense exercise actually burns up less energy than the less intense exercise?

6.51 •• **Magnetar.** On December 27, 2004, astronomers observed the greatest flash of light ever recorded from outside the solar system. It came from the highly magnetic neutron star SGR 1806-20 (a *magnetar*). During 0.20 s, this star released as much energy as our sun does in 250,000 years. If P is the average power output of our sun, what was the average power output (in terms of P) of this magnetar?

6.52 •• A 20.0-kg rock is sliding on a rough, horizontal surface at 8.00 m/s and eventually stops due to friction. The coefficient of kinetic friction between the rock and the surface is 0.200. What average power is produced by friction as the rock stops?

6.53 • A tandem (two-person) bicycle team must overcome a force of 165 N to maintain a speed of 9.00 m/s. Find the power required per rider, assuming that each contributes equally. Express your answer in watts and in horsepower.

6.54 •• When its 75-kW (100-hp) engine is generating full power, a small single-engine airplane with mass 700 kg gains altitude at a rate of 2.5 m/s (150 m/min, or 500 ft/min). What fraction of the engine power is being used to make the airplane climb? (The remainder is used to overcome the effects of air resistance and of inefficiencies in the propeller and engine.)

6.55 •• **Working Like a Horse.** Your job is to lift 30-kg crates a vertical distance of 0.90 m from the ground onto the bed of a truck. (a) How many crates would you have to load onto the truck in 1 minute for the average power output you use to lift the crates to equal 0.50 hp? (b) How many crates for an average power output of 100 W?

6.56 •• An elevator has mass 600 kg, not including passengers. The elevator is designed to ascend, at constant speed, a vertical distance of 20.0 m (five floors) in 16.0 s, and it is driven by a motor that can provide up to 40 hp to the elevator. What is the maximum number of passengers that can ride in the elevator? Assume that an average passenger has mass 65.0 kg.

6.57 •• A ski tow operates on a 15.0° slope of length 300 m. The rope moves at 12.0 km/h and provides power for 50 riders at one time, with an average mass per rider of 70.0 kg. Estimate the power required to operate the tow.

6.58 •• The aircraft carrier *John F. Kennedy* has mass 7.4×10^7 kg. When its engines are developing their full power of 280,000 hp, the *John F. Kennedy* travels at its top speed of 35 knots (65 km/h). If 70% of the power output of the engines is applied to pushing the ship through the water, what is the magnitude of the force of water resistance that opposes the carrier's motion at this speed?

6.59 • **BIO** A typical flying insect applies an average force equal to twice its weight during each downward stroke while hovering. Take the mass of the insect to be 10 g, and assume the wings move an average downward distance of 1.0 cm during each stroke. Assuming 100 downward strokes per second, estimate the average power output of the insect.

PROBLEMS

6.60 ••• **CALC** A balky cow is leaving the barn as you try harder and harder to push her back in. In coordinates with the origin at the barn door, the cow walks from $x = 0$ to $x = 6.9$ m as you apply a force with x-component $F_x = -[20.0 \text{ N} + (3.0 \text{ N/m})x]$. How much work does the force you apply do on the cow during this displacement?

6.61 •• **CALC Rotating Bar.** A thin, uniform 12.0-kg bar that is 2.00 m long rotates uniformly about a pivot at one end, making 5.00 complete revolutions every 3.00 seconds. What is the kinetic energy of this bar? (*Hint:* Different points in the bar have different speeds. Break the bar up into infinitesimal segments of mass dm and integrate to add up the kinetic energies of all these segments.)

6.62 •• **A Near-Earth Asteroid.** On April 13, 2029 (Friday the 13th!), the asteroid 99942 Apophis will pass within 18,600 mi of the earth—about $\frac{1}{13}$ the distance to the moon! It has a density of 2600 kg/m^3, can be modeled as a sphere 320 m in diameter, and will be traveling at 12.6 km/s. (a) If, due to a small disturbance in its orbit, the asteroid were to hit the earth, how much kinetic energy would it deliver? (b) The largest nuclear bomb ever tested by the United States was the "Castle/Bravo" bomb, having a yield of 15 megatons of TNT. (A megaton of TNT releases 4.184×10^{15} J of energy.) How many Castle/Bravo bombs would be equivalent to the energy of Apophis?

6.63 • A luggage handler pulls a 20.0-kg suitcase up a ramp inclined at 25.0° above the horizontal by a force $\vec{F}$ of magnitude 140 N that acts parallel to the ramp. The coefficient of kinetic friction between the ramp and the incline is $\mu_k = 0.300$. If the suitcase travels 3.80 m along the ramp, calculate (a) the work done on the suitcase by the force $\vec{F}$; (b) the work done on the suitcase by the gravitational force; (c) the work done on the suitcase by the normal force; (d) the work done on the suitcase by the friction force; (e) the total work done on the suitcase. (f) If the speed of the suitcase is zero at the bottom of the ramp, what is its speed after it has traveled 3.80 m along the ramp?

6.64 • **BIO Chin-Ups.** While doing a chin-up, a man lifts his body 0.40 m. (a) How much work must the man do per kilogram of body mass? (b) The muscles involved in doing a chin-up can generate about 70 J of work per kilogram of muscle mass. If the man can

just barely do a 0.40-m chin-up, what percentage of his body's mass do these muscles constitute? (For comparison, the *total* percentage of muscle in a typical 70-kg man with 14% body fat is about 43%.) (c) Repeat part (b) for the man's young son, who has arms half as long as his father's but whose muscles can also generate 70 J of work per kilogram of muscle mass. (d) Adults and children have about the same percentage of muscle in their bodies. Explain why children can commonly do chin-ups more easily than their fathers.

6.65 ••• **CP** A 20.0-kg crate sits at rest at the bottom of a 15.0-m-long ramp that is inclined at 34.0° above the horizontal. A constant horizontal force of 290 N is applied to the crate to push it up the ramp. While the crate is moving, the ramp exerts a constant frictional force on it that has magnitude 65.0 N. (a) What is the total work done on the crate during its motion from the bottom to the top of the ramp? (b) How much time does it take the crate to travel to the top of the ramp?

6.66 ••• Consider the blocks in Exercise 6.7 as they move 75.0 cm. Find the total work done on each one (a) if there is no friction between the table and the 20.0-N block, and (b) if $\mu_s = 0.500$ and $\mu_k = 0.325$ between the table and the 20.0-N block.

6.67 • The space shuttle, with mass 86,400 kg, is in a circular orbit of radius 6.66×10^6 m around the earth. It takes 90.1 min for the shuttle to complete each orbit. On a repair mission, the shuttle is cautiously moving 1.00 m closer to a disabled satellite every 3.00 s. Calculate the shuttle's kinetic energy (a) relative to the earth and (b) relative to the satellite.

6.68 •• A 5.00-kg package slides 1.50 m down a long ramp that is inclined at 24.0° below the horizontal. The coefficient of kinetic friction between the package and the ramp is $\mu_k = 0.310$. Calculate (a) the work done on the package by friction; (b) the work done on the package by gravity; (c) the work done on the package by the normal force; (d) the total work done on the package. (e) If the package has a speed of 2.20 m/s at the top of the ramp, what is its speed after sliding 1.50 m down the ramp?

6.69 •• **CP BIO Whiplash Injuries.** When a car is hit from behind, its passengers undergo sudden forward acceleration, which can cause a severe neck injury known as *whiplash*. During normal acceleration, the neck muscles play a large role in accelerating the head so that the bones are not injured. But during a very sudden acceleration, the muscles do not react immediately because they are flexible, so most of the accelerating force is provided by the neck bones. Experimental tests have shown that these bones will fracture if they absorb more than 8.0 J of energy. (a) If a car waiting at a stoplight is rear-ended in a collision that lasts for 10.0 ms, what is the greatest speed this car and its driver can reach without breaking neck bones if the driver's head has a mass of 5.0 kg (which is about right for a 70-kg person)? Express your answer in m/s and in mph. (b) What is the acceleration of the passengers during the collision in part (a), and how large a force is acting to accelerate their heads? Express the acceleration in m/s^2 and in g's.

6.70 •• **CALC** A net force along the x-axis that has x-component $F_x = -12.0 \text{ N} + (0.300 \text{ N/m}^2)x^2$ is applied to a 5.00-kg object that is initially at the origin and moving in the $-x$-direction with a speed of 6.00 m/s. What is the speed of the object when it reaches the point $x = 5.00$ m?

6.71 • **CALC** An object is attracted toward the origin with a force given by $F_x = -k/x^2$. (Gravitational and electrical forces have this distance dependence.) (a) Calculate the work done by the force F_x when the object moves in the x-direction from x_1 to x_2. If $x_2 > x_1$, is the work done by F_x positive or negative? (b) The only other force acting on the object is a force that you exert with your hand to move the object slowly from x_1 to x_2. How much work do you do? If $x_2 > x_1$, is the work you do positive or negative? (c) Explain the similarities and differences between your answers to parts (a) and (b).

6.72 ••• **CALC** The gravitational pull of the earth on an object is inversely proportional to the square of the distance of the object from the center of the earth. At the earth's surface this force is equal to the object's normal weight mg, where $g = 9.8$ m/s^2, and at large distances, the force is zero. If a 20,000-kg asteroid falls to earth from a very great distance away, what will be its minimum speed as it strikes the earth's surface, and how much kinetic energy will it impart to our planet? You can ignore the effects of the earth's atmosphere.

6.73 • **CALC** **Varying Coefficient of Friction.** A box is sliding with a speed of 4.50 m/s on a horizontal surface when, at point P, it encounters a rough section. On the rough section, the coefficient of friction is not constant, but starts at 0.100 at P and increases linearly with distance past P, reaching a value of 0.600 at 12.5 m past point P. (a) Use the work–energy theorem to find how far this box slides before stopping. (b) What is the coefficient of friction at the stopping point? (c) How far would the box have slid if the friction coefficient didn't increase but instead had the constant value of 0.100?

6.74 •• **CALC** Consider a spring that does not obey Hooke's law very faithfully. One end of the spring is fixed. To keep the spring stretched or compressed an amount x, a force along the x-axis with x-component $F_x = kx - bx^2 + cx^3$ must be applied to the free end. Here $k = 100$ N/m, $b = 700$ N/m^2, and $c = 12{,}000$ N/m^3. Note that $x > 0$ when the spring is stretched and $x < 0$ when it is compressed. (a) How much work must be done to stretch this spring by 0.050 m from its unstretched length? (b) How much work must be done to *compress* this spring by 0.050 m from its unstretched length? (c) Is it easier to stretch or compress this spring? Explain why in terms of the dependence of F_x on x. (Many real springs behave qualitatively in the same way.)

6.75 •• **CP** A small block with a mass of 0.0900 kg is attached to a cord passing through a hole in a frictionless, horizontal surface (Fig. P6.75). The block is originally revolving at a distance of 0.40 m from the hole with a speed of 0.70 m/s. The cord is then pulled from below, shortening the radius of the circle in which the block revolves to 0.10 m. At this new distance, the speed of the block is observed to be 2.80 m/s. (a) What is the tension in the cord in the original situation when the block has speed $v = 0.70$ m/s? (b) What is the tension in the cord in the final situation when the block has speed $v = 2.80$ m/s? (c) How much work was done by the person who pulled on the cord?

Figure **P6.75**

6.76 •• **CALC** **Proton Bombardment.** A proton with mass 1.67×10^{-27} kg is propelled at an initial speed of 3.00×10^5 m/s directly toward a uranium nucleus 5.00 m away. The proton is repelled by the uranium nucleus with a force of magnitude $F = \alpha/x^2$, where x is the separation between the two objects and $\alpha = 2.12 \times 10^{-26}$ N·m^2. Assume that the uranium nucleus remains at rest. (a) What is the speed of the proton when it is 8.00×10^{-10} m from the uranium nucleus? (b) As the proton approaches the uranium nucleus, the repulsive force slows down the proton until it comes momentarily to rest, after which the proton moves away from the uranium nucleus. How close to the uranium nucleus does the proton get? (c) What is the speed of the proton when it is again 5.00 m away from the uranium nucleus?

6.77 •• **CP CALC** A block of ice with mass 4.00 kg is initially at rest on a frictionless, horizontal surface. A worker then applies a horizontal force $\vec{F}$ to it. As a result, the block moves along the x-axis such that its position as a function of time is given by $x(t) = \alpha t^2 + \beta t^3$, where $\alpha = 0.200$ m/s^2 and $\beta = 0.0200$ m/s^3. (a) Calculate the velocity of the object when $t = 4.00$ s. (b) Calculate the magnitude of $\vec{F}$ when $t = 4.00$ s. (c) Calculate the work done by the force $\vec{F}$ during the first 4.00 s of the motion.

6.78 •• You and your bicycle have combined mass 80.0 kg. When you reach the base of a bridge, you are traveling along the road at 5.00 m/s (Fig. P6.78). At the top of the bridge, you have climbed a vertical distance of 5.20 m and have slowed to 1.50 m/s. You can ignore work done by friction and any inefficiency in the bike or your legs. (a) What is the total work done on you and your bicycle when you go from the base to the top of the bridge? (b) How much work have you done with the force you apply to the pedals?

Figure **P6.78**

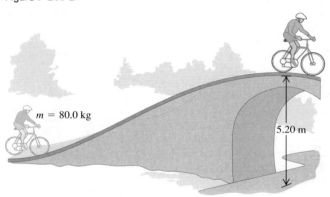

$m = 80.0$ kg

5.20 m

6.79 •• You are asked to design spring bumpers for the walls of a parking garage. A freely rolling 1200-kg car moving at 0.65 m/s is to compress the spring no more than 0.090 m before stopping. What should be the force constant of the spring? Assume that the spring has negligible mass.

6.80 •• The spring of a spring gun has force constant $k = 400$ N/m and negligible mass. The spring is compressed 6.00 cm, and a ball with mass 0.0300 kg is placed in the horizontal barrel against the compressed spring. The spring is then released, and the ball is propelled out the barrel of the gun. The barrel is 6.00 cm long, so the ball leaves the barrel at the same point that it loses contact with the spring. The gun is held so the barrel is horizontal. (a) Calculate the speed with which the ball leaves the barrel if you can ignore friction. (b) Calculate the speed of the ball as it leaves the barrel if a constant resisting force of 6.00 N acts on the ball as it moves along the barrel. (c) For the situation in part (b), at what position along the barrel does the ball have the greatest speed, and what is that speed? (In this case, the maximum speed does not occur at the end of the barrel.)

6.81 ••• A 2.50-kg textbook is forced against a horizontal spring of negligible mass and force constant 250 N/m, compressing the spring a distance of 0.250 m. When released, the textbook slides on a horizontal tabletop with coefficient of kinetic friction

$\mu_k = 0.30$. Use the work–energy theorem to find how far the textbook moves from its initial position before coming to rest.

6.82 ••• **Pushing a Cat.** Your cat "Ms." (mass 7.00 kg) is trying to make it to the top of a frictionless ramp 2.00 m long and inclined upward at 30.0° above the horizontal. Since the poor cat can't get any traction on the ramp, you push her up the entire length of the ramp by exerting a constant 100-N force parallel to the ramp. If Ms. takes a running start so that she is moving at 2.40 m/s at the bottom of the ramp, what is her speed when she reaches the top of the incline? Use the work–energy theorem.

6.83 •• **Crash Barrier.** A student proposes a design for an automobile crash barrier in which a 1700-kg sport utility vehicle moving at 20.0 m/s crashes into a spring of negligible mass that slows it to a stop. So that the passengers are not injured, the acceleration of the vehicle as it slows can be no greater than 5.00g. (a) Find the required spring constant k, and find the distance the spring will compress in slowing the vehicle to a stop. In your calculation, disregard any deformation or crumpling of the vehicle and the friction between the vehicle and the ground. (b) What disadvantages are there to this design?

6.84 ••• A physics professor is pushed up a ramp inclined upward at 30.0° above the horizontal as he sits in his desk chair that slides on frictionless rollers. The combined mass of the professor and chair is 85.0 kg. He is pushed 2.50 m along the incline by a group of students who together exert a constant horizontal force of 600 N. The professor's speed at the bottom of the ramp is 2.00 m/s. Use the work–energy theorem to find his speed at the top of the ramp.

6.85 • A 5.00-kg block is moving at $v_0 = 6.00$ m/s along a frictionless, horizontal surface toward a spring with force constant $k = 500$ N/m that is attached to a wall (Fig. P6.85). The spring has negligible mass. (a) Find the maximum distance the spring will be compressed. (b) If the spring is to compress by no more than 0.150 m, what should be the maximum value of v_0?

Figure **P6.85**

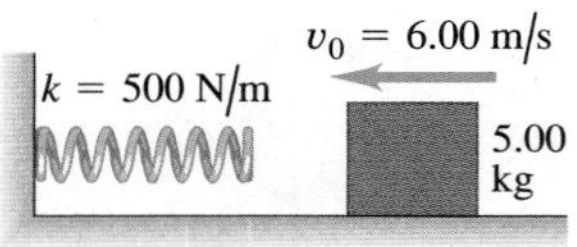

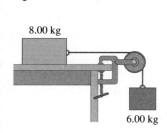

$k = 500$ N/m

$v_0 = 6.00$ m/s

5.00 kg

6.86 •• Consider the system shown in Fig. P6.86. The rope and pulley have negligible mass, and the pulley is frictionless. The coefficient of kinetic friction between the 8.00-kg block and the tabletop is $\mu_k = 0.250$. The blocks are released from rest. Use energy methods to calculate the speed of the 6.00-kg block after it has descended 1.50 m.

Figure **P6.86**

8.00 kg

6.00 kg

6.87 •• Consider the system shown in Fig. P6.86. The rope and pulley have negligible mass, and the pulley is frictionless. Initially the 6.00-kg block is moving downward and the 8.00-kg block is moving to the right, both with a speed of 0.900 m/s. The blocks come to rest after moving 2.00 m. Use the work–energy theorem to calculate the coefficient of kinetic friction between the 8.00-kg block and the tabletop.

6.88 ••• **CALC** **Bow and Arrow.** Figure P6.88 shows how the force exerted by the string of a compound bow on an arrow varies as a function of how far back the arrow is pulled (the

draw length). Assume that the same force is exerted on the arrow as it moves forward after being released. Full draw for this bow is at a draw length of 75.0 cm. If the bow shoots a 0.0250-kg arrow from full draw, what is the speed of the arrow as it leaves the bow?

Figure **P6.88**

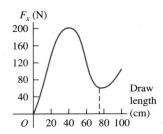

6.89 •• On an essentially frictionless, horizontal ice rink, a skater moving at 3.0 m/s encounters a rough patch that reduces her speed to 1.65 m/s due to a friction force that is 25% of her weight. Use the work–energy theorem to find the length of this rough patch.

6.90 • **Rescue.** Your friend (mass 65.0 kg) is standing on the ice in the middle of a frozen pond. There is very little friction between her feet and the ice, so she is unable to walk. Fortunately, a light rope is tied around her waist and you stand on the bank holding the other end. You pull on the rope for 3.00 s and accelerate your friend from rest to a speed of 6.00 m/s while you remain at rest. What is the average power supplied by the force you applied?

6.91 •• A pump is required to lift 800 kg of water (about 210 gallons) per minute from a well 14.0 m deep and eject it with a speed of 18.0 m/s. (a) How much work is done per minute in lifting the water? (b) How much work is done in giving the water the kinetic energy it has when ejected? (c) What must be the power output of the pump?

6.92 •• **BIO** All birds, independent of their size, must maintain a power output of 10–25 watts per kilogram of body mass in order to fly by flapping their wings. (a) The Andean giant hummingbird *(Patagona gigas)* has mass 70 g and flaps its wings 10 times per second while hovering. Estimate the amount of work done by such a hummingbird in each wingbeat. (b) A 70-kg athlete can maintain a power output of 1.4 kW for no more than a few seconds; the *steady* power output of a typical athlete is only 500 W or so. Is it possible for a human-powered aircraft to fly for extended periods by flapping its wings? Explain.

6.93 ••• A physics student spends part of her day walking between classes or for recreation, during which time she expends energy at an average rate of 280 W. The remainder of the day she is sitting in class, studying, or resting; during these activities, she expends energy at an average rate of 100 W. If she expends a total of 1.1×10^7 J of energy in a 24-hour day, how much of the day did she spend walking?

6.94 ••• The Grand Coulee Dam is 1270 m long and 170 m high. The electrical power output from generators at its base is approximately 2000 MW. How many cubic meters of water must flow from the top of the dam per second to produce this amount of power if 92% of the work done on the water by gravity is converted to electrical energy? (Each cubic meter of water has a mass of 1000 kg.)

6.95 · **BIO** **Power of the Human Heart.** The human heart is a powerful and extremely reliable pump. Each day it takes in and discharges about 7500 L of blood. Assume that the work done by the heart is equal to the work required to lift this amount of blood a height equal to that of the average American woman (1.63 m). The density (mass per unit volume) of blood is 1.05×10^3 kg/m^3. (a) How much work does the heart do in a day? (b) What is the heart's power output in watts?

6.96 ··· Six diesel units in series can provide 13.4 MW of power to the lead car of a freight train. The diesel units have total mass 1.10×10^6 kg. The average car in the train has mass 8.2×10^4 kg and requires a horizontal pull of 2.8 kN to move at a constant 27 m/s on level tracks. (a) How many cars can be in the train under these conditions? (b) This would leave no power for accelerating or climbing hills. Show that the extra force needed to accelerate the train is about the same for a 0.10-m/s^2 acceleration or a 1.0% slope (slope angle $\alpha = $ arctan 0.010). (c) With the 1.0% slope, show that an extra 2.9 MW of power is needed to maintain the 27-m/s speed of the diesel units. (d) With 2.9 MW less power available, how many cars can the six diesel units pull up a 1.0% slope at a constant 27 m/s?

6.97 · It takes a force of 53 kN on the lead car of a 16-car passenger train with mass 9.1×10^5 kg to pull it at a constant 45 m/s (101 mi/h) on level tracks. (a) What power must the locomotive provide to the lead car? (b) How much more power to the lead car than calculated in part (a) would be needed to give the train an acceleration of 1.5 m/s^2, at the instant that the train has a speed of 45 m/s on level tracks? (c) How much more power to the lead car than that calculated in part (a) would be needed to move the train up a 1.5% grade (slope angle $\alpha = $ arctan 0.015) at a constant 45 m/s?

6.98 · **CALC** An object has several forces acting on it. One of these forces is $\vec{F} = \alpha x y \hat{\imath}$, a force in the x-direction whose magnitude depends on the position of the object, with $\alpha = 2.50$ N/m^2. Calculate the work done on the object by this force for the following displacements of the object: (a) The object starts at the point $x = 0$, $y = 3.00$ m and moves parallel to the x-axis to the point $x = 2.00$ m, $y = 3.00$ m. (b) The object starts at the point $x = 2.00$ m, $y = 0$ and moves in the y-direction to the point $x = 2.00$ m, $y = 3.00$ m. (c) The object starts at the origin and moves on the line $y = 1.5x$ to the point $x = 2.00$ m, $y = 3.00$ m.

6.99 ·· **Cycling.** For a touring bicyclist the drag coefficient $C (f_{air} = \frac{1}{2} C A \rho v^2)$ is 1.00, the frontal area A is 0.463 m^2, and the coefficient of rolling friction is 0.0045. The rider has mass 50.0 kg, and her bike has mass 12.0 kg. (a) To maintain a speed of 12.0 m/s (about 27 mi/h) on a level road, what must the rider's power output to the rear wheel be? (b) For racing, the same rider uses a different bike with coefficient of rolling friction 0.0030 and mass 9.00 kg. She also crouches down, reducing her drag coefficient to 0.88 and reducing her frontal area to 0.366 m^2. What must her power output to the rear wheel be then to maintain a speed of 12.0 m/s? (c) For the situation in part (b), what power output is required to maintain a speed of 6.0 m/s? Note the great drop in power requirement when the speed is only halved. (For more on aerodynamic speed limitations for a wide variety of human-powered vehicles, see "The Aerodynamics of Human-Powered Land Vehicles," *Scientific American*, December 1983.)

6.100 ·· **Automotive Power I.** A truck engine transmits 28.0 kW (37.5 hp) to the driving wheels when the truck is traveling at a constant velocity of magnitude 60.0 km/h (37.3 mi/h) on a level road. (a) What is the resisting force acting on the truck? (b) Assume that 65% of the resisting force is due to rolling friction and the remainder is due to air resistance. If the force of rolling friction is independent of speed, and the force of air resistance is proportional to the square of the speed, what power will drive the truck at 30.0 km/h? At 120.0 km/h? Give your answers in kilowatts and in horsepower.

6.101 ·· **Automotive Power II.** (a) If 8.00 hp are required to drive a 1800-kg automobile at 60.0 km/h on a level road, what is the total retarding force due to friction, air resistance, and so on? (b) What power is necessary to drive the car at 60.0 km/h up a 10.0% grade (a hill rising 10.0 m vertically in 100.0 m horizontally)? (c) What power is necessary to drive the car at 60.0 km/h *down* a 1.00% grade? (d) Down what percent grade would the car coast at 60.0 km/h?

CHALLENGE PROBLEMS

6.102 ··· **CALC** On a winter day in Maine, a warehouse worker is shoving boxes up a rough plank inclined at an angle α above the horizontal. The plank is partially covered with ice, with more ice near the bottom of the plank than near the top, so that the coefficient of friction increases with the distance x along the plank: $\mu = Ax$, where A is a positive constant and the bottom of the plank is at $x = 0$. (For this plank the coefficients of kinetic and static friction are equal: $\mu_k = \mu_s = \mu$.) The worker shoves a box up the plank so that it leaves the bottom of the plank moving at speed v_0. Show that when the box first comes to rest, it will remain at rest if

$$v_0{}^2 \geq \frac{3g \sin^2 \alpha}{A \cos \alpha}$$

6.103 ··· **CALC** **A Spring with Mass.** We usually ignore the kinetic energy of the moving coils of a spring, but let's try to get a reasonable approximation to this. Consider a spring of mass M, equilibrium length L_0, and spring constant k. The work done to stretch or compress the spring by a distance L is $\frac{1}{2} k X^2$, where $X = L - L_0$. Consider a spring, as described above, that has one end fixed and the other end moving with speed v. Assume that the speed of points along the length of the spring varies linearly with distance l from the fixed end. Assume also that the mass M of the spring is distributed uniformly along the length of the spring. (a) Calculate the kinetic energy of the spring in terms of M and v. (*Hint:* Divide the spring into pieces of length dl; find the speed of each piece in terms of l, v, and L; find the mass of each piece in terms of dl, M, and L; and integrate from 0 to L. The result is *not* $\frac{1}{2} M v^2$, since not all of the spring moves with the same speed.) In a spring gun, a spring of mass 0.243 kg and force constant 3200 N/m is compressed 2.50 cm from its unstretched length. When the trigger is pulled, the spring pushes horizontally on a 0.053-kg ball. The work done by friction is negligible. Calculate the ball's speed when the spring reaches its uncompressed length (b) ignoring the mass of the spring and (c) including, using the results of part (a), the mass of the spring. (d) In part (c), what is the final kinetic energy of the ball and of the spring?

6.104 ··· **CALC** An airplane in flight is subject to an air resistance force proportional to the square of its speed v. But there is an additional resistive force because the airplane has wings. Air flowing over the wings is pushed down and slightly forward, so from Newton's third law the air exerts a force on the wings and airplane that is up and slightly backward (Fig. P6.104). The upward force is the lift force that keeps the airplane aloft, and the backward force is called *induced drag*. At flying speeds, induced drag is inversely proportional to v^2, so that the total air resistance force can be expressed by $F_{air} = \alpha v^2 + \beta/v^2$, where α and β are positive constants that depend on the shape and size of the airplane and the density of the air. For a Cessna 150, a small single-engine airplane, $\alpha = 0.30$ N·s²/m² and $\beta = 3.5 \times 10^5$ N · m²/s². In steady flight, the engine must provide a forward force that exactly balances the air resistance force. (a) Calculate the speed (in km/h) at which this airplane will have the maximum *range* (that is, travel the greatest distance) for a given quantity of fuel. (b) Calculate the speed (in km/h) for which the airplane will have the maximum *endurance* (that is, remain in the air the longest time).

Figure **P6.104**

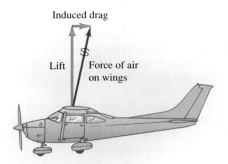

Induced drag

Lift | Force of air on wings

Answers

Chapter Opening Question

The answer is yes. As the ant was exerting an upward force on the piece of cereal, the cereal was exerting a downward force of the same magnitude on the ant (due to Newton's third law). However, because the ant's body had an upward displacement, the work that the cereal did on the ant was *negative* (see Section 6.1).

Test Your Understanding Questions

6.1 Answer: (iii) The electron has constant velocity, so its acceleration is zero and (by Newton's second law) the net force on the electron is also zero. Therefore the total work done by all the forces (equal to the work done by the net force) must be zero as well. The individual forces may do nonzero work, but that's not what the question asks.

6.2 Answer: (iv), (i), (iii), (ii) Body (i) has kinetic energy $K = \frac{1}{2}mv^2 = \frac{1}{2}(2.0 \text{ kg})(5.0 \text{ m/s})^2 = 25$ J. Body (ii) had zero kinetic energy initially and then had 30 J of work done it, so its final kinetic energy is $K_2 = K_1 + W = 0 + 30 \text{ J} = 30$ J. Body (iii) had initial kinetic energy $K_1 = \frac{1}{2}mv_1^2 = \frac{1}{2}(1.0 \text{ kg})(4.0 \text{ m/s})^2 = 8.0$ J and then had 20 J of work done on it, so its final kinetic energy is $K_2 = K_1 + W = 8.0 \text{ J} + 20 \text{ J} = 28$ J. Body (iv) had initial kinetic energy $K_1 = \frac{1}{2}mv_1^2 = \frac{1}{2}(2.0 \text{ kg})(10 \text{ m/s})^2 = 100$ J; when it did 80 J of work on another body, the other body did -80 J of work on body (iv), so the final kinetic energy of body (iv) is $K_2 = K_1 + W = 100 \text{ J} + (-80 \text{ J}) = 20$ J.

6.3 Answers: (a) (iii), (b) (iii) At any point during the pendulum bob's motion, the tension force and the weight both act perpendicular to the motion—that is, perpendicular to an infinitesimal displacement $d\vec{l}$ of the bob. (In Fig. 5.32b, the displacement $d\vec{l}$ would be directed outward from the plane of the free-body diagram.) Hence for either force the scalar product inside the integral in Eq. (6.14) is $\vec{F} \cdot d\vec{l} = 0$, and the work done along any part of the circular path (including a complete circle) is $W = \int \vec{F} \cdot d\vec{l} = 0$.

6.4 Answer: (v) The airliner has a constant horizontal velocity, so the net horizontal force on it must be zero. Hence the backward drag force must have the same magnitude as the forward force due to the combined thrust of the four engines. This means that the drag force must do *negative* work on the airplane at the same rate that the combined thrust force does *positive* work. The combined thrust does work at a rate of $4(108,000 \text{ hp}) = 432,000$ hp, so the drag force must do work at a rate of $-432,000$ hp.

Bridging Problem

Answers: **(a)** $v_1 = \sqrt{\dfrac{2}{m}\left(mgx_1 - \dfrac{1}{3}\alpha x_1^3\right)} = \sqrt{2gx_1 - \dfrac{2\alpha x_1^3}{3m}}$

(b) $P = -F_{\text{spring}-1}v_1 = -\alpha x_1^2 \sqrt{2gx_1 - \dfrac{2\alpha x_1^3}{3m}}$

(c) $x_2 = \sqrt{\dfrac{3mg}{\alpha}}$ **(d)** No

POTENTIAL ENERGY AND ENERGY CONSERVATION

? As this mallard glides in to a landing, it descends along a straight-line path at a constant speed. Does the mallard's mechanical energy increase, decrease, or stay the same during the glide? If it increases, where does the added energy come from? If it decreases, where does the lost energy go?

LEARNING GOALS

By studying this chapter, you will learn:

- How to use the concept of gravitational potential energy in problems that involve vertical motion.

- How to use the concept of elastic potential energy in problems that involve a moving body attached to a stretched or compressed spring.

- The distinction between conservative and nonconservative forces, and how to solve problems in which both kinds of forces act on a moving body.

- How to calculate the properties of a conservative force if you know the corresponding potential-energy function.

- How to use energy diagrams to understand the motion of an object moving in a straight line under the influence of a conservative force.

When a diver jumps off a high board into a swimming pool, he hits the water moving pretty fast, with a lot of kinetic energy. Where does that energy come from? The answer we learned in Chapter 6 was that the gravitational force (his weight) does work on the diver as he falls. The diver's kinetic energy—energy associated with his *motion*—increases by an amount equal to the work done.

However, there is a very useful alternative way to think about work and kinetic energy. This new approach is based on the concept of *potential energy*, which is energy associated with the *position* of a system rather than its motion. In this approach, there is *gravitational potential energy* even while the diver is standing on the high board. Energy is not added to the earth–diver system as the diver falls, but rather a storehouse of energy is *transformed* from one form (potential energy) to another (kinetic energy) as he falls. In this chapter we'll see how the work–energy theorem explains this transformation.

If the diver bounces on the end of the board before he jumps, the bent board stores a second kind of potential energy called *elastic potential energy*. We'll discuss elastic potential energy of simple systems such as a stretched or compressed spring. (An important third kind of potential energy is associated with the positions of electrically charged particles relative to each other. We'll encounter this potential energy in Chapter 23.)

We will prove that in some cases the sum of a system's kinetic and potential energy, called the *total mechanical energy* of the system, is constant during the motion of the system. This will lead us to the general statement of the *law of conservation of energy,* one of the most fundamental and far-reaching principles in all of science.

7.1 Gravitational Potential Energy

We learned in Chapter 6 that a particle gains or loses kinetic energy because it interacts with other objects that exert forces on it. During any interaction, the change in a particle's kinetic energy is equal to the total work done on the particle by the forces that act on it.

In many situations it seems as though energy has been stored in a system, to be recovered later. For example, you must do work to lift a heavy stone over your head. It seems reasonable that in hoisting the stone into the air you are storing energy in the system, energy that is later converted into kinetic energy when you let the stone fall.

This example points to the idea of an energy associated with the *position* of bodies in a system. This kind of energy is a measure of the *potential* or *possibility* for work to be done; when a stone is raised into the air, there is a potential for work to be done on it by the gravitational force, but only if the stone is allowed to fall to the ground. For this reason, energy associated with position is called **potential energy.** Our discussion suggests that there is potential energy associated with a body's weight and its height above the ground. We call this *gravitational potential energy* (Fig. 7.1).

We now have *two* ways to describe what happens when a body falls without air resistance. One way is to say that gravitational potential energy decreases and the falling body's kinetic energy increases. The other way, which we learned in Chapter 6, is that a falling body's kinetic energy increases because the force of the earth's gravity (the body's weight) does work on the body. Later in this section we'll use the work–energy theorem to show that these two descriptions are equivalent.

To begin with, however, let's derive the expression for gravitational potential energy. Suppose a body with mass m moves along the (vertical) y-axis, as in Fig. 7.2. The forces acting on it are its weight, with magnitude $w = mg$, and possibly some other forces; we call the vector sum (resultant) of all the other forces $\vec{F}_{\text{other}}$. We'll assume that the body stays close enough to the earth's surface that the weight is constant. (We'll find in Chapter 13 that weight decreases with altitude.) We want to find the work done by the weight when the body moves downward from a height y_1 above the origin to a lower height y_2 (Fig. 7.2a). The weight and displacement are in the same direction, so the work W_{grav} done on the body by its weight is positive;

$$W_{\text{grav}} = Fs = w(y_1 - y_2) = mgy_1 - mgy_2 \qquad (7.1)$$

This expression also gives the correct work when the body moves *upward* and y_2 is greater than y_1 (Fig. 7.2b). In that case the quantity $(y_1 - y_2)$ is negative, and W_{grav} is negative because the weight and displacement are opposite in direction.

Equation (7.1) shows that we can express W_{grav} in terms of the values of the quantity mgy at the beginning and end of the displacement. This quantity, the product of the weight mg and the height y above the origin of coordinates, is called the **gravitational potential energy,** U_{grav}:

$$U_{\text{grav}} = mgy \qquad \text{(gravitational potential energy)} \qquad (7.2)$$

Its initial value is $U_{\text{grav},1} = mgy_1$ and its final value is $U_{\text{grav},2} = mgy_2$. The change in U_{grav} is the final value minus the initial value, or $\Delta U_{\text{grav}} = U_{\text{grav},2} - U_{\text{grav},1}$. We can express the work W_{grav} done by the gravitational force during the displacement from y_1 to y_2 as

$$W_{\text{grav}} = U_{\text{grav},1} - U_{\text{grav},2} = -(U_{\text{grav},2} - U_{\text{grav},1}) = -\Delta U_{\text{grav}} \qquad (7.3)$$

The negative sign in front of ΔU_{grav} is *essential*. When the body moves up, y increases, the work done by the gravitational force is negative, and the gravitational

7.1 As a basketball descends, gravitational potential energy is converted to kinetic energy and the basketball's speed increases.

7.2 When a body moves vertically from an initial height y_1 to a final height y_2, the gravitational force $\vec{w}$ does work and the gravitational potential energy changes.

(a) A body moves downward

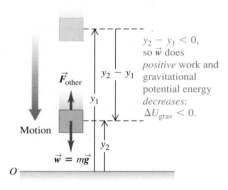

$y_2 - y_1 < 0$, so $\vec{w}$ does *positive* work and gravitational potential energy *decreases:* $\Delta U_{\text{grav}} < 0$.

$\vec{F}_{\text{other}}$

Motion

$\vec{w} = m\vec{g}$

(b) A body moves upward

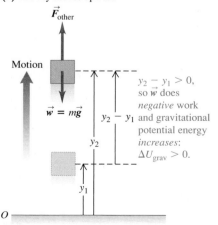

$\vec{F}_{\text{other}}$

Motion

$\vec{w} = m\vec{g}$

$y_2 - y_1 > 0$, so $\vec{w}$ does *negative* work and gravitational potential energy *increases:* $\Delta U_{\text{grav}} > 0$.

potential energy increases ($\Delta U_{\text{grav}} > 0$). When the body moves down, y decreases, the gravitational force does positive work, and the gravitational potential energy decreases ($\Delta U_{\text{grav}} < 0$). It's like drawing money out of the bank (decreasing U_{grav}) and spending it (doing positive work). The unit of potential energy is the joule (J), the same unit as is used for work.

CAUTION **To what body does gravitational potential energy "belong"?** It is *not* correct to call $U_{\text{grav}} = mgy$ the "gravitational potential energy of the body." The reason is that gravitational potential energy U_{grav} is a *shared* property of the body and the earth. The value of U_{grav} increases if the earth stays fixed and the body moves upward, away from the earth; it also increases if the body stays fixed and the earth is moved away from it. Notice that the formula $U_{\text{grav}} = mgy$ involves characteristics of both the body (its mass m) and the earth (the value of g).

Conservation of Mechanical Energy (Gravitational Forces Only)

To see what gravitational potential energy is good for, suppose the body's weight is the *only* force acting on it, so $\vec{F}_{\text{other}} = \mathbf{0}$. The body is then falling freely with no air resistance and can be moving either up or down. Let its speed at point y_1 be v_1 and let its speed at y_2 be v_2. The work–energy theorem, Eq. (6.6), says that the total work done on the body equals the change in the body's kinetic energy: $W_{\text{tot}} = \Delta K = K_2 - K_1$. If gravity is the only force that acts, then from Eq. (7.3), $W_{\text{tot}} = W_{\text{grav}} = -\Delta U_{\text{grav}} = U_{\text{grav},1} - U_{\text{grav},2}$. Putting these together, we get

$$\Delta K = -\Delta U_{\text{grav}} \quad \text{or} \quad K_2 - K_1 = U_{\text{grav},1} - U_{\text{grav},2}$$

which we can rewrite as

$$K_1 + U_{\text{grav},1} = K_2 + U_{\text{grav},2} \quad \text{(if only gravity does work)} \quad (7.4)$$

or

$$\tfrac{1}{2}mv_1^2 + mgy_1 = \tfrac{1}{2}mv_2^2 + mgy_2 \quad \text{(if only gravity does work)} \quad (7.5)$$

Application **Which Egg Has More Mechanical Energy?**
The mechanical energy of each of these identical eggs has the *same* value. The mechanical energy for an egg at rest atop the stone is purely gravitational potential energy. For the falling egg, the gravitational potential energy decreases as the egg descends and the egg's kinetic energy increases. If there is negligible air resistance, the mechanical energy of the falling egg remains constant.

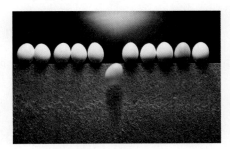

The sum $K + U_{\text{grav}}$ of kinetic and potential energy is called E, the **total mechanical energy of the system.** By "system" we mean the body of mass m and the earth considered together, because gravitational potential energy U is a shared property of both bodies. Then $E_1 = K_1 + U_{\text{grav},1}$ is the total mechanical energy at y_1 and $E_2 = K_2 + U_{\text{grav},2}$ is the total mechanical energy at y_2. Equation (7.4) says that when the body's weight is the only force doing work on it, $E_1 = E_2$. That is, E is constant; it has the same value at y_1 and y_2. But since the positions y_1 and y_2 are arbitrary points in the motion of the body, the total mechanical energy E has the same value at *all* points during the motion:

$$E = K + U_{\text{grav}} = \text{constant} \quad \text{(if only gravity does work)}$$

A quantity that always has the same value is called a *conserved* quantity. *When only the force of gravity does work, the total mechanical energy is constant—that is, it is conserved* (Fig. 7.3). This is our first example of the **conservation of mechanical energy.**

When we throw a ball into the air, its speed decreases on the way up as kinetic energy is converted to potential energy; $\Delta K < 0$ and $\Delta U_{\text{grav}} > 0$. On the way back down, potential energy is converted back to kinetic energy and the ball's speed increases; $\Delta K > 0$ and $\Delta U_{\text{grav}} < 0$. But the *total* mechanical energy (kinetic plus potential) is the same at every point in the motion, provided that no force other than gravity does work on the ball (that is, air resistance must be negligible). It's still true that the gravitational force does work on the body as it

7.3 While this athlete is in midair, only gravity does work on him (if we neglect the minor effects of air resistance). Mechanical energy E—the sum of kinetic and gravitational potential energy—is conserved.

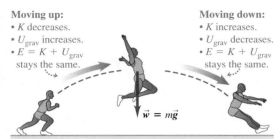

Moving up:
- K decreases.
- U_{grav} increases.
- $E = K + U_{grav}$ stays the same.

Moving down:
- K increases.
- U_{grav} decreases.
- $E = K + U_{grav}$ stays the same.

$\vec{w} = m\vec{g}$

moves up or down, but we no longer have to calculate work directly; keeping track of changes in the value of U_{grav} takes care of this completely.

CAUTION **Choose "zero height" to be wherever you like** When working with gravitational potential energy, we may choose any height to be $y = 0$. If we shift the origin for y, the values of y_1 and y_2 change, as do the values of $U_{grav,1}$ and $U_{grav,2}$. But this shift has no effect on the *difference* in height $y_2 - y_1$ or on the *difference* in gravitational potential energy $U_{grav,2} - U_{grav,1} = mg(y_2 - y_1)$. As the following example shows, the physically significant quantity is not the value of U_{grav} at a particular point, but only the *difference* in U_{grav} between two points. So we can define U_{grav} to be zero at whatever point we choose without affecting the physics.

Example 7.1 | **Height of a baseball from energy conservation**

You throw a 0.145-kg baseball straight up, giving it an initial velocity of magnitude 20.0 m/s. Find how high it goes, ignoring air resistance.

SOLUTION

IDENTIFY and SET UP: After the ball leaves your hand, only gravity does work on it. Hence mechanical energy is conserved, and we can use Eqs. (7.4) and (7.5). We take point 1 to be where the ball leaves your hand and point 2 to be where it reaches its maximum height. As in Fig. 7.2, we take the positive y-direction to be upward. The ball's speed at point 1 is $v_1 = 20.0$ m/s; at its maximum height it is instantaneously at rest, so $v_2 = 0$. We take the origin at point 1, so $y_1 = 0$ (Fig. 7.4). Our target variable, the distance the ball moves vertically between the two points, is the displacement $y_2 - y_1 = y_2 - 0 = y_2$.

EXECUTE: We have $y_1 = 0$, $U_{grav,1} = mgy_1 = 0$, and $K_2 = \frac{1}{2}mv_2^2 = 0$. Then Eq. (7.4), $K_1 + U_{grav,1} = K_2 + U_{grav,2}$, becomes

$$K_1 = U_{grav,2}$$

As the energy bar graphs in Fig. 7.4 show, this equation says that the kinetic energy of the ball at point 1 is completely converted to gravitational potential energy at point 2. We substitute $K_1 = \frac{1}{2}mv_1^2$ and $U_{grav,2} = mgy_2$ and solve for y_2:

$$\frac{1}{2}mv_1^2 = mgy_2$$

$$y_2 = \frac{v_1^2}{2g} = \frac{(20.0 \text{ m/s})^2}{2(9.80 \text{ m/s}^2)} = 20.4 \text{ m}$$

7.4 After a baseball leaves your hand, mechanical energy $E = K + U$ is conserved.

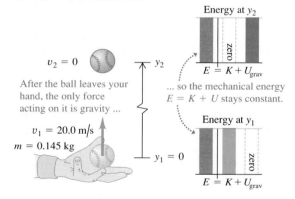

Energy at y_2

$v_2 = 0$

y_2

$E = K + U_{grav}$

After the ball leaves your hand, the only force acting on it is gravity ...

... so the mechanical energy $E = K + U$ stays constant.

$v_1 = 20.0$ m/s
$m = 0.145$ kg

$y_1 = 0$

Energy at y_1

$E = K + U_{grav}$

EVALUATE: As a check on our work, use the given value of v_1 and our result for y_2 to calculate the kinetic energy at point 1 and the gravitational potential energy at point 2. You should find that these are equal: $K_1 = \frac{1}{2}mv_1^2 = 29.0$ J and $U_{grav,2} = mgy_2 = 29.0$ J. Note also that we could have found the result $y_2 = v_1^2/2g$ using Eq. (2.13).

What if we put the origin somewhere else? For example, what if we put it 5.0 m below point 1, so that $y_1 = 5.0$ m? Then the total mechanical energy at point 1 is part kinetic and part potential; at point 2 it's still purely potential because $v_2 = 0$. You'll find that this choice of origin yields $y_2 = 25.4$ m, but again $y_2 - y_1 = 20.4$ m. In problems like this, you are free to choose the height at which $U_{grav} = 0$. The physics doesn't depend on your choice, so don't agonize over it.

When Forces Other Than Gravity Do Work

If other forces act on the body in addition to its weight, then $\vec{F}_{\text{other}}$ in Fig. 7.2 is *not* zero. For the pile driver described in Example 6.4 (Section 6.2), the force applied by the hoisting cable and the friction with the vertical guide rails are examples of forces that might be included in $\vec{F}_{\text{other}}$. The gravitational work W_{grav} is still given by Eq. (7.3), but the total work W_{tot} is then the sum of W_{grav} and the work done by $\vec{F}_{\text{other}}$. We will call this additional work W_{other}, so the total work done by all forces is $W_{\text{tot}} = W_{\text{grav}} + W_{\text{other}}$. Equating this to the change in kinetic energy, we have

$$W_{\text{other}} + W_{\text{grav}} = K_2 - K_1 \qquad (7.6)$$

Also, from Eq. (7.3), $W_{\text{grav}} = U_{\text{grav},1} - U_{\text{grav},2}$, so

$$W_{\text{other}} + U_{\text{grav},1} - U_{\text{grav},2} = K_2 - K_1$$

which we can rearrange in the form

$$K_1 + U_{\text{grav},1} + W_{\text{other}} = K_2 + U_{\text{grav},2} \quad \begin{array}{l}\text{(if forces other than} \\ \text{gravity do work)}\end{array} \qquad (7.7)$$

Finally, using the appropriate expressions for the various energy terms, we obtain

$$\tfrac{1}{2}mv_1^2 + mgy_1 + W_{\text{other}} = \tfrac{1}{2}mv_2^2 + mgy_2 \quad \begin{array}{l}\text{(if forces other than} \\ \text{gravity do work)}\end{array} \qquad (7.8)$$

The meaning of Eqs. (7.7) and (7.8) is this: *The work done by all forces other than the gravitational force equals the change in the total mechanical energy $E = K + U_{\text{grav}}$ of the system, where U_{grav} is the gravitational potential energy.* When W_{other} is positive, E increases and $K_2 + U_{\text{grav},2}$ is greater than $K_1 + U_{\text{grav},1}$. When W_{other} is negative, E decreases (Fig. 7.5). In the special case in which no forces other than the body's weight do work, $W_{\text{other}} = 0$. The total mechanical energy is then constant, and we are back to Eq. (7.4) or (7.5).

7.5 As this skydiver moves downward, the upward force of air resistance does negative work W_{other} on him. Hence the total mechanical energy $E = K + U$ decreases: The skydiver's speed and kinetic energy K stay the same, while the gravitational potential energy U decreases.

Problem-Solving Strategy 7.1   Problems Using Mechanical Energy I

IDENTIFY *the relevant concepts:* Decide whether the problem should be solved by energy methods, by using $\Sigma\vec{F} = m\vec{a}$ directly, or by a combination of these. The energy approach is best when the problem involves varying forces or motion along a curved path (discussed later in this section). If the problem involves elapsed time, the energy approach is usually *not* the best choice because it doesn't involve time directly.

SET UP *the problem* using the following steps:
1. When using the energy approach, first identify the initial and final states (the positions and velocities) of the bodies in question. Use the subscript 1 for the initial state and the subscript 2 for the final state. Draw sketches showing these states.
2. Define a coordinate system, and choose the level at which $y = 0$. Choose the positive y-direction to be upward, as is assumed in Eq. (7.1) and in the equations that follow from it.
3. Identify any forces that do work on each body and that *cannot* be described in terms of potential energy. (So far, this means

any forces other than gravity. In Section 7.2 we'll see that the work done by an ideal spring can also be expressed as a change in potential energy.) Sketch a free-body diagram for each body.
4. List the unknown and known quantities, including the coordinates and velocities at each point. Identify the target variables.

EXECUTE *the solution:* Write expressions for the initial and final kinetic and potential energies K_1, K_2, $U_{\text{grav},1}$, and $U_{\text{grav},2}$. If no other forces do work, use Eq. (7.4). If there are other forces that do work, use Eq. (7.7). Draw bar graphs showing the initial and final values of K, $U_{\text{grav},1}$, and $E = K + U_{\text{grav}}$. Then solve to find your target variables.

EVALUATE *your answer:* Check whether your answer makes physical sense. Remember that the gravitational work is included in ΔU_{grav}, so do not include it in W_{other}.

| Example 7.2 | Work and energy in throwing a baseball |

In Example 7.1 suppose your hand moves upward by 0.50 m while you are throwing the ball. The ball leaves your hand with an upward velocity of 20.0 m/s. (a) Find the magnitude of the force (assumed constant) that your hand exerts on the ball. (b) Find the speed of the ball at a point 15.0 m above the point where it leaves your hand. Ignore air resistance.

SOLUTION

IDENTIFY and SET UP: In Example 7.1 only gravity did work. Here we must include the nongravitational, "other" work done by your hand. Figure 7.6 shows a diagram of the situation, including a free-body diagram for the ball while it is being thrown. We let point 1 be where your hand begins to move, point 2 be where the ball leaves your hand, and point 3 be where the ball is 15.0 m above point 2. The nongravitational force $\vec{F}$ of your hand acts only between points 1 and 2. Using the same coordinate system as in Example 7.1, we have $y_1 = -0.50$ m, $y_2 = 0$, and $y_3 = 15.0$ m. The ball starts at rest at point 1, so $v_1 = 0$, and the ball's speed as it leaves your hand is $v_2 = 20.0$ m/s. Our target variables are (a) the magnitude F of the force of your hand and (b) the ball's velocity v_{3y} at point 3.

EXECUTE: (a) To determine F, we'll first use Eq. (7.7) to calculate the work W_{other} done by this force. We have

$$K_1 = 0$$

$$U_{\text{grav},1} = mgy_1 = (0.145 \text{ kg})(9.80 \text{ m/s}^2)(-0.50 \text{ m}) = -0.71 \text{ J}$$

7.6 (a) Applying energy ideas to a ball thrown vertically upward. (b) Free-body diagram for the ball as you throw it.

(a)

(b)

$$K_2 = \tfrac{1}{2}mv_2^2 = \tfrac{1}{2}(0.145 \text{ kg})(20.0 \text{ m/s})^2 = 29.0 \text{ J}$$

$$U_{\text{grav},2} = mgy_2 = (0.145 \text{ kg})(9.80 \text{ m/s}^2)(0) = 0$$

(Don't worry that $U_{\text{grav},1}$ is less than zero; all that matters is the *difference* in potential energy from one point to another.) From Eq. (7.7),

$$K_1 + U_{\text{grav},1} + W_{\text{other}} = K_2 + U_{\text{grav},2}$$
$$W_{\text{other}} = (K_2 - K_1) + (U_{\text{grav},2} - U_{\text{grav},1})$$
$$= (29.0 \text{ J} - 0) + [0 - (-0.71 \text{ J})] = 29.7 \text{ J}$$

But since $\vec{F}$ is constant and upward, the work done by $\vec{F}$ equals the force magnitude times the displacement: $W_{\text{other}} = F(y_2 - y_1)$. So

$$F = \frac{W_{\text{other}}}{y_2 - y_1} = \frac{29.7 \text{ J}}{0.50 \text{ m}} = 59 \text{ N}$$

This is more than 40 times the weight of the ball (1.42 N).

(b) To find v_{3y}, note that between points 2 and 3 only gravity acts on the ball. So between these points mechanical energy is conserved and $W_{\text{other}} = 0$. From Eq. (7.4), we can solve for K_3 and from that solve for v_{3y}:

$$K_2 + U_{\text{grav},2} = K_3 + U_{\text{grav},3}$$
$$U_{\text{grav},3} = mgy_3 = (0.145 \text{ kg})(9.80 \text{ m/s}^2)(15.0 \text{ m}) = 21.3 \text{ J}$$
$$K_3 = (K_2 + U_{\text{grav},2}) - U_{\text{grav},3}$$
$$= (29.0 \text{ J} + 0 \text{ J}) - 21.3 \text{ J} = 7.7 \text{ J}$$

Since $K_3 = \tfrac{1}{2}mv_{3y}^2$, we find

$$v_{3y} = \pm\sqrt{\frac{2K_3}{m}} = \pm\sqrt{\frac{2(7.7 \text{ J})}{0.145 \text{ kg}}} = \pm 10 \text{ m/s}$$

The plus-or-minus sign reminds us that the ball passes point 3 on the way up and again on the way down. The total mechanical energy E is constant and equal to $K_2 + U_{\text{grav},2} = 29.0 \text{ J}$ while the ball is in free fall, and the potential energy at point 3 is $U_{\text{grav},3} = mgy_3 = 21.3 \text{ J}$ whether the ball is moving up or down. So at point 3, the ball's kinetic energy K_3 (and therefore its speed) don't depend on the direction the ball is moving. The velocity v_{3y} is positive ($+10$ m/s) when the ball is moving up and negative (-10 m/s) when it is moving down; the speed v_3 is 10 m/s in either case.

EVALUATE: In Example 7.1 we found that the ball reaches a maximum height $y = 20.4$ m. At that point all of the kinetic energy it had when it left your hand at $y = 0$ has been converted to gravitational potential energy. At $y = 15.0$ m, the ball is about three-fourths of the way to its maximum height, so about three-fourths of its mechanical energy should be in the form of potential energy. (The energy bar graphs in Fig. 7.6a show this.) Can you show that this is true from our results for K_3 and $U_{\text{grav},3}$?

Gravitational Potential Energy for Motion Along a Curved Path

In our first two examples the body moved along a straight vertical line. What happens when the path is slanted or curved (Fig. 7.7a)? The body is acted on by the gravitational force $\vec{w} = m\vec{g}$ and possibly by other forces whose resultant we call $\vec{F}_{\text{other}}$. To find the work done by the gravitational force during this displacement, we divide the path into small segments $\Delta\vec{s}$; Fig. 7.7b shows a typical segment. The work done by the gravitational force over this segment is the scalar product of the force and the displacement. In terms of unit vectors, the force is $\vec{w} = m\vec{g} = -mg\hat{j}$ and the displacement is $\Delta\vec{s} = \Delta x\hat{i} + \Delta y\hat{j}$, so the work done by the gravitational force is

$$\vec{w} \cdot \Delta\vec{s} = -mg\hat{j} \cdot (\Delta x\hat{i} + \Delta y\hat{j}) = -mg\Delta y$$

The work done by gravity is the same as though the body had been displaced vertically a distance Δy, with no horizontal displacement. This is true for every segment, so the *total* work done by the gravitational force is $-mg$ multiplied by the *total* vertical displacement $(y_2 - y_1)$:

$$W_{\text{grav}} = -mg(y_2 - y_1) = mgy_1 - mgy_2 = U_{\text{grav},1} - U_{\text{grav},2}$$

This is the same as Eq. (7.1) or (7.3), in which we assumed a purely vertical path. So even if the path a body follows between two points is curved, the total work done by the gravitational force depends only on the difference in height between the two points of the path. This work is unaffected by any horizontal motion that may occur. So *we can use the same expression for gravitational potential energy whether the body's path is curved or straight.*

7.7 Calculating the change in gravitational potential energy for a displacement along a curved path.

(a)

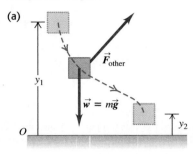

(b)

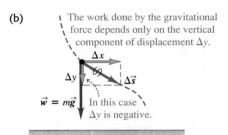

The work done by the gravitational force depends only on the vertical component of displacement Δy.

In this case Δy is negative.

Conceptual Example 7.3 **Energy in projectile motion**

A batter hits two identical baseballs with the same initial speed and from the same initial height but at different initial angles. Prove that both balls have the same speed at any height h if air resistance can be neglected.

SOLUTION

The only force acting on each ball after it is hit is its weight. Hence the total mechanical energy for each ball is constant. Figure 7.8 shows the trajectories of two balls batted at the same height with the same initial speed, and thus the same total mechanical energy, but with different initial angles. At all points at the same height the potential energy is the same. Thus the kinetic energy at this height must be the same for both balls, and the speeds are the same.

7.8 For the same initial speed and initial height, the speed of a projectile at a given elevation h is always the same, neglecting air resistance.

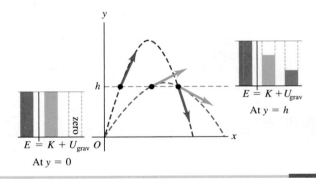

Example 7.4 **Speed at the bottom of a vertical circle**

Your cousin Throckmorton skateboards from rest down a curved, frictionless ramp. If we treat Throcky and his skateboard as a particle, he moves through a quarter-circle with radius $R = 3.00$ m (Fig. 7.9). Throcky and his skateboard have a total mass of 25.0 kg. (a) Find his speed at the bottom of the ramp. (b) Find the normal force that acts on him at the bottom of the curve.

SOLUTION

IDENTIFY: We can't use the constant-acceleration equations of Chapter 2 because Throcky's acceleration isn't constant; the slope decreases as he descends. Instead, we'll use the energy approach. Throcky moves along a circular arc, so we'll also use what we learned about circular motion in Section 5.4.

SET UP: The only forces on Throcky are his weight and the normal force $\vec{n}$ exerted by the ramp (Fig. 7.9b). Although $\vec{n}$ acts all along the path, it does zero work because $\vec{n}$ is perpendicular to Throcky's displacement at every point. Hence $W_{\text{other}} = 0$ and mechanical energy is conserved. We take point 1 at the starting point and point 2 at the bottom of the ramp, and we let $y = 0$ be at the bottom of the ramp (Fig. 7.9a). We take the positive y-direction upward; then $y_1 = R$ and $y_2 = 0$. Throcky starts at rest at the top, so $v_1 = 0$. In part (a) our target variable is his speed v_2 at the bottom; in part (b) the target variable is the magnitude n of the normal force at point 2. To find n, we'll use Newton's second law and the relation $a = v^2/R$.

Continued

EXECUTE: (a) The various energy quantities are

$$K_1 = 0 \qquad U_{grav,1} = mgR$$
$$K_2 = \tfrac{1}{2}mv_2^2 \qquad U_{grav,2} = 0$$

From conservation of mechanical energy, Eq. (7.4),

$$K_1 + U_{grav,1} = K_2 + U_{grav,2}$$
$$0 + mgR = \tfrac{1}{2}mv_2^2 + 0$$
$$v_2 = \sqrt{2gR}$$
$$= \sqrt{2(9.80 \text{ m/s}^2)(3.00 \text{ m})} = 7.67 \text{ m/s}$$

This answer doesn't depend on the ramp being circular; Throcky will have the same speed $v_2 = \sqrt{2gR}$ at the bottom of any ramp of height R, no matter what its shape.

(b) To find n at point 2 using Newton's second law, we need the free-body diagram at that point (Fig. 7.9b). At point 2, Throcky is moving at speed $v_2 = \sqrt{2gR}$ in a circle of radius R; his acceleration is toward the center of the circle and has magnitude

$$a_{rad} = \frac{v_2^2}{R} = \frac{2gR}{R} = 2g$$

The y-component of Newton's second law is

$$\sum F_y = n + (-w) = ma_{rad} = 2mg$$
$$n = w + 2mg = 3mg$$
$$= 3(25.0 \text{ kg})(9.80 \text{ m/s}^2) = 735 \text{ N}$$

At point 2 the normal force is three times Throcky's weight. This result doesn't depend on the radius R of the ramp. We saw in Examples 5.9 and 5.23 that the magnitude of n is the *apparent weight*, so at the bottom of the *curved part* of the ramp Throcky feels as though he weighs three times his true weight mg. But when he reaches the *horizontal* part of the ramp, immediately to the right of point 2, the normal force decreases to $w = mg$ and thereafter Throcky feels his true weight again. Can you see why?

EVALUATE: This example shows a general rule about the role of forces in problems in which we use energy techniques: What matters is not simply whether a force *acts*, but whether that force *does work*. If the force does no work, like the normal force $\vec{n}$ here, then it does not appear in Eqs. (7.4) and (7.7).

7.9 (a) Throcky skateboarding down a frictionless circular ramp. The total mechanical energy is constant. (b) Free-body diagrams for Throcky and his skateboard at various points on the ramp.

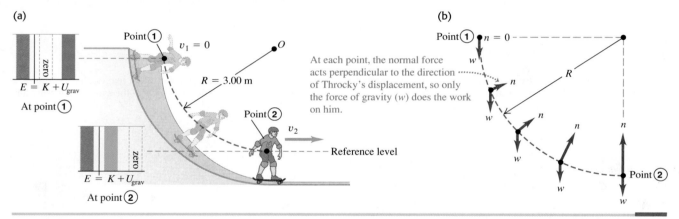

(a)

Point ① $v_1 = 0$ O

$R = 3.00$ m

Point ② v_2

$E = K + U_{grav}$
At point ①

zero

$E = K + U_{grav}$
At point ②

Reference level

(b)

Point ① $n = 0$

At each point, the normal force acts perpendicular to the direction of Throcky's displacement, so only the force of gravity (w) does the work on him.

R

Point ②

Example 7.5   A vertical circle with friction

Suppose that the ramp of Example 7.4 is not frictionless, and that Throcky's speed at the bottom is only 6.00 m/s, not the 7.67 m/s we found there. What work was done on him by the friction force?

SOLUTION

IDENTIFY and SET UP: Figure 7.10 shows that again the normal force does no work, but now there is a friction force $\vec{f}$ that *does* do work W_f. Hence the nongravitational work W_{other} done on Throcky between points 1 and 2 is equal to W_f and is not zero. We use the same coordinate system and the same initial and final points as in Example 7.4. Our target variable is $W_f = W_{other}$, which we'll find using Eq. (7.7).

EXECUTE: The energy quantities are

$$K_1 = 0$$
$$U_{grav,1} = mgR = (25.0 \text{ kg})(9.80 \text{ m/s}^2)(3.00 \text{ m}) = 735 \text{ J}$$
$$K_2 = \tfrac{1}{2}mv_2^2 = \tfrac{1}{2}(25.0 \text{ kg})(6.00 \text{ m/s})^2 = 450 \text{ J}$$
$$U_{grav,2} = 0$$

7.10 Energy bar graphs and free-body diagrams for Throcky skateboarding down a ramp with friction.

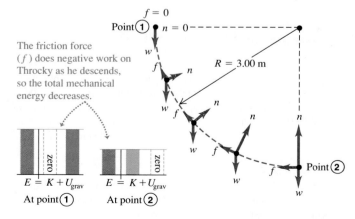

$f = 0$
Point ① $n = 0$

The friction force (f) does negative work on Throcky as he descends, so the total mechanical energy decreases.

$R = 3.00$ m

$E = K + U_{grav}$
At point ①

$E = K + U_{grav}$
At point ②

Point ②

From Eq. (7.7),

$$W_f = W_{\text{other}} = K_2 + U_{\text{grav},2} - K_1 - U_{\text{grav},1}$$
$$= 450 \text{ J} + 0 - 0 - 735 \text{ J} = -285 \text{ J}$$

The work done by the friction force is -285 J, and the total mechanical energy *decreases* by 285 J.

EVALUATE: Our result for W_f is negative. Can you see from the free-body diagrams in Fig. 7.10 why this must be so?

It would be very difficult to apply Newton's second law, $\sum \vec{F} = m\vec{a}$, directly to this problem because the normal and friction forces and the acceleration are continuously changing in both magnitude and direction as Throcky descends. The energy approach, by contrast, relates the motions at the top and bottom of the ramp without involving the details of the motion in between.

Example 7.6 | **An inclined plane with friction**

We want to slide a 12-kg crate up a 2.5-m-long ramp inclined at 30°. A worker, ignoring friction, calculates that he can do this by giving it an initial speed of 5.0 m/s at the bottom and letting it go. But friction is *not* negligible; the crate slides only 1.6 m up the ramp, stops, and slides back down (Fig. 7.11a). (a) Find the magnitude of the friction force acting on the crate, assuming that it is constant. (b) How fast is the crate moving when it reaches the bottom of the ramp?

SOLUTION

IDENTIFY and SET UP: The friction force does work on the crate as it slides. The first part of the motion is from point 1, at the bottom of the ramp, to point 2, where the crate stops instantaneously ($v_2 = 0$). In the second part of the motion, the crate returns to the bottom of the ramp, which we'll also call point 3 (Fig. 7.11a). We take the positive y-direction upward. We take $y = 0$ (and hence $U_{\text{grav}} = 0$) to be at ground level (point 1), so that $y_1 = 0$, $y_2 = (1.6 \text{ m})\sin 30° = 0.80$ m, and $y_3 = 0$. We are given $v_1 = 5.0$ m/s. In part (a) our target variable is f, the magnitude of the friction force as the crate slides up; as in Example 7.2, we'll find this using the energy approach. In part (b) our target variable is v_3, the crate's speed at the bottom of the ramp. We'll calculate the work done by friction as the crate slides back down, then use the energy approach to find v_3.

7.11 (a) A crate slides partway up the ramp, stops, and slides back down. (b) Energy bar graphs for points 1, 2, and 3.

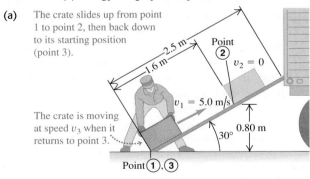

(a) The crate slides up from point 1 to point 2, then back down to its starting position (point 3).

2.5 m
1.6 m
Point ②
$v_2 = 0$
$v_1 = 5.0$ m/s
The crate is moving at speed v_3 when it returns to point 3.
0.80 m
30°
Point ①, ③

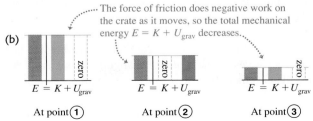

(b)

The force of friction does negative work on the crate as it moves, so the total mechanical energy $E = K + U_{\text{grav}}$ decreases.

zero
$E = K + U_{\text{grav}}$
At point ①

zero
$E = K + U_{\text{grav}}$
At point ②

zero
$E = K + U_{\text{grav}}$
At point ③

EXECUTE: (a) The energy quantities are

$$K_1 = \tfrac{1}{2}(12 \text{ kg})(5.0 \text{ m/s})^2 = 150 \text{ J}$$
$$U_{\text{grav},1} = 0$$
$$K_2 = 0$$
$$U_{\text{grav},2} = (12 \text{ kg})(9.8 \text{ m/s}^2)(0.80 \text{ m}) = 94 \text{ J}$$
$$W_{\text{other}} = -fs$$

Here $s = 1.6$ m. Using Eq. (7.7), we find

$$K_1 + U_{\text{grav},1} + W_{\text{other}} = K_2 + U_{\text{grav},2}$$
$$W_{\text{other}} = -fs = (K_2 + U_{\text{grav},2}) - (K_1 + U_{\text{grav},1})$$
$$= (0 + 94 \text{ J}) - (150 \text{ J} + 0) = -56 \text{ J} = -fs$$
$$f = \frac{W_{\text{other}}}{s} = \frac{56 \text{ J}}{1.6 \text{ m}} = 35 \text{ N}$$

The friction force of 35 N, acting over 1.6 m, causes the mechanical energy of the crate to decrease from 150 J to 94 J (Fig. 7.11b).

(b) As the crate moves from point 2 to point 3, the work done by friction has the same negative value as from point 1 to point 2. (The friction force and the displacement both reverse direction but have the same magnitudes.) The total work done by friction between points 1 and 3 is therefore

$$W_{\text{other}} = W_{\text{fric}} = -2fs = -2(56 \text{ J}) = -112 \text{ J}$$

From part (a), $K_1 = 150$ J and $U_{\text{grav},1} = 0$. Equation (7.7) then gives

$$K_1 + U_{\text{grav},1} + W_{\text{other}} = K_3 + U_{\text{grav},3}$$
$$K_3 = K_1 + U_{\text{grav},1} - U_{\text{grav},3} + W_{\text{other}}$$
$$= 150 \text{ J} + 0 - 0 + (-112 \text{ J}) = 38 \text{ J}$$

The crate returns to the bottom of the ramp with only 38 J of the original 150 J of mechanical energy (Fig. 7.11b). Since $K_3 = \tfrac{1}{2}mv_3^2$,

$$v_3 = \sqrt{\frac{2K_3}{m}} = \sqrt{\frac{2(38 \text{ J})}{12 \text{ kg}}} = 2.5 \text{ m/s}$$

EVALUATE: Energy was lost due to friction, so the crate's speed $v_3 = 2.5$ m/s when it returns to the bottom of the ramp is less than the speed $v_1 = 5.0$ m/s at which it left that point. In part (b) we applied Eq. (7.7) to points 1 and 3, considering the round trip as a whole. Alternatively, we could have considered the second part of the motion by itself and applied Eq. (7.7) to points 2 and 3. Try it; do you get the same result for v_3?

Test Your Understanding of Section 7.1 The figure shows two different frictionless ramps. The heights y_1 and y_2 are the same for both ramps. If a block of mass m is released from rest at the left-hand end of each ramp, which block arrives at the right-hand end with the greater speed? (i) block I; (ii) block II; (iii) the speed is the same for both blocks.

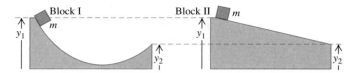

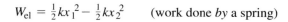

7.2 Elastic Potential Energy

7.12 The Achilles tendon, which runs along the back of the ankle to the heel bone, acts like a natural spring. When it stretches and then relaxes, this tendon stores and then releases elastic potential energy. This spring action reduces the amount of work your leg muscles must do as you run.

MasteringPHYSICS

There are many situations in which we encounter potential energy that is not gravitational in nature. One example is a rubber-band slingshot. Work is done on the rubber band by the force that stretches it, and that work is stored in the rubber band until you let it go. Then the rubber band gives kinetic energy to the projectile.

This is the same pattern we saw with the pile driver in Section 7.1: Do work on the system to store energy, which can later be converted to kinetic energy. We'll describe the process of storing energy in a deformable body such as a spring or rubber band in terms of *elastic potential energy* (Fig. 7.12). A body is called *elastic* if it returns to its original shape and size after being deformed.

To be specific, we'll consider storing energy in an ideal spring, like the ones we discussed in Section 6.3. To keep such an ideal spring stretched by a distance x, we must exert a force $F = kx$, where k is the force constant of the spring. The ideal spring is a useful idealization because many elastic bodies show this same direct proportionality between force $\vec{F}$ and displacement x, provided that x is sufficiently small.

Let's proceed just as we did for gravitational potential energy. We begin with the work done by the elastic (spring) force and then combine this with the work–energy theorem. The difference is that gravitational potential energy is a shared property of a body and the earth, but elastic potential energy is stored just in the spring (or other deformable body).

Figure 7.13 shows the ideal spring from Fig. 6.18, with its left end held stationary and its right end attached to a block with mass m that can move along the x-axis. In Fig. 7.13a the body is at $x = 0$ when the spring is neither stretched nor compressed. We move the block to one side, thereby stretching or compressing the spring, and then let it go. As the block moves from one position x_1 to another position x_2, how much work does the elastic (spring) force do on the block?

We found in Section 6.3 that the work we must do *on* the spring to move one end from an elongation x_1 to a different elongation x_2 is

$$W = \tfrac{1}{2} k x_2^2 - \tfrac{1}{2} k x_1^2 \quad \text{(work done *on* a spring)}$$

where k is the force constant of the spring. If we stretch the spring farther, we do positive work on the spring; if we let the spring relax while holding one end, we do negative work on it. We also saw that this expression for work is still correct if the spring is compressed, not stretched, so that x_1 or x_2 or both are negative. Now we need to find the work done *by* the spring. From Newton's third law the two quantities of work are just negatives of each other. Changing the signs in this equation, we find that in a displacement from x_1 to x_2 the spring does an amount of work W_{el} given by

$$W_{el} = \tfrac{1}{2} k x_1^2 - \tfrac{1}{2} k x_2^2 \quad \text{(work done *by* a spring)}$$

The subscript "el" stands for *elastic*. When x_1 and x_2 are both positive and $x_2 > x_1$ (Fig. 7.13b), the spring does negative work on the block, which moves in the $+x$-direction while the spring pulls on it in the $-x$-direction. The spring stretches farther, and the block slows down. When x_1 and x_2 are both positive and $x_2 < x_1$ (Fig. 7.13c), the spring does positive work as it relaxes and the block speeds up. If the spring can be compressed as well as stretched, x_1 or x_2 or both may be negative, but the expression for W_{el} is still valid. In Fig. 7.13d, both x_1 and x_2 are negative, but x_2 is less negative than x_1; the compressed spring does positive work as it relaxes, speeding the block up.

Just as for gravitational work, we can express the work done by the spring in terms of a given quantity at the beginning and end of the displacement. This quantity is $\frac{1}{2}kx^2$, and we define it to be the **elastic potential energy:**

$$U_{el} = \tfrac{1}{2}kx^2 \qquad \text{(elastic potential energy)} \qquad (7.9)$$

Figure 7.14 is a graph of Eq. (7.9). The unit of U_{el} is the joule (J), the unit used for *all* energy and work quantities; to see this from Eq. (7.9), recall that the units of k are N/m and that $1\ \text{N} \cdot \text{m} = 1\ \text{J}$.

We can use Eq. (7.9) to express the work W_{el} done on the block by the elastic force in terms of the change in elastic potential energy:

$$W_{el} = \tfrac{1}{2}kx_1^{\,2} - \tfrac{1}{2}kx_2^{\,2} = U_{el,1} - U_{el,2} = -\Delta U_{el} \qquad (7.10)$$

When a stretched spring is stretched farther, as in Fig. 7.13b, W_{el} is negative and U_{el} *increases;* a greater amount of elastic potential energy is stored in the spring. When a stretched spring relaxes, as in Fig. 7.13c, x decreases, W_{el} is positive, and U_{el} *decreases;* the spring loses elastic potential energy. Negative values of x refer to a compressed spring. But, as Fig. 7.14 shows, U_{el} is positive for both positive and negative x, and Eqs. (7.9) and (7.10) are valid for both cases. The more a spring is compressed *or* stretched, the greater its elastic potential energy.

CAUTION Gravitational potential energy vs. elastic potential energy An important difference between gravitational potential energy $U_{grav} = mgy$ and elastic potential energy $U_{el} = \frac{1}{2}kx^2$ is that we do *not* have the freedom to choose $x = 0$ to be wherever we wish. To be consistent with Eq. (7.9), $x = 0$ *must* be the position at which the spring is neither stretched nor compressed. At that position, its elastic potential energy and the force that it exerts are both zero.

The work–energy theorem says that $W_{tot} = K_2 - K_1$, no matter what kind of forces are acting on a body. If the elastic force is the *only* force that does work on the body, then

$$W_{tot} = W_{el} = U_{el,1} - U_{el,2}$$

The work–energy theorem, $W_{tot} = K_2 - K_1$, then gives us

$$K_1 + U_{el,1} = K_2 + U_{el,2} \qquad \text{(if only the elastic force does work)} \qquad (7.11)$$

Here U_{el} is given by Eq. (7.9), so

$$\tfrac{1}{2}mv_1^{\,2} + \tfrac{1}{2}kx_1^{\,2} = \tfrac{1}{2}mv_2^{\,2} + \tfrac{1}{2}kx_2^{\,2} \qquad \begin{array}{l}\text{(if only the elastic}\\\text{force does work)}\end{array} \qquad (7.12)$$

In this case the total mechanical energy $E = K + U_{el}$—the sum of kinetic and *elastic* potential energy—is *conserved.* An example of this is the motion of the

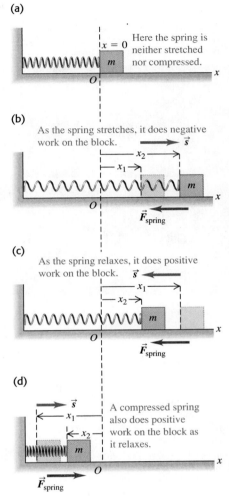

7.13 Calculating the work done by a spring attached to a block on a horizontal surface. The quantity x is the extension or compression of the spring.

(a) Here the spring is neither stretched nor compressed. $x = 0$

(b) As the spring stretches, it does negative work on the block. $\vec{s}$ — x_2, x_1, $\vec{F}_{spring}$

(c) As the spring relaxes, it does positive work on the block. $\vec{s}$ — x_1, x_2, $\vec{F}_{spring}$

(d) A compressed spring also does positive work on the block as it relaxes. $\vec{s}$ — x_1, x_2, $\vec{F}_{spring}$

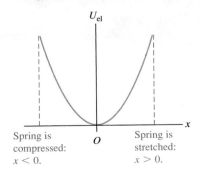

7.14 The graph of elastic potential energy for an ideal spring is a parabola: $U_{el} = \frac{1}{2}kx^2$, where x is the extension or compression of the spring. Elastic potential energy U_{el} is never negative.

U_{el}

Spring is compressed: $x < 0.$ O Spring is stretched: $x > 0.$

Application **Elastic Potential Energy of a Cheetah**

When a cheetah gallops, its back flexes and extends by an exceptional amount. Flexion of the back stretches elastic tendons and muscles along the top of the spine and also compresses the spine, storing mechanical energy. When the cheetah launches into its next bound, this energy helps to extend the spine, enabling the cheetah to run more efficiently.

Difference in nose-to-tail length

7.15 Trampoline jumping involves an interplay among kinetic energy, gravitational potential energy, and elastic potential energy. Due to air resistance and frictional forces within the trampoline, mechanical energy is not conserved. That's why the bouncing eventually stops unless the jumper does work with his or her legs to compensate for the lost energy.

block in Fig. 7.13, provided the horizontal surface is frictionless so that no force does work other than that exerted by the spring.

For Eq. (7.12) to be strictly correct, the ideal spring that we've been discussing must also be *massless*. If the spring has a mass, it also has kinetic energy as the coils of the spring move back and forth. We can neglect the kinetic energy of the spring if its mass is much less than the mass m of the body attached to the spring. For instance, a typical automobile has a mass of 1200 kg or more. The springs in its suspension have masses of only a few kilograms, so their mass can be neglected if we want to study how a car bounces on its suspension.

Situations with Both Gravitational and Elastic Potential Energy

Equations (7.11) and (7.12) are valid when the only potential energy in the system is elastic potential energy. What happens when we have *both* gravitational and elastic forces, such as a block attached to the lower end of a vertically hanging spring? And what if work is also done by other forces that *cannot* be described in terms of potential energy, such as the force of air resistance on a moving block? Then the total work is the sum of the work done by the gravitational force (W_{grav}), the work done by the elastic force (W_{el}), and the work done by other forces (W_{other}): $W_{\text{tot}} = W_{\text{grav}} + W_{\text{el}} + W_{\text{other}}$. Then the work–energy theorem gives

$$W_{\text{grav}} + W_{\text{el}} + W_{\text{other}} = K_2 - K_1$$

The work done by the gravitational force is $W_{\text{grav}} = U_{\text{grav},1} - U_{\text{grav},2}$ and the work done by the spring is $W_{\text{el}} = U_{\text{el},1} - U_{\text{el},2}$. Hence we can rewrite the work–energy theorem for this most general case as

$$K_1 + U_{\text{grav},1} + U_{\text{el},1} + W_{\text{other}} = K_2 + U_{\text{grav},2} + U_{\text{el},2} \quad \begin{array}{c}\text{(valid in}\\\text{general)}\end{array} \quad (7.13)$$

or, equivalently,

$$K_1 + U_1 + W_{\text{other}} = K_2 + U_2 \quad \text{(valid in general)} \quad (7.14)$$

where $U = U_{\text{grav}} + U_{\text{el}} = mgy + \frac{1}{2}kx^2$ is the *sum* of gravitational potential energy and elastic potential energy. For short, we call U simply "the potential energy."

Equation (7.14) is *the most general statement* of the relationship among kinetic energy, potential energy, and work done by other forces. It says:

The work done by all forces other than the gravitational force or elastic force equals the change in the total mechanical energy $E = K + U$ of the system, where $U = U_{\text{grav}} + U_{\text{el}}$ is the sum of the gravitational potential energy and the elastic potential energy.

The "system" is made up of the body of mass m, the earth with which it interacts through the gravitational force, and the spring of force constant k.

If W_{other} is positive, $E = K + U$ increases; if W_{other} is negative, E decreases. If the gravitational and elastic forces are the *only* forces that do work on the body, then $W_{\text{other}} = 0$ and the total mechanical energy (including both gravitational and elastic potential energy) is conserved. (You should compare Eq. (7.14) to Eqs. (7.7) and (7.8), which describe situations in which there is gravitational potential energy but no elastic potential energy.)

Trampoline jumping (Fig. 7.15) involves transformations among kinetic energy, elastic potential energy, and gravitational potential energy. As the jumper descends through the air from the high point of the bounce, gravitational potential energy U_{grav} decreases and kinetic energy K increases. Once the jumper touches the trampoline, some of the mechanical energy goes into elastic potential energy U_{el} stored in the trampoline's springs. Beyond a certain point the jumper's speed and kinetic energy K decrease while U_{grav} continues to decrease and U_{el} continues to increase. At the low point the jumper comes to a momentary halt $(K = 0)$ at the lowest point of the trajectory (U_{grav} is minimum) and the springs are maximally stretched (U_{el} is maximum). The springs then convert their energy back into K and U_{grav}, propelling the jumper upward.

Problem-Solving Strategy 7.2 **Problems Using Mechanical Energy II**

Problem-Solving Strategy 7.1 (Section 7.1) is equally useful in solving problems that involve elastic forces as well as gravitational forces. The only new wrinkle is that the potential energy U now includes the elastic potential energy $U_{el} = \frac{1}{2}kx^2$, where x is the dis-placement of the spring *from its unstretched length*. The work done by the gravitational and elastic forces is accounted for by their potential energies; the work done by other forces, W_{other}, must still be included separately.

Example 7.7 | Motion with elastic potential energy

A glider with mass $m = 0.200$ kg sits on a frictionless horizontal air track, connected to a spring with force constant $k = 5.00$ N/m. You pull on the glider, stretching the spring 0.100 m, and release it from rest. The glider moves back toward its equilibrium position $(x = 0)$. What is its x-velocity when $x = 0.080$ m?

SOLUTION

IDENTIFY and SET UP: As the glider starts to move, elastic potential energy is converted to kinetic energy. The glider remains at the same height throughout the motion, so gravitational potential energy is not a factor and $U = U_{el} = \frac{1}{2}kx^2$. Figure 7.16 shows our sketches. Only the spring force does work on the glider, so $W_{other} = 0$ and we may use Eq. (7.11). We designate the point

7.16 Our sketches and energy bar graphs for this problem.

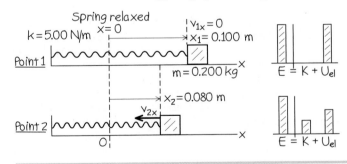

where the glider is released as point 1 (that is, $x_1 = 0.100$ m) and $x_2 = 0.080$ m as point 2. We are given $v_{1x} = 0$; our target variable is v_{2x}.

EXECUTE: The energy quantities are

$$K_1 = \tfrac{1}{2}mv_{1x}^2 = \tfrac{1}{2}(0.200\text{ kg})(0)^2 = 0$$

$$U_1 = \tfrac{1}{2}kx_1^2 = \tfrac{1}{2}(5.00\text{ N/m})(0.100\text{ m})^2 = 0.0250\text{ J}$$

$$K_2 = \tfrac{1}{2}mv_{2x}^2$$

$$U_2 = \tfrac{1}{2}kx_2^2 = \tfrac{1}{2}(5.00\text{ N/m})(0.080\text{ m})^2 = 0.0160\text{ J}$$

We use Eq. (7.11) to solve for K_2 and then find v_{2x}:

$$K_2 = K_1 + U_1 - U_2 = 0 + 0.0250\text{ J} - 0.0160\text{ J} = 0.0090\text{ J}$$

$$v_{2x} = \pm\sqrt{\frac{2K_2}{m}} = \pm\sqrt{\frac{2(0.0090\text{ J})}{0.200\text{ kg}}} = \pm 0.30\text{ m/s}$$

We choose the negative root because the glider is moving in the $-x$-direction. Our answer is $v_{2x} = -0.30$ m/s.

EVALUATE: Eventually the spring will reverse the glider's motion, pushing it back in the $+x$-direction (see Fig. 7.13d). The solution $v_{2x} = +0.30$ m/s tells us that when the glider passes through $x = 0.080$ m on this return trip, its speed will be 0.30 m/s, just as when it passed through this point while moving to the left.

Example 7.8 | Motion with elastic potential energy and work done by other forces

Suppose the glider in Example 7.7 is initially at rest at $x = 0$, with the spring unstretched. You then push on the glider with a constant force $\vec{F}$ (magnitude 0.610 N) in the $+x$-direction. What is the glider's velocity when it has moved to $x = 0.100$ m?

SOLUTION

IDENTIFY and SET UP: Although the force $\vec{F}$ you apply is constant, the spring force isn't, so the acceleration of the glider won't be constant. Total mechanical energy is not conserved because of

Continued

the work done by the force $\vec{F}$, so we must use the generalized energy relationship given by Eq. (7.13). As in Example 7.7, we ignore gravitational potential energy because the glider's height doesn't change. Hence we again have $U = U_{el} = \frac{1}{2}kx^2$. This time, we let point 1 be at $x_1 = 0$, where the velocity is $v_{1x} = 0$, and let point 2 be at $x = 0.100$ m. The glider's displacement is then $\Delta x = x_2 - x_1 = 0.100$ m. Our target variable is v_{2x}, the velocity at point 2.

EXECUTE: The force $\vec{F}$ is constant and in the same direction as the displacement, so the work done by this force is $F\Delta x$. Then the energy quantities are

$$K_1 = 0$$
$$U_1 = \tfrac{1}{2}kx_1^2 = 0$$
$$K_2 = \tfrac{1}{2}mv_{2x}^2$$
$$U_2 = \tfrac{1}{2}kx_2^2 = \tfrac{1}{2}(5.00 \text{ N/m})(0.100 \text{ m})^2 = 0.0250 \text{ J}$$
$$W_{other} = F\Delta x = (0.610 \text{ N})(0.100 \text{ m}) = 0.0610 \text{ J}$$

The initial total mechanical energy is zero; the work done by $\vec{F}$ increases the total mechanical energy to 0.0610 J, of which $U_2 = 0.0250$ J is elastic potential energy. The remainder is kinetic energy. From Eq. (7.13),

$$K_1 + U_1 + W_{other} = K_2 + U_2$$
$$K_2 = K_1 + U_1 + W_{other} - U_2$$

$$= 0 + 0 + 0.0610 \text{ J} - 0.0250 \text{ J} = 0.0360 \text{ J}$$
$$v_{2x} = \sqrt{\frac{2K_2}{m}} = \sqrt{\frac{2(0.0360 \text{ J})}{0.200 \text{ kg}}} = 0.60 \text{ m/s}$$

We choose the positive square root because the glider is moving in the $+x$-direction.

EVALUATE: To test our answer, think what would be different if we disconnected the glider from the spring. Then only $\vec{F}$ would do work, there would be zero elastic potential energy at all times, and Eq. (7.13) would give us

$$K_2 = K_1 + W_{other} = 0 + 0.0610 \text{ J}$$
$$v_{2x} = \sqrt{\frac{2K_2}{m}} = \sqrt{\frac{2(0.0610 \text{ J})}{0.200 \text{ kg}}} = 0.78 \text{ m/s}$$

Our answer $v_{2x} = 0.60$ m/s is less than 0.78 m/s because the spring does negative work on the glider as it stretches (see Fig. 7.13b).

If you stop pushing on the glider when it reaches $x = 0.100$ m, only the spring force does work on it thereafter. Hence for $x > 0.100$ m, the total mechanical energy $E = K + U = 0.0610$ J is constant. As the spring continues to stretch, the glider slows down and the kinetic energy K decreases as the potential energy increases. The glider comes to rest at some point $x = x_3$, at which the kinetic energy is zero and the potential energy $U = U_{el} = \frac{1}{2}kx_3^2$ equals the total mechanical energy 0.0610 J. Can you show that $x_3 = 0.156$ m? (It moves an additional 0.056 m after you stop pushing.) If there is no friction, will the glider remain at rest?

Example 7.9 **Motion with gravitational, elastic, and friction forces**

A 2000-kg (19,600-N) elevator with broken cables in a test rig is falling at 4.00 m/s when it contacts a cushioning spring at the bottom of the shaft. The spring is intended to stop the elevator, compressing 2.00 m as it does so (Fig. 7.17). During the motion a safety clamp applies a constant 17,000-N frictional force to the elevator. What is the necessary force constant k for the spring?

7.17 The fall of an elevator is stopped by a spring and by a constant friction force.

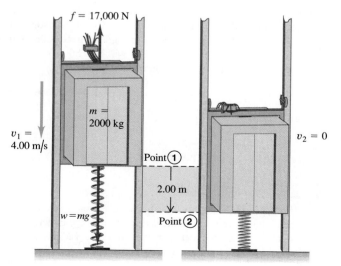

SOLUTION

IDENTIFY and SET UP: We'll use the energy approach to determine k, which appears in the expression for elastic potential energy. This problem involves *both* gravitational and elastic potential energy. Total mechanical energy is not conserved because the friction force does negative work W_{other} on the elevator. We'll therefore use the most general form of the energy relationship, Eq. (7.13). We take point 1 as the position of the bottom of the elevator when it contacts the spring, and point 2 as its position when it stops. We choose the origin to be at point 1, so $y_1 = 0$ and $y_2 = -2.00$ m. With this choice the coordinate of the upper end of the spring after contact is the same as the coordinate of the elevator, so the elastic potential energy at any point between points 1 and 2 is $U_{el} = \frac{1}{2}ky^2$. The gravitational potential energy is $U_{grav} = mgy$ as usual. We know the initial and final speeds of the elevator and the magnitude of the friction force, so the only unknown is the force constant k (our target variable).

EXECUTE: The elevator's initial speed is $v_1 = 4.00$ m/s, so its initial kinetic energy is

$$K_1 = \tfrac{1}{2}mv_1^2 = \tfrac{1}{2}(2000 \text{ kg})(4.00 \text{ m/s})^2 = 16,000 \text{ J}$$

The elevator stops at point 2, so $K_2 = 0$. At point 1 the potential energy $U_1 = U_{grav} + U_{el}$ is zero; U_{grav} is zero because $y_1 = 0$, and $U_{el} = 0$ because the spring is uncompressed. At point 2 there is both gravitational and elastic potential energy, so

$$U_2 = mgy_2 + \tfrac{1}{2}ky_2^2$$

The gravitational potential energy at point 2 is

$$mgy_2 = (2000 \text{ kg})(9.80 \text{ m/s}^2)(-2.00 \text{ m}) = -39,200 \text{ J}$$

The "other" force is the constant 17,000-N friction force. It acts opposite to the 2.00-m displacement, so

$$W_{\text{other}} = -(17,000 \text{ N})(2.00 \text{ m}) = -34,000 \text{ J}$$

We put these terms into Eq. (7.14), $K_1 + U_1 + W_{\text{other}} = K_2 + U_2$:

$$K_1 + 0 + W_{\text{other}} = 0 + (mgy_2 + \tfrac{1}{2}ky_2^2)$$

$$k = \frac{2(K_1 + W_{\text{other}} - mgy_2)}{y_2^2}$$

$$= \frac{2[16,000 \text{ J} + (-34,000 \text{ J}) - (-39,200 \text{ J})]}{(-2.00 \text{ m})^2}$$

$$= 1.06 \times 10^4 \text{ N/m}$$

This is about one-tenth the force constant of a spring in an automobile suspension.

EVALUATE: There might seem to be a paradox here. The elastic potential energy at point 2 is

$$\tfrac{1}{2}ky_2^2 = \tfrac{1}{2}(1.06 \times 10^4 \text{ N/m})(-2.00 \text{ m})^2 = 21,200 \text{ J}$$

This is *more* than the total mechanical energy at point 1:

$$E_1 = K_1 + U_1 = 16,000 \text{ J} + 0 = 16,000 \text{ J}$$

But the friction force *decreased* the mechanical energy of the system by 34,000 J between points 1 and 2. Did energy appear from nowhere? No. At point 2, which is below the origin, there is also *negative* gravitational potential energy $mgy_2 = -39,200 \text{ J}$. The total mechanical energy at point 2 is therefore not 21,200 J but rather

$$E_2 = K_2 + U_2 = 0 + \tfrac{1}{2}ky_2^2 + mgy_2$$

$$= 0 + 21,200 \text{ J} + (-39,200 \text{ J}) = -18,000 \text{ J}$$

This is just the initial mechanical energy of 16,000 J minus 34,000 J lost to friction.

Will the elevator stay at the bottom of the shaft? At point 2 the compressed spring exerts an upward force of magnitude $F_{\text{spring}} = (1.06 \times 10^4 \text{ N/m})(2.00 \text{ m}) = 21,200 \text{ N}$, while the downward force of gravity is only $w = mg = (2000 \text{ kg})(9.80 \text{ m/s}^2) = 19,600 \text{ N}$. If there were no friction, there would be a net upward force of $21,200 \text{ N} - 19,600 \text{ N} = 1600 \text{ N}$, and the elevator would rebound. But the safety clamp can exert a kinetic friction force of 17,000 N, and it can presumably exert a maximum static friction force greater than that. Hence the clamp will keep the elevator from rebounding.

Test Your Understanding of Section 7.2 Consider the situation in Example 7.9 at the instant when the elevator is still moving downward and the spring is compressed by 1.00 m. Which of the energy bar graphs in the figure most accurately shows the kinetic energy K, gravitational potential energy U_{grav}, and elastic potential energy U_{el} at this instant?

(i)

(ii)

(iii)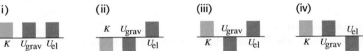

(iv)

7.3 Conservative and Nonconservative Forces

In our discussions of potential energy we have talked about "storing" kinetic energy by converting it to potential energy. We always have in mind that later we may retrieve it again as kinetic energy. For example, when you throw a ball up in the air, it slows down as kinetic energy is converted to gravitational potential energy. But on the way down, the conversion is reversed, and the ball speeds up as potential energy is converted back to kinetic energy. If there is no air resistance, the ball is moving just as fast when you catch it as when you threw it.

Another example is a glider moving on a frictionless horizontal air track that runs into a spring bumper at the end of the track. The glider stops as it compresses the spring and then bounces back. If there is no friction, the glider ends up with the same speed and kinetic energy it had before the collision. Again, there is a two-way conversion from kinetic to potential energy and back. In both cases we can define a potential-energy function so that the total mechanical energy, kinetic plus potential, is constant or *conserved* during the motion.

Conservative Forces

A force that offers this opportunity of two-way conversion between kinetic and potential energies is called a **conservative force.** We have seen two examples of

7.18 The work done by a conservative force such as gravity depends only on the end points of a path, not on the specific path taken between those points.

Because the gravitational force is conservative, the work it does is the same for all three paths.

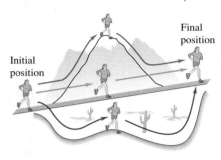

Initial position

Final position

Mastering**PHYSICS**

PhET: The Ramp

conservative forces: the gravitational force and the spring force. (Later in this book we will study another conservative force, the electric force between charged objects.) An essential feature of conservative forces is that their work is always *reversible.* Anything that we deposit in the energy "bank" can later be withdrawn without loss. Another important aspect of conservative forces is that a body may move from point 1 to point 2 by various paths, but the work done by a conservative force is the same for all of these paths (Fig. 7.18). Thus, if a body stays close to the surface of the earth, the gravitational force $m\vec{g}$ is independent of height, and the work done by this force depends only on the change in height. If the body moves around a closed path, ending at the same point where it started, the *total* work done by the gravitational force is always zero.

The work done by a conservative force *always* has four properties:

1. It can be expressed as the difference between the initial and final values of a *potential-energy* function.
2. It is reversible.
3. It is independent of the path of the body and depends only on the starting and ending points.
4. When the starting and ending points are the same, the total work is zero.

When the *only* forces that do work are conservative forces, the total mechanical energy $E = K + U$ is constant.

Nonconservative Forces

Not all forces are conservative. Consider the friction force acting on the crate sliding on a ramp in Example 7.6 (Section 7.1). When the body slides up and then back down to the starting point, the total work done on it by the friction force is *not* zero. When the direction of motion reverses, so does the friction force, and friction does *negative* work in *both* directions. When a car with its brakes locked skids across the pavement with decreasing speed (and decreasing kinetic energy), the lost kinetic energy cannot be recovered by reversing the motion or in any other way, and mechanical energy is *not* conserved. There is *no* potential-energy function for the friction force.

In the same way, the force of fluid resistance (see Section 5.3) is not conservative. If you throw a ball up in the air, air resistance does negative work on the ball while it's rising *and* while it's descending. The ball returns to your hand with less speed and less kinetic energy than when it left, and there is no way to get back the lost mechanical energy.

A force that is not conservative is called a **nonconservative force.** The work done by a nonconservative force *cannot* be represented by a potential-energy function. Some nonconservative forces, like kinetic friction or fluid resistance, cause mechanical energy to be lost or dissipated; a force of this kind is called a **dissipative force.** There are also nonconservative forces that *increase* mechanical energy. The fragments of an exploding firecracker fly off with very large kinetic energy, thanks to a chemical reaction of gunpowder with oxygen. The forces unleashed by this reaction are nonconservative because the process is not reversible. (The fragments never spontaneously reassemble themselves into a complete firecracker!)

Example 7.10 **Frictional work depends on the path**

You are rearranging your furniture and wish to move a 40.0-kg futon 2.50 m across the room. A heavy coffee table, which you don't want to move, blocks this straight-line path. Instead, you slide the futon along a dogleg path; the doglegs are 2.00 m and 1.50 m long. How much more work must you do to push the futon along the dogleg path than along the straight-line path? The coefficient of kinetic friction is $\mu_k = 0.200$.

SOLUTION

IDENTIFY and SET UP: Here both you and friction do work on the futon, so we must use the energy relationship that includes "other" forces. We'll use this relationship to find a connection between the work that *you* do and the work that *friction* does. Figure 7.19 shows our sketch. The futon is at rest at both point 1 and point 2, so

7.19 Our sketch for this problem.

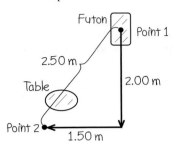

$K_1 = K_2 = 0$. There is no elastic potential energy (there are no springs), and the gravitational potential energy does not change because the futon moves only horizontally, so $U_1 = U_2$. From Eq. (7.14) it follows that $W_{other} = 0$. That "other" work done on the futon is the sum of the positive work you do, W_{you}, and the negative work done by friction, W_{fric}. Since the sum of these is zero, we have

$$W_{you} = -W_{fric}$$

Thus we'll calculate the work done by friction to determine W_{you}.

EXECUTE: The floor is horizontal, so the normal force on the futon equals its weight mg and the magnitude of the friction force is $f_k = \mu_k n = \mu_k mg$. The work you do over each path is then

$$W_{you} = -W_{fric} = -(-f_k s) = +\mu_k mgs$$
$$= (0.200)(40.0 \text{ kg})(9.80 \text{ m/s}^2)(2.50 \text{ m})$$
$$= 196 \text{ J} \quad \text{(straight-line path)}$$

$$W_{you} = -W_{fric} = +\mu_k mgs$$
$$= (0.200)(40.0 \text{ kg})(9.80 \text{ m/s}^2)(2.00 \text{ m} + 1.50 \text{ m})$$
$$= 274 \text{ J} \quad \text{(dogleg path)}$$

The extra work you must do is $274 \text{ J} - 196 \text{ J} = 78 \text{ J}$.

EVALUATE: Friction does different amounts of work on the futon, -196 J and -274 J, on these different paths between points 1 and 2. Hence friction is a *nonconservative* force.

Example 7.11 Conservative or nonconservative?

In a region of space the force on an electron is $\vec{F} = Cx\hat{\jmath}$, where C is a positive constant. The electron moves around a square loop in the xy-plane (Fig. 7.20). Calculate the work done on the electron by the force $\vec{F}$ during a counterclockwise trip around the square. Is this force conservative or nonconservative?

SOLUTION

IDENTIFY and SET UP: The force $\vec{F}$ is not constant, and in general it is not in the same direction as the displacement. To calculate the work done by $\vec{F}$, we'll use the general expression for work, Eq. (6.14):

$$W = \int_{P_1}^{P_2} \vec{F} \cdot d\vec{l}$$

where $d\vec{l}$ is an infinitesimal displacement. We'll calculate the work done on each leg of the square separately, and add the results to find the work done on the round trip. If this round-trip work is zero, force $\vec{F}$ is conservative and can be represented by a potential-energy function.

7.20 An electron moving around a square loop while being acted on by the force $\vec{F} = Cx\hat{\jmath}$.

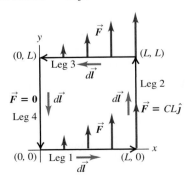

EXECUTE: On the first leg, from $(0, 0)$ to $(L, 0)$, the force is everywhere perpendicular to the displacement. So $\vec{F} \cdot d\vec{l} = 0$, and the work done on the first leg is $W_1 = 0$. The force has the same value $\vec{F} = CL\hat{\jmath}$ everywhere on the second leg, from $(L, 0)$ to (L, L). The displacement on this leg is in the $+y$-direction, so $d\vec{l} = dy\hat{\jmath}$ and

$$\vec{F} \cdot d\vec{l} = CL\hat{\jmath} \cdot dy\hat{\jmath} = CL\, dy$$

The work done on the second leg is then

$$W_2 = \int_{(L, 0)}^{(L, L)} \vec{F} \cdot d\vec{l} = \int_{y=0}^{y=L} CL\, dy = CL \int_0^L dy = CL^2$$

On the third leg, from (L, L) to $(0, L)$, $\vec{F}$ is again perpendicular to the displacement and so $W_3 = 0$. The force is zero on the final leg, from $(0, L)$ to $(0, 0)$, so $W_4 = 0$. The work done by $\vec{F}$ on the round trip is therefore

$$W = W_1 + W_2 + W_3 + W_4 = 0 + CL^2 + 0 + 0 = CL^2$$

The starting and ending points are the same, but the total work done by $\vec{F}$ is not zero. This is a *nonconservative* force; it *cannot* be represented by a potential-energy function.

EVALUATE: Because W is positive, the mechanical energy *increases* as the electron goes around the loop. This is not a mathematical curiosity; it's a much-simplified description of what happens in an electrical generating plant. There, a loop of wire is moved through a magnetic field, which gives rise to a nonconservative force similar to the one here. Electrons in the wire gain energy as they move around the loop, and this energy is carried via transmission lines to the consumer. (We'll discuss how this works in Chapter 29.)

If the electron went *clockwise* around the loop, $\vec{F}$ would be unaffected but the direction of each infinitesimal displacement $d\vec{l}$ would be reversed. Thus the sign of work would also reverse, and the work for a clockwise round trip would be $W = -CL^2$. This is a different behavior than the nonconservative friction force. The work done by friction on a body that slides in any direction over a stationary surface is always negative (see Example 7.6 in Section 7.1).

The Law of Conservation of Energy

Nonconservative forces cannot be represented in terms of potential energy. But we can describe the effects of these forces in terms of kinds of energy other than kinetic and potential energy. When a car with locked brakes skids to a stop, the tires and the road surface both become hotter. The energy associated with this change in the state of the materials is called **internal energy.** Raising the temperature of a body increases its internal energy; lowering the body's temperature decreases its internal energy.

To see the significance of internal energy, let's consider a block sliding on a rough surface. Friction does *negative* work on the block as it slides, and the change in internal energy of the block and surface (both of which get hotter) is *positive.* Careful experiments show that the increase in the internal energy is *exactly* equal to the absolute value of the work done by friction. In other words,

$$\Delta U_{int} = -W_{other}$$

where ΔU_{int} is the change in internal energy. If we substitute this into Eq. (7.7) or (7.14), we find

$$K_1 + U_1 - \Delta U_{int} = K_2 + U_2$$

Writing $\Delta K = K_2 - K_1$ and $\Delta U = U_2 - U_1$, we can finally express this as

$$\Delta K + \Delta U + \Delta U_{int} = 0 \qquad \text{(law of conservation of energy)} \qquad (7.15)$$

This remarkable statement is the general form of the **law of conservation of energy.** In a given process, the kinetic energy, potential energy, and internal energy of a system may all change. But the *sum* of those changes is always zero. If there is a decrease in one form of energy, it is made up for by an increase in the other forms (Fig. 7.21). When we expand our definition of energy to include internal energy, Eq. (7.15) says: *Energy is never created or destroyed; it only changes form.* No exception to this rule has ever been found.

The concept of work has been banished from Eq. (7.15); instead, it suggests that we think purely in terms of the conversion of energy from one form to another. For example, when you throw a baseball straight up, you convert a portion of the internal energy of your molecules to kinetic energy of the baseball. This is converted to gravitational potential energy as the ball climbs and back to kinetic energy as the ball falls. If there is air resistance, part of the energy is used to heat up the air and the ball and increase their internal energy. Energy is converted back to the kinetic form as the ball falls. If you catch the ball in your hand, whatever energy was not lost to the air once again becomes internal energy; the ball and your hand are now warmer than they were at the beginning.

In Chapters 19 and 20, we will study the relationship of internal energy to temperature changes, heat, and work. This is the heart of the area of physics called *thermodynamics.*

7.21 When 1 liter of gasoline is burned in an automotive engine, it releases 3.3×10^7 J of internal energy. Hence $\Delta U_{int} = -3.3 \times 10^7$ J, where the minus sign means that the amount of energy stored in the gasoline has decreased. This energy can be converted to kinetic energy (making the car go faster) or to potential energy (enabling the car to climb uphill).

Conceptual Example 7.12 **Work done by friction**

Let's return to Example 7.5 (Section 7.1), in which Throcky skateboards down a curved ramp. He starts with zero kinetic energy and 735 J of potential energy, and at the bottom he has 450 J of kinetic energy and zero potential energy; hence $\Delta K = +450$ J and $\Delta U = -735$ J. The work $W_{other} = W_{fric}$ done by the friction forces is -285 J, so the change in internal energy is $\Delta U_{int} = -W_{other} = +285$ J. The skateboard wheels and bearings

and the ramp all get a little warmer. In accordance with Eq. (7.15), the sum of the energy changes equals zero:

$$\Delta K + \Delta U + \Delta U_{int} = +450 \text{ J} + (-735 \text{ J}) + 285 \text{ J} = 0$$

The total energy of the system (including internal, nonmechanical forms of energy) is conserved.

Test Your Understanding of Section 7.3 In a hydroelectric generating station, falling water is used to drive turbines ("water wheels"), which in turn run electric generators. Compared to the amount of gravitational potential energy released by the falling water, how much electrical energy is produced? (i) the same; (ii) more; (iii) less.

7.4 Force and Potential Energy

For the two kinds of conservative forces (gravitational and elastic) we have studied, we started with a description of the behavior of the *force* and derived from that an expression for the *potential energy*. For example, for a body with mass m in a uniform gravitational field, the gravitational force is $F_y = -mg$. We found that the corresponding potential energy is $U(y) = mgy$. To stretch an ideal spring by a distance x, we exert a force equal to $+kx$. By Newton's third law the force that an ideal spring exerts on a body is opposite this, or $F_x = -kx$. The corresponding potential energy function is $U(x) = \frac{1}{2}kx^2$.

In studying physics, however, you'll encounter situations in which you are given an expression for the *potential energy* as a function of position and have to find the corresponding *force*. We'll see several examples of this kind when we study electric forces later in this book: It's often far easier to calculate the electric potential energy first and then determine the corresponding electric force afterward.

Here's how we find the force that corresponds to a given potential-energy expression. First let's consider motion along a straight line, with coordinate x. We denote the x-component of force, a function of x, by $F_x(x)$, and the potential energy as $U(x)$. This notation reminds us that both F_x and U are *functions* of x. Now we recall that in any displacement, the work W done by a conservative force equals the negative of the change ΔU in potential energy:

$$W = -\Delta U$$

Let's apply this to a small displacement Δx. The work done by the force $F_x(x)$ during this displacement is approximately equal to $F_x(x)\,\Delta x$. We have to say "approximately" because $F_x(x)$ may vary a little over the interval Δx. But it is at least approximately true that

$$F_x(x)\,\Delta x = -\Delta U \qquad \text{and} \qquad F_x(x) = -\frac{\Delta U}{\Delta x}$$

You can probably see what's coming. We take the limit as $\Delta x \to 0$; in this limit, the variation of F_x becomes negligible, and we have the exact relationship

$$F_x(x) = -\frac{dU(x)}{dx} \qquad \text{(force from potential energy, one dimension)} \quad (7.16)$$

This result makes sense; in regions where $U(x)$ changes most rapidly with x (that is, where $dU(x)/dx$ is large), the greatest amount of work is done during a given displacement, and this corresponds to a large force magnitude. Also, when $F_x(x)$ is in the positive x-direction, $U(x)$ *decreases* with increasing x. So $F_x(x)$ and $dU(x)/dx$ should indeed have opposite signs. The physical meaning of Eq. (7.16) is that *a conservative force always acts to push the system toward lower potential energy.*

As a check, let's consider the function for elastic potential energy, $U(x) = \frac{1}{2}kx^2$. Substituting this into Eq. (7.16) yields

$$F_x(x) = -\frac{d}{dx}\left(\tfrac{1}{2}kx^2\right) = -kx$$

which is the correct expression for the force exerted by an ideal spring (Fig. 7.22a). Similarly, for gravitational potential energy we have $U(y) = mgy$; taking care to change x to y for the choice of axis, we get $F_y = -dU/dy = -d(mgy)/dy = -mg$, which is the correct expression for gravitational force (Fig. 7.22b).

7.22 A conservative force is the negative derivative of the corresponding potential energy.

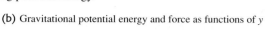

(a) Spring potential energy and force as functions of x

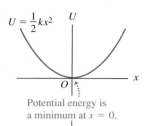

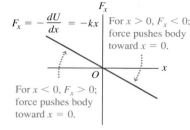

$U = \frac{1}{2}kx^2$

Potential energy is a minimum at $x = 0$.

$F_x = -\dfrac{dU}{dx} = -kx$

For $x > 0$, $F_x < 0$; force pushes body toward $x = 0$.

For $x < 0$, $F_x > 0$; force pushes body toward $x = 0$.

(b) Gravitational potential energy and force as functions of y

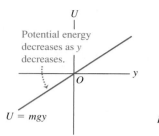

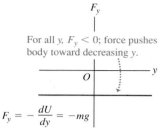

Potential energy decreases as y decreases.

$U = mgy$

For all y, $F_y < 0$; force pushes body toward decreasing y.

$F_y = -\dfrac{dU}{dy} = -mg$

Example 7.13 **An electric force and its potential energy**

An electrically charged particle is held at rest at the point $x = 0$; a second particle with equal charge is free to move along the positive x-axis. The potential energy of the system is $U(x) = C/x$, where C is a positive constant that depends on the magnitude of the charges. Derive an expression for the x-component of force acting on the movable particle as a function of its position.

SOLUTION

IDENTIFY and SET UP: We are given the potential-energy function $U(x)$. We'll find the corresponding force function using Eq. (7.16), $F_x(x) = -dU(x)/dx$.

EXECUTE: The derivative of $1/x$ with respect to x is $-1/x^2$. So for $x > 0$ the force on the movable charged particle $x > 0$ is

$$F_x(x) = -\frac{dU(x)}{dx} = -C\left(-\frac{1}{x^2}\right) = \frac{C}{x^2}$$

EVALUATE: The x-component of force is positive, corresponding to a repulsion between like electric charges. Both the potential energy and the force are very large when the particles are close together (small x), and both get smaller as the particles move farther apart (large x); the force pushes the movable particle toward large positive values of x, where the potential energy is lower. (We'll study electric forces in detail in Chapter 21.)

Force and Potential Energy in Three Dimensions

We can extend this analysis to three dimensions, where the particle may move in the x-, y-, or z-direction, or all at once, under the action of a conservative force that has components F_x, F_y, and F_z. Each component of force may be a function of the coordinates x, y, and z. The potential-energy function U is also a function of all three space coordinates. We can now use Eq. (7.16) to find each component of force. The potential-energy change ΔU when the particle moves a small distance Δx in the x-direction is again given by $-F_x \Delta x$; it doesn't depend on F_y and F_z, which represent force components that are perpendicular to the displacement and do no work. So we again have the approximate relationship

$$F_x = -\frac{\Delta U}{\Delta x}$$

The y- and z-components of force are determined in exactly the same way:

$$F_y = -\frac{\Delta U}{\Delta y} \qquad F_z = -\frac{\Delta U}{\Delta z}$$

To make these relationships exact, we take the limits $\Delta x \to 0$, $\Delta y \to 0$, and $\Delta z \to 0$ so that these ratios become derivatives. Because U may be a function of all three coordinates, we need to remember that when we calculate each of these derivatives, only one coordinate changes at a time. We compute the derivative of U with respect to x by assuming that y and z are constant and only x varies, and so on. Such a derivative is called a *partial derivative*. The usual notation for a partial derivative is $\partial U/\partial x$ and so on; the symbol ∂ is a modified d. So we write

$$F_x = -\frac{\partial U}{\partial x} \qquad F_y = -\frac{\partial U}{\partial y} \qquad F_z = -\frac{\partial U}{\partial z} \qquad \begin{matrix}\text{(force from}\\ \text{potential energy)}\end{matrix} \qquad (7.17)$$

We can use unit vectors to write a single compact vector expression for the force $\vec{F}$:

$$\vec{F} = -\left(\frac{\partial U}{\partial x}\hat{\imath} + \frac{\partial U}{\partial y}\hat{\jmath} + \frac{\partial U}{\partial z}\hat{k}\right) \qquad \text{(force from potential energy)} \qquad (7.18)$$

The expression inside the parentheses represents a particular operation on the function U, in which we take the partial derivative of U with respect to each coordinate, multiply by the corresponding unit vector, and then take the vector sum. This operation is called the **gradient** of U and is often abbreviated as $\vec{\nabla}U$. Thus the force is the negative of the gradient of the potential-energy function:

$$\vec{F} = -\vec{\nabla}U \qquad (7.19)$$

As a check, let's substitute into Eq. (7.19) the function $U = mgy$ for gravitational potential energy:

$$\vec{F} = -\vec{\nabla}(mgy) = -\left(\frac{\partial(mgy)}{\partial x}\hat{\imath} + \frac{\partial(mgy)}{\partial y}\hat{\jmath} + \frac{\partial(mgy)}{\partial z}\hat{k}\right) = (-mg)\hat{\jmath}$$

This is just the familiar expression for the gravitational force.

Application Topography and Potential Energy Gradient

The greater the elevation of a hiker in Canada's Banff National Park, the greater is the gravitational potential energy U_{grav}. Think of an x-axis that runs horizontally from west to east and a y-axis that runs horizontally from south to north. Then the function $U_{\text{grav}}(x, y)$ tells us the elevation as a function of position in the park. Where the mountains have steep slopes, $\vec{F} = -\vec{\nabla}U_{\text{grav}}$ has a large magnitude and there's a strong force pushing you along the mountain's surface toward a region of lower elevation (and hence lower U_{grav}). There's zero force along the surface of the lake, which is all at the same elevation. Hence U_{grav} is constant at all points on the lake surface, and $\vec{F} = -\vec{\nabla}U_{\text{grav}} = \mathbf{0}$.

Example 7.14 Force and potential energy in two dimensions

A puck with coordinates x and y slides on a level, frictionless air-hockey table. It is acted on by a conservative force described by the potential-energy function

$$U(x, y) = \tfrac{1}{2}k(x^2 + y^2)$$

Find a vector expression for the force acting on the puck, and find an expression for the magnitude of the force.

SOLUTION

IDENTIFY and SET UP: Starting with the function $U(x, y)$, we need to find the vector components and magnitude of the corresponding force $\vec{F}$. We'll find the components using Eq. (7.18). The function U doesn't depend on z, so the partial derivative of U with respect to z is $\partial U/\partial z = 0$ and the force has no z-component. We'll determine the magnitude F of the force using $F = \sqrt{F_x^2 + F_y^2}$.

EXECUTE: The x- and y-components of $\vec{F}$ are

$$F_x = -\frac{\partial U}{\partial x} = -kx \qquad F_y = -\frac{\partial U}{\partial y} = -ky$$

From Eq. (7.18), the vector expression for the force is

$$\vec{F} = (-kx)\hat{\imath} + (-ky)\hat{\jmath} = -k(x\hat{\imath} + y\hat{\jmath})$$

The magnitude of the force is

$$F = \sqrt{(-kx)^2 + (-ky)^2} = k\sqrt{x^2 + y^2} = kr$$

EVALUATE: Because $x\hat{\imath} + y\hat{\jmath}$ is just the position vector $\vec{r}$ of the particle, we can rewrite our result as $\vec{F} = -k\vec{r}$. This represents a force that is opposite in direction to the particle's position vector—that is, a force directed toward the origin, $r = 0$. This is the force that would be exerted on the puck if it were attached to one end of a spring that obeys Hooke's law and has a negligibly small unstretched length compared to the other distances in the problem. (The other end is attached to the air-hockey table at $r = 0$.)

To check our result, note that $U = \tfrac{1}{2}kr^2$, where $r^2 = x^2 + y^2$. We can find the force from this expression using Eq. (7.16) with x replaced by r:

$$F_r = -\frac{dU}{dr} = -\frac{d}{dr}\left(\tfrac{1}{2}kr^2\right) = -kr$$

As we found above, the force has magnitude kr; the minus sign indicates that the force is toward the origin (at $r = 0$).

7.23 (a) A glider on an air track. The spring exerts a force $F_x = -kx$. (b) The potential-energy function.

(a)

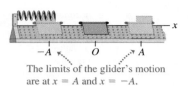

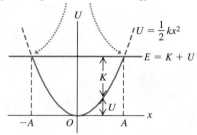

The limits of the glider's motion are at $x = A$ and $x = -A$.

(b)

On the graph, the limits of motion are the points where the U curve intersects the horizontal line representing total mechanical energy E.

$U = \frac{1}{2}kx^2$

$E = K + U$

Application Acrobats in Equilibrium
Each of these acrobats is in *unstable* equilibrium. The gravitational potential energy is lower no matter which way an acrobat tips, so if she begins to fall she will keep on falling. Staying balanced requires the acrobats' constant attention.

Test Your Understanding of Section 7.4 A particle moving along the x-axis is acted on by a conservative force F_x. At a certain point, the force is zero. (a) Which of the following statements about the value of the potential-energy function $U(x)$ at that point is correct? (i) $U(x) = 0$; (ii) $U(x) > 0$; (iii) $U(x) < 0$; (iv) not enough information is given to decide. (b) Which of the following statements about the value of the derivative of $U(x)$ at that point is correct? (i) $dU(x)/dx = 0$; (ii) $dU(x)/dx > 0$; (iii) $dU(x)/dx < 0$; (iv) not enough information is given to decide.

7.5 Energy Diagrams

When a particle moves along a straight line under the action of a conservative force, we can get a lot of insight into its possible motions by looking at the graph of the potential-energy function $U(x)$. Figure 7.23a shows a glider with mass m that moves along the x-axis on an air track. The spring exerts on the glider a force with x-component $F_x = -kx$. Figure 7.23b is a graph of the corresponding potential-energy function $U(x) = \frac{1}{2}kx^2$. If the elastic force of the spring is the *only* horizontal force acting on the glider, the total mechanical energy $E = K + U$ is constant, independent of x. A graph of E as a function of x is thus a straight horizontal line. We use the term **energy diagram** for a graph like this, which shows both the potential-energy function $U(x)$ and the energy of the particle subjected to the force that corresponds to $U(x)$.

The vertical distance between the U and E graphs at each point represents the difference $E - U$, equal to the kinetic energy K at that point. We see that K is greatest at $x = 0$. It is zero at the values of x where the two graphs cross, labeled A and $-A$ in the diagram. Thus the speed v is greatest at $x = 0$, and it is zero at $x = \pm A$, the points of *maximum* possible displacement from $x = 0$ for a given value of the total energy E. The potential energy U can never be greater than the total energy E; if it were, K would be negative, and that's impossible. The motion is a back-and-forth oscillation between the points $x = A$ and $x = -A$.

At each point, the force F_x on the glider is equal to the negative of the slope of the $U(x)$ curve: $F_x = -dU/dx$ (see Fig. 7.22a). When the particle is at $x = 0$, the slope and the force are zero, so this is an *equilibrium* position. When x is positive, the slope of the $U(x)$ curve is positive and the force F_x is negative, directed toward the origin. When x is negative, the slope is negative and F_x is positive, again directed toward the origin. Such a force is called a *restoring force;* when the glider is displaced to either side of $x = 0$, the force tends to "restore" it back to $x = 0$. An analogous situation is a marble rolling around in a round-bottomed bowl. We say that $x = 0$ is a point of **stable equilibrium.** More generally, *any minimum in a potential-energy curve is a stable equilibrium position.*

Figure 7.24a shows a hypothetical but more general potential-energy function $U(x)$. Figure 7.24b shows the corresponding force $F_x = -dU/dx$. Points x_1 and x_3 are stable equilibrium points. At each of these points, F_x is zero because the slope of the $U(x)$ curve is zero. When the particle is displaced to either side, the force pushes back toward the equilibrium point. The slope of the $U(x)$ curve is also zero at points x_2 and x_4, and these are also equilibrium points. But when the particle is displaced a little to the right of either point, the slope of the $U(x)$ curve becomes negative, corresponding to a positive F_x that tends to push the particle still farther from the point. When the particle is displaced a little to the left, F_x is negative, again pushing away from equilibrium. This is analogous to a marble rolling on the top of a bowling ball. Points x_2 and x_4 are called **unstable equilibrium** points; *any maximum in a potential-energy curve is an unstable equilibrium position.*

7.24 The maxima and minima of a potential-energy function $U(x)$ correspond to points where $F_x = 0$.

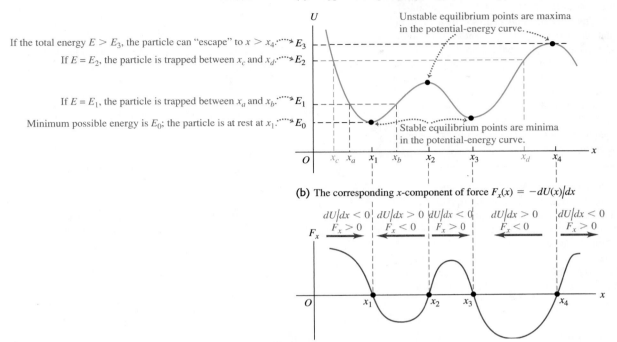

(a) A hypothetical potential-energy function $U(x)$

If the total energy $E > E_3$, the particle can "escape" to $x > x_4$. $\cdots\rightarrow E_3$

If $E = E_2$, the particle is trapped between x_c and x_d. $\cdots\rightarrow E_2$

If $E = E_1$, the particle is trapped between x_a and x_b. $\cdots\rightarrow E_1$

Minimum possible energy is E_0; the particle is at rest at x_1. $\cdots\rightarrow E_0$

Unstable equilibrium points are maxima in the potential-energy curve.

Stable equilibrium points are minima in the potential-energy curve.

(b) The corresponding x-component of force $F_x(x) = -dU(x)/dx$

$dU/dx < 0$ $dU/dx > 0$ $dU/dx < 0$ $dU/dx > 0$ $dU/dx < 0$
$F_x > 0$ $F_x < 0$ $F_x > 0$ $F_x < 0$ $F_x > 0$

CAUTION **Potential energy and the direction of a conservative force** The direction of the force on a body is *not* determined by the sign of the potential energy U. Rather, it's the sign of $F_x = -dU/dx$ that matters. As we discussed in Section 7.1, the physically significant quantity is the *difference* in the values of U between two points, which is just what the derivative $F_x = -dU/dx$ measures. This means that you can always add a constant to the potential-energy function without changing the physics of the situation.

Mastering**PHYSICS**

PhET: Energy Skate Park

If the total energy is E_1 and the particle is initially near x_1, it can move only in the region between x_a and x_b determined by the intersection of the E_1 and U graphs (Fig. 7.24a). Again, U cannot be greater than E_1 because K can't be negative. We speak of the particle as moving in a *potential well*, and x_a and x_b are the *turning points* of the particle's motion (since at these points, the particle stops and reverses direction). If we increase the total energy to the level E_2, the particle can move over a wider range, from x_c to x_d. If the total energy is greater than E_3, the particle can "escape" and move to indefinitely large values of x. At the other extreme, E_0 represents the least possible total energy the system can have.

Test Your Understanding of Section 7.5 The curve in Fig. 7.24b has a maximum at a point between x_2 and x_3. Which statement correctly describes what happens to the particle when it is at this point? (i) The particle's acceleration is zero. (ii) The particle accelerates in the positive x-direction; the magnitude of the acceleration is less than at any other point between x_2 and x_3. (iii) The particle accelerates in the positive x-direction; the magnitude of the acceleration is greater than at any other point between x_2 and x_3. (iv) The particle accelerates in the negative x-direction; the magnitude of the acceleration is less than at any other point between x_2 and x_3. (v) The particle accelerates in the negative x-direction; the magnitude of the acceleration is greater than at any other point between x_2 and x_3.

Gravitational potential energy and elastic potential energy: The work done on a particle by a constant gravitational force can be represented as a change in the gravitational potential energy $U_{\text{grav}} = mgy$. This energy is a shared property of the particle and the earth. A potential energy is also associated with the elastic force $F_x = -kx$ exerted by an ideal spring, where x is the amount of stretch or compression. The work done by this force can be represented as a change in the elastic potential energy of the spring, $U_{\text{el}} = \frac{1}{2}kx^2$.

$$W_{\text{grav}} = mgy_1 - mgy_2$$
$$= U_{\text{grav},1} - U_{\text{grav},2}$$
$$= -\Delta U_{\text{grav}} \qquad (7.1), (7.3)$$

$$W_{\text{el}} = \frac{1}{2}kx_1^2 - \frac{1}{2}kx_2^2$$
$$= U_{\text{el},1} - U_{\text{el},2} = -\Delta U_{\text{el}} \qquad (7.10)$$

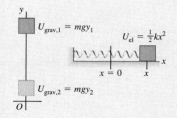

When total mechanical energy is conserved: The total potential energy U is the sum of the gravitational and elastic potential energy: $U = U_{\text{grav}} + U_{\text{el}}$. If no forces other than the gravitational and elastic forces do work on a particle, the sum of kinetic and potential energy is conserved. This sum $E = K + U$ is called the total mechanical energy. (See Examples 7.1, 7.3, 7.4, and 7.7.)

$$K_1 + U_1 = K_2 + U_2 \qquad (7.4), (7.11)$$

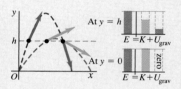

When total mechanical energy is not conserved: When forces other than the gravitational and elastic forces do work on a particle, the work W_{other} done by these other forces equals the change in total mechanical energy (kinetic energy plus total potential energy). (See Examples 7.2, 7.5, 7.6, 7.8, and 7.9.)

$$K_1 + U_1 + W_{\text{other}} = K_2 + U_2 \qquad (7.14)$$

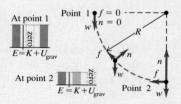

Conservative forces, nonconservative forces, and the law of conservation of energy: All forces are either conservative or nonconservative. A conservative force is one for which the work–kinetic energy relationship is completely reversible. The work of a conservative force can always be represented by a potential-energy function, but the work of a nonconservative force cannot. The work done by nonconservative forces manifests itself as changes in the internal energy of bodies. The sum of kinetic, potential, and internal energy is always conserved. (See Examples 7.10–7.12.)

$$\Delta K + \Delta U + \Delta U_{\text{int}} = 0 \qquad (7.15)$$

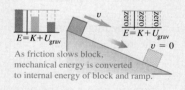

As friction slows block, mechanical energy is converted to internal energy of block and ramp.

Determining force from potential energy: For motion along a straight line, a conservative force $F_x(x)$ is the negative derivative of its associated potential-energy function U. In three dimensions, the components of a conservative force are negative partial derivatives of U. (See Examples 7.13 and 7.14.)

$$F_x(x) = -\frac{dU(x)}{dx} \qquad (7.16)$$

$$F_x = -\frac{\partial U}{\partial x} \qquad F_y = -\frac{\partial U}{\partial y} \qquad (7.17)$$

$$F_z = -\frac{\partial U}{\partial z}$$

$$\vec{F} = -\left(\frac{\partial U}{\partial x}\hat{\imath} + \frac{\partial U}{\partial y}\hat{\jmath} + \frac{\partial U}{\partial z}\hat{k}\right) \qquad (7.18)$$

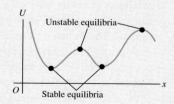

BRIDGING PROBLEM   A Spring and Friction on an Incline

A 2.00-kg package is released on a 53.1° incline, 4.00 m from a long spring with force constant 1.20×10^2 N/m that is attached at the bottom of the incline (Fig. 7.25). The coefficients of friction between the package and incline are $\mu_s = 0.400$ and $\mu_k = 0.200$. The mass of the spring is negligible.

7.25 The initial situation.

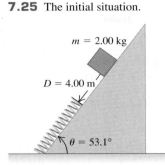

$m = 2.00$ kg

$D = 4.00$ m

$\theta = 53.1°$

(a) What is the maximum compression of the spring? (b) The package rebounds up the incline. How close does it get to its original position? (c) What is the change in the internal energy of the package and incline from when the package is released to when it rebounds to its maximum height?

SOLUTION GUIDE

See MasteringPhysics® study area for a Video Tutor solution.

IDENTIFY and SET UP

1. This problem involves the gravitational force, a spring force, and the friction force, as well as the normal force that acts on the package. Since the spring force isn't constant, you'll have to use energy methods. Is mechanical energy conserved during any part of the motion? Why or why not?
2. Draw free-body diagrams for the package as it is sliding down the incline and sliding back up the incline. Include your choice of coordinate axis. (*Hint:* If you choose $x = 0$ to be at the end of the uncompressed spring, you'll be able to use $U_{el} = \frac{1}{2}kx^2$ for the elastic potential energy of the spring.)
3. Label the three critical points in the package's motion: its starting position, its position when it comes to rest with the spring maximally compressed, and its position when it's rebounded as far as possible up the incline. (*Hint:* You can assume that the

package is no longer in contact with the spring at the last of these positions. If this turns out to be incorrect, you'll calculate a value of x that tells you the spring is still partially compressed at this point.)
4. Make a list of the unknown quantities and decide which of these are the target variables.

EXECUTE

5. Find the magnitude of the friction force that acts on the package. Does the magnitude of this force depend on whether the package is moving up or down the incline, or on whether or not the package is in contact with the spring? Does the *direction* of the normal force depend on any of these?
6. Write the general energy equation for the motion of the package between the first two points you labeled in step 3. Use this equation to solve for the distance that the spring is compressed when the package is at its lowest point. (*Hint:* You'll have to solve a quadratic equation. To decide which of the two solutions of this equation is the correct one, remember that the distance the spring is compressed is positive.)
7. Write the general energy equation for the motion of the package between the second and third points you labeled in step 3. Use this equation to solve for how far the package rebounds.
8. Calculate the change in internal energy for the package's trip down and back up the incline. Remember that the amount the internal energy *increases* is equal to the amount the total mechanical energy *decreases*.

EVALUATE

9. Was it correct to assume in part (b) that the package is no longer in contact with the spring when it reaches it maximum rebound height?
10. Check your result for part (c) by finding the total work done by the force of friction over the entire trip. Is this in accordance with your result from step 8?

Problems

For instructor-assigned homework, go to www.masteringphysics.com

•, ••, •••: Problems of increasing difficulty. **CP**: Cumulative problems incorporating material from earlier chapters. **CALC**: Problems requiring calculus. **BIO**: Biosciences problems.

DISCUSSION QUESTIONS

Q7.1 A baseball is thrown straight up with initial speed v_0. If air resistance cannot be ignored, when the ball returns to its initial height its speed is less than v_0. Explain why, using energy concepts.

Q7.2 A projectile has the same initial kinetic energy no matter what the angle of projection. Why doesn't it rise to the same maximum height in each case?

Q7.3 An object is released from rest at the top of a ramp. If the ramp is frictionless, does the object's speed at the bottom of the ramp depend on the shape of the ramp or just on its height? Explain. What if the ramp is *not* frictionless?

Q7.4 An egg is released from rest from the roof of a building and falls to the ground. Its fall is observed by a student on the roof of the building, who uses coordinates with origin at the roof, and by a student on the ground, who uses coordinates with origin at the ground. Do the two students assign the same or different values to the initial gravitational potential energy, the final gravitational potential energy, the change in gravitational potential energy, and the kinetic energy of the egg just before it strikes the ground? Explain.

Q7.5 A physics teacher had a bowling ball suspended from a very long rope attached to the high ceiling of a large lecture hall. To illustrate his faith in conservation of energy, he would back up to one side of the stage, pull the ball far to one side until the taut rope brought it just to the end of his nose, and then release it. The massive ball would swing in a mighty arc across the stage and then return to stop momentarily just in front of the nose of the stationary, unflinching teacher. However, one day after the demonstration he looked up just in time to see a student at the other side of the stage *push* the ball away from his nose as he tried to duplicate the demonstration. Tell the rest of the story and explain the reason for the potentially tragic outcome.

Q7.6 Lost Energy?   The principle of the conservation of energy tells us that energy is never lost, but only changes from one form to another. Yet in many ordinary situations, energy may appear to be lost. In each case, explain what happens to the "lost" energy. (a) A box sliding on the floor comes to a halt due to friction. How did friction take away its kinetic energy, and what happened to that energy? (b) A car stops when you apply the brakes. What happened to its kinetic energy? (c) Air resistance uses up some of the original gravitational potential energy of a falling object. What type of energy did the "lost" potential energy become? (d) When a returning space shuttle touches down on the runway, it has lost almost all its kinetic energy and gravitational potential energy. Where did all that energy go?

Q7.7 Is it possible for a frictional force to *increase* the mechanical energy of a system? If so, give examples.

Q7.8 A woman bounces on a trampoline, going a little higher with each bounce. Explain how she increases the total mechanical energy.

Q7.9 Fractured Physics.   People often call their electric bill a *power* bill, yet the quantity on which the bill is based is expressed in *kilowatt-hours*. What are people really being billed for?

Q7.10 A rock of mass m and a rock of mass $2m$ are both released from rest at the same height and feel no air resistance as they fall. Which statements about these rocks are true? (There may be more than one correct choice.) (a) Both have the same initial gravitational potential energy. (b) Both have the same kinetic energy when they reach the ground. (c) Both reach the ground with the same speed. (d) When it reaches the ground, the heavier rock has twice the kinetic energy of the lighter one. (e) When it reaches the ground, the heavier rock has four times the kinetic energy of the lighter one.

Q7.11 On a friction-free ice pond, a hockey puck is pressed against (but not attached to) a fixed ideal spring, compressing the spring by a distance x_0. The maximum energy stored in the spring is U_0, the maximum speed the puck gains after being released is v_0, and its maximum kinetic energy is K_0. Now the puck is pressed so it compresses the spring twice as far as before. In this case, (a) what is the maximum potential energy stored in the spring (in terms of U_0), and (b) what are the puck's maximum kinetic energy and speed (in terms of K_0 and x_0)?

Q7.12 When people are cold, they often rub their hands together to warm them up. How does doing this produce heat? Where did the heat come from?

Q7.13 You often hear it said that most of our energy ultimately comes from the sun. Trace each of the following energies back to the sun: (a) the kinetic energy of a jet plane; (b) the potential energy gained by a mountain climber; (c) the electrical energy used to run a computer; (d) the electrical energy from a hydroelectric plant.

Q7.14 A box slides down a ramp and work is done on the box by the forces of gravity and friction. Can the work of each of these forces be expressed in terms of the change in a potential-energy function? For each force explain why or why not.

Q7.15 In physical terms, explain why friction is a nonconservative force. Does it store energy for future use?

Q7.16 A compressed spring is clamped in its compressed position and then is dissolved in acid. What becomes of its potential energy?

Q7.17 Since only changes in potential energy are important in any problem, a student decides to let the elastic potential energy of a spring be zero when the spring is stretched a distance x_1. The student decides, therefore, to let $U = \frac{1}{2}k(x - x_1)^2$. Is this correct? Explain.

Q7.18 Figure 7.22a shows the potential-energy function for the force $F_x = -kx$. Sketch the potential-energy function for the force $F_x = +kx$. For this force, is $x = 0$ a point of equilibrium? Is this equilibrium stable or unstable? Explain.

Q7.19 Figure 7.22b shows the potential-energy function associated with the gravitational force between an object and the earth. Use this graph to explain why objects always fall toward the earth when they are released.

Q7.20 For a system of two particles we often let the potential energy for the force between the particles approach zero as the separation of the particles approaches infinity. If this choice is made, explain why the potential energy at noninfinite separation is positive if the particles repel one another and negative if they attract.

Q7.21 Explain why the points $x = A$ and $x = -A$ in Fig. 7.23b are called *turning points*. How are the values of E and U related at a turning point?

Q7.22 A particle is in *neutral equilibrium* if the net force on it is zero and remains zero if the particle is displaced slightly in any direction. Sketch the potential-energy function near a point of neutral equilibrium for the case of one-dimensional motion. Give an example of an object in neutral equilibrium.

Q7.23 The net force on a particle of mass m has the potential-energy function graphed in Fig. 7.24a. If the total energy is E_1, graph the speed v of the particle versus its position x. At what value of x is the speed greatest? Sketch v versus x if the total energy is E_2.

Q7.24 The potential-energy function for a force $\vec{F}$ is $U = \alpha x^3$, where α is a positive constant. What is the direction of $\vec{F}$?

EXERCISES

Section 7.1 Gravitational Potential Energy

7.1 • In one day, a 75-kg mountain climber ascends from the 1500-m level on a vertical cliff to the top at 2400 m. The next day, she descends from the top to the base of the cliff, which is at an elevation of 1350 m. What is her change in gravitational potential energy (a) on the first day and (b) on the second day?

7.2 • BIO **How High Can We Jump?**   The maximum height a typical human can jump from a crouched start is about 60 cm. By how much does the gravitational potential energy increase for a 72-kg person in such a jump? Where does this energy come from?

7.3 •• CP A 120-kg mail bag hangs by a vertical rope 3.5 m long. A postal worker then displaces the bag to a position 2.0 m sideways from its original position, always keeping the rope taut. (a) What horizontal force is necessary to hold the bag in the new position? (b) As the bag is moved to this position, how much work is done (i) by the rope and (ii) by the worker?

7.4 •• BIO **Food Calories.**   The *food calorie*, equal to 4186 J, is a measure of how much energy is released when food is metabolized by the body. A certain brand of fruit-and-cereal bar contains

140 food calories per bar. (a) If a 65-kg hiker eats one of these bars, how high a mountain must he climb to "work off" the calories, assuming that all the food energy goes only into increasing gravitational potential energy? (b) If, as is typical, only 20% of the food calories go into mechanical energy, what would be the answer to part (a)? (*Note:* In this and all other problems, we are assuming that 100% of the food calories that are eaten are absorbed and used by the body. This is actually not true. A person's "metabolic efficiency" is the percentage of calories eaten that are actually used; the rest are eliminated by the body. Metabolic efficiency varies considerably from person to person.)

7.5 • A baseball is thrown from the roof of a 22.0-m-tall building with an initial velocity of magnitude 12.0 m/s and directed at an angle of 53.1° above the horizontal. (a) What is the speed of the ball just before it strikes the ground? Use energy methods and ignore air resistance. (b) What is the answer for part (a) if the initial velocity is at an angle of 53.1° *below* the horizontal? (c) If the effects of air resistance are included, will part (a) or (b) give the higher speed?

7.6 •• A crate of mass M starts from rest at the top of a frictionless ramp inclined at an angle α above the horizontal. Find its speed at the bottom of the ramp, a distance d from where it started. Do this in two ways: (a) Take the level at which the potential energy is zero to be at the bottom of the ramp with y positive upward. (b) Take the zero level for potential energy to be at the top of the ramp with y positive upward. (c) Why did the normal force not enter into your solution?

7.7 •• **BIO Human Energy vs. Insect Energy.** For its size, the common flea is one of the most accomplished jumpers in the animal world. A 2.0-mm-long, 0.50-mg critter can reach a height of 20 cm in a single leap. (a) Neglecting air drag, what is the takeoff speed of such a flea? (b) Calculate the kinetic energy of this flea at takeoff and its kinetic energy per kilogram of mass. (c) If a 65-kg, 2.0-m-tall human could jump to the same height compared with his length as the flea jumps compared with its length, how high could the human jump, and what takeoff speed would he need? (d) In fact, most humans can jump no more than 60 cm from a crouched start. What is the kinetic energy per kilogram of mass at takeoff for such a 65-kg person? (e) Where does the flea store the energy that allows it to make such a sudden leap?

7.8 •• An empty crate is given an initial push down a ramp, starting with speed v_0, and reaches the bottom with speed v and kinetic energy K. Some books are now placed in the crate, so that the total mass is quadrupled. The coefficient of kinetic friction is constant and air resistance is negligible. Starting again with v_0 at the top of the ramp, what are the speed and kinetic energy at the bottom? Explain the reasoning behind your answers.

7.9 •• **CP** A small rock with mass 0.20 kg is released from rest at point A, which is at the top edge of a large, hemispherical bowl with radius $R = 0.50$ m (Fig. E7.9). Assume that the size of the rock is small compared to R, so that the rock can be treated as a particle, and assume that the rock slides rather than rolls. The work done by friction on the rock when it moves from point A to point B at the bottom of the bowl has magnitude 0.22 J. (a) Between points A and B, how much work is done on the rock by (i) the normal force and (ii) gravity? (b) What is the speed of the rock as it reaches point B? (c) Of the three forces acting on the rock as it slides down the bowl, which (if any) are constant and which

Figure **E7.9**

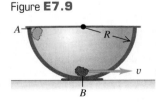

are not? Explain. (d) Just as the rock reaches point B, what is the normal force on it due to the bottom of the bowl?

7.10 •• **BIO Bone Fractures.** The maximum energy that a bone can absorb without breaking depends on its characteristics, such as its cross-sectional area and its elasticity. For healthy human leg bones of approximately 6.0 cm² cross-sectional area, this energy has been experimentally measured to be about 200 J. (a) From approximately what maximum height could a 60-kg person jump and land rigidly upright on both feet without breaking his legs? (b) You are probably surprised at how small the answer to part (a) is. People obviously jump from much greater heights without breaking their legs. How can that be? What else absorbs the energy when they jump from greater heights? (*Hint:* How did the person in part (a) land? How do people normally land when they jump from greater heights?) (c) In light of your answers to parts (a) and (b), what might be some of the reasons that older people are much more prone than younger ones to bone fractures from simple falls (such as a fall in the shower)?

7.11 •• You are testing a new amusement park roller coaster with an empty car of mass 120 kg. One part of the track is a vertical loop with radius 12.0 m. At the bottom of the loop (point A) the car has speed 25.0 m/s, and at the top of the loop (point B) it has speed 8.0 m/s. As the car rolls from point A to point B, how much work is done by friction?

7.12 • **Tarzan and Jane.** Tarzan, in one tree, sights Jane in another tree. He grabs the end of a vine with length 20 m that makes an angle of 45° with the vertical, steps off his tree limb, and swings down and then up to Jane's open arms. When he arrives, his vine makes an angle of 30° with the vertical. Determine whether he gives her a tender embrace or knocks her off her limb by calculating Tarzan's speed just before he reaches Jane. You can ignore air resistance and the mass of the vine.

7.13 •• **CP** A 10.0-kg microwave oven is pushed 8.00 m up the sloping surface of a loading ramp inclined at an angle of 36.9° above the horizontal, by a constant force $\vec{F}$ with a magnitude 110 N and acting parallel to the ramp. The coefficient of kinetic friction between the oven and the ramp is 0.250. (a) What is the work done on the oven by the force $\vec{F}$? (b) What is the work done on the oven by the friction force? (c) Compute the increase in potential energy for the oven. (d) Use your answers to parts (a), (b), and (c) to calculate the increase in the oven's kinetic energy. (e) Use $\Sigma\vec{F} = m\vec{a}$ to calculate the acceleration of the oven. Assuming that the oven is initially at rest, use the acceleration to calculate the oven's speed after traveling 8.00 m. From this, compute the increase in the oven's kinetic energy, and compare it to the answer you got in part (d).

Section 7.2 Elastic Potential Energy

7.14 •• An ideal spring of negligible mass is 12.00 cm long when nothing is attached to it. When you hang a 3.15-kg weight from it, you measure its length to be 13.40 cm. If you wanted to store 10.0 J of potential energy in this spring, what would be its *total* length? Assume that it continues to obey Hooke's law.

7.15 •• A force of 800 N stretches a certain spring a distance of 0.200 m. (a) What is the potential energy of the spring when it is stretched 0.200 m? (b) What is its potential energy when it is compressed 5.00 cm?

7.16 • **BIO Tendons.** Tendons are strong elastic fibers that attach muscles to bones. To a reasonable approximation, they obey Hooke's law. In laboratory tests on a particular tendon, it was found that, when a 250-g object was hung from it, the tendon stretched 1.23 cm. (a) Find the force constant of this tendon in N/m. (b) Because of its thickness, the maximum tension this tendon can support without rupturing is 138 N. By how much can the tendon stretch without rupturing, and how much energy is stored in it at that point?

7.17 • A spring stores potential energy U_0 when it is compressed a distance x_0 from its uncompressed length. (a) In terms of U_0, how much energy does it store when it is compressed (i) twice as much and (ii) half as much? (b) In terms of x_0, how much must it be compressed from its uncompressed length to store (i) twice as much energy and (ii) half as much energy?

7.18 • A slingshot will shoot a 10-g pebble 22.0 m straight up. (a) How much potential energy is stored in the slingshot's rubber band? (b) With the same potential energy stored in the rubber band, how high can the slingshot shoot a 25-g pebble? (c) What physical effects did you ignore in solving this problem?

7.19 •• A spring of negligible mass has force constant $k = 1600$ N/m. (a) How far must the spring be compressed for 3.20 J of potential energy to be stored in it? (b) You place the spring vertically with one end on the floor. You then drop a 1.20-kg book onto it from a height of 0.80 m above the top of the spring. Find the maximum distance the spring will be compressed.

7.20 • A 1.20-kg piece of cheese is placed on a vertical spring of negligible mass and force constant $k = 1800$ N/m that is compressed 15.0 cm. When the spring is released, how high does the cheese rise from this initial position? (The cheese and the spring are *not* attached.)

7.21 •• Consider the glider of Example 7.7 (Section 7.2) and Fig. 7.16. As in the example, the glider is released from rest with the spring stretched 0.100 m. What is the displacement x of the glider from its equilibrium position when its speed is 0.20 m/s? (You should get more than one answer. Explain why.)

7.22 •• Consider the glider of Example 7.7 (Section 7.2) and Fig. 7.16. (a) As in the example, the glider is released from rest with the spring stretched 0.100 m. What is the speed of the glider when it returns to $x = 0$? (b) What must the initial displacement of the glider be if its maximum speed in the subsequent motion is to be 2.50 m/s?

7.23 •• A 2.50-kg mass is pushed against a horizontal spring of force constant 25.0 N/cm on a frictionless air table. The spring is attached to the tabletop, and the mass is not attached to the spring in any way. When the spring has been compressed enough to store 11.5 J of potential energy in it, the mass is suddenly released from rest. (a) Find the greatest speed the mass reaches. When does this occur? (b) What is the greatest acceleration of the mass, and when does it occur?

7.24 •• (a) For the elevator of Example 7.9 (Section 7.2), what is the speed of the elevator after it has moved downward 1:00 m from point 1 in Fig. 7.17? (b) When the elevator is 1.00 m below point 1 in Fig. 7.17, what is its acceleration?

7.25 •• You are asked to design a spring that will give a 1160-kg satellite a speed of 2.50 m/s relative to an orbiting space shuttle. Your spring is to give the satellite a maximum acceleration of $5.00g$. The spring's mass, the recoil kinetic energy of the shuttle, and changes in gravitational potential energy will all be negligible. (a) What must the force constant of the spring be? (b) What distance must the spring be compressed?

7.26 •• A 2.50-kg block on a horizontal floor is attached to a horizontal spring that is initially compressed 0.0300 m. The spring has force constant 840 N/m. The coefficient of kinetic friction between the floor and the block is $\mu_k = 0.40$. The block and spring are released from rest and the block slides along the floor. What is the speed of the block when it has moved a distance of 0.0200 m from its initial position? (At this point the spring is compressed 0.0100 m.)

Section 7.3 Conservative and Nonconservative Forces

7.27 • A 10.0-kg box is pulled by a horizontal wire in a circle on a rough horizontal surface for which the coefficient of kinetic friction is 0.250. Calculate the work done by friction during one complete circular trip if the radius is (a) 2.00 m and (b) 4.00 m. (c) On the basis of the results you just obtained, would you say that friction is a conservative or nonconservative force? Explain.

7.28 • A 75-kg roofer climbs a vertical 7.0-m ladder to the flat roof of a house. He then walks 12 m on the roof, climbs down another vertical 7.0-m ladder, and finally walks on the ground back to his starting point. How much work is done on him by gravity (a) as he climbs up; (b) as he climbs down; (c) as he walks on the roof and on the ground? (d) What is the total work done on him by gravity during this round trip? (e) On the basis of your answer to part (d), would you say that gravity is a conservative or nonconservative force? Explain.

7.29 • A 0.60-kg book slides on a horizontal table. The kinetic friction force on the book has magnitude 1.2 N. (a) How much work is done on the book by friction during a displacement of 3.0 m to the left? (b) The book now slides 3.0 m to the right, returning to its starting point. During this second 3.0-m displacement, how much work is done on the book by friction? (c) What is the total work done on the book by friction during the complete round trip? (d) On the basis of your answer to part (c), would you say that the friction force is conservative or nonconservative? Explain.

7.30 •• **CALC** In an experiment, one of the forces exerted on a proton is $\vec{F} = -\alpha x^2 \hat{\imath}$, where $\alpha = 12$ N/m². (a) How much work does $\vec{F}$ do when the proton moves along the straight-line path from the point $(0.10 \text{ m}, 0)$ to the point $(0.10 \text{ m}, 0.40 \text{ m})$? (b) Along the straight-line path from the point $(0.10 \text{ m}, 0)$ to the point $(0.30 \text{ m}, 0)$? (c) Along the straight-line path from the point $(0.30 \text{ m}, 0)$ to the point $(0.10 \text{ m}, 0)$? (d) Is the force $\vec{F}$ conservative? Explain. If $\vec{F}$ is conservative, what is the potential-energy function for it? Let $U = 0$ when $x = 0$.

7.31 • You and three friends stand at the corners of a square whose sides are 8.0 m long in the middle of the gym floor, as shown in Fig. E7.31. You take your physics book and push it from one person to the other. The book has a mass of 1.5 kg, and the coefficient of kinetic friction between the book and the floor is $\mu_k = 0.25$. (a) The book slides from you to Beth and then from Beth to Carlos, along the lines connecting these people. What is the work done by friction during this displacement? (b) You slide the book from you to Carlos along the diagonal of the square. What is the work done by friction during this displacement? (c) You slide the book to Kim, who then slides it back to you. What is the total work done by friction during this motion of the book? (d) Is the friction force on the book conservative or nonconservative? Explain.

Figure **E7.31**

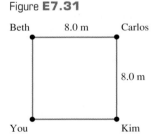

7.32 • While a roofer is working on a roof that slants at 36° above the horizontal, he accidentally nudges his 85.0-N toolbox, causing it to start sliding downward, starting from rest. If it starts 4.25 m from the lower edge of the roof, how fast will the toolbox be moving just as it reaches the edge of the roof if the kinetic friction force on it is 22.0 N?

7.33 •• A 62.0-kg skier is moving at 6.50 m/s on a frictionless, horizontal, snow-covered plateau when she encounters a rough patch 3.50 m long. The coefficient of kinetic friction between this patch and her skis is 0.300. After crossing the rough patch and returning to friction-free snow, she skis down an icy, frictionless hill 2.50 m high. (a) How fast is the skier moving when she gets to the bottom of the hill? (b) How much internal energy was generated in crossing the rough patch?

Section 7.4 Force and Potential Energy

7.34 •• **CALC** The potential energy of a pair of hydrogen atoms separated by a large distance x is given by $U(x) = -C_6/x^6$, where C_6 is a positive constant. What is the force that one atom exerts on the other? Is this force attractive or repulsive?

7.35 •• **CALC** A force parallel to the x-axis acts on a particle moving along the x-axis. This force produces potential energy $U(x)$ given by $U(x) = \alpha x^4$, where $\alpha = 1.20 \text{ J/m}^4$. What is the force (magnitude and direction) when the particle is at $x = -0.800$ m?

7.36 •• **CALC** An object moving in the xy-plane is acted on by a conservative force described by the potential-energy function $U(x, y) = \alpha(1/x^2 + 1/y^2)$, where α is a positive constant. Derive an expression for the force expressed in terms of the unit vectors $\hat{\imath}$ and $\hat{\jmath}$.

7.37 •• **CALC** A small block with mass 0.0400 kg is moving in the xy-plane. The net force on the block is described by the potential-energy function $U(x, y) = (5.80 \text{ J/m}^2)x^2 - (3.60 \text{ J/m}^3)y^3$. What are the magnitude and direction of the acceleration of the block when it is at the point $x = 0.300$ m, $y = 0.600$ m?

Section 7.5 Energy Diagrams

7.38 • A marble moves along the x-axis. The potential-energy function is shown in Fig. E7.38. (a) At which of the labeled x-coordinates is the force on the marble zero? (b) Which of the labeled x-coordinates is a position of stable equilibrium? (c) Which of the labeled x-coordinates is a position of unstable equilibrium?

Figure **E7.38**

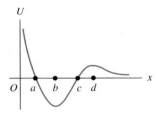

7.39 • **CALC** The potential energy of two atoms in a diatomic molecule is approximated by $U(r) = a/r^{12} - b/r^6$, where r is the spacing between atoms and a and b are positive constants. (a) Find the force $F(r)$ on one atom as a function of r. Draw two graphs: one of $U(r)$ versus r and one of $F(r)$ versus r. (b) Find the equilibrium distance between the two atoms. Is this equilibrium stable? (c) Suppose the distance between the two atoms is equal to the equilibrium distance found in part (b). What minimum energy must be added to the molecule to *dissociate* it—that is, to separate the two atoms to an infinite distance apart? This is called the *dissociation energy* of the molecule. (d) For the molecule CO, the equilibrium distance between the carbon and oxygen atoms is 1.13×10^{-10} m and the dissociation energy is 1.54×10^{-18} J per molecule. Find the values of the constants a and b.

PROBLEMS

7.40 •• Two blocks with different masses are attached to either end of a light rope that passes over a light, frictionless pulley suspended from the ceiling. The masses are released from rest, and the more massive one starts to descend. After this block has descended 1.20 m, its speed is 3.00 m/s. If the total mass of the two blocks is 15.0 kg, what is the mass of each block?

7.41 ••• At a construction site, a 65.0-kg bucket of concrete hangs from a light (but strong) cable that passes over a light, friction-free pulley and is connected to an 80.0-kg box on a horizontal roof (Fig. P7.41). The cable pulls horizontally on the box, and a 50.0-kg bag of gravel rests on top of the box. The coefficients of friction between the box and roof are shown. (a) Find the friction force on the bag of gravel and on the box. (b) Suddenly a worker picks up the bag of gravel. Use energy conservation to find the speed of the bucket after it has descended 2.00 m from rest. (You can check your answer by solving this problem using Newton's laws.)

Figure **P7.41**

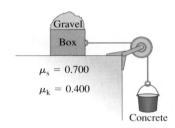

7.42 • A 2.00-kg block is pushed against a spring with negligible mass and force constant $k = 400$ N/m, compressing it 0.220 m. When the block is released, it moves along a frictionless, horizontal surface and then up a frictionless incline with slope 37.0° (Fig. P7.42). (a) What is the speed of the block as it slides along the horizontal surface after having left the spring? (b) How far does the block travel up the incline before starting to slide back down?

Figure **P7.42**

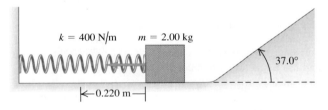

7.43 • A block with mass 0.50 kg is forced against a horizontal spring of negligible mass, compressing the spring a distance of 0.20 m (Fig. P7.43). When released, the block moves on a horizontal tabletop for 1.00 m before coming to rest. The spring constant k is 100 N/m. What is the coefficient of kinetic friction μ_k between the block and the tabletop?

Figure **P7.43**

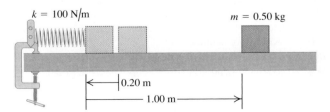

7.44 • On a horizontal surface, a crate with mass 50.0 kg is placed against a spring that stores 360 J of energy. The spring is released, and the crate slides 5.60 m before coming to rest. What is the speed of the crate when it is 2.00 m from its initial position?

7.45 •• A 350-kg roller coaster starts from rest at point A and slides down the frictionless loop-the-loop shown in Fig. P7.45. (a) How fast is this roller coaster moving at point B? (b) How hard does it press against the track at point B?

Figure **P7.45**

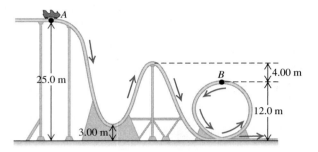

7.46 •• **CP** **Riding a Loop-the-Loop.** A car in an amusement park ride rolls without friction around the track shown in Fig. P7.46. It starts from rest at point A at a height h above the bottom of the loop. Treat the car as a particle. (a) What is the minimum value of h (in terms of R) such that the car moves around the loop without falling off at the top (point B)? (b) If $h = 3.50R$ and $R = 20.0$ m, compute the speed, radial acceleration, and tangential acceleration of the passengers when the car is at point C, which is at the end of a horizontal diameter. Show these acceleration components in a diagram, approximately to scale.

Figure **P7.46**

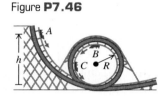

7.47 •• A 2.0-kg piece of wood slides on the surface shown in Fig. P7.47. The curved sides are perfectly smooth, but the rough horizontal bottom is 30 m long and has a kinetic friction coefficient of 0.20 with the wood. The piece of wood starts from rest 4.0 m above the rough bottom. (a) Where will this wood eventually come to rest? (b) For the motion from the initial release until the piece of wood comes to rest, what is the total amount of work done by friction?

Figure **P7.47**

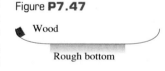

Wood

Rough bottom

7.48 •• **Up and Down the Hill.** A 28-kg rock approaches the foot of a hill with a speed of 15 m/s. This hill slopes upward at a constant angle of 40.0° above the horizontal. The coefficients of static and kinetic friction between the hill and the rock are 0.75 and 0.20, respectively. (a) Use energy conservation to find the maximum height above the foot of the hill reached by the rock. (b) Will the rock remain at rest at its highest point, or will it slide back down the hill? (c) If the rock does slide back down, find its speed when it returns to the bottom of the hill.

7.49 •• A 15.0-kg stone slides down a snow-covered hill (Fig. P7.49), leaving point A with a speed of 10.0 m/s. There is no friction on the hill between points A and B, but there is friction on the level ground at the bottom of the hill, between B and the wall. After entering the rough horizontal

Figure **P7.49**

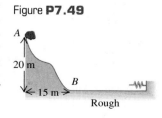

Rough

region, the stone travels 100 m and then runs into a very long, light spring with force constant 2.00 N/m. The coefficients of kinetic and static friction between the stone and the horizontal ground are 0.20 and 0.80, respectively. (a) What is the speed of the stone when it reaches point B? (b) How far will the stone compress the spring? (c) Will the stone move again after it has been stopped by the spring?

7.50 •• **CP** A 2.8-kg block slides over the smooth, icy hill shown in Fig. P7.50. The top of the hill is horizontal and 70 m higher than its base. What minimum speed must the block have at the base of the hill in order for it to pass over the pit at the far side of the hill?

Figure **P7.50**

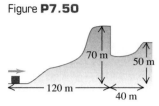

7.51 ••• **Bungee Jump.** A bungee cord is 30.0 m long and, when stretched a distance x, it exerts a restoring force of magnitude kx. Your father-in-law (mass 95.0 kg) stands on a platform 45.0 m above the ground, and one end of the cord is tied securely to his ankle and the other end to the platform. You have promised him that when he steps off the platform he will fall a maximum distance of only 41.0 m before the cord stops him. You had several bungee cords to select from, and you tested them by stretching them out, tying one end to a tree, and pulling on the other end with a force of 380.0 N. When you do this, what distance will the bungee cord that you should select have stretched?

7.52 •• **Ski Jump Ramp.** You are designing a ski jump ramp for the next Winter Olympics. You need to calculate the vertical height h from the starting gate to the bottom of the ramp. The skiers push off hard with their ski poles at the start, just above the starting gate, so they typically have a speed of 2.0 m/s as they reach the gate. For safety, the skiers should have a speed no higher than 30.0 m/s when they reach the bottom of the ramp. You determine that for a 85.0-kg skier with good form, friction and air resistance will do total work of magnitude 4000 J on him during his run down the ramp. What is the maximum height h for which the maximum safe speed will not be exceeded?

7.53 ••• The Great Sandini is a 60-kg circus performer who is shot from a cannon (actually a spring gun). You don't find many men of his caliber, so you help him design a new gun. This new gun has a very large spring with a very small mass and a force constant of 1100 N/m that he will compress with a force of 4400 N. The inside of the gun barrel is coated with Teflon, so the average friction force will be only 40 N during the 4.0 m he moves in the barrel. At what speed will he emerge from the end of the barrel, 2.5 m above his initial rest position?

7.54 ••• You are designing a delivery ramp for crates containing exercise equipment. The 1470-N crates will move at 1.8 m/s at the top of a ramp that slopes downward at 22.0°. The ramp exerts a 550-N kinetic friction force on each crate, and the maximum static friction force also has this value. Each crate will compress a spring at the bottom of the ramp and will come to rest after traveling a total distance of 8.0 m along the ramp. Once stopped, a crate must not rebound back up the ramp. Calculate the force constant of the spring that will be needed in order to meet the design criteria.

7.55 •• A system of two paint buckets connected by a lightweight rope is released from rest with the 12.0-kg bucket 2.00 m above the floor (Fig. P7.55). Use the principle of conservation of energy to find the speed with which this bucket strikes the floor. You can ignore friction and the mass of the pulley.

Figure **P7.55**

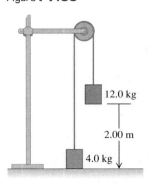

12.0 kg

2.00 m

4.0 kg

7.56 •• A 1500-kg rocket is
to be launched with an initial
upward speed of 50.0 m/s.
In order to assist its engines,
the engineers will start it
from rest on a ramp that
rises 53° above the horizontal (Fig. P7.56). At the bottom, the ramp turns upward
and launches the rocket vertically. The engines provide
a constant forward thrust of
2000 N, and friction with
the ramp surface is a constant 500 N. How far from the base of the
ramp should the rocket start, as measured along the surface of the
ramp?

Figure **P7.56**

Rocket starts
here.

Rocket is
launched
upward.
53°

7.57 • **Legal Physics.** In an auto accident, a car hit a pedestrian
and the driver then slammed on the brakes to stop the car. During
the subsequent trial, the driver's lawyer claimed that he was obeying the posted 35-mph speed limit, but that the legal speed was too
high to allow him to see and react to the pedestrian in time. You
have been called in as the state's expert witness. Your investigation
of the accident found that the skid marks made while the brakes
were applied were 280 ft long, and the tread on the tires produced a
coefficient of kinetic friction of 0.30 with the road. (a) In your
testimony in court, will you say that the driver was obeying the
posted speed? You must be able to back up your conclusion with
clear reasoning because one of the lawyers will surely cross-
examine you. (b) If the driver's speeding ticket were $10 for each
mile per hour he was driving above the posted speed limit, would
he have to pay a fine? If so, how much would it be?

7.58 ••• A wooden rod of negligible mass and length 80.0 cm is
pivoted about a horizontal axis through its center. A white rat with
mass 0.500 kg clings to one end of the stick, and a mouse with
mass 0.200 kg clings to the other end. The system is released from
rest with the rod horizontal. If the animals can manage to hold on,
what are their speeds as the rod swings through a vertical position?

7.59 •• CP A 0.300-kg potato is tied to a string with length 2.50 m,
and the other end of the string is tied to a rigid support. The potato
is held straight out horizontally from the point of support, with the
string pulled taut, and is then released. (a) What is the speed of the
potato at the lowest point of its motion? (b) What is the tension in
the string at this point?

7.60 •• These data are from a computer simulation for a batted
baseball with mass 0.145 kg, including air resistance:

t	x	y	v_x	v_y
0	0	0	30.0 m/s	40.0 m/s
3.05 s	70.2 m	53.6 m	18.6 m/s	0
6.59 s	124.4 m	0	11.9 m/s	−28.7 m/s

(a) How much work was done by the air on the baseball as it
moved from its initial position to its maximum height? (b) How
much work was done by the air on the baseball as it moved from
its maximum height back to the starting elevation? (c) Explain
why the magnitude of the answer in part (b) is smaller than the
magnitude of the answer in part (a).

7.61 •• **Down the Pole.** A fireman of mass m slides a distance d
down a pole. He starts from rest. He moves as fast at the bottom as if
he had stepped off a platform a distance $h \leq d$ above the ground and
descended with negligible air resistance. (a) What average friction
force did the fireman exert on the pole? Does your answer make sense
in the special cases of $h = d$ and $h = 0$? (b) Find a numerical value
for the average friction force a 75-kg fireman exerts, for $d = 2.5$ m
and $h = 1.0$ m. (c) In terms of g, h, and d, what is the speed of the
fireman when he is a distance y above the bottom of the pole?

7.62 •• A 60.0-kg skier starts from rest at the top of a ski slope
65.0 m high. (a) If frictional forces do −10.5 kJ of work on her as
she descends, how fast is she going at the bottom of the slope?
(b) Now moving horizontally, the skier crosses a patch of soft
snow, where $\mu_k = 0.20$. If the patch is 82.0 m wide and the average force of air resistance on the skier is 160 N, how fast is she
going after crossing the patch? (c) The skier hits a snowdrift and
penetrates 2.5 m into it before coming to a stop. What is the average force exerted on her by the snowdrift as it stops her?

7.63 • CP A skier starts at
the top of a very large, frictionless snowball, with a
very small initial speed, and
skis straight down the side
(Fig. P7.63). At what point
does she lose contact with
the snowball and fly off at a
tangent? That is, at the
instant she loses contact with
the snowball, what angle α
does a radial line from the
center of the snowball to the
skier make with the vertical?

Figure **P7.63**

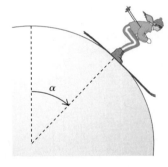

α

7.64 •• A ball is thrown upward with an initial velocity of 15 m/s
at an angle of 60.0° above the horizontal. Use energy conservation
to find the ball's greatest height above the ground.

7.65 •• In a truck-loading station at a post office, a small 0.200-kg
package is released from rest at point A on a track that is one-
quarter of a circle with radius 1.60 m (Fig. P7.65). The size of the
package is much less than 1.60 m, so the package can be treated as
a particle. It slides down the track and reaches point B with a speed
of 4.80 m/s. From point B, it slides on a level surface a distance of

Figure **P7.65**

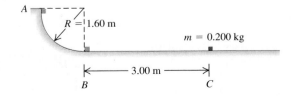

A

R = 1.60 m

m = 0.200 kg

3.00 m

B

C

3.00 m to point *C*, where it comes to rest. (a) What is the coefficient of kinetic friction on the horizontal surface? (b) How much work is done on the package by friction as it slides down the circular arc from *A* to *B*?

7.66 ••• A truck with mass *m* has a brake failure while going down an icy mountain road of constant downward slope angle *α* (Fig. P7.66). Initially the truck is moving downhill at speed v_0. After careening downhill a distance *L* with negligible friction, the truck driver steers the runaway vehicle onto a runaway truck ramp of constant upward slope angle *β*. The truck ramp has a soft sand surface for which the coefficient of rolling friction is μ_r. What is the distance that the truck moves up the ramp before coming to a halt? Solve using energy methods.

Figure **P7.66**

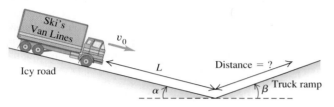

7.67 •• **CALC** A certain spring is found *not* to obey Hooke's law; it exerts a restoring force $F_x(x) = -\alpha x - \beta x^2$ if it is stretched or compressed, where *α* = 60.0 N/m and *β* = 18.0 N/m². The mass of the spring is negligible. (a) Calculate the potential-energy function *U(x)* for this spring. Let *U* = 0 when *x* = 0. (b) An object with mass 0.900 kg on a frictionless, horizontal surface is attached to this spring, pulled a distance 1.00 m to the right (the +*x*-direction) to stretch the spring, and released. What is the speed of the object when it is 0.50 m to the right of the *x* = 0 equilibrium position?

7.68 •• **CP** A sled with rider having a combined mass of 125 kg travels over the perfectly smooth icy hill shown in Fig. 7.68. How far does the sled land from the foot of the cliff?

Figure **P7.68**

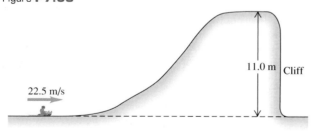

7.69 •• A 0.150-kg block of ice is placed against a horizontal, compressed spring mounted on a horizontal tabletop that is 1.20 m above the floor. The spring has force constant 1900 N/m and is initially compressed 0.045 m. The mass of the spring is negligible. The spring is released, and the block slides along the table, goes off the edge, and travels to the floor. If there is negligible friction between the block of ice and the tabletop, what is the speed of the block of ice when it reaches the floor?

7.70 •• A 3.00-kg block is connected to two ideal horizontal springs having force constants k_1 = 25.0 N/cm   and k_2 = 20.0 N/cm   (Fig. P7.70). The system is initially in equilibrium on a horizontal, frictionless surface. The block is now pushed 15.0 cm to the right and released

Figure **P7.70**

from rest. (a) What is the maximum speed of the block? Where in the motion does the maximum speed occur? (b) What is the maximum compression of spring 1?

7.71 •• An experimental apparatus with mass *m* is placed on a vertical spring of negligible mass and pushed down until the spring is compressed a distance *x*. The apparatus is then released and reaches its maximum height at a distance *h* above the point where it is released. The apparatus is not attached to the spring, and at its maximum height it is no longer in contact with the spring. The maximum magnitude of acceleration the apparatus can have without being damaged is *a*, where *a* > *g*. (a) What should the force constant of the spring be? (b) What distance *x* must the spring be compressed initially?

7.72 •• If a fish is attached to a vertical spring and slowly lowered to its equilibrium position, it is found to stretch the spring by an amount *d*. If the same fish is attached to the end of the unstretched spring and then allowed to fall from rest, through what maximum distance does it stretch the spring? (*Hint:* Calculate the force constant of the spring in terms of the distance *d* and the mass *m* of the fish.)

7.73 ••• **CALC** A 3.00-kg fish is attached to the lower end of a vertical spring that has negligible mass and force constant 900 N/m. The spring initially is neither stretched nor compressed. The fish is released from rest. (a) What is its speed after it has descended 0.0500 m from its initial position? (b) What is the maximum speed of the fish as it descends?

7.74 •• A basket of negligible weight hangs from a vertical spring scale of force constant 1500 N/m. (a) If you suddenly put a 3.0-kg adobe brick in the basket, find the maximum distance that the spring will stretch. (b) If, instead, you release the brick from 1.0 m above the basket, by how much will the spring stretch at its maximum elongation?

7.75 • A 0.500-kg block, attached to a spring with length 0.60 m and force constant 40.0 N/m, is at rest with the back of the block at point *A* on a frictionless, horizontal air table (Fig. P7.75). The mass of the spring is negligible. You move the block to the right along the surface by pulling with a constant 20.0-N horizontal force. (a) What is the block's speed when the back of the block reaches point *B*, which is 0.25 m to the right of point *A*? (b) When the back of the block reaches point *B*, you let go of the block. In the subsequent motion, how close does the block get to the wall where the left end of the spring is attached?

Figure **P7.75**

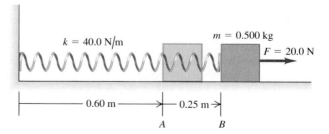

7.76 •• **Fraternity Physics.** The brothers of Iota Eta Pi fraternity build a platform, supported at all four corners by vertical springs, in the basement of their frat house. A brave fraternity brother wearing a football helmet stands in the middle of the platform; his weight compresses the springs by 0.18 m. Then four of his fraternity brothers, pushing down at the corners of the platform, compress the springs another 0.53 m until the top of the brave brother's helmet is 0.90 m below the basement ceiling. They then simultaneously release the platform. You can ignore the masses of the springs and platform. (a) When the dust clears, the fraternity asks you to calculate their fraternity brother's speed just before his helmet hit the flimsy ceiling. (b) Without the ceiling, how high would he have gone? (c) In discussing their probation, the dean of students suggests that the next time they try this, they do it outdoors on another planet. Would the answer to part (b) be the same if this stunt were performed on a planet with a different value of g? Assume that the fraternity brothers push the platform down 0.53 m as before. Explain your reasoning.

7.77 ••• **CP** A small block with mass 0.0500 kg slides in a vertical circle of radius $R = 0.800$ m on the inside of a circular track. There is no friction between the track and the block. At the bottom of the block's path, the normal force the track exerts on the block has magnitude 3.40 N. What is the magnitude of the normal force that the track exerts on the block when it is at the top of its path?

7.78 ••• **CP** A small block with mass 0.0400 kg slides in a vertical circle of radius $R = 0.500$ m on the inside of a circular track. During one of the revolutions of the block, when the block is at the bottom of its path, point A, the magnitude of the normal force exerted on the block by the track has magnitude 3.95 N. In this same revolution, when the block reaches the top of its path, point B, the magnitude of the normal force exerted on the block has magnitude 0.680 N. How much work was done on the block by friction during the motion of the block from point A to point B?

7.79 •• A hydroelectric dam holds back a lake of surface area 3.0×10^6 m^2 that has vertical sides below the water level. The water level in the lake is 150 m above the base of the dam. When the water passes through turbines at the base of the dam, its mechanical energy is converted to electrical energy with 90% efficiency. (a) If gravitational potential energy is taken to be zero at the base of the dam, how much energy is stored in the top meter of the water in the lake? The density of water is 1000 kg/m^3. (b) What volume of water must pass through the dam to produce 1000 kilowatt-hours of electrical energy? What distance does the level of water in the lake fall when this much water passes through the dam?

7.80 •• **CALC** How much total energy is stored in the lake in Problem 7.79? As in that problem, take the gravitational potential energy to be zero at the base of the dam. Express your answer in joules and in kilowatt-hours. (*Hint:* Break the lake up into infinitesimal horizontal layers of thickness dy, and integrate to find the total potential energy.)

7.81 ••• A wooden block with mass 1.50 kg is placed against a compressed spring at the bottom of an incline of slope 30.0° (point A). When the spring is released, it projects the block up the incline. At point B, a distance of 6.00 m up the incline from A, the block is moving up the incline at 7.00 m/s and is no longer in contact with the spring. The coefficient of kinetic friction between the block and the incline is $\mu_k = 0.50$. The mass of the spring is negligible. Calculate the amount of potential energy that was initially stored in the spring.

7.82 •• **CP Pendulum.**   A small rock with mass 0.12 kg is fastened to a massless string with length 0.80 m to form a pendulum. The pendulum is swinging so as to make a maximum angle of 45° with the vertical. Air resistance is negligible. (a) What is the speed of the rock when the string passes through the vertical position? (b) What is the tension in the string when it makes an angle of 45° with the vertical? (c) What is the tension in the string as it passes through the vertical?

7.83 ••• **CALC** A cutting tool under microprocessor control has several forces acting on it. One force is $\vec{F} = -\alpha xy^2\hat{\jmath}$, a force in

the negative y-direction whose magnitude depends on the position of the tool. The constant is $\alpha = 2.50$ N/m^3. Consider the displacement of the tool from the origin to the point $x = 3.00$ m, $y = 3.00$ m. (a) Calculate the work done on the tool by $\vec{F}$ if this displacement is along the straight line $y = x$ that connects these two points. (b) Calculate the work done on the tool by $\vec{F}$ if the tool is first moved out along the x-axis to the point $x = 3.00$ m, $y = 0$ and then moved parallel to the y-axis to the point $x = 3.00$ m, $y = 3.00$ m. (c) Compare the work done by $\vec{F}$ along these two paths. Is $\vec{F}$ conservative or nonconservative? Explain.

7.84 • **CALC** (a) Is the force $\vec{F} = Cy^2\hat{\jmath}$, where C is a negative constant with units of N/m^2, conservative or nonconservative? Justify your answer. (b) Is the force $\vec{F} = Cy^2\hat{\imath}$, where C is a negative constant with units of N/m^2, conservative or nonconservative? Justify your answer.

7.85 •• **CALC** An object has several forces acting on it. One force is $\vec{F} = \alpha xy\hat{\imath}$, a force in the x-direction whose magnitude depends on the position of the object. (See Problem 6.98.) The constant is $\alpha = 2.00$ N/m^2. The object moves along the following path: (1) It starts at the origin and moves along the y-axis to the point $x = 0$, $y = 1.50$ m; (2) it moves parallel to the x-axis to the point $x = 1.50$ m, $y = 1.50$ m; (3) it moves parallel to the y-axis to the point $x = 1.50$ m, $y = 0$; (4) it moves parallel to the x-axis back to the origin. (a) Sketch this path in the xy-plane. (b) Calculate the work done on the object by $\vec{F}$ for each leg of the path and for the complete round trip. (c) Is $\vec{F}$ conservative or nonconservative? Explain.

7.86 • A particle moves along the x-axis while acted on by a single conservative force parallel to the x-axis. The force corresponds to the potential-energy function graphed in Fig. P7.86. The particle is released from rest at point A. (a) What is the direction of the force on the particle when it is at point A? (b) At point B? (c) At what value of x is the kinetic energy of the particle a maximum? (d) What is the force on the particle when it is at point C? (e) What is the largest value of x reached by the particle during its motion? (f) What value or values of x correspond to points of stable equilibrium? (g) Of unstable equilibrium?

Figure **P7.86**

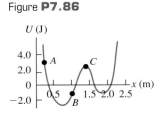

CHALLENGE PROBLEM

7.87 ••• **CALC** A proton with mass m moves in one dimension. The potential-energy function is $U(x) = \alpha/x^2 - \beta/x$, where α and β are positive constants. The proton is released from rest at $x_0 = \alpha/\beta$. (a) Show that $U(x)$ can be written as

$$U(x) = \frac{\alpha}{x_0^2}\left[\left(\frac{x_0}{x}\right)^2 - \frac{x_0}{x}\right]$$

Graph $U(x)$. Calculate $U(x_0)$ and thereby locate the point x_0 on the graph. (b) Calculate $v(x)$, the speed of the proton as a function of position. Graph $v(x)$ and give a qualitative description of the motion. (c) For what value of x is the speed of the proton a maximum? What is the value of that maximum speed? (d) What is the force on the proton at the point in part (c)? (e) Let the proton be released instead at $x_1 = 3\alpha/\beta$. Locate the point x_1 on the graph of $U(x)$. Calculate $v(x)$ and give a qualitative description of the motion. (f) For each release point ($x = x_0$ and $x = x_1$), what are the maximum and minimum values of x reached during the motion?

Answers

Chapter Opening Question ?

The mallard's kinetic energy K remains constant because the speed remains the same, but the gravitational potential energy U_{grav} decreases as the mallard descends. Hence the total mechanical energy $E = K + U_{grav}$ decreases. The lost mechanical energy goes into warming the mallard's skin (that is, an increase in the mallard's internal energy) and stirring up the air through which the mallard passes (an increase in the internal energy of the air). See the discussion in Section 7.3.

Test Your Understanding Questions

7.1 Answer: (iii) The initial kinetic energy $K_1 = 0$, the initial potential energy $U_1 = mgy_1$, and the final potential energy $U_2 = mgy_2$ are the same for both blocks. Mechanical energy is conserved in both cases, so the final kinetic energy $K_2 = \frac{1}{2}mv_2^2$ is also the same for both blocks. Hence the speed at the right-hand end is the *same* in both cases!

7.2 Answer: (iii) The elevator is still moving downward, so the kinetic energy K is positive (remember that K can never be nega-

tive); the elevator is below point 1, so $y < 0$ and $U_{grav} < 0$; and the spring is compressed, so $U_{el} > 0$.

7.3 Answer: (iii) Because of friction in the turbines and between the water and turbines, some of the potential energy goes into raising the temperatures of the water and the mechanism.

7.4 Answers: (a) (iv), (b) (i) If $F_x = 0$ at a point, then the derivative of $U(x)$ must be zero at that point because $F_x = -dU(x)/dx$. However, this tells us absolutely nothing about the *value* of $U(x)$ at that point.

7.5 Answers: (iii) Figure 7.24b shows the x-component of force, F_x. Where this is maximum (most positive), the x-component of force and the x-acceleration have more positive values than at adjacent values of x.

Bridging Problem

Answers: **(a)** 1.06 m
              (b) 1.32 m
              (c) 20.7 J

MOMENTUM, IMPULSE, AND COLLISIONS

? Which could potentially do greater damage to this carrot: a .22-caliber bullet moving at 220 m/s as shown here, or a lightweight bullet of the same length and diameter but half the mass moving at twice the speed?

LEARNING GOALS

By studying this chapter, you will learn:

- The meaning of the momentum of a particle, and how the impulse of the net force acting on a particle causes its momentum to change.

- The conditions under which the total momentum of a system of particles is constant (conserved).

- How to solve problems in which two bodies collide with each other.

- The important distinction among elastic, inelastic, and completely inelastic collisions.

- The definition of the center of mass of a system, and what determines how the center of mass moves.

- How to analyze situations such as rocket propulsion in which the mass of a body changes as it moves.

There are many questions involving forces that cannot be answered by directly applying Newton's second law, $\sum \vec{F} = m\vec{a}$. For example, when a moving van collides head-on with a compact car, what determines which way the wreckage moves after the collision? In playing pool, how do you decide how to aim the cue ball in order to knock the eight ball into the pocket? And when a meteorite collides with the earth, how much of the meteorite's kinetic energy is released in the impact?

A common theme of all these questions is that they involve forces about which we know very little: the forces between the car and the moving van, between the two pool balls, or between the meteorite and the earth. Remarkably, we will find in this chapter that we don't have to know *anything* about these forces to answer questions of this kind!

Our approach uses two new concepts, *momentum* and *impulse,* and a new conservation law, *conservation of momentum.* This conservation law is every bit as important as the law of conservation of energy. The law of conservation of momentum is valid even in situations in which Newton's laws are inadequate, such as bodies moving at very high speeds (near the speed of light) or objects on a very small scale (such as the constituents of atoms). Within the domain of Newtonian mechanics, conservation of momentum enables us to analyze many situations that would be very difficult if we tried to use Newton's laws directly. Among these are *collision* problems, in which two bodies collide and can exert very large forces on each other for a short time.

8.1 Momentum and Impulse

In Chapter 6 we re-expressed Newton's second law for a particle, $\sum \vec{F} = m\vec{a}$, in terms of the work–energy theorem. This theorem helped us tackle a great number of physics problems and led us to the law of conservation of energy. Let's now return to $\sum \vec{F} = m\vec{a}$ and see yet another useful way to restate this fundamental law.

Newton's Second Law in Terms of Momentum

Consider a particle of constant mass m. (Later in this chapter we'll see how to deal with situations in which the mass of a body changes.) Because $\vec{a} = d\vec{v}/dt$, we can write Newton's second law for this particle as

$$\sum \vec{F} = m\frac{d\vec{v}}{dt} = \frac{d}{dt}(m\vec{v}) \qquad (8.1)$$

We can move the mass m inside the derivative because it is constant. Thus Newton's second law says that the net force $\sum \vec{F}$ acting on a particle equals the time rate of change of the combination $m\vec{v}$, the product of the particle's mass and velocity. We'll call this combination the **momentum, or linear momentum,** of the particle. Using the symbol $\vec{p}$ for momentum, we have

$$\vec{p} = m\vec{v} \qquad \text{(definition of momentum)} \qquad (8.2)$$

8.1 The velocity and momentum vectors of a particle.

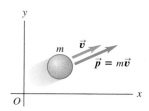

Momentum $\vec{p}$ is a vector quantity; a particle's momentum has the same direction as its velocity $\vec{v}$.

The greater the mass m and speed v of a particle, the greater is its magnitude of momentum mv. Keep in mind, however, that momentum is a *vector* quantity with the same direction as the particle's velocity (Fig. 8.1). Hence a car driving north at 20 m/s and an identical car driving east at 20 m/s have the same *magnitude* of momentum (mv) but different momentum *vectors* ($m\vec{v}$) because their directions are different.

We often express the momentum of a particle in terms of its components. If the particle has velocity components v_x, v_y, and v_z, then its momentum components p_x, p_y, and p_z (which we also call the *x-momentum, y-momentum,* and *z-momentum*) are given by

$$p_x = mv_x \qquad p_y = mv_y \qquad p_z = mv_z \qquad (8.3)$$

These three component equations are equivalent to Eq. (8.2).

The units of the magnitude of momentum are units of mass times speed; the SI units of momentum are kg · m/s. The plural of momentum is "momenta."

If we now substitute the definition of momentum, Eq. (8.2), into Eq. (8.1), we get

$$\sum \vec{F} = \frac{d\vec{p}}{dt} \qquad \text{(Newton's second law in terms of momentum)} \qquad (8.4)$$

8.2 If a fast-moving automobile stops suddenly in a collision, the driver's momentum (mass times velocity) changes from a large value to zero in a short time. An air bag causes the driver to lose momentum more gradually than would an abrupt collision with the steering wheel, reducing the force exerted on the driver as well as the possibility of injury.

The net force (vector sum of all forces) acting on a particle equals the time rate of change of momentum of the particle. This, not $\sum \vec{F} = m\vec{a}$, is the form in which Newton originally stated his second law (although he called momentum the "quantity of motion"). This law is valid only in inertial frames of reference.

According to Eq. (8.4), a rapid change in momentum requires a large net force, while a gradual change in momentum requires less net force. This principle is used in the design of automobile safety devices such as air bags (Fig. 8.2).

The Impulse–Momentum Theorem

A particle's momentum $\vec{p} = m\vec{v}$ and its kinetic energy $K = \frac{1}{2}mv^2$ both depend on the mass and velocity of the particle. What is the fundamental difference between these two quantities? A purely mathematical answer is that momentum is a vector whose magnitude is proportional to speed, while kinetic energy is a scalar proportional to the speed squared. But to see the *physical* difference between momentum and kinetic energy, we must first define a quantity closely related to momentum called *impulse*.

Let's first consider a particle acted on by a *constant* net force $\Sigma \vec{F}$ during a time interval Δt from t_1 to t_2. (We'll look at the case of varying forces shortly.) The **impulse** of the net force, denoted by $\vec{J}$, is defined to be the product of the net force and the time interval:

$$\vec{J} = \Sigma \vec{F}(t_2 - t_1) = \Sigma \vec{F} \, \Delta t \qquad \text{(assuming constant net force)} \qquad (8.5)$$

Impulse is a vector quantity; its direction is the same as the net force $\Sigma \vec{F}$. Its magnitude is the product of the magnitude of the net force and the length of time that the net force acts. The SI unit of impulse is the newton-second ($\text{N} \cdot \text{s}$). Because $1 \, \text{N} = 1 \, \text{kg} \cdot \text{m/s}^2$, an alternative set of units for impulse is $\text{kg} \cdot \text{m/s}$, the same as the units of momentum.

To see what impulse is good for, let's go back to Newton's second law as restated in terms of momentum, Eq. (8.4). If the net force $\Sigma \vec{F}$ is constant, then $d\vec{p}/dt$ is also constant. In that case, $d\vec{p}/dt$ is equal to the *total* change in momentum $\vec{p}_2 - \vec{p}_1$ during the time interval $t_2 - t_1$, divided by the interval:

$$\Sigma \vec{F} = \frac{\vec{p}_2 - \vec{p}_1}{t_2 - t_1}$$

Multiplying this equation by $(t_2 - t_1)$, we have

$$\Sigma \vec{F}(t_2 - t_1) = \vec{p}_2 - \vec{p}_1$$

Comparing with Eq. (8.5), we end up with a result called the **impulse–momentum theorem:**

$$\vec{J} = \vec{p}_2 - \vec{p}_1 \qquad \text{(impulse–momentum theorem)} \qquad (8.6)$$

> **The change in momentum of a particle during a time interval equals the impulse of the net force that acts on the particle during that interval.**

The impulse–momentum theorem also holds when forces are not constant. To see this, we integrate both sides of Newton's second law $\Sigma \vec{F} = d\vec{p}/dt$ over time between the limits t_1 and t_2:

$$\int_{t_1}^{t_2} \Sigma \vec{F} \, dt = \int_{t_1}^{t_2} \frac{d\vec{p}}{dt} \, dt = \int_{\vec{p}_1}^{\vec{p}_2} d\vec{p} = \vec{p}_2 - \vec{p}_1$$

The integral on the left is defined to be the impulse $\vec{J}$ of the net force $\Sigma \vec{F}$ during this interval:

$$\vec{J} = \int_{t_1}^{t_2} \Sigma \vec{F} \, dt \qquad \text{(general definition of impulse)} \qquad (8.7)$$

With this definition, the impulse–momentum theorem $\vec{J} = \vec{p}_2 - \vec{p}_1$, Eq. (8.6), is valid even when the net force $\Sigma \vec{F}$ varies with time.

We can define an *average* net force $\vec{F}_{av}$ such that even when $\Sigma \vec{F}$ is not constant, the impulse $\vec{J}$ is given by

$$\vec{J} = \vec{F}_{av}(t_2 - t_1) \qquad (8.8)$$

When $\Sigma \vec{F}$ is constant, $\Sigma \vec{F} = \vec{F}_{av}$ and Eq. (8.8) reduces to Eq. (8.5).

Figure 8.3a shows the x-component of net force ΣF_x as a function of time during a collision. This might represent the force on a soccer ball that is in contact with a player's foot from time t_1 to t_2. The x-component of impulse during this interval is represented by the red area under the curve between t_1 and t_2. This

Application Woodpecker Impulse
The pileated woodpecker (*Dryocopus pileatus*) has been known to strike its beak against a tree up to 20 times a second and up to 12,000 times a day. The impact force can be as much as 1200 times the weight of the bird's head. Because the impact lasts such a short time, the impulse—the product of the net force during the impact multiplied by the duration of the impact—is relatively small. (The woodpecker has a thick skull of spongy bone as well as shock-absorbing cartilage at the base of the lower jaw, and so avoids injury.)

8.3 The meaning of the area under a graph of ΣF_x versus t.

(a)

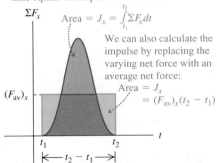

The area under the curve of net force versus time equals the impulse of the net force:

$$\text{Area} = J_x = \int_{t_1}^{t_2} \Sigma F_x \, dt$$

We can also calculate the impulse by replacing the varying net force with an average net force:

$$\text{Area} = J_x = (F_{av})_x(t_2 - t_1)$$

(b)

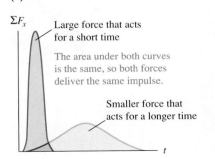

Large force that acts for a short time

The area under both curves is the same, so both forces deliver the same impulse.

Smaller force that acts for a longer time

area is equal to the green rectangular area bounded by t_1, t_2, and $(F_{av})_x$, so $(F_{av})_x(t_2 - t_1)$ is equal to the impulse of the actual time-varying force during the same interval. Note that a large force acting for a short time can have the same impulse as a smaller force acting for a longer time if the areas under the force–time curves are the same (Fig. 8.3b). In this language, an automobile airbag (see Fig. 8.2) provides the same impulse to the driver as would the steering wheel or the dashboard by applying a weaker and less injurious force for a longer time.

Impulse and momentum are both vector quantities, and Eqs. (8.5)–(8.8) are all vector equations. In specific problems, it is often easiest to use them in component form:

$$J_x = \int_{t_1}^{t_2} \Sigma F_x\, dt = (F_{av})_x(t_2 - t_1) = p_{2x} - p_{1x} = mv_{2x} - mv_{1x}$$

$$\tag{8.9}$$

$$J_y = \int_{t_1}^{t_2} \Sigma F_y\, dt = (F_{av})_y(t_2 - t_1) = p_{2y} - p_{1y} = mv_{2y} - mv_{1y}$$

and similarly for the z-component.

Momentum and Kinetic Energy Compared

We can now see the fundamental difference between momentum and kinetic energy. The impulse–momentum theorem $\vec{J} = \vec{p}_2 - \vec{p}_1$ says that changes in a particle's momentum are due to impulse, which depends on the *time* over which the net force acts. By contrast, the work–energy theorem $W_{tot} = K_2 - K_1$ tells us that kinetic energy changes when work is done on a particle; the total work depends on the *distance* over which the net force acts. Consider a particle that starts from rest at t_1 so that $\vec{v}_1 = 0$. Its initial momentum is $\vec{p}_1 = m\vec{v}_1 = 0$, and its initial kinetic energy is $K_1 = \frac{1}{2}mv_1^2 = 0$. Now let a constant net force equal to $\vec{F}$ act on that particle from time t_1 until time t_2. During this interval, the particle moves a distance s in the direction of the force. From Eq. (8.6), the particle's momentum at time t_2 is

$$\vec{p}_2 = \vec{p}_1 + \vec{J} = \vec{J}$$

where $\vec{J} = \vec{F}(t_2 - t_1)$ is the impulse that acts on the particle. So *the momentum of a particle equals the impulse that accelerated it from rest to its present speed;* impulse is the product of the net force that accelerated the particle and the *time* required for the acceleration. By comparison, the kinetic energy of the particle at t_2 is $K_2 = W_{tot} = Fs$, the total *work* done on the particle to accelerate it from rest. The total work is the product of the net force and the *distance* required to accelerate the particle (Fig. 8.4).

Here's an application of the distinction between momentum and kinetic energy. Suppose you have a choice between catching a 0.50-kg ball moving at 4.0 m/s or a 0.10-kg ball moving at 20 m/s. Which will be easier to catch? Both balls have the same magnitude of momentum, $p = mv = (0.50\text{ kg})(4.0\text{ m/s}) = (0.10\text{ kg})(20\text{ m/s}) = 2.0\text{ kg} \cdot \text{m/s}$. However, the two balls have different values of kinetic energy $K = \frac{1}{2}mv^2$; the large, slow-moving ball has $K = 4.0$ J, while the small, fast-moving ball has $K = 20$ J. Since the momentum is the same for both balls, both require the same *impulse* to be brought to rest. But stopping the 0.10-kg ball with your hand requires five times more *work* than stopping the 0.50-kg ball because the smaller ball has five times more kinetic energy. For a given force that you exert with your hand, it takes the same amount of time (the duration of the catch) to stop either ball, but your hand and arm will be pushed back five times farther if you choose to catch the small, fast-moving ball. To minimize arm strain, you should choose to catch the 0.50-kg ball with its lower kinetic energy.

Both the impulse–momentum and work–energy theorems are relationships between force and motion, and both rest on the foundation of Newton's laws. They are *integral* principles, relating the motion at two different times separated

MasteringPHYSICS

ActivPhysics 6.1: Momentum and Energy Change

8.4 The *kinetic energy* of a pitched baseball is equal to the work the pitcher does on it (force multiplied by the distance the ball moves during the throw). The *momentum* of the ball is equal to the impulse the pitcher imparts to it (force multiplied by the time it took to bring the ball up to speed).

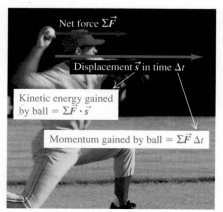

Net force $\Sigma \vec{F}$

Displacement $\vec{s}$ in time Δt

Kinetic energy gained by ball $= \Sigma \vec{F} \cdot \vec{s}$

Momentum gained by ball $= \Sigma \vec{F} \Delta t$

by a finite interval. By contrast, Newton's second law itself (in either of the forms $\sum \vec{F} = m\vec{a}$ or $\sum \vec{F} = d\vec{p}/dt$) is a *differential* principle, relating the forces to the rate of change of velocity or momentum at each instant.

Conceptual Example 8.1   Momentum versus kinetic energy

Consider again the race described in Conceptual Example 6.5 (Section 6.2) between two iceboats on a frictionless frozen lake. The boats have masses m and $2m$, and the wind exerts the same constant horizontal force $\vec{F}$ on each boat (see Fig. 6.14). The boats start from rest and cross the finish line a distance s away. Which boat crosses the finish line with greater momentum?

SOLUTION

In Conceptual Example 6.5 we asked how the *kinetic energies* of the boats compare when they cross the finish line. We answered this by remembering that *a body's kinetic energy equals the total work done to accelerate it from rest*. Both boats started from rest, and the total work done was the same for both boats (because the net force and the displacement were the same for both). Hence both boats had the same kinetic energy at the finish line.

Similarly, to compare the *momenta* of the boats we use the idea that *the momentum of each boat equals the impulse that accelerated*

it from rest. As in Conceptual Example 6.5, the net force on each boat equals the constant horizontal wind force $\vec{F}$. Let Δt be the time a boat takes to reach the finish line, so that the impulse on the boat during that time is $\vec{J} = \vec{F}\,\Delta t$. Since the boat starts from rest, this equals the boat's momentum $\vec{p}$ at the finish line:

$$\vec{p} = \vec{F}\,\Delta t$$

Both boats are subjected to the same force $\vec{F}$, but they take different times Δt to reach the finish line. The boat of mass $2m$ accelerates more slowly and takes a longer time to travel the distance s; thus there is a greater impulse on this boat between the starting and finish lines. So the boat of mass $2m$ crosses the finish line with a greater magnitude of momentum than the boat of mass m (but with the same kinetic energy). Can you show that the boat of mass $2m$ has $\sqrt{2}$ times as much momentum at the finish line as the boat of mass m?

Example 8.2   A ball hits a wall

You throw a ball with a mass of 0.40 kg against a brick wall. It hits the wall moving horizontally to the left at 30 m/s and rebounds horizontally to the right at 20 m/s. (a) Find the impulse of the net force on the ball during its collision with the wall. (b) If the ball is in contact with the wall for 0.010 s, find the average horizontal force that the wall exerts on the ball during the impact.

SOLUTION

IDENTIFY and SET UP: We're given enough information to determine the initial and final values of the ball's momentum, so we can use the impulse–momentum theorem to find the impulse. We'll then use the definition of impulse to determine the average force. Figure 8.5 shows our sketch. We need only a single axis because the motion is purely horizontal. We'll take the positive x-direction to be to the right. In part (a) our target variable is the x-component of impulse, J_x, which we'll find from the x-components of momentum before and after the impact, using Eqs. (8.9). In part (b), our target variable is the average x-component of force $(F_{av})_x$; once we know J_x, we can also find this force by using Eqs. (8.9).

8.5 Our sketch for this problem.

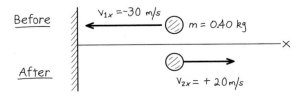

EXECUTE: (a) With our choice of x-axis, the initial and final x-components of momentum of the ball are

$$p_{1x} = mv_{1x} = (0.40\ \text{kg})(-30\ \text{m/s}) = -12\ \text{kg} \cdot \text{m/s}$$
$$p_{2x} = mv_{2x} = (0.40\ \text{kg})(+20\ \text{m/s}) = +8.0\ \text{kg} \cdot \text{m/s}$$

From the x-equation in Eqs. (8.9), the x-component of impulse equals the *change* in the x-momentum:

$$J_x = p_{2x} - p_{1x}$$
$$= 8.0\ \text{kg} \cdot \text{m/s} - (-12\ \text{kg} \cdot \text{m/s}) = 20\ \text{kg} \cdot \text{m/s} = 20\ \text{N} \cdot \text{s}$$

(b) The collision time is $t_2 - t_1 = \Delta t = 0.010$ s. From the x-equation in Eqs. (8.9), $J_x = (F_{av})_x(t_2 - t_1) = (F_{av})_x\,\Delta t$, so

$$(F_{av})_x = \frac{J_x}{\Delta t} = \frac{20\ \text{N} \cdot \text{s}}{0.010\ \text{s}} = 2000\ \text{N}$$

EVALUATE: The x-component of impulse J_x is positive—that is, to the right in Fig. 8.5. This is as it should be: The impulse represents the "kick" that the wall imparts to the ball, and this "kick" is certainly to the right.

CAUTION **Momentum is a vector** Because momentum is a vector, we had to include the negative sign in writing $p_{1x} = -12\ \text{kg} \cdot \text{m/s}$. Had we carelessly omitted it, we would have calculated the impulse to be $8.0\ \text{kg} \cdot \text{m/s} - (12\ \text{kg} \cdot \text{m/s}) = -4\ \text{kg} \cdot \text{m/s}$. This would say that the wall had somehow given the ball a kick to the *left!* Make sure that you account for the *direction* of momentum in your calculations. ▌

The force that the wall exerts on the ball must have such a large magnitude (2000 N, equal to the weight of a 200-kg object) to

Continued

change the ball's momentum in such a short time. Other forces that act on the ball during the collision are comparatively weak; for instance, the gravitational force is only 3.9 N. Thus, during the short time that the collision lasts, we can ignore all other forces on the ball. Figure 8.6 shows the impact of a tennis ball and racket.

Note that the 2000-N value we calculated is the *average* horizontal force that the wall exerts on the ball during the impact. It corresponds to the horizontal line $(F_{av})_x$ in Fig. 8.3a. The horizontal force is zero before impact, rises to a maximum, and then decreases to zero when the ball loses contact with the wall. If the ball is relatively rigid, like a baseball or golf ball, the collision lasts a short time and the maximum force is large, as in the blue curve in Fig. 8.3b. If the ball is softer, like a tennis ball, the collision time is longer and the maximum force is less, as in the orange curve in Fig. 8.3b.

8.6 Typically, a tennis ball is in contact with the racket for approximately 0.01 s. The ball flattens noticeably due to the tremendous force exerted by the racket.

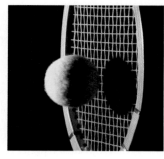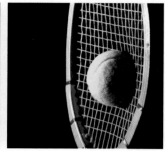

Example 8.3 Kicking a soccer ball

A soccer ball has a mass of 0.40 kg. Initially it is moving to the left at 20 m/s, but then it is kicked. After the kick it is moving at 45° upward and to the right with speed 30 m/s (Fig. 8.7a). Find the impulse of the net force and the average net force, assuming a collision time $\Delta t = 0.010$ s.

SOLUTION

IDENTIFY and SET UP: The ball moves in two dimensions, so we must treat momentum and impulse as vector quantities. We take the x-axis to be horizontally to the right and the y-axis to be vertically upward. Our target variables are the components of the net

8.7 (a) Kicking a soccer ball. (b) Finding the average force on the ball from its components.

(a) Before-and-after diagram

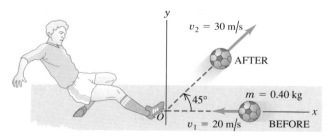

(b) Average force on the ball

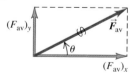

impulse on the ball, J_x and J_y, and the components of the average net force on the ball, $(F_{av})_x$ and $(F_{av})_y$. We'll find them using the impulse–momentum theorem in its component form, Eqs. (8.9).

EXECUTE: Using $\cos 45° = \sin 45° = 0.707$, we find the ball's velocity components before and after the kick:

$$v_{1x} = -20 \text{ m/s} \qquad v_{1y} = 0$$
$$v_{2x} = v_{2y} = (30 \text{ m/s})(0.707) = 21.2 \text{ m/s}$$

From Eqs. (8.9), the impulse components are

$$J_x = p_{2x} - p_{1x} = m(v_{2x} - v_{1x})$$
$$= (0.40 \text{ kg})[21.2 \text{ m/s} - (-20 \text{ m/s})] = 16.5 \text{ kg} \cdot \text{m/s}$$
$$J_y = p_{2y} - p_{1y} = m(v_{2y} - v_{1y})$$
$$= (0.40 \text{ kg})(21.2 \text{ m/s} - 0) = 8.5 \text{ kg} \cdot \text{m/s}$$

From Eq. (8.8), the average net force components are

$$(F_{av})_x = \frac{J_x}{\Delta t} = 1650 \text{ N} \qquad (F_{av})_y = \frac{J_y}{\Delta t} = 850 \text{ N}$$

The magnitude and direction of the average net force $\vec{F}_{av}$ are

$$F_{av} = \sqrt{(1650 \text{ N})^2 + (850 \text{ N})^2} = 1.9 \times 10^3 \text{ N}$$
$$\theta = \arctan \frac{850 \text{ N}}{1650 \text{ N}} = 27°$$

The ball was not initially at rest, so its final velocity does *not* have the same direction as the average force that acted on it.

EVALUATE: $\vec{F}_{av}$ includes the force of gravity, which is very small; the weight of the ball is only 3.9 N. As in Example 8.2, the average force acting during the collision is exerted almost entirely by the object that the ball hit (in this case, the soccer player's foot).

Test Your Understanding of Section 8.1 Rank the following situations according to the magnitude of the impulse of the net force, from largest value to smallest value. In each situation a 1000-kg automobile is moving along a straight east–west road. (i) The automobile is initially moving east at 25 m/s and comes to a stop in 10 s. (ii) The automobile is initially moving east at 25 m/s and comes to a stop in 5 s. (iii) The automobile is initially at rest, and a 2000-N net force toward the east is applied to it for 10 s. (iv) The automobile is initially moving east at 25 m/s, and a 2000-N net force toward the west is applied to it for 10 s. (v) The automobile is initially moving east at 25 m/s. Over a 30-s period, the automobile reverses direction and ends up moving west at 25 m/s. ❙

8.2 Conservation of Momentum

The concept of momentum is particularly important in situations in which we have two or more bodies that *interact*. To see why, let's consider first an idealized system of two bodies that interact with each other but not with anything else—for example, two astronauts who touch each other as they float freely in the zero-gravity environment of outer space (Fig. 8.8). Think of the astronauts as particles. Each particle exerts a force on the other; according to Newton's third law, the two forces are always equal in magnitude and opposite in direction. Hence, the *impulses* that act on the two particles are equal and opposite, and the changes in momentum of the two particles are equal and opposite.

Let's go over that again with some new terminology. For any system, the forces that the particles of the system exert on each other are called **internal forces.** Forces exerted on any part of the system by some object outside it are called **external forces.** For the system shown in Fig. 8.8, the internal forces are $\vec{F}_{B \text{ on } A}$, exerted by particle B on particle A, and $\vec{F}_{A \text{ on } B}$, exerted by particle A on particle B. There are *no* external forces; when this is the case, we have an **isolated system.**

The net force on particle A is $\vec{F}_{B \text{ on } A}$ and the net force on particle B is $\vec{F}_{A \text{ on } B}$, so from Eq. (8.4) the rates of change of the momenta of the two particles are

$$\vec{F}_{B \text{ on } A} = \frac{d\vec{p}_A}{dt} \qquad \vec{F}_{A \text{ on } B} = \frac{d\vec{p}_B}{dt} \qquad (8.10)$$

The momentum of each particle changes, but these changes are related to each other by Newton's third law: The two forces $\vec{F}_{B \text{ on } A}$ and $\vec{F}_{A \text{ on } B}$ are always equal in magnitude and opposite in direction. That is, $\vec{F}_{B \text{ on } A} = -\vec{F}_{A \text{ on } B}$, so $\vec{F}_{B \text{ on } A} + \vec{F}_{A \text{ on } B} = 0$. Adding together the two equations in Eq. (8.10), we have

$$\vec{F}_{B \text{ on } A} + \vec{F}_{A \text{ on } B} = \frac{d\vec{p}_A}{dt} + \frac{d\vec{p}_B}{dt} = \frac{d(\vec{p}_A + \vec{p}_B)}{dt} = 0 \qquad (8.11)$$

The rates of change of the two momenta are equal and opposite, so the rate of change of the vector sum $\vec{p}_A + \vec{p}_B$ is zero. We now define the **total momentum** $\vec{P}$ of the system of two particles as the vector sum of the momenta of the individual particles; that is,

$$\vec{P} = \vec{p}_A + \vec{p}_B \qquad (8.12)$$

Then Eq. (8.11) becomes, finally,

$$\vec{F}_{B \text{ on } A} + \vec{F}_{A \text{ on } B} = \frac{d\vec{P}}{dt} = 0 \qquad (8.13)$$

The time rate of change of the *total* momentum $\vec{P}$ is zero. Hence the total momentum of the system is constant, even though the individual momenta of the particles that make up the system can change.

If external forces are also present, they must be included on the left side of Eq. (8.13) along with the internal forces. Then the total momentum is, in general, not constant. But if the vector sum of the external forces is zero, as in Fig. 8.9, these forces have no effect on the left side of Eq. (8.13), and $d\vec{P}/dt$ is again zero. Thus we have the following general result:

> **If the vector sum of the external forces on a system is zero, the total momentum of the system is constant.**

This is the simplest form of the **principle of conservation of momentum.** This principle is a direct consequence of Newton's third law. What makes this principle useful is that it doesn't depend on the detailed nature of the internal forces that

8.8 Two astronauts push each other as they float freely in the zero-gravity environment of space.

No external forces act on the two-astronaut system, so its total momentum is conserved.

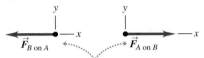

The forces the astronauts exert on each other form an action–reaction pair.

8.9 Two ice skaters push each other as they skate on a frictionless, horizontal surface. (Compare to Fig. 8.8.)

The forces the skaters exert on each other form an action–reaction pair.

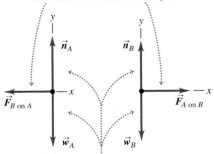

Although the normal and gravitational forces are external, their vector sum is zero, so the total momentum is conserved.

act between members of the system. This means that we can apply conservation of momentum even if (as is often the case) we know very little about the internal forces. We have used Newton's second law to derive this principle, so we have to be careful to use it only in inertial frames of reference.

We can generalize this principle for a system that contains any number of particles A, B, C, ... interacting only with one another. The total momentum of such a system is

$$\vec{P} = \vec{p}_A + \vec{p}_B + \cdots = m_A\vec{v}_A + m_B\vec{v}_B + \cdots \quad \text{(total momentum of a system of particles)} \quad (8.14)$$

We make the same argument as before: The total rate of change of momentum of the system due to each action–reaction pair of internal forces is zero. Thus the total rate of change of momentum of the entire system is zero whenever the vector sum of the external forces acting on it is zero. The internal forces can change the momenta of individual particles in the system but not the *total* momentum of the system.

CAUTION **Conservation of momentum means conservation of its components** When you apply the conservation of momentum to a system, remember that momentum is a *vector* quantity. Hence you must use vector addition to compute the total momentum of a system (Fig. 8.10). Using components is usually the simplest method. If p_{Ax}, p_{Ay}, and p_{Az} are the components of momentum of particle A, and similarly for the other particles, then Eq. (8.14) is equivalent to the component equations

$$P_x = p_{Ax} + p_{Bx} + \cdots$$
$$P_y = p_{Ay} + p_{By} + \cdots \quad (8.15)$$
$$P_z = p_{Az} + p_{Bz} + \cdots$$

If the vector sum of the external forces on the system is zero, then P_x, P_y, and P_z are all constant.

In some ways the principle of conservation of momentum is more general than the principle of conservation of mechanical energy. For example, mechanical energy is conserved only when the internal forces are *conservative*—that is, when the forces allow two-way conversion between kinetic and potential energy—but conservation of momentum is valid even when the internal forces are *not* conservative. In this chapter we will analyze situations in which both momentum and mechanical energy are conserved, and others in which only momentum is conserved. These two principles play a fundamental role in all areas of physics, and we will encounter them throughout our study of physics.

8.10 When applying conservation of momentum, remember that momentum is a vector quantity!

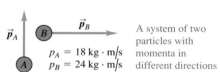

A system of two particles with momenta in different directions

$$p_A = 18 \text{ kg} \cdot \text{m/s}$$
$$p_B = 24 \text{ kg} \cdot \text{m/s}$$

You CANNOT find the magnitude of the total momentum by adding the magnitudes of the individual momenta!

$$P = p_A + p_B = 42 \text{ kg} \cdot \text{m/s} \quad \blacktriangleleft \text{WRONG}$$

Instead, use vector addition:

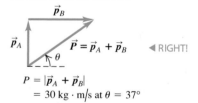

$$\vec{P} = \vec{p}_A + \vec{p}_B \quad \blacktriangleleft \text{RIGHT!}$$

$$P = |\vec{p}_A + \vec{p}_B|$$
$$= 30 \text{ kg} \cdot \text{m/s at } \theta = 37°$$

Problem-Solving Strategy 8.1 **Conservation of Momentum**

IDENTIFY *the relevant concepts:* Confirm that the vector sum of the external forces acting on the system of particles is zero. If it isn't zero, you can't use conservation of momentum.

SET UP *the problem* using the following steps:
1. Treat each body as a particle. Draw "before" and "after" sketches, including velocity vectors. Assign algebraic symbols to each magnitude, angle, and component. Use letters to label each particle and subscripts 1 and 2 for "before" and "after" quantities. Include any given values such as magnitudes, angles, or components.
2. Define a coordinate system and show it in your sketches; define the positive direction for each axis.
3. Identify the target variables.

EXECUTE *the solution:*
1. Write an equation in symbols equating the total initial and final x-components of momentum, using $p_x = mv_x$ for each particle. Write a corresponding equation for the y-components. Velocity components can be positive or negative, so be careful with signs!
2. In some problems, energy considerations (discussed in Section 8.4) give additional equations relating the velocities.
3. Solve your equations to find the target variables.

EVALUATE *your answer:* Does your answer make physical sense? If your target variable is a certain body's momentum, check that the direction of the momentum is reasonable.

Example 8.4 Recoil of a rifle

A marksman holds a rifle of mass $m_R = 3.00$ kg loosely, so it can recoil freely. He fires a bullet of mass $m_B = 5.00$ g horizontally with a velocity relative to the ground of $v_{Bx} = 300$ m/s. What is the recoil velocity v_{Rx} of the rifle? What are the final momentum and kinetic energy of the bullet and rifle?

SOLUTION

IDENTIFY and SET UP: If the marksman exerts negligible horizontal forces on the rifle, then there is no net horizontal force on the system (the bullet and rifle) during the firing, and the total horizontal momentum of the system is conserved. Figure 8.11 shows our sketch. We take the positive x-axis in the direction of aim. The rifle and the bullet are initially at rest, so the initial x-component of total momentum is zero. After the shot is fired, the bullet's x-momentum is $p_{Bx} = m_B v_{Bx}$ and the rifle's x-momentum

8.11 Our sketch for this problem.

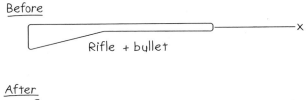

is $p_{Rx} = m_R v_{Rx}$. Our target variables are v_{Rx}, p_{Bx}, p_{Rx}, and the final kinetic energies $K_B = \frac{1}{2} m_B v_{Bx}^2$ and $K_R = \frac{1}{2} m_R v_{Rx}^2$.

EXECUTE: Conservation of the x-component of total momentum gives

$$P_x = 0 = m_B v_{Bx} + m_R v_{Rx}$$

$$v_{Rx} = -\frac{m_B}{m_R} v_{Bx} = -\left(\frac{0.00500 \text{ kg}}{3.00 \text{ kg}}\right)(300 \text{ m/s}) = -0.500 \text{ m/s}$$

The negative sign means that the recoil is in the direction opposite to that of the bullet.

The final momenta and kinetic energies are

$$p_{Bx} = m_B v_{Bx} = (0.00500 \text{ kg})(300 \text{ m/s}) = 1.50 \text{ kg} \cdot \text{m/s}$$

$$K_B = \frac{1}{2} m_B v_{Bx}^2 = \frac{1}{2}(0.00500 \text{ kg})(300 \text{ m/s})^2 = 225 \text{ J}$$

$$p_{Rx} = m_R v_{Rx} = (3.00 \text{ kg})(-0.500 \text{ m/s}) = -1.50 \text{ kg} \cdot \text{m/s}$$

$$K_R = \frac{1}{2} m_R v_{Rx}^2 = \frac{1}{2}(3.00 \text{ kg})(-0.500 \text{ m/s})^2 = 0.375 \text{ J}$$

EVALUATE: The bullet and rifle have equal and opposite final *momenta* thanks to Newton's third law: They experience equal and opposite interaction forces that act for the same *time*, so the impulses are equal and opposite. But the bullet travels a much greater *distance* than the rifle during the interaction. Hence the force on the bullet does more work than the force on the rifle, giving the bullet much greater *kinetic energy* than the rifle. The 600:1 ratio of the two kinetic energies is the inverse of the ratio of the masses; in fact, you can show that this always happens in recoil situations (see Exercise 8.26).

Example 8.5 Collision along a straight line

Two gliders with different masses move toward each other on a frictionless air track (Fig. 8.12a). After they collide (Fig. 8.12b), glider B has a final velocity of $+2.0$ m/s (Fig. 8.12c). What is the final velocity of glider A? How do the changes in momentum and in velocity compare?

SOLUTION

IDENTIFY and SET UP: As for the skaters in Fig. 8.9, the total vertical force on each glider is zero, and the net force on each individual glider is the horizontal force exerted on it by the other glider. The net external force on the *system* of two gliders is zero, so their total momentum is conserved. We take the positive x-axis to be to the right. We are given the masses and initial velocities of both gliders and the final velocity of glider B. Our target variables are v_{A2x}, the final x-component of velocity of glider A, and the changes in momentum and in velocity of the two gliders (the value *after* the collision minus the value *before* the collision).

EXECUTE: The x-component of total momentum before the collision is

$$P_x = m_A v_{A1x} + m_B v_{B1x}$$
$$= (0.50 \text{ kg})(2.0 \text{ m/s}) + (0.30 \text{ kg})(-2.0 \text{ m/s})$$
$$= 0.40 \text{ kg} \cdot \text{m/s}$$

8.12 Two gliders colliding on an air track.

(a) Before collision
$v_{A1x} = 2.0$ m/s $\quad v_{B1x} = -2.0$ m/s
$m_A = 0.50$ kg $\quad m_B = 0.30$ kg

(b) Collision

$v_{A2x} \quad v_{B2x} = 2.0$ m/s
(c) After collision

This is positive (to the right in Fig. 8.12) because A has a greater magnitude of momentum than B. The x-component of total momentum has the same value after the collision, so

$$P_x = m_A v_{A2x} + m_B v_{B2x}$$

Continued

We solve for v_{A2x}:

$$v_{A2x} = \frac{P_x - m_B v_{B2x}}{m_A} = \frac{0.40 \text{ kg} \cdot \text{m/s} - (0.30 \text{ kg})(2.0 \text{ m/s})}{0.50 \text{ kg}}$$

$$= -0.40 \text{ m/s}$$

The changes in the x-momenta are

$$m_A v_{A2x} - m_A v_{A1x} = (0.50 \text{ kg})(-0.40 \text{ m/s})$$
$$- (0.50 \text{ kg})(2.0 \text{ m/s}) = -1.2 \text{ kg} \cdot \text{m/s}$$

$$m_B v_{B2x} - m_B v_{B1x} = (0.30 \text{ kg})(2.0 \text{ m/s})$$
$$- (0.30 \text{ kg})(-2.0 \text{ m/s}) = +1.2 \text{ kg} \cdot \text{m/s}$$

The changes in x-velocities are

$$v_{A2x} - v_{A1x} = (-0.40 \text{ m/s}) - 2.0 \text{ m/s} = -2.4 \text{ m/s}$$
$$v_{B2x} - v_{B1x} = 2.0 \text{ m/s} - (-2.0 \text{ m/s}) = +4.0 \text{ m/s}$$

EVALUATE: The gliders were subjected to equal and opposite interaction forces for the same time during their collision. By the impulse–momentum theorem, they experienced equal and opposite impulses and therefore equal and opposite changes in momentum. But by Newton's second law, the less massive glider (B) had a greater magnitude of acceleration and hence a greater velocity change.

Example 8.6 Collision in a horizontal plane

Figure 8.13a shows two battling robots on a frictionless surface. Robot A, with mass 20 kg, initially moves at 2.0 m/s parallel to the x-axis. It collides with robot B, which has mass 12 kg and is initially at rest. After the collision, robot A moves at 1.0 m/s in a direction that makes an angle $\alpha = 30°$ with its initial direction (Fig. 8.13b). What is the final velocity of robot B?

SOLUTION

IDENTIFY and SET UP: There are no horizontal external forces, so the x- and y-components of the total momentum of the system are both conserved. Momentum conservation requires that the sum of the x-components of momentum *before* the collision (subscript 1) must equal the sum *after* the collision (subscript 2), and similarly for the sums of the y-components. Our target variable is $\vec{v}_{B2}$, the final velocity of robot B.

8.13 Views from above of the velocities (a) before and (b) after the collision.

(a) Before collision

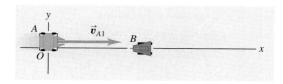

(b) After collision

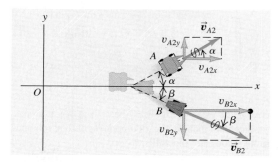

EXECUTE: The momentum-conservation equations and their solutions for v_{B2x} and v_{B2y} are

$$m_A v_{A1x} + m_B v_{B1x} = m_A v_{A2x} + m_B v_{B2x}$$

$$v_{B2x} = \frac{m_A v_{A1x} + m_B v_{B1x} - m_A v_{A2x}}{m_B}$$

$$= \frac{\left[\begin{array}{c} (20 \text{ kg})(2.0 \text{ m/s}) + (12 \text{ kg})(0) \\ - (20 \text{ kg})(1.0 \text{ m/s})(\cos 30°) \end{array} \right]}{12 \text{ kg}}$$

$$= 1.89 \text{ m/s}$$

$$m_A v_{A1y} + m_B v_{B1y} = m_A v_{A2y} + m_B v_{B2y}$$

$$v_{B2y} = \frac{m_A v_{A1y} + m_B v_{B1y} - m_A v_{A2y}}{m_B}$$

$$= \frac{\left[\begin{array}{c} (20 \text{ kg})(0) + (12 \text{ kg})(0) \\ - (20 \text{ kg})(1.0 \text{ m/s})(\sin 30°) \end{array} \right]}{12 \text{ kg}}$$

$$= -0.83 \text{ m/s}$$

Figure 8.13b shows the motion of robot B after the collision. The magnitude of $\vec{v}_{B2}$ is

$$v_{B2} = \sqrt{(1.89 \text{ m/s})^2 + (-0.83 \text{ m/s})^2} = 2.1 \text{ m/s}$$

and the angle of its direction from the positive x-axis is

$$\beta = \arctan\frac{-0.83 \text{ m/s}}{1.89 \text{ m/s}} = -24°$$

EVALUATE: We can check our answer by confirming that the components of total momentum before and after the collision are equal. Initially robot A has x-momentum $m_A v_{A1x} = (20 \text{ kg})(2.0 \text{ m/s}) = 40 \text{ kg} \cdot \text{m/s}$ and zero y-momentum; robot B has zero momentum. After the collision, the momentum components are $m_A v_{A2x} = (20 \text{ kg})(1.0 \text{ m/s})(\cos 30°) = 17 \text{ kg} \cdot \text{m/s}$ and $m_B v_{B2x} = (12 \text{ kg})(1.89 \text{ m/s}) = 23 \text{ kg} \cdot \text{m/s}$; the total x-momentum is 40 kg $\cdot$ m/s, the same as before the collision. The final y-components are $m_A v_{A2y} = (20 \text{ kg})(1.0 \text{ m/s})(\sin 30°) = 10 \text{ kg} \cdot \text{m/s}$ and $m_B v_{B2y} = (12 \text{ kg})(-0.83 \text{ m/s}) = -10 \text{ kg} \cdot \text{m/s}$; the total y-component of momentum is zero, the same as before the collision.

Test Your Understanding of Section 8.2 A spring-loaded toy sits at rest on a horizontal, frictionless surface. When the spring releases, the toy breaks into three equal-mass pieces, *A*, *B*, and *C*, which slide along the surface. Piece *A* moves off in the negative *x*-direction, while piece *B* moves off in the negative *y*-direction. (a) What are the signs of the velocity components of piece *C*? (b) Which of the three pieces is moving the fastest?

8.3 Momentum Conservation and Collisions

To most people the term *collision* is likely to mean some sort of automotive disaster. We'll use it in that sense, but we'll also broaden the meaning to include any strong interaction between bodies that lasts a relatively short time. So we include not only car accidents but also balls colliding on a billiard table, neutrons hitting atomic nuclei in a nuclear reactor, the impact of a meteor on the Arizona desert, and a close encounter of a spacecraft with the planet Saturn.

If the forces between the bodies are much larger than any external forces, as is the case in most collisions, we can neglect the external forces entirely and treat the bodies as an *isolated* system. Then momentum is conserved and the total momentum of the system has the same value before and after the collision. Two cars colliding at an icy intersection provide a good example. Even two cars colliding on dry pavement can be treated as an isolated system during the collision if the forces between the cars are much larger than the friction forces of pavement against tires.

Elastic and Inelastic Collisions

If the forces between the bodies are also *conservative,* so that no mechanical energy is lost or gained in the collision, the total *kinetic* energy of the system is the same after the collision as before. Such a collision is called an **elastic collision.** A collision between two marbles or two billiard balls is almost completely elastic. Figure 8.14 shows a model for an elastic collision. When the gliders collide, their springs are momentarily compressed and some of the original kinetic energy is momentarily converted to elastic potential energy. Then the gliders bounce apart, the springs expand, and this potential energy is converted back to kinetic energy.

A collision in which the total kinetic energy after the collision is *less* than before the collision is called an **inelastic collision.** A meatball landing on a plate of spaghetti and a bullet embedding itself in a block of wood are examples of inelastic collisions. An inelastic collision in which the colliding bodies stick together and move as one body after the collision is often called a **completely inelastic collision.** Figure 8.15 shows an example; we have replaced the spring bumpers in Fig. 8.14 with Velcro®, which sticks the two bodies together.

> **CAUTION** An inelastic collision doesn't have to be *completely* inelastic It's a common misconception that the *only* inelastic collisions are those in which the colliding bodies stick together. In fact, inelastic collisions include many situations in which the bodies do *not* stick. If two cars bounce off each other in a "fender bender," the work done to deform the fenders cannot be recovered as kinetic energy of the cars, so the collision is inelastic (Fig. 8.16).
> Remember this rule: **In any collision in which external forces can be neglected, momentum is conserved and the total momentum before equals the total momentum after; in elastic collisions *only*, the total kinetic energy before equals the total kinetic energy after.**

Completely Inelastic Collisions

Let's look at what happens to momentum and kinetic energy in a *completely* inelastic collision of two bodies (*A* and *B*), as in Fig. 8.15. Because the two bodies stick together after the collision, they have the same final velocity $\vec{v}_2$:

$$\vec{v}_{A2} = \vec{v}_{B2} = \vec{v}_2$$

8.14 Two gliders undergoing an elastic collision on a frictionless surface. Each glider has a steel spring bumper that exerts a conservative force on the other glider.

(a) Before collision

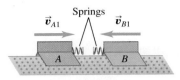

(b) Elastic collision

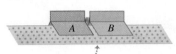

Kinetic energy is stored as potential energy in compressed springs.

(c) After collision

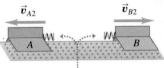

The system of the two gliders has the same kinetic energy after the collision as before it.

8.15 Two gliders undergoing a completely inelastic collision. The spring bumpers on the gliders are replaced by Velcro®, so the gliders stick together after collision.

(a) Before collision

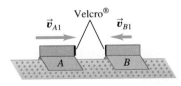

(b) Completely inelastic collision

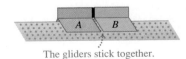

The gliders stick together.

(c) After collision

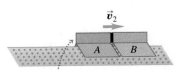

The system of the two gliders has less kinetic energy after the collision than before it.

8.16 Automobile collisions are intended to be inelastic, so that the structure of the car absorbs as much of the energy of the collision as possible. This absorbed energy cannot be recovered, since it goes into a permanent deformation of the car.

Conservation of momentum gives the relationship

$$m_A\vec{v}_{A1} + m_B\vec{v}_{B1} = (m_A + m_B)\vec{v}_2 \quad \text{(completely inelastic collision)} \quad (8.16)$$

If we know the masses and initial velocities, we can compute the common final velocity $\vec{v}_2$.

Suppose, for example, that a body with mass m_A and initial x-component of velocity v_{A1x} collides inelastically with a body with mass m_B that is initially at rest ($v_{B1x} = 0$). From Eq. (8.16) the common x-component of velocity v_{2x} of both bodies after the collision is

$$v_{2x} = \frac{m_A}{m_A + m_B}v_{A1x} \quad \begin{array}{l}\text{(completely inelastic collision,} \\ B \text{ initially at rest)}\end{array} \quad (8.17)$$

Let's verify that the total kinetic energy after this completely inelastic collision is less than before the collision. The motion is purely along the x-axis, so the kinetic energies K_1 and K_2 before and after the collision, respectively, are

$$K_1 = \tfrac{1}{2}m_A v_{A1x}{}^2$$

$$K_2 = \tfrac{1}{2}(m_A + m_B)v_{2x}{}^2 = \tfrac{1}{2}(m_A + m_B)\left(\frac{m_A}{m_A + m_B}\right)^2 v_{A1x}{}^2$$

The ratio of final to initial kinetic energy is

$$\frac{K_2}{K_1} = \frac{m_A}{m_A + m_B} \quad \begin{array}{l}\text{(completely inelastic collision,} \\ B \text{ initially at rest)}\end{array} \quad (8.18)$$

The right side is always less than unity because the denominator is always greater than the numerator. Even when the initial velocity of m_B is not zero, it is not hard to verify that the kinetic energy after a completely inelastic collision is always less than before.

Please note: We don't recommend memorizing Eq. (8.17) or (8.18). We derived them only to prove that kinetic energy is always lost in a completely inelastic collision.

Example 8.7   A completely inelastic collision

We repeat the collision described in Example 8.5 (Section 8.2), but this time equip the gliders so that they stick together when they collide. Find the common final x-velocity, and compare the initial and final kinetic energies of the system.

SOLUTION

IDENTIFY and SET UP: There are no external forces in the x-direction, so the x-component of momentum is conserved. Figure 8.17 shows our sketch. Our target variables are the final x-velocity v_{2x} and the initial and final kinetic energies K_1 and K_2.

8.17 Our sketch for this problem.

Before: $v_{A1x} = 2.0$ m/s $\;\;A \rightarrow$ $\;\;\;\;v_{B1x} = -2.0$ m/s $\;\leftarrow B$ $\;\;\;\;—x$
$m_A = 0.50$ kg $\;\;\;\;\;\;m_B = 0.30$ kg

After: $\;\;\;\;\;\;\boxed{A|B} \rightarrow$ $v_{2x} = ?$ $\;\;\;—x$

EXECUTE: From conservation of momentum,

$$m_A v_{A1x} + m_B v_{B1x} = (m_A + m_B)v_{2x}$$

$$v_{2x} = \frac{m_A v_{A1x} + m_B v_{B1x}}{m_A + m_B}$$

$$= \frac{(0.50\text{ kg})(2.0\text{ m/s}) + (0.30\text{ kg})(-2.0\text{ m/s})}{0.50\text{ kg} + 0.30\text{ kg}}$$

$$= 0.50\text{ m/s}$$

Because v_{2x} is positive, the gliders move together to the right after the collision. Before the collision, the kinetic energies are

$$K_A = \tfrac{1}{2}m_A v_{A1x}{}^2 = \tfrac{1}{2}(0.50\text{ kg})(2.0\text{ m/s})^2 = 1.0\text{ J}$$

$$K_B = \tfrac{1}{2}m_B v_{B1x}{}^2 = \tfrac{1}{2}(0.30\text{ kg})(-2.0\text{ m/s})^2 = 0.60\text{ J}$$

The total kinetic energy before the collision is $K_1 = K_A + K_B = 1.6$ J. The kinetic energy after the collision is

$$K_2 = \tfrac{1}{2}(m_A + m_B)v_{2x}{}^2 = \tfrac{1}{2}(0.50\text{ kg} + 0.30\text{ kg})(0.50\text{ m/s})^2$$

$$= 0.10\text{ J}$$

EVALUATE: The final kinetic energy is only $\frac{1}{16}$ of the original; $\frac{15}{16}$ is converted from mechanical energy to other forms. If there is a wad of chewing gum between the gliders, it squashes and becomes warmer. If there is a spring between the gliders that is compressed as they lock together, the energy is stored as potential energy of the spring. In both cases the *total* energy of the system is conserved, although *kinetic* energy is not. In an isolated system, however, momentum is *always* conserved whether the collision is elastic or not.

Example 8.8 The ballistic pendulum

Figure 8.18 shows a ballistic pendulum, a simple system for measuring the speed of a bullet. A bullet of mass m_B makes a completely inelastic collision with a block of wood of mass m_W, which is suspended like a pendulum. After the impact, the block swings up to a maximum height y. In terms of y, m_B, and m_W, what is the initial speed v_1 of the bullet?

SOLUTION

IDENTIFY: We'll analyze this event in two stages: (1) the embedding of the bullet in the block and (2) the pendulum swing of the block. During the first stage, the bullet embeds itself in the block so quickly that the block does not move appreciably. The supporting strings remain nearly vertical, so negligible external horizontal force acts on the bullet–block system, and the horizontal component of momentum is conserved. Mechanical energy is *not* conserved during this stage, however, because a nonconservative force does work (the force of friction between bullet and block).

In the second stage, the block and bullet move together. The only forces acting on this system are gravity (a conservative force) and the string tensions (which do no work). Thus, as the block swings, *mechanical energy* is conserved. Momentum is *not* conserved during this stage, however, because there is a net external force (the forces of gravity and string tension don't cancel when the strings are inclined).

SET UP: We take the positive x-axis to the right and the positive y-axis upward. Our target variable is v_1. Another unknown quantity is the speed v_2 of the system just after the collision. We'll use momentum conservation in the first stage to relate v_1 to v_2, and we'll use energy conservation in the second stage to relate v_2 to y.

EXECUTE: In the first stage, all velocities are in the $+x$-direction. Momentum conservation gives

$$m_B v_1 = (m_B + m_W)v_2$$
$$v_1 = \frac{m_B + m_W}{m_B}v_2$$

At the beginning of the second stage, the system has kinetic energy $K = \frac{1}{2}(m_B + m_W)v_2^2$. The system swings up and comes to rest for an instant at a height y, where its kinetic energy is zero and the potential energy is $(m_B + m_W)gy$; it then swings back down. Energy conservation gives

$$\tfrac{1}{2}(m_B + m_W)v_2^2 = (m_B + m_W)gy$$
$$v_2 = \sqrt{2gy}$$

We substitute this expression for v_2 into the momentum equation:

$$v_1 = \frac{m_B + m_W}{m_B}\sqrt{2gy}$$

EVALUATE: Let's plug in the realistic numbers $m_B = 5.00\text{ g} = 0.00500\text{ kg}$, $m_W = 2.00\text{ kg}$, and $y = 3.00\text{ cm} = 0.0300\text{ m}$. We then have

$$v_1 = \frac{0.00500\text{ kg} + 2.00\text{ kg}}{0.00500\text{ kg}}\sqrt{2(9.80\text{ m/s}^2)(0.0300\text{ m})}$$

$$= 307\text{ m/s}$$

The speed v_2 of the block just after impact is

$$v_2 = \sqrt{2gy} = \sqrt{2(9.80\text{ m/s}^2)(0.0300\text{ m})}$$

$$= 0.767\text{ m/s}$$

The speeds v_1 and v_2 seem realistic. The kinetic energy of the bullet before impact is $\frac{1}{2}(0.00500\text{ kg})(307\text{ m/s})^2 = 236\text{ J}$. Just after impact the kinetic energy of the system is $\frac{1}{2}(2.005\text{ kg})(0.767\text{ m/s})^2 = 0.590\text{ J}$. Nearly all the kinetic energy disappears as the wood splinters and the bullet and block become warmer.

8.18 A ballistic pendulum.

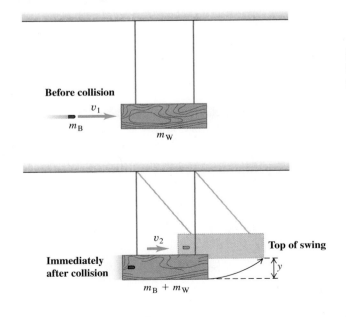

Example 8.9 An automobile collision

A 1000-kg car traveling north at 15 m/s collides with a 2000-kg truck traveling east at 10 m/s. The occupants, wearing seat belts, are uninjured, but the two vehicles move away from the impact point as one. The insurance adjustor asks you to find the velocity of the wreckage just after impact. What is your answer?

SOLUTION

IDENTIFY and SET UP: We'll treat the cars as an isolated system, so that the momentum of the system is conserved. We can do so because (as we show below) the magnitudes of the horizontal forces that the cars exert on each other during the collision are much larger than any external forces such as friction. Figure 8.19 shows our sketch and the coordinate axes. We can find the total momentum $\vec{P}$ before the collision using Eqs. (8.15). The momentum has the same value just after the collision; hence we can find the velocity $\vec{V}$ just after the collision (our target variable) using $\vec{P} = M\vec{V}$, where $M = m_C + m_T = 3000$ kg is the mass of the wreckage.

EXECUTE: From Eqs. (8.15), the components of $\vec{P}$ are

$$P_x = p_{Cx} + p_{Tx} = m_C v_{Cx} + m_T v_{Tx}$$
$$= (1000 \text{ kg})(0) + (2000 \text{ kg})(10 \text{ m/s})$$
$$= 2.0 \times 10^4 \text{ kg} \cdot \text{m/s}$$
$$P_y = p_{Cy} + p_{Ty} = m_C v_{Cy} + m_T v_{Ty}$$
$$= (1000 \text{ kg})(15 \text{ m/s}) + (2000 \text{ kg})(0)$$
$$= 1.5 \times 10^4 \text{ kg} \cdot \text{m/s}$$

The magnitude of $\vec{P}$ is

$$P = \sqrt{(2.0 \times 10^4 \text{ kg} \cdot \text{m/s})^2 + (1.5 \times 10^4 \text{ kg} \cdot \text{m/s})^2}$$
$$= 2.5 \times 10^4 \text{ kg} \cdot \text{m/s}$$

and its direction is given by the angle θ shown in Fig. 8.19:

$$\tan \theta = \frac{P_y}{P_x} = \frac{1.5 \times 10^4 \text{ kg} \cdot \text{m/s}}{2.0 \times 10^4 \text{ kg} \cdot \text{m/s}} = 0.75 \quad \theta = 37°$$

8.19 Our sketch for this problem.

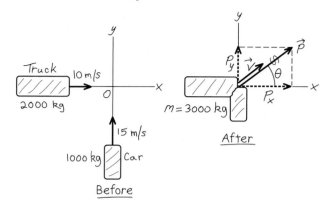

From $\vec{P} = M\vec{V}$, the direction of the velocity $\vec{V}$ just after the collision is also $\theta = 37°$. The velocity magnitude is

$$V = \frac{P}{M} = \frac{2.5 \times 10^4 \text{ kg} \cdot \text{m/s}}{3000 \text{ kg}} = 8.3 \text{ m/s}$$

EVALUATE: This is an inelastic collision, so we expect the total kinetic energy to be less after the collision than before. As you can show, the initial kinetic energy is 2.1×10^5 J and the final value is 1.0×10^5 J.

We can now justify our neglect of the external forces on the vehicles during the collision. The car's weight is about 10,000 N; if the coefficient of kinetic friction is 0.5, the friction force on the car during the impact is about 5000 N. The car's initial kinetic energy is $\frac{1}{2}(1000 \text{ kg})(15 \text{ m/s})^2 = 1.1 \times 10^5$ J, so -1.1×10^5 J of work must be done to stop it. If the car crumples by 0.20 m in stopping, a force of magnitude $(1.1 \times 10^5 \text{ J})/(0.20 \text{ m}) = 5.5 \times 10^5$ N would be needed; that's 110 times the friction force. So it's reasonable to treat the external force of friction as negligible compared with the internal forces the vehicles exert on each other.

Classifying Collisions

It's important to remember that we can classify collisions according to energy considerations (Fig. 8.20). A collision in which kinetic energy is conserved is called *elastic*. (We'll explore these in more depth in the next section.) A collision in which the total kinetic energy decreases is called *inelastic*. When the two bodies have a common final velocity, we say that the collision is *completely inelastic*. There are also cases in which the final kinetic energy is *greater* than the initial value. Rifle recoil, discussed in Example 8.4 (Section 8.2), is an example.

8.20 Collisions are classified according to energy considerations.

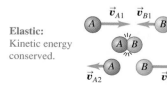

Elastic:
Kinetic energy conserved.

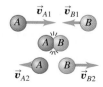

Inelastic:
Some kinetic energy lost.

Completely inelastic:
Bodies have same final velocity.

Finally, we emphasize again that we can sometimes use momentum conservation even when there are external forces acting on the system, if the net external force acting on the colliding bodies is small in comparison with the internal forces during the collision (as in Example 8.9)

Test Your Understanding of Section 8.3   For each situation, state whether the collision is elastic or inelastic. If it is inelastic, state whether it is completely inelastic. (a) You drop a ball from your hand. It collides with the floor and bounces back up so that it just reaches your hand. (b) You drop a different ball from your hand and let it collide with the ground. This ball bounces back up to half the height from which it was dropped. (c) You drop a ball of clay from your hand. When it collides with the ground, it stops.

8.4 Elastic Collisions

We saw in Section 8.3 that an *elastic collision* in an isolated system is one in which kinetic energy (as well as momentum) is conserved. Elastic collisions occur when the forces between the colliding bodies are *conservative*. When two billiard balls collide, they squash a little near the surface of contact, but then they spring back. Some of the kinetic energy is stored temporarily as elastic potential energy, but at the end it is reconverted to kinetic energy (Fig. 8.21).

Let's look at an elastic collision between two bodies A and B. We start with a one-dimensional collision, in which all the velocities lie along the same line; we choose this line to be the x-axis. Each momentum and velocity then has only an x-component. We call the x-velocities before the collision v_{A1x} and v_{B1x}, and those after the collision v_{A2x} and v_{B2x}. From conservation of kinetic energy we have

$$\tfrac{1}{2}m_A v_{A1x}^2 + \tfrac{1}{2}m_B v_{B1x}^2 = \tfrac{1}{2}m_A v_{A2x}^2 + \tfrac{1}{2}m_B v_{B2x}^2$$

and conservation of momentum gives

$$m_A v_{A1x} + m_B v_{B1x} = m_A v_{A2x} + m_B v_{B2x}$$

If the masses m_A and m_B and the initial velocities v_{A1x} and v_{B1x} are known, we can solve these two equations to find the two final velocities v_{A2x} and v_{B2x}.

Elastic Collisions, One Body Initially at Rest

The general solution to the above equations is a little complicated, so we will concentrate on the particular case in which body B is at rest before the collision (so $v_{B1x} = 0$). Think of body B as a target for body A to hit. Then the kinetic energy and momentum conservation equations are, respectively,

$$\tfrac{1}{2}m_A v_{A1x}^2 = \tfrac{1}{2}m_A v_{A2x}^2 + \tfrac{1}{2}m_B v_{B2x}^2 \qquad (8.19)$$

$$m_A v_{A1x} = m_A v_{A2x} + m_B v_{B2x} \qquad (8.20)$$

We can solve for v_{A2x} and v_{B2x} in terms of the masses and the initial velocity v_{A1x}. This involves some fairly strenuous algebra, but it's worth it. No pain, no gain! The simplest approach is somewhat indirect, but along the way it uncovers an additional interesting feature of elastic collisions.

First we rearrange Eqs. (8.19) and (8.20) as follows:

$$m_B v_{B2x}^2 = m_A(v_{A1x}^2 - v_{A2x}^2) = m_A(v_{A1x} - v_{A2x})(v_{A1x} + v_{A2x}) \quad (8.21)$$

$$m_B v_{B2x} = m_A(v_{A1x} - v_{A2x}) \qquad (8.22)$$

Now we divide Eq. (8.21) by Eq. (8.22) to obtain

$$v_{B2x} = v_{A1x} + v_{A2x} \qquad (8.23)$$

8.21 Billiard balls deform very little when they collide, and they quickly spring back from any deformation they do undergo. Hence the force of interaction between the balls is almost perfectly conservative, and the collision is almost perfectly elastic.

Mastering**PHYSICS**

ActivPhysics 6.2: Collisions and Elasticity
ActivPhysics 6.5: Car Collisions: Two Dimensions
ActivPhysics 6.9: Pendulum Bashes Box

8.22 Collisions between (a) a moving Ping-Pong ball and an initially stationary bowling ball, and (b) a moving bowling ball and an initially stationary Ping-Pong ball.

(a) Ping-Pong ball strikes bowling ball.

BEFORE

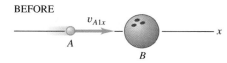

AFTER

(b) Bowling ball strikes Ping-Pong ball.

BEFORE

AFTER

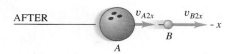

8.23 A one-dimensional elastic collision between bodies of equal mass.

When a moving object A has a 1-D elastic collision with an equal-mass, motionless object B ...

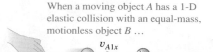

... all of A's momentum and kinetic energy are transferred to B.

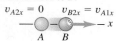

We substitute this expression back into Eq. (8.22) to eliminate v_{B2x} and then solve for v_{A2x}:

$$m_B(v_{A1x} + v_{A2x}) = m_A(v_{A1x} - v_{A2x})$$

$$v_{A2x} = \frac{m_A - m_B}{m_A + m_B} v_{A1x} \tag{8.24}$$

Finally, we substitute this result back into Eq. (8.23) to obtain

$$v_{B2x} = \frac{2m_A}{m_A + m_B} v_{A1x} \tag{8.25}$$

Now we can interpret the results. Suppose body A is a Ping-Pong ball and body B is a bowling ball. Then we expect A to bounce off after the collision with a velocity nearly equal to its original value but in the opposite direction (Fig. 8.22a), and we expect B's velocity to be much less. That's just what the equations predict. When m_A is much smaller than m_B, the fraction in Eq. (8.24) is approximately equal to (-1), so v_{A2x} is approximately equal to $-v_{A1x}$. The fraction in Eq. (8.25) is much smaller than unity, so v_{B2x} is much less than v_{A1x}. Figure 8.22b shows the opposite case, in which A is the bowling ball and B the Ping-Pong ball and m_A is much larger than m_B. What do you expect to happen then? Check your predictions against Eqs. (8.24) and (8.25).

Another interesting case occurs when the masses are equal (Fig. 8.23). If $m_A = m_B$, then Eqs. (8.24) and (8.25) give $v_{A2x} = 0$ and $v_{B2x} = v_{A1x}$. That is, the body that was moving stops dead; it gives all its momentum and kinetic energy to the body that was at rest. This behavior is familiar to all pool players.

Elastic Collisions and Relative Velocity

Let's return to the more general case in which A and B have different masses. Equation (8.23) can be rewritten as

$$v_{A1x} = v_{B2x} - v_{A2x} \tag{8.26}$$

Here $v_{B2x} - v_{A2x}$ is the velocity of B relative to A *after* the collision; from Eq. (8.26), this equals v_{A1x}, which is the *negative* of the velocity of B relative to A *before* the collision. (We discussed relative velocity in Section 3.5.) The relative velocity has the same magnitude, but opposite sign, before and after the collision. The sign changes because A and B are approaching each other before the collision but moving apart after the collision. If we view this collision from a second coordinate system moving with constant velocity relative to the first, the velocities of the bodies are different but the *relative* velocities are the same. Hence our statement about relative velocities holds for *any* straight-line elastic collision, even when neither body is at rest initially. *In a straight-line elastic collision of two bodies, the relative velocities before and after the collision have the same magnitude but opposite sign.* This means that if B is moving before the collision, Eq. (8.26) becomes

$$v_{B2x} - v_{A2x} = -(v_{B1x} - v_{A1x}) \tag{8.27}$$

It turns out that a *vector* relationship similar to Eq. (8.27) is a general property of *all* elastic collisions, even when both bodies are moving initially and the velocities do not all lie along the same line. This result provides an alternative and equivalent definition of an elastic collision: *In an elastic collision, the relative velocity of the two bodies has the same magnitude before and after the collision.* Whenever this condition is satisfied, the total kinetic energy is also conserved.

When an elastic two-body collision isn't head-on, the velocities don't all lie along a single line. If they all lie in a plane, then each final velocity has two unknown components, and there are four unknowns in all. Conservation of energy and conservation of the *x*- and *y*-components of momentum give only three equations. To determine the final velocities uniquely, we need additional information, such as the direction or magnitude of one of the final velocities.

Example 8.10   An elastic straight-line collision

We repeat the air-track collision of Example 8.5 (Section 8.2), but now we add ideal spring bumpers to the gliders so that the collision is elastic. What are the final velocities of the gliders?

SOLUTION

IDENTIFY and SET UP: The net external force on the system is zero, so the momentum of the system is conserved. Figure 8.24 shows our sketch. We'll find our target variables, v_{A2x} and v_{B2x}, using Eq. (8.27), the relative-velocity relationship for an elastic collision, and the momentum-conservation equation.

EXECUTE: From Eq. (8.27),

$$v_{B2x} - v_{A2x} = -(v_{B1x} - v_{A1x})$$
$$= -(-2.0 \text{ m/s} - 2.0 \text{ m/s}) = 4.0 \text{ m/s}$$

From conservation of momentum,

$$m_A v_{A1x} + m_B v_{B1x} = m_A v_{A2x} + m_B v_{B2x}$$
$$(0.50 \text{ kg})(2.0 \text{ m/s}) + (0.30 \text{ kg})(-2.0 \text{ m/s})$$
$$= (0.50 \text{ kg})v_{A2x} + (0.30 \text{ kg})v_{B2x}$$
$$0.50 v_{A2x} + 0.30 v_{B2x} = 0.40 \text{ m/s}$$

(To get the last equation we divided both sides of the equation just above it by the quantity 1 kg. This makes the units the same as in the first equation.) Solving these equations simultaneously, we find

$$v_{A2x} = -1.0 \text{ m/s} \qquad v_{B2x} = 3.0 \text{ m/s}$$

8.24 Our sketch for this problem.

EVALUATE: Both bodies reverse their directions of motion; A moves to the left at 1.0 m/s and B moves to the right at 3.0 m/s. This is unlike the result of Example 8.5 because that collision was *not* elastic. The more massive glider A slows down in the collision and so loses kinetic energy. The less massive glider B speeds up and gains kinetic energy. The total kinetic energy before the collision (which we calculated in Example 8.7) is 1.6 J. The total kinetic energy after the collision is

$$\tfrac{1}{2}(0.50 \text{ kg})(-1.0 \text{ m/s})^2 + \tfrac{1}{2}(0.30 \text{ kg})(3.0 \text{ m/s})^2 = 1.6 \text{ J}$$

As expected, the kinetic energies before and after this elastic collision are equal. Kinetic energy is transferred from A to B, but none of it is lost.

CAUTION   Be careful with the elastic collision equations You could *not* have solved this problem using Eqs. (8.24) and (8.25), which apply only if body B is initially *at rest*. Always be sure that you solve the problem at hand using equations that are applicable!

Example 8.11   Moderating fission neutrons in a nuclear reactor

The fission of uranium nuclei in a nuclear reactor produces high-speed neutrons. Before such neutrons can efficiently cause additional fissions, they must be slowed down by collisions with nuclei in the *moderator* of the reactor. The first nuclear reactor (built in 1942 at the University of Chicago) used carbon (graphite) as the moderator. Suppose a neutron (mass 1.0 u) traveling at 2.6×10^7 m/s undergoes a head-on elastic collision with a carbon nucleus (mass 12 u) initially at rest. Neglecting external forces during the collision, find the velocities after the collision. (1 u is the *atomic mass unit,* equal to 1.66×10^{-27} kg.)

SOLUTION

IDENTIFY and SET UP: We neglect external forces, so momentum is conserved in the collision. The collision is elastic, so kinetic

8.25 Our sketch for this problem.

energy is also conserved. Figure 8.25 shows our sketch. We take the x-axis to be in the direction in which the neutron is moving initially. The collision is head-on, so both particles move along this same axis after the collision. The carbon nucleus is initially at rest, so we can use Eqs. (8.24) and (8.25); we replace A by n (for the neutron) and B by C (for the carbon nucleus). We have $m_n = 1.0$ u, $m_C = 12$ u, and $v_{n1x} = 2.6 \times 10^7$ m/s. The target variables are the final velocities v_{n2x} and v_{C2x}.

EXECUTE: You can do the arithmetic. (*Hint:* There's no reason to convert atomic mass units to kilograms.) The results are

$$v_{n2x} = -2.2 \times 10^7 \text{ m/s} \qquad v_{C2x} = 0.4 \times 10^7 \text{ m/s}$$

EVALUATE: The neutron ends up with $|(m_n - m_C)/(m_n + m_C)| = \frac{11}{13}$ of its initial speed, and the speed of the recoiling carbon nucleus is $|2m_n/(m_n + m_C)| = \frac{2}{13}$ of the neutron's initial speed. Kinetic energy is proportional to speed squared, so the neutron's final kinetic energy is $(\frac{11}{13})^2 \approx 0.72$ of its original value. After a second head-on collision, its kinetic energy is $(0.72)^2$, or about half its original value, and so on. After a few dozen collisions (few of which are head-on), the neutron speed will be low enough that it can efficiently cause a fission reaction in a uranium nucleus.

Example 8.12 **A two-dimensional elastic collision**

Figure 8.26 shows an elastic collision of two pucks (masses $m_A = 0.500$ kg and $m_B = 0.300$ kg) on a frictionless air-hockey table. Puck A has an initial velocity of 4.00 m/s in the positive x-direction and a final velocity of 2.00 m/s in an unknown direction α. Puck B is initially at rest. Find the final speed v_{B2} of puck B and the angles α and β.

SOLUTION

IDENTIFY and SET UP: We'll use the equations for conservation of energy and conservation of x- and y-momentum. These three equations should be enough to solve for the three target variables given in the problem statement.

EXECUTE: The collision is elastic, so the initial and final kinetic energies of the system are equal:

$$\tfrac{1}{2}m_A v_{A1}^2 = \tfrac{1}{2}m_A v_{A2}^2 + \tfrac{1}{2}m_B v_{B2}^2$$

$$v_{B2}^2 = \frac{m_A v_{A1}^2 - m_A v_{A2}^2}{m_B}$$

$$= \frac{(0.500 \text{ kg})(4.00 \text{ m/s})^2 - (0.500 \text{ kg})(2.00 \text{ m/s})^2}{0.300 \text{ kg}}$$

$$v_{B2} = 4.47 \text{ m/s}$$

Conservation of the x- and y-components of total momentum gives

$$m_A v_{A1x} = m_A v_{A2x} + m_B v_{B2x}$$

$$(0.500 \text{ kg})(4.00 \text{ m/s}) = (0.500 \text{ kg})(2.00 \text{ m/s})(\cos\alpha)$$
$$+ (0.300 \text{ kg})(4.47 \text{ m/s})(\cos\beta)$$

$$0 = m_A v_{A2y} + m_B v_{B2y}$$

$$0 = (0.500 \text{ kg})(2.00 \text{ m/s})(\sin\alpha)$$
$$- (0.300 \text{ kg})(4.47 \text{ m/s})(\sin\beta)$$

8.26 An elastic collision that isn't head-on.

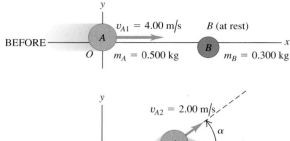

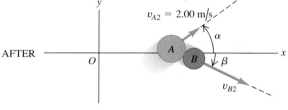

These are two simultaneous equations for α and β. We'll leave it to you to supply the details of the solution. (*Hint:* Solve the first equation for $\cos\beta$ and the second for $\sin\beta$; square each equation and add. Since $\sin^2\beta + \cos^2\beta = 1$, this eliminates β and leaves an equation that you can solve for $\cos\alpha$ and hence for α. Substitute this value into either of the two equations and solve for β.) The results are

$$\alpha = 36.9° \qquad \beta = 26.6°$$

EVALUATE: To check the answers we confirm that the y-momentum, which was zero before the collision, is in fact zero after the collision. The y-momenta are

$$p_{A2y} = (0.500 \text{ kg})(2.00 \text{ m/s})(\sin 36.9°) = +0.600 \text{ kg} \cdot \text{m/s}$$
$$p_{B2y} = -(0.300 \text{ kg})(4.47 \text{ m/s})(\sin 26.6°) = -0.600 \text{ kg} \cdot \text{m/s}$$

and their sum is indeed zero.

Test Your Understanding of Section 8.4 Most present-day nuclear reactors use water as a moderator (see Example 8.11). Are water molecules (mass $m_w = 18.0$ u) a better or worse moderator than carbon atoms? (One advantage of water is that it also acts as a coolant for the reactor's radioactive core.) ❙

8.5 Center of Mass

We can restate the principle of conservation of momentum in a useful way by using the concept of **center of mass.** Suppose we have several particles with masses m_1, m_2, and so on. Let the coordinates of m_1 be (x_1, y_1), those of m_2 be (x_2, y_2), and so on. We define the center of mass of the system as the point that has coordinates (x_{cm}, y_{cm}) given by

$$x_{cm} = \frac{m_1 x_1 + m_2 x_2 + m_3 x_3 + \cdots}{m_1 + m_2 + m_3 + \cdots} = \frac{\sum_i m_i x_i}{\sum_i m_i}$$

$$y_{cm} = \frac{m_1 y_1 + m_2 y_2 + m_3 y_3 + \cdots}{m_1 + m_2 + m_3 + \cdots} = \frac{\sum_i m_i y_i}{\sum_i m_i}$$

(center of mass) (8.28)

The position vector $\vec{r}_{cm}$ of the center of mass can be expressed in terms of the position vectors $\vec{r}_1, \vec{r}_2, \ldots$ of the particles as

$$\vec{r}_{cm} = \frac{m_1\vec{r}_1 + m_2\vec{r}_2 + m_3\vec{r}_3 + \cdots}{m_1 + m_2 + m_3 + \cdots} = \frac{\sum_i m_i\vec{r}_i}{\sum_i m_i} \quad \text{(center of mass)} \quad (8.29)$$

In statistical language, the center of mass is a *mass-weighted average* position of the particles.

Example 8.13   Center of mass of a water molecule

Figure 8.27 shows a simple model of a water molecule. The oxygen-hydrogen separation is $d = 9.57 \times 10^{-11}$ m. Each hydrogen atom has mass 1.0 u, and the oxygen atom has mass 16.0 u. Find the position of the center of mass.

SOLUTION

IDENTIFY and SET UP: Nearly all the mass of each atom is concentrated in its nucleus, whose radius is only about 10^{-5} times the overall radius of the atom. Hence we can safely represent each atom as a point particle. Figure 8.27 shows our coordinate system, with the x-axis chosen to lie along the molecule's symmetry axis. We'll use Eqs. (8.28) to find x_{cm} and y_{cm}.

EXECUTE: The oxygen atom is at $x = 0$, $y = 0$. The x-coordinate of each hydrogen atom is $d\cos(105°/2)$; the y-coordinates are $\pm d\sin(105°/2)$. From Eqs. (8.28),

$$x_{cm} = \frac{\left[\begin{array}{l}(1.0\text{ u})(d\cos 52.5°) + (1.0\text{ u}) \\ \times (d\cos 52.5°) + (16.0\text{ u})(0)\end{array}\right]}{1.0\text{ u} + 1.0\text{ u} + 16.0\text{ u}} = 0.068d$$

$$y_{cm} = \frac{\left[\begin{array}{l}(1.0\text{ u})(d\sin 52.5°) + (1.0\text{ u}) \\ \times (-d\sin 52.5°) + (16.0\text{ u})(0)\end{array}\right]}{1.0\text{ u} + 1.0\text{ u} + 16.0\text{ u}} = 0$$

Substituting $d = 9.57 \times 10^{-11}$ m, we find

$$x_{cm} = (0.068)(9.57 \times 10^{-11}\text{ m}) = 6.5 \times 10^{-12}\text{ m}$$

EVALUATE: The center of mass is much closer to the oxygen atom (located at the origin) than to either hydrogen atom because the oxygen atom is much more massive. The center of mass lies along the molecule's *axis of symmetry*. If the molecule is rotated 180° around this axis, it looks exactly the same as before. The position of the center of mass can't be affected by this rotation, so it *must* lie on the axis of symmetry.

8.27 Where is the center of mass of a water molecule?

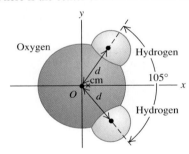

For solid bodies, in which we have (at least on a macroscopic level) a continuous distribution of matter, the sums in Eqs. (8.28) have to be replaced by integrals. The calculations can get quite involved, but we can say three general things about such problems (Fig. 8.28). First, whenever a homogeneous body has a geometric center, such as a billiard ball, a sugar cube, or a can of frozen orange juice, the center of mass is at the geometric center. Second, whenever a body has an axis of symmetry, such as a wheel or a pulley, the center of mass always lies on that axis. Third, there is no law that says the center of mass has to be within the body. For example, the center of mass of a donut is right in the middle of the hole.

We'll talk a little more about locating the center of mass in Chapter 11 in connection with the related concept of *center of gravity*.

8.28 Locating the center of mass of a symmetrical object.

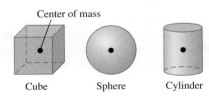

If a homogeneous object has a geometric center, that is where the center of mass is located.

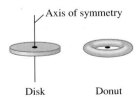

If an object has an axis of symmetry, the center of mass lies along it. As in the case of the donut, the center of mass may not be within the object.

Motion of the Center of Mass

To see the significance of the center of mass of a collection of particles, we must ask what happens to the center of mass when the particles move. The x- and y-components of velocity of the center of mass, $v_{cm\text{-}x}$ and $v_{cm\text{-}y}$, are the time derivatives of x_{cm} and y_{cm}. Also, dx_1/dt is the x-component of velocity of particle 1,

8.29 The center of mass of this wrench is marked with a white dot. The net external force acting on the wrench is almost zero. As the wrench spins on a smooth horizontal surface, the center of mass moves in a straight line with nearly constant velocity.

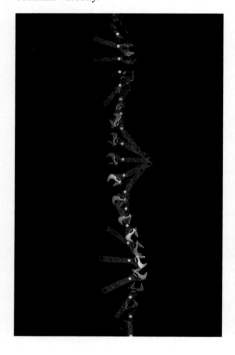

and so on, so $dx_1/dt = v_{1x}$, and so on. Taking time derivatives of Eqs. (8.28), we get

$$v_{\text{cm-}x} = \frac{m_1 v_{1x} + m_2 v_{2x} + m_3 v_{3x} + \cdots}{m_1 + m_2 + m_3 + \cdots}$$

$$v_{\text{cm-}y} = \frac{m_1 v_{1y} + m_2 v_{2y} + m_3 v_{3y} + \cdots}{m_1 + m_2 + m_3 + \cdots} \qquad (8.30)$$

These equations are equivalent to the single vector equation obtained by taking the time derivative of Eq. (8.29):

$$\vec{v}_{\text{cm}} = \frac{m_1 \vec{v}_1 + m_2 \vec{v}_2 + m_3 \vec{v}_3 + \cdots}{m_1 + m_2 + m_3 + \cdots} \qquad (8.31)$$

We denote the *total* mass $m_1 + m_2 + \cdots$ by M. We can then rewrite Eq. (8.31) as

$$M\vec{v}_{\text{cm}} = m_1 \vec{v}_1 + m_2 \vec{v}_2 + m_3 \vec{v}_3 + \cdots = \vec{P} \qquad (8.32)$$

The right side is simply the total momentum $\vec{P}$ of the system. Thus we have proved that *the total momentum is equal to the total mass times the velocity of the center of mass.* When you catch a baseball, you are really catching a collection of a very large number of molecules of masses m_1, m_2, m_3, ... The impulse you feel is due to the total momentum of this entire collection. But this impulse is the same as if you were catching a single particle of mass $M = m_1 + m_2 + m_3 + \cdots$ moving with velocity $\vec{v}_{\text{cm}}$, the velocity of the collection's center of mass. So Eq. (8.32) helps to justify representing an extended body as a particle.

For a system of particles on which the net external force is zero, so that the total momentum $\vec{P}$ is constant, the velocity of the center of mass $\vec{v}_{\text{cm}} = \vec{P}/M$ is also constant. Suppose we mark the center of mass of a wrench and then slide the wrench with a spinning motion across a smooth, horizontal tabletop (Fig. 8.29). The overall motion appears complicated, but the center of mass follows a straight line, as though all the mass were concentrated at that point.

Example 8.14 **A tug-of-war on the ice**

James (mass 90.0 kg) and Ramon (mass 60.0 kg) are 20.0 m apart on a frozen pond. Midway between them is a mug of their favorite beverage. They pull on the ends of a light rope stretched between them. When James has moved 6.0 m toward the mug, how far and in what direction has Ramon moved?

SOLUTION

IDENTIFY and SET UP: The surface is horizontal and (we assume) frictionless, so the net external force on the system of James, Ramon, and the rope is zero; their total momentum is conserved. Initially there is no motion, so the total momentum is zero. The velocity of the center of mass is therefore zero, and it remains at rest. Let's take the origin at the position of the mug and let the $+x$-axis extend from the mug toward Ramon. Figure 8.30 shows

8.30 Our sketch for this problem.

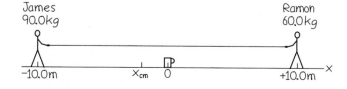

our sketch. We use Eq. (8.28) to calculate the position of the center of mass; we neglect the mass of the light rope.

EXECUTE: The initial x-coordinates of James and Ramon are -10.0 m and $+10.0$ m, respectively, so the x-coordinate of the center of mass is

$$x_{\text{cm}} = \frac{(90.0 \text{ kg})(-10.0 \text{ m}) + (60.0 \text{ kg})(10.0 \text{ m})}{90.0 \text{ kg} + 60.0 \text{ kg}} = -2.0 \text{ m}$$

When James moves 6.0 m toward the mug, his new x-coordinate is -4.0 m; we'll call Ramon's new x-coordinate x_2. The center of mass doesn't move, so

$$x_{\text{cm}} = \frac{(90.0 \text{ kg})(-4.0 \text{ m}) + (60.0 \text{ kg})x_2}{90.0 \text{ kg} + 60.0 \text{ kg}} = -2.0 \text{ m}$$

$$x_2 = 1.0 \text{ m}$$

James has moved 6.0 m and is still 4.0 m from the mug, but Ramon has moved 9.0 m and is only 1.0 m from it.

EVALUATE: The ratio of the distances moved, $(6.0 \text{ m})/(9.0 \text{ m}) = \frac{2}{3}$, is the *inverse* ratio of the masses. Can you see why? Because the surface is frictionless, the two men will keep moving and collide at the center of mass; Ramon will reach the mug first. This is independent of how hard either person pulls; pulling harder just makes them move faster.

External Forces and Center-of-Mass Motion

If the net external force on a system of particles is not zero, then total momentum is not conserved and the velocity of the center of mass changes. Let's look at the relationship between the motion of the center of mass and the forces acting on the system.

Equations (8.31) and (8.32) give the *velocity* of the center of mass in terms of the velocities of the individual particles. We take the time derivatives of these equations to show that the *accelerations* are related in the same way. Let $\vec{a}_{cm} = d\vec{v}_{cm}/dt$ be the acceleration of the center of mass; then we find

$$M\vec{a}_{cm} = m_1\vec{a}_1 + m_2\vec{a}_2 + m_3\vec{a}_3 + \cdots \qquad (8.33)$$

Now $m_1\vec{a}_1$ is equal to the vector sum of forces on the first particle, and so on, so the right side of Eq. (8.33) is equal to the vector sum $\Sigma\vec{F}$ of *all* the forces on *all* the particles. Just as we did in Section 8.2, we can classify each force as *external* or *internal*. The sum of all forces on all the particles is then

$$\Sigma\vec{F} = \Sigma\vec{F}_{ext} + \Sigma\vec{F}_{int} = M\vec{a}_{cm}$$

Because of Newton's third law, the internal forces all cancel in pairs, and $\Sigma\vec{F}_{int} = \mathbf{0}$. What survives on the left side is the sum of only the *external* forces:

$$\Sigma\vec{F}_{ext} = M\vec{a}_{cm} \qquad \text{(body or collection of particles)} \qquad (8.34)$$

When a body or a collection of particles is acted on by external forces, the center of mass moves just as though all the mass were concentrated at that point and it were acted on by a net force equal to the sum of the external forces on the system.

This result may not sound very impressive, but in fact it is central to the whole subject of mechanics. In fact, we've been using this result all along; without it, we would not be able to represent an extended body as a point particle when we apply Newton's laws. It explains why only *external* forces can affect the motion of an extended body. If you pull upward on your belt, your belt exerts an equal downward force on your hands; these are *internal* forces that cancel and have no effect on the overall motion of your body.

Suppose a cannon shell traveling in a parabolic trajectory (neglecting air resistance) explodes in flight, splitting into two fragments with equal mass (Fig. 8.31a). The fragments follow new parabolic paths, but the center of mass continues on the original parabolic trajectory, just as though all the mass were still concentrated at that point. A skyrocket exploding in air (Fig. 8.31b) is a spectacular example of this effect.

8.31 (a) A shell explodes into two fragments in flight. If air resistance is ignored, the center of mass continues on the same trajectory as the shell's path before exploding. (b) The same effect occurs with exploding fireworks.

(a)

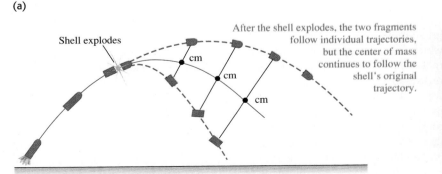

Shell explodes

After the shell explodes, the two fragments follow individual trajectories, but the center of mass continues to follow the shell's original trajectory.

(b)

This property of the center of mass is important when we analyze the motion of rigid bodies. We describe the motion of an extended body as a combination of translational motion of the center of mass and rotational motion about an axis through the center of mass. We will return to this topic in Chapter 10. This property also plays an important role in the motion of astronomical objects. It's not correct to say that the moon orbits the earth; rather, the earth and moon both move in orbits around their center of mass.

There's one more useful way to describe the motion of a system of particles. Using $\vec{a}_{cm} = d\vec{v}_{cm}/dt$, we can rewrite Eq. (8.33) as

$$M\vec{a}_{cm} = M\frac{d\vec{v}_{cm}}{dt} = \frac{d(M\vec{v}_{cm})}{dt} = \frac{d\vec{P}}{dt} \qquad (8.35)$$

The total system mass M is constant, so we're allowed to move it inside the derivative. Substituting Eq. (8.35) into Eq. (8.34), we find

$$\sum\vec{F}_{ext} = \frac{d\vec{P}}{dt} \qquad \text{(extended body or system of particles)} \qquad (8.36)$$

This equation looks like Eq. (8.4). The difference is that Eq. (8.36) describes a *system* of particles, such as an extended body, while Eq. (8.4) describes a single particle. The interactions between the particles that make up the system can change the individual momenta of the particles, but the *total* momentum $\vec{P}$ of the system can be changed only by external forces acting from outside the system.

Finally, we note that if the net external force is zero, Eq. (8.34) shows that the acceleration $\vec{a}_{cm}$ of the center of mass is zero. So the center-of-mass velocity $\vec{v}_{cm}$ is constant, as for the wrench in Fig. 8.29. From Eq. (8.36) the total momentum $\vec{P}$ is also constant. This reaffirms our statement in Section 8.3 of the principle of conservation of momentum.

Test Your Understanding of Section 8.5 Will the center of mass in Fig. 8.31a continue on the same parabolic trajectory even after one of the fragments hits the ground? Why or why not?   ❙

Application Jet Propulsion in Squids
Both a jet engine and a squid use variations in their mass to provide propulsion: They increase their mass by taking in fluid at low speed (air for a jet engine, water for a squid), then decrease their mass by ejecting that fluid at high speed. The net result is a propulsive force.

8.6 Rocket Propulsion

Momentum considerations are particularly useful for analyzing a system in which the masses of parts of the system change with time. In such cases we can't use Newton's second law $\sum\vec{F} = m\vec{a}$ directly because m changes. Rocket propulsion offers a typical and interesting example of this kind of analysis. A rocket is propelled forward by rearward ejection of burned fuel that initially was in the rocket (which is why rocket fuel is also called *propellant*). The forward force on the rocket is the reaction to the backward force on the ejected material. The total mass of the system is constant, but the mass of the rocket itself decreases as material is ejected.

As a simple example, consider a rocket fired in outer space, where there is no gravitational force and no air resistance. Let m denote the mass of the rocket, which will change as it expends fuel. We choose our x-axis to be along the rocket's direction of motion. Figure 8.32a shows the rocket at a time t, when its mass is m and its x-velocity relative to our coordinate system is v. (For simplicity, we will drop the subscript x in this discussion.) The x-component of total momentum at this instant is $P_1 = mv$. In a short time interval dt, the mass of the rocket changes by an amount dm. This is an inherently negative quantity because the rocket's mass m *decreases* with time. During dt, a *positive* mass $-dm$ of burned fuel is ejected from the rocket. Let v_{ex} be the exhaust *speed* of this material *relative to the rocket;* the burned fuel is ejected opposite the direction of motion,

MasteringPHYSICS

ActivPhysics 6.6: Saving an Astronaut

so its x-component of *velocity* relative to the rocket is $-v_{ex}$. The x-velocity v_{fuel} of the burned fuel relative to our coordinate system is then

$$v_{fuel} = v + (-v_{ex}) = v - v_{ex}$$

and the x-component of momentum of the ejected mass $(-dm)$ is

$$(-dm)v_{fuel} = (-dm)(v - v_{ex})$$

Figure 8.32b shows that at the end of the time interval dt, the x-velocity of the rocket and unburned fuel has increased to $v + dv$, and its mass has decreased to $m + dm$ (remember that dm is negative). The rocket's momentum at this time is

$$(m + dm)(v + dv)$$

Thus the *total* x-component of momentum P_2 of the rocket plus ejected fuel at time $t + dt$ is

$$P_2 = (m + dm)(v + dv) + (-dm)(v - v_{ex})$$

According to our initial assumption, the rocket and fuel are an isolated system. Thus momentum is conserved, and the total x-component of momentum of the system must be the same at time t and at time $t + dt$: $P_1 = P_2$. Hence

$$mv = (m + dm)(v + dv) + (-dm)(v - v_{ex})$$

This can be simplified to

$$m\,dv = -dm\,v_{ex} - dm\,dv$$

We can neglect the term $(-dm\,dv)$ because it is a product of two small quantities and thus is much smaller than the other terms. Dropping this term, dividing by dt, and rearranging, we find

$$m\frac{dv}{dt} = -v_{ex}\frac{dm}{dt} \qquad (8.37)$$

Now dv/dt is the acceleration of the rocket, so the left side of this equation (mass times acceleration) equals the net force F, or *thrust,* on the rocket:

$$F = -v_{ex}\frac{dm}{dt} \qquad (8.38)$$

The thrust is proportional both to the relative speed v_{ex} of the ejected fuel and to the mass of fuel ejected per unit time, $-dm/dt$. (Remember that dm/dt is negative because it is the rate of change of the rocket's mass, so F is positive.)

The x-component of acceleration of the rocket is

$$a = \frac{dv}{dt} = -\frac{v_{ex}}{m}\frac{dm}{dt} \qquad (8.39)$$

8.32 A rocket moving in gravity-free outer space at (a) time t and (b) time $t + dt$.

(a)   (b)

At time t, the rocket has mass m and x-component of velocity v.

At time $t + dt$, the rocket has mass $m + dm$ (where dm is inherently *negative*) and x-component of velocity $v + dv$. The burned fuel has x-component of velocity $v_{fuel} = v - v_{ex}$ and mass $-dm$. (The minus sign is needed to make $-dm$ *positive* because dm is negative.)

8.33 To provide enough thrust to lift its payload into space, this Atlas V launch vehicle ejects more than 1000 kg of burned fuel per second at speeds of nearly 4000 m/s.

This is positive because v_{ex} is positive (remember, it's the exhaust *speed*) and dm/dt is negative. The rocket's mass m decreases continuously while the fuel is being consumed. If v_{ex} and dm/dt are constant, the acceleration increases until all the fuel is gone.

Equation (8.38) tells us that an effective rocket burns fuel at a rapid rate (large $-dm/dt$) and ejects the burned fuel at a high relative speed (large v_{ex}), as in Fig. 8.33. In the early days of rocket propulsion, people who didn't understand conservation of momentum thought that a rocket couldn't function in outer space because "it doesn't have anything to push against." On the contrary, rockets work *best* in outer space, where there is no air resistance! The launch vehicle in Fig. 8.33 is *not* "pushing against the ground" to get into the air.

If the exhaust speed v_{ex} is constant, we can integrate Eq. (8.39) to find a relationship between the velocity v at any time and the remaining mass m. At time $t = 0$, let the mass be m_0 and the velocity v_0. Then we rewrite Eq. (8.39) as

$$dv = -v_{\text{ex}} \frac{dm}{m}$$

We change the integration variables to v' and m', so we can use v and m as the upper limits (the final speed and mass). Then we integrate both sides, using limits v_0 to v and m_0 to m, and take the constant v_{ex} outside the integral:

$$\int_{v_0}^{v} dv' = -\int_{m_0}^{m} v_{\text{ex}} \frac{dm'}{m'} = -v_{\text{ex}} \int_{m_0}^{m} \frac{dm'}{m'}$$

$$v - v_0 = -v_{\text{ex}} \ln \frac{m}{m_0} = v_{\text{ex}} \ln \frac{m_0}{m} \qquad (8.40)$$

The ratio m_0/m is the original mass divided by the mass after the fuel has been exhausted. In practical spacecraft this ratio is made as large as possible to maximize the speed gain, which means that the initial mass of the rocket is almost all fuel. The final velocity of the rocket will be greater in magnitude (and is often *much* greater) than the relative speed v_{ex} if $\ln(m_0/m) > 1$—that is, if $m_0/m > e = 2.71828\ldots$.

We've assumed throughout this analysis that the rocket is in gravity-free outer space. However, gravity must be taken into account when a rocket is launched from the surface of a planet, as in Fig. 8.33 (see Problem 8.112).

Example 8.15 | Acceleration of a rocket

The engine of a rocket in outer space, far from any planet, is turned on. The rocket ejects burned fuel at a constant rate; in the first second of firing, it ejects $\frac{1}{120}$ of its initial mass m_0 at a relative speed of 2400 m/s. What is the rocket's initial acceleration?

SOLUTION

IDENTIFY and SET UP: We are given the rocket's exhaust speed v_{ex} and the fraction of the initial mass lost during the first second of firing, from which we can find dm/dt. We'll use Eq. (8.39) to find the acceleration of the rocket.

EXECUTE: The initial rate of change of mass is

$$\frac{dm}{dt} = -\frac{m_0/120}{1 \text{ s}} = -\frac{m_0}{120 \text{ s}}$$

From Eq. (8.39),

$$a = -\frac{v_{\text{ex}}}{m_0} \frac{dm}{dt} = -\frac{2400 \text{ m/s}}{m_0} \left(-\frac{m_0}{120 \text{ s}} \right) = 20 \text{ m/s}^2$$

EVALUATE: The answer doesn't depend on m_0. If v_{ex} is the same, the initial acceleration is the same for a 120,000-kg spacecraft that ejects 1000 kg/s as for a 60-kg astronaut equipped with a small rocket that ejects 0.5 kg/s.

Example 8.16   Speed of a rocket

Suppose that $\frac{3}{4}$ of the initial mass of the rocket in Example 8.15 is fuel, so that the fuel is completely consumed at a constant rate in 90 s. The final mass of the rocket is $m = m_0/4$. If the rocket starts from rest in our coordinate system, find its speed at the end of this time.

SOLUTION

IDENTIFY, SET UP, and EXECUTE: We are given the initial velocity $v_0 = 0$, the exhaust speed $v_{ex} = 2400$ m/s, and the final mass m as a fraction of the initial mass m_0. We'll use Eq. (8.40) to find the final speed v:

$$v = v_0 + v_{ex}\ln\frac{m_0}{m} = 0 + (2400 \text{ m/s})(\ln 4) = 3327 \text{ m/s}$$

EVALUATE: Let's examine what happens as the rocket gains speed. (To illustrate our point, we use more figures than are significant.) At the start of the flight, when the velocity of the rocket is zero, the ejected fuel is moving backward at 2400 m/s relative to our frame of reference. As the rocket moves forward and speeds up, the fuel's speed relative to our system decreases; when the rocket speed reaches 2400 m/s, this relative speed is *zero*. [Knowing the rate of fuel consumption, you can solve Eq. (8.40) to show that this occurs at about $t = 75.6$ s.] After this time the ejected burned fuel moves *forward*, not backward, in our system. Relative to our frame of reference, the last bit of ejected fuel has a forward velocity of 3327 m/s $-$ 2400 m/s $=$ 927 m/s.

Test Your Understanding of Section 8.6   (a) If a rocket in gravity-free outer space has the same thrust at all times, is its acceleration constant, increasing, or decreasing? (b) If the rocket has the same acceleration at all times, is the thrust constant, increasing, or decreasing?

CHAPTER 8 SUMMARY

Momentum of a particle: The momentum $\vec{p}$ of a particle is a vector quantity equal to the product of the particle's mass m and velocity $\vec{v}$. Newton's second law says that the net force on a particle is equal to the rate of change of the particle's momentum.

$$\vec{p} = m\vec{v} \qquad (8.2)$$

$$\sum \vec{F} = \frac{d\vec{p}}{dt} \qquad (8.4)$$

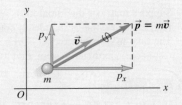

Impulse and momentum: If a constant net force $\sum \vec{F}$ acts on a particle for a time interval Δt from t_1 to t_2, the impulse $\vec{J}$ of the net force is the product of the net force and the time interval. If $\sum \vec{F}$ varies with time, $\vec{J}$ is the integral of the net force over the time interval. In any case, the change in a particle's momentum during a time interval equals the impulse of the net force that acted on the particle during that interval. The momentum of a particle equals the impulse that accelerated it from rest to its present speed. (See Examples 8.1–8.3.)

$$\vec{J} = \sum \vec{F}(t_2 - t_1) = \sum \vec{F}\,\Delta t \qquad (8.5)$$

$$\vec{J} = \int_{t_1}^{t_2} \sum \vec{F}\,dt \qquad (8.7)$$

$$\vec{J} = \vec{p}_2 - \vec{p}_1 \qquad (8.6)$$

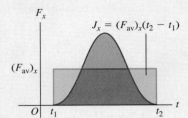

Conservation of momentum: An internal force is a force exerted by one part of a system on another. An external force is a force exerted on any part of a system by something outside the system. If the net external force on a system is zero, the total momentum of the system $\vec{P}$ (the vector sum of the momenta of the individual particles that make up the system) is constant, or conserved. Each component of total momentum is separately conserved. (See Examples 8.4–8.6.)

$$\vec{P} = \vec{p}_A + \vec{p}_B + \cdots$$
$$= m_A\vec{v}_A + m_B\vec{v}_B + \cdots \qquad (8.14)$$

If $\sum \vec{F} = 0$, then $\vec{P}$ = constant.

Collisions: In collisions of all kinds, the initial and final total momenta are equal. In an elastic collision between two bodies, the initial and final total kinetic energies are also equal, and the initial and final relative velocities have the same magnitude. In an inelastic two-body collision, the total kinetic energy is less after the collision than before. If the two bodies have the same final velocity, the collision is completely inelastic. (See Examples 8.7–8.12.)

Center of mass: The position vector of the center of mass of a system of particles, $\vec{r}_{cm}$, is a weighted average of the positions $\vec{r}_1, \vec{r}_2, \ldots$ of the individual particles. The total momentum $\vec{P}$ of a system equals its total mass M multiplied by the velocity of its center of mass, $\vec{v}_{cm}$. The center of mass moves as though all the mass M were concentrated at that point. If the net external force on the system is zero, the center-of-mass velocity $\vec{v}_{cm}$ is constant. If the net external force is not zero, the center of mass accelerates as though it were a particle of mass M being acted on by the same net external force. (See Examples 8.13 and 8.14.)

$$\vec{r}_{cm} = \frac{m_1\vec{r}_1 + m_2\vec{r}_2 + m_3\vec{r}_3 + \cdots}{m_1 + m_2 + m_3 + \cdots}$$

$$= \frac{\sum_i m_i \vec{r}_i}{\sum_i m_i} \qquad (8.29)$$

$$\vec{P} = m_1\vec{v}_1 + m_2\vec{v}_2 + m_3\vec{v}_3 + \cdots$$
$$= M\vec{v}_{cm} \qquad (8.32)$$

$$\sum \vec{F}_{ext} = M\vec{a}_{cm} \qquad (8.34)$$

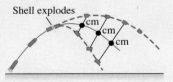

Rocket propulsion: In rocket propulsion, the mass of a rocket changes as the fuel is used up and ejected from the rocket. Analysis of the motion of the rocket must include the momentum carried away by the spent fuel as well as the momentum of the rocket itself. (See Examples 8.15 and 8.16.)

BRIDGING PROBLEM   One Collision After Another

Sphere *A* of mass 0.600 kg is initially moving to the right at 4.00 m/s. Sphere *B*, of mass 1.80 kg, is initially to the right of sphere *A* and moving to the right at 2.00 m/s. After the two spheres collide, sphere *B* is moving at 3.00 m/s in the same direction as before. (a) What is the velocity (magnitude and direction) of sphere *A* after this collision? (b) Is this collision elastic or inelastic? (c) Sphere *B* then has an off-center collision with sphere *C*, which has mass 1.20 kg and is initially at rest. After this collision, sphere *B* is moving at 19.0° to its initial direction at 2.00 m/s. What is the velocity (magnitude and direction) of sphere *C* after this collision? (d) What is the impulse (magnitude and direction) imparted to sphere *B* by sphere *C* when they collide? (e) Is this second collision elastic or inelastic? (f) What is the velocity (magnitude and direction) of the center of mass of the system of three spheres (*A*, *B*, and *C*) after the second collision? No external forces act on any of the spheres in this problem.

SOLUTION GUIDE

See MasteringPhysics® study area for a Video Tutor solution.

IDENTIFY and SET UP

1. Momentum is conserved in these collisions. Can you explain why?
2. Choose the *x*- and *y*-axes, and assign subscripts to values before the first collision, after the first collision but before the second collision, and after the second collision.
3. Make a list of the target variables, and choose the equations that you'll use to solve for these.

EXECUTE

4. Solve for the velocity of sphere *A* after the first collision. Does *A* slow down or speed up in the collision? Does this make sense?
5. Now that you know the velocities of both *A* and *B* after the first collision, decide whether or not this collision is elastic. (How will you do this?)
6. The second collision is two-dimensional, so you'll have to demand that *both* components of momentum are conserved. Use this to find the speed and direction of sphere *C* after the second collision. (*Hint:* After the first collision, sphere *B* maintains the same velocity until it hits sphere *C*.)
7. Use the definition of impulse to find the impulse imparted to sphere *B* by sphere *C*. Remember that impulse is a vector.
8. Use the same technique that you employed in step 5 to decide whether or not the second collision is elastic.
9. Find the velocity of the center of mass after the second collision.

EVALUATE

10. Compare the directions of the vectors you found in steps 6 and 7. Is this a coincidence? Why or why not?
11. Find the velocity of the center of mass before and after the first collision. Compare to your result from step 9. Again, is this a coincidence? Why or why not?

Problems

For instructor-assigned homework, go to www.masteringphysics.com

•, ••, •••: Problems of increasing difficulty. **CP**: Cumulative problems incorporating material from earlier chapters. **CALC**: Problems requiring calculus. **BIO**: Biosciences problems.

DISCUSSION QUESTIONS

Q8.1 In splitting logs with a hammer and wedge, is a heavy hammer more effective than a lighter hammer? Why?

Q8.2 Suppose you catch a baseball and then someone invites you to catch a bowling ball with either the same momentum or the same kinetic energy as the baseball. Which would you choose? Explain.

Q8.3 When rain falls from the sky, what happens to its momentum as it hits the ground? Is your answer also valid for Newton's famous apple?

Q8.4 A car has the same kinetic energy when it is traveling south at 30 m/s as when it is traveling northwest at 30 m/s. Is the momentum of the car the same in both cases? Explain.

Q8.5 A truck is accelerating as it speeds down the highway. One inertial frame of reference is attached to the ground with its origin at a fence post. A second frame of reference is attached to a police car that is traveling down the highway at constant velocity. Is the momentum of the truck the same in these two reference frames? Explain. Is the rate of change of the truck's momentum the same in these two frames? Explain.

Q8.6 (a) When a large car collides with a small car, which one undergoes the greater change in momentum: the large one or the small one? Or is it the same for both? (b) In light of your answer to part (a), why are the occupants of the small car more likely to be hurt than those of the large car, assuming that both cars are equally sturdy?

Q8.7 A woman holding a large rock stands on a frictionless, horizontal sheet of ice. She throws the rock with speed v_0 at an angle α above the horizontal. Consider the system consisting of the woman plus the rock. Is the momentum of the system conserved? Why or why not? Is any component of the momentum of the system conserved? Again, why or why not?

Q8.8 In Example 8.7 (Section 8.3), where the two gliders in Fig. 8.15 stick together after the collision, the collision is inelastic because $K_2 < K_1$. In Example 8.5 (Section 8.2), is the collision inelastic? Explain.

Q8.9 In a completely inelastic collision between two objects, where the objects stick together after the collision, is it possible for the final kinetic energy of the system to be zero? If so, give an example in which this would occur. If the final kinetic energy is zero, what must the initial momentum of the system be? Is the initial kinetic energy of the system zero? Explain.

Q8.10 Since for a particle the kinetic energy is given by $K = \frac{1}{2}mv^2$ and the momentum by $\vec{p} = m\vec{v}$, it is easy to show that $K = p^2/2m$. How, then, is it possible to have an event during which the total momentum of the system is constant but the total kinetic energy changes?

Q8.11 In each of Examples 8.10, 8.11, and 8.12 (Section 8.4), verify that the relative velocity vector of the two bodies has the same magnitude before and after the collision. In each case what happens to the *direction* of the relative velocity vector?

Q8.12 A glass dropped on the floor is more likely to break if the floor is concrete than if it is wood. Why? (Refer to Fig. 8.3b.)

Q8.13 In Fig. 8.22b, the kinetic energy of the Ping-Pong ball is larger after its interaction with the bowling ball than before. From where does the extra energy come? Describe the event in terms of conservation of energy.

Q8.14 A machine gun is fired at a steel plate. Is the average force on the plate from the bullet impact greater if the bullets bounce off or if they are squashed and stick to the plate? Explain.

Q8.15 A net force of 4 N acts on an object initially at rest for 0.25 s and gives it a final speed of 5 m/s. How could a net force of 2 N produce the same final speed?

Q8.16 A net force with x-component $\sum F_x$ acts on an object from time t_1 to time t_2. The x-component of the momentum of the object is the same at t_1 as it is at t_2, but $\sum F_x$ is not zero at all times between t_1 and t_2. What can you say about the graph of $\sum F_x$ versus t?

Q8.17 A tennis player hits a tennis ball with a racket. Consider the system made up of the ball and the racket. Is the total momentum of the system the same just before and just after the hit? Is the total momentum just after the hit the same as 2 s later, when the ball is in midair at the high point of its trajectory? Explain any differences between the two cases.

Q8.18 In Example 8.4 (Section 8.2), consider the system consisting of the rifle plus the bullet. What is the speed of the system's center of mass after the rifle is fired? Explain.

Q8.19 An egg is released from rest from the roof of a building and falls to the ground. As the egg falls, what happens to the momentum of the system of the egg plus the earth?

Q8.20 A woman stands in the middle of a perfectly smooth, frictionless, frozen lake. She can set herself in motion by throwing things, but suppose she has nothing to throw. Can she propel herself to shore *without* throwing anything?

Q8.21 In a zero-gravity environment, can a rocket-propelled spaceship ever attain a speed greater than the relative speed with which the burnt fuel is exhausted?

Q8.22 When an object breaks into two pieces (explosion, radioactive decay, recoil, etc.), the lighter fragment gets more kinetic energy than the heavier one. This is a consequence of momentum conservation, but can you also explain it using Newton's laws of motion?

Q8.23 An apple falls from a tree and feels no air resistance. As it is falling, which of these statements about it are true? (a) Only its momentum is conserved; (b) only its mechanical energy is conserved, (c) both its momentum and its mechanical energy are conserved, (d) its kinetic energy is conserved.

Q8.24 Two pieces of clay collide and stick together. During the collision, which of these statements are true? (a) Only the momentum of the clay is conserved, (b) only the mechanical energy of the clay is conserved, (c) both the momentum and the mechanical energy of the clay are conserved, (d) the kinetic energy of the clay is conserved.

Q8.25 Two marbles are pressed together with a light ideal spring between them, but they are not attached to the spring in any way. They are then released on a frictionless horizontal table and soon move free of the spring. As the marbles are moving away from each other, which of these statements about them are true? (a) Only the momentum of the marbles is conserved, (b) only the mechanical energy of the marbles is conserved, (c) both the momentum and the mechanical energy of the marbles are conserved, (d) the kinetic energy of the marbles is conserved.

Q8.26 A very heavy SUV collides head-on with a very light compact car. Which of these statements about the collision are correct? (a) The amount of kinetic energy lost by the SUV is equal to the amount of kinetic energy gained by the compact, (b) the amount of momentum lost by the SUV is equal to the amount of momentum gained by the compact, (c) the compact feels a considerably greater force during the collision than the SUV does, (d) both cars lose the same amount of kinetic energy.

EXERCISES

Section 8.1 Momentum and Impulse

8.1 • (a) What is the magnitude of the momentum of a 10,000-kg truck whose speed is 12.0 m/s? (b) What speed would a 2000-kg SUV have to attain in order to have (i) the same momentum? (ii) the same kinetic energy?

8.2 • In a certain men's track and field event, the shotput has a mass of 7.30 kg and is released with a speed of 15.0 m/s at 40.0° above the horizontal over a man's straight left leg. What are the initial horizontal and vertical components of the momentum of this shotput?

8.3 •• (a) Show that the kinetic energy K and the momentum magnitude p of a particle with mass m are related by $K = p^2/2m$. (b) A 0.040-kg cardinal (*Richmondena cardinalis*) and a 0.145-kg baseball have the same kinetic energy. Which has the greater magnitude of momentum? What is the ratio of the cardinal's magnitude of momentum to the baseball's? (c) A 700-N man and a 450-N woman have the same momentum. Who has the greater kinetic energy? What is the ratio of the man's kinetic energy to that of the woman?

8.4 • Two vehicles are approaching an intersection. One is a 2500-kg pickup traveling at 14.0 m/s from east to west (the $-x$-direction), and the other is a 1500-kg sedan going from south to north (the $+y$-direction) at 23.0 m/s. (a) Find the x- and y-components of the net momentum of this system. (b) What are the magnitude and direction of the net momentum?

8.5 • One 110-kg football lineman is running to the right at 2.75 m/s while another 125-kg lineman is running directly toward him at 2.60 m/s. What are (a) the magnitude and direction of the net momentum of these two athletes, and (b) their total kinetic energy?

8.6 •• **BIO** **Biomechanics.** The mass of a regulation tennis ball is 57 g (although it can vary slightly), and tests have shown that the ball is in contact with the tennis racket for 30 ms. (This number can also vary, depending on the racket and swing.) We shall assume a 30.0-ms contact time for this exercise. The fastest-known served tennis ball was served by "Big Bill" Tilden in 1931, and its speed was measured to be 73.14 m/s. (a) What impulse and what force did Big Bill exert on the tennis ball in his record serve? (b) If Big Bill's opponent returned his serve with a speed of 55 m/s, what force and what impulse did he exert on the ball, assuming only horizontal motion?

8.7 • **Force of a Golf Swing.** A 0.0450-kg golf ball initially at rest is given a speed of 25.0 m/s when a club strikes. If the club and ball are in contact for 2.00 ms, what average force acts on the ball? Is the effect of the ball's weight during the time of contact significant? Why or why not?

8.8 • **Force of a Baseball Swing.** A baseball has mass 0.145 kg. (a) If the velocity of a pitched ball has a magnitude of 45.0 m/s and the batted ball's velocity is 55.0 m/s in the opposite direction, find the magnitude of the change in momentum of the ball and of the impulse applied to it by the bat. (b) If the ball remains in contact with the bat for 2.00 ms, find the magnitude of the average force applied by the bat.

8.9 • A 0.160-kg hockey puck is moving on an icy, frictionless, horizontal surface. At $t = 0$, the puck is moving to the right at 3.00 m/s. (a) Calculate the velocity of the puck (magnitude and direction) after a force of 25.0 N directed to the right has been applied for 0.050 s. (b) If, instead, a force of 12.0 N directed to the left is applied from $t = 0$ to $t = 0.050$ s, what is the final velocity of the puck?

8.10 • An engine of the orbital maneuvering system (OMS) on a space shuttle exerts a force of $(26{,}700 \text{ N})\hat{\jmath}$ for 3.90 s, exhausting a negligible mass of fuel relative to the 95,000-kg mass of the shuttle. (a) What is the impulse of the force for this 3.90 s? (b) What is the shuttle's change in momentum from this impulse? (c) What is the shuttle's change in velocity from this impulse? (d) Why can't we find the resulting change in the kinetic energy of the shuttle?

8.11 • **CALC** At time $t = 0$, a 2150-kg rocket in outer space fires an engine that exerts an increasing force on it in the $+x$-direction. This force obeys the equation $F_x = At^2$, where t is time, and has a magnitude of 781.25 N when $t = 1.25$ s. (a) Find the SI value of the constant A, including its units. (b) What impulse does the engine exert on the rocket during the 1.50-s interval starting 2.00 s after the engine is fired? (c) By how much does the rocket's velocity change during this interval?

8.12 •• A bat strikes a 0.145-kg baseball. Just before impact, the ball is traveling horizontally to the right at 50.0 m/s, and it leaves the bat traveling to the left at an angle of 30° above horizontal with a speed of 65.0 m/s. If the ball and bat are in contact for 1.75 ms, find the horizontal and vertical components of the average force on the ball.

8.13 • A 2.00-kg stone is sliding to the right on a frictionless horizontal surface at 5.00 m/s when it is suddenly struck by an object that exerts a large horizontal force on it for a short period of time. The graph in Fig. E8.13 shows the magnitude of this force as a function of time. (a) What impulse does this force exert on

Figure **E8.13**

the stone? (b) Just after the force stops acting, find the magnitude and direction of the stone's velocity if the force acts (i) to the right or (ii) to the left.

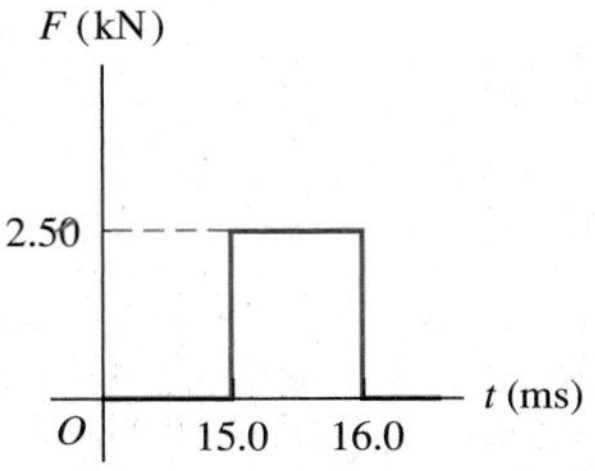

8.14 •• **BIO Bone Fracture.** Experimental tests have shown that bone will rupture if it is subjected to a force density of $1.03 \times 10^8 \text{ N/m}^2$. Suppose a 70.0-kg person carelessly roller-skates into an overhead metal beam that hits his forehead and completely stops his forward motion. If the area of contact with the person's forehead is 1.5 cm², what is the greatest speed with which he can hit the wall without breaking any bone if his head is in contact with the beam for 10.0 ms?

8.15 •• To warm up for a match, a tennis player hits the 57.0-g ball vertically with her racket. If the ball is stationary just

before it is hit and goes 5.50 m high, what impulse did she impart to it?

8.16 •• **CALC** Starting at $t = 0$, a horizontal net force $\vec{F} = (0.280 \text{ N/s})t\hat{\imath} + (-0.450 \text{ N/s}^2)t^2\hat{\jmath}$ is applied to a box that has an initial momentum $\vec{p} = (-3.00 \text{ kg} \cdot \text{m/s})\hat{\imath} + (4.00 \text{ kg} \cdot \text{m/s})\hat{\jmath}$. What is the momentum of the box at $t = 2.00$ s?

Section 8.2 Conservation of Momentum

8.17 •• The expanding gases that leave the muzzle of a rifle also contribute to the recoil. A .30-caliber bullet has mass 0.00720 kg and a speed of 601 m/s relative to the muzzle when fired from a rifle that has mass 2.80 kg. The loosely held rifle recoils at a speed of 1.85 m/s relative to the earth. Find the momentum of the propellant gases in a coordinate system attached to the earth as they leave the muzzle of the rifle.

8.18 • A 68.5-kg astronaut is doing a repair in space on the orbiting space station. She throws a 2.25-kg tool away from her at 3.20 m/s relative to the space station. With what speed and in what direction will she begin to move?

8.19 • **BIO Animal Propulsion.** Squids and octopuses propel themselves by expelling water. They do this by keeping water in a cavity and then suddenly contracting the cavity to force out the water through an opening. A 6.50-kg squid (including the water in the cavity) at rest suddenly sees a dangerous predator. (a) If the squid has 1.75 kg of water in its cavity, at what speed must it expel this water to suddenly achieve a speed of 2.50 m/s to escape the predator? Neglect any drag effects of the surrounding water. (b) How much kinetic energy does the squid create by this maneuver?

8.20 •• You are standing on a sheet of ice that covers the football stadium parking lot in Buffalo; there is negligible friction between your feet and the ice. A friend throws you a 0.400-kg ball that is traveling horizontally at 10.0 m/s. Your mass is 70.0 kg. (a) If you catch the ball, with what speed do you and the ball move afterward? (b) If the ball hits you and bounces off your chest, so afterward it is moving horizontally at 8.0 m/s in the opposite direction, what is your speed after the collision?

8.21 •• On a frictionless, horizontal air table, puck A (with mass 0.250 kg) is moving toward puck B (with mass 0.350 kg), which is initially at rest. After the collision, puck A has a velocity of 0.120 m/s to the left, and puck B has a velocity of 0.650 m/s to the right. (a) What was the speed of puck A before the collision? (b) Calculate the change in the total kinetic energy of the system that occurs during the collision.

8.22 •• When cars are equipped with flexible bumpers, they will bounce off each other during low-speed collisions, thus causing less damage. In one such accident, a 1750-kg car traveling to the right at 1.50 m/s collides with a 1450-kg car going to the left at 1.10 m/s. Measurements show that the heavier car's speed just after the collision was 0.250 m/s in its original direction. You can ignore any road friction during the collision. (a) What was the speed of the lighter car just after the collision? (b) Calculate the change in the combined kinetic energy of the two-car system during this collision.

8.23 •• Two identical 1.50-kg masses are pressed against opposite ends of a light spring of force constant 1.75 N/cm, compressing the spring by 20.0 cm from its normal length. Find the speed of each mass when it has moved free of the spring on a frictionless horizontal table.

8.24 • Block A in Fig. E8.24 has mass 1.00 kg, and block B has mass 3.00 kg. The blocks are forced together, compressing a spring

S between them; then the system is released from rest on a level, frictionless surface. The spring, which has negligible mass, is not fastened to either block and drops to the surface after it has expanded. Block *B* acquires a speed of 1.20 m/s. (a) What is the final speed of block *A*? (b) How much potential energy was stored in the compressed spring?

Figure E8.24

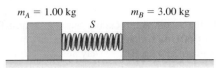

$m_A = 1.00$ kg $m_B = 3.00$ kg

S

8.25 •• A hunter on a frozen, essentially frictionless pond uses a rifle that shoots 4.20-g bullets at 965 m/s. The mass of the hunter (including his gun) is 72.5 kg, and the hunter holds tight to the gun after firing it. Find the recoil velocity of the hunter if he fires the rifle (a) horizontally and (b) at 56.0° above the horizontal.

8.26 • An atomic nucleus suddenly bursts apart (fissions) into two pieces. Piece *A*, of mass m_A, travels off to the left with speed v_A. Piece *B*, of mass m_B, travels off to the right with speed v_B. (a) Use conservation of momentum to solve for v_B in terms of m_A, m_B, and v_A. (b) Use the results of part (a) to show that $K_A/K_B = m_B/m_A$, where K_A and K_B are the kinetic energies of the two pieces.

8.27 •• Two ice skaters, Daniel (mass 65.0 kg) and Rebecca (mass 45.0 kg), are practicing. Daniel stops to tie his shoelace and, while at rest, is struck by Rebecca, who is moving at 13.0 m/s before she collides with him. After the collision, Rebecca has a velocity of magnitude 8.00 m/s at an angle of 53.1° from her initial direction. Both skaters move on the frictionless, horizontal surface of the rink. (a) What are the magnitude and direction of Daniel's velocity after the collision? (b) What is the change in total kinetic energy of the two skaters as a result of the collision?

8.28 •• You are standing on a large sheet of frictionless ice and holding a large rock. In order to get off the ice, you throw the rock so it has velocity 12.0 m/s relative to the earth at an angle of 35.0° above the horizontal. If your mass is 70.0 kg and the rock's mass is 15.0 kg, what is your speed after you throw the rock? (See Discussion Question Q8.7.)

8.29 • **Changing Mass.** An open-topped freight car with mass 24,000 kg is coasting without friction along a level track. It is raining very hard, and the rain is falling vertically downward. Originally, the car is empty and moving with a speed of 4.00 m/s. (a) What is the speed of the car after it has collected 3000 kg of rainwater? (b) Since the rain is falling downward, how is it able to affect the horizontal motion of the car?

8.30 • An astronaut in space cannot use a conventional means, such as a scale or balance, to determine the mass of an object. But she does have devices to measure distance and time accurately. She knows her own mass is 78.4 kg, but she is unsure of the mass of a large gas canister in the airless rocket. When this canister is approaching her at 3.50 m/s, she pushes against it, which slows it down to 1.20 m/s (but does not reverse it) and gives her a speed of 2.40 m/s. What is the mass of this canister?

8.31 •• **Asteroid Collision.** Two asteroids of equal mass in the asteroid belt between Mars and Jupiter collide with a glancing blow. Asteroid *A*, which was initially traveling at 40.0 m/s, is deflected 30.0° from its original direction, while asteroid *B*,

which was initially at rest, travels at 45.0° to the original direction of *A* (Fig. E8.31). (a) Find the speed of each asteroid after the collision. (b) What fraction of the original kinetic energy of asteroid *A* dissipates during this collision?

Figure E8.31

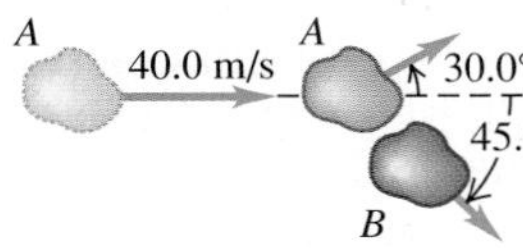

A   40.0 m/s   *A*   30.0°   45.0°   *B*

Section 8.3 Momentum Conservation and Collisions

8.32 • Two skaters collide and grab on to each other on frictionless ice. One of them, of mass 70.0 kg, is moving to the right at 2.00 m/s, while the other, of mass 65.0 kg, is moving to the left at 2.50 m/s. What are the magnitude and direction of the velocity of these skaters just after they collide?

8.33 •• A 15.0-kg fish swimming at 1.10 m/s suddenly gobbles up a 4.50-kg fish that is initially stationary. Neglect any drag effects of the water. (a) Find the speed of the large fish just after it eats the small one. (b) How much mechanical energy was dissipated during this meal?

8.34 • Two fun-loving otters are sliding toward each other on a muddy (and hence frictionless) horizontal surface. One of them, of mass 7.50 kg, is sliding to the left at 5.00 m/s, while the other, of mass 5.75 kg, is slipping to the right at 6.00 m/s. They hold fast to each other after they collide. (a) Find the magnitude and direction of the velocity of these free-spirited otters right after they collide. (b) How much mechanical energy dissipates during this play?

8.35 • **Deep Impact Mission.** In July 2005, NASA's "Deep Impact" mission crashed a 372-kg probe directly onto the surface of the comet Tempel 1, hitting the surface at 37,000 km/h. The original speed of the comet at that time was about 40,000 km/h, and its mass was estimated to be in the range $(0.10 - 2.5) \times 10^{14}$ kg. Use the smallest value of the estimated mass. (a) What change in the comet's velocity did this collision produce? Would this change be noticeable? (b) Suppose this comet were to hit the earth and fuse with it. By how much would it change our planet's velocity? Would this change be noticeable? (The mass of the earth is 5.97×10^{24} kg.)

8.36 • A 1050-kg sports car is moving westbound at 15.0 m/s on a level road when it collides with a 6320-kg truck driving east on the same road at 10.0 m/s. The two vehicles remain locked together after the collision. (a) What is the velocity (magnitude and direction) of the two vehicles just after the collision? (b) At what speed should the truck have been moving so that it and the car are both stopped in the collision? (c) Find the change in kinetic energy of the system of two vehicles for the situations of part (a) and part (b). For which situation is the change in kinetic energy greater in magnitude?

8.37 •• On a very muddy football field, a 110-kg linebacker tackles an 85-kg halfback. Immediately before the collision, the linebacker is slipping with a velocity of 8.8 m/s north and the halfback is sliding with a velocity of 7.2 m/s east. What is the velocity (magnitude and direction) at which the two players move together immediately after the collision?

8.38 •• **Accident Analysis.** Two cars collide at an intersection. Car *A*, with a mass of 2000 kg, is going from west to east, while car *B*, of mass 1500 kg, is going from north to south at 15 m/s. As a result of this collision, the two cars become enmeshed and move as one afterward. In your role as an expert witness, you inspect the scene and determine that, after the collision, the enmeshed cars moved at an angle of 65° south of east from the point of impact.

(a) How fast were the enmeshed cars moving just after the collision? (b) How fast was car *A* going just before the collision?

8.39 • Two cars, one a compact with mass 1200 kg and the other a large gas-guzzler with mass 3000 kg, collide head-on at typical freeway speeds. (a) Which car has a greater magnitude of momentum change? Which car has a greater velocity change? (b) If the larger car changes its velocity by Δv, calculate the change in the velocity of the small car in terms of Δv. (c) Which car's occupants would you expect to sustain greater injuries? Explain.

8.40 •• **BIO** **Bird Defense.** To protect their young in the nest, peregrine falcons will fly into birds of prey (such as ravens) at high speed. In one such episode, a 600-g falcon flying at 20.0 m/s hit a 1.50-kg raven flying at 9.0 m/s. The falcon hit the raven at right angles to its original path and bounced back at 5.0 m/s. (These figures were estimated by the author as he watched this attack occur in northern New Mexico.) (a) By what angle did the falcon change the raven's direction of motion? (b) What was the raven's speed right after the collision?

8.41 • At the intersection of Texas Avenue and University Drive, a yellow subcompact car with mass 950 kg traveling east on University collides with a red pickup truck with mass 1900 kg that is traveling north on Texas and has run a red light (Fig. E8.41). The two vehicles stick together as a result of the collision, and the wreckage slides at 16.0 m/s in the direction 24.0° east of north. Calculate the speed of each vehicle before the collision. The collision occurs during a heavy rainstorm; you can ignore friction forces between the vehicles and the wet road.

Figure **E8.41**

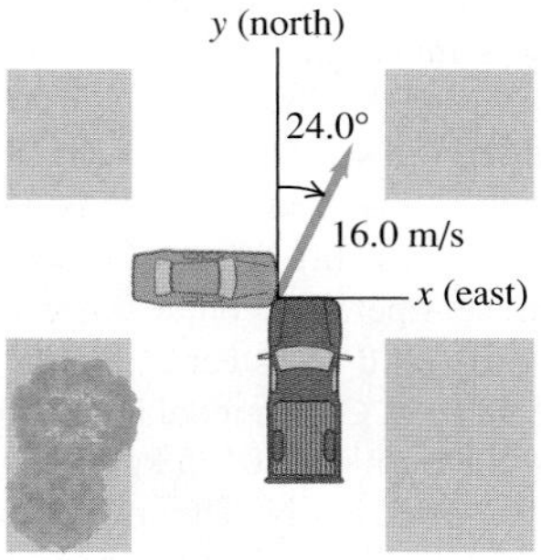

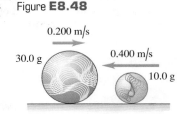

8.42 •• A 5.00-g bullet is fired horizontally into a 1.20-kg wooden block resting on a horizontal surface. The coefficient of kinetic friction between block and surface is 0.20. The bullet remains embedded in the block, which is observed to slide 0.230 m along the surface before stopping. What was the initial speed of the bullet?

8.43 •• **A Ballistic Pendulum.** A 12.0-g rifle bullet is fired with a speed of 380 m/s into a ballistic pendulum with mass 6.00 kg, suspended from a cord 70.0 cm long (see Example 8.8 in Section 8.3). Compute (a) the vertical height through which the pendulum rises, (b) the initial kinetic energy of the bullet, and (c) the kinetic energy of the bullet and pendulum immediately after the bullet becomes embedded in the pendulum.

8.44 •• **Combining Conservation Laws.** A 15.0-kg block is attached to a very light horizontal spring of force constant 500.0 N/m and is resting on a frictionless horizontal table. (Fig. E8.44). Suddenly it is struck by a 3.00-kg stone traveling horizontally at 8.00 m/s to the right, whereupon the stone rebounds at 2.00 m/s horizontally to the left. Find the maximum distance that the block will compress the spring after the collision.

Figure **E8.44**

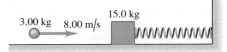

8.45 •• **CP** A 5.00-kg ornament is hanging by a 1.50-m wire when it is suddenly hit by a 3.00-kg missile traveling horizontally at 12.0 m/s. The missile embeds itself in the ornament during the collision. What is the tension in the wire immediately after the collision?

Section 8.4 Elastic Collisions

8.46 •• A 0.150-kg glider is moving to the right on a frictionless, horizontal air track with a speed of 0.80 m/s. It has a head-on collision with a 0.300-kg glider that is moving to the left with a speed of 2.20 m/s. Find the final velocity (magnitude and direction) of each glider if the collision is elastic.

8.47 •• Blocks *A* (mass 2.00 kg) and *B* (mass 10.00 kg) move on a frictionless, horizontal surface. Initially, block *B* is at rest and block *A* is moving toward it at 2.00 m/s. The blocks are equipped with ideal spring bumpers, as in Example 8.10 (Section 8.4). The collision is head-on, so all motion before and after the collision is along a straight line. (a) Find the maximum energy stored in the spring bumpers and the velocity of each block at that time. (b) Find the velocity of each block after they have moved apart.

8.48 • A 10.0-g marble slides to the left with a velocity of magnitude 0.400 m/s on the frictionless, horizontal surface of an icy New York sidewalk and has a head-on, elastic collision with a larger 30.0-g marble sliding to the right with a velocity of magnitude 0.200 m/s (Fig. E8.48). (a) Find the velocity of each marble (magnitude and direction) after the collision. (Since the collision is head-on, all the motion is along a line.) (b) Calculate the *change in momentum* (that is, the momentum after the collision minus the momentum before the collision) for each marble. Compare the values you get for each marble. (c) Calculate the *change in kinetic energy* (that is, the kinetic energy after the collision minus the kinetic energy before the collision) for each marble. Compare the values you get for each marble.

Figure **E8.48**

8.49 •• **Moderators.** Canadian nuclear reactors use *heavy water* moderators in which elastic collisions occur between the neutrons and deuterons of mass 2.0 u (see Example 8.11 in Section 8.4). (a) What is the speed of a neutron, expressed as a fraction of its original speed, after a head-on, elastic collision with a deuteron that is initially at rest? (b) What is its kinetic energy, expressed as a fraction of its original kinetic energy? (c) How many such successive collisions will reduce the speed of a neutron to 1/59,000 of its original value?

8.50 •• You are at the controls of a particle accelerator, sending a beam of 1.50×10^7 m/s protons (mass *m*) at a gas target of an unknown element. Your detector tells you that some protons bounce straight back after a collision with one of the nuclei of the unknown element. All such protons rebound with a speed of 1.20×10^7 m/s. Assume that the initial speed of the target nucleus is negligible and the collision is elastic. (a) Find the mass of one nucleus of the unknown element. Express your answer in terms of the proton mass *m*. (b) What is the speed of the unknown nucleus immediately after such a collision?

Section 8.5 Center of Mass

8.51 • Three odd-shaped blocks of chocolate have the following masses and center-of-mass coordinates: (1) 0.300 kg, (0.200 m,

0.300 m); (2) 0.400 kg, (0.100 m, −0.400 m); (3) 0.200 kg, (−0.300 m, 0.600 m). Find the coordinates of the center of mass of the system of three chocolate blocks.

8.52 • Find the position of the center of mass of the system of the sun and Jupiter. (Since Jupiter is more massive than the rest of the planets combined, this is essentially the position of the center of mass of the solar system.) Does the center of mass lie inside or outside the sun? Use the data in Appendix F.

8.53 •• **Pluto and Charon.** Pluto's diameter is approximately 2370 km, and the diameter of its satellite Charon is 1250 km. Although the distance varies, they are often about 19,700 km apart, center to center. Assuming that both Pluto and Charon have the same composition and hence the same average density, find the location of the center of mass of this system relative to the center of Pluto.

8.54 • A 1200-kg station wagon is moving along a straight highway at 12.0 m/s. Another car, with mass 1800 kg and speed 20.0 m/s, has its center of mass 40.0 m ahead of the center of mass of the station wagon (Fig. E8.54). (a) Find the position of the center of mass of the system consisting of the two automobiles. (b) Find the magnitude of the total momentum of the system from the given data. (c) Find the speed of the center of mass of the system. (d) Find the total momentum of the system, using the speed of the center of mass. Compare your result with that of part (b).

Figure E8.54

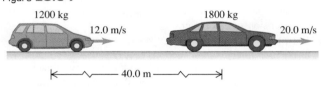

8.55 • A machine part consists of a thin, uniform 4.00-kg bar that is 1.50 m long, hinged perpendicular to a similar vertical bar of mass 3.00 kg and length 1.80 m. The longer bar has a small but dense 2.00-kg ball at one end (Fig. E8.55). By what distance will the center of mass of this part move horizontally and vertically if the vertical bar is pivoted counterclockwise through 90° to make the entire part horizontal?

Figure E8.55

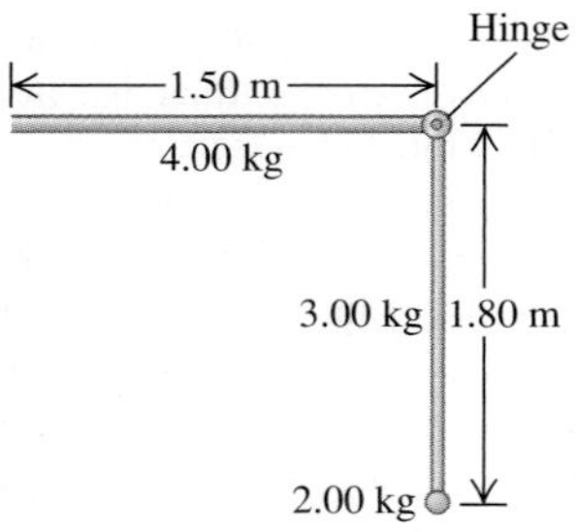

Hinge

1.50 m

4.00 kg

3.00 kg 1.80 m

2.00 kg

8.56 • At one instant, the center of mass of a system of two particles is located on the x-axis at $x = 2.0$ m and has a velocity of $(5.0 \text{ m/s})\hat{\imath}$. One of the particles is at the origin. The other particle has a mass of 0.10 kg and is at rest on the x-axis at $x = 8.0$ m. (a) What is the mass of the particle at the origin? (b) Calculate the total momentum of this system. (c) What is the velocity of the particle at the origin?

8.57 •• In Example 8.14 (Section 8.5), Ramon pulls on the rope to give himself a speed of 0.70 m/s. What is James's speed?

8.58 • **CALC** A system consists of two particles. At $t = 0$ one particle is at the origin; the other, which has a mass of 0.50 kg, is on the y-axis at $y = 6.0$ m. At $t = 0$ the center of mass of the system is on the y-axis at $y = 2.4$ m. The velocity of the center of mass is given by $(0.75 \text{ m/s}^3)t^2\hat{\imath}$. (a) Find the total mass of the system. (b) Find the acceleration of the center of mass at any time t. (c) Find the net external force acting on the system at $t = 3.0$ s.

8.59 • **CALC** A radio-controlled model airplane has a momentum given by $[(-0.75 \text{ kg} \cdot \text{m/s}^3)t^2 + (3.0 \text{ kg} \cdot \text{m/s})]\hat{\imath} + (0.25 \text{ kg} \cdot \text{m/s}^2)t\hat{\jmath}$. What are the x-, y-, and z-components of the net force on the airplane?

8.60 •• **BIO Changing Your Center of Mass.** To keep the calculations fairly simple, but still reasonable, we shall model a human leg that is 92.0 cm long (measured from the hip joint) by assuming that the upper leg and the lower leg (which includes the foot) have equal lengths and that each of them is uniform. For a 70.0-kg person, the mass of the upper leg would be 8.60 kg, while that of the lower leg (including the foot) would be 5.25 kg. Find the location of the center of mass of this leg, relative to the hip joint, if it is (a) stretched out horizontally and (b) bent at the knee to form a right angle with the upper leg remaining horizontal.

Section 8.6 Rocket Propulsion

8.61 •• A 70-kg astronaut floating in space in a 110-kg MMU (manned maneuvering unit) experiences an acceleration of 0.029 m/s² when he fires one of the MMU's thrusters. (a) If the speed of the escaping N_2 gas relative to the astronaut is 490 m/s, how much gas is used by the thruster in 5.0 s? (b) What is the thrust of the thruster?

8.62 • A small rocket burns 0.0500 kg of fuel per second, ejecting it as a gas with a velocity relative to the rocket of magnitude 1600 m/s. (a) What is the thrust of the rocket? (b) Would the rocket operate in outer space where there is no atmosphere? If so, how would you steer it? Could you brake it?

8.63 • A C6-5 model rocket engine has an impulse of 10.0 N · s while burning 0.0125 kg of propellant in 1.70 s. It has a maximum thrust of 13.3 N. The initial mass of the engine plus propellant is 0.0258 kg. (a) What fraction of the maximum thrust is the average thrust? (b) Calculate the relative speed of the exhaust gases, assuming it is constant. (c) Assuming that the relative speed of the exhaust gases is constant, find the final speed of the engine if it was attached to a very light frame and fired from rest in gravity-free outer space.

8.64 •• Obviously, we can make rockets to go very fast, but what is a reasonable top speed? Assume that a rocket is fired from rest at a space station in deep space, where gravity is negligible. (a) If the rocket ejects gas at a relative speed of 2000 m/s and you want the rocket's speed eventually to be $1.00 \times 10^{-3}c$, where c is the speed of light, what fraction of the initial mass of the rocket and fuel is *not* fuel? (b) What is this fraction if the final speed is to be 3000 m/s?

8.65 •• A single-stage rocket is fired from rest from a deep-space platform, where gravity is negligible. If the rocket burns its fuel in 50.0 s and the relative speed of the exhaust gas is $v_{\text{ex}} = 2100$ m/s, what must the mass ratio m_0/m be for a final speed v of 8.00 km/s (about equal to the orbital speed of an earth satellite)?

PROBLEMS

8.66 •• **CP CALC** A young girl with mass 40.0 kg is sliding on a horizontal, frictionless surface with an initial momentum that is due east and that has magnitude 90.0 kg · m/s. Starting at $t = 0$, a net force with magnitude $F = (8.20 \text{ N/s})t$ and direction due west is applied to the girl. (a) At what value of t does the girl have a westward momentum of magnitude 60.0 kg · m/s? (b) How much work has been done on the girl by the force in the time interval from $t = 0$ to the time calculated in part (a)? (c) What is the magnitude of the acceleration of the girl at the time calculated in part (a)?

8.67 •• A steel ball with mass 40.0 g is dropped from a height of 2.00 m onto a horizontal steel slab. The ball rebounds to a height of 1.60 m. (a) Calculate the impulse delivered to the ball during impact. (b) If the ball is in contact with the slab for 2.00 ms, find the average force on the ball during impact.

8.68 • In a volcanic eruption, a 2400-kg boulder is thrown vertically upward into the air. At its highest point, it suddenly explodes (due to trapped gases) into two fragments, one being three times the mass of the other. The lighter fragment starts out with only horizontal velocity and lands 318 m directly north of the point of the explosion. Where will the other fragment land? Neglect any air resistance.

8.69 •• Just before it is struck by a racket, a tennis ball weighing 0.560 N has a velocity of $(20.0 \text{ m/s})\hat{\imath} - (4.0 \text{ m/s})\hat{\jmath}$. During the 3.00 ms that the racket and ball are in contact, the net force on the ball is constant and equal to $-(380 \text{ N})\hat{\imath} + (110 \text{ N})\hat{\jmath}$. (a) What are the x- and y-components of the impulse of the net force applied to the ball? (b) What are the x- and y-components of the final velocity of the ball?

8.70 • Three identical pucks on a horizontal air table have repelling magnets. They are held together and then released simultaneously. Each has the same speed at any instant. One puck moves due west. What is the direction of the velocity of each of the other two pucks?

8.71 •• A 1500-kg blue convertible is traveling south, and a 2000-kg red SUV is traveling west. If the total momentum of the system consisting of the two cars is 7200 kg·m/s directed at 60.0° west of south, what is the speed of each vehicle?

8.72 •• A railroad handcar is moving along straight, frictionless tracks with negligible air resistance. In the following cases, the car initially has a total mass (car and contents) of 200 kg and is traveling east with a velocity of magnitude 5.00 m/s. Find the *final velocity* of the car in each case, assuming that the handcar does not leave the tracks. (a) A 25.0-kg mass is thrown sideways out of the car with a velocity of magnitude 2.00 m/s relative to the car's initial velocity. (b) A 25.0-kg mass is thrown backward out of the car with a velocity of 5.00 m/s relative to the initial motion of the car. (c) A 25.0-kg mass is thrown into the car with a velocity of 6.00 m/s relative to the ground and opposite in direction to the initial velocity of the car.

8.73 • Spheres A (mass 0.020 kg), B (mass 0.030 kg), and C (mass 0.050 kg) are approaching the origin as they slide on a frictionless air table (Fig. P8.73). The initial velocities of A and B are given in the figure. All three spheres arrive at the origin at the same time and stick together. (a) What must the x- and y-components of the initial velocity of C be if all three objects are to end up moving at 0.50 m/s in the $+x$-direction after the collision? (b) If C has the velocity found in part (a), what is the change in the kinetic energy of the system of three spheres as a result of the collision?

8.74 ••• You and your friends are doing physics experiments on a frozen pond that serves as a frictionless, horizontal surface. Sam, with mass 80.0 kg, is given a push and slides eastward. Abigail, with mass 50.0 kg, is sent sliding northward. They collide, and after the collision Sam is moving at 37.0° north of east with a speed of 6.00 m/s and Abigail is moving at 23.0° south of east with a speed of 9.00 m/s. (a) What was the speed of each person before the collision? (b) By how much did the total kinetic energy of the two people decrease during the collision?

8.75 ••• The nucleus of ^{214}Po decays radioactively by emitting an alpha particle (mass 6.65×10^{-27} kg) with kinetic energy 1.23×10^{-12} J, as measured in the laboratory reference frame. Assuming that the Po was initially at rest in this frame, find the recoil velocity of the nucleus that remains after the decay.

8.76 • **CP** At a classic auto show, a 840-kg 1955 Nash Metropolitan motors by at 9.0 m/s, followed by a 1620-kg 1957 Packard Clipper purring past at 5.0 m/s. (a) Which car has the greater kinetic energy? What is the ratio of the kinetic energy of the Nash to that of the Packard? (b) Which car has the greater magnitude of momentum? What is the ratio of the magnitude of momentum of the Nash to that of the Packard? (c) Let F_N be the net force required to stop the Nash in time t, and let F_P be the net force required to stop the Packard in the same time. Which is larger: F_N or F_P? What is the ratio F_N/F_P of these two forces? (d) Now let F_N be the net force required to stop the Nash in a distance d, and let F_P be the net force required to stop the Packard in the same distance. Which is larger: F_N or F_P? What is the ratio F_N/F_P?

8.77 •• **CP** An 8.00-kg block of wood sits at the edge of a frictionless table, 2.20 m above the floor. A 0.500-kg blob of clay slides along the length of the table with a speed of 24.0 m/s, strikes the block of wood, and sticks to it. The combined object leaves the edge of the table and travels to the floor. What horizontal distance has the combined object traveled when it reaches the floor?

8.78 ••• **CP** A small wooden block with mass 0.800 kg is suspended from the lower end of a light cord that is 1.60 m long. The block is initially at rest. A bullet with mass 12.0 g is fired at the block with a horizontal velocity v_0. The bullet strikes the block and becomes embedded in it. After the collision the combined object swings on the end of the cord. When the block has risen a vertical height of 0.800 m, the tension in the cord is 4.80 N. What was the initial speed v_0 of the bullet?

8.79 •• **Combining Conservation Laws.** A 5.00-kg chunk of ice is sliding at 12.0 m/s on the floor of an ice-covered valley when it collides with and sticks to another 5.00-kg chunk of ice that is initially at rest. (Fig. P8.79). Since the valley is icy, there is no friction. After the collision, how high above the valley floor will the combined chunks go?

Figure **P8.73**

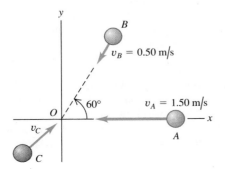

Figure **P8.79**

8.80 •• **Automobile Accident Analysis.** You are called as an expert witness to analyze the following auto accident: Car B, of mass 1900 kg, was stopped at a red light when it was hit from behind by car A, of mass 1500 kg. The cars locked bumpers during the collision and slid to a stop with brakes locked on all wheels. Measurements of the skid marks left by the tires showed them to be 7.15 m long. The coefficient of kinetic friction between the tires and the road was 0.65. (a) What was the speed of car A just before the collision? (b) If the speed limit was 35 mph, was car A speeding, and if so, by how many miles per hour was it *exceeding* the speed limit?

8.81 •• **Accident Analysis.** A 1500-kg sedan goes through a wide intersection traveling from north to south when it is hit by a 2200-kg SUV traveling from east to west. The two cars become enmeshed due to the impact and slide as one thereafter. On-the-scene measurements show that the coefficient of kinetic friction between the tires of these cars and the pavement is 0.75, and the cars slide to a halt at a point 5.39 m west and 6.43 m south of the impact point. How fast was each car traveling just before the collision?

8.82 ••• **CP** A 0.150-kg frame, when suspended from a coil spring, stretches the spring 0.070 m. A 0.200-kg lump of putty is dropped from rest onto the frame from a height of 30.0 cm (Fig. P8.82). Find the maximum distance the frame moves downward from its initial position.

Figure **P8.82**

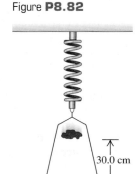

30.0 cm

8.83 • A rifle bullet with mass 8.00 g strikes and embeds itself in a block with mass 0.992 kg that rests on a frictionless, horizontal surface and is attached to a coil spring (Fig. P8.83). The impact compresses the spring 15.0 cm. Calibration of the spring shows that a force of 0.750 N is required to compress the spring 0.250 cm. (a) Find the magnitude of the block's velocity just after impact. (b) What was the initial speed of the bullet?

Figure **P8.83**

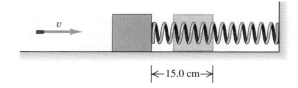

|←15.0 cm→|

8.84 •• **A Ricocheting Bullet.** 0.100-kg stone rests on a frictionless, horizontal surface. A bullet of mass 6.00 g, traveling horizontally at 350 m/s, strikes the stone and rebounds horizontally at right angles to its original direction with a speed of 250 m/s. (a) Compute the magnitude and direction of the velocity of the stone after it is struck. (b) Is the collision perfectly elastic?

8.85 •• A movie stuntman (mass 80.0 kg) stands on a window ledge 5.0 m above the floor (Fig. P8.85). Grabbing a rope attached to a chandelier, he swings down to grapple with the movie's villain (mass 70.0 kg), who is standing directly under the chandelier. (Assume that the stuntman's center of mass moves downward 5.0 m. He releases the rope just as he

Figure **P8.85**

5.0 m $m = 80.0$ kg

$m = 70.0$ kg

reaches the villain.) (a) With what speed do the entwined foes start to slide across the floor? (b) If the coefficient of kinetic friction of their bodies with the floor is $\mu_k = 0.250$, how far do they slide?

8.86 •• **CP** Two identical masses are released from rest in a smooth hemispherical bowl of radius R from the positions shown in Fig. P8.86. You can ignore friction between the masses and the surface of the bowl. If they stick together when they collide, how high above the bottom of the bowl will the masses go after colliding?

Figure **P8.86**

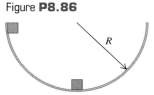

R

8.87 •• A ball with mass M, moving horizontally at 4.00 m/s, collides elastically with a block with mass 3M that is initially hanging at rest from the ceiling on the end of a 50.0-cm wire. Find the maximum angle through which the block swings after it is hit.

8.88 ••• **CP** A 20.00-kg lead sphere is hanging from a hook by a thin wire 3.50 m long and is free to swing in a complete circle. Suddenly it is struck horizontally by a 5.00-kg steel dart that embeds itself in the lead sphere. What must be the minimum initial speed of the dart so that the combination makes a complete circular loop after the collision?

8.89 ••• **CP** An 8.00-kg ball, hanging from the ceiling by a light wire 135 cm long, is struck in an elastic collision by a 2.00-kg ball moving horizontally at 5.00 m/s just before the collision. Find the tension in the wire just after the collision.

8.90 •• A 7.0-kg shell at rest explodes into two fragments, one with a mass of 2.0 kg and the other with a mass of 5.0 kg. If the heavier fragment gains 100 J of kinetic energy from the explosion, how much kinetic energy does the lighter one gain?

8.91 •• A 4.00-g bullet, traveling horizontally with a velocity of magnitude 400 m/s, is fired into a wooden block with mass 0.800 kg, initially at rest on a level surface. The bullet passes through the block and emerges with its speed reduced to 190 m/s. The block slides a distance of 45.0 cm along the surface from its initial position. (a) What is the coefficient of kinetic friction between block and surface? (b) What is the decrease in kinetic energy of the bullet? (c) What is the kinetic energy of the block at the instant after the bullet passes through it?

8.92 •• A 5.00-g bullet is shot *through* a 1.00-kg wood block suspended on a string 2.00 m long. The center of mass of the block rises a distance of 0.38 cm. Find the speed of the bullet as it emerges from the block if its initial speed is 450 m/s.

8.93 •• A neutron with mass m makes a head-on, elastic collision with a nucleus of mass M, which is initially at rest. (a) Show that if the neutron's initial kinetic energy is K_0, the kinetic energy that it loses during the collision is $4mMK_0/(M + m)^2$. (b) For what value of M does the incident neutron lose the most energy? (c) When M has the value calculated in part (b), what is the speed of the neutron after the collision?

8.94 •• **Energy Sharing in Elastic Collisions.** A stationary object with mass m_B is struck head-on by an object with mass m_A that is moving initially at speed v_0. (a) If the collision is elastic, what percentage of the original energy does each object have after the collision? (b) What does your answer in part (a) give for the special cases (i) $m_A = m_B$ and (ii) $m_A = 5m_B$? (c) For what values, if any, of the mass ratio m_A/m_B is the original kinetic energy shared equally by the two objects after the collision?

8.95 •• **CP** In a shipping company distribution center, an open cart of mass 50.0 kg is rolling to the left at a speed of 5.00 m/s

(Fig. P8.95). You can ignore friction between the cart and the floor. A 15.0-kg package slides down a chute that is inclined at 37° from the horizontal and leaves the end of the chute with a speed of 3.00 m/s. The package lands in the cart and they roll off together. If the lower end of the chute is a vertical distance of 4.00 m above the bottom of the cart, what are (a) the speed of the package just before it lands in the cart and (b) the final speed of the cart?

Figure **P8.95**

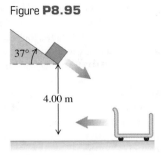

37°

4.00 m

8.96 • A blue puck with mass 0.0400 kg, sliding with a velocity of magnitude 0.200 m/s on a frictionless, horizontal air table, makes a perfectly elastic, head-on collision with a red puck with mass m, initially at rest. After the collision, the velocity of the blue puck is 0.050 m/s in the same direction as its initial velocity. Find (a) the velocity (magnitude and direction) of the red puck after the collision and (b) the mass m of the red puck.

8.97 ••• Jack and Jill are standing on a crate at rest on the frictionless, horizontal surface of a frozen pond. Jack has mass 75.0 kg, Jill has mass 45.0 kg, and the crate has mass 15.0 kg. They remember that they must fetch a pail of water, so each jumps horizontally from the top of the crate. Just after each jumps, that person is moving away from the crate with a speed of 4.00 m/s relative to the crate. (a) What is the final speed of the crate if both Jack and Jill jump simultaneously and in the same direction? (*Hint:* Use an inertial coordinate system attached to the ground.) (b) What is the final speed of the crate if Jack jumps first and then a few seconds later Jill jumps in the same direction? (c) What is the final speed of the crate if Jill jumps first and then Jack, again in the same direction?

8.98 • Suppose you hold a small ball in contact with, and directly over, the center of a large ball. If you then drop the small ball a short time after dropping the large ball, the small ball rebounds with surprising speed. To show the extreme case, ignore air resistance and suppose the large ball makes an elastic collision with the floor and then rebounds to make an elastic collision with the still-descending small ball. Just before the collision between the two balls, the large ball is moving upward with velocity $\vec{v}$ and the small ball has velocity $-\vec{v}$. (Do you see why?) Assume the large ball has a much greater mass than the small ball. (a) What is the velocity of the small ball immediately after its collision with the large ball? (b) From the answer to part (a), what is the ratio of the small ball's rebound distance to the distance it fell before the collision?

8.99 ••• Hockey puck B rests on a smooth ice surface and is struck by a second puck A, which has the same mass. Puck A is initially traveling at 15.0 m/s and is deflected 25.0° from its initial direction. Assume that the collision is perfectly elastic. Find the final speed of each puck and the direction of B's velocity after the collision.

8.100 ••• **Energy Sharing.** An object with mass m, initially at rest, explodes into two fragments, one with mass m_A and the other with mass m_B, where $m_A + m_B = m$. (a) If energy Q is released in the explosion, how much kinetic energy does each fragment have immediately after the explosion? (b) What percentage of the total energy released does each fragment get when one fragment has four times the mass of the other?

8.101 ••• **Neutron Decay.** A neutron at rest decays (breaks up) to a proton and an electron. Energy is released in the decay and appears as kinetic energy of the proton and electron. The mass of a proton is 1836 times the mass of an electron. What fraction of the total energy released goes into the kinetic energy of the proton?

8.102 •• A ^{232}Th (thorium) nucleus at rest decays to a ^{228}Ra (radium) nucleus with the emission of an alpha particle. The total kinetic energy of the decay fragments is 6.54×10^{-13} J. An alpha particle has 1.76% of the mass of a ^{228}Ra nucleus. Calculate the kinetic energy of (a) the recoiling ^{228}Ra nucleus and (b) the alpha particle.

8.103 • **Antineutrino.** In beta decay, a nucleus emits an electron. A ^{210}Bi (bismuth) nucleus at rest undergoes beta decay to ^{210}Po (polonium). Suppose the emitted electron moves to the right with a momentum of 5.60×10^{-22} kg·m/s. The ^{210}Po nucleus, with mass 3.50×10^{-25} kg, recoils to the left at a speed of 1.14×10^{3} m/s. Momentum conservation requires that a second particle, called an antineutrino, must also be emitted. Calculate the magnitude and direction of the momentum of the antineutrino that is emitted in this decay.

8.104 •• Jonathan and Jane are sitting in a sleigh that is at rest on frictionless ice. Jonathan's weight is 800 N, Jane's weight is 600 N, and that of the sleigh is 1000 N. They see a poisonous spider on the floor of the sleigh and immediately jump off. Jonathan jumps to the left with a velocity of 5.00 m/s at 30.0° above the horizontal (relative to the ice), and Jane jumps to the right at 7.00 m/s at 36.9° above the horizontal (relative to the ice). Calculate the sleigh's horizontal velocity (magnitude and direction) after they jump out.

8.105 •• Two friends, Burt and Ernie, are standing at opposite ends of a uniform log that is floating in a lake. The log is 3.0 m long and has mass 20.0 kg. Burt has mass 30.0 kg and Ernie has mass 40.0 kg. Initially the log and the two friends are at rest relative to the shore. Burt then offers Ernie a cookie, and Ernie walks to Burt's end of the log to get it. Relative to the shore, what distance has the log moved by the time Ernie reaches Burt? Neglect any horizontal force that the water exerts on the log and assume that neither Burt nor Ernie falls off the log.

8.106 •• A 45.0-kg woman stands up in a 60.0-kg canoe 5.00 m long. She walks from a point 1.00 m from one end to a point 1.00 m from the other end (Fig. P8.106). If you ignore resistance to motion of the canoe in the water, how far does the canoe move during this process?

Figure **P8.106**

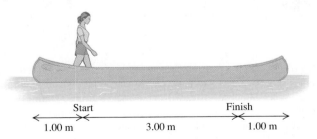

Start   Finish

1.00 m   3.00 m   1.00 m

8.107 •• You are standing on a concrete slab that in turn is resting on a frozen lake. Assume there is no friction between the slab and the ice. The slab has a weight five times your weight. If you begin walking forward at 2.00 m/s relative to the ice, with what speed, relative to the ice, does the slab move?

8.108 •• **CP** A 20.0-kg projectile is fired at an angle of 60.0° above the horizontal with a speed of 80.0 m/s. At the highest point of its trajectory, the projectile explodes into two fragments with equal mass, one of which falls vertically with zero initial speed. You can ignore air resistance. (a) How far from the point of firing does the other fragment strike if the terrain is level? (b) How much energy is released during the explosion?

8.109 ••• **CP** A fireworks rocket is fired vertically upward. At its maximum height of 80.0 m, it explodes and breaks into two pieces: one with mass 1.40 kg and the other with mass 0.28 kg. In the explosion, 860 J of chemical energy is converted to kinetic energy of the two fragments. (a) What is the speed of each fragment just after the explosion? (b) It is observed that the two fragments hit the ground at the same time. What is the distance between the points on the ground where they land? Assume that the ground is level and air resistance can be ignored.

8.110 ••• A 12.0-kg shell is launched at an angle of 55.0° above the horizontal with an initial speed of 150 m/s. When it is at its highest point, the shell explodes into two fragments, one three times heavier than the other. The two fragments reach the ground at the same time. Assume that air resistance can be ignored. If the heavier fragment lands back at the same point from which the shell was launched, where will the lighter fragment land, and how much energy was released in the explosion?

8.111 • **CP** A wagon with two boxes of gold, having total mass 300 kg, is cut loose from the horses by an outlaw when the wagon is at rest 50 m up a 6.0° slope (Fig. P8.111). The outlaw plans to have the wagon roll down the slope and across the level ground, and then fall into a canyon where his confederates wait. But in a tree 40 m from the canyon edge wait the Lone Ranger (mass 75.0 kg) and Tonto (mass 60.0 kg). They drop vertically into the wagon as it passes beneath them. (a) If they require 5.0 s to grab the gold and jump out, will they make it before the wagon goes over the edge? The wagon rolls with negligible friction. (b) When the two heroes drop into the wagon, is the kinetic energy of the system of the heroes plus the wagon conserved? If not, does it increase or decrease, and by how much?

Figure **P8.111**

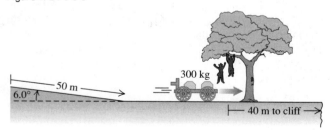

8.112 •• **CALC** In Section 8.6, we considered a rocket fired in outer space where there is no air resistance and where gravity is negligible. Suppose instead that the rocket is accelerating vertically upward from rest on the earth's surface. Continue to ignore air resistance and consider only that part of the motion where the altitude of the rocket is small so that *g* may be assumed to be constant. (a) How is Eq. (8.37) modified by the presence of the gravity force? (b) Derive an expression for the acceleration *a* of the rocket, analogous to Eq. (8.39). (c) What is the acceleration of the rocket in Example 8.15 (Section 8.6) if it is near the earth's surface rather than in outer space? You can ignore air resistance. (d) Find the speed of the rocket in Example 8.16 (Section 8.6) after 90 s if the rocket is fired from the earth's surface rather than in outer space.

You can ignore air resistance. How does your answer compare with the rocket speed calculated in Example 8.16?

8.113 •• **A Multistage Rocket.** Suppose the first stage of a two-stage rocket has total mass 12,000 kg, of which 9000 kg is fuel. The total mass of the second stage is 1000 kg, of which 700 kg is fuel. Assume that the relative speed v_{ex} of ejected material is constant, and ignore any effect of gravity. (The effect of gravity is small during the firing period if the rate of fuel consumption is large.) (a) Suppose the entire fuel supply carried by the two-stage rocket is utilized in a single-stage rocket with the same total mass of 13,000 kg. In terms of v_{ex}, what is the speed of the rocket, starting from rest, when its fuel is exhausted? (b) For the two-stage rocket, what is the speed when the fuel of the first stage is exhausted if the first stage carries the second stage with it to this point? This speed then becomes the initial speed of the second stage. At this point, the second stage separates from the first stage. (c) What is the final speed of the second stage? (d) What value of v_{ex} is required to give the second stage of the rocket a speed of 7.00 km/s?

CHALLENGE PROBLEMS

8.114 • **CALC** **A Variable-Mass Raindrop.** In a rocket-propulsion problem the mass is variable. Another such problem is a raindrop falling through a cloud of small water droplets. Some of these small droplets adhere to the raindrop, thereby *increasing* its mass as it falls. The force on the raindrop is

$$F_{ext} = \frac{dp}{dt} = m\frac{dv}{dt} + v\frac{dm}{dt}$$

Suppose the mass of the raindrop depends on the distance *x* that it has fallen. Then $m = kx$, where *k* is a constant, and $dm/dt = kv$. This gives, since $F_{ext} = mg$,

$$mg = m\frac{dv}{dt} + v(kv)$$

Or, dividing by *k*,

$$xg = x\frac{dv}{dt} + v^2$$

This is a differential equation that has a solution of the form $v = at$, where *a* is the acceleration and is constant. Take the initial velocity of the raindrop to be zero. (a) Using the proposed solution for *v*, find the acceleration *a*. (b) Find the distance the raindrop has fallen in $t = 3.00$ s. (c) Given that $k = 2.00$ g/m, find the mass of the raindrop at $t = 3.00$ s. (For many more intriguing aspects of this problem, see K. S. Krane, *American Journal of Physics,* Vol. 49 (1981), pp. 113–117.)

8.115 •• **CALC** In Section 8.5 we calculated the center of mass by considering objects composed of a *finite* number of point masses or objects that, by symmetry, could be represented by a finite number of point masses. For a solid object whose mass distribution does not allow for a simple determination of the center of mass by symmetry, the sums of Eqs. (8.28) must be generalized to integrals

$$x_{cm} = \frac{1}{M}\int x\,dm \qquad y_{cm} = \frac{1}{M}\int y\,dm$$

where *x* and *y* are the coordinates of the small piece of the object that has mass *dm*. The integration is over the whole of the object.

Consider a thin rod of length L, mass M, and cross-sectional area A. Let the origin of the coordinates be at the left end of the rod and the positive x-axis lie along the rod. (a) If the density $\rho = M/V$ of the object is uniform, perform the integration described above to show that the x-coordinate of the center of mass of the rod is at its geometrical center. (b) If the density of the object varies linearly with x—that is, $\rho = \alpha x$, where α is a positive constant—calculate the x-coordinate of the rod's center of mass.

8.116 •• **CALC** Use the methods of Challenge Problem 8.115 to calculate the x- and y-coordinates of the center of mass of a semicircular metal plate with uniform density ρ and thickness t. Let the radius of the plate be a. The mass of the plate is thus $M = \frac{1}{2}\rho\pi a^2 t$. Use the coordinate system indicated in Fig. P8.116.

Figure **P8.116**

Answers

Chapter Opening Question ?

The two bullets have the same magnitude of momentum $p = mv$ (the product of mass and speed), but the faster, lightweight bullet has twice as much kinetic energy $K = \frac{1}{2}mv^2$. Hence, the lightweight bullet can do twice as much work on the carrot (and twice as much damage) in the process of coming to a halt (see Section 8.1).

Test Your Understanding Questions

8.1 Answer: (v), (i) and (ii) (tied for second place), (iii) and (iv) (tied for third place) We use two interpretations of the impulse of the net force: (1) the net force multiplied by the time that the net force acts, and (2) the change in momentum of the particle on which the net force acts. Which interpretation we use depends on what information we are given. We take the positive x-direction to be to the east. (i) The force is not given, so we use interpretation 2: $J_x = mv_{2x} - mv_{1x} = (1000\text{ kg})(0) - (1000\text{ kg})(25\text{ m/s}) = -25{,}000\text{ kg} \cdot \text{m/s}$, so the magnitude of the impulse is $25{,}000\text{ kg} \cdot \text{m/s} = 25{,}000\text{ N} \cdot \text{s}$. (ii) For the same reason as in (i), we use interpretation 2: $J_x = mv_{2x} - mv_{1x} = (1000\text{ kg})(0) - (1000\text{ kg})(25\text{ m/s}) = -25{,}000\text{ kg} \cdot \text{m/s}$, and the magnitude of the impulse is again $25{,}000\text{ kg} \cdot \text{m/s} = 25{,}000\text{ N} \cdot \text{s}$. (iii) The final velocity is not given, so we use interpretation 1: $J_x = (\Sigma F_x)_{\text{av}}(t_2 - t_1) = (2000\text{ N})(10\text{ s}) = 20{,}000\text{ N} \cdot \text{s}$, so the magnitude of the impulse is $20{,}000\text{ N} \cdot \text{s}$. (iv) For the same reason as in (iii), we use interpretation 1: $J_x = (\Sigma F_x)_{\text{av}}(t_2 - t_1) = (-2000\text{ N})(10\text{ s}) = -20{,}000\text{ N} \cdot \text{s}$, so the magnitude of the impulse is $20{,}000\text{ N} \cdot \text{s}$. (v) The force is not given, so we use interpretation 2: $J_x = mv_{2x} - mv_{1x} = (1000\text{ kg})(-25\text{ m/s}) - (1000\text{ kg})(25\text{ m/s}) = -50{,}000\text{ kg} \cdot \text{m/s}$, so the magnitude of the impulse is $50{,}000\text{ kg} \cdot \text{m/s} = 50{,}000\text{ N} \cdot \text{s}$.

8.2 Answers: (a) $v_{C2x} > 0$, $v_{C2y} > 0$, **(b) piece C** There are no external horizontal forces, so the x- and y-components of the total momentum of the system are both conserved. Both components of the total momentum are zero before the spring releases, so they must be zero after the spring releases. Hence,

$$P_x = 0 = m_A v_{A2x} + m_B v_{B2x} + m_C v_{C2x}$$
$$P_y = 0 = m_A v_{A2y} + m_B v_{B2y} + m_C v_{C2y}$$

We are given that $m_A = m_B = m_C$, $v_{A2x} < 0$, $v_{A2y} = 0$, $v_{B2x} = 0$, and $v_{B2y} < 0$. You can solve the above equations to

show that $v_{C2x} = -v_{A2x} > 0$ and $v_{C2y} = -v_{B2y} > 0$, so the velocity components of piece C are both positive. Piece C has speed $\sqrt{v_{C2x}^2 + v_{C2y}^2} = \sqrt{v_{A2x}^2 + v_{B2y}^2}$, which is greater than the speed of either piece A or piece B.

8.3 Answers: (a) elastic, (b) inelastic, (c) completely inelastic In each case gravitational potential energy is converted to kinetic energy as the ball falls, and the collision is between the ball and the ground. In (a) all of the initial energy is converted back to gravitational potential energy, so no kinetic energy is lost in the bounce and the collision is elastic. In (b) there is less gravitational potential energy at the end than at the beginning, so some kinetic energy was lost in the bounce. Hence the collision is inelastic. In (c) the ball loses all the kinetic energy it has to give, the ball and the ground stick together, and the collision is completely inelastic.

8.4 Answer: worse After a collision with a water molecule initially at rest, the speed of the neutron is $|(m_n - m_w)/(m_n + m_w)| = |(1.0\text{ u} - 18\text{ u})/(1.0\text{ u} + 18\text{ u})| = \frac{17}{19}$ of its initial speed, and its kinetic energy is $\left(\frac{17}{19}\right)^2 = 0.80$ of the initial value. Hence a water molecule is a worse moderator than a carbon atom, for which the corresponding numbers are $\frac{11}{13}$ and $\left(\frac{11}{13}\right)^2 = 0.72$.

8.5 Answer: no If gravity is the only force acting on the system of two fragments, the center of mass will follow the parabolic trajectory of a freely falling object. Once a fragment lands, however, the ground exerts a normal force on that fragment. Hence the net force on the system has changed, and the trajectory of the center of mass changes in response.

8.6 Answers: (a) increasing, (b) decreasing From Eqs. (8.37) and (8.38), the thrust F is equal to $m(dv/dt)$, where m is the rocket's mass and dv/dt is its acceleration. Because m decreases with time, if the thrust F is constant, then the acceleration must increase with time (the same force acts on a smaller mass); if the acceleration dv/dt is constant, then the thrust must decrease with time (a smaller force is all that's needed to accelerate a smaller mass).

Bridging Problem

Answers: **(a)** 1.00 m/s to the right **(b)** Elastic
(c) 1.93 m/s at $-30.4°$
(d) 2.31 kg $\cdot$ m/s at 149.6° **(e)** Inelastic
(f) 1.67 m/s in the positive x-direction

9 ROTATION OF RIGID BODIES

? All segments of a rotating wind turbine blade have the same angular velocity. Compared to a given blade segment, how many times greater is the linear speed of a second segment twice as far from the axis of rotation? How many times greater is the radial acceleration?

LEARNING GOALS

By studying this chapter, you will learn:

- How to describe the rotation of a rigid body in terms of angular coordinate, angular velocity, and angular acceleration.

- How to analyze rigid-body rotation when the angular acceleration is constant.

- How to relate the rotation of a rigid body to the linear velocity and linear acceleration of a point on the body.

- The meaning of a body's moment of inertia about a rotation axis, and how it relates to rotational kinetic energy.

- How to calculate the moment of inertia of various bodies.

What do the motions of a compact disc, a Ferris wheel, a circular saw blade, and a ceiling fan have in common? None of these can be represented adequately as a moving *point;* each involves a body that *rotates* about an axis that is stationary in some inertial frame of reference.

Rotation occurs at all scales, from the motions of electrons in atoms to the motions of entire galaxies. We need to develop some general methods for analyzing the motion of a rotating body. In this chapter and the next we consider bodies that have definite size and definite shape, and that in general can have rotational as well as translational motion.

Real-world bodies can be very complicated; the forces that act on them can deform them—stretching, twisting, and squeezing them. We'll neglect these deformations for now and assume that the body has a perfectly definite and unchanging shape and size. We call this idealized model a **rigid body.** This chapter and the next are mostly about rotational motion of a rigid body.

We begin with kinematic language for *describing* rotational motion. Next we look at the kinetic energy of rotation, the key to using energy methods for rotational motion. Then in Chapter 10 we'll develop dynamic principles that relate the forces on a body to its rotational motion.

9.1 Angular Velocity and Acceleration

In analyzing rotational motion, let's think first about a rigid body that rotates about a *fixed axis*—an axis that is at rest in some inertial frame of reference and does not change direction relative to that frame. The rotating rigid body might be a motor shaft, a chunk of beef on a barbecue skewer, or a merry-go-round.

Figure 9.1 shows a rigid body (in this case, the indicator needle of a speedometer) rotating about a fixed axis. The axis passes through point O and is

9.1 A speedometer needle (an example of a rigid body) rotating counterclockwise about a fixed axis.

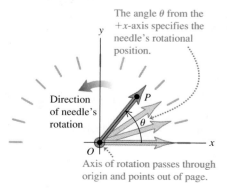

The angle θ from the $+x$-axis specifies the needle's rotational position.

Direction of needle's rotation

Axis of rotation passes through origin and points out of page.

perpendicular to the plane of the diagram, which we choose to call the *xy*-plane. One way to describe the rotation of this body would be to choose a particular point *P* on the body and to keep track of the *x*- and *y*-coordinates of this point. This isn't a terribly convenient method, since it takes two numbers (the two coordinates *x* and *y*) to specify the rotational position of the body. Instead, we notice that the line *OP* is fixed in the body and rotates with it. The angle θ that this line makes with the $+x$-axis describes the rotational position of the body; we will use this single quantity θ as a *coordinate* for rotation.

The angular coordinate θ of a rigid body rotating around a fixed axis can be positive or negative. If we choose positive angles to be measured counterclockwise from the positive *x*-axis, then the angle θ in Fig. 9.1 is positive. If we instead choose the positive rotation direction to be clockwise, then θ in Fig. 9.1 is negative. When we considered the motion of a particle along a straight line, it was essential to specify the direction of positive displacement along that line; when we discuss rotation around a fixed axis, it's just as essential to specify the direction of positive rotation.

To describe rotational motion, the most natural way to measure the angle θ is not in degrees, but in **radians.** As shown in Fig. 9.2a, one radian (1 rad) is the angle subtended at the center of a circle by an arc with a length equal to the radius of the circle. In Fig. 9.2b an angle θ is subtended by an arc of length *s* on a circle of radius *r*. The value of θ (in radians) is equal to *s* divided by *r*:

$$\theta = \frac{s}{r} \quad \text{or} \quad s = r\theta \quad\quad (9.1)$$

An angle in radians is the ratio of two lengths, so it is a pure number, without dimensions. If $s = 3.0$ m and $r = 2.0$ m, then $\theta = 1.5$, but we will often write this as 1.5 rad to distinguish it from an angle measured in degrees or revolutions.

The circumference of a circle (that is, the arc length all the way around the circle) is 2π times the radius, so there are 2π (about 6.283) radians in one complete revolution ($360°$). Therefore

$$1 \text{ rad} = \frac{360°}{2\pi} = 57.3°$$

Similarly, $180° = \pi$ rad, $90° = \pi/2$ rad, and so on. If we had insisted on measuring the angle θ in degrees, we would have needed to include an extra factor of $(2\pi/360)$ on the right-hand side of $s = r\theta$ in Eq. (9.1). By measuring angles in radians, we keep the relationship between angle and distance along an arc as simple as possible.

Angular Velocity

The coordinate θ shown in Fig. 9.1 specifies the rotational position of a rigid body at a given instant. We can describe the rotational *motion* of such a rigid body in terms of the rate of change of θ. We'll do this in an analogous way to our description of straight-line motion in Chapter 2. In Fig. 9.3a, a reference line *OP* in a rotating body makes an angle θ_1 with the $+x$-axis at time t_1. At a later time t_2 the angle has changed to θ_2. We define the **average angular velocity** ω_{av-z} (the Greek letter omega) of the body in the time interval $\Delta t = t_2 - t_1$ as the ratio of the **angular displacement** $\Delta\theta = \theta_2 - \theta_1$ to Δt:

$$\omega_{av-z} = \frac{\theta_2 - \theta_1}{t_2 - t_1} = \frac{\Delta\theta}{\Delta t} \quad\quad (9.2)$$

9.2 Measuring angles in radians.

(a)

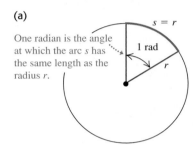

One radian is the angle at which the arc *s* has the same length as the radius *r*.

$s = r$

1 rad

(b)

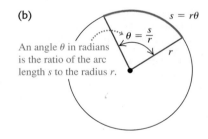

An angle θ in radians is the ratio of the arc length *s* to the radius *r*.

$s = r\theta$

$\theta = \dfrac{s}{r}$

9.3 (a) Angular displacement $\Delta\theta$ of a rotating body. (b) Every part of a rotating rigid body has the same average angular velocity $\Delta\theta/\Delta t$.

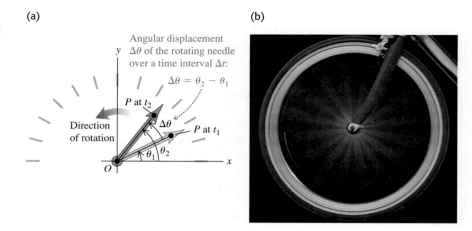

(a)

Angular displacement
$\Delta\theta$ of the rotating needle
over a time interval Δt:

$\Delta\theta = \theta_2 - \theta_1$

P at t_2

Direction of rotation

$\Delta\theta$

P at t_1

θ_1 θ_2

O

(b)

The subscript z indicates that the body in Fig. 9.3a is rotating about the z-axis, which is perpendicular to the plane of the diagram. The **instantaneous angular velocity** ω_z is the limit of $\omega_{\text{av-}z}$ as Δt approaches zero—that is, the derivative of θ with respect to t:

$$\omega_z = \lim_{\Delta t \to 0} \frac{\Delta\theta}{\Delta t} = \frac{d\theta}{dt} \qquad \text{(definition of angular velocity)} \qquad (9.3)$$

When we refer simply to "angular velocity," we mean the instantaneous angular velocity, not the average angular velocity.

The angular velocity ω_z can be positive or negative, depending on the direction in which the rigid body is rotating (Fig. 9.4). The angular *speed* ω, which we will use extensively in Sections 9.3 and 9.4, is the magnitude of angular velocity. Like ordinary (linear) speed v, the angular speed is never negative.

9.4 A rigid body's average angular velocity (shown here) and instantaneous angular velocity can be positive or negative.

Counterclockwise rotation positive:
$\Delta\theta > 0$, so
$\omega_{\text{av-}z} = \Delta\theta/\Delta t > 0$

Clockwise rotation negative:
$\Delta\theta < 0$, so
$\omega_{\text{av-}z} = \Delta\theta/\Delta t < 0$

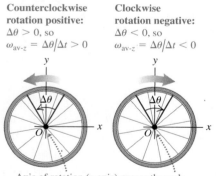

Axis of rotation (z-axis) passes through origin and points out of page.

CAUTION **Angular velocity vs. linear velocity** Keep in mind the distinction between angular velocity ω_z and ordinary velocity, or *linear velocity*, v_x (see Section 2.2). If an object has a velocity v_x, the object as a whole is *moving* along the x-axis. By contrast, if an object has an angular velocity ω_z, then it is *rotating* around the z-axis. We do *not* mean that the object is moving along the z-axis.

Different points on a rotating rigid body move different distances in a given time interval, depending on how far each point lies from the rotation axis. But because the body is rigid, *all* points rotate through the same angle in the same time (Fig. 9.3b). Hence *at any instant, every part of a rotating rigid body has the same angular velocity*. The angular velocity is positive if the body is rotating in the direction of increasing θ and negative if it is rotating in the direction of decreasing θ.

If the angle θ is in radians, the unit of angular velocity is the radian per second (rad/s). Other units, such as the revolution per minute (rev/min or rpm), are often used. Since 1 rev = 2π rad, two useful conversions are

$$1 \text{ rev/s} = 2\pi \text{ rad/s} \qquad \text{and} \qquad 1 \text{ rev/min} = 1 \text{ rpm} = \frac{2\pi}{60} \text{ rad/s}$$

That is, 1 rad/s is about 10 rpm.

Example 9.1 Calculating angular velocity

The angular position θ of a 0.36-m-diameter flywheel is given by

$$\theta = (2.0 \text{ rad/s}^3)t^3$$

(a) Find θ, in radians and in degrees, at $t_1 = 2.0$ s and $t_2 = 5.0$ s. (b) Find the distance that a particle on the flywheel rim moves over the time interval from $t_1 = 2.0$ s to $t_2 = 5.0$ s. (c) Find the average angular velocity, in rad/s and in rev/min, over that interval. (d) Find the instantaneous angular velocities at $t_1 = 2.0$ s and $t_2 = 5.0$ s.

SOLUTION

IDENTIFY and SET UP: We can find the target variables θ_1 (the angular position at time t_1), θ_2 (the angular position at time t_2), and the angular displacement $\Delta\theta = \theta_2 - \theta_1$ from the given expression. Knowing $\Delta\theta$, we'll find the distance traveled and the average angular velocity between t_1 and t_2 using Eqs. (9.1) and (9.2), respectively. To find the instantaneous angular velocities ω_{1z} (at time t_1) and ω_{2z} (at time t_2), we'll take the derivative of the given equation for θ with respect to time, as in Eq. (9.3).

EXECUTE: (a) We substitute the values of t into the equation for θ:

$$\theta_1 = (2.0 \text{ rad/s}^3)(2.0 \text{ s})^3 = 16 \text{ rad}$$

$$= (16 \text{ rad})\frac{360°}{2\pi \text{ rad}} = 920°$$

$$\theta_2 = (2.0 \text{ rad/s}^3)(5.0 \text{ s})^3 = 250 \text{ rad}$$

$$= (250 \text{ rad})\frac{360°}{2\pi \text{ rad}} = 14{,}000°$$

(b) During the interval from t_1 to t_2 the flywheel's angular displacement is $\Delta\theta = \theta_2 - \theta_1 = 250 \text{ rad} - 16 \text{ rad} = 234 \text{ rad}$.

The radius r is half the diameter, or 0.18 m. To use Eq. (9.1), the angles *must* be expressed in radians:

$$s = r\theta_2 - r\theta_1 = r\Delta\theta = (0.18 \text{ m})(234 \text{ rad}) = 42 \text{ m}$$

We can drop "radians" from the unit for s because θ is a pure, dimensionless number; the distance s is measured in meters, the same as r.

(c) From Eq. (9.2),

$$\omega_{av\text{-}z} = \frac{\theta_2 - \theta_1}{t_2 - t_1} = \frac{250 \text{ rad} - 16 \text{ rad}}{5.0 \text{ s} - 2.0 \text{ s}} = 78 \text{ rad/s}$$

$$= \left(78\frac{\text{rad}}{\text{s}}\right)\left(\frac{1 \text{ rev}}{2\pi \text{ rad}}\right)\left(\frac{60 \text{ s}}{1 \text{ min}}\right) = 740 \text{ rev/min}$$

(d) From Eq. (9.3),

$$\omega_z = \frac{d\theta}{dt} = \frac{d}{dt}[(2.0 \text{ rad/s}^3)t^3] = (2.0 \text{ rad/s}^3)(3t^2)$$

$$= (6.0 \text{ rad/s}^3)t^2$$

At times $t_1 = 2.0$ s and $t_2 = 5.0$ s we have

$$\omega_{1z} = (6.0 \text{ rad/s}^3)(2.0 \text{ s})^2 = 24 \text{ rad/s}$$

$$\omega_{2z} = (6.0 \text{ rad/s}^3)(5.0 \text{ s})^2 = 150 \text{ rad/s}$$

EVALUATE: The angular velocity $\omega_z = (6.0 \text{ rad/s}^3)t^2$ increases with time. Our results are consistent with this; the instantaneous angular velocity at the end of the interval ($\omega_{2z} = 150 \text{ rad/s}$) is greater than at the beginning ($\omega_{1z} = 24 \text{ rad/s}$), and the average angular velocity $\omega_{av\text{-}z} = 78 \text{ rad/s}$ over the interval is intermediate between these two values.

Angular Velocity As a Vector

As we have seen, our notation for the angular velocity ω_z about the z-axis is reminiscent of the notation v_x for the ordinary velocity along the x-axis (see Section 2.2). Just as v_x is the x-component of the velocity vector $\vec{v}$, ω_z is the z-component of an angular velocity *vector* $\vec{\omega}$ directed along the axis of rotation. As Fig. 9.5a shows, the direction of $\vec{\omega}$ is given by the right-hand rule that we used to define the vector

(a)

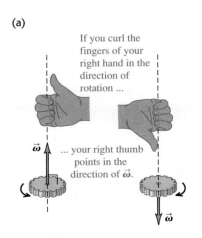

If you curl the fingers of your right hand in the direction of rotation ...

... your right thumb points in the direction of $\vec{\omega}$.

$\vec{\omega}$

(b)

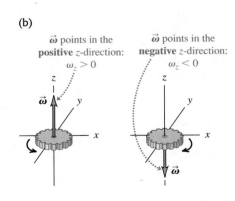

$\vec{\omega}$ points in the **positive** z-direction: $\omega_z > 0$

$\vec{\omega}$ points in the **negative** z-direction: $\omega_z < 0$

9.5 (a) The right-hand rule for the direction of the angular velocity vector $\vec{\omega}$. Reversing the direction of rotation reverses the direction of $\vec{\omega}$. (b) The sign of ω_z for rotation along the z-axis.

product in Section 1.10. If the rotation is about the z-axis, then $\vec{\boldsymbol{\omega}}$ has only a z-component; this component is positive if $\vec{\boldsymbol{\omega}}$ is along the positive z-axis and negative if $\vec{\boldsymbol{\omega}}$ is along the negative z-axis (Fig. 9.5b).

The vector formulation is especially useful in situations in which the direction of the rotation axis *changes*. We'll examine such situations briefly at the end of Chapter 10. In this chapter, however, we'll consider only situations in which the rotation axis is fixed. Hence throughout this chapter we'll use "angular velocity" to refer to ω_z, the component of the angular velocity vector $\vec{\boldsymbol{\omega}}$ along the axis.

Angular Acceleration

When the angular velocity of a rigid body changes, it has an *angular acceleration*. When you pedal your bicycle harder to make the wheels turn faster or apply the brakes to bring the wheels to a stop, you're giving the wheels an angular acceleration. You also impart an angular acceleration whenever you change the rotation speed of a piece of spinning machinery such as an automobile engine's crankshaft.

If ω_{1z} and ω_{2z} are the instantaneous angular velocities at times t_1 and t_2, we define the **average angular acceleration** $\alpha_{\text{av-}z}$ over the interval $\Delta t = t_2 - t_1$ as the change in angular velocity divided by Δt (Fig. 9.6):

$$\alpha_{\text{av-}z} = \frac{\omega_{2z} - \omega_{1z}}{t_2 - t_1} = \frac{\Delta\omega_z}{\Delta t} \tag{9.4}$$

9.6 Calculating the average angular acceleration of a rotating rigid body.

The average angular acceleration is the change in angular velocity divided by the time interval:
$$\alpha_{\text{av-}z} = \frac{\omega_{2z} - \omega_{1z}}{t_2 - t_1} = \frac{\Delta\omega_z}{\Delta t}$$

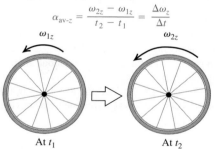

At t_1 At t_2

The **instantaneous angular acceleration** α_z is the limit of $\alpha_{\text{av-}z}$ as $\Delta t \rightarrow 0$:

$$\alpha_z = \lim_{\Delta t \to 0} \frac{\Delta\omega_z}{\Delta t} = \frac{d\omega_z}{dt} \quad \text{(definition of angular acceleration)} \tag{9.5}$$

The usual unit of angular acceleration is the radian per second per second, or rad/s^2. From now on we will use the term "angular acceleration" to mean the instantaneous angular acceleration rather than the average angular acceleration.

Because $\omega_z = d\theta/dt$, we can also express angular acceleration as the second derivative of the angular coordinate:

$$\alpha_z = \frac{d}{dt}\frac{d\theta}{dt} = \frac{d^2\theta}{dt^2} \tag{9.6}$$

You have probably noticed that we are using Greek letters for angular kinematic quantities: θ for angular position, ω_z for angular velocity, and α_z for angular acceleration. These are analogous to x for position, v_x for velocity, and a_x for acceleration, respectively, in straight-line motion. In each case, velocity is the rate of change of position with respect to time and acceleration is the rate of change of velocity with respect to time. We will sometimes use the terms "*linear* velocity" and "*linear* acceleration" for the familiar quantities we defined in Chapters 2 and 3 to distinguish clearly between these and the *angular* quantities introduced in this chapter.

In rotational motion, if the angular acceleration α_z is positive, then the angular velocity ω_z is increasing; if α_z is negative, then ω_z is decreasing. The rotation is speeding up if α_z and ω_z have the same sign and slowing down if α_z and ω_z have opposite signs. (These are exactly the same relationships as those between *linear* acceleration a_x and *linear* velocity v_x for straight-line motion; see Section 2.3.)

Example 9.2	**Calculating angular acceleration**

For the flywheel of Example 9.1, (a) find the average angular acceleration between $t_1 = 2.0$ s and $t_2 = 5.0$ s. (b) Find the instantaneous angular accelerations at $t_1 = 2.0$ s and $t_2 = 5.0$ s.

SOLUTION

IDENTIFY and SET UP: We use Eq. (9.4) for the average angular acceleration $\alpha_{av\text{-}z}$ and Eq. (9.5) for the instantaneous angular acceleration α_z.

EXECUTE: (a) From Example 9.1, the values of ω_z at the two times are

$$\omega_{1z} = 24 \text{ rad/s} \qquad \omega_{2z} = 150 \text{ rad/s}$$

From Eq. (9.4), the average angular acceleration is

$$\alpha_{av\text{-}z} = \frac{150 \text{ rad/s} - 24 \text{ rad/s}}{5.0 \text{ s} - 2.0 \text{ s}} = 42 \text{ rad/s}^2$$

(b) From Eq. (9.5), the value of α_z at any time t is

$$\alpha_z = \frac{d\omega_z}{dt} = \frac{d}{dt}[(6.0 \text{ rad/s}^3)(t^2)] = (6.0 \text{ rad/s}^3)(2t)$$

$$= (12 \text{ rad/s}^3)t$$

Hence

$$\alpha_{1z} = (12 \text{ rad/s}^3)(2.0 \text{ s}) = 24 \text{ rad/s}^2$$

$$\alpha_{2z} = (12 \text{ rad/s}^3)(5.0 \text{ s}) = 60 \text{ rad/s}^2$$

EVALUATE: Note that the angular acceleration is *not* constant in this situation. The angular velocity ω_z is always increasing because α_z is always positive. Furthermore, the rate at which the angular velocity increases is itself increasing, since α_z increases with time.

Angular Acceleration As a Vector

Just as we did for angular velocity, it's useful to define an angular acceleration *vector* $\vec{\alpha}$. Mathematically, $\vec{\alpha}$ is the time derivative of the angular velocity vector $\vec{\omega}$. If the object rotates around the fixed z-axis, then $\vec{\alpha}$ has only a z-component; the quantity α_z is just that component. In this case, $\vec{\alpha}$ is in the same direction as $\vec{\omega}$ if the rotation is speeding up and opposite to $\vec{\omega}$ if the rotation is slowing down (Fig. 9.7).

The angular acceleration vector will be particularly useful in Chapter 10 when we discuss what happens when the rotation axis can change direction. In this chapter, however, the rotation axis will always be fixed and we need use only the z-component α_z.

9.7 When the rotation axis is fixed, the angular acceleration and angular velocity vectors both lie along that axis.

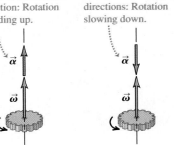

$\vec{\alpha}$ and $\vec{\omega}$ in the **same** direction: Rotation speeding up.

$\vec{\alpha}$ and $\vec{\omega}$ in the **opposite** directions: Rotation slowing down.

Test Your Understanding of Section 9.1
The figure shows a graph of ω_z and α_z versus time for a particular rotating body. (a) During which time intervals is the rotation speeding up? (i) $0 < t < 2$ s; (ii) 2 s $< t < 4$ s; (iii) 4 s $< t < 6$ s. (b) During which time intervals is the rotation slowing down? (i) $0 < t < 2$ s; (ii) 2 s $< t < 4$ s; (iii) 4 s $< t < 6$ s.

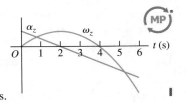

Application **Rotational Motion in Bacteria**
Escherichia coli bacteria (about 2 μm by 0.5 μm) are found in the lower intestines of humans and other warm-blooded animals. The bacteria swim by rotating their long, corkscrew-shaped flagella, which act like the blades of a propeller. Each flagellum is powered by a remarkable protein motor at its base. The motor can rotate the flagellum at angular speeds from 200 to 1000 rev/min (about 20 to 100 rad/s) and can vary its speed to give the flagellum an angular acceleration.

Flagella

9.2 Rotation with Constant Angular Acceleration

In Chapter 2 we found that straight-line motion is particularly simple when the acceleration is constant. This is also true of rotational motion about a fixed axis. When the angular acceleration is constant, we can derive equations for angular velocity and angular position using exactly the same procedure that we used for straight-line motion in Section 2.4. In fact, the equations we are about to derive are identical to Eqs. (2.8), (2.12), (2.13), and (2.14) if we replace x with θ, v_x with ω_z, and a_x with α_z. We suggest that you review Section 2.4 before continuing.

Let ω_{0z} be the angular velocity of a rigid body at time $t = 0$, and let ω_z be its angular velocity at any later time t. The angular acceleration α_z is constant and equal to the average value for any interval. Using Eq. (9.4) with the interval from 0 to t, we find

$$\alpha_z = \frac{\omega_z - \omega_{0z}}{t - 0} \qquad \text{or}$$

$$\omega_z = \omega_{0z} + \alpha_z t \qquad \text{(constant angular acceleration only)} \qquad (9.7)$$

The product $\alpha_z t$ is the total change in ω_z between $t = 0$ and the later time t; the angular velocity ω_z at time t is the sum of the initial value ω_{0z} and this total change.

With constant angular acceleration, the angular velocity changes at a uniform rate, so its average value between 0 and t is the average of the initial and final values:

$$\omega_{\text{av-}z} = \frac{\omega_{0z} + \omega_z}{2} \qquad (9.8)$$

We also know that $\omega_{\text{av-}z}$ is the total angular displacement $(\theta - \theta_0)$ divided by the time interval $(t - 0)$:

$$\omega_{\text{av-}z} = \frac{\theta - \theta_0}{t - 0} \qquad (9.9)$$

When we equate Eqs. (9.8) and (9.9) and multiply the result by t, we get

$$\theta - \theta_0 = \tfrac{1}{2}(\omega_{0z} + \omega_z)t \quad \text{(constant angular acceleration only)} \qquad (9.10)$$

To obtain a relationship between θ and t that doesn't contain ω_z, we substitute Eq. (9.7) into Eq. (9.10):

$$\theta - \theta_0 = \tfrac{1}{2}[\omega_{0z} + (\omega_{0z} + \alpha_z t)]t \quad \text{or}$$

$$\theta = \theta_0 + \omega_{0z}t + \tfrac{1}{2}\alpha_z t^2 \quad \text{(constant angular acceleration only)} \qquad (9.11)$$

That is, if at the initial time $t = 0$ the body is at angular position θ_0 and has angular velocity ω_{0z}, then its angular position θ at any later time t is the sum of three terms: its initial angular position θ_0, plus the rotation $\omega_{0z}t$ it would have if the angular velocity were constant, plus an additional rotation $\tfrac{1}{2}\alpha_z t^2$ caused by the changing angular velocity.

Following the same procedure as for straight-line motion in Section 2.4, we can combine Eqs. (9.7) and (9.11) to obtain a relationship between θ and ω_z that does not contain t. We invite you to work out the details, following the same procedure we used to get Eq. (2.13). (See Exercise 9.12.) In fact, because of the perfect analogy between straight-line and rotational quantities, we can simply take Eq. (2.13) and replace each straight-line quantity by its rotational analog. We get

$$\omega_z^2 = \omega_{0z}^2 + 2\alpha_z(\theta - \theta_0) \quad \text{(constant angular acceleration only)} \qquad (9.12)$$

CAUTION **Constant angular acceleration** Keep in mind that all of these results are valid *only* when the angular acceleration α_z is *constant;* be careful not to try to apply them to problems in which α_z is *not* constant. Table 9.1 shows the analogy between Eqs. (9.7), (9.10), (9.11), and (9.12) for fixed-axis rotation with constant angular acceleration and the corresponding equations for straight-line motion with constant linear acceleration. ▮

Table 9.1 Comparison of Linear and Angular Motion with Constant Acceleration

Straight-Line Motion with Constant Linear Acceleration		Fixed-Axis Rotation with Constant Angular Acceleration	
$a_x = \text{constant}$		$\alpha_z = \text{constant}$	
$v_x = v_{0x} + a_x t$	(2.8)	$\omega_z = \omega_{0z} + \alpha_z t$	(9.7)
$x = x_0 + v_{0x}t + \tfrac{1}{2}a_x t^2$	(2.12)	$\theta = \theta_0 + \omega_{0z}t + \tfrac{1}{2}\alpha_z t^2$	(9.11)
$v_x^2 = v_{0x}^2 + 2a_x(x - x_0)$	(2.13)	$\omega_z^2 = \omega_{0z}^2 + 2\alpha_z(\theta - \theta_0)$	(9.12)
$x - x_0 = \tfrac{1}{2}(v_{0x} + v_x)t$	(2.14)	$\theta - \theta_0 = \tfrac{1}{2}(\omega_{0z} + \omega_z)t$	(9.10)

Example 9.3 **Rotation with constant angular acceleration**

You have finished watching a movie on Blu-ray and the disc is slowing to a stop. The disc's angular velocity at $t = 0$ is 27.5 rad/s, and its angular acceleration is a constant -10.0 rad/s^2. A line PQ on the disc's surface lies along the $+x$-axis at $t = 0$ (Fig. 9.8). (a) What is the disc's angular velocity at $t = 0.300$ s? (b) What angle does the line PQ make with the $+x$-axis at this time?

SOLUTION

IDENTIFY and SET UP: The angular acceleration of the disc is constant, so we can use any of the equations derived in this section (Table 9.1). Our target variables are the angular velocity ω_z and the angular displacement θ at $t = 0.300$ s. Given $\omega_{0z} = 27.5$ rad/s, $\theta_0 = 0$, and $\alpha_z = -10.0$ rad/s^2, it's easiest to use Eqs. (9.7) and (9.11) to find the target variables.

9.8 A line PQ on a rotating Blu-ray disc at $t = 0$.

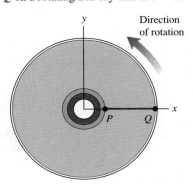

EXECUTE: (a) From Eq. (9.7), at $t = 0.300$ s we have

$$\omega_z = \omega_{0z} + \alpha_z t = 27.5 \text{ rad/s} + (-10.0 \text{ rad/s}^2)(0.300 \text{ s})$$
$$= 24.5 \text{ rad/s}$$

(b) From Eq. (9.11),

$$\theta = \theta_0 + \omega_{0z}t + \tfrac{1}{2}\alpha_z t^2$$
$$= 0 + (27.5 \text{ rad/s})(0.300 \text{ s}) + \tfrac{1}{2}(-10.0 \text{ rad/s}^2)(0.300 \text{ s})^2$$
$$= 7.80 \text{ rad} = 7.80 \text{ rad}\left(\frac{1 \text{ rev}}{2\pi \text{ rad}}\right) = 1.24 \text{ rev}$$

The disc has turned through one complete revolution plus an additional 0.24 revolution—that is, through 360° plus (0.24 rev) (360°/rev) = 87°. Hence the line PQ makes an angle of 87° with the $+x$-axis.

EVALUATE: Our answer to part (a) tells us that the disc's angular velocity has decreased, as it should since $\alpha_z < 0$. We can use our result for ω_z from part (a) with Eq. (9.12) to check our result for θ from part (b). To do so, we solve Eq. (9.12) for θ:

$$\omega_z^2 = \omega_{0z}^2 + 2\alpha_z(\theta - \theta_0)$$
$$\theta = \theta_0 + \left(\frac{\omega_z^2 - \omega_{0z}^2}{2\alpha_z}\right)$$
$$= 0 + \frac{(24.5 \text{ rad/s})^2 - (27.5 \text{ rad/s})^2}{2(-10.0 \text{ rad/s}^2)} = 7.80 \text{ rad}$$

This agrees with our previous result from part (b).

Test Your Understanding of Section 9.2 Suppose the disc in Example 9.3 was initially spinning at twice the rate (55.0 rad/s rather than 27.5 rad/s) and slowed down at twice the rate (-20.0 rad/s^2 rather than -10.0 rad/s^2). (a) Compared to the situation in Example 9.3, how long would it take the disc to come to a stop? (i) the same amount of time; (ii) twice as much time; (iii) 4 times as much time; (iv) $\tfrac{1}{2}$ as much time; (v) $\tfrac{1}{4}$ as much time. (b) Compared to the situation in Example 9.3, through how many revolutions would the disc rotate before coming to a stop? (i) the same number of revolutions; (ii) twice as many revolutions; (iii) 4 times as many revolutions; (iv) $\tfrac{1}{2}$ as many revolutions; (v) $\tfrac{1}{4}$ as many revolutions. ❙

9.3 Relating Linear and Angular Kinematics

How do we find the linear speed and acceleration of a particular point in a rotating rigid body? We need to answer this question to proceed with our study of rotation. For example, to find the kinetic energy of a rotating body, we have to start from $K = \tfrac{1}{2}mv^2$ for a particle, and this requires knowing the speed v for each particle in the body. So it's worthwhile to develop general relationships between the *angular* speed and acceleration of a rigid body rotating about a fixed axis and the *linear* speed and acceleration of a specific point or particle in the body.

Linear Speed in Rigid-Body Rotation

When a rigid body rotates about a fixed axis, every particle in the body moves in a circular path. The circle lies in a plane perpendicular to the axis and is centered on the axis. The speed of a particle is directly proportional to the body's angular

9.9 A rigid body rotating about a fixed axis through point O.

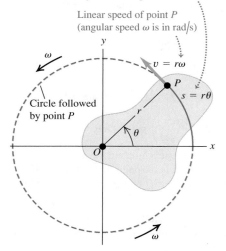

Distance through which point P on the body moves (angle θ is in radians)

Linear speed of point P (angular speed ω is in rad/s)

$v = r\omega$

P

$s = r\theta$

Circle followed by point P

r

θ

O

ω

ω

Mastering**PHYSICS**

PhET: Ladybug Revolution

9.10 A rigid body whose rotation is speeding up. The acceleration of point P has a component a_{rad} toward the rotation axis (perpendicular to $\vec{v}$) and a component a_{tan} along the circle that point P follows (parallel to $\vec{v}$).

Radial and tangential acceleration components:
• $a_{rad} = \omega^2 r$ is point P's centripetal acceleration.
• $a_{tan} = r\alpha$ means that P's rotation is speeding up (the body has angular acceleration).

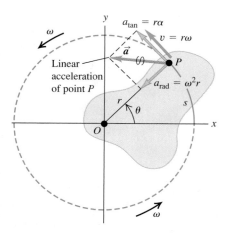

$a_{tan} = r\alpha$

$v = r\omega$

$\vec{a}$

(f)

P

Linear acceleration of point P

$a_{rad} = \omega^2 r$

r

θ

s

O

ω

ω

velocity; the faster the body rotates, the greater the speed of each particle. In Fig. 9.9, point P is a constant distance r from the axis of rotation, so it moves in a circle of radius r. At any time, the angle θ (in radians) and the arc length s are related by

$$s = r\theta$$

We take the time derivative of this, noting that r is constant for any specific particle, and take the absolute value of both sides:

$$\left|\frac{ds}{dt}\right| = r\left|\frac{d\theta}{dt}\right|$$

Now $|ds/dt|$ is the absolute value of the rate of change of arc length, which is equal to the instantaneous *linear* speed v of the particle. Analogously, $|d\theta/dt|$, the absolute value of the rate of change of the angle, is the instantaneous **angular speed** ω—that is, the magnitude of the instantaneous angular velocity in rad/s. Thus

$$v = r\omega \qquad \text{(relationship between linear and angular speeds)} \qquad (9.13)$$

The farther a point is from the axis, the greater its linear speed. The *direction* of the linear velocity *vector* is tangent to its circular path at each point (Fig. 9.9).

CAUTION **Speed vs. velocity** Keep in mind the distinction between the linear and angular *speeds* v and ω, which appear in Eq. (9.13), and the linear and angular *velocities* v_x and ω_z. The quantities without subscripts, v and ω, are never negative; they are the magnitudes of the vectors $\vec{v}$ and $\vec{\omega}$, respectively, and their values tell you only how fast a particle is moving (v) or how fast a body is rotating (ω). The corresponding quantities with subscripts, v_x and ω_z, can be either positive or negative; their signs tell you the direction of the motion.

Linear Acceleration in Rigid-Body Rotation

We can represent the acceleration of a particle moving in a circle in terms of its centripetal and tangential components, a_{rad} and a_{tan} (Fig. 9.10), as we did in Section 3.4. It would be a good idea to review that section now. We found that the **tangential component of acceleration** a_{tan}, the component parallel to the instantaneous velocity, acts to change the *magnitude* of the particle's velocity (i.e., the speed) and is equal to the rate of change of speed. Taking the derivative of Eq. (9.13), we find

$$a_{tan} = \frac{dv}{dt} = r\frac{d\omega}{dt} = r\alpha \qquad \text{(tangential acceleration of a point on a rotating body)} \qquad (9.14)$$

This component of a particle's acceleration is always tangent to the circular path of the particle.

The quantity $\alpha = d\omega/dt$ in Eq. (9.14) is the rate of change of the angular *speed*. It is not quite the same as $\alpha_z = d\omega_z/dt$, which is the rate of change of the angular *velocity*. For example, consider a body rotating so that its angular velocity vector points in the $-z$-direction (see Fig. 9.5b). If the body is gaining angular speed at a rate of 10 rad/s per second, then $\alpha = 10$ rad/s². But ω_z is negative and becoming more negative as the rotation gains speed, so $\alpha_z = -10$ rad/s². The rule for rotation about a fixed axis is that α is equal to α_z if ω_z is positive but equal to $-\alpha_z$ if ω_z is negative.

The component of the particle's acceleration directed toward the rotation axis, the **centripetal component of acceleration** a_{rad}, is associated with the **?**

change of *direction* of the particle's velocity. In Section 3.4 we worked out the relationship $a_{rad} = v^2/r$. We can express this in terms of ω by using Eq. (9.13):

$$a_{rad} = \frac{v^2}{r} = \omega^2 r \qquad \text{(centripetal acceleration of a point on a rotating body)} \qquad (9.15)$$

This is true at each instant, *even when ω and v are not constant.* The centripetal component always points toward the axis of rotation.

The vector sum of the centripetal and tangential components of acceleration of a particle in a rotating body is the linear acceleration $\vec{a}$ (Fig. 9.10).

CAUTION **Use angles in radians in all equations** It's important to remember that Eq. (9.1), $s = r\theta$, is valid *only* when θ is measured in radians. The same is true of any equation derived from this, including Eqs. (9.13), (9.14), and (9.15). When you use these equations, you *must* express the angular quantities in radians, not revolutions or degrees (Fig. 9.11).

Equations (9.1), (9.13), and (9.14) also apply to any particle that has the same tangential velocity as a point in a rotating rigid body. For example, when a rope wound around a circular cylinder unwraps without stretching or slipping, its speed and acceleration at any instant are equal to the speed and tangential acceleration of the point at which it is tangent to the cylinder. The same principle holds for situations such as bicycle chains and sprockets, belts and pulleys that turn without slipping, and so on. We will have several opportunities to use these relationships later in this chapter and in Chapter 10. Note that Eq. (9.15) for the centripetal component a_{rad} is applicable to the rope or chain *only* at points that are in contact with the cylinder or sprocket. Other points do not have the same acceleration toward the center of the circle that points on the cylinder or sprocket have.

9.11 Always use radians when relating linear and angular quantities.

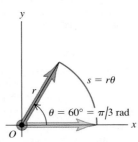

In any equation that relates linear quantities to angular quantities, the angles MUST be expressed in radians ...

RIGHT! ▶ $s = (\pi/3)r$

... never in degrees or revolutions.

WRONG ▶ $s = \cancel{60}r$

Example 9.4   **Throwing a discus**

An athlete whirls a discus in a circle of radius 80.0 cm. At a certain instant, the athlete is rotating at 10.0 rad/s and the angular speed is increasing at 50.0 rad/s². At this instant, find the tangential and centripetal components of the acceleration of the discus and the magnitude of the acceleration.

SOLUTION

IDENTIFY and SET UP: We treat the discus as a particle traveling in a circular path (Fig. 9.12a), so we can use the ideas developed in this section. We are given $r = 0.800$ m, $\omega = 10.0$ rad/s, and $\alpha = 50.0$ rad/s² (Fig. 9.12b). We'll use Eqs. (9.14) and (9.15), respectively, to find the acceleration components a_{tan} and a_{rad}; we'll then find the magnitude a using the Pythagorean theorem.

EXECUTE: From Eqs. (9.14) and (9.15),

$$a_{tan} = r\alpha = (0.800 \text{ m})(50.0 \text{ rad/s}^2) = 40.0 \text{ m/s}^2$$

$$a_{rad} = \omega^2 r = (10.0 \text{ rad/s})^2(0.800 \text{ m}) = 80.0 \text{ m/s}^2$$

Then

$$a = \sqrt{a_{tan}^2 + a_{rad}^2} = 89.4 \text{ m/s}^2$$

EVALUATE: Note that we dropped the unit "radian" from our results for a_{tan}, a_{rad}, and a. We can do this because "radian" is a dimensionless quantity. Can you show that if the angular speed doubles to 20.0 rad/s while α remains the same, the acceleration magnitude a increases to 322 m/s²?

9.12 (a) Whirling a discus in a circle. (b) Our sketch showing the acceleration components for the discus.

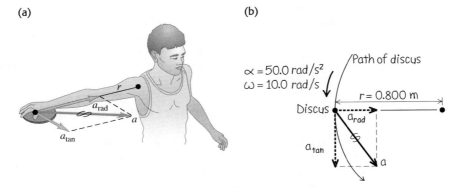

Example 9.5 | **Designing a propeller**

You are designing an airplane propeller that is to turn at 2400 rpm (Fig. 9.13a). The forward airspeed of the plane is to be 75.0 m/s, and the speed of the tips of the propeller blades through the air must not exceed 270 m/s. (This is about 80% of the speed of sound in air. If the speed of the propeller tips were greater than this, they would produce a lot of noise.) (a) What is the maximum possible propeller radius? (b) With this radius, what is the acceleration of the propeller tip?

SOLUTION

IDENTIFY and SET UP: We consider a particle at the tip of the propeller; our target variables are the particle's distance from the axis and its acceleration. The speed of this particle through the air, which cannot exceed 270 m/s, is due to both the propeller's rotation and the forward motion of the airplane. Figure 9.13b shows that the particle's velocity $\vec{v}_{\text{tip}}$ is the vector sum of its tangential velocity due to the propeller's rotation of magnitude $v_{\text{tan}} = \omega r$, given by Eq. (9.13), and the forward velocity of the airplane of magnitude $v_{\text{plane}} = 75.0$ m/s. The propeller rotates in a plane perpendicular to the direction of flight, so $\vec{v}_{\text{tan}}$ and $\vec{v}_{\text{plane}}$ are perpendicular to each other, and we can use the Pythagorean theorem to obtain an expression for v_{tip} from v_{tan} and v_{plane}. We will then set $v_{\text{tip}} = 270$ m/s and solve for the radius r. The angular speed of the propeller is constant, so the acceleration of the propeller tip has only a radial component; we'll find it using Eq. (9.15).

EXECUTE: We first convert ω to rad/s (see Fig. 9.11):

$$\omega = 2400 \text{ rpm} = \left(2400 \frac{\text{rev}}{\text{min}}\right)\left(\frac{2\pi \text{ rad}}{1 \text{ rev}}\right)\left(\frac{1 \text{ min}}{60 \text{ s}}\right)$$
$$= 251 \text{ rad/s}$$

(a) From Fig. 9.13b and Eq. (9.13),

$$v_{\text{tip}}^2 = v_{\text{plane}}^2 + v_{\text{tan}}^2 = v_{\text{plane}}^2 + r^2\omega^2 \quad \text{so}$$

$$r^2 = \frac{v_{\text{tip}}^2 - v_{\text{plane}}^2}{\omega^2} \quad \text{and} \quad r = \frac{\sqrt{v_{\text{tip}}^2 - v_{\text{plane}}^2}}{\omega}$$

If $v_{\text{tip}} = 270$ m/s, the maximum propeller radius is

$$r = \frac{\sqrt{(270 \text{ m/s})^2 - (75.0 \text{ m/s})^2}}{251 \text{ rad/s}} = 1.03 \text{ m}$$

(b) The centripetal acceleration of the particle is

$$a_{\text{rad}} = \omega^2 r = (251 \text{ rad/s})^2(1.03 \text{ m})$$
$$= 6.5 \times 10^4 \text{ m/s}^2 = 6600g$$

The tangential acceleration a_{rad} is zero because the angular speed is constant.

EVALUATE: From $\Sigma\vec{F} = m\vec{a}$, the propeller must exert a force of 6.5×10^4 N on each kilogram of material at its tip! This is why propellers are made out of tough material, usually aluminum alloy.

9.13 (a) A propeller-driven airplane in flight. (b) Our sketch showing the velocity components for the propeller tip.

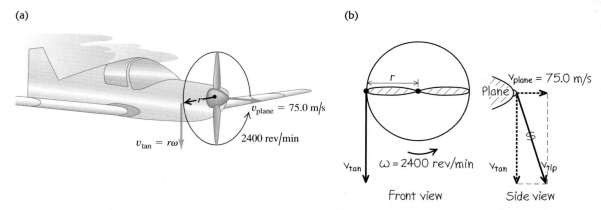

(a)

(b)

Test Your Understanding of Section 9.3 Information is stored on a disc (see Fig. 9.8) in a coded pattern of tiny pits. The pits are arranged in a track that spirals outward toward the rim of the disc. As the disc spins inside a player, the track is scanned at a constant *linear* speed. How must the rotation speed of the disc change as the player's scanning head moves over the track? (i) The rotation speed must increase. (ii) The rotation speed must decrease. (iii) The rotation speed must stay the same.

9.4 Energy in Rotational Motion

A rotating rigid body consists of mass in motion, so it has kinetic energy. As we will see, we can express this kinetic energy in terms of the body's angular speed and a new quantity, called *moment of inertia,* that depends on the body's mass and how the mass is distributed.

To begin, we think of a body as being made up of a large number of particles, with masses $m_1, m_2, \ldots$ at distances $r_1, r_2, \ldots$ from the axis of rotation. We label the particles with the index i: The mass of the ith particle is m_i and its distance from the axis of rotation is r_i. The particles don't necessarily all lie in the same plane, so we specify that r_i is the *perpendicular* distance from the axis to the ith particle.

When a rigid body rotates about a fixed axis, the speed v_i of the ith particle is given by Eq. (9.13), $v_i = r_i\omega$, where ω is the body's angular speed. Different particles have different values of r, but ω is the same for all (otherwise, the body wouldn't be rigid). The kinetic energy of the ith particle can be expressed as

$$\tfrac{1}{2}m_i v_i^2 = \tfrac{1}{2}m_i r_i^2 \omega^2$$

The *total* kinetic energy of the body is the sum of the kinetic energies of all its particles:

$$K = \tfrac{1}{2}m_1 r_1^2 \omega^2 + \tfrac{1}{2}m_2 r_2^2 \omega^2 + \cdots = \sum_i \tfrac{1}{2}m_i r_i^2 \omega^2$$

Taking the common factor $\omega^2/2$ out of this expression, we get

$$K = \tfrac{1}{2}(m_1 r_1^2 + m_2 r_2^2 + \cdots)\omega^2 = \tfrac{1}{2}\Big(\sum_i m_i r_i^2\Big)\omega^2$$

The quantity in parentheses, obtained by multiplying the mass of each particle by the square of its distance from the axis of rotation and adding these products, is denoted by I and is called the **moment of inertia** of the body for this rotation axis:

$$I = m_1 r_1^2 + m_2 r_2^2 + \cdots = \sum_i m_i r_i^2 \qquad \begin{array}{l}\text{(definition of}\\ \text{moment of inertia)}\end{array} \qquad (9.16)$$

The word "moment" means that I depends on how the body's mass is distributed in space; it has nothing to do with a "moment" of time. For a body with a given rotation axis and a given total mass, the greater the distance from the axis to the particles that make up the body, the greater the moment of inertia. In a rigid body, the distances r_i are all constant and I is independent of how the body rotates around the given axis. The SI unit of moment of inertia is the kilogram-meter^2 $(\text{kg} \cdot \text{m}^2)$.

In terms of moment of inertia I, the **rotational kinetic energy** K of a rigid body is

$$K = \tfrac{1}{2}I\omega^2 \qquad \text{(rotational kinetic energy of a rigid body)} \qquad (9.17)$$

The kinetic energy given by Eq. (9.17) is *not* a new form of energy; it's simply the sum of the kinetic energies of the individual particles that make up the rotating rigid body. To use Eq. (9.17), ω *must* be measured in radians per second, not revolutions or degrees per second, to give K in joules. That's because we used $v_i = r_i\omega$ in our derivation.

Equation (9.17) gives a simple physical interpretation of moment of inertia: *The greater the moment of inertia, the greater the kinetic energy of a rigid body rotating with a given angular speed ω.* We learned in Chapter 6 that the kinetic energy of a body equals the amount of work done to accelerate that body from rest. So the greater a body's moment of inertia, the harder it is to start the body rotating if it's at rest and the harder it is to stop its rotation if it's already rotating (Fig. 9.14). For this reason, I is also called the *rotational inertia*.

The next example shows how *changing* the rotation axis can affect the value of I.

Mastering**PHYSICS**

ActivPhysics 7.7: Rotational Inertia

9.14 An apparatus free to rotate around a vertical axis. To vary the moment of inertia, the two equal-mass cylinders can be locked into different positions on the horizontal shaft.

- Mass close to axis
- Small moment of inertia
- Easy to start apparatus rotating

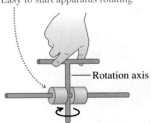

Rotation axis

- Mass farther from axis
- Greater moment of inertia
- Harder to start apparatus rotating

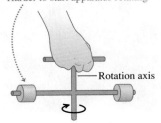

Rotation axis

Example 9.6	**Moments of inertia for different rotation axes**

A machine part (Fig. 9.15) consists of three disks linked by light-weight struts. (a) What is this body's moment of inertia about an axis through the center of disk A, perpendicular to the plane of the diagram? (b) What is its moment of inertia about an axis through the centers of disks B and C? (c) What is the body's kinetic energy if it rotates about the axis through A with angular speed $\omega = 4.0$ rad/s?

SOLUTION

IDENTIFY and SET UP: We'll consider the disks as massive particles located at the centers of the disks, and consider the struts as

9.15 An oddly shaped machine part.

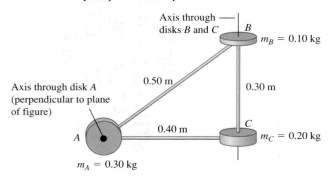

Axis through — disks B and C

B

$m_B = 0.10$ kg

Axis through disk A
(perpendicular to plane
of figure)

0.50 m

0.30 m

0.40 m

C

A

$m_C = 0.20$ kg

$m_A = 0.30$ kg

massless. In parts (a) and (b), we'll use Eq. (9.16) to find the moments of inertia. Given the moment of inertia about axis A, we'll use Eq. (9.17) in part (c) to find the rotational kinetic energy.

EXECUTE: (a) The particle at point A lies *on* the axis through A, so its distance r from the axis is zero and it contributes nothing to the moment of inertia. Hence only B and C contribute, and Eq. (9.16) gives

$$I_A = \sum m_i r_i^2 = (0.10 \text{ kg})(0.50 \text{ m})^2 + (0.20 \text{ kg})(0.40 \text{ m})^2$$
$$= 0.057 \text{ kg} \cdot \text{m}^2$$

(b) The particles at B and C both lie on axis BC, so neither particle contributes to the moment of inertia. Hence only A contributes:

$$I_{BC} = \sum m_i r_i^2 = (0.30 \text{ kg})(0.40 \text{ m})^2 = 0.048 \text{ kg} \cdot \text{m}^2$$

(c) From Eq. (9.17),

$$K_A = \tfrac{1}{2}I_A\omega^2 = \tfrac{1}{2}(0.057 \text{ kg} \cdot \text{m}^2)(4.0 \text{ rad/s})^2 = 0.46 \text{ J}$$

EVALUATE: The moment of inertia about axis A is greater than that about axis BC. Hence of the two axes it's easier to make the machine part rotate about axis BC.

Application Moment of Inertia of a Bird's Wing

When a bird flaps its wings, it rotates the wings up and down around the shoulder. A hummingbird has small wings with a small moment of inertia, so the bird can make its wings move rapidly (up to 70 beats per second). By contrast, the Andean condor (*Vultur gryphus*) has immense wings that are hard to move due to their large moment of inertia. Condors flap their wings at about one beat per second on takeoff, but at most times prefer to soar while holding their wings steady.

CAUTION **Moment of inertia depends on the choice of axis** The results of parts (a) and (b) of Example 9.6 show that the moment of inertia of a body depends on the location and orientation of the axis. It's not enough to just say, "The moment of inertia of this body is 0.048 kg $\cdot$ m^2." We have to be specific and say, "The moment of inertia of this body *about the axis through B and C* is 0.048 kg $\cdot$ m^2." ∎

In Example 9.6 we represented the body as several point masses, and we evaluated the sum in Eq. (9.16) directly. When the body is a *continuous* distribution of matter, such as a solid cylinder or plate, the sum becomes an integral, and we need to use calculus to calculate the moment of inertia. We will give several examples of such calculations in Section 9.6; meanwhile, Table 9.2 gives moments of inertia for several familiar shapes in terms of their masses and dimensions. Each body shown in Table 9.2 is *uniform;* that is, the density has the same value at all points within the solid parts of the body.

CAUTION **Computing the moment of inertia** You may be tempted to try to compute the moment of inertia of a body by assuming that all the mass is concentrated at the center of mass and multiplying the total mass by the square of the distance from the center of mass to the axis. Resist that temptation; it doesn't work! For example, when a uniform thin rod of length L and mass M is pivoted about an axis through one end, perpendicular to the rod, the moment of inertia is $I = ML^2/3$ [case (b) in Table 9.2]. If we took the mass as concentrated at the center, a distance $L/2$ from the axis, we would obtain the *incorrect* result $I = M(L/2)^2 = ML^2/4$. ∎

Now that we know how to calculate the kinetic energy of a rotating rigid body, we can apply the energy principles of Chapter 7 to rotational motion. Here are some points of strategy and some examples.

Table 9.2 Moments of Inertia of Various Bodies

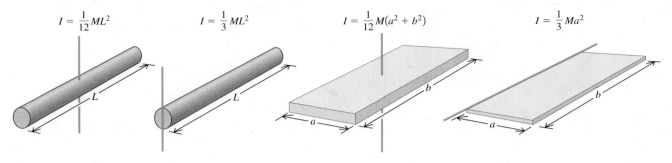

(a) Slender rod, axis through center

$$I = \frac{1}{12}ML^2$$

(b) Slender rod, axis through one end

$$I = \frac{1}{3}ML^2$$

(c) Rectangular plate, axis through center

$$I = \frac{1}{12}M(a^2 + b^2)$$

(d) Thin rectangular plate, axis along edge

$$I = \frac{1}{3}Ma^2$$

(e) Hollow cylinder

$$I = \frac{1}{2}M(R_1^2 + R_2^2)$$

(f) Solid cylinder

$$I = \frac{1}{2}MR^2$$

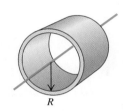

(g) Thin-walled hollow cylinder

$$I = MR^2$$

(h) Solid sphere

$$I = \frac{2}{5}MR^2$$

(i) Thin-walled hollow sphere

$$I = \frac{2}{3}MR^2$$

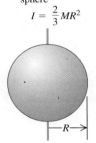

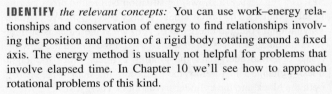

Problem-Solving Strategy 9.1 | Rotational Energy

IDENTIFY *the relevant concepts:* You can use work–energy relationships and conservation of energy to find relationships involving the position and motion of a rigid body rotating around a fixed axis. The energy method is usually not helpful for problems that involve elapsed time. In Chapter 10 we'll see how to approach rotational problems of this kind.

SET UP *the problem* using Problem-Solving Strategy 7.1 (Section 7.1), with the following additions:
5. You can use Eqs. (9.13) and (9.14) in problems involving a rope (or the like) wrapped around a rotating rigid body, if the rope doesn't slip. These equations relate the linear speed and tangential acceleration of a point on the body to the body's angular velocity and angular acceleration. (See Examples 9.7 and 9.8.)
6. Use Table 9.2 to find moments of inertia. Use the parallel-axis theorem, Eq. (9.19) (to be derived in Section 9.5), to find

moments of inertia for rotation about axes parallel to those shown in the table.

EXECUTE *the solution:* Write expressions for the initial and final kinetic and potential energies K_1, K_2, U_1, and U_2 and for the nonconservative work W_{other} (if any), where K_1 and K_2 must now include any rotational kinetic energy $K = \frac{1}{2}I\omega^2$. Substitute these expressions into Eq. (7.14), $K_1 + U_1 + W_{other} = K_2 + U_2$ (if nonconservative work is done), or Eq. (7.11), $K_1 + U_1 = K_2 + U_2$ (if only conservative work is done), and solve for the target variables. It's helpful to draw bar graphs showing the initial and final values of K, U, and $E = K + U$.

EVALUATE *your answer:* Check whether your answer makes physical sense.

Example 9.7 | An unwinding cable I

We wrap a light, nonstretching cable around a solid cylinder of mass 50 kg and diameter 0.120 m, which rotates in frictionless bearings about a stationary horizontal axis (Fig. 9.16). We pull the free end of the cable with a constant 9.0-N force for a distance of 2.0 m; it turns the cylinder as it unwinds without slipping. The cylinder is initially at rest. Find its final angular speed and the final speed of the cable.

SOLUTION

IDENTIFY: We'll solve this problem using energy methods. We'll assume that the cable is massless, so only the cylinder has kinetic energy. There are no changes in gravitational potential energy. There is friction between the cable and the cylinder, but because the cable doesn't slip, there is no motion of the cable relative to the

Continued

9.16 A cable unwinds from a cylinder (side view).

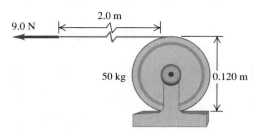

cylinder and no mechanical energy is lost in frictional work. Because the cable is massless, the force that the cable exerts on the cylinder rim is equal to the applied force F.

SET UP: Point 1 is when the cable begins to move. The cylinder starts at rest, so $K_1 = 0$. Point 2 is when the cable has moved a distance $s = 2.0$ m and the cylinder has kinetic energy $K_2 = \frac{1}{2}I\omega^2$. One of our target variables is ω; the other is the speed of the cable at point 2, which is equal to the tangential speed v of the cylinder at that point. We'll use Eq. (9.13) to find v from ω.

EXECUTE: The work done on the cylinder is $W_{\text{other}} = Fs = (9.0 \text{ N})(2.0 \text{ m}) = 18$ J. From Table 9.2 the moment of inertia is

$$I = \tfrac{1}{2}mR^2 = \tfrac{1}{2}(50 \text{ kg})(0.060 \text{ m})^2 = 0.090 \text{ kg} \cdot \text{m}^2$$

(The radius R is half the diameter.) From Eq. (7.14), $K_1 + U_1 + W_{\text{other}} = K_2 + U_2$, so

$$0 + 0 + W_{\text{other}} = \tfrac{1}{2}I\omega^2 + 0$$

$$\omega = \sqrt{\frac{2W_{\text{other}}}{I}} = \sqrt{\frac{2(18 \text{ J})}{0.090 \text{ kg} \cdot \text{m}^2}} = 20 \text{ rad/s}$$

From Eq. (9.13), the final tangential speed of the cylinder, and hence the final speed of the cable, is

$$v = R\omega = (0.060 \text{ m})(20 \text{ rad/s}) = 1.2 \text{ m/s}$$

EVALUATE: If the cable mass is not negligible, some of the 18 J of work would go into the kinetic energy of the cable. Then the cylinder would have less kinetic energy and a lower angular speed than we calculated here.

Example 9.8   An unwinding cable II

We wrap a light, nonstretching cable around a solid cylinder with mass M and radius R. The cylinder rotates with negligible friction about a stationary horizontal axis. We tie the free end of the cable to a block of mass m and release the block from rest at a distance h above the floor. As the block falls, the cable unwinds without stretching or slipping. Find expressions for the speed of the falling block and the angular speed of the cylinder as the block strikes the floor.

SOLUTION

IDENTIFY: As in Example 9.7, the cable doesn't slip and so friction does no work. We assume that the cable is massless, so that the

9.17 Our sketches for this problem.

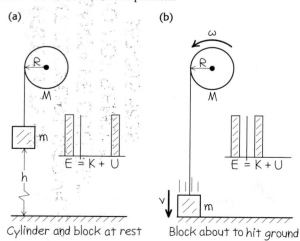

forces it exerts on the cylinder and the block have equal magnitudes. At its upper end the force and displacement are in the same direction, and at its lower end they are in opposite directions, so the cable does no *net* work and $W_{\text{other}} = 0$. Only gravity does work, and mechanical energy is conserved.

SET UP: Figure 9.17a shows the situation before the block begins to fall (point 1). The initial kinetic energy is $K_1 = 0$. We take the gravitational potential energy to be zero when the block is at floor level (point 2), so $U_1 = mgh$ and $U_2 = 0$. (We ignore the gravitational potential energy for the rotating cylinder, since its height doesn't change.) Just before the block hits the floor (Fig. 9.17b), both the block and the cylinder have kinetic energy, so

$$K_2 = \tfrac{1}{2}mv^2 + \tfrac{1}{2}I\omega^2$$

The moment of inertia of the cylinder is $I = \frac{1}{2}MR^2$. Also, $v = R\omega$ since the speed of the falling block must be equal to the tangential speed at the outer surface of the cylinder.

EXECUTE: We use our expressions for K_1, U_1, K_2, and U_2 and the relationship $\omega = v/R$ in Eq. (7.4), $K_1 + U_1 = K_2 + U_2$, and solve for v:

$$0 + mgh = \tfrac{1}{2}mv^2 + \tfrac{1}{2}\left(\tfrac{1}{2}MR^2\right)\left(\frac{v}{R}\right)^2 + 0 = \tfrac{1}{2}\left(m + \tfrac{1}{2}M\right)v^2$$

$$v = \sqrt{\frac{2gh}{1 + M/2m}}$$

The final angular speed of the cylinder is $\omega = v/R$.

EVALUATE: When M is much larger than m, v is very small; when M is much smaller than m, v is nearly equal to $\sqrt{2gh}$, the speed of a body that falls freely from height h. Both of these results are as we would expect.

Gravitational Potential Energy for an Extended Body

In Example 9.8 the cable was of negligible mass, so we could ignore its kinetic energy as well as the gravitational potential energy associated with it. If the mass is *not* negligible, we need to know how to calculate the *gravitational potential energy* associated with such an extended body. If the acceleration of gravity g is the same at all points on the body, the gravitational potential energy is the same as though all the mass were concentrated at the center of mass of the body. Suppose we take the y-axis vertically upward. Then for a body with total mass M, the gravitational potential energy U is simply

$$U = Mgy_{cm} \quad \text{(gravitational potential energy for an extended body)} \quad (9.18)$$

where y_{cm} is the y-coordinate of the center of mass. This expression applies to any extended body, whether it is rigid or not (Fig. 9.18).

To prove Eq. (9.18), we again represent the body as a collection of mass elements m_i. The potential energy for element m_i is m_igy_i, so the total potential energy is

$$U = m_1gy_1 + m_2gy_2 + \cdots = (m_1y_1 + m_2y_2 + \cdots)g$$

But from Eq. (8.28), which defines the coordinates of the center of mass,

$$m_1y_1 + m_2y_2 + \cdots = (m_1 + m_2 + \cdots)y_{cm} = My_{cm}$$

where $M = m_1 + m_2 + \cdots$ is the total mass. Combining this with the above expression for U, we find $U = Mgy_{cm}$ in agreement with Eq. (9.18).

We leave the application of Eq. (9.18) to the problems. We'll make use of this relationship in Chapter 10 in the analysis of rigid-body problems in which the axis of rotation moves.

9.18 In a technique called the "Fosbury flop" after its innovator, this athlete arches her body as she passes over the bar in the high jump. As a result, her center of mass actually passes *under* the bar. This technique requires a smaller increase in gravitational potential energy [Eq. (9.18)] than the older method of straddling the bar.

Test Your Understanding of Section 9.4 Suppose the cylinder and block in Example 9.8 have the same mass, so $m = M$. Just before the block strikes the floor, which statement is correct about the relationship between the kinetic energy of the falling block and the rotational kinetic energy of the cylinder? (i) The block has more kinetic energy than the cylinder. (ii) The block has less kinetic energy than the cylinder. (iii) The block and the cylinder have equal amounts of kinetic energy.

9.5 Parallel-Axis Theorem

We pointed out in Section 9.4 that a body doesn't have just one moment of inertia. In fact, it has infinitely many, because there are infinitely many axes about which it might rotate. But there is a simple relationship between the moment of inertia I_{cm} of a body of mass M about an axis through its center of mass and the moment of inertia I_P about any other axis parallel to the original one but displaced from it by a distance d. This relationship, called the **parallel-axis theorem,** states that

$$I_P = I_{cm} + Md^2 \quad \text{(parallel-axis theorem)} \quad (9.19)$$

To prove this theorem, we consider two axes, both parallel to the z-axis: one through the center of mass and the other through a point P (Fig. 9.19). First we take a very thin slice of the body, parallel to the xy-plane and perpendicular to the z-axis. We take the origin of our coordinate system to be at the center of mass of the body; the coordinates of the center of mass are then $x_{cm} = y_{cm} = z_{cm} = 0$. The axis through the center of mass passes through this thin slice at point O, and the parallel axis passes through point P, whose x- and y-coordinates are (a, b). The distance of this axis from the axis through the center of mass is d, where $d^2 = a^2 + b^2$.

9.19 The mass element m_i has coordinates (x_i, y_i) with respect to an axis of rotation through the center of mass (cm) and coordinates $(x_i - a, y_i - b)$ with respect to the parallel axis through point P.

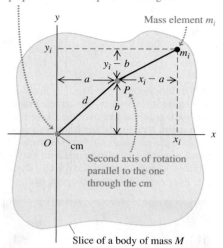

Axis of rotation passing through cm and perpendicular to the plane of the figure

Mass element m_i

Second axis of rotation parallel to the one through the cm

Slice of a body of mass M

We can write an expression for the moment of inertia I_P about the axis through point P. Let m_i be a mass element in our slice, with coordinates (x_i, y_i, z_i). Then the moment of inertia I_{cm} of the slice about the axis through the center of mass (at O) is

$$I_{cm} = \sum_i m_i(x_i^2 + y_i^2)$$

The moment of inertia of the slice about the axis through P is

$$I_P = \sum_i m_i[(x_i - a)^2 + (y_i - b)^2]$$

These expressions don't involve the coordinates z_i measured perpendicular to the slices, so we can extend the sums to include *all* particles in *all* slices. Then I_P becomes the moment of inertia of the *entire* body for an axis through P. We then expand the squared terms and regroup, and obtain

$$I_P = \sum_i m_i(x_i^2 + y_i^2) - 2a\sum_i m_i x_i - 2b\sum_i m_i y_i + (a^2 + b^2)\sum_i m_i$$

The first sum is I_{cm}. From Eq. (8.28), the definition of the center of mass, the second and third sums are proportional to x_{cm} and y_{cm}; these are zero because we have taken our origin to be the center of mass. The final term is d^2 multiplied by the total mass, or Md^2. This completes our proof that $I_P = I_{cm} + Md^2$.

As Eq. (9.19) shows, a rigid body has a lower moment of inertia about an axis through its center of mass than about any other parallel axis. Thus it's easier to start a body rotating if the rotation axis passes through the center of mass. This suggests that it's somehow most natural for a rotating body to rotate about an axis through its center of mass; we'll make this idea more quantitative in Chapter 10.

Example 9.9 **Using the parallel-axis theorem**

A part of a mechanical linkage (Fig. 9.20) has a mass of 3.6 kg. Its moment of inertia I_P about an axis 0.15 m from its center of mass is $I_P = 0.132$ kg·m². What is the moment of inertia I_{cm} about a parallel axis through the center of mass?

9.20 Calculating I_{cm} from a measurement of I_P.

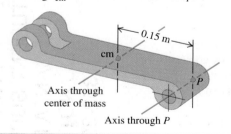

0.15 m

cm

Axis through center of mass

Axis through P

SOLUTION

IDENTIFY, SET UP, and EXECUTE: We'll determine the target variable I_{cm} using the parallel-axis theorem, Eq. (9.19). Rearranging the equation, we obtain

$$I_{cm} = I_P - Md^2 = 0.132 \text{ kg·m}^2 - (3.6 \text{ kg})(0.15 \text{ m})^2$$
$$= 0.051 \text{ kg·m}^2$$

EVALUATE: As we expect, I_{cm} is less than I_P; the moment of inertia for an axis through the center of mass is lower than for any other parallel axis.

Test Your Understanding of Section 9.5 A pool cue is a wooden rod with a uniform composition and tapered with a larger diameter at one end than at the other end. Use the parallel-axis theorem to decide whether a pool cue has a larger moment of inertia (i) for an axis through the thicker end of the rod and perpendicular to the length of the rod, or (ii) for an axis through the thinner end of the rod and perpendicular to the length of the rod. ❙

9.6 Moment-of-Inertia Calculations

If a rigid body is a continuous distribution of mass—like a solid cylinder or a solid sphere—it cannot be represented by a few point masses. In this case the *sum* of masses and distances that defines the moment of inertia [Eq. (9.16)] becomes an *integral*. Imagine dividing the body into elements of mass dm that are very small, so that all points in a particular element are at essentially the same perpendicular distance from the axis of rotation. We call this distance r, as before. Then the moment of inertia is

$$I = \int r^2 \, dm \qquad (9.20)$$

To evaluate the integral, we have to represent r and dm in terms of the same integration variable. When the object is effectively one-dimensional, such as the slender rods (a) and (b) in Table 9.2, we can use a coordinate x along the length and relate dm to an increment dx. For a three-dimensional object it is usually easiest to express dm in terms of an element of volume dV and the *density* ρ of the body. Density is mass per unit volume, $\rho = dm/dV$, so we may also write Eq. (9.20) as

$$I = \int r^2 \rho \, dV$$

This expression tells us that a body's moment of inertia depends on how its density varies within its volume (Fig. 9.21). If the body is uniform in density, then we may take ρ outside the integral:

$$I = \rho \int r^2 \, dV \qquad (9.21)$$

To use this equation, we have to express the volume element dV in terms of the differentials of the integration variables, such as $dV = dx \, dy \, dz$. The element dV must always be chosen so that all points within it are at very nearly the same distance from the axis of rotation. The limits on the integral are determined by the shape and dimensions of the body. For regularly shaped bodies, this integration is often easy to do.

9.21 By measuring small variations in the orbits of satellites, geophysicists can measure the earth's moment of inertia. This tells us how our planet's mass is distributed within its interior. The data show that the earth is far denser at the core than in its outer layers.

Example 9.10   Hollow or solid cylinder, rotating about axis of symmetry

Figure 9.22 shows a hollow cylinder of uniform mass density ρ with length L, inner radius R_1, and outer radius R_2. (It might be a steel cylinder in a printing press.) Using integration, find its moment of inertia about its axis of symmetry.

9.22 Finding the moment of inertia of a hollow cylinder about its symmetry axis.

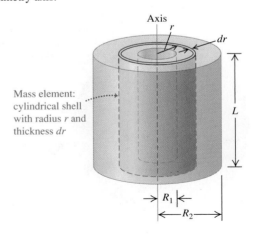

SOLUTION

IDENTIFY and SET UP: We choose as a volume element a thin cylindrical shell of radius r, thickness dr, and length L. All parts of this shell are at very nearly the same distance r from the axis. The volume of the shell is very nearly that of a flat sheet with thickness dr, length L, and width $2\pi r$ (the circumference of the shell). The mass of the shell is

$$dm = \rho \, dV = \rho(2\pi r L \, dr)$$

We'll use this expression in Eq. (9.20), integrating from $r = R_1$ to $r = R_2$.

EXECUTE: From Eq. (9.20), the moment of inertia is

$$I = \int r^2 \, dm = \int_{R_1}^{R_2} r^2 \rho(2\pi r L \, dr)$$

$$= 2\pi\rho L \int_{R_1}^{R_2} r^3 \, dr$$

$$= \frac{2\pi\rho L}{4}(R_2^4 - R_1^4)$$

$$= \frac{\pi\rho L}{2}(R_2^2 - R_1^2)(R_2^2 + R_1^2)$$

(In the last step we used the identity $a^2 - b^2 = (a - b)(a + b)$.) Let's express this result in terms of the total mass M of the body, which is its density ρ multiplied by the total volume V. The cylinder's volume is

$$V = \pi L(R_2^2 - R_1^2)$$

so its total mass M is

$$M = \rho V = \pi L \rho(R_2^2 - R_1^2)$$

Continued

Comparing with the above expression for I, we see that

$$I = \tfrac{1}{2}M(R_1^2 + R_2^2)$$

EVALUATE: Our result agrees with Table 9.2, case (e). If the cylinder is solid, with outer radius $R_2 = R$ and inner radius $R_1 = 0$, its moment of inertia is

$$I = \tfrac{1}{2}MR^2$$

in agreement with case (f). If the cylinder wall is very thin, we have $R_1 \approx R_2 = R$ and the moment of inertia is

$$I = MR^2$$

in agreement with case (g). We could have predicted this last result without calculation; in a thin-walled cylinder, all the mass is at the same distance $r = R$ from the axis, so $I = \int r^2 \, dm = R^2 \int dm = MR^2$.

Example 9.11 Uniform sphere with radius R, axis through center

Find the moment of inertia of a solid sphere of uniform mass density ρ (like a billiard ball) about an axis through its center.

SOLUTION

IDENTIFY and SET UP: We divide the sphere into thin, solid disks of thickness dx (Fig. 9.23), whose moment of inertia we know from Table 9.2, case (f). We'll integrate over these to find the total moment of inertia.

EXECUTE: The radius and hence the volume and mass of a disk depend on its distance x from the center of the sphere. The radius r of the disk shown in Fig. 9.23 is

$$r = \sqrt{R^2 - x^2}$$

Its volume is

$$dV = \pi r^2 \, dx = \pi(R^2 - x^2) \, dx$$

9.23 Finding the moment of inertia of a sphere about an axis through its center.

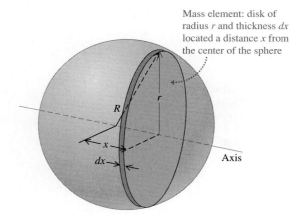

Mass element: disk of radius r and thickness dx located a distance x from the center of the sphere

and so its mass is

$$dm = \rho \, dV = \pi\rho(R^2 - x^2) \, dx$$

From Table 9.2, case (f), the moment of inertia of a disk of radius r and mass dm is

$$dI = \tfrac{1}{2}r^2 \, dm = \tfrac{1}{2}(R^2 - x^2)[\pi\rho(R^2 - x^2) \, dx]$$
$$= \frac{\pi\rho}{2}(R^2 - x^2)^2 \, dx$$

Integrating this expression from $x = 0$ to $x = R$ gives the moment of inertia of the right hemisphere. The total I for the entire sphere, including both hemispheres, is just twice this:

$$I = (2)\frac{\pi\rho}{2} \int_0^R (R^2 - x^2)^2 \, dx$$

Carrying out the integration, we find

$$I = \frac{8\pi\rho R^5}{15}$$

The volume of the sphere is $V = 4\pi R^3/3$, so in terms of its mass M its density is

$$\rho = \frac{M}{V} = \frac{3M}{4\pi R^3}$$

Hence our expression for I becomes

$$I = \left(\frac{8\pi R^5}{15}\right)\left(\frac{3M}{4\pi R^3}\right) = \tfrac{2}{5}MR^2$$

EVALUATE: This is just as in Table 9.2, case (h). Note that the moment of inertia $I = \tfrac{2}{5}MR^2$ of a solid sphere of mass M and radius R is less than the moment of inertia $I = \tfrac{1}{2}MR^2$ of a solid *cylinder* of the same mass and radius, because more of the sphere's mass is located close to the axis.

Test Your Understanding of Section 9.6 Two hollow cylinders have the same inner and outer radii and the same mass, but they have different lengths. One is made of low-density wood and the other of high-density lead. Which cylinder has the greater moment of inertia around its axis of symmetry? (i) the wood cylinder; (ii) the lead cylinder; (iii) the two moments of inertia are equal.

Rotational kinematics: When a rigid body rotates about a stationary axis (usually called the z-axis), its position is described by an angular coordinate θ. The angular velocity ω_z is the time derivative of θ, and the angular acceleration α_z is the time derivative of ω_z or the second derivative of θ. (See Examples 9.1 and 9.2.) If the angular acceleration is constant, then θ, ω_z, and α_z are related by simple kinematic equations analogous to those for straight-line motion with constant linear acceleration. (See Example 9.3.)

$$\omega_z = \lim_{\Delta t \to 0} \frac{\Delta\theta}{\Delta t} = \frac{d\theta}{dt} \tag{9.3}$$

$$\alpha_z = \lim_{\Delta t \to 0} \frac{\Delta\omega_z}{\Delta t} = \frac{d\omega_z}{dt} = \frac{d^2\theta}{dt^2} \tag{9.5}, (9.6)$$

$$\theta = \theta_0 + \omega_{0z}t + \tfrac{1}{2}\alpha_z t^2 \tag{9.11}$$

(constant α_z only)

$$\theta - \theta_0 = \tfrac{1}{2}(\omega_{0z} + \omega_z)t \tag{9.10}$$

(constant α_z only)

$$\omega_z = \omega_{0z} + \alpha_z t \tag{9.7}$$

(constant α_z only)

$$\omega_z^2 = \omega_{0z}^2 + 2\alpha_z(\theta - \theta_0) \tag{9.12}$$

(constant α_z only)

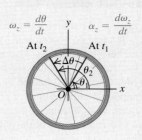

Relating linear and angular kinematics: The angular speed ω of a rigid body is the magnitude of its angular velocity. The rate of change of ω is $\alpha = d\omega/dt$. For a particle in the body a distance r from the rotation axis, the speed v and the components of the acceleration $\vec{a}$ are related to ω and α. (See Examples 9.4 and 9.5.)

$$v = r\omega \tag{9.13}$$

$$a_{\text{tan}} = \frac{dv}{dt} = r\frac{d\omega}{dt} = r\alpha \tag{9.14}$$

$$a_{\text{rad}} = \frac{v^2}{r} = \omega^2 r \tag{9.15}$$

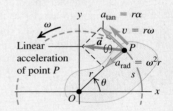

Moment of inertia and rotational kinetic energy: The moment of inertia I of a body about a given axis is a measure of its rotational inertia: The greater the value of I, the more difficult it is to change the state of the body's rotation. The moment of inertia can be expressed as a sum over the particles m_i that make up the body, each of which is at its own perpendicular distance r_i from the axis. The rotational kinetic energy of a rigid body rotating about a fixed axis depends on the angular speed ω and the moment of inertia I for that rotation axis. (See Examples 9.6–9.8.)

$$I = m_1 r_1^2 + m_2 r_2^2 + \cdots$$
$$= \sum_i m_i r_i^2 \tag{9.16}$$

$$K = \tfrac{1}{2}I\omega^2 \tag{9.17}$$

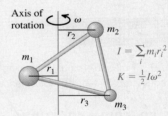

Calculating the moment of inertia: The parallel-axis theorem relates the moments of inertia of a rigid body of mass M about two parallel axes: an axis through the center of mass (moment of inertia I_{cm}) and a parallel axis a distance d from the first axis (moment of inertia I_P). (See Example 9.9.) If the body has a continuous mass distribution, the moment of inertia can be calculated by integration. (See Examples 9.10 and 9.11.)

$$I_P = I_{\text{cm}} + Md^2 \tag{9.19}$$

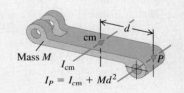

BRIDGING PROBLEM   **A Rotating, Uniform Thin Rod**

Figure 9.24 shows a slender uniform rod with mass M and length L. It might be a baton held by a twirler in a marching band (less the rubber end caps). (a) Use integration to compute its moment of inertia about an axis through O, at an arbitrary distance h from one end. (b) Initially the rod is at rest. It is given a constant angular acceleration of magnitude α around the axis through O. Find how much work is done on the rod in a time t. (c) At time t, what is the *linear* acceleration of the point on the rod farthest from the axis?

9.24 A thin rod with an axis through O.

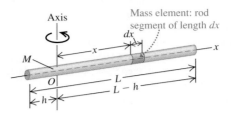

SOLUTION GUIDE

See MasteringPhysics® study area for a Video Tutor solution.

IDENTIFY and SET UP

1. Make a list of the target variables for this problem.
2. To calculate the moment of inertia of the rod, you'll have to divide the rod into infinitesimal elements of mass. If an element has length dx, what is the mass of the element? What are the limits of integration?
3. What is the angular speed of the rod at time t? How does the work required to accelerate the rod from rest to this angular speed compare to the rod's kinetic energy at time t?
4. At time t, does the point on the rod farthest from the axis have a centripetal acceleration? A tangential acceleration? Why or why not?

EXECUTE

5. Do the integration required to find the moment of inertia.
6. Use your result from step 5 to calculate the work done in time t to accelerate the rod from rest.
7. Find the linear acceleration components for the point in question at time t. Use these to find the magnitude of the acceleration.

EVALUATE

8. Check your results for the special cases $h = 0$ (the axis passes through one end of the rod) and $h = L/2$ (the axis passes through the middle of the rod). Are these limits consistent with Table 9.2? With the parallel-axis theorem?
9. Is the acceleration magnitude from step 7 constant? Would you expect it to be?

Problems   For instructor-assigned homework, go to www.masteringphysics.com

•, ••, •••: Problems of increasing difficulty. **CP:** Cumulative problems incorporating material from earlier chapters. **CALC:** Problems requiring calculus. **BIO:** Biosciences problems.

DISCUSSION QUESTIONS

Q9.1 Which of the following formulas is valid if the angular acceleration of an object is *not* constant? Explain your reasoning in each case. (a) $v = r\omega$; (b) $a_{\text{tan}} = r\alpha$; (c) $\omega = \omega_0 + \alpha t$; (d) $a_{\text{tan}} = r\omega^2$; (e) $K = \frac{1}{2}I\omega^2$.

Q9.2 A diatomic molecule can be modeled as two point masses, m_1 and m_2, slightly separated (Fig. Q9.2). If the molecule is oriented along the y-axis, it has kinetic energy K when it spins about the x-axis. What will its kinetic energy (in terms of K) be if it spins at the same angular speed about (a) the z-axis and (b) the y-axis?

Figure **Q9.2**

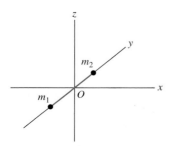

Q9.3 What is the difference between tangential and radial acceleration for a point on a rotating body?

Q9.4 In Fig. Q9.4, all points on the chain have the same linear speed. Is the magnitude of the linear acceleration also the same for all points on the chain? How are the angular accelerations of the two sprockets related? Explain.

Figure **Q9.4**

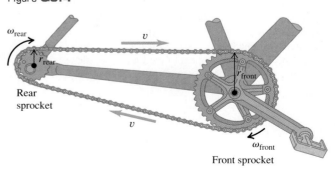

Q9.5 In Fig. Q9.4, how are the radial accelerations of points at the teeth of the two sprockets related? Explain the reasoning behind your answer.

Q9.6 A flywheel rotates with constant angular velocity. Does a point on its rim have a tangential acceleration? A radial acceleration? Are these accelerations constant in magnitude? In direction? In each case give the reasoning behind your answer.

Q9.7 What is the purpose of the spin cycle of a washing machine? Explain in terms of acceleration components.

Q9.8 Although angular velocity and angular acceleration can be treated as vectors, the angular displacement θ, despite having a magnitude and a direction, cannot. This is because θ does not follow the commutative law of vector addition (Eq. 1.3). Prove this to yourself in the following way: Lay your physics textbook flat on the desk in front of you with the cover side up so you can read the writing on it. Rotate it through 90° about a horizontal axis so that the farthest edge comes toward you. Call this angular displacement θ_1. Then rotate it by 90° about a vertical axis so that the left edge comes toward you. Call this angular displacement θ_2. The spine of the book should now face you, with the writing on it oriented so that you can read it. Now start over again but carry out the two rotations in the reverse order. Do you get a different result? That is, does $\theta_1 + \theta_2$ equal $\theta_2 + \theta_1$? Now repeat this experiment but this time with an angle of 1° rather than 90°. Do you think that the infinitesimal displacement $d\vec{\theta}$ obeys the commutative law of addition and hence qualifies as a vector? If so, how is the direction of $d\vec{\theta}$ related to the direction of $\vec{\omega}$?

Q9.9 Can you think of a body that has the same moment of inertia for all possible axes? If so, give an example, and if not, explain why this is not possible. Can you think of a body that has the same moment of inertia for all axes passing through a certain point? If so, give an example and indicate where the point is located.

Q9.10 To maximize the moment of inertia of a flywheel while minimizing its weight, what shape and distribution of mass should it have? Explain.

Q9.11 How might you determine experimentally the moment of inertia of an irregularly shaped body about a given axis?

Q9.12 A cylindrical body has mass M and radius R. Can the mass be distributed within the body in such a way that its moment of inertia about its axis of symmetry is greater than MR^2? Explain.

Q9.13 Describe how you could use part (b) of Table 9.2 to derive the result in part (d).

Q9.14 A hollow spherical shell of radius R that is rotating about an axis through its center has rotational kinetic energy K. If you want to modify this sphere so that it has three times as much kinetic energy at the same angular speed while keeping the same mass, what should be its radius in terms of R?

Q9.15 For the equations for I given in parts (a) and (b) of Table 9.2 to be valid, must the rod have a circular cross section? Is there any restriction on the size of the cross section for these equations to apply? Explain.

Q9.16 In part (d) of Table 9.2, the thickness of the plate must be much less than a for the expression given for I to apply. But in part (c), the expression given for I applies no matter how thick the plate is. Explain.

Q9.17 Two identical balls, A and B, are each attached to very light string, and each string is wrapped around the rim of a frictionless pulley of mass M. The only difference is that the pulley for ball A is a solid disk, while the one for ball B is a hollow disk, like part (e) in Table 9.2. If both balls are released from rest and fall the same distance, which one will have more kinetic energy, or will they have the same kinetic energy? Explain your reasoning.

Q9.18 An elaborate pulley consists of four identical balls at the ends of spokes extending out from a rotating drum (Fig. Q9.18). A box is connected to a light thin rope wound around the rim of the drum. When it is released from rest, the box acquires a speed V after having fallen a distance d. Now the four balls are moved inward closer to the drum, and the box is again released from rest. After it has fallen a distance d, will its speed be equal to V, greater than V, or less than V? Show or explain why.

Figure **Q9.18**

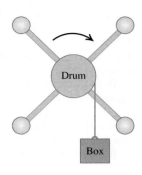

Q9.19 You can use any angular measure—radians, degrees, or revolutions—in some of the equations in Chapter 9, but you can use only radian measure in others. Identify those for which using radians is necessary and those for which it is not, and in each case give the reasoning behind your answer.

Q9.20 When calculating the moment of inertia of an object, can we treat all its mass as if it were concentrated at the center of mass of the object? Justify your answer.

Q9.21 A wheel is rotating about an axis perpendicular to the plane of the wheel and passing through the center of the wheel. The angular speed of the wheel is increasing at a constant rate. Point A is on the rim of the wheel and point B is midway between the rim and center of the wheel. For each of the following quantities, is its magnitude larger at point A or at point B, or is it the same at both points? (a) angular speed; (b) tangential speed; (c) angular acceleration; (d) tangential acceleration; (e) radial acceleration. Justify each of your answers.

Q9.22 Estimate your own moment of inertia about a vertical axis through the center of the top of your head when you are standing up straight with your arms outstretched. Make reasonable approximations and measure or estimate necessary quantities.

EXERCISES

Section 9.1 Angular Velocity and Acceleration

9.1 • (a) What angle in radians is subtended by an arc 1.50 m long on the circumference of a circle of radius 2.50 m? What is this angle in degrees? (b) An arc 14.0 cm long on the circumference of a circle subtends an angle of 128°. What is the radius of the circle? (c) The angle between two radii of a circle with radius 1.50 m is 0.700 rad. What length of arc is intercepted on the circumference of the circle by the two radii?

9.2 • An airplane propeller is rotating at 1900 rpm (rev/min). (a) Compute the propeller's angular velocity in rad/s. (b) How many seconds does it take for the propeller to turn through 35°?

9.3 • CP CALC The angular velocity of a flywheel obeys the equation $\omega_z(t) = A + Bt^2$, where t is in seconds and A and B are constants having numerical values 2.75 (for A) and 1.50 (for B). (a) What are the units of A and B if ω_z is in rad/s? (b) What is the angular acceleration of the wheel at (i) $t = 0.00$ and (ii) $t = 5.00$ s? (c) Through what angle does the flywheel turn during the first 2.00 s? (*Hint:* See Section 2.6.)

9.4 •• CALC A fan blade rotates with angular velocity given by $\omega_z(t) = \gamma - \beta t^2$, where $\gamma = 5.00$ rad/s and $\beta = 0.800$ rad/s³. (a) Calculate the angular acceleration as a function of time. (b) Calculate the instantaneous angular acceleration α_z at $t = 3.00$ s and the average angular acceleration $\alpha_{\text{av-}z}$ for the time interval $t = 0$ to $t = 3.00$ s. How do these two quantities compare? If they are different, why are they different?

9.5 •• **CALC** A child is pushing a merry-go-round. The angle through which the merry-go-round has turned varies with time according to $\theta(t) = \gamma t + \beta t^3$, where $\gamma = 0.400$ rad/s and $\beta = 0.0120$ rad/s³. (a) Calculate the angular velocity of the merry-go-round as a function of time. (b) What is the initial value of the angular velocity? (c) Calculate the instantaneous value of the angular velocity ω_z at $t = 5.00$ s and the average angular velocity $\omega_{\text{av-}z}$ for the time interval $t = 0$ to $t = 5.00$ s. Show that $\omega_{\text{av-}z}$ is *not* equal to the average of the instantaneous angular velocities at $t = 0$ and $t = 5.00$ s, and explain why it is not.

9.6 • **CALC** At $t = 0$ the current to a dc electric motor is reversed, resulting in an angular displacement of the motor shaft given by $\theta(t) = (250 \text{ rad/s})t - (20.0 \text{ rad/s}^2)t^2 - (1.50 \text{ rad/s}^3)t^3$. (a) At what time is the angular velocity of the motor shaft zero? (b) Calculate the angular acceleration at the instant that the motor shaft has zero angular velocity. (c) How many revolutions does the motor shaft turn through between the time when the current is reversed and the instant when the angular velocity is zero? (d) How fast was the motor shaft rotating at $t = 0$, when the current was reversed? (e) Calculate the average angular velocity for the time period from $t = 0$ to the time calculated in part (a).

9.7 • **CALC** The angle θ through which a disk drive turns is given by $\theta(t) = a + bt - ct^3$, where a, b, and c are constants, t is in seconds, and θ is in radians. When $t = 0$, $\theta = \pi/4$ rad and the angular velocity is 2.00 rad/s, and when $t = 1.50$ s, the angular acceleration is 1.25 rad/s². (a) Find a, b, and c, including their units. (b) What is the angular acceleration when $\theta = \pi/4$ rad? (c) What are θ and the angular velocity when the angular acceleration is 3.50 rad/s²?

9.8 • A wheel is rotating about an axis that is in the z-direction. The angular velocity ω_z is -6.00 rad/s at $t = 0$, increases linearly with time, and is $+8.00$ rad/s at $t = 7.00$ s. We have taken counterclockwise rotation to be positive. (a) Is the angular acceleration during this time interval positive or negative? (b) During what time interval is the speed of the wheel increasing? Decreasing? (c) What is the angular displacement of the wheel at $t = 7.00$ s?

Section 9.2 Rotation with Constant Angular Acceleration

9.9 • A bicycle wheel has an initial angular velocity of 1.50 rad/s. (a) If its angular acceleration is constant and equal to 0.300 rad/s², what is its angular velocity at $t = 2.50$ s? (b) Through what angle has the wheel turned between $t = 0$ and $t = 2.50$ s?

9.10 •• An electric fan is turned off, and its angular velocity decreases uniformly from 500 rev/min to 200 rev/min in 4.00 s. (a) Find the angular acceleration in rev/s² and the number of revolutions made by the motor in the 4.00-s interval. (b) How many more seconds are required for the fan to come to rest if the angular acceleration remains constant at the value calculated in part (a)?

9.11 •• The rotating blade of a blender turns with constant angular acceleration 1.50 rad/s². (a) How much time does it take to reach an angular velocity of 36.0 rad/s, starting from rest? (b) Through how many revolutions does the blade turn in this time interval?

9.12 • (a) Derive Eq. (9.12) by combining Eqs. (9.7) and (9.11) to eliminate t. (b) The angular velocity of an airplane propeller increases from 12.0 rad/s to 16.0 rad/s while turning through 7.00 rad. What is the angular acceleration in rad/s²?

9.13 •• A turntable rotates with a constant 2.25 rad/s² angular acceleration. After 4.00 s it has rotated through an angle of 60.0 rad. What was the angular velocity of the wheel at the beginning of the 4.00-s interval?

9.14 • A circular saw blade 0.200 m in diameter starts from rest. In 6.00 s it accelerates with constant angular acceleration to an angular velocity of 140 rad/s. Find the angular acceleration and the angle through which the blade has turned.

9.15 •• A high-speed flywheel in a motor is spinning at 500 rpm when a power failure suddenly occurs. The flywheel has mass 40.0 kg and diameter 75.0 cm. The power is off for 30.0 s, and during this time the flywheel slows due to friction in its axle bearings. During the time the power is off, the flywheel makes 200 complete revolutions. (a) At what rate is the flywheel spinning when the power comes back on? (b) How long after the beginning of the power failure would it have taken the flywheel to stop if the power had not come back on, and how many revolutions would the wheel have made during this time?

9.16 •• At $t = 0$ a grinding wheel has an angular velocity of 24.0 rad/s. It has a constant angular acceleration of 30.0 rad/s² until a circuit breaker trips at $t = 2.00$ s. From then on, it turns through 432 rad as it coasts to a stop at constant angular acceleration. (a) Through what total angle did the wheel turn between $t = 0$ and the time it stopped? (b) At what time did it stop? (c) What was its acceleration as it slowed down?

9.17 •• A safety device brings the blade of a power mower from an initial angular speed of ω_1 to rest in 1.00 revolution. At the same constant acceleration, how many revolutions would it take the blade to come to rest from an initial angular speed ω_3 that was three times as great, $\omega_3 = 3\omega_1$?

Section 9.3 Relating Linear and Angular Kinematics

9.18 • In a charming 19th-century hotel, an old-style elevator is connected to a counterweight by a cable that passes over a rotating disk 2.50 m in diameter (Fig. E9.18). The elevator is raised and lowered by turning the disk, and the cable does not slip on the rim of the disk but turns with it. (a) At how many rpm must the disk turn to raise the elevator at 25.0 cm/s? (b) To start the elevator moving, it must be accelerated at $\frac{1}{8}g$. What must be the angular acceleration of the disk, in rad/s²? (c) Through what angle (in radians and degrees) has the disk turned when it has raised the elevator 3.25 m between floors?

Figure **E9.18**

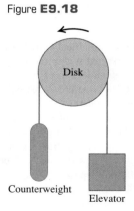

9.19 • Using astronomical data from Appendix F, along with the fact that the earth spins on its axis once per day, calculate (a) the earth's orbital angular speed (in rad/s) due to its motion around the sun, (b) its angular speed (in rad/s) due to its axial spin, (c) the tangential speed of the earth around the sun (assuming a circular orbit), (d) the tangential speed of a point on the earth's equator due to the planet's axial spin, and (e) the radial and tangential acceleration components of the point in part (d).

9.20 • **Compact Disc.** A compact disc (CD) stores music in a coded pattern of tiny pits 10^{-7} m deep. The pits are arranged in a track that spirals outward toward the rim of the disc; the inner and outer radii of this spiral are 25.0 mm and 58.0 mm, respectively. As the disc spins inside a CD player, the track is scanned at a constant *linear* speed of 1.25 m/s. (a) What is the angular speed of the CD when the innermost part of the track is scanned? The outermost part of the track? (b) The maximum playing time of a CD is 74.0 min. What would be the length of the track on such a maximum-duration CD if it were stretched out in a straight line? (c) What is the average angular acceleration of a maximum-duration CD during its 74.0-min playing time? Take the direction of rotation of the disc to be positive.

9.21 •• A wheel of diameter 40.0 cm starts from rest and rotates with a constant angular acceleration of 3.00 rad/s^2. At the instant the wheel has computed its second revolution, compute the radial acceleration of a point on the rim in two ways: (a) using the relationship $a_{rad} = \omega^2 r$ and (b) from the relationship $a_{rad} = v^2/r$.

9.22 •• You are to design a rotating cylindrical axle to lift 800-N buckets of cement from the ground to a rooftop 78.0 m above the ground. The buckets will be attached to a hook on the free end of a cable that wraps around the rim of the axle; as the axle turns, the buckets will rise. (a) What should the diameter of the axle be in order to raise the buckets at a steady 2.00 cm/s when it is turning at 7.5 rpm? (b) If instead the axle must give the buckets an upward acceleration of 0.400 m/s^2, what should the angular acceleration of the axle be?

9.23 • A flywheel with a radius of 0.300 m starts from rest and accelerates with a constant angular acceleration of 0.600 rad/s^2. Compute the magnitude of the tangential acceleration, the radial acceleration, and the resultant acceleration of a point on its rim (a) at the start; (b) after it has turned through 60.0°; (c) after it has turned through 120.0°.

9.24 •• An electric turntable 0.750 m in diameter is rotating about a fixed axis with an initial angular velocity of 0.250 rev/s and a constant angular acceleration of 0.900 rev/s^2. (a) Compute the angular velocity of the turntable after 0.200 s. (b) Through how many revolutions has the turntable spun in this time interval? (c) What is the tangential speed of a point on the rim of the turntable at $t = 0.200$ s? (d) What is the magnitude of the *resultant* acceleration of a point on the rim at $t = 0.200$ s?

9.25 •• **Centrifuge.** An advertisement claims that a centrifuge takes up only 0.127 m of bench space but can produce a radial acceleration of $3000g$ at 5000 rev/min. Calculate the required radius of the centrifuge. Is the claim realistic?

9.26 • (a) Derive an equation for the radial acceleration that includes v and ω, but not r. (b) You are designing a merry-go-round for which a point on the rim will have a radial acceleration of 0.500 m/s^2 when the tangential velocity of that point has magnitude 2.00 m/s. What angular velocity is required to achieve these values?

9.27 • **Electric Drill.** According to the shop manual, when drilling a 12.7-mm-diameter hole in wood, plastic, or aluminum, a drill should have a speed of 1250 rev/min. For a 12.7-mm-diameter drill bit turning at a constant 1250 rev/min, find (a) the maximum linear speed of any part of the bit and (b) the maximum radial acceleration of any part of the bit.

9.28 • At $t = 3.00$ s a point on the rim of a 0.200-m-radius wheel has a tangential speed of 50.0 m/s as the wheel slows down with a tangential acceleration of constant magnitude 10.0 m/s^2. (a) Calculate the wheel's constant angular acceleration. (b) Calculate the angular velocities at $t = 3.00$ s and $t = 0$. (c) Through what angle did the wheel turn between $t = 0$ and $t = 3.00$ s? (d) At what time will the radial acceleration equal g?

9.29 • The spin cycles of a washing machine have two angular speeds, 423 rev/min and 640 rev/min. The internal diameter of the drum is 0.470 m. (a) What is the ratio of the maximum radial force on the laundry for the higher angular speed to that for the lower speed? (b) What is the ratio of the maximum tangential speed of the laundry for the higher angular speed to that for the lower speed? (c) Find the laundry's maximum tangential speed and the maximum radial acceleration, in terms of g.

Section 9.4 Energy in Rotational Motion

9.30 • Four small spheres, each of which you can regard as a point of mass 0.200 kg, are arranged in a square 0.400 m on a side and connected by extremely light rods (Fig. E9.30). Find the moment of inertia of the system about an axis (a) through the center of the square, perpendicular to its plane (an axis through point O in the figure); (b) bisecting two opposite sides of the square (an axis along the line AB in the figure); (c) that passes through the centers of the upper left and lower right spheres and through point O.

Figure **E9.30**

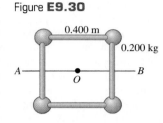

0.400 m

0.200 kg

A ——————— B

O

9.31 • Calculate the moment of inertia of each of the following uniform objects about the axes indicated. Consult Table 9.2 as needed. (a) A thin 2.50-kg rod of length 75.0 cm, about an axis perpendicular to it and passing through (i) one end and (ii) its center, and (iii) about an axis parallel to the rod and passing through it. (b) A 3.00-kg sphere 38.0 cm in diameter, about an axis through its center, if the sphere is (i) solid and (ii) a thin-walled hollow shell. (c) An 8.00-kg cylinder, of length 19.5 cm and diameter 12.0 cm, about the central axis of the cylinder, if the cylinder is (i) thin-walled and hollow, and (ii) solid.

9.32 •• Small blocks, each with mass m, are clamped at the ends and at the center of a rod of length L and negligible mass. Compute the moment of inertia of the system about an axis perpendicular to the rod and passing through (a) the center of the rod and (b) a point one-fourth of the length from one end.

9.33 • A uniform bar has two small balls glued to its ends. The bar is 2.00 m long and has mass 4.00 kg, while the balls each have mass 0.500 kg and can be treated as point masses. Find the moment of inertia of this combination about each of the following axes: (a) an axis perpendicular to the bar through its center; (b) an axis perpendicular to the bar through one of the balls; (c) an axis parallel to the bar through both balls; (d) an axis parallel to the bar and 0.500 m from it.

9.34 • A uniform disk of radius R is cut in half so that the remaining half has mass M (Fig. E9.34a). (a) What is the moment of inertia of this half about an axis perpendicular to its plane through point A? (b) Why did your answer in part (a) come out the same as if this were a complete disk of mass M? (c) What would be the moment of inertia of a quarter disk of mass M and radius R about an axis perpendicular to its plane passing through point B (Fig. E9.34b)?

Figure **E9.34**

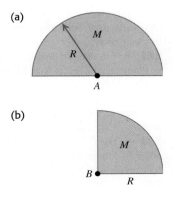

(a)

M

R

A

(b)

M

B

R

9.35 •• A wagon wheel is constructed as shown in Fig. E9.35. The radius of the wheel is 0.300 m, and the rim has mass 1.40 kg. Each of the eight spokes that lie along a diameter and are 0.300 m long has mass 0.280 kg. What is the moment of inertia of the wheel about an axis through its center and perpendicular to the plane of the wheel? (Use the formulas given in Table 9.2.)

Figure **E9.35**

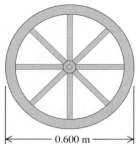

|← 0.600 m →|

9.36 •• An airplane propeller is 2.08 m in length (from tip to tip) with mass 117 kg and is rotating at 2400 rpm (rev/min) about an axis through its center. You can model the propeller as a slender rod. (a) What is its rotational kinetic energy? (b) Suppose that, due to weight constraints, you had to reduce the propeller's mass to 75.0% of its original mass, but you still needed to keep the same size and kinetic energy. What would its angular speed have to be, in rpm?

9.37 •• A compound disk of outside diameter 140.0 cm is made up of a uniform solid disk of radius 50.0 cm and area density 3.00 g/cm² surrounded by a concentric ring of inner radius 50.0 cm, outer radius 70.0 cm, and area density 2.00 g/cm². Find the moment of inertia of this object about an axis perpendicular to the plane of the object and passing through its center.

9.38 • A wheel is turning about an axis through its center with constant angular acceleration. Starting from rest, at $t = 0$, the wheel turns through 8.20 revolutions in 12.0 s. At $t = 12.0$ s the kinetic energy of the wheel is 36.0 J. For an axis through its center, what is the moment of inertia of the wheel?

9.39 • A uniform sphere with mass 28.0 kg and radius 0.380 m is rotating at constant angular velocity about a stationary axis that lies along a diameter of the sphere. If the kinetic energy of the sphere is 176 J, what is the tangential velocity of a point on the rim of the sphere?

9.40 •• A hollow spherical shell has mass 8.20 kg and radius 0.220 m. It is initially at rest and then rotates about a stationary axis that lies along a diameter with a constant acceleration of 0.890 rad/s². What is the kinetic energy of the shell after it has turned through 6.00 rev?

9.41 • **Energy from the Moon?** Suppose that some time in the future we decide to tap the moon's rotational energy for use on earth. In additional to the astronomical data in Appendix F, you may need to know that the moon spins on its axis once every 27.3 days. Assume that the moon is uniform throughout. (a) How much total energy could we get from the moon's rotation? (b) The world presently uses about 4.0×10^{20} J of energy per year. If in the future the world uses five times as much energy yearly, for how many years would the moon's rotation provide us energy? In light of your answer, does this seem like a cost-effective energy source in which to invest?

9.42 •• You need to design an industrial turntable that is 60.0 cm in diameter and has a kinetic energy of 0.250 J when turning at 45.0 rpm (rev/min). (a) What must be the moment of inertia of the turntable about the rotation axis? (b) If your workshop makes this turntable in the shape of a uniform solid disk, what must be its mass?

9.43 •• The flywheel of a gasoline engine is required to give up 500 J of kinetic energy while its angular velocity decreases from 650 rev/min to 520 rev/min. What moment of inertia is required?

9.44 • A light, flexible rope is wrapped several times around a *hollow* cylinder, with a weight of 40.0 N and a radius of 0.25 m, that rotates without friction about a fixed horizontal axis. The cylinder is attached to the axle by spokes of a negligible moment of inertia. The cylinder is initially at rest. The free end of the rope is pulled with a constant force P for a distance of 5.00 m, at which point the end of the rope is moving at 6.00 m/s. If the rope does not slip on the cylinder, what is the value of P?

9.45 •• Energy is to be stored in a 70.0-kg flywheel in the shape of a uniform solid disk with radius $R = 1.20$ m. To prevent structural failure of the flywheel, the maximum allowed radial acceleration of a point on its rim is 3500 m/s². What is the maximum kinetic energy that can be stored in the flywheel?

9.46 •• Suppose the solid cylinder in the apparatus described in Example 9.8 (Section 9.4) is replaced by a thin-walled, hollow cylinder with the same mass M and radius R. The cylinder is attached to the axle by spokes of a negligible moment of inertia. (a) Find the speed of the hanging mass m just as it strikes the floor. (b) Use energy concepts to explain why the answer to part (a) is different from the speed found in Example 9.8.

9.47 •• A frictionless pulley has the shape of a uniform solid disk of mass 2.50 kg and radius 20.0 cm. A 1.50-kg stone is attached to a very light wire that is wrapped around the rim of the pulley (Fig. E9.47), and the system is released from rest. (a) How far must the stone fall so that the pulley has 4.50 J of kinetic energy? (b) What percent of the total kinetic energy does the pulley have?

Figure **E9.47**

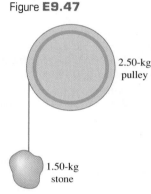

2.50-kg pulley

1.50-kg stone

9.48 •• A bucket of mass m is tied to a massless cable that is wrapped around the outer rim of a frictionless uniform pulley of radius R, similar to the system shown in Fig. E9.47. In terms of the stated variables, what must be the moment of inertia of the pulley so that it always has half as much kinetic energy as the bucket?

9.49 •• **CP** A thin, light wire is wrapped around the rim of a wheel, as shown in Fig. E9.49. The wheel rotates without friction about a stationary horizontal axis that passes through the center of the wheel. The wheel is a uniform disk with radius $R = 0.280$ m. An object of mass $m = 4.20$ kg is suspended from the free end of the wire. The system is released from rest and the suspended object descends with constant acceleration. If the suspended object moves downward a distance of 3.00 m in 2.00 s, what is the mass of the wheel?

Figure **E9.49**

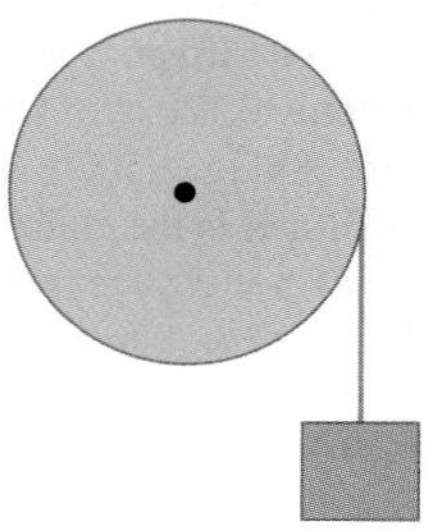

9.50 •• A uniform 2.00-m ladder of mass 9.00 kg is leaning against a vertical wall while making an angle of 53.0° with the floor. A worker pushes the ladder up against the wall until it is vertical. What is the increase in the gravitational potential energy of the ladder?

9.51 •• **How I Scales.** If we multiply all the design dimensions of an object by a scaling factor f, its volume and mass will be multiplied by f^3. (a) By what factor will its moment of inertia be multiplied? (b) If a $\frac{1}{48}$-scale model has a rotational kinetic energy of 2.5 J, what will be the kinetic energy for the full-scale object of the same material rotating at the same angular velocity?

9.52 •• A uniform 3.00-kg rope 24.0 m long lies on the ground at the top of a vertical cliff. A mountain climber at the top lets down half of it to help his partner climb up the cliff. What was the change in potential energy of the rope during this maneuver?

Section 9.5 Parallel-Axis Theorem

9.53 •• About what axis will a uniform, balsa-wood sphere have the same moment of inertia as does a thin-walled, hollow, lead sphere of the same mass and radius, with the axis along a diameter?

9.54 •• Find the moment of inertia of a hoop (a thin-walled, hollow ring) with mass M and radius R about an axis perpendicular to the hoop's plane at an edge.

9.55 •• A thin, rectangular sheet of metal has mass M and sides of length a and b. Use the parallel-axis theorem to calculate the moment of inertia of the sheet for an axis that is perpendicular to the plane of the sheet and that passes through one corner of the sheet.

9.56 • (a) For the thin rectangular plate shown in part (d) of Table 9.2, find the moment of inertia about an axis that lies in the plane of the plate, passes through the center of the plate, and is parallel to the axis shown in the figure. (b) Find the moment of inertia of the plate for an axis that lies in the plane of the plate, passes through the center of the plate, and is perpendicular to the axis in part (a).

9.57 •• A thin uniform rod of mass M and length L is bent at its center so that the two segments are now perpendicular to each other. Find its moment of inertia about an axis perpendicular to its plane and passing through (a) the point where the two segments meet and (b) the midpoint of the line connecting its two ends.

Section 9.6 Moment-of-Inertia Calculations

9.58 • CALC Use Eq. (9.20) to calculate the moment of inertia of a slender, uniform rod with mass M and length L about an axis at one end, perpendicular to the rod.

9.59 •• CALC Use Eq. (9.20) to calculate the moment of inertia of a uniform, solid disk with mass M and radius R for an axis perpendicular to the plane of the disk and passing through its center.

9.60 •• CALC A slender rod with length L has a mass per unit length that varies with distance from the left end, where $x = 0$, according to $dm/dx = \gamma x$, where γ has units of kg/m^2. (a) Calculate the total mass of the rod in terms of γ and L. (b) Use Eq. (9.20) to calculate the moment of inertia of the rod for an axis at the left end, perpendicular to the rod. Use the expression you derived in part (a) to express I in terms of M and L. How does your result compare to that for a uniform rod? Explain this comparison. (c) Repeat part (b) for an axis at the right end of the rod. How do the results for parts (b) and (c) compare? Explain this result.

PROBLEMS

9.61 • CP CALC A flywheel has angular acceleration $\alpha_z(t) = 8.60 \text{ rad/s}^2 - (2.30 \text{ rad/s}^3)t$, where counterclockwise rotation is positive. (a) If the flywheel is at rest at $t = 0$, what is its angular velocity at 5.00 s? (b) Through what angle (in radians) does the flywheel turn in the time interval from $t = 0$ to $t = 5.00$ s?

9.62 •• CALC A uniform disk with radius $R = 0.400$ m and mass 30.0 kg rotates in a horizontal plane on a frictionless vertical axle that passes through the center of the disk. The angle through which the disk has turned varies with time according to $\theta(t) = (1.10 \text{ rad/s})t + (8.60 \text{ rad/s}^2)t^2$. What is the resultant linear acceleration of a point on the rim of the disk at the instant when the disk has turned through 0.100 rev?

9.63 •• CP A circular saw blade with radius 0.120 m starts from rest and turns in a vertical plane with a constant angular acceleration of 3.00 rev/s^2. After the blade has turned through 155 rev, a small piece of the blade breaks loose from the top of the blade. After the piece breaks loose, it travels with a velocity that is initially horizontal and equal to the tangential velocity of the rim of the blade. The piece travels a vertical distance of 0.820 m to the floor. How far does the piece travel horizontally, from where it broke off the blade until it strikes the floor?

9.64 • CALC A roller in a printing press turns through an angle $\theta(t)$ given by $\theta(t) = \gamma t^2 - \beta t^3$, where $\gamma = 3.20 \text{ rad/s}^2$ and $\beta = 0.500 \text{ rad/s}^3$. (a) Calculate the angular velocity of the roller as a function of time. (b) Calculate the angular acceleration of the roller as a function of time. (c) What is the maximum positive angular velocity, and at what value of t does it occur?

9.65 •• CP CALC A disk of radius 25.0 cm is free to turn about an axle perpendicular to it through its center. It has very thin but strong string wrapped around its rim, and the string is attached to a ball that is pulled tangentially away from the rim of the disk (Fig. P9.65). The pull increases in magnitude and produces an acceleration of the ball that obeys the equation $a(t) = At$, where t is in seconds and A is a constant. The cylinder starts from rest, and at the end of the third second, the ball's acceleration is 1.80 m/s^2. (a) Find A. (b) Express the angular acceleration of the disk as a function of time. (c) How much time after the disk has begun to turn does it reach an angular speed of 15.0 rad/s? (d) Through what angle has the disk turned just as it reaches 15.0 rad/s? (*Hint:* See Section 2.6.)

Figure **P9.65**

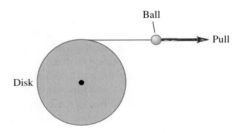

9.66 •• When a toy car is rapidly scooted across the floor, it stores energy in a flywheel. The car has mass 0.180 kg, and its flywheel has moment of inertia $4.00 \times 10^{-5} \text{ kg} \cdot \text{m}^2$. The car is 15.0 cm long. An advertisement claims that the car can travel at a scale speed of up to 700 km/h (440 mi/h). The scale speed is the speed of the toy car multiplied by the ratio of the length of an actual car to the length of the toy. Assume a length of 3.0 m for a real car. (a) For a scale speed of 700 km/h, what is the actual translational speed of the car? (b) If all the kinetic energy that is initially in the flywheel is converted to the translational kinetic energy of the toy, how much energy is originally stored in the flywheel? (c) What initial angular velocity of the flywheel was needed to store the amount of energy calculated in part (b)?

9.67 • A classic 1957 Chevrolet Corvette of mass 1240 kg starts from rest and speeds up with a constant tangential acceleration of 2.00 m/s^2 on a circular test track of radius 60.0 m. Treat the car as a particle. (a) What is its angular acceleration? (b) What is its angular speed 6.00 s after it starts? (c) What is its radial acceleration at this time? (d) Sketch a view from above showing the circular track, the car, the velocity vector, and the acceleration component vectors 6.00 s after the car starts. (e) What are the magnitudes of the total acceleration and net force for the car at this time? (f) What angle do the total acceleration and net force make with the car's velocity at this time?

9.68 •• Engineers are designing a system by which a falling mass m imparts kinetic energy to a rotating uniform drum to which it is attached by thin, very light wire wrapped around the rim of the drum (Fig. P9.68). There is no appreciable friction in the axle of the drum, and everything starts from rest. This system is being tested on earth, but it is to be used on Mars, where the acceleration due to gravity is 3.71 m/s². In the earth tests, when m is set to 15.0 kg and allowed to fall through 5.00 m, it gives 250.0 J of kinetic energy to the drum. (a) If the system is operated on Mars, through what distance would the 15.0-kg mass have to fall to give the same amount of kinetic energy to the drum? (b) How fast would the 15.0-kg mass be moving on Mars just as the drum gained 250.0 J of kinetic energy?

Figure **P9.68**

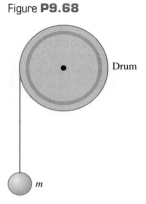

Drum

m

9.69 • A vacuum cleaner belt is looped over a shaft of radius 0.45 cm and a wheel of radius 1.80 cm. The arrangement of the belt, shaft, and wheel is similar to that of the chain and sprockets in Fig. Q9.4. The motor turns the shaft at 60.0 rev/s and the moving belt turns the wheel, which in turn is connected by another shaft to the roller that beats the dirt out of the rug being vacuumed. Assume that the belt doesn't slip on either the shaft or the wheel. (a) What is the speed of a point on the belt? (b) What is the angular velocity of the wheel, in rad/s?

9.70 •• The motor of a table saw is rotating at 3450 rev/min. A pulley attached to the motor shaft drives a second pulley of half the diameter by means of a V-belt. A circular saw blade of diameter 0.208 m is mounted on the same rotating shaft as the second pulley. (a) The operator is careless and the blade catches and throws back a small piece of wood. This piece of wood moves with linear speed equal to the tangential speed of the rim of the blade. What is this speed? (b) Calculate the radial acceleration of points on the outer edge of the blade to see why sawdust doesn't stick to its teeth.

9.71 ••• While riding a multispeed bicycle, the rider can select the radius of the rear sprocket that is fixed to the rear axle. The front sprocket of a bicycle has radius 12.0 cm. If the angular speed of the front sprocket is 0.600 rev/s, what is the radius of the rear sprocket for which the tangential speed of a point on the rim of the rear wheel will be 5.00 m/s? The rear wheel has radius 0.330 m.

9.72 ••• A computer disk drive is turned on starting from rest and has constant angular acceleration. If it took 0.750 s for the drive to make its *second* complete revolution, (a) how long did it take to make the first complete revolution, and (b) what is its angular acceleration, in rad/s²?

9.73 • A wheel changes its angular velocity with a constant angular acceleration while rotating about a fixed axis through its center. (a) Show that the change in the magnitude of the radial acceleration during any time interval of a point on the wheel is twice the product of the angular acceleration, the angular displacement, and the perpendicular distance of the point from the axis. (b) The radial acceleration of a point on the wheel that is 0.250 m from the axis changes from 25.0 m/s² to 85.0 m/s² as the wheel rotates through 20.0 rad. Calculate the tangential acceleration of this point. (c) Show that the change in the wheel's kinetic energy during any time interval is the product of the moment of inertia about the axis, the angular

acceleration, and the angular displacement. (d) During the 20.0-rad angular displacement of part (b), the kinetic energy of the wheel increases from 20.0 J to 45.0 J. What is the moment of inertia of the wheel about the rotation axis?

9.74 •• A sphere consists of a solid wooden ball of uniform density 800 kg/m³ and radius 0.30 m and is covered with a thin coating of lead foil with area density 20 kg/m². Calculate the moment of inertia of this sphere about an axis passing through its center.

9.75 ••• It has been argued that power plants should make use of off-peak hours (such as late at night) to generate mechanical energy and store it until it is needed during peak load times, such as the middle of the day. One suggestion has been to store the energy in large flywheels spinning on nearly frictionless ball bearings. Consider a flywheel made of iron (density 7800 kg/m³) in the shape of a 10.0-cm-thick uniform disk. (a) What would the diameter of such a disk need to be if it is to store 10.0 megajoules of kinetic energy when spinning at 90.0 rpm about an axis perpendicular to the disk at its center? (b) What would be the centripetal acceleration of a point on its rim when spinning at this rate?

9.76 •• While redesigning a rocket engine, you want to reduce its weight by replacing a solid spherical part with a hollow spherical shell of the same size. The parts rotate about an axis through their center. You need to make sure that the new part always has the same rotational kinetic energy as the original part had at any given rate of rotation. If the original part had mass M, what must be the mass of the new part?

9.77 • The earth, which is not a uniform sphere, has a moment of inertia of $0.3308MR^2$ about an axis through its north and south poles. It takes the earth 86,164 s to spin once about this axis. Use Appendix F to calculate (a) the earth's kinetic energy due to its rotation about this axis and (b) the earth's kinetic energy due to its orbital motion around the sun. (c) Explain how the value of the earth's moment of inertia tells us that the mass of the earth is concentrated toward the planet's center.

9.78 ••• A uniform, solid disk with mass m and radius R is pivoted about a horizontal axis through its center. A small object of the same mass m is glued to the rim of the disk. If the disk is released from rest with the small object at the end of a horizontal radius, find the angular speed when the small object is directly below the axis.

9.79 •• **CALC** A metal sign for a car dealership is a thin, uniform right triangle with base length b and height h. The sign has mass M. (a) What is the moment of inertia of the sign for rotation about the side of length h? (b) If $M = 5.40$ kg, $b = 1.60$ m, and $h = 1.20$ m, what is the kinetic energy of the sign when it is rotating about an axis along the 1.20-m side at 2.00 rev/s?

9.80 •• **Measuring I.** As an intern with an engineering firm, you are asked to measure the moment of inertia of a large wheel, for rotation about an axis through its center. Since you were a good physics student, you know what to do. You measure the diameter of the wheel to be 0.740 m and find that it weighs 280 N. You mount the wheel, using frictionless bearings, on a horizontal axis through the wheel's center. You wrap a light rope around the wheel and hang an 8.00-kg mass from the free end of the rope, as shown in Fig. 9.17. You release the mass from rest; the mass descends and the wheel turns as the rope unwinds. You find that the mass has speed 5.00 m/s after it has descended 2.00 m. (a) What is the moment of inertia of the wheel for an axis perpendicular to the wheel at its center? (b) Your boss tells you that a larger I is needed. He asks you to design a wheel of the same mass and radius that has $I = 19.0$ kg·m². How do you reply?

9.81 •• **CP** A meter stick with a mass of 0.180 kg is pivoted about one end so it can rotate without friction about a horizontal axis.

The meter stick is held in a horizontal position and released. As it swings through the vertical, calculate (a) the change in gravitational potential energy that has occurred; (b) the angular speed of the stick; (c) the linear speed of the end of the stick opposite the axis. (d) Compare the answer in part (c) to the speed of a particle that has fallen 1.00 m, starting from rest.

9.82 •• Exactly one turn of a flexible rope with mass m is wrapped around a uniform cylinder with mass M and radius R. The cylinder rotates without friction about a horizontal axle along the cylinder axis. One end of the rope is attached to the cylinder. The cylinder starts with angular speed ω_0. After one revolution of the cylinder the rope has unwrapped and, at this instant, hangs vertically down, tangent to the cylinder. Find the angular speed of the cylinder and the linear speed of the lower end of the rope at this time. You can ignore the thickness of the rope. [*Hint:* Use Eq. (9.18).]

9.83 • The pulley in Fig. P9.83 has radius R and a moment of inertia I. The rope does not slip over the pulley, and the pulley spins on a frictionless axle. The coefficient of kinetic friction between block A and the tabletop is μ_k. The system is released from rest, and block B descends. Block A has mass m_A and block B has mass m_B. Use energy methods to calculate the speed of block B as a function of the distance d that it has descended.

Figure **P9.83**

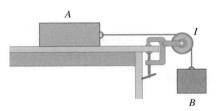

9.84 •• The pulley in Fig. P9.84 has radius 0.160 m and moment of inertia 0.560 kg·m². The rope does not slip on the pulley rim. Use energy methods to calculate the speed of the 4.00-kg block just before it strikes the floor.

Figure **P9.84**

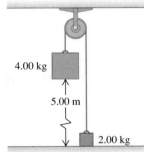

4.00 kg

5.00 m

2.00 kg

9.85 •• You hang a thin hoop with radius R over a nail at the rim of the hoop. You displace it to the side (within the plane of the hoop) through an angle β from its equilibrium position and let it go. What is its angular speed when it returns to its equilibrium position? [*Hint:* Use Eq. (9.18).]

9.86 •• A passenger bus in Zurich, Switzerland, derived its motive power from the energy stored in a large flywheel. The wheel was brought up to speed periodically, when the bus stopped at a station, by an electric motor, which could then be attached to the electric power lines. The flywheel was a solid cylinder with mass 1000 kg and diameter 1.80 m; its top angular speed was 3000 rev/min. (a) At this angular speed, what is the kinetic energy of the flywheel? (b) If the average power required to operate the bus is 1.86×10^4 W, how long could it operate between stops?

9.87 •• Two metal disks, one with radius $R_1 = 2.50$ cm and mass $M_1 = 0.80$ kg and the other with radius $R_2 = 5.00$ cm and mass $M_2 = 1.60$ kg, are welded together and mounted on a frictionless axis through their common center (Fig. P9.87). (a) What is the

total moment of inertia of the two disks? (b) A light string is wrapped around the edge of the smaller disk, and a 1.50-kg block is suspended from the free end of the string. If the block is released from rest at a distance of 2.00 m above the floor, what is its speed just before it strikes the floor? (c) Repeat the calculation of part (b), this time with the string wrapped around the edge of the larger disk. In which case is the final speed of the block greater? Explain why this is so.

Figure **P9.87**

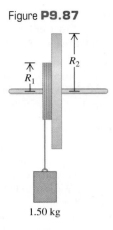

R_2

R_1

1.50 kg

9.88 •• A thin, light wire is wrapped around the rim of a wheel, as shown in Fig. E9.49. The wheel rotates about a stationary horizontal axle that passes through the center of the wheel. The wheel has radius 0.180 m and moment of inertia for rotation about the axle of $I = 0.480$ kg·m². A small block with mass 0.340 kg is suspended from the free end of the wire. When the system is released from rest, the block descends with constant acceleration. The bearings in the wheel at the axle are rusty, so friction there does -6.00 J of work as the block descends 3.00 m. What is the magnitude of the angular velocity of the wheel after the block has descended 3.00 m?

9.89 ••• In the system shown in Fig. 9.17, a 12.0-kg mass is released from rest and falls, causing the uniform 10.0-kg cylinder of diameter 30.0 cm to turn about a frictionless axle through its center. How far will the mass have to descend to give the cylinder 480 J of kinetic energy?

9.90 • In Fig. P9.90, the cylinder and pulley turn without friction about stationary horizontal axles that pass through their centers. A light rope is wrapped around the cylinder, passes over the pulley, and has a 3.00-kg box suspended from its free end. There is no slipping between the rope and the pulley surface. The uniform cylinder has mass 5.00 kg and radius 40.0 cm. The pulley is a uniform disk with mass 2.00 kg and radius 20.0 cm. The box is released from rest and descends as the rope unwraps from the cylinder. Find the speed of the box when it has fallen 2.50 m.

Figure **P9.90**

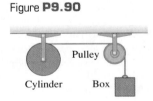

Pulley

Cylinder   Box

9.91 •• A thin, flat, uniform disk has mass M and radius R. A circular hole of radius $R/4$, centered at a point $R/2$ from the disk's center, is then punched in the disk. (a) Find the moment of inertia of the disk with the hole about an axis through the original center of the disk, perpendicular to the plane of the disk. (*Hint:* Find the moment of inertia of the piece punched from the disk.) (b) Find the moment of inertia of the disk with the hole about an axis through the center of the hole, perpendicular to the plane of the disk.

Figure **P9.92**

9.92 •• **BIO Human Rotational Energy.** A dancer is spinning at 72 rpm about an axis through her center with her arms outstretched, as shown in Fig. P9.92. From biomedical measurements, the typical distribution of mass in a human body is as follows:

Head: 7.0%
Arms: 13% (for both)
Trunk and legs: 80.0%

Suppose you are this dancer. Using this information plus length measurements on your own body, calculate (a) your moment of inertia about your spin axis and (b) your rotational kinetic energy. Use the figures in Table 9.2 to model reasonable approximations for the pertinent parts of your body.

9.93 •• **BIO** **The Kinetic Energy of Walking.** If a person of mass M simply moved forward with speed V, his kinetic energy would be $\frac{1}{2}MV^2$. However, in addition to possessing a forward motion, various parts of his body (such as the arms and legs) undergo rotation. Therefore, his total kinetic energy is the sum of the energy from his forward motion plus the rotational kinetic energy of his arms and legs. The purpose of this problem is to see how much this rotational motion contributes to the person's kinetic energy. Biomedical measurements show that the arms and hands together typically make up 13% of a person's mass, while the legs and feet together account for 37%. For a rough (but reasonable) calculation, we can model the arms and legs as thin uniform bars pivoting about the shoulder and hip, respectively. In a brisk walk, the arms and legs each move through an angle of about $\pm 30°$ (a total of 60°) from the vertical in approximately 1 second. We shall assume that they are held straight, rather than being bent, which is not quite true. Let us consider a 75-kg person walking at 5.0 km/h, having arms 70 cm long and legs 90 cm long. (a) What is the average angular velocity of his arms and legs? (b) Using the average angular velocity from part (a), calculate the amount of rotational kinetic energy in this person's arms and legs as he walks. (c) What is the total kinetic energy due to both his forward motion and his rotation? (d) What percentage of his kinetic energy is due to the rotation of his legs and arms?

9.94 •• **BIO** **The Kinetic Energy of Running.** Using Problem 9.93 as a guide, apply it to a person running at 12 km/h, with his arms and legs each swinging through $\pm 30°$ in $\frac{1}{2}$ s. As before, assume that the arms and legs are kept straight.

9.95 •• **Perpendicular-Axis Theorem.** Consider a rigid body that is a thin, plane sheet of arbitrary shape. Take the body to lie in the xy-plane and let the origin O of coordinates be located at any point within or outside the body. Let I_x and I_y be the moments of inertia about the x- and y-axes, and let I_O be the moment of inertia about an axis through O perpendicular to the plane. (a) By considering mass elements m_i with coordinates (x_i, y_i), show that $I_x + I_y = I_O$. This is called the perpendicular-axis theorem. Note that point O does not have to be the center of mass. (b) For a thin washer with mass M and with inner and outer radii R_1 and R_2, use the perpendicular-axis theorem to find the moment of inertia about an axis that is in the plane of the washer and that passes through its center. You may use the information in Table 9.2. (c) Use the perpendicular-axis theorem to show that for a thin, square sheet with mass M and side L, the moment of inertia about *any* axis in the plane of the sheet that passes through the center of the sheet is $\frac{1}{12}ML^2$. You may use the information in Table 9.2.

9.96 ••• A thin, uniform rod is bent into a square of side length a. If the total mass is M, find the moment of inertia about an axis through the center and perpendicular to the plane of the square. (*Hint:* Use the parallel-axis theorem.)

9.97 • **CALC** A cylinder with radius R and mass M has density that increases linearly with distance r from the cylinder axis, $\rho = \alpha r$, where α is a positive constant. (a) Calculate the moment of inertia of the cylinder about a longitudinal axis through its center in terms of M and R. (b) Is your answer greater or smaller than the moment of inertia of a cylinder of the same mass and radius but of uniform density? Explain why this result makes qualitative sense.

9.98 •• **CALC** **Neutron Stars and Supernova Remnants.** The Crab Nebula is a cloud of glowing gas about 10 light-years across, located about 6500 light-years from the earth (Fig. P9.98). It is the remnant of a star that underwent a *supernova explosion,* seen on earth in 1054 A.D. Energy is released by the Crab Nebula at a rate of about 5×10^{31} W, about 10^5 times the rate at which the sun radiates energy. The Crab Nebula obtains its energy from the rotational kinetic energy of a rapidly spinning *neutron star* at its center. This object rotates once every 0.0331 s, and this period is increasing by 4.22×10^{-13} s for each second of time that elapses. (a) If the rate at which energy is lost by the neutron star is equal to the rate at which energy is released by the nebula, find the moment of inertia of the neutron star. (b) Theories of supernovae predict that the neutron star in the Crab Nebula has a mass about 1.4 times that of the sun. Modeling the neutron star as a solid uniform sphere, calculate its radius in kilometers. (c) What is the linear speed of a point on the equator of the neutron star? Compare to the speed of light. (d) Assume that the neutron star is uniform and calculate its density. Compare to the density of ordinary rock (3000 kg/m^3) and to the density of an atomic nucleus (about 10^{17} kg/m^3). Justify the statement that a neutron star is essentially a large atomic nucleus.

Figure **P9.98**

9.99 •• **CALC** A sphere with radius $R = 0.200$ m has density ρ that decreases with distance r from the center of the sphere according to $\rho = 3.00 \times 10^3$ kg/m$^3 - (9.00 \times 10^3$ kg/m$^4)r$. (a) Calculate the total mass of the sphere. (b) Calculate the moment of inertia of the sphere for an axis along a diameter.

CHALLENGE PROBLEMS

9.100 ••• **CALC** Calculate the moment of inertia of a uniform solid cone about an axis through its center (Fig. P9.100). The cone has mass M and altitude h. The radius of its circular base is R.

Figure **P9.100**

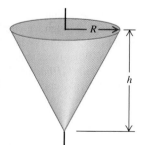

9.101 ••• **CALC** On a compact disc (CD), music is coded in a pattern of tiny pits arranged in a track that spirals outward toward the rim of the disc. As the disc spins inside a CD player, the track is scanned at a constant *linear* speed of $v = 1.25$ m/s. Because the radius of the track varies as it spirals outward, the *angular* speed of the disc must change as the CD is played. (See Exercise 9.20.) Let's see what angular acceleration is required to keep v constant. The equation of a spiral is $r(\theta) = r_0 + \beta\theta$, where r_0 is the radius of the spiral at $\theta = 0$ and β is a constant. On a CD, r_0 is the inner radius of the spiral track. If we take the rotation direction of the CD to be positive, β must be positive so that r increases as the disc turns and θ increases. (a) When the disc rotates through a small angle $d\theta$, the distance scanned along the track is $ds = r\,d\theta$. Using the above expression for $r(\theta)$, integrate ds to find the total distance s scanned along the track as a function of the total angle θ through which the disc has rotated. (b) Since the track is scanned at a constant linear speed v, the distance s found in part (a) is equal to vt. Use this to find θ as a function of time. There will be two solutions for θ; choose the positive one, and explain why this is the solution to choose. (c) Use your expres-

sion for $\theta(t)$ to find the angular velocity ω_z and the angular acceleration α_z as functions of time. Is α_z constant? (d) On a CD, the inner radius of the track is 25.0 mm, the track radius increases by 1.55 μm per revolution, and the playing time is 74.0 min. Find the values of r_0 and β, and find the total number of revolutions made during the playing time. (e) Using your results from parts (c) and (d), make graphs of ω_z (in rad/s) versus t and α_z (in rad/s^2) versus t between $t = 0$ and $t = 74.0$ min.

Answers

Chapter Opening Question

Both segments of the rigid blade have the same angular speed ω. From Eqs. (9.13) and (9.15), doubling the distance r for the same ω doubles the linear speed $v = r\omega$ and doubles the radial acceleration $a_{\text{rad}} = \omega^2 r$.

Test Your Understanding Questions

9.1 Answers: (a) (i) and (iii), (b) (ii) The rotation is speeding up when the angular velocity and angular acceleration have the same sign, and slowing down when they have opposite signs. Hence it is speeding up for $0 < t < 2$ s (ω_z and α_z are both positive) and for 4 s $< t < 6$ s (ω_z and α_z are both negative), but is slowing down for 2 s $< t < 4$ s (ω_z is positive and α_z is negative). Note that the body is rotating in one direction for $t < 4$ s (ω_z is positive) and in the opposite direction for $t > 4$ s (ω_z is negative).

9.2 Answers: (a) (i), (b) (ii) When the disc comes to rest, $\omega_z = 0$. From Eq. (9.7), the *time* when this occurs is $t = (\omega_z - \omega_{0z})/\alpha_z = -\omega_{0z}/\alpha_z$ (this is a positive time because α_z is negative). If we double the initial angular velocity ω_{0z} and also double the angular acceleration α_z, their ratio is unchanged and the rotation stops in the same amount of time. The *angle* through which the disc rotates is given by Eq. (9.10): $\theta - \theta_0 = \frac{1}{2}(\omega_{0z} + \omega_z)t = \frac{1}{2}\omega_{0z}t$ (since the final angular velocity is $\omega_z = 0$). The initial angular velocity ω_{0z} has been doubled but the time t is the same, so the angular displacement $\theta - \theta_0$ (and hence the number of revolutions) has doubled. You can also come to the same conclusion using Eq. (9.12).

9.3 Answer: (ii) From Eq. (9.13), $v = r\omega$. To maintain a constant linear speed v, the angular speed ω must decrease as the scanning head moves outward (greater r).

9.4 Answer: (i) The kinetic energy in the falling block is $\frac{1}{2}mv^2$, and the kinetic energy in the rotating cylinder is $\frac{1}{2}I\omega^2 = \frac{1}{2}\left(\frac{1}{2}mR^2\right)\left(\frac{v}{R}\right)^2 = \frac{1}{4}mv^2$. Hence the total kinetic energy of the system is $\frac{3}{4}mv^2$, of which two-thirds is in the block and one-third is in the cylinder.

9.5 Answer: (ii) More of the mass of the pool cue is concentrated at the thicker end, so the center of mass is closer to that end. The moment of inertia through a point P at either end is $I_P = I_{\text{cm}} + Md^2$; the thinner end is farther from the center of mass, so the distance d and the moment of inertia I_P are greater for the thinner end.

9.6 Answer: (iii) Our result from Example 9.10 does *not* depend on the cylinder length L. The moment of inertia depends only on the *radial* distribution of mass, not on its distribution along the axis.

Bridging Problem

Answers: (a) $I = \left[\dfrac{M}{L}\left(\dfrac{x^3}{3}\right)\right]_{-h}^{L-h} = \frac{1}{3}M(L^2 - 3Lh + 3h^2)$

(b) $W = \frac{1}{6}M(L^2 - 3Lh + 3h^2)\alpha^2 t^2$

(c) $a = (L - h)\alpha\sqrt{1 + \alpha^2 t^4}$

DYNAMICS OF ROTATIONAL MOTION

? If you stand at the north pole, the north star, Polaris, is almost directly overhead, and the other stars appear to trace circles around it. But 5000 years ago a different star, Thuban, was directly above the north pole and was the north star. What caused this change?

We learned in Chapters 4 and 5 that a net force applied to a body gives that body an acceleration. But what does it take to give a body an *angular* acceleration? That is, what does it take to start a stationary body rotating or to bring a spinning body to a halt? A force is required, but it must be applied in a way that gives a twisting or turning action.

In this chapter we will define a new physical quantity, *torque,* that describes the twisting or turning effort of a force. We'll find that the net torque acting on a rigid body determines its angular acceleration, in the same way that the net force on a body determines its linear acceleration. We'll also look at work and power in rotational motion so as to understand such problems as how energy is transmitted by the rotating drive shaft in a car. Finally, we will develop a new conservation principle, *conservation of angular momentum,* that is tremendously useful for understanding the rotational motion of both rigid and nonrigid bodies. We'll finish this chapter by studying *gyroscopes,* rotating devices that seemingly defy common sense and don't fall over when you might think they should—but that actually behave in perfect accordance with the dynamics of rotational motion.

10.1 Torque

We know that forces acting on a body can affect its **translational motion**—that is, the motion of the body as a whole through space. Now we want to learn which aspects of a force determine how effective it is in causing or changing *rotational* motion. The magnitude and direction of the force are important, but so is the point on the body where the force is applied. In Fig. 10.1 a wrench is being used to loosen a tight bolt. Force $\vec{F}_b$, applied near the end of the handle, is more effective than an equal force $\vec{F}_a$ applied near the bolt. Force $\vec{F}_c$ doesn't do any good at all; it's applied at the same point and has the same magnitude as $\vec{F}_b$, but

10.1 Which of these three equal-magnitude forces is most likely to loosen the tight bolt?

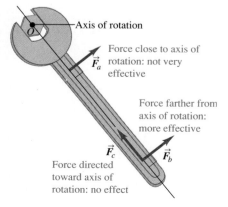

it's directed along the length of the handle. The quantitative measure of the tendency of a force to cause or change a body's rotational motion is called *torque;* we say that $\vec{F}_a$ applies a torque about point O to the wrench in Fig. 10.1, $\vec{F}_b$ applies a greater torque about O, and $\vec{F}_c$ applies zero torque about O.

Figure 10.2 shows three examples of how to calculate torque. The body in the figure can rotate about an axis that is perpendicular to the plane of the figure and passes through point O. Three forces, $\vec{F}_1$, $\vec{F}_2$, and $\vec{F}_3$, act on the body in the plane of the figure. The tendency of the first of these forces, $\vec{F}_1$, to cause a rotation about O depends on its magnitude F_1. It also depends on the *perpendicular* distance l_1 between point O and the **line of action** of the force (that is, the line along which the force vector lies). We call the distance l_1 the **lever arm** (or **moment arm**) of force $\vec{F}_1$ about O. The twisting effort is directly proportional to both F_1 and l_1, so we define the **torque** (or *moment*) of the force $\vec{F}_1$ with respect to O as the product $F_1 l_1$. We use the Greek letter τ (tau) for torque. In general, for a force of magnitude F whose line of action is a perpendicular distance l from O, the torque is

$$\tau = Fl \qquad (10.1)$$

Physicists usually use the term "torque," while engineers usually use "moment" (unless they are talking about a rotating shaft). Both groups use the term "lever arm" or "moment arm" for the distance l.

The lever arm of $\vec{F}_1$ in Fig. 10.2 is the perpendicular distance l_1, and the lever arm of $\vec{F}_2$ is the perpendicular distance l_2. The line of action of $\vec{F}_3$ passes through point O, so the lever arm for $\vec{F}_3$ is zero and its torque with respect to O is zero. In the same way, force $\vec{F}_c$ in Fig. 10.1 has zero torque with respect to point O; $\vec{F}_b$ has a greater torque than $\vec{F}_a$ because its lever arm is greater.

CAUTION **Torque is always measured about a point** Note that torque is *always* defined with reference to a specific point. If we shift the position of this point, the torque of each force may also change. For example, the torque of force $\vec{F}_3$ in Fig. 10.2 is zero with respect to point O, but the torque of $\vec{F}_3$ is *not* zero about point A. It's not enough to refer to "the torque of $\vec{F}$"; you must say "the torque of $\vec{F}$ with respect to point X" or "the torque of $\vec{F}$ about point X." ▮

Force $\vec{F}_1$ in Fig. 10.2 tends to cause *counterclockwise* rotation about O, while $\vec{F}_2$ tends to cause *clockwise* rotation. To distinguish between these two possibilities, we need to choose a positive sense of rotation. With the choice that *counterclockwise torques are positive and clockwise torques are negative,* the torques of $\vec{F}_1$ and $\vec{F}_2$ about O are

$$\tau_1 = +F_1 l_1 \qquad \tau_2 = -F_2 l_2$$

Figure 10.2 shows this choice for the sign of torque. We will often use the symbol $\circlearrowleft_+$ to indicate our choice of the positive sense of rotation.

The SI unit of torque is the newton-meter. In our discussion of work and energy we called this combination the joule. But torque is *not* work or energy, and torque should be expressed in newton-meters, *not* joules.

Figure 10.3 shows a force $\vec{F}$ applied at a point P described by a position vector $\vec{r}$ with respect to the chosen point O. There are three ways to calculate the torque of this force:

1. Find the lever arm l and use $\tau = Fl$.
2. Determine the angle ϕ between the vectors $\vec{r}$ and $\vec{F}$; the lever arm is $r \sin \phi$, so $\tau = rF \sin \phi$.
3. Represent $\vec{F}$ in terms of a radial component F_{rad} along the direction of $\vec{r}$ and a tangential component F_{tan} at right angles, perpendicular to $\vec{r}$. (We call this a tangential component because if the body rotates, the point where the force acts moves in a circle, and this component is tangent to that circle.) Then

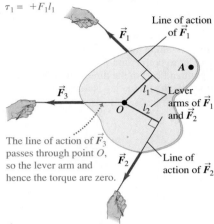

10.2 The torque of a force about a point is the product of the force magnitude and the lever arm of the force.

$\vec{F}_1$ tends to cause *counterclockwise* rotation about point O, so its torque is *positive:* $\tau_1 = +F_1 l_1$

The line of action of $\vec{F}_3$ passes through point O, so the lever arm and hence the torque are zero.

$\vec{F}_2$ tends to cause *clockwise* rotation about point O, so its torque is *negative:* $\tau_2 = -F_2 l_2$

10.3 Three ways to calculate the torque of the force $\vec{F}$ about the point O. In this figure, $\vec{r}$ and $\vec{F}$ are in the plane of the page and the torque vector $\vec{\tau}$ points out of the page toward you.

Three ways to calculate torque:
$$\tau = Fl = rF \sin \phi = F_{tan} r$$

$F_{tan} = F \sin \phi$

$F_{rad} = F \cos \phi$

$\vec{\tau}$ (out of page)

Line of action of $\vec{F}$

$l = r \sin \phi$ = lever arm

$F_{\text{tan}} = F\sin\phi$ and $\tau = r(F\sin\phi) = F_{\text{tan}}r$. The component F_{rad} produces *no* torque with respect to O because its lever arm with respect to that point is zero (compare to forces $\vec{F}_c$ in Fig. 10.1 and $\vec{F}_3$ in Fig. 10.2).

Summarizing these three expressions for torque, we have

$$\tau = Fl = rF\sin\phi = F_{\text{tan}}r \qquad \text{(magnitude of torque)} \qquad (10.2)$$

Torque as a Vector

We saw in Section 9.1 that angular velocity and angular acceleration can be represented as vectors; the same is true for torque. To see how to do this, note that the quantity $rF\sin\phi$ in Eq. (10.2) is the magnitude of the *vector product* $\vec{r} \times \vec{F}$ that we defined in Section 1.10. (You should go back and review that definition.) We now generalize the definition of torque as follows: When a force $\vec{F}$ acts at a point having a position vector $\vec{r}$ with respect to an origin O, as in Fig. 10.3, the torque $\vec{\tau}$ of the force with respect to O is the *vector* quantity

$$\vec{\tau} = \vec{r} \times \vec{F} \qquad \text{(definition of torque vector)} \qquad (10.3)$$

The torque as defined in Eq. (10.2) is just the magnitude of the torque vector $\vec{r} \times \vec{F}$. The direction of $\vec{\tau}$ is perpendicular to both $\vec{r}$ and $\vec{F}$. In particular, if both $\vec{r}$ and $\vec{F}$ lie in a plane perpendicular to the axis of rotation, as in Fig. 10.3, then the torque vector $\vec{\tau} = \vec{r} \times \vec{F}$ is directed along the axis of rotation, with a sense given by the right-hand rule (Fig. 1.29). Figure 10.4 shows the direction relationships.

In diagrams that involve $\vec{r}$, $\vec{F}$, and $\vec{\tau}$, it's common to have one of the vectors oriented perpendicular to the page. (Indeed, by the very nature of the cross product, $\vec{\tau} = \vec{r} \times \vec{F}$ *must* be perpendicular to the plane of the vectors $\vec{r}$ and $\vec{F}$.) We use a dot (•) to represent a vector that points out of the page (see Fig. 10.3) and a cross (✕) to represent a vector that points into the page.

In the following sections we will usually be concerned with rotation of a body about an axis oriented in a specified constant direction. In that case, only the component of torque along that axis is of interest, and we often call that component the torque with respect to the specified *axis*.

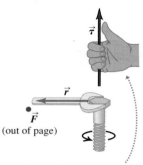

10.4 The torque vector $\vec{\tau} = \vec{r} \times \vec{F}$ is directed along the axis of the bolt, perpendicular to both $\vec{r}$ and $\vec{F}$. The fingers of the right hand curl in the direction of the rotation that the torque tends to cause.

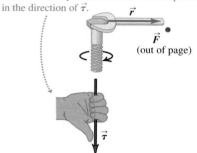

If you point the fingers of your right hand in the direction of $\vec{r}$ and then curl them in the direction of $\vec{F}$, your outstretched thumb points in the direction of $\vec{\tau}$.

Example 10.1 **Applying a torque**

To loosen a pipe fitting, a weekend plumber slips a piece of scrap pipe (a "cheater") over his wrench handle. He stands on the end of the cheater, applying his full 900-N weight at a point 0.80 m from the center of the fitting (Fig. 10.5a). The wrench handle and cheater make an angle of 19° with the horizontal. Find the magnitude and direction of the torque he applies about the center of the fitting.

10.5 (a) A weekend plumber tries to loosen a pipe fitting by standing on a "cheater." (b) Our vector diagram to find the torque about O.

(a) Diagram of situation

(b) Free-body diagram

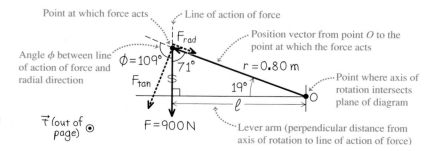

SOLUTION

IDENTIFY and SET UP: Figure 10.5b shows the vectors $\vec{r}$ and $\vec{F}$ and the angle between them ($\phi = 109°$). Equation (10.1) or (10.2) will tell us the magnitude of the torque. The right-hand rule with Eq. (10.3), $\vec{\tau} = \vec{r} \times \vec{F}$, will tell us the direction of the torque.

EXECUTE: To use Eq. (10.1), we first calculate the lever arm l. As Fig. 10.5b shows,

$$l = r\sin\phi = (0.80 \text{ m})\sin 109° = (0.80 \text{ m})\sin 71° = 0.76 \text{ m}$$

Then Eq. (10.1) tells us that the magnitude of the torque is

$$\tau = Fl = (900 \text{ N})(0.76 \text{ m}) = 680 \text{ N} \cdot \text{m}$$

We get the same result from Eq. (10.2):

$$\tau = rF\sin\phi = (0.80 \text{ m})(900 \text{ N})(\sin 109°) = 680 \text{ N} \cdot \text{m}$$

Alternatively, we can find F_{tan}, the tangential component of $\vec{F}$ that acts perpendicular to $\vec{r}$. Figure 10.5b shows that this component is at an angle of $109° - 90° = 19°$ from $\vec{F}$, so $F_{\text{tan}} = F\sin\phi = F(\cos 19°) = (900 \text{ N})(\cos 19°) = 851 \text{ N}$. Then, from Eq. 10.2,

$$\tau = F_{\text{tan}}r = (851 \text{ N})(0.80 \text{ m}) = 680 \text{ N} \cdot \text{m}$$

Curl the fingers of your right hand from the direction of $\vec{r}$ (in the plane of Fig. 10.5b, to the left and up) into the direction of $\vec{F}$ (straight down). Then your right thumb points out of the plane of the figure: This is the direction of $\vec{\tau}$.

EVALUATE: To check the direction of $\vec{\tau}$, note that the force in Fig. 10.5 tends to produce a counterclockwise rotation about O. If you curl the fingers of your right hand in a counterclockwise direction, the thumb points out of the plane of Fig. 10.5, which is indeed the direction of the torque.

Test Your Understanding of Section 10.1 The figure shows a force P being applied to one end of a lever of length L. What is the magnitude of the torque of this force about point A? (i) $PL\sin\theta$; (ii) $PL\cos\theta$; (iii) $PL\tan\theta$.

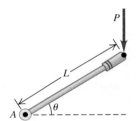

10.2 Torque and Angular Acceleration for a Rigid Body

We are now ready to develop the fundamental relationship for the rotational dynamics of a rigid body. We will show that the angular acceleration of a rotating rigid body is directly proportional to the sum of the torque components along the axis of rotation. The proportionality factor is the moment of inertia.

To develop this relationship, we again imagine the body as being made up of a large number of particles. We choose the axis of rotation to be the z-axis; the first particle has mass m_1 and distance r_1 from this axis (Fig. 10.6). The *net force* $\vec{F}_1$ acting on this particle has a component $F_{1,\text{rad}}$ along the radial direction, a component $F_{1,\text{tan}}$ that is tangent to the circle of radius r_1 in which the particle moves as the body rotates, and a component F_{1z} along the axis of rotation. Newton's second law for the tangential component is

$$F_{1,\text{tan}} = m_1 a_{1,\text{tan}} \qquad (10.4)$$

We can express the tangential acceleration of the first particle in terms of the angular acceleration α_z of the body using Eq. (9.14): $a_{1,\text{tan}} = r_1\alpha_z$. Using this relationship and multiplying both sides of Eq. (10.4) by r_1, we obtain

$$F_{1,\text{tan}}r_1 = m_1 r_1^2 \alpha_z \qquad (10.5)$$

From Eq. (10.2), $F_{1,\text{tan}}r_1$ is just the *torque* of the net force with respect to the rotation axis, equal to the component τ_{1z} of the torque vector along the rotation axis. The subscript z is a reminder that the torque affects rotation around the z-axis, in the same way that the subscript on F_{1z} is a reminder that this force affects the motion of particle 1 along the z-axis.

10.6 As a rigid body rotates around the z-axis, a net force $\vec{F}_1$ acts on one particle of the body. Only the force component $F_{1,\text{tan}}$ can affect the rotation, because only $F_{1,\text{tan}}$ exerts a torque about O with a z-component (along the rotation axis).

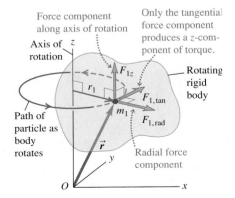

Neither of the components $F_{1,\text{rad}}$ or F_{1z} contributes to the torque about the z-axis, since neither tends to change the particle's rotation about that axis. So $\tau_{1z} = F_{1,\text{tan}}r_1$ is the total torque acting on the particle with respect to the rotation axis. Also, $m_1r_1^2$ is I_1, the moment of inertia of the particle about the rotation axis. Hence we can rewrite Eq. (10.5) as

$$\tau_{1z} = I_1\alpha_z = m_1r_1^2\alpha_z$$

We write an equation like this for every particle in the body and then add all these equations:

$$\tau_{1z} + \tau_{2z} + \cdots = I_1\alpha_z + I_2\alpha_z + \cdots = m_1r_1^2\alpha_z + m_2r_2^2\alpha_z + \cdots$$

or

$$\sum \tau_{iz} = \left(\sum m_ir_i^2\right)\alpha_z \qquad (10.6)$$

The left side of Eq. (10.6) is the sum of all the torques about the rotation axis that act on all the particles. The right side is $I = \sum m_ir_i^2$, the total moment of inertia about the rotation axis, multiplied by the angular acceleration α_z. Note that α_z is the same for every particle because this is a *rigid* body. Thus for the rigid body as a whole, Eq. (10.6) is the *rotational analog of Newton's second law:*

$$\sum \tau_z = I\alpha_z \qquad (10.7)$$

(rotational analog of Newton's second law for a rigid body)

Just as Newton's second law says that the net force on a particle equals the particle's mass times its acceleration, Eq. (10.7) says that the net torque on a rigid body equals the body's moment of inertia about the rotation axis times its angular acceleration (Fig. 10.7).

Note that because our derivation assumed that the angular acceleration α_z is the same for all particles in the body, Eq. (10.7) is valid *only* for *rigid* bodies. Hence this equation doesn't apply to a rotating tank of water or a swirling tornado of air, different parts of which have different angular accelerations. Also note that since our derivation used Eq. (9.14), $a_{\text{tan}} = r\alpha_z$, α_z must be measured in rad/s^2.

The torque on each particle is due to the net force on that particle, which is the vector sum of external and internal forces (see Section 8.2). According to Newton's third law, the *internal* forces that any pair of particles in the rigid body exert on each other are equal and opposite (Fig. 10.8). If these forces act along the line joining the two particles, their lever arms with respect to any axis are also equal. So the torques for each such pair are equal and opposite, and add to zero. Hence *all* the internal torques add to zero, so the sum $\sum \tau_z$ in Eq. (10.7) includes only the torques of the *external* forces.

Often, an important external force acting on a body is its *weight*. This force is not concentrated at a single point; it acts on every particle in the entire body. Nevertheless, it turns out that if $\vec{g}$ has the same value at all points, we always get the correct torque (about any specified axis) if we assume that all the weight is concentrated at the *center of mass* of the body. We will prove this statement in Chapter 11, but meanwhile we will use it for some of the problems in this chapter.

10.7 Loosening or tightening a screw requires giving it an angular acceleration and hence applying a torque. This is made easier by using a screwdriver with a large-radius handle, which provides a large lever arm for the force you apply with your hand.

10.8 Two particles in a rigid body exert equal and opposite forces on each other. If the forces act along the line joining the particles, the lever arms of the forces with respect to an axis through O are the same and the torques due to the two forces are equal and opposite. Only *external* torques affect the body's rotation.

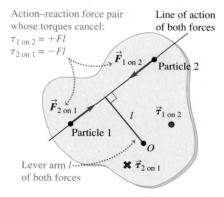

Action–reaction force pair whose torques cancel:
$\tau_{1\text{ on }2} = +Fl$
$\tau_{2\text{ on }1} = -Fl$

$\vec{F}_{1\text{ on }2}$ Particle 2

Line of action of both forces

$\vec{F}_{2\text{ on }1}$ l $\vec{\tau}_{1\text{ on }2}$

Particle 1

O

Lever arm l of both forces $\vec{\tau}_{2\text{ on }1}$

Problem-Solving Strategy 10.1 | Rotational Dynamics for Rigid Bodies

Our strategy for solving problems in rotational dynamics is very similar to Problem-Solving Strategy 5.2 for solving problems involving Newton's second law.

IDENTIFY *the relevant concepts:* Equation (10.7), $\Sigma \tau_z = I\alpha_z$, is useful whenever torques act on a rigid body. Sometimes you can use an energy approach instead, as we did in Section 9.4. However, if the target variable is a force, a torque, an acceleration, an angular acceleration, or an elapsed time, using $\Sigma \tau_z = I\alpha_z$ is almost always best.

SET UP *the problem* using the following steps:
1. Sketch the situation and identify the body or bodies to be analyzed. Indicate the rotation axis.
2. For each body, draw a free-body diagram that shows the *shape* of each body, including all dimensions and angles that you will need for torque calculations. Label pertinent quantities with algebraic symbols.
3. Choose coordinate axes for each body and indicate a positive sense of rotation (clockwise or counterclockwise) for each rotating body. If you know the sense of α_z, pick that as the positive sense of rotation.

EXECUTE *the solution:*
1. For each body, decide whether it undergoes translational motion, rotational motion, or both. Then apply $\Sigma \vec{F} = m\vec{a}$ (as in Section 5.2), $\Sigma \tau_z = I\alpha_z$, or both to the body.
2. Express in algebraic form any *geometrical* relationships between the motions of two or more bodies. An example is a string that unwinds, without slipping, from a pulley or a wheel that rolls without slipping (discussed in Section 10.3). These relationships usually appear as relationships between linear and/or angular accelerations.
3. Ensure that you have as many independent equations as there are unknowns. Solve the equations to find the target variables.

EVALUATE *your answer:* Check that the algebraic signs of your results make sense. As an example, if you are unrolling thread from a spool, your answers should not tell you that the spool is turning in the direction that rolls the thread back on to the spool! Check that any algebraic results are correct for special cases or for extreme values of quantities.

Example 10.2 | An unwinding cable I

Figure 10.9a shows the situation analyzed in Example 9.7 using energy methods. What is the cable's acceleration?

SOLUTION

IDENTIFY and SET UP: We can't use the energy method of Section 9.4, which doesn't involve acceleration. Instead we'll apply rotational dynamics to find the angular acceleration of the cylinder (Fig. 10.9b). We'll then find a relationship between the motion of the cable and the motion of the cylinder rim, and use this to find the acceleration of the cable. The cylinder rotates counterclockwise when the cable is pulled, so we take counterclockwise rotation to be positive. The net force on the cylinder must be zero because its center of mass remains at rest. The force F exerted by the cable produces a torque about the rotation axis. The weight (magnitude Mg) and the normal force (magnitude n) exerted by the cylinder's bearings produce *no* torque about the rotation axis because they both act along lines through that axis.

EXECUTE: The lever arm of F is equal to the radius $R = 0.060$ m of the cylinder, so the torque is $\tau_z = FR$. (This torque is positive, as it tends to cause a counterclockwise rotation.) From Table 9.2, case (f), the moment of inertia of the cylinder about the rotation axis is $I = \frac{1}{2}MR^2$. Then Eq. (10.7) tells us that

$$\alpha_z = \frac{\tau_z}{I} = \frac{FR}{MR^2/2} = \frac{2F}{MR} = \frac{2(9.0 \text{ N})}{(50 \text{ kg})(0.060 \text{ m})} = 6.0 \text{ rad/s}^2$$

(We can add "rad" to our result because radians are dimensionless.)

To get the linear acceleration of the cable, recall from Section 9.3 that the acceleration of a cable unwinding from a cylinder is the same as the tangential acceleration of a point on the surface of the cylinder where the cable is tangent to it. This tangential acceleration is given by Eq. (9.14):

$$a_{\text{tan}} = R\alpha_z = (0.060 \text{ m})(6.0 \text{ rad/s}^2) = 0.36 \text{ m/s}^2$$

EVALUATE: Can you use this result, together with an equation from Chapter 2, to determine the speed of the cable after it has been pulled 2.0 m? Does your result agree with that of Example 9.7?

10.9 (a) Cylinder and cable. (b) Our free-body diagram for the cylinder.

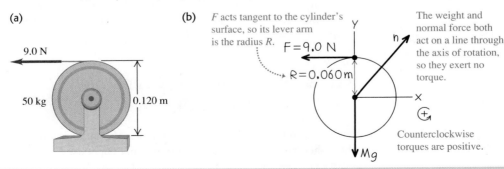

(a)

9.0 N

50 kg   0.120 m

(b) *F* acts tangent to the cylinder's surface, so its lever arm is the radius *R*.

F = 9.0 N

R = 0.060 m

Y

n

The weight and normal force both act on a line through the axis of rotation, so they exert no torque.

X

Counterclockwise torques are positive.

Mg

| Example 10.3 | An unwinding cable II |

In Example 9.8 (Section 9.4), what are the acceleration of the falling block and the tension in the cable?

SOLUTION

IDENTIFY and SET UP: We'll apply translational dynamics to the block and rotational dynamics to the cylinder. As in Example 10.2, we'll relate the linear acceleration of the block (our target variable) to the angular acceleration of the cylinder. Figure 10.10 shows our sketch of the situation and a free-body diagram for each body. We take the positive sense of rotation for the cylinder to be counterclockwise and the positive direction of the y-coordinate for the block to be downward.

10.10 (a) Our diagram of the situation. (b) Our free-body diagrams for the cylinder and the block. We assume the cable has negligible mass.

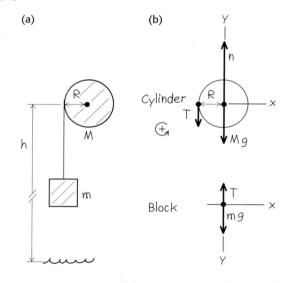

EXECUTE: For the block, Newton's second law gives

$$\sum F_y = mg + (-T) = ma_y$$

For the cylinder, the only torque about its axis is that due to the cable tension T. Hence Eq. (10.7) gives

$$\sum \tau_z = RT = I\alpha_z = \tfrac{1}{2}MR^2\alpha_z$$

As in Example 10.2, the acceleration of the cable is the same as the tangential acceleration of a point on the cylinder rim. From Eq. (9.14), this acceleration is $a_y = a_{\text{tan}} = R\alpha_z$. We use this to replace $R\alpha_z$ with a_y in the cylinder equation above, and then divide by R. The result is $T = \tfrac{1}{2}Ma_y$. Now we substitute this expression for T into Newton's second law for the block and solve for the acceleration a_y:

$$mg - \tfrac{1}{2}Ma_y = ma_y$$

$$a_y = \frac{g}{1 + M/2m}$$

To find the cable tension T, we substitute our expression for a_y into the block equation:

$$T = mg - ma_y = mg - m\left(\frac{g}{1 + M/2m}\right) = \frac{mg}{1 + 2m/M}$$

EVALUATE: The acceleration is positive (in the downward direction) and less than g, as it should be, since the cable is holding back the block. The cable tension is *not* equal to the block's weight mg; if it were, the block could not accelerate.

Let's check some particular cases. When M is much larger than m, the tension is nearly equal to mg and the acceleration is correspondingly much less than g. When M is zero, $T = 0$ and $a_y = g$; the object falls freely. If the object starts from rest $(v_{0y} = 0)$ a height h above the floor, its y-velocity when it strikes the ground is given by $v_y^2 = v_{0y}^2 + 2a_yh = 2a_yh$, so

$$v_y = \sqrt{2a_yh} = \sqrt{\frac{2gh}{1 + M/2m}}$$

We found this same result from energy considerations in Example 9.8.

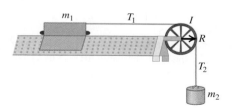

Test Your Understanding of Section 10.2 The figure shows a glider of mass m_1 that can slide without friction on a horizontal air track. It is attached to an object of mass m_2 by a massless string. The pulley has radius R and moment of inertia I about its axis of rotation. When released, the hanging object accelerates downward, the glider accelerates to the right, and the string turns the pulley without slipping or stretching. Rank the magnitudes of the following forces that act during the motion, in order from largest to smallest magnitude. (i) the tension force (magnitude T_1) in the horizontal part of the string; (ii) the tension force (magnitude T_2) in the vertical part of the string; (iii) the weight m_2g of the hanging object. ∣

10.3 Rigid-Body Rotation About a Moving Axis

We can extend our analysis of the dynamics of rotational motion to some cases in which the axis of rotation moves. When that happens, the motion of the body is **combined translation and rotation.** The key to understanding such situations is this: Every possible motion of a rigid body can be represented as a combination of *translational motion of the center of mass* and *rotation about an axis through the center of mass*. This is true even when the center of mass accelerates, so that it is not at rest in any inertial frame. Figure 10.11 illustrates this for the motion of a tossed baton: The center of mass of the baton follows a parabolic curve, as though the baton were a particle located at the center of mass. Other examples of combined translational and rotational motions include a ball rolling down a hill and a yo-yo unwinding at the end of a string.

Combined Translation and Rotation: Energy Relationships

It's beyond the scope of this book to prove that the motion of a rigid body can always be divided into translation of the center of mass and rotation about the center of mass. But we can show that this is true for the *kinetic energy* of a rigid body that has both translational and rotational motions. In this case, the body's kinetic energy is the sum of a part $\frac{1}{2}Mv_{\text{cm}}^2$ associated with motion of the center of mass and a part $\frac{1}{2}I_{\text{cm}}\omega^2$ associated with rotation about an axis through the center of mass:

$$K = \tfrac{1}{2}Mv_{\text{cm}}^2 + \tfrac{1}{2}I_{\text{cm}}\omega^2 \tag{10.8}$$

(rigid body with both translation and rotation)

To prove this relationship, we again imagine the rigid body to be made up of particles. Consider a typical particle with mass m_i as shown in Fig. 10.12. The velocity $\vec{v}_i$ of this particle relative to an inertial frame is the vector sum of the velocity $\vec{v}_{\text{cm}}$ of the center of mass and the velocity $\vec{v}_i'$ of the particle *relative to the center of mass*:

$$\vec{v}_i = \vec{v}_{\text{cm}} + \vec{v}_i' \tag{10.9}$$

The kinetic energy K_i of this particle in the inertial frame is $\frac{1}{2}m_iv_i^2$, which we can also express as $\frac{1}{2}m_i(\vec{v}_i \cdot \vec{v}_i)$. Substituting Eq. (10.9) into this, we get

$$K_i = \tfrac{1}{2}m_i(\vec{v}_{\text{cm}} + \vec{v}_i') \cdot (\vec{v}_{\text{cm}} + \vec{v}_i')$$

$$= \tfrac{1}{2}m_i(\vec{v}_{\text{cm}} \cdot \vec{v}_{\text{cm}} + 2\vec{v}_{\text{cm}} \cdot \vec{v}_i' + \vec{v}_i' \cdot \vec{v}_i')$$

$$= \tfrac{1}{2}m_i(v_{\text{cm}}^2 + 2\vec{v}_{\text{cm}} \cdot \vec{v}_i' + v_i'^2)$$

The total kinetic energy is the sum $\sum K_i$ for all the particles making up the body. Expressing the three terms in this equation as separate sums, we get

$$K = \sum K_i = \sum \left(\tfrac{1}{2}m_iv_{\text{cm}}^2\right) + \sum (m_i\vec{v}_{\text{cm}} \cdot \vec{v}_i') + \sum \left(\tfrac{1}{2}m_iv_i'^2\right)$$

The first and second terms have common factors that can be taken outside the sum:

$$K = \tfrac{1}{2}\left(\sum m_i\right)v_{\text{cm}}^2 + \vec{v}_{\text{cm}} \cdot \left(\sum m_i\vec{v}_i'\right) + \sum \left(\tfrac{1}{2}m_iv_i'^2\right) \tag{10.10}$$

Now comes the reward for our effort. In the first term, $\sum m_i$ is the total mass M. The second term is zero because $\sum m_i\vec{v}_i'$ is M times the velocity of the center of mass *relative to the center of mass*, and this is zero by definition. The last term is the sum of the kinetic energies of the particles computed by using their speeds with respect to the center of mass; this is just the kinetic energy of rotation

10.11 The motion of a rigid body is a combination of translational motion of the center of mass and rotation around the center of mass.

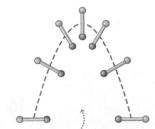

This baton toss can be represented as a combination of ...

... **rotation** about the center of mass ...   ... plus **translation** of the center of mass.

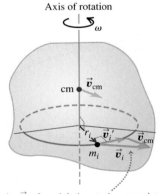

10.12 A rigid body with both translation and rotation.

Velocity $\vec{v}_i$ of particle in rotating, translating rigid body = (velocity $\vec{v}_{\text{cm}}$ of center of mass) + (particle's velocity $\vec{v}_i'$ relative to center of mass)

Application Combined Translation and Rotation

A maple seed consists of a pod attached to a much lighter, flattened wing. Airflow around the wing slows the fall to about 1 m/s and causes the seed to rotate about its center of mass. The seed's slow fall means that a breeze can carry the seed some distance from the parent tree. In the absence of wind, the seed's center of mass falls straight down.

Maple seed

Maple seed falling

ActivPhysics 7.11: Race Between a Block and a Disk

around the center of mass. Using the same steps that led to Eq. (9.17) for the rotational kinetic energy of a rigid body, we can write this last term as $\frac{1}{2}I_{cm}\omega^2$, where I_{cm} is the moment of inertia with respect to the axis through the center of mass and ω is the angular speed. So Eq. (10.10) becomes Eq. (10.8):

$$K = \tfrac{1}{2}Mv_{cm}^2 + \tfrac{1}{2}I_{cm}\omega^2$$

Rolling Without Slipping

An important case of combined translation and rotation is **rolling without slipping,** such as the motion of the wheel shown in Fig. 10.13. The wheel is symmetrical, so its center of mass is at its geometric center. We view the motion in an inertial frame of reference in which the surface on which the wheel rolls is at rest. In this frame, the point on the wheel that contacts the surface must be instantaneously *at rest* so that it does not slip. Hence the velocity $\vec{v}_1'$ of the point of contact relative to the center of mass must have the same magnitude but opposite direction as the center-of-mass velocity $\vec{v}_{cm}$. If the radius of the wheel is R and its angular speed about the center of mass is ω, then the magnitude of $\vec{v}_1'$ is $R\omega$; hence we must have

$$v_{cm} = R\omega \qquad \text{(condition for rolling without slipping)} \qquad (10.11)$$

As Fig. 10.13 shows, the velocity of a point on the wheel is the vector sum of the velocity of the center of mass and the velocity of the point relative to the center of mass. Thus while point 1, the point of contact, is instantaneously at rest, point 3 at the top of the wheel is moving forward *twice as fast* as the center of mass, and points 2 and 4 at the sides have velocities at 45° to the horizontal.

At any instant we can think of the wheel as rotating about an "instantaneous axis" of rotation that passes through the point of contact with the ground. The angular velocity ω is the same for this axis as for an axis through the center of mass; an observer at the center of mass sees the rim make the same number of revolutions per second as does an observer at the rim watching the center of mass spin around him. If we think of the motion of the rolling wheel in Fig. 10.13 in this way, the kinetic energy of the wheel is $K = \frac{1}{2}I_1\omega^2$, where I_1 is the moment of inertia of the wheel about an axis through point 1. But by the parallel-axis theorem, Eq. (9.19), $I_1 = I_{cm} + MR^2$, where M is the total mass of the wheel and I_{cm} is the moment of inertia with respect to an axis through the center of mass. Using Eq. (10.11), the kinetic energy of the wheel is

$$K = \tfrac{1}{2}I_1\omega^2 = \tfrac{1}{2}I_{cm}\omega^2 + \tfrac{1}{2}MR^2\omega^2 = \tfrac{1}{2}I_{cm}\omega^2 + \tfrac{1}{2}Mv_{cm}^2$$

which is the same as Eq. (10.8).

10.13 The motion of a rolling wheel is the sum of the translational motion of the center of mass plus the rotational motion of the wheel around the center of mass.

Translation of center of mass: velocity $\vec{v}_{cm}$

Rotation around center of mass: for rolling without slipping, speed at rim = v_{cm}

Combined motion

Wheel is instantaneously at rest where it contacts the ground.

CAUTION **Rolling without slipping** Note that the relationship $v_{cm} = R\omega$ holds *only* if there is rolling without slipping. When a drag racer first starts to move, the rear tires are spinning very fast even though the racer is hardly moving, so $R\omega$ is greater than v_{cm} (Fig. 10.14). If a driver applies the brakes too heavily so that the car skids, the tires will spin hardly at all and $R\omega$ is less than v_{cm}. ▮

If a rigid body changes height as it moves, we must also consider gravitational potential energy. As we discussed in Section 9.4, the gravitational potential energy associated with any extended body of mass M, rigid or not, is the same as if we replace the body by a particle of mass M located at the body's center of mass. That is,

$$U = Mgy_{cm}$$

10.14 The smoke rising from this drag racer's rear tires shows that the tires are slipping on the road, so v_{cm} is *not* equal to $R\omega$.

Example 10.4 **Speed of a primitive yo-yo**

You make a primitive yo-yo by wrapping a massless string around a solid cylinder with mass M and radius R (Fig. 10.15). You hold the free end of the string stationary and release the cylinder from rest. The string unwinds but does not slip or stretch as the cylinder descends and rotates. Using energy considerations, find the speed v_{cm} of the center of mass of the cylinder after it has descended a distance h.

SOLUTION

IDENTIFY and SET UP: The upper end of the string is held fixed, not pulled upward, so your hand does no work on the string–cylinder system. There is friction between the string and the cylinder, but the string doesn't slip so no mechanical energy is lost. Hence we can use conservation of mechanical energy. The initial kinetic energy of the cylinder is $K_1 = 0$, and its final kinetic energy K_2 is given by

10.15 Calculating the speed of a primitive yo-yo.

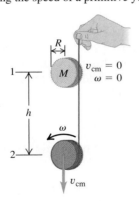

Eq. (10.8); the massless string has no kinetic energy. The moment of inertia is $I = \frac{1}{2}MR^2$, and by Eq. (9.13) $\omega = v_{cm}/R$ because the string doesn't slip. The potential energies are $U_1 = Mgh$ and $U_2 = 0$.

EXECUTE: From Eq. (10.8), the kinetic energy at point 2 is

$$K_2 = \frac{1}{2}Mv_{cm}^2 + \frac{1}{2}\left(\frac{1}{2}MR^2\right)\left(\frac{v_{cm}}{R}\right)^2$$

$$= \frac{3}{4}Mv_{cm}^2$$

The kinetic energy is $1\frac{1}{2}$ times what it would be if the yo-yo were falling at speed v_{cm} without rotating. Two-thirds of the total kinetic energy $\left(\frac{1}{2}Mv_{cm}^2\right)$ is translational and one-third $\left(\frac{1}{4}Mv_{cm}^2\right)$ is rotational. Using conservation of energy,

$$K_1 + U_1 = K_2 + U_2$$

$$0 + Mgh = \frac{3}{4}Mv_{cm}^2 + 0$$

$$v_{cm} = \sqrt{\tfrac{4}{3}gh}$$

EVALUATE: No mechanical energy was lost or gained, so from the energy standpoint the string is merely a way to convert some of the gravitational potential energy (which is released as the cylinder falls) into rotational kinetic energy rather than translational kinetic energy. Because not all of the released energy goes into translation, v_{cm} is less than the speed $\sqrt{2gh}$ of an object dropped from height h with no strings attached.

Example 10.5 **Race of the rolling bodies**

In a physics demonstration, an instructor "races" various bodies that roll without slipping from rest down an inclined plane (Fig. 10.16). What shape should a body have to reach the bottom of the incline first?

SOLUTION

IDENTIFY and SET UP: Kinetic friction does no work if the bodies roll without slipping. We can also ignore the effects of *rolling friction*, introduced in Section 5.3, if the bodies and the surface of the

Continued

incline are rigid. (Later in this section we'll explain why this is so.) We can therefore use conservation of energy. Each body starts from rest at the top of an incline with height h, so $K_1 = 0$, $U_1 = Mgh$, and $U_2 = 0$. Equation (10.8) gives the kinetic energy at the bottom of the incline; since the bodies roll without slipping, $\omega = v_{cm}/R$. We can express the moments of inertia of the four round bodies in Table 9.2, cases (f)–(i), as $I_{cm} = cMR^2$, where c is a number less than or equal to 1 that depends on the shape of the body. Our goal is to find the value of c that gives the body the greatest speed v_{cm} after its center of mass has descended a vertical distance h.

EXECUTE: From conservation of energy,

$$K_1 + U_1 = K_2 + U_2$$

$$0 + Mgh = \tfrac{1}{2}Mv_{cm}^2 + \tfrac{1}{2}cMR^2\left(\frac{v_{cm}}{R}\right)^2 + 0$$

$$Mgh = \tfrac{1}{2}(1 + c)Mv_{cm}^2$$

$$v_{cm} = \sqrt{\frac{2gh}{1 + c}}$$

10.16 Which body rolls down the incline fastest, and why?

EVALUATE: For a given value of c, the speed v_{cm} after descending a distance h is *independent* of the body's mass M and radius R. Hence *all* uniform solid cylinders $\left(c = \tfrac{1}{2}\right)$ have the same speed at the bottom, regardless of their mass and radii. The values of c tell us that the order of finish for uniform bodies will be as follows: (1) any solid sphere $\left(c = \tfrac{2}{5}\right)$, (2) any solid cylinder $\left(c = \tfrac{1}{2}\right)$, (3) any thin-walled, hollow sphere $\left(c = \tfrac{2}{3}\right)$, and (4) any thin-walled, hollow cylinder $(c = 1)$. Small-c bodies always beat large-c bodies because less of their kinetic energy is tied up in rotation and so more is available for translation.

Combined Translation and Rotation: Dynamics

We can also analyze the combined translational and rotational motions of a rigid body from the standpoint of dynamics. We showed in Section 8.5 that for a body with total mass M, the acceleration $\vec{a}_{cm}$ of the center of mass is the same as that of a point mass M acted on by all the external forces on the actual body:

$$\sum \vec{F}_{ext} = M\vec{a}_{cm} \tag{10.12}$$

The rotational motion about the center of mass is described by the rotational analog of Newton's second law, Eq. (10.7):

$$\sum \tau_z = I_{cm}\alpha_z \tag{10.13}$$

10.17 The axle of a bicycle wheel passes through the wheel's center of mass and is an axis of symmetry. Hence the rotation of the wheel is described by Eq. (10.13), provided the bicycle doesn't turn or tilt to one side (which would change the orientation of the axle).

where I_{cm} is the moment of inertia with respect to an axis through the center of mass and the sum $\sum \tau_z$ includes all external torques with respect to this axis. It's not immediately obvious that Eq. (10.13) should apply to the motion of a translating rigid body; after all, our derivation of $\sum \tau_z = I\alpha_z$ in Section 10.2 assumed that the axis of rotation was stationary. But in fact, Eq. (10.13) is valid *even when the axis of rotation moves,* provided the following two conditions are met:

1. The axis through the center of mass must be an axis of symmetry.
2. The axis must not change direction.

These conditions are satisfied for many types of rotation (Fig. 10.17). Note that in general this moving axis of rotation is *not* at rest in an inertial frame of reference.

We can now solve dynamics problems involving a rigid body that undergoes translational and rotational motions at the same time, provided that the rotation axis satisfies the two conditions just mentioned. Problem-Solving Strategy 10.1 (Section 10.2) is equally useful here, and you should review it now. Keep in mind that when a body undergoes translational and rotational motions at the same time, we need two separate equations of motion *for the same body.* One of these, Eq. (10.12), describes the translational motion of the center of mass. The other equation of motion, Eq. (10.13), describes the rotational motion about the axis through the center of mass.

Example 10.6 Acceleration of a primitive yo-yo

For the primitive yo-yo in Example 10.4 (Fig. 10.18a), find the downward acceleration of the cylinder and the tension in the string.

SOLUTION

IDENTIFY and SET UP: Figure 10.18b shows our free-body diagram for the yo-yo, including our choice of positive coordinate directions. Our target variables are $a_{\text{cm-}y}$ and T. We'll use Eq. (10.12) for the

10.18 Dynamics of a primitive yo-yo (see Fig. 10.15).

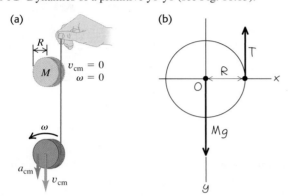

(a)

(b)

translational motion of the center of mass and Eq. (10.13) for rotational motion around the center of mass. We'll also use Eq. (10.11), which says that the string unwinds without slipping. As in Example 10.4, the moment of inertia of the yo-yo for an axis through its center of mass is $I_{\text{cm}} = \frac{1}{2}MR^2$.

EXECUTE: From Eq. (10.12),

$$\sum F_y = Mg + (-T) = Ma_{\text{cm-}y} \qquad (10.14)$$

From Eq. (10.13),

$$\sum \tau_z = TR = I_{\text{cm}}\alpha_z = \frac{1}{2}MR^2\alpha_z \qquad (10.15)$$

From Eq. (10.11), $v_{\text{cm-}z} = R\omega_z$; the derivative of this expression with respect to time gives us

$$a_{\text{cm-}y} = R\alpha_z \qquad (10.16)$$

We now use Eq. (10.16) to eliminate α_z from Eq. (10.15) and then solve Eqs. (10.14) and (10.15) simultaneously for T and $a_{\text{cm-}y}$. The results are

$$a_{\text{cm-}y} = \tfrac{2}{3}g \qquad T = \tfrac{1}{3}Mg$$

EVALUATE: The string slows the fall of the yo-yo, but not enough to stop it completely. Hence $a_{\text{cm-}y}$ is less than the free-fall value g and T is less than the yo-yo weight Mg.

Example 10.7 Acceleration of a rolling sphere

A bowling ball rolls without slipping down a ramp, which is inclined at an angle β to the horizontal (Fig. 10.19a). What are the ball's acceleration and the magnitude of the friction force on the ball? Treat the ball as a uniform solid sphere, ignoring the finger holes.

SOLUTION

IDENTIFY and SET UP: The free-body diagram (Fig. 10.19b) shows that only the friction force exerts a torque about the center of mass. Our target variables are the acceleration $a_{\text{cm-}x}$ of the ball's center of mass and the magnitude f of the friction force. (Because

10.19 A bowling ball rolling down a ramp.

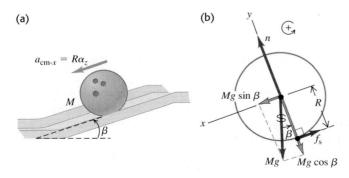

(a)

(b)

the ball does not slip at the instantaneous point of contact with the ramp, this is a *static* friction force; it prevents slipping and gives the ball its angular acceleration.) We use Eqs. (10.12) and (10.13) as in Example 10.6.

EXECUTE: The ball's moment of inertia is $I_{\text{cm}} = \frac{2}{5}MR^2$. The equations of motion are

$$\sum F_x = Mg\sin\beta + (-f) = Ma_{\text{cm-}x} \qquad (10.17)$$

$$\sum \tau_z = fR = I_{\text{cm}}\alpha_z = \left(\tfrac{2}{5}MR^2\right)\alpha_z \qquad (10.18)$$

The ball rolls without slipping, so as in Example 10.6 we use $a_{\text{cm-}x} = R\alpha_z$ to eliminate α_z from Eq. (10.18):

$$fR = \tfrac{2}{5}MRa_{\text{cm-}x}$$

This equation and Eq. (10.17) are two equations for the unknowns $a_{\text{cm-}x}$ and f. We solve Eq. (10.17) for f, substitute that expression into the above equation to eliminate f, and solve for $a_{\text{cm-}x}$:

$$a_{\text{cm-}x} = \tfrac{5}{7}g\sin\beta$$

Finally, we substitute this acceleration into Eq. (10.17) and solve for f:

$$f = \tfrac{2}{7}Mg\sin\beta$$

Continued

EVALUATE: The ball's acceleration is just $\frac{5}{7}$ as large as that of an object *sliding* down the slope without friction. If the ball descends a vertical distance h as it rolls down the ramp, its displacement along the ramp is $h/\sin\beta$. You can show that the speed of the ball at the bottom of the ramp is $v_{cm} = \sqrt{\frac{10}{7}gh}$, the same as our result from Example 10.5 with $c = \frac{2}{5}$.

If the ball were rolling *uphill* without slipping, the force of friction would still be directed uphill as in Fig. 10.19b. Can you see why?

10.20 Rolling down **(a)** a perfectly rigid surface and **(b)** a deformable surface. The deformation in part **(b)** is greatly exaggerated.

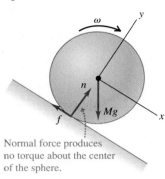

(a) Perfectly rigid sphere rolling on a perfectly rigid surface

Normal force produces no torque about the center of the sphere.

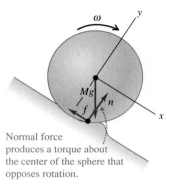

(b) Rigid sphere rolling on a deformable surface

Normal force produces a torque about the center of the sphere that opposes rotation.

Rolling Friction

In Example 10.5 we said that we can ignore rolling friction if both the rolling body and the surface over which it rolls are perfectly rigid. In Fig. 10.20a a perfectly rigid sphere is rolling down a perfectly rigid incline. The line of action of the normal force passes through the center of the sphere, so its torque is zero; there is no sliding at the point of contact, so the friction force does no work. Figure 10.20b shows a more realistic situation, in which the surface "piles up" in front of the sphere and the sphere rides in a shallow trench. Because of these deformations, the contact forces on the sphere no longer act along a single point, but over an area; the forces are concentrated on the front of the sphere as shown. As a result, the normal force now exerts a torque that opposes the rotation. In addition, there is some sliding of the sphere over the surface due to the deformation, causing mechanical energy to be lost. The combination of these two effects is the phenomenon of *rolling friction*. Rolling friction also occurs if the rolling body is deformable, such as an automobile tire. Often the rolling body and the surface are rigid enough that rolling friction can be ignored, as we have assumed in all the examples in this section.

Test Your Understanding of Section 10.3 Suppose the solid cylinder used as a yo-yo in Example 10.6 is replaced by a hollow cylinder of the same mass and radius. (a) Will the acceleration of the yo-yo (i) increase, (ii) decrease, or (iii) remain the same? (b) Will the string tension (i) increase, (ii) decrease, or (iii) remain the same?

10.4 Work and Power in Rotational Motion

When you pedal a bicycle, you apply forces to a rotating body and do work on it. Similar things happen in many other real-life situations, such as a rotating motor shaft driving a power tool or a car engine propelling the vehicle. We can express this work in terms of torque and angular displacement.

Suppose a tangential force $\vec{F}_{tan}$ acts at the rim of a pivoted disk—for example, a child running while pushing on a playground merry-go-round (Fig. 10.21a). The disk rotates through an infinitesimal angle $d\theta$ about a fixed axis during an infinitesimal time interval dt (Fig. 10.21b). The work dW done by the force $\vec{F}_{\text{tan}}$ while a point on the rim moves a distance ds is $dW = F_{\text{tan}}\, ds$. If $d\theta$ is measured in radians, then $ds = R\, d\theta$ and

$$dW = F_{\text{tan}} R\, d\theta$$

Now $F_{\text{tan}}R$ is the *torque* τ_z due to the force $\vec{F}_{\text{tan}}$, so

$$dW = \tau_z\, d\theta \qquad (10.19)$$

The total work W done by the torque during an angular displacement from θ_1 to θ_2 is

$$W = \int_{\theta_1}^{\theta_2} \tau_z\, d\theta \qquad \text{(work done by a torque)} \qquad (10.20)$$

If the torque remains *constant* while the angle changes by a finite amount $\Delta\theta = \theta_2 - \theta_1$, then

$$W = \tau_z(\theta_2 - \theta_1) = \tau_z\Delta\theta \qquad \text{(work done by a constant torque)} \qquad (10.21)$$

The work done by a *constant* torque is the product of torque and the angular displacement. If torque is expressed in newton-meters ($\text{N} \cdot \text{m}$) and angular displacement in radians, the work is in joules. Equation (10.21) is the rotational analog of Eq. (6.1), $W = Fs$, and Eq. (10.20) is the analog of Eq. (6.7), $W = \int F_x\, dx$, for the work done by a force in a straight-line displacement.

If the force in Fig. 10.21 had an axial component (parallel to the rotation axis) or a radial component (directed toward or away from the axis), that component would do no work because the displacement of the point of application has only a tangential component. An axial or radial component of force would also make no contribution to the torque about the axis of rotation. So Eqs. (10.20) and (10.21) are correct for *any* force, no matter what its components.

When a torque does work on a rotating rigid body, the kinetic energy changes by an amount equal to the work done. We can prove this by using exactly the same procedure that we used in Eqs. (6.11) through (6.13) for the translational kinetic energy of a particle. Let τ_z represent the *net* torque on the body so that $\tau_z = I\alpha_z$ from Eq. (10.7), and assume that the body is rigid so that the moment of inertia I is constant. We then transform the integrand in Eq. (10.20) into an integrand with respect to ω_z as follows:

$$\tau_z\, d\theta = (I\alpha_z)\, d\theta = I\frac{d\omega_z}{dt}\, d\theta = I\frac{d\theta}{dt}\, d\omega_z = I\omega_z\, d\omega_z$$

Since τ_z is the net torque, the integral in Eq. (10.20) is the *total* work done on the rotating rigid body. This equation then becomes

$$W_{\text{tot}} = \int_{\omega_1}^{\omega_2} I\omega_z\, d\omega_z = \tfrac{1}{2}I\omega_2^2 - \tfrac{1}{2}I\omega_1^2 \qquad (10.22)$$

The change in the rotational kinetic energy of a *rigid* body equals the work done by forces exerted from outside the body (Fig. 10.22). This equation is analogous to Eq. (6.13), the work–energy theorem for a particle.

What about the *power* associated with work done by a torque acting on a rotating body? When we divide both sides of Eq. (10.19) by the time interval dt during which the angular displacement occurs, we find

$$\frac{dW}{dt} = \tau_z\frac{d\theta}{dt}$$

10.21 A tangential force applied to a rotating body does work.

(a)

Child applies tangential force.

$\vec{F}_{\text{tan}}$

(b) Overhead view of merry-go-round

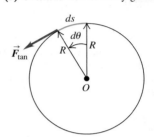

ds

$d\theta$

R

R

$\vec{F}_{\text{tan}}$

O

10.22 The rotational kinetic energy of an airplane propeller is equal to the total work done to set it spinning. When it is spinning at a constant rate, positive work is done on the propeller by the engine and negative work is done on it by air resistance. Hence the net work being done is zero and the kinetic energy remains constant.

But dW/dt is the rate of doing work, or *power P*, and $d\theta/dt$ is angular velocity ω_z, so

$$P = \tau_z\omega_z \qquad (10.23)$$

When a torque τ_z (with respect to the axis of rotation) acts on a body that rotates with angular velocity ω_z, its power (rate of doing work) is the product of τ_z and ω_z. This is the analog of the relationship $P = \vec{F} \cdot \vec{v}$ that we developed in Section 6.4 for particle motion.

Example 10.8 Calculating power from torque

An electric motor exerts a constant 10-N·m torque on a grindstone, which has a moment of inertia of 2.0 kg·m² about its shaft. The system starts from rest. Find the work W done by the motor in 8.0 s and the grindstone kinetic energy K at this time. What average power P_{av} is delivered by the motor?

SOLUTION

IDENTIFY and SET UP: The only torque acting is that due to the motor. Since this torque is constant, the grindstone's angular acceleration α_z is constant. We'll use Eq. (10.7) to find α_z, and then use this in the kinematics equations from Section 9.2 to calculate the angle $\Delta\theta$ through which the grindstone rotates in 8.0 s and its final angular velocity ω_z. From these we'll calculate W, K, and P_{av}.

EXECUTE: We have $\Sigma\tau_z = 10$ N·m and $I = 2.0$ kg·m², so $\Sigma\tau_z = I\alpha_z$ yields $\alpha_z = 5.0$ rad/s². From Eq. (9.11),

$$\Delta\theta = \tfrac{1}{2}\alpha_z t^2 = \tfrac{1}{2}(5.0 \text{ rad/s}^2)(8.0 \text{ s})^2 = 160 \text{ rad}$$

$$W = \tau_z\Delta\theta = (10 \text{ N·m})(160 \text{ rad}) = 1600 \text{ J}$$

From Eqs. (9.7) and (9.17),

$$\omega_z = \alpha_z t = (5.0 \text{ rad/s}^2)(8.0 \text{ s}) = 40 \text{ rad/s}$$

$$K = \tfrac{1}{2}I\omega_z^2 = \tfrac{1}{2}(2.0 \text{ kg·m}^2)(40 \text{ rad/s})^2 = 1600 \text{ J}$$

The average power is the work done divided by the time interval:

$$P_{av} = \frac{1600 \text{ J}}{8.0 \text{ s}} = 200 \text{ J/s} = 200 \text{ W}$$

EVALUATE: The initial kinetic energy was zero, so the work done W must equal the final kinetic energy K [Eq. (10.22)]. This is just as we calculated. We can check our result $P_{av} = 200$ W by considering the *instantaneous* power $P = \tau_z\omega_z$. Because ω_z increases continuously, P increases continuously as well; its value increases from zero at $t = 0$ to $(10 \text{ N·m})(40 \text{ rad/s}) = 400$ W at $t = 8.0$ s. Both ω_z and P increase *uniformly* with time, so the *average* power is just half this maximum value, or 200 W.

Test Your Understanding of Section 10.4 You apply equal torques to two different cylinders, one of which has a moment of inertia twice as large as the other cylinder. Each cylinder is initially at rest. After one complete rotation, which cylinder has the greater kinetic energy? (i) the cylinder with the larger moment of inertia; (ii) the cylinder with the smaller moment of inertia; (iii) both cylinders have the same kinetic energy. ❙

10.5 Angular Momentum

Every rotational quantity that we have encountered in Chapters 9 and 10 is the analog of some quantity in the translational motion of a particle. The analog of *momentum* of a particle is **angular momentum**, a vector quantity denoted as $\vec{L}$. Its relationship to momentum $\vec{p}$ (which we will often call *linear momentum* for clarity) is exactly the same as the relationship of torque to force, $\vec{\tau} = \vec{r} \times \vec{F}$. For a particle with constant mass m, velocity $\vec{v}$, momentum $\vec{p} = m\vec{v}$, and position vector $\vec{r}$ relative to the origin O of an inertial frame, we define angular momentum $\vec{L}$ as

$$\vec{L} = \vec{r} \times \vec{p} = \vec{r} \times m\vec{v} \qquad \text{(angular momentum of a particle)} \quad (10.24)$$

The value of $\vec{L}$ depends on the choice of origin O, since it involves the particle's position vector relative to O. The units of angular momentum are kg $\cdot$ m^2/s.

In Fig. 10.23 a particle moves in the xy-plane; its position vector $\vec{r}$ and momentum $\vec{p} = m\vec{v}$ are shown. The angular momentum vector $\vec{L}$ is perpendicular to the xy-plane. The right-hand rule for vector products shows that its direction is along the $+z$-axis, and its magnitude is

$$L = mvr \sin \phi = mvl \qquad (10.25)$$

where l is the perpendicular distance from the line of $\vec{v}$ to O. This distance plays the role of "lever arm" for the momentum vector.

When a net force $\vec{F}$ acts on a particle, its velocity and momentum change, so its angular momentum may also change. We can show that the *rate of change* of angular momentum is equal to the torque of the net force. We take the time derivative of Eq. (10.24), using the rule for the derivative of a product:

$$\frac{d\vec{L}}{dt} = \left(\frac{d\vec{r}}{dt} \times m\vec{v}\right) + \left(\vec{r} \times m\frac{d\vec{v}}{dt}\right) = (\vec{v} \times m\vec{v}) + (\vec{r} \times m\vec{a})$$

The first term is zero because it contains the vector product of the vector $\vec{v} = d\vec{r}/dt$ with itself. In the second term we replace $m\vec{a}$ with the net force $\vec{F}$, obtaining

$$\frac{d\vec{L}}{dt} = \vec{r} \times \vec{F} = \vec{\tau} \qquad \text{(for a particle acted on by net force } \vec{F}) \quad (10.26)$$

The rate of change of angular momentum of a particle equals the torque of the net force acting on it. Compare this result to Eq. (8.4), which states that the rate of change $d\vec{p}/dt$ of the *linear* momentum of a particle equals the net force that acts on it.

Angular Momentum of a Rigid Body

We can use Eq. (10.25) to find the total angular momentum of a *rigid body* rotating about the z-axis with angular speed ω. First consider a thin slice of the body lying in the xy-plane (Fig. 10.24). Each particle in the slice moves in a circle centered at the origin, and at each instant its velocity $\vec{v}_i$ is perpendicular to its position vector $\vec{r}_i$, as shown. Hence in Eq. (10.25), $\phi = 90°$ for every particle. A particle with mass m_i at a distance r_i from O has a speed v_i equal to $r_i\omega$. From Eq. (10.25) the magnitude L_i of its angular momentum is

$$L_i = m_i(r_i\omega) r_i = m_i r_i^2 \omega \qquad (10.27)$$

The direction of each particle's angular momentum, as given by the right-hand rule for the vector product, is along the $+z$-axis.

The *total* angular momentum of the slice of the body lying in the xy-plane is the sum $\sum L_i$ of the angular momenta L_i of the particles. Summing Eq. (10.27), we have

$$L = \sum L_i = \left(\sum m_i r_i^2\right)\omega = I\omega$$

where I is the moment of inertia of the slice about the z-axis.

We can do this same calculation for the other slices of the body, all parallel to the xy-plane. For points that do not lie in the xy-plane, a complication arises because the $\vec{r}$ vectors have components in the z-direction as well as the x- and y-directions; this gives the angular momentum of each particle a component perpendicular to the z-axis. But *if the z-axis is an axis of symmetry,* the perpendicular components for particles on opposite sides of this axis add up to zero (Fig. 10.25). So when a body rotates about an axis of symmetry, its angular momentum vector $\vec{L}$ lies along the symmetry axis, and its magnitude is $L = I\omega$.

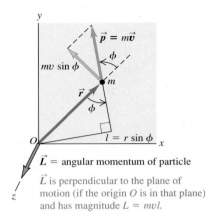

10.23 Calculating the angular momentum $\vec{L} = \vec{r} \times m\vec{v} = \vec{r} \times \vec{p}$ of a particle with mass m moving in the xy-plane.

$\vec{L}$ = angular momentum of particle

$\vec{L}$ is perpendicular to the plane of motion (if the origin O is in that plane) and has magnitude $L = mvl$.

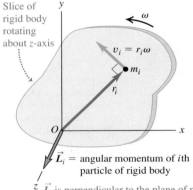

10.24 Calculating the angular momentum of a particle of mass m_i in a rigid body rotating at angular speed ω. (Compare Fig. 10.23.)

$\vec{L}_i$ = angular momentum of ith particle of rigid body

$\vec{L}_i$ is perpendicular to the plane of motion (if the origin O is in that plane) and has magnitude $L_i = m_i v_i r_i = m_i r_i^2 \omega$.

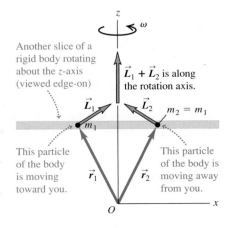

10.25 Two particles of the same mass located symmetrically on either side of the rotation axis of a rigid body. The angular momentum vectors $\vec{L}_1$ and $\vec{L}_2$ of the two particles do not lie along the rotation axis, but their vector sum $\vec{L}_1 + \vec{L}_2$ does.

Another slice of a rigid body rotating about the z-axis (viewed edge-on)

$\vec{L}_1 + \vec{L}_2$ is along the rotation axis.

This particle of the body is moving toward you.

This particle of the body is moving away from you.

10.26 For rotation about an axis of symmetry, $\vec{\omega}$ and $\vec{L}$ are parallel and along the axis. The directions of both vectors are given by the right-hand rule (compare Fig. 9.5).

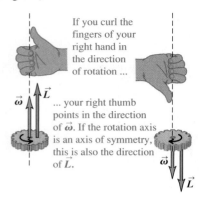

If you curl the fingers of your right hand in the direction of rotation ...

$\vec{\omega}$ $\vec{L}$

... your right thumb points in the direction of $\vec{\omega}$. If the rotation axis is an axis of symmetry, this is also the direction of $\vec{L}$.

$\vec{\omega}$

$\vec{L}$

The angular velocity vector $\vec{\omega}$ also lies along the rotation axis, as we discussed at the end of Section 9.1. Hence for a rigid body rotating around an axis of symmetry, $\vec{L}$ and $\vec{\omega}$ are in the same direction (Fig. 10.26). So we have the *vector* relationship

$$\vec{L} = I\vec{\omega} \qquad \text{(for a rigid body rotating around a symmetry axis)} \quad (10.28)$$

From Eq. (10.26) the rate of change of angular momentum of a particle equals the torque of the net force acting on the particle. For any system of particles (including both rigid and nonrigid bodies), the rate of change of the *total* angular momentum equals the sum of the torques of all forces acting on all the particles. The torques of the *internal* forces add to zero if these forces act along the line from one particle to another, as in Fig. 10.8, and so the sum of the torques includes only the torques of the *external* forces. (A similar cancellation occurred in our discussion of center-of-mass motion in Section 8.5.) If the total angular momentum of the system of particles is $\vec{L}$ and the sum of the external torques is $\sum \vec{\tau}$, then

$$\sum \vec{\tau} = \frac{d\vec{L}}{dt} \qquad \text{(for any system of particles)} \quad (10.29)$$

Finally, if the system of particles is a rigid body rotating about a symmetry axis (the z-axis), then $L_z = I\omega_z$ and I is constant. If this axis has a fixed direction in space, then the vectors $\vec{L}$ and $\vec{\omega}$ change only in magnitude, not in direction. In that case, $dL_z/dt = I \, d\omega_z/dt = I\alpha_z$, or

$$\sum \tau_z = I\alpha_z$$

10.27 If the rotation axis of a rigid body is not a symmetry axis, $\vec{L}$ does not in general lie along the rotation axis. Even if $\vec{\omega}$ is constant, the direction of $\vec{L}$ changes and a net torque is required to maintain rotation.

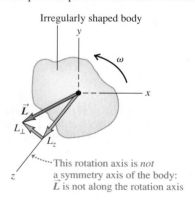

Irregularly shaped body

y

ω

x

$\vec{L}$

$L_\perp$

L_z

z

This rotation axis is *not* a symmetry axis of the body: $\vec{L}$ is not along the rotation axis

which is again our basic relationship for the dynamics of rigid-body rotation. If the body is *not* rigid, I may change, and in that case, L changes even when ω is constant. For a nonrigid body, Eq. (10.29) is still valid, even though Eq. (10.7) is not.

When the axis of rotation is *not* a symmetry axis, the angular momentum is in general *not* parallel to the axis (Fig. 10.27). As the body turns, the angular momentum vector $\vec{L}$ traces out a cone around the rotation axis. Because $\vec{L}$ changes, there must be a net external torque acting on the body even though the angular velocity magnitude ω may be constant. If the body is an unbalanced wheel on a car, this torque is provided by friction in the bearings, which causes the bearings to wear out. "Balancing" a wheel means distributing the mass so that the rotation axis is an axis of symmetry; then $\vec{L}$ points along the rotation axis, and no net torque is required to keep the wheel turning.

In fixed-axis rotation we often use the term "angular momentum of the body" to refer to only the *component* of $\vec{L}$ along the rotation axis of the body (the z-axis in Fig. 10.27), with a positive or negative sign to indicate the sense of rotation just as with angular velocity.

***

Example 10.9   Angular momentum and torque

A turbine fan in a jet engine has a moment of inertia of 2.5 kg $\cdot$ m^2 about its axis of rotation. As the turbine starts up, its angular velocity is given by $\omega_z = (40 \text{ rad/s}^3)t^2$. (a) Find the fan's angular momentum as a function of time, and find its value at $t = 3.0$ s. (b) Find the net torque on the fan as a function of time, and find its value at $t = 3.0$ s.

SOLUTION

IDENTIFY and SET UP: The fan rotates about its axis of symmetry (the z-axis). Hence the angular momentum vector has only a

z-component L_z, which we can determine from the angular velocity ω_z. Since the direction of angular momentum is constant, the net torque likewise has only a component τ_z along the rotation axis. We'll use Eq. (10.28) to find L_z from ω_z and then use Eq. (10.29) to find τ_z.

EXECUTE: (a) From Eq. (10.28),

$$L_z = I\omega_z = (2.5 \text{ kg} \cdot \text{m}^2)(40 \text{ rad/s}^3)t^2 = (100 \text{ kg} \cdot \text{m}^2/\text{s}^3)t^2$$

(We dropped the dimensionless quantity "rad" from the final expression.) At $t = 3.0$ s, $L_z = 900$ kg $\cdot$ m^2/s.

(b) From Eq. (10.29),

$$\tau_z = \frac{dL_z}{dt} = (100 \text{ kg} \cdot \text{m}^2/\text{s}^3)(2t) = (200 \text{ kg} \cdot \text{m}^2/\text{s}^3)t$$

At $t = 3.0$ s,

$$\tau_z = (200 \text{ kg} \cdot \text{m}^2/\text{s}^3)(3.0 \text{ s}) = 600 \text{ kg} \cdot \text{m}^2/\text{s}^2 = 600 \text{ N} \cdot \text{m}$$

EVALUATE: As a check on our expression for τ_z, note that the angular acceleration of the turbine is $\alpha_z = d\omega_z/dt = (40 \text{ rad/s}^3)(2t) = (80 \text{ rad/s}^3)t$. Hence from Eq. (10.7), the torque on the fan is $\tau_z = I\alpha_z = (2.5 \text{ kg} \cdot \text{m}^2)(80 \text{ rad/s}^3)t = (200 \text{ kg} \cdot \text{m}^2/\text{s}^3)t$, just as we calculated.

Test Your Understanding of Section 10.5 A ball is attached to one end of a piece of string. You hold the other end of the string and whirl the ball in a circle around your hand. (a) If the ball moves at a constant speed, is its linear momentum $\vec{p}$ constant? Why or why not? (b) Is its angular momentum $\vec{L}$ constant? Why or why not?

10.28 A falling cat twists different parts of its body in different directions so that it lands feet first. At all times during this process the angular momentum of the cat as a whole remains zero.

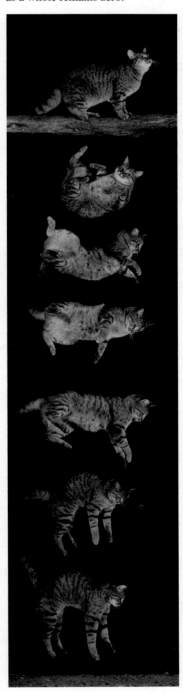

10.6 Conservation of Angular Momentum

We have just seen that angular momentum can be used for an alternative statement of the basic dynamic principle for rotational motion. It also forms the basis for the **principle of conservation of angular momentum.** Like conservation of energy and of linear momentum, this principle is a universal conservation law, valid at all scales from atomic and nuclear systems to the motions of galaxies. This principle follows directly from Eq. (10.29): $\sum \vec{\tau} = d\vec{L}/dt$. If $\sum \vec{\tau} = 0$, then $d\vec{L}/dt = 0$, and $\vec{L}$ is constant.

> **When the net external torque acting on a system is zero, the total angular momentum of the system is constant (conserved).**

A circus acrobat, a diver, and an ice skater pirouetting on the toe of one skate all take advantage of this principle. Suppose an acrobat has just left a swing with arms and legs extended and rotating counterclockwise about her center of mass. When she pulls her arms and legs in, her moment of inertia I_{cm} with respect to her center of mass changes from a large value I_1 to a much smaller value I_2. The only external force acting on her is her weight, which has no torque with respect to an axis through her center of mass. So her angular momentum $L_z = I_{cm}\omega_z$ remains constant, and her angular velocity ω_z increases as I_{cm} decreases. That is,

$$I_1\omega_{1z} = I_2\omega_{2z} \qquad \text{(zero net external torque)} \qquad (10.30)$$

When a skater or ballerina spins with arms outstretched and then pulls her arms in, her angular velocity increases as her moment of inertia decreases. In each case there is conservation of angular momentum in a system in which the net external torque is zero.

When a system has several parts, the internal forces that the parts exert on one another cause changes in the angular momenta of the parts, but the *total* angular momentum doesn't change. Here's an example. Consider two bodies A and B that interact with each other but not with anything else, such as the astronauts we discussed in Section 8.2 (see Fig. 8.8). Suppose body A exerts a force $\vec{F}_{A \text{ on } B}$ on body B; the corresponding torque (with respect to whatever point we choose) is $\vec{\tau}_{A \text{ on } B}$. According to Eq. (10.29), this torque is equal to the rate of change of angular momentum of B:

$$\vec{\tau}_{A \text{ on } B} = \frac{d\vec{L}_B}{dt}$$

At the same time, body B exerts a force $\vec{F}_{B \text{ on } A}$ on body A, with a corresponding torque $\vec{\tau}_{B \text{ on } A}$, and

$$\vec{\tau}_{B \text{ on } A} = \frac{d\vec{L}_A}{dt}$$

From Newton's third law, $\vec{F}_{B\ on\ A} = -\vec{F}_{A\ on\ B}$. Furthermore, if the forces act along the same line, as in Fig. 10.8, their lever arms with respect to the chosen axis are equal. Thus the *torques* of these two forces are equal and opposite, and $\vec{\tau}_{B\ on\ A} = -\vec{\tau}_{A\ on\ B}$. So if we add the two preceding equations, we find

$$\frac{d\vec{L}_A}{dt} + \frac{d\vec{L}_B}{dt} = 0$$

or, because $\vec{L}_A + \vec{L}_B$ is the *total* angular momentum $\vec{L}$ of the system,

$$\frac{d\vec{L}}{dt} = 0 \qquad \text{(zero net external torque)} \qquad (10.31)$$

That is, the total angular momentum of the system is constant. The torques of the internal forces can transfer angular momentum from one body to the other, but they can't change the *total* angular momentum of the system (Fig. 10.28).

MasteringPHYSICS

PhET: Torque
ActivPhysics 7.14: Ball Hits Bat

Example 10.10 Anyone can be a ballerina

A physics professor stands at the center of a frictionless turntable with arms outstretched and a 5.0-kg dumbbell in each hand (Fig. 10.29). He is set rotating about the vertical axis, making one revolution in 2.0 s. Find his final angular velocity if he pulls the dumbbells in to his stomach. His moment of inertia (without the dumbbells) is 3.0 kg · m² with arms outstretched and 2.2 kg · m² with his hands at his stomach. The dumbbells are 1.0 m from the axis initially and 0.20 m at the end.

SOLUTION

IDENTIFY, SET UP, and EXECUTE: No external torques act about the z-axis, so L_z is constant. We'll use Eq. (10.30) to find the final angular velocity ω_{2z}. The moment of inertia of the system is $I = I_{\text{prof}} + I_{\text{dumbbells}}$. We treat each dumbbell as a particle of mass m that contributes mr^2 to $I_{\text{dumbbells}}$, where r is the perpendicular distance from the axis to the dumbbell. Initially we have

$$I_1 = 3.0 \text{ kg} \cdot \text{m}^2 + 2(5.0 \text{ kg})(1.0 \text{ m})^2 = 13 \text{ kg} \cdot \text{m}^2$$

$$\omega_{1z} = \frac{1 \text{ rev}}{2.0 \text{ s}} = 0.50 \text{ rev/s}$$

The final moment of inertia is

$$I_2 = 2.2 \text{ kg} \cdot \text{m}^2 + 2(5.0 \text{ kg})(0.20 \text{ m})^2 = 2.6 \text{ kg} \cdot \text{m}^2$$

From Eq. (10.30), the final angular velocity is

$$\omega_{2z} = \frac{I_1}{I_2}\omega_{1z} = \frac{13 \text{ kg} \cdot \text{m}^2}{2.6 \text{ kg} \cdot \text{m}^2}(0.50 \text{ rev/s}) = 2.5 \text{ rev/s} = 5\omega_{1z}$$

10.29 Fun with conservation of angular momentum.

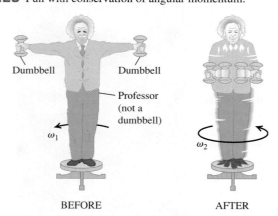

Dumbbell

Dumbbell

Professor (not a dumbbell)

ω_1

ω_2

BEFORE

AFTER

Can you see why we didn't have to change "revolutions" to "radians" in this calculation?

EVALUATE: The angular momentum remained constant, but the angular velocity increased by a factor of 5, from $\omega_{1z} = (0.50 \text{ rev/s})(2\pi \text{ rad/rev}) = 3.14 \text{ rad/s}$ to $\omega_{2z} = (2.5 \text{ rev/s})(2\pi \text{ rad/rev}) = 15.7 \text{ rad/s}$. The initial and final kinetic energies are then

$$K_1 = \tfrac{1}{2}I_1\omega_{1z}^2 = \tfrac{1}{2}(13 \text{ kg} \cdot \text{m}^2)(3.14 \text{ rad/s})^2 = 64 \text{ J}$$

$$K_2 = \tfrac{1}{2}I_2\omega_{2z}^2 = \tfrac{1}{2}(2.6 \text{ kg} \cdot \text{m}^2)(15.7 \text{ rad/s})^2 = 320 \text{ J}$$

The fivefold increase in kinetic energy came from the work that the professor did in pulling his arms and the dumbbells inward.

Example 10.11 A rotational "collision"

Figure 10.30 shows two disks: an engine flywheel (A) and a clutch plate (B) attached to a transmission shaft. Their moments of inertia are I_A and I_B; initially, they are rotating with constant angular speeds ω_A and ω_B, respectively. We push the disks together with forces acting along the axis, so as not to apply any torque on either disk. The disks rub against each other and eventually reach a common angular speed ω. Derive an expression for ω.

SOLUTION

IDENTIFY, SET UP, and EXECUTE: There are no external torques, so the only torque acting on either disk is the torque applied by the other disk. Hence the total angular momentum of the system of two disks is conserved. At the end they rotate together as one body with total moment of inertia $I = I_A + I_B$ and angular speed ω.

10.30 When the net external torque is zero, angular momentum is conserved.

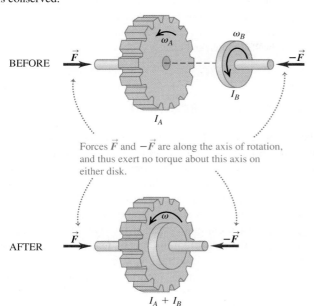

BEFORE

ω_A ω_B

$\vec{F}$ $-\vec{F}$

I_B

I_A

Forces $\vec{F}$ and $-\vec{F}$ are along the axis of rotation, and thus exert no torque about this axis on either disk.

AFTER

ω

$\vec{F}$ $-\vec{F}$

$I_A + I_B$

Figure 10.30 shows that all angular velocities are in the same direction, so we can regard ω_A, ω_B, and ω as components of angular velocity along the rotation axis. Conservation of angular momentum gives

$$I_A\omega_A + I_B\omega_B = (I_A + I_B)\omega$$

$$\omega = \frac{I_A\omega_A + I_B\omega_B}{I_A + I_B}$$

EVALUATE: This "collision" is analogous to a completely inelastic collision (see Section 8.3). When two objects in translational motion along the same axis collide and stick, the linear momentum of the system is conserved. Here two objects in *rotational* motion around the same axis "collide" and stick, and the *angular* momentum of the system is conserved.

The kinetic energy of a system decreases in a completely inelastic collision. Here kinetic energy is lost because nonconservative (frictional) internal forces act while the two disks rub together. Suppose flywheel A has a mass of 2.0 kg, a radius of 0.20 m, and an initial angular speed of 50 rad/s (about 500 rpm), and clutch plate B has a mass of 4.0 kg, a radius of 0.10 m, and an initial angular speed of 200 rad/s. Can you show that the final kinetic energy is only two-thirds of the initial kinetic energy?

Example 10.12 **Angular momentum in a crime bust**

A door 1.00 m wide, of mass 15 kg, can rotate freely about a vertical axis through its hinges. A bullet with a mass of 10 g and a speed of 400 m/s strikes the center of the door, in a direction perpendicular to the plane of the door, and embeds itself there. Find the door's angular speed. Is kinetic energy conserved?

SOLUTION

IDENTIFY and SET UP: We consider the door and bullet as a system. There is no external torque about the hinge axis, so angular momentum about this axis is conserved. Figure 10.31 shows our sketch. The initial angular momentum is that of the bullet, as given by Eq. (10.25). The final angular momentum is that of a rigid body

10.31 Our sketch for this problem.

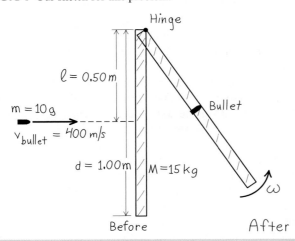

Hinge

$\ell = 0.50\,m$

$m = 10\,g$

$v_{bullet} = 400\,m/s$

$d = 1.00\,m$ $M = 15\,kg$

Bullet

ω

Before   After

composed of the door and the embedded bullet. We'll equate these quantities and solve for the resulting angular speed ω of the door and bullet.

EXECUTE: From Eq. (10.25), the initial angular momentum of the bullet is

$$L = mvl = (0.010 \text{ kg})(400 \text{ m/s})(0.50 \text{ m}) = 2.0 \text{ kg} \cdot \text{m}^2/\text{s}$$

The final angular momentum is $I\omega$, where $I = I_{door} + I_{bullet}$. From Table 9.2, case (d), for a door of width $d = 1.00$ m,

$$I_{door} = \frac{Md^2}{3} = \frac{(15 \text{ kg})(1.00 \text{ m})^2}{3} = 5.0 \text{ kg} \cdot \text{m}^2$$

The moment of inertia of the bullet (with respect to the axis along the hinges) is

$$I_{bullet} = ml^2 = (0.010 \text{ kg})(0.50 \text{ m})^2 = 0.0025 \text{ kg} \cdot \text{m}^2$$

Conservation of angular momentum requires that $mvl = I\omega$, or

$$\omega = \frac{mvl}{I} = \frac{2.0 \text{ kg} \cdot \text{m}^2/\text{s}}{5.0 \text{ kg} \cdot \text{m}^2 + 0.0025 \text{ kg} \cdot \text{m}^2} = 0.40 \text{ rad/s}$$

The initial and final kinetic energies are

$$K_1 = \tfrac{1}{2}mv^2 = \tfrac{1}{2}(0.010 \text{ kg})(400 \text{ m/s})^2 = 800 \text{ J}$$

$$K_2 = \tfrac{1}{2}I\omega^2 = \tfrac{1}{2}(5.0025 \text{ kg} \cdot \text{m}^2)(0.40 \text{ rad/s})^2 = 0.40 \text{ J}$$

EVALUATE: The final kinetic energy is only $\frac{1}{2000}$ of the initial value! We did not expect kinetic energy to be conserved: The collision is inelastic because nonconservative friction forces act during the impact. The door's final angular speed is quite slow: At 0.40 rad/s, it takes 3.9 s to swing through 90° ($\pi/2$ radians).

10.32 A gyroscope supported at one end. The horizontal circular motion of the flywheel and axis is called precession. The angular speed of precession is Ω.

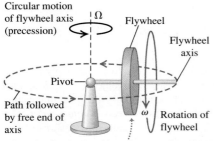

When the flywheel and its axis are stationary, they will fall to the table surface. When the flywheel spins, it and its axis "float" in the air while moving in a circle about the pivot.

10.33 (a) If the flywheel in Fig. 10.32 is initially not spinning, its initial angular momentum is zero (b) In each successive time interval dt, the torque produces a change $d\vec{L} = \vec{\tau}\,dt$ in the angular momentum. The flywheel acquires an angular momentum $\vec{L}$ in the same direction as $\vec{\tau}$, and the flywheel axis falls.

(a) Nonrotating flywheel falls

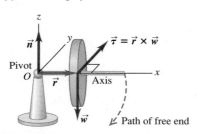

When the flywheel is not rotating, its weight creates a torque around the pivot, causing it to fall along a circular path until its axis rests on the table surface.

(b) View from above as flywheel falls

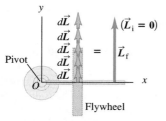

In falling, the flywheel rotates about the pivot and thus acquires an angular momentum $\vec{L}$. The *direction* of $\vec{L}$ stays constant.

Test Your Understanding of Section 10.6 If the polar ice caps were to completely melt due to global warming, the melted ice would redistribute itself over the earth. This change would cause the length of the day (the time needed for the earth to rotate once on its axis) to (i) increase; (ii) decrease; (iii) remain the same. (*Hint:* Use angular momentum ideas. Assume that the sun, moon, and planets exert negligibly small torques on the earth.) ∣

10.7 Gyroscopes and Precession

In all the situations we've looked at so far in this chapter, the axis of rotation either has stayed fixed or has moved and kept the same direction (such as rolling without slipping). But a variety of new physical phenomena, some quite unexpected, can occur when the axis of rotation can change direction. For example, consider a toy gyroscope that's supported at one end (Fig. 10.32). If we hold it with the flywheel axis horizontal and let go, the free end of the axis simply drops owing to gravity—*if* the flywheel isn't spinning. But if the flywheel *is* spinning, what happens is quite different. One possible motion is a steady circular motion of the axis in a horizontal plane, combined with the spin motion of the flywheel about the axis. This surprising, nonintuitive motion of the axis is called **precession.** Precession is found in nature as well as in rotating machines such as gyroscopes. As you read these words, the earth itself is precessing; its spin axis (through the north and south poles) slowly changes direction, going through a complete cycle of precession every 26,000 years.

To study this strange phenomenon of precession, we must remember that angular velocity, angular momentum, and torque are all *vector* quantities. In particular, we need the general relationship between the net torque $\sum \vec{\tau}$ that acts on a body and the rate of change of the body's angular momentum $\vec{L}$, given by Eq. (10.29), $\sum \vec{\tau} = d\vec{L}/dt$. Let's first apply this equation to the case in which the flywheel is *not* spinning (Fig. 10.33a). We take the origin O at the pivot and assume that the flywheel is symmetrical, with mass M and moment of inertia I about the flywheel axis. The flywheel axis is initially along the x-axis. The only external forces on the gyroscope are the normal force $\vec{n}$ acting at the pivot (assumed to be frictionless) and the weight $\vec{w}$ of the flywheel that acts at its center of mass, a distance r from the pivot. The normal force has zero torque with respect to the pivot, and the weight has a torque $\vec{\tau}$ in the y-direction, as shown in Fig. 10.33a. Initially, there is no rotation, and the initial angular momentum $\vec{L}_i$ is zero. From Eq. (10.29) the *change* $d\vec{L}$ in angular momentum in a short time interval dt following this is

$$d\vec{L} = \vec{\tau}\,dt \qquad (10.32)$$

This change is in the y-direction because $\vec{\tau}$ is. As each additional time interval dt elapses, the angular momentum changes by additional increments $d\vec{L}$ in the y-direction because the direction of the torque is constant (Fig. 10.33b). The steadily increasing horizontal angular momentum means that the gyroscope rotates downward faster and faster around the y-axis until it hits either the stand or the table on which it sits.

Now let's see what happens if the flywheel *is* spinning initially, so the initial angular momentum $\vec{L}_i$ is not zero (Fig. 10.34a). Since the flywheel rotates around its symmetry axis, $\vec{L}_i$ lies along the axis. But each change in angular momentum $d\vec{L}$ is perpendicular to the axis because the torque $\vec{\tau} = \vec{r} \times \vec{w}$ is perpendicular to the axis (Fig. 10.34b). This causes the *direction* of $\vec{L}$ to change, but not its magnitude. The changes $d\vec{L}$ are always in the horizontal xy-plane, so the angular momentum vector and the flywheel axis with which it moves are always horizontal. In other words, the axis doesn't fall—it just precesses.

If this still seems mystifying to you, think about a ball attached to a string. If the ball is initially at rest and you pull the string toward you, the ball moves toward you also. But if the ball is initially moving and you continuously pull the

(a) Rotating flywheel

When the flywheel is rotating, the system starts with an angular momentum $\vec{L}_i$ parallel to the flywheel's axis of rotation.

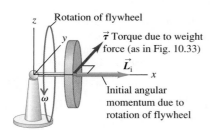

Rotation of flywheel

$\vec{\tau}$ Torque due to weight force (as in Fig. 10.33)

$\vec{L}_i$

Initial angular momentum due to rotation of flywheel

(b) View from above

Now the effect of the torque is to cause the angular momentum to precess around the pivot. The gyroscope circles around its pivot without falling.

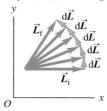

10.34 **(a)** The flywheel is spinning initially with angular momentum $\vec{L}_i$. The forces (not shown) are the same as those in Fig. 10.33a. **(b)** Because the initial angular momentum is not zero, each change $d\vec{L} = \vec{\tau}\,dt$ in angular momentum is perpendicular to $\vec{L}$. As a result, the magnitude of $\vec{L}$ remains the same but its direction changes continuously.

string in a direction perpendicular to the ball's motion, the ball moves in a circle around your hand; it does not approach your hand at all. In the first case the ball has zero linear momentum $\vec{p}$ to start with; when you apply a force $\vec{F}$ toward you for a time dt, the ball acquires a momentum $d\vec{p} = \vec{F}\,dt$, which is also toward you. But if the ball already has linear momentum $\vec{p}$, a change in momentum $d\vec{p}$ that's perpendicular to $\vec{p}$ changes the direction of motion, not the speed. Replace $\vec{p}$ with $\vec{L}$ and $\vec{F}$ with $\vec{\tau}$ in this argument, and you'll see that precession is simply the rotational analog of uniform circular motion.

At the instant shown in Fig. 10.34a, the gyroscope has angular momentum $\vec{L}$. A short time interval dt later, the angular momentum is $\vec{L} + d\vec{L}$; the infinitesimal change in angular momentum is $d\vec{L} = \vec{\tau}\,dt$, which is perpendicular to $\vec{L}$. As the vector diagram in Fig. 10.35 shows, this means that the flywheel axis of the gyroscope has turned through a small angle $d\phi$ given by $d\phi = |d\vec{L}|/|\vec{L}|$. The rate at which the axis moves, $d\phi/dt$, is called the **precession angular speed;** denoting this quantity by Ω, we find

$$\Omega = \frac{d\phi}{dt} = \frac{|d\vec{L}|/|\vec{L}|}{dt} = \frac{\tau_z}{L_z} = \frac{wr}{I\omega} \qquad (10.33)$$

10.35 Detailed view of part of Fig. 10.34b.

In a time dt, the angular momentum vector and the flywheel axis (to which it is parallel) precess together through an angle $d\phi$.

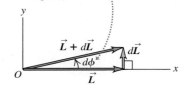

Thus the precession angular speed is *inversely* proportional to the angular speed of spin about the axis. A rapidly spinning gyroscope precesses slowly; if friction in its bearings causes the flywheel to slow down, the precession angular speed *increases!* The precession angular speed of the earth is very slow $(1 \text{ rev}/26{,}000 \text{ yr})$ because its spin angular momentum L_z is large and the torque τ_z, due to the gravitational influences of the moon and sun, is relatively small.

As a gyroscope precesses, its center of mass moves in a circle with radius r in a horizontal plane. Its vertical component of acceleration is zero, so the upward normal force $\vec{n}$ exerted by the pivot must be just equal in magnitude to the weight. The circular motion of the center of mass with angular speed Ω requires a force $\vec{F}$ directed toward the center of the circle, with magnitude $F = M\Omega^2 r$. This force must also be supplied by the pivot.

One key assumption that we made in our analysis of the gyroscope was that the angular momentum vector $\vec{L}$ is associated only with the spin of the flywheel and is purely horizontal. But there will also be a vertical component of angular momentum associated with the precessional motion of the gyroscope. By ignoring this, we've tacitly assumed that the precession is *slow*—that is, that the precession angular speed Ω is very much less than the spin angular speed ω. As Eq. (10.33) shows, a large value of ω automatically gives a small value of Ω, so this approximation is reasonable. When the precession is not slow, additional effects show up, including an up-and-down wobble or *nutation* of the flywheel axis that's superimposed on the precessional motion. You can see nutation occurring in a gyroscope as its spin slows down, so that Ω increases and the vertical component of $\vec{L}$ can no longer be ignored.

Example 10.13 A precessing gyroscope

Figure 10.36a shows a top view of a spinning, cylindrical gyroscope wheel. The pivot is at O, and the mass of the axle is negligible. (a) As seen from above, is the precession clockwise or counterclockwise? (b) If the gyroscope takes 4.0 s for one revolution of precession, what is the angular speed of the wheel?

SOLUTION

IDENTIFY and SET UP: We'll determine the direction of precession using the right-hand rule as in Fig. 10.34, which shows the same kind of gyroscope as Fig. 10.36. We'll use the relationship between precession angular speed Ω and spin angular speed ω, Eq. (10.33), to find ω.

EXECUTE: (a) The right-hand rule shows that $\vec{\omega}$ and $\vec{L}$ are to the left in Fig. 10.36b. The weight $\vec{w}$ points into the page in this top view and acts at the center of mass (denoted by ✕ in the figure). The torque $\vec{\tau} = \vec{r} \times \vec{w}$ is toward the top of the page, so $d\vec{L}/dt$ is

also toward the top of the page. Adding a small $d\vec{L}$ to the initial vector $\vec{L}$ changes the direction of $\vec{L}$ as shown, so the precession is clockwise as seen from above.

(b) Be careful not to confuse ω and Ω! The precession angular speed is $\Omega = (1\ \text{rev})/(4.0\ \text{s}) = (2\pi\ \text{rad})/(4.0\ \text{s}) = 1.57\ \text{rad/s}$. The weight is mg, and if the wheel is a solid, uniform cylinder, its moment of inertia about its symmetry axis is $I = \frac{1}{2}mR^2$. From Eq. (10.33),

$$\omega = \frac{wr}{I\Omega} = \frac{mgr}{(mR^2/2)\Omega} = \frac{2gr}{R^2\Omega}$$

$$= \frac{2(9.8\ \text{m/s}^2)(2.0 \times 10^{-2}\ \text{m})}{(3.0 \times 10^{-2}\ \text{m})^2(1.57\ \text{rad/s})} = 280\ \text{rad/s} = 2600\ \text{rev/min}$$

EVALUATE: The precession angular speed Ω is only about 0.6% of the spin angular speed ω, so this is an example of slow precession.

10.36 In which direction and at what speed does this gyroscope precess?

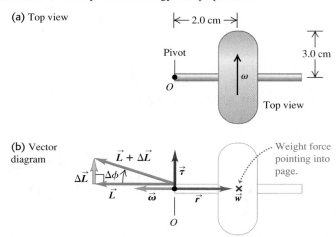

Test Your Understanding of Section 10.7 Suppose the mass of the flywheel in Fig. 10.34 were doubled but all other dimensions and the spin angular speed remained the same. What effect would this change have on the precession angular speed Ω? (i) Ω would increase by a factor of 4; (ii) Ω would double; (iii) Ω would be unaffected; (iv) Ω would be one-half as much; (v) Ω would be one-quarter as much.

Torque: When a force $\vec{F}$ acts on a body, the torque of that force with respect to a point O has a magnitude given by the product of the force magnitude F and the lever arm l. More generally, torque is a vector $\vec{\tau}$ equal to the vector product of $\vec{r}$ (the position vector of the point at which the force acts) and $\vec{F}$. (See Example 10.1.)

$$\tau = Fl \tag{10.2}$$
$$\vec{\tau} = \vec{r} \times \vec{F} \tag{10.3}$$

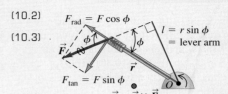

Rotational dynamics: The rotational analog of Newton's second law says that the net torque acting on a body equals the product of the body's moment of inertia and its angular acceleration. (See Examples 10.2 and 10.3.)

$$\sum \tau_z = I\alpha_z \tag{10.7}$$

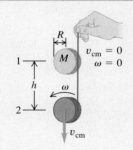

Combined translation and rotation: If a rigid body is both moving through space and rotating, its motion can be regarded as translational motion of the center of mass plus rotational motion about an axis through the center of mass. Thus the kinetic energy is a sum of translational and rotational kinetic energies. For dynamics, Newton's second law describes the motion of the center of mass, and the rotational equivalent of Newton's second law describes rotation about the center of mass. In the case of rolling without slipping, there is a special relationship between the motion of the center of mass and the rotational motion. (See Examples 10.4–10.7.)

$$K = \tfrac{1}{2}Mv_{\text{cm}}^2 + \tfrac{1}{2}I_{\text{cm}}\omega^2 \tag{10.8}$$
$$\sum \vec{F}_{\text{ext}} = M\vec{a}_{\text{cm}} \tag{10.12}$$
$$\sum \tau_z = I_{\text{cm}}\alpha_z \tag{10.13}$$
$$v_{\text{cm}} = R\omega \tag{10.11}$$
(rolling without slipping)

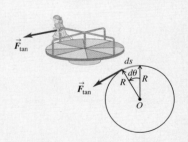

Work done by a torque: A torque that acts on a rigid body as it rotates does work on that body. The work can be expressed as an integral of the torque. The work–energy theorem says that the total rotational work done on a rigid body is equal to the change in rotational kinetic energy. The power, or rate at which the torque does work, is the product of the torque and the angular velocity (See Example 10.8.)

$$W = \int_{\theta_1}^{\theta_2} \tau_z \, d\theta \tag{10.20}$$
$$W = \tau_z(\theta_2 - \theta_1) = \tau_z \Delta\theta \tag{10.21}$$
(constant torque only)
$$W_{\text{tot}} = \tfrac{1}{2}I\omega_2^2 - \tfrac{1}{2}I\omega_1^2 \tag{10.22}$$
$$P = \tau_z\omega_z \tag{10.23}$$

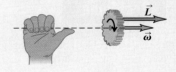

Angular momentum: The angular momentum of a particle with respect to point O is the vector product of the particle's position vector $\vec{r}$ relative to O and its momentum $\vec{p} = m\vec{v}$. When a symmetrical body rotates about a stationary axis of symmetry, its angular momentum is the product of its moment of inertia and its angular velocity vector $\vec{\omega}$. If the body is not symmetrical or the rotation (z) axis is not an axis of symmetry, the component of angular momentum along the rotation axis is $I\omega_z$. (See Example 10.9.)

$$\vec{L} = \vec{r} \times \vec{p} = \vec{r} \times m\vec{v} \tag{10.24}$$
(particle)
$$\vec{L} = I\vec{\omega} \tag{10.28}$$
(rigid body rotating about axis of symmetry)

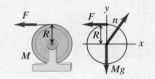

Rotational dynamics and angular momentum: The net external torque on a system is equal to the rate of change of its angular momentum. If the net external torque on a system is zero, the total angular momentum of the system is constant (conserved). (See Examples 10.10–10.13.)

$$\sum \vec{\tau} = \frac{d\vec{L}}{dt} \tag{10.29}$$

BRIDGING PROBLEM **Billiard Physics**

A cue ball (a uniform solid sphere of mass m and radius R) is at rest on a level pool table. Using a pool cue, you give the ball a sharp, horizontal hit of magnitude F at a height h above the center of the ball (Fig. 10.37). The force of the hit is much greater than the friction force f that the table surface exerts on the ball. The hit lasts for a short time Δt. (a) For what value of h will the ball roll without slipping? (b) If you hit the ball dead center ($h = 0$), the ball will slide across the table for a while, but eventually it will roll without slipping. What will the speed of its center of mass be then?

SOLUTION GUIDE

See MasteringPhysics® study area for a Video Tutor solution.

IDENTIFY and SET UP

1. Draw a free-body diagram for the ball for the situation in part (a), including your choice of coordinate axes. Note that the cue exerts both an impulsive force on the ball and an impulsive torque around the center of mass.

2. The cue force applied for a time Δt gives the ball's center of mass a speed v_{cm}, and the cue torque applied for that same time gives the ball an angular speed ω. What must be the relationship between v_{cm} and ω for the ball to roll without slipping?

10.37

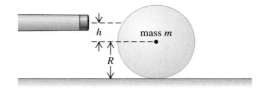

3. Draw two free-body diagrams for the ball in part (b): one showing the forces during the hit and the other showing the forces after the hit but before the ball is rolling without slipping.

4. What is the angular speed of the ball in part (b) just after the hit? While the ball is sliding, does v_{cm} increase or decrease? Does ω increase or decrease? What is the relationship between v_{cm} and ω when the ball is finally rolling without slipping?

EXECUTE

5. In part (a), use the impulse–momentum theorem to find the speed of the ball's center of mass immediately after the hit. Then use the rotational version of the impulse–momentum theorem to find the angular speed immediately after the hit. (*Hint:* To write down the rotational version of the impulse–momentum theorem, remember that the relationship between torque and angular momentum is the same as that between force and linear momentum.)

6. Use your results from step 5 to find the value of h that will cause the ball to roll without slipping immediately after the hit.

7. In part (b), again find the ball's center-of-mass speed and angular speed immediately after the hit. Then write Newton's second law for the translational motion and rotational motion of the ball as it is sliding. Use these equations to write expressions for v_{cm} and ω as functions of the elapsed time t since the hit.

8. Using your results from step 7, find the time t when v_{cm} and ω have the correct relationship for rolling without slipping. Then find the value of v_{cm} at this time.

EVALUATE

9. If you have access to a pool table, test out the results of parts (a) and (b) for yourself!

10. Can you show that if you used a hollow cylinder rather than a solid ball, you would have to hit the top of the cylinder to cause rolling without slipping as in part (a)?

Problems

For instructor-assigned homework, go to www.masteringphysics.com

•, ••, •••: Problems of increasing difficulty. **CP**: Cumulative problems incorporating material from earlier chapters. **CALC**: Problems requiring calculus. **BIO**: Biosciences problems.

DISCUSSION QUESTIONS

Q10.1 When cylinder-head bolts in an automobile engine are tightened, the critical quantity is the *torque* applied to the bolts. Why is the torque more important than the actual *force* applied to the wrench handle?

Q10.2 Can a single force applied to a body change both its translational and rotational motion? Explain.

Q10.3 Suppose you could use wheels of any type in the design of a soapbox-derby racer (an unpowered, four-wheel vehicle that coasts from rest down a hill). To conform to the rules on the total weight of the vehicle and rider, should you design with large massive wheels or small light wheels? Should you use solid wheels or wheels with most of the mass at the rim? Explain.

Q10.4 A four-wheel-drive car is accelerating forward from rest. Show the direction the car's wheels turn and how this causes a friction force due to the pavement that accelerates the car forward.

Q10.5 Serious bicyclists say that if you reduce the weight of a bike, it is more effective if you do so in the wheels rather than in the frame. Why would reducing weight in the wheels make it easier on the bicyclist than reducing the same amount in the frame?

Q10.6 The harder you hit the brakes while driving forward, the more the front end of your car will move down (and the rear end move up). Why? What happens when cars accelerate forward? Why do drag racers not use front-wheel drive only?

Q10.7 When an acrobat walks on a tightrope, she extends her arms straight out from her sides. She does this to make it easier for her to catch herself if she should tip to one side or the other. Explain how this works. [*Hint:* Think about Eq. (10.7).]

Q10.8 When you turn on an electric motor, it takes longer to come up to final speed if a grinding wheel is attached to the shaft. Why?

Q10.9 Experienced cooks can tell whether an egg is raw or hard-boiled by rolling it down a slope (taking care to catch it at the bottom). How is this possible? What are they looking for?

Q10.10 The work done by a force is the product of force and distance. The torque due to a force is the product of force and distance. Does this mean that torque and work are equivalent? Explain.

Q10.11 A valued client brings a treasured ball to your engineering firm, wanting to know whether the ball is solid or hollow. He has tried tapping on it, but that has given insufficient information. Design a simple, inexpensive experiment that you could perform quickly, without injuring the precious ball, to find out whether it is solid or hollow.

Q10.12 You make two versions of the same object out of the same material having uniform density. For one version, all the dimensions are exactly twice as great as for the other one. If the same torque acts on both versions, giving the smaller version angular acceleration α, what will be the angular acceleration of the larger version in terms of α?

Q10.13 Two identical masses are attached to frictionless pulleys by very light strings wrapped around the rim of the pulley and are released from rest. Both pulleys have the same mass and same diameter, but one is solid and the other is a hoop. As the masses fall, in which case is the tension in the string greater, or is it the same in both cases? Justify your answer.

Q10.14 The force of gravity acts on the baton in Fig. 10.11, and forces produce torques that cause a body's angular velocity to change. Why, then, is the angular velocity of the baton in the figure constant?

Q10.15 A certain solid uniform ball reaches a maximum height h_0 when it rolls up a hill without slipping. What maximum height (in terms of h_0) will it reach if you (a) double its diameter, (b) double its mass, (c) double both its diameter and mass, (d) double its angular speed at the bottom of the hill?

Q10.16 A wheel is rolling without slipping on a horizontal surface. In an inertial frame of reference in which the surface is at rest, is there any point on the wheel that has a velocity that is purely vertical? Is there any point that has a horizontal velocity component opposite to the velocity of the center of mass? Explain. Do your answers change if the wheel is slipping as it rolls? Why or why not?

Q10.17 Part of the kinetic energy of a moving automobile is in the rotational motion of its wheels. When the brakes are applied hard on an icy street, the wheels "lock" and the car starts to slide. What becomes of the rotational kinetic energy?

Q10.18 A hoop, a uniform solid cylinder, a spherical shell, and a uniform solid sphere are released from rest at the top of an incline. What is the order in which they arrive at the bottom of the incline? Does it matter whether or not the masses and radii of the objects are all the same? Explain.

Q10.19 A ball is rolling along at speed v without slipping on a horizontal surface when it comes to a hill that rises at a constant angle above the horizontal. In which case will it go higher up the hill: if the hill has enough friction to prevent slipping or if the hill is perfectly smooth? Justify your answers in both cases in terms of energy conservation and in terms of Newton's second law.

Q10.20 You are standing at the center of a large horizontal turntable in a carnival funhouse. The turntable is set rotating on frictionless bearings, and it rotates freely (that is, there is no motor driving the turntable). As you walk toward the edge of the turntable, what happens to the combined angular momentum of you and the turntable? What happens to the rotation speed of the turntable? Explain your answer.

Q10.21 A certain uniform turntable of diameter D_0 has an angular momentum L_0. If you want to redesign it so it retains the same mass but has twice as much angular momentum at the same angular velocity as before, what should be its diameter in terms of D_0?

Q10.22 A point particle travels in a straight line at constant speed, and the closest distance it comes to the origin of coordinates is a distance l. With respect to this origin, does the particle have nonzero angular momentum? As the particle moves along its straight-line path, does its angular momentum with respect to the origin change?

Q10.23 In Example 10.10 (Section 10.6) the angular speed ω changes, and this must mean that there is nonzero angular acceleration. But there is no torque about the rotation axis if the forces the professor applies to the weights are directly, radially inward. Then, by Eq. (10.7), α_z must be zero. Explain what is wrong with this reasoning that leads to this apparent contradiction.

Q10.24 In Example 10.10 (Section 10.6) the rotational kinetic energy of the professor and dumbbells increases. But since there are no external torques, no work is being done to change the rotational kinetic energy. Then, by Eq. (10.22), the kinetic energy must remain the same! Explain what is wrong with this reasoning that leads to this apparent contradiction. Where *does* the extra kinetic energy come from?

Q10.25 As discussed in Section 10.6, the angular momentum of a circus acrobat is conserved as she tumbles through the air. Is her *linear* momentum conserved? Why or why not?

Q10.26 If you stop a spinning raw egg for the shortest possible instant and then release it, the egg will start spinning again. If you do the same to a hard-boiled egg, it will remain stopped. Try it. Explain it.

Q10.27 A helicopter has a large main rotor that rotates in a horizontal plane and provides lift. There is also a small rotor on the tail that rotates in a vertical plane. What is the purpose of the tail rotor? (*Hint:* If there were no tail rotor, what would happen when the pilot changed the angular speed of the main rotor?) Some helicopters have no tail rotor, but instead have two large main rotors that rotate in a horizontal plane. Why is it important that the two main rotors rotate in opposite directions?

Q10.28 In a common design for a gyroscope, the flywheel and flywheel axis are enclosed in a light, spherical frame with the flywheel at the center of the frame. The gyroscope is then balanced on top of a pivot so that the flywheel is directly above the pivot. Does the gyroscope precess if it is released while the flywheel is spinning? Explain.

Q10.29 A gyroscope takes 3.8 s to precess 1.0 revolution about a vertical axis. Two minutes later, it takes only 1.9 s to precess 1.0 revolution. No one has touched the gyroscope. Explain.

Q10.30 A gyroscope is precessing as in Fig. 10.32. What happens if you gently add some weight to the end of the flywheel axis farthest from the pivot?

Q10.31 A bullet spins on its axis as it emerges from a rifle. Explain how this prevents the bullet from tumbling and keeps the streamlined end pointed forward.

EXERCISES

Section 10.1 Torque

10.1 • Calculate the torque (magnitude and direction) about point O due to the force $\vec{F}$ in each of the cases sketched in Fig. E10.1. In each case, the force $\vec{F}$ and the rod both lie in the plane of the page, the rod has length 4.00 m, and the force has magnitude $F = 10.0$ N.

Figure **E10.1**

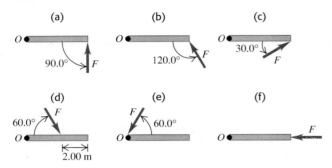

10.2 • Calculate the net torque about point O for the two forces applied as in Fig. E10.2. The rod and both forces are in the plane of the page.

Figure **E10.2**

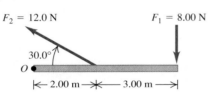

10.3 •• A square metal plate 0.180 m on each side is pivoted about an axis through point O at its center and perpendicular to the plate (Fig. E10.3). Calculate the net torque about this axis due to the three forces shown in the figure if the magnitudes of the forces are $F_1 = 18.0$ N, $F_2 = 26.0$ N, and $F_3 = 14.0$ N. The plate and all forces are in the plane of the page.

Figure **E10.3**   Figure **E10.4**

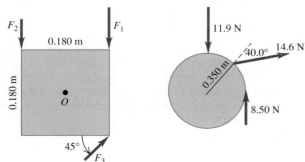

10.4 • Three forces are applied to a wheel of radius 0.350 m, as shown in Fig. E10.4. One force is perpendicular to the rim, one is tangent to it, and the other one makes a 40.0° angle with the radius. What is the net torque on the wheel due to these three forces for an axis perpendicular to the wheel and passing through its center?

10.5 • One force acting on a machine part is $\vec{F} = (-5.00$ N$)\hat{\imath} + (4.00$ N$)\hat{\jmath}$. The vector from the origin to the point where the force is applied is $\vec{r} = (-0.450$ m$)\hat{\imath} + (0.150$ m$)\hat{\jmath}$. (a) In a sketch, show $\vec{r}$, $\vec{F}$, and the origin. (b) Use the right-hand rule to determine the direction of the torque. (c) Calculate the vector torque for an axis at the origin produced by this force. Verify that the direction of the torque is the same as you obtained in part (b).

10.6 • A metal bar is in the xy-plane with one end of the bar at the origin. A force $\vec{F} = (7.00$ N$)\hat{\imath} + (-3.00$ N$)\hat{\jmath}$ is applied to the bar at the point $x = 3.00$ m, $y = 4.00$ m. (a) In terms of unit vectors $\hat{\imath}$ and $\hat{\jmath}$, what is the position vector $\vec{r}$ for the point where the force is applied? (b) What are the magnitude and direction of the torque with respect to the origin produced by $\vec{F}$?

10.7 • In Fig. E10.7, forces $\vec{A}$, $\vec{B}$, $\vec{C}$, and $\vec{D}$ each have magnitude 50 N and act at the same point on the object. (a) What torque (magnitude and direction) does each of these forces exert on the object about point P? (b) What is the total torque about point P?

Figure **E10.7**

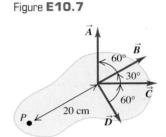

10.8 • A machinist is using a wrench to loosen a nut. The wrench is 25.0 cm long, and he exerts a 17.0-N force at the end of the handle at 37° with the handle (Fig. E10.8). (a) What torque does the machinist exert about the center of the nut? (b) What is the maximum torque he could exert with this force, and how should the force be oriented?

Figure **E10.8**

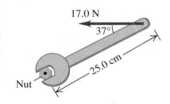

Section 10.2 Torque and Angular Acceleration for a Rigid Body

10.9 •• The flywheel of an engine has moment of inertia 2.50 kg · m² about its rotation axis. What constant torque is required to bring it up to an angular speed of 400 rev/min in 8.00 s, starting from rest?

10.10 •• A uniform disk with mass 40.0 kg and radius 0.200 m is pivoted at its center about a horizontal, frictionless axle that is stationary. The disk is initially at rest, and then a constant force $F = 30.0$ N is applied tangent to the rim of the disk. (a) What is the magnitude v of the tangential velocity of a point on the rim of the disk after the disk has turned through 0.200 revolution? (b) What is the magnitude a of the resultant acceleration of a point on the rim of the disk after the disk has turned through 0.200 revolution?

10.11 •• A machine part has the shape of a solid uniform sphere of mass 225 g and diameter 3.00 cm. It is spinning about a frictionless axle through its center, but at one point on its equator it is scraping against metal, resulting in a friction force of 0.0200 N at that point. (a) Find its angular acceleration. (b) How long will it take to decrease its rotational speed by 22.5 rad/s?

10.12 • A cord is wrapped around the rim of a solid uniform wheel 0.250 m in radius and of mass 9.20 kg. A steady horizontal pull of 40.0 N to the right is exerted on the cord, pulling it off tangentially from the wheel. The wheel is mounted on frictionless bearings on a horizontal axle through its center. (a) Compute the angular acceleration of the wheel and the acceleration of the part of the cord that has already been pulled off the wheel. (b) Find the magnitude and direction of the force that the axle exerts on the wheel. (c) Which of the answers in parts (a) and (b) would change if the pull were upward instead of horizontal?

10.13 ·· **CP** A 2.00-kg textbook rests on a frictionless, horizontal surface. A cord attached to the book passes over a pulley whose diameter is 0.150 m, to a hanging book with mass 3.00 kg. The system is released from rest, and the books are observed to move 1.20 m in 0.800 s. (a) What is the tension in each part of the cord? (b) What is the moment of inertia of the pulley about its rotation axis?

10.14 ·· **CP** A stone is suspended from the free end of a wire that is wrapped around the outer rim of a pulley, similar to what is shown in Fig. 10.10. The pulley is a uniform disk with mass 10.0 kg and radius 50.0 cm and turns on frictionless bearings. You measure that the stone travels 12.6 m in the first 3.00 s starting from rest. Find (a) the mass of the stone and (b) the tension in the wire.

10.15 · A wheel rotates without friction about a stationary horizontal axis at the center of the wheel. A constant tangential force equal to 80.0 N is applied to the rim of the wheel. The wheel has radius 0.120 m. Starting from rest, the wheel has an angular speed of 12.0 rev/s after 2.00 s. What is the moment of inertia of the wheel?

10.16 ·· **CP** A 15.0-kg bucket of water is suspended by a very light rope wrapped around a solid uniform cylinder 0.300 m in diameter with mass 12.0 kg. The cylinder pivots on a frictionless axle through its center. The bucket is released from rest at the top of a well and falls 10.0 m to the water. (a) What is the tension in the rope while the bucket is falling? (b) With what speed does the bucket strike the water? (c) What is the time of fall? (d) While the bucket is falling, what is the force exerted on the cylinder by the axle?

10.17 ·· A 12.0-kg box resting on a horizontal, frictionless surface is attached to a 5.00-kg weight by a thin, light wire that passes over a frictionless pulley (Fig. E10.17). The pulley has the shape of a uniform solid disk of mass 2.00 kg and diameter 0.500 m. After the system is released, find (a) the tension in the wire on both sides of the pulley, (b) the acceleration of the box, and (c) the horizontal and vertical components of the force that the axle exerts on the pulley.

Figure **E10.17**

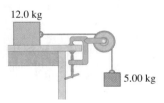

12.0 kg

5.00 kg

Section 10.3 Rigid-Body Rotation About a Moving Axis

10.18 · **BIO** **Gymnastics.** We can roughly model a gymnastic tumbler as a uniform solid cylinder of mass 75 kg and diameter 1.0 m. If this tumbler rolls forward at 0.50 rev/s, (a) how much total kinetic energy does he have, and (b) what percent of his total kinetic energy is rotational?

10.19 · A 2.20-kg hoop 1.20 m in diameter is rolling to the right without slipping on a horizontal floor at a steady 3.00 rad/s. (a) How fast is its center moving? (b) What is the total kinetic energy of the hoop? (c) Find the velocity vector of each of the following points, as viewed by a person at rest on the ground: (i) the highest point on the hoop; (ii) the lowest point on the hoop; (iii) a point on the right side of the hoop, midway between the top and the bottom. (d) Find the velocity vector for each of the points in part (c), but this time as viewed by someone moving along with the same velocity as the hoop.

10.20 ·· A string is wrapped several times around the rim of a small hoop with radius 8.00 cm and mass 0.180 kg. The free end of the string is held in place and the hoop is released from rest (Fig. E10.20). After the hoop has descended 75.0 cm, calculate (a) the angular speed of the rotating hoop and (b) the speed of its center.

Figure **E10.20**

0.0800 m

10.21 · What fraction of the total kinetic energy is rotational for the following objects rolling without slipping on a horizontal surface? (a) a uniform solid cylinder; (b) a uniform sphere; (c) a thin-walled, hollow sphere; (d) a hollow cylinder with outer radius R and inner radius $R/2$.

10.22 ·· A hollow, spherical shell with mass 2.00 kg rolls without slipping down a 38.0° slope. (a) Find the acceleration, the friction force, and the minimum coefficient of friction needed to prevent slipping. (b) How would your answers to part (a) change if the mass were doubled to 4.00 kg?

10.23 ·· A solid ball is released from rest and slides down a hillside that slopes downward at 65.0° from the horizontal. (a) What minimum value must the coefficient of static friction between the hill and ball surfaces have for no slipping to occur? (b) Would the coefficient of friction calculated in part (a) be sufficient to prevent a hollow ball (such as a soccer ball) from slipping? Justify your answer. (c) In part (a), why did we use the coefficient of static friction and not the coefficient of kinetic friction?

10.24 ·· A uniform marble rolls down a symmetrical bowl, starting from rest at the top of the left side. The top of each side is a distance h above the bottom of the bowl. The left half of the bowl is rough enough to cause the marble to roll without slipping, but the right half has no friction because it is coated with oil. (a) How far up the smooth side will the marble go, measured vertically from the bottom? (b) How high would the marble go if both sides were as rough as the left side? (c) How do you account for the fact that the marble goes *higher* with friction on the right side than without friction?

10.25 ·· A 392-N wheel comes off a moving truck and rolls without slipping along a highway. At the bottom of a hill it is rotating at 25.0 rad/s. The radius of the wheel is 0.600 m, and its moment of inertia about its rotation axis is $0.800MR^2$. Friction does work on the wheel as it rolls up the hill to a stop, a height h above the bottom of the hill; this work has absolute value 3500 J. Calculate h.

10.26 ·· **A Ball Rolling Uphill.** A bowling ball rolls without slipping up a ramp that slopes upward at an angle β to the horizontal (see Example 10.7 in Section 10.3). Treat the ball as a uniform solid sphere, ignoring the finger holes. (a) Draw the free-body diagram for the ball. Explain why the friction force must be directed *uphill*. (b) What is the acceleration of the center of mass of the ball? (c) What minimum coefficient of static friction is needed to prevent slipping?

10.27 ·· A thin, light string is wrapped around the outer rim of a uniform hollow cylinder of mass 4.75 kg having inner and outer radii as shown in Fig. E10.27. The cylinder is then released from rest.

Figure **E10.27**

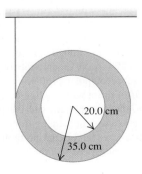

20.0 cm

35.0 cm

(a) How far must the cylinder fall before its center is moving at 6.66 m/s? (b) If you just dropped this cylinder without any string, how fast would its center be moving when it had fallen the distance in part (a)? (c) Why do you get two different answers when the cylinder falls the same distance in both cases?

10.28 •• A bicycle racer is going downhill at 11.0 m/s when, to his horror, one of his 2.25-kg wheels comes off as he is 75.0 m above the foot of the hill. We can model the wheel as a thin-walled cylinder 85.0 cm in diameter and neglect the small mass of the spokes. (a) How fast is the wheel moving when it reaches the foot of the hill if it rolled without slipping all the way down? (b) How much total kinetic energy does the wheel have when it reaches the bottom of the hill?

10.29 •• A size-5 soccer ball of diameter 22.6 cm and mass 426 g rolls up a hill without slipping, reaching a maximum height of 5.00 m above the base of the hill. We can model this ball as a thin-walled hollow sphere. (a) At what rate was it rotating at the base of the hill? (b) How much rotational kinetic energy did it have then?

Section 10.4 Work and Power in Rotational Motion

10.30 • An engine delivers 175 hp to an aircraft propeller at 2400 rev/min. (a) How much torque does the aircraft engine provide? (b) How much work does the engine do in one revolution of the propeller?

10.31 • A playground merry-go-round has radius 2.40 m and moment of inertia $2100 \ \text{kg} \cdot \text{m}^2$ about a vertical axle through its center, and it turns with negligible friction. (a) A child applies an 18.0-N force tangentially to the edge of the merry-go-round for 15.0 s. If the merry-go-round is initially at rest, what is its angular speed after this 15.0-s interval? (b) How much work did the child do on the merry-go-round? (c) What is the average power supplied by the child?

10.32 •• An electric motor consumes 9.00 kJ of electrical energy in 1.00 min. If one-third of this energy goes into heat and other forms of internal energy of the motor, with the rest going to the motor output, how much torque will this engine develop if you run it at 2500 rpm?

10.33 • A 1.50-kg grinding wheel is in the form of a solid cylinder of radius 0.100 m. (a) What constant torque will bring it from rest to an angular speed of 1200 rev/min in 2.5 s? (b) Through what angle has it turned during that time? (c) Use Eq. (10.21) to calculate the work done by the torque. (d) What is the grinding wheel's kinetic energy when it is rotating at 1200 rev/min? Compare your answer to the result in part (c).

10.34 •• An airplane propeller is 2.08 m in length (from tip to tip) and has a mass of 117 kg. When the airplane's engine is first started, it applies a constant torque of 1950 N·m to the propeller, which starts from rest. (a) What is the angular acceleration of the propeller? Model the propeller as a slender rod and see Table 9.2. (b) What is the propeller's angular speed after making 5.00 revolutions? (c) How much work is done by the engine during the first 5.00 revolutions? (d) What is the average power output of the engine during the first 5.00 revolutions? (e) What is the instantaneous power output of the motor at the instant that the propeller has turned through 5.00 revolutions?

10.35 • (a) Compute the torque developed by an industrial motor whose output is 150 kW at an angular speed of 4000 rev/min. (b) A drum with negligible mass, 0.400 m in diameter, is attached to the motor shaft, and the power output of the motor is used to raise a weight hanging from a rope wrapped around the drum. How heavy a weight can the motor lift at constant speed? (c) At what constant speed will the weight rise?

Section 10.5 Angular Momentum

10.36 •• A woman with mass 50 kg is standing on the rim of a large disk that is rotating at 0.50 rev/s about an axis through its center. The disk has mass 110 kg and radius 4.0 m. Calculate the magnitude of the total angular momentum of the woman–disk system. (Assume that you can treat the woman as a point.)

10.37 • A 2.00-kg rock has a horizontal velocity of magnitude 12.0 m/s when it is at point P in Fig. E10.37. (a) At this instant, what are the magnitude and direction of its angular momentum relative to point O? (b) If the only force acting on the rock is its weight, what is the rate of change (magnitude and direction) of its angular momentum at this instant?

Figure **E10.37**

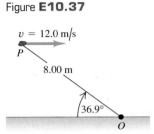

$v = 12.0 \ \text{m/s}$
P
8.00 m
36.9°
O

10.38 •• (a) Calculate the magnitude of the angular momentum of the earth in a circular orbit around the sun. Is it reasonable to model it as a particle? (b) Calculate the magnitude of the angular momentum of the earth due to its rotation around an axis through the north and south poles, modeling it as a uniform sphere. Consult Appendix E and the astronomical data in Appendix F.

10.39 •• Find the magnitude of the angular momentum of the second hand on a clock about an axis through the center of the clock face. The clock hand has a length of 15.0 cm and a mass of 6.00 g. Take the second hand to be a slender rod rotating with constant angular velocity about one end.

10.40 •• **CALC** A hollow, thin-walled sphere of mass 12.0 kg and diameter 48.0 cm is rotating about an axle through its center. The angle (in radians) through which it turns as a function of time (in seconds) is given by $\theta(t) = At^2 + Bt^4$, where A has numerical value 1.50 and B has numerical value 1.10. (a) What are the units of the constants A and B? (b) At the time 3.00 s, find (i) the angular momentum of the sphere and (ii) the net torque on the sphere.

Section 10.6 Conservation of Angular Momentum

10.41 •• Under some circumstances, a star can collapse into an extremely dense object made mostly of neutrons and called a *neutron star*. The density of a neutron star is roughly 10^{14} times as great as that of ordinary solid matter. Suppose we represent the star as a uniform, solid, rigid sphere, both before and after the collapse. The star's initial radius was 7.0×10^5 km (comparable to our sun); its final radius is 16 km. If the original star rotated once in 30 days, find the angular speed of the neutron star.

10.42 • **CP** A small block on a frictionless, horizontal surface has a mass of 0.0250 kg. It is attached to a massless cord passing through a hole in the surface (Fig. E10.42). The block is originally revolving at a distance of 0.300 m from the hole with an angular speed of 1.75 rad/s. The cord is then pulled from below, shortening the radius of the circle in which the block revolves to 0.150 m. Model the block as a particle. (a) Is the angular momentum of the block conserved? Why or why not? (b) What is the new angular speed? (c) Find the change in kinetic energy of the block. (d) How much work was done in pulling the cord?

Figure **E10.42**

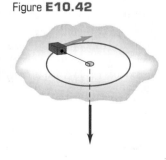

10.43 ·· **The Spinning Figure Skater.** The outstretched hands and arms of a figure skater preparing for a spin can be considered a slender rod pivoting about an axis through its center (Fig. E10.43). When the skater's hands and arms are brought in and wrapped around his body to execute the spin, the hands and arms can be considered a thin-walled, hollow cylinder. His hands and arms have a combined mass of 8.0 kg. When outstretched, they span 1.8 m; when wrapped, they form a cylinder of radius 25 cm. The moment of inertia about the rotation axis of the remainder of his body is constant and equal to 0.40 kg · m². If his original angular speed is 0.40 rev/s, what is his final angular speed?

Figure **E10.43**

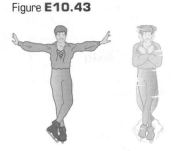

10.44 ·· A diver comes off a board with arms straight up and legs straight down, giving her a moment of inertia about her rotation axis of 18 kg · m². She then tucks into a small ball, decreasing this moment of inertia to 3.6 kg · m². While tucked, she makes two complete revolutions in 1.0 s. If she hadn't tucked at all, how many revolutions would she have made in the 1.5 s from board to water?

10.45 ·· A large wooden turntable in the shape of a flat uniform disk has a radius of 2.00 m and a total mass of 120 kg. The turntable is initially rotating at 3.00 rad/s about a vertical axis through its center. Suddenly, a 70.0-kg parachutist makes a soft landing on the turntable at a point near the outer edge. (a) Find the angular speed of the turntable after the parachutist lands. (Assume that you can treat the parachutist as a particle.) (b) Compute the kinetic energy of the system before and after the parachutist lands. Why are these kinetic energies not equal?

10.46 ·· A solid wood door 1.00 m wide and 2.00 m high is hinged along one side and has a total mass of 40.0 kg. Initially open and at rest, the door is struck at its center by a handful of sticky mud with mass 0.500 kg, traveling perpendicular to the door at 12.0 m/s just before impact. Find the final angular speed of the door. Does the mud make a significant contribution to the moment of inertia?

10.47 ·· A small 10.0-g bug stands at one end of a thin uniform bar that is initially at rest on a smooth horizontal table. The other end of the bar pivots about a nail driven into the table and can rotate freely, without friction. The bar has mass 50.0 g and is 100 cm in length. The bug jumps off in the horizontal direction, perpendicular to the bar, with a speed of 20.0 cm/s relative to the table. (a) What is the angular speed of the bar just after the frisky insect leaps? (b) What is the total kinetic energy of the system just after the bug leaps? (c) Where does this energy come from?

10.48 ·· **Asteroid Collision!** Suppose that an asteroid traveling straight toward the center of the earth were to collide with our planet at the equator and bury itself just below the surface. What would have to be the mass of this asteroid, in terms of the earth's mass *M*, for the day to become 25.0% longer than it presently is as a result of the collision? Assume that the asteroid is very small compared to the earth and that the earth is uniform throughout.

10.49 ·· A thin, uniform metal bar, 2.00 m long and weighing 90.0 N, is hanging vertically from the ceiling by a frictionless pivot. Suddenly it is struck 1.50 m below the ceiling by a small 3.00-kg ball, initially traveling horizontally at 10.0 m/s. The ball rebounds in the opposite direction with a speed of 6.00 m/s. (a) Find the angular speed of the bar just after the collision. (b) During the collision, why is the angular momentum conserved but not the linear momentum?

10.50 ·· A thin uniform rod has a length of 0.500 m and is rotating in a circle on a frictionless table. The axis of rotation is perpendicular to the length of the rod at one end and is stationary. The rod has an angular velocity of 0.400 rad/s and a moment of inertia about the axis of 3.00 × 10⁻³ kg · m². A bug initially standing on the rod at the axis of rotation decides to crawl out to the other end of the rod. When the bug has reached the end of the rod and sits there, its tangential speed is 0.160 m/s. The bug can be treated as a point mass. (a) What is the mass of the rod? (b) What is the mass of the bug?

10.51 ·· A uniform, 4.5-kg, square, solid wooden gate 1.5 m on each side hangs vertically from a frictionless pivot at the center of its upper edge. A 1.1-kg raven flying horizontally at 5.0 m/s flies into this door at its center and bounces back at 2.0 m/s in the opposite direction. (a) What is the angular speed of the gate just after it is struck by the unfortunate raven? (b) During the collision, why is the angular momentum conserved, but not the linear momentum?

10.52 ·· **Sedna.** In November 2003, the now-most-distant-known object in the solar system was discovered by observation with a telescope on Mt. Palomar. This object, known as Sedna, is approximately 1700 km in diameter, takes about 10,500 years to orbit our sun, and reaches a maximum speed of 4.64 km/s. Calculations of its complete path, based on several measurements of its position, indicate that its orbit is highly elliptical, varying from 76 AU to 942 AU in its distance from the sun, where AU is the astronomical unit, which is the average distance of the earth from the sun (1.50 × 10⁸ km). (a) What is Sedna's minimum speed? (b) At what points in its orbit do its maximum and minimum speeds occur? (c) What is the ratio of Sedna's maximum kinetic energy to its minimum kinetic energy?

Section 10.7 Gyroscopes and Precession

10.53 ·· The rotor (flywheel) of a toy gyroscope has mass 0.140 kg. Its moment of inertia about its axis is 1.20 × 10⁻⁴ kg · m². The mass of the frame is 0.0250 kg. The gyroscope is supported on a single pivot (Fig. E10.53) with its center of mass a horizontal distance of 4.00 cm from the pivot. The gyroscope is precessing in a horizontal plane at the rate of one revolution in 2.20 s. (a) Find the upward force exerted by the pivot. (b) Find the angular speed with which the rotor is spinning about its axis, expressed in rev/min. (c) Copy the diagram and draw vectors to show the angular momentum of the rotor and the torque acting on it.

Figure **E10.53**

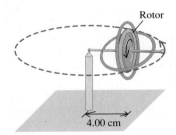

Rotor

4.00 cm

10.54 · **A Gyroscope on the Moon.** A certain gyroscope precesses at a rate of 0.50 rad/s when used on earth. If it were taken to a lunar base, where the acceleration due to gravity is 0.165*g*, what would be its precession rate?

10.55 · A gyroscope is precessing about a vertical axis. Describe what happens to the precession angular speed if the following changes in the variables are made, with all other variables remaining the same: (a) the angular speed of the spinning flywheel is doubled; (b) the total weight is doubled; (c) the moment of inertia about the axis of the spinning flywheel is doubled; (d) the distance from the pivot to the center of gravity is doubled. (e) What happens if all four of the variables in parts (a) through (d) are doubled?

10.56 • **Stabilization of the Hubble Space Telescope.** The Hubble Space Telescope is stabilized to within an angle of about 2-millionths of a degree by means of a series of gyroscopes that spin at 19,200 rpm. Although the structure of these gyroscopes is actually quite complex, we can model each of the gyroscopes as a thin-walled cylinder of mass 2.0 kg and diameter 5.0 cm, spinning about its central axis. How large a torque would it take to cause these gyroscopes to precess through an angle of 1.0×10^{-6} degree during a 5.0-hour exposure of a galaxy?

PROBLEMS

10.57 •• A 50.0-kg grindstone is a solid disk 0.520 m in diameter. You press an ax down on the rim with a normal force of 160 N (Fig. P10.57). The coefficient of kinetic friction between the blade and the stone is 0.60, and there is a constant friction torque of 6.50 N·m between the axle of the stone and its bearings. (a) How much force must be applied tangentially at the end of a crank handle 0.500 m long to bring the stone from rest to 120 rev/min in 9.00 s? (b) After the grindstone attains an angular speed of 120 rev/min, what tangential force at the end of the handle is needed to maintain a constant angular speed of 120 rev/min? (c) How much time does it take the grindstone to come from 120 rev/min to rest if it is acted on by the axle friction alone?

Figure **P10.57**

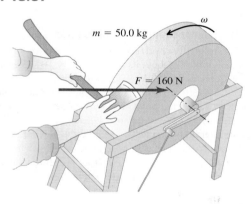

ω
$m = 50.0$ kg
$F = 160$ N

10.58 •• An experimental bicycle wheel is placed on a test stand so that it is free to turn on its axle. If a constant net torque of 7.00 N·m is applied to the tire for 2.00 s, the angular speed of the tire increases from 0 to 100 rev/min. The external torque is then removed, and the wheel is brought to rest by friction in its bearings in 125 s. Compute (a) the moment of inertia of the wheel about the rotation axis; (b) the friction torque; (c) the total number of revolutions made by the wheel in the 125-s time interval.

10.59 ••• A grindstone in the shape of a solid disk with diameter 0.520 m and a mass of 50.0 kg is rotating at 850 rev/min. You press an ax against the rim with a normal force of 160 N (Fig. P10.57), and the grindstone comes to rest in 7.50 s. Find the coefficient of friction between the ax and the grindstone. You can ignore friction in the bearings.

10.60 ••• A uniform, 8.40-kg, spherical shell 50.0 cm in diameter has four small 2.00-kg masses attached to its outer surface and equally spaced around it. This

Figure **P10.60**

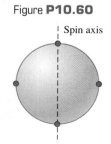

Spin axis

combination is spinning about an axis running through the center of the sphere and two of the small masses (Fig. P10.60). What friction torque is needed to reduce its angular speed from 75.0 rpm to 50.0 rpm in 30.0 s?

10.61 ••• A solid uniform cylinder with mass 8.25 kg and diameter 15.0 cm is spinning at 220 rpm on a thin, frictionless axle that passes along the cylinder axis. You design a simple friction brake to stop the cylinder by pressing the brake against the outer rim with a normal force. The coefficient of kinetic friction between the brake and rim is 0.333. What must the applied normal force be to bring the cylinder to rest after it has turned through 5.25 revolutions?

10.62 ••• A uniform hollow disk has two pieces of thin, light wire wrapped around its outer rim and is supported from the ceiling (Fig. P10.62). Suddenly one of the wires breaks, and the remaining wire does not slip as the disk rolls down. Use energy conservation to find the speed of the center of this disk after it has fallen a distance of 2.20 m.

Figure **P10.62**

30.0 cm

50.0 cm

10.63 ••• A thin, uniform, 3.80-kg bar, 80.0 cm long, has very small 2.50-kg balls glued on at either end (Fig. P10.63). It is supported horizontally by a thin, horizontal, frictionless axle passing through its center and perpendicular to the bar.

Figure **P10.63**

2.50 kg   2.50 kg

Bar   Axle (seen end-on)

Suddenly the right-hand ball becomes detached and falls off, but the other ball remains glued to the bar. (a) Find the angular acceleration of the bar just after the ball falls off. (b) Will the angular acceleration remain constant as the bar continues to swing? If not, will it increase or decrease? (c) Find the angular velocity of the bar just as it swings through its vertical position.

10.64 ••• While exploring a castle, Exena the Exterminator is spotted by a dragon that chases her down a hallway. Exena runs into a room and attempts to swing the heavy door shut before the dragon gets her. The door is initially perpendicular to the wall, so it must be turned through 90° to close. The door is 3.00 m tall and 1.25 m wide, and it weighs 750 N. You can ignore the friction at the hinges. If Exena applies a force of 220 N at the edge of the door and perpendicular to it, how much time does it take her to close the door?

10.65 •• **CALC** You connect a light string to a point on the edge of a uniform vertical disk with radius R and mass M. The disk is free to rotate without friction about a stationary horizontal axis through its center. Initially, the disk is at rest with the string connection at the highest point on the disk. You pull the string with a constant horizontal force $\vec{F}$ until the wheel has made exactly one-quarter revolution about a horizontal axis through its center, and then you let go. (a) Use Eq. (10.20) to find the work done by the string. (b) Use Eq. (6.14) to find the work done by the string. Do you obtain the same result as in part (a)? (c) Find the final angular speed of the disk. (d) Find the maximum tangential acceleration of a point on the disk. (e) Find the maximum radial (centripetal) acceleration of a point on the disk.

10.66 ••• **Balancing Act.** Attached to one end of a long, thin, uniform rod of length L and mass M is a small blob of clay of the same mass M. (a) Locate the position of the center of mass of the system of rod and clay. Note this position on a drawing of the rod.

(b) You carefully balance the rod on a frictionless tabletop so that it is standing vertically, with the end without the clay touching the table. If the rod is now tipped so that it is a small angle θ away from the vertical, determine its angular acceleration at this instant. Assume that the end without the clay remains in contact with the tabletop. (*Hint:* See Table 9.2.) (c) You again balance the rod on the frictionless tabletop so that it is standing vertically, but now the end of the rod *with* the clay is touching the table. If the rod is again tipped so that it is a small angle θ away from the vertical, determine its angular acceleration at this instant. Assume that the end with the clay remains in contact with the tabletop. How does this compare to the angular acceleration in part (b)? (d) A pool cue is a tapered wooden rod that is thick at one end and thin at the other. You can easily balance a pool cue vertically on one finger if the thin end is in contact with your finger; this is quite a bit harder to do if the thick end is in contact with your finger. Explain why there is a difference.

10.67 •• **Atwood's Machine.** Figure P10.67 illustrates an Atwood's machine. Find the linear accelerations of blocks A and B, the angular acceleration of the wheel C, and the tension in each side of the cord if there is no slipping between the cord and the surface of the wheel. Let the masses of blocks A and B be 4.00 kg and 2.00 kg, respectively, the moment of inertia of the wheel about its axis be 0.300 kg·m², and the radius of the wheel be 0.120 m.

Figure **P10.67**

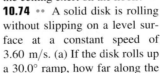

10.68 ••• The mechanism shown in Fig. P10.68 is used to raise a crate of supplies from a ship's hold. The crate has total mass 50 kg. A rope is wrapped around a wooden cylinder that turns on a metal axle. The cylinder has radius 0.25 m and moment of inertia $I = 2.9$ kg·m² about the axle. The crate is suspended from the free end of the rope. One end of the axle pivots on frictionless bearings; a crank handle is attached to the other end. When the crank is turned, the end of the handle rotates about the axle in a vertical circle of radius 0.12 m, the cylinder turns, and the crate is raised. What magnitude of the force $\vec{F}$ applied tangentially to the rotating crank is required to raise the crate with an acceleration of 1.40 m/s²? (You can ignore the mass of the rope as well as the moments of inertia of the axle and the crank.)

Figure **P10.68**

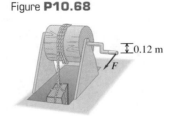

10.69 •• A large 16.0-kg roll of paper with radius $R = 18.0$ cm rests against the wall and is held in place by a bracket attached to a rod through the center of the roll (Fig. P10.69). The rod turns without friction in the bracket, and the moment of inertia of the paper and rod about the axis is 0.260 kg·m². The other end of the bracket is attached by a

Figure **P10.69**

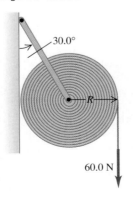

frictionless hinge to the wall such that the bracket makes an angle of 30.0° with the wall. The weight of the bracket is negligible. The coefficient of kinetic friction between the paper and the wall is $\mu_k = 0.25$. A constant vertical force $F = 60.0$ N is applied to the paper, and the paper unrolls. (a) What is the magnitude of the force that the rod exerts on the paper as it unrolls? (b) What is the magnitude of the angular acceleration of the roll?

10.70 •• A block with mass $m = 5.00$ kg slides down a surface inclined 36.9° to the horizontal (Fig. P10.70). The coefficient of kinetic friction is 0.25. A string attached to the block is wrapped around a flywheel on a fixed axis at O. The flywheel has mass 25.0 kg and moment of inertia 0.500 kg·m² with respect to the axis of rotation. The string pulls without slipping at a perpendicular distance of 0.200 m from that axis. (a) What is the acceleration of the block down the plane? (b) What is the tension in the string?

Figure **P10.70**

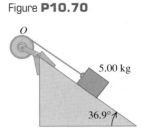

10.71 ••• Two metal disks, one with radius $R_1 = 2.50$ cm and mass $M_1 = 0.80$ kg and the other with radius $R_2 = 5.00$ cm and mass $M_2 = 1.60$ kg, are welded together and mounted on a frictionless axis through their common center, as in Problem 9.87. (a) A light string is wrapped around the edge of the smaller disk, and a 1.50-kg block is suspended from the free end of the string. What is the magnitude of the downward acceleration of the block after it is released? (b) Repeat the calculation of part (a), this time with the string wrapped around the edge of the larger disk. In which case is the acceleration of the block greater? Does your answer make sense?

10.72 •• A lawn roller in the form of a thin-walled, hollow cylinder with mass M is pulled horizontally with a constant horizontal force F applied by a handle attached to the axle. If it rolls without slipping, find the acceleration and the friction force.

10.73 • Two weights are connected by a very light, flexible cord that passes over an 80.0-N frictionless pulley of radius 0.300 m. The pulley is a solid uniform disk and is supported by a hook connected to the ceiling (Fig. P10.73). What force does the ceiling exert on the hook?

Figure **P10.73**

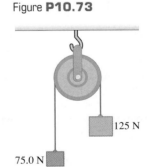

10.74 •• A solid disk is rolling without slipping on a level surface at a constant speed of 3.60 m/s. (a) If the disk rolls up a 30.0° ramp, how far along the ramp will it move before it stops? (b) Explain why your answer in part (a) does not depend on either the mass or the radius of the disk.

10.75 • **The Yo-yo.** A yo-yo is made from two uniform disks, each with mass m and radius R, connected by a light axle of radius b. A light, thin string is wound several times around the axle and then held stationary while the yo-yo is released from rest, dropping as the string unwinds. Find the linear acceleration and angular acceleration of the yo-yo and the tension in the string.

10.76 •• **CP** A thin-walled, hollow spherical shell of mass m and radius r starts from rest and rolls without slipping down the track shown in Fig. P10.76. Points A and B are on a circular part of the track having radius R. The diameter of the shell is very small compared to h_0 and R, and the work done by rolling friction is negligible. (a) What is the minimum height h_0 for which this shell will make a complete loop-the-loop on the circular part of the track? (b) How hard does the track push on the shell at point B, which is at the same level as the center of the circle? (c) Suppose that the track had no friction and the shell was released from the same height h_0 you found in part (a). Would it make a complete loop-the-loop? How do you know? (d) In part (c), how hard does the track push on the shell at point A, the top of the circle? How hard did it push on the shell in part (a)?

Figure **P10.76**

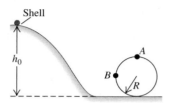

Shell

h_0

A

B

R

10.77 • Starting from rest, a constant force $F = 100$ N is applied to the free end of a 50-m cable wrapped around the outer rim of a uniform solid cylinder, similar to the situation shown in Fig. 10.9(a). The cylinder has mass 4.00 kg and diameter 30.0 cm and is free to turn about a fixed, frictionless axle through its center. (a) How long does it take to unwrap all the cable, and how fast is the cable moving just as the last bit comes off? (b) Now suppose that the cylinder is replaced by a uniform hoop, with all other quantities remaining unchanged. In this case, would the answers in part (a) be larger or smaller? Explain.

10.78 •• As shown in Fig. E10.20, a string is wrapped several times around the rim of a small hoop with radius 0.0800 m and mass 0.180 kg. The free end of the string is pulled upward in just the right way so that the hoop does not move vertically as the string unwinds. (a) Find the tension in the string as the string unwinds. (b) Find the angular acceleration of the hoop as the string unwinds. (c) Find the upward acceleration of the hand that pulls on the free end of the string. (d) How would your answers be different if the hoop were replaced by a solid disk of the same mass and radius?

10.79 •• A basketball (which can be closely modeled as a hollow spherical shell) rolls down a mountainside into a valley and then up the opposite side, starting from rest at a height H_0 above the bottom. In Fig. P10.79, the rough part of the terrain prevents slipping while the smooth part has no friction. (a) How high, in terms of H_0, will the ball go up the other side? (b) Why doesn't the ball return to height H_0? Has it lost any of its original potential energy?

Figure **P10.79**

Rough

Smooth

H_0

10.80 • **CP** A uniform marble rolls without slipping down the path shown in Fig. P10.80, starting from rest. (a) Find the minimum height h required for the marble not to fall into the pit.

(b) The moment of inertia of the marble depends on its radius. Explain why the answer to part (a) does not depend on the radius of the marble. (c) Solve part (a) for a block that slides without friction instead of the rolling marble. How does the minimum h in this case compare to the answer in part (a)?

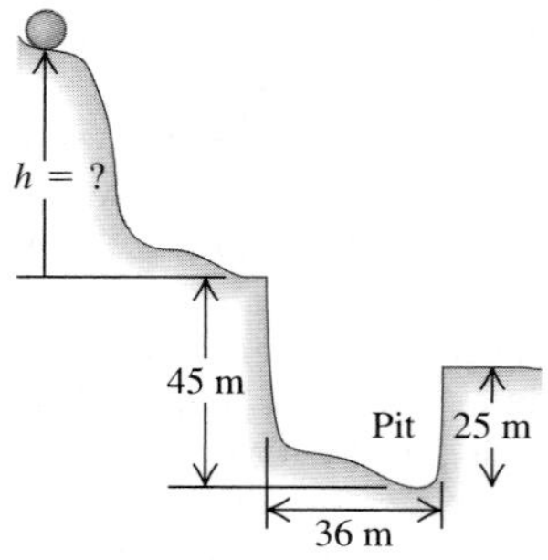

10.81 •• **Rolling Stones.** A solid, uniform, spherical boulder starts from rest and rolls down a 50.0-m-high hill, as shown in Fig. P10.81. The top half of the hill is rough enough to cause the boulder to roll without slipping, but the lower half is covered with ice and there is no friction. What is the translational speed of the boulder when it reaches the bottom of the hill?

10.82 •• **CP** A solid uniform ball rolls without slipping up a hill, as shown in Fig. P10.82. At the top of the hill, it is moving horizontally, and then it goes over the vertical cliff. (a) How far from the foot of the cliff does the ball land, and how fast is it moving just before it lands? (b) Notice that when the balls lands, it has a greater translational speed than when it was at the bottom of the hill. Does this mean that the ball somehow gained energy? Explain!

10.83 •• A 42.0-cm-diameter wheel, consisting of a rim and six spokes, is constructed from a thin, rigid plastic material having a linear mass density of 25.0 g/cm. This wheel is released from rest at the top of a hill 58.0 m high. (a) How fast is it rolling when it reaches the bottom of the hill? (b) How would your answer change if the linear mass density and the diameter of the wheel were each doubled?

10.84 •• A child rolls a 0.600-kg basketball up a long ramp. The basketball can be considered a thin-walled, hollow sphere. When the child releases the basketball at the bottom of the ramp, it has a speed of 8.0 m/s. When the ball returns to her after rolling up the ramp and then rolling back down, it has a speed of 4.0 m/s. Assume the work done by friction on the basketball is the same when the ball moves up or down the ramp and that the basketball rolls without slipping. Find the maximum vertical height increase of the ball as it rolls up the ramp.

10.85 •• **CP** In a lab experiment you let a uniform ball roll down a curved track. The ball starts from rest and rolls without slipping. While on the track, the ball descends a vertical distance h. The lower end of the track is horizontal and extends over the edge of the lab table; the ball leaves the track traveling horizontally. While free-falling after leaving the track, the ball moves a horizontal distance x and a vertical distance y. (a) Calculate x in terms of h and y, ignoring the work done by friction. (b) Would the answer to part (a) be any different on the moon? (c) Although you do the experiment very carefully, your measured value of x is consistently a bit smaller than the value calculated in part (a). Why? (d) What would x be for the same h and y as in part (a) if you let a silver dollar roll down the track? You can ignore the work done by friction.

Figure **P10.80**

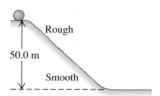

$h = ?$

45 m

Pit | 25 m

36 m

Figure **P10.81**

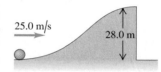

Rough

50.0 m

Smooth

Figure **P10.82**

25.0 m/s

28.0 m

10.86 •• A uniform drawbridge 8.00 m long is attached to the roadway by a frictionless hinge at one end, and it can be raised by a cable attached to the other end. The bridge is at rest, suspended at 60.0° above the horizontal, when the cable suddenly breaks. (a) Find the angular acceleration of the drawbridge just after the cable breaks. (Gravity behaves as though it all acts at the center of mass.) (b) Could you use the equation $\omega = \omega_0 + \alpha t$ to calculate the angular speed of the drawbridge at a later time? Explain why. (c) What is the angular speed of the drawbridge as it becomes horizontal?

10.87 • A uniform solid cylinder with mass M and radius $2R$ rests on a horizontal tabletop. A string is attached by a yoke to a frictionless axle through the center of the cylinder so that the cylinder can rotate about the axle. The string runs over a disk-shaped pulley with mass M and radius R that is mounted on a frictionless axle through its center. A block of mass M is suspended from the free end of the string (Fig. P10.87). The string doesn't slip over the pulley surface, and the cylinder rolls without slipping on the tabletop. Find the magnitude of the acceleration of the block after the system is released from rest.

Figure **P10.87**

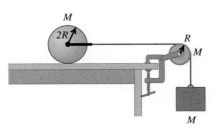

10.88 ••• A uniform, 0.0300-kg rod of length 0.400 m rotates in a horizontal plane about a fixed axis through its center and perpendicular to the rod. Two small rings, each with mass 0.0200 kg, are mounted so that they can slide along the rod. They are initially held by catches at positions 0.0500 m on each side of the center of the rod, and the system is rotating at 30.0 rev/min. With no other changes in the system, the catches are released, and the rings slide outward along the rod and fly off at the ends. (a) What is the angular speed of the system at the instant when the rings reach the ends of the rod? (b) What is the angular speed of the rod after the rings leave it?

10.89 ••• A 5.00-kg ball is dropped from a height of 12.0 m above one end of a uniform bar that pivots at its center. The bar has mass 8.00 kg and is 4.00 m in length. At the other end of the bar sits another 5.00-kg ball, unattached to the bar. The dropped ball sticks to the bar after the collision. How high will the other ball go after the collision?

10.90 •• **Tarzan and Jane in the 21st Century.** Tarzan has foolishly gotten himself into another scrape with the animals and must be rescued once again by Jane. The 60.0-kg Jane starts from rest at a height of 5.00 m in the trees and swings down to the ground using a thin, but very rigid, 30.0-kg vine 8.00 m long. She arrives just in time to snatch the 72.0-kg Tarzan from the jaws of an angry hippopotamus. What is Jane's (and the vine's) angular speed (a) just before she grabs Tarzan and (b) just after she grabs him? (c) How high will Tarzan and Jane go on their first swing after this daring rescue?

10.91 •• A uniform rod of length L rests on a frictionless horizontal surface. The rod pivots about a fixed frictionless axis at one end. The rod is initially at rest. A bullet traveling parallel to the horizontal surface and perpendicular to the rod with speed v strikes the rod at its center and becomes embedded in it. The mass of the bullet is

one-fourth the mass of the rod. (a) What is the final angular speed of the rod? (b) What is the ratio of the kinetic energy of the system after the collision to the kinetic energy of the bullet before the collision?

10.92 •• The solid wood door of a gymnasium is 1.00 m wide and 2.00 m high, has total mass 35.0 kg, and is hinged along one side. The door is open and at rest when a stray basketball hits the center of the door head-on, applying an average force of 1500 N to the door for 8.00 ms. Find the angular speed of the door after the impact. [*Hint:* Integrating Eq. (10.29) yields $\Delta L_z = \int_{t_1}^{t_2}(\Sigma\tau_z)dt = (\Sigma\tau_z)_{\text{av}}\Delta t$. The quantity $\int_{t_1}^{t_2}(\Sigma\tau_z)dt$ is called the angular impulse.]

10.93 ••• A target in a shooting gallery consists of a vertical square wooden board, 0.250 m on a side and with mass 0.750 kg, that pivots on a horizontal axis along its top edge. The board is struck face-on at its center by a bullet with mass 1.90 g that is traveling at 360 m/s and that remains embedded in the board. (a) What is the angular speed of the board just after the bullet's impact? (b) What maximum height above the equilibrium position does the center of the board reach before starting to swing down again? (c) What minimum bullet speed would be required for the board to swing all the way over after impact?

10.94 •• **Neutron Star Glitches.** Occasionally, a rotating neutron star (see Exercise 10.41) undergoes a sudden and unexpected speedup called a *glitch*. One explanation is that a glitch occurs when the crust of the neutron star settles slightly, decreasing the moment of inertia about the rotation axis. A neutron star with angular speed $\omega_0 = 70.4$ rad/s underwent such a glitch in October 1975 that increased its angular speed to $\omega = \omega_0 + \Delta\omega$, where $\Delta\omega/\omega_0 = 2.01 \times 10^{-6}$. If the radius of the neutron star before the glitch was 11 km, by how much did its radius decrease in the starquake? Assume that the neutron star is a uniform sphere.

10.95 ••• A 500.0-g bird is flying horizontally at 2.25 m/s, not paying much attention, when it suddenly flies into a stationary vertical bar, hitting it 25.0 cm below the top (Fig. P10.95). The bar is uniform, 0.750 m long, has a mass of 1.50 kg, and is hinged at its base. The collision stuns the bird so that it just drops to the ground afterward (but soon recovers to fly happily away). What is the angular velocity of the bar (a) just after it is hit by the bird and (b) just as it reaches the ground?

Figure **P10.95**

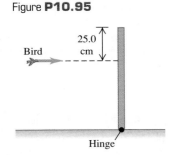

Bird

25.0 cm

Hinge

10.96 ••• **CP** A small block with mass 0.250 kg is attached to a string passing through a hole in a frictionless, horizontal surface (see Fig. E10.42). The block is originally revolving in a circle with a radius of 0.800 m about the hole with a tangential speed of 4.00 m/s. The string is then pulled slowly from below, shortening the radius of the circle in which the block revolves. The breaking strength of the string is 30.0 N. What is the radius of the circle when the string breaks?

10.97 • A horizontal plywood disk with mass 7.00 kg and diameter 1.00 m pivots on frictionless bearings about a vertical axis through its center. You attach a circular model-railroad track of negligible mass and average diameter 0.95 m to the disk. A 1.20-kg, battery-driven model train rests on the tracks. To demonstrate conservation of angular momentum, you switch on the train's engine. The train moves counterclockwise, soon attaining a constant speed of 0.600 m/s relative to the tracks. Find the magnitude and direction of the angular velocity of the disk relative to the earth.

10.98 • A 55-kg runner runs around the edge of a horizontal turntable mounted on a vertical, frictionless axis through its center. The runner's velocity relative to the earth has magnitude 2.8 m/s. The turntable is rotating in the opposite direction with an angular velocity of magnitude 0.20 rad/s relative to the earth. The radius of the turntable is 3.0 m, and its moment of inertia about the axis of rotation is 80 kg · m². Find the final angular velocity of the system if the runner comes to rest relative to the turntable. (You can model the runner as a particle.)

10.99 •• **Center of Percussion.** A baseball bat rests on a frictionless, horizontal surface. The bat has a length of 0.900 m, a mass of 0.800 kg, and its center of mass is 0.600 m from the handle end of the bat (Fig. P10.99). The moment of inertia of the bat about its center of mass is 0.0530 kg · m². The bat is struck by a baseball traveling perpendicular to the bat. The impact applies an impulse $J = \int_{t_1}^{t_2} F\,dt$ at a point a distance x from the handle end of the bat. What must x be so that the handle end of the bat remains at rest as the bat begins to move? [*Hint:* Consider the motion of the center of mass and the rotation about the center of mass. Find x so that these two motions combine to give $v = 0$ for the end of the bat just after the collision. Also, note that integration of Eq. (10.29) gives $\Delta L = \int_{t_1}^{t_2}(\Sigma\tau)\,dt$ (see Problem 10.92).] The point on the bat you have located is called the *center of percussion.* Hitting a pitched ball at the center of percussion of the bat minimizes the "sting" the batter experiences on the hands.

Figure **P10.99**

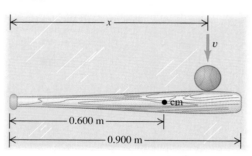

CHALLENGE PROBLEMS

10.100 ••• A uniform ball of radius R rolls without slipping between two rails such that the horizontal distance is d between the two contact points of the rails to the ball. (a) In a sketch, show that at any instant $v_{cm} = \omega\sqrt{R^2 - d^2/4}$. Discuss this expression in the limits $d = 0$ and $d = 2R$. (b) For a uniform ball starting from rest and descending a vertical distance h while rolling without slipping down a ramp, $v_{cm} = \sqrt{10gh/7}$. Replacing the ramp with the two rails, show that

$$v_{cm} = \sqrt{\frac{10gh}{5 + 2/(1 - d^2/4R^2)}}$$

In each case, the work done by friction has been ignored. (c) Which speed in part (b) is smaller? Why? Answer in terms of how the loss of potential energy is shared between the gain in translational and rotational kinetic energies. (d) For which value of the ratio d/R do the two expressions for the speed in part (b) differ by 5.0%? By 0.50%?

10.101 ••• When an object is rolling without slipping, the rolling friction force is much less than the friction force when the object is sliding; a silver dollar will roll on its edge much farther than it will slide on its flat side (see Section 5.3). When an object is rolling without slipping on a horizontal surface, we can approximate the friction force to be zero, so that a_x and α_z are approximately zero and v_x and ω_z are approximately constant. Rolling without slipping means $v_x = r\omega_z$ and $a_x = r\alpha_z$. If an object is set in motion on a surface *without* these equalities, sliding (kinetic) friction will act on the object as it slips until rolling without slipping is established. A solid cylinder with mass M and radius R, rotating with angular speed ω_0 about an axis through its center, is set on a horizontal surface for which the kinetic friction coefficient is μ_k. (a) Draw a free-body diagram for the cylinder on the surface. Think carefully about the direction of the kinetic friction force on the cylinder. Calculate the accelerations a_x of the center of mass and α_z of rotation about the center of mass. (b) The cylinder is initially slipping completely, so initially $\omega_z = \omega_0$ but $v_x = 0$. Rolling without slipping sets in when $v_x = R\omega_z$. Calculate the *distance* the cylinder rolls before slipping stops. (c) Calculate the work done by the friction force on the cylinder as it moves from where it was set down to where it begins to roll without slipping.

10.102 ••• A demonstration gyroscope wheel is constructed by removing the tire from a bicycle wheel 0.650 m in diameter, wrapping lead wire around the rim, and taping it in place. The shaft projects 0.200 m at each side of the wheel, and a woman holds the ends of the shaft in her hands. The mass of the system is 8.00 kg; its entire mass may be assumed to be located at its rim. The shaft is horizontal, and the wheel is spinning about the shaft at 5.00 rev/s. Find the magnitude and direction of the force each hand exerts on the shaft (a) when the shaft is at rest; (b) when the shaft is rotating in a horizontal plane about its center at 0.050 rev/s; (c) when the shaft is rotating in a horizontal plane about its center at 0.300 rev/s. (d) At what rate must the shaft rotate in order that it may be supported at one end only?

10.103 ••• **CP CALC** A block with mass m is revolving with linear speed v_1 in a circle of radius r_1 on a frictionless horizontal surface (see Fig. E10.42). The string is slowly pulled from below until the radius of the circle in which the block is revolving is reduced to r_2. (a) Calculate the tension T in the string as a function of r, the distance of the block from the hole. Your answer will be in terms of the initial velocity v_1 and the radius r_1. (b) Use $W = \int_{r_1}^{r_2} \vec{T}(r) \cdot d\vec{r}$ to calculate the work done by $\vec{T}$ when r changes from r_1 to r_2. (c) Compare the results of part (b) to the change in the kinetic energy of the block.

Answers

Chapter Opening Question

The earth precesses like a top due to torques exerted on it by the sun and moon. As a result, its rotation axis (which passes through the earth's north and south poles) slowly changes its orientation relative to the distant stars, taking 26,000 years for a complete cycle of precession. Today the rotation axis points toward Polaris, but 5000 years ago it pointed toward Thuban, and 12,000 years from now it will point toward the bright star Vega.

Test Your Understanding Questions

10.1 Answer: (ii) The force P acts along a vertical line, so the lever arm is the horizontal distance from A to the line of action. This is the horizontal component of the distance L, which is $L\cos\theta$. Hence the magnitude of the torque is the product of the force magnitude P and the lever arm $L\cos\theta$, or $\tau = PL\cos\theta$.

10.2 Answer: (iii), (ii), (i) In order for the hanging object of mass m_2 to accelerate downward, the net force on it must be downward. Hence the magnitude m_2g of the downward weight force must be greater than the magnitude T_2 of the upward tension force. In order for the pulley to have a clockwise angular acceleration, the net torque on the pulley must be clockwise. The tension T_2 tends to rotate the pulley clockwise, while the tension T_1 tends to rotate the pulley counterclockwise. Both tension forces have the same lever arm R, so there is a clockwise torque T_2R and a counterclockwise torque T_1R. In order for the net torque to be clockwise, T_2 must be greater than T_1. Hence $m_2g > T_2 > T_1$.

10.3 Answers: (a) (ii), (b) (i) If you redo the calculation of Example 10.6 with a hollow cylinder (moment of inertia $I_{cm} = MR^2$) instead of a solid cylinder (moment of inertia $I_{cm} = \frac{1}{2}MR^2$), you will find $a_{cm\text{-}y} = \frac{1}{2}g$ and $T = \frac{1}{2}Mg$ (instead of $a_{cm\text{-}y} = \frac{2}{3}g$ and $T = \frac{1}{3}Mg$ for a solid cylinder). Hence the acceleration is less but the tension is greater. You can come to the same conclusion without doing the calculation. The greater moment of inertia means that the hollow cylinder will rotate more slowly and hence will roll downward more slowly. In order to slow the downward motion, a greater upward tension force is needed to oppose the downward force of gravity.

10.4 Answer: (iii) You apply the same torque over the same angular displacement to both cylinders. Hence, by Eq. (10.21), you do the same amount of work to both cylinders and impart the same kinetic energy to both. (The one with the smaller moment of inertia ends up with a greater angular speed, but that isn't what we are asked. Compare Conceptual Example 6.5 in Section 6.2.)

10.5 Answers: (a) no, (b) yes As the ball goes around the circle, the magnitude of $\vec{p} = m\vec{v}$ remains the same (the speed is constant) but its direction changes, so the linear momentum vector isn't constant. But $\vec{L} = \vec{r} \times \vec{p}$ *is* constant: It has a constant magnitude (the speed and the perpendicular distance from your hand to the ball are both constant) and a constant direction (along the rotation axis, perpendicular to the plane of the ball's motion). The linear momentum changes because there is a net *force* $\vec{F}$ on the ball (toward the center of the circle). The angular momentum remains constant because there is no net *torque;* the vector $\vec{r}$ points from your hand to the ball and the force $\vec{F}$ on the ball is directed toward your hand, so the vector product $\vec{\tau} = \vec{r} \times \vec{F}$ is zero.

10.6 Answer: (i) In the absence of any external torques, the earth's angular momentum $L_z = I\omega_z$ would remain constant. The melted ice would move from the poles toward the equator—that is, away from our planet's rotation axis—and the earth's moment of inertia I would increase slightly. Hence the angular velocity ω_z would decrease slightly and the day would be slightly longer.

10.7 Answer: (iii) Doubling the flywheel mass would double both its moment of inertia I and its weight w, so the ratio I/w would be unchanged. Equation (10.33) shows that the precession angular speed depends on this ratio, so there would be *no* effect on the value of Ω.

Bridging Problem

Answers: (a) $h = \dfrac{2R}{5}$

(b) $\frac{5}{7}$ of the speed it had just after the hit

11

EQUILIBRIUM
AND ELASTICITY

? This Roman aqueduct uses the principle of the arch to sustain the weight of the structure and the water it carries. Are the blocks that make up the arch being compressed, stretched, or a combination?

We've devoted a good deal of effort to understanding why and how bodies accelerate in response to the forces that act on them. But very often we're interested in making sure that bodies *don't* accelerate. Any building, from a multistory skyscraper to the humblest shed, must be designed so that it won't topple over. Similar concerns arise with a suspension bridge, a ladder leaning against a wall, or a crane hoisting a bucket full of concrete.

A body that can be modeled as a *particle* is in equilibrium whenever the vector sum of the forces acting on it is zero. But for the situations we've just described, that condition isn't enough. If forces act at different points on an extended body, an additional requirement must be satisfied to ensure that the body has no tendency to *rotate:* The sum of the *torques* about any point must be zero. This requirement is based on the principles of rotational dynamics developed in Chapter 10. We can compute the torque due to the weight of a body using the concept of center of gravity, which we introduce in this chapter.

Rigid bodies don't bend, stretch, or squash when forces act on them. But the rigid body is an idealization; all real materials are *elastic* and do deform to some extent. Elastic properties of materials are tremendously important. You want the wings of an airplane to be able to bend a little, but you'd rather not have them break off. The steel frame of an earthquake-resistant building has to be able to flex, but not too much. Many of the necessities of everyday life, from rubber bands to suspension bridges, depend on the elastic properties of materials. In this chapter we'll introduce the concepts of *stress, strain,* and *elastic modulus* and a simple principle called *Hooke's law* that helps us predict what deformations will occur when forces are applied to a real (not perfectly rigid) body.

11.1 Conditions for Equilibrium

We learned in Sections 4.2 and 5.1 that a particle is in *equilibrium*—that is, the particle does not accelerate—in an inertial frame of reference if the vector sum of all the forces acting on the particle is zero, $\sum \vec{F} = 0$. For an *extended* body, the equivalent statement is that the center of mass of the body has zero acceleration if the vector sum of all external forces acting on the body is zero, as discussed in Section 8.5. This is often called the **first condition for equilibrium.** In vector and component forms,

$$\sum \vec{F} = 0$$

$$\sum F_x = 0 \qquad \sum F_y = 0 \qquad \sum F_z = 0 \qquad \begin{array}{l}\text{(first condition}\\\text{for equilibrium)}\end{array} \qquad (11.1)$$

A second condition for an extended body to be in equilibrium is that the body must have no tendency to *rotate*. This condition is based on the dynamics of rotational motion in exactly the same way that the first condition is based on Newton's first law. A rigid body that, in an inertial frame, is not rotating about a certain point has zero angular momentum about that point. If it is not to start rotating about that point, the rate of change of angular momentum must *also* be zero. From the discussion in Section 10.5, particularly Eq. (10.29), this means that the sum of torques due to all the external forces acting on the body must be zero. A rigid body in equilibrium can't have any tendency to start rotating about *any* point, so the sum of external torques must be zero about any point. This is the **second condition for equilibrium:**

$$\sum \vec{\tau} = 0 \quad \text{about any point} \qquad \text{(second condition for equilibrium)} \qquad (11.2)$$

The sum of the torques due to all external forces acting on the body, with respect to any specified point, must be zero.

In this chapter we will apply the first and second conditions for equilibrium to situations in which a rigid body is at rest (no translation or rotation). Such a body is said to be in **static equilibrium** (Fig. 11.1). But the same conditions apply to a rigid body in uniform *translational* motion (without rotation), such as an airplane in flight with constant speed, direction, and altitude. Such a body is in equilibrium but is not static.

Test Your Understanding of Section 11.1 Which situation satisfies both the first and second conditions for equilibrium? (i) a seagull gliding at a constant angle below the horizontal and at a constant speed; (ii) an automobile crankshaft turning at an increasing angular speed in the engine of a parked car; (iii) a thrown baseball that does not rotate as it sails through the air.

11.2 Center of Gravity

In most equilibrium problems, one of the forces acting on the body is its weight. We need to be able to calculate the *torque* of this force. The weight doesn't act at a single point; it is distributed over the entire body. But we can always calculate the torque due to the body's weight by assuming that the entire force of gravity (weight) is concentrated at a point called the **center of gravity** (abbreviated "cg"). The acceleration due to gravity decreases with altitude; but if we can ignore this variation over the vertical dimension of the body, then the body's center of gravity is identical to its *center of mass* (abbreviated "cm"), which we defined in Section 8.5. We stated this result without proof in Section 10.2, and now we'll prove it.

11.1 To be in static equilibrium, a body at rest must satisfy *both* conditions for equilibrium: It can have no tendency to accelerate as a whole or to start rotating.

(a) This body is in static equilibrium.

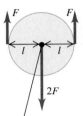

Equilibrium conditions:

First condition satisfied: Net force = 0, so body at rest has no tendency to start moving as a whole.

Second condition satisfied: Net torque about the axis = 0, so body at rest has no tendency to start rotating.

Axis of rotation (perpendicular to figure)

(b) This body has no tendency to accelerate as a whole, but it has a tendency to start rotating.

First condition satisfied: Net force = 0, so body at rest has no tendency to start moving as a whole.

Second condition NOT satisfied: There is a net clockwise torque about the axis, so body at rest will start rotating clockwise.

(c) This body has a tendency to accelerate as a whole but no tendency to start rotating.

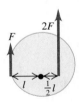

First condition NOT satisfied: There is a net upward force, so body at rest will start moving upward.

Second condition satisfied: Net torque about the axis = 0, so body at rest has no tendency to start rotating.

First let's review the definition of the center of mass. For a collection of particles with masses m_1, m_2, ... and coordinates (x_1, y_1, z_1), (x_2, y_2, z_2), ..., the coordinates x_{cm}, y_{cm}, and z_{cm} of the center of mass are given by

$$x_{cm} = \frac{m_1x_1 + m_2x_2 + m_3x_3 + \cdots}{m_1 + m_2 + m_3 + \cdots} = \frac{\sum\limits_i m_i x_i}{\sum\limits_i m_i}$$

$$y_{cm} = \frac{m_1y_1 + m_2y_2 + m_3y_3 + \cdots}{m_1 + m_2 + m_3 + \cdots} = \frac{\sum\limits_i m_i y_i}{\sum\limits_i m_i} \qquad \text{(center of mass)} \qquad (11.3)$$

$$z_{cm} = \frac{m_1z_1 + m_2z_2 + m_3z_3 + \cdots}{m_1 + m_2 + m_3 + \cdots} = \frac{\sum\limits_i m_i z_i}{\sum\limits_i m_i}$$

Also, x_{cm}, y_{cm}, and z_{cm} are the components of the position vector $\vec{r}_{cm}$ of the center of mass, so Eqs. (11.3) are equivalent to the vector equation

$$\vec{r}_{cm} = \frac{m_1\vec{r}_1 + m_2\vec{r}_2 + m_3\vec{r}_3 + \cdots}{m_1 + m_2 + m_3 + \cdots} = \frac{\sum\limits_i m_i \vec{r}_i}{\sum\limits_i m_i} \qquad (11.4)$$

11.2 The center of gravity (cg) and center of mass (cm) of an extended body.

The gravitational torque about O on a particle of mass m_i within the body is: $\vec{\tau}_i = \vec{r}_i \times \vec{w}_i$.

If $\vec{g}$ has the same value at all points on the body, the cg is identical to the cm.

The net gravitational torque about O on the entire body can be found by assuming that all the weight acts at the cg: $\vec{\tau} = \vec{r}_{cm} \times \vec{w}$.

Now consider the gravitational torque on a body of arbitrary shape (Fig. 11.2). We assume that the acceleration due to gravity $\vec{g}$ is the same at every point in the body. Every particle in the body experiences a gravitational force, and the total weight of the body is the vector sum of a large number of parallel forces. A typical particle has mass m_i and weight $\vec{w}_i = m_i\vec{g}$. If $\vec{r}_i$ is the position vector of this particle with respect to an arbitrary origin O, then the torque vector $\vec{\tau}_i$ of the weight $\vec{w}_i$ with respect to O is, from Eq. (10.3),

$$\vec{\tau}_i = \vec{r}_i \times \vec{w}_i = \vec{r}_i \times m_i\vec{g}$$

The *total* torque due to the gravitational forces on all the particles is

$$\vec{\tau} = \sum_i \vec{\tau}_i = \vec{r}_1 \times m_1\vec{g} + \vec{r}_2 \times m_2\vec{g} + \cdots$$
$$= (m_1\vec{r}_1 + m_2\vec{r}_2 + \cdots) \times \vec{g}$$
$$= \left(\sum_i m_i\vec{r}_i\right) \times \vec{g}$$

When we multiply and divide this by the total mass of the body,

$$M = m_1 + m_2 + \cdots = \sum_i m_i$$

we get

$$\vec{\tau} = \frac{m_1\vec{r}_1 + m_2\vec{r}_2 + \cdots}{m_1 + m_2 + \cdots} \times M\vec{g} = \frac{\sum\limits_i m_i\vec{r}_i}{\sum\limits_i m_i} \times M\vec{g}$$

The fraction in this equation is just the position vector $\vec{r}_{cm}$ of the center of mass, with components x_{cm}, y_{cm}, and z_{cm}, as given by Eq. (11.4), and $M\vec{g}$ is equal to the total weight $\vec{w}$ of the body. Thus

$$\vec{\tau} = \vec{r}_{cm} \times M\vec{g} = \vec{r}_{cm} \times \vec{w} \qquad (11.5)$$

The total gravitational torque, given by Eq. (11.5), is the same as though the total weight $\vec{w}$ were acting on the position $\vec{r}_{cm}$ of the center of mass, which we also call the *center of gravity*. **If $\vec{g}$ has the same value at all points on a body, its center of gravity is identical to its center of mass.** Note, however, that the center of mass is defined independently of any gravitational effect.

While the value of $\vec{g}$ does vary somewhat with elevation, the variation is extremely slight (Fig. 11.3). Hence we will assume throughout this chapter that the center of gravity and center of mass are identical unless explicitly stated otherwise.

Finding and Using the Center of Gravity

We can often use symmetry considerations to locate the center of gravity of a body, just as we did for the center of mass. The center of gravity of a homogeneous sphere, cube, circular sheet, or rectangular plate is at its geometric center. The center of gravity of a right circular cylinder or cone is on its axis of symmetry.

For a body with a more complex shape, we can sometimes locate the center of gravity by thinking of the body as being made of symmetrical pieces. For example, we could approximate the human body as a collection of solid cylinders, with a sphere for the head. Then we can locate the center of gravity of the combination with Eqs. (11.3), letting $m_1, m_2, \ldots$ be the masses of the individual pieces and $(x_1, y_1, z_1), (x_2, y_2, z_2), \ldots$ be the coordinates of their centers of gravity.

When a body acted on by gravity is supported or suspended at a single point, the center of gravity is always at or directly above or below the point of suspension. If it were anywhere else, the weight would have a torque with respect to the point of suspension, and the body could not be in rotational equilibrium. Figure 11.4 shows how to use this fact to determine experimentally the location of the center of gravity of an irregular body.

Using the same reasoning, we can see that a body supported at several points must have its center of gravity somewhere within the area bounded by the supports. This explains why a car can drive on a straight but slanted road if the slant angle is relatively small (Fig. 11.5a) but will tip over if the angle is too steep (Fig. 11.5b). The truck in Fig. 11.5c has a higher center of gravity than the car and will tip over on a shallower incline. When a truck overturns on a highway and blocks traffic for hours, it's the high center of gravity that's to blame.

The lower the center of gravity and the larger the area of support, the more difficult it is to overturn a body. Four-legged animals such as deer and horses have a large area of support bounded by their legs; hence they are naturally stable and need only small feet or hooves. Animals that walk erect on two legs, such as humans and birds, need relatively large feet to give them a reasonable area of

11.3 The acceleration due to gravity at the bottom of the 452-m-tall Petronas Towers in Malaysia is only 0.014% greater than at the top. The center of gravity of the towers is only about 2 cm below the center of mass.

11.4 Finding the center of gravity of an irregularly shaped body—in this case, a coffee mug.

What is the center of gravity of this mug?

① Suspend the mug from any point. A vertical line extending down from the point of suspension passes through the center of gravity.

② Now suspend the mug from a different point. A vertical line extending down from this point intersects the first line at the center of gravity (which is inside the mug).

Center of gravity

11.5 In (a) the center of gravity is within the area bounded by the supports, and the car is in equilibrium. The car in (b) and the truck in (c) will tip over because their centers of gravity lie outside the area of support.

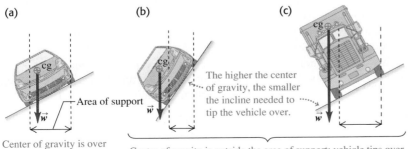

(a)

cg

Area of support

$\vec{w}$

Center of gravity is over the area of support: car is in equilibrium.

(b)

cg

$\vec{w}$

The higher the center of gravity, the smaller the incline needed to tip the vehicle over.

(c)

cg

$\vec{w}$

Center of gravity is outside the area of support: vehicle tips over.

support. If a two-legged animal holds its body approximately horizontal, like a chicken or the dinosaur *Tyrannosaurus rex,* it must perform a delicate balancing act as it walks to keep its center of gravity over the foot that is on the ground. A chicken does this by moving its head; *T. rex* probably did it by moving its massive tail.

Example 11.1   Walking the plank

A uniform plank of length $L = 6.0$ m and mass $M = 90$ kg rests on sawhorses separated by $D = 1.5$ m and equidistant from the center of the plank. Cousin Throckmorton wants to stand on the right-hand end of the plank. If the plank is to remain at rest, how massive can Throckmorton be?

SOLUTION

IDENTIFY and SET UP: To just balance, Throckmorton's mass m must be such that the center of gravity of the plank–Throcky system is directly over the right-hand sawhorse (Fig. 11.6). We take the origin at C, the geometric center and center of gravity of the plank, and take the positive x-axis horizontally to the right. Then the centers of gravity of the plank and Throcky are at $x_P = 0$ and $x_T = L/2 = 3.0$ m, respectively, and the right-hand sawhorse is at

11.6 Our sketch for this problem.

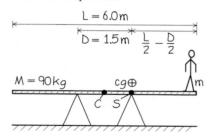

$x_S = D/2$. We'll use Eqs. (11.3) to locate the center of gravity x_{cg} of the plank–Throcky system.

EXECUTE: From the first of Eqs. (11.3),

$$x_{cg} = \frac{M(0) + m(L/2)}{M + m} = \frac{m}{M + m}\frac{L}{2}$$

We set $x_{cg} = x_S$ and solve for m:

$$\frac{m}{M + m}\frac{L}{2} = \frac{D}{2}$$

$$mL = (M + m)D$$

$$m = M\frac{D}{L - D} = (90 \text{ kg})\frac{1.5 \text{ m}}{6.0 \text{ m} - 1.5 \text{ m}} = 30 \text{ kg}$$

EVALUATE: As a check, let's repeat the calculation with the origin at the right-hand sawhorse. Now $x_S = 0$, $x_P = -D/2$, and $x_T = (L/2) - (D/2)$, and we require $x_{cg} = x_S = 0$:

$$x_{cg} = \frac{M(-D/2) + m[(L/2) - (D/2)]}{M + m} = 0$$

$$m = \frac{MD/2}{(L/2) - (D/2)} = M\frac{D}{L - D} = 30 \text{ kg}$$

The result doesn't depend on our choice of origin.

A 60-kg adult could stand only halfway between the right-hand sawhorse and the end of the plank. Can you see why?

11.7 At what point will the meter stick with rock attached be in balance?

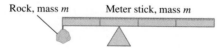

Rock, mass m Meter stick, mass m

Test Your Understanding of Section 11.2 A rock is attached to the left end of a uniform meter stick that has the same mass as the rock. In order for the combination of rock and meter stick to balance atop the triangular object in Fig. 11.7, how far from the left end of the stick should the triangular object be placed? (i) less than 0.25 m; (ii) 0.25 m; (iii) between 0.25 m and 0.50 m; (iv) 0.50 m; (v) more than 0.50 m.

Mastering**PHYSICS**

ActivPhysics 7.4: Two Painters on a Beam
ActivPhysics 7.5: Lecturing from a Beam

11.3 Solving Rigid-Body Equilibrium Problems

There are just two key conditions for rigid-body equilibrium: The vector sum of the forces on the body must be zero, and the sum of the torques about any point must be zero. To keep things simple, we'll restrict our attention to situations in which we can treat all forces as acting in a single plane, which we'll call the xy-plane. Then we can ignore the condition $\sum F_z = 0$ in Eqs. (11.1), and in Eq. (11.2) we need consider only the z-components of torque (perpendicular to the plane). The first and second conditions for equilibrium are then

$$\sum F_x = 0 \quad \text{and} \quad \sum F_y = 0 \qquad \begin{array}{l}\text{(first condition for equilibrium,}\\ \text{forces in } xy\text{-plane)}\end{array}$$

$$\sum \tau_z = 0 \qquad \begin{array}{l}\text{(second condition for equilibrium,}\\ \text{forces in } xy\text{-plane)}\end{array} \qquad (11.6)$$

CAUTION **Choosing the reference point for calculating torques** In equilibrium problems, the choice of reference point for calculating torques in $\sum \tau_z$ is completely arbitrary. But once you make your choice, you must use the *same* point to calculate *all* the torques on a body. Choose the point so as to simplify the calculations as much as possible. ▮

The challenge is to apply these simple conditions to specific problems. Problem-Solving Strategy 11.1 is very similar to the suggestions given in Section 5.2 for the equilibrium of a particle. You should compare it with Problem-Solving Strategy 10.1 (Section 10.2) for rotational dynamics problems.

Problem-Solving Strategy 11.1 **Equilibrium of a Rigid Body**

IDENTIFY *the relevant concepts:* The first and second conditions for equilibrium ($\sum F_x = 0$, $\sum F_y = 0$, and $\sum \tau_z = 0$) are applicable to any rigid body that is not accelerating in space and not rotating.

SET UP *the problem* using the following steps:
1. Sketch the physical situation and identify the body in equilibrium to be analyzed. Sketch the body accurately; do *not* represent it as a point. Include dimensions.
2. Draw a free-body diagram showing all forces acting *on* the body. Show the point on the body at which each force acts.
3. Choose coordinate axes and specify their direction. Specify a positive direction of rotation for torques. Represent forces in terms of their components with respect to the chosen axes.
4. Choose a reference point about which to compute torques. Choose wisely; you can eliminate from your torque equation any force whose line of action goes through the point you

choose. The body doesn't actually have to be pivoted about an axis through the reference point.

EXECUTE *the solution* as follows:
1. Write equations expressing the equilibrium conditions. Remember that $\sum F_x = 0$, $\sum F_y = 0$, and $\sum \tau_z = 0$ are *separate* equations. You can compute the torque of a force by finding the torque of each of its components separately, each with its appropriate lever arm and sign, and adding the results.
2. To obtain as many equations as you have unknowns, you may need to compute torques with respect to two or more reference points; choose them wisely, too.

EVALUATE *your answer:* Check your results by writing $\sum \tau_z = 0$ with respect to a different reference point. You should get the same answers.

Example 11.2 **Weight distribution for a car**

An auto magazine reports that a certain sports car has 53% of its weight on the front wheels and 47% on its rear wheels. (That is, the total normal forces on the front and rear wheels are $0.53w$ and $0.47w$, respectively, where w is the car's weight.) The distance between the axles is 2.46 m. How far in front of the rear axle is the car's center of gravity?

SOLUTION

IDENTIFY and SET UP: We can use the two conditions for equilibrium, Eqs. (11.6), for a car at rest (or traveling in a straight line at constant speed), since the net force and net torque on the car are zero. Figure 11.8 shows our sketch and a free-body diagram, including x- and y-axes and our convention that counterclockwise torques are positive. The weight w acts at the center of gravity. Our target variable is the distance L_{cg}, the lever arm of the weight with respect to the rear axle R, so it is wise to take torques with respect to R. The torque due to the weight is negative because it tends to cause a clockwise rotation about R. The torque due to the upward normal force at the front axle F is positive because it tends to cause a counterclockwise rotation about R.

EXECUTE: The first condition for equilibrium is satisfied (see Fig. 11.8b): $\sum F_x = 0$ because there are no x-components of force and $\sum F_y = 0$ because $0.47w + 0.53w + (-w) = 0$. We write the torque equation and solve for L_{cg}:

$$\sum \tau_R = 0.47w(0) - wL_{cg} + 0.53w(2.46 \text{ m}) = 0$$

$$L_{cg} = 1.30 \text{ m}$$

11.8 Our sketches for this problem.

(a)

(b)

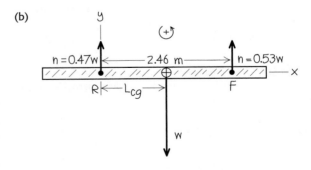

EVALUATE: The center of gravity is between the two supports, as it must be (see Section 11.2). You can check our result by writing the torque equation about the front axle F. You'll find that the center of gravity is 1.16 m behind the front axle, or (2.46 m) − (1.16 m) = 1.30 m in front of the rear axle.

Example 11.3 Will the ladder slip?

Sir Lancelot, who weighs 800 N, is assaulting a castle by climbing a uniform ladder that is 5.0 m long and weighs 180 N (Fig. 11.9a). The bottom of the ladder rests on a ledge and leans across the moat in equilibrium against a frictionless, vertical castle wall. The ladder makes an angle of 53.1° with the horizontal. Lancelot pauses one-third of the way up the ladder. (a) Find the normal and friction forces on the base of the ladder. (b) Find the minimum coefficient of static friction needed to prevent slipping at the base. (c) Find the magnitude and direction of the contact force on the base of the ladder.

SOLUTION

IDENTIFY and SET UP: The ladder–Lancelot system is stationary, so we can use the two conditions for equilibrium to solve part (a). In part (b), we need the relationship among the static friction force, the coefficient of static friction, and the normal force (see Section 5.3). In part (c), the contact force is the vector sum of the normal and friction forces acting at the base of the ladder, found in part (a). Figure 11.9b shows the free-body diagram, with x- and y-directions as shown and with counterclockwise torques taken to be positive. The ladder's center of gravity is at its geometric center. Lancelot's 800-N weight acts at a point one-third of the way up the ladder.

The wall exerts only a normal force n_1 on the top of the ladder. The forces on the base are an upward normal force n_2 and a static friction force f_s, which must point to the right to prevent slipping. The magnitudes n_2 and f_s are the target variables in part (a). From Eq. (5.6), these magnitudes are related by $f_s \leq \mu_s n_2$; the coefficient of static friction μ_s is the target variable in part (b).

EXECUTE: (a) From Eqs. (11.6), the first condition for equilibrium gives

$$\sum F_x = f_s + (-n_1) = 0$$
$$\sum F_y = n_2 + (-800 \text{ N}) + (-180 \text{ N}) = 0$$

These are two equations for the three unknowns n_1, n_2, and f_s. The second equation gives $n_2 = 980$ N. To obtain a third equation, we use the second condition for equilibrium. We take torques about point B, about which n_2 and f_s have no torque. The 53.1° angle creates a 3-4-5 right triangle, so from Fig. 11.9b the lever arm for the ladder's weight is 1.5 m, the lever arm for Lancelot's weight is 1.0 m, and the lever arm for n_1 is 4.0 m. The torque equation for point B is then

$$\sum \tau_B = n_1(4.0 \text{ m}) - (180 \text{ N})(1.5 \text{ m})$$
$$- (800 \text{ N})(1.0 \text{ m}) + n_2(0) + f_s(0) = 0$$

Solving for n_1, we get $n_1 = 268$ N. We substitute this into the $\sum F_x = 0$ equation and get $f_s = 268$ N.

(b) The static friction force f_s cannot exceed $\mu_s n_2$, so the *minimum* coefficient of static friction to prevent slipping is

$$(\mu_s)_{\text{min}} = \frac{f_s}{n_2} = \frac{268 \text{ N}}{980 \text{ N}} = 0.27$$

(c) The components of the contact force $\vec{F}_B$ at the base are the static friction force f_s and the normal force n_2, so

$$\vec{F}_B = f_s \hat{\imath} + n_2 \hat{\jmath} = (268 \text{ N})\hat{\imath} + (980 \text{ N})\hat{\jmath}$$

The magnitude and direction of $\vec{F}_B$ (Fig. 11.9c) are

$$F_B = \sqrt{(268 \text{ N})^2 + (980 \text{ N})^2} = 1020 \text{ N}$$
$$\theta = \arctan \frac{980 \text{ N}}{268 \text{ N}} = 75°$$

EVALUATE: As Fig. 11.9c shows, the contact force $\vec{F}_B$ is *not* directed along the length of the ladder. Can you show that if $\vec{F}_B$ were directed along the ladder, there would be a net counterclockwise torque with respect to the top of the ladder, and equilibrium would be impossible?

As Lancelot climbs higher on the ladder, the lever arm and torque of his weight about B increase. This increases the values of n_1, f_s, and the required friction coefficient $(\mu_s)_{\text{min}}$, so the ladder is more and more likely to slip as he climbs (see Problem 11.10). A simple way to make slipping less likely is to use a larger ladder angle (say, 75° rather than 53.1°). This decreases the lever arms with respect to B of the weights of the ladder and Lancelot and increases the lever arm of n_1, all of which decrease the required friction force.

If we had assumed friction on the wall as well as on the floor, the problem would be impossible to solve by using the equilibrium conditions alone. (Try it!) The difficulty is that it's no longer adequate to treat the body as being perfectly rigid. Another problem of this kind is a four-legged table; there's no way to use the equilibrium conditions alone to find the force on each separate leg.

11.9 (a) Sir Lancelot pauses a third of the way up the ladder, fearing it will slip. (b) Free-body diagram for the system of Sir Lancelot and the ladder. (c) The contact force at B is the superposition of the normal force and the static friction force.

(a)
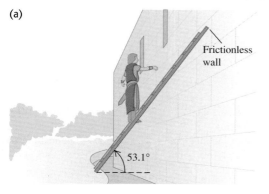

(b)
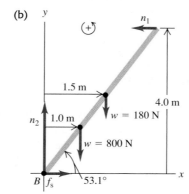

(c)
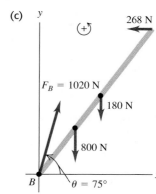

Example 11.4 Equilibrium and pumping iron

Figure 11.10a shows a horizontal human arm lifting a dumbbell. The forearm is in equilibrium under the action of the weight $\vec{w}$ of the dumbbell, the tension $\vec{T}$ in the tendon connected to the biceps muscle, and the force $\vec{E}$ exerted on the forearm by the upper arm at the elbow joint. We neglect the weight of the forearm itself. (For clarity, the point A where the tendon is attached is drawn farther from the elbow than its actual position.) Given the weight w and the angle θ between the tension force and the horizontal, find T and the two components of $\vec{E}$ (three unknown scalar quantities in all).

SOLUTION

IDENTIFY and SET UP: The system is at rest, so we use the conditions for equilibrium. We represent $\vec{T}$ and $\vec{E}$ in terms of their components (Fig. 11.10b). We guess that the directions of E_x and E_y are as shown; the signs of E_x and E_y as given by our solution will tell us the actual directions. Our target variables are T, E_x, and E_y.

EXECUTE: To find T, we take torques about the elbow joint so that the torque equation does not contain E_x, E_y, or T_x:

$$\sum \tau_{elbow} = Lw - DT_y = 0$$

From this we find

$$T_y = \frac{Lw}{D} \quad \text{and} \quad T = \frac{Lw}{D\sin\theta}$$

To find E_x and E_y, we use the first conditions for equilibrium:

$$\sum F_x = T_x + (-E_x) = 0$$

$$E_x = T_x = T\cos\theta = \frac{Lw}{D\sin\theta}\cos\theta$$

$$= \frac{Lw}{D}\cot\theta = \frac{Lw}{D}\frac{D}{h} = \frac{Lw}{h}$$

$$\sum F_y = T_y + E_y + (-w) = 0$$

$$E_y = w - \frac{Lw}{D} = -\frac{(L-D)w}{D}$$

The negative sign for E_y tells us that it should actually point *down* in Fig. 11.10b.

EVALUATE: We can check our results for E_x and E_y by taking torques about points A and B, about both of which T has zero torque:

$$\sum \tau_A = (L-D)w + DE_y = 0 \quad \text{so} \quad E_y = -\frac{(L-D)w}{D}$$

$$\sum \tau_B = Lw - hE_x = 0 \quad \text{so} \quad E_x = \frac{Lw}{h}$$

As a realistic example, take $w = 200$ N, $D = 0.050$ m, $L = 0.30$ m, and $\theta = 80°$, so that $h = D\tan\theta = (0.050 \text{ m})(5.67) = 0.28$ m. Using our results for T, E_x, and E_y, we find

$$T = \frac{Lw}{D\sin\theta} = \frac{(0.30 \text{ m})(200 \text{ N})}{(0.050 \text{ m})(0.98)} = 1220 \text{ N}$$

$$E_y = -\frac{(L-D)w}{D} = -\frac{(0.30 \text{ m} - 0.050 \text{ m})(200 \text{ N})}{0.050 \text{ m}}$$

$$= -1000 \text{ N}$$

$$E_x = \frac{Lw}{h} = \frac{(0.30 \text{ m})(200 \text{ N})}{0.28 \text{ m}} = 210 \text{ N}$$

The magnitude of the force at the elbow is

$$E = \sqrt{E_x^2 + E_y^2} = 1020 \text{ N}$$

The large values of T and E suggest that it was reasonable to neglect the weight of the forearm itself, which may be 20 N or so.

11.10 (a) The situation. (b) Our free-body diagram for the forearm. The weight of the forearm is neglected, and the distance D is greatly exaggerated for clarity.

(a)

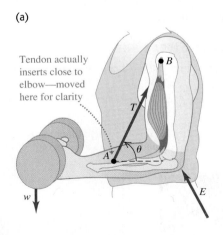

Tendon actually inserts close to elbow—moved here for clarity

(b)

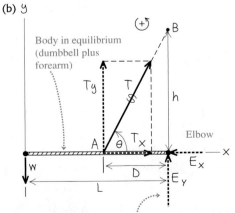

Body in equilibrium (dumbbell plus forearm)

We don't know the sign of this component; we draw it positive for convenience.

11.11 What are the tension in the diagonal cable and the force exerted by the hinge at P?

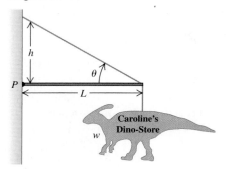

Test Your Understanding of Section 11.3 A metal advertising sign (weight w) for a specialty shop is suspended from the end of a horizontal rod of length L and negligible mass (Fig. 11.11). The rod is supported by a cable at an angle θ from the horizontal and by a hinge at point P. Rank the following force magnitudes in order from greatest to smallest: (i) the weight w of the sign; (ii) the tension in the cable; (iii) the vertical component of force exerted on the rod by the hinge at P.

11.4 Stress, Strain, and Elastic Moduli

The rigid body is a useful idealized model, but the stretching, squeezing, and twisting of real bodies when forces are applied are often too important to ignore. Figure 11.12 shows three examples. We want to study the relationship between the forces and deformations for each case.

For each kind of deformation we will introduce a quantity called **stress** that characterizes the strength of the forces causing the deformation, on a "force per unit area" basis. Another quantity, **strain,** describes the resulting deformation. When the stress and strain are small enough, we often find that the two are directly proportional, and we call the proportionality constant an **elastic modulus.** The harder you pull on something, the more it stretches; the more you squeeze it, the more it compresses. In equation form, this says

$$\frac{\text{Stress}}{\text{Strain}} = \text{Elastic modulus} \qquad \text{(Hooke's law)} \qquad (11.7)$$

The proportionality of stress and strain (under certain conditions) is called **Hooke's law,** after Robert Hooke (1635–1703), a contemporary of Newton. We used one form of Hooke's law in Sections 6.3 and 7.2: The elongation of an ideal spring is proportional to the stretching force. Remember that Hooke's "law" is not really a general law; it is valid over only a limited range. The last section of this chapter discusses what this limited range is.

Tensile and Compressive Stress and Strain

The simplest elastic behavior to understand is the stretching of a bar, rod, or wire when its ends are pulled (Fig. 11.12a). Figure 11.13 shows an object that initially has uniform cross-sectional area A and length l_0. We then apply forces of equal

11.12 Three types of stress. **(a)** Bridge cables under *tensile stress,* being stretched by forces acting at their ends. **(b)** A diver under *bulk stress,* being squeezed from all sides by forces due to water pressure. **(c)** A ribbon under *shear stress,* being deformed and eventually cut by forces exerted by the scissors.

magnitude $F_\perp$ but opposite directions at the ends (this ensures that the object has no tendency to move left or right). We say that the object is in **tension.** We've already talked a lot about tension in ropes and strings; it's the same concept here. The subscript $\perp$ is a reminder that the forces act perpendicular to the cross section.

We define the **tensile stress** at the cross section as the ratio of the force $F_\perp$ to the cross-sectional area A:

$$\text{Tensile stress} = \frac{F_\perp}{A} \qquad (11.8)$$

This is a *scalar* quantity because $F_\perp$ is the *magnitude* of the force. The SI unit of stress is the **pascal** (abbreviated Pa and named for the 17th-century French scientist and philosopher Blaise Pascal). Equation (11.8) shows that 1 pascal equals 1 newton per square meter (N/m^2):

$$1 \text{ pascal} = 1 \text{ Pa} = 1 \text{ N}/\text{m}^2$$

In the British system the logical unit of stress would be the pound per square foot, but the pound per square inch $(\text{lb}/\text{in.}^2$ or psi$)$ is more commonly used. The conversion factors are

$$1 \text{ psi} = 6895 \text{ Pa} \qquad \text{and} \qquad 1 \text{ Pa} = 1.450 \times 10^{-4} \text{ psi}$$

The units of stress are the same as those of *pressure,* which we will encounter often in later chapters. Air pressure in automobile tires is typically around $3 \times 10^5 \text{ Pa} = 300 \text{ kPa}$, and steel cables are commonly required to withstand tensile stresses of the order of 10^8 Pa.

The object shown in Fig. 11.13 stretches to a length $l = l_0 + \Delta l$ when under tension. The elongation Δl does not occur only at the ends; every part of the bar stretches in the same proportion. The **tensile strain** of the object is equal to the fractional change in length, which is the ratio of the elongation Δl to the original length l_0:

$$\text{Tensile strain} = \frac{l - l_0}{l_0} = \frac{\Delta l}{l_0} \qquad (11.9)$$

Tensile strain is stretch per unit length. It is a ratio of two lengths, always measured in the same units, and so is a pure (dimensionless) number with no units.

Experiment shows that for a sufficiently small tensile stress, stress and strain are proportional, as in Eq. (11.7). The corresponding elastic modulus is called **Young's modulus,** denoted by Y:

$$Y = \frac{\text{Tensile stress}}{\text{Tensile strain}} = \frac{F_\perp/A}{\Delta l/l_0} = \frac{F_\perp}{A}\frac{l_0}{\Delta l} \qquad \text{(Young's modulus)} \qquad (11.10)$$

Since strain is a pure number, the units of Young's modulus are the same as those of stress: force per unit area. Some typical values are listed in Table 11.1.

11.13 An object in tension. The net force on the object is zero, but the object deforms. The tensile stress (the ratio of the force to the cross-sectional area) produces a tensile strain (the elongation divided by the initial length). The elongation Δl is exaggerated for clarity.

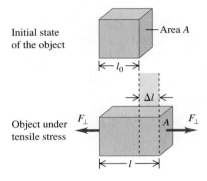

$$\text{Tensile stress} = \frac{F_\perp}{A} \qquad \text{Tensile strain} = \frac{\Delta l}{l_0}$$

Application Young's Modulus of a Tendon
The anterior tibial tendon connects your foot to the large muscle that runs along the side of your shinbone. (You can feel this tendon at the front of your ankle.) Measurements show that this tendon has a Young's modulus of 1.2×10^9 Pa, much less than for the solid materials listed in Table 11.1. Hence this tendon stretches substantially (up to 2.5% of its length) in response to the stresses experienced in walking and running.

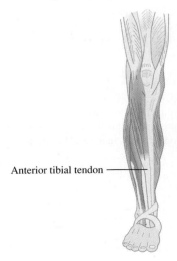

Anterior tibial tendon

Table 11.1 Approximate Elastic Moduli

Material	Young's Modulus, Y (Pa)	Bulk Modulus, B (Pa)	Shear Modulus, S (Pa)
Aluminum	7.0×10^{10}	7.5×10^{10}	2.5×10^{10}
Brass	9.0×10^{10}	6.0×10^{10}	3.5×10^{10}
Copper	11×10^{10}	14×10^{10}	4.4×10^{10}
Crown glass	6.0×10^{10}	5.0×10^{10}	2.5×10^{10}
Iron	21×10^{10}	16×10^{10}	7.7×10^{10}
Lead	1.6×10^{10}	4.1×10^{10}	0.6×10^{10}
Nickel	21×10^{10}	17×10^{10}	7.8×10^{10}
Steel	20×10^{10}	16×10^{10}	7.5×10^{10}

11.14 An object in compression. The compressive stress and compressive strain are defined in the same way as tensile stress and strain (see Fig. 11.13), except that Δl now denotes the distance that the object contracts.

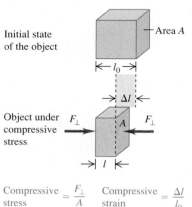

Initial state of the object

Area A

l_0

Δl

Object under compressive stress

$F_\perp$ A $F_\perp$

l

$$\text{Compressive stress} = \frac{F_\perp}{A} \qquad \text{Compressive strain} = \frac{\Delta l}{l_0}$$

(This table also gives values of two other elastic moduli that we will discuss later in this chapter.) A material with a large value of Y is relatively unstretchable; a large stress is required for a given strain. For example, the value of Y for cast steel (2×10^{11} Pa) is much larger than that for rubber (5×10^8 Pa).

When the forces on the ends of a bar are pushes rather than pulls (Fig. 11.14), the bar is in **compression** and the stress is a **compressive stress.** The **compressive strain** of an object in compression is defined in the same way as the tensile strain, but Δl has the opposite direction. Hooke's law and Eq. (11.10) are valid for compression as well as tension if the compressive stress is not too great. For many materials, Young's modulus has the same value for both tensile and compressive stresses. Composite materials such as concrete and stone are an exception; they can withstand compressive stresses but fail under comparable tensile stresses. Stone was the primary building material used by ancient civilizations such as the Babylonians, Assyrians, and Romans, so their structures had to be designed to avoid tensile stresses. Hence they used arches in doorways and bridges, where the weight of the overlying material compresses the stones of the arch together and does not place them under tension.

In many situations, bodies can experience both tensile and compressive stresses at the same time. As an example, a horizontal beam supported at each end sags under its own weight. As a result, the top of the beam is under compression, while the bottom of the beam is under tension (Fig. 11.15a). To minimize the stress and hence the bending strain, the top and bottom of the beam are given a large cross-sectional area. There is neither compression nor tension along the centerline of the beam, so this part can have a small cross section; this helps to keep the weight of the bar to a minimum and further helps to reduce the stress. The result is an I-beam of the familiar shape used in building construction (Fig. 11.15b).

11.15 (a) A beam supported at both ends is under both compression and tension. (b) The cross-sectional shape of an I-beam minimizes both stress and weight.

(a)

Top of beam is under compression.

Beam's centerline is under neither tension nor compression.

Bottom of beam is under tension.

(b)

The top and bottom of an I-beam are broad to minimize the compressive and tensile stresses.

The beam can be narrow near its centerline, which is under neither compression nor tension.

Example 11.5 Tensile stress and strain

A steel rod 2.0 m long has a cross-sectional area of 0.30 cm². It is hung by one end from a support, and a 550-kg milling machine is hung from its other end. Determine the stress on the rod and the resulting strain and elongation.

SOLUTION

IDENTIFY, SET UP, and EXECUTE: The rod is under tension, so we can use Eq. (11.8) to find the tensile stress; Eq. (11.9), with the value of Young's modulus Y for steel from Table 11.1, to find the corresponding strain; and Eq. (11.10) to find the elongation Δl:

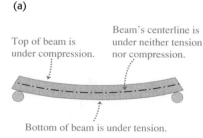

$$\text{Tensile stress} = \frac{F_\perp}{A} = \frac{(550 \text{ kg})(9.8 \text{ m/s}^2)}{3.0 \times 10^{-5} \text{ m}^2} = 1.8 \times 10^8 \text{ Pa}$$

$$\text{Strain} = \frac{\Delta l}{l_0} = \frac{\text{Stress}}{Y} = \frac{1.8 \times 10^8 \text{ Pa}}{20 \times 10^{10} \text{ Pa}} = 9.0 \times 10^{-4}$$

$$\text{Elongation} = \Delta l = (\text{Strain}) \times l_0$$
$$= (9.0 \times 10^{-4})(2.0 \text{ m}) = 0.0018 \text{ m} = 1.8 \text{ mm}$$

EVALUATE: This small elongation, resulting from a load of over half a ton, is a testament to the stiffness of steel.

Bulk Stress and Strain

When a scuba diver plunges deep into the ocean, the water exerts nearly uniform pressure everywhere on his surface and squeezes him to a slightly smaller volume (see Fig. 11.12b). This is a different situation from the tensile and compressive stresses and strains we have discussed. The stress is now a uniform pressure on all sides, and the resulting deformation is a volume change. We use the terms **bulk stress** (or **volume stress**) and **bulk strain** (or **volume strain**) to describe these quantities.

If an object is immersed in a fluid (liquid or gas) at rest, the fluid exerts a force on any part of the object's surface; this force is *perpendicular* to the surface. (If we tried to make the fluid exert a force parallel to the surface, the fluid would slip sideways to counteract the effort.) The force $F_\perp$ per unit area that the fluid exerts on the surface of an immersed object is called the **pressure** p in the fluid:

$$p = \frac{F_\perp}{A} \qquad \text{(pressure in a fluid)} \qquad (11.11)$$

The pressure in a fluid increases with depth. For example, the pressure of the air is about 21% greater at sea level than in Denver (at an elevation of 1.6 km, or 1.0 mi). If an immersed object is relatively small, however, we can ignore pressure differences due to depth for the purpose of calculating bulk stress. Hence we will treat the pressure as having the same value at all points on an immersed object's surface.

Pressure has the same units as stress; commonly used units include 1 Pa ($=1 \text{ N/m}^2$) and 1 lb/in.^{2} (1 psi). Also in common use is the **atmosphere,** abbreviated atm. One atmosphere is the approximate average pressure of the earth's atmosphere at sea level:

$$1 \text{ atmosphere} = 1 \text{ atm} = 1.013 \times 10^5 \text{ Pa} = 14.7 \text{ lb/in.}^2$$

CAUTION **Pressure vs. force** Unlike force, pressure has no intrinsic direction: The pressure on the surface of an immersed object is the same no matter how the surface is oriented. Hence pressure is a *scalar* quantity, not a vector quantity.

Pressure plays the role of stress in a volume deformation. The corresponding strain is the fractional change in volume (Fig. 11.16)—that is, the ratio of the volume change ΔV to the original volume V_0:

$$\text{Bulk (volume) strain} = \frac{\Delta V}{V_0} \qquad (11.12)$$

Volume strain is the change in volume per unit volume. Like tensile or compressive strain, it is a pure number, without units.

When Hooke's law is obeyed, an increase in pressure (bulk stress) produces a *proportional* bulk strain (fractional change in volume). The corresponding elastic modulus (ratio of stress to strain) is called the **bulk modulus,** denoted by B. When the pressure on a body changes by a small amount Δp, from p_0 to $p_0 + \Delta p$, and the resulting bulk strain is $\Delta V/V_0$, Hooke's law takes the form

$$B = \frac{\text{Bulk stress}}{\text{Bulk strain}} = -\frac{\Delta p}{\Delta V/V_0} \qquad \text{(bulk modulus)} \qquad (11.13)$$

We include a minus sign in this equation because an *increase* of pressure always causes a *decrease* in volume. In other words, if Δp is positive, ΔV is negative. The bulk modulus B itself is a positive quantity.

For small pressure changes in a solid or a liquid, we consider B to be constant. The bulk modulus of a *gas,* however, depends on the initial pressure p_0. Table 11.1 includes values of the bulk modulus for several solid materials. Its units, force per unit area, are the same as those of pressure (and of tensile or compressive stress).

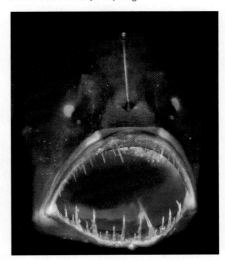

11.16 An object under bulk stress. Without the stress, the cube has volume V_0; when the stress is applied, the cube has a smaller volume V. The volume change ΔV is exaggerated for clarity.

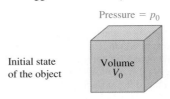

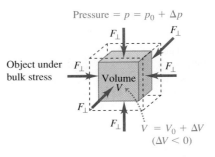

Table 11.2 **Compressibilities of Liquids**

	Compressibility, k	
Liquid	Pa^{-1}	atm^{-1}
Carbon disulfide	93×10^{-11}	94×10^{-6}
Ethyl alcohol	110×10^{-11}	111×10^{-6}
Glycerine	21×10^{-11}	21×10^{-6}
Mercury	3.7×10^{-11}	3.8×10^{-6}
Water	45.8×10^{-11}	46.4×10^{-6}

The reciprocal of the bulk modulus is called the **compressibility** and is denoted by k. From Eq. (11.13),

$$k = \frac{1}{B} = -\frac{\Delta V/V_0}{\Delta p} = -\frac{1}{V_0}\frac{\Delta V}{\Delta p} \quad \text{(compressibility)} \quad (11.14)$$

Compressibility is the fractional decrease in volume, $-\Delta V/V_0$, per unit increase Δp in pressure. The units of compressibility are those of *reciprocal pressure,* Pa^{-1} or atm^{-1}.

Table 11.2 lists the values of compressibility k for several liquids. For example, the compressibility of water is 46.4×10^{-6} atm^{-1}, which means that the volume of water decreases by 46.4 parts per million for each 1-atmosphere increase in pressure. Materials with small bulk modulus and large compressibility are easier to compress.

Example 11.6 Bulk stress and strain

A hydraulic press contains 0.25 m^3 $(250\,L)$ of oil. Find the decrease in the volume of the oil when it is subjected to a pressure increase $\Delta p = 1.6 \times 10^7$ Pa (about 160 atm or 2300 psi). The bulk modulus of the oil is $B = 5.0 \times 10^9$ Pa (about 5.0×10^4 atm) and its compressibility is $k = 1/B = 20 \times 10^{-6}$ atm^{-1}.

SOLUTION

IDENTIFY, SET UP, and EXECUTE: This example uses the ideas of bulk stress and strain. We are given both the bulk modulus and the compressibility, and our target variable is ΔV. Solving Eq. (11.13) for ΔV, we find

$$\Delta V = -\frac{V_0\,\Delta p}{B} = -\frac{(0.25\ m^3)(1.6 \times 10^7\ Pa)}{5.0 \times 10^9\ Pa}$$
$$= -8.0 \times 10^{-4}\ m^3 = -0.80\ L$$

Alternatively, we can use Eq. (11.14) with the approximate unit conversions given above:

$$\Delta V = -kV_0\,\Delta p = -(20 \times 10^{-6}\ atm^{-1})(0.25\ m^3)(160\ atm)$$
$$= -8.0 \times 10^{-4}\ m^3$$

EVALUATE: The negative value of ΔV means that the volume decreases when the pressure increases. Even though the 160-atm pressure increase is large, the *fractional* change in volume is very small:

$$\frac{\Delta V}{V_0} = \frac{-8.0 \times 10^{-4}\ m^3}{0.25\ m^3} = -0.0032 \quad \text{or} \quad -0.32\%$$

Shear Stress and Strain

The third kind of stress-strain situation is called *shear.* The ribbon in Fig. 11.12c is under **shear stress:** One part of the ribbon is being pushed up while an adjacent part is being pushed down, producing a deformation of the ribbon. Figure 11.17 shows a body being deformed by a shear stress. In the figure, forces of equal magnitude but opposite direction act *tangent* to the surfaces of opposite ends of the object. We define the shear stress as the force $F_\parallel$ acting tangent to the surface divided by the area A on which it acts:

$$\text{Shear stress} = \frac{F_\parallel}{A} \quad (11.15)$$

Shear stress, like the other two types of stress, is a force per unit area.

Figure 11.17 shows that one face of the object under shear stress is displaced by a distance x relative to the opposite face. We define **shear strain** as the ratio of the displacement x to the transverse dimension h:

$$\text{Shear strain} = \frac{x}{h} \quad (11.16)$$

In real-life situations, x is nearly always much smaller than h. Like all strains, shear strain is a dimensionless number; it is a ratio of two lengths.

11.17 An object under shear stress. Forces are applied tangent to opposite surfaces of the object (in contrast to the situation in Fig. 11.13, in which the forces act perpendicular to the surfaces). The deformation x is exaggerated for clarity.

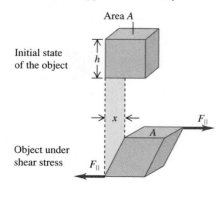

Shear stress $= \dfrac{F_\parallel}{A}$ Shear strain $= \dfrac{x}{h}$

If the forces are small enough that Hooke's law is obeyed, the shear strain is *proportional* to the shear stress. The corresponding elastic modulus (ratio of shear stress to shear strain) is called the **shear modulus,** denoted by S:

$$S = \frac{\text{Shear stress}}{\text{Shear strain}} = \frac{F_{\parallel}/A}{x/h} = \frac{F_{\parallel}}{A}\frac{h}{x} \qquad \text{(shear modulus)} \qquad (11.17)$$

with x and h defined as in Fig. 11.17.

Table 11.1 gives several values of shear modulus. For a given material, S is usually one-third to one-half as large as Young's modulus Y for tensile stress. Keep in mind that the concepts of shear stress, shear strain, and shear modulus apply to *solid* materials only. The reason is that *shear* refers to deforming an object that has a definite shape (see Fig. 11.17). This concept doesn't apply to gases and liquids, which do not have definite shapes.

Example 11.7   Shear stress and strain

Suppose the object in Fig. 11.17 is the brass base plate of an outdoor sculpture that experiences shear forces in an earthquake. The plate is 0.80 m square and 0.50 cm thick. What is the force exerted on each of its edges if the resulting displacement x is 0.16 mm?

SOLUTION

IDENTIFY and SET UP: This example uses the relationship among shear stress, shear strain, and shear modulus. Our target variable is the force $F_{\parallel}$ exerted parallel to each edge, as shown in Fig. 11.17. We'll find the shear strain using Eq. (11.16), the shear stress using Eq. (11.17), and $F_{\parallel}$ using Eq. (11.15). Table 11.1 gives the shear modulus of brass. In Fig. 11.17, h represents the 0.80-m length of each side of the plate. The area A in Eq. (11.15) is the product of the 0.80-m length and the 0.50-cm thickness.

EXECUTE: From Eq. (11.16),

$$\text{Shear strain} = \frac{x}{h} = \frac{1.6 \times 10^{-4}\ \text{m}}{0.80\ \text{m}} = 2.0 \times 10^{-4}$$

From Eq. (11.17),

$$\begin{aligned}\text{Shear stress} &= (\text{Shear strain}) \times S \\ &= (2.0 \times 10^{-4})(3.5 \times 10^{10}\ \text{Pa}) = 7.0 \times 10^{6}\ \text{Pa}\end{aligned}$$

Finally, from Eq. (11.15),

$$\begin{aligned}F_{\parallel} &= (\text{Shear stress}) \times A \\ &= (7.0 \times 10^{6}\ \text{Pa})(0.80\ \text{m})(0.0050\ \text{m}) = 2.8 \times 10^{4}\ \text{N}\end{aligned}$$

EVALUATE: The shear force supplied by the earthquake is more than 3 tons! The large shear modulus of brass makes it hard to deform. Further, the plate is relatively thick (0.50 cm), so the area A is relatively large and a substantial force $F_{\parallel}$ is needed to provide the necessary stress $F_{\parallel}/A$.

Test Your Understanding of Section 11.4   A copper rod of cross-sectional area 0.500 cm^2 and length 1.00 m is elongated by 2.00×10^{-2} mm, and a steel rod of the same cross-sectional area but 0.100 m in length is elongated by 2.00×10^{-3} mm. (a) Which rod has greater tensile *strain*? (i) the copper rod; (ii) the steel rod; (iii) the strain is the same for both. (b) Which rod is under greater tensile *stress*? (i) the copper rod; (ii) the steel rod; (iii) the stress is the same for both.   ❙

11.5 Elasticity and Plasticity

Hooke's law—the proportionality of stress and strain in elastic deformations—has a limited range of validity. In the preceding section we used phrases such as "provided that the forces are small enough that Hooke's law is obeyed." Just what *are* the limitations of Hooke's law? We know that if you pull, squeeze, or twist *anything* hard enough, it will bend or break. Can we be more precise than that?

Let's look at tensile stress and strain again. Suppose we plot a graph of stress as a function of strain. If Hooke's law is obeyed, the graph is a straight line with a

11.18 Typical stress-strain diagram for a ductile metal under tension.

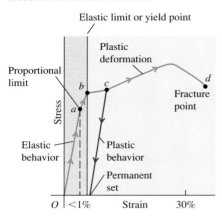

11.19 Typical stress-strain diagram for vulcanized rubber. The curves are different for increasing and decreasing stress, a phenomenon called elastic hysteresis.

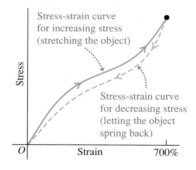

Table 11.3 Approximate Breaking Stresses

Material	Breaking Stress (Pa or N/m^2)
Aluminum	2.2×10^8
Brass	4.7×10^8
Glass	10×10^8
Iron	3.0×10^8
Phosphor bronze	5.6×10^8
Steel	$5–20 \times 10^8$

slope equal to Young's modulus. Figure 11.18 shows a typical stress-strain graph for a metal such as copper or soft iron. The strain is shown as the *percent* elongation; the horizontal scale is not uniform beyond the first portion of the curve, up to a strain of less than 1%. The first portion is a straight line, indicating Hooke's law behavior with stress directly proportional to strain. This straight-line portion ends at point *a*; the stress at this point is called the *proportional limit.*

From *a* to *b*, stress and strain are no longer proportional, and Hooke's law is *not* obeyed. If the load is gradually removed, starting at any point between *O* and *b*, the curve is retraced until the material returns to its original length. The deformation is *reversible,* and the forces are conservative; the energy put into the material to cause the deformation is recovered when the stress is removed. In region *Ob* we say that the material shows *elastic behavior.* Point *b*, the end of this region, is called the *yield point;* the stress at the yield point is called the *elastic limit.*

When we increase the stress beyond point *b*, the strain continues to increase. But now when we remove the load at some point beyond *b*, say *c*, the material does not come back to its original length. Instead, it follows the red line in Fig. 11.18. The length at zero stress is now greater than the original length; the material has undergone an irreversible deformation and has acquired what we call a *permanent set.* Further increase of load beyond *c* produces a large increase in strain for a relatively small increase in stress, until a point *d* is reached at which *fracture* takes place. The behavior of the material from *b* to *d* is called *plastic flow* or *plastic deformation.* A plastic deformation is irreversible; when the stress is removed, the material does not return to its original state.

For some materials, such as the one whose properties are graphed in Fig. 11.18, a large amount of plastic deformation takes place between the elastic limit and the fracture point. Such a material is said to be *ductile.* But if fracture occurs soon after the elastic limit is passed, the material is said to be *brittle.* A soft iron wire that can have considerable permanent stretch without breaking is ductile, while a steel piano string that breaks soon after its elastic limit is reached is brittle.

Something very curious can happen when an object is stretched and then allowed to relax. An example is shown in Fig. 11.19, which is a stress-strain curve for vulcanized rubber that has been stretched by more than seven times its original length. The stress is not proportional to the strain, but the behavior is elastic because when the load is removed, the material returns to its original length. However, the material follows *different* curves for increasing and decreasing stress. This is called *elastic hysteresis.* The work done by the material when it returns to its original shape is less than the work required to deform it; there are nonconservative forces associated with internal friction. Rubber with large elastic hysteresis is very useful for absorbing vibrations, such as in engine mounts and shock-absorber bushings for cars.

The stress required to cause actual fracture of a material is called the *breaking stress,* the *ultimate strength,* or (for tensile stress) the *tensile strength.* Two materials, such as two types of steel, may have very similar elastic constants but vastly different breaking stresses. Table 11.3 gives typical values of breaking stress for several materials in tension. The conversion factor 6.9×10^8 Pa = 100,000 psi may help put these numbers in perspective. For example, if the breaking stress of a particular steel is 6.9×10^8 Pa, then a bar with a 1-in.^{2} cross section has a breaking strength of 100,000 lb.

Test Your Understanding of Section 11.5 While parking your car on a crowded street, you accidentally back into a steel post. You pull forward until the car no longer touches the post and then get out to inspect the damage. What does your rear bumper look like if the strain in the impact was (a) less than at the proportional limit; (b) greater than at the proportional limit, but less than at the yield point; (c) greater than at the yield point, but less than at the fracture point; and (d) greater than at the fracture point?

Conditions for equilibrium: For a rigid body to be in equilibrium, two conditions must be satisfied. First, the vector sum of forces must be zero. Second, the sum of torques about any point must be zero. The torque due to the weight of a body can be found by assuming the entire weight is concentrated at the center of gravity, which is at the same point as the center of mass if $\vec{g}$ has the same value at all points. (See Examples 11.1–11.4.)

$$\sum F_x = 0 \qquad \sum F_y = 0 \qquad \sum F_z = 0$$
$$(11.1)$$

$$\sum \vec{\tau} = \mathbf{0} \quad \text{about } any \text{ point} \qquad (11.2)$$

$$\vec{r}_{cm} = \frac{m_1\vec{r}_1 + m_2\vec{r}_2 + m_3\vec{r}_3 + \cdots}{m_1 + m_2 + m_3 + \cdots}$$
$$(11.4)$$

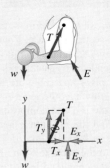

Stress, strain, and Hooke's law: Hooke's law states that in elastic deformations, stress (force per unit area) is proportional to strain (fractional deformation). The proportionality constant is called the elastic modulus.

$$\frac{\text{Stress}}{\text{Strain}} = \text{Elastic modulus} \qquad (11.7)$$

Tensile and compressive stress: Tensile stress is tensile force per unit area, $F_\perp/A$. Tensile strain is fractional change in length, $\Delta l/l_0$. The elastic modulus is called Young's modulus Y. Compressive stress and strain are defined in the same way. (See Example 11.5.)

$$Y = \frac{\text{Tensile stress}}{\text{Tensile strain}} = \frac{F_\perp/A}{\Delta l/l_0} = \frac{F_\perp}{A}\frac{l_0}{\Delta l}$$
$$(11.10)$$

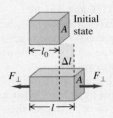

Bulk stress: Pressure in a fluid is force per unit area. Bulk stress is pressure change, Δp, and bulk strain is fractional volume change, $\Delta V/V_0$. The elastic modulus is called the bulk modulus, B. Compressibility, k, is the reciprocal of bulk modulus: $k = 1/B$. (See Example 11.6.)

$$p = \frac{F_\perp}{A} \qquad (11.11)$$

$$B = \frac{\text{Bulk stress}}{\text{Bulk strain}} = -\frac{\Delta p}{\Delta V/V_0} \qquad (11.13)$$

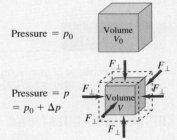

Shear stress: Shear stress is force per unit area, $F_\parallel/A$, for a force applied tangent to a surface. Shear strain is the displacement x of one side divided by the transverse dimension h. The elastic modulus is called the shear modulus, S. (See Example 11.7.)

$$S = \frac{\text{Shear stress}}{\text{Shear strain}} = \frac{F_\parallel/A}{x/h} = \frac{F_\parallel}{A}\frac{h}{x}$$
$$(11.17)$$

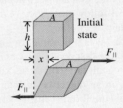

The limits of Hooke's law: The proportional limit is the maximum stress for which stress and strain are proportional. Beyond the proportional limit, Hooke's law is not valid. The elastic limit is the stress beyond which irreversible deformation occurs. The breaking stress, or ultimate strength, is the stress at which the material breaks.

BRIDGING PROBLEM   In Equilibrium and Under Stress

A horizontal, uniform, solid copper rod has an original length l_0, cross-sectional area A, Young's modulus Y, bulk modulus B, shear modulus S, and mass m. It is supported by a frictionless pivot at its right end and by a cable a distance $l_0/4$ from its left end (Fig. 11.20). Both pivot and cable are attached so that they exert their forces uniformly over the rod's cross section. The cable makes an angle θ with the rod and compresses it. (a) Find the tension in the cable. (b) Find the magnitude and direction of the force exerted by the pivot on the right end of the rod. How does this magnitude compare to the cable tension? How does this angle compare to θ? (c) Find the change in length of the rod due to the stresses exerted by the cable and pivot on the rod. (d) By what factor would your answer in part (c) increase if the solid copper rod were twice as long but had the same cross-sectional area?

SOLUTION GUIDE

See MasteringPhysics® study area for a Video Tutor solution.

IDENTIFY and SET UP

1. Draw a free-body diagram for the rod. Be careful to place each force in the correct location.
2. Make a list of the unknown quantities, and decide which are the target variables.
3. What are the conditions that must be met so that the rod remains at rest? What kind of stress (and resulting strain) is involved? Use your answers to select the appropriate equations.

11.20 What are the forces on the rod? What are the stress and strain?

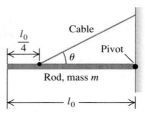

EXECUTE

4. Use your equations to solve for the target variables. (*Hint:* You can make the solution easier by carefully choosing the point around which you calculate torques.)
5. Use your knowledge of trigonometry to decide whether the pivot force or the cable tension has the greater magnitude, as well as to decide whether the angle of the pivot force is greater than, less than, or equal to θ.

EVALUATE

6. Check whether your answers are reasonable. Which force, the cable tension or the pivot force, holds up more of the weight of the rod? Does this make sense?

Problems

For instructor-assigned homework, go to www.masteringphysics.com

•, ••, •••: Problems of increasing difficulty. **CP**: Cumulative problems incorporating material from earlier chapters. **CALC**: Problems requiring calculus. **BIO**: Biosciences problems.

DISCUSSION QUESTIONS

Q11.1 Does a rigid object in uniform rotation about a fixed axis satisfy the first and second conditions for equilibrium? Why? Does it then follow that every particle in this object is in equilibrium? Explain.

Q11.2 (a) Is it possible for an object to be in translational equilibrium (the first condition) but *not* in rotational equilibrium (the second condition)? Illustrate your answer with a simple example. (b) Can an object be in rotational equilibrium yet *not* in translational equilibrium? Justify your answer with a simple example.

Q11.3 Car tires are sometimes "balanced" on a machine that pivots the tire and wheel about the center. Weights are placed around the wheel rim until it does not tip from the horizontal plane. Discuss this procedure in terms of the center of gravity.

Q11.4 Does the center of gravity of a solid body always lie within the material of the body? If not, give a counterexample.

Q11.5 In Section 11.2 we always assumed that the value of g was the same at all points on the body. This is *not* a good approximation if the dimensions of the body are great enough, because the value of g decreases with altitude. If this is taken into account, will the center of gravity of a long, vertical rod be above, below, or at its center of mass? Explain how this can be used to keep the long

axis of an orbiting spacecraft pointed toward the earth. (This would be useful for a weather satellite that must always keep its camera lens trained on the earth.) The moon is not exactly spherical but is somewhat elongated. Explain why this same effect is responsible for keeping the same face of the moon pointed toward the earth at all times.

Q11.6 You are balancing a wrench by suspending it at a single point. Is the equilibrium stable, unstable, or neutral if the point is above, at, or below the wrench's center of gravity? In each case give the reasoning behind your answer. (For rotation, a rigid body is in *stable* equilibrium if a small rotation of the body produces a torque that tends to return the body to equilibrium; it is in *unstable* equilibrium if a small rotation produces a torque that tends to take the body farther from equilibrium; and it is in *neutral* equilibrium if a small rotation produces no torque.)

Q11.7 You can probably stand flatfooted on the floor and then rise up and balance on your tiptoes. Why are you unable do it if your toes are touching the wall of your room? (Try it!)

Q11.8 You freely pivot a horseshoe from a horizontal nail through one of its nail holes. You then hang a long string with a weight at its bottom from the same nail, so that the string hangs vertically in front of the horseshoe without touching it. How do you know that the horseshoe's center of gravity is along the line behind the string? How can you locate the center of gravity by repeating the process at another nail hole? Will the center of gravity be within the solid material of the horseshoe?

Q11.9 An object consists of a ball of weight W glued to the end of a uniform bar also of weight W. If you release it from rest, with the bar horizontal, what will its behavior be as it falls if air resistance is negligible? Will it (a) remain horizontal; (b) rotate about its center of gravity; (c) rotate about the ball; or (d) rotate so that the ball swings downward? Explain your reasoning.

Q11.10 Suppose that the object in Question 11.9 is released from rest with the bar tilted at 60° above the horizontal with the ball at the upper end. As it is falling, will it (a) rotate about its center of gravity until it is horizontal; (b) rotate about its center of gravity until it is vertical with the ball at the bottom; (c) rotate about the ball until it is vertical with the ball at the bottom; or (d) remain at 60° above the horizontal?

Q11.11 Why must a water skier moving with constant velocity lean backward? What determines how far back she must lean? Draw a free-body diagram for the water skier to justify your answers.

Q11.12 In pioneer days, when a Conestoga wagon was stuck in the mud, people would grasp the wheel spokes and try to turn the wheels, rather than simply pushing the wagon. Why?

Q11.13 The mighty Zimbo claims to have leg muscles so strong that he can stand flat on his feet and lean forward to pick up an apple on the floor with his teeth. Should you pay to see him perform, or do you have any suspicions about his claim? Why?

Q11.14 Why is it easier to hold a 10-kg dumbbell in your hand at your side than it is to hold it with your arm extended horizontally?

Q11.15 Certain features of a person, such as height and mass, are fixed (at least over relatively long periods of time). Are the following features also fixed? (a) location of the center of gravity of the body; (b) moment of inertia of the body about an axis through the person's center of mass. Explain your reasoning.

Q11.16 During pregnancy, women often develop back pains from leaning backward while walking. Why do they have to walk this way?

Q11.17 Why is a tapered water glass with a narrow base easier to tip over than a glass with straight sides? Does it matter whether the glass is full or empty?

Q11.18 When a tall, heavy refrigerator is pushed across a rough floor, what factors determine whether it slides or tips?

Q11.19 If a metal wire has its length doubled and its diameter tripled, by what factor does its Young's modulus change?

Q11.20 Why is concrete with steel reinforcing rods embedded in it stronger than plain concrete?

Q11.21 A metal wire of diameter D stretches by 0.100 mm when supporting a weight W. If the same-length wire is used to support a weight three times as heavy, what would its diameter have to be (in terms of D) so it still stretches only 0.100 mm?

Q11.22 Compare the mechanical properties of a steel cable, made by twisting many thin wires together, with the properties of a solid steel rod of the same diameter. What advantages does each have?

Q11.23 The material in human bones and elephant bones is essentially the same, but an elephant has much thicker legs. Explain why, in terms of breaking stress.

Q11.24 There is a small but appreciable amount of elastic hysteresis in the large tendon at the back of a horse's leg. Explain how this can cause damage to the tendon if a horse runs too hard for too long a time.

Q11.25 When rubber mounting blocks are used to absorb machine vibrations through elastic hysteresis, as mentioned in Section 11.5, what becomes of the energy associated with the vibrations?

EXERCISES

Section 11.2 Center of Gravity

11.1 •• A 0.120-kg, 50.0-cm-long uniform bar has a small 0.055-kg mass glued to its left end and a small 0.110-kg mass glued to the other end. The two small masses can each be treated as point masses. You want to balance this system horizontally on a fulcrum placed just under its center of gravity. How far from the left end should the fulcrum be placed?

11.2 •• The center of gravity of a 5.00-kg irregular object is shown in Fig. E11.2. You need to move the center of gravity 2.20 cm to the left by gluing on a 1.50-kg mass, which will then be considered as part of the object. Where should the center of gravity of this additional mass be located?

Figure **E11.2**

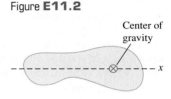

Center of gravity

11.3 • A uniform rod is 2.00 m long and has mass 1.80 kg. A 2.40-kg clamp is attached to the rod. How far should the center of gravity of the clamp be from the left-hand end of the rod in order for the center of gravity of the composite object to be 1.20 m from the left-hand end of the rod?

Section 11.3 Solving Rigid-Body Equilibrium Problems

11.4 • A uniform 300-N trapdoor in a floor is hinged at one side. Find the net upward force needed to begin to open it and the total force exerted on the door by the hinges (a) if the upward force is applied at the center and (b) if the upward force is applied at the center of the edge opposite the hinges.

11.5 •• **Raising a Ladder.** A ladder carried by a fire truck is 20.0 m long. The ladder weighs 2800 N and its center of gravity is at its center. The ladder is pivoted at one end (A) about a pin (Fig. E11.5); you can ignore the friction torque at the pin. The ladder is raised into position by a force applied by a hydraulic piston at C. Point C is 8.0 m from A, and the force $\vec{F}$ exerted by the piston makes an angle of 40° with the ladder. What magnitude must $\vec{F}$ have to just lift the ladder off the support bracket at B? Start with a free-body diagram of the ladder.

Figure **E11.5**

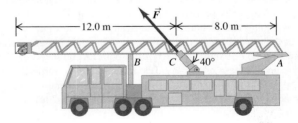

11.6 •• Two people are carrying a uniform wooden board that is 3.00 m long and weighs 160 N. If one person applies an upward force equal to 60 N at one end, at what point does the other person lift? Begin with a free-body diagram of the board.

11.7 •• Two people carry a heavy electric motor by placing it on a light board 2.00 m long. One person lifts at one end with a force of 400 N, and the other lifts the opposite end with a force of 600 N.

(a) What is the weight of the motor, and where along the board is its center of gravity located? (b) Suppose the board is not light but weighs 200 N, with its center of gravity at its center, and the two people each exert the same forces as before. What is the weight of the motor in this case, and where is its center of gravity located?

11.8 •• A 60.0-cm, uniform, 50.0-N shelf is supported horizontally by two vertical wires attached to the sloping ceiling (Fig. E11.8). A very small 25.0-N tool is placed on the shelf midway between the points where the wires are attached to it. Find the ten-

Figure **E11.8**

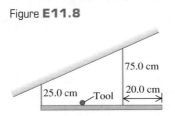

sion in each wire. Begin by making a free-body diagram of the shelf.

11.9 •• A 350-N, uniform, 1.50-m bar is suspended horizontally by two vertical cables at each end. Cable A can support a maximum tension of 500.0 N without breaking, and cable B can support up to 400.0 N. You want to place a small weight on this bar. (a) What is the heaviest weight you can put on without breaking either cable, and (b) where should you put this weight?

11.10 •• A uniform ladder 5.0 m long rests against a frictionless, vertical wall with its lower end 3.0 m from the wall. The ladder weighs 160 N. The coefficient of static friction between the foot of the ladder and the ground is 0.40. A man weighing 740 N climbs slowly up the ladder. Start by drawing a free-body diagram of the ladder. (a) What is the maximum frictional force that the ground can exert on the ladder at its lower end? (b) What is the actual frictional force when the man has climbed 1.0 m along the ladder? (c) How far along the ladder can the man climb before the ladder starts to slip?

11.11 • A diving board 3.00 m long is supported at a point 1.00 m from the end, and a diver weighing 500 N stands at the free end (Fig. E11.11). The diving board is of uniform cross section and weighs 280 N. Find (a) the force at the support point and (b) the force at the left-hand end.

Figure **E11.11**

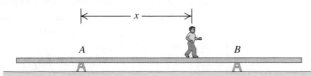

11.12 • A uniform aluminum beam 9.00 m long, weighing 300 N, rests symmetrically on two supports 5.00 m apart (Fig. E11.12). A boy weighing 600 N starts at point A and walks toward the right. (a) In the same diagram construct two graphs showing the upward forces F_A and F_B exerted on the beam at points A and B, as functions of the coordinate x of the boy. Let 1 cm = 100 N vertically, and 1 cm = 1.00 m horizontally. (b) From your diagram, how far beyond point B can the boy walk before the beam tips? (c) How far

Figure **E11.12**

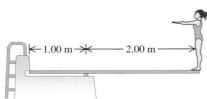

from the right end of the beam should support B be placed so that the boy can walk just to the end of the beam without causing it to tip?

11.13 • Find the tension T in each cable and the magnitude and direction of the force exerted on the strut by the pivot in each of the arrangements in Fig. E11.13. In each case let w be the weight of the suspended crate full of priceless art objects. The strut is uniform and also has weight w. Start each case with a free-body diagram of the strut.

Figure **E11.13**

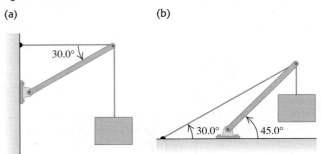

(a)   (b)

11.14 • The horizontal beam in Fig. E11.14 weighs 150 N, and its center of gravity is at its center. Find (a) the tension in the cable and (b) the horizontal and vertical components of the force exerted on the beam at the wall.

Figure **E11.14**

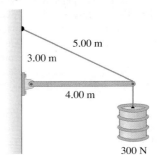

11.15 •• BIO **Push-ups.** To strengthen his arm and chest muscles, an 82-kg athlete who is 2.0 m tall is doing push-ups as shown in Fig. E11.15. His center of mass is 1.15 m from the bottom of his feet, and the centers of

Figure **E11.15**

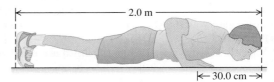

his palms are 30.0 cm from the top of his head. Find the force that the floor exerts on each of his feet and on each hand, assuming that both feet exert the same force and both palms do likewise. Begin with a free-body diagram of the athlete.

11.16 •• Suppose that you can lift no more than 650 N (around 150 lb) unaided. (a) How much can you lift using a 1.4-m-long wheelbarrow that weighs 80.0 N and whose center of gravity is 0.50 m from the center of the wheel (Fig. E11.16)? The center of gravity of the load carried in

Figure **E11.16**

the wheelbarrow is also 0.50 m from the center of the wheel. (b) Where does the force come from to enable you to lift more than 650 N using the wheelbarrow?

11.17 ·· You take your dog Clea to the vet, and the doctor decides he must locate the little beast's center of gravity. It would be awkward to hang the pooch from the ceiling, so the vet must devise another method. He places Clea's front feet on one scale and her hind feet on another. The front scale reads 157 N, while the rear scale reads 89 N. The vet next measures Clea and finds that her rear feet are 0.95 m behind her front feet. How much does Clea weigh, and where is her center of gravity?

11.18 ·· A 15,000-N crane pivots around a friction-free axle at its base and is supported by a cable making a 25° angle with the crane (Fig. E11.18). The crane is 16 m long and is not uniform, its center of gravity being 7.0 m from the axle as measured along the crane. The cable is attached 3.0 m from the upper end of the crane. When the crane is raised to 55° above the horizontal holding an 11,000-N pallet of bricks by a 2.2-m, very light cord, find (a) the tension in the cable and (b) the horizontal and vertical components of the force that the axle exerts on the crane. Start with a free-body diagram of the crane.

Figure **E11.18**

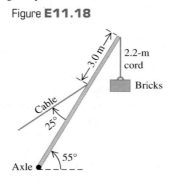

11.19 ·· A 3.00-m-long, 240-N, uniform rod at the zoo is held in a horizontal position by two ropes at its ends (Fig. E11.19). The left rope makes an angle of 150° with the rod and the right rope makes an angle θ with the horizontal. A 90-N howler monkey (*Alouatta seniculus*) hangs motionless 0.50 m from the right end of the rod as he carefully studies you. Calculate the tensions in the two ropes and the angle θ. First make a free-body diagram of the rod.

Figure **E11.19**

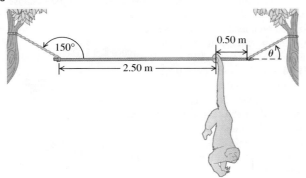

11.20 · A nonuniform beam 4.50 m long and weighing 1.00 kN makes an angle of 25.0° below the horizontal. It is held in position by a frictionless pivot at its upper right end and by a cable 3.00 m farther down the beam and perpendicular to it (Fig. E11.20). The center of gravity of the beam is 2.00 m down the beam from the pivot. Lighting equipment exerts a 5.00-kN downward force on the lower left end of the beam. Find the tension T in the cable and the horizontal and vertical components of the force exerted on the beam by the pivot. Start by sketching a free-body diagram of the beam.

Figure **E11.20**

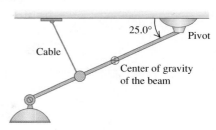

11.21 · **A Couple.** Two forces equal in magnitude and opposite in direction, acting on an object at two different points, form what is called a *couple*. Two antiparallel forces with equal magnitudes $F_1 = F_2 = 8.00$ N are applied to a rod as shown in Fig. E11.21. (a) What should the distance l between the forces be if they are to provide a net torque of 6.40 N·m about the left end of the rod? (b) Is the sense of this torque clockwise or counterclockwise? (c) Repeat parts (a) and (b) for a pivot at the point on the rod where $\vec{F}_2$ is applied.

Figure **E11.21**

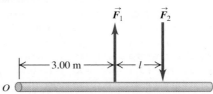

11.22 ·· **BIO A Good Work-out.** You are doing exercises on a Nautilus machine in a gym to strengthen your deltoid (shoulder) muscles. Your arms are raised vertically and can pivot around the shoulder joint, and you grasp the cable of the machine in your hand 64.0 cm from your shoulder joint. The deltoid muscle is attached to the humerus 15.0 cm from the shoulder joint and makes a 12.0° angle with that bone (Fig. E11.22). If you have set the tension in the cable of the machine to 36.0 N on each arm, what is the tension in each deltoid muscle if you simply hold your outstretched arms in place? (*Hint:* Start by making a clear free-body diagram of your arm.)

Figure **E11.22**

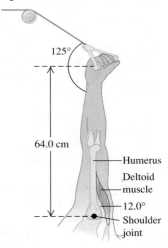

11.23 ·· **BIO Neck Muscles.** A student bends her head at 40.0° from the vertical while intently reading her physics book, pivoting the head around the upper vertebra (point *P* in Fig. E11.23). Her head has a mass of 4.50 kg (which is typical), and its center of mass is 11.0 cm from the pivot point *P*. Her neck muscles are 1.50 cm from point *P*, as measured *perpendicular* to these muscles. The neck itself and the vertebrae are held vertical. (a) Draw a free-body diagram of the student's head. (b) Find the tension in her neck muscles.

Figure **E11.23**

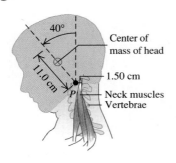

Section 11.4 Stress, Strain, and Elastic Moduli

11.24 • **BIO** **Biceps Muscle.** A relaxed biceps muscle requires a force of 25.0 N for an elongation of 3.0 cm; the same muscle under maximum tension requires a force of 500 N for the same elongation. Find Young's modulus for the muscle tissue under each of these conditions if the muscle is assumed to be a uniform cylinder with length 0.200 m and cross-sectional area 50.0 cm².

11.25 •• A circular steel wire 2.00 m long must stretch no more than 0.25 cm when a tensile force of 400 N is applied to each end of the wire. What minimum diameter is required for the wire?

11.26 •• Two circular rods, one steel and the other copper, are joined end to end. Each rod is 0.750 m long and 1.50 cm in diameter. The combination is subjected to a tensile force with magnitude 4000 N. For each rod, what are (a) the strain and (b) the elongation?

11.27 •• A metal rod that is 4.00 m long and 0.50 cm² in cross-sectional area is found to stretch 0.20 cm under a tension of 5000 N. What is Young's modulus for this metal?

11.28 •• **Stress on a Mountaineer's Rope.** A nylon rope used by mountaineers elongates 1.10 m under the weight of a 65.0-kg climber. If the rope is 45.0 m in length and 7.0 mm in diameter, what is Young's modulus for nylon?

11.29 •• In constructing a large mobile, an artist hangs an aluminum sphere of mass 6.0 kg from a vertical steel wire 0.50 m long and 2.5×10^{-3} cm² in cross-sectional area. On the bottom of the sphere he attaches a similar steel wire, from which he hangs a brass cube of mass 10.0 kg. For each wire, compute (a) the tensile strain and (b) the elongation.

11.30 •• A vertical, solid steel post 25 cm in diameter and 2.50 m long is required to support a load of 8000 kg. You can ignore the weight of the post. What are (a) the stress in the post; (b) the strain in the post; and (c) the change in the post's length when the load is applied?

11.31 •• **BIO** **Compression of Human Bone.** The bulk modulus for bone is 15 GPa. (a) If a diver-in-training is put into a pressurized suit, by how much would the pressure have to be raised (in atmospheres) above atmospheric pressure to compress her bones by 0.10% of their original volume? (b) Given that the pressure in the ocean increases by 1.0×10^4 Pa for every meter of depth below the surface, how deep would this diver have to go for her bones to compress by 0.10%? Does it seem that bone compression is a problem she needs to be concerned with when diving?

11.32 • A solid gold bar is pulled up from the hold of the sunken RMS *Titanic*. (a) What happens to its volume as it goes from the pressure at the ship to the lower pressure at the ocean's surface? (b) The pressure difference is proportional to the depth. How many times greater would the volume change have been had the ship been twice as deep? (c) The bulk modulus of lead is one-fourth that of gold. Find the ratio of the volume change of a solid lead bar to that of a gold bar of equal volume for the same pressure change.

11.33 • **BIO** **Downhill Hiking.** During vigorous downhill hiking, the force on the knee cartilage (the medial and lateral meniscus) can be up to eight times body weight. Depending on the angle of descent, this force can cause a large shear force on the cartilage and deform it. The cartilage has an area of about 10 cm² and a shear modulus of 12 MPa. If the hiker plus his pack have a combined mass of 110 kg (not unreasonable), and if the maximum force at impact is 8 times his body weight (which, of course, includes the weight of his pack) at an angle of 12° with the cartilage (Fig. E11.33), through what angle (in degrees) will his knee cartilage be deformed? (Recall that the bone below the cartilage pushes upward with the same force as the downward force.)

Figure **E11.33**

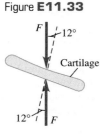

11.34 •• In the Challenger Deep of the Marianas Trench, the depth of seawater is 10.9 km and the pressure is 1.16×10^8 Pa (about 1.15×10^3 atm). (a) If a cubic meter of water is taken from the surface to this depth, what is the change in its volume? (Normal atmospheric pressure is about 1.0×10^5 Pa. Assume that k for seawater is the same as the freshwater value given in Table 11.2.) (b) What is the density of seawater at this depth? (At the surface, seawater has a density of 1.03×10^3 kg/m³.)

11.35 • A specimen of oil having an initial volume of 600 cm³ is subjected to a pressure increase of 3.6×10^6 Pa, and the volume is found to decrease by 0.45 cm³. What is the bulk modulus of the material? The compressibility?

11.36 •• A square steel plate is 10.0 cm on a side and 0.500 cm thick. (a) Find the shear strain that results if a force of magnitude 9.0×10^5 N is applied to each of the four sides, parallel to the side. (b) Find the displacement x in centimeters.

11.37 •• A copper cube measures 6.00 cm on each side. The bottom face is held in place by very strong glue to a flat horizontal surface, while a horizontal force F is applied to the upper face parallel to one of the edges. (Consult Table 11.1.) (a) Show that the glue exerts a force F on the bottom face that is equal but opposite to the force on the top face. (b) How large must F be to cause the cube to deform by 0.250 mm? (c) If the same experiment were performed on a lead cube of the same size as the copper one, by what distance would it deform for the same force as in part (b)?

11.38 • In lab tests on a 9.25-cm cube of a certain material, a force of 1375 N directed at 8.50° to the cube (Fig. E11.38) causes the cube to deform through an angle of 1.24°. What is the shear modulus of the material?

Figure **E11.38**

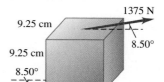

Section 11.5 Elasticity and Plasticity

11.39 •• In a materials testing laboratory, a metal wire made from a new alloy is found to break when a tensile force of 90.8 N is applied perpendicular to each end. If the diameter of the wire is 1.84 mm, what is the breaking stress of the alloy?

11.40 • A 4.0-m-long steel wire has a cross-sectional area of 0.050 cm². Its proportional limit has a value of 0.0016 times its Young's modulus (see Table 11.1). Its breaking stress has a value of 0.0065 times its Young's modulus. The wire is fastened at its upper end and hangs vertically. (a) How great a weight can be hung from the wire without exceeding the proportional limit?

(b) How much will the wire stretch under this load? (c) What is the maximum weight that the wire can support?

11.41 •• **CP** A steel cable with cross-sectional area 3.00 cm² has an elastic limit of 2.40×10^8 Pa. Find the maximum upward acceleration that can be given a 1200-kg elevator supported by the cable if the stress is not to exceed one-third of the elastic limit.

11.42 •• A brass wire is to withstand a tensile force of 350 N without breaking. What minimum diameter must the wire have?

PROBLEMS

11.43 ••• A box of negligible mass rests at the left end of a 2.00-m, 25.0-kg plank (Fig. P11.43). The width of the box is 75.0 cm, and sand is to be distributed uniformly throughout it. The center of gravity of the nonuniform plank is 50.0 cm from the right end. What mass of sand should be put into the box so that the plank balances horizontally on a fulcrum placed just below its midpoint?

Figure **P11.43**

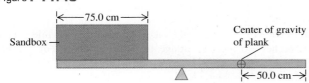

11.44 ••• A door 1.00 m wide and 2.00 m high weighs 280 N and is supported by two hinges, one 0.50 m from the top and the other 0.50 m from the bottom. Each hinge supports half the total weight of the door. Assuming that the door's center of gravity is at its center, find the horizontal components of force exerted on the door by each hinge.

11.45 ••• **Mountain Climbing.** Mountaineers often use a rope to lower themselves down the face of a cliff (this is called *rappelling*). They do this with their body nearly horizontal and their feet pushing against the cliff (Fig. P11.45). Suppose that an 82.0-kg climber, who is 1.90 m tall and has a center of gravity 1.1 m from his feet, rappels down a vertical cliff with his body raised 35.0° above the horizontal. He holds the rope 1.40 m from his feet, and it makes a 25.0° angle with the cliff face. (a) What tension does his rope need to support? (b) Find the horizontal and vertical components of the force that the cliff face exerts on the climber's feet. (c) What minimum coefficient of static friction is needed to prevent the climber's feet from slipping on the cliff face if he has one foot at a time against the cliff?

Figure **P11.45**

11.46 • Sir Lancelot rides slowly out of the castle at Camelot and onto the 12.0-m-long drawbridge that passes over the moat (Fig. P11.46). Unbeknownst to him, his enemies have partially severed the vertical cable holding up the front end of the bridge so that it will break under a tension of 5.80×10^3 N. The bridge has mass 200 kg and its center of gravity is at its center. Lancelot, his lance, his armor, and his horse together have a combined mass of 600 kg. Will the cable break before Lancelot reaches the end of the drawbridge? If so, how far from the castle end of the

bridge will the center of gravity of the horse plus rider be when the cable breaks?

Figure **P11.46**

11.47 • Three vertical forces act on an airplane when it is flying at a constant altitude and with a constant velocity. These are the weight of the airplane, an aerodynamic force on the wing of the airplane, and an aerodynamic force on the airplane's horizontal tail. (The aerodynamic forces are exerted by the surrounding air and are reactions to the forces that the wing and tail exert on the air as the airplane flies through it.) For a particular light airplane with a weight of 6700 N, the center of gravity is 0.30 m in front of the point where the wing's vertical aerodynamic force acts and 3.66 m in front of the point where the tail's vertical aerodynamic force acts. Determine the magnitude and direction (upward or downward) of each of the two vertical aerodynamic forces.

11.48 •• A pickup truck has a wheelbase of 3.00 m. Ordinarily, 10,780 N rests on the front wheels and 8820 N on the rear wheels when the truck is parked on a level road. (a) A box weighing 3600 N is now placed on the tailgate, 1.00 m behind the rear axle. How much total weight now rests on the front wheels? On the rear wheels? (b) How much weight would need to be placed on the tailgate to make the front wheels come off the ground?

11.49 •• A uniform, 255-N rod that is 2.00 m long carries a 225-N weight at its right end and an unknown weight W toward the left end (Fig. P11.49). When W is placed 50.0 cm from the left end of the rod, the system just balances horizontally when the fulcrum is located 75.0 cm from the right end. (a) Find W. (b) If W is now moved 25.0 cm to the right, how far and in what direction must the fulcrum be moved to restore balance?

Figure **P11.49**

11.50 •• A uniform, 8.0-m, 1500-kg beam is hinged to a wall and supported by a thin cable attached 2.0 m from the free end of the beam, (Fig. P11.50). The beam is supported at an angle of 30.0° above the horizontal. (a) Draw a free-body diagram of the beam. (b) Find the tension in the cable. (c) How hard does the beam push inward on the wall?

Figure **P11.50**

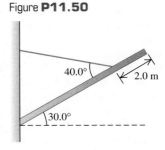

11.51 •• You open a restaurant and hope to entice customers by hanging out a sign (Fig. P11.51). The uniform horizontal beam supporting the sign is 1.50 m long, has a mass of 12.0 kg, and is hinged to the wall. The sign itself is uniform with a mass of 28.0 kg and overall length of 1.20 m. The two wires supporting the sign are each 32.0 cm long, are 90.0 cm apart, and are equally spaced from the middle of the sign. The cable supporting the beam is 2.00 m long. (a) What minimum tension must your cable be able to support without having your sign come crashing down? (b) What minimum vertical force must the hinge be able to support without pulling out of the wall?

Figure **P11.51**

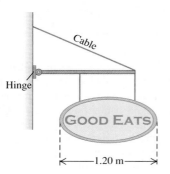

11.52 ••• A claw hammer is used to pull a nail out of a board (Fig. P11.52). The nail is at an angle of 60° to the board, and a force $\vec{F}_1$ of magnitude 400 N applied to the nail is required to pull it from the board. The hammer head contacts the board at point A, which is 0.080 m from where the nail enters the board. A horizontal force $\vec{F}_2$ is applied to the hammer handle at a distance of 0.300 m above the board. What magnitude of force $\vec{F}_2$ is required to apply the required 400-N force (F_1) to the nail? (You can ignore the weight of the hammer.)

Figure **P11.52**

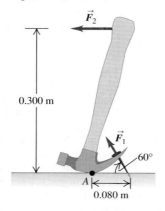

11.53 • End A of the bar AB in Fig. P11.53 rests on a frictionless horizontal surface, and end B is hinged. A horizontal force $\vec{F}$ of magnitude 160 N is exerted on end A. You can ignore the weight of the bar. What are the horizontal and vertical components of the force exerted by the bar on the hinge at B?

Figure **P11.53**

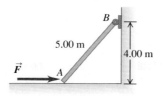

11.54 • A museum of modern art is displaying an irregular 426-N sculpture by hanging it from two thin vertical wires, A and B, that are 1.25 m apart (Fig. P11.54). The center of gravity of this piece of art is located 48.0 cm from its extreme right tip. Find the tension in each wire.

Figure **P11.54**

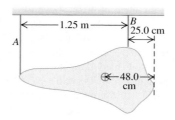

11.55 •• **BIO** **Supporting a Broken Leg.** A therapist tells a 74-kg patient with a broken leg that he must have his leg in a cast suspended horizontally. For minimum discomfort, the leg should be supported by a vertical strap attached at the center of mass of the leg–cast system. (Fig. P11.55). In order to comply with these instructions, the patient consults a table of typical mass distributions and finds that both upper legs (thighs) together typically account for 21.5% of body weight and the center of mass of each thigh is 18.0 cm from the hip joint. The patient also reads that the two lower legs (including the feet) are 14.0% of body weight, with a center of mass 69.0 cm from the hip joint. The cast has a mass of 5.50 kg, and its center of mass is 78.0 cm from the hip joint. How far from the hip joint should the supporting strap be attached to the cast?

Figure **P11.55**

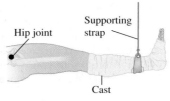

11.56 • **A Truck on a Drawbridge.** A loaded cement mixer drives onto an old drawbridge, where it stalls with its center of gravity three-quarters of the way across the span. The truck driver radios for help, sets the handbrake, and waits. Meanwhile, a boat approaches, so the drawbridge is raised by means of a cable attached to the end opposite the hinge (Fig. P11.56). The drawbridge is 40.0 m long and has a mass of 18,000 kg; its center of gravity is at its midpoint. The cement mixer, with driver, has mass 30,000 kg. When the drawbridge has been raised to an angle of 30° above the horizontal, the cable makes an angle of 70° with the surface of the bridge. (a) What is the tension T in the cable when the drawbridge is held in this position? (b) What are the horizontal and vertical components of the force the hinge exerts on the span?

Figure **P11.56**

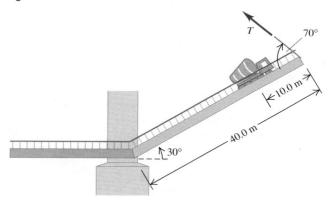

11.57 •• **BIO** **Leg Raises.** In a simplified version of the musculature action in leg raises, the abdominal muscles pull on the femur (thigh bone) to raise the leg by pivoting it about one end (Fig. P11.57). When you are lying horizontally, these muscles make an angle of approximately 5° with the femur,

Figure **P11.57**

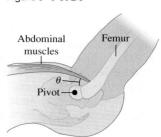

and if you raise your legs, the muscles remain approximately horizontal, so the angle θ increases. We shall assume for simplicity that these muscles attach to the femur in only one place, 10 cm from the hip joint (although, in reality, the situation is more complicated). For a certain 80-kg person having a leg 90 cm long, the mass of the leg is 15 kg and its center of mass is 44 cm from his hip joint as measured along the leg. If the person raises his leg to 60° above the horizontal, the angle between the abdominal muscles and his femur would also be about 60°. (a) With his leg raised to 60°, find the tension in the abdominal muscle on each leg. As usual, begin your solution with a free-body diagram. (b) When is the tension in this muscle greater: when the leg is raised to 60° or when the person just starts to raise it off the ground? Why? (Try this yourself to check your answer.) (c) If the abdominal muscles attached to the femur were perfectly horizontal when a person was lying down, could the person raise his leg? Why or why not?

11.58 • A nonuniform fire escape ladder is 6.0 m long when extended to the icy alley below. It is held at the top by a frictionless pivot, and there is negligible frictional force from the icy surface at the bottom. The ladder weighs 250 N, and its center of gravity is 2.0 m along the ladder from its bottom. A mother and child of total weight 750 N are on the ladder 1.5 m from the pivot. The ladder makes an angle θ with the horizontal. Find the magnitude and direction of (a) the force exerted by the icy alley on the ladder and (b) the force exerted by the ladder on the pivot. (c) Do your answers in parts (a) and (b) depend on the angle θ?

11.59 •• A uniform strut of mass m makes an angle θ with the horizontal. It is supported by a frictionless pivot located at one-third its length from its lower left end and a horizontal rope at its upper right end. A cable and package of total weight w hang from its upper right end. (a) Find the vertical and horizontal components V and H of the pivot's force on the strut as well as the tension T in the rope. (b) If the maximum safe tension in the rope is 700 N and the mass of the strut is 30.0 kg, find the maximum safe weight of the cable and package when the strut makes an angle of 55.0° with the horizontal. (c) For what angle θ can no weight be safely suspended from the right end of the strut?

11.60 • You are asked to design the decorative mobile shown in Fig. P11.60. The strings and rods have negligible weight, and the rods are to hang horizontally. (a) Draw a free-body diagram for each rod. (b) Find the weights of the balls A, B, and C. Find the tensions in the strings S_1, S_2, and S_3. (c) What can you say about the horizontal location of the mobile's center of gravity? Explain.

Figure **P11.60**

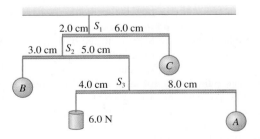

11.61 •• A uniform, 7.5-m-long beam weighing 5860 N is hinged to a wall and supported by a thin cable attached 1.5 m from the free end of the beam. The cable runs between the beam and the wall

and makes a 40° angle with the beam. What is the tension in the cable when the beam is at an angle of 30° above the horizontal?

11.62 •• **CP** A uniform drawbridge must be held at a 37° angle above the horizontal to allow ships to pass underneath. The drawbridge weighs 45,000 N and is 14.0 m long. A cable is connected 3.5 m from the hinge where the bridge pivots (measured along the bridge) and pulls horizontally on the bridge to hold it in place. (a) What is the tension in the cable? (b) Find the magnitude and direction of the force the hinge exerts on the bridge. (c) If the cable suddenly breaks, what is the magnitude of the angular acceleration of the drawbridge just after the cable breaks? (d) What is the angular speed of the drawbridge as it becomes horizontal?

11.63 •• **BIO** **Tendon-Stretch-ing Exercises.** As part of an exercise program, a 75-kg person does toe raises in which he raises his entire body weight on the ball of one foot (Fig. P11.63). The Achilles tendon pulls straight upward on the heel bone of his foot. This tendon is 25 cm long and has a cross-sectional area of 78 mm^2 and a Young's modulus of 1470 MPa. (a) Make a free-body diagram of the person's foot (everything below the ankle joint). You can neglect the weight of the foot. (b) What force does the Achilles tendon exert on the heel during this exercise? Express your answer in newtons and in multiples of his weight. (c) By how many millimeters does the exercise stretch his Achilles tendon?

Figure **P11.63**

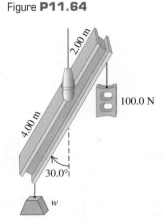

11.64 •• (a) In Fig. P11.64 a 6.00-m-long, uniform beam is hanging from a point 1.00 m to the right of its center. The beam weighs 140 N and makes an angle of 30.0° with the vertical. At the right-hand end of the beam a 100.0-N weight is hung; an unknown weight w hangs at the left end. If the system is in equilibrium, what is w? You can ignore the thickness of the beam. (b) If the beam makes, instead, an angle of 45.0° with the vertical, what is w?

Figure **P11.64**

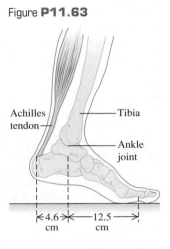

11.65 ••• A uniform, horizontal flagpole 5.00 m long with a weight of 200 N is hinged to a vertical wall at one end. A 600-N stuntwoman hangs from its other end. The flagpole is supported by a guy wire running from its outer end to a point on the wall directly above the pole. (a) If the tension in this wire is not to exceed 1000 N, what is the minimum height above the pole at which it may be fastened to the wall? (b) If the flagpole remains horizontal, by how many newtons would the tension be increased if the wire were fastened 0.50 m below this point?

11.66 • A holiday decoration consists of two shiny glass spheres with masses 0.0240 kg and 0.0360 kg suspended from a uniform rod with mass 0.120 kg and length 1.00 m (Fig. P11.66). The rod is suspended from the ceiling by a vertical cord at each end, so that it is horizontal. Calculate the tension in each of the cords A through F.

Figure **P11.66**

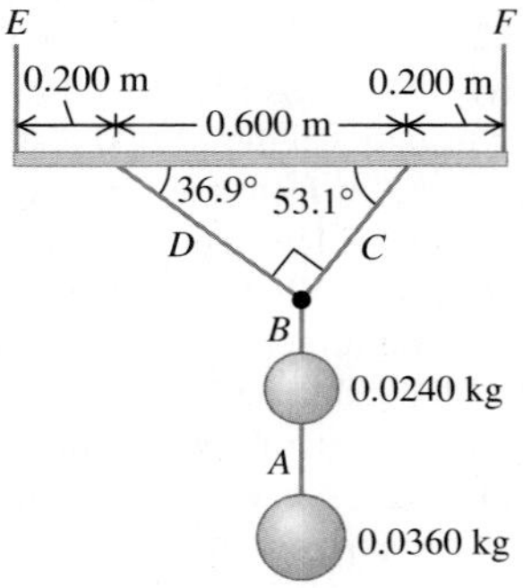

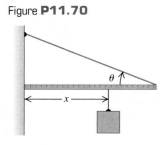

11.67 •• BIO **Downward-Facing Dog.** One yoga exercise, known as the "Downward-Facing Dog," requires stretching your hands straight out above your head and bending down to lean against the floor. This exercise is performed by a 750-N person, as shown in Fig. P11.67. When he bends his body at the hip to a 90° angle between his legs and trunk, his legs, trunk, head, and arms have the dimensions indicated. Furthermore, his legs and feet weigh a total of 277 N, and their center of mass is 41 cm from his hip, measured along his legs. The person's trunk, head, and arms weigh 473 N, and their center of gravity is 65 cm from his hip, measured along the upper body. (a) Find the normal force that the floor exerts on each foot and on each hand, assuming that the person does not favor either hand or either foot. (b) Find the friction force on each foot and on each hand, assuming that it is the same on both feet and on both hands (but not necessarily the same on the feet as on the hands). [*Hint:* First treat his entire body as a system; then isolate his legs (or his upper body).]

Figure **P11.67**

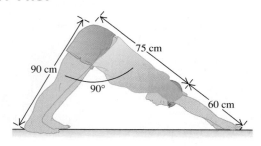

11.68 • When you stretch a wire, rope, or rubber band, it gets thinner as well as longer. When Hooke's law holds, the fractional decrease in width is proportional to the tensile strain. If w_0 is the original width and Δw is the change in width, then $\Delta w/w_0 = -\sigma \Delta l/l_0$, where the minus sign reminds us that width decreases when length increases. The dimensionless constant σ, different for different materials, is called *Poisson's ratio*. (a) If the steel rod of Example 11.5 (Section 11.4) has a circular cross section and a Poisson's ratio of 0.23, what is its change in diameter when the milling machine is hung from it? (b) A cylinder made of nickel (Poisson's ratio = 0.42) has radius 2.0 cm. What tensile force $F_\perp$ must be applied perpendicular to each end of the cylinder to cause its radius to decrease by 0.10 mm? Assume that the breaking stress and proportional limit for the metal are extremely large and are not exceeded.

11.69 • A worker wants to turn over a uniform, 1250-N, rectangular crate by pulling at 53.0° on one of its vertical sides (Fig. P11.69). The floor is rough enough to prevent the crate from slipping. (a) What pull is needed to just start the crate to tip? (b) How hard does the floor push upward on the crate? (c) Find the friction force on the crate. (d) What is the minimum coefficient of static friction needed to prevent the crate from slipping on the floor?

Figure **P11.69**

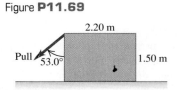

11.70 ••• One end of a uniform meter stick is placed against a vertical wall (Fig. P11.70). The other end is held by a lightweight cord that makes an angle θ with the stick. The coefficient of static friction between the end of the meter stick and the wall is 0.40. (a) What is the maximum value the angle θ can have if the stick is to remain in equilibrium? (b) Let the angle θ be 15°. A block of the same weight as the meter stick is suspended from the stick, as shown, at a distance x from the wall. What is the minimum value of x for which the stick will remain in equilibrium? (c) When $\theta = 15°$, how large must the coefficient of static friction be so that the block can be attached 10 cm from the left end of the stick without causing it to slip?

Figure **P11.70**

11.71 •• Two friends are carrying a 200-kg crate up a flight of stairs. The crate is 1.25 m long and 0.500 m high, and its center of gravity is at its center. The stairs make a 45.0° angle with respect to the floor. The crate also is carried at a 45.0° angle, so that its bottom side is parallel to the slope of the stairs (Fig. P11.71). If the force each person applies is vertical, what is the magnitude of each of these forces? Is it better to be the person above or below on the stairs?

Figure **P11.71**

11.72 •• BIO **Forearm.** In the human arm, the forearm and hand pivot about the elbow joint. Consider a simplified model in which the biceps muscle is attached to the forearm 3.80 cm from the elbow joint. Assume that the person's hand and forearm together weigh 15.0 N and that their center of gravity is 15.0 cm from the elbow (not quite halfway to the hand). The forearm is held horizontally at a right angle to the upper arm, with the biceps muscle exerting its force perpendicular to the forearm. (a) Draw a free-body diagram for the forearm, and find the force exerted by the biceps when the hand is empty. (b) Now the person holds a 80.0-N weight in his hand, with the forearm still horizontal. Assume that the center of gravity of this weight is 33.0 cm from the elbow. Construct a free-body diagram for the forearm, and find the force now exerted by the biceps. Explain why the biceps muscle needs to be very strong. (c) Under the conditions of part (b), find the magnitude and direction of the force that the elbow joint exerts on the forearm. (d) While holding the 80.0-N weight, the person raises his forearm until it is at an angle of 53.0° above the horizontal. If the biceps muscle continues to exert its force perpendicular to the forearm, what is this force when the forearm is in this position? Has the force increased or decreased from its value in part (b)? Explain why this is so, and test your answer by actually doing this with your own arm.

11.73 •• **BIO CALC** Refer to the discussion of holding a dumbbell in Example 11.4 (Section 11.3). The maximum weight that can be held in this way is limited by the maximum allowable tendon tension T (determined by the strength of the tendons) and by the distance D from the elbow to where the tendon attaches to the forearm. (a) Let T_{max} represent the maximum value of the tendon tension. Use the results of Example 11.4 to express w_{max} (the maximum weight that can be held) in terms of T_{max}, L, D, and h. Your expression should *not* include the angle θ. (b) The tendons of different primates are attached to the forearm at different values of D. Calculate the derivative of w_{max} with respect to D, and determine whether the derivative is positive or negative. (c) A chimpanzee tendon is attached to the forearm at a point farther from the elbow than for humans. Use this to explain why chimpanzees have stronger arms than humans. (The disadvantage is that chimpanzees have less flexible arms than do humans.)

11.74 •• A uniform, 90.0-N table is 3.6 m long, 1.0 m high, and 1.2 m wide. A 1500-N weight is placed 0.50 m from one end of the table, a distance of 0.60 m from each side of the table. Draw a free-body diagram for the table and find the force that each of the four legs exerts on the floor.

11.75 ••• **Flying Buttress.** (a) A symmetric building has a roof sloping upward at 35.0° above the horizontal on each side. If each side of the uniform roof weighs 10,000 N, find the horizontal force that this roof exerts at the top of the wall, which tends to push out the walls. Which type of building would be more in danger of collapsing: one with tall walls or one with short walls? Explain. (b) As you saw in part (a), tall walls are in danger of collapsing from the weight of the roof. This problem plagued the ancient builders of large structures. A solution used in the great Gothic cathedrals during the 1200s was the flying buttress, a stone support running between the walls and the ground that helped to hold in the walls. A Gothic church has a uniform roof weighing a total of 20,000 N and rising at 40° above the horizontal at each wall. The walls are 40 m tall, and a flying buttress meets each wall 10 m below the base of the roof. What horizontal force must this flying buttress apply to the wall?

11.76 •• You are trying to raise a bicycle wheel of mass m and radius R up over a curb of height h. To do this, you apply a horizontal force $\vec{F}$ (Fig. P11.76). What is the smallest magnitude of the force $\vec{F}$ that will succeed in raising the wheel onto the curb when the force is applied (a) at the center of the wheel and (b) at the top of the wheel? (c) In which case is less force required?

Figure **P11.76**

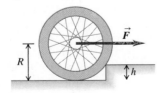

11.77 • **The Farmyard Gate.** A gate 4.00 m wide and 2.00 m high weighs 500 N. Its center of gravity is at its center, and it is hinged at A and B. To relieve the strain on the top hinge, a

Figure **P11.77**

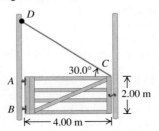

wire CD is connected as shown in Fig. P11.77. The tension in CD is increased until the horizontal force at hinge A is zero. (a) What is the tension in the wire CD? (b) What is the magnitude of the horizontal component of the force at hinge B? (c) What is the combined vertical force exerted by hinges A and B?

11.78 • If you put a uniform block at the edge of a table, the center of the block must be over the table for the block not to fall off. (a) If you stack two identical blocks at the table edge, the center of the top block must be over the bottom block, and the center of gravity of the two blocks together must be over the table. In terms of the length L of each block, what is the maximum overhang possible (Fig. P11.78)? (b) Repeat part (a) for three identical blocks and for four identical blocks. (c) Is it possible to make a stack of blocks such that the uppermost block is not directly over the table at all? How many blocks would it take to do this? (Try this with your friends using copies of this book.)

Figure **P11.78**

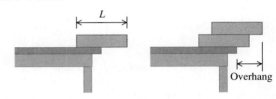

11.79 ••• Two uniform, 75.0-g marbles 2.00 cm in diameter are stacked as shown in Fig. P11.79 in a container that is 3.00 cm wide. (a) Find the force that the container exerts on the marbles at the points of contact A, B, and C. (b) What force does each marble exert on the other?

Figure **P11.79**

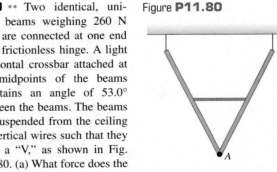

11.80 •• Two identical, uniform beams weighing 260 N each are connected at one end by a frictionless hinge. A light horizontal crossbar attached at the midpoints of the beams maintains an angle of 53.0° between the beams. The beams are suspended from the ceiling by vertical wires such that they form a "V," as shown in Fig. P11.80. (a) What force does the crossbar exert on each beam? (b) Is the crossbar under tension or compression? (c) What force (magnitude and direction) does the hinge at point A exert on each beam?

Figure **P11.80**

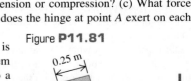

11.81 • An engineer is designing a conveyor system for loading hay bales into a wagon (Fig. P11.81). Each bale is 0.25 m wide, 0.50 m high, and 0.80 m long (the dimension perpendicular to the plane of the figure), with mass 30.0 kg. The center of

Figure **P11.81**

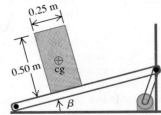

gravity of each bale is at its geometrical center. The coefficient of static friction between a bale and the conveyor belt is 0.60, and the belt moves with constant speed. (a) The angle β of the conveyor is slowly increased. At some critical angle a bale will tip (if it doesn't slip first), and at some different critical angle it will slip (if it doesn't tip first). Find the two critical angles and determine which happens at the smaller angle. (b) Would the outcome of part (a) be different if the coefficient of friction were 0.40?

11.82 • A weight W is supported by attaching it to a vertical uniform metal pole by a thin cord passing over a pulley having negligible mass and friction. The cord is attached to the pole 40.0 cm below the top and pulls horizontally on it (Fig. P11.82). The pole is pivoted about a hinge at its base, is 1.75 m tall, and weighs 55.0 N. A thin wire connects the top of the pole to a vertical wall. The nail that holds this wire to the wall will pull out if an *outward* force greater than 22.0 N acts on it. (a) What is the greatest weight W that can be supported this way without pulling out the nail? (b) What is the *magnitude* of the force that the hinge exerts on the pole?

Figure **P11.82**

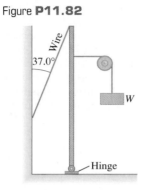

11.83 •• A garage door is mounted on an overhead rail (Fig. P11.83). The wheels at A and B have rusted so that they do not roll, but rather slide along the track. The coefficient of kinetic friction is 0.52. The distance between the wheels is 2.00 m, and each is 0.50 m from the vertical sides of the door. The door is uniform and weighs 950 N. It is pushed to the left at constant speed by a horizontal force $\vec{F}$. (a) If the distance h is 1.60 m, what is the vertical component of the force exerted on each wheel by the track? (b) Find the maximum value h can have without causing one wheel to leave the track.

Figure **P11.83**

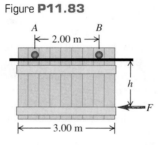

11.84 •• A horizontal boom is supported at its left end by a frictionless pivot. It is held in place by a cable attached to the right-hand end of the boom. A chain and crate of total weight w hang from somewhere along the boom. The boom's weight w_b cannot be ignored and the boom may or may not be uniform. (a) Show that the tension in the cable is the same whether the cable makes an angle θ or an angle $180° - \theta$ with the horizontal, and that the horizontal force component exerted on the boom by the pivot has equal magnitude but opposite direction for the two angles. (b) Show that the cable cannot be horizontal. (c) Show that the tension in the cable is a minimum when the cable is vertical, pulling upward on the right end of the boom. (d) Show that when the cable is vertical, the force exerted by the pivot on the boom is vertical.

11.85 •• Prior to being placed in its hole, a 5700-N, 9.0-m-long, uniform utility pole makes some nonzero angle with the vertical. A vertical cable attached 2.0 m below its upper end holds it in place while its lower end rests on the ground. (a) Find the tension in the cable and the magnitude and direction of the force exerted by the ground on the pole. (b) Why don't we need to know the angle the pole makes with the vertical, as long as it is not zero?

11.86 ••• **Pyramid Builders.** Ancient pyramid builders are balancing a uniform rectangular slab of stone tipped at an angle θ above the horizontal using a rope (Fig. P11.86). The rope is held by five workers who share the force equally. (a) If $\theta = 20.0°$, what force does each worker exert on the rope? (b) As θ increases, does each worker have to exert more or less force than in part (a), assuming they do not change the angle of the rope? Why? (c) At what angle do the workers need to exert *no force* to balance the slab? What happens if θ exceeds this value?

Figure **P11.86**

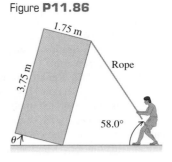

11.87 • You hang a floodlamp from the end of a vertical steel wire. The floodlamp stretches the wire 0.18 mm and the stress is proportional to the strain. How much would it have stretched (a) if the wire were twice as long? (b) if the wire had the same length but twice the diameter? (c) for a copper wire of the original length and diameter?

11.88 •• **Hooke's Law for a Wire.** A wire of length l_0 and cross-sectional area A supports a hanging weight W. (a) Show that if the wire obeys Eq. (11.7), it behaves like a spring of force constant AY/l_0, where Y is Young's modulus for the material of which the wire is made. (b) What would the force constant be for a 75.0-cm length of 16-gauge (diameter = 1.291 mm) copper wire? See Table 11.1. (c) What would W have to be to stretch the wire in part (b) by 1.25 mm?

11.89 ••• **CP** A 12.0-kg mass, fastened to the end of an aluminum wire with an unstretched length of 0.50 m, is whirled in a vertical circle with a constant angular speed of 120 rev/min. The cross-sectional area of the wire is 0.014 cm². Calculate the elongation of the wire when the mass is (a) at the lowest point of the path and (b) at the highest point of its path.

11.90 • A metal wire 3.50 m long and 0.70 mm in diameter was given the following test. A load weighing 20 N was originally hung from the wire to keep it taut. The position of the lower end of the wire was read on a scale as load was added.

Added Load (N)	Scale Reading (cm)
0	3.02
10	3.07
20	3.12
30	3.17
40	3.22
50	3.27
60	3.32
70	4.27

(a) Graph these values, plotting the increase in length horizontally and the added load vertically. (b) Calculate the value of Young's modulus. (c) The proportional limit occurred at a scale reading of 3.34 cm. What was the stress at this point?

11.91 ••• A 1.05-m-long rod of negligible weight is supported at its ends by wires A and B of equal length (Fig. P11.91). The cross-sectional area of A is

Figure **P11.91**

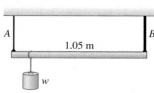

2.00 mm^2 and that of *B* is 4.00 mm^2. Young's modulus for wire *A* is 1.80 × 10^{11} Pa; that for *B* is 1.20 × 10^{11} Pa. At what point along the rod should a weight *w* be suspended to produce (a) equal stresses in *A* and *B* and (b) equal strains in *A* and *B*?

11.92 ••• **CP** An amusement park ride consists of airplane-shaped cars attached to steel rods (Fig. P11.92). Each rod has a length of 15.0 m and a cross-sectional area of 8.00 cm^2. (a) How much is the rod stretched when the ride is at rest? (Assume that each car plus two people seated in it has a total weight of 1900 N.) (b) When operating, the ride has a maximum angular speed of 8.0 rev/min. How much is the rod stretched then?

Figure **P11.92**

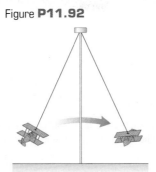

11.93 • A brass rod with a length of 1.40 m and a cross-sectional area of 2.00 cm^2 is fastened end to end to a nickel rod with length *L* and cross-sectional area 1.00 cm^2. The compound rod is subjected to equal and opposite pulls of magnitude 4.00 × 10^4 N at its ends. (a) Find the length *L* of the nickel rod if the elongations of the two rods are equal. (b) What is the stress in each rod? (c) What is the strain in each rod?

11.94 ••• **CP BIO Stress on the Shin Bone.** The compressive strength of our bones is important in everyday life. Young's modulus for bone is about 1.4 × 10^{10} Pa. Bone can take only about a 1.0% change in its length before fracturing. (a) What is the maximum force that can be applied to a bone whose minimum cross-sectional area is 3.0 cm^2? (This is approximately the cross-sectional area of a tibia, or shin bone, at its narrowest point.) (b) Estimate the maximum height from which a 70-kg man could jump and not fracture the tibia. Take the time between when he first touches the floor and when he has stopped to be 0.030 s, and assume that the stress is distributed equally between his legs.

11.95 ••• A moonshiner produces pure ethanol (ethyl alcohol) late at night and stores it in a stainless steel tank in the form of a cylinder 0.300 m in diameter with a tight-fitting piston at the top. The total volume of the tank is 250 L (0.250 m^3). In an attempt to squeeze a little more into the tank, the moonshiner piles 1420 kg of lead bricks on top of the piston. What additional volume of ethanol can the moonshiner squeeze into the tank? (Assume that the wall of the tank is perfectly rigid.)

CHALLENGE PROBLEMS

11.96 ••• Two ladders, 4.00 m and 3.00 m long, are hinged at point *A* and tied together by a horizontal rope 0.90 m above the floor (Fig. P11.96). The ladders weigh 480 N and 360 N, respectively, and the center of gravity of each is at its center. Assume that

Figure **P11.96**

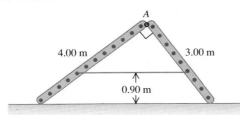

the floor is freshly waxed and frictionless. (a) Find the upward force at the bottom of each ladder. (b) Find the tension in the rope. (c) Find the magnitude of the force one ladder exerts on the other at point *A*. (d) If an 800-N painter stands at point *A*, find the tension in the horizontal rope.

11.97 ••• A bookcase weighing 1500 N rests on a horizontal surface for which the coefficient of static friction is $\mu_s = 0.40$. The bookcase is 1.80 m tall and 2.00 m wide; its center of gravity is at its geometrical center. The bookcase rests on four short legs that are each 0.10 m from the edge of the bookcase. A person pulls on a rope attached to an upper corner of the bookcase with a force $\vec{F}$ that makes an angle θ with the bookcase (Fig. P11.97). (a) If $\theta = 90°$, so $\vec{F}$ is horizontal, show that as *F* is increased from zero, the bookcase will start to slide before it tips, and calculate the magnitude of $\vec{F}$ that will start the bookcase sliding. (b) If $\theta = 0°$, so $\vec{F}$ is vertical, show that the bookcase will tip over rather than slide, and calculate the magnitude of $\vec{F}$ that will cause the bookcase to start to tip. (c) Calculate as a function of θ the magnitude of $\vec{F}$ that will cause the bookcase to start to slide and the magnitude that will cause it to start to tip. What is the smallest value that θ can have so that the bookcase will still start to slide before it starts to tip?

Figure **P11.97**

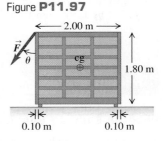

11.98 ••• **Knocking Over a Post.** One end of a post weighing 400 N and with height *h* rests on a rough horizontal surface with $\mu_s = 0.30$. The upper end is held by a rope fastened to the surface and making an angle of 36.9° with the post (Fig. P11.98). A horizontal force $\vec{F}$ is exerted on the post as shown. (a) If the force $\vec{F}$ is applied at the midpoint of the post, what is the largest value it can have without causing the post to slip? (b) How large can the force be without causing the post to slip if its point of application is $\frac{6}{10}$ of the way from the ground to the top of the post? (c) Show that if the point of application of the force is too high, the post cannot be made to slip, no matter how great the force. Find the critical height for the point of application.

Figure **P11.98**

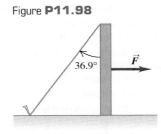

11.99 ••• **CALC Minimizing the Tension.** A heavy horizontal girder of length *L* has several objects suspended from it. It is supported by a frictionless pivot at its left end and a cable of negligible weight that is attached to an I-beam at a point a distance *h* directly above the girder's center. Where should the other end of the cable be attached to the girder so that the cable's tension is a minimum? (*Hint:* In evaluating and presenting your answer, don't forget that the maximum distance of the point of attachment from the pivot is the length *L* of the beam.)

11.100 ••• **Bulk Modulus of an Ideal Gas.** The equation of state (the equation relating pressure, volume, and temperature) for an ideal gas is $pV = nRT$, where *n* and *R* are constants. (a) Show that if the gas is compressed while the temperature *T* is held constant, the bulk modulus is equal to the pressure. (b) When an ideal gas is compressed without the transfer of any heat into or out of it, the pressure and volume are related by pV^γ = constant, where γ is a constant having different values for different gases. Show that, in this case, the bulk modulus is given by $B = \gamma p$.

11.101 ••• **CP** An angler hangs a 4.50-kg fish from a vertical steel wire 1.50 m long and 5.00×10^{-3} cm^2 in cross-sectional area. The upper end of the wire is securely fastened to a support. (a) Calculate the amount the wire is stretched by the hanging fish. The angler now applies a force $\vec{F}$ to the fish, pulling it very slowly downward by 0.500 mm from its equilibrium position. For this downward motion, calculate (b) the work done by gravity; (c) the work done by the force $\vec{F}$; (d) the work done by the force the wire exerts on the fish; and (e) the change in the elastic potential energy (the potential energy associated with the tensile stress in the wire). Compare the answers in parts (d) and (e).

Answers

Chapter Opening Question ?

Each stone in the arch is under compression, not tension. This is because the forces on the stones tend to push them inward toward the center of the arch and thus squeeze them together. Compared to a solid supporting wall, a wall with arches is just as strong yet much more economical to build.

Test Your Understanding Questions

11.1 Answer: (i) Situation (i) satisfies both equilibrium conditions because the seagull has zero acceleration (so $\Sigma \vec{F} = 0$) and no tendency to start rotating (so $\Sigma \vec{\tau} = 0$). Situation (ii) satisfies the first condition because the crankshaft as a whole does not accelerate through space, but it does not satisfy the second condition; the crankshaft has an angular acceleration, so $\Sigma \vec{\tau}$ is not zero. Situation (iii) satisfies the second condition (there is no tendency to rotate) but not the first one; the baseball accelerates in its flight (due to gravity), so $\Sigma \vec{F}$ is not zero.

11.2 Answer: (ii) In equilibrium, the center of gravity must be at the point of support. Since the rock and meter stick have the same mass and hence the same weight, the center of gravity of the system is midway between their respective centers. The center of gravity of the meter stick alone is 0.50 m from the left end (that is, at the middle of the meter stick), so the center of gravity of the combination of rock and meter stick is 0.25 m from the left end.

11.3 Answer: (ii), (i), (iii) This is the same situation described in Example 11.4, with the rod replacing the forearm, the hinge replacing the elbow, and the cable replacing the tendon. The only difference is that the cable attachment point is at the end of the rod, so the distances D and L are identical. From Example 11.4, the tension is

$$T = \frac{Lw}{L \sin \theta} = \frac{w}{\sin \theta}$$

Since $\sin \theta$ is less than 1, the tension T is greater than the weight w. The vertical component of the force exerted by the hinge is

$$E_y = -\frac{(L - L)w}{L} = 0$$

In this situation, the hinge exerts *no* vertical force. You can see this easily if you calculate torques around the right end of the horizontal rod: The only force that exerts a torque around this point is the vertical component of the hinge force, so this force component must be zero.

11.4 Answers: (a) (iii), (b) (ii) In (a), the copper rod has 10 times the elongation Δl of the steel rod, but it also has 10 times the original length l_0. Hence the tensile strain $\Delta l / l_0$ is the same for both rods. In (b), the stress is equal to Young's modulus Y multiplied by the strain. From Table 11.1, steel has a larger value of Y, so a greater stress is required to produce the same strain.

11.5 In (a) and (b), the bumper will have sprung back to its original shape (although the paint may be scratched). In (c), the bumper will have a permanent dent or deformation. In (d), the bumper will be torn or broken.

Bridging Problem

Answers:

(a) $T = \dfrac{2mg}{3 \sin \theta}$

(b) $F = \dfrac{2mg}{3 \sin \theta} \sqrt{\cos^2 \theta + \frac{1}{4} \sin^2 \theta}$, $\phi = \arctan \left(\frac{1}{2} \tan \theta \right)$

(c) $\Delta l = \dfrac{2mgl_0}{3AY \tan \theta}$ (d) 4

FLUID MECHANICS

12

? This shark must swim constantly to keep from sinking to the bottom of the ocean, yet the orange tropical fish can remain at the same level in the water with little effort. Why is there a difference?

LEARNING GOALS

By studying this chapter, you will learn:

• The meaning of the density of a material and the average density of a body.

• What is meant by the pressure in a fluid, and how it is measured.

• How to calculate the buoyant force that a fluid exerts on a body immersed in it.

• The significance of laminar versus turbulent fluid flow, and how the speed of flow in a tube depends on the tube size.

• How to use Bernoulli's equation to relate pressure and flow speed at different points in certain types of flow.

luids play a vital role in many aspects of everyday life. We drink them, breathe them, swim in them. They circulate through our bodies and control our weather. Airplanes fly through them; ships float in them. A fluid is any substance that can flow; we use the term for both liquids and gases. We usually think of a gas as easily compressed and a liquid as nearly incompressible, although there are exceptional cases.

We begin our study with **fluid statics,** the study of fluids at rest in equilibrium situations. Like other equilibrium situations, it is based on Newton's first and third laws. We will explore the key concepts of density, pressure, and buoyancy. **Fluid dynamics,** the study of fluids in motion, is much more complex; indeed, it is one of the most complex branches of mechanics. Fortunately, we can analyze many important situations using simple idealized models and familiar principles such as Newton's laws and conservation of energy. Even so, we will barely scratch the surface of this broad and interesting topic.

12.1 Density

An important property of any material is its **density,** defined as its mass per unit volume. A homogeneous material such as ice or iron has the same density throughout. We use ρ (the Greek letter rho) for density. If a mass m of homogeneous material has volume V, the density ρ is

$$\rho = \frac{m}{V} \qquad \text{(definition of density)} \qquad (12.1)$$

Two objects made of the same material have the same density even though they may have different masses and different volumes. That's because the *ratio* of mass to volume is the same for both objects (Fig. 12.1).

12.1 Two objects with different masses and different volumes but the same density.

Different mass, same density: Because the wrench and nail are both made of steel, they have the same density (mass per unit volume).

Steel wrench   Steel nail

Table 12.1 Densities of Some Common Substances

Material	Density $(kg/m^3)^*$	Material	Density $(kg/m^3)^*$
Air (1 atm, 20°C)	1.20	Iron, steel	7.8×10^3
Ethanol	0.81×10^3	Brass	8.6×10^3
Benzene	0.90×10^3	Copper	8.9×10^3
Ice	0.92×10^3	Silver	10.5×10^3
Water	1.00×10^3	Lead	11.3×10^3
Seawater	1.03×10^3	Mercury	13.6×10^3
Blood	1.06×10^3	Gold	19.3×10^3
Glycerine	1.26×10^3	Platinum	21.4×10^3
Concrete	2×10^3	White dwarf star	10^{10}
Aluminum	2.7×10^3	Neutron star	10^{18}

*To obtain the densities in grams per cubic centimeter, simply divide by 10^3.

The SI unit of density is the kilogram per cubic meter (1 kg/m^3). The cgs unit, the gram per cubic centimeter (1 g/cm^3), is also widely used:

$$1 \text{ g/cm}^3 = 1000 \text{ kg/m}^3$$

The densities of some common substances at ordinary temperatures are given in Table 12.1. Note the wide range of magnitudes. The densest material found on earth is the metal osmium ($\rho = 22{,}500 \text{ kg/m}^3$), but its density pales by comparison to the densities of exotic astronomical objects such as white dwarf stars and neutron stars.

The **specific gravity** of a material is the ratio of its density to the density of water at 4.0°C, 1000 kg/m^3; it is a pure number without units. For example, the specific gravity of aluminum is 2.7. "Specific gravity" is a poor term, since it has nothing to do with gravity; "relative density" would have been better.

The density of some materials varies from point to point within the material. One example is the material of the human body, which includes low-density fat (about 940 kg/m^3) and high-density bone (from 1700 to 2500 kg/m^3). Two others are the earth's atmosphere (which is less dense at high altitudes) and oceans (which are denser at greater depths). For these materials, Eq. (12.1) describes the **average density.** In general, the density of a material depends on environmental factors such as temperature and pressure.

Measuring density is an important analytical technique. For example, we can determine the charge condition of a storage battery by measuring the density of its electrolyte, a sulfuric acid solution. As the battery discharges, the sulfuric acid (H_2SO_4) combines with lead in the battery plates to form insoluble lead sulfate ($PbSO_4$), decreasing the concentration of the solution. The density decreases from about $1.30 \times 10^3 \text{ kg/m}^3$ for a fully charged battery to $1.15 \times 10^3 \text{ kg/m}^3$ for a discharged battery.

Another automotive example is permanent-type antifreeze, which is usually a solution of ethylene glycol ($\rho = 1.12 \times 10^3 \text{ kg/m}^3$) and water. The freezing point of the solution depends on the glycol concentration, which can be determined by measuring the specific gravity. Such measurements can be performed by using a device called a hydrometer, which we'll discuss in Section 12.3.

Example 12.1   The weight of a roomful of air

Find the mass and weight of the air at 20°C in a living room with a 4.0 m × 5.0 m floor and a ceiling 3.0 m high, and the mass and weight of an equal volume of water.

SOLUTION

IDENTIFY and SET UP: We assume that the air density is the same throughout the room. (Air is less dense at high elevations than near sea level, but the density varies negligibly over the room's 3.0-m height; see Section 12.2.) We use Eq. (12.1) to relate the mass m_{air} to the room's volume V (which we'll calculate) and the air density ρ_{air} (given in Table 12.1).

EXECUTE: We have $V = (4.0 \text{ m})(5.0 \text{ m})(3.0 \text{ m}) = 60 \text{ m}^3$, so from Eq. (12.1),

$$m_{air} = \rho_{air}V = (1.20 \text{ kg/m}^3)(60 \text{ m}^3) = 72 \text{ kg}$$

$$w_{air} = m_{air}g = (72 \text{ kg})(9.8 \text{ m/s}^2) = 700 \text{ N} = 160 \text{ lb}$$

The mass and weight of an equal volume of water are

$$m_{water} = \rho_{water}V = (1000 \text{ kg/m}^3)(60 \text{ m}^3) = 6.0 \times 10^4 \text{ kg}$$

$$w_{water} = m_{water}g = (6.0 \times 10^4 \text{ kg})(9.8 \text{ m/s}^2)$$

$$= 5.9 \times 10^5 \text{ N} = 1.3 \times 10^5 \text{ lb} = 66 \text{ tons}$$

EVALUATE: A roomful of air weighs about the same as an average adult. Water is nearly a thousand times denser than air, so its mass and weight are larger by the same factor. The weight of a roomful of water would collapse the floor of an ordinary house.

Test Your Understanding of Section 12.1 Rank the following objects in order from highest to lowest average density: (i) mass 4.00 kg, volume 1.60×10^{-3} m^3; (ii) mass 8.00 kg, volume 1.60×10^{-3} m^3; (iii) mass 8.00 kg, volume 3.20×10^{-3} m^3; (iv) mass 2560 kg, volume 0.640 m^3; (v) mass 2560 kg, volume 1.28 m^3.

12.2 Pressure in a Fluid

When a fluid (either liquid or gas) is at rest, it exerts a force perpendicular to any surface in contact with it, such as a container wall or a body immersed in the fluid. This is the force that you feel pressing on your legs when you dangle them in a swimming pool. While the fluid as a whole is at rest, the molecules that make up the fluid are in motion; the force exerted by the fluid is due to molecules colliding with their surroundings.

If we think of an imaginary surface *within* the fluid, the fluid on the two sides of the surface exerts equal and opposite forces on the surface. (Otherwise, the surface would accelerate and the fluid would not remain at rest.) Consider a small surface of area dA centered on a point in the fluid; the normal force exerted by the fluid on each side is $dF_\perp$ (Fig. 12.2). We define the **pressure** p at that point as the normal force per unit area—that is, the ratio of $dF_\perp$ to dA (Fig. 12.3):

$$p = \frac{dF_\perp}{dA} \quad \text{(definition of pressure)} \tag{12.2}$$

If the pressure is the same at all points of a finite plane surface with area A, then

$$p = \frac{F_\perp}{A} \tag{12.3}$$

where $F_\perp$ is the net normal force on one side of the surface. The SI unit of pressure is the **pascal**, where

$$1 \text{ pascal} = 1 \text{ Pa} = 1 \text{ N/m}^2$$

We introduced the pascal in Chapter 11. Two related units, used principally in meteorology, are the *bar*, equal to 10^5 Pa, and the *millibar*, equal to 100 Pa.

Atmospheric pressure p_a is the pressure of the earth's atmosphere, the pressure at the bottom of this sea of air in which we live. This pressure varies with weather changes and with elevation. Normal atmospheric pressure at sea level (an average value) is 1 *atmosphere* (atm), defined to be exactly 101,325 Pa. To four significant figures,

$$(p_a)_{av} = 1 \text{ atm} = 1.013 \times 10^5 \text{ Pa}$$

$$= 1.013 \text{ bar} = 1013 \text{ millibar} = 14.70 \text{ lb/in.}^2$$

12.2 Forces acting on a small surface within a fluid at rest.

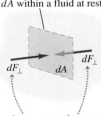

The surface does not accelerate, so the surrounding fluid exerts equal normal forces on both sides of it. (The fluid cannot exert any force parallel to the surface, since that would cause the surface to accelerate.)

12.3 The pressure on either side of a surface is force divided by area. Pressure is a scalar with units of newtons per square meter. By contrast, force is a vector with units of newtons.

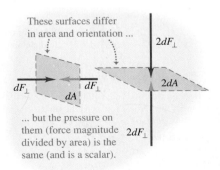

CAUTION **Don't confuse pressure and force** In everyday language the words "pressure" and "force" mean pretty much the same thing. In fluid mechanics, however, these words describe distinct quantities with different characteristics. Fluid pressure acts perpendicular to any surface in the fluid, no matter how that surface is oriented (Fig. 12.3). Hence pressure has no intrinsic direction of its own; it's a scalar. By contrast, force is a vector with a definite direction. Remember, too, that pressure is force per unit area. As Fig. 12.3 shows, a surface with twice the area has twice as much force exerted on it by the fluid, so the pressure is the same.

Example 12.2 The force of air

In the room described in Example 12.1, what is the total downward force on the floor due to an air pressure of 1.00 atm?

SOLUTION

IDENTIFY and SET UP: This example uses the relationship among the pressure p of a fluid (air), the area A subjected to that pressure, and the resulting normal force $F_\perp$ the fluid exerts. The pressure is uniform, so we use Eq. (12.3), $F_\perp = pA$, to determine $F_\perp$. The floor is horizontal, so $F_\perp$ is vertical (downward).

EXECUTE: We have $A = (4.0 \text{ m})(5.0 \text{ m}) = 20 \text{ m}^2$, so from Eq. (12.3),

$$F_\perp = pA = (1.013 \times 10^5 \text{ N/m}^2)(20 \text{ m}^2)$$
$$= 2.0 \times 10^6 \text{ N} = 4.6 \times 10^5 \text{ lb} = 230 \text{ tons}$$

EVALUATE: Unlike the water in Example 12.1, $F_\perp$ will not collapse the floor here, because there is an *upward* force of equal magnitude on the floor's underside. If the house has a basement, this upward force is exerted by the air underneath the floor. In this case, if we neglect the thickness of the floor, the *net* force due to air pressure is zero.

Pressure, Depth, and Pascal's Law

12.4 The forces on an element of fluid in equilibrium.

(a)

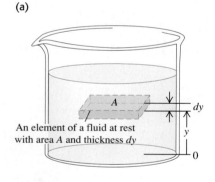

An element of a fluid at rest with area A and thickness dy

(b)

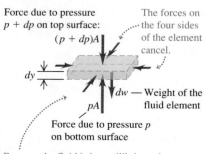

Force due to pressure $p + dp$ on top surface: $(p + dp)A$

The forces on the four sides of the element cancel.

dw — Weight of the fluid element

pA

Force due to pressure p on bottom surface

Because the fluid is in equilibrium, the vector sum of the vertical forces on the fluid element must be zero: $pA - (p + dp)A - dw = 0$.

If the weight of the fluid can be neglected, the pressure in a fluid is the same throughout its volume. We used that approximation in our discussion of bulk stress and strain in Section 11.4. But often the fluid's weight is *not* negligible. Atmospheric pressure is less at high altitude than at sea level, which is why an airplane cabin has to be pressurized when flying at 35,000 feet. When you dive into deep water, your ears tell you that the pressure increases rapidly with increasing depth below the surface.

We can derive a general relationship between the pressure p at any point in a fluid at rest and the elevation y of the point. We'll assume that the density ρ has the same value throughout the fluid (that is, the density is *uniform*), as does the acceleration due to gravity g. If the fluid is in equilibrium, every volume element is in equilibrium. Consider a thin element of fluid with thickness dy (Fig. 12.4a). The bottom and top surfaces each have area A, and they are at elevations y and $y + dy$ above some reference level where $y = 0$. The volume of the fluid element is $dV = A\,dy$, its mass is $dm = \rho\,dV = \rho A\,dy$, and its weight is $dw = dm\,g = \rho g A\,dy$.

What are the other forces on this fluid element (Fig 12.4b)? Let's call the pressure at the bottom surface p; then the total y-component of upward force on this surface is pA. The pressure at the top surface is $p + dp$, and the total y-component of (downward) force on the top surface is $-(p + dp)A$. The fluid element is in equilibrium, so the total y-component of force, including the weight and the forces at the bottom and top surfaces, must be zero:

$$\sum F_y = 0 \qquad \text{so} \qquad pA - (p + dp)A - \rho g A\,dy = 0$$

When we divide out the area A and rearrange, we get

$$\frac{dp}{dy} = -\rho g \tag{12.4}$$

This equation shows that when y increases, p decreases; that is, as we move upward in the fluid, pressure decreases, as we expect. If p_1 and p_2 are the pressures at elevations y_1 and y_2, respectively, and if ρ and g are constant, then

$$p_2 - p_1 = -\rho g(y_2 - y_1) \quad \text{(pressure in a fluid of uniform density)} \quad \text{(12.5)}$$

It's often convenient to express Eq. (12.5) in terms of the *depth* below the surface of a fluid (Fig. 12.5). Take point 1 at any level in the fluid and let p represent the pressure at this point. Take point 2 at the *surface* of the fluid, where the pressure is p_0 (subscript zero for zero depth). The depth of point 1 below the surface is $h = y_2 - y_1$, and Eq. (12.5) becomes

$$p_0 - p = -\rho g(y_2 - y_1) = -\rho g h \qquad \text{or}$$

$$p = p_0 + \rho g h \quad \text{(pressure in a fluid of uniform density)} \quad \text{(12.6)}$$

The pressure p at a depth h is greater than the pressure p_0 at the surface by an amount $\rho g h$. Note that the pressure is the same at any two points at the same level in the fluid. The *shape* of the container does not matter (Fig. 12.6).

Equation (12.6) shows that if we increase the pressure p_0 at the top surface, possibly by using a piston that fits tightly inside the container to push down on the fluid surface, the pressure p at any depth increases by exactly the same amount. This fact was recognized in 1653 by the French scientist Blaise Pascal (1623–1662) and is called *Pascal's law*.

> **Pascal's law: Pressure applied to an enclosed fluid is transmitted undiminished to every portion of the fluid and the walls of the containing vessel.**

The hydraulic lift shown schematically in Fig. 12.7 illustrates Pascal's law. A piston with small cross-sectional area A_1 exerts a force F_1 on the surface of a liquid such as oil. The applied pressure $p = F_1/A_1$ is transmitted through the connecting pipe to a larger piston of area A_2. The applied pressure is the same in both cylinders, so

$$p = \frac{F_1}{A_1} = \frac{F_2}{A_2} \quad \text{and} \quad F_2 = \frac{A_2}{A_1}F_1 \qquad \text{(12.7)}$$

The hydraulic lift is a force-multiplying device with a multiplication factor equal to the ratio of the areas of the two pistons. Dentist's chairs, car lifts and jacks, many elevators, and hydraulic brakes all use this principle.

For gases the assumption that the density ρ is uniform is realistic only over short vertical distances. In a room with a ceiling height of 3.0 m filled with air of uniform density 1.2 kg/m^3, the difference in pressure between floor and ceiling, given by Eq. (12.6), is

$$\rho g h = (1.2 \text{ kg/m}^3)(9.8 \text{ m/s}^2)(3.0 \text{ m}) = 35 \text{ Pa}$$

or about 0.00035 atm, a very small difference. But between sea level and the summit of Mount Everest (8882 m) the density of air changes by nearly a factor of 3, and in this case we cannot use Eq. (12.6). Liquids, by contrast, are nearly incompressible, and it is usually a very good approximation to regard their density as independent of pressure. A pressure of several hundred atmospheres will cause only a few percent increase in the density of most liquids.

Absolute Pressure and Gauge Pressure

If the pressure inside a car tire is equal to atmospheric pressure, the tire is flat. The pressure has to be *greater* than atmospheric to support the car, so the significant quantity is the *difference* between the inside and outside pressures. When we say that the pressure in a car tire is "32 pounds" (actually 32 $lb/in.^2$, equal to 220 kPa or 2.2×10^5 Pa), we mean that it is *greater* than atmospheric pressure

12.5 How pressure varies with depth in a fluid with uniform density.

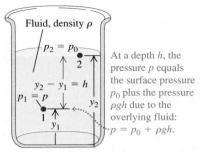

At a depth h, the pressure p equals the surface pressure p_0 plus the pressure $\rho g h$ due to the overlying fluid: $p = p_0 + \rho g h$.

Pressure difference between levels 1 and 2:
$$p_2 - p_1 = -\rho g(y_2 - y_1)$$
The pressure is greater at the lower level.

12.6 Each fluid column has the same height, no matter what its shape.

The pressure at the top of each liquid column is atmospheric pressure, p_0.

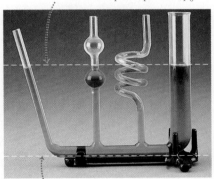

The pressure at the bottom of each liquid column has the same value p.

The difference between p and p_0 is $\rho g h$, where h is the distance from the top to the bottom of the liquid column. Hence all columns have the same height.

12.7 The hydraulic lift is an application of Pascal's law. The size of the fluid-filled container is exaggerated for clarity.

A small force is applied to a small piston.

Because the pressure p is the same at all points at a given height in the fluid ...

F_1

F_2

pA_1

pA_2

... a piston of larger area at the same height experiences a larger force.

($14.7 \, \text{lb/in.}^2$ or $1.01 \times 10^5 \, \text{Pa}$) by this amount. The *total* pressure in the tire is then $47 \, \text{lb/in.}^2$ or $320 \, \text{kPa}$. The excess pressure above atmospheric pressure is usually called **gauge pressure**, and the total pressure is called **absolute pressure**. Engineers use the abbreviations psig and psia for "pounds per square inch gauge" and "pounds per square inch absolute," respectively. If the pressure is *less* than atmospheric, as in a partial vacuum, the gauge pressure is negative.

Example 12.3 Finding absolute and gauge pressures

Water stands 12.0 m deep in a storage tank whose top is open to the atmosphere. What are the absolute and gauge pressures at the bottom of the tank?

SOLUTION

IDENTIFY and SET UP: Table 11.2 indicates that water is nearly incompressible, so we can treat it as having uniform density. The level of the top of the tank corresponds to point 2 in Fig. 12.5, and the level of the bottom of the tank corresponds to point 1. Our target variable is p in Eq. (12.6). We have $h = 12.0 \, \text{m}$ and $p_0 = 1 \, \text{atm} = 1.01 \times 10^5 \, \text{Pa}$.

EXECUTE: From Eq. (12.6), the pressures are

absolute:

$$p = p_0 + \rho g h$$
$$= (1.01 \times 10^5 \, \text{Pa}) + (1000 \, \text{kg/m}^3)(9.80 \, \text{m/s}^2)(12.0 \, \text{m})$$
$$= 2.19 \times 10^5 \, \text{Pa} = 2.16 \, \text{atm} = 31.8 \, \text{lb/in.}^2$$

gauge: $p - p_0 = (2.19 - 1.01) \times 10^5 \, \text{Pa}$
$$= 1.18 \times 10^5 \, \text{Pa} = 1.16 \, \text{atm} = 17.1 \, \text{lb/in.}^2$$

EVALUATE: A pressure gauge at the bottom of such a tank would probably be calibrated to read gauge pressure rather than absolute pressure.

Pressure Gauges

The simplest pressure gauge is the open-tube *manometer* (Fig. 12.8a). The U-shaped tube contains a liquid of density ρ, often mercury or water. The left end of the tube is connected to the container where the pressure p is to be measured, and the right end is open to the atmosphere at pressure $p_0 = p_{atm}$. The pressure at the bottom of the tube due to the fluid in the left column is $p + \rho g y_1$, and the pressure at the bottom due to the fluid in the right column is $p_{atm} + \rho g y_2$. These pressures are measured at the same level, so they must be equal:

$$p + \rho g y_1 = p_{atm} + \rho g y_2$$
$$p - p_{atm} = \rho g (y_2 - y_1) = \rho g h \tag{12.8}$$

In Eq. (12.8), p is the *absolute pressure*, and the difference $p - p_{atm}$ between absolute and atmospheric pressure is the gauge pressure. Thus the gauge pressure is proportional to the difference in height $h = y_2 - y_1$ of the liquid columns.

12.8 Two types of pressure gauge.

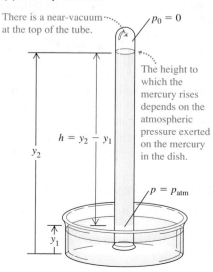

(a) Open-tube manometer

$p_0 = p_{atm}$

$h = y_2 - y_1$

Pressure p

y_2

y_1

$p + \rho g y_1$ $p_{atm} + \rho g y_2$

The pressure is the same at the bottoms of the two tubes.

(b) Mercury barometer

There is a near-vacuum at the top of the tube.

$p_0 = 0$

The height to which the mercury rises depends on the atmospheric pressure exerted on the mercury in the dish.

$h = y_2 - y_1$

y_2

$p = p_{atm}$

y_1

Another common pressure gauge is the **mercury barometer.** It consists of a long glass tube, closed at one end, that has been filled with mercury and then inverted in a dish of mercury (Fig. 12.8b). The space above the mercury column contains only mercury vapor; its pressure is negligibly small, so the pressure p_0 at the top of the mercury column is practically zero. From Eq. (12.6),

$$p_{atm} = p = 0 + \rho g(y_2 - y_1) = \rho g h \qquad (12.9)$$

Thus the mercury barometer reads the atmospheric pressure p_{atm} directly from the height of the mercury column.

Pressures are often described in terms of the height of the corresponding mercury column, as so many "inches of mercury" or "millimeters of mercury" (abbreviated mm Hg). A pressure of 1 mm Hg is called *1 torr,* after Evangelista Torricelli, inventor of the mercury barometer. But these units depend on the density of mercury, which varies with temperature, and on the value of g, which varies with location, so the pascal is the preferred unit of pressure.

Many types of pressure gauges use a flexible sealed tube (Fig. 12.9). A change in the pressure either inside or outside the tube causes a change in its dimensions. This change is detected optically, electrically, or mechanically.

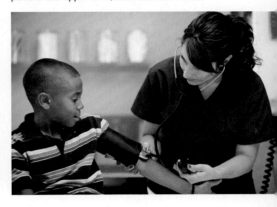

(a)

Changes in the inlet pressure cause the tube to coil or uncoil, which moves the pointer.

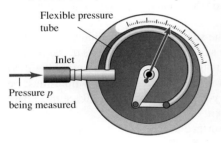
Flexible pressure tube
Inlet
Pressure p being measured

(b)

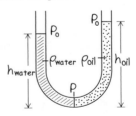

12.9 (a) A Bourdon pressure gauge. When the pressure inside the flexible tube increases, the tube straightens out a little, deflecting the attached pointer. (b) This Bourdon-type pressure gauge is connected to a high-pressure gas line. The gauge pressure shown is just over 5 bars (1 bar = 10^5 Pa).

Example 12.4 | **A tale of two fluids**

A manometer tube is partially filled with water. Oil (which does not mix with water) is poured into the left arm of the tube until the oil–water interface is at the midpoint of the tube as shown. Both arms of the tube are open to the air. Find a relationship between the heights h_{oil} and h_{water}.

SOLUTION

IDENTIFY and SET UP: Figure 12.10 shows our sketch. The relationship between pressure and depth given by Eq. (12.6) applies only to fluids of uniform density; we have two fluids of different densities, so we must write a separate pressure–depth relationship for each. Both fluid columns have pressure p at the bottom (where they are in contact and in equilibrium) and are both at atmospheric pressure p_0 at the top (where both are in contact with and in equilibrium with the air).

EXECUTE: Writing Eq. (12.6) for each fluid gives

$$p = p_0 + \rho_{water}g h_{water}$$
$$p = p_0 + \rho_{oil}g h_{oil}$$

12.10 Our sketch for this problem.

Since the pressure p at the bottom of the tube is the same for both fluids, we set these two expressions equal to each other and solve for h_{oil} in terms of h_{water}. You can show that the result is

$$h_{oil} = \frac{\rho_{water}}{\rho_{oil}}h_{water}$$

EVALUATE: Water ($\rho_{water} = 1000$ kg/m^3) is denser than oil ($\rho_{oil} \approx 850$ kg/m^3), so h_{oil} is greater than h_{water} as Fig. 12.10 shows. It takes a greater height of low-density oil to produce the same pressure p at the bottom of the tube.

Test Your Understanding of Section 12.2 Mercury is less dense at high temperatures than at low temperatures. Suppose you move a mercury barometer from the cold interior of a tightly sealed refrigerator to outdoors on a hot summer day. You find that the column of mercury remains at the same height in the tube. Compared to the air pressure inside the refrigerator, is the air pressure outdoors (i) higher, (ii) lower, or (iii) the same? (Ignore the very small change in the dimensions of the glass tube due to the temperature change.)

12.3 Buoyancy

Buoyancy is a familiar phenomenon: A body immersed in water seems to weigh less than when it is in air. When the body is less dense than the fluid, it floats. The human body usually floats in water, and a helium-filled balloon floats in air.

> **Archimedes's principle: When a body is completely or partially immersed in a fluid, the fluid exerts an upward force on the body equal to the weight of the fluid displaced by the body.**

To prove this principle, we consider an arbitrary element of fluid at rest. In Fig. 12.11a the irregular outline is the surface boundary of this element of fluid. The arrows represent the forces exerted on the boundary surface by the surrounding fluid.

The entire fluid is in equilibrium, so the sum of all the y-components of force on this element of fluid is zero. Hence the sum of the y-components of the *surface* forces must be an upward force equal in magnitude to the weight mg of the fluid inside the surface. Also, the sum of the torques on the element of fluid must be zero, so the line of action of the resultant y-component of surface force must pass through the center of gravity of this element of fluid.

Now we remove the fluid inside the surface and replace it with a solid body having exactly the same shape (Fig. 12.11b). The pressure at every point is exactly the same as before. So the total upward force exerted on the body by the fluid is also the same, again equal in magnitude to the weight mg of the fluid displaced to make way for the body. We call this upward force the **buoyant force** on the solid body. The line of action of the buoyant force again passes through the center of gravity of the displaced fluid (which doesn't necessarily coincide with the center of gravity of the body).

When a balloon floats in equilibrium in air, its weight (including the gas inside it) must be the same as the weight of the air displaced by the balloon. A fish's flesh is denser than water, yet a fish can float while

12.11 Archimedes's principle.

(a) Arbitrary element of fluid in equilibrium

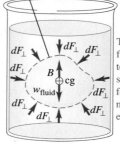

The forces on the fluid element due to pressure must sum to a buoyant force equal in magnitude to the element's weight.

(b) Fluid element replaced with solid body of the same size and shape

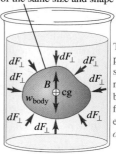

The forces due to pressure are the same, so the body must be acted upon by the same buoyant force as the fluid element, *regardless of the body's weight.*

submerged because it has a gas-filled cavity within its body. This makes the fish's *average* density the same as water's, so its net weight is the same as the weight of the water it displaces. A body whose average density is *less* than that of a liquid can float partially submerged at the free upper surface of the liquid. The greater the density of the liquid, the less of the body is submerged. When you swim in seawater (density 1030 kg/m³), your body floats higher than in fresh water (1000 kg/m³).

A practical example of buoyancy is the hydrometer, used to measure the density of liquids (Fig. 12.12a). The calibrated float sinks into the fluid until the weight of the fluid it displaces is exactly equal to its own weight. The hydrometer floats *higher* in denser liquids than in less dense liquids, and a scale in the top stem permits direct density readings. Figure 12.12b shows a type of hydrometer that is commonly used to measure the density of battery acid or antifreeze. The bottom of the large tube is immersed in the liquid; the bulb is squeezed to expel air and is then released, like a giant medicine dropper. The liquid rises into the outer tube, and the hydrometer floats in this sample of the liquid.

12.12 Measuring the density of a fluid.

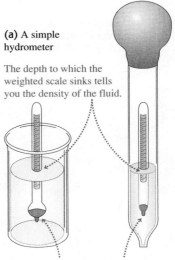

(b) Using a hydrometer to measure the density of battery acid or antifreeze

(a) A simple hydrometer

The depth to which the weighted scale sinks tells you the density of the fluid.

The weight at the bottom makes the scale float upright.

Example 12.5   Buoyancy

A 15.0-kg solid gold statue is raised from the sea bottom (Fig. 12.13a). What is the tension in the hoisting cable (assumed massless) when the statue is (a) at rest and completely underwater and (b) at rest and completely out of the water?

SOLUTION

IDENTIFY and SET UP: In both cases the statue is in equilibrium and experiences three forces: its weight, the cable tension, and a buoyant force equal in magnitude to the weight of the fluid displaced by the statue (seawater in part (a), air in part (b)). Figure 12.13b shows the free-body diagram for the statue. Our target variables are the values of the tension in seawater (T_{sw}) and in air (T_{air}). We are given the mass m_{statue}, and we can calculate the buoyant force in seawater (B_{sw}) and in air (B_{air}) using Archimedes's principle.

EXECUTE: (a) To find B_{sw}, we first find the statue's volume V using the density of gold from Table 12.1:

$$V = \frac{m_{statue}}{\rho_{gold}} = \frac{15.0 \text{ kg}}{19.3 \times 10^3 \text{ kg/m}^3} = 7.77 \times 10^{-4} \text{ m}^3$$

The buoyant force B_{sw} equals the weight of this same volume of seawater. Using Table 12.1 again:

$$B_{sw} = w_{sw} = m_{sw}g = \rho_{sw}Vg$$
$$= (1.03 \times 10^3 \text{ kg/m}^3)(7.77 \times 10^{-4} \text{ m}^3)(9.80 \text{ m/s}^2)$$
$$= 7.84 \text{ N}$$

The statue is at rest, so the net external force acting on it is zero. From Fig. 12.13b,

$$\sum F_y = B_{sw} + T_{sw} + (-m_{statue}g) = 0$$
$$T_{sw} = m_{statue}g - B_{sw} = (15.0 \text{ kg})(9.80 \text{ m/s}^2) - 7.84 \text{ N}$$
$$= 147 \text{ N} - 7.84 \text{ N} = 139 \text{ N}$$

12.13 What is the tension in the cable hoisting the statue?

(a) Immersed statue in equilibrium   **(b)** Free-body diagram of statue

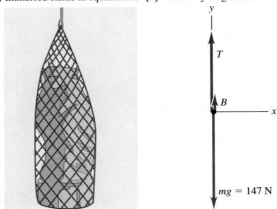

A spring scale attached to the upper end of the cable will indicate a tension 7.84 N less than the statue's actual weight $m_{statue}g = 147$ N.

(b) The density of air is about 1.2 kg/m³, so the buoyant force of air on the statue is

$$B_{air} = \rho_{air}Vg = (1.2 \text{ kg/m}^3)(7.77 \times 10^{-4} \text{ m}^3)(9.80 \text{ m/s}^2)$$
$$= 9.1 \times 10^{-3} \text{ N}$$

This is negligible compared to the statue's actual weight $m_{statue}g = 147$ N. So within the precision of our data, the tension in the cable with the statue in air is $T_{air} = m_{statue}g = 147$ N.

EVALUATE: Note that the buoyant force is proportional to the density of the *fluid* in which the statue is immersed, *not* the density of

Continued

the statue. The denser the fluid, the greater the buoyant force and the smaller the cable tension. If the fluid had the same density as the statue, the buoyant force would be equal to the statue's weight and the tension would be zero (the cable would go slack). If the fluid were denser than the statue, the tension would be *negative:* The buoyant force would be greater than the statue's weight, and a downward force would be required to keep the statue from rising upward.

12.14 The surface of the water acts like a membrane under tension, allowing this water strider to literally "walk on water."

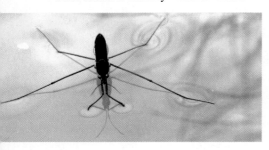

12.15 A molecule at the surface of a liquid is attracted into the bulk liquid, which tends to reduce the liquid's surface area.

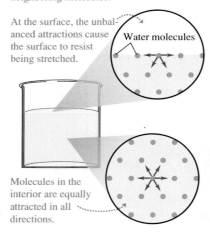

Molecules in a liquid are attracted by neighboring molecules.

At the surface, the unbalanced attractions cause the surface to resist being stretched.

Water molecules

Molecules in the interior are equally attracted in all directions.

12.16 Surface tension makes it difficult to force water through small crevices. The required water pressure p can be reduced by using hot, soapy water, which has less surface tension.

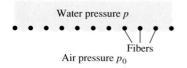

Water pressure p

Fibers

Air pressure p_0

Surface Tension

An object less dense than water, such as an air-filled beach ball, floats with part of its volume below the surface. Conversely, a paper clip can rest *atop* a water surface even though its density is several times that of water. This is an example of **surface tension:** The surface of the liquid behaves like a membrane under tension (Fig. 12.14). Surface tension arises because the molecules of the liquid exert attractive forces on each other. There is zero net force on a molecule inside the volume of the liquid, but a surface molecule is drawn into the volume (Fig. 12.15). Thus the liquid tends to minimize its surface area, just as a stretched membrane does.

Surface tension explains why freely falling raindrops are spherical (*not* teardrop-shaped): A sphere has a smaller surface area for its volume than any other shape. It also explains why hot, soapy water is used for washing. To wash clothing thoroughly, water must be forced through the tiny spaces between the fibers (Fig. 12.16). To do so requires increasing the surface area of the water, which is difficult to achieve because of surface tension. The job is made easier by increasing the temperature of the water and adding soap, both of which decrease the surface tension.

Surface tension is important for a millimeter-sized water drop, which has a relatively large surface area for its volume. (A sphere of radius r has surface area $4\pi r^2$ and volume $(4\pi/3)r^3$. The ratio of surface area to volume is $3/r$, which increases with decreasing radius.) For large quantities of liquid, however, the ratio of surface area to volume is relatively small, and surface tension is negligible compared to pressure forces. For the remainder of this chapter, we will consider only fluids in bulk and hence will ignore the effects of surface tension.

Test Your Understanding of Section 12.3 You place a container of seawater on a scale and note the reading on the scale. You now suspend the statue of Example 12.5 in the water (Fig. 12.17). How does the scale reading change? (i) It increases by 7.84 N; (ii) it decreases by 7.84 N; (iii) it remains the same; (iv) none of these.

12.4 Fluid Flow

We are now ready to consider *motion* of a fluid. Fluid flow can be extremely complex, as shown by the currents in river rapids or the swirling flames of a campfire. But some situations can be represented by relatively simple idealized models. An **ideal fluid** is a fluid that is *incompressible* (that is, its density cannot change) and has no internal friction (called **viscosity**). Liquids are approximately incompressible in most situations, and we may also treat a gas as incompressible if the pressure differences from one region to another are not too great. Internal friction in a fluid causes shear stresses when two adjacent layers of fluid move relative to each other, as when fluid flows inside a tube or around an obstacle. In some cases we can neglect these shear forces in comparison with forces arising from gravitation and pressure differences.

The path of an individual particle in a moving fluid is called a **flow line.** If the overall flow pattern does not change with time, the flow is called **steady flow.** In steady flow, every element passing through a given point follows the same flow line. In this case the "map" of the fluid velocities at various points in space remains constant, although the velocity of a particular particle may change in both magnitude and direction during its motion. A **streamline** is a curve whose tangent at any point is in the direction of the fluid velocity at that point. When the flow pattern changes with time, the streamlines do not coincide with the flow lines. We will consider only steady-flow situations, for which flow lines and streamlines are identical.

The flow lines passing through the edge of an imaginary element of area, such as the area A in Fig. 12.18, form a tube called a **flow tube.** From the definition of a flow line, in steady flow no fluid can cross the side walls of a flow tube; the fluids in different flow tubes cannot mix.

Figure 12.19 shows patterns of fluid flow from left to right around three different obstacles. The photographs were made by injecting dye into water flowing between two closely spaced glass plates. These patterns are typical of **laminar flow,** in which adjacent layers of fluid slide smoothly past each other and the flow is steady. (A *lamina* is a thin sheet.) At sufficiently high flow rates, or when boundary surfaces cause abrupt changes in velocity, the flow can become irregular and chaotic. This is called **turbulent flow** (Fig. 12.20). In turbulent flow there is no steady-state pattern; the flow pattern changes continuously.

The Continuity Equation

The mass of a moving fluid doesn't change as it flows. This leads to an important quantitative relationship called the **continuity equation.** Consider a portion of a flow tube between two stationary cross sections with areas A_1 and A_2 (Fig. 12.21). The fluid speeds at these sections are v_1 and v_2, respectively. No fluid flows in or out across the sides of the tube because the fluid velocity is tangent to the wall at every point on the wall. During a small time interval dt, the fluid at A_1 moves a distance $v_1\, dt$, so a cylinder of fluid with height $v_1\, dt$ and volume $dV_1 = A_1 v_1\, dt$ flows into the tube across A_1. During this same interval, a cylinder of volume $dV_2 = A_2 v_2\, dt$ flows out of the tube across A_2.

Let's first consider the case of an incompressible fluid so that the density ρ has the same value at all points. The mass dm_1 flowing into the tube across A_1 in time dt is $dm_1 = \rho A_1 v_1\, dt$. Similarly, the mass dm_2 that flows out across A_2 in the same time is $dm_2 = \rho A_2 v_2\, dt$. In steady flow the total mass in the tube is constant, so $dm_1 = dm_2$ and

$$\rho A_1 v_1\, dt = \rho A_2 v_2\, dt \qquad \text{or}$$

$$A_1 v_1 = A_2 v_2 \qquad \text{(continuity equation, incompressible fluid)} \qquad (12.10)$$

12.17 How does the scale reading change when the statue is immersed in water?

12.18 A flow tube bounded by flow lines. In steady flow, fluid cannot cross the walls of a flow tube.

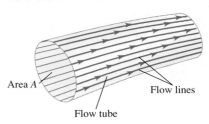

Area A

Flow lines

Flow tube

12.19 Laminar flow around obstacles of different shapes.

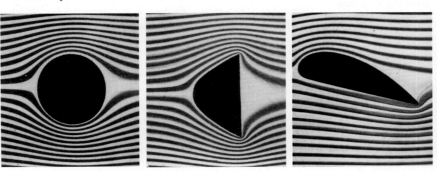

12.20 The flow of smoke rising from these incense sticks is laminar up to a certain point, and then becomes turbulent.

12.21 A flow tube with changing cross-sectional area. If the fluid is incompressible, the product Av has the same value at all points along the tube.

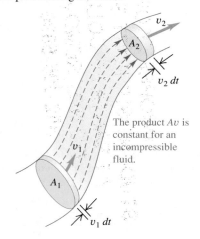

The product Av is constant for an incompressible fluid.

The product Av is the *volume flow rate* dV/dt, the rate at which volume crosses a section of the tube:

$$\frac{dV}{dt} = Av \qquad \text{(volume flow rate)} \qquad (12.11)$$

The *mass* flow rate is the mass flow per unit time through a cross section. This is equal to the density ρ times the volume flow rate dV/dt.

Equation (12.10) shows that the volume flow rate has the same value at all points along any flow tube. When the cross section of a flow tube decreases, the speed increases, and vice versa. A broad, deep part of a river has larger cross section and slower current than a narrow, shallow part, but the volume flow rates are the same in both. This is the essence of the familiar maxim, "Still waters run deep." The stream of water from a faucet narrows as it gains speed during its fall, but dV/dt is the same everywhere along the stream. If a water pipe with 2-cm diameter is connected to a pipe with 1-cm diameter, the flow speed is four times as great in the 1-cm part as in the 2-cm part.

We can generalize Eq. (12.10) for the case in which the fluid is *not* incompressible. If ρ_1 and ρ_2 are the densities at sections 1 and 2, then

$$\rho_1 A_1 v_1 = \rho_2 A_2 v_2 \qquad \text{(continuity equation, compressible fluid)} \qquad (12.12)$$

If the fluid is denser at point 2 than at point 1 ($\rho_2 > \rho_1$), the volume flow rate at point 2 will be less than at point 1 ($A_2 v_2 < A_1 v_1$). We leave the details to you. If the fluid is incompressible so that ρ_1 and ρ_2 are always equal, Eq. (12.12) reduces to Eq. (12.10).

Example 12.6   **Flow of an incompressible fluid**

Incompressible oil of density 850 kg/m^3 is pumped through a cylindrical pipe at a rate of 9.5 liters per second. (a) The first section of the pipe has a diameter of 8.0 cm. What is the flow speed of the oil? What is the mass flow rate? (b) The second section of the pipe has a diameter of 4.0 cm. What are the flow speed and mass flow rate in that section?

SOLUTION

IDENTIFY and SET UP: Since the oil is incompressible, the volume flow rate has the *same* value (9.5 L/s) in both sections of pipe. The mass flow rate (the density times the volume flow rate) also has the same value in both sections. (This is just the statement that no fluid is lost or added anywhere along the pipe.) We use the volume flow rate equation, Eq. (12.11), to determine the speed v_1 in the 8.0-cm-diameter section and the continuity equation for incompressible flow, Eq. (12.10), to find the speed v_2 in the 4.0-cm-diameter section.

EXECUTE: (a) From Eq. (12.11) the volume flow rate in the first section is $dV/dt = A_1 v_1$, where A_1 is the cross-sectional area of the pipe of diameter 8.0 cm and radius 4.0 cm. Hence

$$v_1 = \frac{dV/dt}{A_1} = \frac{(9.5 \text{ L/s})(10^{-3} \text{ m}^3/\text{L})}{\pi(4.0 \times 10^{-2} \text{ m})^2} = 1.9 \text{ m/s}$$

The mass flow rate is $\rho \, dV/dt = (850 \text{ kg/m}^3)(9.5 \times 10^{-3} \text{ m}^3/\text{s}) = 8.1 \text{ kg/s}$.

(b) From the continuity equation, Eq. (12.10),

$$v_2 = \frac{A_1}{A_2} v_1 = \frac{\pi(4.0 \times 10^{-2} \text{ m})^2}{\pi(2.0 \times 10^{-2} \text{ m})^2}(1.9 \text{ m/s}) = 7.6 \text{ m/s} = 4v_1$$

The volume and mass flow rates are the same as in part (a).

EVALUATE: The second section of pipe has one-half the diameter and one-fourth the cross-sectional area of the first section. Hence the speed must be four times greater in the second section, which is just what our result shows.

Test Your Understanding of Section 12.4   A maintenance crew is working on a section of a three-lane highway, leaving only one lane open to traffic. The result is much slower traffic flow (a traffic jam). Do cars on a highway behave like (i) the molecules of an incompressible fluid or (ii) the molecules of a compressible fluid?   (MP)

12.5 Bernoulli's Equation

According to the continuity equation, the speed of fluid flow can vary along the paths of the fluid. The pressure can also vary; it depends on height as in the static situation (see Section 12.2), and it also depends on the speed of flow. We can derive an important relationship called *Bernoulli's equation* that relates the pressure, flow speed, and height for flow of an ideal, incompressible fluid. Bernoulli's equation is an essential tool in analyzing plumbing systems, hydroelectric generating stations, and the flight of airplanes.

The dependence of pressure on speed follows from the continuity equation, Eq. (12.10). When an incompressible fluid flows along a flow tube with varying cross section, its speed *must* change, and so an element of fluid must have an acceleration. If the tube is horizontal, the force that causes this acceleration has to be applied by the surrounding fluid. This means that the pressure *must* be different in regions of different cross section; if it were the same everywhere, the net force on every fluid element would be zero. When a horizontal flow tube narrows and a fluid element speeds up, it must be moving toward a region of lower pressure in order to have a net forward force to accelerate it. If the elevation also changes, this causes an additional pressure difference.

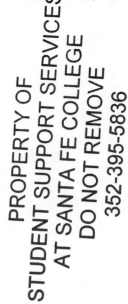

Deriving Bernoulli's Equation

To derive Bernoulli's equation, we apply the work–energy theorem to the fluid in a section of a flow tube. In Fig. 12.22 we consider the element of fluid that at some initial time lies between the two cross sections a and c. The speeds at the lower and upper ends are v_1 and v_2. In a small time interval dt, the fluid that is initially at a moves to b, a distance $ds_1 = v_1\, dt$, and the fluid that is initially at c moves to d, a distance $ds_2 = v_2\, dt$. The cross-sectional areas at the two ends are A_1 and A_2, as shown. The fluid is incompressible; hence by the continuity equation, Eq. (12.10), the volume of fluid dV passing *any* cross section during time dt is the same. That is, $dV = A_1\, ds_1 = A_2\, ds_2$.

Let's compute the *work* done on this fluid element during dt. We assume that there is negligible internal friction in the fluid (i.e., no viscosity), so the only nongravitational forces that do work on the fluid element are due to the pressure of the surrounding fluid. The pressures at the two ends are p_1 and p_2; the force on the cross section at a is p_1A_1, and the force at c is p_2A_2. The net work dW done on the element by the surrounding fluid during this displacement is therefore

$$dW = p_1A_1\, ds_1 - p_2A_2\, ds_2 = (p_1 - p_2)dV \qquad (12.13)$$

The second term has a negative sign because the force at c opposes the displacement of the fluid.

The work dW is due to forces other than the conservative force of gravity, so it equals the change in the total mechanical energy (kinetic energy plus gravitational potential energy) associated with the fluid element. The mechanical energy for the fluid between sections b and c does not change. At the beginning of dt the fluid between a and b has volume $A_1\, ds_1$, mass $\rho A_1\, ds_1$, and kinetic energy $\frac{1}{2}\rho(A_1\, ds_1)v_1^2$. At the end of dt the fluid between c and d has kinetic energy $\frac{1}{2}\rho(A_2\, ds_2)v_2^2$. The net change in kinetic energy dK during time dt is

$$dK = \tfrac{1}{2}\rho\, dV(v_2^2 - v_1^2) \qquad (12.14)$$

What about the change in gravitational potential energy? At the beginning of dt, the potential energy for the mass between a and b is $dm\, gy_1 = \rho\, dV\, gy_1$. At

12.22 Deriving Bernoulli's equation. The net work done on a fluid element by the pressure of the surrounding fluid equals the change in the kinetic energy plus the change in the gravitational potential energy.

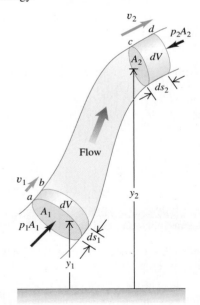

the end of dt, the potential energy for the mass between c and d is $dm\, gy_2 = \rho\, dV\, gy_2$. The net change in potential energy dU during dt is

$$dU = \rho\, dV\, g(y_2 - y_1) \qquad (12.15)$$

Combining Eqs. (12.13), (12.14), and (12.15) in the energy equation $dW = dK + dU$, we obtain

$$(p_1 - p_2)dV = \tfrac{1}{2}\rho\, dV(v_2^2 - v_1^2) + \rho\, dV\, g(y_2 - y_1)$$
$$p_1 - p_2 = \tfrac{1}{2}\rho(v_2^2 - v_1^2) + \rho g(y_2 - y_1) \qquad (12.16)$$

This is **Bernoulli's equation.** It states that the work done on a unit volume of fluid by the surrounding fluid is equal to the sum of the changes in kinetic and potential energies per unit volume that occur during the flow. We may also interpret Eq. (12.16) in terms of pressures. The first term on the right is the pressure difference associated with the change of speed of the fluid. The second term on the right is the additional pressure difference caused by the weight of the fluid and the difference in elevation of the two ends.

We can also express Eq. (12.16) in a more convenient form as

$$p_1 + \rho g y_1 + \tfrac{1}{2}\rho v_1^2 = p_2 + \rho g y_2 + \tfrac{1}{2}\rho v_2^2 \quad \text{(Bernoulli's equation)} \quad (12.17)$$

The subscripts 1 and 2 refer to *any* two points along the flow tube, so we can also write

$$p + \rho g y + \tfrac{1}{2}\rho v^2 = \text{constant} \qquad (12.18)$$

Note that when the fluid is *not* moving (so $v_1 = v_2 = 0$), Eq. (12.17) reduces to the pressure relationship we derived for a fluid at rest, Eq. (12.5).

CAUTION **Bernoulli's principle applies only in certain situations** We stress again that Bernoulli's equation is valid for only incompressible, steady flow of a fluid with no internal friction (no viscosity). It's a simple equation that's easy to use; don't let this tempt you to use it in situations in which it doesn't apply! ▌

Problem-Solving Strategy 12.1 — Bernoulli's Equation

Bernoulli's equation is derived from the work–energy theorem, so much of Problem-Solving Strategy 7.1 (Section 7.1) is applicable here.

IDENTIFY *the relevant concepts:* Bernoulli's equation is applicable to steady flow of an incompressible fluid that has no internal friction (see Section 12.6). It is generally applicable to flows through large pipes and to flows within bulk fluids (e.g., air flowing around an airplane or water flowing around a fish).

SET UP *the problem* using the following steps:
1. Identify the points 1 and 2 referred to in Bernoulli's equation, Eq. (12.17).
2. Define your coordinate system, particularly the level at which $y = 0$. Take the positive y-direction to be upward.

3. Make lists of the unknown and known quantities in Eq. (12.17). Decide which unknowns are the target variables.

EXECUTE *the solution* as follows: Write Bernoulli's equation and solve for the unknowns. You may need the continuity equation, Eq. (12.10), to get a relationship between the two speeds in terms of cross-sectional areas of pipes or containers. You may also need Eq. (12.11) to find the volume flow rate.

EVALUATE *your answer:* Verify that the results make physical sense. Check that you have used consistent units: In SI units, pressure is in pascals, density in kilograms per cubic meter, and speed in meters per second. Also note that the pressures must be either *all* absolute pressures or *all* gauge pressures.

Example 12.7 | Water pressure in the home

Water enters a house (Fig. 12.23) through a pipe with an inside diameter of 2.0 cm at an absolute pressure of 4.0×10^5 Pa (about 4 atm). A 1.0-cm-diameter pipe leads to the second-floor bathroom 5.0 m above. When the flow speed at the inlet pipe is 1.5 m/s, find the flow speed, pressure, and volume flow rate in the bathroom.

SOLUTION

IDENTIFY and SET UP: We assume that the water flows at a steady rate. Water is effectively incompressible, so we can use the continuity equation. It's reasonable to ignore internal friction because the pipe has a relatively large diameter, so we can also use Bernoulli's equation. Let points 1 and 2 be at the inlet pipe and at the bathroom, respectively. We are given the pipe diameters at points 1 and 2, from which we calculate the areas A_1 and A_2, as well as the speed $v_1 = 1.5$ m/s and pressure $p_1 = 4.0 \times 10^5$ Pa at the inlet pipe. We take $y_1 = 0$ and $y_2 = 5.0$ m. We find the speed v_2 using the continuity equation and the pressure p_2 using Bernoulli's equation. Knowing v_2, we calculate the volume flow rate $v_2 A_2$.

EXECUTE: From the continuity equation, Eq. (12.10),

$$v_2 = \frac{A_1}{A_2} v_1 = \frac{\pi(1.0 \text{ cm})^2}{\pi(0.50 \text{ cm})^2}(1.5 \text{ m/s}) = 6.0 \text{ m/s}$$

From Bernoulli's equation, Eq. (12.16),

$$
\begin{aligned}
p_2 &= p_1 - \tfrac{1}{2}\rho(v_2{}^2 - v_1{}^2) - \rho g(y_2 - y_1) \\
&= 4.0 \times 10^5 \text{ Pa} \\
&\quad - \tfrac{1}{2}(1.0 \times 10^3 \text{ kg/m}^3)(36 \text{ m}^2/\text{s}^2 - 2.25 \text{ m}^2/\text{s}^2) \\
&\quad - (1.0 \times 10^3 \text{ kg/m}^3)(9.8 \text{ m/s}^2)(5.0 \text{ m}) \\
&= 4.0 \times 10^5 \text{ Pa} - 0.17 \times 10^5 \text{ Pa} - 0.49 \times 10^5 \text{ Pa} \\
&= 3.3 \times 10^5 \text{ Pa} = 3.3 \text{ atm} = 48 \text{ lb/in.}^2
\end{aligned}
$$

12.23 What is the water pressure in the second-story bathroom of this house?

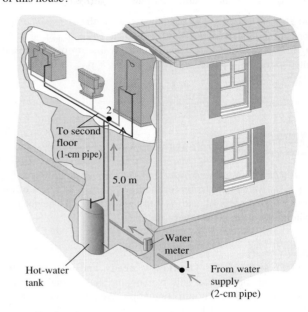

The volume flow rate is

$$
\begin{aligned}
\frac{dV}{dt} &= A_2 v_2 = \pi(0.50 \times 10^{-2} \text{ m})^2(6.0 \text{ m/s}) \\
&= 4.7 \times 10^{-4} \text{ m}^3/\text{s} = 0.47 \text{ L/s}
\end{aligned}
$$

EVALUATE: This is a reasonable flow rate for a bathroom faucet or shower. Note that if the water is turned off, v_1 and v_2 are both zero, the term $\tfrac{1}{2}\rho(v_2{}^2 - v_1{}^2)$ in Bernoulli's equation vanishes, and p_2 rises from 3.3×10^5 Pa to 3.5×10^5 Pa.

Example 12.8 | Speed of efflux

Figure 12.24 shows a gasoline storage tank with cross-sectional area A_1, filled to a depth h. The space above the gasoline contains air at pressure p_0, and the gasoline flows out the bottom of the tank through a short pipe with cross-sectional area A_2. Derive expressions for the flow speed in the pipe and the volume flow rate.

12.24 Calculating the speed of efflux for gasoline flowing out the bottom of a storage tank.

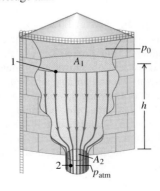

SOLUTION

IDENTIFY and SET UP: We consider the entire volume of moving liquid as a single flow tube of an incompressible fluid with negligible internal friction. Hence, we can use Bernoulli's equation. Points 1 and 2 are at the surface of the gasoline and at the exit pipe, respectively. At point 1 the pressure is p_0, which we assume to be fixed; at point 2 it is atmospheric pressure p_{atm}. We take $y = 0$ at the exit pipe, so $y_1 = h$ and $y_2 = 0$. Because A_1 is very much larger than A_2, the upper surface of the gasoline will drop very slowly and we can regard v_1 as essentially equal to zero. We find v_2 from Eq. (12.17) and the volume flow rate from Eq. (12.11).

EXECUTE: We apply Bernoulli's equation to points 1 and 2:

$$p_0 + \tfrac{1}{2}\rho v_1{}^2 + \rho g h = p_{atm} + \tfrac{1}{2}\rho v_2{}^2 + \rho g(0)$$

$$v_2{}^2 = v_1{}^2 + 2\left(\frac{p_0 - p_{atm}}{\rho}\right) + 2gh$$

Continued

Using $v_1 = 0$, we find

$$v_2 = \sqrt{2\left(\frac{p_0 - p_{atm}}{\rho}\right) + 2gh}$$

From Eq. (12.11), the volume flow rate is $dV/dt = v_2 A_2$.

EVALUATE: The speed v_2, sometimes called the *speed of efflux,* depends on both the pressure difference $(p_0 - p_{atm})$ and the height h of the liquid level in the tank. If the top of the tank is vented to the atmosphere, $p_0 = p_{atm}$ and $p_0 - p_{atm} = 0$. Then

$$v_2 = \sqrt{2gh}$$

That is, the speed of efflux from an opening at a distance h below the top surface of the liquid is the *same* as the speed a body would acquire in falling freely through a height h. This result is called *Torricelli's theorem.* It is valid not only for an opening in the bottom of a container, but also for a hole in a side wall at a depth h below the surface. In this case the volume flow rate is

$$\frac{dV}{dt} = A_2\sqrt{2gh}$$

Example 12.9 The Venturi meter

Figure 12.25 shows a *Venturi meter,* used to measure flow speed in a pipe. Derive an expression for the flow speed v_1 in terms of the cross-sectional areas A_1 and A_2 and the difference in height h of the liquid levels in the two vertical tubes.

SOLUTION

IDENTIFY and SET UP: The flow is steady, and we assume the fluid is incompressible and has negligible internal friction. Hence we can use Bernoulli's equation. We apply that equation to the wide part (point 1) and narrow part (point 2, the *throat*) of the pipe. Equation (12.6) relates h to the pressure difference $p_1 - p_2$.

EXECUTE: Points 1 and 2 have the same vertical coordinate $y_1 = y_2$, so Eq. (12.17) says

$$p_1 + \tfrac{1}{2}\rho v_1^2 = p_2 + \tfrac{1}{2}\rho v_2^2$$

From the continuity equation, $v_2 = (A_1/A_2)v_1$. Substituting this and rearranging, we get

$$p_1 - p_2 = \tfrac{1}{2}\rho v_1^2\left[\left(\frac{A_1}{A_2}\right)^2 - 1\right]$$

12.25 The Venturi meter.

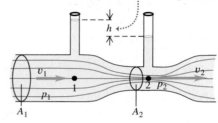

Difference in height results from reduced pressure in throat (point 2).

From Eq. (12.6), the pressure difference $p_1 - p_2$ is also equal to ρgh. Substituting this and solving for v_1, we get

$$v_1 = \sqrt{\frac{2gh}{(A_1/A_2)^2 - 1}}$$

EVALUATE: Because A_1 is greater than A_2, v_2 is greater than v_1 and the pressure p_2 in the throat is *less* than p_1. Those pressure differences produce a net force to the right that makes the fluid speed up as it enters the throat, and a net force to the left that slows it as it leaves.

Conceptual Example 12.10 Lift on an airplane wing

Figure 12.26a shows flow lines around a cross section of an airplane wing. The flow lines crowd together above the wing, corresponding to increased flow speed and reduced pressure, just as in the Venturi throat in Example 12.9. Hence the downward force of the air on the top side of the wing is less than the upward force of the air on the underside of the wing, and there is a net upward force or *lift.* Lift is not simply due to the impulse of air striking the underside of the wing; in fact, the reduced pressure on the upper wing surface makes the greatest contribution to the lift. (This simplified discussion ignores the formation of vortices.)

We can also understand the lift force on the basis of momentum changes. The vector diagram in Fig. 12.26a shows that there is a net *downward* change in the vertical component of momentum of the air flowing past the wing, corresponding to the downward force the wing exerts on the air. The reaction force *on* the wing is *upward,* as we concluded above.

Similar flow patterns and lift forces are found in the vicinity of any humped object in a wind. A moderate wind makes an umbrella

"float"; a strong wind can turn it inside out. At high speed, lift can reduce traction on a car's tires; a "spoiler" at the car's tail, shaped like an upside-down wing, provides a compensating downward force.

CAUTION **A misconception about wings** Some discussions of lift claim that air travels faster over the top of a wing because "it has farther to travel." This claim assumes that air molecules that part company at the front of the wing, one traveling over the wing and one under it, must meet again at the wing's trailing edge. Not so! Figure 12.26b shows a computer simulation of parcels of air flowing around an airplane wing. Parcels that are adjacent at the front of the wing do *not* meet at the trailing edge; the flow over the top of the wing is much faster than if the parcels had to meet. In accordance with Bernoulli's equation, this faster speed means that there is even lower pressure above the wing (and hence greater lift) than the "farther-to-travel" claim would suggest. ▌

12.26 Flow around an airplane wing.

(a) Flow lines around an airplane wing

Flow lines are crowded together above the wing, so flow speed is higher there and pressure is lower.

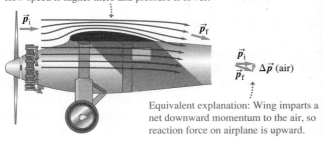

$\vec{p}_i$
$\Delta\vec{p}$ (air)
$\vec{p}_f$

Equivalent explanation: Wing imparts a net downward momentum to the air, so reaction force on airplane is upward.

(b) Computer simulation of air parcels flowing around a wing, showing that air moves much faster over the top than over the bottom.

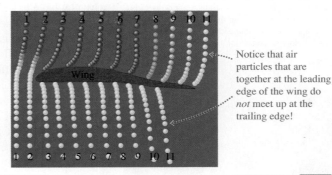

Notice that air particles that are together at the leading edge of the wing do *not* meet up at the trailing edge!

Test Your Understanding of Section 12.5 Which is the most accurate statement of Bernoulli's principle? (i) Fast-moving air causes lower pressure; (ii) lower pressure causes fast-moving air; (iii) both (i) and (ii) are equally accurate.

12.6 Viscosity and Turbulence

In our discussion of fluid flow we assumed that the fluid had no internal friction and that the flow was laminar. While these assumptions are often quite valid, in many important physical situations the effects of viscosity (internal friction) and turbulence (nonlaminar flow) are extremely important. Let's take a brief look at some of these situations.

Viscosity

Viscosity is internal friction in a fluid. Viscous forces oppose the motion of one portion of a fluid relative to another. Viscosity is the reason it takes effort to paddle a canoe through calm water, but it is also the reason the paddle works. Viscous effects are important in the flow of fluids in pipes, the flow of blood, the lubrication of engine parts, and many other situations.

Fluids that flow readily, such as water or gasoline, have smaller viscosities than do "thick" liquids such as honey or motor oil. Viscosities of all fluids are strongly temperature dependent, increasing for gases and decreasing for liquids as the temperature increases (Fig. 12.27). Oils for engine lubrication must flow equally well in cold and warm conditions, and so are designed to have as *little* temperature variation of viscosity as possible.

A viscous fluid always tends to cling to a solid surface in contact with it. There is always a thin *boundary layer* of fluid near the surface, in which the fluid is nearly at rest with respect to the surface. That's why dust particles can cling to a fan blade even when it is rotating rapidly, and why you can't get all the dirt off your car by just squirting a hose at it.

Viscosity has important effects on the flow of liquids through pipes, including the flow of blood in the circulatory system. First think about a fluid with zero viscosity so that we can apply Bernoulli's equation, Eq. (12.17). If the two ends of a long cylindrical pipe are at the same height ($y_1 = y_2$) and the flow speed is the same at both ends (so $v_1 = v_2$), Bernoulli's equation tells us that the pressure is the same at both ends of the pipe. But this result simply isn't true if we take viscosity into account. To see why, consider Fig. 12.28, which shows the flow-speed profile for laminar flow of a viscous fluid in a long cylindrical pipe. Due to viscosity, the speed is *zero* at the pipe walls (to which the fluid clings) and is greatest at the center of the pipe. The motion is like a lot of concentric tubes sliding relative to

12.27 Lava is an example of a viscous fluid. The viscosity decreases with increasing temperature: The hotter the lava, the more easily it can flow.

12.28 Velocity profile for a viscous fluid in a cylindrical pipe.

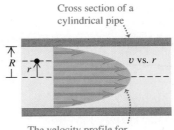

Cross section of a cylindrical pipe

R
r
v vs. r

The velocity profile for viscous fluid flowing in the pipe has a parabolic shape.

one another, with the central tube moving fastest and the outermost tube at rest. Viscous forces between the tubes oppose this sliding, so to keep the flow going we must apply a greater pressure at the back of the flow than at the front. That's why you have to keep squeezing a tube of toothpaste or a packet of ketchup (both viscous fluids) to keep the fluid coming out of its container. Your fingers provide a pressure at the back of the flow that is far greater than the atmospheric pressure at the front of the flow.

The pressure difference required to sustain a given volume flow rate through a cylindrical pipe of length L and radius R turns out to be proportional to L/R^4. If we decrease R by one-half, the required pressure increases by $2^4 = 16$; decreasing R by a factor of 0.90 (a 10% reduction) increases the required pressure difference by a factor of $(1/0.90)^4 = 1.52$ (a 52% increase). This simple relationship explains the connection between a high-cholesterol diet (which tends to narrow the arteries) and high blood pressure. Due to the R^4 dependence, even a small narrowing of the arteries can result in substantially elevated blood pressure and added strain on the heart muscle.

Turbulence

When the speed of a flowing fluid exceeds a certain critical value, the flow is no longer laminar. Instead, the flow pattern becomes extremely irregular and complex, and it changes continuously with time; there is no steady-state pattern. This irregular, chaotic flow is called **turbulence.** Figure 12.20 shows the contrast between laminar and turbulent flow for smoke rising in air. Bernoulli's equation is *not* applicable to regions where there is turbulence because the flow is not steady.

Whether a flow is laminar or turbulent depends in part on the fluid's viscosity. The greater the viscosity, the greater the tendency for the fluid to flow in sheets or lamina and the more likely the flow is to be laminar. (When we discussed Bernoulli's equation in Section 12.5, we assumed that the flow was laminar and that the fluid had zero viscosity. In fact, a *little* viscosity is needed to ensure that the flow is laminar.)

For a fluid of a given viscosity, flow speed is a determining factor for the onset of turbulence. A flow pattern that is stable at low speeds suddenly becomes unstable when a critical speed is reached. Irregularities in the flow pattern can be caused by roughness in the pipe wall, variations in the density of the fluid, and many other factors. At low flow speeds, these disturbances damp out; the flow pattern is *stable* and tends to maintain its laminar nature (Fig. 12.29a). When the critical speed is reached, however, the flow pattern becomes unstable. The disturbances no longer damp out but grow until they destroy the entire laminar-flow pattern (Fig. 12.29b).

12.29 The flow of water from a faucet is **(a)** laminar at low speeds but **(b)** turbulent at sufficiently high speeds.

(a)   (b)

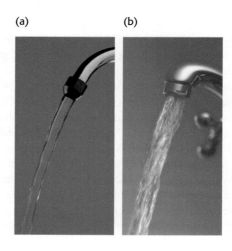

Conceptual Example 12.11 The curve ball

Does a curve ball *really* curve? Yes, it certainly does, and the reason is turbulence. Figure 12.30a shows a nonspinning ball moving through the air from left to right. The flow lines show that to an observer moving with the ball, the air stream appears to move from right to left. Because of the high speeds that are ordinarily involved (near 35 m/s, or 75 mi/h), there is a region of *turbulent* flow behind the ball.

Figure 12.30b shows a *spinning* ball with "top spin." Layers of air near the ball's surface are pulled around in the direction of the spin by friction between the ball and air and by the air's internal friction (viscosity). Hence air moves relative to the ball's surface more slowly at the top of the ball than at the bottom, and turbulence occurs farther forward on the top side than on the bottom. This asymmetry causes a pressure difference; the average pressure at the top of the ball is now greater than that at the bottom. As Fig. 12.30c shows, the resulting net force deflects the ball downward. "Top spin" is used in tennis to keep a fast serve in the court (Fig. 12.30d).

In baseball, a curve ball spins about a nearly *vertical* axis and the resulting deflection is sideways. In that case, Fig. 12.30c is a *top* view of the situation. A curve ball thrown by a left-handed pitcher spins as shown in Fig. 12.30e and will curve *toward* a right-handed batter, making it harder to hit.

A similar effect occurs with golf balls, which acquire "back spin" from impact with the grooved, slanted club face. Figure 12.30f shows the backspin of a golf ball just after impact. The resulting pressure difference between the top and bottom of the ball causes a *lift* force that keeps the ball in the air longer than would be possible without spin. A well-hit drive appears, from the tee, to "float" or even curve *upward* during the initial portion of its flight. This is a real effect, not an illusion. The dimples on the golf ball play an essential role; the viscosity of air gives a dimpled ball a much longer trajectory than an undimpled one with the same initial velocity and spin.

12.30 (a)–(e) Analyzing the motion of a spinning ball through the air. (f) Stroboscopic photograph of a golf ball being struck by a club. The picture was taken at 1000 flashes per second. The ball rotates about once in eight pictures, corresponding to an angular speed of 125 rev/s, or 7500 rpm.

(a) Motion of air relative to a nonspinning ball

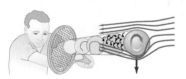

(b) Motion of a spinning ball

This side of the ball moves opposite to the airflow.

This side moves in the direction of the airflow.

(c) Force generated when a spinning ball moves through air

A moving ball drags the adjacent air with it. So, when air moves past a spinning ball:

On one side, the ball **slows the air**, creating a region of **high pressure**.

On the other side, the ball **speeds the air**, creating a region of **low pressure**.

The resultant force points in the direction of the low-pressure side.

(d) Spin pushing a tennis ball downward

(e) Spin causing a curve ball to be deflected sideways

(f) Backspin of a golf ball

Test Your Understanding of Section 12.6 How much more thumb pressure must a nurse use to administer an injection with a hypodermic needle of inside diameter 0.30 mm compared to one with inside diameter 0.60 mm? Assume that the two needles have the same length and that the volume flow rate is the same in both cases. (i) twice as much; (ii) 4 times as much; (iii) 8 times as much; (iv) 16 times as much; (v) 32 times as much.

Density and pressure: Density is mass per unit volume. If a mass m of homogeneous material has volume V, its density ρ is the ratio m/V. Specific gravity is the ratio of the density of a material to the density of water. (See Example 12.1.)

Pressure is normal force per unit area. Pascal's law states that pressure applied to an enclosed fluid is transmitted undiminished to every portion of the fluid. Absolute pressure is the total pressure in a fluid; gauge pressure is the difference between absolute pressure and atmospheric pressure. The SI unit of pressure is the pascal (Pa): $1 \text{ Pa} = 1 \text{ N/m}^2$. (See Example 12.2.)

$$\rho = \frac{m}{V} \qquad (12.1)$$

$$p = \frac{dF_\perp}{dA} \qquad (12.2)$$

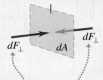

Small area dA within fluid at rest

Equal normal forces exerted on both sides by surrounding fluid

---

Pressures in a fluid at rest: The pressure difference between points 1 and 2 in a static fluid of uniform density ρ (an incompressible fluid) is proportional to the difference between the elevations y_1 and y_2. If the pressure at the surface of an incompressible liquid at rest is p_0, then the pressure at a depth h is greater by an amount $\rho g h$. (See Examples 12.3 and 12.4.)

$$p_2 - p_1 = -\rho g(y_2 - y_1)$$
(pressure in a fluid of uniform density) $\quad (12.5)$

$$p = p_0 + \rho g h$$
(pressure in a fluid of uniform density) $\quad (12.6)$

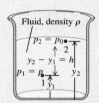

Buoyancy: Archimedes's principle states that when a body is immersed in a fluid, the fluid exerts an upward buoyant force on the body equal to the weight of the fluid that the body displaces. (See Example 12.5.)

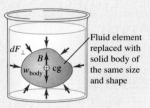

Fluid element replaced with solid body of the same size and shape

---

Fluid flow: An ideal fluid is incompressible and has no viscosity (no internal friction). A flow line is the path of a fluid particle; a streamline is a curve tangent at each point to the velocity vector at that point. A flow tube is a tube bounded at its sides by flow lines. In laminar flow, layers of fluid slide smoothly past each other. In turbulent flow, there is great disorder and a constantly changing flow pattern.

Conservation of mass in an incompressible fluid is expressed by the continuity equation, which relates the flow speeds v_1 and v_2 for two cross sections A_1 and A_2 in a flow tube. The product Av equals the volume flow rate, dV/dt, the rate at which volume crosses a section of the tube. (See Example 12.6.)

Bernoulli's equation relates the pressure p, flow speed v, and elevation y for any two points, assuming steady flow in an ideal fluid. (See Examples 12.7–12.10.)

$$A_1 v_1 = A_2 v_2$$
(continuity equation, incompressible fluid) $\quad (12.10)$

$$\frac{dV}{dt} = Av$$
(volume flow rate) $\quad (12.11)$

$$p_1 + \rho g y_1 + \tfrac{1}{2}\rho v_1^2 = p_2 + \rho g y_2 + \tfrac{1}{2}\rho v_2^2$$
(Bernoulli's equation) $\quad (12.17)$

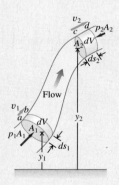

BRIDGING PROBLEM How Long to Drain?

A large cylindrical tank with diameter D is open to the air at the top. The tank contains water to a height H. A small circular hole with diameter d, where d is very much less than D, is then opened at the bottom of the tank. Ignore any effects of viscosity. (a) Find y, the height of water in the tank a time t after the hole is opened, as a function of t. (b) How long does it take to drain the tank completely? (c) If you double the initial height of water in the tank, by what factor does the time to drain the tank increase?

SOLUTION GUIDE

See MasteringPhysics® study area for a Video Tutor solution.

IDENTIFY and SET UP
1. Draw a sketch of the situation that shows all of the relevant dimensions.
2. Make a list of the unknown quantities, and decide which of these are the target variables.

3. What is the speed at which water flows out of the bottom of the tank? How is this related to the volume flow rate of water out of the tank? How is the volume flow rate related to the rate of change of y?

EXECUTE
4. Use your results from step 3 to write an equation for dy/dt.
5. Your result from step 4 is a relatively simple differential equation. With your knowledge of calculus, you can integrate it to find y as a function of t. (*Hint:* Once you've done the integration, you'll still have to do a little algebra.)
6. Use your result from step 5 to find the time when the tank is empty. How does your result depend on the initial height H?

EVALUATE
7. Check whether your answers are reasonable. A good check is to draw a graph of y versus t. According to your graph, what is the algebraic sign of dy/dt at different times? Does this make sense?

Problems

For instructor-assigned homework, go to www.masteringphysics.com

•, ••, •••: Problems of increasing difficulty. **CP**: Cumulative problems incorporating material from earlier chapters. **CALC**: Problems requiring calculus. **BIO**: Biosciences problems.

DISCUSSION QUESTIONS

Q12.1 A cube of oak wood with very smooth faces normally floats in water. Suppose you submerge it completely and press one face flat against the bottom of a tank so that no water is under that face. Will the block float to the surface? Is there a buoyant force on it? Explain.

Q12.2 A rubber hose is attached to a funnel, and the free end is bent around to point upward. When water is poured into the funnel, it rises in the hose to the same level as in the funnel, even though the funnel has a lot more water in it than the hose does. Why? What supports the extra weight of the water in the funnel?

Q12.3 Comparing Example 12.1 (Section 12.1) and Example 12.2 (Section 12.2), it seems that 700 N of air is exerting a downward force of 2.0×10^6 N on the floor. How is this possible?

Q12.4 Equation (12.7) shows that an area ratio of 100 to 1 can give 100 times more output force than input force. Doesn't this violate conservation of energy? Explain.

Q12.5 You have probably noticed that the lower the tire pressure, the larger the contact area between the tire and the road. Why?

Q12.6 In hot-air ballooning, a large balloon is filled with air heated by a gas burner at the bottom. Why must the air be heated? How does the balloonist control ascent and descent?

Q12.7 In describing the size of a large ship, one uses such expressions as "it displaces 20,000 tons." What does this mean? Can the weight of the ship be obtained from this information?

Q12.8 You drop a solid sphere of aluminum in a bucket of water that sits on the ground. The buoyant force equals the weight of water displaced; this is less than the weight of the sphere, so the sphere sinks to the bottom. If you take the bucket with you on an elevator that accelerates upward, the apparent weight of the water increases and the buoyant force on the sphere increases. Could the

acceleration of the elevator be great enough to make the sphere pop up out of the water? Explain.

Q12.9 A rigid, lighter-than-air dirigible filled with helium cannot continue to rise indefinitely. Why? What determines the maximum height it can attain?

Q12.10 Air pressure decreases with increasing altitude. So why is air near the surface not continuously drawn upward toward the lower-pressure regions above?

Q12.11 The purity of gold can be tested by weighing it in air and in water. How? Do you think you could get away with making a fake gold brick by gold-plating some cheaper material?

Q12.12 During the Great Mississippi Flood of 1993, the levees in St. Louis tended to rupture first at the bottom. Why?

Q12.13 A cargo ship travels from the Atlantic Ocean (salt water) to Lake Ontario (freshwater) via the St. Lawrence River. The ship rides several centimeters lower in the water in Lake Ontario than it did in the ocean. Explain why.

Q12.14 You push a piece of wood under the surface of a swimming pool. After it is completely submerged, you keep pushing it deeper and deeper. As you do this, what will happen to the buoyant force on it? Will the force keep increasing, stay the same, or decrease? Why?

Q12.15 An old question is "Which weighs more, a pound of feathers or a pound of lead?" If the weight in pounds is the gravitational force, will a pound of feathers balance a pound of lead on opposite pans of an equal-arm balance? Explain, taking into account buoyant forces.

Q12.16 Suppose the door of a room makes an airtight but frictionless fit in its frame. Do you think you could open the door if the air pressure on one side were standard atmospheric pressure and the air pressure on the other side differed from standard by 1%? Explain.

Q12.17 At a certain depth in an incompressible liquid, the absolute pressure is p. At twice this depth, will the absolute pressure be equal to $2p$, greater than $2p$, or less than $2p$? Justify your answer.

Q12.18 A piece of iron is glued to the top of a block of wood. When the block is placed in a bucket of water with the iron on top, the block floats. The block is now turned over so that the iron is submerged beneath the wood. Does the block float or sink? Does the water level in the bucket rise, drop, or stay the same? Explain your answers.

Q12.19 You take an empty glass jar and push it into a tank of water with the open mouth of the jar downward, so that the air inside the jar is trapped and cannot get out. If you push the jar deeper into the water, does the buoyant force on the jar stay the same? If not, does it increase or decrease? Explain your answer.

Q12.20 You are floating in a canoe in the middle of a swimming pool. Your friend is at the edge of the pool, carefully noting the level of the water on the side of the pool. You have a bowling ball with you in the canoe. If you carefully drop the bowling ball over the side of the canoe and it sinks to the bottom of the pool, does the water level in the pool rise or fall?

Q12.21 You are floating in a canoe in the middle of a swimming pool. A large bird flies up and lights on your shoulder. Does the water level in the pool rise or fall?

Q12.22 At a certain depth in the incompressible ocean the gauge pressure is p_g. At three times this depth, will the gauge pressure be greater than $3p_g$, equal to $3p_g$, or less than $3p_g$? Justify your answer.

Q12.23 An ice cube floats in a glass of water. When the ice melts, will the water level in the glass rise, fall, or remain unchanged? Explain.

Q12.24 You are told, "Bernoulli's equation tells us that where there is higher fluid speed, there is lower fluid pressure, and vice versa." Is this statement always true, even for an idealized fluid? Explain.

Q12.25 If the velocity at each point in space in steady-state fluid flow is constant, how can a fluid particle accelerate?

Q12.26 In a store-window vacuum cleaner display, a table-tennis ball is suspended in midair in a jet of air blown from the outlet hose of a tank-type vacuum cleaner. The ball bounces around a little but always moves back toward the center of the jet, even if the jet is tilted from the vertical. How does this behavior illustrate Bernoulli's equation?

Q12.27 A tornado consists of a rapidly whirling air vortex. Why is the pressure always much lower in the center than at the outside? How does this condition account for the destructive power of a tornado?

Q12.28 Airports at high elevations have longer runways for take-offs and landings than do airports at sea level. One reason is that aircraft engines develop less power in the thin air well above sea level. What is another reason?

Q12.29 When a smooth-flowing stream of water comes out of a faucet, it narrows as it falls. Explain why this happens.

Q12.30 Identical-size lead and aluminum cubes are suspended at different depths by two wires in a large vat of water (Fig. Q12.30). (a) Which cube experiences a greater buoyant force? (b) For which cube is the tension in the wire greater? (c) Which cube experiences a

Figure **Q12.30**

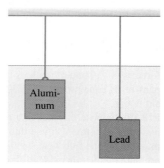

greater force on its lower face? (d) For which cube is the difference in pressure between the upper and lower faces greater?

EXERCISES

Section 12.1 Density

12.1 •• On a part-time job, you are asked to bring a cylindrical iron rod of length 85.8 cm and diameter 2.85 cm from a storage room to a machinist. Will you need a cart? (To answer, calculate the weight of the rod.)

12.2 •• A cube 5.0 cm on each side is made of a metal alloy. After you drill a cylindrical hole 2.0 cm in diameter all the way through and perpendicular to one face, you find that the cube weighs 7.50 N. (a) What is the density of this metal? (b) What did the cube weigh before you drilled the hole in it?

12.3 • You purchase a rectangular piece of metal that has dimensions $5.0 \times 15.0 \times 30.0$ mm and mass 0.0158 kg. The seller tells you that the metal is gold. To check this, you compute the average density of the piece. What value do you get? Were you cheated?

12.4 •• **Gold Brick.** You win the lottery and decide to impress your friends by exhibiting a million-dollar cube of gold. At the time, gold is selling for $426.60 per troy ounce, and 1.0000 troy ounce equals 31.1035 g. How tall would your million-dollar cube be?

12.5 •• A uniform lead sphere and a uniform aluminum sphere have the same mass. What is the ratio of the radius of the aluminum sphere to the radius of the lead sphere?

12.6 • (a) What is the average density of the sun? (b) What is the average density of a neutron star that has the same mass as the sun but a radius of only 20.0 km?

12.7 •• A hollow cylindrical copper pipe is 1.50 m long and has an outside diameter of 3.50 cm and an inside diameter of 2.50 cm. How much does it weigh?

Section 12.2 Pressure in a Fluid

12.8 •• **Black Smokers.** Black smokers are hot volcanic vents that emit smoke deep in the ocean floor. Many of them teem with exotic creatures, and some biologists think that life on earth may have begun around such vents. The vents range in depth from about 1500 m to 3200 m below the surface. What is the gauge pressure at a 3200-m deep vent, assuming that the density of water does not vary? Express your answer in pascals and atmospheres.

12.9 •• **Oceans on Mars.** Scientists have found evidence that Mars may once have had an ocean 0.500 km deep. The acceleration due to gravity on Mars is 3.71 m/s². (a) What would be the gauge pressure at the bottom of such an ocean, assuming it was freshwater? (b) To what depth would you need to go in the earth's ocean to experience the same gauge pressure?

12.10 •• **BIO** (a) Calculate the difference in blood pressure between the feet and top of the head for a person who is 1.65 m tall. (b) Consider a cylindrical segment of a blood vessel 2.00 cm long and 1.50 mm in diameter. What *additional* outward force would such a vessel need to withstand in the person's feet compared to a similar vessel in her head?

12.11 • **BIO** In intravenous feeding, a needle is inserted in a vein in the patient's arm and a tube leads from the needle to a reservoir of fluid (density 1050 kg/m³) located at height h above the arm. The top of the reservoir is open to the air. If the gauge pressure inside the vein is 5980 Pa, what is the minimum value of h that allows fluid to enter the vein? Assume the needle diameter is large enough that you can ignore the viscosity (see Section 12.6) of the fluid.

12.12 • A barrel contains a 0.120-m layer of oil floating on water that is 0.250 m deep. The density of the oil is 600 kg/m³. (a) What is the gauge pressure at the oil–water interface? (b) What is the gauge pressure at the bottom of the barrel?

12.13 • **BIO** **Standing on Your Head.** (a) What is the *difference* between the pressure of the blood in your brain when you stand on your head and the pressure when you stand on your feet? Assume that you are 1.85 m tall. The density of blood is 1060 kg/m³. (b) What effect does the increased pressure have on the blood vessels in your brain?

12.14 •• You are designing a diving bell to withstand the pressure of seawater at a depth of 250 m. (a) What is the gauge pressure at this depth? (You can ignore changes in the density of the water with depth.) (b) At this depth, what is the net force due to the water outside and the air inside the bell on a circular glass window 30.0 cm in diameter if the pressure inside the diving bell equals the pressure at the surface of the water? (You can ignore the small variation of pressure over the surface of the window.)

12.15 •• **BIO** **Ear Damage from Diving.** If the force on the tympanic membrane (eardrum) increases by about 1.5 N above the force from atmospheric pressure, the membrane can be damaged. When you go scuba diving in the ocean, below what depth could damage to your eardrum start to occur? The eardrum is typically 8.2 mm in diameter. (Consult Table 12.1.)

12.16 •• The liquid in the open-tube manometer in Fig. 12.8a is mercury, $y_1 = 3.00$ cm, and $y_2 = 7.00$ cm. Atmospheric pressure is 980 millibars. (a) What is the absolute pressure at the bottom of the U-shaped tube? (b) What is the absolute pressure in the open tube at a depth of 4.00 cm below the free surface? (c) What is the absolute pressure of the gas in the container? (d) What is the gauge pressure of the gas in pascals?

12.17 • **BIO** There is a maximum depth at which a diver can breathe through a snorkel tube (Fig. E12.17) because as the depth increases, so does the pressure difference, which tends to collapse the diver's lungs. Since the snorkel connects the air in the lungs to the atmosphere at the surface, the pressure inside the lungs is atmospheric pressure. What is the external–internal pressure difference when the diver's lungs are at a depth of 6.1 m (about 20 ft)? Assume that the diver is in freshwater. (A scuba diver breathing from compressed air tanks can operate at greater depths than can a snorkeler, since the pressure of the air inside the scuba diver's lungs increases to match the external pressure of the water.)

Figure **E12.17**

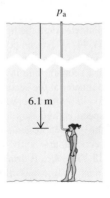

12.18 •• A tall cylinder with a cross-sectional area 12.0 cm² is partially filled with mercury; the surface of the mercury is 5.00 cm above the bottom of the cylinder. Water is slowly poured in on top of the mercury, and the two fluids don't mix. What volume of water must be added to double the gauge pressure at the bottom of the cylinder?

12.19 •• An electrical short cuts off all power to a submersible diving vehicle when it is 30 m below the surface of the ocean. The crew must push out a hatch of area 0.75 m² and weight 300 N on the bottom to escape. If the pressure inside is 1.0 atm, what downward force must the crew exert on the hatch to open it?

12.20 •• A closed container is partially filled with water. Initially, the air above the water is at atmospheric pressure $(1.01 \times 10^5$ Pa)

and the gauge pressure at the bottom of the water is 2500 Pa. Then additional air is pumped in, increasing the pressure of the air above the water by 1500 Pa. (a) What is the gauge pressure at the bottom of the water? (b) By how much must the water level in the container be reduced, by drawing some water out through a valve at the bottom of the container, to return the gauge pressure at the bottom of the water to its original value of 2500 Pa? The pressure of the air above the water is maintained at 1500 Pa above atmospheric pressure.

12.21 •• A cylindrical disk of wood weighing 45.0 N and having a diameter of 30.0 cm floats on a cylinder of oil of density 0.850 g/cm³ (Fig. E12.21). The cylinder of oil is 75.0 cm deep and has a diameter the same as that of the wood. (a) What is the gauge pressure at the top of the oil column? (b) Suppose now that someone puts a weight of 83.0 N on top of the wood, but no oil seeps around the edge of the wood. What is the *change* in pressure at (i) the bottom of the oil and (ii) halfway down in the oil?

Figure **E12.21**

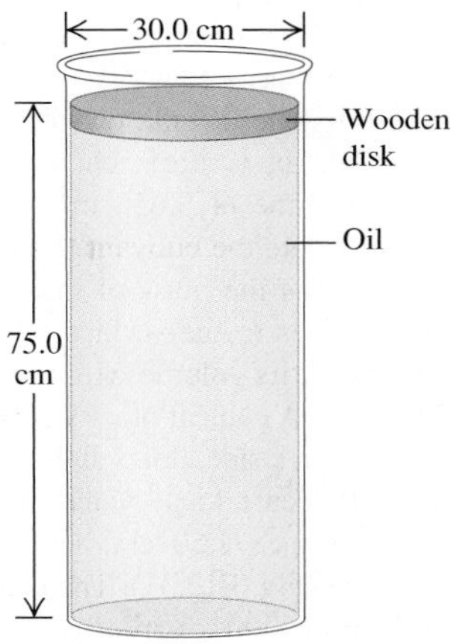

12.22 •• **Exploring   Venus.**
The surface pressure on Venus is 92 atm, and the acceleration due to gravity there is 0.894g. In a future exploratory mission, an upright cylindrical tank of benzene is sealed at the top but still pressurized at 92 atm just above the benzene. The tank has a diameter of 1.72 m, and the benzene column is 11.50 m tall. Ignore any effects due to the very high temperature on Venus. (a) What total force is exerted on the inside surface of the bottom of the tank? (b) What force does the Venusian atmosphere exert on the outside surface of the bottom of the tank? (c) What total inward force does the atmosphere exert on the vertical walls of the tank?

12.23 •• **Hydraulic Lift I.** For the hydraulic lift shown in Fig. 12.7, what must be the ratio of the diameter of the vessel at the car to the diameter of the vessel where the force F_1 is applied so that a 1520-kg car can be lifted with a force F_1 of just 125 N?

12.24 • **Hydraulic Lift II.** The piston of a hydraulic automobile lift is 0.30 m in diameter. What gauge pressure, in pascals, is required to lift a car with a mass of 1200 kg? Also express this pressure in atmospheres.

Section 12.3 Buoyancy

12.25 • A 950-kg cylindrical can buoy floats vertically in salt water. The diameter of the buoy is 0.900 m. Calculate the additional distance the buoy will sink when a 70.0-kg man stands on top of it.

12.26 •• A slab of ice floats on a freshwater lake. What minimum volume must the slab have for a 45.0-kg woman to be able to stand on it without getting her feet wet?

12.27 •• An ore sample weighs 17.50 N in air. When the sample is suspended by a light cord and totally immersed in water, the tension in the cord is 11.20 N. Find the total volume and the density of the sample.

12.28 •• You are preparing some apparatus for a visit to a newly discovered planet Caasi having oceans of glycerine and a surface acceleration due to gravity of 4.15 m/s². If your apparatus floats in the oceans on earth with 25.0% of its volume submerged, what percentage will be submerged in the glycerine oceans of Caasi?

12.29 •• An object of average density ρ floats at the surface of a fluid of density ρ_{fluid}. (a) How must the two densities be related? (b) In view of the answer to part (a), how can steel ships float in water? (c) In terms of ρ and ρ_{fluid}, what fraction of the object is submerged and what fraction is above the fluid? Check that your answers give the correct limiting behavior as $\rho \rightarrow \rho_{\text{fluid}}$ and as $\rho \rightarrow 0$. (d) While on board your yacht, your cousin Throckmorton cuts a rectangular piece (dimensions $5.0 \times 4.0 \times 3.0$ cm) out of a life preserver and throws it into the ocean. The piece has a mass of 42 g. As it floats in the ocean, what percentage of its volume is above the surface?

12.30 • A hollow plastic sphere is held below the surface of a freshwater lake by a cord anchored to the bottom of the lake. The sphere has a volume of 0.650 m^3 and the tension in the cord is 900 N. (a) Calculate the buoyant force exerted by the water on the sphere. (b) What is the mass of the sphere? (c) The cord breaks and the sphere rises to the surface. When the sphere comes to rest, what fraction of its volume will be submerged?

12.31 •• A cubical block of wood, 10.0 cm on a side, floats at the interface between oil and water with its lower surface 1.50 cm below the interface (Fig. E12.31). The density of the oil is 790 kg/m^3. (a) What is the gauge pressure at the upper face of the block? (b) What is the gauge pressure at the lower face of the block? (c) What are the mass and density of the block?

Figure **E12.31**

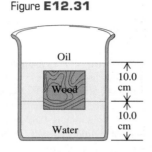

12.32 • A solid aluminum ingot weighs 89 N in air. (a) What is its volume? (b) The ingot is suspended from a rope and totally immersed in water. What is the tension in the rope (the *apparent* weight of the ingot in water)?

12.33 •• A rock is suspended by a light string. When the rock is in air, the tension in the string is 39.2 N. When the rock is totally immersed in water, the tension is 28.4 N. When the rock is totally immersed in an unknown liquid, the tension is 18.6 N. What is the density of the unknown liquid?

Section 12.4 Fluid Flow

12.34 •• Water runs into a fountain, filling all the pipes, at a steady rate of 0.750 m^3/s. (a) How fast will it shoot out of a hole 4.50 cm in diameter? (b) At what speed will it shoot out if the diameter of the hole is three times as large?

12.35 •• A shower head has 20 circular openings, each with radius 1.0 mm. The shower head is connected to a pipe with radius 0.80 cm. If the speed of water in the pipe is 3.0 m/s, what is its speed as it exits the shower-head openings?

12.36 • Water is flowing in a pipe with a varying cross-sectional area, and at all points the water completely fills the pipe. At point 1 the cross-sectional area of the pipe is 0.070 m^2, and the magnitude of the fluid velocity is 3.50 m/s. (a) What is the fluid speed at points in the pipe where the cross-sectional area is (a) 0.105 m^2 and (b) 0.047 m^2? (c) Calculate the volume of water discharged from the open end of the pipe in 1.00 hour.

12.37 • Water is flowing in a pipe with a circular cross section but with varying cross-sectional area, and at all points the water completely fills the pipe. (a) At one point in the pipe the radius is 0.150 m. What is the speed of the water at this point if water is flowing into this pipe at a steady rate of 1.20 m^3/s? (b) At a second point in the

pipe the water speed is 3.80 m/s. What is the radius of the pipe at this point?

12.38 • **Home Repair.** You need to extend a 2.50-inch-diameter pipe, but you have only a 1.00-inch-diameter pipe on hand. You make a fitting to connect these pipes end to end. If the water is flowing at 6.00 cm/s in the wide pipe, how fast will it be flowing through the narrow one?

12.39 • At a point where an irrigation canal having a rectangular cross section is 18.5 m wide and 3.75 m deep, the water flows at 2.50 cm/s. At a point downstream, but on the same level, the canal is 16.5 m wide, but the water flows at 11.0 cm/s. How deep is the canal at this point?

12.40 •• **BIO Artery Blockage.** A medical technician is trying to determine what percentage of a patient's artery is blocked by plaque. To do this, she measures the blood pressure just before the region of blockage and finds that it is 1.20×10^4 Pa, while in the region of blockage it is 1.15×10^4 Pa. Furthermore, she knows that blood flowing through the normal artery just before the point of blockage is traveling at 30.0 cm/s, and the specific gravity of this patient's blood is 1.06. What percentage of the cross-sectional area of the patient's artery is blocked by the plaque?

Section 12.5 Bernoulli's Equation

12.41 •• A sealed tank containing seawater to a height of 11.0 m also contains air above the water at a gauge pressure of 3.00 atm. Water flows out from the bottom through a small hole. How fast is this water moving?

12.42 • A small circular hole 6.00 mm in diameter is cut in the side of a large water tank, 14.0 m below the water level in the tank. The top of the tank is open to the air. Find (a) the speed of efflux of the water and (b) the volume discharged per second.

12.43 • What gauge pressure is required in the city water mains for a stream from a fire hose connected to the mains to reach a vertical height of 15.0 m? (Assume that the mains have a much larger diameter than the fire hose.)

12.44 •• At one point in a pipeline the water's speed is 3.00 m/s and the gauge pressure is 5.00×10^4 Pa. Find the gauge pressure at a second point in the line, 11.0 m lower than the first, if the pipe diameter at the second point is twice that at the first.

12.45 • At a certain point in a horizontal pipeline, the water's speed is 2.50 m/s and the gauge pressure is 1.80×10^4 Pa. Find the gauge pressure at a second point in the line if the cross-sectional area at the second point is twice that at the first.

12.46 • A soft drink (mostly water) flows in a pipe at a beverage plant with a mass flow rate that would fill 220 0.355-L cans per minute. At point 2 in the pipe, the gauge pressure is 152 kPa and the cross-sectional area is 8.00 cm^2. At point 1, 1.35 m above point 2, the cross-sectional area is 2.00 cm^2. Find the (a) mass flow rate; (b) volume flow rate; (c) flow speeds at points 1 and 2; (d) gauge pressure at point 1.

12.47 •• A golf course sprinkler system discharges water from a horizontal pipe at the rate of 7200 cm^3/s. At one point in the pipe, where the radius is 4.00 cm, the water's absolute pressure is 2.40×10^5 Pa. At a second point in the pipe, the water passes through a constriction where the radius is 2.00 cm. What is the water's absolute pressure as it flows through this constriction?

Section 12.6 Viscosity and Turbulence

12.48 • A pressure difference of 6.00×10^4 Pa is required to maintain a volume flow rate of 0.800 m^3/s for a viscous fluid flowing through a section of cylindrical pipe that has radius 0.210 m.

What pressure difference is required to maintain the same volume flow rate if the radius of the pipe is decreased to 0.0700 m?

12.49 •• **BIO Clogged Artery.** Viscous blood is flowing through an artery partially clogged by cholesterol. A surgeon wants to remove enough of the cholesterol to double the flow rate of blood through this artery. If the original diameter of the artery is D, what should be the new diameter (in terms of D) to accomplish this for the same pressure gradient?

PROBLEMS

12.50 •• **CP** The deepest point known in any of the earth's oceans is in the Marianas Trench, 10.92 km deep. (a) Assuming water is incompressible, what is the pressure at this depth? Use the density of seawater. (b) The actual pressure is 1.16×10^8 Pa; your calculated value will be less because the density actually varies with depth. Using the compressibility of water and the actual pressure, find the density of the water at the bottom of the Marianas Trench. What is the percent change in the density of the water?

12.51 ••• In a lecture demonstration, a professor pulls apart two hemispherical steel shells (diameter D) with ease using their attached handles. She then places them together, pumps out the air to an absolute pressure of p, and hands them to a bodybuilder in the back row to pull apart. (a) If atmospheric pressure is p_0, how much force must the bodybuilder exert on each shell? (b) Evaluate your answer for the case $p = 0.025$ atm, $D = 10.0$ cm.

12.52 •• **BIO Fish Navigation.** (a) As you can tell by watching them in an aquarium, fish are able to remain at any depth in water with no effort. What does this ability tell you about their density? (b) Fish are able to inflate themselves using a sac (called the *swim bladder*) located under their spinal column. These sacs can be filled with an oxygen–nitrogen mixture that comes from the blood. If a 2.75-kg fish in freshwater inflates itself and increases its volume by 10%, find the *net* force that the *water* exerts on it. (c) What is the net *external* force on it? Does the fish go up or down when it inflates itself?

12.53 ••• **CALC** A swimming pool is 5.0 m long, 4.0 m wide, and 3.0 m deep. Compute the force exerted by the water against (a) the bottom and (b) either end. (*Hint:* Calculate the force on a thin, horizontal strip at a depth h, and integrate this over the end of the pool.) Do not include the force due to air pressure.

12.54 ••• **CP CALC** The upper edge of a gate in a dam runs along the water surface. The gate is 2.00 m high and 4.00 m wide and is hinged along a horizontal line through its center (Fig. P12.54). Calculate the torque about the hinge arising from the force due to the water.

Figure **P12.54**

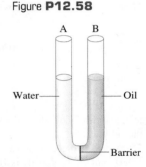

(*Hint:* Use a procedure similar to that used in Problem 12.53; calculate the torque on a thin, horizontal strip at a depth h and integrate this over the gate.)

12.55 ••• **CP CALC Force and Torque on a Dam.** A dam has the shape of a rectangular solid. The side facing the lake has area A and height H. The surface of the freshwater lake behind the dam is at the top of the dam. (a) Show that the net horizontal force exerted by the water on the dam equals $\frac{1}{2}\rho gHA$—that is, the average gauge pressure across the face of the dam times the area (see Problem 12.53). (b) Show that the torque exerted by the water about an axis along the bottom of the dam is $\rho gH^2A/6$. (c) How do the force and torque depend on the size of the lake?

12.56 •• **Ballooning on Mars.** It has been proposed that we could explore Mars using inflated balloons to hover just above the surface. The buoyancy of the atmosphere would keep the balloon aloft. The density of the Martian atmosphere is 0.0154 kg/m³ (although this varies with temperature). Suppose we construct these balloons of a thin but tough plastic having a density such that each square meter has a mass of 5.00 g. We inflate them with a very light gas whose mass we can neglect. (a) What should be the radius and mass of these balloons so they just hover above the surface of Mars? (b) If we released one of the balloons from part (a) on earth, where the atmospheric density is 1.20 kg/m³, what would be its initial acceleration assuming it was the same size as on Mars? Would it go up or down? (c) If on Mars these balloons have five times the radius found in part (a), how heavy an instrument package could they carry?

12.57 •• A 0.180-kg cube of ice (frozen water) is floating in glycerine. The gylcerine is in a tall cylinder that has inside radius 3.50 cm. The level of the glycerine is well below the top of the cylinder. If the ice completely melts, by what distance does the height of liquid in the cylinder change? Does the level of liquid rise or fall? That is, is the surface of the water above or below the original level of the gylcerine before the ice melted?

12.58 •• A narrow, U-shaped glass tube with open ends is filled with 25.0 cm of oil (of specific gravity 0.80) and 25.0 cm of water on opposite sides, with a barrier separating the liquids (Fig. P12.58). (a) Assume that the two liquids do not mix, and find the final heights of the columns of liquid in each side of the tube after the barrier is removed. (b) For the following cases, arrive at your answer by simple physical reasoning, not by calculations: (i) What would be the height on each side if the oil and water had equal densities? (ii) What would the heights be if the oil's density were much less than that of water?

Figure **P12.58**

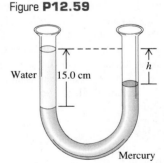

12.59 • A U-shaped tube open to the air at both ends contains some mercury. A quantity of water is carefully poured into the left arm of the U-shaped tube until the vertical height of the water column is 15.0 cm (Fig. P12.59). (a) What is the gauge pressure at the water–mercury interface? (b) Calculate the vertical distance h from the top of the mercury in the right-hand arm of the tube to the top of the water in the left-hand arm.

Figure **P12.59**

12.60 •• **CALC The Great Molasses Flood.** On the afternoon of January 15, 1919, an unusually warm day in Boston, a 17.7-m-high, 27.4-m-diameter cylindrical metal tank used for storing molasses ruptured. Molasses flooded into the streets in a 5-m-deep stream, killing pedestrians and horses and knocking down buildings. The molasses had a density of 1600 kg/m³. If the tank was full before the accident, what was the total outward force the molasses exerted on its sides? (*Hint:* Consider the outward force on a circular ring of the tank wall of width dy and at a depth y below the surface. Integrate to find the total outward force. Assume that before the tank ruptured, the pressure at the surface of the molasses was equal to the air pressure outside the tank.)

12.61 • An open barge has the dimensions shown in Fig. P12.61. If the barge is made out of 4.0-cm-thick steel plate on each of its four sides and its bottom, what mass of coal can the barge carry in freshwater without sinking? Is there enough room in the barge to hold this amount of coal? (The density of coal is about 1500 kg/m³.)

Figure **P12.61**

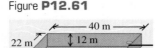

12.62 ••• A hot-air balloon has a volume of 2200 m³. The balloon fabric (the envelope) weighs 900 N. The basket with gear and full propane tanks weighs 1700 N. If the balloon can barely lift an additional 3200 N of passengers, breakfast, and champagne when the outside air density is 1.23 kg/m³, what is the average density of the heated gases in the envelope?

12.63 •• Advertisements for a certain small car claim that it floats in water. (a) If the car's mass is 900 kg and its interior volume is 3.0 m³, what fraction of the car is immersed when it floats? You can ignore the volume of steel and other materials. (b) Water gradually leaks in and displaces the air in the car. What fraction of the interior volume is filled with water when the car sinks?

12.64 • A single ice cube with mass 9.70 g floats in a glass completely full of 420 cm³ of water. You can ignore the water's surface tension and its variation in density with temperature (as long as it remains a liquid). (a) What volume of water does the ice cube displace? (b) When the ice cube has completely melted, has any water overflowed? If so, how much? If not, explain why this is so. (c) Suppose the water in the glass had been very salty water of density 1050 kg/m³. What volume of salt water would the 9.70-g ice cube displace? (d) Redo part (b) for the freshwater ice cube in the salty water.

12.65 ••• A piece of wood is 0.600 m long, 0.250 m wide, and 0.080 m thick. Its density is 700 kg/m³. What volume of lead must be fastened underneath it to sink the wood in calm water so that its top is just even with the water level? What is the mass of this volume of lead?

12.66 •• A hydrometer consists of a spherical bulb and a cylindrical stem with a cross-sectional area of 0.400 cm² (see Fig. 12.12a). The total volume of bulb and stem is 13.2 cm³. When immersed in water, the hydrometer floats with 8.00 cm of the stem above the water surface. When the hydrometer is immersed in an organic fluid, 3.20 cm of the stem is above the surface. Find the density of the organic fluid. (*Note:* This illustrates the precision of such a hydrometer. Relatively small density differences give rise to relatively large differences in hydrometer readings.)

12.67 •• The densities of air, helium, and hydrogen (at $p = 1.0$ atm and $T = 20°C$) are 1.20 kg/m³, 0.166 kg/m³, and 0.0899 kg/m³, respectively. (a) What is the volume in cubic meters displaced by a hydrogen-filled airship that has a total "lift" of 90.0 kN? (The "lift" is the amount by which the buoyant force exceeds the weight of the gas that fills the airship.) (b) What would be the "lift" if helium were used instead of hydrogen? In view of your answer, why is helium used in modern airships like advertising blimps?

12.68 •• When an open-faced boat has a mass of 5750 kg, including its cargo and passengers, it floats with the water just up to the top of its gunwales (sides) on a freshwater lake. (a) What is the volume of this boat? (b) The captain decides that it is too dangerous to float with his boat on the verge of sinking, so he decides to throw some cargo overboard so that 20% of the boat's volume will be above water. How much mass should he throw out?

12.69 •• CP An open cylindrical tank of acid rests at the edge of a table 1.4 m above the floor of the chemistry lab. If this tank springs a small hole in the side at its base, how far from the foot of the table will the acid hit the floor if the acid in the tank is 75 cm deep?

12.70 •• CP A firehose must be able to shoot water to the top of a building 28.0 m tall when aimed straight up. Water enters this hose at a steady rate of 0.500 m³/s and shoots out of a round nozzle. (a) What is the maximum diameter this nozzle can have? (b) If the only nozzle available has a diameter twice as great, what is the highest point the water can reach?

12.71 •• CP You drill a small hole in the side of a vertical cylindrical water tank that is standing on the ground with its top open to the air. (a) If the water level has a height H, at what height above the base should you drill the hole for the water to reach its greatest distance from the base of the cylinder when it hits the ground? (b) What is the greatest distance the water will reach?

12.72 ••• CALC A closed and elevated vertical cylindrical tank with diameter 2.00 m contains water to a depth of 0.800 m. A worker accidently pokes a circular hole with diameter 0.0200 m in the bottom of the tank. As the water drains from the tank, compressed air above the water in the tank maintains a gauge pressure of 5.00×10^3 Pa at the surface of the water. Ignore any effects of viscosity. (a) Just after the hole is made, what is the speed of the water as it emerges from the hole? What is the ratio of this speed to the efflux speed if the top of the tank is open to the air? (b) How much time does it take for all the water to drain from the tank? What is the ratio of this time to the time it takes for the tank to drain if the top of the tank is open to the air?

12.73 •• A block of balsa wood placed in one scale pan of an equal-arm balance is exactly balanced by a 0.115-kg brass mass in the other scale pan. Find the true mass of the balsa wood if its density is 150 kg/m³. Explain why it is accurate to ignore the buoyancy in air of the brass but *not* the buoyancy in air of the balsa wood.

12.74 •• Block A in Fig. P12.74 hangs by a cord from spring balance D and is submerged in a liquid C contained in beaker B. The mass of the beaker is 1.00 kg; the mass of the liquid is 1.80 kg. Balance D reads 3.50 kg, and balance E reads 7.50 kg. The volume of block A is 3.80×10^{-3} m³. (a) What is the density of the liquid? (b) What will each balance read if block A is pulled up out of the liquid?

Figure **P12.74**

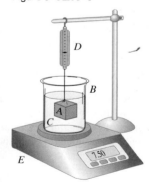

12.75 •• A hunk of aluminum is completely covered with a gold shell to form an ingot of weight 45.0 N. When you suspend the ingot from a spring balance and submerge the ingot in water, the balance reads 39.0 N. What is the weight of the gold in the shell?

12.76 •• A plastic ball has radius 12.0 cm and floats in water with 24.0% of its volume submerged. (a) What force must you apply to the ball to hold it at rest totally below the surface of the water? (b) If you let go of the ball, what is its acceleration the instant you release it?

12.77 •• The weight of a king's solid crown is w. When the crown is suspended by a light rope and completely immersed in water, the tension in the rope (the crown's apparent weight) is fw. (a) Prove that the crown's relative density (specific gravity) is $1/(1 - f)$. Discuss the meaning of the limits as f approaches 0 and 1. (b) If the crown is solid gold and weighs 12.9 N in air, what is its apparent weight when completely immersed in water? (c) Repeat part (b) if the crown is solid lead with a very thin gold plating, but still has a weight in air of 12.9 N.

12.78 •• A piece of steel has a weight w, an apparent weight (see Problem 12.77) w_{water} when completely immersed in water, and an apparent weight w_{fluid} when completely immersed in an unknown fluid. (a) Prove that the fluid's density relative to water (specific gravity) is $(w - w_{\text{fluid}})/(w - w_{\text{water}})$. (b) Is this result reasonable for the three cases of w_{fluid} greater than, equal to, or less than w_{water}? (c) The apparent weight of the piece of steel in water of density 1000 kg/m^3 is 87.2% of its weight. What percentage of its weight will its apparent weight be in formic acid (density 1220 kg/m^3)?

12.79 ••• You cast some metal of density ρ_m in a mold, but you are worried that there might be cavities within the casting. You measure the weight of the casting to be w, and the buoyant force when it is completely surrounded by water to be B. (a) Show that $V_0 = B/(\rho_{\text{water}}g) - w/(\rho_m g)$ is the total volume of any enclosed cavities. (b) If your metal is copper, the casting's weight is 156 N, and the buoyant force is 20 N, what is the total volume of any enclosed cavities in your casting? What fraction is this of the total volume of the casting?

12.80 • A cubical block of wood 0.100 m on a side and with a density of 550 kg/m^3 floats in a jar of water. Oil with a density of 750 kg/m^3 is poured on the water until the top of the oil layer is 0.035 m below the top of the block. (a) How deep is the oil layer? (b) What is the gauge pressure at the block's lower face?

12.81 •• **Dropping Anchor.** An iron anchor with mass 35.0 kg and density 7860 kg/m^3 lies on the deck of a small barge that has vertical sides and floats in a freshwater river. The area of the bottom of the barge is 8.00 m^2. The anchor is thrown overboard but is suspended above the bottom of the river by a rope; the mass and volume of the rope are small enough to ignore. After the anchor is overboard and the barge has finally stopped bobbing up and down, has the barge risen or sunk down in the water? By what vertical distance?

12.82 •• Assume that crude oil from a supertanker has density 750 kg/m^3. The tanker runs aground on a sandbar. To refloat the tanker, its oil cargo is pumped out into steel barrels, each of which has a mass of 15.0 kg when empty and holds 0.120 m^3 of oil. You can ignore the volume occupied by the steel from which the barrel is made. (a) If a salvage worker accidentally drops a filled, sealed barrel overboard, will it float or sink in the seawater? (b) If the barrel floats, what fraction of its volume will be above the water surface? If it sinks, what minimum tension would have to be exerted by a rope to haul the barrel up from the ocean floor? (c) Repeat parts (a) and (b) if the density of the oil is 910 kg/m^3 and the mass of each empty barrel is 32.0 kg.

12.83 ••• A cubical block of density ρ_B and with sides of length L floats in a liquid of greater density ρ_L. (a) What fraction of the block's volume is above the surface of the liquid? (b) The liquid is denser than water (density ρ_W) and does not mix with it. If water is poured on the surface of the liquid, how deep must the water layer be so that the water surface just rises to the top of the block? Express your answer in terms of L, ρ_B, ρ_L, and ρ_W. (c) Find the depth of the water layer in part (b) if the liquid is mercury, the block is made of iron, and the side length is 10.0 cm.

12.84 •• A barge is in a rectangular lock on a freshwater river. The lock is 60.0 m long and 20.0 m wide, and the steel doors on each end are closed. With the barge floating in the lock, a 2.50×10^6 N load of scrap metal is put onto the barge. The metal has density 9000 kg/m^3. (a) When the load of scrap metal, initially on the

bank, is placed onto the barge, what vertical distance does the water in the lock rise? (b) The scrap metal is now pushed overboard into the water. Does the water level in the lock rise, fall, or remain the same? If it rises or falls, by what vertical distance does it change?

12.85 • CP CALC A U-shaped tube with a horizontal portion of length l (Fig. P12.85) contains a liquid. What is the difference in height between the liquid columns in the vertical arms (a) if the tube has an acceleration a toward the right and (b) if the tube is mounted on a horizontal

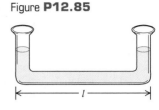

Figure **P12.85**

turntable rotating with an angular speed ω with one of the vertical arms on the axis of rotation? (c) Explain why the difference in height does not depend on the density of the liquid or on the cross-sectional area of the tube. Would it be the same if the vertical tubes did not have equal cross-sectional areas? Would it be the same if the horizontal portion were tapered from one end to the other? Explain.

12.86 • CP CALC A cylindrical container of an incompressible liquid with density ρ rotates with constant angular speed ω about its axis of symmetry, which we take to be the y-axis (Fig. P12.86). (a) Show that the pressure at a given height within the fluid increases in the radial direction (outward from the axis of rotation) according to $\partial p/\partial r = \rho\omega^2 r$. (b) Integrate this partial differential equation to find the pressure as a

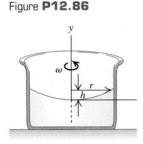

Figure **P12.86**

function of distance from the axis of rotation along a horizontal line at $y = 0$. (c) Combine the result of part (b) with Eq. (12.5) to show that the surface of the rotating liquid has a *parabolic* shape; that is, the height of the liquid is given by $h(r) = \omega^2 r^2/2g$. (This technique is used for making parabolic telescope mirrors; liquid glass is rotated and allowed to solidify while rotating.)

12.87 •• CP CALC An incompressible fluid with density ρ is in a horizontal test tube of inner cross-sectional area A. The test tube spins in a horizontal circle in an ultracentrifuge at an angular speed ω. Gravitational forces are negligible. Consider a volume element of the fluid of area A and thickness dr' a distance r' from the rotation axis. The pressure on its inner surface is p and on its outer surface is $p + dp$. (a) Apply Newton's second law to the volume element to show that $dp = \rho\omega^2 r' dr'$. (b) If the surface of the fluid is at a radius r_0 where the pressure is p_0, show that the pressure p at a distance $r \geq r_0$ is $p = p_0 + \rho\omega^2(r^2 - r_0^2)/2$. (c) An object of volume V and density ρ_{ob} has its center of mass at a distance R_{cmob} from the axis. Show that the net horizontal force on the object is $\rho V\omega^2 R_{\text{cm}}$, where R_{cm} is the distance from the axis to the center of mass of the displaced fluid. (d) Explain why the object will move inward if $\rho R_{\text{cm}} > \rho_{\text{ob}} R_{\text{cmob}}$ and outward if $\rho R_{\text{cm}} < \rho_{\text{ob}} R_{\text{cmob}}$. (e) For small objects of uniform density, $R_{\text{cm}} = R_{\text{cmob}}$. What happens to a mixture of small objects of this kind with different densities in an ultracentrifuge?

12.88 ••• CALC Untethered helium balloons, floating in a car that has all the windows rolled up and outside air vents closed, move in the direction of the car's acceleration, but loose balloons filled with air move in the opposite direction. To show why, consider only the horizontal forces acting on the balloons. Let a be the magnitude of the car's forward acceleration. Consider a horizontal tube of air with a cross-sectional area A that extends from the windshield, where $x = 0$ and $p = p_0$, back along the x-axis. Now consider a volume element of thickness dx in this tube. The pressure on its front surface is p and the pressure on its rear surface is $p + dp$. Assume the air has a constant density ρ. (a) Apply Newton's second law to the volume element to show that $dp = \rho a\, dx$. (b) Integrate the result of part (a) to find the pressure at the front surface in terms of a and x. (c) To show that considering ρ constant is reasonable, calculate the pressure difference in atm for a distance as long as 2.5 m and a large acceleration of 5.0 m/s². (d) Show that the net horizontal force on a balloon of volume V is ρVa. (e) For negligible friction forces, show that the acceleration of the balloon (average density ρ_{bal}) is $(\rho/\rho_{bal})a$, so that the acceleration relative to the car is $a_{rel} = [(\rho/\rho_{bal}) - 1]a$. (f) Use the expression for a_{rel} in part (e) to explain the movement of the balloons.

12.89 • **CP** Water stands at a depth H in a large, open tank whose side walls are vertical (Fig. P12.89). A hole is made in one of the walls at a depth h below the water surface. (a) At what distance R from the foot of the wall does the emerging stream strike the floor? (b) How far above the bottom of the tank could a second hole be cut so that the stream emerging from it could have the same range as for the first hole?

Figure **P12.89**

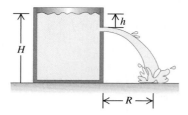

12.90 ••• A cylindrical bucket, open at the top, is 25.0 cm high and 10.0 cm in diameter. A circular hole with a cross-sectional area 1.50 cm² is cut in the center of the bottom of the bucket. Water flows into the bucket from a tube above it at the rate of 2.40×10^{-4} m³/s. How high will the water in the bucket rise?

12.91 • Water flows steadily from an open tank as in Fig. P12.91. The elevation of point 1 is 10.0 m, and the elevation of points 2 and 3 is 2.00 m. The cross-sectional area at point 2 is 0.0480 m²; at point 3 it is 0.0160 m². The area of the tank is very large compared with the cross-sectional area of the pipe. Assuming that Bernoulli's equation applies, compute (a) the discharge rate in cubic meters per second and (b) the gauge pressure at point 2.

Figure **P12.91**

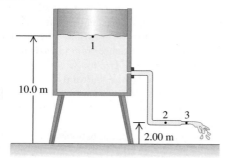

12.92 •• **CP** In 1993 the radius of Hurricane Emily was about 350 km. The wind speed near the center ("eye") of the hurricane, whose radius was about 30 km, reached about 200 km/h. As air swirled in from the rim of the hurricane toward the eye, its angular momentum remained roughly constant. (a) Estimate the wind speed at the rim of the hurricane. (b) Estimate the pressure difference at the earth's surface between the eye and the rim. (*Hint:* See Table 12.1.) Where is the pressure greater? (c) If the kinetic energy of the swirling air in the eye could be converted completely to gravitational potential energy, how high would the air go? (d) In fact, the air in the eye is lifted to heights of several kilometers. How can you reconcile this with your answer to part (c)?

12.93 •• Two very large open tanks A and F (Fig. P12.93) contain the same liquid. A horizontal pipe BCD, having a constriction at C and open to the air at D, leads out of the bottom of tank A, and a vertical pipe E opens into the constriction at C and dips into the liquid in tank F. Assume streamline flow and no viscosity. If the cross-sectional area at C is one-half the area at D and if D is a distance h_1 below the level of the liquid in A, to what height h_2 will liquid rise in pipe E? Express your answer in terms of h_1.

Figure **P12.93**

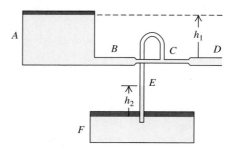

12.94 •• The horizontal pipe shown in Fig. P12.94 has a cross-sectional area of 40.0 cm² at the wider portions and 10.0 cm² at the constriction. Water is flowing in the pipe, and the discharge from the pipe is 6.00×10^{-3} m³/s (6.00 L/s). Find (a) the flow speeds at the wide and the narrow portions; (b) the pressure difference between these portions; (c) the difference in height between the mercury columns in the U-shaped tube.

Figure **P12.94**

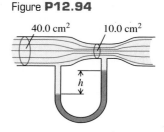

12.95 • A liquid flowing from a vertical pipe has a definite shape as it flows from the pipe. To get the equation for this shape, assume that the liquid is in free fall once it leaves the pipe. Just as it leaves the pipe, the liquid has speed v_0 and the radius of the stream of liquid is r_0. (a) Find an equation for the speed of the liquid as a function of the distance y it has fallen. Combining this with the equation of continuity, find an expression for the radius of the stream as a function of y. (b) If water flows out of a vertical pipe at a speed of 1.20 m/s, how far below the outlet will the radius be one-half the original radius of the stream?

Challenge Problems

12.96 ••• A rock with mass $m = 3.00$ kg is suspended from the roof of an elevator by a light cord. The rock is totally immersed in a bucket of water that sits on the floor of the elevator, but the rock doesn't touch the bottom or sides of the bucket. (a) When the elevator is at rest, the tension in the cord is 21.0 N. Calculate the volume of the rock. (b) Derive an expression for the tension in the cord when the elevator is accelerating *upward* with an acceleration of magnitude a. Calculate the tension when $a = 2.50$ m/s² upward. (c) Derive an expression for the tension in the cord when the elevator is accelerating *downward* with an acceleration of magnitude *a*. Calculate the tension when $a = 2.50$ m/s^2 downward. (d) What is the tension when the elevator is in free fall with a downward acceleration equal to g?

12.97 ••• **CALC** Suppose a piece of styrofoam, $\rho = 180$ kg/m^3, is held completely submerged in water (Fig. P12.97). (a) What is the tension in the cord? Find this using Archimedes's principle. (b) Use $p = p_0 + \rho g h$ to calculate directly the force exerted by the water on the two sloped sides and the bottom of the styrofoam; then show that the vector sum of these forces is the buoyant force.

Figure **P12.97**

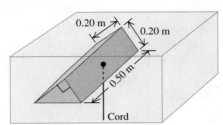

12.98 ••• A *siphon*, as shown in Fig. P12.98, is a convenient device for removing liquids from containers. To establish the flow, the tube must be initially filled with fluid. Let the fluid have density ρ, and let the atmospheric pressure be p_{atm}. Assume that the cross-sectional area of the tube is the same at all points along it. (a) If the lower end of the siphon is at a distance h below the surface of the liquid in the container, what is the speed of the fluid as it flows out the lower end of the siphon? (Assume that the container has a very large diameter, and ignore any effects of viscosity.) (b) A curious feature of a siphon is that the fluid initially flows "uphill." What is the greatest height H that the high point of the tube can have if flow is still to occur?

Figure **P12.98**

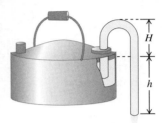

Answers

Chapter Opening Question ?

The flesh of both the shark and the tropical fish is denser than seawater, so left to themselves they would sink. However, a tropical fish has a gas-filled body cavity called a swimbladder, so that the *average* density of the fish's body is the same as that of seawater and the fish neither sinks nor rises. Sharks have no such cavity. Hence they must swim constantly to keep from sinking, using their pectoral fins to provide lift much like the wings of an airplane (see Section 12.5).

Test Your Understanding Questions

12.1 Answer: (ii), (iv), (i) and (iii) (tie), (v) In each case the average density equals the mass divided by the volume. Hence we have
(i) $\rho = (4.00 \text{ kg})/(1.60 \times 10^{-3} \text{ m}^3) = 2.50 \times 10^3$ kg/m^3;
(ii) $\rho = (8.00 \text{ kg})/(1.60 \times 10^{-3} \text{ m}^3) = 5.00 \times 10^3$ kg/m^3;
(iii) $\rho = (8.00 \text{ kg})/(3.20 \times 10^{-3} \text{ m}^3) = 2.50 \times 10^3$ kg/m^3;
(iv) $\rho = (2560 \text{ kg})/(0.640 \text{ m}^3) = 4.00 \times 10^3$ kg/m^3;
(v) $\rho = (2560 \text{ kg})/(1.28 \text{ m}^3) = 2.00 \times 10^3$ kg/m^3. Note that compared to object (i), object (ii) has double the mass but the same volume and so has double the average density. Object (iii) has double the mass and double the volume of object (i), so (i) and (iii) have the same average density. Finally, object (v) has the same mass as object (iv) but double the volume, so (v) has half the average density of (iv).

12.2 Answer: (ii) From Eq. (12.9), the pressure outside the barometer is equal to the product $\rho g h$. When the barometer is taken out of the refrigerator, the density ρ decreases while the height h of the mercury column remains the same. Hence the air pressure must be lower outdoors than inside the refrigerator.

12.3 Answer: (i) Consider the water, the statue, and the container together as a system; the total weight of the system does not depend on whether the statue is immersed. The total supporting force, including the tension T and the upward force F of the scale

on the container (equal to the scale reading), is the same in both cases. But we saw in Example 12.5 that T decreases by 7.84 N when the statue is immersed, so the scale reading F must *increase* by 7.84 N. An alternative viewpoint is that the water exerts an upward buoyant force of 7.84 N on the statue, so the statue must exert an equal downward force on the water, making the scale reading 7.84 N greater than the weight of water and container.

12.4 Answer: (ii) A highway that narrows from three lanes to one is like a pipe whose cross-sectional area narrows to one-third of its value. If cars behaved like the molecules of an incompressible fluid, then as the cars encountered the one-lane section, the spacing between cars (the "density") would stay the same but the cars would triple their speed. This would keep the "volume flow rate" (number of cars per second passing a point on the highway) the same. In real life cars behave like the molecules of a *compressible* fluid: They end up packed closer (the "density" increases) and fewer cars per second pass a point on the highway (the "volume flow rate" decreases).

12.5 Answer: (ii) Newton's second law tells us that a body accelerates (its velocity changes) in response to a net force. In fluid flow, a pressure difference between two points means that fluid particles moving between those two points experience a force, and this force causes the fluid particles to accelerate and change speed.

12.6 Answer: (iv) The required pressure is proportional to $1/R^4$, where R is the inside radius of the needle (half the inside diameter). With the smaller-diameter needle, the pressure is greater by a factor of $[(0.60 \text{ mm})/(0.30 \text{ mm})]^4 = 2^4 = 16$.

Bridging Problem

Answers: (a) $y = H - \left(\dfrac{d}{D}\right)^2 \sqrt{2gH}\, t + \left(\dfrac{d}{D}\right)^4 \dfrac{gt^2}{2}$

(b) $T = \sqrt{\dfrac{2H}{g}} \left(\dfrac{D}{d}\right)^2$ (c) $\sqrt{2}$

13 GRAVITATION

? The rings of Saturn are made of countless individual orbiting particles. Do all the ring particles orbit at the same speed, or do the inner particles orbit faster or slower than the outer ones?

Some of the earliest investigations in physical science started with questions that people asked about the night sky. Why doesn't the moon fall to earth? Why do the planets move across the sky? Why doesn't the earth fly off into space rather than remaining in orbit around the sun? The study of gravitation provides the answers to these and many related questions.

As we remarked in Chapter 5, gravitation is one of the four classes of interactions found in nature, and it was the earliest of the four to be studied extensively. Newton discovered in the 17th century that the same interaction that makes an apple fall out of a tree also keeps the planets in their orbits around the sun. This was the beginning of *celestial mechanics,* the study of the dynamics of objects in space. Today, our knowledge of celestial mechanics allows us to determine how to put a satellite into any desired orbit around the earth or to choose just the right trajectory to send a spacecraft to another planet.

In this chapter you will learn the basic law that governs gravitational interactions. This law is *universal:* Gravity acts in the same fundamental way between the earth and your body, between the sun and a planet, and between a planet and one of its moons. We'll apply the law of gravitation to phenomena such as the variation of weight with altitude, the orbits of satellites around the earth, and the orbits of planets around the sun.

13.1 Newton's Law of Gravitation

The example of gravitational attraction that's probably most familiar to you is your *weight,* the force that attracts you toward the earth. During his study of the motions of the planets and of the moon, Newton discovered the fundamental character of the gravitational attraction between *any* two bodies. Along with his three laws of motion, Newton published the **law of gravitation** in 1687. It may be stated as follows:

> Every particle of matter in the universe attracts every other particle with a force that is directly proportional to the product of the masses of the particles and inversely proportional to the square of the distance between them.

Translating this into an equation, we have

$$F_g = \frac{Gm_1m_2}{r^2} \quad \text{(law of gravitation)} \qquad (13.1)$$

where F_g is the magnitude of the gravitational force on either particle, m_1 and m_2 are their masses, r is the distance between them (Fig. 13.1), and G is a fundamental physical constant called the **gravitational constant.** The numerical value of G depends on the system of units used.

Equation (13.1) tells us that the gravitational force between two particles decreases with increasing distance r: If the distance is doubled, the force is only one-fourth as great, and so on. Although many of the stars in the night sky are far more massive than the sun, they are so far away that their gravitational force on the earth is negligibly small.

CAUTION **Don't confuse g and G** Because the symbols g and G are so similar, it's common to confuse the two very different gravitational quantities that these symbols represent. Lowercase g is the acceleration due to gravity, which relates the weight w of a body to its mass m: $w = mg$. The value of g is different at different locations on the earth's surface and on the surfaces of different planets. By contrast, capital G relates the gravitational force between any two bodies to their masses and the distance between them. We call G a *universal* constant because it has the same value for any two bodies, no matter where in space they are located. In the next section we'll see how the values of g and G are related.

Gravitational forces always act along the line joining the two particles, and they form an action–reaction pair. Even when the masses of the particles are different, the two interaction forces have equal magnitude (Fig. 13.1). The attractive force that your body exerts on the earth has the same magnitude as the force that the earth exerts on you. When you fall from a diving board into a swimming pool, the entire earth rises up to meet you! (You don't notice this because the earth's mass is greater than yours by a factor of about 10^{23}. Hence the earth's acceleration is only 10^{-23} as great as yours.)

Gravitation and Spherically Symmetric Bodies

We have stated the law of gravitation in terms of the interaction between two *particles*. It turns out that the gravitational interaction of any two bodies having *spherically symmetric* mass distributions (such as solid spheres or spherical shells) is the same as though we concentrated all the mass of each at its center, as in Fig. 13.2. Thus, if we model the earth as a spherically symmetric body with mass m_E, the force it exerts on a particle or a spherically symmetric body with mass m, at a distance r between centers, is

$$F_g = \frac{Gm_Em}{r^2} \qquad (13.2)$$

provided that the body lies outside the earth. A force of the same magnitude is exerted *on* the earth by the body. (We will prove these statements in Section 13.6.)

At points *inside* the earth the situation is different. If we could drill a hole to the center of the earth and measure the gravitational force on a body at various depths, we would find that toward the center of the earth the force *decreases,*

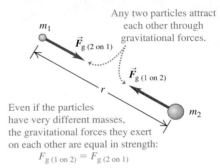

13.1 The gravitational forces between two particles of masses m_1 and m_2.

Any two particles attract each other through gravitational forces.

Even if the particles have very different masses, the gravitational forces they exert on each other are equal in strength:
$$F_{g\,(1\text{ on }2)} = F_{g\,(2\text{ on }1)}$$

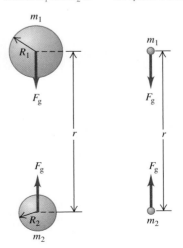

13.2 The gravitational effect *outside* any spherically symmetric mass distribution is the same as though all of the mass were concentrated at its center.

(a) The gravitational force between two spherically symmetric masses m_1 and m_2 ...

(b) ... is the same as if we concentrated all the mass of each sphere at the sphere's center.

13.3 Spherical and nonspherical bodies: the planet Jupiter and one of Jupiter's small moons, Amalthea.

Jupiter's mass is very large (1.90×10^{27} kg), so the mutual gravitational attraction of its parts has pulled it into a nearly spherical shape.

100,000 km

100 km

Amalthea, one of Jupiter's small moons, has a relatively tiny mass (7.17×10^{18} kg, only about 3.8×10^{-9} the mass of Jupiter) and weak mutual gravitation, so it has an irregular shape.

rather than increasing as $1/r^2$. As the body enters the interior of the earth (or other spherical body), some of the earth's mass is on the side of the body opposite from the center and pulls in the opposite direction. Exactly at the center, the earth's gravitational force on the body is zero.

Spherically symmetric bodies are an important case because moons, planets, and stars all tend to be spherical. Since all particles in a body gravitationally attract each other, the particles tend to move to minimize the distance between them. As a result, the body naturally tends to assume a spherical shape, just as a lump of clay forms into a sphere if you squeeze it with equal forces on all sides. This effect is greatly reduced in celestial bodies of low mass, since the gravitational attraction is less, and these bodies tend *not* to be spherical (Fig. 13.3).

Determining the Value of G

To determine the value of the gravitational constant G, we have to *measure* the gravitational force between two bodies of known masses m_1 and m_2 at a known distance r. The force is extremely small for bodies that are small enough to be brought into the laboratory, but it can be measured with an instrument called a *torsion balance*, which Sir Henry Cavendish used in 1798 to determine G.

Figure 13.4 shows a modern version of the Cavendish torsion balance. A light, rigid rod shaped like an inverted T is supported by a very thin, vertical quartz fiber. Two small spheres, each of mass m_1, are mounted at the ends of the horizontal arms of the T. When we bring two large spheres, each of mass m_2, to the positions shown, the attractive gravitational forces twist the T through a small angle. To measure this angle, we shine a beam of light on a mirror fastened to the T. The reflected beam strikes a scale, and as the T twists, the reflected beam moves along the scale.

After calibrating the Cavendish balance, we can measure gravitational forces and thus determine G. The presently accepted value is

$$G = 6.67428(67) \times 10^{-11} \, \text{N} \cdot \text{m}^2/\text{kg}^2$$

To three significant figures, $G = 6.67 \times 10^{-11} \, \text{N} \cdot \text{m}^2/\text{kg}^2$. Because $1 \, \text{N} = 1 \, \text{kg} \cdot \text{m/s}^2$, the units of G can also be expressed as $\text{m}^3/(\text{kg} \cdot \text{s}^2)$.

Gravitational forces combine vectorially. If each of two masses exerts a force on a third, the *total* force on the third mass is the vector sum of the individual forces of the first two. Example 13.3 makes use of this property, which is often called *superposition of forces*.

13.4 The principle of the Cavendish balance, used for determining the value of G. The angle of deflection has been exaggerated here for clarity.

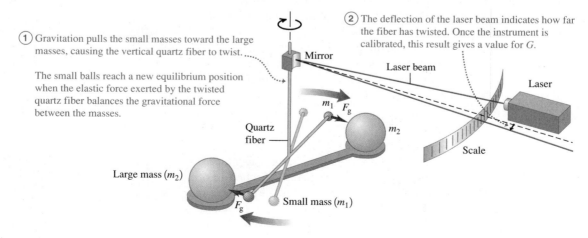

① Gravitation pulls the small masses toward the large masses, causing the vertical quartz fiber to twist.

The small balls reach a new equilibrium position when the elastic force exerted by the twisted quartz fiber balances the gravitational force between the masses.

② The deflection of the laser beam indicates how far the fiber has twisted. Once the instrument is calibrated, this result gives a value for G.

Mirror

Laser beam

Laser

Quartz fiber

m_1 F_g

m_2

Scale

Large mass (m_2)

F_g Small mass (m_1)

Example 13.1 Calculating gravitational force

The mass m_1 of one of the small spheres of a Cavendish balance is 0.0100 kg, the mass m_2 of the nearest large sphere is 0.500 kg, and the center-to-center distance between them is 0.0500 m. Find the gravitational force F_g on each sphere due to the other.

SOLUTION

IDENTIFY, SET UP, and EXECUTE: Because the spheres are spherically symmetric, we can calculate F_g by treating them as *particles* separated by 0.0500 m, as in Fig. 13.2. Each sphere experiences the same magnitude of force from the other sphere. We use Newton's

law of gravitation, Eq. (13.1), to determine F_g:

$$F_g = \frac{(6.67 \times 10^{-11}\,\text{N} \cdot \text{m}^2/\text{kg}^2)(0.0100\,\text{kg})(0.500\,\text{kg})}{(0.0500\,\text{m})^2}$$

$$= 1.33 \times 10^{-10}\,\text{N}$$

EVALUATE: It's remarkable that such a small force could be measured—or even detected—more than 200 years ago. Only a very massive object such as the earth exerts a gravitational force we can feel.

Example 13.2 Acceleration due to gravitational attraction

Suppose the two spheres in Example 13.1 are placed with their centers 0.0500 m apart at a point in space far removed from all other bodies. What is the magnitude of the acceleration of each, relative to an inertial system?

SOLUTION

IDENTIFY, SET UP, and EXECUTE: Each sphere exerts on the other a gravitational force of the same magnitude F_g, which we found in Example 13.1. We can neglect any other forces. The *acceleration* magnitudes a_1 and a_2 are different because the masses are different.

To determine these we'll use Newton's second law:

$$a_1 = \frac{F_g}{m_1} = \frac{1.33 \times 10^{-10}\,\text{N}}{0.0100\,\text{kg}} = 1.33 \times 10^{-8}\,\text{m/s}^2$$

$$a_2 = \frac{F_g}{m_2} = \frac{1.33 \times 10^{-10}\,\text{N}}{0.500\,\text{kg}} = 2.66 \times 10^{-10}\,\text{m/s}^2$$

EVALUATE: The larger sphere has 50 times the mass of the smaller one and hence has $\frac{1}{50}$ the acceleration. These accelerations are *not* constant; the gravitational forces increase as the spheres move toward each other.

Example 13.3 Superposition of gravitational forces

Many stars belong to *systems* of two or more stars held together by their mutual gravitational attraction. Figure 13.5 shows a three-star system at an instant when the stars are at the vertices of a 45° right triangle. Find the total gravitational force exerted on the small star by the two large ones.

SOLUTION

IDENTIFY, SET UP, and EXECUTE: We use the principle of superposition: The total force $\vec{F}$ on the small star is the vector sum of the forces $\vec{F}_1$ and $\vec{F}_2$ due to each large star, as Fig. 13.5 shows. We assume that the stars are spheres as in Fig. 13.2. We first calculate the magnitudes F_1 and F_2 using Eq. (13.1) and then compute the vector sum using components:

$$F_1 = \frac{\left[\begin{array}{l}(6.67 \times 10^{-11}\,\text{N} \cdot \text{m}^2/\text{kg}^2) \\ \times\, (8.00 \times 10^{30}\,\text{kg})(1.00 \times 10^{30}\,\text{kg})\end{array}\right]}{(2.00 \times 10^{12}\,\text{m})^2 + (2.00 \times 10^{12}\,\text{m})^2}$$

$$= 6.67 \times 10^{25}\,\text{N}$$

$$F_2 = \frac{\left[\begin{array}{l}(6.67 \times 10^{-11}\,\text{N} \cdot \text{m}^2/\text{kg}^2) \\ \times\, (8.00 \times 10^{30}\,\text{kg})(1.00 \times 10^{30}\,\text{kg})\end{array}\right]}{(2.00 \times 10^{12}\,\text{m})^2}$$

$$= 1.33 \times 10^{26}\,\text{N}$$

13.5 The total gravitational force on the small star (at O) is the vector sum of the forces exerted on it by the two larger stars. (For comparison, the mass of the sun—a rather ordinary star—is 1.99×10^{30} kg and the earth–sun distance is 1.50×10^{11} m.)

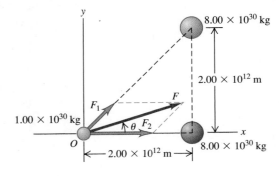

The x- and y-components of these forces are

$$F_{1x} = (6.67 \times 10^{25}\,\text{N})(\cos 45°) = 4.72 \times 10^{25}\,\text{N}$$
$$F_{1y} = (6.67 \times 10^{25}\,\text{N})(\sin 45°) = 4.72 \times 10^{25}\,\text{N}$$
$$F_{2x} = 1.33 \times 10^{26}\,\text{N}$$
$$F_{2y} = 0$$

Continued

The components of the total force $\vec{F}$ on the small star are

$$F_x = F_{1x} + F_{2x} = 1.81 \times 10^{26}\ \text{N}$$
$$F_y = F_{1y} + F_{2y} = 4.72 \times 10^{25}\ \text{N}$$

The magnitude of $\vec{F}$ and its angle θ (see Fig. 13.5) are

$$F = \sqrt{F_x^2 + F_y^2} = \sqrt{(1.81 \times 10^{26}\ \text{N})^2 + (4.72 \times 10^{25}\ \text{N})^2}$$
$$= 1.87 \times 10^{26}\ \text{N}$$
$$\theta = \arctan \frac{F_y}{F_x} = \arctan \frac{4.72 \times 10^{25}\ \text{N}}{1.81 \times 10^{26}\ \text{N}} = 14.6°$$

EVALUATE: While the force magnitude F is tremendous, the magnitude of the resulting acceleration is not: $a = F/m = (1.87 \times 10^{26}\ \text{N})/(1.00 \times 10^{30}\ \text{kg}) = 1.87 \times 10^{-4}\ \text{m/s}^2$. Furthermore, the force $\vec{F}$ is *not* directed toward the center of mass of the two large stars.

13.6 Our solar system is part of a spiral galaxy like this one, which contains roughly 10^{11} stars as well as gas, dust, and other matter. The entire assemblage is held together by the mutual gravitational attraction of all the matter in the galaxy.

Why Gravitational Forces Are Important

Comparing Examples 13.1 and 13.3 shows that gravitational forces are negligible between ordinary household-sized objects, but very substantial between objects that are the size of stars. Indeed, gravitation is *the* most important force on the scale of planets, stars, and galaxies (Fig. 13.6). It is responsible for holding our earth together and for keeping the planets in orbit about the sun. The mutual gravitational attraction between different parts of the sun compresses material at the sun's core to very high densities and temperatures, making it possible for nuclear reactions to take place there. These reactions generate the sun's energy output, which makes it possible for life to exist on earth and for you to read these words.

The gravitational force is so important on the cosmic scale because it acts *at a distance,* without any direct contact between bodies. Electric and magnetic forces have this same remarkable property, but they are less important on astronomical scales because large accumulations of matter are electrically neutral; that is, they contain equal amounts of positive and negative charge. As a result, the electric and magnetic forces between stars or planets are very small or zero. The strong and weak interactions that we discussed in Section 5.5 also act at a distance, but their influence is negligible at distances much greater than the diameter of an atomic nucleus (about $10^{-14}\ \text{m}$).

A useful way to describe forces that act at a distance is in terms of a *field.* One body sets up a disturbance or field at all points in space, and the force that acts on a second body at a particular point is its response to the first body's field at that point. There is a field associated with each force that acts at a distance, and so we refer to gravitational fields, electric fields, magnetic fields, and so on. We won't need the field concept for our study of gravitation in this chapter, so we won't discuss it further here. But in later chapters we'll find that the field concept is an extraordinarily powerful tool for describing electric and magnetic interactions.

Test Your Understanding of Section 13.1 The planet Saturn has about 100 times the mass of the earth and is about 10 times farther from the sun than the earth is. Compared to the acceleration of the earth caused by the sun's gravitational pull, how great is the acceleration of Saturn due to the sun's gravitation? (i) 100 times greater; (ii) 10 times greater; (iii) the same; (iv) $\frac{1}{10}$ as great; (v) $\frac{1}{100}$ as great.

PhET: Lunar Lander

13.2 Weight

We defined the *weight* of a body in Section 4.4 as the attractive gravitational force exerted on it by the earth. We can now broaden our definition:

> **The weight of a body is the total gravitational force exerted on the body by all other bodies in the universe.**

When the body is near the surface of the earth, we can neglect all other gravitational forces and consider the weight as just the earth's gravitational attraction. At the surface of the *moon* we consider a body's weight to be the gravitational attraction of the moon, and so on.

If we again model the earth as a spherically symmetric body with radius R_E and mass m_E, the weight w of a small body of mass m at the earth's surface (a distance R_E from its center) is

$$w = F_g = \frac{Gm_E m}{R_E^2} \qquad \text{(weight of a body of mass } m \text{ at the earth's surface)} \qquad (13.3)$$

But we also know from Section 4.4 that the weight w of a body is the force that causes the acceleration g of free fall, so by Newton's second law, $w = mg$. Equating this with Eq. (13.3) and dividing by m, we find

$$g = \frac{Gm_E}{R_E^2} \qquad \text{(acceleration due to gravity at the earth's surface)} \qquad (13.4)$$

The acceleration due to gravity g is independent of the mass m of the body because m doesn't appear in this equation. We already knew that, but we can now see how it follows from the law of gravitation.

We can *measure* all the quantities in Eq. (13.4) except for m_E, so this relationship allows us to compute the mass of the earth. Solving Eq. (13.4) for m_E and using $R_E = 6380 \text{ km} = 6.38 \times 10^6 \text{ m}$ and $g = 9.80 \text{ m/s}^2$, we find

$$m_E = \frac{gR_E^2}{G} = 5.98 \times 10^{24} \text{ kg}$$

This is very close to the currently accepted value of 5.974×10^{24} kg. Once Cavendish had measured G, he computed the mass of the earth in just this way.

At a point above the earth's surface a distance r from the center of the earth (a distance $r - R_E$ above the surface), the weight of a body is given by Eq. (13.3) with R_E replaced by r:

$$w = F_g = \frac{Gm_E m}{r^2} \qquad (13.5)$$

The weight of a body decreases inversely with the square of its distance from the earth's center (Fig. 13.7). Figure 13.8 shows how the weight varies with height above the earth for an astronaut who weighs 700 N at the earth's surface.

The *apparent* weight of a body on earth differs slightly from the earth's gravitational force because the earth rotates and is therefore not precisely an inertial frame of reference. We have ignored this effect in our earlier discussion and have assumed that the earth *is* an inertial system. We will return to the effect of the earth's rotation in Section 13.7.

While the earth is an approximately spherically symmetric distribution of mass, it is *not* uniform throughout its volume. To demonstrate this, let's first calculate the average *density,* or mass per unit volume, of the earth. If we assume a spherical earth, the volume is

$$V_E = \tfrac{4}{3}\pi R_E^3 = \tfrac{4}{3}\pi (6.38 \times 10^6 \text{ m})^3 = 1.09 \times 10^{21} \text{ m}^3$$

Application Walking and Running on the Moon
You automatically transition from a walk to a run when the vertical force you exert on the ground—which, by Newton's third law, equals the vertical force the ground exerts on you—exceeds your weight. This transition from walking to running happens at much lower speeds on the moon, where objects weigh only 17% as much as on earth. Hence, the Apollo astronauts found themselves running even when moving relatively slowly during their moon "walks."

13.7 In an airliner at high altitude, you are farther from the center of the earth than when on the ground and hence weigh slightly less. Can you show that at an altitude of 10 km above the surface, you weigh 0.3% less than you do on the ground?

13.8 An astronaut who weighs 700 N at the earth's surface experiences less gravitational attraction when above the surface. The relevant distance r is from the astronaut to the *center* of the earth (*not* from the astronaut to the earth's surface).

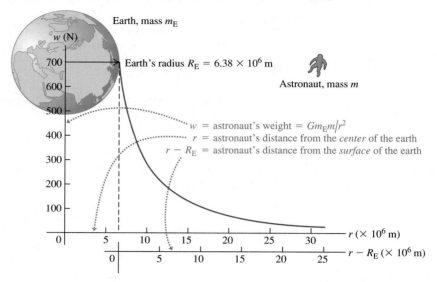

Earth, mass m_E

Earth's radius $R_E = 6.38 \times 10^6$ m

Astronaut, mass m

w = astronaut's weight = $Gm_E m/r^2$
r = astronaut's distance from the *center* of the earth
$r - R_E$ = astronaut's distance from the *surface* of the earth

13.9 The density of the earth decreases with increasing distance from its center.

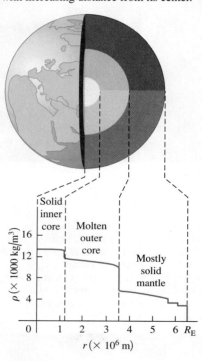

Solid inner core | Molten outer core | Mostly solid mantle

The average density ρ (the Greek letter rho) of the earth is the total mass divided by the total volume:

$$\rho = \frac{m_E}{V_E} = \frac{5.97 \times 10^{24} \text{ kg}}{1.09 \times 10^{21} \text{ m}^3}$$

$$= 5500 \text{ kg/m}^3 = 5.5 \text{ g/cm}^3$$

(For comparison, the density of water is 1000 kg/m³ = 1.00 g/cm³.) If the earth were uniform, we would expect rocks near the earth's surface to have this same density. In fact, the density of surface rocks is substantially lower, ranging from about 2000 kg/m³ for sedimentary rocks to about 3300 kg/m³ for basalt. So the earth *cannot* be uniform, and the interior of the earth must be much more dense than the surface in order that the *average* density be 5500 kg/m³. According to geophysical models of the earth's interior, the maximum density at the center is about 13,000 kg/m³. Figure 13.9 is a graph of density as a function of distance from the center.

Example 13.4 **Gravity on Mars**

A robotic lander with an earth weight of 3430 N is sent to Mars, which has radius $R_M = 3.40 \times 10^6$ m and mass $m_M = 6.42 \times 10^{23}$ kg (see Appendix F). Find the weight F_g of the lander on the Martian surface and the acceleration there due to gravity, g_M.

SOLUTION

IDENTIFY and SET UP: To find F_g we use Eq. (13.3), replacing m_E and R_E with m_M and R_M. We determine the lander mass m from the lander's earth weight w and then find g_M from $F_g = mg_M$.

EXECUTE: The lander's earth weight is $w = mg$, so

$$m = \frac{w}{g} = \frac{3430 \text{ N}}{9.80 \text{ m/s}^2} = 350 \text{ kg}$$

The mass is the same no matter where the lander is. From Eq. (13.3), the lander's weight on Mars is

$$F_g = \frac{Gm_M m}{R_M^2}$$

$$= \frac{(6.67 \times 10^{-11} \text{ N} \cdot \text{m}^2/\text{kg}^2)(6.42 \times 10^{23} \text{ kg})(350 \text{ kg})}{(3.40 \times 10^6 \text{ m})^2}$$

$$= 1.30 \times 10^3 \text{ N}$$

The acceleration due to gravity on Mars is

$$g_M = \frac{F_g}{m} = \frac{1.30 \times 10^3 \, \text{N}}{350 \, \text{kg}} = 3.7 \, \text{m/s}^2$$

EVALUATE: Even though Mars has just 11% of the earth's mass $(6.42 \times 10^{23} \, \text{kg}$ versus $5.98 \times 10^{24} \, \text{kg})$, the acceleration due to gravity g_M (and hence an object's weight F_g) is roughly 40% as large as on earth. That's because g_M is also inversely proportional to the square of the planet's radius, and Mars has only 53% the radius of earth (3.40×10^6 m versus 6.38×10^6 m).

You can check our result for g_M by using Eq. (13.4), with appropriate replacements. Do you get the same answer?

Test Your Understanding of Section 13.2 Rank the following hypothetical planets in order from highest to lowest value of g at the surface: (i) mass = 2 times the mass of the earth, radius = 2 times the radius of the earth; (ii) mass = 4 times the mass of the earth, radius = 4 times the radius of the earth; (iii) mass = 4 times the mass of the earth, radius = 2 times the radius of the earth; (iv) mass = 2 times the mass of the earth, radius = 4 times the radius of the earth.

13.3 Gravitational Potential Energy

When we first introduced gravitational potential energy in Section 7.1, we assumed that the gravitational force on a body is constant in magnitude and direction. This led to the expression $U = mgy$. But the earth's gravitational force on a body of mass m at any point outside the earth is given more generally by Eq. (13.2), $F_g = Gm_Em/r^2$, where m_E is the mass of the earth and r is the distance of the body from the earth's center. For problems in which r changes enough that the gravitational force can't be considered constant, we need a more general expression for gravitational potential energy.

To find this expression, we follow the same steps as in Section 7.1. We consider a body of mass m outside the earth, and first compute the work W_{grav} done by the gravitational force when the body moves directly away from or toward the center of the earth from $r = r_1$ to $r = r_2$, as in Fig. 13.10. This work is given by

$$W_{grav} = \int_{r_1}^{r_2} F_r \, dr \tag{13.6}$$

where F_r is the radial component of the gravitational force $\vec{F}$—that is, the component in the direction *outward* from the center of the earth. Because $\vec{F}$ points directly *inward* toward the center of the earth, F_r is negative. It differs from Eq. (13.2), the magnitude of the gravitational force, by a minus sign:

$$F_r = -\frac{Gm_Em}{r^2} \tag{13.7}$$

Substituting Eq. (13.7) into Eq. (13.6), we see that W_{grav} is given by

$$W_{grav} = -Gm_Em \int_{r_1}^{r_2} \frac{dr}{r^2} = \frac{Gm_Em}{r_2} - \frac{Gm_Em}{r_1} \tag{13.8}$$

The path doesn't have to be a straight line; it could also be a curve like the one in Fig. 13.10. By an argument similar to that in Section 7.1, this work depends only on the initial and final values of r, not on the path taken. This also proves that the gravitational force is always *conservative*.

We now define the corresponding potential energy U so that $W_{grav} = U_1 - U_2$, as in Eq. (7.3). Comparing this with Eq. (13.8), we see that the appropriate definition for **gravitational potential energy** is

$$U = -\frac{Gm_Em}{r} \quad \text{(gravitational potential energy)} \tag{13.9}$$

13.10 Calculating the work done on a body by the gravitational force as the body moves from radial coordinate r_1 to r_2.

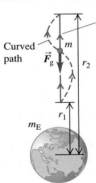

The gravitational force is conservative: The work done by $\vec{F}_g$ does not depend on the path taken from r_1 to r_2.

13.11 A graph of the gravitational potential energy U for the system of the earth (mass m_E) and an astronaut (mass m) versus the astronaut's distance r from the center of the earth.

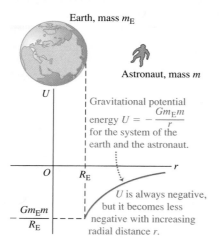

Earth, mass m_E

Astronaut, mass m

U

Gravitational potential energy $U = -\dfrac{Gm_Em}{r}$ for the system of the earth and the astronaut.

O R_E r

$-\dfrac{Gm_Em}{R_E}$

U is always negative, but it becomes less negative with increasing radial distance r.

Figure 13.11 shows how the gravitational potential energy depends on the distance r between the body of mass m and the center of the earth. When the body moves away from the earth, r increases, the gravitational force does negative work, and U increases (i.e., becomes less negative). When the body "falls" toward earth, r decreases, the gravitational work is positive, and the potential energy decreases (i.e., becomes more negative).

You may be troubled by Eq. (13.9) because it states that gravitational potential energy is always negative. But in fact you've seen negative values of U before. In using the formula $U = mgy$ in Section 7.1, we found that U was negative whenever the body of mass m was at a value of y below the arbitrary height we chose to be $y = 0$—that is, whenever the body and the earth were closer together than some certain arbitrary distance. (See, for instance, Example 7.2 in Section 7.1.) In defining U by Eq. (13.9), we have chosen U to be zero when the body of mass m is infinitely far from the earth ($r = \infty$). As the body moves toward the earth, gravitational potential energy decreases and so becomes negative.

If we wanted, we could make $U = 0$ at the surface of the earth, where $r = R_E$, by simply adding the quantity Gm_Em/R_E to Eq. (13.9). This would make U positive when $r > R_E$. We won't do this for two reasons: One, it would make the expression for U more complicated; and two, the added term would not affect the *difference* in potential energy between any two points, which is the only physically significant quantity.

CAUTION Gravitational force vs. gravitational potential energy Be careful not to confuse the expressions for gravitational force, Eq. (13.7), and gravitational potential energy, Eq. (13.9). The force F_r is proportional to $1/r^2$, while potential energy U is proportional to $1/r$. ▮

Armed with Eq. (13.9), we can now use general energy relationships for problems in which the $1/r^2$ behavior of the earth's gravitational force has to be included. If the gravitational force on the body is the only force that does work, the total mechanical energy of the system is constant, or *conserved*. In the following example we'll use this principle to calculate **escape speed,** the speed required for a body to escape completely from a planet.

Example 13.5   **"From the earth to the moon"**

In Jules Verne's 1865 story with this title, three men went to the moon in a shell fired from a giant cannon sunk in the earth in Florida. (a) Find the minimum muzzle speed needed to shoot a shell straight up to a height above the earth equal to the earth's radius R_E. (b) Find the minimum muzzle speed that would allow a shell to escape from the earth completely (the *escape speed*). Neglect air resistance, the earth's rotation, and the gravitational pull of the moon. The earth's radius and mass are $R_E = 6.38 \times 10^6$ m and $m_E = 5.97 \times 10^{24}$ kg.

SOLUTION

IDENTIFY and SET UP: Once the shell leaves the cannon muzzle, only the (conservative) gravitational force does work. Hence we can use conservation of mechanical energy to find the speed at which the shell must leave the muzzle so as to come to a halt (a) at two earth radii from the earth's center and (b) at an infinite distance from earth. The energy-conservation equation is $K_1 + U_1 = K_2 + U_2$, with U given by Eq. (13.9).

Figure 13.12 shows our sketches. Point 1 is at $r_1 = R_E$, where the shell leaves the cannon with speed v_1 (the target variable). Point 2 is where the shell reaches its maximum height; in part

13.12 Our sketches for this problem.

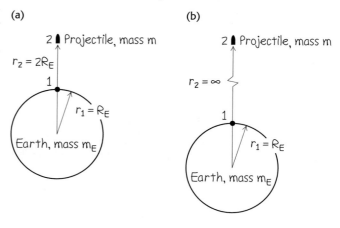

(a)

2 Projectile, mass m

$r_2 = 2R_E$

1

$r_1 = R_E$

Earth, mass m_E

(b)

2 Projectile, mass m

$r_2 = \infty$

1

$r_1 = R_E$

Earth, mass m_E

(a) $r_2 = 2R_E$ (Fig. 13.12a), and in part (b) $r_2 = \infty$ (Fig 13.12b). In both cases $v_2 = 0$ and $K_2 = 0$. Let m be the mass of the shell (with passengers).

EXECUTE: (a) We solve the energy-conservation equation for v_1:

$$K_1 + U_1 = K_2 + U_2$$

$$\frac{1}{2}mv_1^2 + \left(-\frac{Gm_Em}{R_E}\right) = 0 + \left(-\frac{Gm_Em}{2R_E}\right)$$

$$v_1 = \sqrt{\frac{Gm_E}{R_E}} = \sqrt{\frac{(6.67 \times 10^{-11}\ \text{N} \cdot \text{m}^2/\text{kg}^2)(5.97 \times 10^{24}\ \text{kg})}{6.38 \times 10^6\ \text{m}}}$$

$$= 7900\ \text{m/s}\ (= 28{,}400\ \text{km/h} = 17{,}700\ \text{mi/h})$$

(b) Now $r_2 = \infty$ so $U_2 = 0$ (see Fig. 13.11). Since $K_2 = 0$, the total mechanical energy $K_2 + U_2$ is zero in this case. Again we solve the energy-conservation equation for v_1:

$$\frac{1}{2}mv_1^2 + \left(-\frac{Gm_Em}{R_E}\right) = 0 + 0$$

$$v_1 = \sqrt{\frac{2Gm_E}{R_E}}$$

$$= \sqrt{\frac{2(6.67 \times 10^{-11}\ \text{N} \cdot \text{m}^2/\text{kg}^2)(5.97 \times 10^{24}\ \text{kg})}{6.38 \times 10^6\ \text{m}}}$$

$$= 1.12 \times 10^4\ \text{m/s}\ (= 40{,}200\ \text{km/h} = 25{,}000\ \text{mi/h})$$

EVALUATE: Our result in part (b) doesn't depend on the mass of the shell or the direction of launch. A modern spacecraft launched from Florida must attain essentially the speed found in part (b) to escape the earth; however, before launch it's already moving at 410 m/s to the east because of the earth's rotation. Launching to the east takes advantage of this "free" contribution toward escape speed.

To generalize, the initial speed v_1 needed for a body to escape from the surface of a spherical body of mass M and radius R (ignoring air resistance) is $v_1 = \sqrt{2GM/R}$ (escape speed). This equation yields escape speeds of 5.02×10^3 m/s for Mars, 5.95×10^4 m/s for Jupiter, and 6.18×10^5 m/s for the sun.

More on Gravitational Potential Energy

As a final note, let's show that when we are close to the earth's surface, Eq. (13.9) reduces to the familiar $U = mgy$ from Chapter 7. We first rewrite Eq. (13.8) as

$$W_{\text{grav}} = Gm_Em\frac{r_1 - r_2}{r_1r_2}$$

If the body stays close to the earth, then in the denominator we may replace r_1 and r_2 by R_E, the earth's radius, so

$$W_{\text{grav}} = Gm_Em\frac{r_1 - r_2}{R_E^2}$$

According to Eq. (13.4), $g = Gm_E/R_E^2$, so

$$W_{\text{grav}} = mg(r_1 - r_2)$$

If we replace the r's by y's, this is just Eq. (7.1) for the work done by a constant gravitational force. In Section 7.1 we used this equation to derive Eq. (7.2), $U = mgy$, so we may consider Eq. (7.2) for gravitational potential energy to be a special case of the more general Eq. (13.9).

Test Your Understanding of Section 13.3 Is it possible for a planet to have the same surface gravity as the earth (that is, the same value of g at the surface) and yet have a greater escape speed? ❙

13.4 The Motion of Satellites

Artificial satellites orbiting the earth are a familiar part of modern technology (Fig. 13.13). But how do they stay in orbit, and what determines the properties of their orbits? We can use Newton's laws and the law of gravitation to provide the answers. We'll see in the next section that the motion of planets can be analyzed in the same way.

To begin, think back to the discussion of projectile motion in Section 3.3. In Example 3.6 a motorcycle rider rides horizontally off the edge of a cliff, launching himself into a parabolic path that ends on the flat ground at the base of the cliff. If he survives and repeats the experiment with increased launch speed, he will land farther from the starting point. We can imagine him launching himself with great enough speed that the earth's curvature becomes significant. As he falls, the earth curves away beneath him. If he is going fast enough, and if his

13.13 With a length of 13.2 m and a mass of 11,000 kg, the Hubble Space Telescope is among the largest satellites placed in orbit.

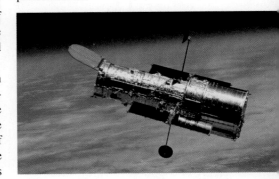

13.14 Trajectories of a projectile launched from a great height (ignoring air resistance). Orbits 1 and 2 would be completed as shown if the earth were a point mass at *C*. (This illustration is based on one in Isaac Newton's *Principia*.)

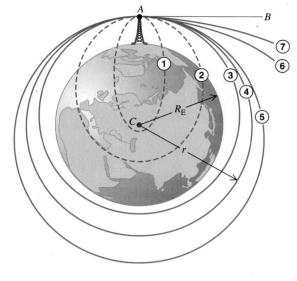

A projectile is launched from *A* toward *B*. Trajectories ① through ⑦ show the effect of increasing initial speed.

launch point is high enough that he clears the mountaintops, he may be able to go right on around the earth without ever landing.

Figure 13.14 shows a variation on this theme. We launch a projectile from point *A* in the direction *AB*, tangent to the earth's surface. Trajectories 1 through 7 show the effect of increasing the initial speed. In trajectories 3 through 5 the projectile misses the earth and becomes a satellite. If there is no retarding force, the projectile's speed when it returns to point *A* is the same as its initial speed and it repeats its motion indefinitely.

Trajectories 1 through 5 close on themselves and are called **closed orbits.** All closed orbits are ellipses or segments of ellipses; trajectory 4 is a circle, a special case of an ellipse. (We'll discuss the properties of an ellipse in Section 13.5.) Trajectories 6 and 7 are **open orbits.** For these paths the projectile never returns to its starting point but travels ever farther away from the earth.

PhET: My Solar System
ActivPhysics 4.6: Satellites Orbit

Satellites: Circular Orbits

A *circular* orbit, like trajectory 4 in Fig. 13.14, is the simplest case. It is also an important case, since many artificial satellites have nearly circular orbits and the orbits of the planets around the sun are also fairly circular. The only force acting on a satellite in circular orbit around the earth is the earth's gravitational attraction, which is directed toward the center of the earth and hence toward the center of the orbit (Fig. 13.15). As we discussed in Section 5.4, this means that the satellite is in *uniform* circular motion and its speed is constant. The satellite isn't falling *toward* the earth; rather, it's constantly falling *around* the earth. In a circular orbit the speed is just right to keep the distance from the satellite to the center of the earth constant.

13.15 The force $\vec{F}_g$ due to the earth's gravitational attraction provides the centripetal acceleration that keeps a satellite in orbit. Compare to Fig. 5.28.

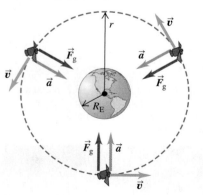

The satellite is in a circular orbit: Its acceleration $\vec{a}$ is always perpendicular to its velocity $\vec{v}$, so its speed v is constant.

Let's see how to find the constant speed v of a satellite in a circular orbit. The radius of the orbit is r, measured from the *center* of the earth; the acceleration of the satellite has magnitude $a_{rad} = v^2/r$ and is always directed toward the center of the circle. By the law of gravitation, the net force (gravitational force) on the satellite of mass m has magnitude $F_g = Gm_E m/r^2$ and is in the same direction as the acceleration. Newton's second law ($\sum \vec{F} = m\vec{a}$) then tells us that

$$\frac{Gm_E m}{r^2} = \frac{mv^2}{r}$$

Solving this for v, we find

$$v = \sqrt{\frac{Gm_E}{r}} \qquad \text{(circular orbit)} \qquad (13.10)$$

This relationship shows that we can't choose the orbit radius r and the speed v independently; for a given radius r, the speed v for a circular orbit is determined.

The satellite's mass m doesn't appear in Eq. (13.10), which shows that the motion of a satellite does not depend on its mass. If we could cut a satellite in half without changing its speed, each half would continue on with the original motion. An astronaut on board a space shuttle is herself a satellite of the earth, held by the earth's gravitational attraction in the same orbit as the shuttle. The astronaut has the same velocity and acceleration as the shuttle, so nothing is pushing her against the floor or walls of the shuttle. She is in a state of *apparent weightlessness,* as in a freely falling elevator; see the discussion following Example 5.9 in Section 5.2. (*True* weightlessness would occur only if the astronaut were infinitely far from any other masses, so that the gravitational force on her would be zero.) Indeed, every part of her body is apparently weightless; she feels nothing pushing her stomach against her intestines or her head against her shoulders (Fig. 13.16).

Apparent weightlessness is not just a feature of circular orbits; it occurs whenever gravity is the only force acting on a spacecraft. Hence it occurs for orbits of any shape, including open orbits such as trajectories 6 and 7 in Fig. 13.14.

We can derive a relationship between the radius r of a circular orbit and the period T, the time for one revolution. The speed v is the distance $2\pi r$ traveled in one revolution, divided by the period:

$$v = \frac{2\pi r}{T} \tag{13.11}$$

To get an expression for T, we solve Eq. (13.11) for T and substitute v from Eq. (13.10):

$$T = \frac{2\pi r}{v} = 2\pi r \sqrt{\frac{r}{Gm_E}} = \frac{2\pi r^{3/2}}{\sqrt{Gm_E}} \quad \text{(circular orbit)} \tag{13.12}$$

Equations (13.10) and (13.12) show that larger orbits correspond to slower speeds and longer periods. As an example, the International Space Station orbits 6800 km from the center of the earth (400 km above the earth's surface) with an orbital speed of 7.7 km/s and an orbital period of 93 minutes. The moon orbits the earth in a much larger orbit of radius 384,000 km, and so has a much slower orbital speed (1.0 km/s) and a much longer orbital period (27.3 days).

It's interesting to compare Eq. (13.10) to the calculation of escape speed in Example 13.5. We see that the escape speed from a spherical body with radius R is $\sqrt{2}$ times greater than the speed of a satellite in a circular orbit at that radius. If our spacecraft is in circular orbit around *any* planet, we have to multiply our speed by a factor of $\sqrt{2}$ to escape to infinity, regardless of the planet's mass.

Since the speed v in a circular orbit is determined by Eq. (13.10) for a given orbit radius r, the total mechanical energy $E = K + U$ is determined as well. Using Eqs. (13.9) and (13.10), we have

$$E = K + U = \tfrac{1}{2}mv^2 + \left(-\frac{Gm_E m}{r}\right) = \tfrac{1}{2}m\left(\frac{Gm_E}{r}\right) - \frac{Gm_E m}{r}$$

$$= -\frac{Gm_E m}{2r} \quad \text{(circular orbit)} \tag{13.13}$$

The total mechanical energy in a circular orbit is negative and equal to one-half the potential energy. Increasing the orbit radius r means increasing the mechanical energy (that is, making E less negative). If the satellite is in a relatively low orbit that encounters the outer fringes of earth's atmosphere, mechanical energy decreases due to negative work done by the force of air resistance; as a result, the orbit radius decreases until the satellite hits the ground or burns up in the atmosphere.

We have talked mostly about earth satellites, but we can apply the same analysis to the circular motion of *any* body under its gravitational attraction to a stationary body. Other examples include the earth's moon and the moons of other worlds (Fig. 13.17).

13.16 These space shuttle astronauts are in a state of apparent weightlessness. Which are right side up and which are upside down?

13.17 The two small satellites of the minor planet Pluto were discovered in 2005. In accordance with Eqs. (13.10) and (13.12), the satellite in the larger orbit has a slower orbital speed and a longer orbital period than the satellite in the smaller orbit.

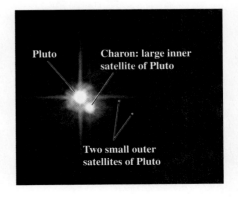

Pluto

Charon: large inner satellite of Pluto

Two small outer satellites of Pluto

Example 13.6 **A satellite orbit**

You wish to put a 1000-kg satellite into a circular orbit 300 km above the earth's surface. (a) What speed, period, and radial acceleration will it have? (b) How much work must be done to the satellite to put it in orbit? (c) How much additional work would have to be done to make the satellite escape the earth? The earth's radius and mass are given in Example 13.5 (Section 13.3).

SOLUTION

IDENTIFY and SET UP: The satellite is in a circular orbit, so we can use the equations derived in this section. In part (a), we first find the radius r of the satellite's orbit from its altitude. We then calculate the speed v and period T using Eqs. (13.10) and (13.12) and the acceleration from $a_{rad} = v^2/r$. In parts (b) and (c), the work required is the difference between the initial and final mechanical energy, which for a circular orbit is given by Eq. (13.13).

EXECUTE: (a) The radius of the satellite's orbit is $r = 6380$ km $+$ 300 km $= 6680$ km $= 6.68 \times 10^6$ m. From Eq. (13.10), the orbital speed is

$$v = \sqrt{\frac{Gm_E}{r}} = \sqrt{\frac{(6.67 \times 10^{-11} \text{ N} \cdot \text{m}^2/\text{kg}^2)(5.97 \times 10^{24} \text{ kg})}{6.68 \times 10^6 \text{ m}}}$$

$$= 7720 \text{ m/s}$$

We find the orbital period from Eq. (13.12):

$$T = \frac{2\pi r}{v} = \frac{2\pi(6.68 \times 10^6 \text{ m})}{7720 \text{ m/s}} = 5440 \text{ s} = 90.6 \text{ min}$$

Finally, the radial acceleration is

$$a_{rad} = \frac{v^2}{r} = \frac{(7720 \text{ m/s})^2}{6.68 \times 10^6 \text{ m}} = 8.92 \text{ m/s}^2$$

This is the value of g at a height of 300 km above the earth's surface; it is about 10% less than the value of g at the surface.

(b) The work required is the difference between E_2, the total mechanical energy when the satellite is in orbit, and E_1, the total mechanical energy when the satellite was at rest on the launch pad. From Eq. (13.13), the energy in orbit is

$$E_2 = -\frac{Gm_E m}{2r}$$

$$= -\frac{(6.67 \times 10^{-11} \text{ N} \cdot \text{m}^2/\text{kg}^2)(5.97 \times 10^{24} \text{ kg})(1000 \text{ kg})}{2(6.68 \times 10^6 \text{ m})}$$

$$= -2.98 \times 10^{10} \text{ J}$$

The satellite's kinetic energy is zero on the launch pad ($r = R_E$), so

$$E_1 = K_1 + U_1 = 0 + \left(-\frac{Gm_E m}{R_E}\right)$$

$$= -\frac{(6.67 \times 10^{-11} \text{ N} \cdot \text{m}^2/\text{kg}^2)(5.97 \times 10^{24} \text{ kg})(1000 \text{ kg})}{6.38 \times 10^6 \text{ m}}$$

$$= -6.24 \times 10^{10} \text{ J}$$

Hence the work required is

$$W_{required} = E_2 - E_1 = (-2.98 \times 10^{10} \text{ J}) - (-6.24 \times 10^{10} \text{ J})$$

$$= 3.26 \times 10^{10} \text{ J}$$

(c) We saw in part (b) of Example 13.5 that the minimum total mechanical energy for a satellite to escape to infinity is zero. Here, the total mechanical energy in the circular orbit is $E_2 = -2.98 \times 10^{10}$ J; to increase this to zero, an amount of work equal to 2.98×10^{10} J would have to be done on the satellite, presumably by rocket engines attached to it.

EVALUATE: In part (b) we ignored the satellite's initial kinetic energy (while it was still on the launch pad) due to the rotation of the earth. How much difference does this make? (See Example 13.5 for useful data.)

Test Your Understanding of Section 13.4 Your personal spacecraft is in a low-altitude circular orbit around the earth. Air resistance from the outer regions of the atmosphere does negative work on the spacecraft, causing the orbital radius to decrease slightly. Does the speed of the spacecraft (i) remain the same, (ii) increase, or (iii) decrease?

13.5 Kepler's Laws and the Motion of Planets

The name *planet* comes from a Greek word meaning "wanderer," and indeed the planets continuously change their positions in the sky relative to the background of stars. One of the great intellectual accomplishments of the 16th and 17th centuries was the threefold realization that the earth is also a planet, that all planets orbit the sun, and that the apparent motions of the planets as seen from the earth can be used to precisely determine their orbits.

The first and second of these ideas were published by Nicolaus Copernicus in Poland in 1543. The nature of planetary orbits was deduced between 1601 and 1619 by the German astronomer and mathematician Johannes Kepler, using a voluminous set of precise data on apparent planetary motions compiled by his mentor, the Danish astronomer Tycho Brahe. By trial and error, Kepler discovered three empirical laws that accurately described the motions of the planets:

1. **Each planet moves in an elliptical orbit, with the sun at one focus of the ellipse.**
2. **A line from the sun to a given planet sweeps out equal areas in equal times.**
3. **The periods of the planets are proportional to the $\frac{3}{2}$ powers of the major axis lengths of their orbits.**

Kepler did not know *why* the planets moved in this way. Three generations later, when Newton turned his attention to the motion of the planets, he discovered that each of Kepler's laws can be *derived;* they are consequences of Newton's laws of motion and the law of gravitation. Let's see how each of Kepler's laws arises.

Kepler's First Law

First consider the elliptical orbits described in Kepler's first law. Figure 13.18 shows the geometry of an ellipse. The longest dimension is the *major axis,* with half-length a; this half-length is called the **semi-major axis.** The sum of the distances from S to P and from S' to P is the same for all points on the curve. S and S' are the *foci* (plural of *focus*). The sun is at S, and the planet is at P; we think of them both as points because the size of each is very small in comparison to the distance between them. There is nothing at the other focus S'.

The distance of each focus from the center of the ellipse is ea, where e is a dimensionless number between 0 and 1 called the **eccentricity.** If $e = 0$, the ellipse is a circle. The actual orbits of the planets are fairly circular; their eccentricities range from 0.007 for Venus to 0.206 for Mercury. (The earth's orbit has $e = 0.017$.) The point in the planet's orbit closest to the sun is the *perihelion,* and the point most distant from the sun is the *aphelion.*

Newton was able to show that for a body acted on by an attractive force proportional to $1/r^2$, the only possible closed orbits are a circle or an ellipse; he also showed that open orbits (trajectories 6 and 7 in Fig. 13.14) must be parabolas or hyperbolas. These results can be derived by a straightforward application of Newton's laws and the law of gravitation, together with a lot more differential equations than we're ready for.

Kepler's Second Law

Figure 13.19 shows Kepler's second law. In a small time interval dt, the line from the sun S to the planet P turns through an angle $d\theta$. The area swept out is the colored triangle with height r, base length $r\,d\theta$, and area $dA = \frac{1}{2}r^2\,d\theta$ in Fig. 13.19b. The rate at which area is swept out, dA/dt, is called the *sector velocity:*

$$\frac{dA}{dt} = \frac{1}{2}r^2\frac{d\theta}{dt} \tag{13.14}$$

The essence of Kepler's second law is that the sector velocity has the same value at all points in the orbit. When the planet is close to the sun, r is small and $d\theta/dt$ is large; when the planet is far from the sun, r is large and $d\theta/dt$ is small.

To see how Kepler's second law follows from Newton's laws, we express dA/dt in terms of the velocity vector $\vec{v}$ of the planet P. The component of $\vec{v}$ perpendicular to the radial line is $v_\perp = v\sin\phi$. From Fig. 13.19b the displacement along the direction of $v_\perp$ during time dt is $r\,d\theta$, so we also have $v_\perp = r\,d\theta/dt$. Using this relationship in Eq. (13.14), we find

$$\frac{dA}{dt} = \frac{1}{2}rv\sin\phi \quad \text{(sector velocity)} \tag{13.15}$$

13.18 Geometry of an ellipse. The sum of the distances SP and $S'P$ is the same for every point on the curve. The sizes of the sun (S) and planet (P) are exaggerated for clarity.

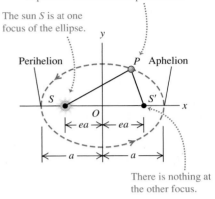

13.19 (a) The planet (P) moves about the sun (S) in an elliptical orbit. (b) In a time dt the line SP sweeps out an area $dA = \frac{1}{2}(r\,d\theta)r = \frac{1}{2}r^2\,d\theta$. (c) The planet's speed varies so that the line SP sweeps out the same area A in a given time t regardless of the planet's position in its orbit.

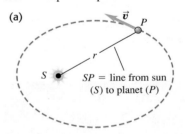

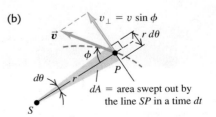

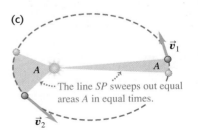

Now $rv\sin\phi$ is the magnitude of the vector product $\vec{r} \times \vec{v}$, which in turn is $1/m$ times the angular momentum $\vec{L} = \vec{r} \times m\vec{v}$ of the planet with respect to the sun. So we have

$$\frac{dA}{dt} = \frac{1}{2m}|\vec{r} \times m\vec{v}| = \frac{L}{2m} \qquad (13.16)$$

Thus Kepler's second law—that sector velocity is constant—means that angular momentum is constant!

It is easy to see why the angular momentum of the planet *must* be constant. According to Eq. (10.26), the rate of change of $\vec{L}$ equals the torque of the gravitational force $\vec{F}$ acting on the planet:

$$\frac{d\vec{L}}{dt} = \vec{\tau} = \vec{r} \times \vec{F}$$

In our situation, $\vec{r}$ is the vector from the sun to the planet, and the force $\vec{F}$ is directed from the planet to the sun. So these vectors always lie along the same line, and their vector product $\vec{r} \times \vec{F}$ is zero. Hence $d\vec{L}/dt = \mathbf{0}$. This conclusion does not depend on the $1/r^2$ behavior of the force; angular momentum is conserved for *any* force that acts always along the line joining the particle to a fixed point. Such a force is called a *central force*. (Kepler's first and third laws are valid *only* for a $1/r^2$ force.)

Conservation of angular momentum also explains why the orbit lies in a plane. The vector $\vec{L} = \vec{r} \times m\vec{v}$ is always perpendicular to the plane of the vectors $\vec{r}$ and $\vec{v}$; since $\vec{L}$ is constant in magnitude *and* direction, $\vec{r}$ and $\vec{v}$ always lie in the same plane, which is just the plane of the planet's orbit.

Kepler's Third Law

We have already derived Kepler's third law for the particular case of circular orbits. Equation (13.12) shows that the period of a satellite or planet in a circular orbit is proportional to the $\frac{3}{2}$ power of the orbit radius. Newton was able to show that this same relationship holds for an *elliptical* orbit, with the orbit radius r replaced by the semi-major axis a:

$$T = \frac{2\pi a^{3/2}}{\sqrt{Gm_\text{S}}} \qquad \text{(elliptical orbit around the sun)} \qquad (13.17)$$

Since the planet orbits the sun, not the earth, we have replaced the earth's mass m_E in Eq. (13.12) with the sun's mass m_S. Note that the period does not depend on the eccentricity e. An asteroid in an elongated elliptical orbit with semi-major axis a will have the same orbital period as a planet in a circular orbit of radius a. The key difference is that the asteroid moves at different speeds at different points in its elliptical orbit (Fig. 13.19c), while the planet's speed is constant around its circular orbit.

Application Biological Hazards of Interplanetary Travel
A spacecraft sent from earth to another planet spends most of its journey coasting along an elliptical orbit with the sun at one focus. Rockets are used only at the start and end of the journey, and even the trip to a nearby planet like Mars takes several months. During its journey, the spacecraft is exposed to cosmic rays—radiation that emanates from elsewhere in our galaxy. (On earth we're shielded from this radiation by our planet's magnetic field, as we'll describe in Chapter 27.) This poses no problem for a robotic spacecraft, but would be a severe medical hazard for astronauts undertaking such a voyage.

Conceptual Example 13.7 **Orbital speeds**

At what point in an elliptical orbit (see Fig. 13.19) does a planet move the fastest? The slowest?

SOLUTION

Mechanical energy is conserved as a planet moves in its orbit. The planet's kinetic energy $K = \frac{1}{2}mv^2$ is maximum when the potential energy $U = -Gm_\text{S}m/r$ is minimum (that is, most negative; see

Fig. 13.11), which occurs when the sun–planet distance r is a minimum. Hence the speed v is greatest at perihelion. Similarly, K is minimum when r is maximum, so the speed is slowest at aphelion.

Your intuition about falling bodies is helpful here. As the planet falls inward toward the sun, it picks up speed, and its speed is maximum when closest to the sun. The planet slows down as it moves away from the sun, and its speed is minimum at aphelion.

Example 13.8   Kepler's third law

The asteroid Pallas has an orbital period of 4.62 years and an orbital eccentricity of 0.233. Find the semi-major axis of its orbit.

SOLUTION

IDENTIFY and SET UP: This example uses Kepler's third law, which relates the period T and the semi-major axis a for an orbiting object (such as an asteroid). We use Eq. (13.17) to determine a; from Appendix F we have $m_S = 1.99 \times 10^{30}$ kg, and a conversion factor from Appendix E gives $T = (4.62 \text{ yr})(3.156 \times 10^7 \text{ s/yr}) = 1.46 \times 10^8$ s. Note that we don't need the value of the eccentricity.

EXECUTE: From Eq. (13.17), $a^{3/2} = [(Gm_S)^{1/2}T]/2\pi$. To solve for a, we raise both sides of this expression to the $\frac{2}{3}$ power and then substitute the values of G, m_S, and T:

$$a = \left(\frac{Gm_S T^2}{4\pi^2}\right)^{1/3} = 4.15 \times 10^{11} \text{ m}$$

(Plug in the numbers yourself to check.)

EVALUATE: Our result is intermediate between the semi-major axes of Mars and Jupiter (see Appendix F). Most known asteroids orbit in an "asteroid belt" between the orbits of these two planets.

Example 13.9   Comet Halley

Comet Halley moves in an elongated elliptical orbit around the sun (Fig. 13.20). Its distances from the sun at perihelion and aphelion are 8.75×10^7 km and 5.26×10^9 km, respectively. Find the orbital semi-major axis, eccentricity, and period.

SOLUTION

IDENTIFY and SET UP: We are to find the semi-major axis a, eccentricity e, and orbital period T. We can use Fig. 13.18 to find a and e from the given perihelion and aphelion distances. Knowing a, we can find T from Kepler's third law, Eq. (13.17).

EXECUTE: From Fig. 13.18, the length $2a$ of the major axis equals the sum of the comet–sun distance at perihelion and the comet–sun distance at aphelion. Hence

$$a = \frac{(8.75 \times 10^7 \text{ km}) + (5.26 \times 10^9 \text{ km})}{2} = 2.67 \times 10^9 \text{ km}$$

Figure 13.19 also shows that the comet–sun distance at perihelion is $a - ea = a(1 - e)$. This distance is 8.75×10^7 km, so

$$e = 1 - \frac{8.75 \times 10^7 \text{ km}}{a} = 1 - \frac{8.75 \times 10^7 \text{ km}}{2.67 \times 10^9 \text{ km}} = 0.967$$

From Eq. (13.17), the period is

$$T = \frac{2\pi a^{3/2}}{\sqrt{Gm_S}} = \frac{2\pi(2.67 \times 10^{12} \text{ m})^{3/2}}{\sqrt{(6.67 \times 10^{-11} \text{ N} \cdot \text{m}^2/\text{kg}^2)(1.99 \times 10^{30} \text{ kg})}}$$

$$= 2.38 \times 10^9 \text{ s} = 75.5 \text{ years}$$

EVALUATE: The eccentricity is close to 1, so the orbit is very elongated (see Fig. 13.20a). Comet Halley was at perihelion in early 1986 (Fig. 13.20b); it will next reach perihelion one period later, in 2061.

13.20 (a) The orbit of Comet Halley. (b) Comet Halley as it appeared in 1986. At the heart of the comet is an icy body, called the nucleus, that is about 10 km across. When the comet's orbit carries it close to the sun, the heat of sunlight causes the nucleus to partially evaporate. The evaporated material forms the tail, which can be tens of millions of kilometers long.

(a)

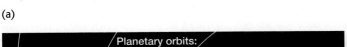
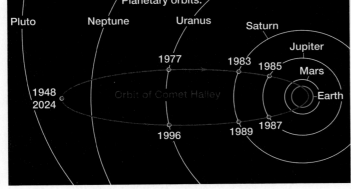

(b)

Planetary Motions and the Center of Mass

We have assumed that as a planet or comet orbits the sun, the sun remains absolutely stationary. Of course, this can't be correct; because the sun exerts a

13.21 A star and its planet both orbit about their common center of mass.

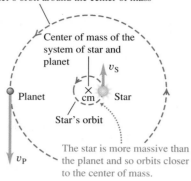

Planet's orbit around the center of mass

Center of mass of the system of star and planet

v_S

Planet

cm

Star

Star's orbit

v_P

The star is more massive than the planet and so orbits closer to the center of mass.

The planet and star are always on opposite sides of the center of mass.

gravitational force on the planet, the planet exerts a gravitational force on the sun of the same magnitude but opposite direction. In fact, *both* the sun and the planet orbit around their common center of mass (Fig. 13.21). We've made only a small error by ignoring this effect, however; the sun's mass is about 750 times the total mass of all the planets combined, so the center of mass of the solar system is not far from the center of the sun. Remarkably, astronomers have used this effect to detect the presence of planets orbiting other stars. Sensitive telescopes are able to detect the apparent "wobble" of a star as it orbits the common center of mass of the star and an unseen companion planet. (The planets are too faint to observe directly.) By analyzing these "wobbles," astronomers have discovered planets in orbit around hundreds of other stars.

Newton's analysis of planetary motions is used on a daily basis by modern-day astronomers. But the most remarkable result of Newton's work is that the motions of bodies in the heavens obey the *same* laws of motion as do bodies on the earth. This *Newtonian synthesis*, as it has come to be called, is one of the great unifying principles of science. It has had profound effects on the way that humanity looks at the universe—not as a realm of impenetrable mystery, but as a direct extension of our everyday world, subject to scientific study and calculation.

13.22 Calculating the gravitational potential energy of interaction between a point mass m outside a spherical shell and a ring on the surface of the shell.

(a) Geometry of the situation

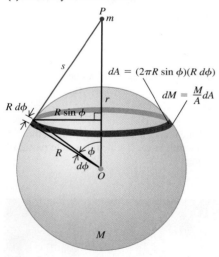

P
m

s

$dA = (2\pi R \sin\phi)(R\, d\phi)$

$dM = \dfrac{M}{A}dA$

r

$R\, d\phi$

$R \sin\phi$

R

ϕ

$d\phi$

O

M

(b) The distance s is the hypotenuse of a right triangle with sides $(r - R\cos\phi)$ and $R\sin\phi$.

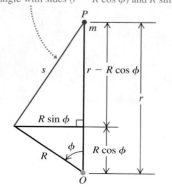

P
m

s

$r - R\cos\phi$

r

$R\sin\phi$

R

ϕ

$R\cos\phi$

O

Test Your Understanding of Section 13.5 The orbit of Comet X has a semi-major axis that is four times longer than the semi-major axis of Comet Y. What is the ratio of the orbital period of X to the orbital period of Y? (i) 2; (ii) 4; (iii) 8; (iv) 16; (v) 32; (vi) 64.

13.6 Spherical Mass Distributions

We have stated without proof that the gravitational interaction between two spherically symmetric mass distributions is the same as though all the mass of each were concentrated at its center. Now we're ready to prove this statement. Newton searched for a proof for several years, and he delayed publication of the law of gravitation until he found one.

Here's our program. Rather than starting with two spherically symmetric masses, we'll tackle the simpler problem of a point mass m interacting with a thin spherical shell with total mass M. We will show that when m is outside the sphere, the *potential energy* associated with this gravitational interaction is the same as though M were all concentrated at the center of the sphere. We learned in Section 7.4 that the force is the negative derivative of the potential energy, so the *force* on m is also the same as for a point mass M. Any spherically symmetric mass distribution can be thought of as being made up of many concentric spherical shells, so our result will also hold for *any* spherically symmetric M.

A Point Mass Outside a Spherical Shell

We start by considering a ring on the surface of the shell (Fig. 13.22a), centered on the line from the center of the shell to m. We do this because all of the particles that make up the ring are the same distance s from the point mass m. From Eq. (13.9) the potential energy of interaction between the earth (mass m_E) and a point mass m, separated by a distance r, is $U = -Gm_Em/r$. By changing notation in this expression, we see that the potential energy of interaction between the point mass m and a particle of mass m_i within the ring is given by

$$U_i = -\frac{Gmm_i}{s}$$

To find the potential energy of interaction between m and the entire ring of mass $dM = \Sigma_i m_i$, we sum this expression for U_i over all particles in the ring. Calling this potential energy dU, we find

$$dU = \sum_i U_i = \sum_i \left(-\frac{Gmm_i}{s} \right) = -\frac{Gm}{s} \sum_i m_i = -\frac{Gm\, dM}{s} \qquad \text{(13.18)}$$

To proceed, we need to know the mass dM of the ring. We can find this with the aid of a little geometry. The radius of the shell is R, so in terms of the angle ϕ shown in the figure, the radius of the ring is $R\sin\phi$, and its circumference is $2\pi R \sin\phi$. The width of the ring is $R\, d\phi$, and its area dA is approximately equal to its width times its circumference:

$$dA = 2\pi R^2 \sin\phi\, d\phi$$

The ratio of the ring mass dM to the total mass M of the shell is equal to the ratio of the area dA of the ring to the total area $A = 4\pi R^2$ of the shell:

$$\frac{dM}{M} = \frac{2\pi R^2 \sin\phi\, d\phi}{4\pi R^2} = \tfrac{1}{2} \sin\phi\, d\phi \qquad \text{(13.19)}$$

Now we solve Eq. (13.19) for dM and substitute the result into Eq. (13.18) to find the potential energy of interaction between the point mass m and the ring:

$$dU = -\frac{GMm \sin\phi\, d\phi}{2s} \qquad \text{(13.20)}$$

The total potential energy of interaction between the point mass and the *shell* is the integral of Eq. (13.20) over the whole sphere as ϕ varies from 0 to π (*not* 2π!) and s varies from $r - R$ to $r + R$. To carry out the integration, we have to express the integrand in terms of a single variable; we choose s. To express ϕ and $d\phi$ in terms of s, we have to do a little more geometry. Figure 13.22b shows that s is the hypotenuse of a right triangle with sides $(r - R\cos\phi)$ and $R\sin\phi$, so the Pythagorean theorem gives

$$s^2 = (r - R\cos\phi)^2 + (R\sin\phi)^2$$
$$= r^2 - 2rR\cos\phi + R^2 \qquad \text{(13.21)}$$

We take differentials of both sides:

$$2s\, ds = 2rR \sin\phi\, d\phi$$

Next we divide this by $2rR$ and substitute the result into Eq. (13.20):

$$dU = -\frac{GMm}{2s} \frac{s\, ds}{rR} = -\frac{GMm}{2rR}\, ds \qquad \text{(13.22)}$$

We can now integrate Eq. (13.22), recalling that s varies from $r - R$ to $r + R$:

$$U = -\frac{GMm}{2rR} \int_{r-R}^{r+R} ds = -\frac{GMm}{2rR}[(r + R) - (r - R)] \qquad \text{(13.23)}$$

Finally, we have

$$U = -\frac{GMm}{r} \qquad \text{(point mass m outside spherical shell M)} \qquad \text{(13.24)}$$

This is equal to the potential energy of two point masses m and M at a distance r. So we have proved that the gravitational potential energy of the spherical shell M and the point mass m at any distance r is the same as though they were point masses. Because the force is given by $F_r = -dU/dr$, the force is also the same.

The Gravitational Force Between Spherical Mass Distributions

Any spherically symmetric mass distribution can be thought of as a combination of concentric spherical shells. Because of the principle of superposition of forces, what is true of one shell is also true of the combination. So we have proved half of what we set out to prove: that the gravitational interaction between any spherically symmetric mass distribution and a point mass is the same as though all the mass of the spherically symmetric distribution were concentrated at its center.

The other half is to prove that *two* spherically symmetric mass distributions interact as though they were both points. That's easier. In Fig. 13.22a the forces the two bodies exert on each other are an action–reaction pair, and they obey Newton's third law. So we have also proved that the force that m exerts *on* the sphere M is the same as though M were a point. But now if we replace m with a spherically symmetric mass distribution centered at m's location, the resulting gravitational force on any part of M is the same as before, and so is the total force. This completes our proof.

A Point Mass Inside a Spherical Shell

We assumed at the beginning that the point mass m was outside the spherical shell, so our proof is valid only when m is outside a spherically symmetric mass distribution. When m is *inside* a spherical shell, the geometry is as shown in Fig. 13.23. The entire analysis goes just as before; Eqs. (13.18) through (13.22) are still valid. But when we get to Eq. (13.23), the limits of integration have to be changed to $R - r$ and $R + r$. We then have

$$U = -\frac{GMm}{2rR} \int_{R-r}^{R+r} ds = -\frac{GMm}{2rR}[(R + r) - (R - r)] \quad \text{(13.25)}$$

and the final result is

$$U = -\frac{GMm}{R} \qquad \text{(point mass } m \text{ inside spherical shell } M\text{)} \quad \text{(13.26)}$$

Compare this result to Eq. (13.24): Instead of having r, the distance between m and the center of M, in the denominator, we have R, the radius of the shell. This means that U in Eq. (13.26) doesn't depend on r and thus has the same value everywhere inside the shell. When m moves around inside the shell, no work is done on it, so the force on m at any point inside the shell must be zero.

More generally, at any point in the interior of any spherically symmetric mass distribution (not necessarily a shell), at a distance r from its center, the gravitational force on a point mass m is the same as though we removed all the mass at points farther than r from the center and concentrated all the remaining mass at the center.

13.23 When a point mass m is *inside* a uniform spherical shell of mass M, the potential energy is the same no matter where inside the shell the point mass is located. The force from the masses' mutual gravitational interaction is zero.

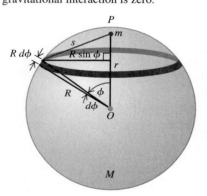

Example 13.10   "Journey to the center of the earth"

Imagine that we drill a hole through the earth along a diameter and drop a mail pouch down the hole. Derive an expression for the gravitational force F_g on the pouch as a function of its distance from the earth's center. Assume that the earth's density is uniform (not a very realistic model; see Fig. 13.9).

SOLUTION

IDENTIFY and SET UP: From the discussion immediately above, the value of F_g at a distance r from the earth's center is determined only by the mass M within a spherical region of radius r

(Fig. 13.24). Hence F_g is the same as if all the mass within radius r were concentrated at the center of the earth. The mass of a uniform sphere is proportional to the volume of the sphere, which is $\frac{4}{3}\pi r^3$ for a sphere of arbitrary radius r and $\frac{4}{3}\pi R_E^3$ for the entire earth.

EXECUTE: The ratio of the mass M of the sphere of radius r to the mass m_E of the earth is

$$\frac{M}{m_E} = \frac{\frac{4}{3}\pi r^3}{\frac{4}{3}\pi R_E^3} = \frac{r^3}{R_E^3} \qquad \text{so} \qquad M = m_E\frac{r^3}{R_E^3}$$

The magnitude of the gravitational force on m is then

$$F_g = \frac{GMm}{r^2} = \frac{Gm}{r^2}\left(m_E \frac{r^3}{R_E^3}\right) = \frac{Gm_E m}{R_E^3} r$$

EVALUATE: Inside this uniform-density sphere, F_g is *directly proportional* to the distance r from the center, rather than to $1/r^2$ as it is outside the sphere. At the surface $r = R_E$, we have $F_g = Gm_E m/R_E^2$, as we should. In the next chapter we'll learn how to compute the time it would take for the mail pouch to emerge on the other side of the earth.

13.24 A hole through the center of the earth (assumed to be uniform). When an object is a distance r from the center, only the mass inside a sphere of radius r exerts a net gravitational force on it.

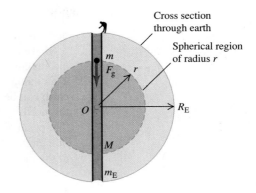

Test Your Understanding of Section 13.6 In the classic 1913 science-fiction novel *At the Earth's Core* by Edgar Rice Burroughs, explorers discover that the earth is a hollow sphere and that an entire civilization lives on the inside of the sphere. Would it be possible to stand and walk on the inner surface of a hollow, nonrotating planet? ❚

13.7 Apparent Weight and the Earth's Rotation

Because the earth rotates on its axis, it is not precisely an inertial frame of reference. For this reason the apparent weight of a body on earth is not precisely equal to the earth's gravitational attraction, which we will call the **true weight** $\vec{w}_0$ of the body. Figure 13.25 is a cutaway view of the earth, showing three observers. Each one holds a spring scale with a body of mass m hanging from it. Each scale applies a tension force $\vec{F}$ to the body hanging from it, and the reading on each scale is the magnitude F of this force. If the observers are unaware of the earth's

13.25 Except at the poles, the reading for an object being weighed on a scale (the *apparent weight*) is less than the gravitational force of attraction on the object (the *true weight*). The reason is that a net force is needed to provide a centripetal acceleration as the object rotates with the earth. For clarity, the illustration greatly exaggerates the angle β between the true and apparent weight vectors.

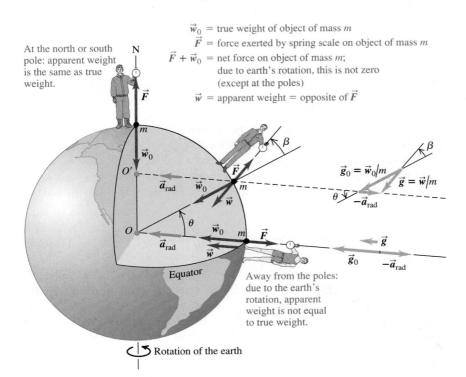

rotation, each one *thinks* that the scale reading equals the weight of the body because he thinks the body on his spring scale is in equilibrium. So each observer thinks that the tension $\vec{F}$ must be opposed by an equal and opposite force $\vec{w}$, which we call the **apparent weight.** But if the bodies are rotating with the earth, they are *not* precisely in equilibrium. Our problem is to find the relationship between the apparent weight $\vec{w}$ and the true weight $\vec{w}_0$.

If we assume that the earth is spherically symmetric, then the true weight $\vec{w}_0$ has magnitude $Gm_\mathrm{E}m/R_\mathrm{E}^2$, where m_E and R_E are the mass and radius of the earth. This value is the same for all points on the earth's surface. If the center of the earth can be taken as the origin of an inertial coordinate system, then the body at the north pole really *is* in equilibrium in an inertial system, and the reading on that observer's spring scale is equal to w_0. But the body at the equator is moving in a circle of radius R_E with speed v, and there must be a net inward force equal to the mass times the centripetal acceleration:

$$w_0 - F = \frac{mv^2}{R_\mathrm{E}}$$

So the magnitude of the apparent weight (equal to the magnitude of F) is

$$w = w_0 - \frac{mv^2}{R_\mathrm{E}} \quad \text{(at the equator)} \tag{13.27}$$

If the earth were not rotating, the body when released would have a free-fall acceleration $g_0 = w_0/m$. Since the earth *is* rotating, the falling body's actual acceleration relative to the observer at the equator is $g = w/m$. Dividing Eq. (13.27) by m and using these relationships, we find

$$g = g_0 - \frac{v^2}{R_\mathrm{E}} \quad \text{(at the equator)}$$

To evaluate v^2/R_E, we note that in 86,164 s a point on the equator moves a distance equal to the earth's circumference, $2\pi R_\mathrm{E} = 2\pi(6.38 \times 10^6 \text{ m})$. (The solar day, 86,400 s, is $\frac{1}{365}$ longer than this because in one day the earth also completes $\frac{1}{365}$ of its orbit around the sun.) Thus we find

$$v = \frac{2\pi(6.38 \times 10^6 \text{ m})}{86,164 \text{ s}} = 465 \text{ m/s}$$

$$\frac{v^2}{R_\mathrm{E}} = \frac{(465 \text{ m/s})^2}{6.38 \times 10^6 \text{ m}} = 0.0339 \text{ m/s}^2$$

So for a spherically symmetric earth the acceleration due to gravity should be about 0.03 m/s^2 less at the equator than at the poles.

At locations intermediate between the equator and the poles, the true weight $\vec{w}_0$ and the centripetal acceleration are not along the same line, and we need to write a vector equation corresponding to Eq. (13.27). From Fig. 13.25 we see that the appropriate equation is

$$\vec{w} = \vec{w}_0 - m\vec{a}_\mathrm{rad} = m\vec{g}_0 - m\vec{a}_\mathrm{rad} \tag{13.28}$$

The difference in the magnitudes of g and g_0 lies between zero and 0.0339 m/s^2. As shown in Fig. 13.25, the *direction* of the apparent weight differs from the direction toward the center of the earth by a small angle β, which is 0.1° or less.

Table 13.1 gives the values of g at several locations, showing variations with latitude. There are also small additional variations due to the lack of perfect spherical symmetry of the earth, local variations in density, and differences in elevation.

Table 13.1 Variations of g with Latitude and Elevation

Station	North Latitude	Elevation (m)	$g(\text{m/s}^2)$
Canal Zone	09°	0	9.78243
Jamaica	18°	0	9.78591
Bermuda	32°	0	9.79806
Denver, CO	40°	1638	9.79609
Pittsburgh, PA	40.5°	235	9.80118
Cambridge, MA	42°	0	9.80398
Greenland	70°	0	9.82534

Test Your Understanding of Section 13.7   Imagine a planet that has
the same mass and radius as the earth, but that makes 10 rotations during the time
the earth makes one rotation. What would be the difference between the acceleration due to gravity at the planet's equator and the acceleration due to gravity at its poles?
(i) 0.00339 m/s^2; (ii) 0.0339 m/s^2; (iii) 0.339 m/s^2; (iv) 3.39 m/s^2.

13.8 Black Holes

The concept of a black hole is one of the most interesting and startling products
of modern gravitational theory, yet the basic idea can be understood on the basis
of Newtonian principles.

The Escape Speed from a Star

Think first about the properties of our own sun. Its mass $M = 1.99 \times 10^{30}$ kg and
radius $R = 6.96 \times 10^8$ m are much larger than those of any planet, but compared
to other stars, our sun is not exceptionally massive. You can find the sun's average
density ρ in the same way we found the average density of the earth in Section 13.2:

$$\rho = \frac{M}{V} = \frac{M}{\frac{4}{3}\pi R^3} = \frac{1.99 \times 10^{30} \text{ kg}}{\frac{4}{3}\pi(6.96 \times 10^8 \text{ m})^3} = 1410 \text{ kg/m}^3$$

The sun's temperatures range from 5800 K (about 5500°C or 10,000°F) at the
surface up to 1.5×10^7 K (about 2.7×10^7°F) in the interior, so it surely contains no solids or liquids. Yet gravitational attraction pulls the sun's gas atoms
together until the sun is, on average, 41% denser than water and about 1200 times
as dense as the air we breathe.

Now think about the escape speed for a body at the surface of the sun. In
Example 13.5 (Section 13.3) we found that the escape speed from the surface of a
spherical mass M with radius R is $v = \sqrt{2GM/R}$. We can relate this to the average
density. Substituting $M = \rho V = \rho(\frac{4}{3}\pi R^3)$ into the expression for escape speed
gives

$$v = \sqrt{\frac{2GM}{R}} = \sqrt{\frac{8\pi G\rho}{3}}R \qquad (13.29)$$

Using either form of this equation, you can show that the escape speed for a body
at the surface of our sun is $v = 6.18 \times 10^5$ m/s (about 2.2 million km/h, or
1.4 million mi/h). This value, roughly $\frac{1}{500}$ the speed of light, is independent of the
mass of the escaping body; it depends on only the mass and radius (or average
density and radius) of the sun.

Now consider various stars with the same average density ρ and different radii R.
Equation (13.29) shows that for a given value of density ρ, the escape speed v is
directly proportional to R. In 1783 the Rev. John Mitchell, an amateur
astronomer, noted that if a body with the same average density as the sun had
about 500 times the radius of the sun, its escape speed would be greater than the
speed of light c. With his statement that "all light emitted from such a body
would be made to return toward it," Mitchell became the first person to suggest
the existence of what we now call a **black hole**—an object that exerts a gravitational force on other bodies but cannot emit any light of its own.

Black Holes, the Schwarzschild Radius,
and the Event Horizon

The first expression for escape speed in Eq. (13.29) suggests that a body of mass
M will act as a black hole if its radius R is less than or equal to a certain critical
radius. How can we determine this critical radius? You might think that you can
find the answer by simply setting $v = c$ in Eq. (13.29). As a matter of fact, this
does give the correct result, but only because of two compensating errors.

13.26 (a) A body with a radius R greater than the Schwarzschild radius R_S. (b) If the body collapses to a radius smaller than R_S, it is a black hole with an escape speed greater than the speed of light. The surface of the sphere of radius R_S is called the event horizon of the black hole.

(a) When the radius R of a body is greater than the Schwarzschild radius R_S, light can escape from the surface of the body.

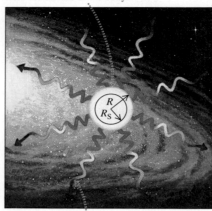

Gravity acting on the escaping light "red shifts" it to longer wavelengths.

(b) If all the mass of the body lies inside radius R_S, the body is a black hole: No light can escape from it.

The kinetic energy of light is *not* $mc^2/2$, and the gravitational potential energy near a black hole is *not* given by Eq. (13.9). In 1916, Karl Schwarzschild used Einstein's general theory of relativity (in part a generalization and extension of Newtonian gravitation theory) to derive an expression for the critical radius R_S, now called the **Schwarzschild radius.** The result turns out to be the same as though we had set $v = c$ in Eq. (13.29), so

$$c = \sqrt{\frac{2GM}{R_S}}$$

Solving for the Schwarzschild radius R_S, we find

$$R_S = \frac{2GM}{c^2} \qquad \text{(Schwarzschild radius)} \qquad (13.30)$$

If a spherical, nonrotating body with mass M has a radius less than R_S, then *nothing* (not even light) can escape from the surface of the body, and the body is a black hole (Fig. 13.26). In this case, any other body within a distance R_S of the center of the black hole is trapped by the gravitational attraction of the black hole and cannot escape from it.

The surface of the sphere with radius R_S surrounding a black hole is called the **event horizon:** Since light can't escape from within that sphere, we can't see events occurring inside. All that an observer outside the event horizon can know about a black hole is its mass (from its gravitational effects on other bodies), its electric charge (from the electric forces it exerts on other charged bodies), and its angular momentum (because a rotating black hole tends to drag space—and everything in that space—around with it). All other information about the body is irretrievably lost when it collapses inside its event horizon.

Example 13.11 | **Black hole calculations**

Astrophysical theory suggests that a burned-out star whose mass is at least three solar masses will collapse under its own gravity to form a black hole. If it does, what is the radius of its event horizon?

SOLUTION

IDENTIFY, SET UP, and EXECUTE: The radius in question is the Schwarzschild radius. We use Eq. (13.30) with a value of M

equal to three solar masses, or $M = 3(1.99 \times 10^{30} \text{ kg}) = 6.0 \times 10^{30}$ kg:

$$R_S = \frac{2GM}{c^2} = \frac{2(6.67 \times 10^{-11} \text{ N} \cdot \text{m}^2/\text{kg}^2)(6.0 \times 10^{30} \text{ kg})}{(3.00 \times 10^8 \text{ m/s})^2}$$

$$= 8.9 \times 10^3 \text{ m} = 8.9 \text{ km} = 5.5 \text{ mi}$$

EVALUATE: The average density of such an object is

$$\rho = \frac{M}{\frac{4}{3}\pi R^3} = \frac{6.0 \times 10^{30}\,\text{kg}}{\frac{4}{3}\pi(8.9 \times 10^3\,\text{m})^3} = 2.0 \times 10^{18}\,\text{kg/m}^3$$

This is about 10^{15} times as great as the density of familiar matter on earth and is comparable to the densities of atomic nuclei.

In fact, once the body collapses to a radius of R_{S}, nothing can prevent it from collapsing further. All of the mass ends up being crushed down to a single point called a *singularity* at the center of the event horizon. This point has zero volume and so has *infinite* density.

A Visit to a Black Hole

At points far from a black hole, its gravitational effects are the same as those of any normal body with the same mass. If the sun collapsed to form a black hole, the orbits of the planets would be unaffected. But things get dramatically different close to the black hole. If you decided to become a martyr for science and jump into a black hole, the friends you left behind would notice several odd effects as you moved toward the event horizon, most of them associated with effects of general relativity.

If you carried a radio transmitter to send back your comments on what was happening, your friends would have to retune their receiver continuously to lower and lower frequencies, an effect called the *gravitational red shift*. Consistent with this shift, they would observe that your clocks (electronic or biological) would appear to run more and more slowly, an effect called *time dilation*. In fact, during their lifetimes they would never see you make it to the event horizon.

In your frame of reference, you would make it to the event horizon in a rather short time but in a rather disquieting way. As you fell feet first into the black hole, the gravitational pull on your feet would be greater than that on your head, which would be slightly farther away from the black hole. The *differences* in gravitational force on different parts of your body would be great enough to stretch you along the direction toward the black hole and compress you perpendicular to it. These effects (called *tidal forces*) would rip you to atoms, and then rip your atoms apart, before you reached the event horizon.

Detecting Black Holes

If light cannot escape from a black hole and if black holes are as small as Example 13.11 suggests, how can we know that such things exist? The answer is that any gas or dust near the black hole tends to be pulled into an *accretion disk* that swirls around and into the black hole, rather like a whirlpool (Fig. 13.27). Friction within the accretion disk's material causes it to lose mechanical energy

13.27 A binary star system in which an ordinary star and a black hole orbit each other. The black hole itself cannot be seen, but the x rays from its accretion disk can be detected.

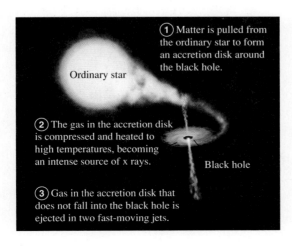

① Matter is pulled from the ordinary star to form an accretion disk around the black hole.

Ordinary star

② The gas in the accretion disk is compressed and heated to high temperatures, becoming an intense source of x rays.

Black hole

③ Gas in the accretion disk that does not fall into the black hole is ejected in two fast-moving jets.

and spiral into the black hole; as it moves inward, it is compressed together. This causes heating of the material, just as air compressed in a bicycle pump gets hotter. Temperatures in excess of 10^6 K can occur in the accretion disk, so hot that the disk emits not just visible light (as do bodies that are "red-hot" or "white-hot") but x rays. Astronomers look for these x rays (emitted by the material *before* it crosses the event horizon) to signal the presence of a black hole. Several promising candidates have been found, and astronomers now express considerable confidence in the existence of black holes.

Black holes in binary star systems like the one depicted in Fig. 13.27 have masses a few times greater than the sun's mass. There is also mounting evidence for the existence of much larger *supermassive black holes.* One example is thought to lie at the center of our Milky Way galaxy, some 26,000 light-years from earth in the direction of the constellation Sagittarius. High-resolution images of the galactic center reveal stars moving at speeds greater than 1500 km/s about an unseen object that lies at the position of a source of radio waves called Sgr A* (Fig. 13.28). By analyzing these motions, astronomers can infer the period T and semi-major axis a of each star's orbit. The mass m_X of the unseen object can then be calculated using Kepler's third law in the form given in Eq. (13.17), with the mass of the sun m_S replaced by m_X:

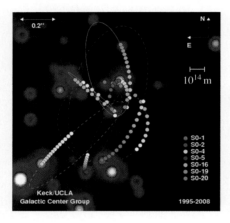

13.28 This false-color image shows the motions of stars at the center of our galaxy over a 13-year period. Analyzing these orbits using Kepler's third law indicates that the stars are moving about an unseen object that is some 4.1×10^6 times the mass of the sun. The scale bar indicates a length of 10^{14} m (670 times the distance from the earth to the sun) at the distance of the galactic center.

$$T = \frac{2\pi a^{3/2}}{\sqrt{Gm_X}} \qquad \text{so} \qquad m_X = \frac{4\pi^2 a^3}{GT^2}$$

The conclusion is that the mysterious dark object at the galactic center has a mass of 8.2×10^{36} kg, or 4.1 *million* times the mass of the sun. Yet observations with radio telescopes show that it has a radius no more than 4.4×10^{10} m, about one-third of the distance from the earth to the sun. These observations suggest that this massive, compact object is a black hole with a Schwarzschild radius of 1.1×10^{10} m. Astronomers hope to improve the resolution of their observations so that they can actually see the event horizon of this black hole.

Other lines of research suggest that even larger black holes, in excess of 10^9 times the mass of the sun, lie at the centers of other galaxies. Observational and theoretical studies of black holes of all sizes continue to be an exciting area of research in both physics and astronomy.

Test Your Understanding of Section 13.8 If the sun somehow collapsed to form a black hole, what effect would this event have on the orbit of the earth? (i) The orbit would shrink; (ii) the orbit would expand; (iii) the orbit would remain the same size.

Newton's law of gravitation: *Any* two bodies with masses m_1 and m_2, a distance r apart, attract each other with forces inversely proportional to r^2. These forces form an action–reaction pair and obey Newton's third law. When two or more bodies exert gravitational forces on a particular body, the total gravitational force on that individual body is the vector sum of the forces exerted by the other bodies. The gravitational interaction between spherical mass distributions, such as planets or stars, is the same as if all the mass of each distribution were concentrated at the center. (See Examples 13.1–13.3 and 13.10.)

$$F_g = \frac{Gm_1m_2}{r^2} \qquad (13.1)$$

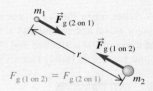

Gravitational force, weight, and gravitational potential energy: The weight w of a body is the total gravitational force exerted on it by all other bodies in the universe. Near the surface of the earth (mass m_E and radius R_E), the weight is essentially equal to the gravitational force of the earth alone. The gravitational potential energy U of two masses m and m_E separated by a distance r is inversely proportional to r. The potential energy is never positive; it is zero only when the two bodies are infinitely far apart. (See Examples 13.4 and 13.5.)

$$w = F_g = \frac{Gm_Em}{R_E^2} \qquad (13.3)$$
(weight at earth's surface)

$$g = \frac{Gm_E}{R_E^2} \qquad (13.4)$$
(acceleration due to gravity at earth's surface)

$$U = -\frac{Gm_Em}{r} \qquad (13.9)$$

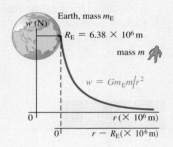

Orbits: When a satellite moves in a circular orbit, the centripetal acceleration is provided by the gravitational attraction of the earth. Kepler's three laws describe the more general case: an elliptical orbit of a planet around the sun or a satellite around a planet. (See Examples 13.6–13.9.)

$$v = \sqrt{\frac{Gm_E}{r}}$$
(speed in circular orbit) $\qquad (13.10)$

$$T = \frac{2\pi r}{v} = 2\pi r\sqrt{\frac{r}{Gm_E}} = \frac{2\pi r^{3/2}}{\sqrt{Gm_E}}$$
(period in circular orbit) $\qquad (13.12)$

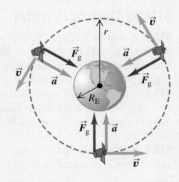

Black holes: If a nonrotating spherical mass distribution with total mass M has a radius less than its Schwarzschild radius R_S, it is called a black hole. The gravitational interaction prevents anything, including light, from escaping from within a sphere with radius R_S. (See Example 13.11.)

$$R_S = \frac{2GM}{c^2} \qquad (13.30)$$
(Schwarzschild radius)

If all of the body is inside its Schwarzschild radius $R_S = 2GM/c^2$, the body is a black hole.

BRIDGING PROBLEM | **Speeds in an Elliptical Orbit**

A comet orbits the sun (mass m_S) in an elliptical orbit of semimajor axis a and eccentricity e. (a) Find expressions for the speeds of the comet at perihelion and aphelion. (b) Evaluate these expressions for Comet Halley (see Example 13.9).

SOLUTION GUIDE

See MasteringPhysics® study area for a Video Tutor solution.

IDENTIFY and SET UP

1. Sketch the situation; show all relevant dimensions. Label the perihelion and aphelion.
2. List the unknown quantities, and identify the target variables.
3. Just as for a satellite orbiting the earth, the mechanical energy is conserved for a comet orbiting the sun. (Why?) What other quantity is conserved as the comet moves in its orbit? (*Hint:* See Section 13.5.)

EXECUTE

4. You'll need at least two equations that involve the two unknown speeds, and you'll need expressions for the sun–comet distances at perihelion and aphelion. (*Hint:* See Fig. 13.18.)
5. Solve the equations for your target variables. Compare your expressions: Which speed is lower? Does this make sense?
6. Use your expressions from step 5 to find the perihelion and aphelion speeds for Comet Halley. (*Hint:* See Appendix F.)

EVALUATE

7. Check whether your results make sense for the special case of a circular orbit ($e = 0$).

Problems

For instructor-assigned homework, go to www.masteringphysics.com

•, ••, •••: Problems of increasing difficulty. **CP**: Cumulative problems incorporating material from earlier chapters. **CALC**: Problems requiring calculus. **BIO**: Biosciences problems.

DISCUSSION QUESTIONS

Q13.1 A student wrote: "The only reason an apple falls downward to meet the earth instead of the earth rising upward to meet the apple is that the earth is much more massive and so exerts a much greater pull." Please comment.

Q13.2 A planet makes a circular orbit with period T around a star. If it were to orbit, at the same distance, a star with three times the mass of the original star, would the new period (in terms of T) be (a) $3T$, (b) $T\sqrt{3}$, (c) T, (d) $T/\sqrt{3}$, or (e) $T/3$?

Q13.3 If all planets had the same average density, how would the acceleration due to gravity at the surface of a planet depend on its radius?

Q13.4 Is a pound of butter on the earth the same amount as a pound of butter on Mars? What about a kilogram of butter? Explain.

Q13.5 Example 13.2 (Section 13.1) shows that the acceleration of each sphere caused by the gravitational force is inversely proportional to the mass of that sphere. So why does the force of gravity give all masses the same acceleration when they are dropped near the surface of the earth?

Q13.6 When will you attract the sun more: today at noon, or tonight at midnight? Explain.

Q13.7 Since the moon is constantly attracted toward the earth by the gravitational interaction, why doesn't it crash into the earth?

Q13.8 A planet makes a circular orbit with period T around a star. If the planet were to orbit at the same distance around this star, but had three times as much mass, what would the new period (in terms of T) be: (a) $3T$, (b) $T\sqrt{3}$, (c) T, (d) $T/\sqrt{3}$, or (e) $T/3$?

Q13.9 The sun pulls on the moon with a force that is more than twice the magnitude of the force with which the earth attracts the moon. Why, then, doesn't the sun take the moon away from the earth?

Q13.10 As defined in Chapter 7, gravitational potential energy is $U = mgy$ and is positive for a body of mass m above the earth's surface (which is at $y = 0$). But in this chapter, gravitational potential energy is $U = -Gm_E m/r$, which is *negative* for a body of mass m above the earth's surface (which is at $r = R_E$). How can you reconcile these seemingly incompatible descriptions of gravitational potential energy?

Q13.11 A planet is moving at constant speed in a circular orbit around a star. In one complete orbit, what is the net amount of work done on the planet by the star's gravitational force: positive, negative, or zero? What if the planet's orbit is an ellipse, so that the speed is not constant? Explain your answers.

Q13.12 Does the escape speed for an object at the earth's surface depend on the direction in which it is launched? Explain. Does your answer depend on whether or not you include the effects of air resistance?

Q13.13 If a projectile is fired straight up from the earth's surface, what would happen if the total mechanical energy (kinetic plus potential) is (a) less than zero, and (b) greater than zero? In each case, ignore air resistance and the gravitational effects of the sun, the moon, and the other planets.

Q13.14 Discuss whether this statement is correct: "In the absence of air resistance, the trajectory of a projectile thrown near the earth's surface is an *ellipse,* not a parabola."

Q13.15 The earth is closer to the sun in November than in May. In which of these months does it move faster in its orbit? Explain why.

Q13.16 A communications firm wants to place a satellite in orbit so that it is always directly above the earth's 45th parallel (latitude 45° north). This means that the plane of the orbit will not pass through the center of the earth. Is such an orbit possible? Why or why not?

Q13.17 At what point in an elliptical orbit is the acceleration maximum? At what point is it minimum? Justify your answers.

Q13.18 Which takes more fuel: a voyage from the earth to the moon or from the moon to the earth? Explain.

Q13.19 What would Kepler's third law be for circular orbits if an amendment to Newton's law of gravitation made the gravitational force inversely proportional to r^3? Would this change affect Kepler's other two laws? Explain.

Q13.20 In the elliptical orbit of Comet Halley shown in Fig. 13.20a, the sun's gravity is responsible for making the comet fall inward from aphelion to perihelion. But what is responsible for making the comet move from perihelion back outward to aphelion?

Q13.21 Many people believe that orbiting astronauts feel weightless because they are "beyond the pull of the earth's gravity." How far from the earth would a spacecraft have to travel to be truly beyond the earth's gravitational influence? If a spacecraft were really unaffected by the earth's gravity, would it remain in orbit? Explain. What is the real reason astronauts in orbit feel weightless?

Q13.22 As part of their training before going into orbit, astronauts ride in an airliner that is flown along the same parabolic trajectory as a freely falling projectile. Explain why this gives the same experience of apparent weightlessness as being in orbit.

EXERCISES

Section 13.1 Newton's Law of Gravitation

13.1 • What is the ratio of the gravitational pull of the sun on the moon to that of the earth on the moon? (Assume the distance of the moon from the sun can be approximated by the distance of the earth from the sun.) Use the data in Appendix F. Is it more accurate to say that the moon orbits the earth, or that the moon orbits the sun?

13.2 •• **CP Cavendish Experiment.** In the Cavendish balance apparatus shown in Fig. 13.4, suppose that $m_1 = 1.10$ kg, $m_2 = 25.0$ kg, and the rod connecting the m_1 pairs is 30.0 cm long. If, in each pair, m_1 and m_2 are 12.0 cm apart center to center, find (a) the net force and (b) the net torque (about the rotation axis) on the rotating part of the apparatus. (c) Does it seem that the torque in part (b) would be enough to easily rotate the rod? Suggest some ways to improve the sensitivity of this experiment.

13.3 • **Rendezvous in Space!** A couple of astronauts agree to rendezvous in space after hours. Their plan is to let gravity bring them together. One of them has a mass of 65 kg and the other a mass of 72 kg, and they start from rest 20.0 m apart. (a) Make a free-body diagram of each astronaut, and use it to find his or her initial acceleration. As a rough approximation, we can model the astronauts as uniform spheres. (b) If the astronauts' acceleration remained constant, how many days would they have to wait before reaching each other? (Careful! They *both* have acceleration toward each other.) (c) Would their acceleration, in fact, remain constant? If not, would it increase or decrease? Why?

13.4 •• Two uniform spheres, each with mass M and radius R, touch each other. What is the magnitude of their gravitational force of attraction?

13.5 • Two uniform spheres, each of mass 0.260 kg, are fixed at points A and B (Fig. E13.5). Find the magnitude and direction of the initial acceleration of a uniform sphere with mass 0.010 kg if released from rest at

Figure **E13.5**

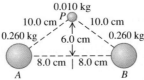

0.010 kg
10.0 cm P 10.0 cm
0.260 kg   0.260 kg
6.0 cm
8.0 cm | 8.0 cm
A B

point P and acted on only by forces of gravitational attraction of the spheres at A and B.

13.6 •• Find the magnitude and direction of the net gravitational force on mass A due to masses B and C in Fig. E13.6. Each mass is 2.00 kg.

Figure **E13.6**

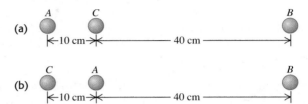

(a)
A C B
|←10 cm→|←———————— 40 cm ————————→|

(b)
C A B
|←10 cm→|←———————— 40 cm ————————→|

13.7 • A typical adult human has a mass of about 70 kg. (a) What force does a full moon exert on such a human when it is directly overhead with its center 378,000 km away? (b) Compare this force with the force exerted on the human by the earth.

13.8 •• An 8.00-kg point mass and a 15.0-kg point mass are held in place 50.0 cm apart. A particle of mass m is released from a point between the two masses 20.0 cm from the 8.00-kg mass along the line connecting the two fixed masses. Find the magnitude and direction of the acceleration of the particle.

13.9 •• A particle of mass $3m$ is located 1.00 m from a particle of mass m. (a) Where should you put a third mass M so that the net gravitational force on M due to the two masses is exactly zero? (b) Is the equilibrium of M at this point stable or unstable (i) for points along the line connecting m and $3m$, and (ii) for points along the line passing through M and perpendicular to the line connecting m and $3m$?

13.10 •• The point masses m and $2m$ lie along the x-axis, with m at the origin and $2m$ at $x = L$. A third point mass M is moved along the x-axis. (a) At what point is the net gravitational force on M due to the other two masses equal to zero? (b) Sketch the x-component of the net force on M due to m and $2m$, taking quantities to the right as positive. Include the regions $x < 0$, $0 < x < L$, and $x > L$. Be especially careful to show the behavior of the graph on either side of $x = 0$ and $x = L$.

Section 13.2 Weight

13.11 •• At what distance above the surface of the earth is the acceleration due to the earth's gravity 0.980 m/s^2 if the acceleration due to gravity at the surface has magnitude 9.80 m/s^2?

13.12 • The mass of Venus is 81.5% that of the earth, and its radius is 94.9% that of the earth. (a) Compute the acceleration due to gravity on the surface of Venus from these data. (b) If a rock weighs 75.0 N on earth, what would it weigh at the surface of Venus?

13.13 • Titania, the largest moon of the planet Uranus, has $\frac{1}{8}$ the radius of the earth and $\frac{1}{1700}$ the mass of the earth. (a) What is the acceleration due to gravity at the surface of Titania? (b) What is the average density of Titania? (This is less than the density of rock, which is one piece of evidence that Titania is made primarily of ice.)

13.14 • Rhea, one of Saturn's moons, has a radius of 765 km and an acceleration due to gravity of 0.278 m/s^2 at its surface. Calculate its mass and average density.

13.15 •• Calculate the earth's gravity force on a 75-kg astronaut who is repairing the Hubble Space Telescope 600 km above the earth's surface, and then compare this value with his weight at the earth's surface. In view of your result, explain why we say astronauts are weightless when they orbit the earth in a satellite such as a space shuttle. Is it because the gravitational pull of the earth is negligibly small?

Section 13.3 Gravitational Potential Energy

13.16 •• **Volcanoes on Io.** Jupiter's moon Io has active volcanoes (in fact, it is the most volcanically active body in the solar system) that eject material as high as 500 km (or even higher) above the surface. Io has a mass of 8.94×10^{22} kg and a radius of 1815 km. Ignore any variation in gravity over the 500-km range of the debris. How high would this material go on earth if it were ejected with the same speed as on Io?

13.17 • Use the results of Example 13.5 (Section 13.3) to calculate the escape speed for a spacecraft (a) from the surface of Mars and (b) from the surface of Jupiter. Use the data in Appendix F. (c) Why is the escape speed for a spacecraft independent of the spacecraft's mass?

13.18 •• Ten days after it was launched toward Mars in December 1998, the *Mars Climate Orbiter* spacecraft (mass 629 kg) was 2.87×10^6 km from the earth and traveling at 1.20×10^4 km/h relative to the earth. At this time, what were (a) the spacecraft's kinetic energy relative to the earth and (b) the potential energy of the earth–spacecraft system?

Section 13.4 The Motion of Satellites

13.19 • For a satellite to be in a circular orbit 780 km above the surface of the earth, (a) what orbital speed must it be given, and (b) what is the period of the orbit (in hours)?

13.20 •• **Aura Mission.** On July 15, 2004, NASA launched the Aura spacecraft to study the earth's climate and atmosphere. This satellite was injected into an orbit 705 km above the earth's surface. Assume a circular orbit. (a) How many hours does it take this satellite to make one orbit? (b) How fast (in km/s) is the Aura spacecraft moving?

13.21 •• Two satellites are in circular orbits around a planet that has radius 9.00×10^6 m. One satellite has mass 68.0 kg, orbital radius 5.00×10^7 m, and orbital speed 4800 m/s. The second satellite has mass 84.0 kg and orbital radius 3.00×10^7 m. What is the orbital speed of this second satellite?

13.22 •• **International Space Station.** The International Space Station makes 15.65 revolutions per day in its orbit around the earth. Assuming a circular orbit, how high is this satellite above the surface of the earth?

13.23 • Deimos, a moon of Mars, is about 12 km in diameter with mass 2.0×10^{15} kg. Suppose you are stranded alone on Deimos and want to play a one-person game of baseball. You would be the pitcher, and you would be the batter! (a) With what speed would you have to throw a baseball so that it would go into a circular orbit just above the surface and return to you so you could hit it? Do you think you could actually throw it at this speed? (b) How long (in hours) after throwing the ball should you be ready to hit it? Would this be an action-packed baseball game?

Section 13.5 Kepler's Laws and the Motion of Planets

13.24 •• **Planet Vulcan.** Suppose that a planet were discovered between the sun and Mercury, with a circular orbit of radius equal to $\frac{2}{3}$ of the average orbit radius of Mercury. What would be the orbital period of such a planet? (Such a planet was once postulated, in part to explain the precession of Mercury's orbit. It was even given the name Vulcan, although we now have no evidence that it actually exists. Mercury's precession has been explained by general relativity.)

13.25 •• The star Rho[1] Cancri is 57 light-years from the earth and has a mass 0.85 times that of our sun. A planet has been detected in a circular orbit around Rho[1] Cancri with an orbital radius equal to 0.11 times the radius of the earth's orbit around the sun. What are (a) the orbital speed and (b) the orbital period of the planet of Rho[1] Cancri?

13.26 •• In March 2006, two small satellites were discovered orbiting Pluto, one at a distance of 48,000 km and the other at 64,000 km. Pluto already was known to have a large satellite Charon, orbiting at 19,600 km with an orbital period of 6.39 days. Assuming that the satellites do not affect each other, find the orbital periods of the two small satellites *without* using the mass of Pluto.

13.27 • (a) Use Fig. 13.18 to show that the sun–planet distance at perihelion is $(1 - e)a$, the sun–planet distance at aphelion is $(1 + e)a$, and therefore the sum of these two distances is $2a$. (b) When the dwarf planet Pluto was at perihelion in 1989, it was almost 100 million km closer to the sun than Neptune. The semi-major axes of the orbits of Pluto and Neptune are 5.92×10^{12} m and 4.50×10^{12} m, respectively, and the eccentricities are 0.248 and 0.010. Find Pluto's closest distance and Neptune's farthest distance from the sun. (c) How many years after being at perihelion in 1989 will Pluto again be at perihelion?

13.28 •• **Hot Jupiters.** In 2004 astronomers reported the discovery of a large Jupiter-sized planet orbiting very close to the star HD 179949 (hence the term "hot Jupiter"). The orbit was just $\frac{1}{9}$ the distance of Mercury from our sun, and it takes the planet only 3.09 days to make one orbit (assumed to be circular). (a) What is the mass of the star? Express your answer in kilograms and as a multiple of our sun's mass. (b) How fast (in km/s) is this planet moving?

13.29 •• **Planets Beyond the Solar System.** On October 15, 2001, a planet was discovered orbiting around the star HD 68988. Its orbital distance was measured to be 10.5 million kilometers from the center of the star, and its orbital period was estimated at 6.3 days. What is the mass of HD 68988? Express your answer in kilograms and in terms of our sun's mass. (Consult Appendix F.)

Section 13.6 Spherical Mass Distributions

13.30 • A uniform, spherical, 1000.0-kg shell has a radius of 5.00 m. (a) Find the gravitational force this shell exerts on a 2.00-kg point mass placed at the following distances from the center of the shell: (i) 5.01 m, (ii) 4.99 m, (iii) 2.72 m. (b) Sketch a qualitative graph of the magnitude of the gravitational force this sphere exerts on a point mass m as a function of the distance r of m from the center of the sphere. Include the region from $r = 0$ to $r \to \infty$.

13.31 •• A uniform, solid, 1000.0-kg sphere has a radius of 5.00 m. (a) Find the gravitational force this sphere exerts on a 2.00-kg point mass placed at the following distances from the center of the sphere: (i) 5.01 m, (ii) 2.50 m. (b) Sketch a qualitative graph of the magnitude of the gravitational force this sphere exerts on a point mass m as a function of the distance r of m from the center of the sphere. Include the region from $r = 0$ to $r \to \infty$.

13.32 • **CALC** A thin, uniform rod has length L and mass M. A small uniform sphere of mass m is placed a distance x from one end of the rod, along the axis of the rod (Fig. E13.32). (a) Calculate

Figure **E13.32**

the gravitational potential energy of the rod–sphere system. Take the potential energy to be zero when the rod and sphere are infinitely far apart. Show that your answer reduces to the expected result when x is much larger than L. (*Hint:* Use the power series expansion for $\ln(1 + x)$ given in Appendix B.) (b) Use $F_x = -dU/dx$ to find the magnitude and direction of the gravitational force exerted on the sphere by the rod (see Section 7.4). Show that your answer reduces to the expected result when x is much larger than L.

13.33 • **CALC** Consider the ring-shaped body of Fig. E13.33. A particle with mass m is placed a distance x from the center of the ring, along the line through the center of the ring and perpendicular to its plane. (a) Calculate the gravitational potential energy U of this system. Take the potential energy to be zero when the two objects are far apart. (b) Show that your answer to part (a) reduces to the expected result when x is much larger than the radius a of the ring. (c) Use $F_x = -dU/dx$ to find the magnitude and direction of the force on the particle (see Section 7.4). (d) Show that your answer to part (c) reduces to the expected result when x is much larger than a. (e) What are the values of U and F_x when $x = 0$? Explain why these results make sense.

Figure **E13.33**

Section 13.7 Apparent Weight and the Earth's Rotation

13.34 •• **A Visit to Santa.** You decide to visit Santa Claus at the north pole to put in a good word about your splendid behavior throughout the year. While there, you notice that the elf Sneezy, when hanging from a rope, produces a tension of 475.0 N in the rope. If Sneezy hangs from a similar rope while delivering presents at the earth's equator, what will the tension in it be? (Recall that the earth is rotating about an axis through its north and south poles.) Consult Appendix F and start with a free-body diagram of Sneezy at the equator.

13.35 • The acceleration due to gravity at the north pole of Neptune is approximately 10.7 m/s^2. Neptune has mass 1.0×10^{26} kg and radius 2.5×10^4 km and rotates once around its axis in about 16 h. (a) What is the gravitational force on a 5.0-kg object at the north pole of Neptune? (b) What is the apparent weight of this same object at Neptune's equator? (Note that Neptune's "surface" is gaseous, not solid, so it is impossible to stand on it.)

Section 13.8 Black Holes

13.36 •• **Mini Black Holes.** Cosmologists have speculated that black holes the size of a proton could have formed during the early days of the Big Bang when the universe began. If we take the diameter of a proton to be 1.0×10^{-15} m, what would be the mass of a mini black hole?

13.37 •• **At the Galaxy's Core.** Astronomers have observed a small, massive object at the center of our Milky Way galaxy (see Section 13.8). A ring of material orbits this massive object; the ring has a diameter of about 15 light-years and an orbital speed of about 200 km/s. (a) Determine the mass of the object at the center of the Milky Way galaxy. Give your answer both in kilograms and in solar masses (one solar mass is the mass of the sun). (b) Observations of stars, as well as theories of the structure of stars, suggest that it

is impossible for a single star to have a mass of more than about 50 solar masses. Can this massive object be a single, ordinary star? (c) Many astronomers believe that the massive object at the center of the Milky Way galaxy is a black hole. If so, what must the Schwarzschild radius of this black hole be? Would a black hole of this size fit inside the earth's orbit around the sun?

13.38 • (a) Show that a black hole attracts an object of mass m with a force of $mc^2 R_S/(2r^2)$, where r is the distance between the object and the center of the black hole. (b) Calculate the magnitude of the gravitational force exerted by a black hole of Schwarzschild radius 14.0 mm on a 5.00-kg mass 3000 km from it. (c) What is the mass of this black hole?

13.39 • In 2005 astronomers announced the discovery of a large black hole in the galaxy Markarian 766 having clumps of matter orbiting around once every 27 hours and moving at 30,000 km/s. (a) How far are these clumps from the center of the black hole? (b) What is the mass of this black hole, assuming circular orbits? Express your answer in kilograms and as a multiple of our sun's mass. (c) What is the radius of its event horizon?

PROBLEMS

13.40 ••• Four identical masses of 800 kg each are placed at the corners of a square whose side length is 10.0 cm. What is the net gravitational force (magnitude and direction) on one of the masses, due to the other three?

13.41 ••• Neutron stars, such as the one at the center of the Crab Nebula, have about the same mass as our sun but have a *much* smaller diameter. If you weigh 675 N on the earth, what would you weigh at the surface of a neutron star that has the same mass as our sun and a diameter of 20 km?

13.42 ••• **CP Exploring Europa.** There is strong evidence that Europa, a satellite of Jupiter, has a liquid ocean beneath its icy surface. Many scientists think we should land a vehicle there to search for life. Before launching it, we would want to test such a lander under the gravity conditions at the surface of Europa. One way to do this is to put the lander at the end of a rotating arm in an orbiting earth satellite. If the arm is 4.25 m long and pivots about one end, at what angular speed (in rpm) should it spin so that the acceleration of the lander is the same as the acceleration due to gravity at the surface of Europa? The mass of Europa is 4.8×10^{22} kg and its diameter is 3138 km.

13.43 • Three uniform spheres are fixed at the positions shown in Fig. P13.43. (a) What are the magnitude and direction of the force on a 0.0150-kg particle placed at P? (b) If the spheres are in deep outer space and a 0.0150-kg particle is released from rest 300 m from the origin along a line 45° below the $-x$-axis, what will the particle's speed be when it reaches the origin?

Figure **P13.43**

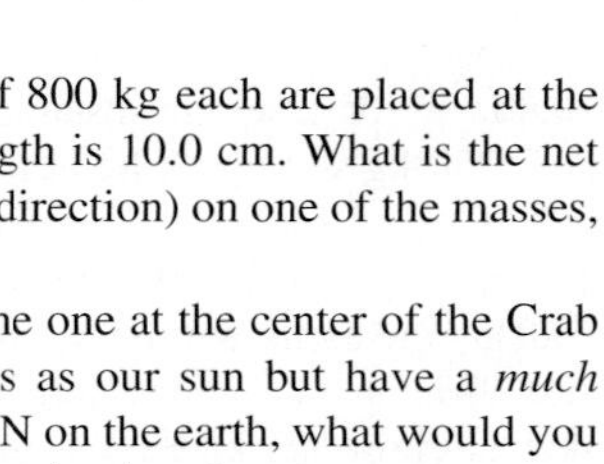

13.44 •• A uniform sphere with mass 60.0 kg is held with its center at the origin, and a second uniform sphere with mass 80.0 kg is held with its center at the point $x = 0$, $y = 3.00$ m. (a) What are the magnitude and direction of the net gravitational force due to these objects on a third uniform sphere with mass 0.500 kg placed at the point $x = 4.00$ m, $y = 0$? (b) Where, other than infinitely far away, could the third sphere be placed such that the net gravitational force acting on it from the other two spheres is equal to zero?

13.45 •• **CP BIO Hip Wear on the Moon.** (a) Use data from Appendix F to calculate the acceleration due to gravity on the moon. (b) Calculate the friction force on a walking 65-kg astronaut carrying a 43-kg instrument pack on the moon if the coefficient of kinetic friction at her hip joint is 0.0050. (c) What would be the friction force on earth for this astronaut?

13.46 •• **Mission to Titan.** On December 25, 2004, the *Huygens* probe separated from the *Cassini* spacecraft orbiting Saturn and began a 22-day journey to Saturn's giant moon Titan, on whose surface it landed. Besides the data in Appendix F, it is useful to know that Titan is 1.22×10^6 km from the center of Saturn and has a mass of 1.35×10^{23} kg and a diameter of 5150 km. At what distance from Titan should the gravitational pull of Titan just balance the gravitational pull of Saturn?

13.47 •• The asteroid Toro has a radius of about 5.0 km. Consult Appendix F as necessary. (a) Assuming that the density of Toro is the same as that of the earth (5.5 g/cm^3), find its total mass and find the acceleration due to gravity at its surface. (b) Suppose an object is to be placed in a circular orbit around Toro, with a radius just slightly larger than the asteroid's radius. What is the speed of the object? Could you launch yourself into orbit around Toro by running?

13.48 ••• At a certain instant, the earth, the moon, and a stationary 1250-kg spacecraft lie at the vertices of an equilateral triangle whose sides are 3.84×10^5 km in length. (a) Find the magnitude and direction of the net gravitational force exerted on the spacecraft by the earth and moon. State the direction as an angle measured from a line connecting the earth and the spacecraft. In a sketch, show the earth, the moon, the spacecraft, and the force vector. (b) What is the minimum amount of work that you would have to do to move the spacecraft to a point far from the earth and moon? You can ignore any gravitational effects due to the other planets or the sun.

13.49 ••• **CP** An experiment is performed in deep space with two uniform spheres, one with mass 50.0 kg and the other with mass 100.0 kg. They have equal radii, $r = 0.20$ m. The spheres are released from rest with their centers 40.0 m apart. They accelerate toward each other because of their mutual gravitational attraction. You can ignore all gravitational forces other than that between the two spheres. (a) Explain why linear momentum is conserved. (b) When their centers are 20.0 m apart, find (i) the speed of each sphere and (ii) the magnitude of the relative velocity with which one sphere is approaching the other. (c) How far from the initial position of the center of the 50.0-kg sphere do the surfaces of the two spheres collide?

13.50 •• **CP Submarines on Europa.** Some scientists are eager to send a remote-controlled submarine to Jupiter's moon Europa to search for life in its oceans below an icy crust. Europa's mass has been measured to be 4.8×10^{22} kg, its diameter is 3138 km, and it has no appreciable atmosphere. Assume that the layer of ice at the surface is not thick enough to exert substantial force on the water. If the windows of the submarine you are designing are 25.0 cm square and can stand a maximum inward force of 9750 N per window, what is the greatest depth to which this submarine can safely dive?

13.51 • **Geosynchronous Satellites.** Many satellites are moving in a circle in the earth's equatorial plane. They are at such a height above the earth's surface that they always remain above the same point. (a) Find the altitude of these satellites above the earth's surface. (Such an orbit is said to be *geosynchronous*.) (b) Explain, with a sketch, why the radio signals from these satellites cannot directly reach receivers on earth that are north of 81.3° N latitude.

13.52 ••• A landing craft with mass 12,500 kg is in a circular orbit 5.75×10^5 m above the surface of a planet. The period of the orbit is 5800 s. The astronauts in the lander measure the diameter of the planet to be 9.60×10^6 m. The lander sets down at the north pole of the planet. What is the weight of an 85.6-kg astronaut as he steps out onto the planet's surface?

13.53 ••• What is the escape speed from a 300-km-diameter asteroid with a density of 2500 kg/m^3?

13.54 •• (a) Asteroids have average densities of about 2500 kg/m^3 and radii from 470 km down to less than a kilometer. Assuming that the asteroid has a spherically symmetric mass distribution, estimate the radius of the largest asteroid from which you could escape simply by jumping off. (*Hint:* You can estimate your jump speed by relating it to the maximum height that you can jump on earth.) (b) Europa, one of Jupiter's four large moons, has a radius of 1570 km. The acceleration due to gravity at its surface is 1.33 m/s^2. Calculate its average density.

13.55 ••• (a) Suppose you are at the earth's equator and observe a satellite passing directly overhead and moving from west to east in the sky. Exactly 12.0 hours later, you again observe this satellite to be directly overhead. How far above the earth's surface is the satellite's orbit? (b) You observe another satellite directly overhead and traveling east to west. This satellite is again overhead in 12.0 hours. How far is this satellite's orbit above the surface of the earth?

13.56 •• Planet X rotates in the same manner as the earth, around an axis through its north and south poles, and is perfectly spherical. An astronaut who weighs 943.0 N on the earth weighs 915.0 N at the north pole of Planet X and only 850.0 N at its equator. The distance from the north pole to the equator is 18,850 km, measured along the surface of Planet X. (a) How long is the day on Planet X? (b) If a 45,000-kg satellite is placed in a circular orbit 2000 km above the surface of Planet X, what will be its orbital period?

13.57 •• There are two equations from which a change in the gravitational potential energy U of the system of a mass m and the earth can be calculated. One is $U = mgy$ (Eq. 7.2). The other is $U = -Gm_Em/r$ (Eq. 13.9). As shown in Section 13.3, the first equation is correct only if the gravitational force is a constant over the change in height Δy. The second is always correct. Actually, the gravitational force is never exactly constant over any change in height, but if the variation is small, we can ignore it. Consider the difference in U between a mass at the earth's surface and a distance h above it using both equations, and find the value of h for which Eq. (7.2) is in error by 1%. Express this value of h as a fraction of the earth's radius, and also obtain a numerical value for it.

13.58 ••• **CP** Your starship, the *Aimless Wanderer*, lands on the mysterious planet Mongo. As chief scientist-engineer, you make the following measurements: A 2.50-kg stone thrown upward from the ground at 12.0 m/s returns to the ground in 6.00 s; the circumference of Mongo at the equator is 2.00×10^5 km; and there is no appreciable atmosphere on Mongo. The starship commander, Captain Confusion, asks for the following information: (a) What is the mass of Mongo? (b) If the *Aimless Wanderer* goes into a circular orbit 30,000 km above the surface of Mongo, how many hours will it take the ship to complete one orbit?

13.59 •• **CP** An astronaut, whose mission is to go where no one has gone before, lands on a spherical planet in a distant galaxy. As she stands on the surface of the planet, she releases a small rock from rest and finds that it takes the rock 0.480 s to fall 1.90 m. If the radius of the planet is 8.60×10^7 m, what is the mass of the planet?

13.60 •• In Example 13.5 (Section 13.3) we ignored the gravitational effects of the moon on a spacecraft en route from the earth to the moon. In fact, we must include the gravitational potential energy due to the moon as well. For this problem, you can ignore the motion of the earth and moon. (a) If the moon has radius R_M and the distance between the centers of the earth and the moon is R_{EM}, find the total gravitational potential energy of the particle–earth and particle–moon systems when a particle with mass m is between the earth and the moon, and a distance r from the center of the earth. Take the gravitational potential energy to be zero when the objects are far from each other. (b) There is a point along a line between the earth and the moon where the net gravitational force is zero. Use the expression derived in part (a) and numerical values from Appendix F to find the distance of this point from the center of the earth. With what speed must a spacecraft be launched from the surface of the earth just barely to reach this point? (c) If a spacecraft were launched from the earth's surface toward the moon with an initial speed of 11.2 km/s, with what speed would it impact the moon?

13.61 •• Calculate the percent difference between your weight in Sacramento, near sea level, and at the top of Mount Everest, which is 8800 m above sea level.

13.62 •• The 0.100-kg sphere in Fig. P13.62 is released from rest at the position shown in the sketch, with its center 0.400 m from the center of the 5.00-kg mass. Assume that the only forces on the 0.100-kg sphere are the gravitational forces exerted by the other two spheres and that the 5.00-kg and 10.0-kg spheres are held in place at their initial positions. What is the speed of the 0.100-kg sphere when it has moved 0.400 m to the right from its initial position?

Figure **P13.62**

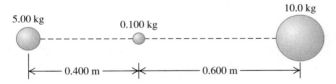

13.63 ••• An unmanned spacecraft is in a circular orbit around the moon, observing the lunar surface from an altitude of 50.0 km (see Appendix F). To the dismay of scientists on earth, an electrical fault causes an on-board thruster to fire, decreasing the speed of the spacecraft by 20.0 m/s. If nothing is done to correct its orbit, with what speed (in km/h) will the spacecraft crash into the lunar surface?

13.64 ••• **Mass of a Comet.** On July 4, 2005, the NASA spacecraft Deep Impact fired a projectile onto the surface of Comet Tempel 1. This comet is about 9.0 km across. Observations of surface debris released by the impact showed that dust with a speed as low as 1.0 m/s was able to escape the comet. (a) Assuming a spherical shape, what is the mass of this comet? (*Hint:* See Example 13.5 in Section 13.3.) (b) How far from the comet's center will this debris be when it has lost (i) 90.0% of its initial kinetic energy at the surface and (ii) all of its kinetic energy at the surface?

13.65 • **Falling Hammer.** A hammer with mass m is dropped from rest from a height h above the earth's surface. This height is not necessarily small compared with the radius R_E of the earth. If you ignore air resistance, derive an expression for the speed v of the hammer when it reaches the surface of the earth. Your expression should involve h, R_E, and m_E, the mass of the earth.

13.66 • (a) Calculate how much work is required to launch a spacecraft of mass m from the surface of the earth (mass m_E, radius R_E) and place it in a circular *low earth orbit*—that is, an orbit whose altitude above the earth's surface is much less than R_E. (As an example, the International Space Station is in low earth orbit at an altitude of about 400 km, much less than $R_E = 6380$ km.) You can ignore the kinetic energy that the spacecraft has on the ground due to the earth's rotation. (b) Calculate the minimum amount of additional work required to move the spacecraft from low earth orbit to a very great distance from the earth. You can ignore the gravitational effects of the sun, the moon, and the other planets. (c) Justify the statement: "In terms of energy, low earth orbit is halfway to the edge of the universe."

13.67 • A spacecraft is to be launched from the surface of the earth so that it will escape from the solar system altogether. (a) Find the speed relative to the center of the earth with which the spacecraft must be launched. Take into consideration the gravitational effects of both the earth and the sun, and include the effects of the earth's orbital speed, but ignore air resistance. (b) The rotation of the earth can help this spacecraft achieve escape speed. Find the speed that the spacecraft must have relative to the earth's *surface* if the spacecraft is launched from Florida at the point shown in Fig. P13.67. The rotation and orbital motions of the earth are in the same direction. The launch facilities in Florida are 28.5° north of the equator. (c) The European Space Agency (ESA) uses launch facilities in French Guiana (immediately north of Brazil), 5.15° north of the equator. What speed relative to the earth's surface would a spacecraft need to escape the solar system if launched from French Guiana?

Figure **P13.67**

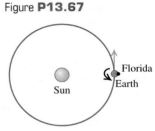

13.68 • **Gravity Inside the Earth.** Find the gravitational force that the earth exerts on a 10.0-kg mass if it is placed at the following locations. Consult Fig. 13.9, and assume a constant density through each of the interior regions (mantle, outer core, inner core), but *not* the same density in each of these regions. Use the graph to estimate the average density for each region: (a) at the surface of the earth; (b) at the outer surface of the molten outer core; (c) at the surface of the solid inner core; (d) at the center of the earth.

13.69 • **Kirkwood Gaps.** Hundreds of thousands of asteroids orbit the sun within the *asteroid belt*, which extends from about 3×10^8 km to about 5×10^8 km from the sun. (a) Find the orbital period (in years) of (i) an asteroid at the inside of the belt and (ii) an asteroid at the outside of the belt. Assume circular orbits. (b) In 1867 the American astronomer Daniel Kirkwood pointed out that several gaps exist in the asteroid belt where relatively few asteroids are found. It is now understood that these *Kirkwood gaps* are caused by the gravitational attraction of Jupiter, the largest planet, which orbits the sun once every 11.86 years. As an example, if an asteroid has an orbital period half that of Jupiter, or 5.93 years, on every other orbit this asteroid would be at its closest to Jupiter and feel a strong attraction toward the planet. This attraction, acting over and over on successive orbits, could sweep asteroids out of the Kirkwood gap. Use this hypothesis to determine the orbital radius for this Kirkwood gap. (c) One of several other Kirkwood gaps appears at a distance from the sun where the orbital period is 0.400 that of Jupiter. Explain why this happens, and find the orbital radius for this Kirkwood gap.

13.70 ••• If a satellite is in a sufficiently low orbit, it will encounter air drag from the earth's atmosphere. Since air drag does negative work (the force of air drag is directed opposite the motion), the mechanical energy will decrease. According to Eq. (13.13), if E decreases (becomes more negative), the radius r of the orbit will decrease. If air drag is relatively small, the satellite can be considered to be in a circular orbit of continually decreasing radius. (a) According to Eq. (13.10), if the radius of a satellite's circular orbit decreases, the satellite's orbital speed v *increases*. How can you reconcile this with the statement that the mechanical energy *decreases*? (*Hint*: Is air drag the only force that does work on the satellite as the orbital radius decreases?) (b) Due to air drag, the radius of a satellite's circular orbit decreases from r to $r - \Delta r$, where the positive quantity Δr is much less than r. The mass of the satellite is m. Show that the increase in orbital speed is $\Delta v = +(\Delta r/2) \sqrt{Gm_E/r^3}$; that the change in kinetic energy is $\Delta K = +(Gm_E m/2r^2) \Delta r$; that the change in gravitational potential energy is $\Delta U = -2 \Delta K = -(Gm_E m/r^2) \Delta r$; and that the amount of work done by the force of air drag is $W = -(Gm_E m/2r^2) \Delta r$. Interpret these results in light of your comments in part (a). (c) A satellite with mass 3000 kg is initially in a circular orbit 300 km above the earth's surface. Due to air drag, the satellite's altitude decreases to 250 km. Calculate the initial orbital speed; the increase in orbital speed; the initial mechanical energy; the change in kinetic energy; the change in gravitational potential energy; the change in mechanical energy; and the work done by the force of air drag. (d) Eventually a satellite will descend to a low enough altitude in the atmosphere that the satellite burns up and the debris falls to the earth. What becomes of the initial mechanical energy?

13.71 • **Binary Star—Equal Masses.** Two identical stars with mass M orbit around their center of mass. Each orbit is circular and has radius R, so that the two stars are always on opposite sides of the circle. (a) Find the gravitational force of one star on the other. (b) Find the orbital speed of each star and the period of the orbit. (c) How much energy would be required to separate the two stars to infinity?

13.72 •• **CP Binary Star—Different Masses.** Two stars, with masses M_1 and M_2, are in circular orbits around their center of mass. The star with mass M_1 has an orbit of radius R_1; the star with mass M_2 has an orbit of radius R_2. (a) Show that the ratio of the orbital radii of the two stars equals the reciprocal of the ratio of their masses—that is, $R_1/R_2 = M_2/M_1$. (b) Explain why the two stars have the same orbital period, and show that the period T is given by $T = 2\pi(R_1 + R_2)^{3/2}/\sqrt{G(M_1 + M_2)}$. (c) The two stars in a certain binary star system move in circular orbits. The first star, Alpha, has an orbital speed of 36.0 km/s. The second star, Beta, has an orbital speed of 12.0 km/s. The orbital period is 137 d. What are the masses of each of the two stars? (d) One of the best candidates for a black hole is found in the binary system called A0620-0090. The two objects in the binary system are an orange star, V616 Monocerotis, and a compact object believed to be a black hole (see Fig. 13.27). The orbital period of A0620-0090 is 7.75 hours, the mass of V616 Monocerotis is estimated to be 0.67 times the mass of the sun, and the mass of the black hole is estimated to be 3.8 times the mass of the sun. Assuming that the orbits are circular, find the radius of each object's orbit and the orbital speed of each object. Compare these answers to the orbital radius and orbital speed of the earth in its orbit around the sun.

13.73 ••• Comets travel around the sun in elliptical orbits with large eccentricities. If a comet has speed 2.0×10^4 m/s when at a distance of 2.5×10^{11} m from the center of the sun, what is its speed when at a distance of 5.0×10^{10} m?

13.74 •• **CP** An astronaut is standing at the north pole of a newly discovered, spherically symmetric planet of radius R. In his hands he holds a container full of a liquid with mass m and volume V. At the surface of the liquid, the pressure is p_0; at a depth d below the surface, the pressure has a greater value p. From this information, determine the mass of the planet.

13.75 •• **CALC** The earth does not have a uniform density; it is most dense at its center and least dense at its surface. An approximation of its density is $\rho(r) = A - Br$, where $A = 12{,}700$ kg/m^3 and $B = 1.50 \times 10^{-3}$ kg/m^4. Use $R = 6.37 \times 10^6$ m for the radius of the earth approximated as a sphere. (a) Geological evidence indicates that the densities are 13,100 kg/m^3 and 2400 kg/m^3 at the earth's center and surface, respectively. What values does the linear approximation model give for the densities at these two locations? (b) Imagine dividing the earth into concentric, spherical shells. Each shell has radius r, thickness dr, volume $dV = 4\pi r^2 \, dr$, and mass $dm = \rho(r)dV$. By integrating from $r = 0$ to $r = R$, show that the mass of the earth in this model is $M = \frac{4}{3}\pi R^3(A - \frac{3}{4}BR)$. (c) Show that the given values of A and B give the correct mass of the earth to within 0.4%. (d) We saw in Section 13.6 that a uniform spherical shell gives no contribution to g inside it. Show that $g(r) = \frac{4}{3}\pi Gr(A - \frac{3}{4}Br)$ inside the earth in this model. (e) Verify that the expression of part (d) gives $g = 0$ at the center of the earth and $g = 9.85$ m/s^2 at the surface. (f) Show that in this model g does *not* decrease uniformly with depth but rather has a maximum of $4\pi GA^2/9B = 10.01$ m/s^2 at $r = 2A/3B = 5640$ km.

13.76 •• **CP CALC** In Example 13.10 (Section 13.6) we saw that inside a planet of uniform density (not a realistic assumption for the earth) the acceleration due to gravity increases uniformly with distance from the center of the planet. That is, $g(r) = g_s r/R$, where g_s is the acceleration due to gravity at the surface, r is the distance from the center of the planet, and R is the radius of the planet. The interior of the planet can be treated approximately as an incompressible fluid of density ρ. (a) Replace the height y in Eq. (12.4) with the radial coordinate r and integrate to find the pressure inside a uniform planet as a function of r. Let the pressure at the surface be zero. (This means ignoring the pressure of the planet's atmosphere.) (b) Using this model, calculate the pressure at the center of the earth. (Use a value of ρ equal to the average density of the earth, calculated from the mass and radius given in Appendix F.) (c) Geologists estimate the pressure at the center of the earth to be approximately 4×10^{11} Pa. Does this agree with your calculation for the pressure at $r = 0$? What might account for any differences?

13.77 ••• **CP** Consider a spacecraft in an elliptical orbit around the earth. At the low point, or perigee, of its orbit, it is 400 km above the earth's surface; at the high point, or apogee, it is 4000 km above the earth's surface. (a) What is the period of the spacecraft's orbit? (b) Using conservation of angular momentum, find the ratio of the spacecraft's speed at perigee to its speed at apogee. (c) Using conservation of energy, find the speed at perigee and the speed at apogee. (d) It is necessary to have the spacecraft escape from the earth completely. If the spacecraft's rockets are fired at perigee, by how much would the speed have to be increased to achieve this? What if the rockets were fired at apogee? Which point in the orbit is more efficient to use?

13.78 • The planet Uranus has a radius of 25,560 km and a surface acceleration due to gravity of 11.1 m/s^2 at its poles. Its moon Miranda (discovered by Kuiper in 1948) is in a circular orbit about Uranus at an altitude of 104,000 km above the planet's surface. Miranda has a mass of 6.6×10^{19} kg and a radius of 235 km. (a) Calculate the mass of Uranus from the given data. (b) Calculate the magnitude of Miranda's acceleration due to its orbital motion about Uranus. (c) Calculate the acceleration due to Miranda's gravity at the surface of Miranda. (d) Do the answers to parts (b) and (c) mean that an object released 1 m above Miranda's surface on the side toward Uranus will fall *up* relative to Miranda? Explain.

13.79 ••• A 5000-kg spacecraft is in a circular orbit 2000 km above the surface of Mars. How much work must the spacecraft engines perform to move the spacecraft to a circular orbit that is 4000 km above the surface?

13.80 •• One of the brightest comets of the 20th century was Comet Hyakutake, which passed close to the sun in early 1996. The orbital period of this comet is estimated to be about 30,000 years. Find the semi-major axis of this comet's orbit. Compare it to the average sun–Pluto distance and to the distance to Alpha Centauri, the nearest star to the sun, which is 4.3 light-years distant.

13.81 ••• **CALC** Planets are not uniform inside. Normally, they are densest at the center and have decreasing density outward toward the surface. Model a spherically symmetric planet, with the same radius as the earth, as having a density that decreases linearly with distance from the center. Let the density be 15.0×10^3 kg/m³ at the center and 2.0×10^3 kg/m³ at the surface. What is the acceleration due to gravity at the surface of this planet?

13.82 •• **CALC** A uniform wire with mass M and length L is bent into a semicircle. Find the magnitude and direction of the gravitational force this wire exerts on a point with mass m placed at the center of curvature of the semicircle.

13.83 ••• **CALC** An object in the shape of a thin ring has radius a and mass M. A uniform sphere with mass m and radius R is placed with its center at a distance x to the right of the center of the ring, along a line through the center of the ring, and perpendicular to its plane (see Fig. E13.33). What is the gravitational force that the sphere exerts on the ring-shaped object? Show that your result reduces to the expected result when x is much larger than a.

13.84 ••• **CALC** A thin, uniform rod has length L and mass M. Calculate the magnitude of the gravitational force the rod exerts on a particle with mass m that is at a point along the axis of the rod a distance x from one end (see Fig. E13.32). Show that your result reduces to the expected result when x is much larger than L.

13.85 • **CALC** A shaft is drilled from the surface to the center of the earth (see Fig. 13.24). As in Example 13.10 (Section 13.6), make the unrealistic assumption that the density of the earth is uniform. With this approximation, the gravitational force on an object with mass m, that is inside the earth at a distance r from the center, has magnitude $F_g = Gm_E mr/R_E^3$ (as shown in Example 13.10) and points toward the center of the earth. (a) Derive an expression for the gravitational potential energy $U(r)$ of the object–earth system as a function of the object's distance from the center of the earth. Take the potential energy to be zero when the object is at the center of the earth. (b) If an object is released in the shaft at the earth's surface, what speed will it have when it reaches the center of the earth?

CHALLENGE PROBLEMS

13.86 ••• (a) When an object is in a circular orbit of radius r around the earth (mass m_E), the period of the orbit is T, given by Eq. (13.12), and the orbital speed is v, given by Eq. (13.10). Show that when the object is moved into a circular orbit of slightly larger radius $r + \Delta r$, where $\Delta r \ll r$, its new period is $T + \Delta T$ and its new orbital speed is $v - \Delta v$, where Δr, ΔT, and Δv are all positive quantities and

$$\Delta T = \frac{3\pi \, \Delta r}{v} \quad \text{and} \quad \Delta v = \frac{\pi \, \Delta r}{T}$$

[*Hint:* Use the expression $(1 + x)^n \approx 1 + nx$, valid for $|x| \ll 1$.] (b) The International Space Station (ISS) is in a nearly circular orbit at an altitude of 398.00 km above the surface of the earth. A maintenance crew is about to arrive on the space shuttle that is also in a circular orbit in the same orbital plane as the ISS, but with an altitude of 398.10 km. The crew has come to remove a faulty 125-m electrical cable, one end of which is attached to the ISS and the other end of which is floating free in space. The plan is for the shuttle to snag the free end just at the moment that the shuttle, the ISS, and the center of the earth all lie along the same line. The cable will then break free from the ISS when it becomes taut. How long after the free end is caught by the space shuttle will it detach from the ISS? Give your answer in minutes. (c) If the shuttle misses catching the cable, show that the crew must wait a time $t \approx T^2/\Delta T$ before they have a second chance. Find the numerical value of t and explain whether it would be worth the wait.

13.87 ••• **Interplanetary Navigation.** The most efficient way to send a spacecraft from the earth to another planet is by using a *Hohmann transfer orbit* (Fig. P13.87). If the orbits of the departure and destination planets are circular, the Hohmann transfer orbit is an elliptical orbit whose perihelion and aphelion are tangent to the orbits of the two planets. The rockets are fired briefly at the departure planet to put the spacecraft into the transfer orbit; the spacecraft then coasts until it reaches the destination planet. The rockets are then fired again to put the spacecraft into the same orbit about the sun as the destination planet. (a) For a flight from earth to Mars, in what direction must the rockets be fired at the earth and at Mars: in the direction of motion, or opposite the direction of motion? What about for a flight from Mars to the earth? (b) How long does a one-way trip from the the earth to Mars take, between the firings of the rockets? (c) To reach Mars from the earth, the launch must be timed so that Mars will be at the right spot when the spacecraft reaches Mars's orbit around the sun. At launch, what must the angle between a sun–Mars line and a sun–earth line be? Use data from Appendix F.

Figure **P13.87**

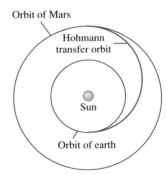

Orbit of Mars

Hohmann transfer orbit

Sun

Orbit of earth

13.88 ••• **CP Tidal Forces near a Black Hole.** An astronaut inside a spacecraft, which protects her from harmful radiation, is orbiting a black hole at a distance of 120 km from its center. The black hole is 5.00 times the mass of the sun and has a Schwarzschild radius of 15.0 km. The astronaut is positioned inside the spaceship such that one of her 0.030-kg ears is 6.0 cm farther from the black hole than the center of mass of the spacecraft and the other ear is 6.0 cm closer. (a) What is the tension between her ears? Would the astronaut find it difficult to keep from being torn apart by the gravitational forces? (Since her whole body orbits with the same angular velocity, one ear is moving too slowly for the radius of its orbit and the other is moving too fast. Hence her head must exert forces on her ears to keep them in their orbits.) (b) Is the center of gravity of her head at the same point as the center of mass? Explain.

13.89 ••• **CALC** Mass M is distributed uniformly over a disk of radius a. Find the gravitational force (magnitude and direction) between this disk-shaped mass and a particle with mass m located a distance x above the center of the disk (Fig. P13.89). Does your result reduce to the correct expression as x becomes very large? (*Hint:* Divide the disk into infinitesimally thin concentric rings, use

Figure **P13.89**

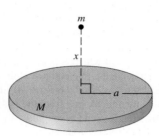

the expression derived in Exercise 13.33 for the gravitational force due to each ring, and integrate to find the total force.)

13.90 ••• **CALC** Mass M is distributed uniformly along a line of length $2L$. A particle with mass m is at a point that is a distance a above the center of the line on its perpendicular bisector (point P in Fig. P13.90). For the gravitational force that the line exerts on the particle, calculate the components perpendicular and parallel to the line. Does your result reduce to the correct expression as a becomes very large?

Figure **P13.90**

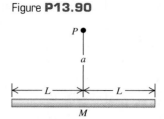

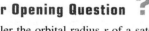

Answers

Chapter Opening Question **?**

The smaller the orbital radius r of a satellite, the faster its orbital speed v [see Eq. (13.10)]. Hence a particle near the inner edge of Saturn's rings has a faster speed than a particle near the outer edge of the rings.

Test Your Understanding Questions

13.1 Answer: (v) From Eq. (13.1), the gravitational force of the sun (mass m_1) on a planet (mass m_2) a distance r away has magnitude $F_g = Gm_1m_2/r^2$. Compared to the earth, Saturn has a value of r^2 that is $10^2 = 100$ times greater and a value of m_2 that is also 100 times greater. Hence the *force* that the sun exerts on Saturn has the same magnitude as the force that the sun exerts on earth. The *acceleration* of a planet equals the net force divided by the planet's mass: Since Saturn has 100 times more mass than the earth, its acceleration is $\frac{1}{100}$ as great as that of the earth.

13.2 Answer: (iii), (i), (ii), (iv) From Eq. (13.4), the acceleration due to gravity at the surface of a planet of mass m_P and radius R_P is $g_P = Gm_P/R_P^2$. That is, g_P is directly proportional to the planet's mass and inversely proportional to the square of its radius. It follows that compared to the value of g at the earth's surface, the value of g_P on each planet is (i) $2/2^2 = \frac{1}{2}$ as great; (ii) $4/4^2 = \frac{1}{4}$ as great; (iii) $4/2^2 = 1$ time as great—that is, the same as on earth; and (iv) $2/4^2 = \frac{1}{8}$ as great.

13.3 Answer: yes This is possible because surface gravity and escape speed depend in different ways on the planet's mass m_P and radius R_P: The value of g at the surface is Gm_P/R_P^2, while the escape speed is $\sqrt{2Gm_P/R_P}$. For the planet Saturn, for example, m_P is about 100 times the earth's mass and R_P is about 10 times the earth's radius. The value of g is different than on earth by a factor of $(100)/(10)^2 = 1$ (i.e., it is the same as on earth), while the escape speed is greater by a factor of $\sqrt{100/10} = 3.2$. It may help to remember that the surface gravity tells you about conditions right next to the planet's surface, while the escape speed (which tells you how fast you must travel to escape to infinity) depends on conditions at *all* points between the planet's surface and infinity.

13.4 Answer: (ii) Equation (13.10) shows that in a smaller-radius orbit, the spacecraft has a faster speed. The negative work

done by air resistance decreases the *total* mechanical energy $E = K + U$; the kinetic energy K increases (becomes more positive), but the gravitational potential energy U decreases (becomes more negative) by a greater amount.

13.5 Answer: (iii) Equation (13.17) shows that the orbital period T is proportional to the $\frac{3}{2}$ power of the semi-major axis a. Hence the orbital period of Comet X is longer than that of Comet Y by a factor of $4^{3/2} = 8$.

13.6 Answer: no Our analysis shows that there is *zero* gravitational force inside a hollow spherical shell. Hence visitors to the interior of a hollow planet would find themselves weightless, and they could not stand or walk on the planet's inner surface.

13.7 Answer: (iv) The discussion following Eq. (13.27) shows that the difference between the acceleration due to gravity at the equator and at the poles is v^2/R_E. Since this planet has the same radius and hence the same circumference as the earth, the speed v at its equator must be 10 times the speed of the earth's equator. Hence v^2/R_E is $10^2 = 100$ times greater than for the earth, or $100(0.0339 \text{ m/s}^2) = 3.39 \text{ m/s}^2$. The acceleration due to gravity at the poles is 9.80 m/s^2, while at the equator it is dramatically less, $9.80 \text{ m/s}^2 - 3.39 \text{ m/s}^2 = 6.41 \text{ m/s}^2$. You can show that if this planet were to rotate 17.0 times faster than the earth, the acceleration due to gravity at the equator would be *zero* and loose objects would fly off the equator's surface!

13.8 Answer: (iii) If the sun collapsed into a black hole (which, according to our understanding of stars, it cannot do), the sun would have the same mass but a much smaller radius. Because the gravitational attraction of the sun on the earth does not depend on the sun's radius, the earth's orbit would be unaffected.

Bridging Problem

Answers: (a) Perihelion: $v_P = \sqrt{\dfrac{Gm_S}{a}\dfrac{(1+e)}{(1-e)}}$

aphelion: $v_A = \sqrt{\dfrac{Gm_S}{a}\dfrac{(1-e)}{(1+e)}}$

(b) $v_P = 54.4 \text{ km/s}, v_A = 0.913 \text{ km/s}$

PERIODIC MOTION

Dogs walk with much quicker strides than do humans. Is this primarily because dogs' legs are shorter than human legs, less massive than human legs, or both?

LEARNING GOALS

By studying this chapter, you will learn:

- How to describe oscillations in terms of amplitude, period, frequency, and angular frequency.
- How to do calculations with simple harmonic motion, an important type of oscillation.
- How to use energy concepts to analyze simple harmonic motion.
- How to apply the ideas of simple harmonic motion to different physical situations.
- How to analyze the motions of a simple pendulum.
- What a physical pendulum is, and how to calculate the properties of its motion.
- What determines how rapidly an oscillation dies out.
- How a driving force applied to an oscillator at the right frequency can cause a very large response, or resonance.

Many kinds of motion repeat themselves over and over: the vibration of a quartz crystal in a watch, the swinging pendulum of a grandfather clock, the sound vibrations produced by a clarinet or an organ pipe, and the back-and-forth motion of the pistons in a car engine. This kind of motion, called **periodic motion** or **oscillation,** is the subject of this chapter. Understanding periodic motion will be essential for our later study of waves, sound, alternating electric currents, and light.

A body that undergoes periodic motion always has a stable equilibrium position. When it is moved away from this position and released, a force or torque comes into play to pull it back toward equilibrium. But by the time it gets there, it has picked up some kinetic energy, so it overshoots, stopping somewhere on the other side, and is again pulled back toward equilibrium. Picture a ball rolling back and forth in a round bowl or a pendulum that swings back and forth past its straight-down position.

In this chapter we will concentrate on two simple examples of systems that can undergo periodic motions: spring-mass systems and pendulums. We will also study why oscillations often tend to die out with time and why some oscillations can build up to greater and greater displacements from equilibrium when periodically varying forces act.

14.1 Describing Oscillation

Figure 14.1 shows one of the simplest systems that can have periodic motion. A body with mass m rests on a frictionless horizontal guide system, such as a linear air track, so it can move only along the x-axis. The body is attached to a spring of negligible mass that can be either stretched or compressed. The left end of the spring is held fixed and the right end is attached to the body. The spring force is the only horizontal force acting on the body; the vertical normal and gravitational forces always add to zero.

14.1 A system that can have periodic motion.

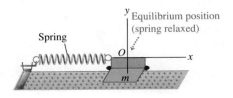

14.2 Model for periodic motion. When the body is displaced from its equilibrium position at $x = 0$, the spring exerts a restoring force back toward the equilibrium position.

(a)

$x > 0$: glider displaced to the right from the equilibrium position.

$F_x < 0$, so $a_x < 0$: stretched spring pulls glider toward equilibrium position.

(b)

$x = 0$: The relaxed spring exerts no force on the glider, so the glider has zero acceleration.

(c)

$x < 0$: glider displaced to the left from the equilibrium position.

$F_x > 0$, so $a_x > 0$: compressed spring pushes glider toward equilibrium position.

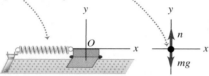

Application **Wing Frequencies**
The ruby-throated hummingbird (*Archilochus colubris*) normally flaps its wings at about 50 Hz, producing the characteristic sound that gives hummingbirds their name. Insects can flap their wings at even faster rates, from 330 Hz for a house fly and 600 Hz for a mosquito to an amazing 1040 Hz for the tiny biting midge.

It's simplest to define our coordinate system so that the origin O is at the equilibrium position, where the spring is neither stretched nor compressed. Then x is the x-component of the **displacement** of the body from equilibrium and is also the change in the length of the spring. The x-component of the force that the spring exerts on the body is F_x, and the x-component of acceleration a_x is given by $a_x = F_x/m$.

Figure 14.2 shows the body for three different displacements of the spring. Whenever the body is displaced from its equilibrium position, the spring force tends to restore it to the equilibrium position. We call a force with this character a **restoring force.** Oscillation can occur only when there is a restoring force tending to return the system to equilibrium.

Let's analyze how oscillation occurs in this system. If we displace the body to the right to $x = A$ and then let go, the net force and the acceleration are to the left (Fig. 14.2a). The speed increases as the body approaches the equilibrium position O. When the body is at O, the net force acting on it is zero (Fig. 14.2b), but because of its motion it *overshoots* the equilibrium position. On the other side of the equilibrium position the body is still moving to the left, but the net force and the acceleration are to the right (Fig. 14.2c); hence the speed decreases until the body comes to a stop. We will show later that with an ideal spring, the stopping point is at $x = -A$. The body then accelerates to the right, overshoots equilibrium again, and stops at the starting point $x = A$, ready to repeat the whole process. The body is oscillating! If there is no friction or other force to remove mechanical energy from the system, this motion repeats forever; the restoring force perpetually draws the body back toward the equilibrium position, only to have the body overshoot time after time.

In different situations the force may depend on the displacement x from equilibrium in different ways. But oscillation *always* occurs if the force is a *restoring* force that tends to return the system to equilibrium.

Amplitude, Period, Frequency, and Angular Frequency

Here are some terms that we'll use in discussing periodic motions of all kinds:

The **amplitude** of the motion, denoted by A, is the maximum magnitude of displacement from equilibrium—that is, the maximum value of $|x|$. It is always positive. If the spring in Fig. 14.2 is an ideal one, the total overall range of the motion is $2A$. The SI unit of A is the meter. A complete vibration, or **cycle,** is one complete round trip—say, from A to $-A$ and back to A, or from O to A, back through O to $-A$, and back to O. Note that motion from one side to the other (say, $-A$ to A) is a half-cycle, not a whole cycle.

The **period,** T, is the time for one cycle. It is always positive. The SI unit is the second, but it is sometimes expressed as "seconds per cycle."

The **frequency,** f, is the number of cycles in a unit of time. It is always positive. The SI unit of frequency is the hertz:

$$1 \text{ hertz} = 1 \text{ Hz} = 1 \text{ cycle/s} = 1 \text{ s}^{-1}$$

This unit is named in honor of the German physicist Heinrich Hertz (1857–1894), a pioneer in investigating electromagnetic waves.

The **angular frequency,** ω, is 2π times the frequency:

$$\omega = 2\pi f$$

We'll learn shortly why ω is a useful quantity. It represents the rate of change of an angular quantity (not necessarily related to a rotational motion) that is always measured in radians, so its units are rad/s. Since f is in cycle/s, we may regard the number 2π as having units rad/cycle.

From the definitions of period T and frequency f we see that each is the reciprocal of the other:

$$f = \frac{1}{T} \qquad T = \frac{1}{f} \qquad \text{(relationships between frequency and period)} \qquad \text{(14.1)}$$

Also, from the definition of ω,

$$\omega = 2\pi f = \frac{2\pi}{T} \quad \text{(angular frequency)} \quad (14.2)$$

Example 14.1 Period, frequency, and angular frequency

An ultrasonic transducer used for medical diagnosis oscillates at $6.7 \text{ MHz} = 6.7 \times 10^6 \text{ Hz}$. How long does each oscillation take, and what is the angular frequency?

SOLUTION

IDENTIFY and SET UP: The target variables are the period T and the angular frequency ω. We can find these using the given frequency f in Eqs. (14.1) and (14.2).

EXECUTE: From Eqs. (14.1) and (14.2),

$$T = \frac{1}{f} = \frac{1}{6.7 \times 10^6 \text{ Hz}} = 1.5 \times 10^{-7} \text{ s} = 0.15 \ \mu\text{s}$$

$$\omega = 2\pi f = 2\pi(6.7 \times 10^6 \text{ Hz})$$
$$= (2\pi \text{ rad/cycle})(6.7 \times 10^6 \text{ cycle/s})$$
$$= 4.2 \times 10^7 \text{ rad/s}$$

EVALUATE: This is a very rapid vibration, with large f and ω and small T. A slow vibration has small f and ω and large T.

Test Your Understanding of Section 14.1 A body like that shown in Fig. 14.2 oscillates back and forth. For each of the following values of the body's x-velocity v_x and x-acceleration a_x, state whether its displacement x is positive, negative, or zero. (a) $v_x > 0$ and $a_x > 0$; (b) $v_x > 0$ and $a_x < 0$; (c) $v_x < 0$ and $a_x > 0$; (d) $v_x < 0$ and $a_x < 0$; (e) $v_x = 0$ and $a_x < 0$; (f) $v_x > 0$ and $a_x = 0$.

14.2 Simple Harmonic Motion

The simplest kind of oscillation occurs when the restoring force F_x is *directly proportional* to the displacement from equilibrium x. This happens if the spring in Figs. 14.1 and 14.2 is an ideal one that obeys Hooke's law. The constant of proportionality between F_x and x is the force constant k. (You may want to review Hooke's law and the definition of the force constant in Section 6.3.) On either side of the equilibrium position, F_x and x always have opposite signs. In Section 6.3 we represented the force acting *on* a stretched ideal spring as $F_x = kx$. The x-component of force the spring exerts *on the body* is the negative of this, so the x-component of force F_x on the body is

$$F_x = -kx \quad \text{(restoring force exerted by an ideal spring)} \quad (14.3)$$

This equation gives the correct magnitude and sign of the force, whether x is positive, negative, or zero (Fig. 14.3). The force constant k is always positive and has units of N/m (a useful alternative set of units is kg/s^2). We are assuming that there is no friction, so Eq. (14.3) gives the *net* force on the body.

When the restoring force is directly proportional to the displacement from equilibrium, as given by Eq. (14.3), the oscillation is called **simple harmonic motion,** *abbreviated* **SHM.** The acceleration $a_x = d^2x/dt^2 = F_x/m$ of a body in SHM is given by

$$a_x = \frac{d^2x}{dt^2} = -\frac{k}{m}x \quad \text{(simple harmonic motion)} \quad (14.4)$$

The minus sign means the acceleration and displacement always have opposite signs. This acceleration is *not* constant, so don't even think of using the constant-acceleration equations from Chapter 2. We'll see shortly how to solve this equation to find the displacement x as a function of time. A body that undergoes simple harmonic motion is called a **harmonic oscillator.**

14.3 An idealized spring exerts a restoring force that obeys Hooke's law, $F_x = -kx$. Oscillation with such a restoring force is called simple harmonic motion.

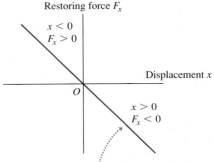

The restoring force exerted by an idealized spring is directly proportional to the displacement (Hooke's law, $F_x = -kx$): the graph of F_x versus x is a straight line.

14.4 In most real oscillations Hooke's law applies provided the body doesn't move too far from equilibrium. In such a case small-amplitude oscillations are approximately simple harmonic.

Ideal case: The restoring force obeys Hooke's law ($F_x = -kx$), so the graph of F_x versus x is a straight line.

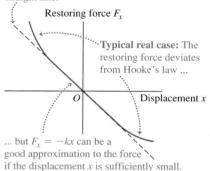

Restoring force F_x

Typical real case: The restoring force deviates from Hooke's law ...

O

Displacement x

... but $F_x = -kx$ can be a good approximation to the force if the displacement x is sufficiently small.

Why is simple harmonic motion important? Keep in mind that not all periodic motions are simple harmonic; in periodic motion in general, the restoring force depends on displacement in a more complicated way than in Eq. (14.3). But in many systems the restoring force is *approximately* proportional to displacement if the displacement is sufficiently small (Fig. 14.4). That is, if the amplitude is small enough, the oscillations of such systems are approximately simple harmonic and therefore approximately described by Eq. (14.4). Thus we can use SHM as an approximate model for many different periodic motions, such as the vibration of the quartz crystal in a watch, the motion of a tuning fork, the electric current in an alternating-current circuit, and the oscillations of atoms in molecules and solids.

Circular Motion and the Equations of SHM

To explore the properties of simple harmonic motion, we must express the displacement x of the oscillating body as a function of time, $x(t)$. The second derivative of this function, d^2x/dt^2, must be equal to $(-k/m)$ times the function itself, as required by Eq. (14.4). As we mentioned, the formulas for constant acceleration from Section 2.4 are no help because the acceleration changes constantly as the displacement x changes. Instead, we'll find $x(t)$ by noticing a striking similarity between SHM and another form of motion that we've already studied.

Figure 14.5a shows a top view of a horizontal disk of radius A with a ball attached to its rim at point Q. The disk rotates with constant angular speed ω (measured in rad/s), so the ball moves in uniform circular motion. A horizontal light beam shines on the rotating disk and casts a shadow of the ball on a screen. The shadow at point P oscillates back and forth as the ball moves in a circle. We then arrange a body attached to an ideal spring, like the combination shown in Figs. 14.1 and 14.2, so that the body oscillates parallel to the shadow. We will prove that the motion of the body and the motion of the ball's shadow are *identical* if the amplitude of the body's oscillation is equal to the disk radius A, and if the angular frequency $2\pi f$ of the oscillating body is equal to the angular speed ω of the rotating disk. That is, *simple harmonic motion is the projection of uniform circular motion onto a diameter.*

We can verify this remarkable statement by finding the acceleration of the shadow at P and comparing it to the acceleration of a body undergoing SHM, given by Eq. (14.4). The circle in which the ball moves so that its projection matches the motion of the oscillating body is called the **reference circle**; we will call the point Q the *reference point.* We take the reference circle to lie in the

14.5 (a) Relating uniform circular motion and simple harmonic motion. (b) The ball's shadow moves exactly like a body oscillating on an ideal spring.

(a) Apparatus for creating the reference circle

While the ball Q on the turntable moves in uniform circular motion, its *shadow P* moves back and forth on the screen in simple harmonic motion.

Illuminated vertical screen

$-A$ $\quad O$ $\quad P$ $\quad A$

Shadow of ball on screen

Ball's shadow

Ball on rotating turntable

Illumination

Table

Light beam

(b) An abstract representation of the motion in **(a)**

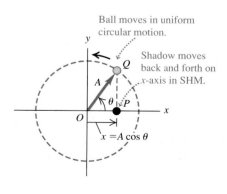

Ball moves in uniform circular motion.

Shadow moves back and forth on x-axis in SHM.

$x = A \cos \theta$

xy-plane, with the origin *O* at the center of the circle (Fig. 14.5b). At time *t* the vector *OQ* from the origin to the reference point *Q* makes an angle θ with the positive *x*-axis. As the point *Q* moves around the reference circle with constant angular speed ω, the vector *OQ* rotates with the same angular speed. Such a rotating vector is called a **phasor.** (This term was in use long before the invention of the Star Trek stun gun with a similar name. The phasor method for analyzing oscillations is useful in many areas of physics. We'll use phasors when we study alternating-current circuits in Chapter 31 and the interference of light in Chapters 35 and 36.)

The *x*-component of the phasor at time *t* is just the *x*-coordinate of the point *Q*:

$$x = A\cos\theta \qquad (14.5)$$

This is also the *x*-coordinate of the shadow *P*, which is the *projection* of *Q* onto the *x*-axis. Hence the *x*-velocity of the shadow *P* along the *x*-axis is equal to the *x*-component of the velocity vector of point *Q* (Fig. 14.6a), and the *x*-acceleration of *P* is equal to the *x*-component of the acceleration vector of *Q* (Fig. 14.6b). Since point *Q* is in uniform circular motion, its acceleration vector $\vec{a}_Q$ is always directed toward *O*. Furthermore, the magnitude of $\vec{a}_Q$ is constant and given by the angular speed squared times the radius of the circle (see Section 9.3):

$$a_Q = \omega^2 A \qquad (14.6)$$

Figure 14.6b shows that the *x*-component of $\vec{a}_Q$ is $a_x = -a_Q\cos\theta$. Combining this with Eqs. (14.5) and (14.6), we get that the acceleration of point *P* is

$$a_x = -a_Q\cos\theta = -\omega^2 A\cos\theta \qquad \text{or} \qquad (14.7)$$

$$a_x = -\omega^2 x \qquad (14.8)$$

The acceleration of point *P* is directly proportional to the displacement *x* and always has the opposite sign. These are precisely the hallmarks of simple harmonic motion.

Equation (14.8) is *exactly* the same as Eq. (14.4) for the acceleration of a harmonic oscillator, provided that the angular speed ω of the reference point *Q* is related to the force constant *k* and mass *m* of the oscillating body by

$$\omega^2 = \frac{k}{m} \qquad \text{or} \qquad \omega = \sqrt{\frac{k}{m}} \qquad (14.9)$$

We have been using the same symbol ω for the angular *speed* of the reference point *Q* and the angular *frequency* of the oscillating point *P*. The reason is that these quantities are equal! If point *Q* makes one complete revolution in time *T*, then point *P* goes through one complete cycle of oscillation in the same time; hence *T* is the period of the oscillation. During time *T* the point *Q* moves through 2π radians, so its angular speed is $\omega = 2\pi/T$. But this is just the same as Eq. (14.2) for the angular frequency of the point *P*, which verifies our statement about the two interpretations of ω. This is why we introduced angular frequency in Section 14.1; this quantity makes the connection between oscillation and circular motion. So we reinterpret Eq. (14.9) as an expression for the angular frequency of simple harmonic motion for a body of mass *m*, acted on by a restoring force with force constant *k*:

$$\omega = \sqrt{\frac{k}{m}} \qquad \text{(simple harmonic motion)} \qquad (14.10)$$

When you start a body oscillating in SHM, the value of ω is not yours to choose; it is predetermined by the values of *k* and *m*. The units of *k* are N/m or kg/s², so k/m is in $(\text{kg/s}^2)/\text{kg} = \text{s}^{-2}$. When we take the square root in Eq. (14.10), we get s^{-1}, or more properly rad/s because this is an *angular* frequency (recall that a radian is not a true unit).

14.6 The (a) *x*-velocity and (b) *x*-acceleration of the ball's shadow *P* (see Fig. 14.5) are the *x*-components of the velocity and acceleration vectors, respectively, of the ball *Q*.

(a) Using the reference circle to determine the *x*-velocity of point *P*

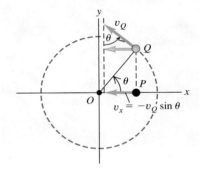

(b) Using the reference circle to determine the *x*-acceleration of point *P*

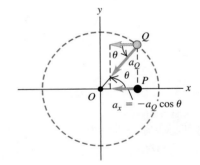

According to Eqs. (14.1) and (14.2), the frequency f and period T are

$$f = \frac{\omega}{2\pi} = \frac{1}{2\pi}\sqrt{\frac{k}{m}} \quad \text{(simple harmonic motion)} \qquad (14.11)$$

$$T = \frac{1}{f} = \frac{2\pi}{\omega} = 2\pi\sqrt{\frac{m}{k}} \quad \text{(simple harmonic motion)} \qquad (14.12)$$

14.7 The greater the mass m in a tuning fork's tines, the lower the frequency of oscillation $f = (1/2\pi)\sqrt{k/m}$ and the lower the pitch of the sound that the tuning fork produces.

Tines with large mass m:
low frequency $f = 128$ Hz

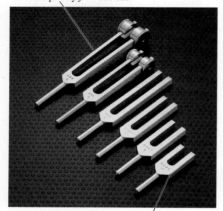

Tines with small mass m:
high frequency $f = 4096$ Hz

We see from Eq. (14.12) that a larger mass m, with its greater inertia, will have less acceleration, move more slowly, and take a longer time for a complete cycle (Fig. 14.7). In contrast, a stiffer spring (one with a larger force constant k) exerts a greater force at a given deformation x, causing greater acceleration, higher speeds, and a shorter time T per cycle.

CAUTION **Don't confuse frequency and angular frequency** You can run into trouble if you don't make the distinction between frequency f and angular frequency $\omega = 2\pi f$. Frequency tells you how many cycles of oscillation occur per second, while angular frequency tells you how many radians per second this corresponds to on the reference circle. In solving problems, pay careful attention to whether the goal is to find f or ω.

Period and Amplitude in SHM

Equations (14.11) and (14.12) show that the period and frequency of simple harmonic motion are completely determined by the mass m and the force constant k. *In simple harmonic motion the period and frequency do not depend on the amplitude A.* For given values of m and k, the time of one complete oscillation is the same whether the amplitude is large or small. Equation (14.3) shows why we should expect this. Larger A means that the body reaches larger values of $|x|$ and is subjected to larger restoring forces. This increases the average speed of the body over a complete cycle; this exactly compensates for having to travel a larger distance, so the same total time is involved.

The oscillations of a tuning fork are essentially simple harmonic motion, which means that it always vibrates with the same frequency, independent of amplitude. This is why a tuning fork can be used as a standard for musical pitch. If it were not for this characteristic of simple harmonic motion, it would be impossible to make familiar types of mechanical and electronic clocks run accurately or to play most musical instruments in tune. If you encounter an oscillating body with a period that *does* depend on the amplitude, the oscillation is *not* simple harmonic motion.

Example 14.2   **Angular frequency, frequency, and period in SHM**

A spring is mounted horizontally, with its left end fixed. A spring balance attached to the free end and pulled toward the right (Fig. 14.8a) indicates that the stretching force is proportional to the displacement, and a force of 6.0 N causes a displacement of 0.030 m. We replace the spring balance with a 0.50-kg glider, pull it 0.020 m to the right along a frictionless air track, and release it from rest (Fig. 14.8b). (a) Find the force constant k of the spring. (b) Find the angular frequency ω, frequency f, and period T of the resulting oscillation.

14.8 (a) The force exerted *on* the spring (shown by the vector F) has x-component $F_x = +6.0$ N. The force exerted *by* the spring has x-component $F_x = -6.0$ N. (b) A glider is attached to the same spring and allowed to oscillate.

(a)

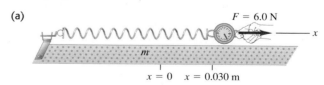

(b)

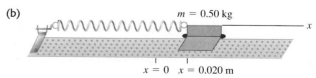

SOLUTION

IDENTIFY and SET UP: Because the spring force (equal in magnitude to the stretching force) is proportional to the displacement, the motion is simple harmonic. We find k using Hooke's law, Eq. (14.3), and ω, f, and T using Eqs. (14.10), (14.11), and (14.12), respectively.

EXECUTE: (a) When $x = 0.030$ m, the force the spring exerts on the spring balance is $F_x = -6.0$ N. From Eq. (14.3),

$$k = -\frac{F_x}{x} = -\frac{-6.0 \text{ N}}{0.030 \text{ m}} = 200 \text{ N/m} = 200 \text{ kg/s}^2$$

(b) From Eq. (14.10), with $m = 0.50$ kg,

$$\omega = \sqrt{\frac{k}{m}} = \sqrt{\frac{200 \text{ kg/s}^2}{0.50 \text{ kg}}} = 20 \text{ rad/s}$$

$$f = \frac{\omega}{2\pi} = \frac{20 \text{ rad/s}}{2\pi \text{ rad/cycle}} = 3.2 \text{ cycle/s} = 3.2 \text{ Hz}$$

$$T = \frac{1}{f} = \frac{1}{3.2 \text{ cycle/s}} = 0.31 \text{ s}$$

EVALUATE: The amplitude of the oscillation is 0.020 m, the distance that we pulled the glider before releasing it. In SHM the angular frequency, frequency, and period are all independent of the amplitude. Note that a period is usually stated in "seconds" rather than "seconds per cycle."

Displacement, Velocity, and Acceleration in SHM

We still need to find the displacement x as a function of time for a harmonic oscillator. Equation (14.4) for a body in simple harmonic motion along the x-axis is identical to Eq. (14.8) for the x-coordinate of the reference point in uniform circular motion with constant angular speed $\omega = \sqrt{k/m}$. Hence Eq. (14.5), $x = A \cos \theta$, describes the x-coordinate for both of these situations. If at $t = 0$ the phasor OQ makes an angle ϕ (the Greek letter phi) with the positive x-axis, then at any later time t this angle is $\theta = \omega t + \phi$. We substitute this into Eq. (14.5) to obtain

$$x = A\cos(\omega t + \phi) \quad \text{(displacement in SHM)} \quad (14.13)$$

where $\omega = \sqrt{k/m}$. Figure 14.9 shows a graph of Eq. (14.13) for the particular case $\phi = 0$. The displacement x is a periodic function of time, as expected for SHM. We could also have written Eq. (14.13) in terms of a sine function rather than a cosine by using the identity $\cos \alpha = \sin(\alpha + \pi/2)$. *In simple harmonic motion the position is a periodic, sinusoidal function of time.* There are many other periodic functions, but none so simple as a sine or cosine function.

The value of the cosine function is always between -1 and 1, so in Eq. (14.13), x is always between $-A$ and A. This confirms that A is the amplitude of the motion.

The period T is the time for one complete cycle of oscillation, as Fig. 14.9 shows. The cosine function repeats itself whenever the quantity in parentheses in Eq. (14.13) increases by 2π radians. Thus, if we start at time $t = 0$, the time T to complete one cycle is given by

$$\omega T = \sqrt{\frac{k}{m}} T = 2\pi \quad \text{or} \quad T = 2\pi \sqrt{\frac{m}{k}}$$

which is just Eq. (14.12). Changing either m or k changes the period of oscillation, as shown in Figs. 14.10a and 14.10b. The period does not depend on the amplitude A (Fig. 14.10c).

MasteringPHYSICS

PhET: Motion in 2D
ActivPhysics 9.1: Position Graphs and Equations
ActivPhysics 9.2: Describing Vibrational Motion
ActivPhysics 9.5: Age Drops Tarzan

14.9 Graph of x versus t [see Eq. (14.13)] for simple harmonic motion. The case shown has $\phi = 0$.

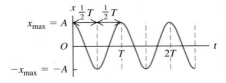

14.10 Variations of simple harmonic motion. All cases shown have $\phi = 0$ [see Eq. (14.13)].

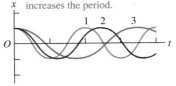

(a) Increasing m; same A and k

Mass m increases from curve 1 to 2 to 3. Increasing m alone increases the period.

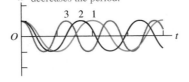

(b) Increasing k; same A and m

Force constant k increases from curve 1 to 2 to 3. Increasing k alone decreases the period.

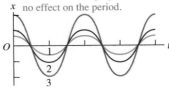

(c) Increasing A; same k and m

Amplitude A increases from curve 1 to 2 to 3. Changing A alone has no effect on the period.

14.11 Variations of SHM: displacement versus time for the same harmonic oscillator with different phase angles ϕ.

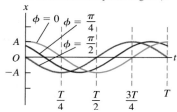

These three curves show SHM with the same period T and amplitude A but with different phase angles ϕ.

14.12 Graphs of (a) x versus t, (b) v_x versus t, and (c) a_x versus t for a body in SHM. For the motion depicted in these graphs, $\phi = \pi/3$.

(a) Displacement x as a function of time t

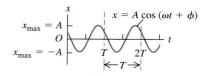

(b) Velocity v_x as a function of time t

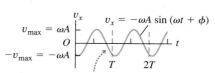

The v_x-t graph is shifted by $\frac{1}{4}$ cycle from the x-t graph.

(c) Acceleration a_x as a function of time t

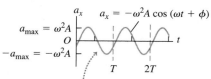

The a_x-t graph is shifted by $\frac{1}{4}$ cycle from the v_x-t graph and by $\frac{1}{2}$ cycle from the x-t graph.

The constant ϕ in Eq. (14.13) is called the **phase angle.** It tells us at what point in the cycle the motion was at $t = 0$ (equivalent to where around the circle the point Q was at $t = 0$). We denote the position at $t = 0$ by x_0. Putting $t = 0$ and $x = x_0$ in Eq. (14.13), we get

$$x_0 = A \cos \phi \qquad (14.14)$$

If $\phi = 0$, then $x_0 = A \cos 0 = A$, and the body starts at its maximum positive displacement. If $\phi = \pi$, then $x_0 = A \cos \pi = -A$, and the particle starts at its maximum *negative* displacement. If $\phi = \pi/2$, then $x_0 = A \cos(\pi/2) = 0$, and the particle is initially at the origin. Figure 14.11 shows the displacement x versus time for three different phase angles.

We find the velocity v_x and acceleration a_x as functions of time for a harmonic oscillator by taking derivatives of Eq. (14.13) with respect to time:

$$v_x = \frac{dx}{dt} = -\omega A \sin(\omega t + \phi) \qquad \text{(velocity in SHM)} \qquad (14.15)$$

$$a_x = \frac{dv_x}{dt} = \frac{d^2x}{dt^2} = -\omega^2 A \cos(\omega t + \phi) \qquad \text{(acceleration in SHM)} \qquad (14.16)$$

The velocity v_x oscillates between $v_{max} = +\omega A$ and $-v_{max} = -\omega A$, and the acceleration a_x oscillates between $a_{max} = +\omega^2 A$ and $-a_{max} = -\omega^2 A$ (Fig. 14.12). Comparing Eq. (14.16) with Eq. (14.13) and recalling that $\omega^2 = k/m$ from Eq. (14.9), we see that

$$a_x = -\omega^2 x = -\frac{k}{m} x$$

which is just Eq. (14.4) for simple harmonic motion. This confirms that Eq. (14.13) for x as a function of time is correct.

We actually derived Eq. (14.16) earlier in a geometrical way by taking the x-component of the acceleration vector of the reference point Q. This was done in Fig. 14.6b and Eq. (14.7) (recall that $\theta = \omega t + \phi$). In the same way, we could have derived Eq. (14.15) by taking the x-component of the velocity vector of Q, as shown in Fig. 14.6b. We'll leave the details for you to work out.

Note that the sinusoidal graph of displacement versus time (Fig. 14.12a) is shifted by one-quarter period from the graph of velocity versus time (Fig. 14.12b) and by one-half period from the graph of acceleration versus time (Fig. 14.12c). Figure 14.13 shows why this is so. When the body is passing through the equilibrium position so that the displacement is zero, the velocity equals either v_{max} or $-v_{max}$ (depending on which way the body is moving) and the acceleration is zero. When the body is at either its maximum positive displacement, $x = +A$, or its maximum negative displacement, $x = -A$, the velocity is zero and the body is instantaneously at rest. At these points, the restoring force $F_x = -kx$ and the acceleration of the body have their maximum magnitudes. At $x = +A$ the acceleration is negative and equal to $-a_{max}$. At $x = -A$ the acceleration is positive: $a_x = +a_{max}$.

If we are given the initial position x_0 and initial velocity v_{0x} for the oscillating body, we can determine the amplitude A and the phase angle ϕ. Here's how to do it. The initial velocity v_{0x} is the velocity at time $t = 0$; putting $v_x = v_{0x}$ and $t = 0$ in Eq. (14.15), we find

$$v_{0x} = -\omega A \sin \phi \qquad (14.17)$$

To find ϕ, we divide Eq. (14.17) by Eq. (14.14). This eliminates A and gives an equation that we can solve for ϕ:

$$\frac{v_{0x}}{x_0} = \frac{-\omega A \sin\phi}{A\cos\phi} = -\omega\tan\phi$$

$$\phi = \arctan\left(-\frac{v_{0x}}{\omega x_0}\right) \qquad \text{(phase angle in SHM)} \qquad (14.18)$$

It is also easy to find the amplitude A if we are given x_0 and v_{0x}. We'll sketch the derivation, and you can fill in the details. Square Eq. (14.14); then divide Eq. (14.17) by ω, square it, and add to the square of Eq. (14.14). The right side will be $A^2(\sin^2\phi + \cos^2\phi)$, which is equal to A^2. The final result is

$$A = \sqrt{x_0^2 + \frac{v_{0x}^2}{\omega^2}} \qquad \text{(amplitude in SHM)} \qquad (14.19)$$

Note that when the body has both an initial displacement x_0 and a nonzero initial velocity v_{0x}, the amplitude A is *not* equal to the initial displacement. That's reasonable; if you start the body at a positive x_0 but give it a positive velocity v_{0x}, it will go *farther* than x_0 before it turns and comes back.

14.13 How x-velocity v_x and x-acceleration a_x vary during one cycle of SHM.

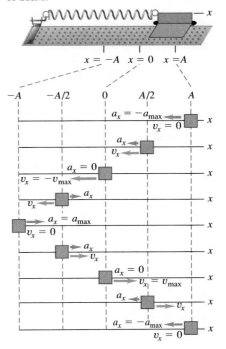

Simple Harmonic Motion I: Describing Motion

IDENTIFY *the relevant concepts:* An oscillating system undergoes simple harmonic motion (SHM) *only* if the restoring force is directly proportional to the displacement.

SET UP *the problem* using the following steps:
1. Identify the known and unknown quantities, and determine which are the target variables.
2. Distinguish between two kinds of quantities. *Properties of the system* include the mass m, the force constant k, and quantities derived from m and k, such as the period T, frequency f, and angular frequency ω. These are independent of *properties of the motion*, which describe how the system behaves when it is set into motion in a particular way; they include the amplitude A, maximum velocity v_{max}, and phase angle ϕ, and values of x, v_x, and a_x at particular times.
3. If necessary, define an x-axis as in Fig. 14.13, with the equilibrium position at $x = 0$.

EXECUTE *the solution* as follows:
1. Use the equations given in Sections 14.1 and 14.2 to solve for the target variables.
2. To find the values of x, v_x, and a_x at particular times, use Eqs. (14.13), (14.15), and (14.16), respectively. If the initial position x_0 and initial velocity v_{0x} are both given, determine ϕ and A from Eqs. (14.18) and (14.19). If the body has an initial positive displacement x_0 but zero initial velocity ($v_{0x} = 0$), then the amplitude is $A = x_0$ and the phase angle is $\phi = 0$. If it has an initial positive velocity v_{0x} but no initial displacement ($x_0 = 0$), the amplitude is $A = v_{0x}/\omega$ and the phase angle is $\phi = -\pi/2$. Express all phase angles in *radians*.

EVALUATE *your answer:* Make sure that your results are consistent. For example, suppose you used x_0 and v_{0x} to find general expressions for x and v_x at time t. If you substitute $t = 0$ into these expressions, you should get back the given values of x_0 and v_{0x}.

Example 14.3 **Describing SHM**

We give the glider of Example 14.2 an initial displacement $x_0 = +0.015$ m and an initial velocity $v_{0x} = +0.40$ m/s. (a) Find the period, amplitude, and phase angle of the resulting motion. (b) Write equations for the displacement, velocity, and acceleration as functions of time.

SOLUTION

IDENTIFY and SET UP: As in Example 14.2, the oscillations are SHM. We use equations from this section and the given values $k = 200$ N/m, $m = 0.50$ kg, x_0, and v_{0x} to calculate the target variables A and ϕ and to obtain expressions for x, v_x, and a_x.

Continued

EXECUTE: (a) In SHM the period and angular frequency are *properties of the system* that depend only on k and m, not on the amplitude, and so are the same as in Example 14.2 ($T = 0.31$ s and $\omega = 20$ rad/s). From Eq. (14.19), the amplitude is

$$A = \sqrt{x_0^2 + \frac{v_{0x}^2}{\omega^2}} = \sqrt{(0.015 \text{ m})^2 + \frac{(0.40 \text{ m/s})^2}{(20 \text{ rad/s})^2}} = 0.025 \text{ m}$$

We use Eq. (14.18) to find the phase angle:

$$\phi = \arctan\left(-\frac{v_{0x}}{\omega x_0}\right)$$

$$= \arctan\left(-\frac{0.40 \text{ m/s}}{(20 \text{ rad/s})(0.015 \text{ m})}\right) = -53° = -0.93 \text{ rad}$$

(b) The displacement, velocity, and acceleration at any time are given by Eqs. (14.13), (14.15), and (14.16), respectively. We substitute the values of A, ω, and ϕ into these equations:

$$x = (0.025 \text{ m})\cos[(20 \text{ rad/s})t - 0.93 \text{ rad}]$$
$$v_x = -(0.50 \text{ m/s})\sin[(20 \text{ rad/s})t - 0.93 \text{ rad}]$$
$$a_x = -(10 \text{ m/s}^2)\cos[(20 \text{ rad/s})t - 0.93 \text{ rad}]$$

EVALUATE: You can check the expressions for x and v_x by confirming that if you substitute $t = 0$, they yield $x = x_0 = 0.015$ m and $v_x = v_{0x} = 0.40$ m/s.

Test Your Understanding of Section 14.2 A glider is attached to a spring as shown in Fig. 14.13. If the glider is moved to $x = 0.10$ m and released from rest at time $t = 0$, it will oscillate with amplitude $A = 0.10$ m and phase angle $\phi = 0$. (a) Suppose instead that at $t = 0$ the glider is at $x = 0.10$ m and is moving to the right in Fig. 14.13. In this situation is the amplitude greater than, less than, or equal to 0.10 m? Is the phase angle greater than, less than, or equal to zero? (b) Suppose instead that at $t = 0$ the glider is at $x = 0.10$ m and is moving to the left in Fig. 14.13. In this situation is the amplitude greater than, less than, or equal to 0.10 m? Is the phase angle greater than, less than, or equal to zero? ∎

PhET: Masses & Springs
ActivPhysics 9.3: Vibrational Energy
ActivPhysics 9.4: Two Ways to Weigh Young Tarzan
ActivPhysics 9.6: Releasing a Vibrating Skier I
ActivPhysics 9.7: Releasing a Vibrating Skier II
ActivPhysics 9.8: One- and Two-Spring Vibrating Systems
ActivPhysics 9.9: Vibro-Ride

14.3 Energy in Simple Harmonic Motion

We can learn even more about simple harmonic motion by using energy considerations. Take another look at the body oscillating on the end of a spring in Figs. 14.2 and 14.13. We've already noted that the spring force is the only horizontal force on the body. The force exerted by an ideal spring is a conservative force, and the vertical forces do no work, so the total mechanical energy of the system is *conserved.* We also assume that the mass of the spring itself is negligible.

The kinetic energy of the body is $K = \frac{1}{2}mv^2$ and the potential energy of the spring is $U = \frac{1}{2}kx^2$, just as in Section 7.2. (You'll find it helpful to review that section.) There are no nonconservative forces that do work, so the total mechanical energy $E = K + U$ is conserved:

$$E = \tfrac{1}{2}mv_x^2 + \tfrac{1}{2}kx^2 = \text{constant} \qquad (14.20)$$

(Since the motion is one-dimensional, $v^2 = v_x^2$.)

The total mechanical energy E is also directly related to the amplitude A of the motion. When the body reaches the point $x = A$, its maximum displacement from equilibrium, it momentarily stops as it turns back toward the equilibrium position. That is, when $x = A$ (or $-A$), $v_x = 0$. At this point the energy is entirely potential, and $E = \frac{1}{2}kA^2$. Because E is constant, it is equal to $\frac{1}{2}kA^2$ at any other point. Combining this expression with Eq. (14.20), we get

$$E = \tfrac{1}{2}mv_x^2 + \tfrac{1}{2}kx^2 = \tfrac{1}{2}kA^2 = \text{constant} \qquad \begin{array}{l}\text{(total mechanical}\\ \text{energy in SHM)}\end{array} \qquad (14.21)$$

We can verify this equation by substituting x and v_x from Eqs. (14.13) and (14.15) and using $\omega^2 = k/m$ from Eq. (14.9):

$$E = \tfrac{1}{2}mv_x^2 + \tfrac{1}{2}kx^2 = \tfrac{1}{2}m[-\omega A\sin(\omega t + \phi)]^2 + \tfrac{1}{2}k[A\cos(\omega t + \phi)]^2$$
$$= \tfrac{1}{2}kA^2\sin^2(\omega t + \phi) + \tfrac{1}{2}kA^2\cos^2(\omega t + \phi)$$
$$= \tfrac{1}{2}kA^2$$

14.14 Graphs of E, K, and U versus displacement in SHM. The velocity of the body is *not* constant, so these images of the body at equally spaced positions are *not* equally spaced in time.

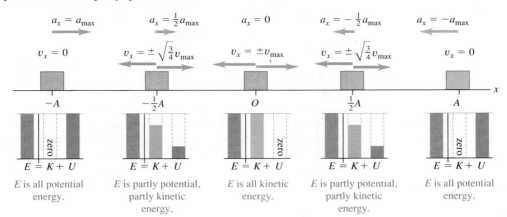

(Recall that $\sin^2 \alpha + \cos^2 \alpha = 1$.) Hence our expressions for displacement and velocity in SHM are consistent with energy conservation, as they must be.

We can use Eq. (14.21) to solve for the velocity v_x of the body at a given displacement x:

$$v_x = \pm \sqrt{\frac{k}{m}} \sqrt{A^2 - x^2} \qquad (14.22)$$

The $\pm$ sign means that at a given value of x the body can be moving in either direction. For example, when $x = \pm A/2$,

$$v_x = \pm \sqrt{\frac{k}{m}} \sqrt{A^2 - \left(\pm \frac{A}{2}\right)^2} = \pm \sqrt{\frac{3}{4}} \sqrt{\frac{k}{m}} A$$

Equation (14.22) also shows that the *maximum* speed v_{max} occurs at $x = 0$. Using Eq. (14.10), $\omega = \sqrt{k/m}$, we find that

$$v_{max} = \sqrt{\frac{k}{m}} A = \omega A \qquad (14.23)$$

This agrees with Eq. (14.15): v_x oscillates between $-\omega A$ and $+\omega A$.

Interpreting E, K, and U in SHM

Figure 14.14 shows the energy quantities E, K, and U at $x = 0$, $x = \pm A/2$, and $x = \pm A$. Figure 14.15 is a graphical display of Eq. (14.21); energy (kinetic, potential, and total) is plotted vertically and the coordinate x is plotted horizontally.

(a) The potential energy U and total mechanical energy E for a body in SHM as a function of displacement x

The total mechanical energy E is constant.

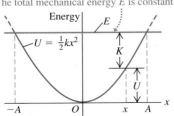

(b) The same graph as in (a), showing kinetic energy K as well

At $x = \pm A$ the energy is all potential; the kinetic energy is zero.

At $x = 0$ the energy is all kinetic; the potential energy is zero.

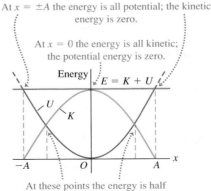

At these points the energy is half kinetic and half potential.

14.15 Kinetic energy K, potential energy U, and total mechanical energy E as functions of position for SHM. At each value of x the sum of the values of K and U equals the constant value of E. Can you show that the energy is half kinetic and half potential at $x = \pm \sqrt{\frac{1}{2}} A$?

The parabolic curve in Fig. 14.15a represents the potential energy $U = \frac{1}{2}kx^2$. The horizontal line represents the total mechanical energy E, which is constant and does not vary with x. At any value of x between $-A$ and A, the vertical distance from the x-axis to the parabola is U; since $E = K + U$, the remaining vertical distance up to the horizontal line is K. Figure 14.15b shows both K and U as functions of x. The horizontal line for E intersects the potential-energy curve at $x = -A$ and $x = A$, so at these points the energy is entirely potential, the kinetic energy is zero, and the body comes momentarily to rest before reversing direction. As the body oscillates between $-A$ and A, the energy is continuously transformed from potential to kinetic and back again.

Figure 14.15a shows the connection between the amplitude A and the corresponding total mechanical energy $E = \frac{1}{2}kA^2$. If we tried to make x greater than A (or less than $-A$), U would be greater than E, and K would have to be negative. But K can never be negative, so x can't be greater than A or less than $-A$.

Problem-Solving Strategy 14.2 **Simple Harmonic Motion II: Energy**

The SHM energy equation, Eq. (14.21), is a useful relationship among velocity, position, and total mechanical energy. If the problem requires you to relate position, velocity, and acceleration without reference to time, consider using Eq. (14.4) (from Newton's second law) or Eq. (14.21) (from energy conservation). Because Eq. (14.21) involves x^2 and v_x^2, you must infer the *signs* of x and v_x from the situation. For instance, if the body is moving from the equilibrium position toward the point of greatest positive displacement, then x is positive and v_x is positive.

Example 14.4 **Velocity, acceleration, and energy in SHM**

(a) Find the maximum and minimum velocities attained by the oscillating glider of Example 14.2. (b) Find the maximum and minimum accelerations. (c) Find the velocity v_x and acceleration a_x when the glider is halfway from its initial position to the equilibrium position $x = 0$. (d) Find the total energy, potential energy, and kinetic energy at this position.

SOLUTION

IDENTIFY and SET UP: The problem concerns properties of the motion at specified *positions*, not at specified *times*, so we can use the energy relationships of this section. Figure 14.13 shows our choice of x-axis. The maximum displacement from equilibrium is $A = 0.020$ m. We use Eqs. (14.22) and (14.4) to find v_x and a_x for a given x. We then use Eq. (14.21) for given x and v_x to find the total, potential, and kinetic energies E, U, and K.

EXECUTE: (a) From Eq. (14.22), the velocity v_x at any displacement x is

$$v_x = \pm\sqrt{\frac{k}{m}}\sqrt{A^2 - x^2}$$

The glider's maximum *speed* occurs when it is moving through $x = 0$:

$$v_{\text{max}} = \sqrt{\frac{k}{m}}\,A = \sqrt{\frac{200 \text{ N/m}}{0.50 \text{ kg}}}(0.020 \text{ m}) = 0.40 \text{ m/s}$$

Its maximum and minimum (most negative) *velocities* are $+0.40$ m/s and -0.40 m/s, which occur when it is moving through $x = 0$ to the right and left, respectively.

(b) From Eq. (14.4), $a_x = -(k/m)x$. The glider's maximum (most positive) acceleration occurs at the most negative value of x, $x = -A$:

$$a_{\text{max}} = -\frac{k}{m}(-A) = -\frac{200 \text{ N/m}}{0.50 \text{ kg}}(-0.020 \text{ m}) = 8.0 \text{ m/s}^2$$

The minimum (most negative) acceleration is $a_{\text{min}} = -8.0$ m/s^2, which occurs at $x = +A = +0.020$ m.

(c) The point halfway from $x = x_0 = A$ to $x = 0$ is $x = A/2 = 0.010$ m. From Eq. (14.22), at this point

$$v_x = -\sqrt{\frac{200 \text{ N/m}}{0.50 \text{ kg}}}\sqrt{(0.020 \text{ m})^2 - (0.010 \text{ m})^2} = -0.35 \text{ m/s}$$

We choose the negative square root because the glider is moving from $x = A$ toward $x = 0$. From Eq. (14.4),

$$a_x = -\frac{200 \text{ N/m}}{0.50 \text{ kg}}(0.010 \text{ m}) = -4.0 \text{ m/s}^2$$

Figure 14.14 shows the conditions at $x = 0$, $\pm A/2$, and $\pm A$.

(d) The energies are

$$E = \tfrac{1}{2}kA^2 = \tfrac{1}{2}(200 \text{ N/m})(0.020 \text{ m})^2 = 0.040 \text{ J}$$

$$U = \tfrac{1}{2}kx^2 = \tfrac{1}{2}(200 \text{ N/m})(0.010 \text{ m})^2 = 0.010 \text{ J}$$

$$K = \tfrac{1}{2}mv_x^2 = \tfrac{1}{2}(0.50 \text{ kg})(-0.35 \text{ m/s})^2 = 0.030 \text{ J}$$

EVALUATE: At $x = A/2$, the total energy is one-fourth potential energy and three-fourths kinetic energy. You can confirm this by inspecting Fig. 14.15b.

Example 14.5 Energy and momentum in SHM

A block of mass M attached to a horizontal spring with force constant k is moving in SHM with amplitude A_1. As the block passes through its equilibrium position, a lump of putty of mass m is dropped from a small height and sticks to it. (a) Find the new amplitude and period of the motion. (b) Repeat part (a) if the putty is dropped onto the block when it is at one end of its path.

SOLUTION

IDENTIFY and SET UP: The problem involves the motion at a given position, not a given time, so we can use energy methods. Figure 14.16 shows our sketches. Before the putty falls, the mechanical energy of the block–spring system is constant. In part (a), the putty–block collision is completely inelastic: The horizontal component of momentum is conserved, kinetic energy decreases, and the amount of mass that's oscillating increases. After the collision, the mechanical energy remains constant at its new value. In part (b) the oscillating mass also increases, but the block isn't moving when the putty is added; there is effectively no collision at all, and no mechanical energy is lost. We find the amplitude A_2 after each collision from the final energy of the system using Eq. (14.21) and conservation of momentum. The period T_2 after the collision is a *property of the system*, so it is the same in both parts (a) and (b); we find it using Eq. (14.12).

EXECUTE: (a) Before the collision the total mechanical energy of the block and spring is $E_1 = \frac{1}{2}kA_1^2$. The block is at $x = 0$, so $U = 0$ and the energy is purely kinetic (Fig. 14.16a). If we let v_1 be the speed of the block at this point, then $E_1 = \frac{1}{2}kA_1^2 = \frac{1}{2}Mv_1^2$ and

$$v_1 = \sqrt{\frac{k}{M}}A_1$$

During the collision the x-component of momentum of the block–putty system is conserved. (Why?) Just before the collision this component is the sum of Mv_1 (for the block) and zero (for the putty). Just after the collision the block and putty move together with speed v_2, so their combined x-component of momentum is $(M + m)v_2$. From conservation of momentum,

$$Mv_1 + 0 = (M + m)v_2 \quad \text{so} \quad v_2 = \frac{M}{M + m}v_1$$

We assume that the collision lasts a very short time, so that the block and putty are still at the equilibrium position just after the collision. The energy is still purely kinetic but is *less* than before the collision:

$$E_2 = \frac{1}{2}(M + m)v_2^2 = \frac{1}{2}\frac{M^2}{M + m}v_1^2$$

$$= \frac{M}{M + m}\left(\frac{1}{2}Mv_1^2\right) = \left(\frac{M}{M + m}\right)E_1$$

14.16 Our sketches for this problem.

(a)

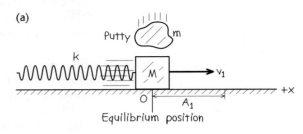

(b)

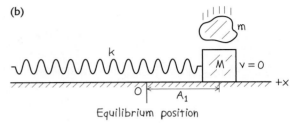

Since $E_2 = \frac{1}{2}kA_2^2$, where A_2 is the amplitude after the collision, we have

$$\frac{1}{2}kA_2^2 = \left(\frac{M}{M + m}\right)\frac{1}{2}kA_1^2$$

$$A_2 = A_1\sqrt{\frac{M}{M + m}}$$

From Eq. (14.12), the period of oscillation after the collision is

$$T_2 = 2\pi\sqrt{\frac{M + m}{k}}$$

(b) When the putty falls, the block is instantaneously at rest (Fig. 14.16b). The x-component of momentum is zero both before and after the collision. The block and putty have zero kinetic energy just before and just after the collision. The energy is all potential energy stored in the spring, so adding the putty has *no effect* on the mechanical energy. That is, $E_2 = E_1 = \frac{1}{2}kA_1^2$, and the amplitude is unchanged: $A_2 = A_1$. The period is again $T_2 = 2\pi\sqrt{(M + m)/k}$.

EVALUATE: Energy is lost in part (a) because the putty slides against the moving block during the collision, and energy is dissipated by kinetic friction. No energy is lost in part (b), because there is no sliding during the collision.

Test Your Understanding of Section 14.3 (a) To double the total energy for a mass-spring system oscillating in SHM, by what factor must the amplitude increase? (i) 4; (ii) 2; (iii) $\sqrt{2} = 1.414$; (iv) $\sqrt[4]{2} = 1.189$. (b) By what factor will the frequency change due to this amplitude increase? (i) 4; (ii) 2; (iii) $\sqrt{2} = 1.414$; (iv) $\sqrt[4]{2} = 1.189$; (v) it does not change.

14.4 Applications of Simple Harmonic Motion

So far, we've looked at a grand total of *one* situation in which simple harmonic motion (SHM) occurs: a body attached to an ideal horizontal spring. But SHM can occur in any system in which there is a restoring force that is directly proportional to the displacement from equilibrium, as given by Eq. (14.3), $F_x = -kx$. The restoring force will originate in different ways in different situations, so the force constant k has to be found for each case by examining the net force on the system. Once this is done, it's straightforward to find the angular frequency ω, frequency f, and period T; we just substitute the value of k into Eqs. (14.10), (14.11), and (14.12), respectively. Let's use these ideas to examine several examples of simple harmonic motion.

Vertical SHM

Suppose we hang a spring with force constant k (Fig. 14.17a) and suspend from it a body with mass m. Oscillations will now be vertical; will they still be SHM? In Fig. 14.17b the body hangs at rest, in equilibrium. In this position the spring is stretched an amount Δl just great enough that the spring's upward vertical force $k\,\Delta l$ on the body balances its weight mg:

$$k\,\Delta l = mg$$

Take $x = 0$ to be this equilibrium position and take the positive x-direction to be upward. When the body is a distance x *above* its equilibrium position (Fig. 14.17c), the extension of the spring is $\Delta l - x$. The upward force it exerts on the body is then $k(\Delta l - x)$, and the net x-component of force on the body is

$$F_{\text{net}} = k(\Delta l - x) + (-mg) = -kx$$

that is, a net downward force of magnitude kx. Similarly, when the body is *below* the equilibrium position, there is a net upward force with magnitude kx. In either case there is a restoring force with magnitude kx. If the body is set in vertical motion, it oscillates in SHM with the same angular frequency as though it were horizontal, $\omega = \sqrt{k/m}$. So vertical SHM doesn't differ in any essential way from horizontal SHM. The only real change is that the equilibrium position $x = 0$ no longer corresponds to the point at which the spring is unstretched. The same ideas hold if a body with weight mg is placed atop a compressible spring (Fig. 14.18) and compresses it a distance Δl.

14.17 A body attached to a hanging spring.

(a)

A hanging spring that obeys Hooke's law

(b) A body is suspended from the spring. It is in equilibrium when the upward force exerted by the stretched spring equals the body's weight.

(c) If the body is displaced from equilibrium, the net force on the body is proportional to its displacement. The oscillations are SHM.

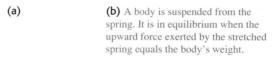

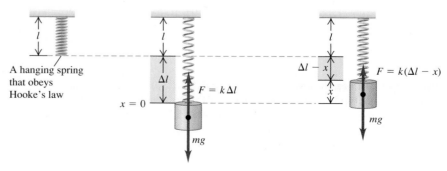

Example 14.6   Vertical SHM in an old car

The shock absorbers in an old car with mass 1000 kg are completely worn out. When a 980-N person climbs slowly into the car at its center of gravity, the car sinks 2.8 cm. The car (with the person aboard) hits a bump, and the car starts oscillating up and down in SHM. Model the car and person as a single body on a single spring, and find the period and frequency of the oscillation.

SOLUTION

IDENTIFY and SET UP: The situation is like that shown in Fig. 14.18. The compression of the spring when the person's weight is added tells us the force constant, which we can use to find the period and frequency (the target variables).

EXECUTE: When the force increases by 980 N, the spring compresses an additional 0.028 m, and the x-coordinate of the car

changes by -0.028 m. Hence the effective force constant (including the effect of the entire suspension) is

$$k = -\frac{F_x}{x} = -\frac{980 \text{ N}}{-0.028 \text{ m}} = 3.5 \times 10^4 \text{ kg/s}^2$$

The person's mass is $w/g = (980 \text{ N})/(9.8 \text{ m/s}^2) = 100$ kg. The *total* oscillating mass is $m = 1000$ kg $+ 100$ kg $= 1100$ kg. The period T is

$$T = 2\pi\sqrt{\frac{m}{k}} = 2\pi\sqrt{\frac{1100 \text{ kg}}{3.5 \times 10^4 \text{ kg/s}^2}} = 1.11 \text{ s}$$

The frequency is $f = 1/T = 1/(1.11 \text{ s}) = 0.90$ Hz.

EVALUATE: A persistent oscillation with a period of about 1 second makes for a very unpleasant ride. The purpose of shock absorbers is to make such oscillations die out (see Section 14.7).

Angular SHM

A mechanical watch keeps time based on the oscillations of a balance wheel (Fig. 14.19). The wheel has a moment of inertia I about its axis. A coil spring exerts a restoring torque τ_z that is proportional to the angular displacement θ from the equilibrium position. We write $\tau_z = -\kappa\theta$, where κ (the Greek letter kappa) is a constant called the *torsion constant*. Using the rotational analog of Newton's second law for a rigid body, $\Sigma\tau_z = I\alpha_z = I\,d^2\theta/dt^2$, we can find the equation of motion:

$$-\kappa\theta = I\alpha \quad \text{or} \quad \frac{d^2\theta}{dt^2} = -\frac{\kappa}{I}\theta$$

The form of this equation is exactly the same as Eq. (14.4) for the acceleration in simple harmonic motion, with x replaced by θ and k/m replaced by κ/I. So we are dealing with a form of *angular* simple harmonic motion. The angular frequency ω and frequency f are given by Eqs. (14.10) and (14.11), respectively, with the same replacement:

$$\omega = \sqrt{\frac{\kappa}{I}} \quad \text{and} \quad f = \frac{1}{2\pi}\sqrt{\frac{\kappa}{I}} \qquad \text{(angular SHM)} \qquad (14.24)$$

The motion is described by the function

$$\theta = \Theta\cos(\omega t + \phi)$$

where Θ (the Greek letter theta) plays the role of an angular amplitude.

It's a good thing that the motion of a balance wheel *is* simple harmonic. If it weren't, the frequency might depend on the amplitude, and the watch would run too fast or too slow as the spring ran down.

Vibrations of Molecules

The following discussion of the vibrations of molecules uses the binomial theorem. If you aren't familiar with this theorem, you should read about it in the appropriate section of a math textbook.

When two atoms are separated from each other by a few atomic diameters, they can exert attractive forces on each other. But if the atoms are so close to each other that their electron shells overlap, the forces between the atoms are repulsive. Between these limits, there can be an equilibrium separation distance at which two atoms form a *molecule*. If these atoms are displaced slightly from equilibrium, they will oscillate.

14.18 If the weight mg compresses the spring a distance Δl, the force constant is $k = mg/\Delta l$ and the angular frequency for vertical SHM is $\omega = \sqrt{k/m}$—the same as if the body were suspended from the spring (see Fig. 14.17).

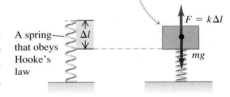

A body is placed atop the spring. It is in equilibrium when the upward force exerted by the compressed spring equals the body's weight.

14.19 The balance wheel of a mechanical watch. The spring exerts a restoring torque that is proportional to the angular displacement θ, so the motion is angular SHM.

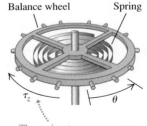

The spring torque τ_z opposes the angular displacement θ.

14.20 (a) Two atoms with centers separated by r. (b) Potential energy U in the van der Waals interaction as a function of r. (c) Force F_r on the right-hand atom as a function of r.

(a) Two-atom system

Distance between atom centers

Atoms

F_r = the force exerted by the left-hand atom on the right-hand atom

(b) Potential energy U of the two-atom system as a function of r

Near equilibrium, U can be approximated by a parabola.

The equilibrium point is at $r = R_0$ (where U is minimum).

(c) The force F_r as a function of r

Near equilibrium, F_r can be approximated by a straight line.

The equilibrium point is at $r = R_0$ (where F_r is zero).

As an example, we'll consider one type of interaction between atoms called the *van der Waals interaction*. Our immediate task here is to study oscillations, so we won't go into the details of how this interaction arises. Let the center of one atom be at the origin and let the center of the other atom be a distance r away (Fig. 14.20a); the equilibrium distance between centers is $r = R_0$. Experiment shows that the van der Waals interaction can be described by the potential-energy function

$$U = U_0\left[\left(\frac{R_0}{r}\right)^{12} - 2\left(\frac{R_0}{r}\right)^6\right] \qquad (14.25)$$

where U_0 is a positive constant with units of joules. When the two atoms are very far apart, $U = 0$; when they are separated by the equilibrium distance $r = R_0$, $U = -U_0$. The force on the second atom is the negative derivative of Eq. (14.25):

$$F_r = -\frac{dU}{dr} = U_0\left[\frac{12R_0^{12}}{r^{13}} - 2\frac{6R_0^6}{r^7}\right] = 12\frac{U_0}{R_0}\left[\left(\frac{R_0}{r}\right)^{13} - \left(\frac{R_0}{r}\right)^7\right] \qquad (14.26)$$

Figures 14.20b and 14.20c plot the potential energy and force, respectively. The force is positive for $r < R_0$ and negative for $r > R_0$, so it is a *restoring* force.

Let's examine the restoring force F_r in Eq. (14.26). We let x represent the displacement from equilibrium:

$$x = r - R_0 \qquad \text{so} \qquad r = R_0 + x$$

In terms of x, the force F_r in Eq. (14.26) becomes

$$F_r = 12\frac{U_0}{R_0}\left[\left(\frac{R_0}{R_0 + x}\right)^{13} - \left(\frac{R_0}{R_0 + x}\right)^7\right]$$

$$= 12\frac{U_0}{R_0}\left[\frac{1}{(1 + x/R_0)^{13}} - \frac{1}{(1 + x/R_0)^7}\right] \qquad (14.27)$$

This looks nothing like Hooke's law, $F_x = -kx$, so we might be tempted to conclude that molecular oscillations cannot be SHM. But let us restrict ourselves to *small-amplitude* oscillations so that the absolute value of the displacement x is small in comparison to R_0 and the absolute value of the ratio x/R_0 is much less than 1. We can then simplify Eq. (14.27) by using the *binomial theorem*:

$$(1 + u)^n = 1 + nu + \frac{n(n - 1)}{2!}u^2 + \frac{n(n - 1)(n - 2)}{3!}u^3 + \cdots \qquad (14.28)$$

If $|u|$ is much less than 1, each successive term in Eq. (14.28) is much smaller than the one it follows, and we can safely approximate $(1 + u)^n$ by just the first two terms. In Eq. (14.27), u is replaced by x/R_0 and n equals -13 or -7, so

$$\frac{1}{(1 + x/R_0)^{13}} = (1 + x/R_0)^{-13} \approx 1 + (-13)\frac{x}{R_0}$$

$$\frac{1}{(1 + x/R_0)^7} = (1 + x/R_0)^{-7} \approx 1 + (-7)\frac{x}{R_0}$$

$$F_r \approx 12\frac{U_0}{R_0}\left[\left(1 + (-13)\frac{x}{R_0}\right) - \left(1 + (-7)\frac{x}{R_0}\right)\right] = -\left(\frac{72U_0}{R_0^2}\right)x \quad \text{(14.29)}$$

This is just Hooke's law, with force constant $k = 72U_0/R_0^2$. (Note that k has the correct units, J/m^2 or N/m.) So oscillations of molecules bound by the van der Waals interaction can be simple harmonic motion, provided that the amplitude is small in comparison to R_0 so that the approximation $|x/R_0| \ll 1$ used in the derivation of Eq. (14.29) is valid.

You can also use the binomial theorem to show that the potential energy U in Eq. (14.25) can be written as $U \approx \frac{1}{2}kx^2 + C$, where $C = -U_0$ and k is again equal to $72U_0/R_0^2$. Adding a constant to the potential energy has no effect on the physics, so the system of two atoms is fundamentally no different from a mass attached to a horizontal spring for which $U = \frac{1}{2}kx^2$.

Example 14.7 | Molecular vibration

Two argon atoms form the molecule Ar_2 as a result of a van der Waals interaction with $U_0 = 1.68 \times 10^{-21}$ J and $R_0 = 3.82 \times 10^{-10}$ m. Find the frequency of small oscillations of one Ar atom about its equilibrium position.

SOLUTION

IDENTIFY and SET UP This is just the situation shown in Fig. 14.20. Because the oscillations are small, we can use Eq. (14.29) to find the force constant k and Eq. (14.11) to find the frequency f of SHM.

EXECUTE: From Eq. (14.29),

$$k = \frac{72U_0}{R_0^2} = \frac{72(1.68 \times 10^{-21} \text{ J})}{(3.82 \times 10^{-10} \text{ m})^2} = 0.829 \text{ J/m}^2 = 0.829 \text{ N/m}$$

(This force constant is comparable to that of a loose toy spring like a Slinky™.) From Appendix D, the average atomic mass of argon is $(39.948 \text{ u})(1.66 \times 10^{-27} \text{ kg/1 u}) = 6.63 \times 10^{-26}$ kg.

From Eq. (14.11), if one atom is fixed and the other oscillates,

$$f = \frac{1}{2\pi}\sqrt{\frac{k}{m}} = \frac{1}{2\pi}\sqrt{\frac{0.829 \text{ N/m}}{6.63 \times 10^{-26} \text{ kg}}} = 5.63 \times 10^{11} \text{ Hz}$$

EVALUATE: Our answer for f isn't quite right. If no net external force acts on the molecule, its center of mass (halfway between the atoms) doesn't accelerate, so *both* atoms must oscillate with the same amplitude in opposite directions. It turns out that we can account for this by replacing m with $m/2$ in our expression for f. This makes f larger by a factor of $\sqrt{2}$, so the correct frequency is $f = \sqrt{2}(5.63 \times 10^{11} \text{ Hz}) = 7.96 \times 10^{11} \text{ Hz}$. What's more, on the atomic scale we must use *quantum mechanics* rather than Newtonian mechanics to describe motion; happily, quantum mechanics also yields $f = 7.96 \times 10^{11} \text{ Hz}$.

Test Your Understanding of Section 14.4 A block attached to a hanging ideal spring oscillates up and down with a period of 10 s on earth. If you take the block and spring to Mars, where the acceleration due to gravity is only about 40% as large as on earth, what will be the new period of oscillation? (i) 10 s; (ii) more than 10 s; (iii) less than 10 s.

14.5 The Simple Pendulum

A **simple pendulum** is an idealized model consisting of a point mass suspended by a massless, unstretchable string. When the point mass is pulled to one side of its straight-down equilibrium position and released, it oscillates about the equilibrium position. Familiar situations such as a wrecking ball on a crane's cable or a person on a swing (Fig. 14.21a) can be modeled as simple pendulums.

Mastering**PHYSICS**

PhET: Pendulum Lab
ActivPhysics 9.10: Pendulum Frequency
ActivPhysics 9.11: Risky Pendulum Walk
ActivPhysics 9.12: Physical Pendulum

14.21 The dynamics of a simple pendulum.

(a) A real pendulum

(b) An idealized simple pendulum

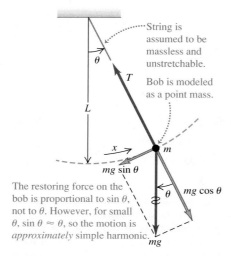

String is assumed to be massless and unstretchable.

Bob is modeled as a point mass.

The restoring force on the bob is proportional to sin θ, not to θ. However, for small θ, sin $\theta \approx \theta$, so the motion is *approximately* simple harmonic.

14.22 For small angular displacements θ, the restoring force $F_\theta = -mg \sin\theta$ on a simple pendulum is approximately equal to $-mg\theta$; that is, it is approximately proportional to the displacement θ. Hence for small angles the oscillations are simple harmonic.

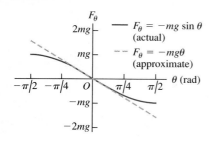

The path of the point mass (sometimes called a pendulum bob) is not a straight line but the arc of a circle with radius L equal to the length of the string (Fig. 14.21b). We use as our coordinate the distance x measured along the arc. If the motion is simple harmonic, the restoring force must be directly proportional to x or (because $x = L\theta$) to θ. Is it?

In Fig. 14.21b we represent the forces on the mass in terms of tangential and radial components. The restoring force F_θ is the tangential component of the net force:

$$F_\theta = -mg \sin\theta \qquad (14.30)$$

The restoring force is provided by gravity; the tension T merely acts to make the point mass move in an arc. The restoring force is proportional *not* to θ but to $\sin\theta$, so the motion is *not* simple harmonic. However, if the angle θ is *small,* $\sin\theta$ is very nearly equal to θ in radians (Fig. 14.22). For example, when $\theta = 0.1$ rad (about 6°), $\sin\theta = 0.0998$, a difference of only 0.2%. With this approximation, Eq. (14.30) becomes

$$F_\theta = -mg\theta = -mg\frac{x}{L} \qquad \text{or}$$

$$F_\theta = -\frac{mg}{L}x \qquad (14.31)$$

The restoring force is then proportional to the coordinate for small displacements, and the force constant is $k = mg/L$. From Eq. (14.10) the angular frequency ω of a simple pendulum with small amplitude is

$$\omega = \sqrt{\frac{k}{m}} = \sqrt{\frac{mg/L}{m}} = \sqrt{\frac{g}{L}} \qquad \text{(simple pendulum, small amplitude)} \qquad (14.32)$$

The corresponding frequency and period relationships are

$$f = \frac{\omega}{2\pi} = \frac{1}{2\pi}\sqrt{\frac{g}{L}} \qquad \text{(simple pendulum, small amplitude)} \qquad (14.33)$$

$$T = \frac{2\pi}{\omega} = \frac{1}{f} = 2\pi\sqrt{\frac{L}{g}} \qquad \text{(simple pendulum, small amplitude)} \qquad (14.34)$$

Note that these expressions do not involve the *mass* of the particle. This is because the restoring force, a component of the particle's weight, is proportional to m. Thus the mass appears on *both* sides of $\Sigma\vec{F} = m\vec{a}$ and cancels out. (This is the same physics that explains why bodies of different masses fall with the same acceleration in a vacuum.) For small oscillations, the period of a pendulum for a given value of g is determined entirely by its length.

The dependence on L and g in Eqs. (14.32) through (14.34) is just what we should expect. A long pendulum has a longer period than a shorter one. Increasing g increases the restoring force, causing the frequency to increase and the period to decrease.

We emphasize again that the motion of a pendulum is only *approximately* simple harmonic. When the amplitude is not small, the departures from simple harmonic motion can be substantial. But how small is "small"? The period can be expressed by an infinite series; when the maximum angular displacement is Θ, the period T is given by

$$T = 2\pi\sqrt{\frac{L}{g}}\left(1 + \frac{1^2}{2^2}\sin^2\frac{\Theta}{2} + \frac{1^2 \cdot 3^2}{2^2 \cdot 4^2}\sin^4\frac{\Theta}{2} + \cdots\right) \qquad (14.35)$$

We can compute the period to any desired degree of precision by taking enough terms in the series. We invite you to check that when $\Theta = 15°$ (on either side of the central position), the true period is longer than that given by the approximate Eq. (14.34) by less than 0.5%.

The usefulness of the pendulum as a timekeeper depends on the period being *very nearly* independent of amplitude, provided that the amplitude is small. Thus, as a pendulum clock runs down and the amplitude of the swings decreases a little, the clock still keeps very nearly correct time.

Example 14.8 **A simple pendulum**

Find the period and frequency of a simple pendulum 1.000 m long at a location where $g = 9.800$ m/s^2.

SOLUTION

IDENTIFY and SET UP: This is a simple pendulum, so we can use the ideas of this section. We use Eq. (14.34) to determine the pendulum's period T from its length, and Eq. (14.1) to find the frequency f from T.

EXECUTE: From Eqs. (14.34) and (14.1),

$$T = 2\pi\sqrt{\frac{L}{g}} = 2\pi\sqrt{\frac{1.000 \text{ m}}{9.800 \text{ m/s}^2}} = 2.007 \text{ s}$$

$$f = \frac{1}{T} = \frac{1}{2.007 \text{ s}} = 0.4983 \text{ Hz}$$

EVALUATE: The period is almost exactly 2 s. When the metric system was established, the second was *defined* as half the period of a 1-m simple pendulum. This was a poor standard, however, because the value of g varies from place to place. We discussed more modern time standards in Section 1.3.

Test Your Understanding of Section 14.5 When a body oscillating on a horizontal spring passes through its equilibrium position, its acceleration is zero (see Fig. 14.2b). When the bob of an oscillating simple pendulum passes through its equilibrium position, is its acceleration zero?

14.6 **The Physical Pendulum**

A **physical pendulum** is any *real* pendulum that uses an extended body, as contrasted to the idealized model of the *simple* pendulum with all the mass concentrated at a single point. For small oscillations, analyzing the motion of a real, physical pendulum is almost as easy as for a simple pendulum. Figure 14.23 shows a body of irregular shape pivoted so that it can turn without friction about an axis through point O. In the equilibrium position the center of gravity is directly below the pivot; in the position shown in the figure, the body is displaced from equilibrium by an angle θ, which we use as a coordinate for the system. The distance from O to the center of gravity is d, the moment of inertia of the body about the axis of rotation through O is I, and the total mass is m. When the body is displaced as shown, the weight mg causes a restoring torque

$$\tau_z = -(mg)(d\sin\theta) \tag{14.36}$$

The negative sign shows that the restoring torque is clockwise when the displacement is counterclockwise, and vice versa.

When the body is released, it oscillates about its equilibrium position. The motion is not simple harmonic because the torque τ_z is proportional to $\sin\theta$ rather than to θ itself. However, if θ is small, we can approximate $\sin\theta$ by θ in radians, just as we did in analyzing the simple pendulum. Then the motion is *approximately* simple harmonic. With this approximation,

$$\tau_z = -(mgd)\theta$$

The equation of motion is $\Sigma\tau_z = I\alpha_z$, so

$$-(mgd)\theta = I\alpha_z = I\frac{d^2\theta}{dt^2}$$

$$\frac{d^2\theta}{dt^2} = -\frac{mgd}{I}\theta \tag{14.37}$$

14.23 Dynamics of a physical pendulum.

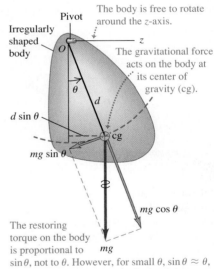

The body is free to rotate around the *z*-axis.

The gravitational force acts on the body at its center of gravity (cg).

The restoring torque on the body is proportional to $\sin\theta$, not to θ. However, for small θ, $\sin\theta \approx \theta$, so the motion is *approximately* simple harmonic.

Comparing this with Eq. (14.4), we see that the role of (k/m) for the spring-mass system is played here by the quantity (mgd/I). Thus the angular frequency is

$$\omega = \sqrt{\frac{mgd}{I}} \qquad \text{(physical pendulum, small amplitude)} \qquad (14.38)$$

The frequency f is $1/2\pi$ times this, and the period T is

$$T = 2\pi\sqrt{\frac{I}{mgd}} \qquad \text{(physical pendulum, small amplitude)} \qquad (14.39)$$

Equation (14.39) is the basis of a common method for experimentally determining the moment of inertia of a body with a complicated shape. First locate the center of gravity of the body by balancing. Then suspend the body so that it is free to oscillate about an axis, and measure the period T of small-amplitude oscillations. Finally, use Eq. (14.39) to calculate the moment of inertia I of the body about this axis from T, the body's mass m, and the distance d from the axis to the center of gravity (see Exercise 14.53). Biomechanics researchers use this method to find the moments of inertia of an animal's limbs. This information is important for analyzing how an animal walks, as we'll see in the second of the two following examples.

Example 14.9 Physical pendulum versus simple pendulum

If the body in Fig. 14.23 is a uniform rod with length L, pivoted at one end, what is the period of its motion as a pendulum?

SOLUTION

IDENTIFY and SET UP: Our target variable is the oscillation period T of a rod that acts as a physical pendulum. We find the rod's moment of inertia in Table 9.2, and then determine T using Eq. (14.39).

EXECUTE: The moment of inertia of a uniform rod about an axis through one end is $I = \frac{1}{3}ML^2$. The distance from the pivot to the rod's center of gravity is $d = L/2$. Then from Eq. (14.39),

$$T = 2\pi\sqrt{\frac{I}{mgd}} = 2\pi\sqrt{\frac{\frac{1}{3}ML^2}{MgL/2}} = 2\pi\sqrt{\frac{2L}{3g}}$$

EVALUATE: If the rod is a meter stick $(L = 1.00 \text{ m})$ and $g = 9.80 \text{ m/s}^2$, then

$$T = 2\pi\sqrt{\frac{2(1.00 \text{ m})}{3(9.80 \text{ m/s}^2)}} = 1.64 \text{ s}$$

The period is smaller by a factor of $\sqrt{\frac{2}{3}} = 0.816$ than that of a simple pendulum of the same length (see Example 14.8). The rod's moment of inertia around one end, $I = \frac{1}{3}ML^2$, is one-third that of the simple pendulum, and the rod's cg is half as far from the pivot as that of the simple pendulum. You can show that, taken together in Eq. (14.39), these two differences account for the factor $\sqrt{\frac{2}{3}}$ by which the periods differ.

Example 14.10 *Tyrannosaurus rex* and the physical pendulum

All walking animals, including humans, have a natural walking pace—a number of steps per minute that is more comfortable than a faster or slower pace. Suppose that this pace corresponds to the oscillation of the leg as a physical pendulum. (a) How does this pace depend on the length L of the leg from hip to foot? Treat the leg as a uniform rod pivoted at the hip joint. (b) Fossil evidence shows that *T. rex*, a two-legged dinosaur that lived about 65 million years ago, had a leg length $L = 3.1$ m and a stride length $S = 4.0$ m (the distance from one footprint to the next print of the same foot; see Fig. 14.24). Estimate the walking speed of *T. rex*.

SOLUTION

IDENTIFY and SET UP: Our target variables are (a) the relationship between walking pace and leg length L and (b) the walking speed of *T. rex*. We treat the leg as a physical pendulum, with a period of

14.24 The walking speed of *Tyrannosaurus rex* can be estimated from leg length L and stride length S.

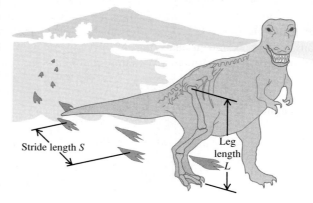

Stride length S

Leg length L

oscillation as found in Example 14.9. We can find the walking speed from the period and the stride length.

EXECUTE: (a) From Example 14.9 the period of oscillation of the leg is $T = 2\pi \sqrt{2L/3g}$, which is proportional to $\sqrt{L}$. Each step takes one-half a period, so the walking pace (in steps per second) is twice the oscillation frequency $f = 1/T$, which is proportional to $1/\sqrt{L}$. The greater the leg length L, the slower the walking pace.

(b) According to our model, *T. rex* traveled one stride length S in a time

$$T = 2\pi\sqrt{\frac{2L}{3g}} = 2\pi\sqrt{\frac{2(3.1 \text{ m})}{3(9.8 \text{ m/s}^2)}} = 2.9 \text{ s}$$

so its walking speed was

$$v = \frac{S}{T} = \frac{4.0 \text{ m}}{2.9 \text{ s}} = 1.4 \text{ m/s} = 5.0 \text{ km/h} = 3.1 \text{ mi/h}$$

This is roughly the walking speed of an adult human.

EVALUATE: A uniform rod isn't a very good model for a leg. The legs of many animals, including both *T. rex* and humans, are tapered; there is more mass between hip and knee than between knee and foot. The center of mass is therefore less than $L/2$ from the hip; a reasonable guess would be about $L/4$. The moment of inertia is therefore *considerably* less than $ML^2/3$—say, $ML^2/15$. Use the analysis of Example 14.9 with these corrections; you'll get a shorter oscillation period and an even greater walking speed for *T. rex*.

Test Your Understanding of Section 14.6   The center of gravity of a simple pendulum of mass m and length L is located at the position of the pendulum bob, a distance L from the pivot point. The center of gravity of a uniform rod of the same mass m and length $2L$ pivoted at one end is also a distance L from the pivot point. How does the period of this uniform rod compare to the period of the simple pendulum? (i) The rod has a longer period; (ii) the rod has a shorter period; (iii) the rod has the same period. ❚

14.7 Damped Oscillations

The idealized oscillating systems we have discussed so far are frictionless. There are no nonconservative forces, the total mechanical energy is constant, and a system set into motion continues oscillating forever with no decrease in amplitude.

Real-world systems always have some dissipative forces, however, and oscillations die out with time unless we replace the dissipated mechanical energy (Fig. 14.25). A mechanical pendulum clock continues to run because potential energy stored in the spring or a hanging weight system replaces the mechanical energy lost due to friction in the pivot and the gears. But eventually the spring runs down or the weights reach the bottom of their travel. Then no more energy is available, and the pendulum swings decrease in amplitude and stop.

The decrease in amplitude caused by dissipative forces is called **damping,** and the corresponding motion is called **damped oscillation.** The simplest case to analyze in detail is a simple harmonic oscillator with a frictional damping force that is directly proportional to the *velocity* of the oscillating body. This behavior occurs in friction involving viscous fluid flow, such as in shock absorbers or sliding between oil-lubricated surfaces. We then have an additional force on the body due to friction, $F_x = -bv_x$, where $v_x = dx/dt$ is the velocity and b is a constant that describes the strength of the damping force. The negative sign shows that the force is always opposite in direction to the velocity. The *net* force on the body is then

$$\sum F_x = -kx - bv_x \tag{14.40}$$

and Newton's second law for the system is

$$-kx - bv_x = ma_x \quad \text{or} \quad -kx - b\frac{dx}{dt} = m\frac{d^2x}{dt^2} \tag{14.41}$$

Equation (14.41) is a differential equation for x; it would be the same as Eq. (14.4), the equation for the acceleration in SHM, except for the added term $-b\,dx/dt$. Solving this equation is a straightforward problem in differential equations, but we won't go into the details here. If the damping force is relatively small, the motion is described by

$$x = Ae^{-(b/2m)t}\cos(\omega' t + \phi) \quad \text{(oscillator with little damping)} \tag{14.42}$$

14.25 A swinging bell left to itself will eventually stop oscillating due to damping forces (air resistance and friction at the point of suspension).

14.26 Graph of displacement versus time for an oscillator with little damping [see Eq. (14.42)] and with phase angle $\phi = 0$. The curves are for two values of the damping constant b.

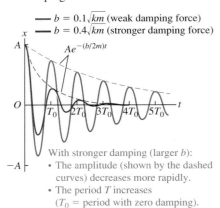

With stronger damping (larger b):
• The amplitude (shown by the dashed curves) decreases more rapidly.
• The period T increases (T_0 = period with zero damping).

14.27 An automobile shock absorber. The viscous fluid causes a damping force that depends on the relative velocity of the two ends of the unit.

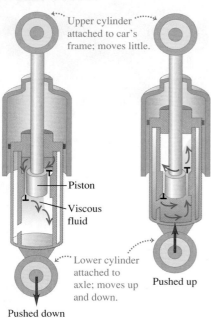

Upper cylinder attached to car's frame; moves little.

Piston

Viscous fluid

Lower cylinder attached to axle; moves up and down.

Pushed up

Pushed down

The angular frequency of oscillation ω' is given by

$$\omega' = \sqrt{\frac{k}{m} - \frac{b^2}{4m^2}} \quad \text{(oscillator with little damping)} \quad (14.43)$$

You can verify that Eq. (14.42) is a solution of Eq. (14.41) by calculating the first and second derivatives of x, substituting them into Eq. (14.41), and checking whether the left and right sides are equal. This is a straightforward but slightly tedious procedure.

The motion described by Eq. (14.42) differs from the undamped case in two ways. First, the amplitude $Ae^{-(b/2m)t}$ is not constant but decreases with time because of the decreasing exponential factor $e^{-(b/2m)t}$. Figure 14.26 is a graph of Eq. (14.42) for the case $\phi = 0$; it shows that the larger the value of b, the more quickly the amplitude decreases.

Second, the angular frequency ω', given by Eq. (14.43), is no longer equal to $\omega = \sqrt{k/m}$ but is somewhat smaller. It becomes zero when b becomes so large that

$$\frac{k}{m} - \frac{b^2}{4m^2} = 0 \quad \text{or} \quad b = 2\sqrt{km} \quad (14.44)$$

When Eq. (14.44) is satisfied, the condition is called **critical damping.** The system no longer oscillates but returns to its equilibrium position without oscillation when it is displaced and released.

If b is greater than $2\sqrt{km}$, the condition is called **overdamping.** Again there is no oscillation, but the system returns to equilibrium more slowly than with critical damping. For the overdamped case the solutions of Eq. (14.41) have the form

$$x = C_1 e^{-a_1 t} + C_2 e^{-a_2 t}$$

where C_1 and C_2 are constants that depend on the initial conditions and a_1 and a_2 are constants determined by m, k, and b.

When b is less than the critical value, as in Eq. (14.42), the condition is called **underdamping.** The system oscillates with steadily decreasing amplitude.

In a vibrating tuning fork or guitar string, it is usually desirable to have as little damping as possible. By contrast, damping plays a beneficial role in the oscillations of an automobile's suspension system. The shock absorbers provide a velocity-dependent damping force so that when the car goes over a bump, it doesn't continue bouncing forever (Fig. 14.27). For optimal passenger comfort, the system should be critically damped or slightly underdamped. Too much damping would be counterproductive; if the suspension is overdamped and the car hits a second bump just after the first one, the springs in the suspension will still be compressed somewhat from the first bump and will not be able to fully absorb the impact.

Energy in Damped Oscillations

In damped oscillations the damping force is nonconservative; the mechanical energy of the system is not constant but decreases continuously, approaching zero after a long time. To derive an expression for the rate of change of energy, we first write an expression for the total mechanical energy E at any instant:

$$E = \tfrac{1}{2}mv_x^2 + \tfrac{1}{2}kx^2$$

To find the rate of change of this quantity, we take its time derivative:

$$\frac{dE}{dt} = mv_x\frac{dv_x}{dt} + kx\frac{dx}{dt}$$

But $dv_x/dt = a_x$ and $dx/dt = v_x$, so

$$\frac{dE}{dt} = v_x(ma_x + kx)$$

From Eq. (14.41), $ma_x + kx = -bdx/dt = -bv_x$, so

$$\frac{dE}{dt} = v_x(-bv_x) = -bv_x{}^2 \quad \text{(damped oscillations)} \qquad (14.45)$$

The right side of Eq. (14.45) is **negative** whenever the oscillating body is in motion, whether the x-velocity v_x is positive or negative. This shows that as the body moves, the energy decreases, though not at a uniform rate. The term $-bv_x{}^2 = (-bv_x)v_x$ (force times velocity) is the rate at which the damping force does (negative) work on the system (that is, the damping *power*). This equals the rate of change of the total mechanical energy of the system.

Similar behavior occurs in electric circuits containing inductance, capacitance, and resistance. There is a natural frequency of oscillation, and the resistance plays the role of the damping constant b. We will study these circuits in detail in Chapters 30 and 31.

Test Your Understanding of Section 14.7   An airplane is flying in a straight line at a constant altitude. If a wind gust strikes and raises the nose of the airplane, the nose will bob up and down until the airplane eventually returns to its original attitude. Are these oscillations (i) undamped, (ii) underdamped, (iii) critically damped, or (iv) overdamped?

14.8 Forced Oscillations and Resonance

A damped oscillator left to itself will eventually stop moving altogether. But we can maintain a constant-amplitude oscillation by applying a force that varies with time in a periodic or cyclic way, with a definite period and frequency. As an example, consider your cousin Throckmorton on a playground swing. You can keep him swinging with constant amplitude by giving him a little push once each cycle. We call this additional force a **driving force.**

Damped Oscillation with a Periodic Driving Force

If we apply a periodically varying driving force with angular frequency ω_d to a damped harmonic oscillator, the motion that results is called a **forced oscillation** or a *driven oscillation*. It is different from the motion that occurs when the system is simply displaced from equilibrium and then left alone, in which case the system oscillates with a **natural angular frequency** ω' determined by m, k, and b, as in Eq. (14.43). In a forced oscillation, however, the angular frequency with which the mass oscillates is equal to the driving angular frequency ω_d. This does *not* have to be equal to the angular frequency ω' with which the system would oscillate without a driving force. If you grab the ropes of Throckmorton's swing, you can force the swing to oscillate with any frequency you like.

Suppose we force the oscillator to vibrate with an angular frequency ω_d that is nearly *equal* to the angular frequency ω' it would have with no driving force. What happens? The oscillator is naturally disposed to oscillate at $\omega = \omega'$, so we expect the amplitude of the resulting oscillation to be larger than when the two frequencies are very different. Detailed analysis and experiment show that this is just what happens. The easiest case to analyze is a *sinusoidally* varying force— say, $F(t) = F_{max} \cos \omega_d t$. If we vary the frequency ω_d of the driving force, the amplitude of the resulting forced oscillation varies in an interesting way (Fig. 14.28). When there is very little damping (small b), the amplitude goes through a sharp peak as the driving angular frequency ω_d nears the natural oscillation angular frequency ω'. When the damping is increased (larger b), the peak becomes broader and smaller in height and shifts toward lower frequencies.

We could work out an expression that shows how the amplitude A of the forced oscillation depends on the frequency of a sinusoidal driving force, with

14.28 Graph of the amplitude A of forced oscillation as a function of the angular frequency ω_d of the driving force. The horizontal axis shows the ratio of ω_d to the angular frequency $\omega = \sqrt{k/m}$ of an undamped oscillator. Each curve has a different value of the damping constant b.

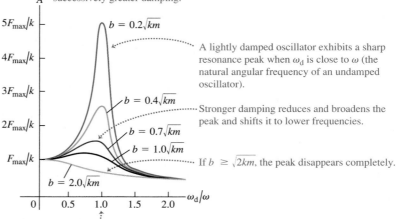

Each curve shows the amplitude A for an oscillator subjected to a driving force at various angular frequencies ω_d. Successive curves from blue to gold represent successively greater damping.

A lightly damped oscillator exhibits a sharp resonance peak when ω_d is close to ω (the natural angular frequency of an undamped oscillator).

Stronger damping reduces and broadens the peak and shifts it to lower frequencies.

If $b \geq \sqrt{2km}$, the peak disappears completely.

Driving frequency ω_d equals natural angular frequency ω of an undamped oscillator.

maximum value F_{max}. That would involve more differential equations than we're ready for, but here is the result:

$$A = \frac{F_{max}}{\sqrt{(k - m\omega_d^2)^2 + b^2\omega_d^2}} \qquad \text{(amplitude of a driven oscillator)} \qquad (14.46)$$

When $k - m\omega_d^2 = 0$, the first term under the radical is zero, so A has a maximum near $\omega_d = \sqrt{k/m}$. The height of the curve at this point is proportional to $1/b$; the less damping, the higher the peak. At the low-frequency extreme, when $\omega_d = 0$, we get $A = F_{max}/k$. This corresponds to a *constant* force F_{max} and a constant displacement $A = F_{max}/k$ from equilibrium, as we might expect.

Resonance and Its Consequences

The fact that there is an amplitude peak at driving frequencies close to the natural frequency of the system is called **resonance.** Physics is full of examples of resonance; building up the oscillations of a child on a swing by pushing with a frequency equal to the swing's natural frequency is one. A vibrating rattle in a car that occurs only at a certain engine speed or wheel-rotation speed is an all-too-familiar example. Inexpensive loudspeakers often have an annoying boom or buzz when a musical note happens to coincide with the resonant frequency of the speaker cone or the speaker housing. In Chapter 16 we will study other examples of resonance that involve sound. Resonance also occurs in electric circuits, as we will see in Chapter 31; a tuned circuit in a radio or television receiver responds strongly to waves having frequencies near its resonant frequency, and this fact is used to select a particular station and reject the others.

Resonance in mechanical systems can be destructive. A company of soldiers once destroyed a bridge by marching across it in step; the frequency of their steps was close to a natural vibration frequency of the bridge, and the resulting oscillation had large enough amplitude to tear the bridge apart. Ever since, marching soldiers have been ordered to break step before crossing a bridge. Some years ago, vibrations of the engines of a particular airplane had just the right frequency to resonate with the natural frequencies of its wings. Large oscillations built up, and occasionally the wings fell off.

Application Canine Resonance
Unlike humans, dogs have no sweat glands and so must pant in order to cool down. The frequency at which a dog pants is very close to the resonant frequency of its respiratory system. This causes the maximum amount of air to move in and out of the dog and so minimizes the effort that the dog must exert to cool itself.

Test Your Understanding of Section 14.8 When driven at a frequency near its natural frequency, an oscillator with very little damping has a much greater response than the same oscillator with more damping. When driven at a frequency that is much higher or lower than the natural frequency, which oscillator will have the greater response: (i) the one with very little damping or (ii) the one with more damping? ∎

Periodic motion: Periodic motion is motion that repeats itself in a definite cycle. It occurs whenever a body has a stable equilibrium position and a restoring force that acts when it is displaced from equilibrium. Period T is the time for one cycle. Frequency f is the number of cycles per unit time. Angular frequency ω is 2π times the frequency. (See Example 14.1.)

$$f = \frac{1}{T} \qquad T = \frac{1}{f} \qquad (14.1)$$

$$\omega = 2\pi f = \frac{2\pi}{T} \qquad (14.2)$$

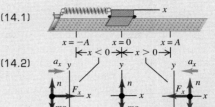

Simple harmonic motion: If the restoring force F_x in periodic motion is directly proportional to the displacement x, the motion is called simple harmonic motion (SHM). In many cases this condition is satisfied if the displacement from equilibrium is small. The angular frequency, frequency, and period in SHM do not depend on the amplitude, but only on the mass m and force constant k. The displacement, velocity, and acceleration in SHM are sinusoidal functions of time; the amplitude A and phase angle ϕ of the oscillation are determined by the initial position and velocity of the body. (See Examples 14.2, 14.3, 14.6, and 14.7.)

$$F_x = -kx \qquad (14.3)$$

$$a_x = \frac{F_x}{m} = -\frac{k}{m}x \qquad (14.4)$$

$$\omega = \sqrt{\frac{k}{m}} \qquad (14.10)$$

$$f = \frac{\omega}{2\pi} = \frac{1}{2\pi}\sqrt{\frac{k}{m}} \qquad (14.11)$$

$$T = \frac{1}{f} = 2\pi\sqrt{\frac{m}{k}} \qquad (14.12)$$

$$x = A\cos(\omega t + \phi) \qquad (14.13)$$

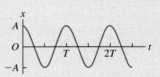

Energy in simple harmonic motion: Energy is conserved in SHM. The total energy can be expressed in terms of the force constant k and amplitude A. (See Examples 14.4 and 14.5.)

$$E = \tfrac{1}{2}mv_x^2 + \tfrac{1}{2}kx^2 = \tfrac{1}{2}kA^2 = \text{constant} \qquad (14.21)$$

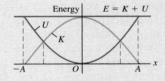

Angular simple harmonic motion: In angular SHM, the frequency and angular frequency are related to the moment of inertia I and the torsion constant κ.

$$\omega = \sqrt{\frac{\kappa}{I}} \quad \text{and} \quad f = \frac{1}{2\pi}\sqrt{\frac{\kappa}{I}} \qquad (14.24)$$

Spring torque τ_z opposes angular displacement θ.

Simple pendulum: A simple pendulum consists of a point mass m at the end of a massless string of length L. Its motion is approximately simple harmonic for sufficiently small amplitude; the angular frequency, frequency, and period then depend only on g and L, not on the mass or amplitude. (See Example 14.8.)

$$\omega = \sqrt{\frac{g}{L}} \qquad (14.32)$$

$$f = \frac{\omega}{2\pi} = \frac{1}{2\pi}\sqrt{\frac{g}{L}} \qquad (14.33)$$

$$T = \frac{2\pi}{\omega} = \frac{1}{f} = 2\pi\sqrt{\frac{L}{g}} \qquad (14.34)$$

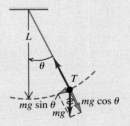

Physical pendulum: A physical pendulum is any body suspended from an axis of rotation. The angular frequency and period for small-amplitude oscillations are independent of amplitude, but depend on the mass m, distance d from the axis of rotation to the center of gravity, and moment of inertia I about the axis. (See Examples 14.9 and 14.10.)

$$\omega = \sqrt{\frac{mgd}{I}} \qquad (14.38)$$

$$T = 2\pi\sqrt{\frac{I}{mgd}} \qquad (14.39)$$

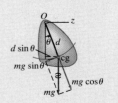

Damped oscillations: When a force $F_x = -bv_x$ proportional to velocity is added to a simple harmonic oscillator, the motion is called a damped oscillation. If $b < 2\sqrt{km}$ (called underdamping), the system oscillates with a decaying amplitude and an angular frequency ω' that is lower than it would be without damping. If $b = 2\sqrt{km}$ (called critical damping) or $b > 2\sqrt{km}$ (called overdamping), when the system is displaced it returns to equilibrium without oscillating.

$$x = Ae^{-(b/2m)t}\cos(\omega' t + \phi) \quad (14.42)$$

$$\omega' = \sqrt{\frac{k}{m} - \frac{b^2}{4m^2}} \quad (14.43)$$

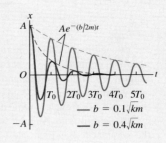

Driven oscillations and resonance: When a sinusoidally varying driving force is added to a damped harmonic oscillator, the resulting motion is called a forced oscillation. The amplitude is a function of the driving frequency ω_d and reaches a peak at a driving frequency close to the natural frequency of the system. This behavior is called resonance.

$$A = \frac{F_{max}}{\sqrt{(k - m\omega_d^2)^2 + b^2\omega_d^2}} \quad (14.46)$$

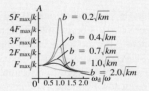

BRIDGING PROBLEM **Oscillating and Rolling**

Two uniform, solid cylinders of radius R and total mass M are connected along their common axis by a short, light rod and rest on a horizontal tabletop (Fig. 14.29). A frictionless ring at the center of the rod is attached to a spring with force constant k; the other end of the spring is fixed. The cylinders are pulled to the left a distance x, stretching the spring, and then released from rest. Due to friction between the tabletop and the cylinders, the cylinders roll without slipping as they oscillate. Show that the motion of the center of mass of the cylinders is simple harmonic, and find its period.

14.29

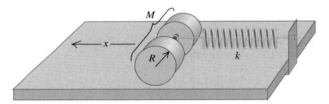

SOLUTION GUIDE

See MasteringPhysics® study area for a Video Tutor solution. (MP)

IDENTIFY and SET UP

1. What condition must be satisfied for the motion of the center of mass of the cylinders to be simple harmonic? (*Hint:* See Section 14.2.)
2. Which equations should you use to describe the translational and rotational motions of the cylinders? Which equation should you use to describe the condition that the cylinders roll without slipping? (*Hint:* See Section 10.3.)
3. Sketch the situation and choose a coordinate system. Make a list of the unknown quantities and decide which is the target variable.

EXECUTE

4. Draw a free-body diagram for the cylinders when they are displaced a distance x from equilibrium.
5. Solve the equations to find an expression for the acceleration of the center of mass of the cylinders. What does this expression tell you?
6. Use your result from step 5 to find the period of oscillation of the center of mass of the cylinders.

EVALUATE

7. What would be the period of oscillation if there were no friction and the cylinders didn't roll? Is this period larger or smaller than your result from step 6? Is this reasonable?

•, ••, •••: Problems of increasing difficulty. **CP**: Cumulative problems incorporating material from earlier chapters. **CALC**: Problems requiring calculus. **BIO**: Biosciences problems.

DISCUSSION QUESTIONS

Q14.1 An object is moving with SHM of amplitude A on the end of a spring. If the amplitude is doubled, what happens to the total distance the object travels in one period? What happens to the period? What happens to the maximum speed of the object? Discuss how these answers are related.

Q14.2 Think of several examples in everyday life of motions that are, at least approximately, simple harmonic. In what respects does each differ from SHM?

Q14.3 Does a tuning fork or similar tuning instrument undergo SHM? Why is this a crucial question for musicians?

Q14.4 A box containing a pebble is attached to an ideal horizontal spring and is oscillating on a friction-free air table. When the box has reached its maximum distance from the equilibrium point, the pebble is suddenly lifted out vertically without disturbing the box. Will the following characteristics of the motion increase, decrease, or remain the same in the subsequent motion of the box? Justify each answer. (a) frequency; (b) period; (c) amplitude; (d) the maximum kinetic energy of the box; (e) the maximum speed of the box.

Q14.5 If a uniform spring is cut in half, what is the force constant of each half? Justify your answer. How would the frequency of SHM using a half-spring differ from the frequency using the same mass and the entire spring?

Q14.6 The analysis of SHM in this chapter ignored the mass of the spring. How does the spring's mass change the characteristics of the motion?

Q14.7 Two identical gliders on an air track are connected by an ideal spring. Could such a system undergo SHM? Explain. How would the period compare with that of a single glider attached to a spring whose other end is rigidly attached to a stationary object? Explain.

Q14.8 You are captured by Martians, taken into their ship, and put to sleep. You awake some time later and find yourself locked in a small room with no windows. All the Martians have left you with is your digital watch, your school ring, and your long silver-chain necklace. Explain how you can determine whether you are still on earth or have been transported to Mars.

Q14.9 The system shown in Fig. 14.17 is mounted in an elevator. What happens to the period of the motion (does it increase, decrease, or remain the same) if the elevator (a) accelerates upward at 5.0 m/s²; (b) moves upward at a steady 5.0 m/s; (c) accelerates downward at 5.0 m/s²? Justify your answers.

Q14.10 If a pendulum has a period of 2.5 s on earth, what would be its period in a space station orbiting the earth? If a mass hung from a vertical spring has a period of 5.0 s on earth, what would its period be in the space station? Justify each of your answers.

Q14.11 A simple pendulum is mounted in an elevator. What happens to the period of the pendulum (does it increase, decrease, or remain the same) if the elevator (a) accelerates upward at 5.0 m/s²; (b) moves upward at a steady 5.0 m/s; (c) accelerates downward at 5.0 m/s²; (d) accelerates downward at 9.8 m/s²? Justify your answers.

Q14.12 What should you do to the length of the string of a simple pendulum to (a) double its frequency; (b) double its period; (c) double its angular frequency?

Q14.13 If a pendulum clock is taken to a mountaintop, does it gain or lose time, assuming it is correct at a lower elevation? Explain your answer.

Q14.14 When the amplitude of a simple pendulum increases, should its period increase or decrease? Give a qualitative argument; do not rely on Eq. (14.35). Is your argument also valid for a physical pendulum?

Q14.15 Why do short dogs (like Chihuahuas) walk with quicker strides than do tall dogs (like Great Danes)?

Q14.16 At what point in the motion of a simple pendulum is the string tension greatest? Least? In each case give the reasoning behind your answer.

Q14.17 Could a standard of time be based on the period of a certain standard pendulum? What advantages and disadvantages would such a standard have compared to the actual present-day standard discussed in Section 1.3?

Q14.18 For a simple pendulum, clearly distinguish between ω (the angular velocity) and ω (the angular frequency). Which is constant and which is variable?

Q14.19 A glider is attached to a fixed ideal spring and oscillates on a horizontal, friction-free air track. A coin is atop the glider and oscillating with it. At what points in the motion is the friction force on the coin greatest? At what points is it least? Justify your answers.

Q14.20 In designing structures in an earthquake-prone region, how should the natural frequencies of oscillation of a structure relate to typical earthquake frequencies? Why? Should the structure have a large or small amount of damping?

EXERCISES

Section 14.1 Describing Oscillation

14.1 • **BIO** (a) **Music.** When a person sings, his or her vocal cords vibrate in a repetitive pattern that has the same frequency as the note that is sung. If someone sings the note B flat, which has a frequency of 466 Hz, how much time does it take the person's vocal cords to vibrate through one complete cycle, and what is the angular frequency of the cords? (b) **Hearing.** When sound waves strike the eardrum, this membrane vibrates with the same frequency as the sound. The highest pitch that typical humans can hear has a period of 50.0 μs. What are the frequency and angular frequency of the vibrating eardrum for this sound? (c) **Vision.** When light having vibrations with angular frequency ranging from 2.7×10^{15} rad/s to 4.7×10^{15} rad/s strikes the retina of the eye, it stimulates the receptor cells there and is perceived as visible light. What are the limits of the period and frequency of this light? (d) **Ultrasound.** High-frequency sound waves (ultrasound) are used to probe the interior of the body, much as x rays do. To detect small objects such as tumors, a frequency of around 5.0 MHz is used. What are the period and angular frequency of the molecular vibrations caused by this pulse of sound?

14.2 • If an object on a horizontal, frictionless surface is attached to a spring, displaced, and then released, it will oscillate. If it is displaced 0.120 m from its equilibrium position and released with zero initial speed, then after 0.800 s its displacement is found to be

0.120 m on the opposite side, and it has passed the equilibrium position once during this interval. Find (a) the amplitude; (b) the period; (c) the frequency.

14.3 • The tip of a tuning fork goes through 440 complete vibrations in 0.500 s. Find the angular frequency and the period of the motion.

14.4 • The displacement of an oscillating object as a function of time is shown in Fig. E14.4. What are (a) the frequency; (b) the amplitude; (c) the period; (d) the angular frequency of this motion?

Figure **E14.4**

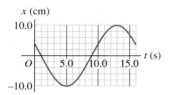

14.5 •• A machine part is undergoing SHM with a frequency of 5.00 Hz and amplitude 1.80 cm. How long does it take the part to go from $x = 0$ to $x = -1.80$ cm?

Section 14.2 Simple Harmonic Motion

14.6 •• In a physics lab, you attach a 0.200-kg air-track glider to the end of an ideal spring of negligible mass and start it oscillating. The elapsed time from when the glider first moves through the equilibrium point to the second time it moves through that point is 2.60 s. Find the spring's force constant.

14.7 • When a body of unknown mass is attached to an ideal spring with force constant 120 N/m, it is found to vibrate with a frequency of 6.00 Hz. Find (a) the period of the motion; (b) the angular frequency; (c) the mass of the body.

14.8 • When a 0.750-kg mass oscillates on an ideal spring, the frequency is 1.33 Hz. What will the frequency be if 0.220 kg are (a) added to the original mass and (b) subtracted from the original mass? Try to solve this problem *without* finding the force constant of the spring.

14.9 •• An object is undergoing SHM with period 0.900 s and amplitude 0.320 m. At $t = 0$ the object is at $x = 0.320$ m and is instantaneously at rest. Calculate the time it takes the object to go (a) from $x = 0.320$ m to $x = 0.160$ m and (b) from $x = 0.160$ m to $x = 0$.

14.10 • A small block is attached to an ideal spring and is moving in SHM on a horizontal, frictionless surface. When the block is at $x = 0.280$ m, the acceleration of the block is -5.30 m/s². What is the frequency of the motion?

14.11 • A 2.00-kg, frictionless block is attached to an ideal spring with force constant 300 N/m. At $t = 0$ the spring is neither stretched nor compressed and the block is moving in the negative direction at 12.0 m/s. Find (a) the amplitude and (b) the phase angle. (c) Write an equation for the position as a function of time.

14.12 •• Repeat Exercise 14.11, but assume that at $t = 0$ the block has velocity -4.00 m/s and displacement $+0.200$ m.

14.13 • The point of the needle of a sewing machine moves in SHM along the x-axis with a frequency of 2.5 Hz. At $t = 0$ its position and velocity components are $+1.1$ cm and -15 cm/s, respectively. (a) Find the acceleration component of the needle at $t = 0$. (b) Write equations giving the position, velocity, and acceleration components of the point as a function of time.

14.14 •• A small block is attached to an ideal spring and is moving in SHM on a horizontal, frictionless surface. When the ampli-

tude of the motion is 0.090 m, it takes the block 2.70 s to travel from $x = 0.090$ m to $x = -0.090$ m. If the amplitude is doubled, to 0.180 m, how long does it take the block to travel (a) from $x = 0.180$ m to $x = -0.180$ m and (b) from $x = 0.090$ m to $x = -0.090$ m?

14.15 • **BIO** **Weighing Astronauts.** This procedure has actually been used to "weigh" astronauts in space. A 42.5-kg chair is attached to a spring and allowed to oscillate. When it is empty, the chair takes 1.30 s to make one complete vibration. But with an astronaut sitting in it, with her feet off the floor, the chair takes 2.54 s for one cycle. What is the mass of the astronaut?

14.16 • A 0.400-kg object undergoing SHM has $a_x = -2.70$ m/s² when $x = 0.300$ m. What is the time for one oscillation?

14.17 • On a frictionless, horizontal air track, a glider oscillates at the end of an ideal spring of force constant 2.50 N/cm. The graph in Fig. E14.17 shows the acceleration of the glider as a function of time. Find (a) the mass of the glider; (b) the maximum displacement of the glider from the equilibrium point; (c) the maximum force the spring exerts on the glider.

Figure **E14.17**

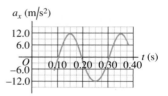

14.18 • A 0.500-kg mass on a spring has velocity as a function of time given by $v_x(t) = -(3.60$ cm/s$) \sin[(4.71$ s$^{-1})t - \pi/2]$. What are (a) the period; (b) the amplitude; (c) the maximum acceleration of the mass; (d) the force constant of the spring?

14.19 • A 1.50-kg mass on a spring has displacement as a function of time given by the equation

$$x(t) = (7.40 \text{ cm}) \cos[(4.16 \text{ s}^{-1})t - 2.42]$$

Find (a) the time for one complete vibration; (b) the force constant of the spring; (c) the maximum speed of the mass; (d) the maximum force on the mass; (e) the position, speed, and acceleration of the mass at $t = 1.00$ s; (f) the force on the mass at that time.

14.20 • **BIO** **Weighing a Virus.** In February 2004, scientists at Purdue University used a highly sensitive technique to measure the mass of a vaccinia virus (the kind used in smallpox vaccine). The procedure involved measuring the frequency of oscillation of a tiny sliver of silicon (just 30 nm long) with a laser, first without the virus and then after the virus had attached itself to the silicon. The difference in mass caused a change in the frequency. We can model such a process as a mass on a spring. (a) Show that the ratio of the frequency with the virus attached (f_{S+V}) to the frequency without the virus (f_S) is given by the formula $\dfrac{f_{S+V}}{f_S} = \dfrac{1}{\sqrt{1 + (m_V/m_S)}}$, where m_V is the mass of the virus and m_S is the mass of the silicon sliver. Notice that it is *not* necessary to know or measure the force constant of the spring. (b) In some data, the silicon sliver has a mass of 2.10×10^{-16} g and a frequency of 2.00×10^{15} Hz without the virus and 2.87×10^{14} Hz with the virus. What is the mass of the virus, in grams and in femtograms?

14.21 •• **CALC** **Jerk.** A guitar string vibrates at a frequency of 440 Hz. A point at its center moves in SHM with an amplitude of

3.0 mm and a phase angle of zero. (a) Write an equation for the position of the center of the string as a function of time. (b) What are the maximum values of the magnitudes of the velocity and acceleration of the center of the string? (c) The derivative of the acceleration with respect to time is a quantity called the *jerk*. Write an equation for the jerk of the center of the string as a function of time, and find the maximum value of the magnitude of the jerk.

Section 14.3 Energy in Simple Harmonic Motion

14.22 •• For the oscillating object in Fig. E14.4, what are (a) its maximum speed and (b) its maximum acceleration?

14.23 • A small block is attached to an ideal spring and is moving in SHM on a horizontal, frictionless surface. The amplitude of the motion is 0.120 m. The maximum speed of the block is 3.90 m/s. What is the maximum magnitude of the acceleration of the block?

14.24 • A small block is attached to an ideal spring and is moving in SHM on a horizontal, frictionless surface. The amplitude of the motion is 0.250 m and the period is 3.20 s. What are the speed and acceleration of the block when $x = 0.160$ m?

14.25 •• A tuning fork labeled 392 Hz has the tip of each of its two prongs vibrating with an amplitude of 0.600 mm. (a) What is the maximum speed of the tip of a prong? (b) A housefly (*Musca domestica*) with mass 0.0270 g is holding onto the tip of one of the prongs. As the prong vibrates, what is the fly's maximum kinetic energy? Assume that the fly's mass has a negligible effect on the frequency of oscillation.

14.26 •• A harmonic oscillator has angular frequency ω and amplitude A. (a) What are the magnitudes of the displacement and velocity when the elastic potential energy is equal to the kinetic energy? (Assume that $U = 0$ at equilibrium.) (b) How often does this occur in each cycle? What is the time between occurrences? (c) At an instant when the displacement is equal to $A/2$, what fraction of the total energy of the system is kinetic and what fraction is potential?

14.27 • A 0.500-kg glider, attached to the end of an ideal spring with force constant $k = 450$ N/m, undergoes SHM with an amplitude of 0.040 m. Compute (a) the maximum speed of the glider; (b) the speed of the glider when it is at $x = -0.015$ m; (c) the magnitude of the maximum acceleration of the glider; (d) the acceleration of the glider at $x = -0.015$ m; (e) the total mechanical energy of the glider at any point in its motion.

14.28 •• A cheerleader waves her pom-pom in SHM with an amplitude of 18.0 cm and a frequency of 0.850 Hz. Find (a) the maximum magnitude of the acceleration and of the velocity; (b) the acceleration and speed when the pom-pom's coordinate is $x = +9.0$ cm; (c) the time required to move from the equilibrium position directly to a point 12.0 cm away. (d) Which of the quantities asked for in parts (a), (b), and (c) can be found using the energy approach used in Section 14.3, and which cannot? Explain.

14.29 • **CP** For the situation described in part (a) of Example 14.5, what should be the value of the putty mass m so that the amplitude after the collision is one-half the original amplitude? For this value of m, what fraction of the original mechanical energy is converted into heat?

14.30 • A 0.150-kg toy is undergoing SHM on the end of a horizontal spring with force constant $k = 300$ N/m. When the object is 0.0120 m from its equilibrium position, it is observed to have a speed of 0.300 m/s. What are (a) the total energy of the object at any point of its motion; (b) the amplitude of the motion; (c) the maximum speed attained by the object during its motion?

14.31 •• You are watching an object that is moving in SHM. When the object is displaced 0.600 m to the right of its equilibrium position, it has a velocity of 2.20 m/s to the right and an acceleration of 8.40 m/s² to the left. How much farther from this point will the object move before it stops momentarily and then starts to move back to the left?

14.32 •• On a horizontal, frictionless table, an open-topped 5.20-kg box is attached to an ideal horizontal spring having force constant 375 N/m. Inside the box is a 3.44-kg stone. The system is oscillating with an amplitude of 7.50 cm. When the box has reached its maximum speed, the stone is suddenly plucked vertically out of the box without touching the box. Find (a) the period and (b) the amplitude of the resulting motion of the box. (c) Without doing any calculations, is the new period greater or smaller than the original period? How do you know?

14.33 •• A mass is oscillating with amplitude A at the end of a spring. How far (in terms of A) is this mass from the equilibrium position of the spring when the elastic potential energy equals the kinetic energy?

14.34 •• A mass m is attached to a spring of force constant 75 N/m and allowed to oscillate. Figure E14.34 shows a graph of its velocity v_x as a function of time t. Find (a) the period, (b) the frequency, and (c) the angular frequency of this motion. (d) What is the amplitude (in cm), and at what times does the mass reach this position? (e) Find the maximum acceleration of the mass and the times at which it occurs. (f) What is the mass m?

Figure **E14.34**

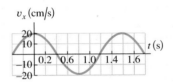

14.35 • Inside a NASA test vehicle, a 3.50-kg ball is pulled along by a horizontal ideal spring fixed to a friction-free table. The force constant of the spring is 225 N/m. The vehicle has a steady acceleration of 5.00 m/s², and the ball is not oscillating. Suddenly, when the vehicle's speed has reached 45.0 m/s, its engines turn off, thus eliminating its acceleration but not its velocity. Find (a) the amplitude and (b) the frequency of the resulting oscillations of the ball. (c) What will be the ball's maximum speed relative to the vehicle?

Section 14.4 Applications of Simple Harmonic Motion

14.36 • A proud deep-sea fisherman hangs a 65.0-kg fish from an ideal spring having negligible mass. The fish stretches the spring 0.120 m. (a) Find the force constant of the spring. The fish is now pulled down 5.00 cm and released. (b) What is the period of oscillation of the fish? (c) What is the maximum speed it will reach?

14.37 • A 175-g glider on a horizontal, frictionless air track is attached to a fixed ideal spring with force constant 155 N/m. At the instant you make measurements on the glider, it is moving at 0.815 m/s and is 3.00 cm from its equilibrium point. Use *energy conservation* to find (a) the amplitude of the motion and (b) the maximum speed of the glider. (c) What is the angular frequency of the oscillations?

14.38 • A thrill-seeking cat with mass 4.00 kg is attached by a harness to an ideal spring of negligible mass and oscillates vertically in SHM. The amplitude is 0.050 m, and at the highest point of the motion the spring has its natural unstretched length. Calculate the elastic potential energy of the spring (take it to be zero for the unstretched spring), the kinetic energy of the cat, the gravitational potential energy of the system relative to the lowest point of the motion, and the sum of these three energies when the cat is (a) at its highest point; (b) at its lowest point; (c) at its equilibrium position.

14.39 •• A 1.50-kg ball and a 2.00-kg ball are glued together with the lighter one below the heavier one. The upper ball is attached to a vertical ideal spring of force constant 165 N/m, and the system is vibrating vertically with amplitude 15.0 cm. The glue connecting the balls is old and weak, and it suddenly comes loose when the balls are at the lowest position in their motion. (a) Why is the glue more likely to fail at the *lowest* point than at any other point in the motion? (b) Find the amplitude and frequency of the vibrations after the lower ball has come loose.

14.40 •• A uniform, solid metal disk of mass 6.50 kg and diameter 24.0 cm hangs in a horizontal plane, supported at its center by a vertical metal wire. You find that it requires a horizontal force of 4.23 N tangent to the rim of the disk to turn it by 3.34°, thus twisting the wire. You now remove this force and release the disk from rest. (a) What is the torsion constant for the metal wire? (b) What are the frequency and period of the torsional oscillations of the disk? (c) Write the equation of motion for $\theta(t)$ for the disk.

14.41 •• A certain alarm clock ticks four times each second, with each tick representing half a period. The balance wheel consists of a thin rim with radius 0.55 cm, connected to the balance staff by thin spokes of negligible mass. The total mass of the balance wheel is 0.90 g. (a) What is the moment of inertia of the balance wheel about its shaft? (b) What is the torsion constant of the coil spring (Fig. 14.19)?

14.42 • A thin metal disk with mass 2.00×10^{-3} kg and radius 2.20 cm is attached at its center to a long fiber (Fig. E14.42). The disk, when twisted and released, oscillates with a period of 1.00 s. Find the torsion constant of the fiber.

Figure **E14.42**

14.43 •• You want to find the moment of inertia of a complicated machine part about an axis through its center of mass. You suspend it from a wire along this axis. The wire has a torsion constant of 0.450 N·m/rad. You twist the part a small amount about this axis and let it go, timing 125 oscillations in 265 s. What is the moment of inertia you want to find?

14.44 •• **CALC** The balance wheel of a watch vibrates with an angular amplitude Θ, angular frequency ω, and phase angle $\phi = 0$. (a) Find expressions for the angular velocity $d\theta/dt$ and angular acceleration $d^2\theta/dt^2$ as functions of time. (b) Find the balance wheel's angular velocity and angular acceleration when its angular displacement is Θ, and when its angular displacement is $\Theta/2$ and θ is decreasing. (*Hint:* Sketch a graph of θ versus t.)

Section 14.5 The Simple Pendulum

14.45 •• You pull a simple pendulum 0.240 m long to the side through an angle of 3.50° and release it. (a) How much time does it take the pendulum bob to reach its highest speed? (b) How much time does it take if the pendulum is released at an angle of 1.75° instead of 3.50°?

14.46 • An 85.0-kg mountain climber plans to swing down, starting from rest, from a ledge using a light rope 6.50 m long. He holds one end of the rope, and the other end is tied higher up on a rock face. Since the ledge is not very far from the rock face, the rope makes a small angle with the vertical. At the lowest point of his swing, he plans to let go and drop a short distance to the ground. (a) How long after he begins his swing will the climber first reach his lowest point? (b) If he missed the first chance to drop off, how long after first beginning his swing will the climber reach his lowest point for the second time?

14.47 • A building in San Francisco has light fixtures consisting of small 2.35-kg bulbs with shades hanging from the ceiling at the end of light, thin cords 1.50 m long. If a minor earthquake occurs, how many swings per second will these fixtures make?

14.48 • **A Pendulum on Mars.** A certain simple pendulum has a period on the earth of 1.60 s. What is its period on the surface of Mars, where $g = 3.71$ m/s^2?

14.49 • After landing on an unfamiliar planet, a space explorer constructs a simple pendulum of length 50.0 cm. She finds that the pendulum makes 100 complete swings in 136 s. What is the value of g on this planet?

14.50 •• A small sphere with mass m is attached to a massless rod of length L that is pivoted at the top, forming a simple pendulum. The pendulum is pulled to one side so that the rod is at an angle Θ from the vertical, and released from rest. (a) In a diagram, show the pendulum just after it is released. Draw vectors representing the *forces* acting on the small sphere and the *acceleration* of the sphere. Accuracy counts! At this point, what is the linear acceleration of the sphere? (b) Repeat part (a) for the instant when the pendulum rod is at an angle $\Theta/2$ from the vertical. (c) Repeat part (a) for the instant when the pendulum rod is vertical. At this point, what is the linear speed of the sphere?

14.51 • A simple pendulum 2.00 m long swings through a maximum angle of 30.0° with the vertical. Calculate its period (a) assuming a small amplitude, and (b) using the first three terms of Eq. (14.35). (c) Which of the answers in parts (a) and (b) is more accurate? For the one that is less accurate, by what percent is it in error from the more accurate answer?

Section 14.6 The Physical Pendulum

14.52 •• We want to hang a thin hoop on a horizontal nail and have the hoop make one complete small-angle oscillation each 2.0 s. What must the hoop's radius be?

14.53 • A 1.80-kg connecting rod from a car engine is pivoted about a horizontal knife edge as shown in Fig. E14.53. The center of gravity of the rod was located by balancing and is 0.200 m from the pivot. When the rod is set into small-amplitude oscillation, it makes 100 complete swings in 120 s. Calculate the moment of inertia of the rod about the rotation axis through the pivot.

Figure **E14.53**

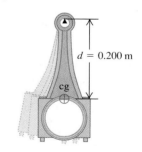

$d = 0.200$ m

cg

14.54 •• A 1.80-kg monkey wrench is pivoted 0.250 m from its center of mass and allowed to swing as a physical pendulum. The period for small-angle oscillations is 0.940 s. (a) What is the moment of inertia of the wrench about an axis through the pivot? (b) If the wrench is initially displaced 0.400 rad from its equilibrium position, what is the angular speed of the wrench as it passes through the equilibrium position?

14.55 • Two pendulums have the same dimensions (length L) and total mass (m). Pendulum A is a very small ball swinging at the end of a uniform massless bar. In pendulum B, half the mass is in the ball and half is in the uniform bar. Find the period of each pendulum for small oscillations. Which one takes longer for a swing?

14.56 •• **CP** A holiday ornament in the shape of a hollow sphere with mass $M = 0.015$ kg and radius $R = 0.050$ m is hung from a tree limb by a small loop of wire attached to the surface of the sphere. If the ornament is displaced a small distance and released, it swings back and forth as a physical pendulum with negligible friction. Calculate its period. (*Hint:* Use the parallel-axis theorem to find the moment of inertia of the sphere about the pivot at the tree limb.)

14.57 •• The two pendulums shown in Fig. E14.57 each consist of a uniform solid ball of mass M supported by a rigid massless rod, but the ball for pendulum A is very tiny while the ball for pendulum B is much larger. Find the period of each pendulum for small displacements. Which ball takes longer to complete a swing?

Figure **E14.57**

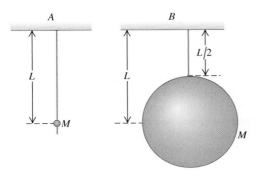

Section 14.7 Damped Oscillations

14.58 • A 2.50-kg rock is attached at the end of a thin, very light rope 1.45 m long. You start it swinging by releasing it when the rope makes an 11° angle with the vertical. You record the observation that it rises only to an angle of 4.5° with the vertical after $10\frac{1}{2}$ swings. (a) How much energy has this system lost during that time? (b) What happened to the "lost" energy? Explain *how* it could have been "lost."

14.59 • An unhappy 0.300-kg rodent, moving on the end of a spring with force constant $k = 2.50$ N/m, is acted on by a damping force $F_x = -bv_x$. (a) If the constant b has the value 0.900 kg/s, what is the frequency of oscillation of the rodent? (b) For what value of the constant b will the motion be critically damped?

14.60 •• A 50.0-g hard-boiled egg moves on the end of a spring with force constant $k = 25.0$ N/m. Its initial displacement is 0.300 m. A damping force $F_x = -bv_x$ acts on the egg, and the amplitude of the motion decreases to 0.100 m in 5.00 s. Calculate the magnitude of the damping constant b.

14.61 •• **CALC** The motion of an underdamped oscillator is described by Eq. (14.42). Let the phase angle ϕ be zero. (a) According to this equation, what is the value of x at $t = 0$? (b) What are the magnitude and direction of the velocity at $t = 0$? What does the result tell you about the slope of the graph of x versus t near $t = 0$? (c) Obtain an expression for the acceleration a_x at $t = 0$. For what value or range of values of the damping constant b (in terms of k and m) is the acceleration at $t = 0$ negative, zero, and positive? Discuss each case in terms of the shape of the graph of x versus t near $t = 0$.

14.62 •• A mass is vibrating at the end of a spring of force constant 225 N/m. Figure E14.62 shows a graph of its position x as a function of time t. (a) At what times is the mass not moving? (b) How much energy did this system originally contain? (c) How much energy did the system lose between $t = 1.0$ s and $t = 4.0$ s? Where did this energy go?

Figure **E14.62**

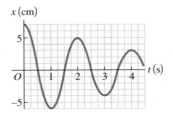

Section 14.8 Forced Oscillations and Resonance

14.63 • A sinusoidally varying driving force is applied to a damped harmonic oscillator. (a) What are the units of the damping constant b? (b) Show that the quantity $\sqrt{km}$ has the same units as b. (c) In terms of F_{max} and k, what is the amplitude for $\omega_d = \sqrt{k/m}$ when (i) $b = 0.2\sqrt{km}$ and (ii) $b = 0.4\sqrt{km}$? Compare your results to Fig. 14.28.

14.64 • A sinusoidally varying driving force is applied to a damped harmonic oscillator of force constant k and mass m. If the damping constant has a value b_1, the amplitude is A_1 when the driving angular frequency equals $\sqrt{k/m}$. In terms of A_1, what is the amplitude for the same driving frequency and the same driving force amplitude F_{max}, if the damping constant is (a) $3b_1$ and (b) $b_1/2$?

PROBLEMS

14.65 •• An object is undergoing SHM with period 1.200 s and amplitude 0.600 m. At $t = 0$ the object is at $x = 0$ and is moving in the negative x-direction. How far is the object from the equilibrium position when $t = 0.480$ s?

14.66 ••• An object is undergoing SHM with period 0.300 s and amplitude 6.00 cm. At $t = 0$ the object is instantaneously at rest at $x = 6.00$ cm. Calculate the time it takes the object to go from $x = 6.00$ cm to $x = -1.50$ cm.

14.67 • **CP SHM in a Car Engine.** The motion of the piston of an automobile engine is approximately simple harmonic. (a) If the stroke of an engine (twice the amplitude) is 0.100 m and the engine runs at 4500 rev/min, compute the acceleration of the piston at the endpoint of its stroke. (b) If the piston has mass 0.450 kg, what net force must be exerted on it at this point? (c) What are the speed and kinetic energy of the piston at the midpoint of its stroke? (d) What average power is required to accelerate the piston from rest to the speed found in part (c)? (e) If the engine runs at 7000 rev/min, what are the answers to parts (b), (c), and (d)?

14.68 • Four passengers with combined mass 250 kg compress the springs of a car with worn-out shock absorbers by 4.00 cm when they get in. Model the car and passengers as a single body on a single ideal spring. If the loaded car has a period of vibration of 1.92 s, what is the period of vibration of the empty car?

14.69 • A glider is oscillating in SHM on an air track with an amplitude A_1. You slow it so that its amplitude is halved. What happens to its (a) period, frequency, and angular frequency;

(b) total mechanical energy; (c) maximum speed; (d) speed at $x = \pm A_1/4$; (e) potential and kinetic energies at $x = \pm A_1/4$?

14.70 ••• **CP** A child with poor table manners is sliding his 250-g dinner plate back and forth in SHM with an amplitude of 0.100 m on a horizontal surface. At a point 0.060 m away from equilibrium, the speed of the plate is 0.400 m/s. (a) What is the period? (b) What is the displacement when the speed is 0.160 m/s? (c) In the center of the dinner plate is a 10.0-g carrot slice. If the carrot slice is just on the verge of slipping at the endpoint of the path, what is the coefficient of static friction between the carrot slice and the plate?

14.71 ••• A 1.50-kg, horizontal, uniform tray is attached to a vertical ideal spring of force constant 185 N/m and a 275-g metal ball is in the tray. The spring is below the tray, so it can oscillate up and down. The tray is then pushed down to point A, which is 15.0 cm below the equilibrium point, and released from rest. (a) How high above point A will the tray be when the metal ball leaves the tray? (*Hint:* This does *not* occur when the ball and tray reach their maximum speeds.) (b) How much time elapses between releasing the system at point A and the ball leaving the tray? (c) How fast is the ball moving just as it leaves the tray?

14.72 •• **CP** A block with mass M rests on a frictionless surface and is connected to a horizontal spring of force constant k. The other end of the spring is attached to a wall (Fig. P14.72). A second block with mass m rests on top of the first block. The coefficient of static friction between the blocks is μ_s. Find the *maximum* amplitude of oscillation such that the top block will not slip on the bottom block.

Figure **P14.72**

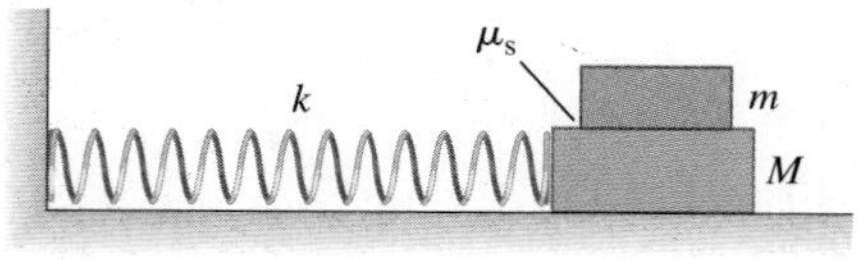

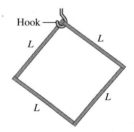

14.73 • **CP** A 10.0-kg mass is traveling to the right with a speed of 2.00 m/s on a smooth horizontal surface when it collides with and sticks to a second 10.0-kg mass that is initially at rest but is attached to a light spring with force constant 110.0 N/m. (a) Find the frequency, amplitude, and period of the subsequent oscillations. (b) How long does it take the system to return the first time to the position it had immediately after the collision?

14.74 • **CP** A rocket is accelerating upward at 4.00 m/s² from the launchpad on the earth. Inside a small, 1.50-kg ball hangs from the ceiling by a light, 1.10-m wire. If the ball is displaced 8.50° from the vertical and released, find the amplitude and period of the resulting swings of this pendulum.

14.75 ••• An apple weighs 1.00 N. When you hang it from the end of a long spring of force constant 1.50 N/m and negligible mass, it bounces up and down in SHM. If you stop the bouncing and let the apple swing from side to side through a small angle, the frequency of this simple pendulum is half the bounce frequency. (Because the angle is small, the back-and-forth swings do not cause any appreciable change in the length of the spring.) What is the unstretched length of the spring (with the apple removed)?

14.76 ••• **CP SHM of a Floating Object.** An object with height h, mass M, and a uniform cross-sectional area A floats upright in a liquid with density ρ. (a) Calculate the vertical distance from the surface of the liquid to the bottom of the floating object at equilibrium. (b) A downward force with magnitude F is applied to the top of the object. At the new equilibrium position, how much farther below the surface of the liquid is the bottom of the object than it was in part (a)? (Assume that some of the object remains above the surface of the liquid.) (c) Your result in part (b) shows that if the force is suddenly removed, the object will oscillate up and down in SHM. Calculate the period of this motion in terms of the density ρ of the liquid, the mass M, and the cross-sectional area A of the object. You can ignore the damping due to fluid friction (see Section 14.7).

14.77 •• **CP** A 950-kg, cylindrical can buoy floats vertically in salt water. The diameter of the buoy is 0.900 m. (a) Calculate the additional distance the buoy will sink when a 70.0-kg man stands on top of it. (Use the expression derived in part (b) of Problem 14.76.) (b) Calculate the period of the resulting vertical SHM when the man dives off. (Use the expression derived in part (c) of Problem 14.76, and as in that problem, you can ignore the damping due to fluid friction.)

14.78 ••• **CP Tarzan to the Rescue!** Tarzan spies a 35-kg chimpanzee in severe danger, so he swings to the rescue. He adjusts his strong, but very light, vine so that he will first come to rest 4.0 s after beginning his swing, at which time his vine makes a 12° angle with the vertical. (a) How long is Tarzan's vine, assuming that he swings at the bottom end of it? (b) What are the frequency and amplitude (in degrees) of Tarzan's swing? (c) Just as he passes through the lowest point in his swing, Tarzan nabs the chimp from the ground and sweeps him out of the jaws of danger. If Tarzan's mass is 65 kg, find the frequency and amplitude (in degrees) of the swing with Tarzan holding onto the grateful chimp.

14.79 •• **CP** A square object of mass m is constructed of four identical uniform thin sticks, each of length L, attached together. This object is hung on a hook at its upper corner (Fig. P14.79). If it is rotated slightly to the left and then released, at what frequency will it swing back and forth?

Figure **P14.79**

14.80 ••• An object with mass 0.200 kg is acted on by an elastic restoring force with force constant 10.0 N/m. (a) Graph elastic potential energy U as a function of displacement x over a range of x from −0.300 m to +0.300 m. On your graph, let 1 cm = 0.05 J vertically and 1 cm = 0.05 m horizontally. The object is set into oscillation with an initial potential energy of 0.140 J and an initial kinetic energy of 0.060 J. Answer the following questions by referring to the graph. (b) What is the amplitude of oscillation? (c) What is the potential energy when the displacement is one-half the amplitude? (d) At what displacement are the kinetic and potential energies equal? (e) What is the value of the phase angle ϕ if the initial velocity is positive and the initial displacement is negative?

14.81 • **CALC** A 2.00-kg bucket containing 10.0 kg of water is hanging from a vertical ideal spring of force constant 125 N/m and oscillating up and down with an amplitude of 3.00 cm. Suddenly the bucket springs a leak in the bottom such that water drops out at a steady rate of 2.00 g/s. When the bucket is half full, find

(a) the period of oscillation and (b) the rate at which the period is changing with respect to time. Is the period getting longer or shorter? (c) What is the shortest period this system can have?

14.82 •• **CP** A hanging wire is 1.80 m long. When a 60.0-kg steel ball is suspended from the wire, the wire stretches by 2.00 mm. If the ball is pulled down a small additional distance and released, at what frequency will it vibrate? Assume that the stress on the wire is less than the proportional limit (see Section 11.5).

14.83 •• A 5.00-kg partridge is suspended from a pear tree by an ideal spring of negligible mass. When the partridge is pulled down 0.100 m below its equilibrium position and released, it vibrates with a period of 4.20 s. (a) What is its speed as it passes through the equilibrium position? (b) What is its acceleration when it is 0.050 m above the equilibrium position? (c) When it is moving upward, how much time is required for it to move from a point 0.050 m below its equilibrium position to a point 0.050 m above it? (d) The motion of the partridge is stopped, and then it is removed from the spring. How much does the spring shorten?

14.84 •• A 0.0200-kg bolt moves with SHM that has an amplitude of 0.240 m and a period of 1.500 s. The displacement of the bolt is $+0.240$ m when $t = 0$. Compute (a) the displacement of the bolt when $t = 0.500$ s; (b) the magnitude and direction of the force acting on the bolt when $t = 0.500$ s; (c) the minimum time required for the bolt to move from its initial position to the point where $x = -0.180$ m; (d) the speed of the bolt when $x = -0.180$ m.

14.85 •• **CP** **SHM of a Butcher's Scale.** A spring of negligible mass and force constant $k = 400$ N/m is hung vertically, and a 0.200-kg pan is suspended from its lower end. A butcher drops a 2.2-kg steak onto the pan from a height of 0.40 m. The steak makes a totally inelastic collision with the pan and sets the system into vertical SHM. What are (a) the speed of the pan and steak immediately after the collision; (b) the amplitude of the subsequent motion; (c) the period of that motion?

14.86 •• A uniform beam is suspended horizontally by two identical vertical springs that are attached between the ceiling and each end of the beam. The beam has mass 225 kg, and a 175-kg sack of gravel sits on the middle of it. The beam is oscillating in SHM, with an amplitude of 40.0 cm and a frequency of 0.600 cycle/s. (a) The sack of gravel falls off the beam when the beam has its maximum upward displacement. What are the frequency and amplitude of the subsequent SHM of the beam? (b) If the gravel instead falls off when the beam has its maximum speed, what are the frequency and amplitude of the subsequent SHM of the beam?

14.87 ••• **CP** On the planet Newtonia, a simple pendulum having a bob with mass 1.25 kg and a length of 185.0 cm takes 1.42 s, when released from rest, to swing through an angle of 12.5°, where it again has zero speed. The circumference of Newtonia is measured to be 51,400 km. What is the mass of the planet Newtonia?

14.88 •• A 40.0-N force stretches a vertical spring 0.250 m. (a) What mass must be suspended from the spring so that the system will oscillate with a period of 1.00 s? (b) If the amplitude of the motion is 0.050 m and the period is that specified in part (a), where is the object and in what direction is it moving 0.35 s after it has passed the equilibrium position, moving downward? (c) What force (magnitude and direction) does the spring exert on the object when it is 0.030 m below the equilibrium position, moving upward?

14.89 •• **Don't Miss the Boat.** While on a visit to Minnesota ("Land of 10,000 Lakes"), you sign up to take an excursion around one of the larger lakes. When you go to the dock where the 1500-kg boat is tied, you find that the boat is bobbing up and down in the waves, executing simple harmonic motion with amplitude 20 cm. The boat takes 3.5 s to make one complete up-and-down cycle.

When the boat is at its highest point, its deck is at the same height as the stationary dock. As you watch the boat bob up and down, you (mass 60 kg) begin to feel a bit woozy, due in part to the previous night's dinner of lutefisk. As a result, you refuse to board the boat unless the level of the boat's deck is within 10 cm of the dock level. How much time do you have to board the boat comfortably during each cycle of up-and-down motion?

14.90 • **CP** An interesting, though highly impractical example of oscillation is the motion of an object dropped down a hole that extends from one side of the earth, through its center, to the other side. With the assumption (not realistic) that the earth is a sphere of uniform density, prove that the motion is simple harmonic and find the period. [*Note:* The gravitational force on the object as a function of the object's distance r from the center of the earth was derived in Example 13.10 (Section 13.6). The motion is simple harmonic if the acceleration a_x and the displacement from equilibrium x are related by Eq. (14.8), and the period is then $T = 2\pi/\omega$.]

14.91 ••• **CP** A rifle bullet with mass 8.00 g and initial horizontal velocity 280 m/s strikes and embeds itself in a block with mass 0.992 kg that rests on a frictionless surface and is attached to one end of an ideal spring. The other end of the spring is attached to the wall. The impact compresses the spring a maximum distance of 18.0 cm. After the impact, the block moves in SHM. Calculate the period of this motion.

14.92 •• **CP CALC** For a certain oscillator the net force on the body with mass m is given by $F_x = -cx^3$. (a) What is the potential energy function for this oscillator if we take $U = 0$ at $x = 0$? (b) One-quarter of a period is the time for the body to move from $x = 0$ to $x = A$. Calculate this time and hence the period. [*Hint:* Begin with Eq. (14.20), modified to include the potential-energy function you found in part (a), and solve for the velocity v_x as a function of x. Then replace v_x with dx/dt. Separate the variable by writing all factors containing x on one side and all factors containing t on the other side so that each side can be integrated. In the x-integral make the change of variable $u = x/A$. The resulting integral can be evaluated by numerical methods on a computer and has the value $\int_0^1 du/\sqrt{1 - u^4} = 1.31$.] (c) According to the result you obtained in part (b), does the period depend on the amplitude A of the motion? Are the oscillations simple harmonic?

14.93 • **CP CALC** An approximation for the potential energy of a KCl molecule is $U = A[(R_0^7/8r^8) - 1/r]$, where $R_0 = 2.67 \times 10^{-10}$ m, $A = 2.31 \times 10^{-28}$ J $\cdot$ m, and r is the distance between the two atoms. Using this approximation: (a) Show that the radial component of the force on each atom is $F_r = A[(R_0^7/r^9) - 1/r^2]$. (b) Show that R_0 is the equilibrium separation. (c) Find the minimum potential energy. (d) Use $r = R_0 + x$ and the first two terms of the binomial theorem (Eq. 14.28) to show that $F_r \approx -(7A/R_0^3)x$, so that the molecule's force constant is $k = 7A/R_0^3$. (e) With both the K and Cl atoms vibrating in opposite directions on opposite sides of the molecule's center of mass, $m_1 m_2/(m_1 + m_2) = 3.06 \times 10^{-26}$ kg is the mass to use in calculating the frequency. Calculate the frequency of small-amplitude vibrations.

14.94 ••• **CP** Two uniform solid spheres, each with mass $M = 0.800$ kg and radius $R = 0.0800$ m, are connected by a short, light rod that is along a diameter of each sphere and are at rest on a horizontal tabletop. A spring with force constant $k = 160$ N/m has one end attached to the wall and the other end attached to a frictionless ring that passes over the rod at the center of mass of the spheres, which is midway between the centers of the two spheres. The spheres are each pulled the same distance from the wall, stretching the spring, and released. There is sufficient friction between the tabletop and the spheres for the spheres to roll without slipping as they move back and forth on the end of the spring. Show that the motion of the center of mass of the spheres is simple harmonic and calculate the period.

14.95 • **CP** In Fig. P14.95 the upper ball is released from rest, collides with the stationary lower ball, and sticks to it. The strings are both 50.0 cm long. The upper ball has mass 2.00 kg, and it is initially 10.0 cm higher than the lower ball, which has mass 3.00 kg. Find the frequency and maximum angular displacement of the motion after the collision.

Figure **P14.95**

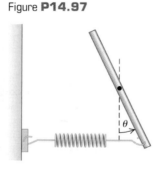

10.0 cm

14.96 •• **CP BIO** *T. rex.* Model the leg of the *T. rex* in Example 14.10 (Section 14.6) as two uniform rods, each 1.55 m long, joined rigidly end to end. Let the lower rod have mass M and the upper rod mass $2M$. The composite object is pivoted about the top of the upper rod. Compute the oscillation period of this object for small-amplitude oscillations. Compare your result to that of Example 14.10.

14.97 •• **CALC** A slender, uniform, metal rod with mass M is pivoted without friction about an axis through its midpoint and perpendicular to the rod. A horizontal spring with force constant k is attached to the lower end of the rod, with the other end of the spring attached to a rigid support. If the rod is displaced by a small angle Θ from the vertical (Fig. P14.97) and released, show that it moves in angular SHM and calculate the period. (*Hint:* Assume that the angle Θ is small enough for the approximations $\sin \Theta \approx \Theta$ and $\cos \Theta \approx 1$ to be valid. The motion is simple harmonic if $d^2\theta/dt^2 = -\omega^2\theta$, and the period is then $T = 2\pi/\omega$.)

Figure **P14.97**

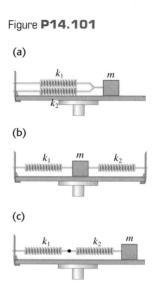

14.98 •• **The Silently Ringing Bell Problem.** A large bell is hung from a wooden beam so it can swing back and forth with negligible friction. The center of mass of the bell is 0.60 m below the pivot, the bell has mass 34.0 kg, and the moment of inertia of the bell about an axis at the pivot is 18.0 kg·m². The clapper is a small, 1.8-kg mass attached to one end of a slender rod that has length L and negligible mass. The other end of the rod is attached to the inside of the bell so it can swing freely about the same axis as the bell. What should be the length L of the clapper rod for the bell to ring silently—that is, for the period of oscillation for the bell to equal that for the clapper?

14.99 ••• Two identical thin rods, each with mass m and length L, are joined at right angles to form an L-shaped object. This object is balanced on top of a sharp edge (Fig. P14.99). If the L-shaped object is deflected slightly, it oscillates. Find the frequency of oscillation.

Figure **P14.99**

14.100 • **CP CALC** A uniform rod of length L oscillates through small angles about a point a distance x from its center. (a) Prove that its angular frequency is $\sqrt{gx/[(L^2/12) + x^2]}$. (b) Show that its maximum angular frequency occurs when $x = L/\sqrt{12}$. (c) What is the length of the rod if the maximum angular frequency is 2π rad/s?

CHALLENGE PROBLEMS

14.101 ••• **The Effective Force Constant of Two Springs.** Two springs with the same unstretched length but different force constants k_1 and k_2 are attached to a block with mass m on a level, frictionless surface. Calculate the effective force constant k_{eff} in each of the three cases (a), (b), and (c) depicted in Fig. P14.101. (The effective force constant is defined by $\Sigma F_x = -k_{\text{eff}}x$.) (d) An object with mass m, suspended from a uniform spring with a force constant k, vibrates with a frequency f_1. When the spring is cut in half and the same object is suspended from one of the halves, the frequency is f_2. What is the ratio f_2/f_1?

Figure **P14.101**

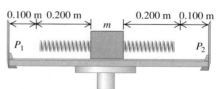

14.102 ••• Two springs, each with unstretched length 0.200 m but with different force constants k_1 and k_2, are attached to opposite ends of a block with mass m on a level, frictionless surface. The outer ends of the springs are now attached to two pins P_1 and P_2, 0.100 m from the original positions of the ends of the springs (Fig. P14.102). Let $k_1 = 2.00$ N/m, $k_2 = 6.00$ N/m, and $m = 0.100$ kg. (a) Find the length of each spring when the block is in its new equilibrium position after the springs have been attached to the pins. (b) Find the period of vibration of the block if it is slightly displaced from its new equilibrium position and released.

Figure **P14.102**

0.100 m 0.200 m   0.200 m 0.100 m

P_1 m P_2

14.103 ••• **CALC A Spring with Mass.** The preceding problems in this chapter have assumed that the springs had negligible mass. But of course no spring is completely massless. To find the effect of the spring's mass, consider a spring with mass M, equilibrium length L_0, and spring constant k. When stretched or compressed to a length L, the potential energy is $\frac{1}{2}kx^2$, where $x = L - L_0$. (a) Consider a spring, as described above, that has one end fixed and the other end moving with speed v. Assume that the speed of points along the length of the spring varies linearly with distance l from the fixed end. Assume also that the mass M of the spring is distributed uniformly along the length of the spring. Calculate the kinetic energy of the spring in terms of M and v. (*Hint:* Divide the spring into pieces of length dl; find the speed of each piece in terms of l, v, and L; find the mass of each piece in terms of dl, M, and L; and integrate from 0 to L. The result is *not* $\frac{1}{2}Mv^2$, since not all of the spring moves with the same speed.) (b) Take the time derivative of the conservation of energy equation, Eq. (14.21), for a mass m moving on the end of a *massless* spring. By comparing your results to Eq. (14.8), which defines ω, show that the angular frequency of oscillation is $\omega = \sqrt{k/m}$. (c) Apply the procedure of part (b) to obtain the angular frequency of oscillation ω of the spring considered in part (a). If the *effective mass* M' of the spring is defined by $\omega = \sqrt{k/M'}$, what is M' in terms of M?

Answers

Chapter Opening Question

The length of the leg is more important. The back-and-forth motion of a leg during walking is like a physical pendulum, for which the oscillation period is $T = 2\pi\sqrt{I/mgd}$ [see Eq. (14.39)]. In this expression I is the moment of inertia of the pendulum, m is its mass, and d is the distance from the rotation axis to the pendulum center of mass. The moment of inertia I is proportional to the mass m, so the mass cancels out of this expression for the period T. Hence only the dimensions of the leg matter. (See Examples 14.9 and 14.10.)

Test Your Understanding Questions

14.1 Answers: (a) $x < 0$, **(b)** $x > 0$, **(c)** $x < 0$, **(d)** $x > 0$, **(e)** $x > 0$, **(f)** $x = 0$ Figure 14.2 shows that the net x-component of force F_x and the x-acceleration a_x are both positive when $x < 0$ (so the body is displaced to the left and the spring is compressed), while F_x and a_x are both negative when $x > 0$ (so the body is displaced to the right and the spring is stretched). Hence x and a_x always have *opposite* signs. This is true whether the object is moving to the right ($v_x > 0$), to the left ($v_x < 0$), or not at all ($v_x = 0$), since the force exerted by the spring depends only on whether it is compressed or stretched and by what distance. This explains the answers to (a) through (e). If the acceleration is zero as in (f), the net force must also be zero and so the spring must be relaxed; hence $x = 0$.

14.2 Answers: (a) $A > 0.10$ m, $\phi < 0$; **(b)** $A > 0.10$ m, $\phi > 0$ In both situations the initial ($t = 0$) x-velocity v_{0x} is nonzero, so from Eq. (14.19) the amplitude $A = \sqrt{x_0^2 + (v_{0x}^2/\omega^2)}$ is greater than the initial x-coordinate $x_0 = 0.10$ m. From Eq. (14.18) the phase angle is $\phi = \arctan(-v_{0x}/\omega x_0)$, which is positive if the quantity $-v_{0x}/\omega x_0$ (the argument of the arctangent function) is positive and negative if $-v_{0x}/\omega x_0$ is negative. In part (a) x_0 and v_{0x} are both positive, so $-v_{0x}/\omega x_0 < 0$ and $\phi < 0$. In part (b) x_0 is positive and v_{0x} is negative, so $-v_{0x}/\omega x_0 > 0$ and $\phi > 0$.

14.3 Answers: (a) (iii), **(b)** (v) To increase the total energy $E = \frac{1}{2}kA^2$ by a factor of 2, the amplitude A must increase by a factor of $\sqrt{2}$. Because the motion is SHM, changing the amplitude has no effect on the frequency.

14.4 Answer: (i) The oscillation period of a body of mass m attached to a hanging spring of force constant k is given by $T = 2\pi\sqrt{m/k}$, the same expression as for a body attached to a horizontal spring. Neither m nor k changes when the apparatus is taken to Mars, so the period is unchanged. The only difference is that in equilibrium, the spring will stretch a shorter distance on Mars than on earth due to the weaker gravity.

14.5 Answer: no Just as for an object oscillating on a spring, at the equilibrium position the *speed* of the pendulum bob is instantaneously not changing (this is where the speed is maximum, so its derivative at this time is zero). But the *direction* of motion is changing because the pendulum bob follows a circular path. Hence the bob must have a component of acceleration perpendicular to the path and toward the center of the circle (see Section 3.4). To cause this acceleration at the equilibrium position when the string is vertical, the upward tension force at this position must be greater than the weight of the bob. This causes a net upward force on the bob and an upward acceleration toward the center of the circular path.

14.6 Answer: (i) The period of a physical pendulum is given by Eq. (14.39), $T = 2\pi\sqrt{I/mgd}$. The distance $d = L$ from the pivot to the center of gravity is the same for both the rod and the simple pendulum, as is the mass m. This means that for any displacement angle θ the same restoring torque acts on both the rod and the simple pendulum. However, the rod has a greater moment of inertia: $I_{rod} = \frac{1}{3}m(2L)^2 = \frac{4}{3}mL^2$ and $I_{simple} = mL^2$ (all the mass of the pendulum is a distance L from the pivot). Hence the rod has a longer period.

14.7 Answer: (ii) The oscillations are underdamped with a decreasing amplitude on each cycle of oscillation, like those graphed in Fig. 14.26. If the oscillations were undamped, they would continue indefinitely with the same amplitude. If they were critically damped or overdamped, the nose would not bob up and down but would return smoothly to the original equilibrium attitude without overshooting.

14.8 Answer: (i) Figure 14.28 shows that the curve of amplitude versus driving frequency moves upward at *all* frequencies as the value of the damping constant b is decreased. Hence for fixed values of k and m, the oscillator with the least damping (smallest value of b) will have the greatest response at any driving frequency.

Bridging Problem

Answer: $T = 2\pi\sqrt{3M/2k}$

15

MECHANICAL WAVES

? When an earthquake strikes, the news of the event travels through the body of the earth in the form of seismic waves. Which aspects of a seismic wave determine how much power is carried by the wave?

LEARNING GOALS

By studying this chapter, you will learn:

- What is meant by a mechanical wave, and the different varieties of mechanical waves.

- How to use the relationship among speed, frequency, and wavelength for a periodic wave.

- How to interpret and use the mathematical expression for a sinusoidal periodic wave.

- How to calculate the speed of waves on a rope or string.

- How to calculate the rate at which a mechanical wave transports energy.

- What happens when mechanical waves overlap and interfere.

- The properties of standing waves on a string, and how to analyze these waves.

- How stringed instruments produce sounds of specific frequencies.

Ripples on a pond, musical sounds, seismic tremors triggered by an earthquake—all these are *wave* phenomena. Waves can occur whenever a system is disturbed from equilibrium and when the disturbance can travel, or *propagate,* from one region of the system to another. As a wave propagates, it carries energy. The energy in light waves from the sun warms the surface of our planet; the energy in seismic waves can crack our planet's crust.

This chapter and the next are about mechanical waves—waves that travel within some material called a *medium.* (Chapter 16 is concerned with sound, an important type of mechanical wave.) We'll begin this chapter by deriving the basic equations for describing waves, including the important special case of *sinusoidal* waves in which the wave pattern is a repeating sine or cosine function. To help us understand waves in general, we'll look at the simple case of waves that travel on a stretched string or rope.

Waves on a string play an important role in music. When a musician strums a guitar or bows a violin, she makes waves that travel in opposite directions along the instrument's strings. What happens when these oppositely directed waves overlap is called *interference.* We'll discover that sinusoidal waves can occur on a guitar or violin string only for certain special frequencies, called *normal-mode frequencies,* determined by the properties of the string. The normal-mode frequencies of a stringed instrument determine the pitch of the musical sounds that the instrument produces. (In the next chapter we'll find that interference also helps explain the pitches of *wind* instruments such as flutes and pipe organs.)

Not all waves are mechanical in nature. *Electromagnetic* waves—including light, radio waves, infrared and ultraviolet radiation, and x rays—can propagate even in empty space, where there is *no* medium. We'll explore these and other nonmechanical waves in later chapters.

15.1 Types of Mechanical Waves

A **mechanical wave** is a disturbance that travels through some material or substance called the **medium** for the wave. As the wave travels through the medium, the particles that make up the medium undergo displacements of various kinds, depending on the nature of the wave.

Figure 15.1 shows three varieties of mechanical waves. In Fig. 15.1a the medium is a string or rope under tension. If we give the left end a small upward shake or wiggle, the wiggle travels along the length of the string. Successive sections of string go through the same motion that we gave to the end, but at successively later times. Because the displacements of the medium are perpendicular or *transverse* to the direction of travel of the wave along the medium, this is called a **transverse wave.**

In Fig. 15.1b the medium is a liquid or gas in a tube with a rigid wall at the right end and a movable piston at the left end. If we give the piston a single back-and-forth motion, displacement and pressure fluctuations travel down the length of the medium. This time the motions of the particles of the medium are back and forth along the *same* direction that the wave travels. We call this a **longitudinal wave.**

In Fig. 15.1c the medium is a liquid in a channel, such as water in an irrigation ditch or canal. When we move the flat board at the left end forward and back once, a wave disturbance travels down the length of the channel. In this case the displacements of the water have *both* longitudinal and transverse components.

Each of these systems has an equilibrium state. For the stretched string it is the state in which the system is at rest, stretched out along a straight line. For the fluid in a tube it is a state in which the fluid is at rest with uniform pressure. And for the liquid in a trough it is a smooth, level water surface. In each case the wave motion is a disturbance from the equilibrium state that travels from one region of the medium to another. And in each case there are forces that tend to restore the system to its equilibrium position when it is displaced, just as the force of gravity tends to pull a pendulum toward its straight-down equilibrium position when it is displaced.

Application Waves on a Snake's Body
A snake moves itself along the ground by producing waves that travel backward along its body from its head to its tail. The waves remain stationary with respect to the ground as they push against the ground, so the snake moves forward.

MasteringPHYSICS

ActivPhysics 10.1: Properties of Mechanical Waves

15.1 Three ways to make a wave that moves to the right. (a) The hand moves the string up and then returns, producing a transverse wave. (b) The piston moves to the right, compressing the gas or liquid, and then returns, producing a longitudinal wave. (c) The board moves to the right and then returns, producing a combination of longitudinal and transverse waves.

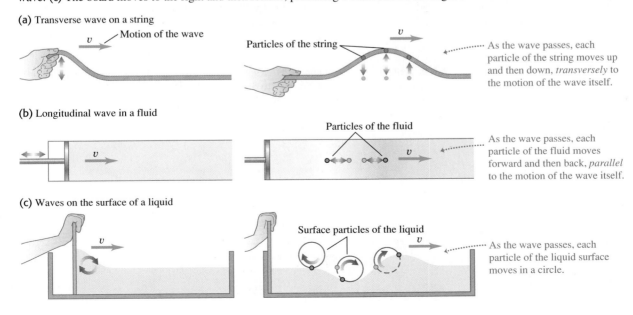

(a) Transverse wave on a string

Motion of the wave

Particles of the string

As the wave passes, each particle of the string moves up and then down, *transversely* to the motion of the wave itself.

(b) Longitudinal wave in a fluid

Particles of the fluid

As the wave passes, each particle of the fluid moves forward and then back, *parallel* to the motion of the wave itself.

(c) Waves on the surface of a liquid

Surface particles of the liquid

As the wave passes, each particle of the liquid surface moves in a circle.

15.2 "Doing the wave" at a sports stadium is an example of a mechanical wave: The disturbance propagates through the crowd, but there is no transport of matter (none of the spectators moves from one seat to another).

These examples have three things in common. First, in each case the disturbance travels or *propagates* with a definite speed through the medium. This speed is called the speed of propagation, or simply the **wave speed.** Its value is determined in each case by the mechanical properties of the medium. We will use the symbol v for wave speed. (The wave speed is *not* the same as the speed with which particles move when they are disturbed by the wave. We'll return to this point in Section 15.3.) Second, the medium itself does not travel through space; its individual particles undergo back-and-forth or up-and-down motions around their equilibrium positions. The overall pattern of the wave disturbance is what travels. Third, to set any of these systems into motion, we have to put in energy by doing mechanical work on the system. The wave motion transports this energy from one region of the medium to another. *Waves transport energy, but not matter, from one region to another* (Fig. 15.2).

Test Your Understanding of Section 15.1 What type of wave is "the wave" shown in Fig. 15.2? (i) transverse; (ii) longitudinal; (iii) a combination of transverse and longitudinal. ❙

15.2 Periodic Waves

The transverse wave on a stretched string in Fig. 15.1a is an example of a *wave pulse.* The hand shakes the string up and down just once, exerting a transverse force on it as it does so. The result is a single "wiggle," or pulse, that travels along the length of the string. The tension in the string restores its straight-line shape once the pulse has passed.

A more interesting situation develops when we give the free end of the string a repetitive, or *periodic,* motion. (You may want to review the discussion of periodic motion in Chapter 14 before going ahead.) Then each particle in the string also undergoes periodic motion as the wave propagates, and we have a **periodic wave.**

Periodic Transverse Waves

In particular, suppose we move the string up and down with *simple harmonic motion* (SHM) with amplitude A, frequency f, angular frequency $\omega = 2\pi f$, and period $T = 1/f = 2\pi/\omega$. Figure 15.3 shows one way to do this. The wave that results is a symmetrical sequence of *crests* and *troughs.* As we will see, periodic

15.3 A block of mass m attached to a spring undergoes simple harmonic motion, producing a sinusoidal wave that travels to the right on the string. (In a real-life system a driving force would have to be applied to the block to replace the energy carried away by the wave.)

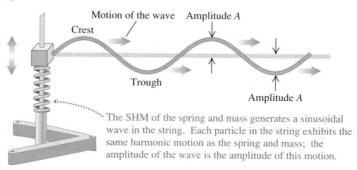

Motion of the wave   Amplitude A

Crest

Trough

Amplitude A

The SHM of the spring and mass generates a sinusoidal wave in the string. Each particle in the string exhibits the same harmonic motion as the spring and mass; the amplitude of the wave is the amplitude of this motion.

waves with simple harmonic motion are particularly easy to analyze; we call them **sinusoidal waves.** It also turns out that *any* periodic wave can be represented as a combination of sinusoidal waves. So this particular kind of wave motion is worth special attention.

In Fig. 15.3 the wave that advances along the string is a *continuous succession* of transverse sinusoidal disturbances. Figure 15.4 shows the shape of a part of the string near the left end at time intervals of $\frac{1}{8}$ of a period, for a total time of one period. The wave shape advances steadily toward the right, as indicated by the highlighted area. As the wave moves, any point on the string (any of the red dots, for example) oscillates up and down about its equilibrium position with simple harmonic motion. *When a sinusoidal wave passes through a medium, every particle in the medium undergoes simple harmonic motion with the same frequency.*

CAUTION **Wave motion vs. particle motion** Be very careful to distinguish between the motion of the *transverse wave* along the string and the motion of a *particle* of the string. The wave moves with constant speed v *along* the length of the string, while the motion of the particle is simple harmonic and *transverse* (perpendicular) to the length of the string.

For a periodic wave, the shape of the string at any instant is a repeating pattern. The length of one complete wave pattern is the distance from one crest to the next, or from one trough to the next, or from any point to the corresponding point on the next repetition of the wave shape. We call this distance the **wavelength** of the wave, denoted by λ (the Greek letter lambda). The wave pattern travels with constant speed v and advances a distance of one wavelength λ in a time interval of one period T. So the wave speed v is given by $v = \lambda/T$ or, because $f = 1/T$,

$$v = \lambda f \qquad \text{(periodic wave)} \qquad (15.1)$$

The speed of propagation equals the product of wavelength and frequency. The frequency is a property of the *entire* periodic wave because all points on the string oscillate with the same frequency f.

Waves on a string propagate in just one dimension (in Fig. 15.4, along the *x*-axis). But the ideas of frequency, wavelength, and amplitude apply equally well to waves that propagate in two or three dimensions. Figure 15.5 shows a wave propagating in two dimensions on the surface of a tank of water. As with waves on a string, the wavelength is the distance from one crest to the next, and the amplitude is the height of a crest above the equilibrium level.

In many important situations including waves on a string, the wave speed v is determined entirely by the mechanical properties of the medium. In this case, increasing f causes λ to decrease so that the product $v = \lambda f$ remains the same, and waves of *all* frequencies propagate with the same wave speed. In this chapter we will consider *only* waves of this kind. (In later chapters we will study the propagation of light waves in matter for which the wave speed depends on frequency; this turns out to be the reason prisms break white light into a spectrum and raindrops create a rainbow.)

Periodic Longitudinal Waves

To understand the mechanics of a periodic *longitudinal* wave, we consider a long tube filled with a fluid, with a piston at the left end as in Fig. 15.1b. If we push the piston in, we compress the fluid near the piston, increasing the pressure in this

15.4 A sinusoidal transverse wave traveling to the right along a string. The vertical scale is exaggerated.

The string is shown at time intervals of $\frac{1}{8}$ period for a total of one period T. The highlighting shows the motion of one wavelength of the wave.

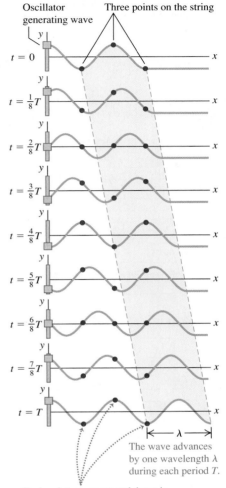

The wave advances by one wavelength λ during each period T.

Each point moves up and down in place. Particles one wavelength apart move in phase with each other.

15.5 A series of drops falling into water produces a periodic wave that spreads radially outward. The wave crests and troughs are concentric circles. The wavelength λ is the radial distance between adjacent crests or adjacent troughs.

15.6 Using an oscillating piston to make a sinusoidal longitudinal wave in a fluid.

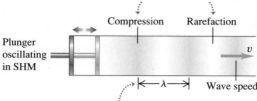

Forward motion of the plunger creates a compression (a zone of high density); backward motion creates a rarefaction (a zone of low density).

Wavelength λ is the distance between corresponding points on successive cycles.

15.7 A sinusoidal longitudinal wave traveling to the right in a fluid. The wave has the same amplitude A and period T as the oscillation of the piston.

Longitudinal waves are shown at intervals of $\frac{1}{8}T$ for one period T.

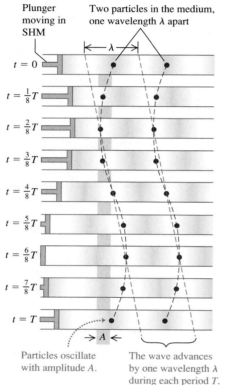

Particles oscillate with amplitude A.

The wave advances by one wavelength λ during each period T.

region. This region then pushes against the neighboring region of fluid, and so on, and a wave pulse moves along the tube.

Now suppose we move the piston back and forth with simple harmonic motion, along a line parallel to the axis of the tube (Fig. 15.6). This motion forms regions in the fluid where the pressure and density are greater or less than the equilibrium values. We call a region of increased density a *compression;* a region of reduced density is a *rarefaction.* Figure 15.6 shows compressions as darkly shaded areas and rarefactions as lightly shaded areas. The wavelength is the distance from one compression to the next or from one rarefaction to the next.

Figure 15.7 shows the wave propagating in the fluid-filled tube at time intervals of $\frac{1}{8}$ of a period, for a total time of one period. The pattern of compressions and rarefactions moves steadily to the right, just like the pattern of crests and troughs in a sinusoidal transverse wave (compare Fig. 15.4). Each particle in the fluid oscillates in SHM parallel to the direction of wave propagation (that is, left and right) with the same amplitude A and period T as the piston. The particles shown by the two red dots in Fig. 15.7 are one wavelength apart, and so oscillate in phase with each other.

Just like the sinusoidal transverse wave shown in Fig. 15.4, in one period T the longitudinal wave in Fig. 15.7 travels one wavelength λ to the right. Hence the fundamental equation $v = \lambda f$ holds for longitudinal waves as well as for transverse waves, and indeed for *all* types of periodic waves. Just as for transverse waves, in this chapter and the next we will consider only situations in which the speed of longitudinal waves does not depend on the frequency.

Example 15.1 **Wavelength of a musical sound**

Sound waves are longitudinal waves in air. The speed of sound depends on temperature; at 20°C it is 344 m/s (1130 ft/s). What is the wavelength of a sound wave in air at 20°C if the frequency is 262 Hz (the approximate frequency of middle C on a piano)?

SOLUTION

IDENTIFY and SET UP: This problem involves Eq. (15.1), $v = \lambda f$, which relates wave speed v, wavelength λ, and frequency f for a periodic wave. The target variable is the wavelength λ. We are given $v = 344$ m/s and $f = 262$ Hz $= 262$ s^{-1}.

EXECUTE: We solve Eq. (15.1) for λ:

$$\lambda = \frac{v}{f} = \frac{344 \text{ m/s}}{262 \text{ Hz}} = \frac{344 \text{ m/s}}{262 \text{ s}^{-1}} = 1.31 \text{ m}$$

EVALUATE: The speed v of sound waves does *not* depend on the frequency. Hence $\lambda = v/f$ says that wavelength changes in inverse proportion to frequency. As an example, high (soprano) C is two octaves above middle C. Each octave corresponds to a factor of 2 in frequency, so the frequency of high C is four times that of middle C: $f = 4(262 \text{ Hz}) = 1048$ Hz. Hence the *wavelength* of high C is *one-fourth* as large: $\lambda = (1.31 \text{ m})/4 = 0.328$ m.

Test Your Understanding of Section 15.2   If you double the wave-length of a wave on a particular string, what happens to the wave speed v and the frequency f? (i) v doubles and f is unchanged; (ii) v is unchanged and f doubles; (iii) v becomes one-half as great and f is unchanged; (iv) v is unchanged and f becomes one-half as great; (v) none of these.

15.3 Mathematical Description of a Wave

Many characteristics of periodic waves can be described by using the concepts of wave speed, amplitude, period, frequency, and wavelength. Often, though, we need a more detailed description of the positions and motions of individual particles of the medium at particular times during wave propagation.

As a specific example, let's look at waves on a stretched string. If we ignore the sag of the string due to gravity, the equilibrium position of the string is along a straight line. We take this to be the x-axis of a coordinate system. Waves on a string are *transverse;* during wave motion a particle with equilibrium position x is displaced some distance y in the direction perpendicular to the x-axis. The value of y depends on which particle we are talking about (that is, y depends on x) and also on the time t when we look at it. Thus y is a *function* of both x and t; $y = y(x, t)$. We call $y(x, t)$ the **wave function** that describes the wave. If we know this function for a particular wave motion, we can use it to find the displacement (from equilibrium) of any particle at any time. From this we can find the velocity and acceleration of any particle, the shape of the string, and anything else we want to know about the behavior of the string at any time.

Wave Function for a Sinusoidal Wave

Let's see how to determine the form of the wave function for a sinusoidal wave. Suppose a sinusoidal wave travels from left to right (the direction of increasing x) along the string, as in Fig. 15.8. Every particle of the string oscillates with simple harmonic motion with the same amplitude and frequency. But the oscillations of particles at different points on the string are *not* all in step with each other. The particle at point B in Fig. 15.8 is at its maximum positive value of y at $t = 0$ and returns to $y = 0$ at $t = \frac{2}{8}T$; these same events occur for a particle at point A or point C at $t = \frac{4}{8}T$ and $t = \frac{6}{8}T$, exactly one half-period later. For any two particles of the string, the motion of the particle on the right (in terms of the wave, the "downstream" particle) lags behind the motion of the particle on the left by an amount proportional to the distance between the particles.

Hence the cyclic motions of various points on the string are out of step with each other by various fractions of a cycle. We call these differences *phase differences,* and we say that the *phase* of the motion is different for different points. For example, if one point has its maximum positive displacement at the same time that another has its maximum negative displacement, the two are a half-cycle out of phase. (This is the case for points A and B, or points B and C.)

Suppose that the displacement of a particle at the left end of the string ($x = 0$), where the wave originates, is given by

$$y(x = 0, t) = A\cos\omega t = A\cos 2\pi ft \qquad (15.2)$$

That is, the particle oscillates in simple harmonic motion with amplitude A, frequency f, and angular frequency $\omega = 2\pi f$. The notation $y(x = 0, t)$ reminds us that the motion of this particle is a special case of the wave function $y(x, t)$ that describes the entire wave. At $t = 0$ the particle at $x = 0$ is at its maximum positive displacement ($y = A$) and is instantaneously at rest (because the value of y is a maximum).

The wave disturbance travels from $x = 0$ to some point x to the right of the origin in an amount of time given by x/v, where v is the wave speed. So the motion of point x at time t is the same as the motion of point $x = 0$ at the earlier time $t - x/v$. Hence we can find the displacement of point x at time t by simply

15.8 Tracking the oscillations of three points on a string as a sinusoidal wave propagates along it.

The string is shown at time intervals of $\frac{1}{8}$ period for a total of one period T.

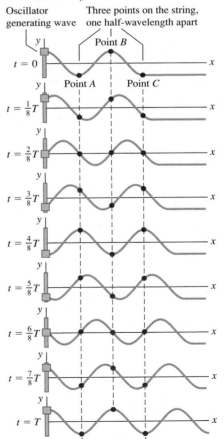

replacing t in Eq. (15.2) by $(t - x/v)$. When we do that, we find the following expression for the wave function:

$$y(x, t) = A \cos\left[\omega\left(t - \frac{x}{v}\right)\right]$$

Because $\cos(-\theta) = \cos\theta$, we can rewrite the wave function as

$$y(x, t) = A \cos\left[\omega\left(\frac{x}{v} - t\right)\right] = A \cos\left[2\pi f\left(\frac{x}{v} - t\right)\right] \quad \begin{array}{l}\text{(sinusoidal wave} \\ \text{moving in} \\ \text{+x-direction)}\end{array} \quad (15.3)$$

The displacement $y(x, t)$ is a function of both the location x of the point and the time t. We could make Eq. (15.3) more general by allowing for different values of the phase angle, as we did for simple harmonic motion in Section 14.2, but for now we omit this.

We can rewrite the wave function given by Eq. (15.3) in several different but useful forms. We can express it in terms of the period $T = 1/f$ and the wavelength $\lambda = v/f$:

$$y(x, t) = A \cos\left[2\pi\left(\frac{x}{\lambda} - \frac{t}{T}\right)\right] \quad \begin{array}{l}\text{(sinusoidal wave moving} \\ \text{in +x-direction)}\end{array} \quad (15.4)$$

It's convenient to define a quantity k, called the **wave number:**

$$k = \frac{2\pi}{\lambda} \quad \text{(wave number)} \quad (15.5)$$

Substituting $\lambda = 2\pi/k$ and $f = \omega/2\pi$ into the wavelength–frequency relationship $v = \lambda f$ gives

$$\omega = vk \quad \text{(periodic wave)} \quad (15.6)$$

We can then rewrite Eq. (15.4) as

$$y(x, t) = A \cos(kx - \omega t) \quad \begin{array}{l}\text{(sinusoidal wave moving} \\ \text{in +x-direction)}\end{array} \quad (15.7)$$

Which of these various forms for the wave function $y(x, t)$ we use in any specific problem is a matter of convenience. Note that ω has units rad/s, so for unit consistency in Eqs. (15.6) and (15.7) the wave number k must have the units rad/m. (Some physicists define the wave number as $1/\lambda$ rather than $2\pi/\lambda$. When reading other texts, be sure to determine how this term is defined.)

Graphing the Wave Function

Figure 15.9a graphs the wave function $y(x, t)$ as a function of x for a specific time t. This graph gives the displacement y of a particle from its equilibrium position as a function of the coordinate x of the particle. If the wave is a transverse wave on a string, the graph in Fig. 15.9a represents the shape of the string at that instant, like a flash photograph of the string. In particular, at time $t = 0$,

$$y(x, t = 0) = A \cos kx = A \cos 2\pi\frac{x}{\lambda}$$

Figure 15.9b is a graph of the wave function versus time t for a specific coordinate x. This graph gives the displacement y of the particle at that coordinate as a function of time; that is, it describes the motion of that particle. In particular, at the position $x = 0$,

$$y(x = 0, t) = A \cos(-\omega t) = A \cos\omega t = A \cos 2\pi\frac{t}{T}$$

This is consistent with our original statement about the motion at $x = 0$, Eq. (15.2).

15.9 Two graphs of the wave function $y(x, t)$ in Eq. (15.7). **(a)** Graph of displacement y versus coordinate x at time $t = 0$. **(b)** Graph of displacement y versus time t at coordinate $x = 0$. The vertical scale is exaggerated in both (a) and (b).

(a) If we use Eq. (15.7) to plot y as a function of x for time $t = 0$, the curve shows the *shape* of the string at $t = 0$.

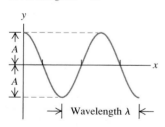

(b) If we use Eq. (15.7) to plot y as a function of t for position $x = 0$, the curve shows the *displacement* y of the particle at $x = 0$ as a function of time.

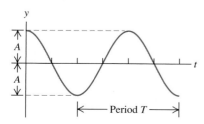

CAUTION **Wave graphs** Although they may look the same at first glance, Figs. 15.9a and 15.9b are *not* identical. Figure 15.9a is a picture of the shape of the string at $t = 0$, while Fig. 15.9b is a graph of the displacement y of a particle at $x = 0$ as a function of time.

More on the Wave Function

We can modify Eqs. (15.3) through (15.7) to represent a wave traveling in the *negative x*-direction. In this case the displacement of point x at time t is the same as the motion of point $x = 0$ at the *later* time $(t + x/v)$, so in Eq. (15.2) we replace t by $(t + x/v)$. For a wave traveling in the negative x-direction,

$$y(x, t) = A\cos\left[2\pi f\left(\frac{x}{v} + t\right)\right] = A\cos\left[2\pi\left(\frac{x}{\lambda} + \frac{t}{T}\right)\right] = A\cos(kx + \omega t) \quad (15.8)$$

(sinusoidal wave moving in $-x$-direction)

In the expression $y(x, t) = A\cos(kx \pm \omega t)$ for a wave traveling in the $-x$- or $+x$-direction, the quantity $(kx \pm \omega t)$ is called the **phase.** It plays the role of an angular quantity (always measured in radians) in Eq. (15.7) or (15.8), and its value for any values of x and t determines what part of the sinusoidal cycle is occurring at a particular point and time. For a crest (where $y = A$ and the cosine function has the value 1), the phase could be 0, 2π, 4π, and so on; for a trough (where $y = -A$ and the cosine has the value -1), it could be π, 3π, 5π, and so on.

The wave speed is the speed with which we have to move along with the wave to keep alongside a point of a given phase, such as a particular crest of a wave on a string. For a wave traveling in the $+x$-direction, that means $kx - \omega t = $ constant. Taking the derivative with respect to t, we find $k\, dx/dt = \omega$, or

$$\frac{dx}{dt} = \frac{\omega}{k}$$

Comparing this with Eq. (15.6), we see that dx/dt is equal to the speed v of the wave. Because of this relationship, v is sometimes called the *phase velocity* of the wave. (*Phase speed* would be a better term.)

Problem-Solving Strategy 15.1 | **Mechanical Waves**

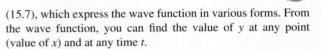

IDENTIFY *the relevant concepts:* As always, identify the target variables; these may include mathematical *expressions* (for example, the wave function for a given situation). Note that wave problems fall into two categories. *Kinematics* problems, concerned with describing wave motion, involve wave speed v, wavelength λ (or wave number k), frequency f (or angular frequency ω), and amplitude A. They may also involve the position, velocity, and acceleration of individual particles in the medium. *Dynamics* problems also use concepts from Newton's laws. Later in this chapter we'll encounter problems that involve the relationship of wave speed to the mechanical properties of the medium.

SET UP *the problem* using the following steps:
1. List the given quantities. Sketch graphs of y versus x (like Fig. 15.9a) and of y versus t (like Fig. 15.9b), and label them with known values.
2. Identify useful equations. These may include Eq. (15.1) $(v = \lambda f)$, Eq. (15.6) $(\omega = vk)$, and Eqs. (15.3), (15.4), and

(15.7), which express the wave function in various forms. From the wave function, you can find the value of y at any point (value of x) and at any time t.
3. If you need to determine the wave speed v and don't know both λ and f, you may be able to use a relationship between v and the mechanical properties of the system. (In the next section we'll develop this relationship for waves on a string.)

EXECUTE *the solution:* Solve for the unknown quantities using the equations you've identified. To determine the wave function from Eq. (15.3), (15.4), or (15.7), you must know A and any two of v, λ, and f (or v, k, and ω).

EVALUATE *your answer:* Confirm that the values of v, f, and λ (or v, ω, and k) agree with the relationships given in Eq. (15.1) or (15.6). If you've calculated the wave function, check one or more special cases for which you can predict the results.

Example 15.2 — Wave on a clothesline

Cousin Throckmorton holds one end of the clothesline taut and wiggles it up and down sinusoidally with frequency 2.00 Hz and amplitude 0.075 m. The wave speed on the clothesline is $v = 12.0$ m/s. At $t = 0$ Throcky's end has maximum positive displacement and is instantaneously at rest. Assume that no wave bounces back from the far end. (a) Find the wave amplitude A, angular frequency ω, period T, wavelength λ, and wave number k. (b) Write a wave function describing the wave. (c) Write equations for the displacement, as a function of time, of Throcky's end of the clothesline and of a point 3.00 m from that end.

SOLUTION

IDENTIFY and SET UP: This is a kinematics problem about the clothesline's wave motion. Throcky produces a sinusoidal wave that propagates along the clothesline, so we can use all of the expressions of this section. In part (a) our target variables are A, ω, T, λ, and k. We use the relationships $\omega = 2\pi f$, $f = 1/T$, $v = \lambda f$, and $k = 2\pi/\lambda$. In parts (b) and (c) our target "variables" are expressions for displacement, which we'll obtain from an appropriate equation for the wave function. We take the positive x-direction to be the direction in which the wave propagates, so either Eq. (15.4) or (15.7) will yield the desired expression. A photograph of the clothesline at time $t = 0$ would look like Fig. 15.9a, with the maximum displacement at $x = 0$ (the end that Throcky holds).

EXECUTE: (a) The wave amplitude and frequency are the same as for the oscillations of Throcky's end of the clothesline, $A = 0.075$ m and $f = 2.00$ Hz. Hence

$$\omega = 2\pi f = \left(2\pi\frac{\text{rad}}{\text{cycle}}\right)\left(2.00\frac{\text{cycles}}{\text{s}}\right)$$

$$= 4.00\pi \text{ rad/s} = 12.6 \text{ rad/s}$$

The period is $T = 1/f = 0.500$ s, and from Eq. (15.1),

$$\lambda = \frac{v}{f} = \frac{12.0 \text{ m/s}}{2.00 \text{ s}^{-1}} = 6.00 \text{ m}$$

We find the wave number from Eq. (15.5) or (15.6):

$$k = \frac{2\pi}{\lambda} = \frac{2\pi \text{ rad}}{6.00 \text{ m}} = 1.05 \text{ rad/m}$$

or

$$k = \frac{\omega}{v} = \frac{4.00\pi \text{ rad/s}}{12.0 \text{ m/s}} = 1.05 \text{ rad/m}$$

(b) We write the wave function using Eq. (15.4) and the values of A, T, and λ from part (a):

$$y(x,t) = A\cos 2\pi\left(\frac{x}{\lambda} - \frac{t}{T}\right)$$

$$= (0.075 \text{ m})\cos 2\pi\left(\frac{x}{6.00 \text{ m}} - \frac{t}{0.500 \text{ s}}\right)$$

$$= (0.075 \text{ m})\cos[(1.05 \text{ rad/m})x - (12.6 \text{ rad/s})t]$$

We can also get this same expression from Eq. (15.7) by using the values of ω and k from part (a).

(c) We can find the displacement as a function of time at $x = 0$ and $x = +3.00$ m by substituting these values into the wave function from part (b):

$$y(x = 0, t) = (0.075 \text{ m})\cos 2\pi\left(\frac{0}{6.00 \text{ m}} - \frac{t}{0.500 \text{ s}}\right)$$

$$= (0.075 \text{ m})\cos(12.6 \text{ rad/s})t$$

$$y(x = +3.00 \text{ m}, t) = (0.075 \text{ m})\cos 2\pi\left(\frac{3.00 \text{ m}}{6.00 \text{ m}} - \frac{t}{0.500 \text{ s}}\right)$$

$$= (0.075 \text{ m})\cos[\pi - (12.6 \text{ rad/s})t]$$

$$= -(0.075 \text{ m})\cos(12.6 \text{ rad/s})t$$

EVALUATE: In part (b), the quantity $(1.05 \text{ rad/m})x - (12.6 \text{ rad/s})t$ is the *phase* of a point x on the string at time t. The two points in part (c) oscillate in SHM with the same frequency and amplitude, but their oscillations differ in phase by $(1.05 \text{ rad/m})(3.00 \text{ m}) = 3.15$ rad $= \pi$ radians—that is, one-half cycle—because the points are separated by one half-wavelength: $\lambda/2 = (6.00 \text{ m})/2 = 3.00$ m. Thus, while a graph of y versus t for the point at $x = 0$ is a cosine curve (like Fig. 15.9b), a graph of y versus t for the point $x = 3.00$ m is a *negative* cosine curve (the same as a cosine curve shifted by one half-cycle).

Using the expression for $y(x = 0, t)$ in part (c), can you show that the end of the string at $x = 0$ is instantaneously at rest at $t = 0$, as stated at the beginning of this example? (*Hint:* Calculate the y-velocity at this point by taking the derivative of y with respect to t.)

Particle Velocity and Acceleration in a Sinusoidal Wave

From the wave function we can get an expression for the transverse velocity of any *particle* in a transverse wave. We call this v_y to distinguish it from the wave propagation speed v. To find the transverse velocity v_y at a particular point x, we take the derivative of the wave function $y(x, t)$ with respect to t, keeping x constant. If the wave function is

$$y(x, t) = A\cos(kx - \omega t)$$

then

$$v_y(x, t) = \frac{\partial y(x, t)}{\partial t} = \omega A\sin(kx - \omega t) \tag{15.9}$$

The ∂ in this expression is a modified d, used to remind us that $y(x, t)$ is a function of *two* variables and that we are allowing only one (t) to vary. The other (x) is constant because we are looking at a particular point on the string. This derivative is called a *partial derivative*. If you haven't reached this point yet in your study of calculus, don't fret; it's a simple idea.

Equation (15.9) shows that the transverse velocity of a particle varies with time, as we expect for simple harmonic motion. The maximum particle speed is ωA; this can be greater than, less than, or equal to the wave speed v, depending on the amplitude and frequency of the wave.

The *acceleration* of any particle is the *second* partial derivative of $y(x, t)$ with respect to t:

$$a_y(x, t) = \frac{\partial^2 y(x, t)}{\partial t^2} = -\omega^2 A \cos(kx - \omega t) = -\omega^2 y(x, t) \quad \text{(15.10)}$$

The acceleration of a particle equals $-\omega^2$ times its displacement, which is the result we obtained in Section 14.2 for simple harmonic motion.

We can also compute partial derivatives of $y(x, t)$ with respect to x, holding t constant. The first derivative $\partial y(x, t)/\partial x$ is the *slope* of the string at point x and at time t. The second partial derivative with respect to x is the *curvature* of the string:

$$\frac{\partial^2 y(x, t)}{\partial x^2} = -k^2 A \cos(kx - \omega t) = -k^2 y(x, t) \quad \text{(15.11)}$$

From Eqs. (15.10) and (15.11) and the relationship $\omega = vk$ we see that

$$\frac{\partial^2 y(x, t)/\partial t^2}{\partial^2 y(x, t)/\partial x^2} = \frac{\omega^2}{k^2} = v^2 \quad \text{and}$$

$$\frac{\partial^2 y(x, t)}{\partial x^2} = \frac{1}{v^2}\frac{\partial^2 y(x, t)}{\partial t^2} \quad \text{(wave equation)} \quad \text{(15.12)}$$

We've derived Eq. (15.12) for a wave traveling in the positive x-direction. You can use the same steps to show that the wave function for a sinusoidal wave propagating in the *negative* x-direction, $y(x, t) = A \cos(kx + \omega t)$, also satisfies this equation.

Equation (15.12), called the **wave equation,** is one of the most important equations in all of physics. Whenever it occurs, we know that a disturbance can propagate as a wave along the x-axis with wave speed v. The disturbance need not be a sinusoidal wave; we'll see in the next section that *any* wave on a string obeys Eq. (15.12), whether the wave is periodic or not (see also Problem 15.65). In Chapter 32 we will find that electric and magnetic fields satisfy the wave equation; the wave speed will turn out to be the speed of light, which will lead us to the conclusion that light is an electromagnetic wave.

Figure 15.10a shows the transverse velocity v_y and transverse acceleration a_y, given by Eqs. (15.9) and (15.10), for several points on a string as a sinusoidal wave passes along it. Note that at points where the string has an upward curvature $(\partial^2 y/\partial x^2 > 0)$, the acceleration of that point is positive $(a_y = \partial^2 y/\partial t^2 > 0)$; this follows from the wave equation, Eq. (15.12). For the same reason the acceleration is negative $(a_y = \partial^2 y/\partial t^2 < 0)$ at points where the string has a downward curvature $(\partial^2 y/\partial x^2 < 0)$, and the acceleration is zero $(a_y = \partial^2 y/\partial t^2 = 0)$ at points of inflection where the curvature is zero $(\partial^2 y/\partial x^2 = 0)$. We emphasize again that v_y and a_y are the *transverse* velocity and acceleration of points on the string; these points move along the y-direction, not along the propagation direction of the

15.10 (a) Another view of the wave at $t = 0$ in Fig. 15.9a. The vectors show the transverse velocity v_y and transverse acceleration a_y at several points on the string. (b) From $t = 0$ to $t = 0.05T$, a particle at point 1 is displaced to point $1'$, a particle at point 2 is displaced to point $2'$, and so on.

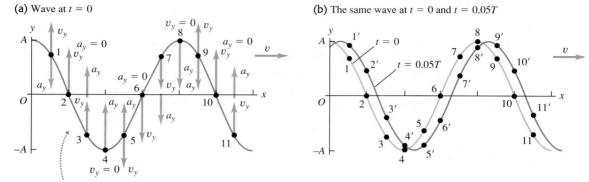

(a) Wave at $t = 0$

(b) The same wave at $t = 0$ and $t = 0.05T$

- Acceleration a_y at each point on the string is proportional to displacement y at that point.
- Acceleration is upward where string curves upward, downward where string curves downward.

wave. Figure 15.10b shows the transverse motions of several points on the string.

The concept of wave function is equally useful with *longitudinal* waves. The quantity y still measures the displacement of a particle of the medium from its equilibrium position; the difference is that for a longitudinal wave, this displacement is *parallel* to the x-axis instead of perpendicular to it. We'll discuss longitudinal waves in detail in Chapter 16.

Test Your Understanding of Section 15.3 Figure 15.8 shows a sinusoidal wave of period T on a string at times $0, \frac{1}{8}T, \frac{2}{8}T, \frac{3}{8}T, \frac{4}{8}T, \frac{5}{8}T, \frac{6}{8}T, \frac{7}{8}T$, and T. (a) At which time is point A on the string moving upward with maximum speed? (b) At which time does point B on the string have the greatest upward acceleration? (c) At which time does point C on the string have a downward acceleration but an upward velocity?

MasteringPHYSICS

ActivPhysics 10.2: Speed of Waves on a String

15.4 Speed of a Transverse Wave

One of the key properties of any wave is the wave *speed*. Light waves in air have a much greater speed of propagation than do sound waves in air (3.00×10^8 m/s versus 344 m/s); that's why you see the flash from a bolt of lightning before you hear the clap of thunder. In this section we'll see what determines the speed of propagation of one particular kind of wave: transverse waves on a string. The speed of these waves is important to understand because it is an essential part of analyzing stringed musical instruments, as we'll discuss later in this chapter. Furthermore, the speeds of many kinds of mechanical waves turn out to have the same basic mathematical expression as does the speed of waves on a string.

The physical quantities that determine the speed of transverse waves on a string are the *tension* in the string and its *mass per unit length* (also called *linear mass density*). We might guess that increasing the tension should increase the restoring forces that tend to straighten the string when it is disturbed, thus increasing the wave speed. We might also guess that increasing the mass should make the motion more sluggish and decrease the speed. Both these guesses turn out to be right. We'll develop the exact relationship among wave speed, tension, and mass per unit length by two different methods. The first is simple in concept and considers a specific wave shape; the second is more general but also more formal. Choose whichever you like better.

15.11 Propagation of a transverse wave on a string.

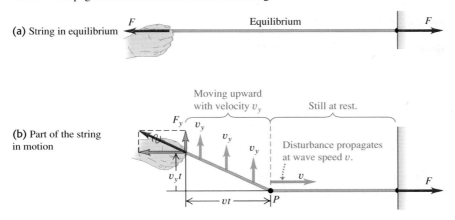

(a) String in equilibrium

(b) Part of the string in motion

Wave Speed on a String: First Method

We consider a perfectly flexible string (Fig. 15.11). In the equilibrium position the tension is F and the linear mass density (mass per unit length) is μ. (When portions of the string are displaced from equilibrium, the mass per unit length decreases a little, and the tension increases a little.) We ignore the weight of the string so that when the string is at rest in the equilibrium position, the string forms a perfectly straight line as in Fig. 15.11a.

Starting at time $t = 0$, we apply a constant upward force F_y at the left end of the string. We might expect that the end would move with constant acceleration; that would happen if the force were applied to a *point* mass. But here the effect of the force F_y is to set successively more and more mass in motion. The wave travels with constant speed v, so the division point P between moving and nonmoving portions moves with the same constant speed v (Fig. 15.11b).

Figure 15.11b shows that all particles in the moving portion of the string move upward with constant *velocity* v_y, not constant acceleration. To see why this is so, we note that the *impulse* of the force F_y up to time t is $F_y t$. According to the impulse–momentum theorem (see Section 8.1), the impulse is equal to the change in the total transverse component of momentum $(mv_y - 0)$ of the moving part of the string. Because the system started with *no* transverse momentum, this is equal to the total momentum at time t:

$$F_y t = mv_y$$

The total momentum thus must increase proportionately with time. But since the division point P moves with constant speed, the length of string that is in motion and hence the total mass m in motion are also proportional to the time t that the force has been acting. So the *change* of momentum must be associated entirely with the increasing amount of mass in motion, not with an increasing velocity of an individual mass element. That is, mv_y changes because m, not v_y, changes.

At time t, the left end of the string has moved up a distance $v_y t$, and the boundary point P has advanced a distance vt. The total force at the left end of the string has components F and F_y. Why F? There is no motion in the direction along the length of the string, so there is no unbalanced horizontal force. Therefore F, the magnitude of the horizontal component, does not change when the string is displaced. In the displaced position the tension is $(F^2 + F_y{}^2)^{1/2}$ (greater than F), and the string stretches somewhat.

To derive an expression for the wave speed v, we again apply the impulse–momentum theorem to the portion of the string in motion at time t—that is, the portion to the left of P in Fig. 15.11b. The transverse *impulse* (transverse force times time) is equal to the change of transverse *momentum* of the moving portion (mass times transverse component of velocity). The impulse of the transverse force F_y in time t is $F_y t$. In Fig. 15.11b the right triangle whose vertex is at P, with sides $v_y t$ and vt, is similar to the right triangle whose vertex is at the position of the hand, with sides F_y and F. Hence

$$\frac{F_y}{F} = \frac{v_y t}{vt} \qquad F_y = F\frac{v_y}{v}$$

and

$$\text{Transverse impulse} = F_y t = F\frac{v_y}{v}t$$

The mass of the moving portion of the string is the product of the mass per unit length μ and the length vt, or μvt. The transverse momentum is the product of this mass and the transverse velocity v_y:

$$\text{Transverse momentum} = (\mu vt)v_y$$

We note again that the momentum increases with time *not* because mass is moving faster, as was usually the case in Chapter 8, but because *more mass* is brought into motion. But the impulse of the force F_y is still equal to the total change in momentum of the system. Applying this relationship, we obtain

$$F\frac{v_y}{v}t = \mu vt v_y$$

Solving this for v, we find

$$v = \sqrt{\frac{F}{\mu}} \qquad \text{(speed of a transverse wave on a string)} \qquad (15.13)$$

Equation (15.13) confirms our prediction that the wave speed v should increase when the tension F increases but decrease when the mass per unit length μ increases (Fig. 15.12).

Note that v_y does not appear in Eq. (15.13); thus the wave speed doesn't depend on v_y. Our calculation considered only a very special kind of pulse, but we can consider *any* shape of wave disturbance as a series of pulses with different values of v_y. So even though we derived Eq. (15.13) for a special case, it is valid for *any* transverse wave motion on a string, including the sinusoidal and other periodic waves we discussed in Section 15.3. Note also that the wave speed doesn't depend on the amplitude or frequency of the wave, in accordance with our assumptions in Section 15.3.

Wave Speed on a String: Second Method

Here is an alternative derivation of Eq. (15.13). If you aren't comfortable with partial derivatives, it can be omitted. We apply Newton's second law, $\sum \vec{F} = m\vec{a}$, to a small segment of string whose length in the equilibrium position is Δx (Fig. 15.13). The mass of the segment is $m = \mu\,\Delta x$; the forces at the ends are represented in terms of their x- and y-components. The x-components have equal magnitude F and add to zero because the motion is transverse and there is no component of acceleration in the x-direction. To obtain F_{1y} and F_{2y}, we note that the ratio F_{1y}/F is equal in magnitude to the *slope* of the string at point x and that F_{2y}/F is equal to the slope at point $x + \Delta x$. Taking proper account of signs, we find

$$\frac{F_{1y}}{F} = -\left(\frac{\partial y}{\partial x}\right)_x \qquad \frac{F_{2y}}{F} = \left(\frac{\partial y}{\partial x}\right)_{x+\Delta x} \qquad (15.14)$$

15.12 These cables have a relatively large amount of mass per unit length (μ) and a low tension (F). If the cables are disturbed—say, by a bird landing on them—transverse waves will travel along them at a slow speed $v = \sqrt{F/\mu}$.

15.13 Free-body diagram for a segment of string. The force at each end of the string is tangent to the string at the point of application.

The string to the right of the segment (not shown) exerts a force $\vec{F}_2$ on the segment.

There can be a net vertical force on the segment, but the net horizontal force is zero (the motion is transverse).

F_2

F_{2y}

F

Equilibrium length of this segment of the string

F

F_{1y}

F_1

Δx

x

$x + \Delta x$

The string to the left of the segment (not shown) exerts a force $\vec{F}_1$ on the segment.

The notation reminds us that the derivatives are evaluated at points x and $x + \Delta x$, respectively. From Eq. (15.14) we find that the net y-component of force is

$$F_y = F_{1y} + F_{2y} = F\left[\left(\frac{\partial y}{\partial x}\right)_{x+\Delta x} - \left(\frac{\partial y}{\partial x}\right)_x\right] \qquad (15.15)$$

We now equate F_y from Eq. (15.15) to the mass $\mu\,\Delta x$ times the y-component of acceleration $\partial^2 y / \partial t^2$. We obtain

$$F\left[\left(\frac{\partial y}{\partial x}\right)_{x+\Delta x} - \left(\frac{\partial y}{\partial x}\right)_x\right] = \mu\,\Delta x \frac{\partial^2 y}{\partial t^2} \qquad (15.16)$$

or, dividing by $F\Delta x$,

$$\frac{\left(\dfrac{\partial y}{\partial x}\right)_{x+\Delta x} - \left(\dfrac{\partial y}{\partial x}\right)_x}{\Delta x} = \frac{\mu}{F}\frac{\partial^2 y}{\partial t^2} \qquad (15.17)$$

We now take the limit as $\Delta x \to 0$. In this limit, the left side of Eq. (15.17) becomes the derivative of $\partial y / \partial x$ with respect to x (at constant t)—that is, the *second* (partial) derivative of y with respect to x:

$$\frac{\partial^2 y}{\partial x^2} = \frac{\mu}{F}\frac{\partial^2 y}{\partial t^2} \qquad (15.18)$$

Now, finally, comes the punch line of our story. Equation (15.18) has exactly the same form as the *wave equation,* Eq. (15.12), that we derived at the end of Section 15.3. That equation and Eq. (15.18) describe the very same wave motion, so they must be identical. Comparing the two equations, we see that for this to be so, we must have

$$v = \sqrt{\frac{F}{\mu}} \qquad (15.19)$$

which is the same expression as Eq. (15.13).

In going through this derivation, we didn't make any special assumptions about the shape of the wave. Since our derivation led us to rediscover Eq. (15.12), the wave equation, we conclude that the wave equation is valid for waves on a string that have *any* shape.

The Speed of Mechanical Waves

Equation (15.13) or (15.19) gives the wave speed for only the special case of mechanical waves on a stretched string or rope. Remarkably, it turns out that for many types of mechanical waves, including waves on a string, the expression for wave speed has the same general form:

$$v = \sqrt{\frac{\text{Restoring force returning the system to equilibrium}}{\text{Inertia resisting the return to equilibrium}}}$$

To interpret this expression, let's look at the now-familiar case of waves on a string. The tension F in the string plays the role of the restoring force; it tends to bring the string back to its undisturbed, equilibrium configuration. The mass of the string—or, more properly, the linear mass density μ—provides the inertia that prevents the string from returning instantaneously to equilibrium. Hence we have $v = \sqrt{F/\mu}$ for the speed of waves on a string.

In Chapter 16 we'll see a similar expression for the speed of sound waves in a gas. Roughly speaking, the gas pressure provides the force that tends to return the gas to its undisturbed state when a sound wave passes through. The inertia is provided by the density, or mass per unit volume, of the gas.

Example 15.3 Calculating wave speed

One end of a 2.00-kg rope is tied to a support at the top of a mine shaft 80.0 m deep (Fig. 15.14). The rope is stretched taut by a 20.0-kg box of rocks attached at the bottom. (a) The geologist at the bottom of the shaft signals to a colleague at the top by jerking the rope sideways. What is the speed of a transverse wave on the rope? (b) If a point on the rope is in transverse SHM with $f = 2.00$ Hz, how many cycles of the wave are there in the rope's length?

SOLUTION

IDENTIFY and SET UP: In part (a) we can find the wave speed (our target variable) using the *dynamic* relationship $v = \sqrt{F/\mu}$

15.14 Sending signals along a vertical rope using transverse waves.

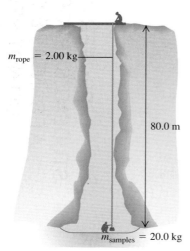

$m_{rope} = 2.00$ kg

80.0 m

$m_{samples} = 20.0$ kg

[Eq. (15.13)]. In part (b) we find the wavelength from the *kinematic* relationship $v = f\lambda$; from that we can find the target variable, the number of wavelengths that fit into the rope's 80.0-m length. We'll assume that the rope is massless (even though its weight is 10% that of the box), so that the box alone provides the tension in the rope.

EXECUTE: (a) The tension in the rope due to the box is

$$F = m_{box}g = (20.0 \text{ kg})(9.80 \text{ m/s}^2) = 196 \text{ N}$$

and the rope's linear mass density is

$$\mu = \frac{m_{rope}}{L} = \frac{2.00 \text{ kg}}{80.0 \text{ m}} = 0.0250 \text{ kg/m}$$

Hence, from Eq. (15.13), the wave speed is

$$v = \sqrt{\frac{F}{\mu}} = \sqrt{\frac{196 \text{ N}}{0.0250 \text{ kg/m}}} = 88.5 \text{ m/s}$$

(b) From Eq. (15.1), the wavelength is

$$\lambda = \frac{v}{f} = \frac{88.5 \text{ m/s}}{2.00 \text{ s}^{-1}} = 44.3 \text{ m}$$

There are $(80.0 \text{ m})/(44.3 \text{ m}) = 1.81$ wavelengths (that is, cycles of the wave) in the rope.

EVALUATE: Because of the rope's weight, its tension is greater at the top than at the bottom. Hence both the wave speed and the wavelength increase as a wave travels up the rope. If you take account of this, can you verify that the wave speed at the top of the rope is 92.9 m/s?

Test Your Understanding of Section 15.4 The six strings of a guitar are the same length and under nearly the same tension, but they have different thicknesses. On which string do waves travel the fastest? (i) the thickest string; (ii) the thinnest string; (iii) the wave speed is the same on all strings.

15.5 Energy in Wave Motion

Every wave motion has *energy* associated with it. The energy we receive from sunlight and the destructive effects of ocean surf and earthquakes bear this out. To produce any of the wave motions we have discussed in this chapter, we have to apply a force to a portion of the wave medium; the point where the force is applied moves, so we do *work* on the system. As the wave propagates, each portion of the medium exerts a force and does work on the adjoining portion. In this way a wave can transport energy from one region of space to another.

As an example of energy considerations in wave motion, let's look again at transverse waves on a string. How is energy transferred from one portion of string to another? Picture a wave traveling from left to right (the positive x-direction) on the string, and consider a particular point a on the string (Fig. 15.15a). The string to the left of point a exerts a force on the string to the right of it, and vice versa. In Fig. 15.15b the string to the left of a has been removed, and the force it exerts at a is represented by the components F and F_y, as we did in Figs. 15.11 and

Application Surface Waves and the Swimming Speed of Ducks
When a duck swims, it necessarily produces waves on the surface of the water. The faster the duck swims, the larger the wave amplitude and the more power the duck must supply to produce these waves. The maximum power available from their leg muscles limits the maximum swimming speed of ducks to only about 0.7 m/s (2.5 km/h = 1.6 mi/h).

15.13. We note again that F_y/F is equal to the negative of the *slope* of the string at a, which is also given by $\partial y/\partial x$. Putting these together, we have

$$F_y(x, t) = -F\frac{\partial y(x, t)}{\partial x} \qquad (15.20)$$

We need the negative sign because F_y is negative when the slope is positive. We write the vertical force as $F_y(x, t)$ as a reminder that its value may be different at different points along the string and at different times.

When point a moves in the y-direction, the force F_y does *work* on this point and therefore transfers energy into the part of the string to the right of a. The corresponding power P (rate of doing work) at the point a is the transverse force $F_y(x, t)$ at a times the transverse velocity $v_y(x, t) = \partial y(x, t)/\partial t$ of that point:

$$P(x, t) = F_y(x, t)v_y(x, t) = -F\frac{\partial y(x, t)}{\partial x}\frac{\partial y(x, t)}{\partial t} \qquad (15.21)$$

This power is the *instantaneous* rate at which energy is transferred along the string. Its value depends on the position x on the string and on the time t. Note that energy is being transferred only at points where the string has a nonzero slope ($\partial y/\partial x$ is nonzero), so that there is a transverse component of the tension force, and where the string has a nonzero transverse velocity ($\partial y/\partial t$ is nonzero) so that the transverse force can do work.

Equation (15.21) is valid for *any* wave on a string, sinusoidal or not. For a sinusoidal wave with wave function given by Eq. (15.7), we have

$$y(x, t) = A\cos(kx - \omega t)$$
$$\frac{\partial y(x, t)}{\partial x} = -kA\sin(kx - \omega t)$$
$$\frac{\partial y(x, t)}{\partial t} = \omega A\sin(kx - \omega t)$$
$$P(x, t) = Fk\omega A^2\sin^2(kx - \omega t) \qquad (15.22)$$

By using the relationships $\omega = vk$ and $v^2 = F/\mu$, we can also express Eq. (15.22) in the alternative form

$$P(x, t) = \sqrt{\mu F}\,\omega^2 A^2\sin^2(kx - \omega t) \qquad (15.23)$$

The $\sin^2$ function is never negative, so the instantaneous power in a sinusoidal wave is either positive (so that energy flows in the positive x-direction) or zero (at points where there is no energy transfer). Energy is never transferred in the direction opposite to the direction of wave propagation (Fig. 15.16).

The maximum value of the instantaneous power $P(x, t)$ occurs when the $\sin^2$ function has the value unity:

$$P_{max} = \sqrt{\mu F}\,\omega^2 A^2 \qquad (15.24)$$

To obtain the *average* power from Eq. (15.23), we note that the *average* value of the $\sin^2$ function, averaged over any whole number of cycles, is $\frac{1}{2}$. Hence the average power is

$$P_{av} = \tfrac{1}{2}\sqrt{\mu F}\,\omega^2 A^2 \quad \text{(average power, sinusoidal wave on a string)} \quad (15.25)$$

The average power is just one-half of the maximum instantaneous power (see Fig. 15.16).

The average rate of energy transfer is proportional to the square of the amplitude and to the square of the frequency. This proportionality is a general result for mechanical waves of all types, including seismic waves (see the photo that opens this chapter). For a mechanical wave, the rate of energy transfer

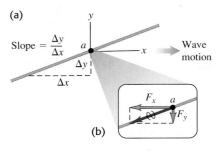

15.15 (a) Point a on a string carrying a wave from left to right. (b) The components of the force exerted on the part of the string to the right of point a by the part of the string to the left of point a.

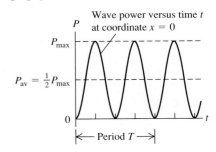

15.16 The instantaneous power $P(x, t)$ in a sinusoidal wave as given by Eq. (15.23), shown as a function of time at coordinate $x = 0$. The power is never negative, which means that energy never flows opposite to the direction of wave propagation.

quadruples if the frequency is doubled (for the same amplitude) or if the amplitude is doubled (for the same frequency).

Electromagnetic waves turn out to be a bit different. While the average rate of energy transfer in an electromagnetic wave is proportional to the square of the amplitude, just as for mechanical waves, it is independent of the value of ω.

Example 15.4 **Power in a wave**

(a) In Example 15.2 (Section 15.3), at what maximum rate does Throcky put energy into the clothesline? That is, what is his maximum instantaneous power? The linear mass density of the clothesline is $\mu = 0.250$ kg/m, and Throcky applies tension $F = 36.0$ N. (b) What is his average power? (c) As Throcky tires, the amplitude decreases. What is the average power when the amplitude is 7.50 mm?

SOLUTION

IDENTIFY and SET UP: In part (a) our target variable is the *maximum instantaneous power* P_{max}, while in parts (b) and (c) it is the *average* power. For part (a) we'll use Eq. (15.24), and for parts (b) and (c) we'll use Eq. (15.25); Example 15.2 gives us all the needed quantities.

EXECUTE: (a) From Eq. (15.24),

$$P_{max} = \sqrt{\mu F}\omega^2 A^2$$
$$= \sqrt{(0.250 \text{ kg/m})(36.0 \text{ N})}(4.00\pi \text{ rad/s})^2(0.075 \text{ m})^2$$
$$= 2.66 \text{ W}$$

(b) From Eqs. (15.24) and (15.25), the average power is one-half of the maximum instantaneous power, so

$$P_{av} = \tfrac{1}{2}P_{max} = \tfrac{1}{2}(2.66 \text{ W}) = 1.33 \text{ W}$$

(c) The new amplitude is $\frac{1}{10}$ of the value we used in parts (a) and (b). From Eq. (15.25), the average power is proportional to A^2, so the new average power is

$$P_{av} = \left(\tfrac{1}{10}\right)^2(1.33 \text{ W}) = 0.0133 \text{ W} = 13.3 \text{ mW}$$

EVALUATE: Equation (15.23) shows that P_{max} occurs when $\sin^2(kx - \omega t) = 1$. At any given position x, this happens twice per period of the wave—once when the sine function is equal to $+1$, and once when it's equal to -1. The *minimum* instantaneous power is zero; this occurs when $\sin^2(kx - \omega t) = 0$, which also happens twice per period.

Can you confirm that the given values of μ and F give the wave speed mentioned in Example 15.2?

Wave Intensity

Waves on a string carry energy in just one dimension of space (along the direction of the string). But other types of waves, including sound waves in air and seismic waves in the body of the earth, carry energy across all three dimensions of space. For waves that travel in three dimensions, we define the **intensity** (denoted by I) to be *the time average rate at which energy is transported by the wave, per unit area,* across a surface perpendicular to the direction of propagation. That is, intensity I is average power per unit area. It is usually measured in watts per square meter (W/m^2).

If waves spread out equally in all directions from a source, the intensity at a distance r from the source is inversely proportional to r^2 (Fig. 15.17). This follows directly from energy conservation. If the power output of the source is P, then the average intensity I_1 through a sphere with radius r_1 and surface area $4\pi r_1^2$ is

$$I_1 = \frac{P}{4\pi r_1^2}$$

The average intensity I_2 through a sphere with a different radius r_2 is given by a similar expression. If no energy is absorbed between the two spheres, the power P must be the same for both, and

$$4\pi r_1^2 I_1 = 4\pi r_2^2 I_2$$

$$\frac{I_1}{I_2} = \frac{r_2^2}{r_1^2} \qquad \text{(inverse-square law for intensity)} \qquad (15.26)$$

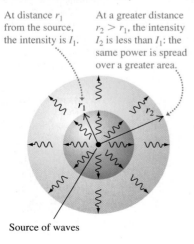

15.17 The greater the distance from a wave source, the greater the area over which the wave power is distributed and the smaller the wave intensity.

At distance r_1 from the source, the intensity is I_1.

At a greater distance $r_2 > r_1$, the intensity I_2 is less than I_1: the same power is spread over a greater area.

Source of waves

The intensity I at any distance r is therefore inversely proportional to r^2. This relationship is called the *inverse-square law* for intensity.

Example 15.5 The inverse-square law

A siren on a tall pole radiates sound waves uniformly in all directions. At a distance of 15.0 m from the siren, the sound intensity is 0.250 W/m². At what distance is the intensity 0.010 W/m²?

SOLUTION

IDENTIFY and SET UP: Because sound is radiated uniformly in all directions, we can use the inverse-square law, Eq. (15.26). At $r_1 = 15.0$ m the intensity is $I_1 = 0.250$ W/m², and the target variable is the distance r_2 at which the intensity is $I_2 = 0.010$ W/m².

EXECUTE: We solve Eq. (15.26) for r_2:

$$r_2 = r_1\sqrt{\frac{I_1}{I_2}} = (15.0 \text{ m})\sqrt{\frac{0.250 \text{ W/m}^2}{0.010 \text{ W/m}^2}} = 75.0 \text{ m}$$

EVALUATE: As a check on our answer, note that r_2 is five times greater than r_1. By the inverse-square law, the intensity I_2 should be $1/5^2 = 1/25$ as great as I_1, and indeed it is.

By using the inverse-square law, we've assumed that the sound waves travel in straight lines away from the siren. A more realistic solution, which is beyond our scope, would account for the reflection of sound waves from the ground.

Test Your Understanding of Section 15.5 Four identical strings each carry a sinusoidal wave of frequency 10 Hz. The string tension and wave amplitude are different for different strings. Rank the following strings in order from highest to lowest value of the average wave power: (i) tension 10 N, amplitude 1.0 mm; (ii) tension 40 N, amplitude 1.0 mm; (iii) tension 10 N, amplitude 4.0 mm; (iv) tension 20 N, amplitude 2.0 mm.

15.6 Wave Interference, Boundary Conditions, and Superposition

Up to this point we've been discussing waves that propagate continuously in the same direction. But when a wave strikes the boundaries of its medium, all or part of the wave is *reflected*. When you yell at a building wall or a cliff face some distance away, the sound wave is reflected from the rigid surface and you hear an echo. When you flip the end of a rope whose far end is tied to a rigid support, a pulse travels the length of the rope and is reflected back to you. In both cases, the initial and reflected waves overlap in the same region of the medium. This overlapping of waves is called **interference.** (In general, the term "interference" refers to what happens when two or more waves pass through the same region at the same time.)

As a simple example of wave reflections and the role of the boundary of a wave medium, let's look again at transverse waves on a stretched string. What happens when a wave pulse or a sinusoidal wave arrives at the *end* of the string?

If the end is fastened to a rigid support, it is a *fixed* end that cannot move. The arriving wave exerts a force on the support; the reaction to this force, exerted *by* the support *on* the string, "kicks back" on the string and sets up a *reflected* pulse or wave traveling in the reverse direction. Figure 15.18 is a series of photographs showing the reflection of a pulse at the fixed end of a long coiled spring. The reflected pulse moves in the opposite direction from the initial, or *incident*, pulse, and its displacement is also opposite. Figure 15.19a illustrates this situation for a wave pulse on a string.

The opposite situation from an end that is held stationary is a *free* end, one that is perfectly free to move in the direction perpendicular to the length of the string. For example, the string might be tied to a light ring that slides on a frictionless rod perpendicular to the string, as in Fig. 15.19b. The ring and rod maintain the tension but exert no transverse force. When a wave arrives at this free end, the ring slides along the rod. The ring reaches a maximum displacement, and both it and the string come momentarily to rest, as in drawing 4 in Fig. 15.19b. But the string is now stretched, giving increased tension, so the free end of the string is

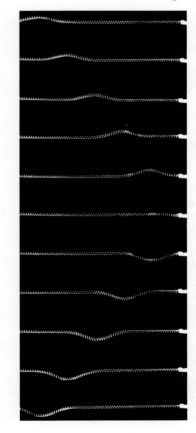

15.18 A series of images of a wave pulse, equally spaced in time from top to bottom. The pulse starts at the left in the top image, travels to the right, and is reflected from the fixed end at the right.

15.19 Reflection of a wave pulse **(a)** at a fixed end of a string and **(b)** at a free end. Time increases from top to bottom in each figure.

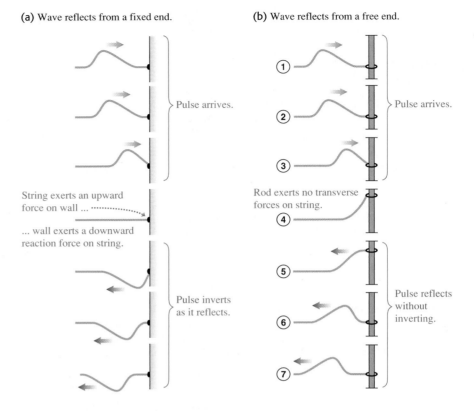

(a) Wave reflects from a fixed end.

Pulse arrives.

String exerts an upward force on wall ...

... wall exerts a downward reaction force on string.

Pulse inverts as it reflects.

(b) Wave reflects from a free end.

Pulse arrives.

Rod exerts no transverse forces on string.

Pulse reflects without inverting.

15.20 Overlap of two wave pulses—one right side up, one inverted—traveling in opposite directions. Time increases from top to bottom.

As the pulses overlap, the displacement of the string at any point is the algebraic sum of the displacements due to the individual pulses.

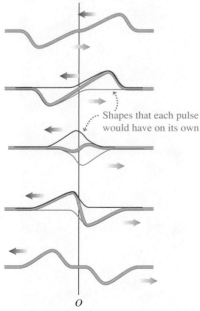

Shapes that each pulse would have on its own

O

pulled back down, and again a reflected pulse is produced (drawing 7). As for a fixed end, the reflected pulse moves in the opposite direction from the initial pulse, but now the direction of the displacement is the same as for the initial pulse. The conditions at the end of the string, such as a rigid support or the complete absence of transverse force, are called **boundary conditions.**

The formation of the reflected pulse is similar to the overlap of two pulses traveling in opposite directions. Figure 15.20 shows two pulses with the same shape, one inverted with respect to the other, traveling in opposite directions. As the pulses overlap and pass each other, the total displacement of the string is the *algebraic sum* of the displacements at that point in the individual pulses. Because these two pulses have the same shape, the total displacement at point *O* in the middle of the figure is zero at all times. Thus the motion of the left half of the string would be the same if we cut the string at point *O*, threw away the right side, and held the end at *O* fixed. The two pulses on the left side then correspond to the incident and reflected pulses, combining so that the total displacement at *O* is *always* zero. For this to occur, the reflected pulse must be inverted relative to the incident pulse.

Figure 15.21 shows two pulses with the same shape, traveling in opposite directions but *not* inverted relative to each other. The displacement at point *O* in the middle of the figure is not zero, but the slope of the string at this point is always zero. According to Eq. (15.20), this corresponds to the absence of any transverse force at this point. In this case the motion of the left half of the string would be the same as if we cut the string at point *O* and attached the end to a frictionless sliding ring (Fig. 15.19b) that maintains tension without exerting any transverse force. In other words, this situation corresponds to reflection of a pulse at a free end of a string at point *O*. In this case the reflected pulse is *not* inverted.

The Principle of Superposition

Combining the displacements of the separate pulses at each point to obtain the actual displacement is an example of the **principle of superposition:** When two waves overlap, the actual displacement of any point on the string at any time is obtained by adding the displacement the point would have if only the first wave were present and the displacement it would have if only the second wave were present. In other words, the wave function $y(x, t)$ that describes the resulting motion in this situation is obtained by *adding* the two wave functions for the two separate waves:

$$y(x, t) = y_1(x, t) + y_2(x, t) \quad \text{(principle of superposition)} \quad (15.27)$$

Mathematically, this additive property of wave functions follows from the form of the wave equation, Eq. (15.12) or (15.18), which every physically possible wave function must satisfy. Specifically, the wave equation is *linear;* that is, it contains the function $y(x, t)$ only to the first power (there are no terms involving $y(x, t)^2$, $y(x, t)^{1/2}$, etc.). As a result, if any two functions $y_1(x, t)$ and $y_2(x, t)$ satisfy the wave equation separately, their sum $y_1(x, t) + y_2(x, t)$ also satisfies it and is therefore a physically possible motion. Because this principle depends on the linearity of the wave equation and the corresponding linear-combination property of its solutions, it is also called the *principle of linear superposition.* For some physical systems, such as a medium that does not obey Hooke's law, the wave equation is *not* linear; this principle does not hold for such systems.

The principle of superposition is of central importance in all types of waves. When a friend talks to you while you are listening to music, you can distinguish the sound of speech and the sound of music from each other. This is precisely because the total sound wave reaching your ears is the algebraic sum of the wave produced by your friend's voice and the wave produced by the speakers of your stereo. If two sound waves did *not* combine in this simple linear way, the sound you would hear in this situation would be a hopeless jumble. Superposition also applies to electromagnetic waves (such as light) and many other types of waves.

Test Your Understanding of Section 15.6   Figure 15.22 shows two wave pulses with different shapes traveling in different directions along a string. Make a series of sketches like Fig. 15.21 showing the shape of the string as the two pulses approach, overlap, and then pass each other.

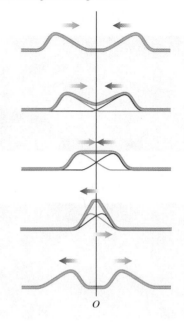

15.21 Overlap of two wave pulses—both right side up—traveling in opposite directions. Time increases from top to bottom. Compare to Fig. 15.20.

O

15.22 Two wave pulses with different shapes.

15.7 Standing Waves on a String

We have talked about the reflection of a wave *pulse* on a string when it arrives at a boundary point (either a fixed end or a free end). Now let's look at what happens when a *sinusoidal* wave is reflected by a fixed end of a string. We'll again approach the problem by considering the superposition of two waves propagating through the string, one representing the original or incident wave and the other representing the wave reflected at the fixed end.

Figure 15.23 shows a string that is fixed at its left end. Its right end is moved up and down in simple harmonic motion to produce a wave that travels to the left; the wave reflected from the fixed end travels to the right. The resulting motion when the two waves combine no longer looks like two waves traveling in opposite directions. The string appears to be subdivided into a number of segments, as in the time-exposure photographs of Figs. 15.23a, 15.23b, 15.23c, and 15.23d. Figure 15.23e shows two instantaneous shapes of the string in Fig. 15.23b. Let's compare this behavior with the waves we studied in Sections 15.1 through 15.5. In a wave that travels along the string, the amplitude is constant and the wave pattern moves with a speed equal to the wave speed. Here, instead, the wave pattern remains in the same position along the string and its amplitude

15.23 (a)–(d) Time exposures of standing waves in a stretched string. From (a) to (d), the frequency of oscillation of the right-hand end increases and the wavelength of the standing wave decreases. (e) The extremes of the motion of the standing wave in part (b), with nodes at the center and at the ends. The right-hand end of the string moves very little compared to the antinodes and so is essentially a node.

(a) String is one-half wavelength long.

(b) String is one wavelength long.

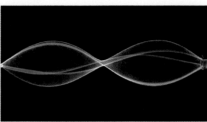

(c) String is one and a half wavelengths long.

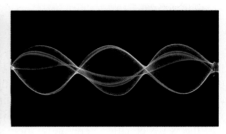

(d) String is two wavelengths long.

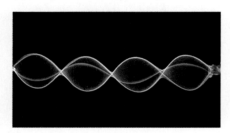

(e) The shape of the string in (b) at two different instants

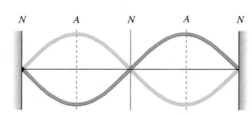

N = **nodes:** points at which the string never moves

A = **antinodes:** points at which the amplitude of string motion is greatest

fluctuates. There are particular points called **nodes** (labeled N in Fig. 15.23e) that never move at all. Midway between the nodes are points called **antinodes** (labeled A in Fig. 15.23e) where the amplitude of motion is greatest. Because the wave pattern doesn't appear to be moving in either direction along the string, it is called a **standing wave.** (To emphasize the difference, a wave that *does* move along the string is called a **traveling wave.**)

The principle of superposition explains how the incident and reflected waves combine to form a standing wave. In Fig. 15.24 the red curves show a wave traveling to the left. The blue curves show a wave traveling to the right with the same propagation speed, wavelength, and amplitude. The waves are shown at nine instants, $\frac{1}{16}$ of a period apart. At each point along the string, we add the displacements (the values of y) for the two separate waves; the result is the total wave on the string, shown in brown.

At certain instants, such as $t = \frac{1}{4}T$, the two wave patterns are exactly in phase with each other, and the shape of the string is a sine curve with twice the amplitude of either individual wave. At other instants, such as $t = \frac{1}{2}T$, the two waves are exactly out of phase with each other, and the total wave at that instant is zero. The resultant displacement is *always* zero at those places marked N at the bottom of Fig. 15.24. These are the *nodes.* At a node the displacements of the two waves in red and blue are always equal and opposite and cancel each other out. This cancellation is called **destructive interference.** Midway between the nodes are the points of *greatest* amplitude, or the *antinodes,* marked A. At the antinodes the displacements of the two waves in red and blue are always identical, giving a large resultant displacement; this phenomenon is called **constructive interference.** We can see from the figure that the distance between successive nodes or between successive antinodes is one half-wavelength, or $\lambda/2$.

We can derive a wave function for the standing wave of Fig. 15.24 by adding the wave functions $y_1(x, t)$ and $y_2(x, t)$ for two waves with equal amplitude, period, and wavelength traveling in opposite directions. Here $y_1(x, t)$ (the red curves in Fig. 15.24) represents an incoming, or *incident,* wave traveling to the

15.24 Formation of a standing wave. A wave traveling to the left (red curves) combines with a wave traveling to the right (blue curves) to form a standing wave (brown curves).

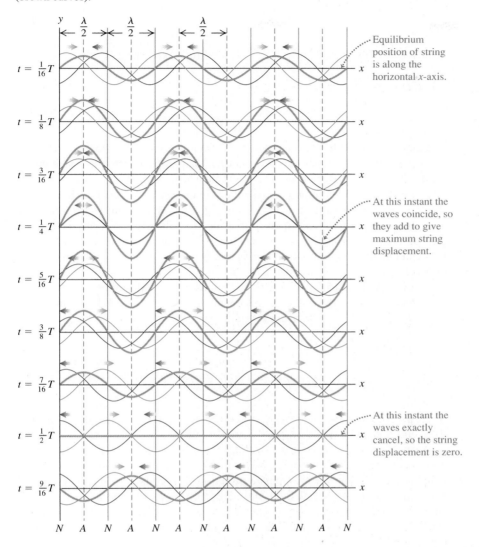

Equilibrium position of string is along the horizontal x-axis.

At this instant the waves coincide, so they add to give maximum string displacement.

At this instant the waves exactly cancel, so the string displacement is zero.

left along the $+x$-axis, arriving at the point $x = 0$ and being reflected; $y_2(x, t)$ (the blue curves in Fig. 15.24) represents the *reflected* wave traveling to the right from $x = 0$. We noted in Section 15.6 that the wave reflected from a fixed end of a string is inverted, so we give a negative sign to one of the waves:

$$y_1(x, t) = -A\cos(kx + \omega t) \quad \text{(incident wave traveling to the left)}$$
$$y_2(x, t) = A\cos(kx - \omega t) \quad \text{(reflected wave traveling to the right)}$$

Note also that the change in sign corresponds to a shift in *phase* of 180° or π radians. At $x = 0$ the motion from the reflected wave is $A\cos\omega t$ and the motion from the incident wave is $-A\cos\omega t$, which we can also write as $A\cos(\omega t + \pi)$. From Eq. (15.27), the wave function for the standing wave is the sum of the individual wave functions:

$$y(x, t) = y_1(x, t) + y_2(x, t) = A[-\cos(kx + \omega t) + \cos(kx - \omega t)]$$

We can rewrite each of the cosine terms by using the identities for the cosine of the sum and difference of two angles: $\cos(a \mp b) = \cos a\cos b \mp \sin a\sin b$.

Applying these and combining terms, we obtain the wave function for the standing wave:

$$y(x, t) = y_1(x, t) + y_2(x, t) = (2A \sin kx) \sin \omega t \quad \text{or}$$

$$y(x, t) = (A_{SW} \sin kx) \sin \omega t \quad \begin{array}{l}\text{(standing wave on a} \\ \text{string, fixed end at } x = 0)\end{array} \quad \text{(15.28)}$$

The standing-wave amplitude A_{SW} is twice the amplitude A of either of the original traveling waves:

$$A_{SW} = 2A$$

Equation (15.28) has two factors: a function of x and a function of t. The factor $A_{SW} \sin kx$ shows that at each instant the shape of the string is a sine curve. But unlike a wave traveling along a string, the wave shape stays in the same position, oscillating up and down as described by the $\sin \omega t$ factor. This behavior is shown graphically by the brown curves in Fig. 15.24. Each point in the string still undergoes simple harmonic motion, but all the points between any successive pair of nodes oscillate *in phase*. This is in contrast to the phase differences between oscillations of adjacent points that we see with a wave traveling in one direction.

We can use Eq. (15.28) to find the positions of the nodes; these are the points for which $\sin kx = 0$, so the displacement is *always* zero. This occurs when $kx = 0, \pi, 2\pi, 3\pi, \ldots$, or, using $k = 2\pi/\lambda$,

$$x = 0, \frac{\pi}{k}, \frac{2\pi}{k}, \frac{3\pi}{k}, \ldots$$

$$= 0, \frac{\lambda}{2}, \frac{2\lambda}{2}, \frac{3\lambda}{2}, \ldots \quad \begin{array}{l}\text{(nodes of a standing wave on} \\ \text{a string, fixed end at } x = 0)\end{array} \quad \text{(15.29)}$$

In particular, there is a node at $x = 0$, as there should be, since this point is a fixed end of the string.

A standing wave, unlike a traveling wave, *does not* transfer energy from one end to the other. The two waves that form it would individually carry equal amounts of power in opposite directions. There is a local flow of energy from each node to the adjacent antinodes and back, but the *average* rate of energy transfer is zero at every point. If you evaluate the wave power given by Eq. (15.21) using the wave function of Eq. (15.28), you will find that the average power is zero.

Problem-Solving Strategy 15.2 | **Standing Waves**

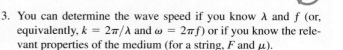

IDENTIFY *the relevant concepts:* Identify the target variables. Then determine whether the problem is purely *kinematic* (involving only such quantities as wave speed v, wavelength λ, and frequency f) or whether *dynamic* properties of the medium (such as F and μ for transverse waves on a string) are also involved.

SET UP *the problem* using the following steps:
1. Sketch the shape of the standing wave at a particular instant. This will help you visualize the nodes (label them N) and antinodes (A). The distance between adjacent nodes (or antinodes) is $\lambda/2$; the distance between a node and the adjacent antinode is $\lambda/4$.
2. Choose the equations you'll use. The wave function for the standing wave, like Eq. (15.28), is often useful.

3. You can determine the wave speed if you know λ and f (or, equivalently, $k = 2\pi/\lambda$ and $\omega = 2\pi f$) or if you know the relevant properties of the medium (for a string, F and μ).

EXECUTE *the solution:* Solve for the target variables. Once you've found the wave function, you can find the displacement y at any point x and at any time t. You can find the velocity and acceleration of a particle in the medium by taking the first and second partial derivatives of y with respect to time.

EVALUATE *your answer:* Compare your numerical answers with your sketch. Check that the wave function satisfies the boundary conditions (for example, the displacement should be zero at a fixed end).

Standing waves on a guitar string

A guitar string lies along the x-axis when in equilibrium. The end of the string at $x = 0$ (the bridge of the guitar) is fixed. A sinusoidal wave with amplitude $A = 0.750$ mm $= 7.50 \times 10^{-4}$ m and frequency $f = 440$ Hz, corresponding to the red curves in Fig. 15.24, travels along the string in the $-x$-direction at 143 m/s. It is reflected from the fixed end, and the superposition of the incident and reflected waves forms a standing wave. (a) Find the equation giving the displacement of a point on the string as a function of position and time. (b) Locate the nodes. (c) Find the amplitude of the standing wave and the maximum transverse velocity and acceleration.

SOLUTION

IDENTIFY and SET UP: This is a *kinematics* problem (see Problem-Solving Strategy 15.1 in Section 15.3). The target variables are: in part (a), the wave function of the standing wave; in part (b), the locations of the nodes; and in part (c), the maximum displacement y, transverse velocity v_y, and transverse acceleration a_y. Since there is a fixed end at $x = 0$, we can use Eqs. (15.28) and (15.29) to describe this standing wave. We will need the relationships $\omega = 2\pi f$, $v = \omega/k$, and $v = \lambda f$.

EXECUTE: (a) The standing-wave amplitude is $A_{SW} = 2A = 1.50 \times 10^{-3}$ m (twice the amplitude of either the incident or reflected wave). The angular frequency and wave number are

$$\omega = 2\pi f = (2\pi \text{ rad})(440 \text{ s}^{-1}) = 2760 \text{ rad/s}$$

$$k = \frac{\omega}{v} = \frac{2760 \text{ rad/s}}{143 \text{ m/s}} = 19.3 \text{ rad/m}$$

Equation (15.28) then gives

$$y(x, t) = (A_{SW} \sin kx) \sin \omega t$$
$$= [(1.50 \times 10^{-3} \text{ m}) \sin(19.3 \text{ rad/m})x] \sin(2760 \text{ rad/s})t$$

(b) From Eq. (15.29), the positions of the nodes are $x = 0$, $\lambda/2$, λ, $3\lambda/2, \ldots$. The wavelength is $\lambda = v/f = (143 \text{ m/s})/(440 \text{ Hz})$

$= 0.325$ m, so the nodes are at $x = 0$, 0.163 m, 0.325 m, 0.488 m, ...

(c) From the expression for $y(x, t)$ in part (a), the maximum displacement from equilibrium is $A_{SW} = 1.50 \times 10^{-3}$ m $= 1.50$ mm. This occurs at the *antinodes,* which are midway between adjacent nodes (that is, at $x = 0.081$ m, 0.244 m, 0.406 m, . . .).

For a particle on the string at any point x, the transverse (y-) velocity is

$$v_y(x, t) = \frac{\partial y(x, t)}{\partial t}$$
$$= [(1.50 \times 10^{-3} \text{ m}) \sin(19.3 \text{ rad/m})x]$$
$$\times [(2760 \text{ rad/s}) \cos(2760 \text{ rad/s})t]$$
$$= [(4.15 \text{ m/s}) \sin(19.3 \text{ rad/m})x] \cos(2760 \text{ rad/s})t$$

At an antinode, $\sin(19.3 \text{ rad/m})x = \pm 1$ and the transverse velocity varies between $+4.15$ m/s and -4.15 m/s. As is always the case in SHM, the maximum velocity occurs when the particle is passing through the equilibrium position ($y = 0$).

The transverse acceleration $a_y(x, t)$ is the *second* partial derivative of $y(x, t)$ with respect to time. You can show that

$$a_y(x, t) = \frac{\partial v_y(x, t)}{\partial t} = \frac{\partial^2 y(x, t)}{\partial t^2}$$
$$= [(-1.15 \times 10^4 \text{ m/s}^2) \sin(19.3 \text{ rad/m})x]$$
$$\times \sin(2760 \text{ rad/s})t$$

At the antinodes, the transverse acceleration varies between $+1.15 \times 10^4$ m/s^2 and -1.15×10^4 m/s^2.

EVALUATE: The maximum transverse velocity at an antinode is quite respectable (about 15 km/h, or 9.3 mi/h). But the maximum transverse acceleration is tremendous, 1170 times the acceleration due to gravity! Guitar strings are actually fixed at *both* ends; we'll see the consequences of this in the next section.

Test Your Understanding of Section 15.7 Suppose the frequency of the standing wave in Example 15.6 were doubled from 440 Hz to 880 Hz. Would all of the nodes for $f = 440$ Hz also be nodes for $f = 880$ Hz? If so, would there be additional nodes for $f = 880$ Hz? If not, which nodes are absent for $f = 880$ Hz?   ❙

15.8 Normal Modes of a String

When we described standing waves on a string rigidly held at one end, as in Fig. 15.23, we made no assumptions about the length of the string or about what was happening at the other end. Let's now consider a string of a definite length L, rigidly held at *both* ends. Such strings are found in many musical instruments, including pianos, violins, and guitars. When a guitar string is plucked, a wave is produced in the string; this wave is reflected and re-reflected from the ends of the string, making a standing wave. This standing wave on the string in turn produces a sound wave in the air, with a frequency determined by the properties of the string. This is what makes stringed instruments so useful in making music.

To understand these properties of standing waves on a string fixed at both ends, let's first examine what happens when we set up a sinusoidal wave on such a string. The standing wave that results must have a node at *both* ends of the string. We saw in the preceding section that adjacent nodes are one half-wavelength

MasteringPHYSICS

PhET: Fourier: Making Waves
PhET: Waves on a String
ActivPhysics 10.4: Standing Waves on Strings
ActivPhysics 10.5: Tuning a Stringed Instrument: Standing Waves
ActivPhysics 10.6: String Mass and Standing Waves

$(\lambda/2)$ apart, so the length of the string must be $\lambda/2$, or $2(\lambda/2)$, or $3(\lambda/2)$, or in general some integer number of half-wavelengths:

$$L = n\frac{\lambda}{2} \quad (n = 1, 2, 3, \ldots) \quad \text{(string fixed at both ends)} \quad (15.30)$$

That is, if a string with length L is fixed at both ends, a standing wave can exist only if its wavelength satisfies Eq. (15.30).

Solving this equation for λ and labeling the possible values of λ as λ_n, we find

$$\lambda_n = \frac{2L}{n} \quad (n = 1, 2, 3, \ldots) \quad \text{(string fixed at both ends)} \quad (15.31)$$

Waves can exist on the string if the wavelength is *not* equal to one of these values, but there cannot be a steady wave pattern with nodes and antinodes, and the total wave cannot be a standing wave. Equation (15.31) is illustrated by the standing waves shown in Figs. 15.23a, 15.23b, 15.23c, and 15.23d; these represent $n = 1, 2, 3,$ and 4, respectively.

Corresponding to the series of possible standing-wave wavelengths λ_n is a series of possible standing-wave frequencies f_n, each related to its corresponding wavelength by $f_n = v/\lambda_n$. The smallest frequency f_1 corresponds to the largest wavelength (the $n = 1$ case), $\lambda_1 = 2L$:

$$f_1 = \frac{v}{2L} \quad \text{(string fixed at both ends)} \quad (15.32)$$

This is called the **fundamental frequency.** The other standing-wave frequencies are $f_2 = 2v/2L$, $f_3 = 3v/2L$, and so on. These are all integer multiples of the fundamental frequency f_1, such as $2f_1, 3f_1, 4f_1,$ and so on, and we can express *all* the frequencies as

$$f_n = n\frac{v}{2L} = nf_1 \quad (n = 1, 2, 3, \ldots) \quad \text{(string fixed at both ends)} \quad (15.33)$$

These frequencies are called **harmonics,** and the series is called a **harmonic series.** Musicians sometimes call $f_2, f_3,$ and so on **overtones;** f_2 is the second harmonic or the first overtone, f_3 is the third harmonic or the second overtone, and so on. The first harmonic is the same as the fundamental frequency (Fig. 15.25).

For a string with fixed ends at $x = 0$ and $x = L$, the wave function $y(x, t)$ of the nth standing wave is given by Eq. (15.28) (which satisfies the condition that there is a node at $x = 0$), with $\omega = \omega_n = 2\pi f_n$ and $k = k_n = 2\pi/\lambda_n$:

$$y_n(x, t) = A_{SW} \sin k_n x \sin \omega_n t \quad (15.34)$$

You can easily show that this wave function has nodes at both $x = 0$ and $x = L$, as it must.

A **normal mode** of an oscillating system is a motion in which all particles of the system move sinusoidally with the same frequency. For a system made up of a string of length L fixed at both ends, each of the wavelengths given by Eq. (15.31) corresponds to a possible normal-mode pattern and frequency. There are infinitely many normal modes, each with its characteristic frequency and vibration pattern. Figure 15.26 shows the first four normal-mode patterns and their associated frequencies and wavelengths; these correspond to Eq. (15.34) with $n = 1, 2, 3,$ and 4. By contrast, a harmonic oscillator, which has only one oscillating particle, has only one normal mode and one characteristic frequency. The string fixed at both ends has infinitely many normal modes because it is made up of a very large (effectively infinite) number of particles. More complicated oscillating systems also have infinite numbers of normal modes, though with more complex normal-mode patterns than a string (Fig. 15.27).

15.25 Each string of a violin naturally oscillates at one or more of its harmonic frequencies, producing sound waves in the air with the same frequencies.

15.26 The first four normal modes of a string fixed at both ends. (Compare these to the photographs in Fig. 15.23.)

(a) $n = 1$: fundamental frequency, f_1

N A N

$\frac{\lambda}{2} = L$

(b) $n = 2$: second harmonic, f_2 (first overtone)

N A N A N

$2\frac{\lambda}{2} = L$

(c) $n = 3$: third harmonic, f_3 (second overtone)

N A N A N A N

$3\frac{\lambda}{2} = L$

(d) $n = 4$: fourth harmonic, f_4 (third overtone)

N A N A N A N A N

$4\frac{\lambda}{2} = L$

Complex Standing Waves

If we could displace a string so that its shape is the same as one of the normal-mode patterns and then release it, it would vibrate with the frequency of that mode. Such a vibrating string would displace the surrounding air with the same frequency, producing a traveling sinusoidal sound wave that your ears would perceive as a pure tone. But when a string is struck (as in a piano) or plucked (as is done to guitar strings), the shape of the displaced string is *not* as simple as one of the patterns in Fig. 15.26. The fundamental as well as many overtones are present in the resulting vibration. This motion is therefore a combination or *superposition* of many normal modes. Several simple harmonic motions of different frequencies are present simultaneously, and the displacement of any point on the string is the sum (or superposition) of the displacements associated with the individual modes. The sound produced by the vibrating string is likewise a superposition of traveling sinusoidal sound waves, which you perceive as a rich, complex tone with the fundamental frequency f_1. The standing wave on the string and the traveling sound wave in the air have similar **harmonic content** (the extent to which frequencies higher than the fundamental are present). The harmonic content depends on how the string is initially set into motion. If you pluck the strings of an acoustic guitar in the normal location over the sound hole, the sound that you hear has a different harmonic content than if you pluck the strings next to the fixed end on the guitar body.

It is possible to represent every possible motion of the string as some superposition of normal-mode motions. Finding this representation for a given vibration pattern is called *harmonic analysis*. The sum of sinusoidal functions that represents a complex wave is called a *Fourier series*. Figure 15.28 shows how a standing wave that is produced by plucking a guitar string of length L at a point $L/4$ from one end can be represented as a combination of sinusoidal functions.

Standing Waves and String Instruments

As we have seen, the fundamental frequency of a vibrating string is $f_1 = v/2L$. The speed v of waves on the string is determined by Eq. (15.13), $v = \sqrt{F/\mu}$. Combining these equations, we find

$$f_1 = \frac{1}{2L}\sqrt{\frac{F}{\mu}} \quad \text{(string fixed at both ends)} \quad (15.35)$$

This is also the fundamental frequency of the sound wave created in the surrounding air by the vibrating string. Familiar musical instruments show how f_1 depends on the properties of the string. The inverse dependence of frequency on length L is illustrated by the long strings of the bass (low-frequency) section of the piano or the bass viol compared with the shorter strings of the treble section of the piano or the violin (Fig. 15.29). The pitch of a violin or guitar is usually

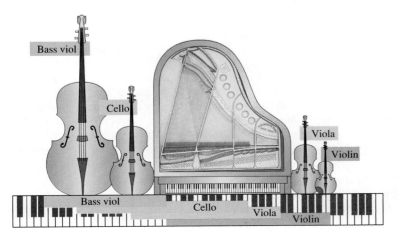

15.27 Astronomers have discovered that the sun oscillates in several different normal modes. This computer simulation shows one such mode.

Cross section of the sun's interior

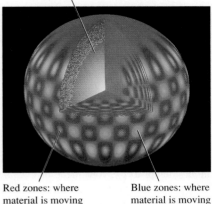

Red zones: where material is moving outward

Blue zones: where material is moving inward

Mastering**PHYSICS**

ActivPhysics 10.10: Complex Waves: Fourier Analysis

15.28 When a guitar string is plucked (pulled into a triangular shape) and released, a standing wave results. The standing wave is well represented (except at the sharp maximum point) by the sum of just three sinusoidal functions. Including additional sinusoidal functions further improves the representation.

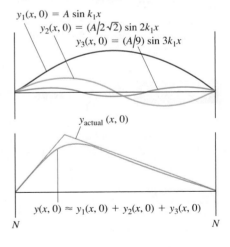

15.29 Comparing the range of a concert grand piano to the ranges of a bass viol, a cello, a viola, and a violin. In all cases, longer strings produce bass notes and shorter strings produce treble notes.

varied by pressing a string against the fingerboard with the fingers to change the length L of the vibrating portion of the string. Increasing the tension F increases the wave speed v and thus increases the frequency (and the pitch). All string instruments are "tuned" to the correct frequencies by varying the tension; you tighten the string to raise the pitch. Finally, increasing the mass per unit length μ decreases the wave speed and thus the frequency. The lower notes on a steel guitar are produced by thicker strings, and one reason for winding the bass strings of a piano with wire is to obtain the desired low frequency from a relatively short string.

Wind instruments such as saxophones and trombones also have normal modes. As for stringed instruments, the frequencies of these normal modes determine the pitch of the musical tones that these instruments produce. We'll discuss these instruments and many other aspects of sound in Chapter 16.

Example 15.7 A giant bass viol

In an attempt to get your name in *Guinness World Records,* you build a bass viol with strings of length 5.00 m between fixed points. One string, with linear mass density 40.0 g/m, is tuned to a 20.0-Hz fundamental frequency (the lowest frequency that the human ear can hear). Calculate (a) the tension of this string, (b) the frequency and wavelength on the string of the second harmonic, and (c) the frequency and wavelength on the string of the second overtone.

SOLUTION

IDENTIFY and SET UP: In part (a) the target variable is the string tension F; we'll use Eq. (15.35), which relates F to the known values $f_1 = 20.0$ Hz, $L = 5.00$ m, and $\mu = 40.0$ g/m. In parts (b) and (c) the target variables are the frequency and wavelength of a given harmonic and a given overtone. We determine these from the given length of the string and the fundamental frequency, using Eqs. (15.31) and (15.33).

EXECUTE: (a) We solve Eq. (15.35) for F:

$$F = 4\mu L^2 f_1^2 = 4(40.0 \times 10^{-3}\text{ kg/m})(5.00\text{ m})^2(20.0\text{ s}^{-1})^2$$
$$= 1600\text{ N} = 360\text{ lb}$$

(b) From Eqs. (15.33) and (15.31), the frequency and wavelength of the second harmonic $(n = 2)$ are

$$f_2 = 2f_1 = 2(20.0\text{ Hz}) = 40.0\text{ Hz}$$
$$\lambda_2 = \frac{2L}{2} = \frac{2(5.00\text{ m})}{2} = 5.00\text{ m}$$

(c) The second overtone is the "second tone over" (above) the fundamental—that is, $n = 3$. Its frequency and wavelength are

$$f_3 = 3f_1 = 3(20.0\text{ Hz}) = 60.0\text{ Hz}$$
$$\lambda_3 = \frac{2L}{3} = \frac{2(5.00\text{ m})}{3} = 3.33\text{ m}$$

EVALUATE: The string tension in a real bass viol is typically a few hundred newtons; the tension in part (a) is a bit higher than that. The wavelengths in parts (b) and (c) are equal to the length of the string and two-thirds the length of the string, respectively, which agrees with the drawings of standing waves in Fig. 15.26.

Example 15.8 From waves on a string to sound waves in air

What are the frequency and wavelength of the sound waves produced in the air when the string in Example 15.7 is vibrating at its fundamental frequency? The speed of sound in air at 20°C is 344 m/s.

SOLUTION

IDENTIFY and SET UP: Our target variables are the frequency and wavelength for the *sound wave* produced by the bass viol string. The frequency of the sound wave is the same as the fundamental frequency f_1 of the standing wave, because the string forces the surrounding air to vibrate at the same frequency. The wavelength of the sound wave is $\lambda_{1(\text{sound})} = v_{\text{sound}}/f_1$.

EXECUTE: We have $f = f_1 = 20.0$ Hz, so

$$\lambda_{1(\text{sound})} = \frac{v_{\text{sound}}}{f_1} = \frac{344\text{ m/s}}{20.0\text{ Hz}} = 17.2\text{ m}$$

EVALUATE: In Example 15.7, the wavelength of the fundamental on the string was $\lambda_{1(\text{string})} = 2L = 2(5.00\text{ m}) = 10.0$ m. Here $\lambda_{1(\text{sound})} = 17.2$ m is greater than that by the factor of $17.2/10.0 = 1.72$. This is as it should be: Because the frequencies of the sound wave and the standing wave are equal, $\lambda = v/f$ says that the wavelengths in air and on the string are in the same ratio as the corresponding wave speeds; here $v_{\text{sound}} = 344$ m/s is greater than $v_{\text{string}} = (10.0\text{ m})(20.0\text{ Hz}) = 200$ m/s by just the factor 1.72.

Test Your Understanding of Section 15.8 While a guitar string is vibrating, you gently touch the midpoint of the string to ensure that the string does not vibrate at that point. Which normal modes *cannot* be present on the string while you are touching it in this way? ▌

Waves and their properties: A wave is any disturbance that propagates from one region to another. A mechanical wave travels within some material called the medium. The wave speed v depends on the type of wave and the properties of the medium.

In a periodic wave, the motion of each point of the medium is periodic with frequency f and period T. The wavelength λ is the distance over which the wave pattern repeats, and the amplitude A is the maximum displacement of a particle in the medium. The product of λ and f equals the wave speed. A sinusoidal wave is a special periodic wave in which each point moves in simple harmonic motion. (See Example 15.1.)

$$v = \lambda f \qquad (15.1)$$

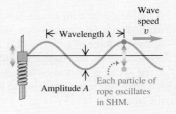

Wave functions and wave dynamics: The wave function $y(x, t)$ describes the displacements of individual particles in the medium. Equations (15.3), (15.4), and (15.7) give the wave equation for a sinusoidal wave traveling in the $+x$-direction. If the wave is moving in the $-x$-direction, the minus signs in the cosine functions are replaced by plus signs. (See Example 15.2.)

The wave function obeys a partial differential equation called the wave equation, Eq. (15.12).

The speed of transverse waves on a string depends on the tension F and mass per unit length μ. (See Example 15.3.)

$$y(x, t) = A \cos\left[\omega\left(\frac{x}{v} - t\right)\right]$$
$$= A \cos 2\pi f\left(\frac{x}{v} - t\right) \qquad (15.3)$$
$$y(x, t) = A \cos 2\pi\left(\frac{x}{\lambda} - \frac{t}{T}\right) \qquad (15.4)$$
$$y(x, t) = A \cos(kx - \omega t) \qquad (15.7)$$
where $k = 2\pi/\lambda$ and $\omega = 2\pi f = vk$
$$\frac{\partial^2 y(x, t)}{\partial x^2} = \frac{1}{v^2} \frac{\partial^2 y(x, t)}{\partial t^2} \qquad (15.12)$$
$$v = \sqrt{\frac{F}{\mu}} \quad \text{(waves on a string)} \qquad (15.13)$$

Wave power: Wave motion conveys energy from one region to another. For a sinusoidal mechanical wave, the average power P_{av} is proportional to the square of the wave amplitude and the square of the frequency. For waves that spread out in three dimensions, the wave intensity I is inversely proportional to the square of the distance from the source. (See Examples 15.4 and 15.5.)

$$P_{av} = \frac{1}{2}\sqrt{\mu F}\,\omega^2 A^2 \qquad (15.25)$$
(average power, sinusoidal wave)

$$\frac{I_1}{I_2} = \frac{r_2^2}{r_1^2} \qquad (15.26)$$
(inverse-square law for intensity)

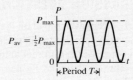
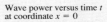

Wave power versus time t at coordinate $x = 0$

Wave superposition: A wave reflects when it reaches a boundary of its medium. At any point where two or more waves overlap, the total displacement is the sum of the displacements of the individual waves (principle of superposition).

$$y(x, t) = y_1(x, t) + y_2(x, t) \qquad (15.27)$$
(principle of superposition)

Standing waves on a string: When a sinusoidal wave is reflected from a fixed or free end of a stretched string, the incident and reflected waves combine to form a standing sinusoidal wave with nodes and antinodes. Adjacent nodes are spaced a distance $\lambda/2$ apart, as are adjacent antinodes. (See Example 15.6.)

When both ends of a string with length L are held fixed, standing waves can occur only when L is an integer multiple of $\lambda/2$. Each frequency with its associated vibration pattern is called a normal mode. (See Examples 15.7 and 15.8.)

$$y(x, t) = (A_{SW} \sin kx) \sin \omega t \qquad (15.28)$$
(standing wave on a string, fixed end at $x = 0$)

$$f_n = n\frac{v}{2L} = nf_1 \ (n = 1, 2, 3, \dots) \qquad (15.33)$$

$$f_1 = \frac{1}{2L}\sqrt{\frac{F}{\mu}} \qquad (15.35)$$
(string fixed at both ends)

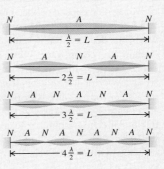

499

BRIDGING PROBLEM Waves on a Rotating Rope

A uniform rope with length L and mass m is held at one end and whirled in a horizontal circle with angular velocity ω. You can ignore the force of gravity on the rope. (a) At a point on the rope a distance r from the end that is held, what is the tension F? (b) What is the speed of transverse waves at this point? (c) Find the time required for a transverse wave to travel from one end of the rope to the other.

SOLUTION GUIDE

See MasteringPhysics® Study Area for a Video Tutor solution.

IDENTIFY and SET UP

1. Draw a sketch of the situation and label the distances r and L. The tension in the rope will be different at different values of r. Do you see why? Where on the rope do you expect the tension to be greatest? Where do you expect it will be least?

2. Where on the rope do you expect the wave speed to be greatest? Where do you expect it will be least?

3. Think about the portion of the rope that is farther out than r from the end that is held. What forces act on this portion? (Remember that you can ignore gravity.) What is the mass of this portion? How far is its center of mass from the rotation axis?

4. Make a list of the unknown quantities and decide which are your target variables.

EXECUTE

5. Draw a free-body diagram for the portion of the rope that is farther out than r from the end that is held.

6. Use your free-body diagram to help you determine the tension in the rope at distance r.

7. Use your result from step 6 to find the wave speed at distance r.

8. Use your result from step 7 to find the time for a wave to travel from one end to the other. (*Hint:* The wave speed is $v = dr/dt$, so the time for the wave to travel a distance dr along the rope is $dt = dr/v$. Integrate this to find the total time. See Appendix B.)

EVALUATE

9. Do your results for parts (a) and (b) agree with your expectations from steps 1 and 2? Are the units correct?

10. Check your result for part (a) by considering the net force on a small segment of the rope at distance r with length dr and mass $dm = (m/L)dr$. [*Hint:* The tension forces on this segment are $F(r)$ on one side and $F(r + dr)$ on the other side. You will get an equation for dF/dr that you can integrate to find F as a function of r.]

Problems

For instructor-assigned homework, go to www.masteringphysics.com (MP)

•, ••, •••: Problems of increasing difficulty. **CP**: Cumulative problems incorporating material from earlier chapters. **CALC**: Problems requiring calculus. **BIO**: Biosciences problems.

DISCUSSION QUESTIONS

Q15.1 Two waves travel on the same string. Is it possible for them to have (a) different frequencies; (b) different wavelengths; (c) different speeds; (d) different amplitudes; (e) the same frequency but different wavelengths? Explain your reasoning.

Q15.2 Under a tension F, it takes 2.00 s for a pulse to travel the length of a taut wire. What tension is required (in terms of F) for the pulse to take 6.00 s instead?

Q15.3 What kinds of energy are associated with waves on a stretched string? How could you detect such energy experimentally?

Q15.4 The amplitude of a wave decreases gradually as the wave travels down a long, stretched string. What happens to the energy of the wave when this happens?

Q15.5 For the wave motions discussed in this chapter, does the speed of propagation depend on the amplitude? What makes you say this?

Q15.6 The speed of ocean waves depends on the depth of the water; the deeper the water, the faster the wave travels. Use this to explain why ocean waves crest and "break" as they near the shore.

Q15.7 Is it possible to have a longitudinal wave on a stretched string? Why or why not? Is it possible to have a transverse wave on a steel rod? Again, why or why not? If your answer is yes in either case, explain how you would create such a wave.

Q15.8 An echo is sound reflected from a distant object, such as a wall or a cliff. Explain how you can determine how far away the object is by timing the echo.

Q15.9 Why do you see lightning before you hear the thunder? A familiar rule of thumb is to start counting slowly, once per second, when you see the lightning; when you hear the thunder, divide the number you have reached by 3 to obtain your distance from the lightning in kilometers (or divide by 5 to obtain your distance in miles). Why does this work, or does it?

Q15.10 For transverse waves on a string, is the wave speed the same as the speed of any part of the string? Explain the difference between these two speeds. Which one is constant?

Q15.11 Children make toy telephones by sticking each end of a long string through a hole in the bottom of a paper cup and knotting it so it will not pull out. When the spring is pulled taut, sound can be transmitted from one cup to the other. How does this work? Why is the transmitted sound louder than the sound traveling through air for the same distance?

Q15.12 The four strings on a violin have different thicknesses, but are all under approximately the same tension. Do waves travel faster on the thick strings or the thin strings? Why? How does the fundamental vibration frequency compare for the thick versus the thin strings?

Q15.13 A sinusoidal wave can be described by a cosine function, which is negative just as often as positive. So why isn't the average power delivered by this wave zero?

Q15.14 Two strings of different mass per unit length μ_1 and μ_2 are tied together and stretched with a tension F. A wave travels along the string and passes the discontinuity in μ. Which of the following wave properties will be the same on both sides of the discontinuity, and which ones will change? speed of the wave; frequency; wavelength. Explain the physical reasoning behind each of your answers.

Q15.15 A long rope with mass m is suspended from the ceiling and hangs vertically. A wave pulse is produced at the lower end of the rope, and the pulse travels up the rope. Does the speed of the wave pulse change as it moves up the rope, and if so, does it increase or decrease?

Q15.16 In a transverse wave on a string, the motion of the string is perpendicular to the length of the string. How, then, is it possible for energy to move along the length of the string?

Q15.17 Both wave intensity and gravitation obey inverse-square laws. Do they do so for the same reason? Discuss the reason for each of these inverse-square laws as well as you can.

Q15.18 Energy can be transferred along a string by wave motion. However, in a standing wave on a string, no energy can ever be transferred past a node. Why not?

Q15.19 Can a standing wave be produced on a string by superposing two waves traveling in opposite directions with the same frequency but different amplitudes? Why or why not? Can a standing wave be produced by superposing two waves traveling in opposite directions with different frequencies but the same amplitude? Why or why not?

Q15.20 If you stretch a rubber band and pluck it, you hear a (somewhat) musical tone. How does the frequency of this tone change as you stretch the rubber band further? (Try it!) Does this agree with Eq. (15.35) for a string fixed at both ends? Explain.

Q15.21 A musical interval of an *octave* corresponds to a factor of 2 in frequency. By what factor must the tension in a guitar or violin string be increased to raise its pitch one octave? To raise it two octaves? Explain your reasoning. Is there any danger in attempting these changes in pitch?

Q15.22 By touching a string lightly at its center while bowing, a violinist can produce a note exactly one octave above the note to which the string is tuned—that is, a note with exactly twice the frequency. Why is this possible?

Q15.23 As we discussed in Section 15.1, water waves are a combination of longitudinal and transverse waves. Defend the following statement: "When water waves hit a vertical wall, the wall is a node of the longitudinal displacement but an antinode of the transverse displacement."

Q15.24 Violins are short instruments, while cellos and basses are long. In terms of the frequency of the waves they produce, explain why this is so.

Q15.25 What is the purpose of the frets on a guitar? In terms of frequency of the vibration of the strings, explain their use.

EXERCISES

Section 15.2 Periodic Waves

15.1 • The speed of sound in air at 20°C is 344 m/s. (a) What is the wavelength of a sound wave with a frequency of 784 Hz, corresponding to the note G_5 on a piano, and how many milliseconds does each vibration take? (b) What is the wavelength of a sound wave one octave higher than the note in part (a)?

15.2 • **BIO** **Audible Sound.** Provided the amplitude is sufficiently great, the human ear can respond to longitudinal waves over a range of frequencies from about 20.0 Hz to about 20.0 kHz. (a) If you were to mark the beginning of each complete wave pattern with a red dot for the long-wavelength sound and a blue dot

for the short-wavelength sound, how far apart would the red dots be, and how far apart would the blue dots be? (b) In reality would adjacent dots in each set be far enough apart for you to easily measure their separation with a meter stick? (c) Suppose you repeated part (a) in water, where sound travels at 1480 m/s. How far apart would the dots be in each set? Could you readily measure their separation with a meter stick?

15.3 • **Tsunami!** On December 26, 2004, a great earthquake occurred off the coast of Sumatra and triggered immense waves (tsunami) that killed some 200,000 people. Satellites observing these waves from space measured 800 km from one wave crest to the next and a period between waves of 1.0 hour. What was the speed of these waves in m/s and in km/h? Does your answer help you understand why the waves caused such devastation?

15.4 • **BIO** **Ultrasound Imaging.** Sound having frequencies above the range of human hearing (about 20,000 Hz) is called *ultrasound*. Waves above this frequency can be used to penetrate the body and to produce images by reflecting from surfaces. In a typical ultrasound scan, the waves travel through body tissue with a speed of 1500 m/s. For a good, detailed image, the wavelength should be no more than 1.0 mm. What frequency sound is required for a good scan?

15.5 • **BIO** (a) **Audible wavelengths.** The range of audible frequencies is from about 20 Hz to 20,000 Hz. What is the range of the wavelengths of audible sound in air? (b) **Visible light.** The range of visible light extends from 400 nm to 700 nm. What is the range of visible frequencies of light? (c) **Brain surgery.** Surgeons can remove brain tumors by using a cavitron ultrasonic surgical aspirator, which produces sound waves of frequency 23 kHz. What is the wavelength of these waves in air? (d) **Sound in the body.** What would be the wavelength of the sound in part (c) in bodily fluids in which the speed of sound is 1480 m/s but the frequency is unchanged?

15.6 •• A fisherman notices that his boat is moving up and down periodically, owing to waves on the surface of the water. It takes 2.5 s for the boat to travel from its highest point to its lowest, a total distance of 0.62 m. The fisherman sees that the wave crests are spaced 6.0 m apart. (a) How fast are the waves traveling? (b) What is the amplitude of each wave? (c) If the total vertical distance traveled by the boat were 0.30 m but the other data remained the same, how would the answers to parts (a) and (b) be affected?

Section 15.3 Mathematical Description of a Wave

15.7 • Transverse waves on a string have wave speed 8.00 m/s, amplitude 0.0700 m, and wavelength 0.320 m. The waves travel in the $-x$-direction, and at $t = 0$ the $x = 0$ end of the string has its maximum upward displacement. (a) Find the frequency, period, and wave number of these waves. (b) Write a wave function describing the wave. (c) Find the transverse displacement of a particle at $x = 0.360$ m at time $t = 0.150$ s. (d) How much time must elapse from the instant in part (c) until the particle at $x = 0.360$ m next has maximum upward displacement?

15.8 • A certain transverse wave is described by

$$y(x, t) = (6.50 \text{ mm}) \cos 2\pi \left(\frac{x}{28.0 \text{ cm}} - \frac{t}{0.0360 \text{ s}} \right)$$

Determine the wave's (a) amplitude; (b) wavelength; (c) frequency; (d) speed of propagation; (e) direction of propagation.

15.9 • **CALC** Which of the following wave functions satisfies the wave equation, Eq. (15.12)? (a) $y(x, t) = A \cos(kx + \omega t)$; (b) $y(x, t) = A \sin(kx + \omega t)$; (c) $y(x, t) = A(\cos kx + \cos \omega t)$. (d) For the wave of part (b), write the equations for the transverse velocity and transverse acceleration of a particle at point x.

15.10 • A water wave traveling in a straight line on a lake is described by the equation

$$y(x, t) = (3.75 \text{ cm}) \cos(0.450 \text{ cm}^{-1} x + 5.40 \text{ s}^{-1} t)$$

where y is the displacement perpendicular to the undisturbed surface of the lake. (a) How much time does it take for one complete wave pattern to go past a fisherman in a boat at anchor, and what horizontal distance does the wave crest travel in that time? (b) What are the wave number and the number of waves per second that pass the fisherman? (c) How fast does a wave crest travel past the fisherman, and what is the maximum speed of his cork floater as the wave causes it to bob up and down?

15.11 • A sinusoidal wave is propagating along a stretched string that lies along the x-axis. The displacement of the string as a function of time is graphed in Fig. E15.11 for particles at $x = 0$ and at $x = 0.0900$ m. (a) What is the amplitude of the wave? (b) What is the period of the wave? (c) You are told that the two points $x = 0$ and $x = 0.0900$ m are within one wavelength of each other. If the wave is moving in the $+x$-direction, determine the wavelength and the wave speed. (d) If instead the wave is moving in the $-x$-direction, determine the wavelength and the wave speed. (e) Would it be possible to determine definitively the wavelength in parts (c) and (d) if you were not told that the two points were within one wavelength of each other? Why or why not?

Figure **E15.11**

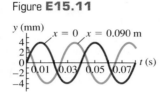

15.12 •• CALC **Speed of Propagation vs. Particle Speed.** (a) Show that Eq. (15.3) may be written as

$$y(x, t) = A \cos\left[\frac{2\pi}{\lambda}(x - vt)\right]$$

(b) Use $y(x, t)$ to find an expression for the transverse velocity v_y of a particle in the string on which the wave travels. (c) Find the maximum speed of a particle of the string. Under what circumstances is this equal to the propagation speed v? Less than v? Greater than v?

15.13 •• A transverse wave on a string has amplitude 0.300 cm, wavelength 12.0 cm, and speed 6.00 cm/s. It is represented by $y(x, t)$ as given in Exercise 15.12. (a) At time $t = 0$, compute y at 1.5-cm intervals of x (that is, at $x = 0, x = 1.5$ cm, $x = 3.0$ cm, and so on) from $x = 0$ to $x = 12.0$ cm. Graph the results. This is the shape of the string at time $t = 0$. (b) Repeat the calculations for the same values of x at times $t = 0.400$ s and $t = 0.800$ s. Graph the shape of the string at these instants. In what direction is the wave traveling?

15.14 • A wave on a string is described by $y(x, t) = A \cos(kx - \omega t)$. (a) Graph y, v_y, and a_y as functions of x for time $t = 0$. (b) Consider the following points on the string: (i) $x = 0$; (ii) $x = \pi/4k$; (iii) $x = \pi/2k$; (iv) $x = 3\pi/4k$; (v) $x = \pi/k$; (vi) $x = 5\pi/4k$; (vii) $x = 3\pi/2k$; (viii) $x = 7\pi/4k$. For a particle at each of these points at $t = 0$, describe in words whether the particle is moving and in what direction, and whether the particle is speeding up, slowing down, or instantaneously not accelerating.

Section 15.4 Speed of a Transverse Wave

15.15 • One end of a horizontal rope is attached to a prong of an electrically driven tuning fork that vibrates the rope transversely at 120 Hz. The other end passes over a pulley and supports a 1.50-kg mass. The linear mass density of the rope is 0.0550 kg/m.

(a) What is the speed of a transverse wave on the rope? (b) What is the wavelength? (c) How would your answers to parts (a) and (b) change if the mass were increased to 3.00 kg?

15.16 • With what tension must a rope with length 2.50 m and mass 0.120 kg be stretched for transverse waves of frequency 40.0 Hz to have a wavelength of 0.750 m?

15.17 •• The upper end of a 3.80-m-long steel wire is fastened to the ceiling, and a 54.0-kg object is suspended from the lower end of the wire. You observe that it takes a transverse pulse 0.0492 s to travel from the bottom to the top of the wire. What is the mass of the wire?

15.18 •• A 1.50-m string of weight 0.0125 N is tied to the ceiling at its upper end, and the lower end supports a weight W. Neglect the very small variation in tension along the length of the string that is produced by the weight of the string. When you pluck the string slightly, the waves traveling up the string obey the equation

$$y(x, t) = (8.50 \text{ mm}) \cos(172 \text{ m}^{-1} x - 4830 \text{ s}^{-1} t)$$

Assume that the tension of the string is constant and equal to W. (a) How much time does it take a pulse to travel the full length of the string? (b) What is the weight W? (c) How many wavelengths are on the string at any instant of time? (d) What is the equation for waves traveling *down* the string?

15.19 • A thin, 75.0-cm wire has a mass of 16.5 g. One end is tied to a nail, and the other end is attached to a screw that can be adjusted to vary the tension in the wire. (a) To what tension (in newtons) must you adjust the screw so that a transverse wave of wavelength 3.33 cm makes 875 vibrations per second? (b) How fast would this wave travel?

15.20 • **Weighty Rope.** If in Example 15.3 (Section 15.4) we do *not* neglect the weight of the rope, what is the wave speed (a) at the bottom of the rope; (b) at the middle of the rope; (c) at the top of the rope?

15.21 • A simple harmonic oscillator at the point $x = 0$ generates a wave on a rope. The oscillator operates at a frequency of 40.0 Hz and with an amplitude of 3.00 cm. The rope has a linear mass density of 50.0 g/m and is stretched with a tension of 5.00 N. (a) Determine the speed of the wave. (b) Find the wavelength. (c) Write the wave function $y(x, t)$ for the wave. Assume that the oscillator has its maximum upward displacement at time $t = 0$. (d) Find the maximum transverse acceleration of points on the rope. (e) In the discussion of transverse waves in this chapter, the force of gravity was ignored. Is that a reasonable approximation for this wave? Explain.

Section 15.5 Energy in Wave Motion

15.22 •• A piano wire with mass 3.00 g and length 80.0 cm is stretched with a tension of 25.0 N. A wave with frequency 120.0 Hz and amplitude 1.6 mm travels along the wire. (a) Calculate the average power carried by the wave. (b) What happens to the average power if the wave amplitude is halved?

15.23 • A horizontal wire is stretched with a tension of 94.0 N, and the speed of transverse waves for the wire is 492 m/s. What must the amplitude of a traveling wave of frequency 69.0 Hz be in order for the average power carried by the wave to be 0.365 W?

15.24 •• A light wire is tightly stretched with tension F. Transverse traveling waves of amplitude A and wavelength λ_1 carry average power $P_{av,1} = 0.400$ W. If the wavelength of the waves is doubled, so $\lambda_2 = 2\lambda_1$, while the tension F and amplitude A are not altered, what then is the average power $P_{av,2}$ carried by the waves?

15.25 •• A jet plane at takeoff can produce sound of intensity 10.0 W/m² at 30.0 m away. But you prefer the tranquil sound of normal conversation, which is 1.0 μW/m^2. Assume that the plane behaves like a point source of sound. (a) What is the closest distance you should live from the airport runway to preserve your peace of mind? (b) What intensity from the jet does your friend experience if she lives twice as far from the runway as you do? (c) What power of sound does the jet produce at takeoff?

15.26 ·· Threshold of Pain.   You are investigating the report of a UFO landing in an isolated portion of New Mexico, and you encounter a strange object that is radiating sound waves uniformly in all directions. Assume that the sound comes from a point source and that you can ignore reflections. You are slowly walking toward the source. When you are 7.5 m from it, you measure its intensity to be 0.11 W/m^2. An intensity of 1.0 W/m^2 is often used as the "threshold of pain." How much closer to the source can you move before the sound intensity reaches this threshold?

15.27 · Energy Output.   By measurement you determine that sound waves are spreading out equally in all directions. from a point source and that the intensity is 0.026 W/m^2 at a distance of 4.3 m from the source. (a) What is the intensity at a distance of 3.1 m from the source? (b) How much sound energy does the source emit in one hour if its power output remains constant?

15.28 ·   A fellow student with a mathematical bent tells you that the wave function of a traveling wave on a thin rope is $y(x, t) = 2.30$ mm $\cos[(6.98 \text{ rad/m})x + (742 \text{ rad/s})t]$. Being more practical, you measure the rope to have a length of 1.35 m and a mass of 0.00338 kg. You are then asked to determine the following: (a) amplitude; (b) frequency; (c) wavelength; (d) wave speed; (e) direction the wave is traveling; (f) tension in the rope; (g) average power transmitted by the wave.

15.29 ·   At a distance of 7.00×10^{12} m from a star, the intensity of the radiation from the star is 15.4 W/m^2. Assuming that the star radiates uniformly in all directions, what is the total power output of the star?

Section 15.6 Wave Interference, Boundary Conditions, and Superposition

15.30 · Reflection.   A wave pulse on a string has the dimensions shown in Fig. E15.30 at $t = 0$. The wave speed is 40 cm/s. (a) If point O is a fixed end, draw the total wave on the string at $t = 15$ ms, 20 ms, 25 ms, 30 ms, 35 ms, 40 ms, and 45 ms. (b) Repeat part (a) for the case in which point O is a free end.

Figure **E15.30**

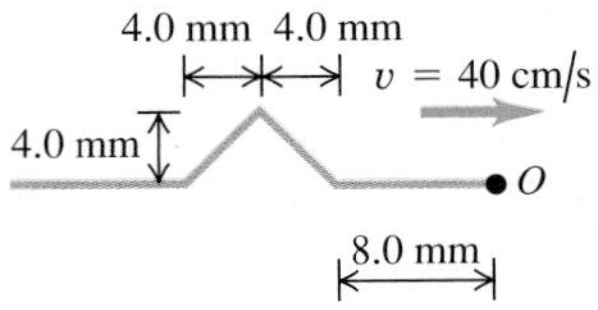

Figure **E15.31**

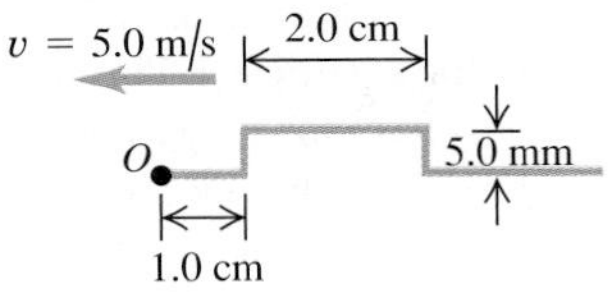

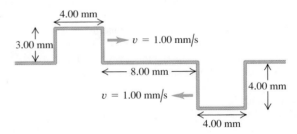

15.31 · Reflection.   A wave pulse on a string has the dimensions shown in Fig. E15.31 at $t = 0$. The wave speed is 5.0 m/s. (a) If point O is a fixed end, draw the total wave on the string at $t = 1.0$ ms, 2.0 ms, 3.0 ms, 4.0 ms, 5.0 ms, 6.0 ms, and 7.0 ms. (b) Repeat part (a) for the case in which point O is a free end.

15.32 · Interference of Triangular Pulses.   Two triangular wave pulses are traveling toward each other on a stretched string as shown in Fig. E15.32. Each pulse is identical to the other and travels at 2.00 cm/s. The leading edges of the pulses are 1.00 cm apart at $t = 0$. Sketch the shape of the string at $t = 0.250$ s, $t = 0.500$ s, $t = 0.750$ s, $t = 1.000$ s, and $t = 1.250$ s.

Figure **E15.32**

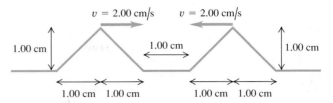

15.33 ·   Suppose that the left-traveling pulse in Exercise 15.32 is *below* the level of the unstretched string instead of above it. Make the same sketches that you did in that exercise.

15.34 ··   Two pulses are moving in opposite directions at 1.0 cm/s on a taut string, as shown in Fig. E15.34. Each square is 1.0 cm. Sketch the shape of the string at the end of (a) 6.0 s; (b) 7.0 s; (c) 8.0 s.

Figure **E15.34**

15.35 ·· Interference of Rectangular Pulses.   Figure E15.35 shows two rectangular wave pulses on a stretched string traveling toward each other. Each pulse is traveling with a speed of 1.00 mm/s and has the height and width shown in the figure. If the leading edges of the pulses are 8.00 mm apart at $t = 0$, sketch the shape of the string at $t = 4.00$ s, $t = 6.00$ s, and $t = 10.0$ s.

Figure **E15.35**

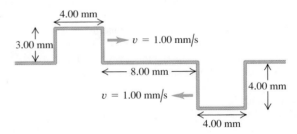

Section 15.7 Standing Waves on a String
Section 15.8 Normal Modes of a String

15.36 ·· CALC   Adjacent antinodes of a standing wave on a string are 15.0 cm apart. A particle at an antinode oscillates in simple harmonic motion with amplitude 0.850 cm and period 0.0750 s. The string lies along the $+x$-axis and is fixed at $x = 0$. (a) How far apart are the adjacent nodes? (b) What are the wavelength, amplitude, and speed of the two traveling waves that form this pattern? (c) Find the maximum and minimum transverse speeds of a point at an antinode. (d) What is the shortest distance along the string between a node and an antinode?

15.37 ·   Standing waves on a wire are described by Eq. (15.28), with $A_{SW} = 2.50$ mm, $\omega = 942$ rad/s, and $k = 0.750\pi$ rad/m. The left end of the wire is at $x = 0$. At what distances from the left end are (a) the nodes of the standing wave and (b) the antinodes of the standing wave?

15.38 · CALC Wave Equation and Standing Waves.   (a) Prove by direct substitution that $y(x, t) = (A_{SW} \sin kx) \sin \omega t$ is a solution of the wave equation, Eq. (15.12), for $v = \omega/k$. (b) Explain why the relationship $v = \omega/k$ for *traveling* waves also applies to *standing* waves.

15.39 • **CALC** Let $y_1(x, t) = A \cos(k_1 x - \omega_1 t)$ and $y_2(x, t) = A \cos(k_2 x - \omega_2 t)$ be two solutions to the wave equation, Eq. (15.12), for the same v. Show that $y(x, t) = y_1(x, t) + y_2(x, t)$ is also a solution to the wave equation.

15.40 • A 1.50-m-long rope is stretched between two supports with a tension that makes the speed of transverse waves 48.0 m/s. What are the wavelength and frequency of (a) the fundamental; (b) the second overtone; (c) the fourth harmonic?

15.41 • A wire with mass 40.0 g is stretched so that its ends are tied down at points 80.0 cm apart. The wire vibrates in its fundamental mode with frequency 60.0 Hz and with an amplitude at the antinodes of 0.300 cm. (a) What is the speed of propagation of transverse waves in the wire? (b) Compute the tension in the wire. (c) Find the maximum transverse velocity and acceleration of particles in the wire.

15.42 • A piano tuner stretches a steel piano wire with a tension of 800 N. The steel wire is 0.400 m long and has a mass of 3.00 g. (a) What is the frequency of its fundamental mode of vibration? (b) What is the number of the highest harmonic that could be heard by a person who is capable of hearing frequencies up to 10,000 Hz?

15.43 • **CALC** A thin, taut string tied at both ends and oscillating in its third harmonic has its shape described by the equation $y(x, t) = (5.60 \text{ cm}) \sin[(0.0340 \text{ rad/cm})x] \sin[(50.0 \text{ rad/s})t]$, where the origin is at the left end of the string, the x-axis is along the string, and the y-axis is perpendicular to the string. (a) Draw a sketch that shows the standing-wave pattern. (b) Find the amplitude of the two traveling waves that make up this standing wave. (c) What is the length of the string? (d) Find the wavelength, frequency, period, and speed of the traveling waves. (e) Find the maximum transverse speed of a point on the string. (f) What would be the equation $y(x, t)$ for this string if it were vibrating in its eighth harmonic?

15.44 • The wave function of a standing wave is $y(x, t) = 4.44 \text{ mm} \sin[(32.5 \text{ rad/m})x] \sin[(754 \text{ rad/s})t]$. For the two traveling waves that make up this standing wave, find the (a) amplitude; (b) wavelength; (c) frequency; (d) wave speed; (e) wave functions. (f) From the information given, can you determine which harmonic this is? Explain.

15.45 •• Consider again the rope and traveling wave of Exercise 15.28. Assume that the ends of the rope are held fixed and that this traveling wave and the reflected wave are traveling in the opposite direction. (a) What is the wave function $y(x, t)$ for the standing wave that is produced? (b) In which harmonic is the standing wave oscillating? (c) What is the frequency of the fundamental oscillation?

15.46 •• One string of a certain musical instrument is 75.0 cm long and has a mass of 8.75 g. It is being played in a room where the speed of sound is 344 m/s. (a) To what tension must you adjust the string so that, when vibrating in its second overtone, it produces sound of wavelength 0.765 m? (Assume that the breaking stress of the wire is very large and isn't exceeded.) (b) What frequency sound does this string produce in its fundamental mode of vibration?

15.47 • The portion of the string of a certain musical instrument between the bridge and upper end of the finger board (that part of the string that is free to vibrate) is 60.0 cm long, and this length of the string has mass 2.00 g. The string sounds an A_4 note (440 Hz) when played. (a) Where must the player put a finger (what distance x from the bridge) to play a D_5 note (587 Hz)? (See Fig. E15.47.)

For both the A_4 and D_5 notes, the string vibrates in its fundamental mode. (b) Without retuning, is it possible to play a G_4 note (392 Hz) on this string? Why or why not?

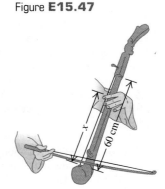

Figure **E15.47**

15.48 •• (a) A horizontal string tied at both ends is vibrating in its fundamental mode. The traveling waves have speed v, frequency f, amplitude A, and wavelength λ. Calculate the maximum transverse velocity and maximum transverse acceleration of points located at (i) $x = \lambda/2$, (ii) $x = \lambda/4$, and (iii) $x = \lambda/8$ from the left-hand end of the string. (b) At each of the points in part (a), what is the amplitude of the motion? (c) At each of the points in part (a), how much time does it take the string to go from its largest upward displacement to its largest downward displacement?

15.49 • **Guitar String.** One of the 63.5-cm-long strings of an ordinary guitar is tuned to produce the note B_3 (frequency 245 Hz) when vibrating in its fundamental mode. (a) Find the speed of transverse waves on this string. (b) If the tension in this string is increased by 1.0%, what will be the new fundamental frequency of the string? (c) If the speed of sound in the surrounding air is 344 m/s, find the frequency and wavelength of the sound wave produced in the air by the vibration of the B_3 string. How do these compare to the frequency and wavelength of the standing wave on the string?

15.50 • **Waves on a Stick.** A flexible stick 2.0 m long is not fixed in any way and is free to vibrate. Make clear drawings of this stick vibrating in its first three harmonics, and then use your drawings to find the wavelengths of each of these harmonics. (*Hint:* Should the ends be nodes or antinodes?)

PROBLEMS

15.51 • **CALC** A transverse sine wave with an amplitude of 2.50 mm and a wavelength of 1.80 m travels from left to right along a long, horizontal, stretched string with a speed of 36.0 m/s. Take the origin at the left end of the undisturbed string. At time $t = 0$ the left end of the string has its maximum upward displacement. (a) What are the frequency, angular frequency, and wave number of the wave? (b) What is the function $y(x, t)$ that describes the wave? (c) What is $y(t)$ for a particle at the left end of the string? (d) What is $y(t)$ for a particle 1.35 m to the right of the origin? (e) What is the maximum magnitude of transverse velocity of any particle of the string? (f) Find the transverse displacement and the transverse velocity of a particle 1.35 m to the right of the origin at time $t = 0.0625$ s.

15.52 • A transverse wave on a rope is given by

$$y(x, t) = (0.750 \text{ cm}) \cos \pi[(0.400 \text{ cm}^{-1})x + (250 \text{ s}^{-1})t]$$

(a) Find the amplitude, period, frequency, wavelength, and speed of propagation. (b) Sketch the shape of the rope at these values of t: 0, 0.0005 s, 0.0010 s. (c) Is the wave traveling in the $+x$- or $-x$-direction? (d) The mass per unit length of the rope is 0.0500 kg/m. Find the tension. (e) Find the average power of this wave.

15.53 •• Three pieces of string, each of length L, are joined together end to end, to make a combined string of length $3L$. The first piece of string has mass per unit length μ_1, the second piece has mass per unit length $\mu_2 = 4\mu_1$, and the third piece has mass per unit length $\mu_3 = \mu_1/4$. (a) If the combined string is under tension F, how much time does it take a transverse wave to travel the entire length $3L$? Give your answer in terms of L, F, and μ_1. (b) Does your answer to part (a) depend on the order in which the three pieces are joined together? Explain.

15.54 •• **CP** A 1750-N irregular beam is hanging horizontally by its ends from the ceiling by two vertical wires (A and B), each 1.25 m long and weighing 0.360 N. The center of gravity of this beam is one-third of the way along the beam from the end where wire A is attached. If you pluck both strings at the same time at the beam, what is the time delay between the arrival of the two pulses at the ceiling? Which pulse arrives first? (Neglect the effect of the weight of the wires on the tension in the wires.)

15.55 • **CALC Ant Joy Ride.** You place your pet ant Klyde (mass m) on top of a horizontal, stretched rope, where he holds on tightly. The rope has mass M and length L and is under tension F. You start a sinusoidal transverse wave of wavelength λ and amplitude A propagating along the rope. The motion of the rope is in a vertical plane. Klyde's mass is so small that his presence has no effect on the propagation of the wave. (a) What is Klyde's top speed as he oscillates up and down? (b) Klyde enjoys the ride and begs for more. You decide to double his top speed by changing the tension while keeping the wavelength and amplitude the same. Should the tension be increased or decreased, and by what factor?

15.56 •• **Weightless Ant.** An ant with mass m is standing peacefully on top of a horizontal, stretched rope. The rope has mass per unit length μ and is under tension F. Without warning, Cousin Throckmorton starts a sinusoidal transverse wave of wavelength λ propagating along the rope. The motion of the rope is in a vertical plane. What minimum wave amplitude will make the ant become momentarily weightless? Assume that m is so small that the presence of the ant has no effect on the propagation of the wave.

15.57 • **CP** When a transverse sinusoidal wave is present on a string, the particles of the string undergo SHM. This is the same motion as that of a mass m attached to an ideal spring of force constant k', for which the angular frequency of oscillation was found in Chapter 14 to be $\omega = \sqrt{k'/m}$. Consider a string with tension F and mass per unit length μ, along which is propagating a sinusoidal wave with amplitude A and wavelength λ. (a) Find the "force constant" k' of the restoring force that acts on a short segment of the string of length Δx (where $\Delta x \ll \lambda$). (b) How does the "force constant" calculated in part (b) depend on F, μ, A, and λ? Explain the physical reasons this should be so.

15.58 •• **Music.** You are designing a two-string instrument with metal strings 35.0 cm long, as shown in Fig. P15.58. Both strings are under the *same tension*. String S_1 has a mass of 8.00 g and produces the note middle C (frequency 262 Hz) in its fundamental mode. (a) What should be the tension in the string? (b) What should be the mass of string S_2 so that it will produce A-sharp (frequency 466 Hz) as its fundamental? (c) To extend the range of your instrument, you include a fret located just under the strings but not normally touching them. How far from the upper end should you put this fret so that when you press S_1 tightly against it, this string will produce C-sharp (frequency 277 Hz) in its fundamental? That is, what is x in the figure?

Figure **P15.58**

(d) If you press S_2 against the fret, what frequency of sound will it produce in its fundamental?

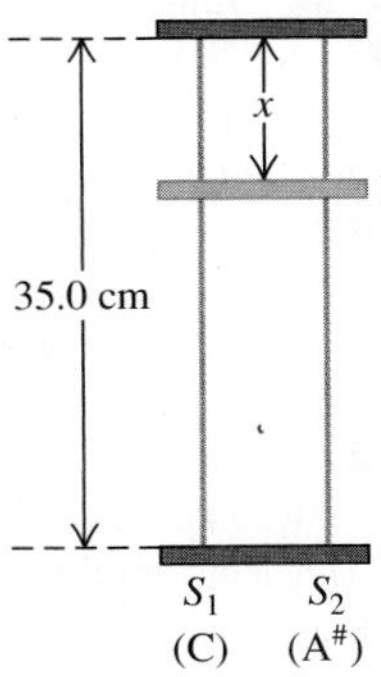

15.59 ••• **CP** The lower end of a uniform bar of mass 45.0 kg is attached to a wall by a frictionless hinge. The bar is held by a horizontal wire attached at its upper end so that the bar makes an angle of 30.0° with the wall. The wire has length 0.330 m and mass 0.0920 kg. What is the frequency of the fundamental standing wave for transverse waves on the wire?

15.60 ••• **CP** You are exploring a newly discovered planet. The radius of the planet is 7.20×10^7 m. You suspend a lead weight from the lower end of a light string that is 4.00 m long and has mass 0.0280 kg. You measure that it takes 0.0600 s for a transverse pulse to travel from the lower end to the upper end of the string. On earth, for the same string and lead weight, it takes 0.0390 s for a transverse pulse to travel the length of the string. The weight of the string is small enough that its effect on the tension in the string can be neglected. Assuming that the mass of the planet is distributed with spherical symmetry, what is its mass?

15.61 •• For a string stretched between two supports, two successive standing-wave frequencies are 525 Hz and 630 Hz. There are other standing-wave frequencies lower than 525 Hz and higher than 630 Hz. If the speed of transverse waves on the string is 384 m/s, what is the length of the string? Assume that the mass of the wire is small enough for its effect on the tension in the wire to be neglected.

15.62 ••• **CP** A 5.00-m, 0.732-kg wire is used to support two uniform 235-N posts of equal length (Fig. P15.62). Assume that the wire is essentially horizontal and that the speed of sound is 344 m/s. A strong wind is blowing, causing the wire to vibrate in its 5th overtone. What are the frequency and wavelength of the sound this wire produces?

Figure **P15.62**

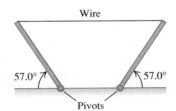

15.63 ••• **CP** A 1.80-m-long uniform bar that weighs 536 N is suspended in a horizontal position by two vertical wires that are attached to the ceiling. One wire is aluminum and the other is copper. The aluminum wire is attached to the left-hand end of the bar, and the copper wire is attached 0.40 m to the left of the right-hand end. Each wire has length 0.600 m and a circular cross section with radius 0.280 mm. What is the fundamental frequency of transverse standing waves for each wire?

15.64 •• A continuous succession of sinusoidal wave pulses are produced at one end of a very long string and travel along the length of the string. The wave has frequency 70.0 Hz, amplitude 5.00 mm, and wavelength 0.600 m. (a) How long does it take the wave to travel a distance of 8.00 m along the length of the string? (b) How long does it take a point on the string to travel a distance of 8.00 m, once the wave train has reached the point and set it into motion? (c) In parts (a) and (b), how does the time change if the amplitude is doubled?

15.65 ••• **CALC Waves of Arbitrary Shape.** (a) Explain why *any* wave described by a function of the form $y(x, t) = f(x - vt)$ moves in the $+x$-direction with speed v. (b) Show that $y(x, t) = f(x - vt)$ satisfies the wave equation, no matter what the functional form of f. To do this, write $y(x, t) = f(u)$, where

$u = x - vt$. Then, to take partial derivatives of $y(x, t)$, use the chain rule:

$$\frac{\partial y(x, t)}{\partial t} = \frac{df(u)}{du} \frac{\partial u}{\partial t} = \frac{df(u)}{du}(-v)$$

$$\frac{\partial y(x, t)}{\partial t} = \frac{df(u)}{du} \frac{\partial u}{\partial x} = \frac{df(u)}{du}$$

(c) A wave pulse is described by the function $y(x, t) = De^{-(Bx - Ct)^2}$, where B, C, and D are all positive constants. What is the speed of this wave?

15.66 ••• **CP** A vertical, 1.20-m length of 18-gauge (diameter of 1.024 mm) copper wire has a 100.0-N ball hanging from it. (a) What is the wavelength of the third harmonic for this wire? (b) A 500.0-N ball now *replaces* the original ball. What is the change in the wavelength of the third harmonic caused by replacing the light ball with the heavy one? (*Hint:* See Table 11.1 for Young's modulus.)

15.67 • (a) Show that Eq. (15.25) can also be written as $P_{av} = \frac{1}{2}Fk\omega A^2$, where k is the wave number of the wave. (b) If the tension F in the string is quadrupled while the amplitude A is kept the same, how must k and ω each change to keep the average power constant? [*Hint:* Recall Eq. (15.6).]

15.68 ••• **CALC** Equation (15.7) for a sinusoidal wave can be made more general by including a phase angle ϕ, where $0 \le \phi \le 2\pi$ (in radians). Then the wave function $y(x, t)$ becomes

$$y(x, t) = A \cos(kx - \omega t + \phi)$$

(a) Sketch the wave as a function of x at $t = 0$ for $\phi = 0$, $\phi = \pi/4$, $\phi = \pi/2$, $\phi = 3\pi/4$, and $\phi = 3\pi/2$. (b) Calculate the transverse velocity $v_y = \partial y/\partial t$. (c) At $t = 0$, a particle on the string at $x = 0$ has displacement $y = A/\sqrt{2}$. Is this enough information to determine the value of ϕ? In addition, if you are told that a particle at $x = 0$ is moving toward $y = 0$ at $t = 0$, what is the value of ϕ? (d) Explain in general what you must know about the wave's behavior at a given instant to determine the value of ϕ.

15.69 ••• A sinusoidal transverse wave travels on a string. The string has length 8.00 m and mass 6.00 g. The wave speed is 30.0 m/s, and the wavelength is 0.200 m. (a) If the wave is to have an average power of 50.0 W, what must be the amplitude of the wave? (b) For this same string, if the amplitude and wavelength are the same as in part (a), what is the average power for the wave if the tension is increased such that the wave speed is doubled?

15.70 ••• **CALC** **Energy in a Triangular Pulse.** A triangular wave pulse on a taut string travels in the positive x-direction with speed v. The tension in the string is F, and the linear mass density of the string is μ. At $t = 0$, the shape of the pulse is given by

$$y(x, 0) = \begin{cases} 0 & \text{if } x < -L \\ h(L + x)/L & \text{for } -L < x < 0 \\ h(L - x)/L & \text{for } 0 < x < L \\ 0 & \text{for } x > L \end{cases}$$

(a) Draw the pulse at $t = 0$. (b) Determine the wave function $y(x, t)$ at all times t. (c) Find the instantaneous power in the wave. Show that the power is zero except for $-L < (x - vt) < L$ and that in this interval the power is constant. Find the value of this constant power.

15.71 ••• **CALC** **Instantaneous Power in a Wave.** (a) Graph $y(x, t)$ as given by Eq. (15.7) as a function of x for a given time t (say, $t = 0$). On the same axes, make a graph of the instantaneous power $P(x, t)$ as given by Eq. (15.23). (b) Explain the connection between the slope of the graph of $y(x, t)$ versus x and the value of $P(x, t)$. In particular, explain what is happening at points where $P = 0$, where there is no instantaneous energy transfer. (c) The quantity $P(x, t)$ always has the same sign. What does this imply about the direction of energy flow? (d) Consider a wave moving in the $-x$-direction, for which $y(x, t) = A \cos(kx + \omega t)$. Calculate $P(x, t)$ for this wave, and make a graph of $y(x, t)$ and $P(x, t)$ as functions of x for a given time t (say, $t = 0$). What differences arise from reversing the direction of the wave?

15.72 •• A vibrating string 50.0 cm long is under a tension of 1.00 N. The results from five successive stroboscopic pictures are shown in Fig. P15.72. The strobe rate is set at 5000 flashes per minute, and observations reveal that the maximum displacement occurred at flashes 1 and 5 with no other maxima in between. (a) Find the period, frequency, and wavelength for the traveling waves on this string. (b) In what normal mode (harmonic) is the string vibrating? (c) What is the speed of the traveling waves on the string? (d) How fast is point P moving when the string is in (i) position 1 and (ii) position 3? (e) What is the mass of this string? (See Section 15.3.)

Figure **P15.72**

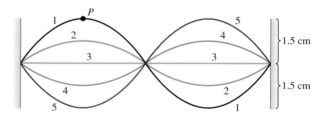

15.73 • **Clothesline Nodes.** Cousin Throckmorton is once again playing with the clothesline in Example 15.2 (Section 15.3). One end of the clothesline is attached to a vertical post. Throcky holds the other end loosely in his hand, so that the speed of waves on the clothesline is a relatively slow 0.720 m/s. He finds several frequencies at which he can oscillate his end of the clothesline so that a light clothespin 45.0 cm from the post doesn't move. What are these frequencies?

15.74 ••• **CALC** A guitar string is vibrating in its fundamental mode, with nodes at each end. The length of the segment of the string that is free to vibrate is 0.386 m. The maximum transverse acceleration of a point at the middle of the segment is 8.40×10^3 m/s^2 and the maximum transverse velocity is 3.80 m/s. (a) What is the amplitude of this standing wave? (b) What is the wave speed for the transverse traveling waves on this string?

15.75 •• **CALC** A string that lies along the $+x$-axis has a free end at $x = 0$. (a) By using steps similar to those used to derive Eq. (15.28), show that an incident traveling wave $y_1(x, t) = A \cos(kx + \omega t)$ gives rise to a standing wave $y(x, t) = 2A \cos\omega t \cos kx$. (b) Show that the standing wave has an antinode at its free end ($x = 0$). (c) Find the maximum displacement, maximum speed, and maximum acceleration of the free end of the string.

15.76 •• A string with both ends held fixed is vibrating in its third harmonic. The waves have a speed of 192 m/s and a frequency of 240 Hz. The amplitude of the standing wave at an antinode is 0.400 cm. (a) Calculate the amplitude at points on the string a distance of (i) 40.0 cm; (ii) 20.0 cm; and (iii) 10.0 cm from the left end of the string. (b) At each point in part (a), how much time does it take the string to go from its largest upward displacement to its largest downward displacement? (c) Calculate the maximum transverse velocity and the maximum transverse acceleration of the string at each of the points in part (a).

15.77 ••• A uniform cylindrical steel wire, 55.0 cm long and 1.14 mm in diameter, is fixed at both ends. To what tension must it be adjusted so that, when vibrating in its first overtone, it produces the note D-sharp of frequency 311 Hz? Assume that it stretches an insignificant amount. (*Hint:* See Table 12.1.)

15.78 • **Holding Up Under Stress.** A string or rope will break apart if it is placed under too much tensile stress [Eq. (11.8)]. Thicker ropes can withstand more tension without breaking because the thicker the rope, the greater the cross-sectional area and the smaller the stress. One type of steel has density 7800 kg/m³ and will break if the tensile stress exceeds 7.0×10^8 N/m². You want to make a guitar string from 4.0 g of this type of steel. In use, the guitar string must be able to withstand a tension of 900 N without breaking. Your job is the following: (a) Determine the maximum length and minimum radius the string can have. (b) Determine the highest possible fundamental frequency of standing waves on this string, if the entire length of the string is free to vibrate.

15.79 ••• **Combining Standing Waves.** A guitar string of length L is plucked in such a way that the total wave produced is the sum of the fundamental and the second harmonic. That is, the standing wave is given by

$$y(x, t) = y_1(x, t) + y_2(x, t)$$

where

$$y_1(x, t) = C \sin \omega_1 t \sin k_1 x$$
$$y_2(x, t) = C \sin \omega_2 t \sin k_2 x$$

with $\omega_1 = vk_1$ and $\omega_2 = vk_2$. (a) At what values of x are the nodes of y_1? (b) At what values of x are the nodes of y_2? (c) Graph the total wave at $t = 0$, $t = \frac{1}{8}f_1$, $t = \frac{1}{4}f_1$, $t = \frac{3}{8}f_1$, and $t = \frac{1}{2}f_1$. (d) Does the sum of the two standing waves y_1 and y_2 produce a standing wave? Explain.

15.80 •• **CP** When a massive aluminum sculpture is hung from a steel wire, the fundamental frequency for transverse standing waves on the wire is 250.0 Hz. The sculpture (but not the wire) is then completely submerged in water. (a) What is the new fundamental frequency? (*Hint:* See Table 12.1.) (b) Why is it a good approximation to treat the wire as being fixed at both ends?

15.81 ••• **CP** A large rock that weighs 164.0 N is suspended from the lower end of a thin wire that is 3.00 m long. The density of the rock is 3200 kg/m³. The mass of the wire is small enough that its effect on the tension in the wire can be neglected. The upper end of the wire is held fixed. When the rock is in air, the fundamental frequency for transverse standing waves on the wire is 42.0 Hz. When the rock is totally submerged in a liquid, with the top of the rock just below the surface, the fundamental frequency for the wire is 28.0 Hz. What is the density of the liquid?

15.82 •• **Tuning an Instrument.** A musician tunes the C-string of her instrument to a fundamental frequency of 65.4 Hz. The vibrating portion of the string is 0.600 m long and has a mass of 14.4 g. (a) With what tension must the musician stretch it? (b) What percent increase in tension is needed to increase the frequency from 65.4 Hz to 73.4 Hz, corresponding to a rise in pitch from C to D?

15.83 ••• One type of steel has a density of 7.8×10^3 kg/m³ and a breaking stress of 7.0×10^8 N/m². A cylindrical guitar string is to be made of 4.00 g of this steel. (a) What are the length and radius of the longest and thinnest string that can be placed under a tension of 900 N without breaking? (b) What is the highest fundamental frequency that this string could have?

CHALLENGE PROBLEMS

15.84 ••• **CP CALC** A deep-sea diver is suspended beneath the surface of Loch Ness by a 100-m-long cable that is attached to a boat on the surface (Fig. P15.84). The diver and his suit have a total mass of 120 kg and a volume of 0.0800 m³. The cable has a diameter of 2.00 cm and a linear mass density of $\mu =$ 1.10 kg/m. The diver thinks he sees something moving in the murky depths and jerks the end of the cable back and forth to send transverse waves up the cable as a signal to his companions in the boat. (a) What is the tension in the cable at its lower end, where it is attached to the diver? Do not forget to include the buoyant force that the water (density 1000 kg/m³) exerts on him. (b) Calculate the tension in the cable a distance x above the diver. The buoyant force on the cable must be included in your calculation. (c) The speed of transverse waves on the cable is given by $v = \sqrt{F/\mu}$ (Eq. 15.13). The speed therefore varies along the cable, since the tension is not constant. (This expression neglects the damping force that the water exerts on the moving cable.) Integrate to find the time required for the first signal to reach the surface.

Figure **P15.84**

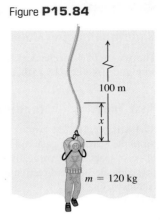

15.85 ••• **CALC** (a) Show that for a wave on a string, the kinetic energy *per unit length of string* is

$$u_k(x, t) = \tfrac{1}{2}\mu v_y^2(x, t) = \tfrac{1}{2}\mu \left(\frac{\partial y(x, t)}{\partial t} \right)^2$$

where μ is the mass per unit length. (b) Calculate $u_k(x, t)$ for a sinusoidal wave given by Eq. (15.7). (c) There is also elastic potential energy in the string, associated with the work required to deform and stretch the string. Consider a short segment of string at position x that has unstretched length Δx, as in Fig. 15.13. Ignoring the (small) curvature of the segment, its slope is $\partial y(x, t)/\partial x$. Assume that the displacement of the string from equilibrium is small, so that $\partial y/\partial x$ has a magnitude much less than unity. Show that the stretched length of the segment is approximately

$$\Delta x \left[1 + \tfrac{1}{2}\left(\frac{\partial y(x, t)}{\partial x} \right)^2 \right]$$

(*Hint:* Use the relationship $\sqrt{1 + u} \approx 1 + \tfrac{1}{2}u$, valid for $|u| \ll 1$.) (d) The potential energy stored in the segment equals the work done by the string tension F (which acts along the string) to stretch the segment from its unstretched length Δx to the length calculated in part (c). Calculate this work and show that the potential energy *per unit length of string* is

$$u_p(x, t) = \tfrac{1}{2}F\left(\frac{\partial y(x, t)}{\partial x} \right)^2$$

(e) Calculate $u_p(x, t)$ for a sinusoidal wave given by Eq. (15.7). (f) Show that $u_k(x, t) = u_p(x, t)$, for all x and t. (g) Show $y(x, t)$, $u_k(x, t)$, and $u_p(x, t)$ as functions of x for $t = 0$ in one graph with all three functions on the same axes. Explain why u_k and u_p are maximum where y is zero, and vice versa. (h) Show that the instantaneous power in the wave, given by Eq. (15.22), is equal to the total energy per unit length multiplied by the wave speed v. Explain why this result is reasonable.

Answers

Chapter Opening Question

The power of a mechanical wave depends on its frequency and amplitude [see Eq. (15.25)].

Test Your Understanding Questions

15.1 Answer: (i) The "wave" travels horizontally from one spectator to the next along each row of the stadium, but the displacement of each spectator is vertically upward. Since the displacement is perpendicular to the direction in which the wave travels, the wave is transverse.

15.2 Answer: (iv) The speed of waves on a string, v, does not depend on the wavelength. We can rewrite the relationship $v = \lambda f$ as $f = v/\lambda$, which tells us that if the wavelength λ doubles, the frequency f becomes one-half as great.

15.3 Answers: (a) $\frac{2}{8}T$, **(b)** $\frac{4}{8}T$, **(c)** $\frac{5}{8}T$ Since the wave is sinusoidal, each point on the string oscillates in simple harmonic motion (SHM). Hence we can apply all of the ideas from Chapter 14 about SHM to the wave depicted in Fig. 15.8. (a) A particle in SHM has its maximum speed when it is passing through the equilibrium position ($y = 0$ in Fig. 15.8). The particle at point A is moving upward through this position at $t = \frac{2}{8}T$. (b) In vertical SHM the greatest *upward* acceleration occurs when a particle is at its maximum *downward* displacement. This occurs for the particle at point B at $t = \frac{4}{8}T$. (c) A particle in vertical SHM has a *downward* acceleration when its displacement is *upward*. The particle at C has an upward displacement and is moving downward at $t = \frac{5}{8}T$.

15.4 Answer: (ii) The relationship $v = \sqrt{F/\mu}$ [Eq. (15.13)] says that the wave speed is greatest on the string with the smallest linear mass density. This is the thinnest string, which has the smallest amount of mass m and hence the smallest linear mass density $\mu = m/L$ (all strings are the same length).

15.5 Answer: (iii), (iv), (ii), (i) Equation (15.25) says that the average power in a sinusoidal wave on a string is $P_{av} = \frac{1}{2}\sqrt{\mu F}\,\omega^2 A^2$. All four strings are identical, so all have the same mass, the same length, and the same linear mass density μ. The frequency f is the same for each wave, as is the angular frequency $\omega = 2\pi f$. Hence the average wave power for each string is proportional to the square root of the string tension F and the square of the amplitude A. Compared to string (i), the average power in each string is (ii) $\sqrt{4} = 2$ times greater; (iii) $4^2 = 16$ times greater; and (iv) $\sqrt{2}\,(2)^2 = 4\sqrt{2}$ times greater.

15.6 Answer:

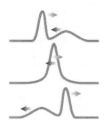

15.7 Answers: yes, yes Doubling the frequency makes the wavelength half as large. Hence the spacing between nodes (equal to $\lambda/2$) is also half as large. There are nodes at all of the previous positions, but there is also a new node between every pair of old nodes.

15.8 Answers: $n = 1, 3, 5, \ldots$ When you touch the string at its center, you are demanding that there be a node at the center. Hence only standing waves with a node at $x = L/2$ are allowed. From Figure 15.26 you can see that the normal modes $n = 1, 3, 5, \ldots$ cannot be present.

Bridging Problem

Answers: (a) $F(r) = \dfrac{m\omega^2}{2L}(L^2 - r^2)$

(b) $v(r) = \omega\sqrt{\dfrac{L^2 - r^2}{2}}$

(c) $\dfrac{\pi}{\omega\sqrt{2}}$

SOUND AND HEARING

16

? Most people like to listen to music, but hardly anyone likes to listen to noise. What is the physical difference between musical sound and noise?

LEARNING GOALS

By studying this chapter, you will learn:

• How to describe a sound wave in terms of either particle displacements or pressure fluctuations.

• How to calculate the speed of sound waves in different materials.

• How to calculate the intensity of a sound wave.

• What determines the particular frequencies of sound produced by an organ or a flute.

• How resonance occurs in musical instruments.

• What happens when sound waves from different sources overlap.

• How to describe what happens when two sound waves of slightly different frequencies are combined.

• Why the pitch of a siren changes as it moves past you.

O f all the mechanical waves that occur in nature, the most important in our everyday lives are longitudinal waves in a medium—usually air—called *sound* waves. The reason is that the human ear is tremendously sensitive and can detect sound waves even of very low intensity. Besides their use in spoken communication, our ears allow us to pick up a myriad of cues about our environment, from the welcome sound of a meal being prepared to the warning sound of an approaching car. The ability to hear an unseen nocturnal predator was essential to the survival of our ancestors, so it is no exaggeration to say that we humans owe our existence to our highly evolved sense of hearing.

Up to this point we have described mechanical waves primarily in terms of displacement; however, a description of sound waves in terms of *pressure* fluctuations is often more appropriate, largely because the ear is primarily sensitive to changes in pressure. We'll study the relationships among displacement, pressure fluctuation, and intensity and the connections between these quantities and human sound perception.

When a source of sound or a listener moves through the air, the listener may hear a frequency different from the one emitted by the source. This is the Doppler effect, which has important applications in medicine and technology.

16.1 Sound Waves

The most general definition of **sound** is a longitudinal wave in a medium. Our main concern in this chapter is with sound waves in air, but sound can travel through any gas, liquid, or solid. You may be all too familiar with the propagation of sound through a solid if your neighbor's stereo speakers are right next to your wall.

The simplest sound waves are sinusoidal waves, which have definite frequency, amplitude, and wavelength. The human ear is sensitive to waves in the frequency range from about 20 to 20,000 Hz, called the **audible range,** but we also use the

16.1 A sinusoidal longitudinal wave traveling to the right in a fluid. (Compare to Fig. 15.7.)

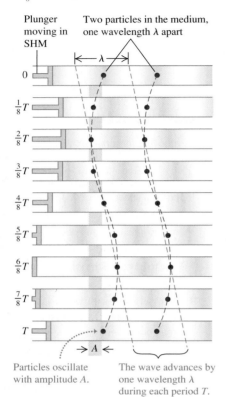

Longitudinal waves are shown at intervals of $\frac{1}{8}T$ for one period T.

Plunger moving in SHM

Two particles in the medium, one wavelength λ apart

Particles oscillate with amplitude A.

The wave advances by one wavelength λ during each period T.

term "sound" for similar waves with frequencies above (**ultrasonic**) and below (**infrasonic**) the range of human hearing.

Sound waves usually travel out in all directions from the source of sound, with an amplitude that depends on the direction and distance from the source. We'll return to this point in the next section. For now, we concentrate on the idealized case of a sound wave that propagates in the positive x-direction only. As we discussed in Section 15.3, such a wave is described by a wave function $y(x, t)$, which gives the instantaneous displacement y of a particle in the medium at position x at time t. If the wave is sinusoidal, we can express it using Eq. (15.7):

$$y(x, t) = A\cos(kx - \omega t)$$

(sound wave propagating in the +x-direction) (16.1)

Remember that in a longitudinal wave the displacements are *parallel* to the direction of travel of the wave, so distances x and y are measured parallel to each other, not perpendicular as in a transverse wave. The amplitude A is the maximum displacement of a particle in the medium from its equilibrium position (Fig. 16.1). Hence A is also called the **displacement amplitude.**

Sound Waves As Pressure Fluctuations

Sound waves may also be described in terms of variations of *pressure* at various points. In a sinusoidal sound wave in air, the pressure fluctuates above and below atmospheric pressure p_a in a sinusoidal variation with the same frequency as the motions of the air particles. The human ear operates by sensing such pressure variations. A sound wave entering the ear canal exerts a fluctuating pressure on one side of the eardrum; the air on the other side of the eardrum, vented to the outside by the Eustachian tube, is at atmospheric pressure. The pressure difference on the two sides of the eardrum sets it into motion. Microphones and similar devices also usually sense pressure differences, not displacements, so it is very useful to develop a relationship between these two descriptions.

Let $p(x, t)$ be the instantaneous pressure fluctuation in a sound wave at any point x at time t. That is, $p(x, t)$ is the amount by which the pressure *differs* from normal atmospheric pressure p_a. Think of $p(x, t)$ as the *gauge pressure* defined in Section 12.2; it can be either positive or negative. The *absolute* pressure at a point is then $p_a + p(x, t)$.

To see the connection between the pressure fluctuation $p(x, t)$ and the displacement $y(x, t)$ in a sound wave propagating in the +x-direction, consider an imaginary cylinder of a wave medium (gas, liquid, or solid) with cross-sectional area S and axis along the direction of propagation (Fig. 16.2). When no sound wave is present, the cylinder has length Δx and volume $V = S\,\Delta x$, as shown by the shaded volume in Fig. 16.2. When a wave is present, at time t the end of the cylinder that is initially at x is displaced by $y_1 = y(x, t)$, and the end that is initially at $x + \Delta x$ is displaced by $y_2 = y(x + \Delta x, t)$; this is shown by the red lines. If $y_2 > y_1$, as shown in Fig. 16.2, the cylinder's volume has increased, which causes a decrease in pressure. If $y_2 < y_1$, the cylinder's volume has decreased and the pressure has increased. If $y_2 = y_1$, the cylinder is simply shifted to the left or right; there is no volume change and no pressure fluctuation. The pressure fluctuation depends on the *difference* between the displacements at neighboring points in the medium.

Quantitatively, the change in volume ΔV of the cylinder is

$$\Delta V = S(y_2 - y_1) = S[y(x + \Delta x, t) - y(x, t)]$$

In the limit as $\Delta x \rightarrow 0$, the fractional change in volume dV/V (volume change divided by original volume) is

$$\frac{dV}{V} = \lim_{\Delta x \to 0} \frac{S[y(x + \Delta x, t) - y(x, t)]}{S\,\Delta x} = \frac{\partial y(x, t)}{\partial x}$$ (16.2)

16.2 As a sound wave propagates along the x-axis, the left and right ends undergo different displacements y_1 and y_2.

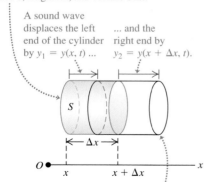

Undisturbed cylinder of fluid has cross-sectional area S, length Δx, and volume $S\Delta x$.

A sound wave displaces the left end of the cylinder by $y_1 = y(x, t)$ ...

... and the right end by $y_2 = y(x + \Delta x, t)$.

The change in volume of the disturbed cylinder of fluid is $S(y_2 - y_1)$.

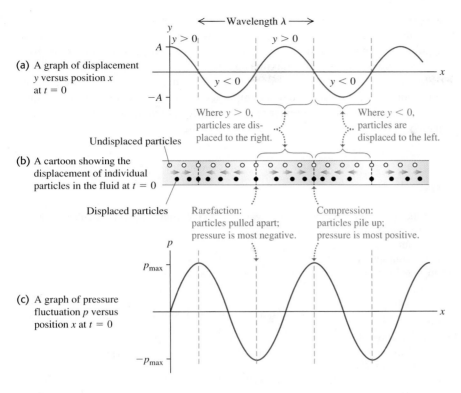

(a) A graph of displacement y versus position x at $t = 0$

Undisplaced particles

(b) A cartoon showing the displacement of individual particles in the fluid at $t = 0$

Displaced particles

(c) A graph of pressure fluctuation p versus position x at $t = 0$

The fractional volume change is related to the pressure fluctuation by the bulk modulus B, which by definition [Eq. (11.13)] is $B = -p(x, t)/(dV/V)$ (see Section 11.4). Solving for $p(x, t)$, we have

$$p(x, t) = -B\frac{\partial y(x, t)}{\partial x} \tag{16.3}$$

The negative sign arises because when $\partial y(x, t)/\partial x$ is positive, the displacement is greater at $x + \Delta x$ than at x, corresponding to an increase in volume and a *decrease* in pressure.

When we evaluate $\partial y(x, t)/\partial x$ for the sinusoidal wave of Eq. (16.1), we find

$$p(x, t) = BkA \sin(kx - \omega t) \tag{16.4}$$

Figure 16.3 shows $y(x, t)$ and $p(x, t)$ for a sinusoidal sound wave at $t = 0$. It also shows how individual particles of the wave are displaced at this time. While $y(x, t)$ and $p(x, t)$ describe the same wave, these two functions are one-quarter cycle out of phase: At any time, the displacement is greatest where the pressure fluctuation is zero, and vice versa. In particular, note that the compressions (points of greatest pressure and density) and rarefactions (points of lowest pressure and density) are points of *zero* displacement.

CAUTION **Graphs of a sound wave** Keep in mind that the graphs in Fig. 16.3 show the wave at only *one* instant of time. Because the wave is propagating in the $+x$-direction, as time goes by the wave patterns in the functions $y(x, t)$ and $p(x, t)$ move to the right at the wave speed $v = \omega/k$. Hence the positions of the compressions and rarefactions also move to the right at this same speed. The particles, by contrast, simply oscillate back and forth in simple harmonic motion as shown in Fig. 16.1. ▮

Equation (16.4) shows that the quantity BkA represents the maximum pressure fluctuation. We call this the **pressure amplitude,** denoted by p_{max}:

$$p_{max} = BkA \quad \text{(sinusoidal sound wave)} \tag{16.5}$$

The pressure amplitude is directly proportional to the displacement amplitude A, as we might expect, and it also depends on wavelength. Waves of shorter wavelength λ (larger wave number $k = 2\pi/\lambda$) have greater pressure variations for a given amplitude because the maxima and minima are squeezed closer together. A medium with a large value of bulk modulus B requires a relatively large pressure amplitude for a given displacement amplitude because large B means a less compressible medium; that is, greater pressure change is required for a given volume change.

Example 16.1 Amplitude of a sound wave in air

In a sinusoidal sound wave of moderate loudness, the maximum pressure variations are about 3.0×10^{-2} Pa above and below atmospheric pressure. Find the corresponding maximum displacement if the frequency is 1000 Hz. In air at normal atmospheric pressure and density, the speed of sound is 344 m/s and the bulk modulus is 1.42×10^5 Pa.

SOLUTION

IDENTIFY and SET UP: This problem involves the relationship between two ways of describing a sound wave: in terms of displacement and in terms of pressure. The target variable is the displacement amplitude A. We are given the pressure amplitude p_{max}, wave speed v, frequency f, and bulk modulus B. Equation (16.5) relates the target variable A to p_{max}. We use $\omega = vk$ [Eq. (15.6)] to determine the wave number k from v and the angular frequency $\omega = 2\pi f$.

EXECUTE: From Eq. (15.6),

$$k = \frac{\omega}{v} = \frac{2\pi f}{v} = \frac{(2\pi \text{ rad})(1000 \text{ Hz})}{344 \text{ m/s}} = 18.3 \text{ rad/m}$$

Then from Eq. (16.5), the maximum displacement is

$$A = \frac{p_{max}}{Bk} = \frac{3.0 \times 10^{-2} \text{ Pa}}{(1.42 \times 10^5 \text{ Pa})(18.3 \text{ rad/m})} = 1.2 \times 10^{-8} \text{ m}$$

EVALUATE: This displacement amplitude is only about $\frac{1}{100}$ the size of a human cell. The ear actually senses pressure fluctuations; it detects these minuscule displacements only indirectly.

Example 16.2 Amplitude of a sound wave in the inner ear

A sound wave that enters the human ear sets the eardrum into oscillation, which in turn causes oscillation of the *ossicles*, a chain of three tiny bones in the middle ear (Fig. 16.4). The ossicles transmit this oscillation to the fluid (mostly water) in the inner ear; there the fluid motion disturbs hair cells that send nerve impulses to the brain with information about the sound. The area of the moving part of the eardrum is about 43 mm^2, and that of the stapes (the smallest of the ossicles) where it connects to the inner ear is about 3.2 mm^2. For the sound in Example 16.1, determine (a) the pressure amplitude and (b) the displacement amplitude of the wave in the fluid of the inner ear, in which the speed of sound is 1500 m/s.

SOLUTION

IDENTIFY and SET UP: Although the sound wave here travels in liquid rather than air, the same principles and relationships among the properties of the wave apply. We can neglect the mass of the tiny ossicles (about 58 mg $= 5.8 \times 10^{-5}$ kg), so the force they exert on the inner-ear fluid is the same as that exerted on the eardrum and ossicles by the incident sound wave. (In Chapters 4 and 5 we used the same idea to say that the tension is the same at either end of a massless rope.) Hence the pressure amplitude in the inner ear, $p_{max(\text{inner ear})}$, is greater than in the outside air, $p_{max(\text{air})}$, because the same force is exerted on a smaller area (the area of the stapes versus the area of the eardrum). Given $p_{max(\text{inner ear})}$, we find the displacement amplitude $A_{\text{inner ear}}$ using Eq. (16.5).

16.4 The anatomy of the human ear. The middle ear is the size of a small marble; the ossicles (incus, malleus, and stapes) are the smallest bones in the human body.

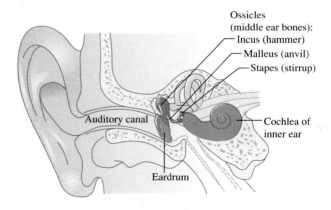

EXECUTE: (a) From the area of the eardrum and the pressure amplitude in air found in Example 16.1, the maximum force exerted by the sound wave in air on the eardrum is $F_{max} = p_{max(\text{air})} S_{\text{eardrum}}$. Then

$$p_{max(\text{inner ear})} = \frac{F_{max}}{S_{\text{stapes}}} = p_{max(\text{air})} \frac{S_{\text{eardrum}}}{S_{\text{stapes}}}$$

$$= (3.0 \times 10^{-2} \text{ Pa})\frac{43 \text{ mm}^2}{3.2 \text{ mm}^2} = 0.40 \text{ Pa}$$

(b) To find the maximum displacement $A_{\text{inner ear}}$, we use $A = p_{\text{max}}/Bk$ as in Example 16.1. The inner-ear fluid is mostly water, which has a much greater bulk modulus B than air. From Table 11.2 the compressibility of water (unfortunately also called k) is 45.8×10^{-11} Pa^{-1}, so $B_{\text{fluid}} = 1/(45.8 \times 10^{-11} \text{ Pa}^{-1}) = 2.18 \times 10^9$ Pa.

The wave in the inner ear has the same angular frequency ω as the wave in the air because the air, eardrum, ossicles, and inner-ear fluid all oscillate together (see Example 15.8 in Section 15.8). But because the wave speed v is greater in the inner ear than in the air (1500 m/s versus 344 m/s), the wave number $k = \omega/v$ is smaller. Using the value of ω from Example 16.1,

$$k_{\text{inner ear}} = \frac{\omega}{v_{\text{inner ear}}} = \frac{(2\pi \text{ rad})(1000 \text{ Hz})}{1500 \text{ m/s}} = 4.2 \text{ rad/m}$$

Putting everything together, we have

$$A_{\text{inner ear}} = \frac{p_{\text{max (inner ear)}}}{B_{\text{fluid}} k_{\text{inner ear}}} = \frac{0.40 \text{ Pa}}{(2.18 \times 10^9 \text{ Pa})(4.2 \text{ rad/m})}$$
$$= 4.4 \times 10^{-11} \text{ m}$$

EVALUATE: In part (a) we see that the ossicles increase the pressure amplitude by a factor of $(43 \text{ mm}^2)/(3.2 \text{ mm}^2) = 13$. This amplification helps give the human ear its great sensitivity.

The displacement amplitude in the inner ear is even smaller than in the air. But *pressure* variations within the inner-ear fluid are what set the hair cells into motion, so what matters is that the pressure amplitude is larger in the inner ear than in the air.

Perception of Sound Waves

The physical characteristics of a sound wave are directly related to the perception of that sound by a listener. For a given frequency, the greater the pressure amplitude of a sinusoidal sound wave, the greater the perceived **loudness.** The relationship between pressure amplitude and loudness is not a simple one, and it varies from one person to another. One important factor is that the ear is not equally sensitive to all frequencies in the audible range. A sound at one frequency may seem louder than one of equal pressure amplitude at a different frequency. At 1000 Hz the minimum pressure amplitude that can be perceived with normal hearing is about 3×10^{-5} Pa; to produce the same loudness at 200 Hz or 15,000 Hz requires about 3×10^{-4} Pa. Perceived loudness also depends on the health of the ear. A loss of sensitivity at the high-frequency end usually happens naturally with age but can be further aggravated by excessive noise levels.

The frequency of a sound wave is the primary factor in determining the **pitch** of a sound, the quality that lets us classify the sound as "high" or "low." The higher the frequency of a sound (within the audible range), the higher the pitch that a listener will perceive. Pressure amplitude also plays a role in determining pitch. When a listener compares two sinusoidal sound waves with the same frequency but different pressure amplitudes, the one with the greater pressure amplitude is usually perceived as louder but also as slightly lower in pitch.

Musical sounds have wave functions that are more complicated than a simple sine function. The pressure fluctuation in the sound wave produced by a clarinet is shown in Fig. 16.5a. The pattern is so complex because the column of air in a wind instrument like a clarinet vibrates at a fundamental frequency and at many harmonics at the same time. (In Section 15.8, we described this same behavior for a string that has been plucked, bowed, or struck. We'll examine the physics of wind instruments in Section 16.5.) The sound wave produced in the surrounding air has a similar amount of each harmonic—that is, a similar *harmonic content.* Figure 16.5b shows the harmonic content of the sound of a clarinet. The mathematical process of translating a pressure–time graph like Fig. 16.5a into a graph of harmonic content like Fig. 16.5b is called *Fourier analysis.*

Two tones produced by different instruments might have the same fundamental frequency (and thus the same pitch) but sound different because of different harmonic content. The difference in sound is called *tone color, quality,* or **timbre** and is often described in subjective terms such as reedy, golden, round, mellow, and tinny. A tone that is rich in harmonics, like the clarinet tone in Figs. 16.5a and 16.5b, usually sounds thin and "stringy" or "reedy," while a tone containing mostly a fundamental, like the alto recorder tone in Figs. 16.5c and 16.5d, is more mellow and flutelike. The same principle applies to the human voice, which is another example of a wind instrument; the vowels "a" and "e" sound different because of differences in harmonic content.

Application Hearing Loss from Amplified Sound
Due to exposure to highly amplified music, many young popular musicians have suffered permanent ear damage and have hearing typical of persons 65 years of age. Headphones for personal music players used at high volume pose similar threats to hearing. Be careful!

16.5 Different representations of the sound of (a), (b) a clarinet and (c), (d) an alto recorder. (Graphs adapted from R.E. Berg and D.G. Stork, *The Physics of Sound,* Prentice-Hall, 1982.)

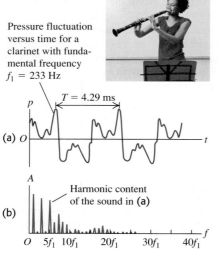

Pressure fluctuation versus time for a clarinet with fundamental frequency $f_1 = 233$ Hz

(a)

$T = 4.29$ ms

(b) Harmonic content of the sound in (a)

$O \quad 5f_1 \ 10f_1 \quad 20f_1 \quad 30f_1 \quad 40f_1$

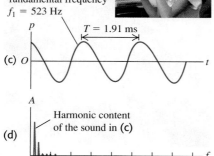

Pressure fluctuation versus time for an alto recorder with fundamental frequency $f_1 = 523$ Hz

(c)

$T = 1.91$ ms

(d) Harmonic content of the sound in (c)

$O \quad 5f_1 \ 10f_1 \quad 20f_1 \quad 30f_1 \quad 40f_1$

Another factor in determining tone quality is the behavior at the beginning (*attack*) and end (*decay*) of a tone. A piano tone begins with a thump and then dies away gradually. A harpsichord tone, in addition to having different harmonic content, begins much more quickly with a click, and the higher harmonics begin before the lower ones. When the key is released, the sound also dies away much more rapidly with a harpsichord than with a piano. Similar effects are present in other musical instruments. With wind and string instruments the player has considerable control over the attack and decay of the tone, and these characteristics help to define the unique characteristics of each instrument.

Unlike the tones made by musical instruments or the vowels in human speech, **noise** is a combination of *all* frequencies, not just frequencies that are integer multiples of a fundamental frequency. (An extreme case is "white noise," which contains equal amounts of all frequencies across the audible range.) Examples include the sound of the wind and the hissing sound you make in saying the consonant "s."

Test Your Understanding of Section 16.1   You use an electronic signal generator to produce a sinusoidal sound wave in air. You then increase the frequency of the wave from 100 Hz to 400 Hz while keeping the pressure amplitude constant. What effect does this have on the displacement amplitude of the sound wave? (i) It becomes four times greater; (ii) it becomes twice as great; (iii) it is unchanged; (iv) it becomes $\frac{1}{2}$ as great; (v) it becomes $\frac{1}{4}$ as great.

16.2 Speed of Sound Waves

We found in Section 15.4 that the speed v of a transverse wave on a string depends on the string tension F and the linear mass density μ: $v = \sqrt{F/\mu}$. What, we may ask, is the corresponding expression for the speed of sound waves in a gas or liquid? On what properties of the medium does the speed depend?

We can make an educated guess about these questions by remembering a claim that we made in Section 15.4: For mechanical waves in general, the expression for the wave speed is of the form

$$v = \sqrt{\frac{\text{Restoring force returning the system to equilibrium}}{\text{Inertia resisting the return to equilibrium}}}$$

A sound wave in a bulk fluid causes compressions and rarefactions of the fluid, so the restoring-force term in the above expression must be related to how easy or difficult it is to compress the fluid. This is precisely what the bulk modulus B of the medium tells us. According to Newton's second law, inertia is related to mass. The "massiveness" of a bulk fluid is described by its density, or mass per unit volume, ρ. (The corresponding quantity for a string is the mass per unit length, μ.) Hence we expect that the speed of sound waves should be of the form $v = \sqrt{B/\rho}$.

To check our guess, we'll derive the speed of sound waves in a fluid in a pipe. This is a situation of some importance, since all musical wind instruments are fundamentally pipes in which a longitudinal wave (sound) propagates in a fluid (air) (Fig. 16.6). Human speech works on the same principle; sound waves propagate in your vocal tract, which is basically an air-filled pipe connected to the lungs at one end (your larynx) and to the outside air at the other end (your mouth). The steps in our derivation are completely parallel to those we used in Section 15.4 to find the speed of transverse waves, so you'll find it useful to review that section.

Speed of Sound in a Fluid

Figure 16.7 shows a fluid (either liquid or gas) with density ρ in a pipe with cross-sectional area A. In the equilibrium state, the fluid is under a uniform pressure p. In Fig. 16.7a the fluid is at rest. We take the x-axis along the length of the pipe. This is also the direction in which we make a longitudinal wave propagate, so the displacement y is also measured along the pipe, just as in Section 16.1 (see Fig. 16.2).

At time $t = 0$ we start the piston at the left end moving toward the right with constant speed v_y. This initiates a wave motion that travels to the right along the length of the pipe, in which successive sections of fluid begin to move and become compressed at successively later times.

Figure 16.7b shows the fluid at time t. All portions of fluid to the left of point P are moving to the right with speed v_y, and all portions to the right of P are still at rest. The boundary between the moving and stationary portions travels to the right with a speed equal to the speed of propagation or wave speed v. At time t the piston has moved a distance $v_y t$, and the boundary has advanced a distance vt. As with a transverse disturbance in a string, we can compute the speed of propagation from the impulse–momentum theorem.

The quantity of fluid set in motion in time t is the amount that originally occupied a section of the cylinder with length vt, cross-sectional area A, and volume vtA. The mass of this fluid is ρvtA, and its longitudinal momentum (that is, momentum along the length of the pipe) is

$$\text{Longitudinal momentum} = (\rho vtA)v_y$$

Next we compute the increase of pressure, Δp, in the moving fluid. The original volume of the moving fluid, Avt, has decreased by an amount $Av_y t$. From the definition of the bulk modulus B, Eq. (11.13) in Section 11.5,

$$B = \frac{-\text{Pressure change}}{\text{Fractional volume change}} = \frac{-\Delta p}{-Av_y t/Avt}$$

$$\Delta p = B\frac{v_y}{v}$$

The pressure in the moving fluid is $p + \Delta p$ and the force exerted on it by the piston is $(p + \Delta p)A$. The net force on the moving fluid (see Fig. 16.7b) is ΔpA, and the longitudinal impulse is

$$\text{Longitudinal impulse} = \Delta pAt = B\frac{v_y}{v}At$$

Because the fluid was at rest at time $t = 0$, the change in momentum up to time t is equal to the momentum at that time. Applying the impulse–momentum theorem (see Section 8.1), we find

$$B\frac{v_y}{v}At = \rho vtAv_y \qquad (16.6)$$

When we solve this expression for v, we get

$$v = \sqrt{\frac{B}{\rho}} \qquad \text{(speed of a longitudinal wave in a fluid)} \qquad (16.7)$$

which agrees with our educated guess. Thus the speed of propagation of a longitudinal pulse in a fluid depends only on the bulk modulus B and the density ρ of the medium.

While we derived Eq. (16.7) for waves in a pipe, it also applies to longitudinal waves in a bulk fluid. Thus the speed of sound waves traveling in air or water is determined by this equation.

Speed of Sound in a Solid

When a longitudinal wave propagates in a *solid* rod or bar, the situation is somewhat different. The rod expands sideways slightly when it is compressed

16.6 When a wind instrument like this French horn is played, sound waves propagate through the air within the instrument's pipes. The properties of the sound that emerges from the large bell depend on the speed of these waves.

16.7 A sound wave propagating in a fluid confined to a tube. (a) Fluid in equilibrium. (b) A time t after the piston begins moving to the right at speed v_y, the fluid between the piston and point P is in motion. The speed of sound waves is v.

longitudinally, while a fluid in a pipe with constant cross section cannot move sideways. Using the same kind of reasoning that led us to Eq. (16.7), we can show that the speed of a longitudinal pulse in the rod is given by

$$v = \sqrt{\frac{Y}{\rho}} \qquad \text{(speed of a longitudinal wave in a solid rod)} \qquad (16.8)$$

where Y is Young's modulus, defined in Section 11.4.

CAUTION **Solid rods vs. bulk solids** Equation (16.8) applies only to a rod or bar whose sides are free to bulge and shrink a little as the wave travels. It does not apply to longitudinal waves in a *bulk* solid, since in these materials, sideways motion in any element of material is prevented by the surrounding material. The speed of longitudinal waves in a bulk solid depends on the density, the bulk modulus, and the *shear* modulus; a full discussion is beyond the scope of this book. ▮

As with the derivation for a transverse wave on a string, Eqs. (16.7) and (16.8) are valid for sinusoidal and other periodic waves, not just for the special case discussed here.

Table 16.1 lists the speed of sound in several bulk materials. Sound waves travel more slowly in lead than in aluminum or steel because lead has a lower bulk modulus and shear modulus and a higher density.

Table 16.1 Speed of Sound in Various Bulk Materials

Material	Speed of Sound (m/s)
Gases	
Air (20°C)	344
Helium (20°C)	999
Hydrogen (20°C)	1330
Liquids	
Liquid helium (4 K)	211
Mercury (20°C)	1451
Water (0°C)	1402
Water (20°C)	1482
Water (100°C)	1543
Solids	
Aluminum	6420
Lead	1960
Steel	5941

Example 16.3 Wavelength of sonar waves

A ship uses a sonar system (Fig. 16.8) to locate underwater objects. Find the speed of sound waves in water using Eq. (16.7), and find the wavelength of a 262-Hz wave.

SOLUTION

IDENTIFY and SET UP: Our target variables are the speed and wavelength of a sound wave in water. In Eq. (16.7), we use the density of water, $\rho = 1.00 \times 10^3 \text{ kg/m}^3$, and the bulk modulus of water, which we find from the compressibility (see Table 11.2). Given the speed and the frequency $f = 262$ Hz, we find the wavelength from $v = f\lambda$.

EXECUTE: In Example 16.2, we used Table 11.2 to find $B = 2.18 \times 10^9$ Pa. Then

$$v = \sqrt{\frac{B}{\rho}} = \sqrt{\frac{2.18 \times 10^9 \text{ Pa}}{1.00 \times 10^3 \text{ kg/m}^3}} = 1480 \text{ m/s}$$

and

$$\lambda = \frac{v}{f} = \frac{1480 \text{ m/s}}{262 \text{ s}^{-1}} = 5.65 \text{ m}$$

EVALUATE: The calculated value of v agrees well with the value in Table 16.1. Water is denser than air (ρ is larger) but is also much

16.8 A sonar system uses underwater sound waves to detect and locate submerged objects.

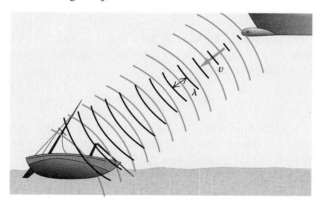

more incompressible (B is much larger), and so the speed $v = \sqrt{B/\rho}$ is greater than the 344-m/s speed of sound in air at ordinary temperatures. The relationship $\lambda = v/f$ then says that a sound wave in water must have a longer wavelength than a wave of the same frequency in air. Indeed, we found in Example 15.1 (Section 15.2) that a 262-Hz sound wave in air has a wavelength of only 1.31 m.

Dolphins emit high-frequency sound waves (typically 100,000 Hz) and use the echoes for guidance and for hunting. The corresponding wavelength in water is 1.48 cm. With this high-frequency "sonar" system they can sense objects that are roughly as small as the wavelength (but not much smaller). *Ultrasonic imaging* is a medical technique that uses exactly the same physical principle; sound waves of very high frequency and very short wavelength, called *ultrasound*, are scanned over the human body, and the "echoes" from interior organs are used to create an image. With ultrasound of frequency 5 MHz $= 5 \times 10^6$ Hz, the wavelength in water (the primary constituent of the body) is 0.3 mm, and features as small as this can be discerned in the image. Ultrasound is used for the study of heart-valve action, detection of tumors, and prenatal examinations (Fig. 16.9). Ultrasound is more sensitive than x rays in distinguishing various kinds of tissues and does not have the radiation hazards associated with x rays.

16.9 This three-dimensional image of a fetus in the womb was made using a sequence of ultrasound scans. Each individual scan reveals a two-dimensional "slice" through the fetus; many such slices were then combined digitally.

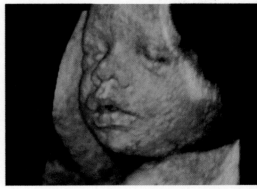

Speed of Sound in a Gas

Most of the sound waves that we encounter on a daily basis propagate in air. To use Eq. (16.7) to find the speed of sound waves in air, we must keep in mind that the bulk modulus of a gas depends on the pressure of the gas: The greater the pressure applied to a gas to compress it, the more it resists further compression and hence the greater the bulk modulus. (That's why specific values of the bulk modulus for gases are not given in Table 11.1.) The expression for the bulk modulus of a gas for use in Eq. (16.7) is

$$B = \gamma p_0 \tag{16.9}$$

where p_0 is the equilibrium pressure of the gas. The quantity γ (the Greek letter gamma) is called the *ratio of heat capacities*. It is a dimensionless number that characterizes the thermal properties of the gas. (We'll learn more about this quantity in Chapter 19.) As an example, the ratio of heat capacities for air is $\gamma = 1.40$. At normal atmospheric pressure $p_0 = 1.013 \times 10^5$ Pa, so $B = (1.40)(1.013 \times 10^5 \text{ Pa}) = 1.42 \times 10^5$ Pa. This value is minuscule compared to the bulk modulus of a typical solid (see Table 11.1), which is approximately 10^{10} to 10^{11} Pa. This shouldn't be surprising: It's simply a statement that air is far easier to compress than steel.

The density ρ of a gas also depends on the pressure, which in turn depends on the temperature. It turns out that the ratio B/ρ for a given type of ideal gas does *not* depend on the pressure at all, only the temperature. From Eq. (16.7), this means that the speed of sound in a gas is fundamentally a function of temperature T:

$$v = \sqrt{\frac{\gamma RT}{M}} \qquad \text{(speed of sound in an ideal gas)} \tag{16.10}$$

This expression incorporates several quantities that you may recognize from your study of ideal gases in chemistry and that we will study in Chapters 17, 18, and 19. The temperature T is the *absolute* temperature in kelvins (K), equal to the Celsius temperature plus 273.15; thus 20.00°C corresponds to $T = 293.15$ K. The quantity M is the *molar mass,* or mass per mole of the substance of which the gas is composed. The *gas constant R* has the same value for all gases. The current best numerical value of R is

$$R = 8.314472(15) \text{ J/mol} \cdot \text{K}$$

which for practical calculations we can write as 8.314 J/mol·K.

For any particular gas, γ, R, and M are constants, and the wave speed is proportional to the square root of the absolute temperature. We will see in Chapter 18 that Eq. (16.10) is almost identical to the expression for the average speed of molecules in an ideal gas. This shows that sound speeds and molecular speeds are closely related.

Example 16.4   **Speed of sound in air**

Find the speed of sound in air at $T = 20$°C, and find the range of wavelengths in air to which the human ear (which can hear frequencies in the range of 20–20,000 Hz) is sensitive. The mean molar mass for air (a mixture of mostly nitrogen and oxygen) is $M = 28.8 \times 10^{-3}$ kg/mol and the ratio of heat capacities is $\gamma = 1.40$.

Continued

SOLUTION

IDENTIFY and SET UP: We use Eq. (16.10) to find the sound speed from γ, T, and M, and we use $v = f\lambda$ to find the wavelengths corresponding to the frequency limits. Note that in Eq. (16.10) temperature T *must* be expressed in kelvins, not Celsius degrees.

EXECUTE: At $T = 20°C = 293$ K, we find

$$v = \sqrt{\frac{\gamma RT}{M}} = \sqrt{\frac{(1.40)(8.314 \text{ J/mol} \cdot \text{K})(293 \text{ K})}{28.8 \times 10^{-3} \text{ kg/mol}}} = 344 \text{ m/s}$$

Using this value of v in $\lambda = v/f$, we find that at 20°C the frequency $f = 20$ Hz corresponds to $\lambda = 17$ m and $f = 20{,}000$ Hz to $\lambda = 1.7$ cm.

EVALUATE: Our calculated value of v agrees with the measured sound speed at $T = 20°C$ to within 0.3%.

In this discussion we have treated a gas as a continuous medium. A gas is actually composed of molecules in random motion, separated by distances that are large in comparison with their diameters. The vibrations that constitute a wave in a gas are superposed on the random thermal motion. At atmospheric pressure, a molecule travels an average distance of about 10^{-7} m between collisions, while the displacement amplitude of a faint sound may be only 10^{-9} m. We can think of a gas with a sound wave passing through as being comparable to a swarm of bees; the swarm as a whole oscillates slightly while individual insects move about through the swarm, apparently at random.

Test Your Understanding of Section 16.2   Mercury is 13.6 times denser than water. Based on Table 16.1, at 20°C which of these liquids has the greater bulk modulus? (i) mercury; (ii) water; (iii) both are about the same; (iv) not enough information is given to decide.

16.3 Sound Intensity

Traveling sound waves, like all other traveling waves, transfer energy from one region of space to another. We saw in Section 15.5 that a useful way to describe the energy carried by a sound wave is through the *wave intensity I*, equal to the time average rate at which energy is transported per unit area across a surface perpendicular to the direction of propagation. Let's see how to express the intensity of a sound wave in terms of the displacement amplitude A or pressure amplitude $p_{\max}$.

Intensity and Displacement Amplitude

For simplicity, let us consider a sound wave propagating in the $+x$-direction so that we can use our expressions from Section 16.1 for the displacement $y(x, t)$ and pressure fluctuation $p(x, t)$—Eqs. (16.1) and (16.4), respectively. In Section 6.4 we saw that power equals the product of force and velocity [see Eq. (6.18)]. So the power per unit area in this sound wave equals the product of $p(x, t)$ (force per unit area) and the *particle* velocity $v_y(x, t)$. The particle velocity $v_y(x, t)$ is the velocity at time t of that portion of the wave medium at coordinate x. Using Eqs. (16.1) and (16.4), we find

$$v_y(x, t) = \frac{\partial y(x, t)}{\partial t} = \omega A \sin(kx - \omega t)$$

$$p(x, t)v_y(x, t) = [BkA\sin(kx - \omega t)][\omega A \sin(kx - \omega t)]$$

$$= B\omega k A^2 \sin^2(kx - \omega t)$$

CAUTION **Wave velocity vs. particle velocity** Remember that the velocity of the wave as a whole is *not* the same as the particle velocity. While the wave continues to move in the direction of propagation, individual particles in the wave medium merely slosh back and forth, as shown in Fig. 16.1. Furthermore, the maximum speed of a particle of the medium can be very different from the wave speed. ▌

The intensity is, by definition, the time average value of $p(x, t)v_y(x, t)$. For any value of x the average value of the function $\sin^2(kx - \omega t)$ over one period $T = 2\pi/\omega$ is $\frac{1}{2}$, so

$$I = \tfrac{1}{2} B\omega k A^2 \qquad (16.11)$$

By using the relationships $\omega = vk$ and $v^2 = B/\rho$, we can transform Eq. (16.11) into the form

$$I = \tfrac{1}{2}\sqrt{\rho B}\,\omega^2 A^2 \qquad \text{(intensity of a sinusoidal sound wave)} \qquad (16.12)$$

This equation shows why in a stereo system, a low-frequency woofer has to vibrate with much larger amplitude than a high-frequency tweeter to produce the same sound intensity.

Intensity and Pressure Amplitude

It is usually more useful to express I in terms of the pressure amplitude p_{max}. Using Eq. (16.5) and the relationship $\omega = vk$, we find

$$I = \frac{\omega p_{max}^2}{2Bk} = \frac{v p_{max}^2}{2B} \qquad (16.13)$$

By using the wave speed relationship $v^2 = B/\rho$, we can also write Eq. (16.13) in the alternative forms

$$I = \frac{p_{max}^2}{2\rho v} = \frac{p_{max}^2}{2\sqrt{\rho B}} \qquad \text{(intensity of a sinusoidal sound wave)} \qquad (16.14)$$

You should verify these expressions. Comparison of Eqs. (16.12) and (16.14) shows that sinusoidal sound waves of the same intensity but different frequency have different displacement amplitudes A but the *same* pressure amplitude p_{max}. This is another reason it is usually more convenient to describe a sound wave in terms of pressure fluctuations, not displacement.

The *total* average power carried across a surface by a sound wave equals the product of the intensity at the surface and the surface area, if the intensity over the surface is uniform. The average total sound power emitted by a person speaking in an ordinary conversational tone is about 10^{-5} W, while a loud shout corresponds to about 3×10^{-2} W. If all the residents of New York City were to talk at the same time, the total sound power would be about 100 W, equivalent to the electric power requirement of a medium-sized light bulb. On the other hand, the power required to fill a large auditorium or stadium with loud sound is considerable (see Example 16.7.)

If the sound source emits waves in all directions equally, the intensity decreases with increasing distance r from the source according to the inverse-square law: The intensity is proportional to $1/r^2$. We discussed this law and its consequences in Section 15.5. If the sound goes predominantly in one direction, the inverse-square law does not apply and the intensity decreases with distance more slowly than $1/r^2$ (Fig. 16.10).

16.10 By cupping your hands like this, you direct the sound waves emerging from your mouth so that they don't propagate to the sides. Hence the intensity decreases with distance more slowly than the inverse-square law would predict, and you can be heard at greater distances.

The inverse-square relationship also does not apply indoors because sound energy can reach a listener by reflection from the walls and ceiling. Indeed, part of the architect's job in designing an auditorium is to tailor these reflections so that the intensity is as nearly uniform as possible over the entire auditorium.

Problem-Solving Strategy 16.1　Sound Intensity

IDENTIFY *the relevant concepts:* The relationships between the intensity and amplitude of a sound wave are straightforward. Other quantities are involved in these relationships, however, so it's particularly important to decide which is your target variable.

SET UP *the problem* using the following steps:
1. Sort the physical quantities into categories. Wave properties include the displacement and pressure amplitudes A and p_{max} and the frequency f, which can be determined from the angular frequency ω, the wave number k, or the wavelength λ. These quantities are related through the wave speed v, which is determined by properties of the medium (B and ρ for a liquid, and γ, T, and M for a gas).

2. List the given quantities and identify the target variables. Find relationships that take you where you want to go.

EXECUTE *the solution:* Use your selected equations to solve for the target variables. Express the temperature in kelvins (Celsius temperature plus 273.15) to calculate the speed of sound in a gas.

EVALUATE *your answer:* If possible, use an alternative relationship to check your results.

Example 16.5　Intensity of a sound wave in air

Find the intensity of the sound wave in Example 16.1, with $p_{max} = 3.0 \times 10^{-2}$ Pa. Assume the temperature is 20°C so that the density of air is $\rho = 1.20$ kg/m^3 and the speed of sound is $v = 344$ m/s.

SOLUTION

IDENTIFY and SET UP: Our target variable is the intensity I of the sound wave. We are given the pressure amplitude p_{max} of the wave as well as the density ρ and wave speed v for the medium. We can determine I from p_{max}, ρ, and v using Eq. (16.14).

EXECUTE: From Eq. (16.14),

$$I = \frac{p_{max}^2}{2\rho v} = \frac{(3.0 \times 10^{-2} \text{ Pa})^2}{2(1.20 \text{ kg/m}^3)(344 \text{ m/s})}$$

$$= 1.1 \times 10^{-6} \text{ J/(s} \cdot \text{m}^2) = 1.1 \times 10^{-6} \text{ W/m}^2$$

EVALUATE: This seems like a very low intensity, but it is well within the range of sound intensities encountered on a daily basis. A very loud sound wave at the threshold of pain has a pressure amplitude of about 30 Pa and an intensity of about 1 W/m^2. The pressure amplitude of the faintest sound wave that can be heard is about 3×10^{-5} Pa, and the corresponding intensity is about 10^{-12} W/m^2. (Try these values of p_{max} in Eq. (16.14) to check that the corresponding intensities are as we have stated.)

Example 16.6　Same intensity, different frequencies

What are the pressure and displacement amplitudes of a 20-Hz sound wave with the same intensity as the 1000-Hz sound wave of Examples 16.1 and 16.5?

SOLUTION

IDENTIFY and SET UP: In Examples 16.1 and 16.5 we found that for a 1000-Hz sound wave with $p_{max} = 3.0 \times 10^{-2}$ Pa, $A = 1.2 \times 10^{-8}$ m and $I = 1.1 \times 10^{-6}$ W/m^2. Our target variables are p_{max} and A for a 20-Hz sound wave of the same intensity I. We can find these using Eqs. (16.14) and (16.12), respectively.

EXECUTE: We can rearrange Eqs. (16.14) and (16.12) as $p_{max}^2 = 2I\sqrt{\rho B}$ and $\omega^2 A^2 = 2I/\sqrt{\rho B}$, respectively. These tell us that for a given sound intensity I in a given medium (constant ρ and B), the

quantities p_{max} and ωA (or, equivalently, fA) are both *constants* that don't depend on frequency. From the first result we immediately have $p_{max} = 3.0 \times 10^{-2}$ Pa for $f = 20$ Hz, the same as for $f = 1000$ Hz. If write the second result as $f_{20}A_{20} = f_{1000}A_{1000}$, we have

$$A_{20} = \left(\frac{f_{1000}}{f_{20}}\right)A_{1000}$$

$$= \left(\frac{1000 \text{ Hz}}{20 \text{ Hz}}\right)(1.2 \times 10^{-8} \text{ m}) = 6.0 \times 10^{-7} \text{ m} = 0.60 \text{ }\mu\text{m}$$

EVALUATE: Our result reinforces the idea that pressure amplitude is a more convenient description of a sound wave and its intensity than displacement amplitude.

Example 16.7   **"Play it loud!"**

For an outdoor concert we want the sound intensity to be 1 W/m^2 at a distance of 20 m from the speaker array. If the sound intensity is uniform in all directions, what is the required acoustic power output of the array?

SOLUTION

IDENTIFY, SET UP, and EXECUTE: This example uses the definition of sound intensity as power per unit area. The total power is the target variable; the area in question is a hemisphere centered on the speaker array. We assume that the speakers are on the ground and

that none of the acoustic power is directed into the ground, so the acoustic power is uniform over a hemisphere 20 m in radius. The surface area of this hemisphere is $(\frac{1}{2})(4\pi)(20 \text{ m})^2$, or about 2500 m^2. The required power is the product of this area and the intensity: $(1 \text{ W/m}^2)(2500 \text{ m}^2) = 2500 \text{ W} = 2.5 \text{ kW}$.

EVALUATE: The electrical power input to the speaker would need to be considerably greater than 2.5 kW, because speaker efficiency is not very high (typically a few percent for ordinary speakers, and up to 25% for horn-type speakers).

The Decibel Scale

Because the ear is sensitive over a broad range of intensities, a *logarithmic* intensity scale is usually used. The **sound intensity level** β of a sound wave is defined by the equation

$$\beta = (10 \text{ dB}) \log \frac{I}{I_0} \quad \text{(definition of sound intensity level)} \quad (16.15)$$

In this equation, I_0 is a reference intensity, chosen to be 10^{-12} W/m^2, approximately the threshold of human hearing at 1000 Hz. Recall that "log" means the logarithm to base 10. Sound intensity levels are expressed in **decibels,** abbreviated dB. A decibel is $\frac{1}{10}$ of a *bel,* a unit named for Alexander Graham Bell (the inventor of the telephone). The bel is inconveniently large for most purposes, and the decibel is the usual unit of sound intensity level.

If the intensity of a sound wave equals I_0 or 10^{-12} W/m^2, its sound intensity level is 0 dB. An intensity of 1 W/m^2 corresponds to 120 dB. Table 16.2 gives the sound intensity levels in decibels of some familiar sounds. You can use Eq. (16.15) to check the value of sound intensity level β given for each intensity in the table.

Because the ear is not equally sensitive to all frequencies in the audible range, some sound-level meters weight the various frequencies unequally. One such scheme leads to the so-called dBA scale; this scale deemphasizes the low and very high frequencies, where the ear is less sensitive than at midrange frequencies.

Table 16.2 Sound Intensity Levels from Various Sources (Representative Values)

Source or Description of Sound	Sound Intensity Level, β (dB)	Intensity, I (W/m^2)
Military jet aircraft 30 m away	140	10^2
Threshold of pain	120	1
Riveter	95	3.2×10^{-3}
Elevated train	90	10^{-3}
Busy street traffic	70	10^{-5}
Ordinary conversation	65	3.2×10^{-6}
Quiet automobile	50	10^{-7}
Quiet radio in home	40	10^{-8}
Average whisper	20	10^{-10}
Rustle of leaves	10	10^{-11}
Threshold of hearing at 1000 Hz	0	10^{-12}

Example 16.8 **Temporary—or permanent—hearing loss**

A 10-min exposure to 120-dB sound will temporarily shift your threshold of hearing at 1000 Hz from 0 dB up to 28 dB. Ten years of exposure to 92-dB sound will cause a *permanent* shift to 28 dB. What sound intensities correspond to 28 dB and 92 dB?

SOLUTION

IDENTIFY and SET UP: We are given two sound intensity levels β; our target variables are the corresponding intensities. We can solve Eq. (16.15) to find the intensity I that corresponds to each value of β.

EXECUTE: We solve Eq. (16.15) for I by dividing both sides by 10 dB and using the relationship $10^{\log x} = x$:

$$I = I_0 10^{\beta/(10 \text{ dB})}$$

For $\beta = 28$ dB and $\beta = 92$ dB, the exponents are $\beta/(10 \text{ dB}) = 2.8$ and 9.2, respectively, so that

$$I_{28 \text{ dB}} = (10^{-12} \text{ W/m}^2)10^{2.8} = 6.3 \times 10^{-10} \text{ W/m}^2$$
$$I_{92 \text{ dB}} = (10^{-12} \text{ W/m}^2)10^{9.2} = 1.6 \times 10^{-3} \text{ W/m}^2$$

EVALUATE: If your answers are a factor of 10 too large, you may have entered 10×10^{-12} in your calculator instead of 1×10^{-12}. Be careful!

Example 16.9 **A bird sings in a meadow**

Consider an idealized bird (treated as a point source) that emits constant sound power, with intensity obeying the inverse-square law (Fig. 16.11). If you move twice the distance from the bird, by how many decibels does the sound intensity level drop?

SOLUTION

IDENTIFY and SET UP: The decibel scale is logarithmic, so the *difference* between two sound intensity levels (the target variable) corresponds to the *ratio* of the corresponding intensities, which is determined by the inverse-square law. We label the two points P_1 and P_2 (Fig. 16.11). We use Eq. (16.15), the definition of sound intensity level, at each point. We use Eq. (15.26), the inverse-square law, to relate the intensities at the two points.

EXECUTE: The difference $\beta_2 - \beta_1$ between any two sound intensity levels is related to the corresponding intensities by

$$\beta_2 - \beta_1 = (10 \text{ dB})\left(\log\frac{I_2}{I_0} - \log\frac{I_1}{I_0}\right)$$
$$= (10 \text{ dB})[(\log I_2 - \log I_0) - (\log I_1 - \log I_0)]$$
$$= (10 \text{ dB})\log\frac{I_2}{I_1}$$

For this inverse-square-law source, Eq. (15.26) yields $I_2/I_1 = r_1^2/r_2^2 = \frac{1}{4}$, so

$$\beta_2 - \beta_1 = (10 \text{ dB})\log\frac{I_1}{I_2} = (10 \text{ dB})\log\frac{1}{4} = -6.0 \text{ dB}$$

16.11 When you double your distance from a point source of sound, by how much does the sound intensity level decrease?

Point source

P_2 P_1

EVALUATE: Our result is negative, which tells us (correctly) that the sound intensity level is less at P_2 than at P_1. The 6-dB difference doesn't depend on the sound intensity level at P_1; *any* doubling of the distance from an inverse-square-law source reduces the sound intensity level by 6 dB.

Note that the perceived *loudness* of a sound is not directly proportional to its intensity. For example, most people interpret an increase of 8 dB to 10 dB in sound intensity level (corresponding to increasing intensity by a factor of 6 to 10) as a doubling of loudness.

Test Your Understanding of Section 16.3 You double the intensity of a sound wave in air while leaving the frequency unchanged. (The pressure, density, and temperature of the air remain unchanged as well.) What effect does this have on the displacement amplitude, pressure amplitude, bulk modulus, sound speed, and sound intensity level?

16.4 Standing Sound Waves and Normal Modes

When longitudinal (sound) waves propagate in a fluid in a pipe with finite length, the waves are reflected from the ends in the same way that transverse waves on a string are reflected at its ends. The superposition of the waves traveling in opposite directions again forms a standing wave. Just as for transverse standing waves on a string (see Section 15.7), standing sound waves (normal modes) in a pipe can

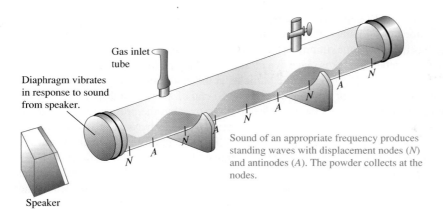

Gas inlet tube

Diaphragm vibrates in response to sound from speaker.

Speaker

Sound of an appropriate frequency produces standing waves with displacement nodes (N) and antinodes (A). The powder collects at the nodes.

16.12 Demonstrating standing sound waves using a Kundt's tube. The blue shading represents the density of the gas at an instant when the gas pressure at the displacement nodes is a maximum or a minimum.

be used to create sound waves in the surrounding air. This is the operating principle of the human voice as well as many musical instruments, including woodwinds, brasses, and pipe organs.

Transverse waves on a string, including standing waves, are usually described only in terms of the displacement of the string. But, as we have seen, sound waves in a fluid may be described either in terms of the displacement of the fluid or in terms of the pressure variation in the fluid. To avoid confusion, we'll use the terms **displacement node** and **displacement antinode** to refer to points where particles of the fluid have zero displacement and maximum displacement, respectively.

We can demonstrate standing sound waves in a column of gas using an apparatus called a Kundt's tube (Fig. 16.12). A horizontal glass tube a meter or so long is closed at one end and has a flexible diaphragm at the other end that can transmit vibrations. A nearby loudspeaker is driven by an audio oscillator and amplifier; this produces sound waves that force the diaphragm to vibrate sinusoidally with a frequency that we can vary. The sound waves within the tube are reflected at the other, closed end of the tube. We spread a small amount of light powder uniformly along the bottom of the tube. As we vary the frequency of the sound, we pass through frequencies at which the amplitude of the standing waves becomes large enough for the powder to be swept along the tube at those points where the gas is in motion. The powder therefore collects at the displacement nodes (where the gas is not moving). Adjacent nodes are separated by a distance equal to $\lambda/2$, and we can measure this distance. Given the wavelength, we can use this experiment to determine the wave speed: We read the frequency f from the oscillator dial, and we can then calculate the speed v of the waves from the relationship $v = \lambda f$.

Figure 16.13 shows the motions of nine different particles within a gas-filled tube in which there is a standing sound wave. A particle at a displacement node (N) does not move, while a particle at a displacement antinode (A) oscillates with maximum amplitude. Note that particles on opposite sides of a displacement node vibrate in opposite phase. When these particles approach each other, the gas between them is compressed and the pressure rises; when they recede from each other, there is an expansion and the pressure drops. Hence at a displacement *node* the gas undergoes the maximum amount of compression and expansion, and the variations in pressure and density above and below the average have their maximum value. By contrast, particles on opposite sides of a displacement *antinode* vibrate *in phase;* the distance between the particles is nearly constant, and there is *no* variation in pressure or density at a displacement antinode.

We use the term **pressure node** to describe a point in a standing sound wave at which the pressure and density do not vary and the term **pressure antinode** to describe a point at which the variations in pressure and density are greatest. Using these terms, we can summarize our observations about standing sound waves as follows:

A pressure node is always a displacement antinode, and a pressure antinode is always a displacement node.

16.13 In a standing sound wave, a displacement node N is a pressure antinode (a point where the pressure fluctuates the most) and a displacement antinode A is a pressure node (a point where the pressure does not fluctuate at all).

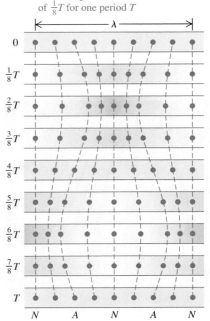

A standing wave shown at intervals of $\frac{1}{8}T$ for one period T

λ

0

$\frac{1}{8}T$

$\frac{2}{8}T$

$\frac{3}{8}T$

$\frac{4}{8}T$

$\frac{5}{8}T$

$\frac{6}{8}T$

$\frac{7}{8}T$

T

N A N A N

N = a displacement node = a pressure antinode
A = a displacement antinode = a pressure node

Figure 16.12 depicts a standing sound wave at an instant at which the pressure variations are greatest; the blue shading shows that the density and pressure of the gas have their maximum and minimum values at the displacement nodes.

When reflection takes place at a *closed* end of a pipe (an end with a rigid barrier or plug), the displacement of the particles at this end must always be zero, analogous to a fixed end of a string. Thus a closed end of a pipe is a displacement node and a pressure antinode; the particles do not move, but the pressure variations are maximum. An *open* end of a pipe is a pressure node because it is open to the atmosphere, where the pressure is constant. Because of this, an open end is always a displacement *antinode,* in analogy to a free end of a string; the particles oscillate with maximum amplitude, but the pressure does not vary. (Strictly speaking, the pressure node actually occurs somewhat beyond an open end of a pipe. But if the diameter of the pipe is small in comparison to the wavelength, which is true for most musical instruments, this effect can safely be neglected.) Thus longitudinal waves in a column of fluid are reflected at the closed and open ends of a pipe in the same way that transverse waves in a string are reflected at fixed and free ends, respectively.

Conceptual Example 16.10	The sound of silence

A directional loudspeaker directs a sound wave of wavelength λ at a wall (Fig. 16.14). At what distances from the wall could you stand and hear no sound at all?

SOLUTION

Your ear detects pressure variations in the air; you will therefore hear no sound if your ear is at a *pressure node,* which is a displacement antinode. The wall is at a displacement node; the distance from any node to an adjacent antinode is $\lambda/4$, and the distance from one antinode to the next is $\lambda/2$ (Fig. 16.14). Hence the displacement antinodes (pressure nodes), at which no sound will be heard, are at distances $d = \lambda/4$, $d = \lambda/4 + \lambda/2 = 3\lambda/4$, $d = 3\lambda/4 + \lambda/2 = 5\lambda/4$, . . . from the wall. If the loudspeaker is not highly directional, this effect is hard to notice because of reflections of sound waves from the floor, ceiling, and other walls.

16.14 When a sound wave is directed at a wall, it interferes with the reflected wave to create a standing wave. The N's and A's are *displacement* nodes and antinodes.

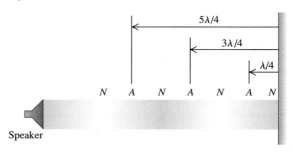

Organ Pipes and Wind Instruments

The most important application of standing sound waves is the production of musical tones by wind instruments. Organ pipes are one of the simplest examples (Fig. 16.15). Air is supplied by a blower, at a gauge pressure typically of the order of 10^3 Pa (10^{-2} atm), to the bottom end of the pipe (Fig. 16.16). A stream of air emerges from the narrow opening at the edge of the horizontal surface and is directed against the top edge of the opening, which is called the *mouth* of the pipe. The column of air in the pipe is set into vibration, and there is a series of possible normal modes, just as with the stretched string. The mouth always acts as an open end; thus it is a pressure node and a displacement antinode. The other end of the pipe (at the top in Fig. 16.16) may be either open or closed.

In Fig. 16.17, both ends of the pipe are open, so both ends are pressure nodes and displacement antinodes. An organ pipe that is open at both ends is called an *open pipe.* The fundamental frequency f_1 corresponds to a standing-wave pattern with a displacement antinode at each end and a displacement node in the middle (Fig. 16.17a). The distance between adjacent antinodes is always equal to one

16.15 Organ pipes of different sizes produce tones with different frequencies.

half-wavelength, and in this case that is equal to the length L of the pipe; $\lambda/2 = L$. The corresponding frequency, obtained from the relationship $f = v/\lambda$, is

$$f_1 = \frac{v}{2L} \quad \text{(open pipe)} \tag{16.16}$$

Figures 16.17b and 16.17c show the second and third harmonics (first and second overtones); their vibration patterns have two and three displacement nodes, respectively. For these, a half-wavelength is equal to $L/2$ and $L/3$, respectively, and the frequencies are twice and three times the fundamental, respectively. That is, $f_2 = 2f_1$ and $f_3 = 3f_1$. For *every* normal mode of an open pipe the length L must be an integer number of half-wavelengths, and the possible wavelengths λ_n are given by

$$L = n\frac{\lambda_n}{2} \quad \text{or} \quad \lambda_n = \frac{2L}{n} \quad (n = 1, 2, 3, \dots) \quad \text{(open pipe)} \tag{16.17}$$

The corresponding frequencies f_n are given by $f_n = v/\lambda_n$, so all the normal-mode frequencies for a pipe that is open at both ends are given by

$$f_n = \frac{nv}{2L} \quad (n = 1, 2, 3, \dots) \quad \text{(open pipe)} \tag{16.18}$$

The value $n = 1$ corresponds to the fundamental frequency, $n = 2$ to the second harmonic (or first overtone), and so on. Alternatively, we can say

$$f_n = nf_1 \quad (n = 1, 2, 3, \dots) \quad \text{(open pipe)} \tag{16.19}$$

with f_1 given by Eq. (16.16).

Figure 16.18 shows a pipe that is open at the left end but closed at the right end. This is called a *stopped pipe*. The left (open) end is a displacement antinode (pressure node), but the right (closed) end is a displacement node (pressure antinode). The distance between a node and the adjacent antinode is always one quarter-wavelength. Figure 16.18a shows the lowest-frequency mode; the length

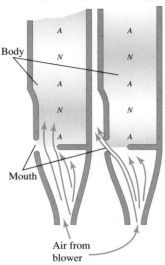

16.16 Cross sections of an organ pipe at two instants one half-period apart. The N's and A's are *displacement* nodes and antinodes; as the blue shading shows, these are points of maximum pressure variation and zero pressure variation, respectively.

16.17 A cross section of an open pipe showing the first three normal modes. The shading indicates the pressure variations. The red curves are graphs of the displacement along the pipe axis at two instants separated in time by one half-period. The N's and A's are the *displacement* nodes and antinodes; interchange these to show the *pressure* nodes and antinodes.

(a) Fundamental: $f_1 = \dfrac{v}{2L}$

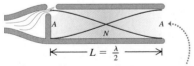

Open end is always a displacement antinode.

(b) Second harmonic: $f_2 = 2\dfrac{v}{2L} = 2f_1$

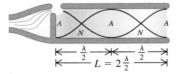

(c) Third harmonic: $f_3 = 3\dfrac{v}{2L} = 3f_1$

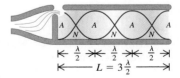

16.18 A cross section of a stopped pipe showing the first three normal modes as well as the *displacement* nodes and antinodes. Only odd harmonics are possible.

(a) Fundamental: $f_1 = \dfrac{v}{4L}$

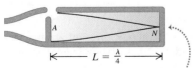

Closed end is always a displacement node.

(b) Third harmonic: $f_3 = 3\dfrac{v}{4L} = 3f_1$

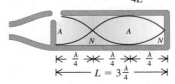

(c) Fifth harmonic: $f_5 = 5\dfrac{v}{4L} = 5f_1$

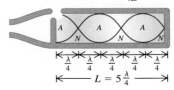

of the pipe is a quarter-wavelength $(L = \lambda_1/4)$. The fundamental frequency is $f_1 = v/\lambda_1$, or

$$f_1 = \frac{v}{4L} \quad \text{(stopped pipe)} \tag{16.20}$$

This is one-half the fundamental frequency for an *open* pipe of the same length. In musical language, the *pitch* of a closed pipe is one octave lower (a factor of 2 in frequency) than that of an open pipe of the same length. Figure 16.18b shows the next mode, for which the length of the pipe is *three-quarters* of a wavelength, corresponding to a frequency $3f_1$. For Fig. 16.18c, $L = 5\lambda/4$ and the frequency is $5f_1$. The possible wavelengths are given by

$$L = n\frac{\lambda_n}{4} \quad \text{or} \quad \lambda_n = \frac{4L}{n} \quad (n = 1, 3, 5, \ldots) \quad \text{(stopped pipe)} \tag{16.21}$$

The normal-mode frequencies are given by $f_n = v/\lambda_n$, or

$$f_n = \frac{nv}{4L} \quad (n = 1, 3, 5, \ldots) \quad \text{(stopped pipe)} \tag{16.22}$$

or

$$f_n = nf_1 \quad (n = 1, 3, 5, \ldots) \quad \text{(stopped pipe)} \tag{16.23}$$

with f_1 given by Eq. (16.20). We see that the second, fourth, and all *even* harmonics are missing. In a pipe that is closed at one end, the fundamental frequency is $f_1 = v/4L$, and only the odd harmonics in the series $(3f_1, 5f_1, \ldots)$ are possible.

A final possibility is a pipe that is closed at *both* ends, with displacement nodes and pressure antinodes at both ends. This wouldn't be of much use as a musical instrument because there would be no way for the vibrations to get out of the pipe.

Example 16.11 A tale of two pipes

On a day when the speed of sound is 345 m/s, the fundamental frequency of a particular stopped organ pipe is 220 Hz. (a) How long is this pipe? (b) The second *overtone* of this pipe has the same wavelength as the third *harmonic* of an *open* pipe. How long is the open pipe?

SOLUTION

IDENTIFY and SET UP: This problem uses the relationship between the length and normal-mode frequencies of open pipes (Fig. 16.17) and stopped pipes (Fig. 16.18). In part (a), we determine the length of the stopped pipe from Eq. (16.22). In part (b), we must determine the length of an open pipe, for which Eq. (16.18) gives the frequencies.

EXECUTE: (a) For a stopped pipe $f_1 = v/4L$, so

$$L_{\text{stopped}} = \frac{v}{4f_1} = \frac{345 \text{ m/s}}{4(220 \text{ s}^{-1})} = 0.392 \text{ m}$$

(b) The frequency of the second overtone of a stopped pipe (the *third* possible frequency) is $f_5 = 5f_1 = 5(220 \text{ Hz}) = 1100 \text{ Hz}$. If the wavelengths for the two pipes are the same, the frequencies are also the same. Hence the frequency of the third harmonic of the open pipe, which is at $3f_1 = 3(v/2L)$, equals 1100 Hz. Then

$$1100 \text{ Hz} = 3\left(\frac{345 \text{ m/s}}{2L_{\text{open}}}\right) \quad \text{and} \quad L_{\text{open}} = 0.470 \text{ m}$$

EVALUATE: The 0.392-m stopped pipe has a fundamental frequency of 220 Hz; the *longer* (0.470-m) open pipe has a *higher* fundamental frequency, $(1100 \text{ Hz})/3 = 367 \text{ Hz}$. This is not a contradiction, as you can see if you compare Figs. 16.17a and 16.18a.

In an organ pipe in actual use, several modes are always present at once; the motion of the air is a superposition of these modes. This situation is analogous to a string that is struck or plucked, as in Fig. 15.28. Just as for a vibrating string, a complex standing wave in the pipe produces a traveling sound wave in the surrounding air with a harmonic content similar to that of the standing wave. A very narrow pipe produces a sound wave rich in higher harmonics, which we hear as a thin and "stringy" tone; a fatter pipe produces mostly the fundamental mode, heard as a softer, more flutelike tone. The harmonic content also depends on the shape of the pipe's mouth.

We have talked about organ pipes, but this discussion is also applicable to other wind instruments. The flute and the recorder are directly analogous. The most significant difference is that those instruments have holes along the pipe. Opening and closing the holes with the fingers changes the effective length L of the air column and thus changes the pitch. Any individual organ pipe, by comparison, can play only a single note. The flute and recorder behave as *open* pipes, while the clarinet acts as a *stopped* pipe (closed at the reed end, open at the bell).

Equations (16.18) and (16.22) show that the frequencies of any wind instrument are proportional to the speed of sound v in the air column inside the instrument. As Eq. (16.10) shows, v depends on temperature; it increases when temperature increases. Thus the pitch of all wind instruments rises with increasing temperature. An organ that has some of its pipes at one temperature and others at a different temperature is bound to sound out of tune.

Test Your Understanding of Section 16.4 If you connect a hose to one end of a metal pipe and blow compressed air into it, the pipe produces a musical tone. If instead you blow compressed helium into the pipe at the same pressure and temperature, will the pipe produce (i) the same tone, (ii) a higher-pitch tone, or (iii) a lower-pitch tone?

16.5 Resonance and Sound

Many mechanical systems have normal modes of oscillation. As we have seen, these include columns of air (as in an organ pipe) and stretched strings (as in a guitar; see Section 15.8). In each mode, every particle of the system oscillates with simple harmonic motion at the same frequency as the mode. Air columns and stretched strings have an infinite series of normal modes, but the basic concept is closely related to the simple harmonic oscillator, discussed in Chapter 14, which has only a single normal mode (that is, only one frequency at which it oscillates after being disturbed).

Suppose we apply a periodically varying force to a system that can oscillate. The system is then forced to oscillate with a frequency equal to the frequency of the applied force (called the *driving frequency*). This motion is called a *forced oscillation*. We talked about forced oscillations of the harmonic oscillator in Section 14.8, and we suggest that you review that discussion. In particular, we described the phenomenon of mechanical **resonance**. A simple example of resonance is pushing Cousin Throckmorton on a swing. The swing is a pendulum; it has only a single normal mode, with a frequency determined by its length. If we push the swing periodically with this frequency, we can build up the amplitude of the motion. But if we push with a very different frequency, the swing hardly moves at all.

Resonance also occurs when a periodically varying force is applied to a system with many normal modes. An example is shown in Fig. 16.19a. An open organ pipe is placed next to a loudspeaker that is driven by an amplifier and emits pure sinusoidal sound waves of frequency f, which can be varied by adjusting the amplifier. The air in the pipe is forced to vibrate with the same frequency f as the *driving force* provided by the loudspeaker. In general the amplitude of this motion is relatively small, and the air inside the pipe will not move in any of the normal-mode patterns shown in Fig. 16.17. But if the frequency f of the force is close to one of the normal-mode frequencies, the air in the pipe moves in the normal-mode pattern for that frequency, and the amplitude can become quite large. Figure 16.19b shows the amplitude of oscillation of the air

16.19 (a) The air in an open pipe is forced to oscillate at the same frequency as the sinusoidal sound waves coming from the loudspeaker. (b) The resonance curve of the open pipe graphs the amplitude of the standing sound wave in the pipe as a function of the driving frequency.

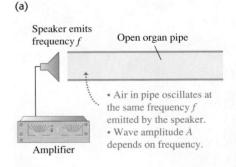

(a)

Speaker emits frequency f Open organ pipe

• Air in pipe oscillates at the same frequency f emitted by the speaker.
• Wave amplitude A depends on frequency.

Amplifier

(b) Resonance curve: graph of amplitude A versus driving frequency f. Peaks occur at normal-mode frequencies of the pipe: f_1, $f_2 = 2f_1$, $f_3 = 3f_1$, ...

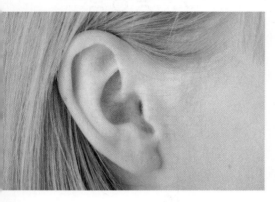

16.20 The frequency of the sound from this trumpet exactly matches one of the normal-mode frequencies of the goblet. The resonant vibrations of the goblet have such large amplitude that the goblet tears itself apart.

in the pipe as a function of the driving frequency f. The shape of this graph is called the **resonance curve** of the pipe; it has peaks where f equals the normal-mode frequencies of the pipe. The detailed shape of the resonance curve depends on the geometry of the pipe.

If the frequency of the force is precisely *equal* to a normal-mode frequency, the system is in resonance, and the amplitude of the forced oscillation is maximum. If there were no friction or other energy-dissipating mechanism, a driving force at a normal-mode frequency would continue to add energy to the system, and the amplitude would increase indefinitely. In such an idealized case the peaks in the resonance curve of Fig. 16.19b would be infinitely high. But in any real system there is always some dissipation of energy, or damping, as we discussed in Section 14.8; the amplitude of oscillation in resonance may be large, but it cannot be infinite.

The "sound of the ocean" you hear when you put your ear next to a large seashell is due to resonance. The noise of the outside air moving past the seashell is a mixture of sound waves of almost all audible frequencies, which forces the air inside the seashell to oscillate. The seashell behaves like an organ pipe, with a set of normal-mode frequencies; hence the inside air oscillates most strongly at those frequencies, producing the seashell's characteristic sound. To hear a similar phenomenon, uncap a full bottle of your favorite beverage and blow across the open top. The noise is provided by your breath blowing across the top, and the "organ pipe" is the column of air inside the bottle above the surface of the liquid. If you take a drink and repeat the experiment, you will hear a lower tone because the "pipe" is longer and the normal-mode frequencies are lower.

Resonance also occurs when a stretched string is forced to oscillate (see Section 15.8). Suppose that one end of a stretched string is held fixed while the other is given a transverse sinusoidal motion with small amplitude, setting up standing waves. If the frequency of the driving mechanism is *not* equal to one of the normal-mode frequencies of the string, the amplitude at the antinodes is fairly small. However, if the frequency is equal to any one of the normal-mode frequencies, the string is in resonance, and the amplitude at the antinodes is very much larger than that at the driven end. The driven end is not precisely a node, but it lies much closer to a node than to an antinode when the string is in resonance. The photographs in Fig. 15.23 were made this way, with the left end of the string fixed and the right end oscillating vertically with small amplitude; large-amplitude standing waves resulted when the frequency of oscillation of the right end was equal to the fundamental frequency or to one of the first three overtones.

It is easy to demonstrate resonance with a piano. Push down the damper pedal (the right-hand pedal) so that the dampers are lifted and the strings are free to vibrate, and then sing a steady tone into the piano. When you stop singing, the piano seems to continue to sing the same note. The sound waves from your voice excite vibrations in the strings that have natural frequencies close to the frequencies (fundamental and harmonics) present in the note you sang.

A more spectacular example is a singer breaking a wine glass with her amplified voice. A good-quality wine glass has normal-mode frequencies that you can hear by tapping it. If the singer emits a loud note with a frequency corresponding exactly to one of these normal-mode frequencies, large-amplitude oscillations can build up and break the glass (Fig. 16.20).

Example 16.12   **An organ–guitar duet**

A stopped organ pipe is sounded near a guitar, causing one of the strings to vibrate with large amplitude. We vary the string tension until we find the maximum amplitude. The string is 80% as long as the pipe. If both pipe and string vibrate at their fundamental frequency, calculate the ratio of the wave speed on the string to the speed of sound in air.

SOLUTION

IDENTIFY and SET UP: The large response of the string is an example of resonance. It occurs because the organ pipe and the guitar string have the same fundamental frequency. Letting the subscripts a and s stand for the air in the pipe and the string, respectively, the condition for resonance is $f_{1a} = f_{1s}$. Equation (16.20) gives the fundamental frequency for a stopped pipe, and Eq. (15.32) gives the fundamental frequency for a guitar string held at both ends. These expressions involve the wave speed in air (v_a) and on the string (v_s) and the lengths of the pipe and string. We are given that $L_s = 0.80L_a$; our target variable is the ratio v_s/v_a.

EXECUTE: From Eqs. (16.20) and (15.32), $f_{1a} = v_a/4L_a$ and $f_{1s} = v_s/2L_s$. These frequencies are equal, so

$$\frac{v_a}{4L_a} = \frac{v_s}{2L_s}$$

Substituting $L_s = 0.80L_a$ and rearranging, we get $v_s/v_a = 0.40$.

EVALUATE: As an example, if the speed of sound in air is 345 m/s, the wave speed on the string is $(0.40)(345 \text{ m/s}) = 138$ m/s. Note that while the standing waves in the pipe and on the string have the same frequency, they have different *wavelengths* $\lambda = v/f$ because the two media have different wave speeds v. Which standing wave has the greater wavelength?

Test Your Understanding of Section 16.5 A stopped organ pipe of length L has a fundamental frequency of 220 Hz. For which of the following organ pipes will there be a resonance if a tuning fork of frequency 660 Hz is sounded next to the pipe? (There may be more than one correct answer.) (i) a stopped organ pipe of length L; (ii) a stopped organ pipe of length $2L$; (iii) an open organ pipe of length L; (iii) an open organ pipe of length $2L$.

16.6 Interference of Waves

Wave phenomena that occur when two or more waves overlap in the same region of space are grouped under the heading *interference*. As we have seen, standing waves are a simple example of an interference effect: Two waves traveling in opposite directions in a medium combine to produce a standing wave pattern with nodes and antinodes that do not move.

Figure 16.21 shows an example of another type of interference that involves waves that spread out in space. Two speakers, driven in phase by the same amplifier, emit identical sinusoidal sound waves with the same constant frequency. We place a microphone at point P in the figure, equidistant from the speakers. Wave crests emitted from the two speakers at the same time travel equal distances and arrive at point P at the same time; hence the waves arrive in phase, and there is constructive interference. The total wave amplitude at P is twice the amplitude from each individual wave, and we can measure this combined amplitude with the microphone.

Now let's move the microphone to point Q, where the distances from the two speakers to the microphone differ by a half-wavelength. Then the two waves arrive a half-cycle out of step, or *out of phase;* a positive crest from one speaker arrives at the same time as a negative crest from the other. Destructive interference takes place, and the amplitude measured by the microphone is much *smaller* than when only one speaker is present. If the amplitudes from the two speakers are equal, the two waves cancel each other out completely at point Q, and the total amplitude there is zero.

> **CAUTION** **Interference and traveling waves** Although this situation bears some resemblance to standing waves in a pipe, the total wave in Fig. 16.21 is a *traveling* wave, not a standing wave. To see why, recall that in a standing wave there is no net flow of energy in any direction. By contrast, in Fig. 16.21 there is an overall flow of energy from the speakers into the surrounding air; this is characteristic of a traveling wave. The interference between the waves from the two speakers simply causes the energy flow to be *channeled* into certain directions (for example, toward P) and away from other directions (for example, away from Q). You can see another difference between Fig. 16.21 and a standing wave by considering a point, such as Q, where destructive interference occurs. Such a point is *both* a displacement node *and* a pressure node because there is no wave at all at this point. Compare this to a standing wave, in which a pressure node is a displacement antinode, and vice versa.

16.21 Two speakers driven by the same amplifier. Constructive interference occurs at point P, and destructive interference occurs at point Q.

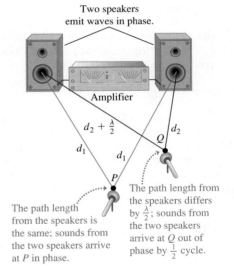

Two speakers emit waves in phase.

Amplifier

$d_2 + \frac{\lambda}{2}$ d_2

d_1 d_1 Q

P

The path length from the speakers is the same; sounds from the two speakers arrive at P in phase.

The path length from the speakers differs by $\frac{\lambda}{2}$; sounds from the two speakers arrive at Q out of phase by $\frac{1}{2}$ cycle.

Constructive interference occurs wherever the distances traveled by the two waves differ by a whole number of wavelengths, 0, λ, 2λ, 3λ, ...; in all these cases the waves arrive at the microphone in phase (Fig. 16.22a). If the distances from the two speakers to the microphone differ by any half-integer number of wavelengths, $\lambda/2$, $3\lambda/2$, $5\lambda/2$, ..., the waves arrive at the microphone out of phase and there will be destructive interference (Fig. 16.22b). In this case, little or no sound energy flows toward the microphone directly in front of the speakers. The energy is instead directed to the sides, where constructive interference occurs.

16.22 Two speakers driven by the same amplifier, emitting waves in phase. Only the waves directed toward the microphone are shown, and they are separated for clarity. (a) Constructive interference occurs when the path difference is 0, λ, 2λ, 3λ, ... (b) Destructive interference occurs when the path difference is $\lambda/2$, $3\lambda/2$, $5\lambda/2$, ...

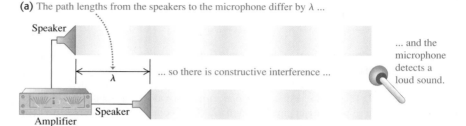

(a) The path lengths from the speakers to the microphone differ by λ ...

... so there is constructive interference ...

... and the microphone detects a loud sound.

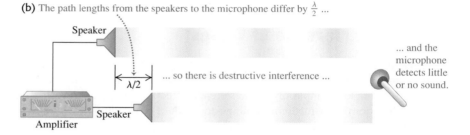

(b) The path lengths from the speakers to the microphone differ by $\frac{\lambda}{2}$ ...

... so there is destructive interference ...

... and the microphone detects little or no sound.

Example 16.13 Loudspeaker interference

Two small loudspeakers, A and B (Fig. 16.23), are driven by the same amplifier and emit pure sinusoidal waves in phase. (a) For what frequencies does constructive interference occur at point P? (b) For what frequencies does destructive interference occur? The speed of sound is 350 m/s.

SOLUTION

IDENTIFY and SET UP: The nature of the interference at P depends on the difference d in path lengths from point A to P and from point B to P. We calculate the path lengths using the Pythagorean theorem. Constructive interference occurs when d equals a whole number of wavelengths, while destructive interference occurs

16.23 What sort of interference occurs at P?

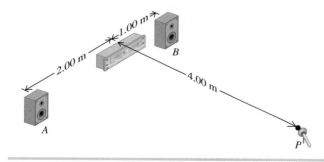

when d is a half-integer number of wavelengths. To find the corresponding frequencies, we use $v = f\lambda$.

EXECUTE: The distance from A to P is $[(2.00 \text{ m})^2 + (4.00 \text{ m})^2]^{1/2} = 4.47$ m, and the distance from B to P is $[(1.00 \text{ m})^2 + (4.00 \text{ m})^2]^{1/2} = 4.12$ m. The path difference is $d = 4.47 \text{ m} - 4.12 \text{ m} = 0.35$ m.

(a) Constructive interference occurs when $d = 0$, λ, 2λ, ... or $d = 0$, v/f, $2v/f$, ... $= nv/f$. So the possible frequencies are

$$f_n = \frac{nv}{d} = n\frac{350 \text{ m/s}}{0.35 \text{ m}} \qquad (n = 1, 2, 3, ...)$$

$$= 1000 \text{ Hz}, 2000 \text{ Hz}, 3000 \text{ Hz}, ...$$

(b) Destructive interference occurs when $d = \lambda/2$, $3\lambda/2$, $5\lambda/2$, ... or $d = v/2f$, $3v/2f$, $5v/2f$, ... The possible frequencies are

$$f_n = \frac{nv}{2d} = n\frac{350 \text{ m/s}}{2(0.35 \text{ m})} \qquad (n = 1, 3, 5, ...)$$

$$= 500 \text{ Hz}, 1500 \text{ Hz}, 2500 \text{ Hz}, ...$$

EVALUATE: As we increase the frequency, the sound at point P alternates between large and small (near zero) amplitudes, with maxima and minima at the frequencies given above. This effect may not be strong in an ordinary room because of reflections from the walls, floor, and ceiling. It is stronger outdoors and best in an anechoic chamber, which has walls that absorb almost all sound and thereby eliminate reflections.

Interference effects are used to control noise from very loud sound sources such as gas-turbine power plants or jet engine test cells. The idea is to use additional sound sources that in some regions of space interfere destructively with the unwanted sound and cancel it out. Microphones in the controlled area feed signals back to the sound sources, which are continuously adjusted for optimum cancellation of noise in the controlled area.

Test Your Understanding of Section 16.6 Suppose that speaker A in Fig. 16.23 emits a sinusoidal sound wave of frequency 500 Hz and speaker B emits a sinusoidal sound wave of frequency 1000 Hz. What sort of interference will there be between these two waves? (i) constructive interference at various points, including point P, and destructive interference at various other points; (ii) destructive interference at various points, including point P, and constructive interference at various points; (iii) neither (i) nor (ii).

16.7 Beats

In Section 16.6 we talked about *interference* effects that occur when two different waves with the same frequency overlap in the same region of space. Now let's look at what happens when we have two waves with equal amplitude but slightly different frequencies. This occurs, for example, when two tuning forks with slightly different frequencies are sounded together, or when two organ pipes that are supposed to have exactly the same frequency are slightly "out of tune."

Consider a particular point in space where the two waves overlap. The displacements of the individual waves at this point are plotted as functions of time in Fig. 16.24a. The total length of the time axis represents 1 second, and the frequencies are 16 Hz (blue graph) and 18 Hz (red graph). Applying the principle of superposition, we add the two displacements at each instant of time to find the total displacement at that time. The result is the graph of Fig. 16.24b. At certain times the two waves are in phase; their maxima coincide and their amplitudes add. But because of their slightly different frequencies, the two waves cannot be in phase at all times. Indeed, at certain times (like $t = 0.50$ s in Fig. 16.24) the two waves are exactly *out* of phase. The two waves then cancel each other, and the total amplitude is zero.

The resultant wave in Fig. 16.24b looks like a single sinusoidal wave with a varying amplitude that goes from a maximum to zero and back. In this example the amplitude goes through two maxima and two minima in 1 second, so the frequency of this amplitude variation is 2 Hz. The amplitude variation causes variations of loudness called **beats,** and the frequency with which the loudness varies is called the **beat frequency.** In this example the beat frequency is the *difference*

MasteringPHYSICS

ActivPhysics 10.7: Beats and Beat Frequency

16.24 Beats are fluctuations in amplitude produced by two sound waves of slightly different frequency, here 16 Hz and 18 Hz. (a) Individual waves. (b) Resultant wave formed by superposition of the two waves. The beat frequency is 18 Hz − 16 Hz = 2 Hz.

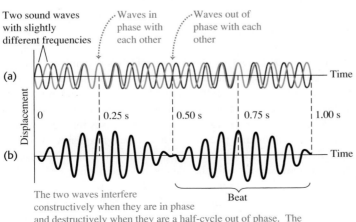

Two sound waves with slightly different frequencies

Waves in phase with each other

Waves out of phase with each other

(a)

Displacement

0   0.25 s   0.50 s   0.75 s   1.00 s

(b)   Time

The two waves interfere constructively when they are in phase and destructively when they are a half-cycle out of phase. The resultant wave rises and falls in intensity, forming beats.

Beat

of the two frequencies. If the beat frequency is a few hertz, we hear it as a waver or pulsation in the tone.

We can prove that the beat frequency is *always* the difference of the two frequencies f_a and f_b. Suppose f_a is larger than f_b; the corresponding periods are T_a and T_b, with $T_a < T_b$. If the two waves start out in phase at time $t = 0$, they are again in phase when the first wave has gone through exactly one more cycle than the second. This happens at a value of t equal to T_{beat}, the *period* of the beat. Let n be the number of cycles of the first wave in time T_{beat}; then the number of cycles of the second wave in the same time is $(n - 1)$, and we have the relationships

$$T_{\text{beat}} = nT_a \qquad \text{and} \qquad T_{\text{beat}} = (n - 1)T_b$$

Eliminating n between these two equations, we find

$$T_{\text{beat}} = \frac{T_a T_b}{T_b - T_a}$$

The reciprocal of the beat period is the beat *frequency*, $f_{\text{beat}} = 1/T_{\text{beat}}$, so

$$f_{\text{beat}} = \frac{T_b - T_a}{T_a T_b} = \frac{1}{T_a} - \frac{1}{T_b}$$

and finally

$$f_{\text{beat}} = f_a - f_b \qquad \text{(beat frequency)} \tag{16.24}$$

As claimed, the beat frequency is the difference of the two frequencies. In using Eq. (16.24), remember that f_a is the higher frequency.

An alternative way to derive Eq. (16.24) is to write functions to describe the curves in Fig. 16.24a and then add them. Suppose that at a certain position the two waves are given by $y_a(t) = A \sin 2\pi f_a t$ and $y_b(t) = -A \sin 2\pi f_b t$. We use the trigonometric identity

$$\sin a - \sin b = 2 \sin \tfrac{1}{2}(a - b) \cos \tfrac{1}{2}(a + b)$$

We can then express the total wave $y(t) = y_a(t) + y_b(t)$ as

$$y_a(t) + y_b(t) = \left[2A \sin \tfrac{1}{2}(2\pi)(f_a - f_b)t \right] \cos \tfrac{1}{2}(2\pi)(f_a + f_b)t$$

The amplitude factor (the quantity in brackets) varies slowly with frequency $\tfrac{1}{2}(f_a - f_b)$. The cosine factor varies with a frequency equal to the *average* frequency $\tfrac{1}{2}(f_a + f_b)$. The *square* of the amplitude factor, which is proportional to the intensity that the ear hears, goes through two maxima and two minima per cycle. So the beat frequency f_{beat} that is heard is twice the quantity $\tfrac{1}{2}(f_a - f_b)$, or just $f_a - f_b$, in agreement with Eq. (16.24).

Beats between two tones can be heard up to a beat frequency of about 6 or 7 Hz. Two piano strings or two organ pipes differing in frequency by 2 or 3 Hz sound wavery and "out of tune," although some organ stops contain two sets of pipes deliberately tuned to beat frequencies of about 1 to 2 Hz for a gently undulating effect. Listening for beats is an important technique in tuning all musical instruments.

At frequency differences greater than about 6 or 7 Hz, we no longer hear individual beats, and the sensation merges into one of *consonance* or *dissonance,* depending on the frequency ratio of the two tones. In some cases the ear perceives a tone called a *difference tone,* with a pitch equal to the beat frequency of the two tones. For example, if you listen to a whistle that produces sounds at 1800 Hz and 1900 Hz when blown, you will hear not only these tones but also a much lower 100-Hz tone.

The engines on multiengine propeller aircraft have to be synchronized so that the propeller sounds don't cause annoying beats, which are heard as loud throbbing sounds (Fig. 16.25). On some planes this is done electronically; on others the pilot does it by ear, just like tuning a piano.

16.25 If the two propellers on this airplane are not precisely synchronized, the pilots, passengers, and listeners on the ground will hear beats.

Test Your Understanding of Section 16.7   One tuning fork vibrates at 440 Hz, while a second tuning fork vibrates at an unknown frequency. When both tuning forks are sounded simultaneously, you hear a tone that rises and falls in intensity three times per second. What is the frequency of the second tuning fork? (i) 434 Hz; (ii) 437 Hz; (iii) 443 Hz; (iv) 446 Hz; (v) either 434 Hz or 446 Hz; (vi) either 437 Hz or 443 Hz.   ❙

16.8 The Doppler Effect

MasteringPHYSICS

ActivPhysics 10.8: Doppler Effect: Conceptual Introduction
ActivPhysics 10.9: Doppler Effect: Problems

You've probably noticed that when a car approaches you with its horn sounding, the pitch seems to drop as the car passes. This phenomenon, first described by the 19th-century Austrian scientist Christian Doppler, is called the **Doppler effect.** When a source of sound and a listener are in motion relative to each other, the frequency of the sound heard by the listener is not the same as the source frequency. A similar effect occurs for light and radio waves; we'll return to this later in this section.

To analyze the Doppler effect for sound, we'll work out a relationship between the frequency shift and the velocities of source and listener relative to the medium (usually air) through which the sound waves propagate. To keep things simple, we consider only the special case in which the velocities of both source and listener lie along the line joining them. Let v_S and v_L be the velocity components along this line for the source and the listener, respectively, relative to the medium. We choose the positive direction for both v_S and v_L to be the direction from the listener L to the source S. The speed of sound relative to the medium, v, is always considered positive.

Moving Listener and Stationary Source

Let's think first about a listener L moving with velocity v_L toward a stationary source S (Fig. 16.26). The source emits a sound wave with frequency f_S and wavelength $\lambda = v/f_S$. The figure shows four wave crests, separated by equal distances λ. The wave crests approaching the moving listener have a speed of propagation *relative to the listener* of $(v + v_L)$. So the frequency f_L with which the crests arrive at the listener's position (that is, the frequency the listener hears) is

$$f_L = \frac{v + v_L}{\lambda} = \frac{v + v_L}{v/f_S} \tag{16.25}$$

or

$$f_L = \left(\frac{v + v_L}{v}\right)f_S = \left(1 + \frac{v_L}{v}\right)f_S \qquad \begin{array}{l}\text{(moving listener,}\\\text{stationary source)}\end{array} \tag{16.26}$$

16.26 A listener moving toward a stationary source hears a frequency that is higher than the source frequency. This is because the relative speed of listener and wave is greater than the wave speed v.

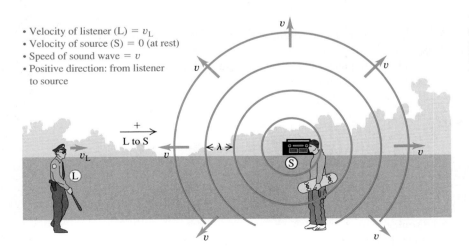

- Velocity of listener (L) = v_L
- Velocity of source (S) = 0 (at rest)
- Speed of sound wave = v
- Positive direction: from listener to source

+
L to S

16.27 Wave crests emitted by a moving source are crowded together in front of the source (to the right of this source) and stretched out behind it (to the left of this source).

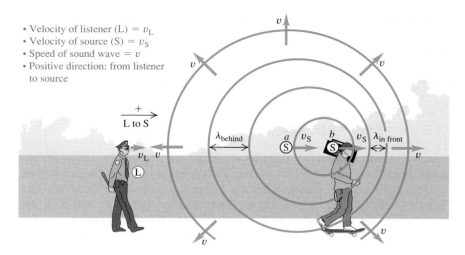

- Velocity of listener (L) = v_L
- Velocity of source (S) = v_S
- Speed of sound wave = v
- Positive direction: from listener to source

So a listener moving toward a source ($v_L > 0$), as in Fig. 16.26, hears a higher frequency (higher pitch) than does a stationary listener. A listener moving away from the source ($v_L < 0$) hears a lower frequency (lower pitch).

Moving Source and Moving Listener

Now suppose the source is also moving, with velocity v_S (Fig. 16.27). The wave speed relative to the wave medium (air) is still v; it is determined by the properties of the medium and is not changed by the motion of the source. But the wavelength is no longer equal to v/f_S. Here's why. The time for emission of one cycle of the wave is the period $T = 1/f_S$. During this time, the wave travels a distance $vT = v/f_S$ and the source moves a distance $v_S T = v_S/f_S$. The wavelength is the distance between successive wave crests, and this is determined by the *relative* displacement of source and wave. As Fig. 16.27 shows, this is different in front of and behind the source. In the region to the right of the source in Fig. 16.27 (that is, in front of the source), the wavelength is

$$\lambda_{\text{in front}} = \frac{v}{f_S} - \frac{v_S}{f_S} = \frac{v - v_S}{f_S} \qquad \text{(wavelength in front of a moving source)} \qquad \text{(16.27)}$$

In the region to the left of the source (that is, behind the source), it is

$$\lambda_{\text{behind}} = \frac{v + v_S}{f_S} \qquad \text{(wavelength behind a moving source)} \qquad \text{(16.28)}$$

The waves in front of and behind the source are compressed and stretched out, respectively, by the motion of the source.

To find the frequency heard by the listener behind the source, we substitute Eq. (16.28) into the first form of Eq. (16.25):

$$f_L = \frac{v + v_L}{\lambda_{\text{behind}}} = \frac{v + v_L}{(v + v_S)/f_S}$$

$$f_L = \frac{v + v_L}{v + v_S} f_S \qquad \text{(Doppler effect, moving source and moving listener)} \qquad \text{(16.29)}$$

This expresses the frequency f_L heard by the listener in terms of the frequency f_S of the source.

Although we derived it for the particular situation shown in Fig. 16.27, Eq. (16.29) includes *all* possibilities for motion of source and listener (relative to the medium) along the line joining them. If the listener happens to be at rest in the medium, v_L is zero. When both source and listener are at rest or have the same velocity relative to the medium, $v_L = v_S$ and $f_L = f_S$. Whenever the direction of the source or listener velocity is opposite to the direction from the listener toward the source (which we have defined as positive), the corresponding velocity to be used in Eq. (16.29) is negative.

As an example, the frequency heard by a listener at rest $(v_L = 0)$ is $f_L = [v/(v + v_S)]f_S$. If the source is moving toward the listener (in the negative direction), then $v_S < 0$, $f_L > f_S$, and the listener hears a higher frequency than that emitted by the source. If instead the source is moving away from the listener (in the positive direction), then $v_S > 0$, $f_L < f_S$, and the listener hears a lower frequency. This explains the change in pitch that you hear from the siren of an ambulance as it passes you (Fig. 16.28).

16.28 The Doppler effect explains why the siren on a fire engine or ambulance has a high pitch $(f_L > f_S)$ when it is approaching you $(v_S < 0)$ and a low pitch $(f_L < f_S)$ when it is moving away $(v_S > 0)$.

Problem-Solving Strategy 16.2 | **Doppler Effect**

(MP)

IDENTIFY *the relevant concepts:* The Doppler effect occurs whenever the source of waves, the wave detector (listener), or both are in motion.

SET UP *the problem* using the following steps:
1. Establish a coordinate system, with the positive direction from the listener toward the source. Carefully determine the signs of all relevant velocities. A velocity in the direction from the listener toward the source is positive; a velocity in the opposite direction is negative. All velocities must be measured relative to the air in which the sound travels.
2. Use consistent subscripts to identify the various quantities: S for source and L for listener.
3. Identify which unknown quantities are the target variables.

EXECUTE *the solution* as follows:
1. Use Eq. (16.29) to relate the frequencies at the source and the listener, the sound speed, and the velocities of the source and

the listener according to the sign convention of step 1. If the source is moving, you can find the wavelength measured by the listener using Eq. (16.27) or (16.28).
2. When a wave is reflected from a stationary or moving surface, solve the problem in two steps. In the first, the surface is the "listener"; the frequency with which the wave crests arrive at the surface is f_L. In the second, the surface is the "source," emitting waves with this same frequency f_L. Finally, determine the frequency heard by a listener detecting this new wave.

EVALUATE *your answer:* Is the *direction* of the frequency shift reasonable? If the source and the listener are moving toward each other, $f_L > f_S$; if they are moving apart, $f_L < f_S$. If the source and the listener have no relative motion, $f_L = f_S$.

Example 16.14 | **Doppler effect I: Wavelengths**

A police car's siren emits a sinusoidal wave with frequency $f_S = 300$ Hz. The speed of sound is 340 m/s and the air is still. (a) Find the wavelength of the waves if the siren is at rest. (b) Find the wavelengths of the waves in front of and behind the siren if it is moving at 30 m/s.

SOLUTION

IDENTIFY and SET UP: In part (a) there is no Doppler effect because neither source nor listener is moving with respect to the air; $v = \lambda f$ gives the wavelength. Figure 16.29 shows the situation in part (b): The source is in motion, so we find the wavelengths using Eqs. (16.27) and (16.28) for the Doppler effect.

EXECUTE: (a) When the source is at rest,

$$\lambda = \frac{v}{f_S} = \frac{340 \text{ m/s}}{300 \text{ Hz}} = 1.13 \text{ m}$$

16.29 Our sketch for this problem.

Police car

$$\lambda_{\text{behind}} = ? \qquad \textcircled{S} \; v_S = 30 \text{ m/s} \qquad \lambda_{\text{in front}} = ?$$

(b) From Eq. (16.27), in front of the siren

$$\lambda_{\text{in front}} = \frac{v - v_S}{f_S} = \frac{340 \text{ m/s} - 30 \text{ m/s}}{300 \text{ Hz}} = 1.03 \text{ m}$$

From Eq. (16.28), behind the siren

$$\lambda_{\text{behind}} = \frac{v + v_S}{f_S} = \frac{340 \text{ m/s} + 30 \text{ m/s}}{300 \text{ Hz}} = 1.23 \text{ m}$$

EVALUATE: The wavelength is shorter in front of the siren and longer behind it, as we expect.

Example 16.15 Doppler effect II: Frequencies

If a listener L is at rest and the siren in Example 16.14 is moving away from L at 30 m/s, what frequency does the listener hear?

SOLUTION

IDENTIFY and SET UP: Our target variable is the frequency f_L heard by a listener behind the moving source. Figure 16.30 shows the situation. We have $v_L = 0$ and $v_S = +30$ m/s (positive, since the velocity of the source is in the direction from listener to source).

EXECUTE: From Eq. (16.29),

$$f_L = \frac{v}{v + v_S} f_S = \frac{340 \text{ m/s}}{340 \text{ m/s} + 30 \text{ m/s}} (300 \text{ Hz}) = 276 \text{ Hz}$$

EVALUATE: The source and listener are moving apart, so $f_L < f_S$. Here's a check on our numerical result. From Example 16.14, the

16.30 Our sketch for this problem.

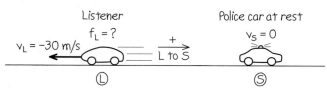

Listener at rest   Police car

wavelength behind the source (where the listener in Fig. 16.30 is located) is 1.23 m. The wave speed relative to the stationary listener is $v = 340$ m/s even though the source is moving, so

$$f_L = \frac{v}{\lambda} = \frac{340 \text{ m/s}}{1.23 \text{ m}} = 276 \text{ Hz}$$

Example 16.16 Doppler effect III: A moving listener

If the siren is at rest and the listener is moving away from it at 30 m/s, what frequency does the listener hear?

SOLUTION

IDENTIFY and SET UP: Again our target variable is f_L, but now L is in motion and S is at rest. Figure 16.31 shows the situation. The velocity of the listener is $v_L = -30$ m/s (negative, since the motion is in the direction from source to listener).

EXECUTE: From Eq. (16.29),

$$f_L = \frac{v + v_L}{v} f_S = \frac{340 \text{ m/s} + (-30 \text{ m/s})}{340 \text{ m/s}} (300 \text{ Hz}) = 274 \text{ Hz}$$

16.31 Our sketch for this problem.

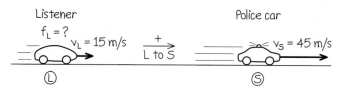

Listener   Police car at rest

EVALUATE: Again the source and listener are moving apart, so $f_L < f_S$. Note that the *relative velocity* of source and listener is the same as in Example 16.15, but the Doppler shift is different because v_S and v_L are different.

Example 16.17 Doppler effect IV: Moving source, moving listener

The siren is moving away from the listener with a speed of 45 m/s relative to the air, and the listener is moving toward the siren with a speed of 15 m/s relative to the air. What frequency does the listener hear?

SOLUTION

IDENTIFY and SET UP: Now *both* L and S are in motion. Again our target variable is f_L. Both the source velocity $v_S = +45$ m/s and the listener's velocity $v_L = +15$ m/s are positive because both velocities are in the direction from listener to source.

EXECUTE: From Eq. (16.29),

$$f_L = \frac{v + v_L}{v + v_S} f_S = \frac{340 \text{ m/s} + 15 \text{ m/s}}{340 \text{ m/s} + 45 \text{ m/s}} (300 \text{ Hz}) = 277 \text{ Hz}$$

16.32 Our sketch for this problem.

Listener   Police car

EVALUATE: As in Examples 16.15 and 16.16, the source and listener again move away from each other at 30 m/s, so again $f_L < f_S$. But f_L is different in all three cases because the Doppler effect for sound depends on how the source and listener are moving relative to the *air*, not simply on how they move relative to each other.

Example 16.18 Doppler effect V: A double Doppler shift

The police car is moving toward a warehouse at 30 m/s. What frequency does the driver hear reflected from the warehouse?

SOLUTION

IDENTIFY: In this situation there are *two* Doppler shifts (Fig. 16.33). In the first shift, the warehouse is the stationary "listener."

16.33 Two stages of the sound wave's motion from the police car to the warehouse and back to the police car.

(a) Sound travels from police car's siren (source S) to warehouse ("listener" L).

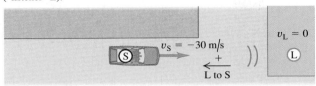

(b) Reflected sound travels from warehouse (source S) to police car (listener L).

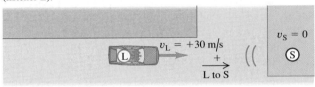

The frequency of sound reaching the warehouse, which we call f_W, is greater than 300 Hz because the source is approaching. In the second shift, the warehouse acts as a source of sound with frequency f_W, and the listener is the driver of the police car; she hears a frequency greater than f_W because she is approaching the source.

SET UP: To determine f_W, we use Eq. (16.29) with f_L replaced by f_W. For this part of the problem, $v_L = v_W = 0$ (the warehouse is at rest) and $v_S = -30$ m/s (the siren is moving in the negative direction from source to listener).

To determine the frequency heard by the driver (our target variable), we again use Eq. (16.29) but now with f_S replaced by f_W. For this second part of the problem, $v_S = 0$ because the stationary warehouse is the source and the velocity of the listener (the driver) is $v_L = +30$ m/s. (The listener's velocity is positive because it is in the direction from listener to source.)

EXECUTE: The frequency reaching the warehouse is

$$f_W = \frac{v}{v + v_S} f_S = \frac{340 \text{ m/s}}{340 \text{ m/s} + (-30 \text{ m/s})} (300 \text{ Hz}) = 329 \text{ Hz}$$

Then the frequency heard by the driver is

$$f_L = \frac{v + v_L}{v} f_W = \frac{340 \text{ m/s} + 30 \text{ m/s}}{340 \text{ m/s}} (329 \text{ Hz}) = 358 \text{ Hz}$$

EVALUATE: Because there are two Doppler shifts, the reflected sound heard by the driver has an even higher frequency than the sound heard by a stationary listener in the warehouse.

Doppler Effect for Electromagnetic Waves

In the Doppler effect for sound, the velocities v_L and v_S are always measured relative to the *air* or whatever medium we are considering. There is also a Doppler effect for *electromagnetic* waves in empty space, such as light waves or radio waves. In this case there is no medium that we can use as a reference to measure velocities, and all that matters is the *relative* velocity of source and receiver. (By contrast, the Doppler effect for sound does not depend simply on this relative velocity, as discussed in Example 16.17.)

To derive the expression for the Doppler frequency shift for light, we have to use the special theory of relativity. We will discuss this in Chapter 37, but for now we quote the result without derivation. The wave speed is the speed of light, usually denoted by c, and it is the same for both source and receiver. In the frame of reference in which the receiver is at rest, the source is moving away from the receiver with velocity v. (If the source is *approaching* the receiver, v is negative.) The source frequency is again f_S. The frequency f_R measured by the receiver R (the frequency of arrival of the waves at the receiver) is then

$$f_R = \sqrt{\frac{c - v}{c + v}} f_S \qquad \text{(Doppler effect for light)} \qquad (16.30)$$

When v is positive, the source is moving directly *away* from the receiver and f_R is always *less* than f_S; when v is negative, the source is moving directly *toward* the receiver and f_R is *greater* than f_S. The qualitative effect is the same as for sound, but the quantitative relationship is different.

A familiar application of the Doppler effect for radio waves is the radar device mounted on the side window of a police car to check other cars' speeds. The electromagnetic wave emitted by the device is reflected from a moving car, which acts as a moving source, and the wave reflected back to the device is Doppler-shifted in frequency. The transmitted and reflected signals are combined to produce beats, and the speed can be computed from the frequency of the beats. Similar techniques ("Doppler radar") are used to measure wind velocities in the atmosphere.

16.34 Change of velocity component along the line of sight of a satellite passing a tracking station. The frequency received at the tracking station changes from high to low as the satellite passes overhead.

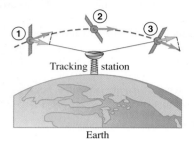

The Doppler effect is also used to track satellites and other space vehicles. In Fig. 16.34 a satellite emits a radio signal with constant frequency f_S. As the satellite orbits past, it first approaches and then moves away from the receiver; the frequency f_R of the signal received on earth changes from a value greater than f_S to a value less than f_S as the satellite passes overhead.

Test Your Understanding of Section 16.8 You are at an outdoor concert with a wind blowing at 10 m/s from the performers toward you. Is the sound you hear Doppler-shifted? If so, is it shifted to lower or higher frequencies? ❙

16.9 Shock Waves

You may have experienced "sonic booms" caused by an airplane flying overhead faster than the speed of sound. We can see qualitatively why this happens from Fig. 16.35. Let v_S denote the *speed* of the airplane relative to the air, so that it is always positive. The motion of the airplane through the air produces sound; if v_S is less than the speed of sound v, the waves in front of the airplane are crowded together with a wavelength given by Eq. (16.27):

$$\lambda_{\text{in front}} = \frac{v - v_S}{f_S}$$

As the speed v_S of the airplane approaches the speed of sound v, the wavelength approaches zero and the wave crests pile up on each other (Fig. 16.35a). The airplane must exert a large force to compress the air in front of it; by Newton's third law, the air exerts an equally large force back on the airplane. Hence there is a large increase in aerodynamic drag (air resistance) as the airplane approaches the speed of sound, a phenomenon known as the "sound barrier."

When v_S is greater in magnitude than v, the source of sound is **supersonic,** and Eqs. (16.27) and (16.29) for the Doppler effect no longer describe the sound wave in front of the source. Figure 16.35b shows a cross section of what happens. As the airplane moves, it displaces the surrounding air and produces sound. A series of wave crests is emitted from the nose of the airplane; each spreads out in a circle centered at the position of the airplane when it emitted the crest. After a time t the crest emitted from point S_1 has spread to a circle with radius vt, and the airplane has moved a greater distance $v_S t$ to position S_2. You can see that the circular crests interfere constructively at points along the blue line that makes an angle α with

16.35 Wave crests around a sound source S moving **(a)** slightly slower than the speed of sound v and **(b)** faster than the sound speed v. **(c)** This photograph shows a T-38 jet airplane moving at 1.1 times the speed of sound. Separate shock waves are produced by the nose, wings, and tail. The angles of these waves vary because the air speeds up and slows down as it moves around the airplane, so the relative speed v_S of the airplane and air is different for shock waves produced at different points.

(a) Sound source S (airplane) moving at nearly the speed of sound

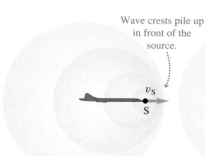

(b) Sound source moving faster than the speed of sound

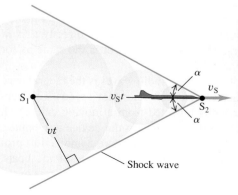

(c) Shock waves around a supersonic airplane

the direction of the airplane velocity, leading to a very-large-amplitude wave crest along this line. This large-amplitude crest is called a **shock wave** (Fig. 16.35c).

From the right triangle in Fig. 16.35b we can see that the angle α is given by

$$\sin\alpha = \frac{vt}{v_St} = \frac{v}{v_S} \qquad \text{(shock wave)} \qquad (16.31)$$

In this relationship, v_S is the *speed* of the source (the magnitude of its velocity) relative to the air and is always positive. The ratio v_S/v is called the **Mach number.** It is greater than unity for all supersonic speeds, and $\sin\alpha$ in Eq. (16.31) is the reciprocal of the Mach number. The first person to break the sound barrier was Capt. Chuck Yeager of the U.S. Air Force, flying the Bell X-1 at Mach 1.06 on October 14, 1947 (Fig. 16.36).

Shock waves are actually three-dimensional; a shock wave forms a *cone* around the direction of motion of the source. If the source (possibly a supersonic jet airplane or a rifle bullet) moves with constant velocity, the angle α is constant, and the shock-wave cone moves along with the source. It's the arrival of this shock wave that causes the sonic boom you hear after a supersonic airplane has passed by. The larger the airplane, the stronger the sonic boom; the shock wave produced at ground level by the (now retired) Concorde supersonic airliner flying at 12,000 m (40,000 ft) caused a sudden jump in air pressure of about 20 Pa. In front of the shock-wave cone, there is no sound. Inside the cone a stationary listener hears the Doppler-shifted sound of the airplane moving away.

CAUTION **Shock waves** We emphasize that a shock wave is produced *continuously* by any object that moves through the air at supersonic speed, not only at the instant that it "breaks the sound barrier." The sound waves that combine to form the shock wave, as in Fig. 16.35b, are created by the motion of the object itself, not by any sound source that the object may carry. The cracking noises of a bullet and of the tip of a circus whip are due to their supersonic motion. A supersonic jet airplane may have very loud engines, but these do not cause the shock wave. Indeed, a space shuttle makes a very loud sonic boom when coming in for a landing; its engines are out of fuel at this point, so it is a supersonic glider. ▮

Shock waves have applications outside of aviation. They are used to break up kidney stones and gallstones without invasive surgery, using a technique with the impressive name *extracorporeal shock-wave lithotripsy.* A shock wave produced outside the body is focused by a reflector or acoustic lens so that as much of it as possible converges on the stone. When the resulting stresses in the stone exceed its tensile strength, it breaks into small pieces and can be eliminated. This technique requires accurate determination of the location of the stone, which may be done using ultrasonic imaging techniques (see Fig. 16.9).

16.36 The first supersonic airplane, the Bell X-1, was shaped much like a 50-caliber bullet—which was known to be able to travel faster than sound.

Example 16.19   **Sonic boom from a supersonic airplane**

An airplane is flying at Mach 1.75 at an altitude of 8000 m, where the speed of sound is 320 m/s. How long after the plane passes directly overhead will you hear the sonic boom?

SOLUTION

IDENTIFY and SET UP: The shock wave forms a cone trailing backward from the airplane, so the problem is really asking for how much time elapses from when the airplane flies overhead to when the shock wave reaches you at point L (Fig. 16.37). During the time t (our target variable) since the airplane traveling at speed v_S passed overhead, it has traveled a distance v_St. Equation (16.31) gives the shock cone angle α; we use trigonometry to solve for t.

EXECUTE: From Eq. (16.31) the angle α of the shock cone is

$$\alpha = \arcsin\frac{1}{1.75} = 34.8°$$

The speed of the plane is the speed of sound multiplied by the Mach number:

$$v_S = (1.75)(320 \text{ m/s}) = 560 \text{ m/s}$$

Continued

16.37 You hear a sonic boom when the shock wave reaches you at L (*not* just when the plane breaks the sound barrier). A listener to the right of L has not yet heard the sonic boom but will shortly; a listener to the left of L has already heard the sonic boom.

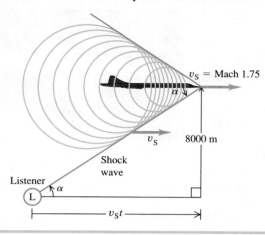

v_S = Mach 1.75

v_S 8000 m

Shock wave

Listener

L α

$v_S t$

From Fig. 16.37 we have

$$\tan \alpha = \frac{8000 \text{ m}}{v_S t}$$

$$t = \frac{8000 \text{ m}}{(560 \text{ m/s})(\tan 34.8°)} = 20.5 \text{ s}$$

EVALUATE: You hear the boom 20.5 s after the airplane passes overhead, at which time it has traveled $(560 \text{ m/s})(20.5 \text{ s}) =$ 11.5 km since it passed overhead. We have assumed that the speed of sound is the same at all altitudes, so that $\alpha = \arcsin v/v_S$ is constant and the shock wave forms a perfect cone. In fact, the speed of sound decreases with increasing altitude. How would this affect the value of t?

Test Your Understanding of Section 16.9   What would you hear if you were directly behind (to the left of) the supersonic airplane in Fig. 16.37? (i) a sonic boom; (ii) the sound of the airplane, Doppler-shifted to higher frequencies; (iii) the sound of the airplane, Doppler-shifted to lower frequencies; (iv) nothing.   ❙

Sound waves: Sound consists of longitudinal waves in a medium. A sinusoidal sound wave is characterized by its frequency f and wavelength λ (or angular frequency ω and wave number k) and by its displacement amplitude A. The pressure amplitude p_{max} is directly proportional to the displacement amplitude, the wave number, and the bulk modulus B of the wave medium. (See Examples 16.1 and 16.2.)

The speed of a sound wave in a fluid depends on the bulk modulus B and density ρ. If the fluid is an ideal gas, the speed can be expressed in terms of the temperature T, molar mass M, and ratio of heat capacities γ of the gas. The speed of longitudinal waves in a solid rod depends on the density and Young's modulus Y. (See Examples 16.3 and 16.4.)

$$p_{max} = BkA \qquad (16.5)$$
(sinusoidal sound wave)

$$v = \sqrt{\frac{B}{\rho}} \qquad (16.7)$$
(longitudinal wave in a fluid)

$$v = \sqrt{\frac{\gamma RT}{M}} \qquad (16.10)$$
(sound wave in an ideal gas)

$$v = \sqrt{\frac{Y}{\rho}} \qquad (16.8)$$
(longitudinal wave in a solid rod)

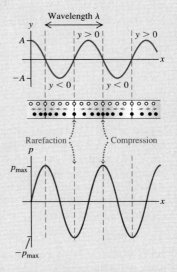

Intensity and sound intensity level: The intensity I of a sound wave is the time average rate at which energy is transported by the wave, per unit area. For a sinusoidal wave, the intensity can be expressed in terms of the displacement amplitude A or the pressure amplitude p_{max}. (See Examples 16.5–16.7.)

The sound intensity level β of a sound wave is a logarithmic measure of its intensity. It is measured relative to I_0, an arbitrary intensity defined to be 10^{-12} W/m^2. Sound intensity levels are expressed in decibels (dB). (See Examples 16.8 and 16.9.)

$$I = \frac{1}{2}\sqrt{\rho B}\,\omega^2 A^2 = \frac{p_{max}^2}{2\rho v}$$
$$= \frac{p_{max}^2}{2\sqrt{\rho B}} \qquad (16.12), (16.14)$$
(intensity of a sinusoidal sound wave)

$$\beta = (10\text{ dB})\log\frac{I}{I_0} \qquad (16.15)$$
(definition of sound intensity level)

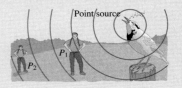

Standing sound waves: Standing sound waves can be set up in a pipe or tube. A closed end is a displacement node and a pressure antinode; an open end is a displacement antinode and a pressure node. For a pipe of length L open at both ends, the normal-mode frequencies are integer multiples of the sound speed divided by $2L$. For a stopped pipe (one that is open at only one end), the normal-mode frequencies are the odd multiples of the sound speed divided by $4L$. (See Examples 16.10 and 16.11.)

A pipe or other system with normal-mode frequencies can be driven to oscillate at any frequency. A maximum response, or resonance, occurs if the driving frequency is close to one of the normal-mode frequencies of the system. (See Example 16.12.)

$$f_n = \frac{nv}{2L} \quad (n = 1, 2, 3, \ldots) \quad (16.18)$$
(open pipe)

$$f_n = \frac{nv}{4L} \quad (n = 1, 3, 5, \ldots) \quad (16.22)$$
(stopped pipe)

Interference: When two or more waves overlap in the same region of space, the resulting effects are called interference. The resulting amplitude can be either larger or smaller than the amplitude of each individual wave, depending on whether the waves are in phase (constructive interference) or out of phase (destructive interference). (See Example 16.13.)

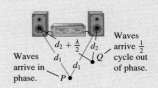

Beats: Beats are heard when two tones with slightly different frequencies f_a and f_b are sounded together. The beat frequency f_{beat} is the difference between f_a and f_b.

$$f_{\text{beat}} = f_a - f_b \qquad (16.24)$$
(beat frequency)

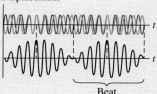

Displacement

Beat

Doppler effect: The Doppler effect for sound is the frequency shift that occurs when there is motion of a source of sound, a listener, or both, relative to the medium. The source and listener frequencies f_S and f_L are related by the source and listener velocities v_S and v_L relative to the medium and to the speed of sound v. (See Examples 16.14–16.18.)

$$f_L = \frac{v + v_L}{v + v_S} f_S \qquad (16.29)$$
(Doppler effect, moving source and moving listener)

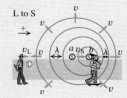

L to S

Shock waves: A sound source moving with a speed v_S greater than the speed of sound v creates a shock wave. The wave front is a cone with angle α. (See Example 16.19.)

$$\sin \alpha = \frac{v}{v_S} \qquad \text{(shock wave)} \qquad (16.31)$$

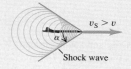

$v_S > v$

Shock wave

BRIDGING PROBLEM   Loudspeaker Interference

Loudspeakers A and B are 7.00 m apart and vibrate in phase at 172 Hz. They radiate sound uniformly in all directions. Their acoustic power outputs are 8.00×10^{-4} W and 6.00×10^{-5} W, respectively. The air temperature is 20°C. (a) Determine the difference in phase of the two signals at a point C along the line joining A and B, 3.00 m from B and 4.00 m from A. (b) Determine the intensity and sound intensity level at C from speaker A alone (with B turned off) and from speaker B alone (with A turned off). (c) Determine the intensity and sound intensity level at C from both speakers together.

SOLUTION GUIDE

See MasteringPhysics® study area for a Video Tutor solution.

IDENTIFY and SET UP

1. Sketch the situation and label the distances between A, B, and C.
2. Choose the equations that relate power, distance from the source, intensity, pressure amplitude, and sound intensity level.
3. Decide how you will determine the phase difference in part (a). Once you have found the phase difference, how can you use it to find the amplitude of the combined wave at C due to both sources?

4. List the unknown quantities for each part of the problem and identify your target variables.

EXECUTE

5. Determine the phase difference at point C.
6. Find the intensity, sound intensity level, and pressure amplitude at C due to each speaker alone.
7. Use your results from steps 5 and 6 to find the pressure amplitude at C due to both loudspeakers together.
8. Use your result from step 7 to find the intensity and sound intensity level at C due to both loudspeakers together.

EVALUATE

9. How do your results from part (c) for intensity and sound intensity level at C compare to those from part (b)? Does this make sense?
10. What result would you have gotten in part (c) if you had (incorrectly) combined the *intensities* from A and B directly, rather than (correctly) combining the *pressure amplitudes* as you did in step 7?

Problems

For instructor-assigned homework, go to www.masteringphysics.com (MP)

•, ••, •••: Problems of increasing difficulty. **CP**: Cumulative problems incorporating material from earlier chapters. **CALC**: Problems requiring calculus. **BIO**: Biosciences problems.

DISCUSSION QUESTIONS

Q16.1 When sound travels from air into water, does the frequency of the wave change? The speed? The wavelength? Explain your reasoning.

Q16.2 The hero of a western movie listens for an oncoming train by putting his ear to the track. Why does this method give an earlier warning of the approach of a train than just listening in the usual way?

Q16.3 Would you expect the pitch (or frequency) of an organ pipe to increase or decrease with increasing temperature? Explain.

Q16.4 In most modern wind instruments the pitch is changed by using keys or valves to change the length of the vibrating air column. The bugle, however, has no valves or keys, yet it can play many notes. How might this be possible? Are there restrictions on what notes a bugle can play?

Q16.5 Symphonic musicians always "warm up" their wind instruments by blowing into them before a performance. What purpose does this serve?

Q16.6 In a popular and amusing science demonstration, a person inhales helium and then his voice becomes high and squeaky. Why does this happen? (*Warning:* Inhaling too much helium can cause unconsciousness or death.)

Q16.7 Lane dividers on highways sometimes have regularly spaced ridges or ripples. When the tires of a moving car roll along such a divider, a musical note is produced. Why? Explain how this phenomenon could be used to measure the car's speed.

Q16.8 The tone quality of an acoustic guitar is different when the strings are plucked near the bridge (the lower end of the strings) than when they are plucked near the sound hole (close to the center of the strings). Why?

Q16.9 Which has a more direct influence on the loudness of a sound wave: the *displacement* amplitude or the *pressure* amplitude? Explain your reasoning.

Q16.10 If the pressure amplitude of a sound wave is halved, by what factor does the intensity of the wave decrease? By what factor must the pressure amplitude of a sound wave be increased in order to increase the intensity by a factor of 16? Explain.

Q16.11 Does the sound intensity level β obey the inverse-square law? Why?

Q16.12 A small fraction of the energy in a sound wave is absorbed by the air through which the sound passes. How does this modify the inverse-square relationship between intensity and distance from the source? Explain your reasoning.

Q16.13 A wire under tension and vibrating in its first overtone produces sound of wavelength λ. What is the new wavelength of the sound (in terms of λ) if the tension is doubled?

Q16.14 A small metal band is slipped onto one of the tines of a tuning fork. As this band is moved closer and closer to the end of the tine, what effect does this have on the wavelength and frequency of the sound the tine produces? Why?

Q16.15 An organist in a cathedral plays a loud chord and then releases the keys. The sound persists for a few seconds and gradually dies away. Why does it persist? What happens to the sound energy when the sound dies away?

Q16.16 Two vibrating tuning forks have identical frequencies, but one is stationary and the other is mounted at the rim of a rotating platform. What does a listener hear? Explain.

Q16.17 A large church has part of the organ in the front of the church and part in the back. A person walking rapidly down the aisle while both segments are playing at once reports that the two segments sound out of tune. Why?

Q16.18 A sound source and a listener are both at rest on the earth, but a strong wind is blowing from the source toward the listener. Is there a Doppler effect? Why or why not?

Q16.19 Can you think of circumstances in which a Doppler effect would be observed for surface waves in water? For elastic waves propagating in a body of water deep below the surface? If so, describe the circumstances and explain your reasoning. If not, explain why not.

Q16.20 Stars other than our sun normally appear featureless when viewed through telescopes. Yet astronomers can readily use the light from these stars to determine that they are rotating and even measure the speed of their surface. How do you think they can do this?

Q16.21 If you wait at a railroad crossing as a train approaches and passes, you hear a Doppler shift in its sound. But if you listen closely, you hear that the change in frequency is continuous; it does not suddenly go from one high frequency to another low frequency. Instead the frequency *smoothly* (but rather quickly) changes from high to low as the train passes. Why does this smooth change occur?

Q16.22 In case 1, a source of sound approaches a stationary observer at speed v. In case 2, the observer moves toward the stationary source at the same speed v. If the source is always producing the same frequency sound, will the observer hear the same frequency in both cases, since the relative speed is the same each time? Why or why not?

Q16.23 Does an aircraft make a sonic boom only at the instant its speed exceeds Mach 1? Explain your reasoning.

Q16.24 If you are riding in a supersonic aircraft, what do you hear? Explain your reasoning. In particular, do you hear a continuous sonic boom? Why or why not?

Q16.25 A jet airplane is flying at a constant altitude at a steady speed v_S greater than the speed of sound. Describe what observers at points A, B, and C hear at the instant shown in Fig. Q16.25, when the shock wave has just reached point B. Explain your reasoning.

Figure **Q16.25**

EXERCISES

Unless indicated otherwise, assume the speed of sound in air to be $v = 344$ m/s.

Section 16.1 Sound Waves

16.1 • Example 16.1 (Section 16.1) showed that for sound waves in air with frequency 1000 Hz, a displacement amplitude of

1.2×10^{-8} m produces a pressure amplitude of 3.0×10^{-2} Pa. (a) What is the wavelength of these waves? (b) For 1000-Hz waves in air, what displacement amplitude would be needed for the pressure amplitude to be at the pain threshold, which is 30 Pa? (c) For what wavelength and frequency will waves with a displacement amplitude of 1.2×10^{-8} m produce a pressure amplitude of 1.5×10^{-3} Pa?

16.2 • Example 16.1 (Section 16.1) showed that for sound waves in air with frequency 1000 Hz, a displacement amplitude of 1.2×10^{-8} m produces a pressure amplitude of 3.0×10^{-2} Pa. Water at 20°C has a bulk modulus of 2.2×10^{9} Pa, and the speed of sound in water at this temperature is 1480 m/s. For 1000-Hz sound waves in 20°C water, what displacement amplitude is produced if the pressure amplitude is 3.0×10^{-2} Pa? Explain why your answer is much less than 1.2×10^{-8} m.

16.3 • Consider a sound wave in air that has displacement amplitude 0.0200 mm. Calculate the pressure amplitude for frequencies of (a) 150 Hz; (b) 1500 Hz; (c) 15,000 Hz. In each case compare the result to the pain threshold, which is 30 Pa.

16.4 • A loud factory machine produces sound having a displacement amplitude of 1.00 μm, but the frequency of this sound can be adjusted. In order to prevent ear damage to the workers, the maximum pressure amplitude of the sound waves is limited to 10.0 Pa. Under the conditions of this factory, the bulk modulus of air is 1.42×10^{5} Pa. What is the highest-frequency sound to which this machine can be adjusted without exceeding the prescribed limit? Is this frequency audible to the workers?

16.5 • **BIO Ultrasound and Infrasound.** (a) **Whale communication.** Blue whales apparently communicate with each other using sound of frequency 17 Hz, which can be heard nearly 1000 km away in the ocean. What is the wavelength of such a sound in seawater, where the speed of sound is 1531 m/s? (b) **Dolphin clicks.** One type of sound that dolphins emit is a sharp click of wavelength 1.5 cm in the ocean. What is the frequency of such clicks? (c) **Dog whistles.** One brand of dog whistles claims a frequency of 25 kHz for its product. What is the wavelength of this sound? (d) **Bats.** While bats emit a wide variety of sounds, one type emits pulses of sound having a frequency between 39 kHz and 78 kHz. What is the range of wavelengths of this sound? (e) **Sonograms.** Ultrasound is used to view the interior of the body, much as x rays are utilized. For sharp imagery, the wavelength of the sound should be around one-fourth (or less) the size of the objects to be viewed. Approximately what frequency of sound is needed to produce a clear image of a tumor that is 1.0 mm across if the speed of sound in the tissue is 1550 m/s?

Section 16.2 Speed of Sound Waves

16.6 • (a) In a liquid with density 1300 kg/m³, longitudinal waves with frequency 400 Hz are found to have wavelength 8.00 m. Calculate the bulk modulus of the liquid. (b) A metal bar with a length of 1.50 m has density 6400 kg/m³. Longitudinal sound waves take 3.90×10^{-4} s to travel from one end of the bar to the other. What is Young's modulus for this metal?

Figure E16.7

16.7 • A submerged scuba diver hears the sound of a boat horn directly above her on the surface of the lake. At the same time, a friend on dry land 22.0 m from the boat also hears the horn (Fig. E16.7). The horn is 1.2 m above the surface of the water.

22.0 m

?

What is the distance (labeled by "?" in Fig. E16.7) from the horn to the diver? Both air and water are at 20°C.

16.8 • At a temperature of 27.0°C, what is the speed of longitudinal waves in (a) hydrogen (molar mass 2.02 g/mol); (b) helium (molar mass 4.00 g/mol); (c) argon (molar mass 39.9 g/mol)? See Table 19.1 for values of γ. (d) Compare your answers for parts (a), (b), and (c) with the speed in air at the same temperature.

16.9 • An oscillator vibrating at 1250 Hz produces a sound wave that travels through an ideal gas at 325 m/s when the gas temperature is 22.0°C. For a certain experiment, you need to have the same oscillator produce sound of wavelength 28.5 cm in this gas. What should the gas temperature be to achieve this wavelength?

16.10 •• **CALC** (a) Show that the fractional change in the speed of sound (dv/v) due to a very small temperature change dT is given by $dv/v = \frac{1}{2}dT/T$. (*Hint:* Start with Eq. 16.10.) (b) The speed of sound in air at 20°C is found to be 344 m/s. Use the result in part (a) to find the change in the speed of sound for a 1.0°C change in air temperature.

16.11 •• An 80.0-m-long brass rod is struck at one end. A person at the other end hears two sounds as a result of two longitudinal waves, one traveling in the metal rod and the other traveling in the air. What is the time interval between the two sounds? (The speed of sound in air is 344 m/s; relevant information about brass can be found in Table 11.1 and Table 12.1.)

16.12 •• What must be the stress (F/A) in a stretched wire of a material whose Young's modulus is Y for the speed of longitudinal waves to equal 30 times the speed of transverse waves?

Section 16.3 Sound Intensity

16.13 •• **BIO Energy Delivered to the Ear.** Sound is detected when a sound wave causes the tympanic membrane (the eardrum) to vibrate. Typically, the diameter of this membrane is about 8.4 mm in humans. (a) How much energy is delivered to the eardrum each second when someone whispers (20 dB) a secret in your ear? (b) To comprehend how sensitive the ear is to very small amounts of energy, calculate how fast a typical 2.0-mg mosquito would have to fly (in mm/s) to have this amount of kinetic energy.

16.14 • Use information from Table 16.2 to answer the following questions about sound in air. At 20°C the bulk modulus for air is 1.42×10^{5} Pa and its density is 1.20 kg/m³. At this temperature, what are the pressure amplitude (in Pa and atm) and the displacement amplitude (in m and nm) (a) for the softest sound a person can normally hear at 1000 Hz and (b) for the sound from a riveter at the same frequency? (c) How much energy per second does each wave deliver to a square 5.00 mm on a side?

16.15 •• **Longitudinal Waves in Different Fluids.** (a) A longitudinal wave propagating in a water-filled pipe has intensity 3.00×10^{-6} W/m² and frequency 3400 Hz. Find the amplitude A and wavelength λ of the wave. Water has density 1000 kg/m³ and bulk modulus 2.18×10^{9} Pa. (b) If the pipe is filled with air at pressure 1.00×10^{5} Pa and density 1.20 kg/m³, what will be the amplitude A and wavelength λ of a longitudinal wave with the same intensity and frequency as in part (a)? (c) In which fluid is the amplitude larger, water or air? What is the ratio of the two amplitudes? Why is this ratio so different from 1.00?

16.16 •• **BIO Human Hearing.** A fan at a rock concert is 30 m from the stage, and at this point the sound intensity level is 110 dB. (a) How much energy is transferred to her eardrums each second? (b) How fast would a 2.0-mg mosquito have to fly (in mm/s) to have this much kinetic energy? Compare the mosquito's speed with that found for the whisper in part (a) of Exercise 16.13.

16.17 • A sound wave in air at 20°C has a frequency of 150 Hz and a displacement amplitude of 5.00×10^{-3} mm. For this sound wave calculate the (a) pressure amplitude (in Pa); (b) intensity (in W/m^2); (c) sound intensity level (in decibels).

16.18 •• You live on a busy street, but as a music lover, you want to reduce the traffic noise. (a) If you install special sound-reflecting windows that reduce the sound intensity level (in dB) by 30 dB, by what fraction have you lowered the sound intensity (in W/m^2)? (b) If, instead, you reduce the intensity by half, what change (in dB) do you make in the sound intensity level?

16.19 • BIO For a person with normal hearing, the faintest sound that can be heard at a frequency of 400 Hz has a pressure amplitude of about 6.0×10^{-5} Pa. Calculate the (a) intensity; (b) sound intensity level; (c) displacement amplitude of this sound wave at 20°C.

16.20 •• The intensity due to a number of independent sound sources is the sum of the individual intensities. (a) When four quadruplets cry simultaneously, how many decibels greater is the sound intensity level than when a single one cries? (b) To increase the sound intensity level again by the same number of decibels as in part (a), how many more crying babies are required?

16.21 • CP A baby's mouth is 30 cm from her father's ear and 1.50 m from her mother's ear. What is the difference between the sound intensity levels heard by the father and by the mother?

16.22 •• The Sacramento City Council adopted a law to reduce the allowed sound intensity level of the much-despised leaf blowers from their current level of about 95 dB to 70 dB. With the new law, what is the ratio of the new allowed intensity to the previously allowed intensity?

16.23 •• CP At point A, 3.0 m from a small source of sound that is emitting uniformly in all directions, the sound intensity level is 53 dB. (a) What is the intensity of the sound at A? (b) How far from the source must you go so that the intensity is one-fourth of what it was at A? (c) How far must you go so that the sound intensity level is one-fourth of what it was at A? (d) Does intensity obey the inverse-square law? What about sound intensity level?

16.24 •• (a) If two sounds differ by 5.00 dB, find the ratio of the intensity of the louder sound to that of the softer one. (b) If one sound is 100 times as intense as another, by how much do they differ in sound intensity level (in decibels)? (c) If you increase the volume of your stereo so that the intensity doubles, by how much does the sound intensity level increase?

Section 16.4 Standing Sound Waves and Normal Modes

16.25 • Standing sound waves are produced in a pipe that is 1.20 m long. For the fundamental and first two overtones, determine the locations along the pipe (measured from the left end) of the displacement nodes and the pressure nodes if (a) the pipe is open at both ends and (b) the pipe is closed at the left end and open at the right end.

16.26 • The fundamental frequency of a pipe that is open at both ends is 594 Hz. (a) How long is this pipe? If one end is now closed, find (b) the wavelength and (c) the frequency of the new fundamental.

16.27 • BIO **The Human Voice.** The human vocal tract is a pipe that extends about 17 cm from the lips to the vocal folds (also called "vocal cords") near the middle of your throat. The vocal folds behave rather like the reed of a clarinet, and the vocal tract acts like a stopped pipe. Estimate the first three standing-wave frequencies of the vocal tract. Use $v = 344$ m/s. (The answers are only an estimate, since the position of lips and tongue affects the motion of air in the vocal tract.)

16.28 •• BIO **The Vocal Tract.** Many opera singers (and some pop singers) have a range of about $2\frac{1}{2}$ octaves or even greater. Suppose a soprano's range extends from A below middle C (frequency 220 Hz) up to E^b-flat above high C (frequency 1244 Hz). Although the vocal tract is quite complicated, we can model it as a resonating air column, like an organ pipe, that is open at the top and closed at the bottom. The column extends from the mouth down to the diaphragm in the chest cavity, and we can also assume that the lowest note is the fundamental. How long is this column of air if $v = 354$ m/s? Does your result seem reasonable, on the basis of observations of your own body?

16.29 •• A certain pipe produces a fundamental frequency of 262 Hz in air. (a) If the pipe is filled with helium at the same temperature, what fundamental frequency does it produce? (The molar mass of air is 28.8 g/mol, and the molar mass of helium is 4.00 g/mol.) (b) Does your answer to part (a) depend on whether the pipe is open or stopped? Why or why not?

16.30 • **Singing in the Shower.** A pipe closed at both ends can have standing waves inside of it, but you normally don't hear them because little of the sound can get out. But you *can* hear them if you are *inside* the pipe, such as someone singing in the shower. (a) Show that the wavelengths of standing waves in a pipe of length L that is closed at both ends are $\lambda_n = 2L/n$ and the frequencies are given by $f_n = nv/2L = nf_1$, where $n = 1, 2, 3, \ldots$. (b) Modeling it as a pipe, find the frequency of the fundamental and the first two overtones for a shower 2.50 m tall. Are these frequencies audible?

Section 16.5 Resonance and Sound

16.31 • You blow across the open mouth of an empty test tube and produce the fundamental standing wave of the air column inside the test tube. The speed of sound in air is 344 m/s and the test tube acts as a stopped pipe. (a) If the length of the air column in the test tube is 14.0 cm, what is the frequency of this standing wave? (b) What is the frequency of the fundamental standing wave in the air column if the test tube is half filled with water?

16.32 •• CP You have a stopped pipe of adjustable length close to a taut 85.0-cm, 7.25-g wire under a tension of 4110 N. You want to adjust the length of the pipe so that, when it produces sound at its fundamental frequency, this sound causes the wire to vibrate in its second *overtone* with very large amplitude. How long should the pipe be?

Section 16.6 Interference of Waves

16.33 • Two loudspeakers, A and B (Fig. E16.33), are driven by the same amplifier and emit sinusoidal waves in phase. Speaker B is 2.00 m to the right of speaker A. Consider point Q along the extension of the line connecting the speakers, 1.00 m to the right of speaker B. Both speakers emit sound waves that travel directly from the speaker to point Q. (a) What is the lowest frequency for which *constructive* interference occurs at point Q? (b) What is the lowest frequency for which *destructive* interference occurs at point Q?

Figure **E16.33**

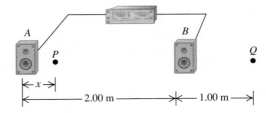

16.34 •• Two loudspeakers, *A* and *B* (see Fig. E16.33), are driven by the same amplifier and emit sinusoidal waves in phase. Speaker *B* is 2.00 m to the right of speaker *A*. The frequency of the sound waves produced by the loudspeakers is 206 Hz. Consider point *P* between the speakers and along the line connecting them, a distance *x* to the right of speaker *A*. Both speakers emit sound waves that travel directly from the speaker to point *P*. (a) For what values of *x* will *destructive* interference occur at point *P*? (b) For what values of *x* will *constructive* interference occur at point *P*? (c) Interference effects like those in parts (a) and (b) are almost never a factor in listening to home stereo equipment. Why not?

16.35 •• Two loudspeakers, *A* and *B*, are driven by the same amplifier and emit sinusoidal waves in phase. Speaker *B* is 12.0 m to the right of speaker *A*. The frequency of the waves emitted by each speaker is 688 Hz. You are standing between the speakers, along the line connecting them, and are at a point of constructive interference. How far must you walk toward speaker *B* to move to a point of destructive interference?

16.36 • Two loudspeakers, *A* and *B*, are driven by the same amplifier and emit sinusoidal waves in phase. The frequency of the waves emitted by each speaker is 172 Hz. You are 8.00 m from *A*. What is the closest you can be to *B* and be at a point of destructive interference?

16.37 • Two loudspeakers, *A* and *B*, are driven by the same amplifier and emit sinusoidal waves in phase. The frequency of the waves emitted by each speaker is 860 Hz. Point *P* is 12.0 m from *A* and 13.4 m from *B*. Is the interference at *P* constructive or destructive? Give the reasoning behind your answer.

16.38 •• Two small stereo speakers are driven in step by the same variable-frequency oscillator. Their sound is picked up by a microphone arranged as shown in Fig. E16.38. For what frequencies does their sound at the speakers produce (a) constructive interference and (b) destructive interference?

Figure **E16.38**

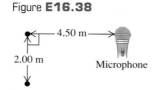

Section 16.7 Beats

16.39 •• **Tuning a Violin.** A violinist is tuning her instrument to concert A (440 Hz). She plays the note while listening to an electronically generated tone of exactly that frequency and hears a beat of frequency 3 Hz, which increases to 4 Hz when she tightens her violin string slightly. (a) What was the frequency of the note played by her violin when she heard the 3-Hz beat? (b) To get her violin perfectly tuned to concert A, should she tighten or loosen her string from what it was when she heard the 3-Hz beat?

16.40 •• Two guitarists attempt to play the same note of wavelength 6.50 cm at the same time, but one of the instruments is slightly out of tune and plays a note of wavelength 6.52 cm instead. What is the frequency of the beat these musicians hear when they play together?

16.41 •• Two organ pipes, open at one end but closed at the other, are each 1.14 m long. One is now lengthened by 2.00 cm. Find the frequency of the beat they produce when playing together in their fundamental.

16.42 •• **Adjusting Airplane Motors.** The motors that drive airplane propellers are, in some cases, tuned by using beats. The whirring motor produces a sound wave having the same frequency as the propeller. (a) If one single-bladed propeller is turning at 575 rpm and you hear a 2.0-Hz beat when you run the second propeller, what are the two possible frequencies (in rpm) of the second propeller? (b) Suppose you increase the speed of the second propeller slightly and find that the beat frequency changes to 2.1 Hz. In part (a), which of the two answers was the correct one for the frequency of the second single-bladed propeller? How do you know?

Section 16.8 The Doppler Effect

16.43 •• On the planet Arrakis a male ornithoid is flying toward his mate at 25.0 m/s while singing at a frequency of 1200 Hz. If the stationary female hears a tone of 1240 Hz, what is the speed of sound in the atmosphere of Arrakis?

16.44 •• In Example 16.18 (Section 16.8), suppose the police car is moving away from the warehouse at 20 m/s. What frequency does the driver of the police car hear reflected from the warehouse?

16.45 • Two train whistles, *A* and *B*, each have a frequency of 392 Hz. *A* is stationary and *B* is moving toward the right (away from *A*) at a speed of 35.0 m/s. A listener is between the two whistles and is moving toward the right with a speed of 15.0 m/s (Fig. E16.45). No wind is blowing. (a) What is the frequency from *A* as heard by the listener? (b) What is the frequency from *B* as heard by the listener? (c) What is the beat frequency detected by the listener?

Figure **E16.45**

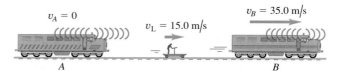

16.46 • A railroad train is traveling at 25.0 m/s in still air. The frequency of the note emitted by the locomotive whistle is 400 Hz. What is the wavelength of the sound waves (a) in front of the locomotive and (b) behind the locomotive? What is the frequency of the sound heard by a stationary listener (c) in front of the locomotive and (d) behind the locomotive?

16.47 • A swimming duck paddles the water with its feet once every 1.6 s, producing surface waves with this period. The duck is moving at constant speed in a pond where the speed of surface waves is 0.32 m/s, and the crests of the waves ahead of the duck are spaced 0.12 m apart. (a) What is the duck's speed? (b) How far apart are the crests behind the duck?

16.48 • **Moving Source vs. Moving Listener.** (a) A sound source producing 1.00-kHz waves moves toward a stationary listener at one-half the speed of sound. What frequency will the listener hear? (b) Suppose instead that the source is stationary and the listener moves toward the source at one-half the speed of sound. What frequency does the listener hear? How does your answer compare to that in part (a)? Explain on physical grounds why the two answers differ.

16.49 • A car alarm is emitting sound waves of frequency 520 Hz. You are on a motorcycle, traveling directly away from the car. How fast must you be traveling if you detect a frequency of 490 Hz?

16.50 • A railroad train is traveling at 30.0 m/s in still air. The frequency of the note emitted by the train whistle is 262 Hz. What frequency is heard by a passenger on a train moving in the opposite direction to the first at 18.0 m/s and (a) approaching the first and (b) receding from the first?

16.51 • Two swift canaries fly toward each other, each moving at 15.0 m/s relative to the ground, each warbling a note of frequency 1750 Hz. (a) What frequency note does each bird hear from the other one? (b) What wavelength will each canary measure for the note from the other one?

16.52 •• The siren of a fire engine that is driving northward at 30.0 m/s emits a sound of frequency 2000 Hz. A truck in front of this fire engine is moving northward at 20.0 m/s. (a) What is the frequency of the siren's sound that the fire engine's driver hears reflected from the back of the truck? (b) What wavelength would this driver measure for these reflected sound waves?

16.53 •• How fast (as a percentage of light speed) would a star have to be moving so that the frequency of the light we receive from it is 10.0% higher than the frequency of the light it is emitting? Would it be moving away from us or toward us? (Assume it is moving either directly away from us or directly toward us.)

16.54 • **Extrasolar Planets.** In the not-too-distant future, it should be possible to detect the presence of planets moving around other stars by measuring the Doppler shift in the infrared light they emit. If a planet is going around its star at 50.00 km/s while emitting infrared light of frequency 3.330×10^{14} Hz, what frequency light will be received from this planet when it is moving directly away from us? (*Note:* Infrared light is light having wavelengths longer than those of visible light.)

Section 16.9 Shock Waves

16.55 •• A jet plane flies overhead at Mach 1.70 and at a constant altitude of 950 m. (a) What is the angle α of the shock-wave cone? (b) How much time after the plane passes directly overhead do you hear the sonic boom? Neglect the variation of the speed of sound with altitude.

16.56 • The shock-wave cone created by the space shuttle at one instant during its reentry into the atmosphere makes an angle of 58.0° with its direction of motion. The speed of sound at this altitude is 331 m/s. (a) What is the Mach number of the shuttle at this instant, and (b) how fast (in m/s and in mi/h) is it traveling relative to the atmosphere? (c) What would be its Mach number and the angle of its shock-wave cone if it flew at the same speed but at low altitude where the speed of sound is 344 m/s?

PROBLEMS

16.57 ••• **CP** Two identical taut strings under the same tension F produce a note of the same fundamental frequency f_0. The tension in one of them is now increased by a very small amount ΔF. (a) If they are played together in their fundamental, show that the frequency of the beat produced is $f_{\text{beat}} = f_0 (\Delta F/2F)$. (b) Two identical violin strings, when in tune and stretched with the same tension, have a fundamental frequency of 440.0 Hz. One of the strings is retuned by increasing its tension. When this is done, 1.5 beats per second are heard when both strings are plucked simultaneously at their centers. By what percentage was the string tension changed?

16.58 •• **CALC** (a) Defend the following statement: "In a sinusoidal sound wave, the pressure variation given by Eq. (16.4) is greatest where the displacement given by Eq. (16.1) is zero." (b) For a sinusoidal sound wave given by Eq. (16.1) with amplitude $A = 10.0 \ \mu\text{m}$ and wavelength $\lambda = 0.250$ m, graph the displacement y and pressure fluctuation p as functions of x at time $t = 0$. Show at least two wavelengths of the wave on your graphs. (c) The displacement y in a *non*sinusoidal sound wave is shown in Fig. P16.58 as a function of x for $t = 0$. Draw a graph showing the pressure fluctuation p in this wave as a function of x at $t = 0$. This sound wave has the same 10.0-μm amplitude as the wave in part (b). Does it have the same pressure amplitude? Why or why not? (d) Is the statement in part (a) necessarily true if the sound wave is *not* sinusoidal? Explain your reasoning.

Figure **P16.58**

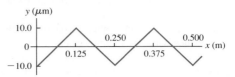

16.59 •• A soprano and a bass are singing a duet. While the soprano sings an A-sharp at 932 Hz, the bass sings an A-sharp but three octaves lower. In this concert hall, the density of air is 1.20 kg/m^3 and its bulk modulus is 1.42×10^5 Pa. In order for their notes to have the same sound intensity level, what must be (a) the ratio of the pressure amplitude of the bass to that of the soprano and (b) the ratio of the displacement amplitude of the bass to that of the soprano? (c) What displacement amplitude (in m and in nm) does the soprano produce to sing her A-sharp at 72.0 dB?

16.60 •• **CP** The sound from a trumpet radiates uniformly in all directions in 20°C air. At a distance of 5.00 m from the trumpet the sound intensity level is 52.0 dB. The frequency is 587 Hz. (a) What is the pressure amplitude at this distance? (b) What is the displacement amplitude? (c) At what distance is the sound intensity level 30.0 dB?

16.61 ••• **A Thermometer.** Suppose you have a tube of length L containing a gas whose temperature you want to take, but you cannot get inside the tube. One end is closed, and the other end is open but a small speaker producing sound of variable frequency is at that end. You gradually increase the frequency of the speaker until the sound from the tube first becomes very loud. With further increase of the frequency, the loudness decreases but then gets very loud again at still higher frequencies. Call f_0 the lowest frequency at which the sound is very loud. (a) Show that the absolute temperature of this gas is given by $T = 16ML^2f_0^2/\gamma R$, where M is the molar mass of the gas, γ is the ratio of its heat capacities, and R is the ideal gas constant. (b) At what frequency above f_0 will the sound from the tube next reach a maximum in loudness? (c) How could you determine the speed of sound in this tube at temperature T?

16.62 •• **CP** A uniform 165-N bar is supported horizontally by two identical wires A and B (Fig. P16.62). A small 185-N cube of lead is placed three-fourths of the way from A to B. The wires are each 75.0 cm long and have a mass of 5.50 g. If both of them are simultaneously plucked at the center, what is the frequency of the beats that they will produce when vibrating in their fundamental?

Figure **P16.62**

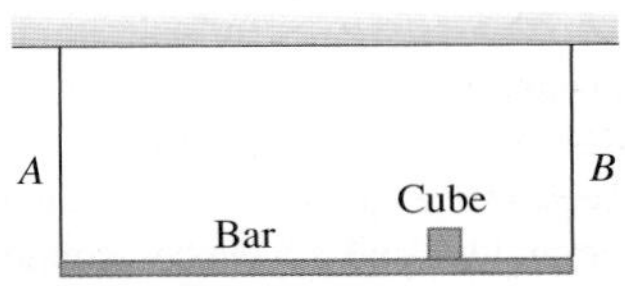

16.63 • **CP** A person is playing a small flute 10.75 cm long, open at one end and closed at the other, near a taut string having a fundamental frequency of 600.0 Hz. If the speed of sound is 344.0 m/s, for which harmonics of the flute will the string resonate? In each case, which harmonic of the string is in resonance?

16.64 ••• **CP** **A New Musical Instrument.** You have designed a new musical instrument of very simple construction. Your design consists of a metal tube with length L and diameter $L/10$. You have stretched a string of mass per unit length μ across the open end of the tube. The other end of the tube is closed. To produce the musical effect you're looking for, you want the frequency of the third-harmonic standing wave on the string to be the same as the fundamental frequency for sound waves in the air column in the tube. The speed of sound waves in this air column is v_s. (a) What must be the tension of the string to produce the desired effect? (b) What happens to the sound produced by the instrument if the tension is changed to twice the value calculated in part (a)? (c) For the tension calculated in part (a), what other harmonics of the string, if any, are in resonance with standing waves in the air column?

16.65 • An organ pipe has two successive harmonics with frequencies 1372 and 1764 Hz. (a) Is this an open or a stopped pipe? Explain. (b) What two harmonics are these? (c) What is the length of the pipe?

16.66 • **Longitudinal Standing Waves in a Solid.** Longitudinal standing waves can be produced in a solid rod by holding it at some point between the fingers of one hand and stroking it with the other hand. The rod oscillates with antinodes at both ends. (a) Why are the ends antinodes and not nodes? (b) The fundamental frequency can be obtained by stroking the rod while it is held at its center. Explain why this is the *only* place to hold the rod to obtain the fundamental. (c) Calculate the fundamental frequency of a steel rod of length 1.50 m (see Table 16.1). (d) What is the next possible standing-wave frequency of this rod? Where should the rod be held to excite a standing wave of this frequency?

16.67 •• A long tube contains air at a pressure of 1.00 atm and a temperature of 77.0°C. The tube is open at one end and closed at the other by a movable piston. A tuning fork near the open end is vibrating with a frequency of 500 Hz. Resonance is produced when the piston is at distances 18.0, 55.5, and 93.0 cm from the open end. (a) From these measurements, what is the speed of sound in air at 77.0°C? (b) From the result of part (a), what is the value of γ? (c) These data show that a displacement antinode is slightly outside of the open end of the tube. How far outside is it?

16.68 ••• The frequency of the note F_4 is 349 Hz. (a) If an organ pipe is open at one end and closed at the other, what length must it have for its fundamental mode to produce this note at 20.0°C? (b) At what air temperature will the frequency be 370 Hz, corresponding to a rise in pitch from F to F-sharp? (Ignore the change in length of the pipe due to the temperature change.)

16.69 • A standing wave with a frequency of 1100 Hz in a column of methane (CH_4) at 20.0°C produces nodes that are 0.200 m apart. What is the value of γ for methane? (The molar mass of methane is 16.0 g/mol.)

16.70 •• Two identical loudspeakers are located at points A and B, 2.00 m apart. The loudspeakers are driven by the same amplifier and produce sound waves with a frequency of 784 Hz. Take the speed of sound in air to be 344 m/s. A small microphone is moved out from point B along a line perpendicular to the line connecting A and B (line BC in Fig. P16.70). (a) At what distances from B will there be *destructive* interference? (b) At what distances from B will there be *constructive* interference? (c) If the frequency is made low enough, there will be no positions along the line BC at which destructive interference occurs. How low must the frequency be for this to be the case?

Figure **P16.70**

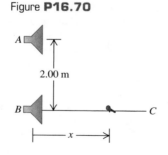

16.71 • **Wagnerian Opera.** A man marries a great Wagnerian soprano but, alas, he discovers he cannot stand Wagnerian opera. In order to save his eardrums, the unhappy man decides he must silence his larklike wife for good. His plan is to tie her to the front of his car and send car and soprano speeding toward a brick wall. This soprano is quite shrewd, however, having studied physics in her student days at the music conservatory. She realizes that this wall has a resonant frequency of 600 Hz, which means that if a continuous sound wave of this frequency hits the wall, it will fall down, and she will be saved to sing more Isoldes. The car is heading toward the wall at a high speed of 30 m/s. (a) At what frequency must the soprano sing so that the wall will crumble? (b) What frequency will the soprano hear reflected from the wall just before it crumbles?

16.72 •• A bat flies toward a wall, emitting a steady sound of frequency 1.70 kHz. This bat hears its own sound plus the sound reflected by the wall. How fast should the bat fly in order to hear a beat frequency of 10.0 Hz?

16.73 •• **CP** A person leaning over a 125-m-deep well accidentally drops a siren emitting sound of frequency 2500 Hz. Just before this siren hits the bottom of the well, find the frequency and wavelength of the sound the person hears (a) coming directly from the siren and (b) reflected off the bottom of the well. (c) What beat frequency does this person perceive?

16.74 ••• **BIO** **Ultrasound in Medicine.** A 2.00-MHz sound wave travels through a pregnant woman's abdomen and is reflected from the fetal heart wall of her unborn baby. The heart wall is moving toward the sound receiver as the heart beats. The reflected sound is then mixed with the transmitted sound, and 72 beats per second are detected. The speed of sound in body tissue is 1500 m/s. Calculate the speed of the fetal heart wall at the instant this measurement is made.

16.75 •• The sound source of a ship's sonar system operates at a frequency of 22.0 kHz. The speed of sound in water (assumed to be at a uniform 20°C) is 1482 m/s. (a) What is the wavelength of the waves emitted by the source? (b) What is the difference in frequency between the directly radiated waves and the waves reflected from a whale traveling directly toward the ship at 4.95 m/s? The ship is at rest in the water.

16.76 • **CP** A police siren of frequency f_{siren} is attached to a vibrating platform. The platform and siren oscillate up and down in simple harmonic motion with amplitude A_p and frequency f_p. (a) Find the maximum and minimum sound frequencies that you would hear at a position directly above the siren. (b) At what point in the motion of the platform is the maximum frequency heard? The minimum frequency? Explain.

16.77 ••• **BIO** Horseshoe bats (genus *Rhinolophus*) emit sounds from their nostrils and then listen to the frequency of the sound reflected from their prey to determine the prey's speed. (The "horseshoe" that gives the bat its name is a depression around the nostrils that acts like a focusing mirror, so that the bat emits sound in a narrow beam like a flashlight.) A *Rhinolophus* flying at speed v_{bat} emits sound of frequency f_{bat}; the sound it hears reflected from an insect flying toward it has a higher frequency f_{refl}. (a) Show that the speed of the insect is

$$v_{insect} = v \left[\frac{f_{refl}(v - v_{bat}) - f_{bat}(v + v_{bat})}{f_{refl}(v - v_{bat}) + f_{bat}(v + v_{bat})} \right]$$

where v is the speed of sound. (b) If $f_{bat} = 80.7$ kHz, $f_{refl} = 83.5$ kHz, and $v_{bat} = 3.9$ m/s, calculate the speed of the insect.

16.78 •• (a) Show that Eq. (16.30) can be written as

$$f_R = f_S \left(1 - \frac{v}{c} \right)^{1/2} \left(1 + \frac{v}{c} \right)^{-1/2}$$

(b) Use the binomial theorem to show that if $v \ll c$, this is approximately equal to

$$f_R = f_S\left(1 - \frac{v}{c}\right)$$

(c) A pilotless reconnaissance aircraft emits a radio signal with a frequency of 243 MHz. It is flying directly toward a test engineer on the ground. The engineer detects beats between the received signal and a local signal also of frequency 243 MHz. The beat frequency is 46.0 Hz. What is the speed of the aircraft? (Radio waves travel at the speed of light, $c = 3.00 \times 10^8$ m/s.)

16.79 •• **Supernova!** The gas cloud known as the Crab Nebula can be seen with even a small telescope. It is the remnant of a *supernova,* a cataclysmic explosion of a star. The explosion was seen on the earth on July 4, 1054 C.E. The streamers glow with the characteristic red color of heated hydrogen gas. In a laboratory on the earth, heated hydrogen produces red light with frequency 4.568×10^{14} Hz; the red light received from streamers in the Crab Nebula pointed toward the earth has frequency 4.586×10^{14} Hz. (a) Estimate the speed with which the outer edges of the Crab Nebula are expanding. Assume that the speed of the center of the nebula relative to the earth is negligible. (You may use the formulas derived in Problem 16.78. The speed of light is 3.00×10^8 m/s.) (b) Assuming that the expansion speed has been constant since the supernova explosion, estimate the diameter of the Crab Nebula. Give your answer in meters and in light-years. (c) The angular diameter of the Crab Nebula as seen from earth is about 5 arc minutes (1 arc minute = $\frac{1}{60}$ degree). Estimate the distance (in light-years) to the Crab Nebula, and estimate the year in which the supernova explosion actually took place.

16.80 •• **CP** A turntable 1.50 m in diameter rotates at 75 rpm. Two speakers, each giving off sound of wavelength 31.3 cm, are attached to the rim of the table at opposite ends of a diameter. A listener stands in front of the turntable. (a) What is the greatest beat frequency the listener will receive from this system? (b) Will the listener be able to distinguish individual beats?

16.81 •• A woman stands at rest in front of a large, smooth wall. She holds a vibrating tuning fork of frequency f_0 directly in front of her (between her and the wall). (a) The woman now runs toward the wall with speed v_W. She detects beats due to the interference between the sound waves reaching her directly from the fork and those reaching her after being reflected from the wall. How many beats per second will she detect? (*Note:* If the beat frequency is too large, the woman may have to use some instrumentation other than

her ears to detect and count the beats.) (b) If the woman instead runs away from the wall, holding the tuning fork at her back so it is between her and the wall, how many beats per second will she detect?

16.82 •• On a clear day you see a jet plane flying overhead. From the apparent size of the plane, you determine that it is flying at a constant altitude h. You hear the sonic boom at time T after the plane passes directly overhead. Show that if the speed of sound v is the same at all altitudes, the speed of the plane is

$$v_S = \frac{hv}{\sqrt{h^2 - v^2 T^2}}$$

(*Hint:* Trigonometric identities will be useful.)

CHALLENGE PROBLEMS

16.83 ••• **CALC** Figure P16.83 shows the pressure fluctuation p of a nonsinusoidal sound wave as a function of x for $t = 0$. The wave is traveling in the $+x$-direction. (a) Graph the pressure fluctuation p as a function of t for $x = 0$. Show at least two cycles of oscillation. (b) Graph the displacement y in this sound wave as a function of x at $t = 0$. At $x = 0$, the displacement at $t = 0$ is zero. Show at least two wavelengths of the wave. (c) Graph the displacement y as a function of t for $x = 0$. Show at least two cycles of oscillation. (d) Calculate the maximum velocity and the maximum acceleration of an element of the air through which this sound wave is traveling. (e) Describe how the cone of a loudspeaker must move as a function of time to produce the sound wave in this problem.

Figure **P16.83**

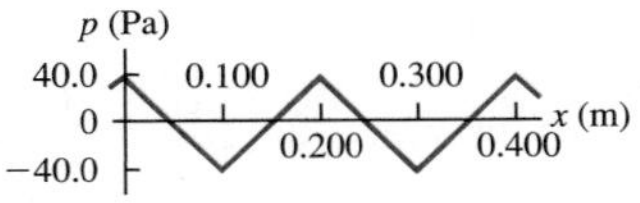

16.84 ••• **CP** **Longitudinal Waves on a Spring.** A long spring such as a Slinky™ is often used to demonstrate longitudinal waves. (a) Show that if a spring that obeys Hooke's law has mass m, length L, and force constant k', the speed of longitudinal waves on the spring is $v = L\sqrt{k'/m}$. (see Section 16.2). (b) Evaluate v for a spring with $m = 0.250$ kg, $L = 2.00$ m, and $k' = 1.50$ N/m.

Answers

Chapter Opening Question ?

Both musical sound and noise are made up of a combination of sinusoidal sound waves. The difference is that the frequencies of the sine waves in musical sound are all integer multiples of a fundamental frequency, while *all* frequencies are present in noise.

Test Your Understanding Questions

16.1 Answer: (v) From Eq. (16.5), the displacement amplitude is $A = p_{max}/Bk$. The pressure amplitude p_{max} and bulk modulus B remain the same, but the frequency f increases by a factor of 4. Hence the wave number $k = \omega/v = 2\pi f/v$ also increases by a factor of 4. Since A is inversely proportional to k, the displacement amplitude becomes $\frac{1}{4}$ as great. In other words, at higher frequency

a smaller maximum displacement is required to produce the same maximum pressure fluctuation.

16.2 Answer: (i) From Eq. (16.7), the speed of longitudinal waves (sound) in a fluid is $v = \sqrt{B/\rho}$. We can rewrite this to give an expression for the bulk modulus B in terms of the fluid density ρ and the sound speed v: $B = \rho v^2$. At 20°C the speed of sound in mercury is slightly less than in water (1451 m/s versus 1482 m/s), but the density of mercury is greater than that of water by a large factor (13.6). Hence the bulk modulus of mercury is greater than that of water by a factor of $(13.6)(1451/1482)^2 = 13.0$.

16.3 Answer: A and p_{max} increase by a factor of $\sqrt{2}$, B and v are unchanged, β increases by 3.0 dB Equations (16.9) and (16.10) show that the bulk modulus B and sound speed v remain the same because the physical properties of the air are unchanged. From Eqs. (16.12) and (16.14), the intensity is proportional to the square of the displacement amplitude or the square of the pressure amplitude. Hence doubling the intensity means that A and p_{max} both increase by a factor of $\sqrt{2}$. Example 16.9 shows that *multiplying* the intensity by a factor of $2(I_2/I_1 = 2)$ corresponds to *adding* to the sound intensity level by $(10\text{ dB})\log(I_2/I_1) = (10\text{ dB})\log 2 = 3.0\text{ dB}$.

16.4 Answer: (ii) Helium is less dense and has a lower molar mass than air, so sound travels faster in helium than in air. The normal-mode frequencies for a pipe are proportional to the sound speed v, so the frequency and hence the pitch increase when the air in the pipe is replaced with helium.

16.5 Answer: (i) and (iv) There will be a resonance if 660 Hz is one of the pipe's normal-mode frequencies. A stopped organ pipe has normal-mode frequencies that are odd multiples of its fundamental frequency [see Eq. (16.22) and Fig. 16.18]. Hence pipe (i), which has fundamental frequency 220 Hz, also has a normal-mode frequency of $3(220\text{ Hz}) = 660\text{ Hz}$. Pipe (ii) has twice the length of pipe (i); from Eq. (16.20), the fundamental frequency of a stopped pipe is inversely proportional to the length, so pipe (ii) has a fundamental frequency of $\left(\frac{1}{2}\right)(220\text{ Hz}) = 110\text{ Hz}$. Its other normal-mode frequencies are 330 Hz, 550 Hz, 770 Hz, ..., so a 660-Hz tuning fork will not cause resonance. Pipe (iii) is an open pipe of the same length as pipe (i), so its fundamental frequency is twice as great as for pipe (i) [compare Eqs. (16.16) and (16.20)], or $2(220\text{ Hz}) = 440\text{ Hz}$. Its other normal-mode frequencies are integer multiples of the fundamental frequency [see Eq. (16.19)], or 880 Hz, 1320 Hz, ..., none of which match the 660-Hz frequency of the tuning fork. Pipe (iv) is also an open pipe but with twice the length of pipe (iii) [see Eq. (16.18)], so its normal-mode frequencies are one-half those of pipe (iii): 220 Hz, 440 Hz, 660 Hz, ..., so the third harmonic will resonate with the tuning fork.

16.6 Answer: (iii) Constructive and destructive interference between two waves can occur only if the two waves have the same frequency. In this case the frequencies are different, so there are no points where the two waves always reinforce each other (constructive interference) or always cancel each other (destructive interference).

16.7 Answer: (vi) The beat frequency is 3 Hz, so the difference between the two tuning fork frequencies is also 3 Hz. Hence the second tuning fork vibrates at a frequency of either 443 Hz or 437 Hz. You can distinguish between the two possibilities by comparing the pitches of the two tuning forks sounded one at a time: The frequency is 437 Hz if the second tuning fork has a lower pitch and 443 Hz if it has a higher pitch.

16.8 Answer: no The air (the medium for sound waves) is moving from the source toward the listener. Hence, relative to the air, both the source and the listener are moving in the direction from listener to source. So both velocities are positive and $v_S = v_L = +10\text{ m/s}$. The equality of these two velocities means that the numerator and the denominator in Eq. (16.29) are the same, so $f_L = f_S$ and there is *no* Doppler shift.

16.9 Answer: (iii) Figure 16.37 shows that there are sound waves inside the cone of the shock wave. Behind the airplane the wave crests are spread apart, just as they are behind the moving source in Fig. 16.27. Hence the waves that reach you have an increased wavelength and a lower frequency.

Bridging Problem

Answers: (a) $180° = \pi$ rad

(b) A alone: $I = 3.98 \times 10^{-6}\text{ W/m}^2$, $\beta = 66.0\text{ dB}$;
B alone: $I = 5.31 \times 10^{-7}\text{ W/m}^2$, $\beta = 57.2\text{ dB}$

(c) $I = 1.60 \times 10^{-6}\text{ W/m}^2$, $\beta = 62.1\text{ dB}$

TEMPERATURE
AND HEAT

? At a steelworks, molten iron is heated to 1500° Celsius to remove impurities.
Is it accurate to say that the molten iron contains heat?

LEARNING GOALS

*By studying this chapter, you will
learn:*

* The meaning of thermal equilibrium,
  and what thermometers really
  measure.

* How different types of thermometers
  function.

* The physics behind the absolute, or
  Kelvin, temperature scale.

* How the dimensions of an object
  change as a result of a temperature
  change.

* The meaning of heat, and how it
  differs from temperature.

* How to do calculations that involve
  heat flow, temperature changes, and
  changes of phase.

* How heat is transferred by conduction, convection, and radiation.

Whether it's a sweltering summer day or a frozen midwinter night, your body needs to be kept at a nearly constant temperature. It has effective temperature-control mechanisms, but sometimes it needs help. On a hot day you wear less clothing to improve heat transfer from your body to the air and for better cooling by evaporation of perspiration. You drink cold beverages and may sit near a fan or in an air-conditioned room. On a cold day you wear more clothes or stay indoors where it's warm. When you're outside, you keep active and drink hot liquids to stay warm. The concepts in this chapter will help you understand the basic physics of keeping warm or cool.

The terms "temperature" and "heat" are often used interchangeably in everyday language. In physics, however, these two terms have very different meanings. In this chapter we'll define temperature in terms of how it's measured and see how temperature changes affect the dimensions of objects. We'll see that heat refers to energy transfer caused by temperature differences and learn how to calculate and control such energy transfers.

Our emphasis in this chapter is on the concepts of temperature and heat as they relate to *macroscopic* objects such as cylinders of gas, ice cubes, and the human body. In Chapter 18 we'll look at these same concepts from a *microscopic* viewpoint in terms of the behavior of individual atoms and molecules. These two chapters lay the groundwork for the subject of **thermodynamics,** the study of energy transformations involving heat, mechanical work, and other aspects of energy and how these transformations relate to the properties of matter. Thermodynamics forms an indispensable part of the foundation of physics, chemistry, and the life sciences, and its applications turn up in such places as car engines, refrigerators, biochemical processes, and the structure of stars. We'll explore the key ideas of thermodynamics in Chapters 19 and 20.

551

17.1 Temperature and Thermal Equilibrium

The concept of **temperature** is rooted in qualitative ideas of "hot" and "cold" based on our sense of touch. A body that feels hot usually has a higher temperature than a similar body that feels cold. That's pretty vague, and the senses can be deceived. But many properties of matter that we can *measure* depend on temperature. The length of a metal rod, steam pressure in a boiler, the ability of a wire to conduct an electric current, and the color of a very hot glowing object—all these depend on temperature.

Temperature is also related to the kinetic energies of the molecules of a material. In general this relationship is fairly complex, so it's not a good place to start in *defining* temperature. In Chapter 18 we will look at the relationship between temperature and the energy of molecular motion for an ideal gas. It is important to understand, however, that temperature and heat can be defined independently of any detailed molecular picture. In this section we'll develop a *macroscopic* definition of temperature.

To use temperature as a measure of hotness or coldness, we need to construct a temperature scale. To do this, we can use any measurable property of a system that varies with its "hotness" or "coldness." Figure 17.1a shows a familiar system that is used to measure temperature. When the system becomes hotter, the colored liquid (usually mercury or ethanol) expands and rises in the tube, and the value of L increases. Another simple system is a quantity of gas in a constant-volume container (Fig. 17.1b). The pressure p, measured by the gauge, increases or decreases as the gas becomes hotter or colder. A third example is the electrical resistance R of a conducting wire, which also varies when the wire becomes hotter or colder. Each of these properties gives us a number (L, p, or R) that varies with hotness and coldness, so each property can be used to make a **thermometer.**

To measure the temperature of a body, you place the thermometer in contact with the body. If you want to know the temperature of a cup of hot coffee, you stick the thermometer in the coffee; as the two interact, the thermometer becomes hotter and the coffee cools off a little. After the thermometer settles down to a steady value, you read the temperature. The system has reached an *equilibrium* condition, in which the interaction between the thermometer and the coffee causes no further change in the system. We call this a state of **thermal equilibrium.**

If two systems are separated by an insulating material or **insulator** such as wood, plastic foam, or fiberglass, they influence each other more slowly. Camping coolers are made with insulating materials to delay the ice and cold food inside from warming up and attaining thermal equilibrium with the hot summer air outside. An *ideal insulator* is a material that permits no interaction at all between the two systems. It prevents the systems from attaining thermal equilibrium if they aren't in thermal equilibrium at the start. An ideal insulator is just that, an idealization; real insulators, like those in camping coolers, aren't ideal, so the contents of the cooler will warm up eventually.

The Zeroth Law of Thermodynamics

We can discover an important property of thermal equilibrium by considering three systems, A, B, and C, that initially are not in thermal equilibrium (Fig. 17.2). We surround them with an ideal insulating box so that they cannot interact with anything except each other. We separate systems A and B with an ideal insulating wall (the green slab in Fig. 17.2a), but we let system C interact with both systems A and B. This interaction is shown in the figure by a yellow slab representing a thermal **conductor,** a material that *permits* thermal interactions through it. We wait until thermal equilibrium is attained; then A and B are each in thermal equilibrium with C. But are they in thermal equilibrium *with each other?*

To find out, we separate system C from systems A and B with an ideal insulating wall (Fig. 17.2b), and then we replace the insulating wall between A and B

17.1 Two devices for measuring temperature.

(a) Changes in temperature cause the liquid's volume to change.

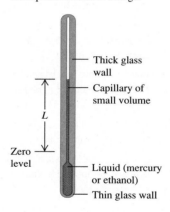

- Thick glass wall
- Capillary of small volume
- L
- Zero level
- Liquid (mercury or ethanol)
- Thin glass wall

(b) Changes in temperature cause the pressure of the gas to change.

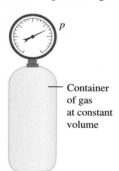

- p
- Container of gas at constant volume

with a *conducting* wall that lets *A* and *B* interact. What happens? Experiment shows that *nothing* happens; there are no additional changes to *A* or *B*. We conclude:

> If *C* is initially in thermal equilibrium with both *A* and *B*, then *A* and *B* are also in thermal equilibrium with each other. This result is called the **zeroth law of thermodynamics.**

(The importance of this law was recognized only after the first, second, and third laws of thermodynamics had been named. Since it is fundamental to all of them, the name "zeroth" seemed appropriate.)

Now suppose system *C* is a thermometer, such as the liquid-in-tube system of Fig. 17.1a. In Fig. 17.2a the thermometer *C* is in contact with both *A* and *B*. In thermal equilibrium, when the thermometer reading reaches a stable value, the thermometer measures the temperature of both *A* and *B*; hence *A* and *B* both have the *same* temperature. Experiment shows that thermal equilibrium isn't affected by adding or removing insulators, so the reading of thermometer *C* wouldn't change if it were in contact only with *A* or only with *B*. We conclude:

> Two systems are in thermal equilibrium if and only if they have the same temperature.

This is what makes a thermometer useful; a thermometer actually measures *its own* temperature, but when a thermometer is in thermal equilibrium with another body, the temperatures must be equal. When the temperatures of two systems are different, they *cannot* be in thermal equilibrium.

Test Your Understanding of Section 17.1   You put a thermometer in a pot of hot water and record the reading. What temperature have you recorded? (i) the temperature of the water; (ii) the temperature of the thermometer; (iii) an equal average of the temperatures of the water and thermometer; (iv) a weighted average of the temperatures of the water and thermometer, with more emphasis on the temperature of the water; (v) a weighted average of the water and thermometer, with more emphasis on the temperature of the thermometer.   ❙

17.2 **Thermometers and Temperature Scales**

To make the liquid-in-tube device shown in Fig. 17.1a into a useful thermometer, we need to mark a scale on the tube wall with numbers on it. These numbers are arbitrary, and historically many different schemes have been used. Suppose we label the thermometer's liquid level at the freezing temperature of pure water "zero" and the level at the boiling temperature "100," and divide the distance between these two points into 100 equal intervals called *degrees*. The result is the **Celsius temperature scale** (formerly called the *centigrade* scale in English-speaking countries). The Celsius temperature for a state colder than freezing water is a negative number. The Celsius scale is used, both in everyday life and in science and industry, almost everywhere in the world.

Another common type of thermometer uses a *bimetallic strip,* made by bonding strips of two different metals together (Fig. 17.3a). When the temperature of the composite strip increases, one metal expands more than the other and the strip bends (Fig. 17.3b). This strip is usually formed into a spiral, with the outer end anchored to the thermometer case and the inner end attached to a pointer (Fig. 17.3c). The pointer rotates in response to temperature changes.

In a *resistance thermometer* the changing electrical resistance of a coil of fine wire, a carbon cylinder, or a germanium crystal is measured. Resistance thermometers are usually more precise than most other types.

17.2 The zeroth law of thermodynamics.

(a) If systems *A* and *B* are each in thermal equilibrium with system *C* …

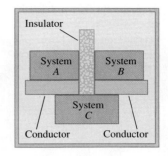

(b) … then systems *A* and *B* are in thermal equilibrium with each other.

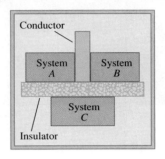

17.3 Use of a bimetallic strip as a thermometer.

(a) A bimetallic strip

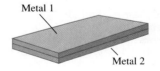

(b) The strip bends when its temperature is raised.

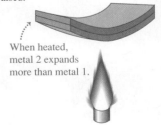

When heated, metal 2 expands more than metal 1.

(c) A bimetallic strip used in a thermometer

17.4 A temporal artery thermometer measures infrared radiation from the skin that overlies one of the important arteries in the head. Although the thermometer cover touches the skin, the infrared detector inside the cover does not.

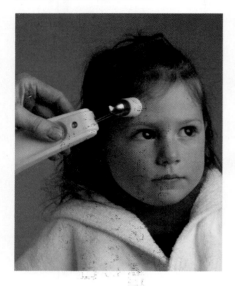

Application Mammalian Body Temperatures

Most mammals maintain body temperatures in the range from 36°C to 40°C (309 K to 313 K). A high metabolic rate warms the animal from within, and insulation (such as fur, feathers, and body fat) slows heat loss.

Some thermometers work by detecting the amount of infrared radiation emitted by an object. (We'll see in Section 17.7 that *all* objects emit electromagnetic radiation, including infrared, as a consequence of their temperature.) A modern example is a *temporal artery thermometer* (Fig. 17.4). A nurse runs this over a patient's forehead in the vicinity of the temporal artery, and an infrared sensor in the thermometer measures the radiation from the skin. Tests show that this device gives more accurate values of body temperature than do oral or ear thermometers.

In the **Fahrenheit temperature scale,** still used in everyday life in the United States, the freezing temperature of water is 32°F (thirty-two degrees Fahrenheit) and the boiling temperature is 212°F, both at standard atmospheric pressure. There are 180 degrees between freezing and boiling, compared to 100 on the Celsius scale, so one Fahrenheit degree represents only $\frac{100}{180}$, or $\frac{5}{9}$, as great a temperature change as one Celsius degree.

To convert temperatures from Celsius to Fahrenheit, note that a Celsius temperature T_C is the number of Celsius degrees above freezing; the number of Fahrenheit degrees above freezing is $\frac{9}{5}$ of this. But freezing on the Fahrenheit scale is at 32°F, so to obtain the actual Fahrenheit temperature T_F, multiply the Celsius value by $\frac{9}{5}$ and then add 32°. Symbolically,

$$T_F = \tfrac{9}{5}T_C + 32° \tag{17.1}$$

To convert Fahrenheit to Celsius, solve this equation for T_C:

$$T_C = \tfrac{5}{9}(T_F - 32°) \tag{17.2}$$

In words, subtract 32° to get the number of Fahrenheit degrees above freezing, and then multiply by $\frac{5}{9}$ to obtain the number of Celsius degrees above freezing—that is, the Celsius temperature.

We don't recommend memorizing Eqs. (17.1) and (17.2). Instead, try to understand the reasoning that led to them so that you can derive them on the spot when you need them, checking your reasoning with the relationship 100°C = 212°F.

It is useful to distinguish between an actual temperature and a temperature *interval* (a difference or change in temperature). An actual temperature of 20° is stated as 20°C (twenty degrees Celsius), and a temperature *interval* of 10° is 10 C° (ten Celsius degrees). A beaker of water heated from 20°C to 30°C undergoes a temperature change of 10 C°.

Test Your Understanding of Section 17.2 Which of the following types of thermometers have to be in thermal equilibrium with the object being measured in order to give accurate readings? (i) a bimetallic strip; (ii) a resistance thermometer; (iii) a temporal artery thermometer; (iv) both (i) and (ii); (v) all of (i), (ii), and (iii).

17.3 Gas Thermometers and the Kelvin Scale

When we calibrate two thermometers, such as a liquid-in-tube system and a resistance thermometer, so that they agree at 0°C and 100°C, they may not agree exactly at intermediate temperatures. Any temperature scale defined in this way always depends somewhat on the specific properties of the material used. Ideally, we would like to define a temperature scale that *doesn't* depend on the properties of a particular material. To establish a truly material-independent scale, we first need to develop some principles of thermodynamics. We'll return to this fundamental problem in Chapter 20. Here we'll discuss a thermometer that comes close to the ideal, the *gas thermometer*.

17.5 (a) Using a constant-volume gas thermometer to measure temperature. (b) The greater the amount of gas in the thermometer, the higher the graph of pressure p versus temperature T.

(a) A constant-volume gas thermometer

(b) Graphs of pressure versus temperature at constant volume for three different types and quantities of gas

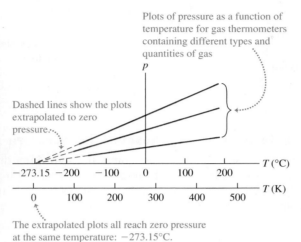

Plots of pressure as a function of temperature for gas thermometers containing different types and quantities of gas

Dashed lines show the plots extrapolated to zero pressure.

The extrapolated plots all reach zero pressure at the same temperature: −273.15°C.

The principle of a gas thermometer is that the pressure of a gas at constant volume increases with temperature. A quantity of gas is placed in a constant-volume container (Fig. 17.5a), and its pressure is measured by one of the devices described in Section 12.2. To calibrate a constant-volume gas thermometer, we measure the pressure at two temperatures, say 0°C and 100°C, plot these points on a graph, and draw a straight line between them. Then we can read from the graph the temperature corresponding to any other pressure. Figure 17.5b shows the results of three such experiments, each using a different type and quantity of gas.

By extrapolating this graph, we see that there is a hypothetical temperature, −273.15°C, at which the absolute pressure of the gas would become zero. We might expect that this temperature would be different for different gases, but it turns out to be the *same* for many different gases (at least in the limit of very low gas density). We can't actually observe this zero-pressure condition. Gases liquefy and solidify at very low temperatures, and the proportionality of pressure to temperature no longer holds.

We use this extrapolated zero-pressure temperature as the basis for a temperature scale with its zero at this temperature. This is the **Kelvin temperature scale,** named for the British physicist Lord Kelvin (1824–1907). The units are the same size as those on the Celsius scale, but the zero is shifted so that 0 K = −273.15°C and 273.15 K = 0°C; that is,

$$T_K = T_C + 273.15 \qquad (17.3)$$

Figure 17.5b shows both the Celsius and Kelvin scales. A common room temperature, 20°C (= 68°F), is 20 + 273.15, or about 293 K.

CAUTION *Never say "degrees kelvin"* In SI nomenclature, "degree" is not used with the Kelvin scale; the temperature mentioned above is read "293 kelvins," not "degrees kelvin" (Fig. 17.6). We capitalize Kelvin when it refers to the temperature scale; however, the *unit* of temperature is the *kelvin,* which is not capitalized (but is nonetheless abbreviated as a capital K). ▌

17.6 Correct and incorrect uses of the Kelvin scale.

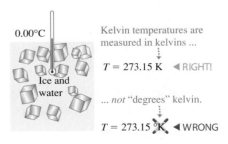

0.00°C

Ice and water

Kelvin temperatures are measured in kelvins ...

$T = 273.15$ K ◀ RIGHT!

... *not* "degrees" kelvin.

$T = 273.15$ °K ◀ WRONG

Example 17.1 **Body temperature**

You place a small piece of ice in your mouth. Eventually, the water all converts from ice at $T_1 = 32.00°F$ to body temperature, $T_2 = 98.60°F$. Express these temperatures in both Celsius degrees and kelvins, and find $\Delta T = T_2 - T_1$ in both cases.

SOLUTION

IDENTIFY and SET UP: Our target variables are stated above. We convert Fahrenheit temperatures to Celsius using Eq. (17.2), and Celsius temperatures to Kelvin using Eq. (17.3).

EXECUTE: From Eq. (17.2), $T_1 = 0.00°C$ and $T_2 = 37.00°C$; then $\Delta T = T_2 - T_1 = 37.00$ C°. To get the Kelvin temperatures, just add 273.15 to each Celsius temperature: $T_1 = 273.15$ K and $T_2 = 310.15$ K. The temperature difference is $\Delta T = T_2 - T_1 = 37.00$ K.

EVALUATE: The Celsius and Kelvin scales have different zero points but the same size degrees. Therefore *any* temperature difference ΔT is the *same* on the Celsius and Kelvin scales. However, ΔT is *not* the same on the Fahrenheit scale; here, for example, $\Delta T = 66.60$ F°.

The Kelvin Scale and Absolute Temperature

The Celsius scale has two fixed points: the normal freezing and boiling temperatures of water. But we can define the Kelvin scale using a gas thermometer with only a single reference temperature. We define the ratio of any two temperatures T_1 and T_2 on the Kelvin scale as the ratio of the corresponding gas-thermometer pressures p_1 and p_2:

$$\frac{T_2}{T_1} = \frac{p_2}{p_1} \quad \text{(constant-volume gas thermometer, } T \text{ in kelvins)} \quad (17.4)$$

The pressure p is directly proportional to the Kelvin temperature, as shown in Fig. 17.5b. To complete the definition of T, we need only specify the Kelvin temperature of a single specific state. For reasons of precision and reproducibility, the state chosen is the *triple point* of water. This is the unique combination of temperature and pressure at which solid water (ice), liquid water, and water vapor can all coexist. It occurs at a temperature of 0.01°C and a water-vapor pressure of 610 Pa (about 0.006 atm). (This is the pressure of the *water;* it has nothing to do directly with the gas pressure in the *thermometer.*) The triple-point temperature T_{triple} of water is *defined* to have the value $T_{triple} = 273.16$ K, corresponding to 0.01°C. From Eq. (17.4), if p_{triple} is the pressure in a gas thermometer at temperature T_{triple} and p is the pressure at some other temperature T, then T is given on the Kelvin scale by

$$T = T_{triple}\frac{p}{p_{triple}} = (273.16 \text{ K})\frac{p}{p_{triple}} \quad (17.5)$$

Low-pressure gas thermometers using various gases are found to agree very closely, but they are large, bulky, and very slow to come to thermal equilibrium. They are used principally to establish high-precision standards and to calibrate other thermometers.

Figure 17.7 shows the relationships among the three temperature scales we have discussed. The Kelvin scale is called an **absolute temperature scale,** and its zero point ($T = 0$ K $= -273.15°C$, the temperature at which $p = 0$ in Eq. (17.5)) is called **absolute zero.** At absolute zero a system of molecules (such as a quantity of a gas, a liquid, or a solid) has its *minimum* possible total energy (kinetic plus potential); because of quantum effects, however, it is *not* correct to say that all molecular motion ceases at absolute zero. To define more completely what we mean by absolute zero, we need to use the thermodynamic principles developed in the next several chapters. We will return to this concept in Chapter 20.

17.7 Relationships among Kelvin (K), Celsius (C), and Fahrenheit (F) temperature scales. Temperatures have been rounded off to the nearest degree.

	K	C	F
Water boils	373	100°	212°
	100 K	100 C°	180 F°
Water freezes	273	0°	32°
CO_2 solidifies	195	−78°	−109°
Oxygen liquefies	90	−183°	−298°
Absolute zero	0	−273°	−460°

Test Your Understanding of Section 17.3 Rank the following temperatures from highest to lowest: (i) 0.00°C; (ii) 0.00°F; (iii) 260.00 K; (iv) 77.00 K; (v) −180.00°C.

17.4 Thermal Expansion

Most materials expand when their temperatures increase. Rising temperatures make the liquid expand in a liquid-in-tube thermometer (Fig. 17.1a) and bend bimetallic strips (Fig. 17.3b). The decks of bridges need special joints and supports to allow for expansion. A completely filled and tightly capped bottle of water cracks when it is heated, but you can loosen a metal jar lid by running hot water over it. These are all examples of *thermal expansion.*

Linear Expansion

Suppose a rod of material has a length L_0 at some initial temperature T_0. When the temperature changes by ΔT, the length changes by ΔL. Experiments show that if ΔT is not too large (say, less than 100 C° or so), ΔL is *directly proportional* to ΔT (Fig. 17.8a). If two rods made of the same material have the same temperature change, but one is twice as long as the other, then the *change* in its length is also twice as great. Therefore ΔL must also be proportional to L_0 (Fig. 17.8b). Introducing a proportionality constant α (which is different for different materials), we may express these relationships in an equation:

$$\Delta L = \alpha L_0 \Delta T \quad \text{(linear thermal expansion)} \quad [17.6]$$

If a body has length L_0 at temperature T_0, then its length L at a temperature $T = T_0 + \Delta T$ is

$$L = L_0 + \Delta L = L_0 + \alpha L_0 \Delta T = L_0(1 + \alpha \Delta T) \quad [17.7]$$

The constant α, which describes the thermal expansion properties of a particular material, is called the **coefficient of linear expansion.** The units of α are K^{-1} or $(C°)^{-1}$. (Remember that a temperature *interval* is the same in the Kelvin and Celsius scales.) For many materials, every linear dimension changes according to Eq. (17.6) or (17.7). Thus L could be the thickness of a rod, the side length of a square sheet, or the diameter of a hole. Some materials, such as wood or single crystals, expand differently in different directions. We won't consider this complication.

We can understand thermal expansion qualitatively on a molecular basis. Picture the interatomic forces in a solid as springs, as in Fig. 17.9a. (We explored the analogy between spring forces and interatomic forces in Section 14.4.) Each atom vibrates about its equilibrium position. When the temperature increases, the energy and amplitude of the vibration also increase. The interatomic spring forces are not symmetrical about the equilibrium position; they usually behave like a spring that is easier to stretch than to compress. As a result, when the amplitude of vibration increases, the *average* distance between atoms also increases (Fig. 17.9b). As the atoms get farther apart, every dimension increases.

17.8 How the length of a rod changes with a change in temperature. (Length changes are exaggerated for clarity.)

(a) For moderate temperature changes, ΔL is directly proportional to ΔT.

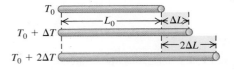

(b) ΔL is also directly proportional to L_0.

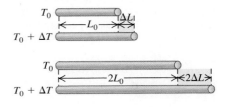

(a) A model of the forces between neighboring atoms in a solid

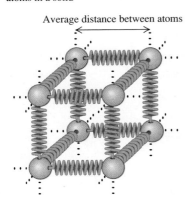

Average distance between atoms

(b) A graph of the "spring" potential energy $U(x)$

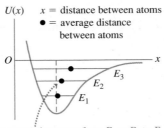

As energy increases from E_1 to E_2 to E_3, average distance between atoms increases.

17.9 (a) We can model atoms in a solid as being held together by "springs" that are easier to stretch than to compress. (b) A graph of the "spring" potential energy $U(x)$ versus distance x between neighboring atoms is *not* symmetrical (compare Fig. 14.20b). As the energy increases and the atoms oscillate with greater amplitude, the average distance increases.

17.10 When an object undergoes thermal expansion, any holes in the object expand as well. (The expansion is exaggerated.)

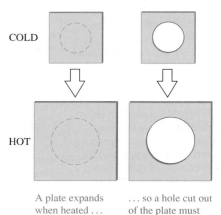

COLD

HOT

A plate expands when heated ...

... so a hole cut out of the plate must expand, too.

17.11 When this SR-71 aircraft is sitting on the ground, its wing panels fit together so loosely that fuel leaks out of the wings onto the ground. But once it is in flight at over three times the speed of sound, air friction heats the panels so much that they expand to make a perfect fit. (In-flight refueling makes up for the lost fuel.)

CAUTION **Heating an object with a hole** If a solid object has a hole in it, what happens to the size of the hole when the temperature of the object increases? A common misconception is that if the object expands, the hole will shrink because material expands into the hole. But the truth of the matter is that if the object expands, the hole will expand too (Fig. 17.10); as we stated above, *every* linear dimension of an object changes in the same way when the temperature changes. If you're not convinced, think of the atoms in Fig. 17.9a as outlining a cubical hole. When the object expands, the atoms move apart and the hole increases in size. The only situation in which a "hole" will fill in due to thermal expansion is when two separate objects expand and close the gap between them (Fig. 17.11).

The direct proportionality expressed by Eq. (17.6) is not exact; it is *approximately* correct only for sufficiently small temperature changes. For a given material, α varies somewhat with the initial temperature T_0 and the size of the temperature interval. We'll ignore this complication here, however. Average values of α for several materials are listed in Table 17.1. Within the precision of these values we don't need to worry whether T_0 is 0°C or 20°C or some other temperature. Note that typical values of α are very small; even for a temperature change of 100 C°, the fractional length change $\Delta L/L_0$ is only of the order of $\frac{1}{1000}$ for the metals in the table.

Volume Expansion

Increasing temperature usually causes increases in *volume* for both solid and liquid materials. Just as with linear expansion, experiments show that if the temperature change ΔT is not too great (less than 100 C° or so), the increase in volume ΔV is approximately proportional to both the temperature change ΔT and the initial volume V_0:

$$\Delta V = \beta V_0 \, \Delta T \quad \text{(volume thermal expansion)} \quad (17.8)$$

The constant β characterizes the volume expansion properties of a particular material; it is called the **coefficient of volume expansion.** The units of β are K^{-1} or $(\text{C}°)^{-1}$. As with linear expansion, β varies somewhat with temperature, and Eq. (17.8) is an approximate relationship that is valid only for small temperature changes. For many substances, β decreases at low temperatures. Several values of β in the neighborhood of room temperature are listed in Table 17.2. Note that the values for liquids are generally much larger than those for solids.

For solid materials there is a simple relationship between the volume expansion coefficient β and the linear expansion coefficient α. To derive this relationship, we consider a cube of material with side length L and volume $V = L^3$. At the initial temperature the values are L_0 and V_0. When the temperature increases by dT, the side length increases by dL and the volume increases by an amount dV given by

$$dV = \frac{dV}{dL} \, dL = 3L^2 \, dL$$

Table 17.1 Coefficients of Linear Expansion

Material	$\alpha \, [\text{K}^{-1} \text{ or } (\text{C}°)^{-1}]$
Aluminum	2.4×10^{-5}
Brass	2.0×10^{-5}
Copper	1.7×10^{-5}
Glass	0.4–0.9×10^{-5}
Invar (nickel–iron alloy)	0.09×10^{-5}
Quartz (fused)	0.04×10^{-5}
Steel	1.2×10^{-5}

Table 17.2 Coefficients of Volume Expansion

Solids	$\beta \, [\text{K}^{-1} \text{ or } (\text{C}°)^{-1}]$	Liquids	$\beta \, [\text{K}^{-1} \text{ or } (\text{C}°)^{-1}]$
Aluminum	7.2×10^{-5}	Ethanol	75×10^{-5}
Brass	6.0×10^{-5}	Carbon disulfide	115×10^{-5}
Copper	5.1×10^{-5}	Glycerin	49×10^{-5}
Glass	1.2–2.7×10^{-5}	Mercury	18×10^{-5}
Invar	0.27×10^{-5}		
Quartz (fused)	0.12×10^{-5}		
Steel	3.6×10^{-5}		

Now we replace L and V by the initial values L_0 and V_0. From Eq. (17.6), dL is

$$dL = \alpha L_0 \, dT$$

Since $V_0 = L_0^3$, this means that dV can also be expressed as

$$dV = 3L_0^2 \alpha L_0 \, dT = 3\alpha V_0 \, dT$$

This is consistent with the infinitesimal form of Eq. (17.8), $dV = \beta V_0 \, dT$, only if

$$\beta = 3\alpha \qquad (17.9)$$

You should check this relationship for some of the materials listed in Tables 17.1 and 17.2.

Problem-Solving Strategy 17.1   Thermal Expansion

IDENTIFY *the relevant concepts:* Decide whether the problem involves changes in length (linear thermal expansion) or in volume (volume thermal expansion).

SET UP *the problem* using the following steps:
1. List the known and unknown quantities and identify the target variables.
2. Choose Eq. (17.6) for linear expansion and Eq. (17.8) for volume expansion.

EXECUTE *the solution* as follows:
1. Solve for the target variables. If you are given an initial temperature T_0 and must find a final temperature T corresponding to a given length or volume change, find ΔT and calculate $T = T_0 + \Delta T$. Remember that the size of a hole in a material varies with temperature just as any other linear dimension, and that the volume of a hole (such as the interior of a container) varies just as that of the corresponding solid shape.
2. Maintain unit consistency. Both L_0 and ΔL (or V_0 and ΔV) must have the same units. If you use a value of α or β in K^{-1} or $(C°)^{-1}$, then ΔT must be in either kelvins or Celsius degrees; from Example 17.1, the two scales are equivalent *for temperature differences.*

EVALUATE *your answer:* Check whether your results make sense.

Example 17.2   Length change due to temperature change

A surveyor uses a steel measuring tape that is exactly 50.000 m long at a temperature of 20°C. The markings on the tape are calibrated for this temperature. (a) What is the length of the tape when the temperature is 35°C? (b) When it is 35°C, the surveyor uses the tape to measure a distance. The value that she reads off the tape is 35.794 m. What is the actual distance?

SOLUTION

IDENTIFY and SET UP: This problem concerns the linear expansion of a measuring tape. We are given the tape's initial length $L_0 = 50.000$ m at $T_0 = 20$°C. In part (a) we use Eq. (17.6) to find the change ΔL in the tape's length at $T = 35$°C, and use Eq. (17.7) to find L. (Table 17.1 gives the value of α for steel.) Since the tape expands, at 35°C the distance between two successive meter marks is greater than 1 m. Hence the actual distance in part (b) is *larger* than the distance read off the tape by a factor equal to the ratio of the tape's length L at 35°C to its length L_0 at 20°C.

EXECUTE: (a) The temperature change is $\Delta T = T - T_0 = 15$ C°; from Eqs. (17.6) and (17.7),

$$\Delta L = \alpha L_0 \, \Delta T = (1.2 \times 10^{-5} \text{ K}^{-1})(50 \text{ m})(15 \text{ K})$$
$$= 9.0 \times 10^{-3} \text{ m} = 9.0 \text{ mm}$$
$$L = L_0 + \Delta L = 50.000 \text{ m} + 0.009 \text{ m} = 50.009 \text{ m}$$

(b) Our result from part (a) shows that at 35°C, the slightly expanded tape reads a distance of 50.000 m when the true distance is 50.009 m. We can rewrite the algebra of part (a) as $L = L_0(1 + \alpha \Delta T)$; at 35°C, *any* true distance will be greater than the reading by the factor $50.009/50.000 = 1 + \alpha \Delta T = 1 + 1.8 \times 10^{-4}$. The true distance is therefore

$$(1 + 1.8 \times 10^{-4})(35.794 \text{ m}) = 35.800 \text{ m}$$

EVALUATE: Note that in part (a) we needed only two of the five significant figures of L_0 to compute ΔL to the same number of decimal places as L_0. Our result shows that metals expand very little under moderate temperature changes. However, even the small difference 0.009 m = 9 mm found in part (b) between the scale reading and the true distance can be important in precision work.

Example 17.3   Volume change due to temperature change

A 200-cm³ glass flask is filled to the brim with mercury at 20°C. How much mercury overflows when the temperature of the system is raised to 100°C? The coefficient of *linear* expansion of the glass is 0.40×10^{-5} K^{-1}.

Continued

SOLUTION

IDENTIFY and SET UP: This problem involves the volume expansion of the glass and of the mercury. The amount of overflow depends on the *difference* between the volume changes ΔV for these two materials, both given by Eq. (17.8). The mercury will overflow if its coefficient of volume expansion β (given in Table 17.2) is greater than that of glass, which we find from Eq. (17.9) using the given value of α.

EXECUTE: From Table 17.2, $\beta_{Hg} = 18 \times 10^{-5}\,\text{K}^{-1}$. That is indeed greater than β_{glass}: From Eq. (17.9), $\beta_{glass} = 3\alpha_{glass} = 3(0.40 \times 10^{-5}\,\text{K}^{-1}) = 1.2 \times 10^{-5}\,\text{K}^{-1}$. The volume overflow is then

$$\Delta V_{Hg} - \Delta V_{glass} = \beta_{Hg}V_0\Delta T - \beta_{glass}V_0\Delta T$$
$$= V_0\Delta T(\beta_{Hg} - \beta_{glass})$$
$$= (200\ \text{cm}^3)(80\ \text{C}°)(18 \times 10^{-5} - 1.2 \times 10^{-5})$$
$$= 2.7\ \text{cm}^3$$

EVALUATE: This is basically how a mercury-in-glass thermometer works; the column of mercury inside a sealed tube rises as T increases because mercury expands faster than glass.

As Tables 17.1 and 17.2 show, glass has smaller coefficients of expansion α and β than do most metals. This is why you can use hot water to loosen a metal lid on a glass jar; the metal expands more than the glass does.

Thermal Expansion of Water

17.12 The volume of 1 gram of water in the temperature range from 0°C to 100°C. By 100°C the volume has increased to 1.034 cm³. If the coefficient of volume expansion were constant, the curve would be a straight line.

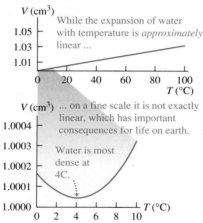

Water, in the temperature range from 0°C to 4°C, *decreases* in volume with increasing temperature. In this range its coefficient of volume expansion is *negative*. Above 4°C, water expands when heated (Fig. 17.12). Hence water has its greatest density at 4°C. Water also expands when it freezes, which is why ice humps up in the middle of the compartments in an ice cube tray. By contrast, most materials contract when they freeze.

This anomalous behavior of water has an important effect on plant and animal life in lakes. A lake cools from the surface down; above 4°C, the cooled water at the surface flows to the bottom because of its greater density. But when the surface temperature drops below 4°C, the water near the surface is less dense than the warmer water below. Hence the downward flow ceases, and the water near the surface remains colder than that at the bottom. As the surface freezes, the ice floats because it is less dense than water. The water at the bottom remains at 4°C until nearly the entire lake is frozen. If water behaved like most substances, contracting continuously on cooling and freezing, lakes would freeze from the bottom up. Circulation due to density differences would continuously carry warmer water to the surface for efficient cooling, and lakes would freeze solid much more easily. This would destroy all plant and animal life that cannot withstand freezing. If water did not have this special property, the evolution of life would have taken a very different course.

Thermal Stress

If we clamp the ends of a rod rigidly to prevent expansion or contraction and then change the temperature, **thermal stresses** develop. The rod would like to expand or contract, but the clamps won't let it. The resulting stresses may become large enough to strain the rod irreversibly or even break it. (You may want to review the discussion of stress and strain in Section 11.4).

Engineers must account for thermal stress when designing structures. Concrete highways and bridge decks usually have gaps between sections, filled with a flexible material or bridged by interlocking teeth (Fig. 17.13), to permit expansion and contraction of the concrete. Long steam pipes have expansion joints or U-shaped sections to prevent buckling or stretching with temperature changes. If one end of a steel bridge is rigidly fastened to its abutment, the other end usually rests on rollers.

To calculate the thermal stress in a clamped rod, we compute the amount the rod *would* expand (or contract) if not held and then find the stress needed to compress (or stretch) it back to its original length. Suppose that a rod with length L_0 and cross-sectional area A is held at constant length while the temperature is reduced (negative ΔT), causing a tensile stress. The fractional change in length if the rod were free to contract would be

$$\left(\frac{\Delta L}{L_0}\right)_{thermal} = \alpha\,\Delta T \tag{17.10}$$

17.13 Expansion joints on bridges are needed to accommodate changes in length that result from thermal expansion.

Both ΔL and ΔT are negative. The tension must increase by an amount F that is just enough to produce an equal and opposite fractional change in length $(\Delta L/L_0)_{\text{tension}}$. From the definition of Young's modulus, Eq. (11.10),

$$Y = \frac{F/A}{\Delta L/L_0} \qquad \text{so} \qquad \left(\frac{\Delta L}{L_0}\right)_{\text{tension}} = \frac{F}{AY} \qquad (17.11)$$

If the length is to be constant, the *total* fractional change in length must be zero. From Eqs. (17.10) and (17.11), this means that

$$\left(\frac{\Delta L}{L_0}\right)_{\text{thermal}} + \left(\frac{\Delta L}{L_0}\right)_{\text{tension}} = \alpha \, \Delta T + \frac{F}{AY} = 0$$

Solving for the tensile stress F/A required to keep the rod's length constant, we find

$$\frac{F}{A} = -Y\alpha \, \Delta T \qquad \text{(thermal stress)} \qquad (17.12)$$

For a decrease in temperature, ΔT is negative, so F and F/A are positive; this means that a *tensile* force and stress are needed to maintain the length. If ΔT is positive, F and F/A are negative, and the required force and stress are *compressive.*

If there are temperature differences within a body, nonuniform expansion or contraction will result and thermal stresses can be induced. You can break a glass bowl by pouring very hot water into it; the thermal stress between the hot and cold parts of the bowl exceeds the breaking stress of the glass, causing cracks. The same phenomenon makes ice cubes crack when dropped into warm water. Heat-resistant glasses such as Pyrex™ have exceptionally low expansion coefficients and high strength.

Example 17.4 Thermal stress

An aluminum cylinder 10 cm long, with a cross-sectional area of 20 cm², is used as a spacer between two steel walls. At 17.2°C it just slips between the walls. Calculate the stress in the cylinder and the total force it exerts on each wall when it warms to 22.3°C, assuming that the walls are perfectly rigid and a constant distance apart.

SOLUTION

IDENTIFY and SET UP: Figure 17.14 shows our sketch of the situation. Our target variables are the thermal stress F/A in the cylinder, whose cross-sectional area A is given, and the associated force F it

17.14 Our sketch for this problem.

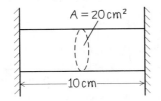

A = 20 cm²

10 cm

exerts on the walls. We use Eq. (17.12) to relate F/A to the temperature change ΔT, and from that calculate F. (The length of the cylinder is irrelevant.) We find Young's modulus Y_{Al} and the coefficient of linear expansion α_{Al} from Tables 11.1 and 17.1, respectively.

EXECUTE: We have $Y_{\text{Al}} = 7.0 \times 10^{10}$ Pa and $\alpha_{\text{Al}} = 2.4 \times 10^{-5} \, \text{K}^{-1}$, and $\Delta T = 22.3°\text{C} - 17.2°\text{C} = 5.1 \, \text{C}° = 5.1$ K. From Eq. (17.12), the stress is

$$\frac{F}{A} = -Y_{\text{Al}}\alpha_{\text{Al}}\Delta T$$
$$= -(7.0 \times 10^{10} \text{ Pa})(2.4 \times 10^{-5} \text{ K}^{-1})(5.1 \text{ K})$$
$$= -8.6 \times 10^6 \text{ Pa} = -1200 \text{ lb/in.}^2$$

The total force is the cross-sectional area times the stress:

$$F = A\left(\frac{F}{A}\right) = (20 \times 10^{-4} \text{ m}^2)(-8.6 \times 10^6 \text{ Pa})$$
$$= -1.7 \times 10^4 \text{ N} = 1.9 \text{ tons}$$

EVALUATE: The stress on the cylinder and the force it exerts on each wall are immense. Such thermal stresses must be accounted for in engineering.

Test Your Understanding of Section 17.4 In the bimetallic strip shown in Fig. 17.3a, metal 1 is copper. Which of the following materials could be used for metal 2? (There may be more than one correct answer). (i) steel; (ii) brass; (iii) aluminum. ❙

17.5 Quantity of Heat

17.15 The same temperature change of the same system may be accomplished by (a) doing work on it or (b) adding heat to it.

(a) Raising the temperature of water by doing work on it

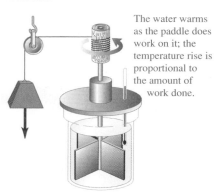

The water warms as the paddle does work on it; the temperature rise is proportional to the amount of work done.

(b) Raising the temperature of water by direct heating

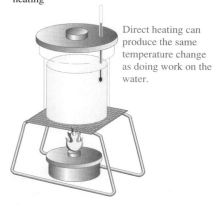

Direct heating can produce the same temperature change as doing work on the water.

17.16 The word "energy" is of Greek origin. This label on a can of Greek coffee shows that 100 milliliters of prepared coffee have an energy content ($\varepsilon\nu\acute{\varepsilon}\rho\gamma\varepsilon\iota\alpha$) of 9.6 kilojoules or 2.3 kilocalories.

When you put a cold spoon into a cup of hot coffee, the spoon warms up and the coffee cools down as they approach thermal equilibrium. The interaction that causes these temperature changes is fundamentally a transfer of *energy* from one substance to another. Energy transfer that takes place solely because of a temperature difference is called *heat flow* or *heat transfer,* and energy transferred in this way is called **heat.**

An understanding of the relationship between heat and other forms of energy emerged during the 18th and 19th centuries. Sir James Joule (1818–1889) studied how water can be warmed by vigorous stirring with a paddle wheel (Fig. 17.15a). The paddle wheel adds energy to the water by doing *work* on it, and Joule found that *the temperature rise is directly proportional to the amount of work done.* The same temperature change can also be caused by putting the water in contact with some hotter body (Fig. 17.15b); hence this interaction must also involve an energy exchange. We will explore the relationship between heat and mechanical energy in Chapters 19 and 20.

CAUTION **Temperature vs. heat** It is absolutely essential for you to distinguish between *temperature* and *heat.* Temperature depends on the physical state of a material and is a quantitative description of its hotness or coldness. In physics the term "heat" always refers to energy in transit from one body or system to another because of a temperature difference, never to the amount of energy contained within a particular system. We can change the temperature of a body by adding heat to it or taking heat away, or by adding or subtracting energy in other ways, such as mechanical work (Fig. 17.15a). If we cut a body in half, each half has the same temperature as the whole; but to raise the temperature of each half by a given interval, we add *half* as much heat as for the whole. ▮

We can define a *unit* of quantity of heat based on temperature changes of some specific material. The **calorie** (abbreviated cal) is defined as *the amount of heat required to raise the temperature of 1 gram of water from 14.5°C to 15.5°C.* The kilocalorie (kcal), equal to 1000 cal, is also used; a food-value calorie is actually a kilocalorie (Fig. 17.16). A corresponding unit of heat using Fahrenheit degrees and British units is the **British thermal unit,** or Btu. One Btu is the quantity of heat required to raise the temperature of 1 pound (weight) of water 1 F° from 63°F to 64°F.

Because heat is energy in transit, there must be a definite relationship between these units and the familiar mechanical energy units such as the joule. Experiments similar in concept to Joule's have shown that

$$1 \text{ cal} = 4.186 \text{ J}$$
$$1 \text{ kcal} = 1000 \text{ cal} = 4186 \text{ J}$$
$$1 \text{ Btu} = 778 \text{ ft} \cdot \text{lb} = 252 \text{ cal} = 1055 \text{ J}$$

The calorie is not a fundamental SI unit. The International Committee on Weights and Measures recommends using the joule as the basic unit of energy in all forms, including heat. We will follow that recommendation in this book.

Specific Heat

We use the symbol Q for quantity of heat. When it is associated with an infinitesimal temperature change dT, we call it dQ. The quantity of heat Q required to increase the temperature of a mass m of a certain material from T_1 to T_2 is found to be approximately proportional to the temperature change $\Delta T = T_2 - T_1$. It is also proportional to the mass m of material. When you're heating water to make tea, you need twice as much heat for two cups as for one if the temperature change is the same. The quantity of heat needed also depends on the nature of the material; raising the temperature of 1 kilogram of water by 1 C° requires 4190 J of heat, but only 910 J is needed to raise the temperature of 1 kilogram of aluminum by 1 C°.

Putting all these relationships together, we have

$$Q = mc\ \Delta T \quad \text{(heat required for temperature change } \Delta T \text{ of mass } m)} \quad \text{(17.13)}$$

where c is a quantity, different for different materials, called the **specific heat** of the material. For an infinitesimal temperature change dT and corresponding quantity of heat dQ,

$$dQ = mc\ dT \qquad (17.14)$$

$$c = \frac{1}{m}\frac{dQ}{dT} \quad \text{(specific heat)} \qquad (17.15)$$

In Eqs. (17.13), (17.14), and (17.15), Q (or dQ) and ΔT (or dT) can be either positive or negative. When they are positive, heat enters the body and its temperature increases; when they are negative, heat leaves the body and its temperature decreases.

CAUTION   **The definition of heat** Remember that dQ does not represent a change in the amount of heat *contained* in a body; this is a meaningless concept. Heat is always energy *in transit* as a result of a temperature difference. There is no such thing as "the amount of heat in a body."

The specific heat of water is approximately

$$4190\ \text{J/kg} \cdot \text{K} \qquad 1\ \text{cal/g} \cdot \text{C}^{\circ} \qquad \text{or} \qquad 1\ \text{Btu/lb} \cdot \text{F}^{\circ}$$

The specific heat of a material always depends somewhat on the initial temperature and the temperature interval. Figure 17.17 shows this dependence for water. In the problems and examples in this chapter we will usually ignore this small variation.

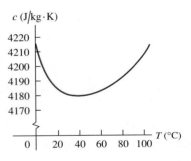

17.17 Specific heat of water as a function of temperature. The value of c varies by less than 1% between 0°C and 100°C.

Example 17.5   **Feed a cold, starve a fever**

During a bout with the flu an 80-kg man ran a fever of 39.0°C (102.2°F) instead of the normal body temperature of 37.0°C (98.6°F). Assuming that the human body is mostly water, how much heat is required to raise his temperature by that amount?

SOLUTION

IDENTIFY and SET UP: This problem uses the relationship among heat (the target variable), mass, specific heat, and temperature change. We use Eq. (17.13) to determine the required heat Q, with $m = 80$ kg, $c = 4190$ J/kg $\cdot$ K (for water), and $\Delta T = 39.0°C - 37.0°C = 2.0\ \text{C}^{\circ} = 2.0$ K.

EXECUTE: From Eq. (17.13),

$$Q = mc\ \Delta T = (80\ \text{kg})(4190\ \text{J/kg}\cdot\text{K})(2.0\ \text{K}) = 6.7 \times 10^5\ \text{J}$$

EVALUATE: This corresponds to 160 kcal. In fact, the specific heat of the human body is about 3480 J/kg $\cdot$ K, 83% that of water, because protein, fat, and minerals have lower specific heats. Hence a more accurate answer is $Q = 5.6 \times 10^5$ J $= 133$ kcal. Either result shows us that were it not for the body's temperature-regulating systems, taking in energy in the form of food would produce measurable changes in body temperature. (The elevated temperature of a person with the flu results from the body's extra activity in response to infection.)

Example 17.6   **Overheating electronics**

You are designing an electronic circuit element made of 23 mg of silicon. The electric current through it adds energy at the rate of 7.4 mW $= 7.4 \times 10^{-3}$ J/s. If your design doesn't allow any heat transfer out of the element, at what rate does its temperature increase? The specific heat of silicon is 705 J/kg $\cdot$ K.

SOLUTION

IDENTIFY and SET UP: The energy added to the circuit element gives rise to a temperature increase, just as if heat were flowing into the element at the rate $dQ/dt = 7.4 \times 10^{-3}$ J/s. Our target variable is the rate of temperature change dT/dt. We can use Eq. (17.14),

which relates infinitesimal temperature changes dT to the corresponding heat dQ, to obtain an expression for dQ/dt in terms of dT/dt.

EXECUTE: We divide both sides of Eq. (17.14) by dt and rearrange:

$$\frac{dT}{dt} = \frac{dQ/dt}{mc} = \frac{7.4 \times 10^{-3}\ \text{J/s}}{(23 \times 10^{-6}\ \text{kg})(705\ \text{J/kg}\cdot\text{K})} = 0.46\ \text{K/s}$$

EVALUATE: At this rate of temperature rise (27 K/min), the circuit element would soon self-destruct. Heat transfer is an important design consideration in electronic circuit elements.

Molar Heat Capacity

Sometimes it's more convenient to describe a quantity of substance in terms of the number of *moles n* rather than the *mass m* of material. Recall from your study of chemistry that a mole of any pure substance always contains the same number of molecules. (We will discuss this point in more detail in Chapter 18.) The *molar mass* of any substance, denoted by *M*, is the mass per mole. (The quantity *M* is sometimes called *molecular weight,* but *molar mass* is preferable; the quantity depends on the mass of a molecule, not its weight.) For example, the molar mass of water is $18.0 \text{ g/mol} = 18.0 \times 10^{-3} \text{ kg/mol}$; 1 mole of water has a mass of $18.0 \text{ g} = 0.0180 \text{ kg}$. The total mass *m* of material is equal to the mass per mole *M* times the number of moles *n*:

$$m = nM \tag{17.16}$$

Replacing the mass *m* in Eq. (17.13) by the product *nM*, we find

$$Q = nMc \, \Delta T \tag{17.17}$$

The product *Mc* is called the **molar heat capacity** (or *molar specific heat*) and is denoted by *C* (capitalized). With this notation we rewrite Eq. (17.17) as

$$Q = nC \, \Delta T \quad \text{(heat required for temperature change of } n \text{ moles)} \tag{17.18}$$

Comparing to Eq. (17.15), we can express the molar heat capacity *C* (heat per mole per temperature change) in terms of the specific heat *c* (heat per mass per temperature change) and the molar mass *M* (mass per mole):

$$C = \frac{1}{n}\frac{dQ}{dT} = Mc \quad \text{(molar heat capacity)} \tag{17.19}$$

For example, the molar heat capacity of water is

$$C = Mc = (0.0180 \text{ kg/mol})(4190 \text{ J/kg} \cdot \text{K}) = 75.4 \text{ J/mol} \cdot \text{K}$$

Values of specific heat and molar heat capacity for several substances are given in Table 17.3. Note the remarkably large specific heat for water (Fig. 17.18).

> **CAUTION** **The meaning of "heat capacity"** The term "heat capacity" is unfortunate because it gives the erroneous impression that a body *contains* a certain amount of heat. Remember, heat is energy in transit to or from a body, not the energy residing in the body.

17.18 Water has a much higher specific heat than the glass or metals used to make cookware. This helps explain why it takes several minutes to boil water on a stove, even though the pot or kettle reaches a high temperature very quickly.

Table 17.3 Approximate Specific Heats and Molar Heat Capacities (Constant Pressure)

Substance	Specific Heat, c (J/kg · K)	Molar Mass, M (kg/mol)	Molar Heat Capacity, C (J/mol · K)
Aluminum	910	0.0270	24.6
Beryllium	1970	0.00901	17.7
Copper	390	0.0635	24.8
Ethanol	2428	0.0461	111.9
Ethylene glycol	2386	0.0620	148.0
Ice (near 0°C)	2100	0.0180	37.8
Iron	470	0.0559	26.3
Lead	130	0.207	26.9
Marble ($CaCO_3$)	879	0.100	87.9
Mercury	138	0.201	27.7
Salt (NaCl)	879	0.0585	51.4
Silver	234	0.108	25.3
Water (liquid)	4190	0.0180	75.4

Precise measurements of specific heats and molar heat capacities require great experimental skill. Usually, a measured quantity of energy is supplied by an electric current in a heater wire wound around the specimen. The temperature change ΔT is measured with a resistance thermometer or thermocouple embedded in the specimen. This sounds simple, but great care is needed to avoid or compensate for unwanted heat transfer between the sample and its surroundings. Measurements for solid materials are usually made at constant atmospheric pressure; the corresponding values are called the *specific heat* and *molar heat capacity at constant pressure,* denoted by c_p and C_p. For a gas it is usually easier to keep the substance in a container with constant *volume;* the corresponding values are called the *specific heat* and *molar heat capacity at constant volume,* denoted by c_V and C_V. For a given substance, C_V and C_p are different. If the system can expand while heat is added, there is additional energy exchange through the performance of *work* by the system on its surroundings. If the volume is constant, the system does no work. For gases the difference between C_p and C_V is substantial. We will study heat capacities of gases in detail in Section 19.7.

The last column of Table 17.3 shows something interesting. The molar heat capacities for most elemental solids are about the same: about 25 J/mol · K. This correlation, named the *rule of Dulong and Petit* (for its discoverers), forms the basis for a very important idea. The number of atoms in 1 mole is the same for all elemental substances. This means that on a *per atom* basis, about the same amount of heat is required to raise the temperature of each of these elements by a given amount, even though the *masses* of the atoms are very different. The heat required for a given temperature increase depends only on *how many* atoms the sample contains, not on the mass of an individual atom. We will see the reason the rule of Dulong and Petit works so well when we study the molecular basis of heat capacities in greater detail in Chapter 18.

Test Your Understanding of Section 17.5 You wish to raise the temperature of each of the following samples from 20°C to 21°C. Rank these in order of the amount of heat needed to do this, from highest to lowest. (i) 1 kilogram of mercury; (ii) 1 kilogram of ethanol; (iii) 1 mole of mercury; (iv) 1 mole of ethanol.

17.6 Calorimetry and Phase Changes

Calorimetry means "measuring heat." We have discussed the energy transfer (heat) involved in temperature changes. Heat is also involved in *phase changes,* such as the melting of ice or boiling of water. Once we understand these additional heat relationships, we can analyze a variety of problems involving quantity of heat.

Phase Changes

We use the term **phase** to describe a specific state of matter, such as a solid, liquid, or gas. The compound H_2O exists in the *solid phase* as ice, in the *liquid phase* as water, and in the *gaseous phase* as steam. (These are also referred to as **states of matter:** the solid state, the liquid state, and the gaseous state.) A transition from one phase to another is called a **phase change** or *phase transition.* For any given pressure a phase change takes place at a definite temperature, usually accompanied by absorption or emission of heat and a change of volume and density.

A familiar example of a phase change is the melting of ice. When we add heat to ice at 0°C and normal atmospheric pressure, the temperature of the ice *does not* increase. Instead, some of it melts to form liquid water. If we add the heat slowly, to maintain the system very close to thermal equilibrium, the temperature remains at 0°C until all the ice is melted (Fig. 17.19). The effect of adding heat to this system is not to raise its temperature but to change its *phase* from solid to liquid.

To change 1 kg of ice at 0°C to 1 kg of liquid water at 0°C and normal atmospheric pressure requires 3.34×10^5 J of heat. The heat required per unit mass is

17.19 The surrounding air is at room temperature, but this ice–water mixture remains at 0°C until all of the ice has melted and the phase change is complete.

called the **heat of fusion** (or sometimes *latent heat of fusion*), denoted by L_f. For water at normal atmospheric pressure the heat of fusion is

$$L_f = 3.34 \times 10^5 \text{ J/kg} = 79.6 \text{ cal/g} = 143 \text{ Btu/lb}$$

More generally, to melt a mass m of material that has a heat of fusion L_f requires a quantity of heat Q given by

$$Q = mL_f$$

This process is *reversible.* To freeze liquid water to ice at 0°C, we have to *remove* heat; the magnitude is the same, but in this case, Q is negative because heat is removed rather than added. To cover both possibilities and to include other kinds of phase changes, we write

$$Q = \pm mL \qquad \text{(heat transfer in a phase change)} \qquad (17.20)$$

The plus sign (heat entering) is used when the material melts; the minus sign (heat leaving) is used when it freezes. The heat of fusion is different for different materials, and it also varies somewhat with pressure.

For any given material at any given pressure, the freezing temperature is the same as the melting temperature. At this unique temperature the liquid and solid phases (liquid water and ice, for example) can coexist in a condition called **phase equilibrium.**

We can go through this whole story again for *boiling* or *evaporation,* a phase transition between liquid and gaseous phases. The corresponding heat (per unit mass) is called the **heat of vaporization** L_v. At normal atmospheric pressure the heat of vaporization L_v for water is

$$L_v = 2.256 \times 10^6 \text{ J/kg} = 539 \text{ cal/g} = 970 \text{ Btu/lb}$$

That is, it takes 2.256×10^6 J to change 1 kg of liquid water at 100°C to 1 kg of water vapor at 100°C. By comparison, to raise the temperature of 1 kg of water from 0°C to 100°C requires $Q = mc \, \Delta T = (1.00 \text{ kg})(4190 \text{ J/kg} \cdot \text{C}°) \times (100 \text{ C}°) = 4.19 \times 10^5$ J, less than one-fifth as much heat as is required for vaporization at 100°C. This agrees with everyday kitchen experience; a pot of water may reach boiling temperature in a few minutes, but it takes a much longer time to completely evaporate all the water away.

Like melting, boiling is a reversible transition. When heat is removed from a gas at the boiling temperature, the gas returns to the liquid phase, or *condenses,* giving up to its surroundings the same quantity of heat (heat of vaporization) that was needed to vaporize it. At a given pressure the boiling and condensation temperatures are always the same; at this temperature the liquid and gaseous phases can coexist in phase equilibrium.

Both L_v and the boiling temperature of a material depend on pressure. Water boils at a lower temperature (about 95°C) in Denver than in Pittsburgh because Denver is at higher elevation and the average atmospheric pressure is lower. The heat of vaporization is somewhat greater at this lower pressure, about 2.27×10^6 J/kg.

Table 17.4 lists heats of fusion and vaporization for some materials and their melting and boiling temperatures at normal atmospheric pressure. Very few *elements* have melting temperatures in the vicinity of ordinary room temperatures; one of the few is the metal gallium, shown in Fig. 17.20.

Figure 17.21 shows how the temperature varies when we add heat continuously to a specimen of ice with an initial temperature below 0°C (point *a*). The temperature rises until we reach the melting point (point *b*). As more heat is added, the temperature remains constant until all the ice has melted (point *c*). Then the temperature rises again until the boiling temperature is reached (point *d*). At that point the temperature again is constant until all the water is transformed into the vapor phase (point *e*). If the rate of heat input is constant, the line for the solid phase (ice) has a steeper slope than does the line for the liquid phase (water). Do you see why? (See Table 17.3.)

17.20 The metal gallium, shown here melting in a person's hand, is one of the few elements that melt in the vicinity of room temperature. Its melting temperature is 29.8°C, and its heat of fusion is 8.04×10^4 J/kg.

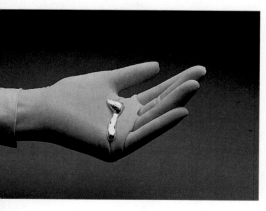

Table 17.4 Heats of Fusion and Vaporization

Substance	Normal Melting Point		Heat of Fusion, L_f (J/kg)	Normal Boiling Point		Heat of Vaporization, L_v (J/kg)
	K	°C		K	°C	
Helium	*	*	*	4.216	−268.93	20.9×10^3
Hydrogen	13.84	−259.31	58.6×10^3	20.26	−252.89	452×10^3
Nitrogen	63.18	−209.97	25.5×10^3	77.34	−195.8	201×10^3
Oxygen	54.36	−218.79	13.8×10^3	90.18	−183.0	213×10^3
Ethanol	159	−114	104.2×10^3	351	78	854×10^3
Mercury	234	−39	11.8×10^3	630	357	272×10^3
Water	273.15	0.00	334×10^3	373.15	100.00	2256×10^3
Sulfur	392	119	38.1×10^3	717.75	444.60	326×10^3
Lead	600.5	327.3	24.5×10^3	2023	1750	871×10^3
Antimony	903.65	630.50	165×10^3	1713	1440	561×10^3
Silver	1233.95	960.80	88.3×10^3	2466	2193	2336×10^3
Gold	1336.15	1063.00	64.5×10^3	2933	2660	1578×10^3
Copper	1356	1083	134×10^3	1460	1187	5069×10^3

*A pressure in excess of 25 atmospheres is required to make helium solidify. At 1 atmosphere pressure, helium remains a liquid down to absolute zero.

A substance can sometimes change directly from the solid to the gaseous phase. This process is called *sublimation,* and the solid is said to *sublime.* The corresponding heat is called the *heat of sublimation,* L_s. Liquid carbon dioxide cannot exist at a pressure lower than about 5×10^5 Pa (about 5 atm), and "dry ice" (solid carbon dioxide) sublimes at atmospheric pressure. Sublimation of water from frozen food causes freezer burn. The reverse process, a phase change from gas to solid, occurs when frost forms on cold bodies such as refrigerator cooling coils.

Very pure water can be cooled several degrees below the freezing temperature without freezing; the resulting unstable state is described as *supercooled.* When a small ice crystal is dropped in or the water is agitated, it crystallizes within a second or less. Supercooled water *vapor* condenses quickly into fog droplets when a disturbance, such as dust particles or ionizing radiation, is introduced. This principle is used in "seeding" clouds, which often contain supercooled water vapor, to cause condensation and rain.

A liquid can sometimes be *superheated* above its normal boiling temperature. Any small disturbance such as agitation causes local boiling with bubble formation.

Steam heating systems for buildings use a boiling–condensing process to transfer heat from the furnace to the radiators. Each kilogram of water that is

17.21 Graph of temperature versus time for a specimen of water initially in the solid phase (ice). Heat is added to the specimen at a constant rate. The temperature remains constant during each change of phase, provided that the pressure remains constant.

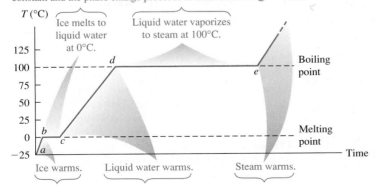

Phase of water changes. During these periods, temperature stays constant and the phase change proceeds as heat is added: $Q = +mL$.

Ice melts to liquid water at 0°C.

Liquid water vaporizes to steam at 100°C.

Boiling point

Melting point

Ice warms.　Liquid water warms.　Steam warms.

Temperature of water changes. During these periods, temperature rises as heat is added: $Q = mc\,\Delta T$.

turned to steam in the boiler absorbs over 2×10^6 J (the heat of vaporization L_v of water) from the boiler and gives it up when it condenses in the radiators. Boiling–condensing processes are also used in refrigerators, air conditioners, and heat pumps. We will discuss these systems in Chapter 20.

The temperature-control mechanisms of many warm-blooded animals make use of heat of vaporization, removing heat from the body by using it to evaporate water from the tongue (panting) or from the skin (sweating). Evaporative cooling enables humans to maintain normal body temperature in hot, dry desert climates where the air temperature may reach 55°C (about 130°F). The skin temperature may be as much as 30°C cooler than the surrounding air. Under these conditions a normal person may perspire several liters per day, and this lost water must be replaced. Old-time desert rats (such as one of the authors) state that in the desert, any canteen that holds less than a gallon should be viewed as a toy! Evaporative cooling also explains why you feel cold when you first step out of a swimming pool (Fig. 17.22).

Evaporative cooling is also used to cool buildings in hot, dry climates and to condense and recirculate "used" steam in coal-fired or nuclear-powered electric-generating plants. That's what goes on in the large, tapered concrete towers that you see at such plants.

Chemical reactions such as combustion are analogous to phase changes in that they involve definite quantities of heat. Complete combustion of 1 gram of gasoline produces about 46,000 J or about 11,000 cal, so the **heat of combustion** L_c of gasoline is

$$L_c = 46{,}000 \text{ J/g} = 4.6 \times 10^7 \text{ J/kg}$$

Energy values of foods are defined similarly. When we say that a gram of peanut butter "contains 6 calories," we mean that 6 kcal of heat (6000 cal or 25,000 J) is released when the carbon and hydrogen atoms in the peanut butter react with oxygen (with the help of enzymes) and are completely converted to CO_2 and H_2O. Not all of this energy is directly useful for mechanical work. We will study the *efficiency* of energy utilization in Chapter 20.

Heat Calculations

Let's look at some examples of calorimetry calculations (calculations with heat). The basic principle is very simple: When heat flow occurs between two bodies that are isolated from their surroundings, the amount of heat lost by one body must equal the amount gained by the other. Heat is energy in transit, so this principle is really just conservation of energy. Calorimetry, dealing entirely with one conserved quantity, is in many ways the simplest of all physical theories!

17.22 The water may be warm and it may be a hot day, but these children will feel cold when they first step out of the swimming pool. That's because as water evaporates from their skin, it removes the heat of vaporization from their bodies. To stay warm, they will need to dry off immediately.

Problem-Solving Strategy 17.2 | **Calorimetry Problems**

IDENTIFY *the relevant concepts:* When heat flow occurs between two or more bodies that are isolated from their surroundings, the *algebraic sum* of the quantities of heat transferred to all the bodies is zero. We take a quantity of heat *added* to a body as *positive* and a quantity *leaving* a body as *negative*.

SET UP *the problem* using the following steps:
1. Identify the objects that exchange heat.
2. Each object may undergo a temperature change only, a phase change at constant temperature, or both. Use Eq. (17.13) for the heat transferred in a temperature change and Eq. (17.20) for the heat transferred in a phase change.
3. Consult Table 17.3 for values of specific heat or molar heat capacity and Table 17.4 for heats of fusion or vaporization.
4. List the known and unknown quantities and identify the target variables.

EXECUTE *the solution* as follows:
1. Use Eq. (17.13) and/or Eq. (17.20) and the energy-conservation relation $\Sigma Q = 0$ to solve for the target variables. Ensure that you use the correct algebraic signs for Q and ΔT terms, and that you correctly write $\Delta T = T_{\text{final}} - T_{\text{initial}}$ and not the reverse.
2. If a phase change occurs, you may not know in advance whether all, or only part, of the material undergoes a phase change. Make a reasonable guess; if that leads to an unreasonable result (such as a final temperature higher or lower than any initial temperature), the guess was wrong. Try again!

EVALUATE *your answer:* Double-check your calculations, and ensure that the results are physically sensible.

Example 17.7 A temperature change with no phase change

A camper pours 0.300 kg of coffee, initially in a pot at 70.0°C, into a 0.120-kg aluminum cup initially at 20.0°C. What is the equilibrium temperature? Assume that coffee has the same specific heat as water and that no heat is exchanged with the surroundings.

SOLUTION

IDENTIFY and SET UP: The target variable is the common final temperature T of the cup and coffee. No phase changes occur, so we need only Eq. (17.13). With subscripts C for coffee, W for water, and Al for aluminum, we have $T_{0C} = 70.0°$ and $T_{0Al} = 20.0°$; Table 17.3 gives $c_W = 4190 \text{ J/kg} \cdot \text{K}$ and $c_{Al} = 910 \text{ J/kg} \cdot \text{K}$.

EXECUTE: The (negative) heat gained by the coffee is $Q_C = m_C c_W \Delta T_C$. The (positive) heat gained by the cup is $Q_{Al} = m_{Al} c_{Al} \Delta T_{Al}$. We set $Q_C + Q_{Al} = 0$ (see Problem-Solving Strategy 17.2) and substitute $\Delta T_C = T - T_{0C}$ and $\Delta T_{Al} = T - T_{0Al}$:

$$Q_C + Q_{Al} = m_C c_W \Delta T_C + m_{Al} c_{Al} \Delta T_{Al} = 0$$
$$m_C c_W (T - T_{0C}) + m_{Al} c_{Al} (T - T_{0Al}) = 0$$

Then we solve this expression for the final temperature T. A little algebra gives

$$T = \frac{m_C c_W T_{0C} + m_{Al} c_{Al} T_{0Al}}{m_C c_W + m_{Al} c_{Al}} = 66.0°C$$

EVALUATE: The final temperature is much closer to the initial temperature of the coffee than to that of the cup; water has a much higher specific heat than aluminum, and we have more than twice as much mass of water. We can also find the quantities of heat by substituting the value $T = 66.0°C$ back into the original equations. We find $Q_C = -5.0 \times 10^3$ J and $Q_{Al} = +5.0 \times 10^3$ J. As expected, Q_C is negative: The coffee loses heat to the cup.

Example 17.8 Changes in both temperature and phase

A glass contains 0.25 kg of Omni-Cola (mostly water) initially at 25°C. How much ice, initially at −20°C, must you add to obtain a final temperature of 0°C with all the ice melted? Neglect the heat capacity of the glass.

SOLUTION

IDENTIFY and SET UP: The Omni-Cola and ice exchange heat. The cola undergoes a temperature change; the ice undergoes both a temperature change and a phase change from solid to liquid. We use subscripts C for cola, I for ice, and W for water. The target variable is the mass of ice, m_I. We use Eq. (17.13) to obtain an expression for the amount of heat involved in cooling the drink to $T = 0°C$ and warming the ice to $T = 0°C$, and Eq. (17.20) to obtain an expression for the heat required to melt the ice at 0°C. We have $T_{0C} = 25°C$ and $T_{0I} = -20°C$, Table 17.3 gives $c_W = 4190 \text{ J/kg} \cdot \text{K}$ and $c_I = 2100 \text{ J/kg} \cdot \text{K}$, and Table 17.4 gives $L_f = 3.34 \times 10^5$ J/kg.

EXECUTE: From Eq. (17.13), the (negative) heat gained by the Omni-Cola is $Q_C = m_C c_W \Delta T_C$. The (positive) heat gained by the ice in warming is $Q_I = m_I c_I \Delta T_I$. The (positive) heat required to melt the ice is $Q_2 = m_I L_f$. We set $Q_C + Q_I + Q_2 = 0$, insert $\Delta T_C = T - T_{0C}$ and $\Delta T_I = T - T_{0I}$, and solve for m_I:

$$m_C c_W \Delta T_C + m_I c_I \Delta T_I + m_I L_f = 0$$
$$m_C c_W (T - T_{0C}) + m_I c_I (T - T_{0I}) + m_I L_f = 0$$
$$m_I [c_I (T - T_{0I}) + L_f] = -m_C c_W (T - T_{0C})$$
$$m_I = m_C \frac{c_W (T_{0C} - T)}{c_I (T - T_{0I}) + L_f}$$

Substituting numerical values, we find that $m_I = 0.070$ kg = 70 g.

EVALUATE: Three or four medium-size ice cubes would make about 70 g, which seems reasonable given the 250 g of Omni-Cola to be cooled.

Example 17.9 What's cooking?

A hot copper pot of mass 2.0 kg (including its copper lid) is at a temperature of 150°C. You pour 0.10 kg of cool water at 25°C into the pot, then quickly replace the lid so no steam can escape. Find the final temperature of the pot and its contents, and determine the phase of the water (liquid, gas, or a mixture). Assume that no heat is lost to the surroundings.

SOLUTION

IDENTIFY and SET UP: The water and the pot exchange heat. Three outcomes are possible: (1) No water boils, and the final temperature T is less than 100°C; (2) some water boils, giving a mixture of water and steam at 100°C; or (3) all the water boils, giving 0.10 kg of steam at 100°C or greater. We use Eq. (17.13) for the heat transferred in a temperature change and Eq. (17.20) for the heat transferred in a phase change.

EXECUTE: First consider case (1), which parallels Example 17.8 exactly. The equation that states that the heat flow into the water equals the heat flow out of the pot is

$$Q_W + Q_{Cu} = m_W c_W (T - T_{0W}) + m_{Cu} c_{Cu} (T - T_{0Cu}) = 0$$

Here we use subscripts W for water and Cu for copper, with $m_W = 0.10$ kg, $m_{Cu} = 2.0$ kg, $T_{0W} = 25°C$, and $T_{0Cu} = 150°C$. From Table 17.3, $c_W = 4190 \text{ J/kg} \cdot \text{K}$ and $c_{Cu} = 390 \text{ J/kg} \cdot \text{K}$. Solving for the final temperature T and substituting these values, we get

$$T = \frac{m_W c_W T_{0W} + m_{Cu} c_{Cu} T_{0Cu}}{m_W c_W + m_{Cu} c_{Cu}} = 106°C$$

But this is above the boiling point of water, which contradicts our assumption that no water boils! So at least some of the water boils.

Continued

So consider case (2), in which the final temperature is $T = 100°C$ and some unknown fraction x of the water boils, where (if this case is correct) x is greater than zero and less than or equal to 1. The (positive) amount of heat needed to vaporize this water is xm_WL_v. The energy-conservation condition $Q_W + Q_{Cu} = 0$ is then

$$m_Wc_W(100°C - T_{0W}) + xm_WL_v + m_{Cu}c_{Cu}(100°C - T_{0Cu}) = 0$$

We solve for the target variable x:

$$x = \frac{-m_{Cu}c_{Cu}(100°C - T_{0Cu}) - m_Wc_W(100°C - T_{0W})}{m_WL_v}$$

With $L_v = 2.256 \times 10^6$ J from Table 17.4, this yields $x = 0.034$. We conclude that the final temperature of the water and copper is 100°C and that $0.034(0.10 \text{ kg}) = 0.0034 \text{ kg} = 3.4$ g of the water is converted to steam at 100°C.

EVALUATE: Had x turned out to be greater than 1, case (3) would have held; all the water would have vaporized, and the final temperature would have been greater than 100°C. Can you show that this would have been the case if we had originally poured less than 15 g of 25°C water into the pot?

Example 17.10 Combustion, temperature change, and phase change

In a particular camp stove, only 30% of the energy released in burning gasoline goes to heating the water in a pot on the stove. How much gasoline must we burn to heat 1.00 L (1.00 kg) of water from 20°C to 100°C and boil away 0.25 kg of it?

SOLUTION

IDENTIFY and SET UP: All of the water undergoes a temperature change and part of it undergoes a phase change, from liquid to gas. We determine the heat required to cause both of these changes, and then use the 30% combustion efficiency to determine the amount of gasoline that must be burned (the target variable). We use Eqs. (17.13) and (17.20) and the idea of heat of combustion.

EXECUTE: To raise the temperature of the water from 20°C to 100°C requires

$$Q_1 = mc \, \Delta T = (1.00 \text{ kg})(4190 \text{ J/kg} \cdot \text{K})(80 \text{ K})$$
$$= 3.35 \times 10^5 \text{ J}$$

To boil 0.25 kg of water at 100°C requires

$$Q_2 = mL_v = (0.25 \text{ kg})(2.256 \times 10^6 \text{ J/kg}) = 5.64 \times 10^5 \text{ J}$$

The total energy needed is $Q_1 + Q_2 = 8.99 \times 10^5$ J. This is 30% = 0.30 of the total heat of combustion, which is therefore $(8.99 \times 10^5 \text{ J})/0.30 = 3.00 \times 10^6$ J. As we mentioned earlier, the combustion of 1 g of gasoline releases 46,000 J, so the mass of gasoline required is $(3.00 \times 10^6 \text{ J})/(46,000 \text{ J/g}) = 65$ g, or a volume of about 0.09 L of gasoline.

EVALUATE: This result suggests the tremendous amount of energy released in burning even a small quantity of gasoline. Another 123 g of gasoline would be required to boil away the remaining water; can you prove this?

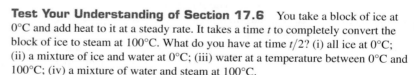

Test Your Understanding of Section 17.6 You take a block of ice at 0°C and add heat to it at a steady rate. It takes a time t to completely convert the block of ice to steam at 100°C. What do you have at time $t/2$? (i) all ice at 0°C; (ii) a mixture of ice and water at 0°C; (iii) water at a temperature between 0°C and 100°C; (iv) a mixture of water and steam at 100°C.

PhET: The Greenhouse Effect

17.7 Mechanisms of Heat Transfer

We have talked about *conductors* and *insulators*, materials that permit or prevent heat transfer between bodies. Now let's look in more detail at *rates* of energy transfer. In the kitchen you use a metal or glass pot for good heat transfer from the stove to whatever you're cooking, but your refrigerator is insulated with a material that *prevents* heat from flowing into the food inside the refrigerator. How do we describe the difference between these two materials?

The three mechanisms of heat transfer are conduction, convection, and radiation. *Conduction* occurs within a body or between two bodies in contact. *Convection* depends on motion of mass from one region of space to another. *Radiation* is heat transfer by electromagnetic radiation, such as sunshine, with no need for matter to be present in the space between bodies.

Conduction

If you hold one end of a copper rod and place the other end in a flame, the end you are holding gets hotter and hotter, even though it is not in direct contact with the flame. Heat reaches the cooler end by **conduction** through the material.

On the atomic level, the atoms in the hotter regions have more kinetic energy, on the average, than their cooler neighbors. They jostle their neighbors, giving them some of their energy. The neighbors jostle *their* neighbors, and so on through the material. The atoms themselves do not move from one region of material to another, but their energy does.

Most metals also use another, more effective mechanism to conduct heat. Within the metal, some electrons can leave their parent atoms and wander through the crystal lattice. These "free" electrons can rapidly carry energy from the hotter to the cooler regions of the metal, so metals are generally good conductors of heat. A metal rod at 20°C feels colder than a piece of wood at 20°C because heat can flow more easily from your hand into the metal. The presence of "free" electrons also causes most metals to be good electrical conductors.

Heat transfer occurs only between regions that are at different temperatures, and the direction of heat flow is always from higher to lower temperature. Figure 17.23a shows a rod of conducting material with cross-sectional area A and length L. The left end of the rod is kept at a temperature T_H and the right end at a lower temperature T_C, so heat flows from left to right. The sides of the rod are covered by an ideal insulator, so no heat transfer occurs at the sides.

When a quantity of heat dQ is transferred through the rod in a time dt, the rate of heat flow is dQ/dt. We call this rate the **heat current,** denoted by H. That is, $H = dQ/dt$. Experiments show that the heat current is proportional to the cross-sectional area A of the rod (Fig. 17.23b) and to the temperature difference $(T_H - T_C)$ and is inversely proportional to the rod length L (Fig. 17.23c). Introducing a proportionality constant k called the **thermal conductivity** of the material, we have

$$H = \frac{dQ}{dt} = kA\frac{T_H - T_C}{L} \quad \text{(heat current in conduction)} \quad (17.21)$$

The quantity $(T_H - T_C)/L$ is the temperature difference *per unit length;* it is called the magnitude of the **temperature gradient.** The numerical value of k depends on the material of the rod. Materials with large k are good conductors of heat; materials with small k are poor conductors, or insulators. Equation (17.21) also gives the heat current through a slab or through *any* homogeneous body with uniform cross section A perpendicular to the direction of flow; L is the length of the heat-flow path.

The units of heat current H are units of energy per time, or power; the SI unit of heat current is the watt (1 W = 1 J/s). We can find the units of k by solving Eq. (17.21) for k; you can show that the SI units are W/m·K. Some numerical values of k are given in Table 17.5.

The thermal conductivity of "dead" (that is, nonmoving) air is very small. A wool sweater keeps you warm because it traps air between the fibers. In fact, many insulating materials such as Styrofoam and fiberglass are mostly dead air.

If the temperature varies in a nonuniform way along the length of the conducting rod, we introduce a coordinate x along the length and generalize the temperature gradient to be dT/dx. The corresponding generalization of Eq. (17.21) is

$$H = \frac{dQ}{dt} = -kA\frac{dT}{dx} \quad (17.22)$$

The negative sign shows that heat always flows in the direction of *decreasing* temperature.

17.23 Steady-state heat flow due to conduction in a uniform rod.

(a) Heat current H

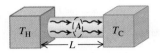

(b) Doubling the cross-sectional area of the conductor doubles the heat current (H is proportional to A).

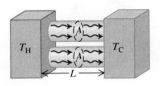

(c) Doubling the length of the conductor halves the heat current (H is inversely proportional to L).

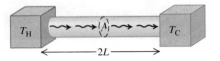

Table 17.5 Thermal Conductivities

Substance	k (W/m · K)
Metals	
Aluminum	205.0
Brass	109.0
Copper	385.0
Lead	34.7
Mercury	8.3
Silver	406.0
Steel	50.2
Solids (representative values)	
Brick, insulating	0.15
Brick, red	0.6
Concrete	0.8
Cork	0.04
Felt	0.04
Fiberglass	0.04
Glass	0.8
Ice	1.6
Rock wool	0.04
Styrofoam	0.027
Wood	0.12–0.04
Gases	
Air	0.024
Argon	0.016
Helium	0.14
Hydrogen	0.14
Oxygen	0.023

Application **Fur Versus Blubber**
The fur of an arctic fox is a good thermal insulator because it traps air, which has a low thermal conductivity k. (The value $k = 0.04$ W/m·K for fur is higher than for air, $k = 0.024$ W/m·K, because fur also includes solid hairs.) The layer of fat beneath a bowhead whale's skin, called blubber, has six times the thermal conductivity of fur $(k = 0.24$ W/m·K). So a 6-cm thickness of blubber $(L = 6$ cm) is required to give the same insulation as 1 cm of fur.

For thermal insulation in buildings, engineers use the concept of **thermal resistance,** denoted by R. The thermal resistance R of a slab of material with area A is defined so that the heat current H through the slab is

$$H = \frac{A(T_H - T_C)}{R} \qquad (17.23)$$

where T_H and T_C are the temperatures on the two sides of the slab. Comparing this with Eq. (17.21), we see that R is given by

$$R = \frac{L}{k} \qquad (17.24)$$

where L is the thickness of the slab. The SI unit of R is 1 m²·K/W. In the units used for commercial insulating materials in the United States, H is expressed in Btu/h, A is in ft², and $T_H - T_C$ in F°. (1 Btu/h = 0.293 W.) The units of R are then ft²·F°·h/Btu, though values of R are usually quoted without units; a 6-inch-thick layer of fiberglass has an R value of 19 (that is, $R = 19$ ft²·F°·h/Btu), a 2-inch-thick slab of polyurethane foam has an R value of 12, and so on. Doubling the thickness doubles the R value. Common practice in new construction in severe northern climates is to specify R values of around 30 for exterior walls and ceilings. When the insulating material is in layers, such as a plastered wall, fiberglass insulation, and wood exterior siding, the R values are additive. Do you see why? (See Problem 17.108.)

Problem-Solving Strategy 17.3 | **Heat Conduction**

IDENTIFY *the relevant concepts:* Heat conduction occurs whenever two objects at different temperatures are placed in contact.

SET UP *the problem* using the following steps:
1. Identify the direction of heat flow (from hot to cold). In Eq. (17.21), L is measured along this direction, and A is an area perpendicular to this direction. You can often approximate an irregular-shaped container with uniform wall thickness as a flat slab with the same thickness and total wall area.

2. List the known and unknown quantities and identify the target variable.

EXECUTE *the solution* as follows:
1. If heat flows through a single object, use Eq. (17.21) to solve for the target variable.
2. If the heat flows through two different materials in succession (in *series*), the temperature T at the interface between them is

intermediate between T_H and T_C, so that the temperature differences across the two materials are $(T_H - T)$ and $(T - T_C)$. In steady-state heat flow, the same heat must pass through both materials, so the heat current H must be the *same* in both materials.

3. If heat flows through two or more *parallel* paths, then the total heat current H is the sum of the currents H_1, H_2, ... for the separate paths. An example is heat flow from inside a room to outside, both through the glass in a window and through the surrounding wall. In parallel heat flow the temperature difference is the same for each path, but L, A, and k may be different for each path.

4. Use consistent units. If k is expressed in W/m·K, for example, use distances in meters, heat in joules, and T in kelvins.

EVALUATE *your answer:* Are the results physically reasonable?

Example 17.11 | **Conduction into a picnic cooler**

A Styrofoam cooler (Fig. 17.24a) has total wall area (including the lid) of 0.80 m² and wall thickness 2.0 cm. It is filled with ice, water, and cans of Omni-Cola, all at 0°C. What is the rate of heat flow into the cooler if the temperature of the outside wall is 30°C? How much ice melts in 3 hours?

SOLUTION

IDENTIFY and SET UP: The target variables are the heat current H and the mass m of ice melted. We use Eq. (17.21) to determine H and Eq. (17.20) to determine m.

EXECUTE: We assume that the total heat flow is the same as it would be through a flat Styrofoam slab of area 0.80 m² and thickness 2.0 cm = 0.020 m (Fig. 17.24b). We find k from Table 17.5. From Eq. (17.21),

$$H = kA\frac{T_H - T_C}{L} = (0.027 \text{ W/m·K})(0.80 \text{ m}^2)\frac{30°C - 0°C}{0.020 \text{ m}}$$
$$= 32.4 \text{ W} = 32.4 \text{ J/s}$$

The total heat flow is $Q = Ht$, with $t = 3$ h $= 10,800$ s. From Table 17.4, the heat of fusion of ice is $L_f = 3.34 \times 10^5$ J/kg, so from Eq. (17.20) the mass of ice that melts is

$$m = \frac{Q}{L_f} = \frac{(32.4 \text{ J/s})(10,800 \text{ s})}{3.34 \times 10^5 \text{ J/kg}} = 1.0 \text{ kg}$$

EVALUATE: The low heat current is a result of the low thermal conductivity of Styrofoam.

17.24 Conduction of heat across the walls of a Styrofoam cooler.

(a) A cooler at the beach

(b) Our sketch for this problem

Example 17.12 **Conduction through two bars I**

A steel bar 10.0 cm long is welded end to end to a copper bar 20.0 cm long. Each bar has a square cross section, 2.00 cm on a side. The free end of the steel bar is kept at 100°C by placing it in contact with steam, and the free end of the copper bar is kept at 0°C by placing it in contact with ice. Both bars are perfectly insulated on their sides. Find the steady-state temperature at the junction of the two bars and the total rate of heat flow through the bars.

17.25 Our sketch for this problem.

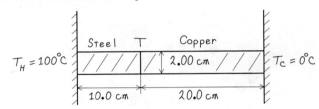

SOLUTION

IDENTIFY and SET UP: Figure 17.25 shows the situation. The heat currents in these end-to-end bars must be the same (see Problem-Solving Strategy 17.3). We are given "hot" and "cold" temperatures $T_H = 100$°C and $T_C = 0$°C. With subscripts S for steel and Cu for copper, we write Eq. (17.21) separately for the heat currents H_S and H_{Cu} and set the resulting expressions equal to each other.

EXECUTE: Setting $H_S = H_{Cu}$, we have from Eq. (17.21)

$$H_S = k_S A \frac{T_H - T}{L_S} = H_{Cu} = k_{Cu} A \frac{T - T_C}{L_{Cu}}$$

We divide out the equal cross-sectional areas A and solve for T:

$$T = \frac{\dfrac{k_S}{L_S} T_H + \dfrac{k_{Cu}}{L_{Cu}} T_C}{\left(\dfrac{k_S}{L_S} + \dfrac{k_{Cu}}{L_{Cu}}\right)}$$

Substituting $L_S = 10.0$ cm and $L_{Cu} = 20.0$ cm, the given values of T_H and T_C, and the values of k_S and k_{Cu} from Table 17.5, we find $T = 20.7$°C.

We can find the total heat current by substituting this value of T into either the expression for H_S or the one for H_{Cu}:

$$H_S = (50.2 \text{ W/m} \cdot \text{K})(0.0200 \text{ m})^2 \frac{100°\text{C} - 20.7°\text{C}}{0.100 \text{ m}}$$
$$= 15.9 \text{ W}$$

$$H_{Cu} = (385 \text{ W/m} \cdot \text{K})(0.0200 \text{ m})^2 \frac{20.7°\text{C}}{0.200 \text{ m}} = 15.9 \text{ W}$$

EVALUATE: Even though the steel bar is shorter, the temperature drop across it is much greater (from 100°C to 20.7°C) than across the copper bar (from 20.7°C to 0°C). That's because steel is a much poorer conductor than copper.

Example 17.13 **Conduction through two bars II**

Suppose the two bars of Example 17.12 are separated. One end of each bar is kept at 100°C and the other end of each bar is kept at 0°C. What is the *total* heat current in the two bars?

17.26 Our sketch for this problem.

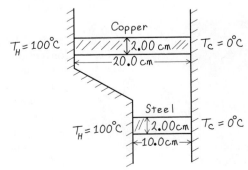

SOLUTION

IDENTIFY and SET UP: Figure 17.26 shows the situation. For each bar, $T_H - T_C = 100$°C $- 0$°C $= 100$ K. The total heat current is the sum of the currents in the two bars, $H_S + H_{Cu}$.

Continued

EXECUTE: We write the heat currents for the two rods individually, and then add them to get the total heat current:

$$H = H_S + H_{Cu} = k_S A \frac{T_H - T_C}{L_S} + k_{Cu} A \frac{T_H - T_C}{L_{Cu}}$$

$$= (50.2 \text{ W/m} \cdot \text{K})(0.0200 \text{ m})^2 \frac{100 \text{ K}}{0.100 \text{ m}}$$

$$+ (385 \text{ W/m} \cdot \text{K})(0.0200 \text{ m})^2 \frac{100 \text{ K}}{0.200 \text{ m}}$$

$$= 20.1 \text{ W} + 77.0 \text{ W} = 97.1 \text{ W}$$

EVALUATE: The heat flow in the copper bar is much greater than that in the steel bar, even though it is longer, because the thermal conductivity of copper is much larger. The total heat flow is greater than in Example 17.12 because the total cross section for heat flow is greater and because the full 100-K temperature difference appears across each bar.

Convection

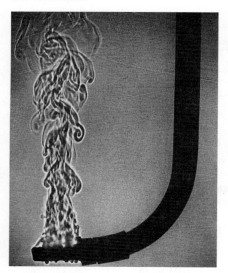

17.27 A heating element in the tip of this submerged tube warms the surrounding water, producing a complex pattern of free convection.

Convection is the transfer of heat by mass motion of a fluid from one region of space to another. Familiar examples include hot-air and hot-water home heating systems, the cooling system of an automobile engine, and the flow of blood in the body. If the fluid is circulated by a blower or pump, the process is called *forced convection;* if the flow is caused by differences in density due to thermal expansion, such as hot air rising, the process is called *natural convection* or *free convection* (Fig. 17.27).

Free convection in the atmosphere plays a dominant role in determining the daily weather, and convection in the oceans is an important global heat-transfer mechanism. On a smaller scale, soaring hawks and glider pilots make use of thermal updrafts from the warm earth. The most important mechanism for heat transfer within the human body (needed to maintain nearly constant temperature in various environments) is *forced* convection of blood, with the heart serving as the pump.

Convective heat transfer is a very complex process, and there is no simple equation to describe it. Here are a few experimental facts:

1. The heat current due to convection is directly proportional to the surface area. This is the reason for the large surface areas of radiators and cooling fins.
2. The viscosity of fluids slows natural convection near a stationary surface, giving a surface film that on a vertical surface typically has about the same insulating value as 1.3 cm of plywood (R value $= 0.7$). Forced convection decreases the thickness of this film, increasing the rate of heat transfer. This is the reason for the "wind-chill factor"; you get cold faster in a cold wind than in still air with the same temperature.
3. The heat current due to convection is found to be approximately proportional to the $\frac{5}{4}$ power of the temperature difference between the surface and the main body of fluid.

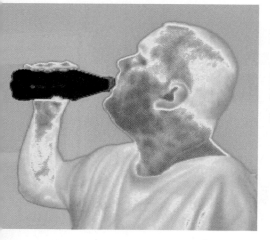

17.28 This false-color infrared photograph reveals radiation emitted by various parts of the man's body. The strongest emission (colored red) comes from the warmest areas, while there is very little emission from the bottle of cold beverage.

Radiation

Radiation is the transfer of heat by electromagnetic waves such as visible light, infrared, and ultraviolet radiation. Everyone has felt the warmth of the sun's radiation and the intense heat from a charcoal grill or the glowing coals in a fireplace. Most of the heat from these very hot bodies reaches you not by conduction or convection in the intervening air but by *radiation*. This heat transfer would occur even if there were nothing but vacuum between you and the source of heat.

Every body, even at ordinary temperatures, emits energy in the form of electromagnetic radiation. Around 20°C, nearly all the energy is carried by infrared waves with wavelengths much longer than those of visible light (see Figs. 17.4 and 17.28). As the temperature rises, the wavelengths shift to shorter values. At 800°C, a body emits enough visible radiation to appear "red-hot," although even at this temperature most of the energy is carried by infrared waves. At 3000°C, the temperature of an incandescent lamp filament, the radiation contains enough visible light that the body appears "white-hot."

The rate of energy radiation from a surface is proportional to the surface area A and to the fourth power of the absolute (Kelvin) temperature T. The rate also depends on the nature of the surface; this dependence is described by a quantity e called the **emissivity.** A dimensionless number between 0 and 1, e represents the ratio of the rate of radiation from a particular surface to the rate of radiation from an equal area of an ideal radiating surface at the same temperature. Emissivity also depends somewhat on temperature. Thus the heat current $H = dQ/dt$ due to radiation from a surface area A with emissivity e at absolute temperature T can be expressed as

$$H = Ae\sigma T^4 \qquad \text{(heat current in radiation)} \qquad (17.25)$$

where σ is a fundamental physical constant called the **Stefan–Boltzmann constant.** This relationship is called the **Stefan–Boltzmann law** in honor of its late-19th-century discoverers. The current best numerical value of σ is

$$\sigma = 5.670400(40) \times 10^{-8} \text{ W/m}^2 \cdot \text{K}^4$$

We invite you to check unit consistency in Eq. (17.25). Emissivity (e) is often larger for dark surfaces than for light ones. The emissivity of a smooth copper surface is about 0.3, but e for a dull black surface can be close to unity.

Example 17.14 Heat transfer by radiation

A thin, square steel plate, 10 cm on a side, is heated in a blacksmith's forge to 800°C. If the emissivity is 0.60, what is the total rate of radiation of energy from the plate?

SOLUTION

IDENTIFY and SET UP: The target variable is H, the rate of emission of energy from the plate's two surfaces. We use Eq. (17.25) to calculate H.

EXECUTE: The total surface area is $2(0.10 \text{ m})^2 = 0.020 \text{ m}^2$, and $T = 800°C = 1073 \text{ K}$. Then Eq. (17.25) gives

$$H = Ae\sigma T^4$$
$$= (0.020 \text{ m}^2)(0.60)(5.67 \times 10^{-8} \text{ W/m}^2 \cdot \text{K}^4)(1073 \text{ K})^4$$
$$= 900 \text{ W}$$

EVALUATE: The nearby blacksmith will easily feel the heat radiated from this plate.

Radiation and Absorption

While a body at absolute temperature T is radiating, its surroundings at temperature T_s are also radiating, and the body *absorbs* some of this radiation. If it is in thermal equilibrium with its surroundings, $T = T_s$ and the rates of radiation and absorption must be equal. For this to be true, the rate of absorption must be given in general by $H = Ae\sigma T_s^4$. Then the *net* rate of radiation from a body at temperature T with surroundings at temperature T_s is

$$H_{\text{net}} = Ae\sigma T^4 - Ae\sigma T_s^4 = Ae\sigma(T^4 - T_s^4) \qquad (17.26)$$

In this equation a positive value of H means a net heat flow *out of* the body. Equation (17.26) shows that for radiation, as for conduction and convection, the heat current depends on the temperature *difference* between two bodies.

Example 17.15 Radiation from the human body

What is the total rate of radiation of energy from a human body with surface area 1.20 m^2 and surface temperature 30°C = 303 K? If the surroundings are at a temperature of 20°C, what is the *net* rate of radiative heat loss from the body? The emissivity of the human body is very close to unity, irrespective of skin pigmentation.

SOLUTION

IDENTIFY and SET UP: We must consider both the radiation that the body emits and the radiation that it absorbs from its surroundings. Equation (17.25) gives the rate of radiation of energy from the body, and Eq. (17.26) gives the net rate of heat loss.

Continued

EXECUTE: Taking $e = 1$ in Eq. (17.25), we find that the body radiates at a rate

$$H = Ae\sigma T^4$$
$$= (1.20 \text{ m}^2)(1)(5.67 \times 10^{-8} \text{ W/m}^2 \cdot \text{K}^4)(303 \text{ K})^4 = 574 \text{ W}$$

This loss is partly offset by absorption of radiation, which depends on the temperature of the surroundings. From Eq. (17.26), the *net* rate of radiative energy transfer is

$$H_{\text{net}} = Ae\sigma(T^4 - T_s{}^4)$$
$$= (1.20 \text{ m}^2)(1)(5.67 \times 10^{-8} \text{ W/m}^2 \cdot \text{K}^4)[(303 \text{ K})^4$$
$$- (293 \text{ K})^4] = 72 \text{ W}$$

EVALUATE: The value of H_{net} is positive because the body is losing heat to its colder surroundings.

Applications of Radiation

Heat transfer by radiation is important in some surprising places. A premature baby in an incubator can be cooled dangerously by radiation if the walls of the incubator happen to be cold, even when the *air* in the incubator is warm. Some incubators regulate the air temperature by measuring the baby's skin temperature.

A body that is a good absorber must also be a good emitter. An ideal radiator, with an emissivity of unity, is also an ideal absorber, absorbing *all* of the radiation that strikes it. Such an ideal surface is called an ideal black body or simply a **blackbody.** Conversely, an ideal *reflector,* which absorbs *no* radiation at all, is also a very ineffective radiator.

This is the reason for the silver coatings on vacuum ("Thermos") bottles, invented by Sir James Dewar (1842–1923). A vacuum bottle has double glass walls. The air is pumped out of the spaces between the walls; this eliminates nearly all heat transfer by conduction and convection. The silver coating on the walls reflects most of the radiation from the contents back into the container, and the wall itself is a very poor emitter. Thus a vacuum bottle can keep coffee or soup hot for several hours. The Dewar flask, used to store very cold liquefied gases, is exactly the same in principle.

Radiation, Climate, and Climate Change

Our planet constantly absorbs radiation coming from the sun. In thermal equilibrium, the rate at which our planet absorbs solar radiation must equal the rate at which it emits radiation into space. The presence of an atmosphere on our planet has a significant effect on this equilibrium.

Most of the radiation emitted by the sun (which has a surface temperature of 5800 K) is in the visible part of the spectrum, to which our atmosphere is transparent. But the average surface temperature of the earth is only 287 K (14°C). Hence most of the radiation that our planet emits into space is infrared radiation, just like the radiation from the person shown in Fig. 17.28. However, our atmosphere is *not* completely transparent to infrared radiation. This is because our atmosphere contains carbon dioxide (CO_2), which is its fourth most abundant constituent (after nitrogen, oxygen, and argon). Molecules of CO_2 in the atmosphere have the property that they *absorb* some of the infrared radiation coming upward from the surface. They then re-radiate the absorbed energy, but some of the re-radiated energy is directed back down toward the surface instead of escaping into space. In order to maintain thermal equilibrium, the earth's surface must compensate for this by increasing its temperature T and hence its total rate of radiating energy (which is proportional to T^4). This phenomenon, called the **greenhouse effect,** makes our planet's surface temperature about 33°C higher than it would be if there were no atmospheric CO_2. If CO_2 were absent, the earth's average surface temperature would be below the freezing point of water, and life as we know it would be impossible.

While atmospheric CO_2 has a beneficial effect, too much of it can have extremely negative consequences. Measurements of air trapped in ancient Antarctic ice show that over the past 650,000 years CO_2 has constituted less than 300 parts per million of our atmosphere. Since the beginning of the industrial age, however, the burning of fossil fuels such as coal and petroleum has elevated the atmospheric CO_2 concentration to unprecedented levels (Fig. 17.29a). As a consequence, since the 1950s the global average surface temperature has increased by 0.6°C and the earth has experienced the hottest years ever recorded (Fig. 17.29b). If we continue to consume fossil fuels at the same rate, by 2050 the atmospheric CO_2 concentration will reach 600 parts per million, well off the scale of Fig. 17.29a. The resulting temperature increase will have dramatic effects on climate around the world. In the polar regions massive quantities of ice will melt and run from solid land to the sea, thus raising ocean levels worldwide and threatening the homes and lives of hundreds of millions of people who live near the coast. Coping with these threats is one of the greatest challenges facing 21st-century civilization.

17.29 (a) The concentration of atmospheric CO_2 has increased by 22% since continuous measurements began in 1958. (The yearly variations are due to increased intake of CO_2 by plants in spring and summer.) (b) The increase in global average temperature since the beginning of the industrial era is a result of the increase in CO_2 concentration.

(a)　　　　　　　　　　　　　　　　　　　　　　　　(b)

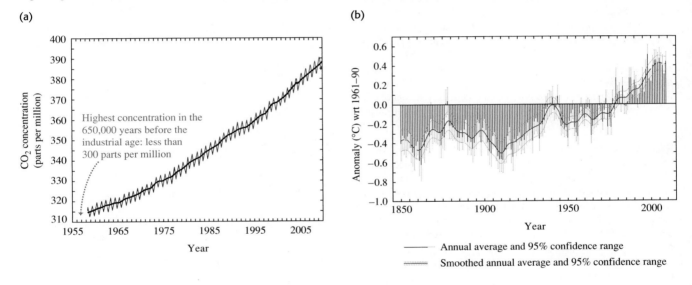

Test Your Understanding of Section 17.7 A room has one wall made of concrete, one wall made of copper, and one wall made of steel. All of the walls are the same size and at the same temperature of 20°C. Which wall feels coldest to the touch? (i) the concrete wall; (ii) the copper wall; (iii) the steel wall; (iv) all three walls feel equally cold.

Temperature and temperature scales: Two bodies in thermal equilibrium must have the same temperature. A conducting material between two bodies permits them to interact and come to thermal equilibrium; an insulating material impedes this interaction.

The Celsius and Fahrenheit temperature scales are based on the freezing ($0°C = 32°F$) and boiling ($100°C = 212°F$) temperatures of water. One Celsius degree equals $\frac{9}{5}$ Fahrenheit degrees. (See Example 17.1.)

The Kelvin scale has its zero at the extrapolated zero-pressure temperature for a gas thermometer, $-273.15°C = 0$ K. In the gas-thermometer scale, the ratio of two temperatures T_1 and T_2 is defined to be equal to the ratio of the two corresponding gas-thermometer pressures p_1 and p_2.

$$T_F = \frac{9}{5}T_C + 32° \quad (17.1)$$

$$T_C = \frac{5}{9}(T_F - 32°) \quad (17.2)$$

$$T_K = T_C + 273.15 \quad (17.3)$$

$$\frac{T_2}{T_1} = \frac{p_2}{p_1} \quad (17.4)$$

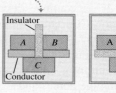

If systems A and B are each in thermal equilibrium with system C ...

... then systems A and B are in thermal equilibrium with each other.

Thermal expansion and thermal stress: A temperature change ΔT causes a change in any linear dimension L_0 of a solid body. The change ΔL is approximately proportional to L_0 and ΔT. Similarly, a temperature change causes a change ΔV in the volume V_0 of any solid or liquid; ΔV is approximately proportional to V_0 and ΔT. The quantities α and β are the coefficients of linear expansion and volume expansion, respectively. For solids, $\beta = 3\alpha$. (See Examples 17.2 and 17.3.)

When a material is cooled or heated and held so it cannot contract or expand, it is under a tensile stress F/A. (See Example 17.4.)

$$\Delta L = \alpha L_0 \Delta T \quad (17.6)$$

$$\Delta V = \beta V_0 \Delta T \quad (17.8)$$

$$\frac{F}{A} = -Y\alpha \Delta T \quad (17.12)$$

$$L = L_0 + \Delta L$$
$$= L_0(1 + \alpha \Delta T)$$

Heat, phase changes, and calorimetry: Heat is energy in transit from one body to another as a result of a temperature difference. Equations (17.13) and (17.18) give the quantity of heat Q required to cause a temperature change ΔT in a quantity of material with mass m and specific heat c (alternatively, with number of moles n and molar heat capacity $C = Mc$, where M is the molar mass and $m = nM$). When heat is added to a body, Q is positive; when it is removed, Q is negative. (See Examples 17.5 and 17.6.)

To change a mass m of a material to a different phase at the same temperature (such as liquid to vapor), a quantity of heat given by Eq. (17.20) must be added or subtracted. Here L is the heat of fusion, vaporization, or sublimation.

In an isolated system whose parts interact by heat exchange, the algebraic sum of the Q's for all parts of the system must be zero. (See Examples 17.7–17.10.)

$$Q = mc \Delta T \quad (17.13)$$

$$Q = nC \Delta T \quad (17.18)$$

$$Q = \pm mL \quad (17.20)$$

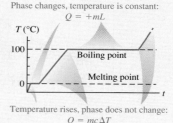

Phase changes, temperature is constant:
$Q = +mL$

Temperature rises, phase does not change:
$Q = mc\Delta T$

Conduction, convection, and radiation: Conduction is the transfer of heat within materials without bulk motion of the materials. The heat current H depends on the area A through which the heat flows, the length L of the heat-flow path, the temperature difference ($T_H - T_C$), and the thermal conductivity k of the material. (See Examples 17.11–17.13.)

Convection is a complex heat-transfer process that involves mass motion from one region to another.

Radiation is energy transfer through electromagnetic radiation. The radiation heat current H depends on the surface area A, the emissivity e of the surface (a pure number between 0 and 1), and the Kelvin temperature T. Here σ is the Stefan–Boltzmann constant. The net radiation heat current H_{net} from a body at temperature T to its surroundings at temperature T_s depends on both T and T_s. (See Examples 17.14 and 17.15.)

$$H = \frac{dQ}{dt} = kA\frac{T_H - T_C}{L} \quad (17.21)$$

$$H = Ae\sigma T^4 \quad (17.25)$$

$$H_{net} = Ae\sigma(T^4 - T_s^4) \quad (17.26)$$

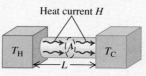

Heat current H

Heat current $H = kA\dfrac{T_H - T_C}{L}$

BRIDGING PROBLEM | Steady-State Heat Flow: Radiation and Conduction

One end of a solid cylindrical copper rod 0.200 m long and 0.0250 m in radius is inserted into a large block of solid hydrogen at its melting temperature, 13.84 K. The other end is blackened and exposed to thermal radiation from surrounding walls at 500.0 K. The sides of the rod are insulated, so no energy is lost or gained except at the ends of the rod. (a) When equilibrium is reached, what is the temperature of the blackened end? The thermal conductivity of copper at temperatures near 20 K is 1670 Wm · K. (b) At what rate (in kg/h) does the solid hydrogen melt?

SOLUTION GUIDE

See MasteringPhysics® study area for a Video Tutor solution.

IDENTIFY and SET UP

1. Draw a sketch of the situation, showing all relevant dimensions.
2. List the known and unknown quantities, and identify the target variables.
3. In order for the rod to be in equilibrium, how must the radiation heat current from the walls into the blackened end of the rod compare to the conduction heat current from this end to the

other end and into the solid hydrogen? Use your answers to select the appropriate equations for part (a).
4. How does the heat current from the rod into the hydrogen determine the rate at which the hydrogen melts? (*Hint:* See Table 17.4.) Use your answer to select the appropriate equations for part (b).

EXECUTE

5. Solve for the temperature of the blackened end of the rod. (*Hint:* Since copper is an excellent conductor of heat at low temperature, you can assume that the temperature of the blackened end is only slightly higher than 13.84 K.)
6. Use your result from step 5 to find the rate at which the hydrogen melts.

EVALUATE

7. Is your result from step 5 consistent with the hint given in that step?
8. How would your results from steps 5 and 6 be affected if the rod had twice the radius?

Problems

For instructor-assigned homework, go to www.masteringphysics.com

•, ••, •••: Problems of increasing difficulty. **CP**: Cumulative problems incorporating material from earlier chapters. **CALC**: Problems requiring calculus. **BIO**: Biosciences problems.

DISCUSSION QUESTIONS

Q17.1 Explain why it would not make sense to use a full-size glass thermometer to measure the temperature of a thimbleful of hot water.

Q17.2 If you heat the air inside a rigid, sealed container until its Kelvin temperature doubles, the air pressure in the container will also double. Is the same thing true if you double the Celsius temperature of the air in the container? Explain.

Q17.3 Many automobile engines have cast-iron cylinders and aluminum pistons. What kinds of problems could occur if the engine gets too hot? (The coefficient of volume expansion of cast iron is approximately the same as that of steel.)

Q17.4 Why do frozen water pipes burst? Would a mercury thermometer break if the temperature went below the freezing temperature of mercury? Why or why not?

Q17.5 Two bodies made of the same material have the same external dimensions and appearance, but one is solid and the other is hollow. When their temperature is increased, is the overall volume expansion the same or different? Why?

Q17.6 The inside of an oven is at a temperature of 200°C (392°F). You can put your hand in the oven without injury as long as you don't touch anything. But since the air inside the oven is also at 200°C, why isn't your hand burned just the same?

Q17.7 A newspaper article about the weather states that "the temperature of a body measures how much heat the body contains." Is this description correct? Why or why not?

Q17.8 To raise the temperature of an object, must you add heat to it? If you add heat to an object, must you raise its temperature? Explain.

Q17.9 A student asserts that a suitable unit for specific heat is $1 \text{ m}^2/\text{s}^2 \cdot \text{C}°$. Is she correct? Why or why not?

Q17.10 In some household air conditioners used in dry climates, air is cooled by blowing it through a water-soaked filter, evaporating some of the water. How does this cool the air? Would such a system work well in a high-humidity climate? Why or why not?

Q17.11 The units of specific heat c are $\text{J/kg} \cdot \text{K}$, but the units of heat of fusion L_f or heat of vaporization L_v are simply J/kg. Why do the units of L_f and L_v not include a factor of $(\text{K})^{-1}$ to account for a temperature change?

Q17.12 Why is a hot, humid day in the tropics generally more uncomfortable for human beings than a hot, dry day in the desert?

Q17.13 A piece of aluminum foil used to wrap a potato for baking in a hot oven can usually be handled safely within a few seconds after the potato is removed from the oven. The same is not true of the potato, however! Give two reasons for this difference.

Q17.14 Desert travelers sometimes keep water in a canvas bag. Some water seeps through the bag and evaporates. How does this cool the water inside the bag?

Q17.15 When you first step out of the shower, you feel cold. But as soon as you are dry you feel warmer, even though the room temperature does not change. Why?

Q17.16 The climate of regions adjacent to large bodies of water (like the Pacific and Atlantic coasts) usually features a narrower range of temperature than the climate of regions far from large bodies of water (like the prairies). Why?

Q17.17 When water is placed in ice-cube trays in a freezer, why doesn't the water freeze all at once when the temperature has reached 0°C? In fact, the water freezes first in a layer adjacent to the sides of the tray. Why?

Q17.18 Before giving you an injection, a physician swabs your arm with isopropyl alcohol at room temperature. Why does this make your arm feel cold? (*Hint:* The reason is *not* the fear of the injection! The boiling point of isopropyl alcohol is 82.4°C.)

Q17.19 A cold block of metal feels colder than a block of wood at the same temperature. Why? A *hot* block of metal feels hotter than a block of wood at the same temperature. Again, why? Is there any temperature at which the two blocks feel equally hot or cold? What temperature is this?

Q17.20 A person pours a cup of hot coffee, intending to drink it five minutes later. To keep the coffee as hot as possible, should she put cream in it now or wait until just before she drinks it? Explain.

Q17.21 When a freshly baked apple pie has just been removed from the oven, the crust and filling are both at the same temperature. Yet if you sample the pie, the filling will burn your tongue but the crust will not. Why is there a difference? (*Hint:* The filling is moist while the crust is dry.)

Q17.22 Old-time kitchen lore suggests that things cook better (evenly and without burning) in heavy cast-iron pots. What desirable characteristics do such pots have?

Q17.23 In coastal regions in the winter, the temperature over the land is generally colder than the temperature over the nearby ocean; in the summer, the reverse is usually true. Explain. (*Hint:* The specific heat of soil is only 0.2–0.8 times as great as that of water.)

Q17.24 It is well known that a potato bakes faster if a large nail is stuck through it. Why? Does an aluminum nail work better than a steel one? Why or why not? (*Note:* Don't try this in a microwave oven!) There is also a gadget on the market to hasten the roasting of meat; it consists of a hollow metal tube containing a wick and some water. This is claimed to work much better than a solid metal rod. How does it work?

Q17.25 Glider pilots in the Midwest know that thermal updrafts are likely to occur in the vicinity of freshly plowed fields. Why?

Q17.26 Some folks claim that ice cubes freeze faster if the trays are filled with hot water, because hot water cools off faster than cold water. What do you think?

Q17.27 We're lucky that the earth isn't in thermal equilibrium with the sun (which has a surface temperature of 5800 K). But why aren't the two bodies in thermal equilibrium?

Q17.28 When energy shortages occur, magazine articles sometimes urge us to keep our homes at a constant temperature day and night to conserve fuel. They argue that when we turn down the heat at night, the walls, ceilings, and other areas cool off and must be reheated in the morning. So if we keep the temperature constant, these parts of the house will not cool off and will not have to be reheated. Does this argument make sense? Would we really save energy by following this advice?

EXERCISES

Section 17.2 Thermometers and Temperature Scales

17.1 • Convert the following Celsius temperatures to Fahrenheit: (a) −62.8°C, the lowest temperature ever recorded in North America (February 3, 1947, Snag, Yukon); (b) 56.7°C, the highest temperature ever recorded in the United States (July 10, 1913, Death Valley, California); (c) 31.1°C, the world's highest average annual temperature (Lugh Ferrandi, Somalia).

17.2 • BIO **Temperatures in Biomedicine.** (a) **Normal body temperature.** The average normal body temperature measured in the mouth is 310 K. What would Celsius and Fahrenheit thermometers read for this temperature? (b) **Elevated body temperature.** During very vigorous exercise, the body's temperature can go as high as 40°C. What would Kelvin and Fahrenheit thermometers read for this temperature? (c) **Temperature difference in the body.** The surface temperature of the body is normally about 7 C° lower than the internal temperature. Express this temperature difference in kelvins and in Fahrenheit degrees. (d) **Blood storage.** Blood stored at 4.0°C lasts safely for about 3 weeks, whereas blood stored at −160°C lasts for 5 years. Express both temperatures on the Fahrenheit and Kelvin scales. (e) **Heat stroke.** If the body's temperature is above 105°F for a prolonged period, heat stroke can result. Express this temperature on the Celsius and Kelvin scales.

17.3 • (a) On January 22, 1943, the temperature in Spearfish, South Dakota, rose from −4.0°F to 45.0°F in just 2 minutes. What was the temperature change in Celsius degrees? (b) The temperature in Browning, Montana, was 44.0°F on January 23, 1916. The next day the temperature plummeted to −56°F. What was the temperature change in Celsius degrees?

Section 17.3 Gas Thermometers and the Kelvin Scale

17.4 • (a) Calculate the one temperature at which Fahrenheit and Celsius thermometers agree with each other. (b) Calculate the one temperature at which Fahrenheit and Kelvin thermometers agree with each other.

17.5 •• You put a bottle of soft drink in a refrigerator and leave it until its temperature has dropped 10.0 K. What is its temperature change in (a) F° and (b) C°?

17.6 • Convert the following Kelvin temperatures to the Celsius and Fahrenheit scales: (a) the midday temperature at the surface of the moon (400 K); (b) the temperature at the tops of the clouds in the atmosphere of Saturn (95 K); (c) the temperature at the center of the sun (1.55×10^7 K).

17.7 • The pressure of a gas at the triple point of water is 1.35 atm. If its volume remains unchanged, what will its pressure be at the temperature at which CO_2 solidifies?

17.8 •• A gas thermometer registers an absolute pressure corresponding to 325 mm of mercury when in contact with water at the triple point. What pressure does it read when in contact with water at the normal boiling point?

17.9 •• **A Constant-Volume Gas Thermometer.** An experimenter using a gas thermometer found the pressure at the triple point of water (0.01°C) to be 4.80×10^4 Pa and the pressure at the normal boiling point (100°C) to be 6.50×10^4 Pa. (a) Assuming that the pressure varies linearly with temperature, use these two data points to find the Celsius temperature at which the gas pressure would be zero (that is, find the Celsius temperature of absolute zero). (b) Does the gas in this thermometer obey Eq. (17.4) precisely? If that equation were precisely obeyed and the pressure at 100°C were 6.50×10^4 Pa, what pressure would the experimenter have measured at 0.01°C? (As we will learn in Section 18.1, Eq. (17.4) is accurate only for gases at very low density.)

17.10 • Like the Kelvin scale, the *Rankine scale* is an absolute temperature scale: Absolute zero is zero degrees Rankine (0°R). However, the units of this scale are the same size as those of the Fahrenheit scale rather than the Celsius scale. What is the numerical value of the triple-point temperature of water on the Rankine scale?

Section 17.4 Thermal Expansion

17.11 • The Humber Bridge in England has the world's longest single span, 1410 m. Calculate the change in length of the steel deck of the span when the temperature increases from $-5.0°C$ to $18.0°C$.

17.12 • One of the tallest buildings in the world is the Taipei 101 in Taiwan, at a height of 1671 feet. Assume that this height was measured on a cool spring day when the temperature was $15.5°C$. You could use the building as a sort of giant thermometer on a hot summer day by carefully measuring its height. Suppose you do this and discover that the Taipei 101 is 0.471 foot taller than its official height. What is the temperature, assuming that the building is in thermal equilibrium with the air and that its entire frame is made of steel?

17.13 • A U.S. penny has a diameter of 1.9000 cm at $20.0°C$. The coin is made of a metal alloy (mostly zinc) for which the coefficient of linear expansion is $2.6 \times 10^{-5} \text{ K}^{-1}$. What would its diameter be on a hot day in Death Valley ($48.0°C$)? On a cold night in the mountains of Greenland ($-53°C$)?

17.14 • **Ensuring a Tight Fit.** Aluminum rivets used in airplane construction are made slightly larger than the rivet holes and cooled by "dry ice" (solid CO_2) before being driven. If the diameter of a hole is 4.500 mm, what should be the diameter of a rivet at $23.0°C$ if its diameter is to equal that of the hole when the rivet is cooled to $-78.0°C$, the temperature of dry ice? Assume that the expansion coefficient remains constant at the value given in Table 17.1.

17.15 •• The outer diameter of a glass jar and the inner diameter of its iron lid are both 725 mm at room temperature ($20.0°C$). What will be the size of the difference in these diameters if the lid is briefly held under hot water until its temperature rises to $50.0°C$, without changing the temperature of the glass?

17.16 •• A geodesic dome constructed with an aluminum framework is a nearly perfect hemisphere; its diameter measures 55.0 m on a winter day at a temperature of $-15°C$. How much more interior space does the dome have in the summer, when the temperature is $35°C$?

17.17 •• A copper cylinder is initially at $20.0°C$. At what temperature will its volume be 0.150% larger than it is at $20.0°C$?

17.18 •• A steel tank is completely filled with 2.80 m³ of ethanol when both the tank and the ethanol are at a temperature of $32.0°C$. When the tank and its contents have cooled to $18.0°C$, what additional volume of ethanol can be put into the tank?

17.19 •• A glass flask whose volume is 1000.00 cm³ at $0.0°C$ is completely filled with mercury at this temperature. When flask and mercury are warmed to $55.0°C$, 8.95 cm³ of mercury overflow. If the coefficient of volume expansion of mercury is $18.0 \times 10^{-5} \text{ K}^{-1}$, compute the coefficient of volume expansion of the glass.

17.20 •• (a) If an area measured on the surface of a solid body is A_0 at some initial temperature and then changes by ΔA when the temperature changes by ΔT, show that

$$\Delta A = (2\alpha)A_0\Delta T$$

where α is the coefficient of linear expansion. (b) A circular sheet of aluminum is 55.0 cm in diameter at $15.0°C$. By how much does the area of one side of the sheet change when the temperature increases to $27.5°C$?

17.21 •• A machinist bores a hole of diameter 1.35 cm in a steel plate at a temperature of $25.0°C$. What is the cross-sectional area of the hole (a) at $25.0°C$ and (b) when the temperature of the plate is increased to $175°C$? Assume that the coefficient of linear expansion remains constant over this temperature range. (*Hint:* See Exercise 17.20.)

17.22 •• As a new mechanical engineer for Engines Inc., you have been assigned to design brass pistons to slide inside steel cylinders. The engines in which these pistons will be used will operate between $20.0°C$ and $150.0°C$. Assume that the coefficients of expansion are constant over this temperature range. (a) If the piston just fits inside the chamber at $20.0°C$, will the engines be able to run at higher temperatures? Explain. (b) If the cylindrical pistons are 25.000 cm in diameter at $20.0°C$, what should be the minimum diameter of the cylinders at that temperature so the pistons will operate at $150.0°C$?

17.23 • (a) A wire that is 1.50 m long at $20.0°C$ is found to increase in length by 1.90 cm when warmed to $420.0°C$. Compute its average coefficient of linear expansion for this temperature range. (b) The wire is stretched just taut (zero tension) at $420.0°C$. Find the stress in the wire if it is cooled to $20.0°C$ without being allowed to contract. Young's modulus for the wire is 2.0×10^{11} Pa.

17.24 •• A brass rod is 185 cm long and 1.60 cm in diameter. What force must be applied to each end of the rod to prevent it from contracting when it is cooled from $120.0°C$ to $10.0°C$?

17.25 •• Steel train rails are laid in 12.0-m-long segments placed end to end. The rails are laid on a winter day when their temperature is $-2.0°C$. (a) How much space must be left between adjacent rails if they are just to touch on a summer day when their temperature is $33.0°C$? (b) If the rails are originally laid in contact, what is the stress in them on a summer day when their temperature is $33.0°C$?

Section 17.5 Quantity of Heat

17.26 • In an effort to stay awake for an all-night study session, a student makes a cup of coffee by first placing a 200-W electric immersion heater in 0.320 kg of water. (a) How much heat must be added to the water to raise its temperature from $20.0°C$ to $80.0°C$? (b) How much time is required? Assume that all of the heater's power goes into heating the water.

17.27 •• An aluminum tea kettle with mass 1.50 kg and containing 1.80 kg of water is placed on a stove. If no heat is lost to the surroundings, how much heat must be added to raise the temperature from $20.0°C$ to $85.0°C$?

17.28 • **BIO Heat Loss During Breathing.** In very cold weather a significant mechanism for heat loss by the human body is energy expended in warming the air taken into the lungs with each breath. (a) On a cold winter day when the temperature is $-20°C$, what amount of heat is needed to warm to body temperature ($37°C$) the 0.50 L of air exchanged with each breath? Assume that the specific heat of air is 1020 J/kg·K and that 1.0 L of air has mass 1.3×10^{-3} kg. (b) How much heat is lost per hour if the respiration rate is 20 breaths per minute?

17.29 • You are given a sample of metal and asked to determine its specific heat. You weigh the sample and find that its weight is 28.4 N. You carefully add 1.25×10^4 J of heat energy to the sample and find that its temperature rises 18.0 C°. What is the sample's specific heat?

17.30 •• **On-Demand Water Heaters.** Conventional hot-water heaters consist of a tank of water maintained at a fixed temperature. The hot water is to be used when needed. The drawbacks are that energy is wasted because the tank loses heat when it is not in use and that you can run out of hot water if you use too much. Some utility companies are encouraging the use of *on-demand* water heaters (also known as *flash heaters*), which consist of heating units to heat the water as you use it. No water tank is involved, so no heat is wasted. A typical household shower flow rate is 2.5 gal/min

(9.46 L/min) with the tap water being heated from 50°F (10°C) to 120°F (49°C) by the on-demand heater. What rate of heat input (either electrical or from gas) is required to operate such a unit, assuming that all the heat goes into the water?

17.31 • BIO While running, a 70-kg student generates thermal energy at a rate of 1200 W. For the runner to maintain a constant body temperature of 37°C, this energy must be removed by perspiration or other mechanisms. If these mechanisms failed and the heat could not flow out of the student's body, for what amount of time could a student run before irreversible body damage occurred? (*Note:* Protein structures in the body are irreversibly damaged if body temperature rises to 44°C or higher. The specific heat of a typical human body is 3480 J/kg·K, slightly less than that of water. The difference is due to the presence of protein, fat, and minerals, which have lower specific heats.)

17.32 • CP While painting the top of an antenna 225 m in height, a worker accidentally lets a 1.00-L water bottle fall from his lunchbox. The bottle lands in some bushes at ground level and does not break. If a quantity of heat equal to the magnitude of the change in mechanical energy of the water goes into the water, what is its increase in temperature?

17.33 •• CP A crate of fruit with mass 35.0 kg and specific heat 3650 J/kg·K slides down a ramp inclined at 36.9° below the horizontal. The ramp is 8.00 m long. (a) If the crate was at rest at the top of the incline and has a speed of 2.50 m/s at the bottom, how much work was done on the crate by friction? (b) If an amount of heat equal to the magnitude of the work done by friction goes into the crate of fruit and the fruit reaches a uniform final temperature, what is its temperature change?

17.34 • CP A 25,000-kg subway train initially traveling at 15.5 m/s slows to a stop in a station and then stays there long enough for its brakes to cool. The station's dimensions are 65.0 m long by 20.0 m wide by 12.0 m high. Assuming all the work done by the brakes in stopping the train is transferred as heat uniformly to all the air in the station, by how much does the air temperature in the station rise? Take the density of the air to be 1.20 kg/m³ and its specific heat to be 1020 J/kg·K.

17.35 • CP A nail driven into a board increases in temperature. If we assume that 60% of the kinetic energy delivered by a 1.80-kg hammer with a speed of 7.80 m/s is transformed into heat that flows into the nail and does not flow out, what is the temperature increase of an 8.00-g aluminum nail after it is struck ten times?

17.36 • A technician measures the specific heat of an unidentified liquid by immersing an electrical resistor in it. Electrical energy is converted to heat transferred to the liquid for 120 s at a constant rate of 65.0 W. The mass of the liquid is 0.780 kg, and its temperature increases from 18.55°C to 22.54°C. (a) Find the average specific heat of the liquid in this temperature range. Assume that negligible heat is transferred to the container that holds the liquid and that no heat is lost to the surroundings. (b) Suppose that in this experiment heat transfer from the liquid to the container or surroundings cannot be ignored. Is the result calculated in part (a) an *overestimate* or an *underestimate* of the average specific heat? Explain.

17.37 •• CP A 15.0-g bullet traveling horizontally at 865 m/s passes through a tank containing 13.5 kg of water and emerges with a speed of 534 m/s. What is the maximum temperature increase that the water could have as a result of this event?

Section 17.6 Calorimetry and Phase Changes

17.38 •• As a physicist, you put heat into a 500.0-g solid sample at the rate of 10.0 kJ/min, while recording its temperature as a function of time. You plot your data and obtain the graph shown in Fig. E17.38. (a) What is the latent heat of fusion for this solid? (b) What are the specific heats of the liquid and solid states of the material?

Figure **E17.38**

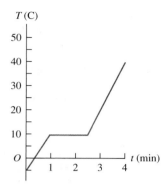

17.39 •• A 500.0-g chunk of an unknown metal, which has been in boiling water for several minutes, is quickly dropped into an insulating Styrofoam beaker containing 1.00 kg of water at room temperature (20.0°C). After waiting and gently stirring for 5.00 minutes, you observe that the water's temperature has reached a constant value of 22.0°C. (a) Assuming that the Styrofoam absorbs a negligibly small amount of heat and that no heat was lost to the surroundings, what is the specific heat of the metal? (b) Which is more useful for storing thermal energy: this metal or an equal weight of water? Explain. (c) What if the heat absorbed by the Styrofoam actually is not negligible. How would the specific heat you calculated in part (a) be in error? Would it be too large, too small, or still correct? Explain.

17.40 • BIO **Treatment for a Stroke.** One suggested treatment for a person who has suffered a stroke is immersion in an ice-water bath at 0°C to lower the body temperature, which prevents damage to the brain. In one set of tests, patients were cooled until their internal temperature reached 32.0°C. To treat a 70.0-kg patient, what is the minimum amount of ice (at 0°C) you need in the bath so that its temperature remains at 0°C? The specific heat of the human body is 3480 J/kg·C°, and recall that normal body temperature is 37.0°C.

17.41 •• A copper pot with a mass of 0.500 kg contains 0.170 kg of water, and both are at a temperature of 20.0°C. A 0.250-kg block of iron at 85.0°C is dropped into the pot. Find the final temperature of the system, assuming no heat loss to the surroundings.

17.42 •• BIO **Bicycling on a Warm Day.** If the air temperature is the same as the temperature of your skin (about 30°C), your body cannot get rid of heat by transferring it to the air. In that case, it gets rid of the heat by evaporating water (sweat). During bicycling, a typical 70-kg person's body produces energy at a rate of about 500 W due to metabolism, 80% of which is converted to heat. (a) How many kilograms of water must the person's body evaporate in an hour to get rid of this heat? The heat of vaporization of water at body temperature is 2.42×10^6 J/kg. (b) The evaporated water must, of course, be replenished, or the person will dehydrate. How many 750-mL bottles of water must the bicyclist drink per hour to replenish the lost water? (Recall that the mass of a liter of water is 1.0 kg.)

17.43 •• BIO **Overheating.** (a) By how much would the body temperature of the bicyclist in the preceeding problem increase in an hour if he were unable to get rid of the excess heat? (b) Is this temperature increase large enough to be serious? To find out, how high a fever would it be equivalent to, in °F? (Recall that the normal internal body temperature is 98.6°F and the specific heat of the body is 3480 J/kg·C°.)

17.44 • In a container of negligible mass, 0.200 kg of ice at an initial temperature of −40.0°C is mixed with a mass m of water that has an initial temperature of 80.0°C. No heat is lost to the surroundings. If the final temperature of the system is 20.0°C, what is the mass m of the water that was initially at 80.0°C?

17.45 • A 6.00-kg piece of solid copper metal at an initial temperature T is placed with 2.00 kg of ice that is initially at −20.0°C. The ice is in an insulated container of negligible mass and no heat is exchanged with the surroundings. After thermal equilibrium is reached, there is 1.20 kg of ice and 0.80 kg of liquid water. What was the initial temperature of the piece of copper?

17.46 • BIO Before going in for his annual physical, a 70.0-kg man whose body temperature is 37.0°C consumes an entire 0.355-L can of a soft drink (mostly water) at 12.0°C. (a) What will his body temperature be after equilibrium is attained? Ignore any heating by the man's metabolism. The specific heat of the man's body is 3480 J/kg·K. (b) Is the change in his body temperature great enough to be measured by a medical thermometer?

17.47 •• BIO In the situation described in Exercise 17.46, the man's metabolism will eventually return the temperature of his body (and of the soft drink that he consumed) to 37.0°C. If his body releases energy at a rate of 7.00×10^3 kJ/day (the *basal metabolic rate,* or BMR), how long does this take? Assume that all of the released energy goes into raising the temperature.

17.48 •• An ice-cube tray of negligible mass contains 0.350 kg of water at 18.0°C. How much heat must be removed to cool the water to 0.00°C and freeze it? Express your answer in joules, calories, and Btu.

17.49 • How much heat is required to convert 12.0 g of ice at −10.0°C to steam at 100.0°C? Express your answer in joules, calories, and Btu.

17.50 •• An open container holds 0.550 kg of ice at −15.0°C. The mass of the container can be ignored. Heat is supplied to the container at the constant rate of 800.0 J/min for 500.0 min. (a) After how many minutes does the ice *start* to melt? (b) After how many minutes, from the time when the heating is first started, does the temperature begin to rise above 0.0°C? (c) Plot a curve showing the temperature as a function of the elapsed time.

17.51 • CP What must the initial speed of a lead bullet be at a temperature of 25.0°C so that the heat developed when it is brought to rest will be just sufficient to melt it? Assume that all the initial mechanical energy of the bullet is converted to heat and that no heat flows from the bullet to its surroundings. (Typical rifles have muzzle speeds that exceed the speed of sound in air, which is 347 m/s at 25.0°C.)

17.52 •• BIO **Steam Burns Versus Water Burns.** What is the amount of heat input to your skin when it receives the heat released (a) by 25.0 g of steam initially at 100.0°C, when it is cooled to skin temperature (34.0°C)? (b) By 25.0 g of water initially at 100.0°C, when it is cooled to 34.0°C? (c) What does this tell you about the relative severity of steam and hot water burns?

17.53 • BIO **"The Ship of the Desert."** Camels require very little water because they are able to tolerate relatively large changes in their body temperature. While humans keep their body temperatures constant to within one or two Celsius degrees, a dehydrated camel permits its body temperature to drop to 34.0°C overnight and rise to 40.0°C during the day. To see how effective this mechanism is for saving water, calculate how many liters of water a 400-kg

camel would have to drink if it attempted to keep its body temperature at a constant 34.0°C by evaporation of sweat during the day (12 hours) instead of letting it rise to 40.0°C. (*Note:* The specific heat of a camel or other mammal is about the same as that of a typical human, 3480 J/kg·K. The heat of vaporization of water at 34°C is 2.42×10^6 J/kg.)

17.54 • BIO Evaporation of sweat is an important mechanism for temperature control in some warm-blooded animals. (a) What mass of water must evaporate from the skin of a 70.0-kg man to cool his body 1.00 C°? The heat of vaporization of water at body temperature (37°C) is 2.42×10^6 J/kg. The specific heat of a typical human body is 3480 J/kg·K (see Exercise 17.31). (b) What volume of water must the man drink to replenish the evaporated water? Compare to the volume of a soft-drink can (355 cm³).

17.55 •• CP An asteroid with a diameter of 10 km and a mass of 2.60×10^{15} kg impacts the earth at a speed of 32.0 km/s, landing in the Pacific Ocean. If 1.00% of the asteroid's kinetic energy goes to boiling the ocean water (assume an initial water temperature of 10.0°C), what mass of water will be boiled away by the collision? (For comparison, the mass of water contained in Lake Superior is about 2×10^{15} kg.)

17.56 • A laboratory technician drops a 0.0850-kg sample of unknown solid material, at a temperature of 100.0°C, into a calorimeter. The calorimeter can, initially at 19.0°C, is made of 0.150 kg of copper and contains 0.200 kg of water. The final temperature of the calorimeter can and contents is 26.1°C. Compute the specific heat of the sample.

17.57 •• An insulated beaker with negligible mass contains 0.250 kg of water at a temperature of 75.0°C. How many kilograms of ice at a temperature of −20.0°C must be dropped into the water to make the final temperature of the system 40.0°C?

17.58 •• A glass vial containing a 16.0-g sample of an enzyme is cooled in an ice bath. The bath contains water and 0.120 kg of ice. The sample has specific heat 2250 J/kg·K; the glass vial has mass 6.00 g and specific heat 2800 J/kg·K. How much ice melts in cooling the enzyme sample from room temperature (19.5°C) to the temperature of the ice bath?

17.59 • A 4.00-kg silver ingot is taken from a furnace, where its temperature is 750.0°C, and placed on a large block of ice at 0.0°C. Assuming that all the heat given up by the silver is used to melt the ice, how much ice is melted?

17.60 •• A copper calorimeter can with mass 0.100 kg contains 0.160 kg of water and 0.0180 kg of ice in thermal equilibrium at atmospheric pressure. If 0.750 kg of lead at a temperature of 255°C is dropped into the calorimeter can, what is the final temperature? Assume that no heat is lost to the surroundings.

17.61 •• A vessel whose walls are thermally insulated contains 2.40 kg of water and 0.450 kg of ice, all at a temperature of 0.0°C. The outlet of a tube leading from a boiler in which water is boiling at atmospheric pressure is inserted into the water. How many grams of steam must condense inside the vessel (also at atmospheric pressure) to raise the temperature of the system to 28.0°C? You can ignore the heat transferred to the container.

Section 17.7 Mechanisms of Heat Transfer

17.62 •• Two rods, one made of brass and the other made of copper, are joined end to end. The length of the brass section is 0.200 m and the length of the copper section is 0.800 m. Each segment has cross-sectional area 0.00500 m². The free end of the brass segment is in boiling water and the free end of the copper segment is in an ice and water mixture, in both cases under normal atmospheric pressure. The sides of the rods are insulated so there is no heat loss to the surroundings. (a) What is the temperature of the point where the brass and copper segments are joined? (b) What mass of ice is melted in 5.00 min by the heat conducted by the composite rod?

17.63 • Suppose that the rod in Fig. 17.23a is made of copper, is 45.0 cm long, and has a cross-sectional area of 1.25 cm². Let $T_H = 100.0°C$ and $T_C = 0.0°C$. (a) What is the final steady-state temperature gradient along the rod? (b) What is the heat current in the rod in the final steady state? (c) What is the final steady-state temperature at a point in the rod 12.0 cm from its left end?

17.64 •• One end of an insulated metal rod is maintained at 100.0°C, and the other end is maintained at 0.00°C by an ice–water mixture. The rod is 60.0 cm long and has a cross-sectional area of 1.25 cm². The heat conducted by the rod melts 8.50 g of ice in 10.0 min. Find the thermal conductivity k of the metal.

17.65 •• A carpenter builds an exterior house wall with a layer of wood 3.0 cm thick on the outside and a layer of Styrofoam insulation 2.2 cm thick on the inside wall surface. The wood has $k = 0.080$ W/m·K, and the Styrofoam has $k = 0.010$ W/m·K. The interior surface temperature is 19.0°C, and the exterior surface temperature is −10.0°C. (a) What is the temperature at the plane where the wood meets the Styrofoam? (b) What is the rate of heat flow per square meter through this wall?

17.66 • An electric kitchen range has a total wall area of 1.40 m² and is insulated with a layer of fiberglass 4.00 cm thick. The inside surface of the fiberglass has a temperature of 175°C, and its outside surface is at 35.0°C. The fiberglass has a thermal conductivity of 0.040 W/m·K. (a) What is the heat current through the insulation, assuming it may be treated as a flat slab with an area of 1.40 m²? (b) What electric-power input to the heating element is required to maintain this temperature?

17.67 • **BIO** **Conduction Through the Skin.** The blood plays an important role in removing heat from the body by bringing this heat directly to the surface where it can radiate away. Nevertheless, this heat must still travel through the skin before it can radiate away. We shall assume that the blood is brought to the bottom layer of skin at a temperature of 37.0°C and that the outer surface of the skin is at 30.0°C. Skin varies in thickness from 0.50 mm to a few millimeters on the palms and soles, so we shall assume an average thickness of 0.75 mm. A 165-lb, 6-ft-tall person has a surface area of about 2.0 m² and loses heat at a net rate of 75 W while resting. On the basis of our assumptions, what is the thermal conductivity of this person's skin?

17.68 • A long rod, insulated to prevent heat loss along its sides, is in perfect thermal contact with boiling water (at atmospheric pressure) at one end and with an ice–water mixture at the other (Fig. E17.68). The rod consists of a 1.00-m section of copper (one end in boiling water) joined end to end to a length L_2 of steel (one end in the ice–water mixture). Both sections of the rod have cross-sectional areas of 4.00 cm². The temperature of the copper–steel junction is 65.0°C after a steady state has been set up. (a) How much heat per second flows from the boiling water to the ice–water mixture? (b) What is the length L_2 of the steel section?

Figure **E17.68**

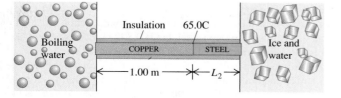

17.69 • A pot with a steel bottom 8.50 mm thick rests on a hot stove. The area of the bottom of the pot is 0.150 m². The water inside the pot is at 100.0°C, and 0.390 kg are evaporated every 3.00 min. Find the temperature of the lower surface of the pot, which is in contact with the stove.

17.70 •• You are asked to design a cylindrical steel rod 50.0 cm long, with a circular cross section, that will conduct 150.0 J/s from a furnace at 400.0°C to a container of boiling water under 1 atmosphere. What must the rod's diameter be?

17.71 •• A picture window has dimensions of 1.40 m × 2.50 m and is made of glass 5.20 mm thick. On a winter day, the outside temperature is −20.0°C, while the inside temperature is a comfortable 19.5°C. (a) At what rate is heat being lost through the window by conduction? (b) At what rate would heat be lost through the window if you covered it with a 0.750-mm-thick layer of paper (thermal conductivity 0.0500 W/m·K)?

17.72 • What is the rate of energy radiation per unit area of a blackbody at a temperature of (a) 273 K and (b) 2730 K?

17.73 • **Size of a Light-Bulb Filament.** The operating temperature of a tungsten filament in an incandescent light bulb is 2450 K, and its emissivity is 0.350. Find the surface area of the filament of a 150-W bulb if all the electrical energy consumed by the bulb is radiated by the filament as electromagnetic waves. (Only a fraction of the radiation appears as visible light.)

17.74 •• The emissivity of tungsten is 0.350. A tungsten sphere with radius 1.50 cm is suspended within a large evacuated enclosure whose walls are at 290.0 K. What power input is required to maintain the sphere at a temperature of 3000.0 K if heat conduction along the supports is neglected?

17.75 • **The Sizes of Stars.** The hot glowing surfaces of stars emit energy in the form of electromagnetic radiation. It is a good approximation to assume $e = 1$ for these surfaces. Find the radii of the following stars (assumed to be spherical): (a) Rigel, the bright blue star in the constellation Orion, which radiates energy at a rate of 2.7×10^{32} W and has surface temperature 11,000 K; (b) Procyon B (visible only using a telescope), which radiates energy at a rate of 2.1×10^{23} W and has surface temperature 10,000 K. (c) Compare your answers to the radius of the earth, the radius of the sun, and the distance between the earth and the sun. (Rigel is an example of a *supergiant* star, and Procyon B is an example of a *white dwarf* star.)

PROBLEMS

17.76 •• Suppose that a steel hoop could be constructed to fit just around the earth's equator at a temperature of 20.0°C. What would be the thickness of space between the hoop and the earth if the temperature of the hoop were increased by 0.500 C°?

17.77 ••• You propose a new temperature scale with temperatures given in °M. You define 0.0°M to be the normal melting point of mercury and 100.0° to be the normal boiling point of mercury. (a) What is the normal boiling point of water in °M? (b) A temperature change of 10.0 M° corresponds to how many C°?

17.78 • **CP, CALC** A 250-kg weight is hanging from the ceiling by a thin copper wire. In its fundamental mode, this wire vibrates at the frequency of concert A (440 Hz). You then increase the temperature of the wire by 40 C°. (a) By how much will the fundamental frequency change? Will it increase or decrease? (b) By what percentage will the speed of a wave on the wire change? (c) By what percentage will the wavelength of the fundamental standing wave change? Will it increase or decrease?

17.79 ••• You are making pesto for your pasta and have a cylindrical measuring cup 10.0 cm high made of ordinary glass $[\beta = 2.7 \times 10^{-5} \ (\text{C}°)^{-1}]$ that is filled with olive oil $[\beta = 6.8 \times 10^{-4} \ (\text{C}°)^{-1}]$ to a height of 2.00 mm below the top of the cup. Initially, the cup and oil are at room temperature (22.0°C). You get a phone call and forget about the olive oil, which you inadvertently leave on the hot stove. The cup and oil heat up slowly and have a common temperature. At what temperature will the olive oil start to spill out of the cup?

17.80 •• A surveyor's 30.0-m steel tape is correct at a temperature of 20.0°C. The distance between two points, as measured by this tape on a day when its temperature is 5.00°C, is 25.970 m. What is the true distance between the points?

17.81 •• **CP** A Foucault pendulum consists of a brass sphere with a diameter of 35.0 cm suspended from a steel cable 10.5 m long (both measurements made at 20.0°C). Due to a design oversight, the swinging sphere clears the floor by a distance of only 2.00 mm when the temperature is 20.0°C. At what temperature will the sphere begin to brush the floor?

17.82 •• You pour 108 cm³ of ethanol, at a temperature of −10.0°C, into a graduated cylinder initially at 20.0°C, filling it to the very top. The cylinder is made of glass with a specific heat of 840 J/kg·K and a coefficient of volume expansion of $1.2 \times 10^{-5} \ \text{K}^{-1}$; its mass is 0.110 kg. The mass of the ethanol is 0.0873 kg. (a) What will be the final temperature of the ethanol, once thermal equilibrium is reached? (b) How much ethanol will overflow the cylinder before thermal equilibrium is reached?

17.83 •• A metal rod that is 30.0 cm long expands by 0.0650 cm when its temperature is raised from 0.0°C to 100.0°C. A rod of a different metal and of the same length expands by 0.0350 cm for the same rise in temperature. A third rod, also 30.0 cm long, is made up of pieces of each of the above metals placed end to end and expands 0.0580 cm between 0.0°C and 100.0°C. Find the length of each portion of the composite rod.

17.84 •• On a cool (4.0°C) Saturday morning, a pilot fills the fuel tanks of her Pitts S-2C (a two-seat aerobatic airplane) to their full capacity of 106.0 L. Before flying on Sunday morning, when the temperature is again 4.0°C, she checks the fuel level and finds only 103.4 L of gasoline in the tanks. She realizes that it was hot on Saturday afternoon, and that thermal expansion of the gasoline caused the missing fuel to empty out of the tank's vent. (a) What was the maximum temperature (in °C) reached by the fuel and the tank on Saturday afternoon? The coefficient of volume expansion of gasoline is $9.5 \times 10^{-4} \ \text{K}^{-1}$, and the tank is made of aluminum. (b) In order to have the maximum amount of fuel available for flight, when should the pilot have filled the fuel tanks?

17.85 ••• (a) Equation (17.12) gives the stress required to keep the length of a rod constant as its temperature changes. Show that if the length is permitted to change by an amount ΔL when its temperature changes by ΔT, the stress is equal to

$$\frac{F}{A} = Y\left(\frac{\Delta L}{L_0} - \alpha\Delta T\right)$$

where F is the tension on the rod, L_0 is the original length of the rod, A its cross-sectional area, α its coefficient of linear expansion, and Y its Young's modulus. (b) A heavy brass bar has projections at its ends, as in Fig. P17.85. Two fine steel wires, fastened between the pro-

Figure **P17.85**

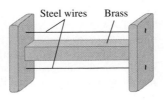

Steel wires   Brass

jections, are just taut (zero tension) when the whole system is at 20°C. What is the tensile stress in the steel wires when the temperature of the system is raised to 140°C? Make any simplifying assumptions you think are justified, but state what they are.

17.86 •• **CP** A metal wire, with density ρ and Young's modulus Y, is stretched between rigid supports. At temperature T, the speed of a transverse wave is found to be v_1. When the temperature is increased to $T + \Delta T$, the speed decreases to $v_2 < v_1$. Determine the coefficient of linear expansion of the wire.

17.87 •• **CP** **Out of Tune.** The B-string of a guitar is made of steel (density 7800 kg/m³), is 63.5 cm long, and has diameter 0.406 mm. The fundamental frequency is $f = 247.0$ Hz. (a) Find the string tension. (b) If the tension F is changed by a small amount ΔF, the frequency f changes by a small amount Δf. Show that

$$\frac{\Delta f}{f} = \frac{\Delta F}{2F}$$

(c) The string is tuned to a fundamental frequency of 247.0 Hz when its temperature is 18.5°C. Strenuous playing can make the temperature of the string rise, changing its vibration frequency. Find Δf if the temperature of the string rises to 29.5°C. The steel string has a Young's modulus of 2.00×10^{11} Pa and a coefficient of linear expansion of $1.20 \times 10^{-5} \ (\text{C}°)^{-1}$. Assume that the temperature of the body of the guitar remains constant. Will the vibration frequency rise or fall?

17.88 ••• A steel rod 0.450 m long and an aluminum rod 0.250 m long, both with the same diameter, are placed end to end between rigid supports with no initial stress in the rods. The temperature of the rods is now raised by 60.0 C°. What is the stress in each rod? (*Hint:* The length of the combined rod remains the same, but the lengths of the individual rods do not. See Problem 17.85.)

17.89 •• A steel ring with a 2.5000-in. inside diameter at 20.0°C is to be warmed and slipped over a brass shaft with a 2.5020-in. outside diameter at 20.0°C. (a) To what temperature should the ring be warmed? (b) If the ring and the shaft together are cooled by some means such as liquid air, at what temperature will the ring just slip off the shaft?

17.90 •• **Bulk Stress Due to a Temperature Increase.** (a) Prove that, if an object under pressure has its temperature raised but is not allowed to expand, the increase in pressure is

$$\Delta p = B\beta\Delta T$$

where the bulk modulus B and the average coefficient of volume expansion β are both assumed positive and constant. (b) What pressure is necessary to prevent a steel block from expanding when its temperature is increased from 20.0°C to 35.0°C?

17.91 •• A liquid is enclosed in a metal cylinder that is provided with a piston of the same metal. The system is originally at a pressure of 1.00 atm (1.013×10^5 Pa) and at a temperature of 30.0°C. The piston is forced down until the pressure on the liquid is increased by 50.0 atm, and then clamped in this position. Find the new temperature at which the pressure of the liquid is again 1.00 atm. Assume that the cylinder is sufficiently strong so that its volume is not altered by changes in pressure, but only by changes in temperature. Use the result derived in Problem 17.90. (*Hint:* See Section 11.4.)
Compressibility of liquid: $k = 8.50 \times 10^{-10} \ \text{Pa}^{-1}$
Coefficient of volume expansion of liquid: $\beta = 4.80 \times 10^{-4} \ \text{K}^{-1}$
Coefficient of volume expansion of metal: $\beta = 3.90 \times 10^{-5} \ \text{K}^{-1}$

17.92 •• You cool a 100.0-g slug of red-hot iron (temperature 745°C) by dropping it into an insulated cup of negligible mass containing 85.0 g of water at 20.0°C. Assuming no heat exchange with the surroundings, (a) what is the final temperature of the water and (b) what is the final mass of the iron and the remaining water?

17.93 • **CP Spacecraft Reentry.** A spacecraft made of aluminum circles the earth at a speed of 7700 m/s. (a) Find the ratio of its kinetic energy to the energy required to raise its temperature from 0°C to 600°C. (The melting point of aluminum is 660°C. Assume a constant specific heat of 910 J/kg · K.) (b) Discuss the bearing of your answer on the problem of the reentry of a manned space vehicle into the earth's atmosphere.

17.94 • **CP** A capstan is a rotating drum or cylinder over which a rope or cord slides in order to provide a great amplification of the rope's tension while keeping both ends free (Fig. P17.94). Since the added tension in the rope is due to friction, the capstan generates thermal energy. (a) If the difference in tension between the two ends of the rope is 520.0 N and the capstan has a diameter of 10.0 cm and turns once in 0.900 s, find the rate at which thermal energy is generated. Why does the number of turns not matter? (b) If the capstan is made of iron and has mass 6.00 kg, at what rate does its temperature rise? Assume that the temperature in the capstan is uniform and that all the thermal energy generated flows into it.

Figure **P17.94**

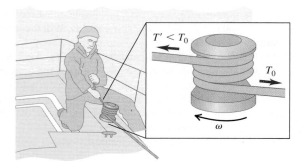

17.95 •• **CALC Debye's T^3 Law.** At very low temperatures the molar heat capacity of rock salt varies with temperature according to Debye's T^3 law:

$$C = k\frac{T^3}{\Theta^3}$$

where $k = 1940$ J/mol · K and $\Theta = 281$ K. (a) How much heat is required to raise the temperature of 1.50 mol of rock salt from 10.0 K to 40.0 K? (*Hint:* Use Eq. (17.18) in the form $dQ = nC\ dT$ and integrate.) (b) What is the average molar heat capacity in this range? (c) What is the true molar heat capacity at 40.0 K?

17.96 •• **CP** A person of mass 70.0 kg is sitting in the bathtub. The bathtub is 190.0 cm by 80.0 cm; before the person got in, the water was 16.0 cm deep. The water is at a temperature of 37.0°C. Suppose that the water were to cool down spontaneously to form ice at 0.0°C, and that all the energy released was used to launch the hapless bather vertically into the air. How high would the bather go? (As you will see in Chapter 20, this event is allowed by energy conservation but is prohibited by the second law of thermodynamics.)

17.97 • **Hot Air in a Physics Lecture.** (a) A typical student listening attentively to a physics lecture has a heat output of 100 W. How much heat energy does a class of 90 physics students release into a lecture hall over the course of a 50-min lecture? (b) Assume that all the heat energy in part (a) is transferred to the 3200 m³ of air in the room. The air has specific heat 1020 J/kg · K and density 1.20 kg/m³. If none of the heat escapes and the air conditioning system is off, how much will the temperature of the air in the room rise during the 50-min lecture? (c) If the class is taking an exam,

the heat output per student rises to 280 W. What is the temperature rise during 50 min in this case?

17.98 ••• **CALC** The molar heat capacity of a certain substance varies with temperature according to the empirical equation

$$C = 29.5 \text{ J/mol·K} + (8.20 \times 10^{-3} \text{ J/mol·K}^2)T$$

How much heat is necessary to change the temperature of 3.00 mol of this substance from 27°C to 227°C? (*Hint:* Use Eq. (17.18) in the form $dQ = nC\ dT$ and integrate.)

17.99 ••• For your cabin in the wilderness, you decide to build a primitive refrigerator out of Styrofoam, planning to keep the interior cool with a block of ice that has an initial mass of 24.0 kg. The box has dimensions of 0.500 m × 0.800 m × 0.500 m. Water from melting ice collects in the bottom of the box. Suppose the ice block is at 0.00°C and the outside temperature is 21.0°C. If the top of the empty box is never opened and you want the interior of the box to remain at 5.00°C for exactly one week, until all the ice melts, what must be the thickness of the Styrofoam?

17.100 •• **Hot Water Versus Steam Heating.** In a household hot-water heating system, water is delivered to the radiators at 70.0°C (158.0°F) and leaves at 28.0°C (82.4°F). The system is to be replaced by a steam system in which steam at atmospheric pressure condenses in the radiators and the condensed steam leaves the radiators at 35.0°C (95.0°F). How many kilograms of steam will supply the same heat as was supplied by 1.00 kg of hot water in the first system?

17.101 ••• A copper calorimeter can with mass 0.446 kg contains 0.0950 kg of ice. The system is initially at 0.0°C. (a) If 0.0350 kg of steam at 100.0°C and 1.00 atm pressure is added to the can, what is the final temperature of the calorimeter can and its contents? (b) At the final temperature, how many kilograms are there of ice, how many of liquid water, and how many of steam?

17.102 • A Styrofoam bucket of negligible mass contains 1.75 kg of water and 0.450 kg of ice. More ice, from a refrigerator at −15.0°C, is added to the mixture in the bucket, and when thermal equilibrium has been reached, the total mass of ice in the bucket is 0.868 kg. Assuming no heat exchange with the surroundings, what mass of ice was added?

17.103 ••• In a container of negligible mass, 0.0400 kg of steam at 100°C and atmospheric pressure is added to 0.200 kg of water at 50.0°C. (a) If no heat is lost to the surroundings, what is the final temperature of the system? (b) At the final temperature, how many kilograms are there of steam and how many of liquid water?

17.104 •• **BIO Mammal Insulation.** Animals in cold climates often depend on *two* layers of insulation: a layer of body fat (of thermal conductivity 0.20 W/m · K) surrounded by a layer of air trapped inside fur or down. We can model a black bear (*Ursus americanus*) as a sphere 1.5 m in diameter having a layer of fat 4.0 cm thick. (Actually, the thickness varies with the season, but we are interested in hibernation, when the fat layer is thickest.) In studies of bear hibernation, it was found that the outer surface layer of the fur is at 2.7°C during hibernation, while the inner surface of the fat layer is at 31.0°C. (a) What is the temperature at the fat–inner fur boundary? (b) How thick should the air layer (contained within the fur) be so that the bear loses heat at a rate of 50.0 W?

17.105 ••• A worker pours 1.250 kg of molten lead at a temperature of 327.3°C into 0.5000 kg of water at a temperature of 75.00°C in an insulated bucket of negligible mass. Assuming no heat loss to the surroundings, calculate the mass of lead and water remaining in the bucket when the materials have reached thermal equilibrium.

17.106 •• One experimental method of measuring an insulating material's thermal conductivity is to construct a box of the material and measure the power input to an electric heater inside the box that maintains the interior at a measured temperature above the outside surface. Suppose that in such an apparatus a power input of 180 W is required to keep the interior surface of the box 65.0 C° (about 120 F°) above the temperature of the outer surface. The total area of the box is 2.18 m², and the wall thickness is 3.90 cm. Find the thermal conductivity of the material in SI units.

17.107 •• **Effect of a Window in a Door.** A carpenter builds a solid wood door with dimensions 2.00 m × 0.95 m × 5.0 cm. Its thermal conductivity is $k = 0.120$ W/m·K. The air films on the inner and outer surfaces of the door have the same combined thermal resistance as an additional 1.8-cm thickness of solid wood. The inside air temperature is 20.0°C, and the outside air temperature is −8.0°C. (a) What is the rate of heat flow through the door? (b) By what factor is the heat flow increased if a window 0.500 m on a side is inserted in the door? The glass is 0.450 cm thick, and the glass has a thermal conductivity of 0.80 W/m·K. The air films on the two sides of the glass have a total thermal resistance that is the same as an additional 12.0 cm of glass.

17.108 • A wood ceiling with thermal resistance R_1 is covered with a layer of insulation with thermal resistance R_2. Prove that the effective thermal resistance of the combination is $R = R_1 + R_2$.

17.109 •• Compute the ratio of the rate of heat loss through a single-pane window with area 0.15 m² to that for a double-pane window with the same area. The glass of a single pane is 4.2 mm thick, and the air space between the two panes of the double-pane window is 7.0 mm thick. The glass has thermal conductivity 0.80 W/m·K. The air films on the room and outdoor surfaces of either window have a combined thermal resistance of 0.15 m²·K/W.

17.110 • Rods of copper, brass, and steel are welded together to form a Y-shaped figure. The cross-sectional area of each rod is 2.00 cm². The free end of the copper rod is maintained at 100.0°C, and the free ends of the brass and steel rods at 0.0°C. Assume there is no heat loss from the surfaces of the rods. The lengths of the rods are: copper, 13.0 cm; brass, 18.0 cm; steel, 24.0 cm. (a) What is the temperature of the junction point? (b) What is the heat current in each of the three rods?

17.111 ••• **CALC** **Time Needed for a Lake to Freeze Over.** (a) When the air temperature is below 0°C, the water at the surface of a lake freezes to form an ice sheet. Why doesn't freezing occur throughout the entire volume of the lake? (b) Show that the thickness of the ice sheet formed on the surface of a lake is proportional to the square root of the time if the heat of fusion of the water freezing on the underside of the ice sheet is conducted through the sheet. (c) Assuming that the upper surface of the ice sheet is at −10°C and the bottom surface is at 0°C, calculate the time it will take to form an ice sheet 25 cm thick. (d) If the lake in part (c) is uniformly 40 m deep, how long would it take to freeze all the water in the lake? Is this likely to occur?

17.112 •• A rod is initially at a uniform temperature of 0°C throughout. One end is kept at 0°C, and the other is brought into contact with a steam bath at 100°C. The surface of the rod is insulated so that heat can flow only lengthwise along the rod. The cross-sectional area of the rod is 2.50 cm², its length is 120 cm, its thermal conductivity is 380 W/m·K, its density is 1.00×10^4 kg/m³, and its specific heat is 520 J/kg·K. Consider a short cylindrical element of the rod 1.00 cm in length. (a) If the temperature gradient at the cooler end of this element is 140 C°/m, how many joules of heat energy flow across this end per second? (b) If the average temperature of the element is

increasing at the rate of what is the temperature gradient at the other end of the element?

17.113 •• A rustic cabin has a floor area of 3.50 m × 3.00 m. Its walls, which are 2.50 m tall, are made of wood (thermal conductivity 0.0600 W/m·K) 1.80 cm thick and are further insulated with 1.50 cm of a synthetic material. When the outside temperature is 2.00°C, it is found necessary to heat the room at a rate of 1.25 kW to maintain its temperature at 19.0°C. Calculate the thermal conductivity of the insulating material. Neglect the heat lost through the ceiling and floor. Assume the inner and outer surfaces of the wall have the same termperature as the air inside and outside the cabin.

17.114 • The rate at which radiant energy from the sun reaches the earth's upper atmosphere is about 1.50 kW/m². The distance from the earth to the sun is 1.50×10^{11} m, and the radius of the sun is 6.96×10^8 m. (a) What is the rate of radiation of energy per unit area from the sun's surface? (b) If the sun radiates as an ideal blackbody, what is the temperature of its surface?

17.115 ••• **A Thermos for Liquid Helium.** A physicist uses a cylindrical metal can 0.250 m high and 0.090 m in diameter to store liquid helium at 4.22 K; at that temperature the heat of vaporization of helium is 2.09×10^4 J/kg. Completely surrounding the metal can are walls maintained at the temperature of liquid nitrogen, 77.3 K, with vacuum between the can and the surrounding walls. How much helium is lost per hour? The emissivity of the metal can is 0.200. The only heat transfer between the metal can and the surrounding walls is by radiation.

17.116 •• **BIO** **Basal Metabolic Rate.** The basal metabolic rate is the rate at which energy is produced in the body when a person is at rest. A 75-kg (165-lb) person of height 1.83 m (6 ft) has a body surface area of approximately 2.0 m². (a) What is the net amount of heat this person could radiate per second into a room at 18°C (about 65°F) if his skin's surface temperature is 30°C? (At such temperatures, nearly all the heat is infrared radiation, for which the body's emissivity is 1.0, regardless of the amount of pigment.) (b) Normally, 80% of the energy produced by metabolism goes into heat, while the rest goes into things like pumping blood and repairing cells. Also normally, a person at rest can get rid of this excess heat just through radiation. Use your answer to part (a) to find this person's basal metabolic rate.

17.117 •• **BIO** **Jogging in the Heat of the Day.** You have probably seen people jogging in extremely hot weather and wondered Why? As we shall see, there are good reasons not to do this! When jogging strenuously, an average runner of mass 68 kg and surface area 1.85 m² produces energy at a rate of up to 1300 W, 80% of which is converted to heat. The jogger radiates heat, but actually absorbs more from the hot air than he radiates away. At such high levels of activity, the skin's temperature can be elevated to around 33°C instead of the usual 30°C. (We shall neglect conduction, which would bring even more heat into his body.) The only way for the body to get rid of this extra heat is by evaporating water (sweating). (a) How much heat per second is produced just by the act of jogging? (b) How much *net* heat per second does the runner gain just from radiation if the air temperature is 40.0°C (104°F)? (Remember that he radiates out, but the environment radiates back in.) (c) What is the *total* amount of excess heat this runner's body must get rid of per second? (d) How much water must the jogger's body evaporate every minute due to his activity? The heat of vaporization of water at body temperature is 2.42×10^6 J/kg. (e) How many 750-mL bottles of water must he drink after (or preferably before!) jogging for a half hour? Recall that a liter of water has a mass of 1.0 kg.

17.118 •• BIO **Overheating While Jogging.** (a) If the jogger in the preceding problem were not able to get rid of the excess heat, by how much would his body temperature increase above the normal 37°C in a half hour of jogging? The specific heat for a human is about 3500 J/kg · K. (b) How high a fever (in °F) would this temperature increase be equivalent to? Is the increase large enough to be of concern? (Recall that normal body temperature is 98.6°F.)

17.119 •• An engineer is developing an electric water heater to provide a continuous supply of hot water. One trial design is shown in Fig. P17.119. Water is flowing at the rate of 0.500 kg/min, the inlet thermometer registers 18.0°C, the voltmeter reads 120 V, and the ammeter reads 15.0 A [corresponding to a power input of $(120 \text{ V}) \times (15.0 \text{ A}) = 1800 \text{ W}$]. (a) When a steady state is finally reached, what is the reading of the outlet thermometer? (b) Why is it unnecessary to take into account the heat capacity mc of the apparatus itself?

Figure **P17.119**

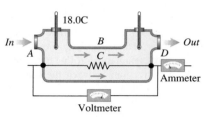

17.120 • **Food Intake of a Hamster.** The energy output of an animal engaged in an activity is called the basal metabolic rate (BMR) and is a measure of the conversion of food energy into other forms of energy. A simple calorimeter to measure the BMR consists of an insulated box with a thermometer to measure the temperature of the air. The air has density 1.20 kg/m³ and specific heat 1020 J/kg · K. A 50.0-g hamster is placed in a calorimeter that contains 0.0500 m³ of air at room temperature. (a) When the hamster is running in a wheel, the temperature of the air in the calorimeter rises 1.60 C° per hour. How much heat does the running hamster generate in an hour? Assume that all this heat goes into the air in the calorimeter. You can ignore the heat that goes into the walls of the box and into the thermometer, and assume that no heat is lost to the surroundings. (b) Assuming that the hamster converts seed into heat with an efficiency of 10% and that hamster seed has a food energy value of 24 J/g, how many grams of seed must the hamster eat per hour to supply this energy?

17.121 •• The icecaps of Greenland and Antarctica contain about 1.75% of the total water (by mass) on the earth's surface; the oceans contain about 97.5%, and the other 0.75% is mainly groundwater. Suppose the icecaps, currently at an average temperature of about −30°C, somehow slid into the ocean and melted. What would be the resulting temperature decrease of the ocean? Assume that the average temperature of ocean water is currently 5.00°C.

17.122 •• **Why Do the Seasons Lag?** In the northern hemisphere, June 21 (the summer solstice) is both the longest day of the year and the day on which the sun's rays strike the earth most vertically, hence delivering the greatest amount of heat to the surface. Yet the hottest summer weather usually occurs about a month or so later. Let us see why this is the case. Because of the large specific heat of water, the oceans are slower to warm up than the land (and also slower to cool off in winter). In addition to perusing pertinent information in the tables included in this book, it is useful to know

that approximately two-thirds of the earth's surface is ocean composed of salt water having a specific heat of 3890 J/kg · K and that the oceans, on the average, are 4000 m deep. Typically, an average of 1050 W/m² of solar energy falls on the earth's surface, and the oceans absorb essentially all of the light that strikes them. However, most of that light is absorbed in the upper 100 m of the surface. Depths below that do not change temperature seasonally. Assume that the sunlight falls on the surface for only 12 hours per day and that the ocean retains all the heat it absorbs. What will be the rise in temperature of the upper 100 m of the oceans during the month following the summer solstice? Does this seem to be large enough to be perceptible?

CHALLENGE PROBLEMS

17.123 ••• CALC Suppose that both ends of the rod in Fig. 17.23a are kept at a temperature of 0°C, and that the initial temperature distribution along the rod is given by $T = (100°C) \sin \pi x/L$, where x is measured from the left end of the rod. Let the rod be copper, with length $L = 0.100$ m and cross-sectional area 1.00 cm². (a) Show the initial temperature distribution in a diagram. (b) What is the final temperature distribution after a very long time has elapsed? (c) Sketch curves that you think would represent the temperature distribution at intermediate times. (d) What is the initial temperature gradient at the ends of the rod? (e) What is the initial heat current from the ends of the rod into the bodies making contact with its ends? (f) What is the initial heat current at the center of the rod? Explain. What is the heat current at this point at any later time? (g) What is the value of the *thermal diffusivity* $k/\rho c$ for copper, and in what unit is it expressed? (Here k is the thermal conductivity, $\rho = 8.9 \times 10^3$ kg/m³ is the density, and c is the specific heat.) (h) What is the initial time rate of change of temperature at the center of the rod? (i) How much time would be required for the center of the rod to reach its final temperature if the temperature continued to decrease at this rate? (This time is called the *relaxation time* of the rod.) (j) From the graphs in part (c), would you expect the magnitude of the rate of temperature change at the midpoint to remain constant, increase, or decrease as a function of time? (k) What is the initial rate of change of temperature at a point in the rod 2.5 cm from its left end?

17.124 ••• CALC (a) A spherical shell has inner and outer radii a and b, respectively, and the temperatures at the inner and outer surfaces are T_2 and T_1. The thermal conductivity of the material of which the shell is made is k. Derive an equation for the total heat current through the shell. (b) Derive an equation for the temperature variation within the shell in part (a); that is, calculate T as a function of r, the distance from the center of the shell. (c) A hollow cylinder has length L, inner radius a, and outer radius b, and the temperatures at the inner and outer surfaces are T_2 and T_1. (The cylinder could represent an insulated hot-water pipe, for example.) The thermal conductivity of the material of which the cylinder is made is k. Derive an equation for the total heat current through the walls of the cylinder. (d) For the cylinder of part (c), derive an equation for the temperature variation inside the cylinder walls. (e) For the spherical shell of part (a) and the hollow cylinder of part (c), show that the equation for the total heat current in each case reduces to Eq. (17.21) for linear heat flow when the shell or cylinder is very thin.

17.125 ••• A steam pipe with a radius of 2.00 cm, carrying steam at 140°C, is surrounded by a cylindrical jacket with inner and outer radii 2.00 cm and 4.00 cm and made of a type of cork with thermal conductivity 4.00×10^{-2} W/m · K. This in turn is surrounded by a cylindrical jacket made of a brand of Styrofoam with thermal conductivity 1.00×10^{-2} W/m·K and having inner and outer radii 4.00 cm and 6.00 cm (Fig. P17.125). The outer surface of the Styrofoam is in contact with air at 15°C. Assume that this outer surface has a temperature of 15°C. (a) What is the temperature at a radius of 4.00 cm, where the two insulating layers meet? (b) What is the total rate of transfer of heat out of a 2.00-m length of pipe? (*Hint:* Use the expression derived in part (c) of Challenge Problem 17.124.)

Figure **P17.125**

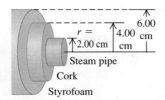

17.126 ••• **CP Temperature Change in a Clock.** A pendulum clock is designed to tick off one second on each side-to-side swing of the pendulum (two ticks per complete period). (a) Will a pendulum clock gain time in hot weather and lose it in cold, or the reverse? Explain your reasoning. (b) A particular pendulum clock keeps correct time at 20.0°C. The pendulum shaft is steel, and its mass can be ignored compared with that of the bob. What is the fractional change in the length of the shaft when it is cooled to 10.0°C? (c) How many seconds per day will the clock gain or lose at 10.0°C? (d) How closely must the temperature be controlled if the clock is not to gain or lose more than 1.00 s a day? Does the answer depend on the period of the pendulum?

17.127 ••• **BIO A Walk in the Sun.** Consider a poor lost soul walking at 5 km/h on a hot day in the desert, wearing only a bathing suit. This person's skin temperature tends to rise due to four mechanisms: (i) energy is generated by metabolic reactions in the body at a rate of 280 W, and almost all of this energy is converted to heat that flows to the skin; (ii) heat is delivered to the skin by convection from the outside air at a rate equal to $k'A_{\text{skin}}(T_{\text{air}} - T_{\text{skin}})$, where k' is 54 J/h·C°·m^2, the exposed skin area A_{skin} is 1.5 m^2, the air temperature T_{air} is 47°C, and the skin temperature T_{skin} is 36°C; (iii) the skin absorbs radiant energy from the sun at a rate of 1400 W/m^2; (iv) the skin absorbs radiant energy from the environment, which has temperature 47°C. (a) Calculate the net rate (in watts) at which the person's skin is heated by all four of these mechanisms. Assume that the emissivity of the skin is $e = 1$ and that the skin temperature is initially 36°C. Which mechanism is the most important? (b) At what rate (in L/h) must perspiration evaporate from this person's skin to maintain a constant skin temperature? (The heat of vaporization of water at 36°C is 2.42×10^6 J/kg.) (c) Suppose instead the person is protected by light-colored clothing ($e \approx 0$) so that the exposed skin area is only 0.45 m^2. What rate of perspiration is required now? Discuss the usefulness of the traditional clothing worn by desert peoples.

Answers

Chapter Opening Question

No. By "heat" we mean energy that is in transit from one body to another as a result of temperature difference between the bodies. Bodies do not *contain* heat.

Test Your Understanding Questions

17.1 Answer: (ii) A liquid-in-tube thermometer actually measures its own temperature. If the thermometer stays in the hot water long enough, it will come to thermal equilibrium with the water and its temperature will be the same as that of the water.

17.2 Answer: (iv) Both a bimetallic strip and a resistance thermometer measure their own temperature. For this to be equal to the temperature of the object being measured, the thermometer and object must be in contact and in thermal equilibrium. A temporal artery thermometer detects the infrared radiation from a person's skin, so there is no need for the detector and skin to be at the same temperature.

17.3 Answer: (i), (iii), (ii), (v), (iv) To compare these temperatures, convert them all to the Kelvin scale. For (i), the Kelvin temperature is $T_K = T_C + 273.15 = 0.00 + 273.15 = 273.15$ K; for (ii), $T_C = \frac{5}{9}(T_F - 32°) = \frac{5}{9}(0.00° - 32°) = -17.78°$C and $T_K = T_C + 273.15 = -17.78 + 273.15 = 255.37$ K; for (iii), $T_K = 260.00$ K; for (iv), $T_K = 77.00$ K; and for (v), $T_K = T_C + 273.15 = -180.00 + 273.15 = 93.15$ K.

17.4 Answer: (ii) and (iii) Metal 2 must expand more than metal 1 when heated and so must have a larger coefficient of linear expansion

α. From Table 17.1, brass and aluminum have larger values of α than copper, but steel does not.

17.5 Answer: (ii), (i), (iv), (iii) For (i) and (ii), the relevant quantity is the specific heat c of the substance, which is the amount of heat required to raise the temperature of 1 *kilogram* of that substance by 1 K (1 C°). From Table 17.3, these values are (i) 138 J for mercury and (ii) 2428 J for ethanol. For (iii) and (iv) we need the molar heat capacity C, which is the amount of heat required to raise the temperature of 1 *mole* of that substance by 1 C°. Again from Table 17.3, these values are (iii) 27.7 J for mercury and (iv) 111.9 J for ethanol. (The ratio of molar heat capacities is different from the ratio of the specific heats because a mole of mercury and a mole of ethanol have different masses.)

17.6 Answer: (iv) In time t the system goes from point b to point e in Fig. 17.21. According to this figure, at time $t/2$ (halfway along the horizontal axis from b to e), the system is at 100°C and is still boiling; that is, it is a mixture of liquid and gas. This says that most of the heat added goes into boiling the water.

17.7 Answer: (ii) When you touch one of the walls, heat flows from your hand to the lower-temperature wall. The more rapidly heat flows from your hand, the colder you will feel. Equation (17.21) shows that the rate of heat flow is proportional to the thermal conductivity k. From Table 17.5, copper has a much higher thermal conductivity (385.0 W/m·K) than steel (50.2 W/m·K) or concrete (0.8 W/m·K), and so the copper wall feels the coldest.

Bridging Problem

Answers: (a) 14.26 K **(b)** 0.427 kg/h

18 THERMAL PROPERTIES OF MATTER

LEARNING GOALS

By studying this chapter, you will learn:

- How to relate the pressure, volume, and temperature of a gas.

- How the interactions between the molecules of a substance determine the properties of the substance.

- How the pressure and temperature of a gas are related to the kinetic energy of its molecules.

- How the heat capacities of a gas reveal whether its molecules are rotating or vibrating.

- What determines whether a substance is a gas, a liquid, or a solid.

? The higher the temperature of a gas, the greater the average kinetic energy of its molecules. How much faster are molecules moving in the air above a frying pan (100°C) than in the surrounding kitchen air (25°C)?

The kitchen is a great place to learn about how the properties of matter depend on temperature. When you boil water in a tea kettle, the increase in temperature produces steam that whistles out of the spout at high pressure. If you forget to poke holes in a potato before baking it, the high-pressure steam produced inside the potato can cause it to explode messily. Water vapor in the air can condense into droplets of liquid on the sides of a glass of ice water; if the glass is just out of the freezer, frost will form on the sides as water vapor changes to a solid.

These examples show the relationships among the large-scale or *macroscopic* properties of a substance, such as pressure, volume, temperature, and mass. But we can also describe a substance using a *microscopic* perspective. This means investigating small-scale quantities such as the masses, speeds, kinetic energies, and momenta of the individual molecules that make up a substance.

The macroscopic and microscopic descriptions are intimately related. For example, the (microscopic) forces that occur when air molecules strike a solid surface (such as your skin) cause (macroscopic) atmospheric pressure. To produce standard atmospheric pressure of 1.01×10^5 Pa, 10^{32} molecules strike your skin every day with an average speed of over 1700 km/h (1000 mi/h)!

In this chapter we'll begin our study of the thermal properties of matter by looking at some macroscopic aspects of matter in general. We'll pay special attention to the *ideal gas*, one of the simplest types of matter to understand. Using our knowledge of momentum and kinetic energy, we'll relate the macroscopic properties of an ideal gas to the microscopic behavior of its individual molecules. We'll also use microscopic ideas to understand the heat capacities of both gases and solids. Finally, we'll take a look at the various phases of matter—gas, liquid, and solid—and the conditions under which each occurs.

18.1 Equations of State

The conditions in which a particular material exists are described by physical quantities such as pressure, volume, temperature, and amount of substance. For example, a tank of oxygen in a welding outfit has a pressure gauge and a label stating its volume. We could add a thermometer and place the tank on a scale to determine its mass. These variables describe the *state* of the material and are called **state variables.**

The volume V of a substance is usually determined by its pressure p, temperature T, and amount of substance, described by the mass m_{total} or number of moles n. (We are calling the total mass of a substance m_{total} because later in the chapter we will use m for the mass of one molecule.) Ordinarily, we can't change one of these variables without causing a change in another. When the tank of oxygen gets hotter, the pressure increases. If the tank gets too hot, it explodes.

In a few cases the relationship among p, V, T, and m (or n) is simple enough that we can express it as an equation called the **equation of state.** When it's too complicated for that, we can use graphs or numerical tables. Even then, the relationship among the variables still exists; we call it an equation of state even when we don't know the actual equation.

Here's a simple (though approximate) equation of state for a solid material. The temperature coefficient of volume expansion β (see Section 17.4) is the fractional volume change $\Delta V / V_0$ per unit temperature change, and the compressibility k (see Section 11.4) is the negative of the fractional volume change $\Delta V / V_0$ per unit pressure change. If a certain amount of material has volume V_0 when the pressure is p_0 and the temperature is T_0, the volume V at slightly differing pressure p and temperature T is approximately

$$V = V_0[1 + \beta(T - T_0) - k(p - p_0)] \tag{18.1}$$

(There is a negative sign in front of the term $k(p - p_0)$ because an *increase* in pressure causes a *decrease* in volume.)

The Ideal-Gas Equation

Another simple equation of state is the one for an *ideal gas.* Figure 18.1 shows an experimental setup to study the behavior of a gas. The cylinder has a movable piston to vary the volume, the temperature can be varied by heating, and we can pump any desired amount of any gas into the cylinder. We then measure the pressure, volume, temperature, and amount of gas. Note that *pressure* refers both to the force per unit area exerted by the cylinder on the gas and to the force per unit area exerted by the gas on the cylinder; by Newton's third law, these must be equal.

It is usually easiest to describe the amount of gas in terms of the number of moles n, rather than the mass. (We did this when we defined molar heat capacity in Section 17.5.) The **molar mass** M of a compound (sometimes called *molecular weight*) is the mass per mole, and the total mass m_{total} of a given quantity of that compound is the number of moles n times the mass per mole M:

$$m_{total} = nM \quad \text{(total mass, number of moles, and molar mass)} \tag{18.2}$$

Hence if we know the number of moles of gas in the cylinder, we can determine the mass of gas using Eq. (18.2).

Measurements of the behavior of various gases lead to three conclusions:

1. The volume V is proportional to the number of moles n. If we double the number of moles, keeping pressure and temperature constant, the volume doubles.

ActivPhysics 8.4: State Variables and Ideal Gas Law

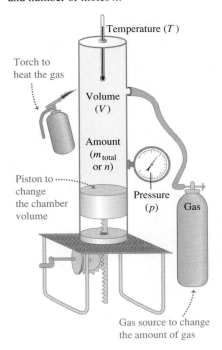

18.1 A hypothetical setup for studying the behavior of gases. By heating the gas, varying the volume with a movable piston, and adding more gas, we can control the gas pressure p, volume V, temperature T, and number of moles n.

Temperature (T)

Torch to heat the gas

Volume (V)

Amount (m_{total} or n)

Piston to change the chamber volume

Pressure (p)

Gas

Gas source to change the amount of gas

2. The volume varies *inversely* with the absolute pressure p. If we double the pressure while holding the temperature T and number of moles n constant, the gas compresses to one-half of its initial volume. In other words, pV = constant when n and T are constant.

3. The pressure is proportional to the *absolute* temperature. If we double the absolute temperature, keeping the volume and number of moles constant, the pressure doubles. In other words, p = (constant)T when n and V are constant.

18.2 The ideal-gas equation $pV = nRT$ gives a good description of the air inside an inflated vehicle tire, where the pressure is about 3 atmospheres and the temperature is much too high for nitrogen or oxygen to liquefy. As the tire warms (T increases), the volume V changes only slightly but the pressure p increases.

These three relationships can be combined neatly into a single equation, called the **ideal-gas equation:**

$$pV = nRT \qquad \text{(ideal-gas equation)} \qquad (18.3)$$

where R is a proportionality constant. An **ideal gas** is one for which Eq. (18.3) holds precisely for *all* pressures and temperatures. This is an idealized model; it works best at very low pressures and high temperatures, when the gas molecules are far apart and in rapid motion. It is reasonably good (within a few percent) at moderate pressures (such as a few atmospheres) and at temperatures well above those at which the gas liquefies (Fig. 18.2).

We might expect that the constant R in the ideal-gas equation would have different values for different gases, but it turns out to have the same value for *all* gases, at least at sufficiently high temperature and low pressure. It is called the **gas constant** (or *ideal-gas constant*). The numerical value of R depends on the units of p, V, and T. In SI units, in which the unit of p is Pa $(1 \text{ Pa} = 1 \text{ N/m}^2)$ and the unit of V is m^3, the current best numerical value of R is

$$R = 8.314472(15) \text{ J/mol} \cdot \text{K}$$

or $R = 8.314 \text{ J/mol} \cdot \text{K}$ to four significant figures. Note that the units of pressure times volume are the same as the units of work or energy (for example, N/m^2 times m^3); that's why R has units of energy per mole per unit of absolute temperature. In chemical calculations, volumes are often expressed in liters (L) and pressures in atmospheres (atm). In this system, to four significant figures,

$$R = 0.08206 \frac{\text{L} \cdot \text{atm}}{\text{mol} \cdot \text{K}}$$

Application **Respiration and the Ideal-Gas Equation**
To breathe, you rely on the ideal-gas equation $pV = nRT$. Contraction of the dome-shaped diaphragm muscle increases the volume V of the thoracic cavity (which encloses the lungs), decreasing its pressure p. The lowered pressure causes the lungs to expand and fill with air. (The temperature T is kept constant.) When you exhale, the diaphragm relaxes, allowing the lungs to contract and expel the air.

We can express the ideal-gas equation, Eq. (18.3), in terms of the mass m_{total} of gas, using $m_{\text{total}} = nM$ from Eq. (18.2):

$$pV = \frac{m_{\text{total}}}{M} RT \qquad (18.4)$$

From this we can get an expression for the density $\rho = m_{\text{total}}/V$ of the gas:

$$\rho = \frac{pM}{RT} \qquad (18.5)$$

CAUTION **Density vs. pressure** When using Eq. (18.5), be certain that you distinguish between the Greek letter ρ (rho) for density and the letter p for pressure. ▮

For a *constant mass* (or constant number of moles) of an ideal gas the product nR is constant, so the quantity pV/T is also constant. If the subscripts 1 and 2 refer to any two states of the same mass of a gas, then

$$\frac{p_1 V_1}{T_1} = \frac{p_2 V_2}{T_2} = \text{constant} \qquad \text{(ideal gas, constant mass)} \qquad (18.6)$$

Notice that you don't need the value of R to use this equation.

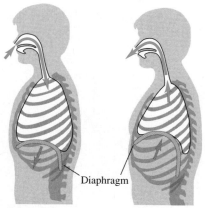

Inhalation:
Diaphragm contracts; lungs expand

Exhalation:
Diaphragm relaxes; lungs contract

Diaphragm

We used the proportionality of pressure to absolute temperature in Chapter 17 to define a temperature scale in terms of pressure in a constant-volume gas thermometer. That may make it seem that the pressure–temperature relationship in the ideal-gas equation, Eq. (18.3), is just a result of the way we define temperature. But the equation also tells us what happens when we change the volume or the amount of substance. Also, we'll see in Chapter 20 that the gas-thermometer scale corresponds closely to a temperature scale that does *not* depend on the properties of any particular material. For now, consider Eq. (18.6) as being based on this genuinely material-independent temperature scale.

Problem-Solving Strategy 18.1   Ideal Gases

IDENTIFY *the relevant concepts:* Unless the problem states otherwise, you can use the ideal-gas equation to find quantities related to the state of a gas, such as pressure p, volume V, temperature T, and/or number of moles n.

SET UP *the problem* using the following steps:
1. List the known and unknown quantities. Identify the target variables.
2. If the problem concerns only one state of the system, use Eq. (18.3), $pV = nRT$.
3. Use Eq. (18.5), $\rho = pM/RT$, as an alternative to Eq. (18.3) if the problem involves the density ρ rather than n and V.
4. In problems that concern two states (call them 1 and 2) of the same amount of gas, if all but one of the six quantities p_1, p_2, V_1, V_2, T_1, and T_2 are known, use Eq. (18.6), $p_1V_1/T_1 = p_2V_2/T_2 = $ constant. Otherwise, use Eq. (18.3) or Eq. (18.5) as appropriate.

EXECUTE *the solution* as follows:
1. Use consistent units. (SI units are entirely consistent.) The problem statement may make one system of units more con-

venient than others. Make appropriate unit conversions, such as from atmospheres to pascals or from liters to cubic meters.
2. You may have to convert between mass m_{total} and number of moles n, using $m_{total} = Mn$, where M is the molar mass. If you use Eq. (18.4), you *must* use the same mass units for m_{total} and M. So if M is in grams per mole (the usual units for molar mass), then m_{total} must also be in grams. To use m_{total} in kilograms, you must convert M to kg/mol. For example, the molar mass of oxygen is 32 g/mol or 32×10^{-3} kg/mol.
3. Remember that in the ideal-gas equations, T is always an *absolute* (Kelvin) temperature and p is always an absolute (not gauge) pressure.
4. Solve for the target variables.

EVALUATE *your answer:* Do your results make physical sense? Use benchmarks, such as the result of Example 18.1 below that a mole of an ideal gas at 1 atmosphere pressure occupies a volume of 22.4 liters.

Example 18.1   Volume of an ideal gas at STP

What is the volume of a container that holds exactly 1 mole of an ideal gas at *standard temperature and pressure* (STP), defined as $T = 0°C = 273.15$ K and $p = 1$ atm $= 1.013 \times 10^5$ Pa?

SOLUTION

IDENTIFY and SET UP: This problem involves the properties of a single state of an ideal gas, so we use Eq. (18.3). We are given the pressure p, temperature T, and number of moles n; our target variable is the corresponding volume V.

EXECUTE: From Eq. (18.3), using R in $\text{J/mol} \cdot \text{K}$,

$$V = \frac{nRT}{p} = \frac{(1 \text{ mol})(8.314 \text{ J/mol} \cdot \text{K})(273.15 \text{ K})}{1.013 \times 10^5 \text{ Pa}}$$

$$= 0.0224 \text{ m}^3 = 22.4 \text{ L}$$

EVALUATE: At STP, 1 mole of an ideal gas occupies 22.4 L. This is the volume of a cube 0.282 m (11.1 in.) on a side, or of a sphere 0.350 m (13.8 in.) in diameter.

Example 18.2   Compressing gas in an automobile engine

In an automobile engine, a mixture of air and vaporized gasoline is compressed in the cylinders before being ignited. A typical engine has a compression ratio of 9.00 to 1; that is, the gas in the cylinders is compressed to $\frac{1}{9.00}$ of its original volume (Fig. 18.3). The intake

and exhaust valves are closed during the compression, so the quantity of gas is constant. What is the final temperature of the compressed gas if its initial temperature is 27°C and the initial and final pressures are 1.00 atm and 21.7 atm, respectively?

Continued

SOLUTION

IDENTIFY and SET UP: We must compare two states of the same quantity of ideal gas, so we use Eq. (18.6). In the uncompressed state, $p_1 = 1.00$ atm and $T_1 = 27°C = 300$ K. In the compressed state 2, $p_2 = 21.7$ atm. The cylinder volumes are not given, but we have $V_1 = 9.00V_2$. The temperature T_2 of the compressed gas is the target variable.

EXECUTE: We solve Eq. (18.6) for T_2:

$$T_2 = T_1 \frac{p_2 V_2}{p_1 V_1} = (300 \text{ K}) \frac{(21.7 \text{ atm})V_2}{(1.00 \text{ atm})(9.00V_2)} = 723 \text{ K} = 450°C$$

EVALUATE: This is the temperature of the air–gasoline mixture *before* the mixture is ignited; when burning starts, the temperature becomes higher still.

18.3 Cutaway of an automobile engine. While the air–gasoline mixture is being compressed prior to ignition, the intake and exhaust valves are both in the closed (up) position.

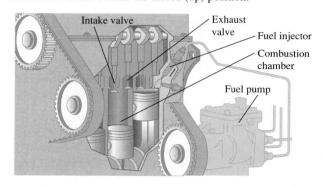

Example 18.3   **Mass of air in a scuba tank**

An "empty" aluminum scuba tank contains 11.0 L of air at 21°C and 1 atm. When the tank is filled rapidly from a compressor, the air temperature is 42°C and the gauge pressure is 2.10×10^7 Pa. What mass of air was added? (Air is about 78% nitrogen, 21% oxygen, and 1% miscellaneous; its average molar mass is 28.8 g/mol = 28.8×10^{-3} kg/mol.)

SOLUTION

IDENTIFY and SET UP: Our target variable is the difference $m_2 - m_1$ between the masses present at the end (state 2) and at the beginning (state 1). We are given the molar mass M of air, so we can use Eq. (18.2) to find the target variable if we know the number of moles present in states 1 and 2. We determine n_1 and n_2 by applying Eq. (18.3) to each state individually.

EXECUTE: We convert temperatures to the Kelvin scale by adding 273 and convert the pressure to absolute by adding 1.013×10^5 Pa.

The tank's volume is hardly affected by the increased temperature and pressure, so $V_2 = V_1$. From Eq. (18.3), the numbers of moles in the empty tank (n_1) and the full tank (n_2) are

$$n_1 = \frac{p_1 V_1}{RT_1} = \frac{(1.013 \times 10^5 \text{ Pa})(11.0 \times 10^{-3} \text{ m}^3)}{(8.314 \text{ J/mol} \cdot \text{K})(294 \text{ K})} = 0.46 \text{ mol}$$

$$n_2 = \frac{p_2 V_2}{RT_2} = \frac{(2.11 \times 10^7 \text{ Pa})(11.0 \times 10^{-3} \text{ m}^3)}{(8.314 \text{ J/mol} \cdot \text{K})(315 \text{ K})} = 88.6 \text{ mol}$$

We added $n_2 - n_1 = 88.6$ mol − 0.46 mol = 88.1 mol to the tank. From Eq. (18.2), the added mass is $M(n_2 - n_1) = (28.8 \times 10^{-3} \text{ kg/mol})(88.1 \text{ mol}) = 2.54$ kg.

EVALUATE: The added mass is not insubstantial: You could certainly use a scale to determine whether the tank was empty or full.

Example 18.4   **Variation of atmospheric pressure with elevation**

Find the variation of atmospheric pressure with elevation in the earth's atmosphere. Assume that at all elevations, $T = 0°C$ and $g = 9.80$ m/s^2.

SOLUTION

IDENTIFY and SET UP: As the elevation y increases, both the atmospheric pressure p and the density ρ decrease. Hence we have *two* unknown functions of y; to solve for them, we need two independent equations. One is the ideal-gas equation, Eq. (18.5), which is expressed in terms of p and ρ. The other is Eq. (12.4), the relationship that we found in Section 12.2 among p, ρ, and y in a fluid in equilibrium: $dp/dy = -\rho g$. We are told to assume that g and T are the same at all elevations; we also assume that the atmosphere has the same chemical composition, and hence the same molar mass M, at all heights. We combine the two equations and solve for $p(y)$.

EXECUTE: We substitute $\rho = pM/RT$ into $dp/dy = -\rho g$, separate variables, and integrate, letting p_1 be the pressure at elevation y_1 and p_2 be the pressure at y_2:

$$\frac{dp}{dy} = -\frac{pM}{RT}g$$

$$\int_{p_1}^{p_2} \frac{dp}{p} = -\frac{Mg}{RT} \int_{y_1}^{y_2} dy$$

$$\ln\frac{p_2}{p_1} = -\frac{Mg}{RT}(y_2 - y_1)$$

$$\frac{p_2}{p_1} = e^{-Mg(y_2 - y_1)/RT}$$

Now let $y_1 = 0$ be at sea level and let the pressure at that point be $p_0 = 1.013 \times 10^5$ Pa. Then the pressure p at any height y is

$$p = p_0 e^{-Mgy/RT}$$

EVALUATE: According to our calculation, the pressure decreases exponentially with elevation. The graph in Fig. 18.4 shows that the slope dp/dy becomes less negative with greater elevation. That result makes sense, since $dp/dy = -\rho g$ and the density also

18.4 The variation of atmospheric pressure p with elevation y, assuming a constant temperature T.

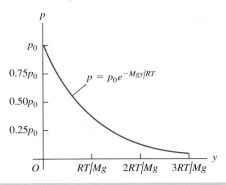

decreases with elevation. At the summit of Mount Everest, where $y = 8863$ m,

$$\frac{Mgy}{RT} = \frac{(28.8 \times 10^{-3}\text{ kg/mol})(9.80\text{ m/s}^2)(8863\text{ m})}{(8.314\text{ J/mol} \cdot \text{K})(273\text{ K})} = 1.10$$

$$p = (1.013 \times 10^5\text{ Pa})e^{-1.10} = 0.337 \times 10^5\text{ Pa} = 0.33\text{ atm}$$

The assumption of constant temperature isn't realistic, and g decreases a little with increasing elevation (see Challenge Problem 18.92). Even so, this example shows why mountaineers need to carry oxygen on Mount Everest. It also shows why jet airliners, which typically fly at altitudes of 8000 to 12,000 m, *must* have pressurized cabins for passenger comfort and health.

The van der Waals Equation

The ideal-gas equation, Eq. (18.3), can be obtained from a simple molecular model that ignores the volumes of the molecules themselves and the attractive forces between them (Fig. 18.5a). We'll examine that model in Section 18.3. Meanwhile, we mention another equation of state, the **van der Waals equation,** that makes approximate corrections for these two omissions (Fig. 18.5b). This equation was developed by the 19th-century Dutch physicist J. D. van der Waals; the interaction between atoms that we discussed in Section 14.4 was named the *van der Waals interaction* after him. The van der Waals equation is

$$\left(p + \frac{an^2}{V^2}\right)(V - nb) = nRT \tag{18.7}$$

The constants a and b are empirical constants, different for different gases. Roughly speaking, b represents the volume of a mole of molecules; the total volume of the molecules is then nb, and the volume remaining in which the molecules can move is $V - nb$. The constant a depends on the attractive intermolecular forces, which reduce the pressure of the gas for given values of n, V, and T by *pulling* the molecules together as they *push* on the walls of the container. The decrease in pressure is proportional to the number of molecules per unit volume in a layer near the wall (which are exerting the pressure on the wall) and is also proportional to the number per unit volume in the next layer beyond the wall (which are doing the attracting). Hence the decrease in pressure due to intermolecular forces is proportional to n^2/V^2.

When n/V is small (that is, when the gas is *dilute*), the average distance between molecules is large, the corrections in the van der Waals equation become insignificant, and Eq. (18.7) reduces to the ideal-gas equation. As an example, for carbon dioxide gas (CO_2) the constants in the van der Waals equation are $a = 0.364\text{ J} \cdot \text{m}^3/\text{mol}^2$ and $b = 4.27 \times 10^{-5}\text{ m}^3/\text{mol}$. We found in Example 18.1 that 1 mole of an ideal gas at $T = 0°\text{C} = 273.15\text{ K}$ and $p = 1\text{ atm} = 1.013 \times 10^5\text{ Pa}$ occupies a volume $V = 0.0224\text{ m}^3$; according to Eq. (18.7),

(a) An idealized model of a gas

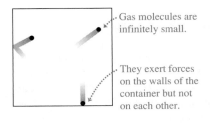

· Gas molecules are infinitely small.

· They exert forces on the walls of the container but not on each other.

(b) A more realistic model of a gas

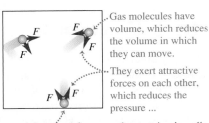

· Gas molecules have volume, which reduces the volume in which they can move.

· They exert attractive forces on each other, which reduces the pressure ...

... and they exert forces on the container's walls.

18.5 A gas as modeled by **(a)** the ideal-gas equation and **(b)** the van der Waals equation.

1 mole of CO_2 occupying this volume at this temperature would be at a pressure 532 Pa less than 1 atm, a difference of only 0.5% from the ideal-gas value.

pV-Diagrams

We could in principle represent the p-V-T relationship graphically as a *surface* in a three-dimensional space with coordinates p, V, and T. This representation sometimes helps us grasp the overall behavior of the substance, but ordinary two-dimensional graphs are usually more convenient. One of the most useful of these is a set of graphs of pressure as a function of volume, each for a particular constant temperature. Such a diagram is called a **pV-diagram.** Each curve, representing behavior at a specific temperature, is called an **isotherm,** or a *pV-isotherm.*

Figure 18.6 shows pV-isotherms for a constant amount of an ideal gas. Since $p = nRT/V$ from Eq. (18.3), along an isotherm (constant T) the pressure p is inversely proportional to the volume V and the isotherms are hyperbolic curves.

Figure 18.7 shows a pV-diagram for a material that *does not* obey the ideal-gas equation. At temperatures below T_c the isotherms develop flat regions in which we can compress the material (that is, reduce the volume V) without increasing the pressure p. Observation shows that the gas is *condensing* from the vapor (gas) to the liquid phase. The flat parts of the isotherms in the shaded area of Fig. 18.7 represent conditions of liquid-vapor *phase equilibrium.* As the volume decreases, more and more material goes from vapor to liquid, but the pressure does not change. (To keep the temperature constant during condensation, we have to remove the heat of vaporization, discussed in Section 17.6.)

When we compress such a gas at a constant temperature T_2 in Fig. 18.7, it is vapor until point a is reached. Then it begins to liquefy; as the volume decreases further, more material liquefies, and *both* the pressure and the temperature remain constant. At point b, all the material is in the liquid state. After this, any further compression requires a very rapid rise of pressure, because liquids are in general much less compressible than gases. At a lower constant temperature T_1, similar behavior occurs, but the condensation begins at lower pressure and greater volume than at the constant temperature T_2. At temperatures greater than T_c, *no* phase transition occurs as the material is compressed; at the highest temperatures, such as T_4, the curves resemble the ideal-gas curves of Fig. 18.6. We call T_c the *critical temperature* for this material. In Section 18.6 we'll discuss what happens to the phase of the gas above the critical temperature.

We will use pV-diagrams often in the next two chapters. We will show that the *area* under a pV-curve (whether or not it is an isotherm) represents the *work* done by the system during a volume change. This work, in turn, is directly related to heat transfer and changes in the *internal energy* of the system.

18.6 Isotherms, or constant-temperature curves, for a constant amount of an ideal gas. The highest temperature is T_4; the lowest is T_1. This is a graphical representation of the ideal-gas equation of state.

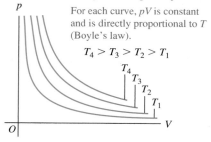

Each curve represents pressure as a function of volume for an ideal gas at a single temperature.

For each curve, pV is constant and is directly proportional to T (Boyle's law).

$T_4 > T_3 > T_2 > T_1$

18.7 A pV-diagram for a nonideal gas, showing isotherms for temperatures above and below the critical temperature T_c. The liquid–vapor equilibrium region is shown as a green shaded area. At still lower temperatures the material might undergo phase transitions from liquid to solid or from gas to solid; these are not shown in this diagram.

$T_4 > T_3 > T_c > T_2 > T_1$

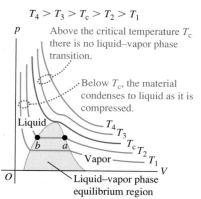

Above the critical temperature T_c there is no liquid–vapor phase transition.

Below T_c, the material condenses to liquid as it is compressed.

Liquid–vapor phase equilibrium region

Test Your Understanding of Section 18.1 Rank the following ideal gases in order from highest to lowest number of moles: (i) pressure 1 atm, volume 1 L, and temperature 300 K; (ii) pressure 2 atm, volume 1 L, and temperature 300 K; (iii) pressure 1 atm, volume 2 L, and temperature 300 K; (iv) pressure 1 atm, volume 1 L, and temperature 600 K; (v) pressure 2 atm, volume 1 L, and temperature 600 K.

18.2 Molecular Properties of Matter

We have studied several properties of matter in bulk, including elasticity, density, surface tension, heat capacities, and equations of state. Now we want to look in more detail at the relationship of bulk behavior to *molecular* structure. We begin with a general discussion of the molecular structure of matter. Then in the next two sections we develop the kinetic-molecular model of an ideal gas, obtaining from this molecular model the equation of state and an expression for heat capacity.

Molecules and Intermolecular Forces

Any specific chemical compound is made up of identical **molecules.** The smallest molecules contain one atom each and are of the order of 10^{-10} m in size; the largest contain many atoms and are at least 10,000 times larger. In gases the molecules move nearly independently; in liquids and solids they are held together by intermolecular forces. These forces arise from interactions among the electrically charged particles that make up the molecules. Gravitational forces between molecules are negligible in comparison with electrical forces.

The interaction of two *point* electric charges is described by a force (repulsive for like charges, attractive for unlike charges) with a magnitude proportional to $1/r^2$, where r is the distance between the points. We will study this relationship, called *Coulomb's law,* in Chapter 21. Molecules are *not* point charges but complex structures containing both positive and negative charge, and their interactions are more complex. The force between molecules in a gas varies with the distance r between molecules somewhat as shown in Fig. 18.8, where a positive F_r corresponds to a repulsive force and a negative F_r to an attractive force. When molecules are far apart, the intermolecular forces are very small and usually attractive. As a gas is compressed and its molecules are brought closer together, the attractive forces increase. The intermolecular force becomes zero at an equilibrium spacing r_0, corresponding roughly to the spacing between molecules in the liquid and solid states. In liquids and solids, relatively large pressures are needed to compress the substance appreciably. This shows that at molecular distances slightly *less* than the equilibrium spacing, the forces become *repulsive* and relatively large.

Figure 18.8 also shows the potential energy as a function of r. This function has a *minimum* at r_0, where the force is zero. The two curves are related by $F_r(r) = -dU/dr$, as we showed in Section 7.4. Such a potential-energy function is often called a **potential well.** A molecule at rest at a distance r_0 from a second molecule would need an additional energy $|U_0|$, the "depth" of the potential well, to "escape" to an indefinitely large value of r.

Molecules are always in motion; their kinetic energies usually increase with temperature. At very low temperatures the average kinetic energy of a molecule may be much *less* than the depth of the potential well. The molecules then condense into the liquid or solid phase with average intermolecular spacings of about r_0. But at higher temperatures the average kinetic energy becomes larger than the depth $|U_0|$ of the potential well. Molecules can then escape the intermolecular force and become free to move independently, as in the gaseous phase of matter.

In *solids,* molecules vibrate about more or less fixed points. In a crystalline solid these points are arranged in a *crystal lattice.* Figure 18.9 shows the cubic crystal structure of sodium chloride, and Fig. 18.10 shows a scanning tunneling microscope image of individual silicon atoms on the surface of a crystal.

The vibration of molecules in a solid about their equilibrium positions may be nearly simple harmonic if the potential well is approximately parabolic in shape at distances close to r_0. (We discussed this kind of simple harmonic motion in Section 14.4.) But if the potential-energy curve rises more gradually for $r > r_0$ than for $r < r_0$, as in Fig. 18.8, the average position shifts to larger r with increasing amplitude. As we pointed out in Section 17.4, this is the basis of thermal expansion.

In a *liquid,* the intermolecular distances are usually only slightly greater than in the solid phase of the same substance, but the molecules have much greater freedom of movement. Liquids show regularity of structure only in the immediate neighborhood of a few molecules.

The molecules of a *gas* are usually widely separated and so have only very small attractive forces. A gas molecule moves in a straight line until it collides with another molecule or with a wall of the container. In molecular terms, an *ideal gas* is a gas whose molecules exert *no* attractive forces on each other (see Fig. 18.5a) and therefore have no *potential* energy.

18.8 How the force between molecules and their potential energy of interaction depend on their separation r.

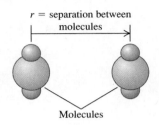

r = separation between molecules

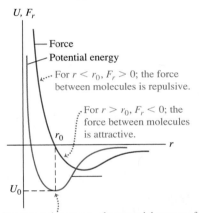

U, F_r

— Force
— Potential energy

··· For $r < r_0$, $F_r > 0$; the force between molecules is repulsive.

For $r > r_0$, $F_r < 0$; the force between molecules is attractive.

r_0

r

U_0

At a separation $r = r_0$, the potential energy of the two molecules is minimum and the force between the molecules is zero.

18.9 Schematic representation of the cubic crystal structure of sodium chloride (ordinary salt).

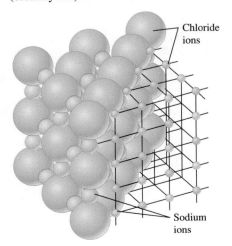

Chloride ions

Sodium ions

18.10 A scanning tunneling microscope image of the surface of a silicon crystal. The area shown is only 9.0 nm (9.0×10^{-9} m) across. Each blue "bead" is an individual silicon atom; you can clearly see how these atoms are arranged in a (nearly) perfect array of hexagons.

At low temperatures, most common substances are in the solid phase. As the temperature rises, a substance melts and then vaporizes. From a molecular point of view, these transitions are in the direction of increasing molecular kinetic energy. Thus temperature and molecular kinetic energy are closely related.

Moles and Avogadro's Number

We have used the mole as a measure of quantity of substance. One **mole** of any pure chemical element or compound contains a definite number of molecules, the same number for all elements and compounds. The official SI definition is:

> **One mole is the amount of substance that contains as many elementary entities as there are atoms in 0.012 kilogram of carbon-12.**

In our discussion, the "elementary entities" are molecules. (In a monatomic substance such as carbon or helium, each molecule is a single atom.) Atoms of a given element may occur in any of several isotopes, which are chemically identical but have different atomic masses; "carbon-12" is a specific isotope of carbon.

The number of molecules in a mole is called **Avogadro's number,** denoted by N_A. The current best numerical value of N_A is

$$N_A = 6.02214179(30) \times 10^{23} \text{ molecules/mol} \quad \text{(Avogadro's number)}$$

The *molar mass M* of a compound is the mass of 1 mole. It is equal to the mass m of a single molecule multiplied by Avogadro's number:

$$M = N_A m \qquad \begin{array}{l}\text{(molar mass, Avogadro's number,} \\ \text{and mass of a molecule)}\end{array} \qquad (18.8)$$

When the molecule consists of a single atom, the term *atomic mass* is often used instead of molar mass or molecular weight.

Example 18.5 | **Atomic and molecular mass**

Find the mass of a single hydrogen atom and of a single oxygen molecule.

SOLUTION

IDENTIFY and SET UP: This problem involves the relationship between the mass of a molecule or atom (our target variable) and the corresponding molar mass M. We use Eq. (18.8) in the form $m = M/N_A$ and the values of the atomic masses from the periodic table of the elements (see Appendix D).

EXECUTE: For atomic hydrogen the atomic mass (molar mass) is $M_H = 1.008$ g/mol, so the mass m_H of a single hydrogen atom is

$$m_H = \frac{1.008 \text{ g/mol}}{6.022 \times 10^{23} \text{ atoms/mol}} = 1.674 \times 10^{-24} \text{ g/atom}$$

For oxygen the atomic mass is 16.0 g/mol, so for the diatomic (two-atom) oxygen molecule the molar mass is 32.0 g/mol. Then the mass of a single oxygen molecule is

$$m_{O_2} = \frac{32.0 \text{ g/mol}}{6.022 \times 10^{23} \text{ molecules/mol}} = 53.1 \times 10^{-24} \text{ g/molecule}$$

EVALUATE: We note that the values in Appendix D are for the *average* atomic masses of a natural sample of each element. Such a sample may contain several *isotopes* of the element, each with a different atomic mass. Natural samples of hydrogen and oxygen are almost entirely made up of just one isotope.

Test Your Understanding of Section 18.2 Suppose you could adjust the value of r_0 for the molecules of a certain chemical compound (Fig. 18.8) by turning a dial. If you doubled the value of r_0, the density of the solid form of this compound would become (i) twice as great; (ii) four times as great; (iii) eight times as great; (iv) $\frac{1}{2}$ as great; (v) $\frac{1}{4}$ as great; (vi) $\frac{1}{8}$ as great.

18.3 Kinetic-Molecular Model of an Ideal Gas

Mastering**PHYSICS**

PhET: Balloons & Buoyancy
PhET: Friction
PhET: Gas Properties
ActivPhysics 8.1: Characteristics of a Gas

The goal of any molecular theory of matter is to understand the *macroscopic* properties of matter in terms of its atomic or molecular structure and behavior. Once we have this understanding, we can design materials to have specific desired properties. Theories have led to the development of high-strength steels, semiconductor materials for electronic devices, and countless other materials essential to contemporary technology.

In this and the following sections we will consider a simple molecular model of an ideal gas. This *kinetic-molecular model* represents the gas as a large number of particles bouncing around in a closed container. In this section we use the kinetic-molecular model to understand how the ideal-gas equation of state, Eq. (18.3), is related to Newton's laws. In the following section we use the kinetic-molecular model to predict the molar heat capacity of an ideal gas. We'll go on to elaborate the model to include "particles" that are not points but have a finite size.

Our discussion of the kinetic-molecular model has several steps, and you may need to go over them several times. Don't get discouraged!

Here are the assumptions of our model:

1. A container with volume V contains a very large number N of identical molecules, each with mass m.
2. The molecules behave as point particles that are small compared to the size of the container and to the average distance between molecules.
3. The molecules are in constant motion. Each molecule collides occasionally with a wall of the container. These collisions are perfectly elastic.
4. The container walls are rigid and infinitely massive and do not move.

CAUTION **Molecules vs. moles** Make sure you don't confuse N, the number of *molecules* in the gas, with n, the number of *moles*. The number of molecules is equal to the number of moles multiplied by Avogadro's number: $N = nN_A$. ▮

Collisions and Gas Pressure

During collisions the molecules exert *forces* on the walls of the container; this is the origin of the *pressure* that the gas exerts. In a typical collision (Fig. 18.11) the velocity component parallel to the wall is unchanged, and the component perpendicular to the wall reverses direction but does not change in magnitude.

Our program is first to determine the *number* of collisions that occur per unit time for a certain area A of wall. Then we find the total momentum change associated with these collisions and the force needed to cause this momentum change. From this we can determine the pressure, which is force per unit area, and compare the result to the ideal-gas equation. We'll find a direct connection between the temperature of the gas and the kinetic energy of the gas molecules.

To begin, we will assume that all molecules in the gas have the same *magnitude* of x-velocity, $|v_x|$. This isn't right, but making this temporary assumption helps to clarify the basic ideas. We will show later that this assumption isn't really necessary.

As shown in Fig. 18.11, for each collision the x-component of velocity changes from $-|v_x|$ to $+|v_x|$. So the x-component of momentum changes from $-m|v_x|$ to $+m|v_x|$, and the *change* in the x-component of momentum is $m|v_x| - (-m|v_x|) = 2m|v_x|$.

If a molecule is going to collide with a given wall area A during a small time interval dt, then at the beginning of dt it must be within a distance $|v_x| \, dt$ from the wall (Fig. 18.12) and it must be headed toward the wall. So the number of molecules that collide with A during dt is equal to the number of molecules within a cylinder with base area A and length $|v_x| \, dt$ that have their x-velocity aimed

18.11 Elastic collision of a molecule with an idealized container wall.

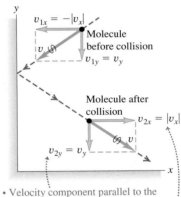

- Velocity component parallel to the wall (y-component) does not change.
- Velocity component perpendicular to the wall (x-component) reverses direction.
- Speed v does not change.

18.12 For a molecule to strike the wall in area A during a time interval dt, the molecule must be headed for the wall and be within the shaded cylinder of length $|v_x| \, dt$ at the beginning of the interval.

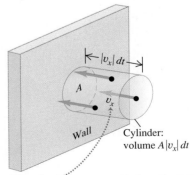

All molecules are assumed to have the same magnitude $|v_x|$ of x-velocity.

toward the wall. The volume of such a cylinder is $A|v_x|\,dt$. Assuming that the number of molecules per unit volume (N/V) is uniform, the *number* of molecules in this cylinder is $(N/V)(A|v_x|\,dt)$. On the average, half of these molecules are moving toward the wall and half are moving away from it. So the number of collisions with A during dt is

$$\frac{1}{2}\left(\frac{N}{V}\right)(A|v_x|\,dt)$$

For the system of all molecules in the gas, the total momentum change dP_x during dt is the *number* of collisions multiplied by $2m|v_x|$:

$$dP_x = \frac{1}{2}\left(\frac{N}{V}\right)(A|v_x|\,dt)(2m|v_x|) = \frac{NAmv_x^2\,dt}{V} \tag{18.9}$$

(We are using capital P for total momentum and small p for pressure. Be careful!) We wrote v_x^2 rather than $|v_x|^2$ in the final expression because the square of the absolute value of a number is equal to the square of that number. The *rate* of change of momentum component P_x is

$$\frac{dP_x}{dt} = \frac{NAmv_x^2}{V} \tag{18.10}$$

According to Newton's second law, this rate of change of momentum equals the force exerted by the wall area A on the gas molecules. From Newton's *third* law this is equal and opposite to the force exerted *on* the wall *by* the molecules. Pressure p is the magnitude of the force exerted on the wall per unit area, and we obtain

$$p = \frac{F}{A} = \frac{Nmv_x^2}{V} \tag{18.11}$$

The pressure exerted by the gas depends on the number of molecules per volume (N/V), the mass m per molecule, and the speed of the molecules.

Pressure and Molecular Kinetic Energies

We mentioned that $|v_x|$ is really *not* the same for all the molecules. But we could have sorted the molecules into groups having the same $|v_x|$ within each group, then added up the resulting contributions to the pressure. The net effect of all this is just to replace v_x^2 in Eq. (18.11) by the *average* value of v_x^2 which we denote by $(v_x^2)_{av}$. We can relate $(v_x^2)_{av}$ to the *speeds* of the molecules. The speed v of a molecule is related to the velocity components v_x, v_y, and v_z by

$$v^2 = v_x^2 + v_y^2 + v_z^2$$

We can average this relation over all molecules:

$$(v^2)_{av} = (v_x^2)_{av} + (v_y^2)_{av} + (v_z^2)_{av}$$

But there is no real difference in our model between the x-, y-, and z-directions. (Molecular speeds are very fast in a typical gas, so the effects of gravity are negligibly small.) It follows that $(v_x^2)_{av}$, $(v_y^2)_{av}$, and $(v_z^2)_{av}$ must all be *equal*. Hence $(v^2)_{av}$ is equal to $3(v_x^2)_{av}$ and

$$(v_x^2)_{av} = \frac{1}{3}(v^2)_{av}$$

so Eq. (18.11) becomes

$$pV = \frac{1}{3}Nm(v^2)_{av} = \frac{2}{3}N\left[\frac{1}{2}m(v^2)_{av}\right] \tag{18.12}$$

We notice that $\frac{1}{2}m(v^2)_{av}$ is the average translational kinetic energy of a single molecule. The product of this and the total number of molecules N equals the total random kinetic energy K_{tr} of translational motion of all the molecules. (The notation K_{tr} reminds us that this is the energy of *translational* motion. There may also be energies associated with molecular rotation and vibration.) The product pV equals two-thirds of the total translational kinetic energy:

$$pV = \tfrac{2}{3}K_{tr} \qquad (18.13)$$

Now we compare this with the ideal-gas equation,

$$pV = nRT$$

which is based on experimental studies of gas behavior. For the two equations to agree, we must have

$$K_{tr} = \tfrac{3}{2}nRT \qquad \begin{array}{l}\text{(average translational kinetic} \\ \text{energy of } n \text{ moles of ideal gas)}\end{array} \qquad (18.14)$$

This remarkably simple result shows that K_{tr} is *directly proportional* to the absolute temperature T (Fig. 18.13).

The average translational kinetic energy of a single molecule is the total translational kinetic energy K_{tr} of all molecules divided by the number of molecules, N:

$$\frac{K_{tr}}{N} = \tfrac{1}{2}m(v^2)_{av} = \frac{3nRT}{2N}$$

Also, the total number of molecules N is the number of moles n multiplied by Avogadro's number N_A, so

$$N = nN_A \qquad \frac{n}{N} = \frac{1}{N_A}$$

and

$$\frac{K_{tr}}{N} = \tfrac{1}{2}m(v^2)_{av} = \tfrac{3}{2}\left(\frac{R}{N_A}\right)T \qquad (18.15)$$

The ratio R/N_A occurs frequently in molecular theory. It is called the **Boltzmann constant,** k:

$$k = \frac{R}{N_A} = \frac{8.314 \text{ J/mol} \cdot \text{K}}{6.022 \times 10^{23} \text{ molecules/mol}}$$

$$= 1.381 \times 10^{-23} \text{ J/molecule} \cdot \text{K}$$

(The current best numerical value of k is $1.3806504(24) \times 10^{-23}$ J/molecule $\cdot$ K). In terms of k we can rewrite Eq. (18.15) as

$$\tfrac{1}{2}m(v^2)_{av} = \tfrac{3}{2}kT \qquad \begin{array}{l}\text{(average translational kinetic} \\ \text{energy of a gas molecule)}\end{array} \qquad (18.16)$$

This shows that the average translational kinetic energy *per molecule* depends only on the temperature, not on the pressure, volume, or kind of molecule. We can obtain the average translational kinetic energy *per mole* by multiplying Eq. (18.16) by Avogadro's number and using the relation $M = N_A m$:

$$N_A\tfrac{1}{2}m(v^2)_{av} = \tfrac{1}{2}M(v^2)_{av} = \tfrac{3}{2}RT \qquad \begin{array}{l}\text{(average translational kinetic} \\ \text{energy per mole of gas)}\end{array} \qquad (18.17)$$

The translational kinetic energy of a mole of an ideal gas depends only on T.

18.13 Summer air (top) is warmer than winter air (bottom); that is, the average translational kinetic energy of air molecules is greater in summer.

Finally, it is sometimes convenient to rewrite the ideal-gas equation on a molecular basis. We use $N = N_A n$ and $R = N_A k$ to obtain this alternative form:

$$pV = NkT \qquad (18.18)$$

This shows that we can think of the Boltzmann constant k as a gas constant on a "per-molecule" basis instead of the usual "per-mole" basis for R.

Molecular Speeds

From Eqs. (18.16) and (18.17) we can obtain expressions for the square root of $(v^2)_{av}$, called the **root-mean-square speed** (or **rms speed**) v_{rms}:

$$v_{rms} = \sqrt{(v^2)_{av}} = \sqrt{\frac{3kT}{m}} = \sqrt{\frac{3RT}{M}} \quad \begin{array}{l}\text{(root-mean-square speed} \\ \text{of a gas molecule)}\end{array} \qquad (18.19)$$

It might seem more natural to characterize molecular speeds by their *average* value rather than by v_{rms}, but we see that v_{rms} follows more directly from Eqs. (18.16) and (18.17). To compute the rms speed, we square each molecular speed, add, divide by the number of molecules, and take the square root; v_{rms} is the *root* of the *mean* of the *squares*. Example 18.7 illustrates this procedure.

Equations (18.16) and (18.19) show that at a given temperature T, gas molecules of different mass m have the same average kinetic energy but different root-mean-square speeds. On average, the nitrogen molecules ($M = 28$ g/mol) in the air around you are moving faster than are the oxygen molecules ($M = 32$ g/mol). Hydrogen molecules ($M = 2$ g/mol) are fastest of all; this is why there is hardly any hydrogen in the earth's atmosphere, despite its being the most common element in the universe (Fig. 18.14). A sizable fraction of any H_2 molecules in the atmosphere would have speeds greater than the earth's escape speed of 1.12×10^4 m/s (calculated in Example 13.5 in Section 13.3) and would escape into space. The heavier, slower-moving gases cannot escape so easily, which is why they predominate in our atmosphere.

The assumption that individual molecules undergo perfectly elastic collisions with the container wall is actually a little too simple. More detailed investigation has shown that in most cases, molecules actually adhere to the wall for a short time and then leave again with speeds that are characteristic of the temperature *of the wall*. However, the gas and the wall are ordinarily in thermal equilibrium and have the same temperature. So there is no net energy transfer between gas and wall, and this discovery does not alter the validity of our conclusions.

18.14 While hydrogen is a desirable fuel for vehicles, it is only a trace constituent of our atmosphere (0.00005% by volume). Hence hydrogen fuel has to be generated by electrolysis of water, which is itself an energy-intensive process.

Problem-Solving Strategy 18.2 | Kinetic-Molecular Theory

IDENTIFY *the relevant concepts:* Use the results of the kinetic-molecular model to relate the macroscopic properties of a gas, such as temperature and pressure, to microscopic properties, such as molecular speeds.

SET UP *the problem* using the following steps:
1. List knowns and unknowns; identify the target variables.
2. Choose appropriate equation(s) from among Eqs. (18.14), (18.16), and (18.19).

EXECUTE *the solution* as follows: Maintain consistency in units. Note especially the following:
1. The usual units for molar mass M are grams per mole; these units are often omitted in tables. In equations such as Eq. (18.19), when you use SI units you must express M in kilograms per

mole. For example, for oxygen $M_{O_2} = 32$ g/mol $= 32 \times 10^{-3}$ kg/mol.
2. Are you working on a "per-molecule" basis (with m, N, and k) or a "per-mole" basis (with M, n, and R)? To check units, think of N as having units of "molecules"; then m has units of mass per molecule, and k has units of joules per molecule per kelvin. Similarly, n has units of moles; then M has units of mass per mole and R has units of joules per mole per kelvin.
3. Remember that T is always *absolute* (Kelvin) temperature.

EVALUATE *your answer:* Are your answers reasonable? Here's a benchmark: Typical molecular speeds at room temperature are several hundred meters per second.

Example 18.6 **Molecular kinetic energy and v_{rms}**

(a) What is the average translational kinetic energy of an ideal-gas molecule at 27°C? (b) What is the total random translational kinetic energy of the molecules in 1 mole of this gas? (c) What is the root-mean-square speed of oxygen molecules at this temperature?

SOLUTION

IDENTIFY and SET UP: This problem involves the translational kinetic energy of an ideal gas on a per-molecule and per-mole basis, as well as the root-mean-square molecular speed v_{rms}. We are given $T = 27°C = 300$ K and $n = 1$ mol; we use the molecular mass m for oxygen. We use Eq. (18.16) to determine the average kinetic energy of a molecule, Eq. (18.14) to find the total molecular kinetic energy K_{tr} of 1 mole, and Eq. (18.19) to find v_{rms}.

EXECUTE: (a) From Eq. (18.16),

$$\tfrac{1}{2}m(v^2)_{av} = \tfrac{3}{2}kT = \tfrac{3}{2}(1.38 \times 10^{-23} \text{ J/K})(300 \text{ K})$$
$$= 6.21 \times 10^{-21} \text{ J}$$

(b) From Eq. (18.14), the kinetic energy of one mole is

$$K_{tr} = \tfrac{3}{2}nRT = \tfrac{3}{2}(1 \text{ mol})(8.314 \text{ J/mol}\cdot\text{K})(300 \text{ K}) = 3740 \text{ J}$$

(c) We found the mass per molecule m and molar mass M of molecular oxygen in Example 18.5. Using Eq. (18.19), we can calculate v_{rms} in two ways:

$$v_{rms} = \sqrt{\frac{3kT}{m}} = \sqrt{\frac{3(1.38 \times 10^{-23} \text{ J/K})(300 \text{ K})}{5.31 \times 10^{-26} \text{ kg}}}$$
$$= 484 \text{ m/s} = 1740 \text{ km/h} = 1080 \text{ mi/h}$$

$$v_{rms} = \sqrt{\frac{3RT}{M}} = \sqrt{\frac{3(8.314 \text{ J/mol}\cdot\text{K})(300 \text{ K})}{32.0 \times 10^{-3} \text{ kg/mol}}} = 484 \text{ m/s}$$

EVALUATE: The answer in part (a) does not depend on the mass of the molecule. We can check our result in part (b) by noting that the translational kinetic energy per mole must be equal to the product of the average translational kinetic energy per molecule from part (a) and Avogadro's number N_A: $K_{tr} = (6.022 \times 10^{23} \text{ molecules})$ $(6.21 \times 10^{-21} \text{ J/molecule}) = 3740 \text{ J}$.

Example 18.7 **Calculating rms and average speeds**

Five gas molecules chosen at random are found to have speeds of 500, 600, 700, 800, and 900 m/s. What is the rms speed? What is the *average* speed?

SOLUTION

IDENTIFY and SET UP: We use the definitions of the root mean square and the average of a collection of numbers. To find v_{rms}, we square each speed, find the average (mean) of the squares, and take the square root of the result. We find v_{av} as usual.

EXECUTE: The average value of v^2 and the resulting v_{rms} for the five molecules are

$$(v^2)_{av} = \frac{500^2 + 600^2 + 700^2 + 800^2 + 900^2}{5} \text{ m}^2/\text{s}^2$$
$$= 5.10 \times 10^5 \text{ m}^2/\text{s}^2$$

$$v_{rms} = \sqrt{(v^2)_{av}} = 714 \text{ m/s}$$

The average speed v_{av} is

$$v_{av} = \frac{500 + 600 + 700 + 800 + 900}{5} \text{ m/s} = 700 \text{ m/s}$$

EVALUATE: In general v_{rms} and v_{av} are *not* the same. Roughly speaking, v_{rms} gives greater weight to the higher speeds than does v_{av}.

Collisions Between Molecules

We have ignored the possibility that two gas molecules might collide. If they are really points, they *never* collide. But consider a more realistic model in which the molecules are rigid spheres with radius r. How often do they collide with other molecules? How far do they travel, on average, between collisions? We can get approximate answers from the following rather primitive model.

Consider N spherical molecules with radius r in a volume V. Suppose only one molecule is moving. When it collides with another molecule, the distance between centers is $2r$. Suppose we draw a cylinder with radius $2r$, with its axis parallel to the velocity of the molecule (Fig. 18.15). The moving molecule collides with any other molecule whose center is inside this cylinder. In a short time dt a molecule with speed v travels a distance $v\, dt$; during this time it collides with any molecule that is in the cylindrical volume of radius $2r$ and length $v\, dt$. The volume of the cylinder is $4\pi r^2 v\, dt$. There are N/V molecules per unit volume, so the number dN with centers in this cylinder is

$$dN = 4\pi r^2 v\, dt\, N/V$$

18.15 In a time dt a molecule with radius r will collide with any other molecule within a cylindrical volume of radius $2r$ and length $v\, dt$.

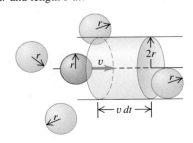

Thus the number of collisions *per unit time* is

$$\frac{dN}{dt} = \frac{4\pi r^2 vN}{V}$$

This result assumes that only one molecule is moving. The analysis is quite a bit more involved when all the molecules move at once. It turns out that in this case the collisions are more frequent, and the above equation has to be multiplied by a factor of $\sqrt{2}$:

$$\frac{dN}{dt} = \frac{4\pi \sqrt{2} r^2 vN}{V}$$

The average time t_{mean} between collisions, called the *mean free time,* is the reciprocal of this expression:

$$t_{mean} = \frac{V}{4\pi \sqrt{2} r^2 vN} \qquad (18.20)$$

The average distance traveled between collisions is called the **mean free path,** denoted by λ (the Greek letter lambda). In our simple model, this is just the molecule's speed v multiplied by t_{mean}:

18.16 If you try to walk through a crowd, your mean free path—the distance you can travel on average without running into another person—depends on how large the people are and how closely they are spaced.

$$\lambda = vt_{mean} = \frac{V}{4\pi \sqrt{2} r^2 N} \qquad \text{(mean free path of a gas molecule)} \qquad (18.21)$$

The mean free path is inversely proportional to the number of molecules per unit volume (N/V) and inversely proportional to the cross-sectional area πr^2 of a molecule; the more molecules there are and the larger the molecule, the shorter the mean distance between collisions (Fig. 18.16). Note that the mean free path *does not* depend on the speed of the molecule.

We can express Eq. (18.21) in terms of macroscopic properties of the gas, using the ideal-gas equation in the form of Eq. (18.18), $pV = NkT$. We find

$$\lambda = \frac{kT}{4\pi \sqrt{2} r^2 p} \qquad (18.22)$$

If the temperature is increased at constant pressure, the gas expands, the average distance between molecules increases, and λ increases. If the pressure is increased at constant temperature, the gas compresses and λ decreases.

Example 18.8 **Calculating mean free path**

(a) Estimate the mean free path of a molecule of air at 27°C and 1 atm. Model the molecules as spheres with radius $r = 2.0 \times 10^{-10}$ m. (b) Estimate the mean free time of an oxygen molecule with $v = v_{rms}$ at 27°C and 1 atm.

SOLUTION

IDENTIFY and SET UP: This problem uses the concepts of mean free path and mean free time (our target variables). We use Eq. (18.22) to determine the mean free path λ. We then use the basic relationship $\lambda = vt_{mean}$ in Eq. (18.21), with $v = v_{rms}$, to find the mean free time t_{mean}.

EXECUTE: (a) From Eq. (18.22),

$$\lambda = \frac{kT}{4\pi \sqrt{2} r^2 p} = \frac{(1.38 \times 10^{-23} \text{ J/K})(300 \text{ K})}{4\pi \sqrt{2}(2.0 \times 10^{-10} \text{ m})^2(1.01 \times 10^5 \text{ Pa})}$$

$$= 5.8 \times 10^{-8} \text{ m}$$

(b) From Example 18.6, for oxygen at 27°C the root-mean-square speed is $v_{rms} = 484$ m/s, so the mean free time for a molecule with this speed is

$$t_{mean} = \frac{\lambda}{v} = \frac{5.8 \times 10^{-8} \text{ m}}{484 \text{ m/s}} = 1.2 \times 10^{-10} \text{ s}$$

This molecule undergoes about 10^{10} collisions per second!

EVALUATE: Note that from Eqs. (18.21) and (18.22) the mean free *path* doesn't depend on the molecule's speed, but the mean free *time* does. Slower molecules have a longer average time interval t_{mean} between collisions than do fast ones, but the average *distance* λ between collisions is the same no matter what the molecule's speed. Our answer to part (a) says that the molecule doesn't go far between collisions, but the mean free path is still several hundred times the molecular radius r.

Test Your Understanding of Section 18.3 Rank the following gases in order from (a) highest to lowest rms speed of molecules and (b) highest to lowest average translational kinetic energy of a molecule: (i) oxygen ($M = 32.0$ g/mol) at 300 K; (ii) nitrogen ($M = 28.0$ g/mol) at 300 K; (iii) oxygen at 330 K; (iv) nitrogen at 330 K.

18.4 Heat Capacities

When we introduced the concept of heat capacity in Section 17.5, we talked about ways to *measure* the specific heat or molar heat capacity of a particular material. Now we'll see how to *predict* these on theoretical grounds.

Heat Capacities of Gases

The basis of our analysis is that heat is *energy* in transit. When we add heat to a substance, we are increasing its molecular energy. In this discussion the volume of the gas will remain constant so that we don't have to worry about energy transfer through mechanical work. If we were to let the gas expand, it would do work by pushing on moving walls of its container, and this additional energy transfer would have to be included in our calculations. We'll return to this more general case in Chapter 19. For now, with the volume held constant, we are concerned with C_V, the molar heat capacity *at constant volume*.

In the simple kinetic-molecular model of Section 18.3 the molecular energy consists only of the translational kinetic energy K_{tr} of the pointlike molecules. This energy is directly proportional to the absolute temperature T, as shown by Eq. (18.14), $K_{tr} = \frac{3}{2}nRT$. When the temperature changes by a small amount dT, the corresponding change in kinetic energy is

$$dK_{tr} = \frac{3}{2} nR \, dT \tag{18.23}$$

From the definition of molar heat capacity at constant volume, C_V (see Section 17.5), we also have

$$dQ = nC_V \, dT \tag{18.24}$$

where dQ is the heat input needed for a temperature change dT. Now if K_{tr} represents the total molecular energy, as we have assumed, then dQ and dK_{tr} must be *equal* (Fig. 18.17). From Eqs. (18.23) and (18.24), this says

$$nC_V \, dT = \frac{3}{2} nR \, dT$$

$$C_V = \frac{3}{2}R \qquad \text{(ideal gas of point particles)} \tag{18.25}$$

This surprisingly simple result says that the molar heat capacity at constant volume is $3R/2$ for *any* gas whose molecules can be represented as points.

Does Eq. (18.25) agree with measured values of molar heat capacities? In SI units, Eq. (18.25) gives

$$C_V = \frac{3}{2}(8.314 \text{ J/mol} \cdot \text{K}) = 12.47 \text{ J/mol} \cdot \text{K}$$

For comparison, Table 18.1 gives measured values of C_V for several gases. We see that for *monatomic* gases our prediction is right on the money, but that it is way off for diatomic and polyatomic gases.

This comparison tells us that our point-molecule model is good enough for monatomic gases but that for diatomic and polyatomic molecules we need something more sophisticated. For example, we can picture a diatomic molecule as

18.17 (a) A fixed volume V of a monatomic ideal gas. (b) When an amount of heat dQ is added to the gas, the total translational kinetic energy increases by $dK_{tr} = dQ$ and the temperature increases by $dT = dQ/nC_V$.

(a)

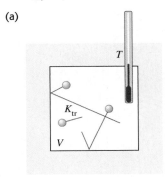

(b)

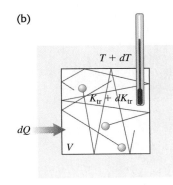

Table 18.1 Molar Heat Capacities of Gases

Type of Gas	Gas	C_V (J/mol · K)
Monatomic	He	12.47
	Ar	12.47
Diatomic	H_2	20.42
	N_2	20.76
	O_2	20.85
	CO	20.85
Polyatomic	CO_2	28.46
	SO_2	31.39
	H_2S	25.95

18.18 Motions of a diatomic molecule.

(a) Translational motion. The molecule moves as a whole; its velocity may be described as the *x-*, *y-*, and *z*-velocity components of its center of mass.

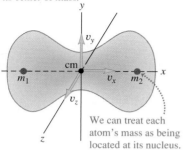

We can treat each atom's mass as being located at its nucleus.

(b) Rotational motion. The molecule rotates about its center of mass. This molecule has two independent axes of rotation.

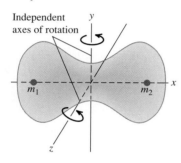

Independent axes of rotation

(c) Vibrational motion. The molecule oscillates as though the nuclei were connected by a spring.

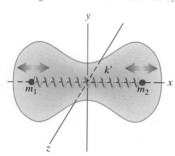

two point masses, like a little elastic dumbbell, with an interaction force between the atoms of the kind shown in Fig. 18.8. Such a molecule can have additional kinetic energy associated with *rotation* about axes through its center of mass. The atoms may also have *vibrating* motion along the line joining them, with additional kinetic and potential energies. Figure 18.18 shows these possibilities.

When heat flows into a *monatomic* gas at constant volume, *all* of the added energy goes into an increase in random *translational* molecular kinetic energy. Equation (18.23) shows that this gives rise to an increase in temperature. But when the temperature is increased by the same amount in a *diatomic* or *polyatomic* gas, additional heat is needed to supply the increased rotational and vibrational energies. Thus polyatomic gases have *larger* molar heat capacities than monatomic gases, as Table 18.1 shows.

But how do we know how much energy is associated with each additional kind of motion of a complex molecule, compared to the translational kinetic energy? The new principle that we need is called the principle of **equipartition of energy.** It can be derived from sophisticated statistical-mechanics considerations; that derivation is beyond our scope, and we will treat the principle as an axiom.

The principle of equipartition of energy states that each velocity component (either linear or angular) has, on average, an associated kinetic energy per molecule of $\frac{1}{2}kT$, or one-half the product of the Boltzmann constant and the absolute temperature. The number of velocity components needed to describe the motion of a molecule completely is called the number of **degrees of freedom.** For a monatomic gas, there are three degrees of freedom (for the velocity components v_x, v_y, and v_z); this gives a total average kinetic energy per molecule of $3(\frac{1}{2}kT)$, consistent with Eq. (18.16).

For a *diatomic* molecule there are two possible axes of rotation, perpendicular to each other and to the molecule's axis. (We don't include rotation about the molecule's own axis because in ordinary collisions there is no way for this rotational motion to change.) If we assign five degrees of freedom to a diatomic molecule, the average total kinetic energy per molecule is $\frac{5}{2}kT$ instead of $\frac{3}{2}kT$. The total kinetic energy of n moles is $K_{\text{total}} = nN_A(\frac{5}{2}kT) = \frac{5}{2}n(kN_A)T = \frac{5}{2}nRT$, and the molar heat capacity (at constant volume) is

$$C_V = \tfrac{5}{2}R \qquad \text{(diatomic gas, including rotation)} \qquad (18.26)$$

In SI units,

$$C_V = \tfrac{5}{2}(8.314 \text{ J/mol} \cdot \text{K}) = 20.79 \text{ J/mol} \cdot \text{K}$$

This agrees within a few percent with the measured values for diatomic gases given in Table 18.1.

Vibrational motion can also contribute to the heat capacities of gases. Molecular bonds are not rigid; they can stretch and bend, and the resulting vibrations lead to additional degrees of freedom and additional energies. For most diatomic gases, however, vibrational motion does *not* contribute appreciably to heat capacity. The reason for this is a little subtle and involves some concepts of quantum mechanics. Briefly, vibrational energy can change only in finite steps. If the energy change of the first step is much larger than the energy possessed by most molecules, then nearly all the molecules remain in the minimum-energy state of motion. In that case, changing the temperature does not change their average vibrational energy appreciably, and the vibrational degrees of freedom are said to be "frozen out." In more complex molecules the gaps between permitted energy levels are sometimes much smaller, and then vibration *does* contribute to heat capacity. The rotational energy of a molecule also changes by finite steps, but they are usually much

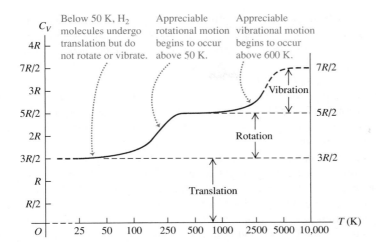

Below 50 K, H_2 molecules undergo translation but do not rotate or vibrate.

Appreciable rotational motion begins to occur above 50 K.

Appreciable vibrational motion begins to occur above 600 K.

18.19 Experimental values of C_V, the molar heat capacity at constant volume, for hydrogen gas (H_2). The temperature is plotted on a logarithmic scale.

smaller; the "freezing out" of rotational degrees of freedom occurs only in rare instances, such as for the hydrogen molecule below about 100 K.

In Table 18.1 the large values of C_V for some polyatomic molecules show the contributions of vibrational energy. In addition, a molecule with three or more atoms that are not in a straight line has three, not two, rotational degrees of freedom.

From this discussion we expect heat capacities to be temperature-dependent, generally increasing with increasing temperature. Figure 18.19 is a graph of the temperature dependence of C_V for hydrogen gas (H_2), showing the temperatures at which the rotational and vibrational energies begin to contribute.

Heat Capacities of Solids

We can carry out a similar heat-capacity analysis for a crystalline solid. Consider a crystal consisting of N identical atoms (a *monatomic solid*). Each atom is bound to an equilibrium position by interatomic forces. The elasticity of solid materials shows us that these forces must permit stretching and bending of the bonds. We can think of a crystal as an array of atoms connected by little springs (Fig. 18.20). Each atom can *vibrate* about its equilibrium position.

Each atom has three degrees of freedom, corresponding to its three components of velocity. According to the equipartition principle, each atom has an average kinetic energy of $\frac{1}{2}kT$ for each degree of freedom. In addition, each atom has *potential* energy associated with the elastic deformation. For a simple harmonic oscillator (discussed in Chapter 14) it is not hard to show that the average kinetic energy of an atom is *equal* to its average potential energy. In our model of a crystal, each atom is essentially a three-dimensional harmonic oscillator; it can be shown that the equality of average kinetic and potential energies also holds here, provided that the "spring" forces obey Hooke's law.

Thus we expect each atom to have an average kinetic energy $\frac{3}{2}kT$ and an average potential energy $\frac{3}{2}kT$, or an average total energy $3kT$ per atom. If the crystal contains N atoms or n moles, its total energy is

$$E_{total} = 3NkT = 3nRT \qquad (18.27)$$

18.20 To visualize the forces between neighboring atoms in a crystal, envision every atom as being attached to its neighbors by springs.

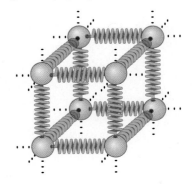

From this we conclude that the molar heat capacity of a crystal should be

$$C_V = 3R \qquad \text{(ideal monatomic solid)} \qquad (18.28)$$

In SI units,

$$C_V = (3)(8.314 \text{ J/mol} \cdot \text{K}) = 24.9 \text{ J/mol} \cdot \text{K}$$

This is the **rule of Dulong and Petit,** which we encountered as an *empirical* finding in Section 17.5: Elemental solids all have molar heat capacities of about 25 J/mol · K. Now we have *derived* this rule from kinetic theory. The agreement is only approximate, to be sure, but considering the very simple nature of our model, it is quite significant.

At low temperatures, the heat capacities of most solids *decrease* with decreasing temperature (Fig. 18.21) for the same reason that vibrational degrees of freedom of molecules are frozen out at low temperatures. At very low temperatures the quantity kT is much *smaller* than the smallest energy step the vibrating atoms can take. Hence most of the atoms remain in their lowest energy states because the next higher energy level is out of reach. The average vibrational energy per atom is then *less* than $3kT$, and the heat capacity per molecule is *less* than $3k$. At higher temperatures when kT is *large* in comparison to the minimum energy step, the equipartition principle holds, and the total heat capacity is $3k$ per molecule or $3R$ per mole as the Dulong and Petit rule predicts. Quantitative understanding of the temperature variation of heat capacities was one of the triumphs of quantum mechanics during its initial development in the 1920s.

Test Your Understanding of Section 18.4 A cylinder with a fixed volume contains hydrogen gas (H_2) at 25 K. You then add heat to the gas at a constant rate until its temperature reaches 500 K. Does the temperature of the gas increase at a constant rate? Why or why not? If not, does the temperature increase most rapidly near the beginning or near the end of this process? ❙

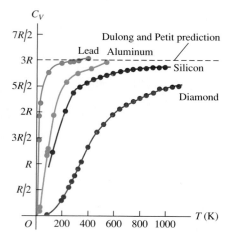

18.21 Experimental values of C_V for lead, aluminum, silicon, and diamond. At high temperatures, C_V for each solid approaches about $3R$, in agreement with the rule of Dulong and Petit. At low temperatures, C_V is much less than $3R$.

18.5 Molecular Speeds

As we mentioned in Section 18.3, the molecules in a gas don't all have the same speed. Figure 18.22 shows one experimental scheme for measuring the distribution of molecular speeds. A substance is vaporized in a hot oven; molecules of the vapor escape through an aperture in the oven wall and into a vacuum chamber. A series of slits blocks all molecules except those in a narrow beam, which is aimed at a pair of rotating disks. A molecule passing through the slit in the first disk is blocked by the second disk unless it arrives just as the slit in the second disk is lined up with the beam. The disks function as a speed selector that passes only molecules within a certain narrow speed range. This range can be varied by changing the disk rotation speed, and we can measure how many molecules lie within each of various speed ranges.

To describe the results of such measurements, we define a function $f(v)$ called a *distribution function*. If we observe a total of N molecules, the number dN having speeds in the range between v and $v + dv$ is given by

$$dN = Nf(v)\,dv \qquad (18.29)$$

18.22 A molecule with a speed v passes through the slit in the first rotating disk. When the molecule reaches the second rotating disk, the disks have rotated through the offset angle θ. If $v = \omega x/\theta$, the molecule passes through the slit in the second rotating disk and reaches the detector.

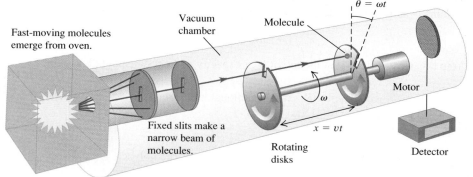

We can also say that the *probability* that a randomly chosen molecule will have a speed in the interval v to $v + dv$ is $f(v)\,dv$. Hence $f(v)$ is the probability per unit speed *interval;* it is *not* equal to the probability that a molecule has speed exactly equal to v. Since a probability is a pure number, $f(v)$ has units of reciprocal speed (s/m).

Figure 18.23a shows distribution functions for three different temperatures. At each temperature the height of the curve for any value of v is proportional to the number of molecules with speeds near v. The peak of the curve represents the *most probable speed* v_{mp} for the corresponding temperature. As the temperature increases, the average molecular kinetic energy increases, and so the peak of $f(v)$ shifts to higher and higher speeds.

Figure 18.23b shows that the area under a curve between any two values of v represents the fraction of all the molecules having speeds in that range. Every molecule must have *some* value of v, so the integral of $f(v)$ over all v must be unity for any T.

If we know $f(v)$, we can calculate the most probable speed v_{mp}, the average speed v_{av}, and the rms speed v_{rms}. To find v_{mp}, we simply find the point where $df/dv = 0$; this gives the value of the speed where the curve has its peak. To find v_{av}, we take the number $Nf(v)\,dv$ having speeds in each interval dv, multiply each number by the corresponding speed v, add all these products (by integrating over all v from zero to infinity), and finally divide by N. That is,

$$v_{av} = \int_0^\infty v f(v)\,dv \tag{18.30}$$

The rms speed is obtained similarly; the average of v^2 is given by

$$(v^2)_{av} = \int_0^\infty v^2 f(v)\,dv \tag{18.31}$$

and v_{rms} is the square root of this.

The Maxwell–Boltzmann Distribution

The function $f(v)$ describing the actual distribution of molecular speeds is called the **Maxwell–Boltzmann distribution.** It can be derived from statistical-mechanics considerations, but that derivation is beyond our scope. Here is the result:

$$f(v) = 4\pi\left(\frac{m}{2\pi kT}\right)^{3/2} v^2 e^{-mv^2/2kT} \quad \text{(Maxwell–Boltzmann distribution)} \tag{18.32}$$

We can also express this function in terms of the translational kinetic energy of a molecule, which we denote by ϵ; that is, $\epsilon = \frac{1}{2}mv^2$. We invite you to verify that when this is substituted into Eq. (18.32), the result is

$$f(v) = \frac{8\pi}{m}\left(\frac{m}{2\pi kT}\right)^{3/2} \epsilon e^{-\epsilon/kT} \tag{18.33}$$

This form shows that the exponent in the Maxwell–Boltzmann distribution function is $-\epsilon/kT$ and that the shape of the curve is determined by the relative magnitude of ϵ and kT at any point. We leave it to you (see Exercise 18.48) to prove that the *peak* of each curve occurs where $\epsilon = kT$, corresponding to a most probable speed v_{mp} given by

$$v_{mp} = \sqrt{\frac{2kT}{m}} \tag{18.34}$$

To find the average speed, we substitute Eq. (18.32) into Eq. (18.30) and carry out the integration, making a change of variable $v^2 = x$ and then integrating by parts. The result is

$$v_{av} = \sqrt{\frac{8kT}{\pi m}} \tag{18.35}$$

18.23 (a) Curves of the Maxwell–Boltzmann distribution function $f(v)$ for three temperatures. (b) The shaded areas under the curve represent the fractions of molecules within certain speed ranges. The most probable speed v_{mp} for a given temperature is at the peak of the curve.

(a)

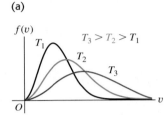

As temperature increases:
• the curve flattens.
• the maximum shifts to higher speeds.

(b)

Fraction of molecules with speeds from v_1 to v_2

Fraction of molecules with speeds greater than v_A

MasteringPHYSICS

ActivPhysics 8.2: Maxwell-Boltzmann Distribution—Conceptual Analysis
ActivPhysics 8.3: Maxwell-Boltzmann Distribution—Quantitative Analysis

Finally, to find the rms speed, we substitute Eq. (18.32) into Eq. (18.31). Evaluating the resulting integral takes some mathematical acrobatics, but we can find it in a table of integrals. The result is

$$v_{rms} = \sqrt{\frac{3kT}{m}} \qquad (18.36)$$

This result agrees with Eq. (18.19); it *must* agree if the Maxwell–Boltzmann distribution is to be consistent with the equipartition principle and our other kinetic-theory calculations.

Table 18.2 shows the fraction of all the molecules in an ideal gas that have speeds *less than* various multiples of v_{rms}. These numbers were obtained by numerical integration; they are the same for all ideal gases.

The distribution of molecular speeds in liquids is similar, although not identical, to that for gases. We can understand the vapor pressure of a liquid and the phenomenon of boiling on this basis. Suppose a molecule must have a speed at least as great as v_A in Fig. 18.23b to escape from the surface of a liquid into the adjacent vapor. The number of such molecules, represented by the area under the "tail" of each curve (to the right of v_A), increases rapidly with temperature. Thus the rate at which molecules can escape is strongly temperature-dependent. This process is balanced by another one in which molecules in the vapor phase collide inelastically with the surface and are trapped back into the liquid phase. The number of molecules suffering this fate per unit time is proportional to the pressure in the vapor phase. Phase equilibrium between liquid and vapor occurs when these two competing processes proceed at exactly the same rate. So if the molecular speed distributions are known for various temperatures, we can make a theoretical prediction of vapor pressure as a function of temperature. When liquid evaporates, it's the high-speed molecules that escape from the surface. The ones that are left have less energy on average; this gives us a molecular view of evaporative cooling.

Rates of chemical reactions are often strongly temperature-dependent, and the reason is contained in the Maxwell–Boltzmann distribution. When two reacting molecules collide, the reaction can occur only when the molecules are close enough for the electric-charge distributions of their electrons to interact strongly. This requires a minimum energy, called the *activation energy,* and thus a certain minimum molecular speed. Figure 18.23a shows that the number of molecules in the high-speed tail of the curve increases rapidly with temperature. Thus we expect the rate of any reaction that depends on an activation energy to increase rapidly with temperature.

Table 18.2 Fractions of Molecules in an Ideal Gas with Speeds Less than Various Multiples of v/v_{rms}

v/v_{rms}	Fraction
0.20	0.011
0.40	0.077
0.60	0.218
0.80	0.411
1.00	0.608
1.20	0.771
1.40	0.882
1.60	0.947
1.80	0.979
2.00	0.993

Application Activation Energy and Moth Activity

This hawkmoth of genus *Manduca* cannot fly if the temperature of its muscles is below 29°C. The reason is that the enzyme-catalyzed reactions that power aerobic metabolism and enable muscle action require a minimum molecular energy (activation energy). Just like the molecules in an ideal gas, at low temperatures very few of the molecules involved in these reactions have high energy. As the temperature increases, more molecules have the required minimum energy and the reactions take place at a greater rate. Above 29°C, enough power is generated to allow the hawkmoth to fly.

Test Your Understanding of Section 18.5 A quantity of gas containing N molecules has a speed distribution function $f(v)$. How many molecules have speeds between v_1 and $v_2 > v_1$? (i) $\int_0^{v_2} f(v)\, dv - \int_0^{v_1} f(v)\, dv$; (ii) $N[\int_0^{v_2} f(v)\, dv - \int_0^{v_1} f(v)\, dv]$; (iii) $\int_0^{v_1} f(v)\, dv - \int_0^{v_2} f(v)\, dv$; (iv) $N[\int_0^{v_1} f(v)\, dv - \int_0^{v_2} f(v)\, dv]$; (v) none of these.

18.6 Phases of Matter

An ideal gas is the simplest system to analyze from a molecular viewpoint because we ignore the interactions between molecules. But those interactions are the very thing that makes matter condense into the liquid and solid phases under some conditions. So it's not surprising that theoretical analysis of liquid and solid structure and behavior is a lot more complicated than that for gases. We won't try to go far here with a microscopic picture, but we can talk in general about phases of matter, phase equilibrium, and phase transitions.

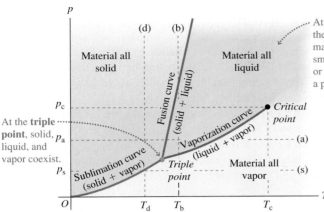

18.24 A typical pT phase diagram, showing regions of temperature and pressure at which the various phases exist and where phase changes occur.

In Section 17.6 we learned that each phase is stable only in certain ranges of temperature and pressure. A transition from one phase to another ordinarily requires **phase equilibrium** between the two phases, and for a given pressure this occurs at only one specific temperature. We can represent these conditions on a graph with axes p and T, called a **phase diagram;** Fig. 18.24 shows an example. Each point on the diagram represents a pair of values of p and T.

Only a single phase can exist at each point in Fig. 18.24, except for points on the solid lines, where two phases can coexist in phase equilibrium. The fusion curve separates the solid and liquid areas and represents possible conditions of solid-liquid phase equilibrium. The vaporization curve separates the liquid and vapor areas, and the sublimation curve separates the solid and vapor areas. All three curves meet at the **triple point,** the only condition under which all three phases can coexist (Fig. 18.25). In Section 17.3 we used the triple-point temperature of water to define the Kelvin temperature scale. Table 18.3 gives triple-point data for several substances.

If we add heat to a substance at a constant pressure p_a, it goes through a series of states represented by the horizontal line (a) in Fig. 18.24. The melting and boiling temperatures at this pressure are the temperatures at which the line intersects the fusion and vaporization curves, respectively. When the pressure is p_s, constant-pressure heating transforms a substance from solid directly to vapor. This process is called *sublimation;* the intersection of line (s) with the sublimation curve gives the temperature T_s at which it occurs for a pressure p_s. At any pressure less than the triple-point pressure, no liquid phase is possible. The triple-point pressure for carbon dioxide is 5.1 atm. At normal atmospheric pressure, solid carbon dioxide ("dry ice") undergoes sublimation; there is no liquid phase at this pressure.

Line (b) in Fig. 18.24 represents compression at a constant temperature T_b. The material passes from vapor to liquid and then to solid at the points where line (b) crosses the vaporization curve and fusion curve, respectively. Line (d) shows constant-temperature compression at a lower temperature T_d; the material passes from vapor to solid at the point where line (d) crosses the sublimation curve.

We saw in the pV-diagram of Fig. 18.7 that a liquid-vapor phase transition occurs only when the temperature and pressure are less than those at the point lying at the top of the green shaded area labeled "Liquid-vapor phase equilibrium region." This point corresponds to the endpoint at the top of the vaporization curve in Fig. 18.24. It is called the **critical point,** and the corresponding values of p and T are called the critical pressure and temperature, p_c and T_c. A gas at a pressure *above* the critical pressure does not separate into two phases when it is cooled at constant pressure (along a horizontal line above the critical point in Fig. 18.24). Instead, its properties change gradually and continuously from those we ordinarily associate with a gas (low density, large compressibility) to those of a liquid (high density, small compressibility) *without a phase transition*.

18.25 Atmospheric pressure on earth is higher than the triple-point pressure of water (see line (a) in Fig. 18.24). Depending on the temperature, water can exist as a vapor (in the atmosphere), as a liquid (in the ocean), or as a solid (like the iceberg shown here).

Table 18.3 Triple-Point Data

Substance	Temperature (K)	Pressure (Pa)
Hydrogen	13.80	0.0704×10^5
Deuterium	18.63	0.171×10^5
Neon	24.56	0.432×10^5
Nitrogen	63.18	0.125×10^5
Oxygen	54.36	0.00152×10^5
Ammonia	195.40	0.0607×10^5
Carbon dioxide	216.55	5.17×10^5
Sulfur dioxide	197.68	0.00167×10^5
Water	273.16	0.00610×10^5

You can understand this by thinking about liquid-phase transitions at successively higher points on the vaporization curve. As we approach the critical point, the *differences* in physical properties (such as density and compressibility) between the liquid and vapor phases become smaller. Exactly *at* the critical point they all become zero, and at this point the distinction between liquid and vapor disappears. The heat of vaporization also grows smaller as we approach the critical point, and it too becomes zero at the critical point.

For nearly all familiar materials the critical pressures are much greater than atmospheric pressure, so we don't observe this behavior in everyday life. For example, the critical point for water is at 647.4 K and 221.2×10^5 Pa (about 218 atm or 3210 psi). But high-pressure steam boilers in electric generating plants regularly run at pressures and temperatures well above the critical point.

Many substances can exist in more than one solid phase. A familiar example is carbon, which exists as noncrystalline soot and crystalline graphite and diamond. Water is another example; at least eight types of ice, differing in crystal structure and physical properties, have been observed at very high pressures.

pVT-Surfaces

We remarked in Section 18.1 that the equation of state of any material can be represented graphically as a surface in a three-dimensional space with coordinates p, V, and T. Visualizing such a surface can add to our understanding of the behavior of materials at various temperatures and pressures. Figure 18.26 shows a typical *pVT*-surface. The light lines represent *pV*-isotherms; projecting them onto the *pV*-plane gives a diagram similar to Fig. 18.7. The *pV*-isotherms represent contour lines on the *pVT*-surface, just as contour lines on a topographic map represent the elevation (the third dimension) at each point. The projections of the edges of the surface onto the *pT*-plane give the *pT* phase diagram of Fig. 18.24.

Line *abcdef* in Fig. 18.26 represents constant-pressure heating, with melting along *bc* and vaporization along *de*. Note the volume changes that occur as *T* increases along this line. Line *ghjklm* corresponds to an isothermal (constant temperature) compression, with liquefaction along *hj* and solidification along *kl*. Between these, segments *gh* and *jk* represent isothermal compression with increase in pressure; the pressure increases are much greater in the liquid region

18.26 A *pVT*-surface for a substance that expands on melting. Projections of the boundaries on the surface onto the *pT*- and *pV*-planes are also shown.

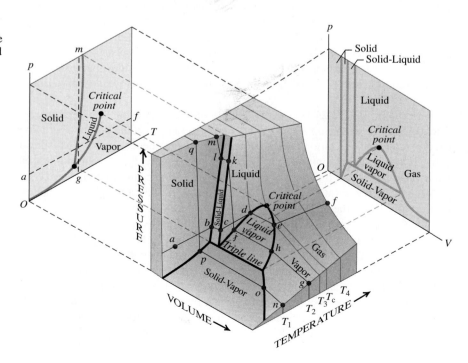

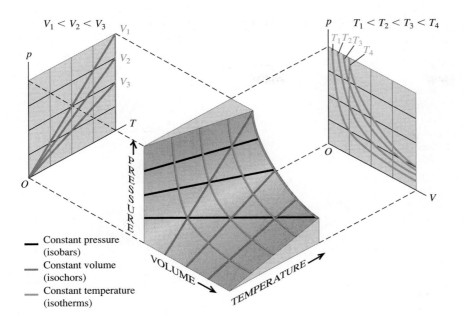

$V_1 < V_2 < V_3$

$T_1 < T_2 < T_3 < T_4$

— Constant pressure (isobars)
— Constant volume (isochors)
— Constant temperature (isotherms)

18.27 A pVT-surface for an ideal gas. At the left, each red line corresponds to a certain constant volume; at the right, each green line corresponds to a certain constant temperature.

jk and the solid region *lm* than in the vapor region *gh*. Finally, line *nopq* represents isothermal solidification directly from vapor, as in the formation of snowflakes or frost.

Figure 18.27 shows the much simpler pVT-surface for a substance that obeys the ideal-gas equation of state under all conditions. The projections of the constant-temperature curves onto the pV-plane correspond to the curves of Fig. 18.6, and the projections of the constant-volume curves onto the pT-plane show that pressure is directly proportional to absolute temperature.

Test Your Understanding of Section 18.6 The average atmospheric pressure on Mars is 6.0×10^2 Pa. Could there be lakes or rivers of liquid water on Mars today? What about in the past, when the atmospheric pressure is thought to have been substantially greater than today? ❙

Equations of state: The pressure p, volume V, and absolute temperature T of a given quantity of a substance are related by an equation of state. This relationship applies only for equilibrium states, in which p and T are uniform throughout the system. The ideal-gas equation of state, Eq. (18.3), involves the number of moles n and a constant R that is the same for all gases. (See Examples 18.1–18.4.)

$$pV = nRT \qquad (18.3)$$

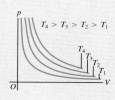

Molecular properties of matter: The molar mass M of a pure substance is the mass per mole. The mass m_{total} of a quantity of substance equals M multiplied by the number of moles n. Avogadro's number N_A is the number of molecules in a mole. The mass m of an individual molecule is M divided by N_A. (See Example 18.5.)

$$m_{total} = nM \qquad (18.2)$$

$$M = N_A m \qquad (18.8)$$

Kinetic-molecular model of an ideal gas: In an ideal gas, the total translational kinetic energy of the gas as a whole (K_{tr}) and the average translational kinetic energy per molecule $[\frac{1}{2}m(v^2)_{av}]$ are proportional to the absolute temperature T, and the root-mean-square speed of molecules is proportional to the square root of T. These expressions involve the Boltzmann constant $k = R/N_A$. (See Examples 18.6 and 18.7.) The mean free path λ of molecules in an ideal gas depends on the number of molecules per volume (N/V) and the molecular radius r. (See Example 18.8.)

$$K_{tr} = \tfrac{3}{2}nRT \qquad (18.14)$$

$$\tfrac{1}{2}m(v^2)_{av} = \tfrac{3}{2}kT \qquad (18.16)$$

$$v_{rms} = \sqrt{(v^2)_{av}} = \sqrt{\frac{3kT}{m}} \qquad (18.19)$$

$$\lambda = vt_{mean} = \frac{V}{4\pi\sqrt{2}\,r^2 N} \qquad (18.21)$$

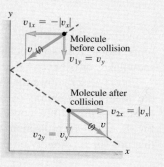

Heat capacities: The molar heat capacity at constant volume C_V is a simple multiple of the gas constant R for certain idealized cases: an ideal monatomic gas [Eq. (18.25)]; an ideal diatomic gas including rotational energy [Eq. (18.26)]; and an ideal monatomic solid [Eq. (18.28)]. Many real systems are approximated well by these idealizations.

$$C_V = \tfrac{3}{2}R \quad \text{(monatomic gas)} \qquad (18.25)$$

$$C_V = \tfrac{5}{2}R \quad \text{(diatomic gas)} \qquad (18.26)$$

$$C_V = 3R \quad \text{(monatomic solid)} \qquad (18.28)$$

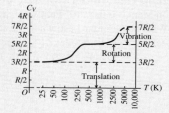

Molecular speeds: The speeds of molecules in an ideal gas are distributed according to the Maxwell–Boltzmann distribution $f(v)$. The quantity $f(v)\,dv$ describes what fraction of the molecules have speeds between v and $v + dv$.

$$f(v) = 4\pi\left(\frac{m}{2\pi kT}\right)^{3/2} v^2 e^{-mv^2/2kT} \qquad (18.32)$$

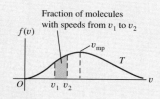

Phases of matter: Ordinary matter exists in the solid, liquid, and gas phases. A phase diagram shows conditions under which two phases can coexist in phase equilibrium. All three phases can coexist at the triple point. The vaporization curve ends at the critical point, above which the distinction between the liquid and gas phases disappears.

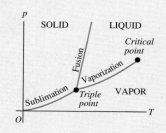

BRIDGING PROBLEM   Gas on Jupiter's Moon Europa

An astronaut visiting Jupiter's satellite Europa leaves a canister of 1.20 mol of nitrogen gas (28.0 g/mol) at 25.0°C on the satellite's surface. Europa has no significant atmosphere, and the acceleration due to gravity at its surface is 1.30 m/s^2. The canister springs a leak, allowing molecules to escape from a small hole. (a) What is the maximum height (in km) above Europa's surface that is reached by a nitrogen molecule whose speed equals the rms speed? Assume that the molecule is shot straight up out of the hole in the canister, and ignore the variation in g with altitude. (b) The escape speed from Europa is 2025 m/s. Can any of the nitrogen molecules escape from Europa and into space?

SOLUTION GUIDE

See MasteringPhysics® study area for a Video Tutor solution.

IDENTIFY and SET UP

1. Draw a sketch of the situation, showing all relevant dimensions.
2. Make a list of the unknown quantities, and decide which are the target variables.

3. How will you find the rms speed of the nitrogen molecules? What principle will you use to find the maximum height that a molecule with this speed can reach?
4. Does the rms speed of molecules in an ideal gas represent the maximum speed of the molecules? If not, what is the maximum speed?

EXECUTE

5. Solve for the rms speed. Use this to calculate the maximum height that a molecule with this speed can reach.
6. Use your result from step 5 to answer the question in part (b).

EVALUATE

7. Do your results depend on the amount of gas in the container? Why or why not?
8. How would your results from steps 5 and 6 be affected if the gas cylinder were instead left on Jupiter's satellite Ganymede, which has higher surface gravity than Europa and a higher escape speed? Like Europa, Ganymede has no significant atmosphere.

Problems   For instructor-assigned homework, go to www.masteringphysics.com MP

•, ••, •••: Problems of increasing difficulty. **CP**: Cumulative problems incorporating material from earlier chapters. **CALC**: Problems requiring calculus. **BIO**: Biosciences problems.

DISCUSSION QUESTIONS

Q18.1 Section 18.1 states that ordinarily, pressure, volume, and temperature cannot change individually without one affecting the others. Yet when a liquid evaporates, its volume changes, even though its pressure and temperature are constant. Is this inconsistent? Why or why not?

Q18.2 In the ideal-gas equation, could an equivalent Celsius temperature be used instead of the Kelvin one if an appropriate numerical value of the constant R is used? Why or why not?

Q18.3 On a chilly morning you can "see your breath." Can you really? What are you actually seeing? Does this phenomenon depend on the temperature of the air, the humidity, or both? Explain.

Q18.4 When a car is driven some distance, the air pressure in the tires increases. Why? Should you let out some air to reduce the pressure? Why or why not?

Q18.5 The coolant in an automobile radiator is kept at a pressure higher than atmospheric pressure. Why is this desirable? The radiator cap will release coolant when the gauge pressure of the coolant reaches a certain value, typically 15 lb/in.^{2} or so. Why not just seal the system completely?

Q18.6 Unwrapped food placed in a freezer experiences dehydration, known as "freezer burn." Why?

Q18.7 "Freeze-drying" food involves the same process as "freezer burn," referred to in Discussion Question Q18.6. For freeze-drying, the food is usually frozen first, and then placed in a vacuum chamber and irradiated with infrared radiation. What is the purpose of the vacuum? The radiation? What advantages might freeze-drying have in comparison to ordinary drying?

Q18.8 A group of students drove from their university (near sea level) up into the mountains for a skiing weekend. Upon arriving at the slopes, they discovered that the bags of potato chips they had brought for snacks had all burst open. What caused this to happen?

Q18.9 How does evaporation of perspiration from your skin cool your body?

Q18.10 A rigid, perfectly insulated container has a membrane dividing its volume in half. One side contains a gas at an absolute temperature T_0 and pressure p_0, while the other half is completely empty. Suddenly a small hole develops in the membrane, allowing the gas to leak out into the other half until it eventually occupies twice its original volume. In terms of T_0 and p_0, what will be the new temperature and pressure of the gas when it is distributed equally in both halves of the container? Explain your reasoning.

Q18.11 (a) Which has more atoms: a kilogram of hydrogen or a kilogram of lead? Which has more mass? (b) Which has more atoms: a mole of hydrogen or a mole of lead? Which has more mass? Explain your reasoning.

Q18.12 Use the concepts of the kinetic-molecular model to explain: (a) why the pressure of a gas in a rigid container increases as heat is added to the gas and (b) why the pressure of a gas increases as we compress it, even if we do not change its temperature.

Q18.13 The proportions of various gases in the earth's atmosphere change somewhat with altitude. Would you expect the proportion of oxygen at high altitude to be greater or less than at sea level compared to the proportion of nitrogen? Why?

Q18.14 Comment on the following statement: *When two gases are mixed, if they are to be in thermal equilibrium, they must have the* *same average molecular speed.* Is the statement correct? Why or why not?

Q18.15 The kinetic-molecular model contains a hidden assumption about the temperature of the container walls. What is this assumption? What would happen if this assumption were not valid?

Q18.16 The temperature of an ideal gas is directly proportional to the average kinetic energy of its molecules. If a container of ideal gas is moving past you at 2000 m/s, is the temperature of the gas higher than if the container was at rest? Explain your reasoning.

Q18.17 If the pressure of an ideal monatomic gas is increased while the number of moles is kept constant, what happens to the average translational kinetic energy of one atom of the gas? Is it possible to change *both* the volume and the pressure of an ideal gas and keep the average translational kinetic energy of the atoms constant? Explain.

Q18.18 In deriving the ideal-gas equation from the kinetic-molecular model, we ignored potential energy due to the earth's gravity. Is this omission justified? Why or why not?

Q18.19 The derivation of the ideal-gas equation included the assumption that the number of molecules is very large, so that we could compute the average force due to many collisions. However, the ideal-gas equation holds accurately only at low pressures, where the molecules are few and far between. Is this inconsistent? Why or why not?

Q18.20 A gas storage tank has a small leak. The pressure in the tank drops more quickly if the gas is hydrogen or helium than if it is oxygen. Why?

Q18.21 Consider two specimens of ideal gas at the same temperature. Specimen A has the same total mass as specimen B, but the molecules in specimen A have greater molar mass than they do in specimen B. In which specimen is the total kinetic energy of the gas greater? Does your answer depend on the molecular structure of the gases? Why or why not?

Q18.22 The temperature of an ideal monatomic gas is increased from 25°C to 50°C. Does the average translational kinetic energy of each gas atom double? Explain. If your answer is no, what would the final temperature be if the average translational kinetic energy was doubled?

Q18.23 If the root-mean-square speed of the atoms of an ideal gas is to be doubled, by what factor must the Kelvin temperature of the gas be increased? Explain.

Q18.24 (a) If you apply the same amount of heat to 1.00 mol of an ideal monatomic gas and 1.00 mol of an ideal diatomic gas, which one (if any) will increase more in temperature? (b) Physically, *why* do diatomic gases have a greater molar heat capacity than monatomic gases?

Q18.25 The discussion in Section 18.4 concluded that all ideal diatomic gases have the same heat capacity C_V. Does this mean that it takes the same amount of heat to raise the temperature of 1.0 g of each one by 1.0 K? Explain your reasoning.

Q18.26 In a gas that contains N molecules, is it accurate to say that the number of molecules with speed v is equal to $f(v)$? Is it accurate to say that this number is given by $Nf(v)$? Explain your answers.

Q18.27 Imagine a special air filter placed in a window of a house. The tiny holes in the filter allow only air molecules moving faster than a certain speed to exit the house, and allow only air molecules moving slower than that speed to enter the house from outside. What effect would this filter have on the temperature inside the house? (It turns out that the second law of thermodynamics—which we will discuss in Chapter 20—tells us that such a wonderful air filter would be impossible to make.)

Q18.28 A beaker of water at room temperature is placed in an enclosure, and the air pressure in the enclosure is slowly reduced. When the air pressure is reduced sufficiently, the water begins to boil. The temperature of the water does not rise when it boils; in fact, the temperature *drops* slightly. Explain these phenomena.

Q18.29 Ice is slippery to walk on, and especially slippery if you wear ice skates. What does this tell you about how the melting temperature of ice depends on pressure? Explain.

Q18.30 Hydrothermal vents are openings in the ocean floor that discharge very hot water. The water emerging from one such vent off the Oregon coast, 2400 m below the surface, has a temperature of 279°C. Despite its high temperature, the water doesn't boil. Why not?

Q18.31 The dark areas on the moon's surface are called *maria,* Latin for "seas," and were once thought to be bodies of water. In fact, the maria are not "seas" at all, but plains of solidified lava. Given that there is no atmosphere on the moon, how can you explain the absence of liquid water on the moon's surface?

Q18.32 In addition to the normal cooking directions printed on the back of a box of rice, there are also "high-altitude directions." The only difference is that the "high-altitude directions" suggest increasing the cooking time and using a greater volume of boiling water in which to cook the rice. Why should the directions depend on the altitude in this way?

EXERCISES

Section 18.1 Equations of State

18.1 • A 20.0-L tank contains 4.86×10^{-4} kg of helium at 18.0°C. The molar mass of helium is 4.00 g/mol. (a) How many moles of helium are in the tank? (b) What is the pressure in the tank, in pascals and in atmospheres?

18.2 •• Helium gas with a volume of 2.60 L, under a pressure of 0.180 atm and at a temperature of 41.0°C, is warmed until both pressure and volume are doubled. (a) What is the final temperature? (b) How many grams of helium are there? The molar mass of helium is 4.00 g/mol.

18.3 • A cylindrical tank has a tight-fitting piston that allows the volume of the tank to be changed. The tank originally contains 0.110 m³ of air at a pressure of 0.355 atm. The piston is slowly pulled out until the volume of the gas is increased to 0.390 m³. If the temperature remains constant, what is the final value of the pressure?

18.4 • A 3.00-L tank contains air at 3.00 atm and 20.0°C. The tank is sealed and cooled until the pressure is 1.00 atm. (a) What is the temperature then in degrees Celsius? Assume that the volume of the tank is constant. (b) If the temperature is kept at the value found in part (a) and the gas is compressed, what is the volume when the pressure again becomes 3.00 atm?

18.5 • **Planetary Atmospheres.** (a) Calculate the density of the atmosphere at the surface of Mars (where the pressure is 650 Pa and the temperature is typically 253 K, with a CO_2 atmosphere), Venus (with an average temperature of 730 K and pressure of 92 atm, with a CO_2 atmosphere), and Saturn's moon Titan (where the pressure is 1.5 atm and the temperature is −178°C, with a N_2 atmosphere). (b) Compare each of these densities with that of the earth's atmosphere, which is 1.20 kg/m³. Consult the periodic chart in Appendix D to determine molar masses.

18.6 •• You have several identical balloons. You experimentally determine that a balloon will break if its volume exceeds 0.900 L. The pressure of the gas inside the balloon equals air pressure (1.00 atm). (a) If the air inside the balloon is at a constant temperature of

22.0°C and behaves as an ideal gas, what mass of air can you blow into one of the balloons before it bursts? (b) Repeat part (a) if the gas is helium rather than air.

18.7 •• A Jaguar XK8 convertible has an eight-cylinder engine. At the beginning of its compression stroke, one of the cylinders contains 499 cm³ of air at atmospheric pressure (1.01×10^5 Pa) and a temperature of 27.0°C. At the end of the stroke, the air has been compressed to a volume of 46.2 cm³ and the gauge pressure has increased to 2.72×10^6 Pa. Compute the final temperature.

18.8 •• A welder using a tank of volume 0.0750 m³ fills it with oxygen (molar mass 32.0 g/mol) at a gauge pressure of 3.00×10^5 Pa and temperature of 37.0°C. The tank has a small leak, and in time some of the oxygen leaks out. On a day when the temperature is 22.0°C, the gauge pressure of the oxygen in the tank is 1.80×10^5 Pa. Find (a) the initial mass of oxygen and (b) the mass of oxygen that has leaked out.

18.9 •• A large cylindrical tank contains 0.750 m³ of nitrogen gas at 27°C and 7.50×10^3 Pa (absolute pressure). The tank has a tight-fitting piston that allows the volume to be changed. What will be the pressure if the volume is decreased to 0.480 m³ and the temperature is increased to 157°C?

18.10 • An empty cylindrical canister 1.50 m long and 90.0 cm in diameter is to be filled with pure oxygen at 22.0°C to store in a space station. To hold as much gas as possible, the absolute pressure of the oxygen will be 21.0 atm. The molar mass of oxygen is 32.0 g/mol. (a) How many moles of oxygen does this canister hold? (b) For someone lifting this canister, by how many kilograms does this gas increase the mass to be lifted?

18.11 • The gas inside a balloon will always have a pressure nearly equal to atmospheric pressure, since that is the pressure applied to the outside of the balloon. You fill a balloon with helium (a nearly ideal gas) to a volume of 0.600 L at a temperature of 19.0°C. What is the volume of the balloon if you cool it to the boiling point of liquid nitrogen (77.3 K)?

18.12 • **Deviations from the Ideal-Gas Equation.** For carbon dioxide gas (CO_2), the constants in the van der Waals equation are $a = 0.364$ J·m³/mol² and $b = 4.27 \times 10^{-5}$ m³/mol. (a) If 1.00 mol of CO_2 gas at 350 K is confined to a volume of 400 cm³, find the pressure of the gas using the ideal-gas equation and the van der Waals equation. (b) Which equation gives a lower pressure? Why? What is the percentage difference of the van der Waals equation result from the ideal-gas equation result? (c) The gas is kept at the same temperature as it expands to a volume of 4000 cm³. Repeat the calculations of parts (a) and (b). (d) Explain how your calculations show that the van der Waals equation is equivalent to the ideal-gas equation if n/V is small.

18.13 •• If a certain amount of ideal gas occupies a volume V at STP on earth, what would be its volume (in terms of V) on Venus, where the temperature is 1003°C and the pressure is 92 atm?

18.14 • A diver observes a bubble of air rising from the bottom of a lake (where the absolute pressure is 3.50 atm) to the surface (where the pressure is 1.00 atm). The temperature at the bottom is 4.0°C, and the temperature at the surface is 23.0°C. (a) What is the ratio of the volume of the bubble as it reaches the surface to its volume at the bottom? (b) Would it be safe for the diver to hold his breath while ascending from the bottom of the lake to the surface? Why or why not?

18.15 • A metal tank with volume 3.10 L will burst if the absolute pressure of the gas it contains exceeds 100 atm. (a) If 11.0 mol of an ideal gas is put into the tank at a temperature of 23.0°C, to what temperature can the gas be warmed before the tank ruptures? You can ignore the thermal expansion of the tank. (b) Based on your

answer to part (a), is it reasonable to ignore the thermal expansion of the tank? Explain.

18.16 • Three moles of an ideal gas are in a rigid cubical box with sides of length 0.200 m. (a) What is the force that the gas exerts on each of the six sides of the box when the gas temperature is 20.0°C? (b) What is the force when the temperature of the gas is increased to 100.0°C?

18.17 • With the assumptions of Example 18.4 (Section 18.1), at what altitude above sea level is air pressure 90% of the pressure at sea level?

18.18 • Make the same assumptions as in Example 18.4 (Section 18.1). How does the percentage decrease in air pressure in going from sea level to an altitude of 100 m compare to that when going from sea level to an altitude of 1000 m? If your second answer is not 10 times your first answer, explain why.

18.19 •• (a) Calculate the mass of nitrogen present in a volume of 3000 cm³ if the temperature of the gas is 22.0°C and the absolute pressure of 2.00×10^{-13} atm is a partial vacuum easily obtained in laboratories. (b) What is the density (in kg/m³) of the N_2?

18.20 •• With the assumption that the air temperature is a uniform 0.0°C (as in Example 18.4), what is the density of the air at an altitude of 1.00 km as a percentage of the density at the surface?

18.21 • At an altitude of 11,000 m (a typical cruising altitude for a jet airliner), the air temperature is -56.5°C and the air density is 0.364 kg/m³. What is the pressure of the atmosphere at that altitude? (*Note:* The temperature at this altitude is not the same as at the surface of the earth, so the calculation of Example 18.4 in Section 18.1 doesn't apply.)

Section 18.2 Molecular Properties of Matter

18.22 • A large organic molecule has a mass of 1.41×10^{-21} kg. What is the molar mass of this compound?

18.23 •• Suppose you inherit 3.00 mol of gold from your uncle (an eccentric chemist) at a time when this metal is selling for $14.75 per gram. Consult the periodic table in Appendix D and Table 12.1. (a) To the nearest dollar, what is this gold worth? (b) If you have your gold formed into a spherical nugget, what is its diameter?

18.24 •• Modern vacuum pumps make it easy to attain pressures of the order of 10^{-13} atm in the laboratory. Consider a volume of air and treat the air as an ideal gas. (a) At a pressure of 9.00×10^{-14} atm and an ordinary temperature of 300.0 K, how many molecules are present in a volume of 1.00 cm³? (b) How many molecules would be present at the same temperature but at 1.00 atm instead?

18.25 •• The Lagoon Nebula (Fig. E18.25) is a cloud of hydrogen gas located 3900 light-years from the earth. The cloud is about 45 light-years in diameter and glows because of its high temperature of 7500 K. (The gas is raised to this temperature by the stars that

Figure **E18.25**

lie within the nebula.) The cloud is also very thin; there are only 80 molecules per cubic centimeter. (a) Find the gas pressure (in atmospheres) in the Lagoon Nebula. Compare it to the laboratory pressure referred to in Exercise 18.24. (b) Science-fiction films sometimes show starships being buffeted by turbulence as they fly through gas clouds such as the Lagoon Nebula. Does this seem realistic? Why or why not?

18.26 •• In a gas at standard conditions, what is the length of the side of a cube that contains a number of molecules equal to the population of the earth (about 6×10^9 people)?

18.27 • How many moles are in a 1.00-kg bottle of water? How many molecules? The molar mass of water is 18.0 g/mol.

18.28 •• **How Close Together Are Gas Molecules?** Consider an ideal gas at 27°C and 1.00 atm pressure. To get some idea how close these molecules are to each other, on the average, imagine them to be uniformly spaced, with each molecule at the center of a small cube. (a) What is the length of an edge of each cube if adjacent cubes touch but do not overlap? (b) How does this distance compare with the diameter of a typical molecule? (c) How does their separation compare with the spacing of atoms in solids, which typically are about 0.3 nm apart?

18.29 •• Consider 5.00 mol of liquid water. (a) What volume is occupied by this amount of water? The molar mass of water is 18.0 g/mol. (b) Imagine the molecules to be, on average, uniformly spaced, with each molecule at the center of a small cube. What is the length of an edge of each small cube if adjacent cubes touch but don't overlap? (c) How does this distance compare with the diameter of a molecule?

Section 18.3 Kinetic-Molecular Model of an Ideal Gas

18.30 • A flask contains a mixture of neon (Ne), krypton (Kr), and radon (Rn) gases. Compare (a) the average kinetic energies of the three types of atoms and (b) the root-mean-square speeds. (*Hint:* The periodic table in Appendix D shows the molar mass (in g/mol) of each element under the chemical symbol for that element.)

18.31 • **Gaseous Diffusion of Uranium.** (a) A process called *gaseous diffusion* is often used to separate isotopes of uranium—that is, atoms of the elements that have different masses, such as ^{235}U and ^{238}U. The only gaseous compound of uranium at ordinary temperatures is uranium hexafluoride, UF_6. Speculate on how $^{235}UF_6$ and $^{238}UF_6$ molecules might be separated by diffusion. (b) The molar masses for $^{235}UF_6$ and $^{238}UF_6$ molecules are 0.349 kg/mol and 0.352 kg/mol, respectively. If uranium hexafluoride acts as an ideal gas, what is the ratio of the root-mean-square speed of $^{235}UF_6$ molecules to that of $^{238}UF_6$ molecules if the temperature is uniform?

18.32 • The ideas of average and root-mean-square value can be applied to any distribution. A class of 150 students had the following scores on a 100-point quiz:

Score	Number of Students
10	11
20	12
30	24
40	15
50	19
60	10
70	12
80	20
90	17
100	10

(a) Find the average score for the class. (b) Find the root-mean-square score for the class.

18.33 • We have two equal-size boxes, A and B. Each box contains gas that behaves as an ideal gas. We insert a thermometer into each box and find that the gas in box A is at a temperature of 50°C while the gas in box B is at 10°C. This is all we know about the gas in the boxes. Which of the following statements *must* be true? Which *could* be true? (a) The pressure in A is higher than in B. (b) There are more molecules in A than in B. (c) A and B do not contain the same type of gas. (d) The molecules in A have more average kinetic energy per molecule than those in B. (e) The molecules in A are moving faster than those in B. Explain the reasoning behind your answers.

18.34 • A container with volume 1.48 L is initially evacuated. Then it is filled with 0.226 g of N_2. Assume that the pressure of the gas is low enough for the gas to obey the ideal-gas law to a high degree of accuracy. If the root-mean-square speed of the gas molecules is 182 m/s, what is the pressure of the gas?

18.35 •• (a) A deuteron, ^2_1H, is the nucleus of a hydrogen isotope and consists of one proton and one neutron. The plasma of deuterons in a nuclear fusion reactor must be heated to about 300 million K. What is the rms speed of the deuterons? Is this a significant fraction of the speed of light ($c = 3.0 \times 10^8$ m/s)? (b) What would the temperature of the plasma be if the deuterons had an rms speed equal to $0.10c$?

18.36 • **Martian Climate.** The atmosphere of Mars is mostly CO_2 (molar mass 44.0 g/mol) under a pressure of 650 Pa, which we shall assume remains constant. In many places the temperature varies from 0.0°C in summer to −100°C in winter. Over the course of a Martian year, what are the ranges of (a) the rms speeds of the CO_2 molecules and (b) the density (in mol/m³) of the atmosphere?

18.37 •• (a) Oxygen (O_2) has a molar mass of 32.0 g/mol. What is the average translational kinetic energy of an oxygen molecule at a temperature of 300 K? (b) What is the average value of the square of its speed? (c) What is the root-mean-square speed? (d) What is the momentum of an oxygen molecule traveling at this speed? (e) Suppose an oxygen molecule traveling at this speed bounces back and forth between opposite sides of a cubical vessel 0.10 m on a side. What is the average force the molecule exerts on one of the walls of the container? (Assume that the molecule's velocity is perpendicular to the two sides that it strikes.) (f) What is the average force per unit area? (g) How many oxygen molecules traveling at this speed are necessary to produce an average pressure of 1 atm? (h) Compute the number of oxygen molecules that are actually contained in a vessel of this size at 300 K and atmospheric pressure. (i) Your answer for part (h) should be three times as large as the answer for part (g). Where does this discrepancy arise?

18.38 •• Calculate the mean free path of air molecules at a pressure of 3.50×10^{-13} atm and a temperature of 300 K. (This pressure is readily attainable in the laboratory; see Exercise 18.24.) As in Example 18.8, model the air molecules as spheres of radius 2.0×10^{-10} m.

18.39 •• At what temperature is the root-mean-square speed of nitrogen molecules equal to the root-mean-square speed of hydrogen molecules at 20.0°C? (*Hint:* The periodic table in Appendix D shows the molar mass (in g/mol) of each element under the chemical symbol for that element. The molar mass of H_2 is twice the molar mass of hydrogen atoms, and similarly for N_2.)

18.40 • Smoke particles in the air typically have masses of the order of 10^{-16} kg. The Brownian motion (rapid, irregular movement) of these particles, resulting from collisions with air molecules, can be observed with a microscope. (a) Find the root-mean-square speed of Brownian motion for a particle with a mass of 3.00×10^{-16} kg in air at 300 K. (b) Would the root-mean-square speed be different if the particle were in hydrogen gas at the same temperature? Explain.

Section 18.4 Heat Capacities

18.41 • (a) How much heat does it take to increase the temperature of 2.50 mol of a diatomic ideal gas by 50.0 K near room temperature if the gas is held at constant volume? (b) What is the answer to the question in part (a) if the gas is monatomic rather than diatomic?

18.42 •• Perfectly rigid containers each hold n moles of ideal gas, one being hydrogen (H_2) and other being neon (Ne). If it takes 300 J of heat to increase the temperature of the hydrogen by 2.50°C, by how many degrees will the same amount of heat raise the temperature of the neon?

18.43 •• (a) Compute the specific heat at constant volume of nitrogen (N_2) gas, and compare it with the specific heat of liquid water. The molar mass of N_2 is 28.0 g/mol. (b) You warm 1.00 kg of water at a constant volume of 1.00 L from 20.0°C to 30.0°C in a kettle. For the same amount of heat, how many kilograms of 20.0°C air would you be able to warm to 30.0°C? What volume (in liters) would this air occupy at 20.0°C and a pressure of 1.00 atm? Make the simplifying assumption that air is 100% N_2.

18.44 •• (a) Calculate the specific heat at constant volume of water vapor, assuming the nonlinear triatomic molecule has three translational and three rotational degrees of freedom and that vibrational motion does not contribute. The molar mass of water is 18.0 g/mol. (b) The actual specific heat of water vapor at low pressures is about 2000 J/kg·K. Compare this with your calculation and comment on the actual role of vibrational motion.

18.45 •• (a) Use Eq. 18.28 to calculate the specific heat at constant volume of aluminum in units of J/kg·K. Consult the periodic table in Appendix D. (b) Compare the answer in part (a) with the value given in Table 17.3. Try to explain any disagreement between these two values.

Section 18.5 Molecular Speeds

18.46 • For a gas of nitrogen molecules (N_2), what must the temperature be if 94.7% of all the molecules have speeds less than (a) 1500 m/s; (b) 1000 m/s; (c) 500 m/s? Use Table 18.2. The molar mass of N_2 is 28.0 g/mol.

18.47 • For diatomic carbon dioxide gas (CO_2, molar mass 44.0 g/mol) at $T = 300$ K, calculate (a) the most probable speed v_{mp}; (b) the average speed v_{av}; (c) the root-mean-square speed v_{rms}.

18.48 • CALC Prove that $f(v)$ as given by Eq. (18.33) is maximum for $\epsilon = kT$. Use this result to obtain Eq. (18.34).

Section 18.6 Phases of Matter

18.49 • Solid water (ice) is slowly warmed from a very low temperature. (a) What minimum external pressure p_1 must be applied to the solid if a melting phase transition is to be observed? Describe the sequence of phase transitions that occur if the applied pressure p is such that $p < p_1$. (b) Above a certain maximum pressure p_2, no boiling transition is observed. What is this pressure? Describe the sequence of phase transitions that occur if $p_1 < p < p_2$.

18.50 • Puffy cumulus clouds, which are made of water droplets, occur at lower altitudes in the atmosphere. Wispy cirrus clouds,

which are made of ice crystals, occur only at higher altitudes. Find the altitude y (measured from sea level) above which only cirrus clouds can occur. On a typical day and at altitudes less than 11 km, the temperature at an altitude y is given by $T = T_0 - \alpha y$, where $T_0 = 15.0$°C and $\alpha = 6.0$ C°/1000 m.

18.51 • The atmosphere of the planet Mars is 95.3% carbon dioxide (CO_2) and about 0.03% water vapor. The atmospheric pressure is only about 600 Pa, and the surface temperature varies from -30°C to -100°C. The polar ice caps contain both CO_2 ice and water ice. Could there be *liquid* CO_2 on the surface of Mars? Could there be liquid water? Why or why not?

18.52 • A physics lecture room has a volume of 216 m³. (a) For a pressure of 1.00 atm and a temperature of 27.0°C, use the ideal-gas law to estimate the number of air molecules in the room. Assume all the air is N_2. (b) Calculate the particle density—that is, the number of N_2 molecules per cubic centimeter. (c) Calculate the mass of the air in the room.

PROBLEMS

18.53 •• CP BIO **The Effect of Altitude on the Lungs.** (a) Calculate the *change* in air pressure you will experience if you climb a 1000-m mountain, assuming that the temperature and air density do not change over this distance and that they were 22°C and 1.2 kg/m³, respectively, at the bottom of the mountain. (Note that the result of Example 18.4 doesn't apply, since the expression derived in that example accounts for the variation of air density with altitude and we are told to ignore that in this problem.) (b) If you took a 0.50-L breath at the foot of the mountain and managed to hold it until you reached the top, what would be the volume of this breath when you exhaled it there?

18.54 •• CP BIO **The Bends.** If deep-sea divers rise to the surface too quickly, nitrogen bubbles in their blood can expand and prove fatal. This phenomenon is known as the *bends*. If a scuba diver rises quickly from a depth of 25 m in Lake Michigan (which is fresh water), what will be the volume at the surface of an N_2 bubble that occupied 1.0 mm³ in his blood at the lower depth? Does it seem that this difference is large enough to be a problem? (Assume that the pressure difference is due only to the changing water pressure, not to any temperature difference, an assumption that is reasonable, since we are warm-blooded creatures.)

18.55 ••• CP A hot-air balloon stays aloft because hot air at atmospheric pressure is less dense than cooler air at the same pressure. If the volume of the balloon is 500.0 m³ and the surrounding air is at 15.0°C, what must the temperature of the air in the balloon be for it to lift a total load of 290 kg (in addition to the mass of the hot air)? The density of air at 15.0°C and atmospheric pressure is 1.23 kg/m³.

18.56 •• (a) Use Eq. (18.1) to estimate the change in the volume of a solid steel sphere of volume 11 L when the temperature and pressure increase from 21°C and 1.013×10^5 Pa to 42°C and 2.10×10^7 Pa. (*Hint:* Consult Chapters 11 and 17 to determine the values of β and k.) (b) In Example 18.3 the change in volume of an 11-L steel scuba tank was ignored. Was this a good approximation? Explain.

18.57 ••• A cylinder 1.00 m tall with inside diameter 0.120 m is used to hold propane gas (molar mass 44.1 g/mol) for use in a barbecue. It is initially filled with gas until the gauge pressure is 1.30×10^6 Pa and the temperature is 22.0°C. The temperature of the gas remains constant as it is partially emptied out of the tank, until the gauge pressure is 2.50×10^5 Pa. Calculate the mass of propane that has been used.

18.58 • **CP** During a test dive in 1939, prior to being accepted by the U.S. Navy, the submarine *Squalus* sank at a point where the depth of water was 73.0 m. The temperature at the surface was 27.0°C, and at the bottom it was 7.0°C. The density of seawater is 1030 kg/m^3. (a) A diving bell was used to rescue 33 trapped crewmen from the *Squalus*. The diving bell was in the form of a circular cylinder 2.30 m high, open at the bottom and closed at the top. When the diving bell was lowered to the bottom of the sea, to what height did water rise within the diving bell? (*Hint:* You may ignore the relatively small variation in water pressure between the bottom of the bell and the surface of the water within the bell.) (b) At what gauge pressure must compressed air have been supplied to the bell while on the bottom to expel all the water from it?

18.59 • **Atmosphere of Titan.** Titan, the largest satellite of Saturn, has a thick nitrogen atmosphere. At its surface, the pressure is 1.5 earth-atmospheres and the temperature is 94 K. (a) What is the surface temperature in °C? (b) Calculate the surface density in Titan's atmosphere in molecules per cubic meter. (c) Compare the density of Titan's surface atmosphere to the density of earth's atmosphere at 22°C. Which body has denser atmosphere?

18.60 • **Pressure on Venus.** At the surface of Venus the average temperature is a balmy 460°C due to the greenhouse effect (global warming!), the pressure is 92 earth-atmospheres, and the acceleration due to gravity is 0.894g_{earth}. The atmosphere is nearly all CO_2 (molar mass 44.0 g/mol) and the temperature remains remarkably constant. We shall assume that the temperature does not change at all with altitude. (a) What is the atmospheric pressure 1.00 km above the surface of Venus? Express your answer in Venus-atmospheres and earth-atmospheres. (b) What is the root-mean-square speed of the CO_2 molecules at the surface of Venus and at an altitude of 1.00 km?

18.61 •• An automobile tire has a volume of 0.0150 m^3 on a cold day when the temperature of the air in the tire is 5.0°C and atmospheric pressure is 1.02 atm. Under these conditions the gauge pressure is measured to be 1.70 atm (about 25 lb/in.^{2}). After the car is driven on the highway for 30 min, the temperature of the air in the tires has risen to 45.0°C and the volume has risen to 0.0159 m^3. What then is the gauge pressure?

18.62 •• A flask with a volume of 1.50 L, provided with a stopcock, contains ethane gas (C_2H_6) at 300 K and atmospheric pressure (1.013 × 10^5 Pa). The molar mass of ethane is 30.1 g/mol. The system is warmed to a temperature of 490 K, with the stopcock open to the atmosphere. The stopcock is then closed, and the flask is cooled to its original temperature. (a) What is the final pressure of the ethane in the flask? (b) How many grams of ethane remain in the flask?

18.63 •• **CP** A balloon whose volume is 750 m^3 is to be filled with hydrogen at atmospheric pressure (1.01 × 10^5 Pa). (a) If the hydrogen is stored in cylinders with volumes of 1.90 m^3 at a gauge pressure of 1.20 × 10^6 Pa, how many cylinders are required? Assume that the temperature of the hydrogen remains constant. (b) What is the total weight (in addition to the weight of the gas) that can be supported by the balloon if the gas in the balloon and the surrounding air are both at 15.0°C? The molar mass of hydrogen (H_2) is 2.02 g/mol. The density of air at 15.0°C and atmospheric pressure is 1.23 kg/m^3. See Chapter 12 for a discussion of buoyancy. (c) What weight could be supported if the balloon were filled with helium (molar mass 4.00 g/mol) instead of hydrogen, again at 15.0°C?

18.64 •• A vertical cylindrical tank contains 1.80 mol of an ideal gas under a pressure of 0.500 atm at 20.0°C. The round part of the tank has a radius of 10.0 cm, and the gas is supporting a piston that can move up and down in the cylinder without friction. There is a vacuum above the piston. (a) What is the mass of this piston? (b) How tall is the column of gas that is supporting the piston?

18.65 •• **CP** A large tank of water has a hose connected to it, as shown in Fig. P18.65. The tank is sealed at the top and has compressed air between the water surface and the top. When the water height h has the value 3.50 m, the absolute pressure p of the compressed air is 4.20 × 10^5 Pa. Assume that the air above the water expands at constant temperature, and take the atmospheric pressure to be 1.00 × 10^5 Pa. (a) What is the speed with which water flows out of the hose when h = 3.50 m? (b) As water flows out of the tank, h decreases. Calculate the speed of flow for h = 3.00 m and for h = 2.00 m. (c) At what value of h does the flow stop?

Figure **P18.65**

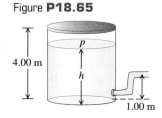

18.66 •• **BIO** A person at rest inhales 0.50 L of air with each breath at a pressure of 1.00 atm and a temperature of 20.0°C. The inhaled air is 21.0% oxygen. (a) How many oxygen molecules does this person inhale with each breath? (b) Suppose this person is now resting at an elevation of 2000 m but the temperature is still 20.0°C. Assuming that the oxygen percentage and volume per inhalation are the same as stated above, how many oxygen molecules does this person now inhale with each breath? (c) Given that the body still requires the same number of oxygen molecules per second as at sea level to maintain its functions, explain why some people report "shortness of breath" at high elevations.

18.67 •• **BIO How Many Atoms Are You?** Estimate the number of atoms in the body of a 50-kg physics student. Note that the human body is mostly water, which has molar mass 18.0 g/mol, and that each water molecule contains three atoms.

18.68 • The size of an oxygen molecule is about 2.0 × 10^{-10} m. Make a rough estimate of the pressure at which the finite volume of the molecules should cause noticeable deviations from ideal-gas behavior at ordinary temperatures (T = 300 K).

18.69 •• You have two identical containers, one containing gas A and the other gas B. The masses of these molecules are m_A = 3.34 × 10^{-27} kg and m_B = 5.34 × 10^{-26} kg. Both gases are under the same pressure and are at 10.0°C. (a) Which molecules (A or B) have greater translational kinetic energy per molecule and rms speeds? (b) Now you want to raise the temperature of only one of these containers so that both gases will have the same rms speed. For which gas should you raise the temperature? (c) At what temperature will you accomplish your goal? (d) Once you have accomplished your goal, which molecules (A or B) now have greater average translational kinetic energy per molecule?

18.70 • **Insect Collisions.** A cubical cage 1.25 m on each side contains 2500 angry bees, each flying randomly at 1.10 m/s. We can model these insects as spheres 1.50 cm in diameter. On the average, (a) how far does a typical bee travel between collisions, (b) what is the average time between collisions, and (c) how many collisions per second does a bee make?

18.71 •• You blow up a spherical balloon to a diameter of 50.0 cm until the absolute pressure inside is 1.25 atm and the temperature is 22.0°C. Assume that all the gas is N_2, of molar mass 28.0 g/mol. (a) Find the mass of a single N_2 molecule. (b) How much translational kinetic energy does an average N_2 molecule have? (c) How many N_2 molecules are in this balloon? (d) What is the *total* translational kinetic energy of all the molecules in the balloon?

18.72 • **CP** (a) Compute the increase in gravitational potential energy for a nitrogen molecule (molar mass 28.0 g/mol) for an increase in elevation of 400 m near the earth's surface. (b) At what temperature is this equal to the average kinetic energy of a nitrogen molecule? (c) Is it possible that a nitrogen molecule near sea level where $T = 15.0°C$ could rise to an altitude of 400 m? Is it likely that it could do so without hitting any other molecules along the way? Explain.

18.73 •• **CP, CALC** **The Lennard-Jones Potential.** A commonly used potential-energy function for the interaction of two molecules (see Fig. 18.8) is the Lennard-Jones 6-12 potential:

$$U(r) = U_0\left[\left(\frac{R_0}{r}\right)^{12} - 2\left(\frac{R_0}{r}\right)^6\right]$$

where r is the distance between the centers of the molecules and U_0 and R_0 are positive constants. The corresponding force $F(r)$ is given in Eq. (14.26). (a) Graph $U(r)$ and $F(r)$ versus r. (b) Let r_1 be the value of r at which $U(r) = 0$, and let r_2 be the value of r at which $F(r) = 0$. Show the locations of r_1 and r_2 on your graphs of $U(r)$ and $F(r)$. Which of these values represents the equilibrium separation between the molecules? (c) Find the values of r_1 and r_2 in terms of R_0, and find the ratio r_1/r_2. (d) If the molecules are located a distance r_2 apart [as calculated in part (c)], how much work must be done to pull them apart so that $r \rightarrow \infty$?

18.74 • (a) What is the total random translational kinetic energy of 5.00 L of hydrogen gas (molar mass 2.016 g/mol) with pressure 1.01×10^5 Pa and temperature 300 K? (*Hint:* Use the procedure of Problem 18.71 as a guide.) (b) If the tank containing the gas is placed on a swift jet moving at 300.0 m/s, by what percentage is the *total* kinetic energy of the gas increased? (c) Since the kinetic energy of the gas molecules is greater when it is on the jet, does this mean that its temperature has gone up? Explain.

18.75 • The speed of propagation of a sound wave in air at 27°C is about 350 m/s. Calculate, for comparison, (a) v_{rms} for nitrogen molecules and (b) the rms value of v_x at this temperature. The molar mass of nitrogen (N_2) is 28.0 g/mol.

18.76 • **Hydrogen on the Sun.** The surface of the sun has a temperature of about 5800 K and consists largely of hydrogen atoms. (a) Find the rms speed of a hydrogen atom at this temperature. (The mass of a single hydrogen atom is 1.67×10^{-27} kg.) (b) The escape speed for a particle to leave the gravitational influence of the sun is given by $(2GM/R)^{1/2}$, where M is the sun's mass, R its radius, and G the gravitational constant (see Example 13.5 of Section 13.3). Use the data in Appendix F to calculate this escape speed. (c) Can appreciable quantities of hydrogen escape from the sun? Can *any* hydrogen escape? Explain.

18.77 •• **CP** (a) Show that a projectile with mass m can "escape" from the surface of a planet if it is launched vertically upward with a kinetic energy greater than mgR_p, where g is the acceleration due to gravity at the planet's surface and R_p is the planet's radius. Ignore air resistance. (See Problem 18.76.) (b) If the planet in question is the earth, at what temperature does the average translational kinetic energy of a nitrogen molecule (molar mass 28.0 g/mol) equal that required to escape? What about a hydrogen molecule (molar mass 2.02 g/mol)? (c) Repeat part (b) for the moon, for which $g = 1.63$ m/s^2 and $R_p = 1740$ km. (d) While the earth and the moon have similar average surface temperatures, the moon has essentially no atmosphere. Use your results from parts (b) and (c) to explain why.

18.78 • **Planetary Atmospheres.** (a) The temperature near the top of Jupiter's multicolored cloud layer is about 140 K. The temperature at the top of the earth's troposphere, at an altitude of about

20 km, is about 220 K. Calculate the rms speed of hydrogen molecules in both these environments. Give your answers in m/s and as a fraction of the escape speed from the respective planet (see Problem 18.76). (b) Hydrogen gas (H_2) is a rare element in the earth's atmosphere. In the atmosphere of Jupiter, by contrast, 89% of all molecules are H_2. Explain why, using your results from part (a). (c) Suppose an astronomer claims to have discovered an oxygen (O_2) atmosphere on the asteroid Ceres. How likely is this? Ceres has a mass equal to 0.014 times the mass of the moon, a density of 2400 kg/m^3, and a surface temperature of about 200 K.

18.79 •• (a) For what mass of molecule or particle is v_{rms} equal to 1.00 mm/s at 300 K? (b) If the particle is an ice crystal, how many molecules does it contain? The molar mass of water is 18.0 g/mol. (c) Calculate the diameter of the particle if it is a spherical piece of ice. Would it be visible to the naked eye?

18.80 •• In describing the heat capacities of solids in Section 18.4, we stated that the potential energy $U = \frac{1}{2}kx^2$ of a harmonic oscillator averaged over one period of the motion is equal to the kinetic energy $K = \frac{1}{2}mv^2$ averaged over one period. Prove this result using Eqs. (14.13) and (14.15) for the position and velocity of a simple harmonic oscillator. For simplicity, assume that the initial position and velocity make the phase angle ϕ equal to zero. (*Hint:* Use the trigonometric identities $\cos^2(\theta) = [1 + \cos(2\theta)]/2$ and $\sin^2(\theta) = [1 - \cos(2\theta)]/2$. What is the average value of $\cos(2\omega t)$ over one period?)

18.81 •• It is possible to make crystalline solids that are only one layer of atoms thick. Such "two-dimensional" crystals can be created by depositing atoms on a very flat surface. (a) If the atoms in such a two-dimensional crystal can move only within the plane of the crystal, what will be its molar heat capacity near room temperature? Give your answer as a multiple of R and in J/mol · K. (b) At very low temperatures, will the molar heat capacity of a two-dimensional crystal be greater than, less than, or equal to the result you found in part (a)? Explain why.

18.82 •• (a) Calculate the total *rotational* kinetic energy of the molecules in 1.00 mol of a diatomic gas at 300 K. (b) Calculate the moment of inertia of an oxygen molecule (O_2) for rotation about either the y- or z-axis shown in Fig. 18.18b. Treat the molecule as two massive points (representing the oxygen atoms) separated by a distance of 1.21×10^{-10} m. The molar mass of oxygen *atoms* is 16.0 g/mol. (c) Find the rms angular velocity of rotation of an oxygen molecule about either the y- or z-axis shown in Fig. 18.18b. How does your answer compare to the angular velocity of a typical piece of rapidly rotating machinery (10,000 rev/min)?

18.83 • For each polyatomic gas in Table 18.1, compute the value of the molar heat capacity at constant volume, C_V, on the assumption that there is no vibrational energy. Compare with the measured values in the table, and compute the fraction of the total heat capacity that is due to vibration for each of the three gases. (*Note:* CO_2 is linear; SO_2 and H_2S are not. Recall that a linear polyatomic molecule has two rotational degrees of freedom, and a nonlinear molecule has three.)

18.84 •• **CALC** (a) Show that $\int_0^\infty f(v)\,dv = 1$, where $f(v)$ is the Maxwell–Boltzmann distribution of Eq. (18.32). (b) In terms of the physical definition of $f(v)$, explain why the integral in part (a) *must* have this value.

18.85 •• **CALC** Calculate the integral in Eq. (18.31), $\int_0^\infty v^2 f(v)\,dv$, and compare this result to $(v^2)_{av}$ as given by Eq. (18.16). (*Hint:* You may use the tabulated integral

$$\int_0^\infty x^{2n}e^{-\alpha x^2}\,dx = \frac{1 \cdot 3 \cdot 5 \cdots (2n-1)}{2^{n+1}\alpha^n}\sqrt{\frac{\pi}{\alpha}}$$

where n is a positive integer and α is a positive constant.)

18.86 •• **CALC** Calculate the integral in Eq. (18.30), $\int_0^\infty v f(v)\, dv$, and compare this result to v_{av} as given by Eq. (18.35). (*Hint:* Make the change of variable $v^2 = x$ and use the tabulated integral

$$\int_0^\infty x^n e^{-\alpha x}\, dx = \frac{n!}{\alpha^{n+1}}$$

where n is a positive integer and α is a positive constant.)

18.87 •• **CALC** (a) Explain why in a gas of N molecules, the number of molecules having speeds in the *finite* interval v to $v + \Delta v$ is $\Delta N = N \int_v^{v+\Delta v} f(v)\, dv$. (b) If Δv is small, then $f(v)$ is approximately constant over the interval and $\Delta N \approx N f(v) \Delta v$. For oxygen gas ($O_2$, molar mass 32.0 g/mol) at $T = 300$ K, use this approximation to calculate the number of molecules with speeds within $\Delta v = 20$ m/s of v_{mp}. Express your answer as a multiple of N. (c) Repeat part (b) for speeds within $\Delta v = 20$ m/s of $7 v_{mp}$. (d) Repeat parts (b) and (c) for a temperature of 600 K. (e) Repeat parts (b) and (c) for a temperature of 150 K. (f) What do your results tell you about the shape of the distribution as a function of temperature? Do your conclusions agree with what is shown in Fig. 18.23?

18.88 • **Meteorology.** The *vapor pressure* is the pressure of the vapor phase of a substance when it is in equilibrium with the solid or liquid phase of the substance. The *relative humidity* is the partial pressure of water vapor in the air divided by the vapor pressure of water at that same temperature, expressed as a percentage. The air is saturated when the humidity is 100%. (a) The vapor pressure of water at 20.0°C is 2.34×10^3 Pa. If the air temperature is 20.0°C and the relative humidity is 60%, what is the partial pressure of water vapor in the atmosphere (that is, the pressure due to water vapor alone)? (b) Under the conditions of part (a), what is the mass of water in 1.00 m³ of air? (The molar mass of water is 18.0 g/mol. Assume that water vapor can be treated as an ideal gas.)

18.89 • **The Dew Point.** The vapor pressure of water (see Problem 18.88) decreases as the temperature decreases. If the amount of water vapor in the air is kept constant as the air is cooled, a temperature is reached, called the *dew point,* at which the partial pressure and vapor pressure coincide and the vapor is saturated. If the air is cooled further, vapor condenses to liquid until the partial pressure again equals the vapor pressure at that temperature. The temperature in a room is 30.0°C. A meteorologist cools a metal can by gradually adding cold water. When the can temperature reaches 16.0°C, water droplets form on its outside surface. What is the relative humidity of the 30.0°C air in the room? The table lists the vapor pressure of water at various temperatures:

Temperature (°C)	Vapor Pressure (Pa)
10.0	1.23×10^3
12.0	1.40×10^3
14.0	1.60×10^3
16.0	1.81×10^3
18.0	2.06×10^3
20.0	2.34×10^3
22.0	2.65×10^3
24.0	2.99×10^3
26.0	3.36×10^3
28.0	3.78×10^3
30.0	4.25×10^3

18.90 ••• **Altitude at Which Clouds Form.** On a spring day in the midwestern United States, the air temperature at the surface is 28.0°C. Puffy cumulus clouds form at an altitude where the air temperature equals the dew point (see Problem 18.89). If the air temperature decreases with altitude at a rate of 0.6 C°/100 m, at approximately what height above the ground will clouds form if the relative humidity at the surface is 35% and 80%? (*Hint:* Use the table in Problem 18.89.)

CHALLENGE PROBLEMS

18.91 ••• **CP Dark Nebulae and the Interstellar Medium.** The dark area in Fig. P18.91 that appears devoid of stars is a *dark nebula,* a cold gas cloud in interstellar space that contains enough material to block out light from the stars behind it. A typical dark nebula is about 20 light-years in diameter and contains about 50 hydrogen atoms per cubic centimeter (monatomic hydrogen, *not* H_2) at a temperature of about 20 K. (A light-year is the distance light travels in vacuum in one year and is equal to 9.46×10^{15} m.) (a) Estimate the mean free path for a hydrogen atom in a dark nebula. The radius of a hydrogen atom is 5.0×10^{-11} m. (b) Estimate the rms speed of a hydrogen atom and the mean free time (the average time between collisions for a given atom). Based on this result, do you think that atomic collisions, such as those leading to H_2 molecule formation, are very important in determining the composition of the nebula? (c) Estimate the pressure inside a dark nebula. (d) Compare the rms speed of a hydrogen atom to the escape speed at the surface of the nebula (assumed spherical). If the space around the nebula were a vacuum, would such a cloud be stable or would it tend to evaporate? (e) The stability of dark nebulae is explained by the presence of the *interstellar medium* (ISM), an even thinner gas that permeates space and in which the dark nebulae are embedded. Show that for dark nebulae to be in equilibrium with the ISM, the numbers of atoms per volume (N/V) and the temperatures (T) of dark nebulae and the ISM must be related by

$$\frac{(N/V)_{\text{nebula}}}{(N/V)_{\text{ISM}}} = \frac{T_{\text{ISM}}}{T_{\text{nebula}}}$$

(f) In the vicinity of the sun, the ISM contains about 1 hydrogen atom per 200 cm³. Estimate the temperature of the ISM in the vicinity of the sun. Compare to the temperature of the sun's surface, about 5800 K. Would a spacecraft coasting through interstellar space burn up? Why or why not?

Figure **P18.91**

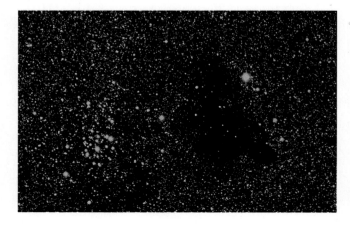

18.92 ••• **CALC Earth's Atmosphere.** In the *troposphere,* the part of the atmosphere that extends from earth's surface to an altitude of about 11 km, the temperature is not uniform but decreases with increasing elevation. (a) Show that if the temperature variation is approximated by the linear relationship

$$T = T_0 - \alpha y$$

where T_0 is the temperature at the earth's surface and T is the temperature at height y, the pressure p at height y is given by

$$\ln\left(\frac{p}{p_0}\right) = \frac{Mg}{R\alpha} \ln\left(\frac{T_0 - \alpha y}{T_0}\right)$$

where p_0 is the pressure at the earth's surface and M is the molar mass for air. The coefficient α is called the lapse rate of temperature. It varies with atmospheric conditions, but an average value is about 0.6 C°/100 m. (b) Show that the above result reduces to the result of Example 18.4 (Section 18.1) in the limit that $\alpha \to 0$. (c) With $\alpha = 0.6$ C°/100 m, calculate p for $y = 8863$ m and compare your answer to the result of Example 18.4. Take $T_0 = 288$ K and $p_0 = 1.00$ atm.

18.93 ••• In Example 18.7 (Section 18.3) we saw that $v_{rms} > v_{av}$. It is not difficult to show that this is *always* the case. (The only exception is when the particles have the same speed, in which case $v_{rms} = v_{av}$.) (a) For two particles with speeds v_1 and v_2, show that $v_{rms} \geq v_{av}$, regardless of the numerical values of v_1 and v_2. Then show that $v_{rms} > v_{av}$ if $v_1 \neq v_2$. (b) Suppose that for a collection of N particles you know that $v_{rms} > v_{av}$. Another particle, with speed u, is added to the collection of particles. If the new rms and average speeds are denoted as v'_{rms} and v'_{av}, show that

$$v'_{rms} = \sqrt{\frac{N v_{rms}^2 + u^2}{N+1}} \quad \text{and} \quad v'_{av} = \frac{N v_{av} + u}{N+1}$$

(c) Use the expressions in part (b) to show that $v'_{rms} > v'_{av}$ regardless of the numerical value of u. (d) Explain why your results for (a) and (c) together show that $v_{rms} > v_{av}$ for any collection of particles if the particles do not all have the same speed.

Answers

Chapter Opening Question

From Eq. (18.19), the root-mean-square speed of a gas molecule is proportional to the square root of the absolute temperature T. The temperature range we're considering is from $(25 + 273.15)$ K = 298 K to $(100 + 273.15)$ K = 373 K. Hence the speeds increase by a factor of $\sqrt{(373 \text{ K})/(298 \text{ K})} = 1.12$; that is, there is a 12% increase. While 100°C feels far warmer than 25°C, the difference in molecular speeds is relatively small.

Test Your Understanding Questions

18.1 Answer: (ii) and (iii) (tie), (i) and (v) (tie), (iv) We can rewrite the ideal-gas equation, Eq. (18.3), as $n = pV/RT$. This tells us that the number of moles n is proportional to the pressure and volume and inversely proportional to the absolute temperature. Hence, compared to (i), the number of moles in each case is (ii) $(2)(1)/(1) = 2$ times as much, (iii) $(1)(2)/(1) = 2$ times as much, (iv) $(1)(1)/(2) = \frac{1}{2}$ as much, and (v) $(2)(1)/(2) = 1$ time as much (that is, equal).

18.2 Answer: (vi) The value of r_0 determines the equilibrium separation of the molecules in the solid phase, so doubling r_0 means that the separation doubles as well. Hence a solid cube of this compound might grow from 1 cm on a side to 2 cm on a side. The volume would then be $2^3 = 8$ times larger, and the density (mass divided by volume) would be $\frac{1}{8}$ as great.

18.3 Answers: (a) (iv), (ii), (iii), (i); (b) (iii) and (iv) (tie), (i) and (ii) (tie) (a) Equation (18.19) tells us that $v_{rms} = \sqrt{3RT/M}$, so the rms speed is proportional to the square root of the ratio of absolute temperature T to molar mass M. Compared to (i) oxygen at 300 K, v_{rms} in the other cases is (ii) $\sqrt{(32.0 \text{ g/mol})/(28.0 \text{ g/mol})} = 1.07$ times faster, (iii) $\sqrt{(330 \text{ K})/(300 \text{ K})} = 1.05$ times faster, and (iv) $\sqrt{(330 \text{ K})(32.0 \text{ g/mol})/(300 \text{ K})(28.0 \text{ g/mol})} = 1.12$ times faster. (b) From Eq. (18.16), the average translational kinetic energy per molecule is $\frac{1}{2}m(v^2)_{av} = \frac{3}{2}kT$, which is directly proportional to T and independent of M. We have $T = 300$ K for cases (i) and (ii) and $T = 330$ K for cases (iii) and (iv), so $\frac{1}{2}m(v^2)_{av}$ has equal values for cases (iii) and (iv) and equal (but smaller) values for cases (i) and (ii).

18.4 Answers: no, near the beginning Adding a small amount of heat dQ to the gas changes the temperature by dT, where $dQ = nC_V \, dT$ from Eq. (18.24). Figure 18.19 shows that C_V for H_2 varies with temperature between 25 K and 500 K, so a given amount of heat gives rise to different amounts of temperature change during the process. Hence the temperature will *not* increase at a constant rate. The temperature change $dT = dQ/nC_V$ is inversely proportional to C_V, so the temperature increases most rapidly at the beginning of the process when the temperature is lowest and C_V is smallest (see Fig. 18.19).

18.5 Answer: (ii) Figure 18.23b shows that the *fraction* of molecules with speeds between v_1 and v_2 equals the area under the curve of $f(v)$ versus v from $v = v_1$ to $v = v_2$. This is equal to the integral $\int_{v_1}^{v_2} f(v) \, dv$, which in turn is equal to the difference between the integrals $\int_0^{v_2} f(v) \, dv$ (the fraction of molecules with speeds from 0 to v_2) and $\int_0^{v_1} f(v) \, dv$ (the fraction of molecules with speeds from 0 to the slower speed v_1). The *number* of molecules with speeds from v_1 to v_2 equals the fraction of molecules in this speed range multiplied by N, the total number of molecules.

18.6 Answers: no, yes The triple-point pressure of water from Table 18.3 is 6.10×10^2 Pa. The present-day pressure on Mars is just less than this value, corresponding to the line labeled p_s in Fig. 18.24. Hence liquid water cannot exist on the present-day Martian surface, and there are no rivers or lakes. Planetary scientists conclude that liquid water could have and almost certainly did exist on Mars in the past, when the atmosphere was thicker.

Bridging Problem

Answers: **(a)** 102 km **(b)** yes

19 THE FIRST LAW OF THERMODYNAMICS

? A steam locomotive operates using the first law of thermodynamics: Water is heated and boils, and the expanding steam does work to propel the locomotive. Would it be possible for the steam to propel the locomotive by doing work as it *condenses?*

Every time you drive a car, turn on an air conditioner, or cook a meal, you reap the practical benefits of *thermodynamics,* the study of relationships involving heat, mechanical work, and other aspects of energy and energy transfer. For example, in a car engine heat is generated by the chemical reaction of oxygen and vaporized gasoline in the engine's cylinders. The heated gas pushes on the pistons within the cylinders, doing mechanical work that is used to propel the car. This is an example of a *thermodynamic process.*

The first law of thermodynamics, central to the understanding of such processes, is an extension of the principle of conservation of energy. It broadens this principle to include energy exchange by both heat transfer and mechanical work and introduces the concept of the *internal energy* of a system. Conservation of energy plays a vital role in every area of physical science, and the first law has extremely broad usefulness. To state energy relationships precisely, we need the concept of a *thermodynamic system.* We'll discuss *heat* and *work* as two means of transferring energy into or out of such a system.

19.1 Thermodynamic Systems

We have studied energy transfer through mechanical work (Chapter 6) and through heat transfer (Chapters 17 and 18). Now we are ready to combine and generalize these principles.

We always talk about energy transfer to or from some specific *system.* The system might be a mechanical device, a biological organism, or a specified quantity of material, such as the refrigerant in an air conditioner or steam expanding in a turbine. In general, a **thermodynamic system** is any collection of objects that is convenient to regard as a unit, and that may have the potential to exchange energy with its surroundings. A familiar example is a quantity of popcorn kernels in a pot with a lid. When the pot is placed on a stove, energy is added to the popcorn

19.1 The popcorn in the pot is a thermodynamic system. In the thermodynamic process shown here, heat is added to the system, and the system does work on its surroundings to lift the lid of the pot.

by conduction of heat. As the popcorn pops and expands, it does work as it exerts an upward force on the lid and moves it through a displacement (Fig. 19.1). The *state* of the popcorn changes in this process, since the volume, temperature, and pressure of the popcorn all change as it pops. A process such as this one, in which there are changes in the state of a thermodynamic system, is called a **thermodynamic process.**

In mechanics we used the concept of *system* with free-body diagrams and with conservation of energy and momentum. For *thermodynamic* systems, as for all others, it is essential to define clearly at the start exactly what is and is not included in the system. Only then can we describe unambiguously the energy transfers into and out of that system. For instance, in our popcorn example we defined the system to include the popcorn but not the pot, lid, or stove.

Thermodynamics has its roots in many practical problems other than popping popcorn (Fig. 19.2). The gasoline engine in an automobile, the jet engines in an airplane, and the rocket engines in a launch vehicle use the heat of combustion of their fuel to perform mechanical work in propelling the vehicle. Muscle tissue in living organisms metabolizes chemical energy in food and performs mechanical work on the organism's surroundings. A steam engine or steam turbine uses the heat of combustion of coal or other fuel to perform mechanical work such as driving an electric generator or pulling a train.

Signs for Heat and Work in Thermodynamics

We describe the energy relationships in any thermodynamic process in terms of the quantity of heat Q added *to* the system and the work W done *by* the system. Both Q and W may be positive, negative, or zero (Fig. 19.3). A positive value of Q represents heat flow *into* the system, with a corresponding input of energy to it; negative Q represents heat flow *out of* the system. A positive value of W represents work done *by* the system against its surroundings, such as work done by an expanding gas, and hence corresponds to energy *leaving* the system. Negative W, such as work done during compression of a gas in which work is done *on the gas* by its surroundings, represents energy *entering* the system. We will use these conventions consistently in the examples in this chapter and the next.

CAUTION **Be careful with the sign of work** W Note that our sign rule for work is *opposite* to the one we used in mechanics, in which we always spoke of the work done by the forces acting *on* a body. In thermodynamics it is usually more convenient to call W the work done *by* the system so that when a system expands, the pressure, volume change, and work are all positive. Take care to use the sign rules for work and heat consistently!

Test Your Understanding of Section 19.1 In Example 17.8 (Section 17.6), what is the sign of Q for the coffee? For the aluminum cup? If a block slides along a horizontal surface with friction, what is the sign of W for the block? ∣

19.2 Work Done During Volume Changes

A simple but common example of a thermodynamic system is a quantity of gas enclosed in a cylinder with a movable piston. Internal-combustion engines, steam engines, and compressors in refrigerators and air conditioners all use some version of such a system. In the next several sections we will use the gas-in-cylinder system to explore several kinds of processes involving energy transformations.

We'll use a microscopic viewpoint, based on the kinetic and potential energies of individual molecules in a material, to develop intuition about thermodynamic quantities. But it is important to understand that the central principles of thermodynamics can be treated in a completely *macroscopic* way, without reference to microscopic models. Indeed, part of the great power and generality of thermodynamics is that it does *not* depend on details of the structure of matter.

19.2 (a) A rocket engine uses the heat of combustion of its fuel to do work propelling the launch vehicle. (b) Humans and other biological organisms are more complicated systems than we can analyze fully in this book, but the same basic principles of thermodynamics apply to them.

(a) (b)

19.3 A thermodynamic system may exchange energy with its surroundings (environment) by means of heat, work, or both. Note the sign conventions for Q and W.

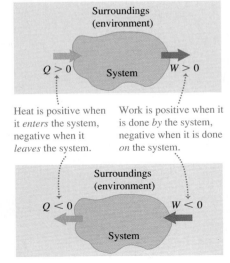

Heat is positive when it *enters* the system, negative when it *leaves* the system.

Work is positive when it is done *by* the system, negative when it is done *on* the system.

19.4 A molecule striking a piston (a) does positive work if the piston is moving away from the molecule and (b) does negative work if the piston is moving toward the molecule. Hence a gas does positive work when it expands as in (a) but does negative work when it compresses as in (b).

(a)

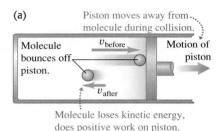

Piston moves away from molecule during collision.

Molecule bounces off piston.

v_{before}

Motion of piston

v_{after}

Molecule loses kinetic energy, does positive work on piston.

(b)

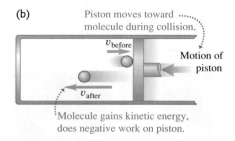

Piston moves toward molecule during collision.

v_{before}

Motion of piston

v_{after}

Molecule gains kinetic energy, does negative work on piston.

19.5 The infinitesimal work done by the system during the small expansion dx is $dW = pA\ dx$.

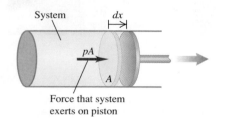

System

dx

pA

A

Force that system exerts on piston

First we consider the *work* done by the system during a volume change. When a gas expands, it pushes outward on its boundary surfaces as they move outward. Hence an expanding gas always does positive work. The same thing is true of any solid or fluid material that expands under pressure, such as the popcorn in Fig. 19.1.

We can understand the work done by a gas in a volume change by considering the molecules that make up the gas. When one such molecule collides with a stationary surface, it exerts a momentary force on the wall but does no work because the wall does not move. But if the surface is moving, like a piston in a gasoline engine, the molecule *does* do work on the surface during the collision. If the piston in Fig. 19.4a moves to the right, so that the volume of the gas increases, the molecules that strike the piston exert a force through a distance and do *positive* work on the piston. If the piston moves toward the left as in Fig. 19.4b, so the volume of the gas decreases, then positive work is done *on* the molecule during the collision. Hence the gas molecules do *negative* work on the piston.

Figure 19.5 shows a system whose volume can change (a gas, liquid, or solid) in a cylinder with a movable piston. Suppose that the cylinder has cross-sectional area A and that the pressure exerted by the system at the piston face is p. The total force F exerted by the system on the piston is $F = pA$. When the piston moves out an infinitesimal distance dx, the work dW done by this force is

$$dW = F\ dx = pA\ dx$$

But

$$A\ dx = dV$$

where dV is the infinitesimal change of volume of the system. Thus we can express the work done by the system in this infinitesimal volume change as

$$dW = p\ dV \qquad (19.1)$$

In a finite change of volume from V_1 to V_2,

$$W = \int_{V_1}^{V_2} p\ dV \qquad \text{(work done in a volume change)} \qquad (19.2)$$

In general, the pressure of the system may vary during the volume change. For example, this is the case in the cylinders of an automobile engine as the pistons move back and forth. To evaluate the integral in Eq. (19.2), we have to know how the pressure varies as a function of volume. We can represent this relationship as a graph of p as a function of V (a pV-diagram, described at the end of Section 18.1). Figure 19.6 a shows a simple example. In this figure, Eq. (19.2) is represented

19.6 The work done equals the area under the curve on a pV-diagram.

(a) pV-diagram for a system undergoing an expansion with varying pressure

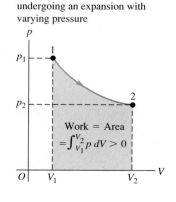

p

p_1 ⸱ 1

p_2 ⸱⸱⸱⸱⸱⸱⸱⸱ 2

Work = Area
$= \int_{V_1}^{V_2} p\ dV > 0$

O | V_1 | V_2 | V

(b) pV-diagram for a system undergoing a compression with varying pressure

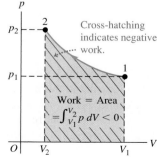

p

p_2 ⸱ 2

Cross-hatching indicates negative work.

p_1 ⸱⸱⸱⸱⸱⸱⸱⸱ 1

Work = Area
$= \int_{V_1}^{V_2} p\ dV < 0$

O | V_2 | V_1 | V

(c) pV-diagram for a system undergoing an expansion with constant pressure

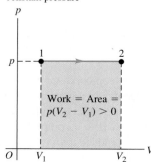

p

p ⸱ 1 ⟶ 2

Work = Area =
$p(V_2 - V_1) > 0$

O | V_1 | V_2 | V

graphically as the *area* under the curve of p versus V between the limits V_1 and V_2. (In Section 6.3 we used a similar interpretation of the work done by a force F as the area under the curve of F versus x between the limits x_1 and x_2.)

According to the rule we stated in Section 19.1, work is *positive* when a system *expands*. In an expansion from state 1 to state 2 in Fig. 19.6a, the area under the curve and the work are positive. A *compression* from 1 to 2 in Fig. 19.6b gives a *negative* area; when a system is compressed, its volume decreases and it does *negative* work on its surroundings (see also Fig. 19.4b).

CAUTION **Be careful with subscripts 1 and 2** When using Eq. (19.2), always remember that V_1 is the *initial* volume and V_2 is the *final* volume. That's why the labels 1 and 2 are reversed in Fig. 19.6b compared to Fig. 19.6a, even though both processes move between the same two thermodynamic states.

If the pressure p remains constant while the volume changes from V_1 to V_2 (Fig. 19.6c), the work done by the system is

$$W = p(V_2 - V_1) \quad \text{(work done in a volume change at constant pressure)} \quad (19.3)$$

MasteringPHYSICS

ActivPhysics 8.5: Work Done By a Gas

In any process in which the volume is *constant,* the system does no work because there is no displacement.

Example 19.1 **Isothermal expansion of an ideal gas**

As an ideal gas undergoes an *isothermal* (constant-temperature) expansion at temperature T, its volume changes from V_1 to V_2. How much work does the gas do?

SOLUTION

IDENTIFY and SET UP: The ideal-gas equation, Eq. (18.3), tells us that if the temperature T of n moles of an ideal gas is constant, the quantity $pV = nRT$ is also constant: p and V are inversely related. If V changes, p changes as well, so we *cannot* use Eq. (19.3) to calculate the work done. Instead we must use Eq. (19.2). To evaluate the integral in Eq. (19.2) we must know p as a function of V; for this we use Eq. (18.3).

EXECUTE: From Eq. (18.3),

$$p = \frac{nRT}{V}$$

We substitute this into the integral of Eq. (19.2), take the constant factor nRT outside, and evaluate the integral:

$$W = \int_{V_1}^{V_2} p \, dV$$

$$= nRT \int_{V_1}^{V_2} \frac{dV}{V} = nRT \ln \frac{V_2}{V_1} \quad \text{(ideal gas, isothermal process)}$$

We can rewrite this expression for W in terms of p_1 and p_2. Because $pV = nRT$ is constant,

$$p_1 V_1 = p_2 V_2 \quad \text{or} \quad \frac{V_2}{V_1} = \frac{p_1}{p_2}$$

so

$$W = nRT \ln \frac{p_1}{p_2} \quad \text{(ideal gas, isothermal process)}$$

EVALUATE: We check our result by noting that in an expansion $V_2 > V_1$ and the ratio V_2/V_1 is greater than 1. The logarithm of a number greater than 1 is positive, so $W > 0$, as it should be. As an additional check, look at our second expression for W: In an isothermal expansion the volume increases and the pressure drops, so $p_2 < p_1$, the ratio $p_1/p_2 > 1$, and $W = nRT \ln(p_1/p_2)$ is again positive.

These results also apply to an isothermal *compression* of a gas, for which $V_2 < V_1$ and $p_2 > p_1$.

Test Your Understanding of Section 19.2 A quantity of ideal gas undergoes an expansion that increases its volume from V_1 to $V_2 = 2V_1$. The final pressure of the gas is p_2. Does the gas do more work on its surroundings if the expansion is at constant *pressure* or at constant *temperature*? (i) constant pressure; (ii) constant temperature; (iii) the same amount of work is done in both cases; (iv) not enough information is given to decide.

MP

19.3 Paths Between Thermodynamic States

We've seen that if a thermodynamic process involves a change in volume, the system undergoing the process does work (either positive or negative) on its surroundings. Heat also flows into or out of the system during the process if there is a temperature difference between the system and its surroundings. Let's now examine how the work done by and the heat added to the system during a thermodynamic process depend on the details of how the process takes place.

Work Done in a Thermodynamic Process

When a thermodynamic system changes from an initial state to a final state, it passes through a series of intermediate states. We call this series of states a **path.** There are always infinitely many different possibilities for these intermediate states. When they are all equilibrium states, the path can be plotted on a pV-diagram (Fig. 19.7a). Point 1 represents an initial state with pressure p_1 and volume V_1, and point 2 represents a final state with pressure p_2 and volume V_2. To pass from state 1 to state 2, we could keep the pressure constant at p_1 while the system expands to volume V_2 (point 3 in Fig. 19.7b), then reduce the pressure to p_2 (probably by decreasing the temperature) while keeping the volume constant at V_2 (to point 2 on the diagram). The work done by the system during this process is the area under the line $1 \rightarrow 3$; no work is done during the constant-volume process $3 \rightarrow 2$. Or the system might traverse the path $1 \rightarrow 4 \rightarrow 2$ (Fig. 19.7c); in that case the work is the area under the line $4 \rightarrow 2$, since no work is done during the constant-volume process $1 \rightarrow 4$. The smooth curve from 1 to 2 is another possibility (Fig. 19.7d), and the work for this path is different from that for either of the other paths.

We conclude that *the work done by the system depends not only on the initial and final states, but also on the intermediate states—that is, on the path.* Furthermore, we can take the system through a series of states forming a closed loop, such as $1 \rightarrow 3 \rightarrow 2 \rightarrow 4 \rightarrow 1$. In this case the final state is the same as the initial state, but the total work done by the system is *not* zero. (In fact, it is represented on the graph by the area enclosed by the loop; can you prove that? See Exercise 19.7.) It follows that it doesn't make sense to talk about the amount of work *contained in* a system. In a particular state, a system may have definite values of the state coordinates p, V, and T, but it wouldn't make sense to say that it has a definite value of W.

Heat Added in a Thermodynamic Process

Like work, the *heat* added to a thermodynamic system when it undergoes a change of state depends on the path from the initial state to the final state. Here's an example. Suppose we want to change the volume of a certain quantity of an ideal gas from 2.0 L to 5.0 L while keeping the temperature constant at $T = 300$ K. Figure 19.8 shows two different ways in which we can do this. In Fig. 19.8a the gas is contained in a cylinder with a piston, with an initial volume of 2.0 L. We let the gas expand slowly, supplying heat from the electric heater to keep the temperature at 300 K. After expanding in this slow, controlled, isothermal manner, the gas reaches its final volume of 5.0 L; it absorbs a definite amount of heat in the process.

Figure 19.8b shows a different process leading to the same final state. The container is surrounded by insulating walls and is divided by a thin, breakable partition into two compartments. The lower part has volume 2.0 L and the upper part has volume 3.0 L. In the lower compartment we place the same amount of the same gas as in Fig. 19.8a, again at $T = 300$ K. The initial state is the same as before. Now we break the partition; the gas undergoes a rapid, uncontrolled expansion, with no heat passing through the insulating walls. The final volume is 5.0 L, the same as in Fig. 19.8a. The gas does no work during this expansion

19.7 The work done by a system during a transition between two states depends on the path chosen.

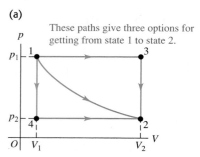

(a)

These paths give three options for getting from state 1 to state 2.

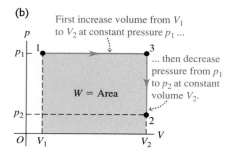

(b)

First increase volume from V_1 to V_2 at constant pressure p_1 ...

... then decrease pressure from p_1 to p_2 at constant volume V_2.

W = Area

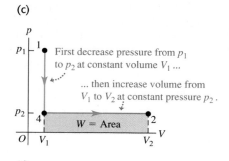

(c)

First decrease pressure from p_1 to p_2 at constant volume V_1 ...

... then increase volume from V_1 to V_2 at constant pressure p_2 .

W = Area

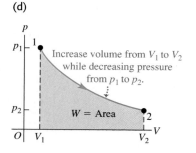

(d)

Increase volume from V_1 to V_2 while decreasing pressure from p_1 to p_2.

W = Area

because it doesn't push against anything that moves. This uncontrolled expansion of a gas into vacuum is called a **free expansion;** we will discuss it further in Section 19.6.

Experiments have shown that when an ideal gas undergoes a free expansion, there is no temperature change. Therefore the final state of the gas is the same as in Fig. 19.8a. The intermediate states (pressures and volumes) during the transition from state 1 to state 2 are entirely different in the two cases; Figs. 19.8a and 19.8b represent *two different paths* connecting the *same states* 1 and 2. For the path in Fig. 19.8b, *no heat is transferred into the system, and the system does no work.* Like work, *heat depends not only on the initial and final states but also on the path.*

Because of this path dependence, it would not make sense to say that a system "contains" a certain quantity of heat. To see this, suppose we assign an arbitrary value to the "heat in a body" in some reference state. Then presumably the "heat in the body" in some other state would equal the heat in the reference state plus the heat added when the body goes to the second state. But that's ambiguous, as we have just seen; the heat added depends on the *path* we take from the reference state to the second state. We are forced to conclude that there is *no* consistent way to define "heat in a body"; it is not a useful concept.

While it doesn't make sense to talk about "work in a body" or "heat in a body," it *does* make sense to speak of the amount of *internal energy* in a body. This important concept is our next topic.

Test Your Understanding of Section 19.3 The system described in Fig. 19.7a undergoes four different thermodynamic processes. Each process is represented in a *pV*-diagram as a straight line from the initial state to the final state. (These processes are different from those shown in the *pV*-diagrams of Fig. 19.7.) Rank the processes in order of the amount of work done by the system, from the most positive to the most negative. (i) 1 → 2; (ii) 2 → 1; (iii) 3 → 4; (iv) 4 → 3.

19.4 Internal Energy and the First Law of Thermodynamics

Internal energy is one of the most important concepts in thermodynamics. In Section 7.3, when we discussed energy changes for a body sliding with friction, we stated that warming a body increased its internal energy and that cooling the body decreased its internal energy. But what *is* internal energy? We can look at it in various ways; let's start with one based on the ideas of mechanics. Matter consists of atoms and molecules, and these are made up of particles having kinetic and potential energies. We *tentatively* define the **internal energy** of a system as the sum of the kinetic energies of all of its constituent particles, plus the sum of all the potential energies of interaction among these particles.

CAUTION **Is it internal?** Note that internal energy does *not* include potential energy arising from the interaction between the system and its surroundings. If the system is a glass of water, placing it on a high shelf increases the gravitational potential energy arising from the interaction between the glass and the earth. But this has no effect on the interaction between the molecules of the water, and so the internal energy of the water does not change.

We use the symbol U for internal energy. (We used this same symbol in our study of mechanics to represent potential energy. You may have to remind yourself occasionally that U has a different meaning in thermodynamics.) During a change of state of the system, the internal energy may change from an initial value U_1 to a final value U_2. We denote the change in internal energy as $\Delta U = U_2 - U_1$.

19.8 (a) Slow, controlled isothermal expansion of a gas from an initial state 1 to a final state 2 with the same temperature but lower pressure. (b) Rapid, uncontrolled expansion of the same gas starting at the same state 1 and ending at the same state 2.

(a) System does work on piston; hot plate adds heat to system ($W > 0$ and $Q > 0$).

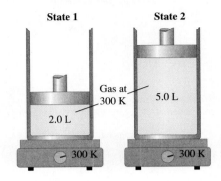

(b) System does no work; no heat enters or leaves system ($W = 0$ and $Q = 0$).

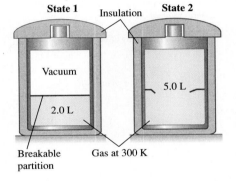

19.9 In a thermodynamic process, the internal energy U of a system may (a) increase ($\Delta U > 0$), (b) decrease ($\Delta U < 0$), or (c) remain the same ($\Delta U = 0$).

(a) More heat is added to system than system does work: Internal energy of system increases.

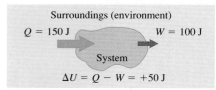

Surroundings (environment)

$Q = 150\ \text{J}$ $W = 100\ \text{J}$

System

$\Delta U = Q - W = +50\ \text{J}$

(b) More heat flows out of system than work is done: Internal energy of system decreases.

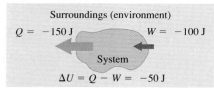

Surroundings (environment)

$Q = -150\ \text{J}$ $W = -100\ \text{J}$

System

$\Delta U = Q - W = -50\ \text{J}$

(c) Heat added to system equals work done by system: Internal energy of system unchanged.

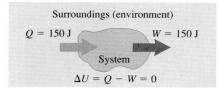

Surroundings (environment)

$Q = 150\ \text{J}$ $W = 150\ \text{J}$

System

$\Delta U = Q - W = 0$

Application The First Law of Exercise Thermodynamics
Your body is a thermodynamic system. When you exercise, your body does work (such as the work done to lift your body as a whole in a push-up). Hence $W > 0$. Your body also warms up during exercise; by perspiration and other means the body rids itself of this heat, so $Q < 0$. Since Q is negative and W is positive, $\Delta U = Q - W < 0$ and the body's internal energy decreases. That's why exercise helps you lose weight: It uses up some of the internal energy stored in your body in the form of fat.

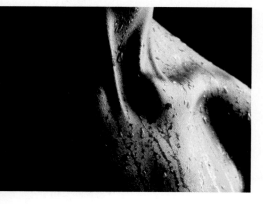

When we add a quantity of heat Q to a system and the system does no work during the process (so $W = 0$), the internal energy increases by an amount equal to Q; that is, $\Delta U = Q$. When a system does work W by expanding against its surroundings and no heat is added during the process, energy leaves the system and the internal energy decreases: W is positive, Q is zero, and $\Delta U = -W$. When *both* heat transfer and work occur, the *total* change in internal energy is

$$U_2 - U_1 = \Delta U = Q - W \quad \text{(first law of thermodynamics)} \quad (19.4)$$

We can rearrange this to the form

$$Q = \Delta U + W \quad (19.5)$$

The message of Eq. (19.5) is that in general, when heat Q is added to a system, some of this added energy remains within the system, changing its internal energy by an amount ΔU; the remainder leaves the system again as the system does work W against its surroundings. Because W and Q may be positive, negative, or zero, ΔU can be positive, negative, or zero for different processes (Fig. 19.9).

Equation (19.4) or (19.5) is the **first law of thermodynamics.** It is a generalization of the principle of conservation of energy to include energy transfer through heat as well as mechanical work. As you will see in later chapters, this principle can be extended to ever-broader classes of phenomena by identifying additional forms of energy and energy transfer. In every situation in which it seems that the total energy in all known forms is not conserved, it has been possible to identify a new form of energy such that the total energy, including the new form, *is* conserved. There is energy associated with electric fields, with magnetic fields, and, according to the theory of relativity, even with mass itself.

Understanding the First Law of Thermodynamics

At the beginning of this discussion we tentatively defined internal energy in terms of microscopic kinetic and potential energies. This has drawbacks, however. Actually *calculating* internal energy in this way for any real system would be hopelessly complicated. Furthermore, this definition isn't an *operational* one because it doesn't describe how to determine internal energy from physical quantities that we can measure directly.

So let's look at internal energy in another way. Starting over, we define the *change* in internal energy ΔU during any change of a system as the quantity given by Eq. (19.4), $\Delta U = Q - W$. This *is* an operational definition because we can measure Q and W. It does not define U itself, only ΔU. This is not a shortcoming because we can *define* the internal energy of a system to have a specified value in some reference state, and then use Eq. (19.4) to define the internal energy in any other state. This is analogous to our treatment of potential energy in Chapter 7, in which we arbitrarily defined the potential energy of a mechanical system to be zero at a certain position.

This new definition trades one difficulty for another. If we define ΔU by Eq. (19.4), then when the system goes from state 1 to state 2 by two different paths, how do we know that ΔU is the same for the two paths? We have already seen that Q and W are, in general, *not* the same for different paths. If ΔU, which equals $Q - W$, is also path dependent, then ΔU is ambiguous. If so, the concept of internal energy of a system is subject to the same criticism as the erroneous concept of quantity of heat in a system, as we discussed at the end of Section 19.3.

The only way to answer this question is through *experiment*. For various materials we measure Q and W for various changes of state and various paths to learn whether ΔU is or is not path dependent. The results of many such investigations are clear and unambiguous: While Q and W depend on the path, $\Delta U = Q - W$ *is independent of path. The change in internal energy of a system during any thermodynamic process depends only on the initial and final states, not on the path leading from one to the other.*

Experiment, then, is the ultimate justification for believing that a thermodynamic system in a specific state has a unique internal energy that depends only on that state. An equivalent statement is that the internal energy U of a system is a function of the state coordinates p, V, and T (actually, any two of these, since the three variables are related by the equation of state).

To say that the first law of thermodynamics, given by Eq. (19.4) or (19.5), represents conservation of energy for thermodynamic processes is correct, as far as it goes. But an important *additional* aspect of the first law is the fact that internal energy depends only on the state of a system (Fig. 19.10). In changes of state, the change in internal energy is independent of the path.

All this may seem a little abstract if you are satisfied to think of internal energy as microscopic mechanical energy. There's nothing wrong with that view, and we will make use of it at various times during our discussion. But in the interest of precise *operational* definitions, internal energy, like heat, can and must be defined in a way that is independent of the detailed microscopic structure of the material.

Cyclic Processes and Isolated Systems

Two special cases of the first law of thermodynamics are worth mentioning. A process that eventually returns a system to its initial state is called a *cyclic process.* For such a process, the final state is the same as the initial state, and so the *total* internal energy change must be zero. Then

$$U_2 = U_1 \quad \text{and} \quad Q = W$$

If a net quantity of work W is done by the system during this process, an equal amount of energy must have flowed into the system as heat Q. But there is no reason either Q or W individually has to be zero (Fig. 19.11).

Another special case occurs in an *isolated system,* one that does no work on its surroundings and has no heat flow to or from its surroundings. For any process taking place in an isolated system,

$$W = Q = 0$$

and therefore

$$U_2 = U_1 = \Delta U = 0$$

In other words, *the internal energy of an isolated system is constant.*

19.10 The internal energy of a cup of coffee depends on just its thermodynamic state—how much water and ground coffee it contains, and what its temperature is. It does not depend on the history of how the coffee was prepared—that is, the thermodynamic path that led to its current state.

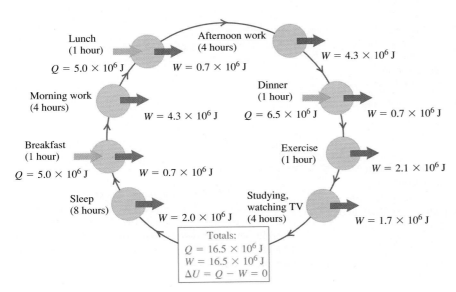

19.11 Every day, your body (a thermodynamic system) goes through a cyclic thermodynamic process like this one. Heat Q is added by metabolizing food, and your body does work W in breathing, walking, and other activities. If you return to the same state at the end of the day, $Q = W$ and the net change in your internal energy is zero.

Lunch (1 hour)
$Q = 5.0 \times 10^6$ J

Afternoon work (4 hours)
$W = 0.7 \times 10^6$ J
$W = 4.3 \times 10^6$ J

Morning work (4 hours)
$W = 4.3 \times 10^6$ J

Dinner (1 hour)
$Q = 6.5 \times 10^6$ J
$W = 0.7 \times 10^6$ J

Breakfast (1 hour)
$Q = 5.0 \times 10^6$ J
$W = 0.7 \times 10^6$ J

Exercise (1 hour)
$W = 2.1 \times 10^6$ J

Sleep (8 hours)
$W = 2.0 \times 10^6$ J

Studying, watching TV (4 hours)
$W = 1.7 \times 10^6$ J

Totals:
$Q = 16.5 \times 10^6$ J
$W = 16.5 \times 10^6$ J
$\Delta U = Q - W = 0$

Problem-Solving Strategy 19.1 · The First Law of Thermodynamics

IDENTIFY *the relevant concepts:* The first law of thermodynamics is the statement of the law of conservation of energy in its most general form. You can apply it to *any* thermodynamic process in which the internal energy of a system changes, heat flows into or out of the system, and/or work is done by or on the system.

SET UP *the problem* using the following steps:
1. Define the thermodynamic system to be considered.
2. If the thermodynamic process has more than one step, identify the initial and final states for each step.
3. List the known and unknown quantities and identify the target variables.
4. Confirm that you have enough equations. You can apply the first law, $\Delta U = Q - W$, just once to each step in a thermodynamic process, so you will often need additional equations. These may include Eq. (19.2), $W = \int_{V_1}^{V_2} p\, dV$, which gives the work W done in a volume change, and the equation of state of the material that makes up the thermodynamic system (for an ideal gas, $pV = nRT$).

EXECUTE *the solution* as follows:
1. Be sure to use consistent units. If p is in Pa and V in m³, then W is in joules. If a heat capacity is given in terms of calories,

convert it to joules. When you use $n = m_{total}/M$ to relate total mass m_{total} to number of moles n, remember that if m_{total} is in kilograms, M must be in *kilograms* per mole; M is usually tabulated in *grams* per mole.
2. The internal energy change ΔU in any thermodynamic process or series of processes is independent of the path, whether the substance is an ideal gas or not. If you can calculate ΔU for *any* path between given initial and final states, you know ΔU for *every possible path* between those states; you can then relate the various energy quantities for any of those other paths.
3. In a process comprising several steps, tabulate Q, W, and ΔU for each step, with one line per step and with the Q's, W's, and ΔU's forming columns (see Example 19.4). You can apply the first law to each line, and you can add each column and apply the first law to the sums. Do you see why?
4. Using steps 1–3, solve for the target variables.

EVALUATE *your answer:* Check your results for reasonableness. Ensure that each of your answers has the correct algebraic sign. A positive Q means that heat flows *into* the system; a negative Q means that heat flows *out of* the system. A positive W means that work is done *by* the system on its environment; a negative W means that work is done *on* the system by its environment.

Example 19.2 · Working off your dessert

You propose to climb several flights of stairs to work off the energy you took in by eating a 900-calorie hot fudge sundae. How high must you climb? Assume that your mass is 60.0 kg.

SOLUTION

IDENTIFY and SET UP: The thermodynamic system is your body. You climb the stairs to make the final state of the system the same as the initial state (no fatter, no leaner). There is therefore no net change in internal energy: $\Delta U = 0$. Eating the hot fudge sundae corresponds to a heat flow into your body, and you do work climbing the stairs. We can relate these quantities using the first law of thermodynamics. We are given that $Q = 900$ food calories (900 kcal) of heat flow into your body. The work you must do to raise your mass m a height h is $W = mgh$; our target variable is h.

EXECUTE: From the first law of thermodynamics, $\Delta U = 0 = Q - W$, so $W = mgh = Q$. Hence you must climb to height $h = Q/mg$. First convert units: $Q = (900 \text{ kcal})(4186 \text{ J}/1 \text{ kcal}) = 3.77 \times 10^6$ J. Then

$$h = \frac{Q}{mg} = \frac{3.77 \times 10^6 \text{ J}}{(60.0 \text{ kg})(9.80 \text{ m/s}^2)} = 6410 \text{ m}$$

EVALUATE: We have unrealistically assumed 100% efficiency in the conversion of food energy into mechanical work. You would in fact have to climb considerably *less* than 6140 m (about 21,000 ft).

Example 19.3 · A cyclic process

Figure 19.12 shows a *pV*-diagram for a *cyclic* process in which the initial and final states of some thermodynamic system are the same. As shown, the state of the system starts at point a and proceeds counterclockwise in the *pV*-diagram to point b, then back to a; the total work is $W = -500$ J. (a) Why is the work negative? (b) Find the change in internal energy and the heat added during this process.

SOLUTION

IDENTIFY and SET UP: We must relate the change in internal energy, the heat added, and the work done in a thermodynamic process. Hence we can apply the first law of thermodynamics. The process is cyclic, and it has two steps: $a \to b$ via the lower curve in Fig. 19.12 and $b \to a$ via the upper curve. We are asked only about the *entire* cyclic process $a \to b \to a$.

19.12 The net work done by the system in the process *aba* is -500 J. What would it have been if the process had proceeded clockwise in this pV-diagram?

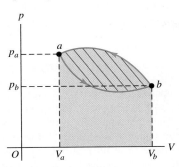

EXECUTE: (a) The work done in any step equals the area under the curve in the pV-diagram, with the area taken as positive if $V_2 > V_1$

and negative if $V_2 < V_1$; this rule yields the signs that result from the actual integrations in Eq. (19.2), $W = \int_{V_1}^{V_2} p \, dV$. The area under the lower curve $a \rightarrow b$ is therefore positive, but it is smaller than the absolute value of the (negative) area under the upper curve $b \rightarrow a$. Therefore the net area (the area enclosed by the path, shown with red stripes) and the net work W are negative. In other words, 500 J more work is done *on* the system than *by* the system in the complete process.

(b) In any cyclic process, $\Delta U = 0$, so $Q = W$. Here, that means $Q = -500$ J; that is, 500 J of heat flows *out of* the system.

EVALUATE:

In cyclic processes, the total work is positive if the process goes clockwise around the pV-diagram representing the cycle, and negative if the process goes counterclockwise (as here).

Example 19.4 | Comparing thermodynamic processes

The pV-diagram of Fig. 19.13 shows a series of thermodynamic processes. In process *ab*, 150 J of heat is added to the system; in process *bd*, 600 J of heat is added. Find (a) the internal energy change in process *ab*; (b) the internal energy change in process *abd* (shown in light blue); and (c) the total heat added in process *acd* (shown in dark blue).

SOLUTION

IDENTIFY and SET UP: In each process we use $\Delta U = Q - W$ to determine the desired quantity. We are given $Q_{ab} = +150$ J and $Q_{bd} = +600$ J (both values are positive because heat is *added* to the system). Our target variables are (a) ΔU_{ab}, (b) ΔU_{abd}, and (c) Q_{acd}.

EXECUTE: (a) No volume change occurs during process *ab*, so the system does no work: $W_{ab} = 0$ and so $\Delta U_{ab} = Q_{ab} = 150$ J.

(b) Process *bd* is an expansion at constant pressure, so from Eq. (19.3),

$$W_{bd} = p(V_2 - V_1)$$
$$= (8.0 \times 10^4 \text{ Pa})(5.0 \times 10^{-3} \text{ m}^3 - 2.0 \times 10^{-3} \text{ m}^3)$$
$$= 240 \text{ J}$$

The total work for the two-step process *abd* is then

$$W_{abd} = W_{ab} + W_{bd} = 0 + 240 \text{ J} = 240 \text{ J}$$

and the total heat is

$$Q_{abd} = Q_{ab} + Q_{bd} = 150 \text{ J} + 600 \text{ J} = 750 \text{ J}$$

Applying Eq. (19.4) to *abd*, we then have

$$\Delta U_{abd} = Q_{abd} - W_{abd} = 750 \text{ J} - 240 \text{ J} = 510 \text{ J}$$

(c) Because ΔU is *independent of the path* from *a* to *d*, the internal energy change is the same for path *acd* as for path *abd*:

$$\Delta U_{acd} = \Delta U_{abd} = 510 \text{ J}$$

19.13 A pV-diagram showing the various thermodynamic processes.

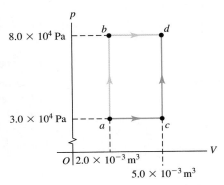

The total work for path *acd* is

$$W_{acd} = W_{ac} + W_{cd} = p(V_2 - V_1) + 0$$
$$= (3.0 \times 10^4 \text{ Pa})(5.0 \times 10^{-3} \text{ m}^3 - 2.0 \times 10^{-3} \text{ m}^3)$$
$$= 90 \text{ J}$$

Now we apply Eq. (19.5) to process *acd*:

$$Q_{acd} = \Delta U_{acd} + W_{acd} = 510 \text{ J} + 90 \text{ J} = 600 \text{ J}$$

We tabulate the quantities above:

Step	Q	W	$\Delta U = Q - W$	Step	Q	W	$\Delta U = Q - W$
ab	150 J	0 J	150 J	*ac*	?	90 J	?
bd	600 J	240 J	360 J	*cd*	?	0 J	?
abd	750 J	240 J	510 J	*acd*	600 J	90 J	510 J

EVALUATE: Be sure that you understand how each entry in the table above was determined. Although ΔU is the same (510 J) for *abd* and *acd*, W (240 J versus 90 J) and Q (750 J versus 600 J) are quite different. Although we couldn't find Q or ΔU for processes *ac* and *cd*, we could analyze the composite process *acd* by comparing it with process *abd*, which has the same initial and final states and for which we have more information.

Example 19.5 Thermodynamics of boiling water

One gram of water (1 cm^3) becomes 1671 cm^3 of steam when boiled at a constant pressure of 1 atm $(1.013 \times 10^5 \text{ Pa})$. The heat of vaporization at this pressure is $L_v = 2.256 \times 10^6 \text{ J/kg}$. Compute (a) the work done by the water when it vaporizes and (b) its increase in internal energy.

SOLUTION

IDENTIFY and SET UP: The heat added causes the system (water) to change phase from liquid to vapor. We can analyze this process using the first law of thermodynamics, which holds for thermodynamic processes of *all* kinds. The water is boiled at constant pressure, so we can use Eq. (19.3) to calculate the work W done by the vaporizing water as it expands. We are given the mass of water and the heat of vaporization, so we can use Eq. (17.20), $Q = mL_v$, to calculate the heat Q added to the water. We can then find the internal energy change using Eq. (19.4), $\Delta U = Q - W$.

EXECUTE: (a) From Eq. (19.3), the water does work

$$W = p(V_2 - V_1)$$
$$= (1.013 \times 10^5 \text{ Pa})(1671 \times 10^{-6} \text{ m}^3 - 1 \times 10^{-6} \text{ m}^3)$$
$$= 169 \text{ J}$$

(b) From Eq. (17.20), the heat added to the water is

$$Q = mL_v = (10^{-3} \text{ kg})(2.256 \times 10^6 \text{ J/kg}) = 2256 \text{ J}$$

Then from Eq. (19.4),

$$\Delta U = Q - W = 2256 \text{ J} - 169 \text{ J} = 2087 \text{ J}$$

EVALUATE: To vaporize 1 g of water, we must add 2256 J of heat, most of which (2087 J) remains in the system as an increase in internal energy. The remaining 169 J leaves the system as the system expands from liquid to vapor and does work against the surroundings. (The increase in internal energy is associated mostly with the attractive intermolecular forces that hold the molecules together in the liquid state. The associated potential energies are greater after work has been done to pull the molecules apart, forming the vapor state. It's like increasing gravitational potential energy by pulling an elevator farther from the center of the earth.)

Infinitesimal Changes of State

In the preceding examples the initial and final states differ by a finite amount. Later we will consider *infinitesimal* changes of state in which a small amount of heat dQ is added to the system, the system does a small amount of work dW, and its internal energy changes by an amount dU. For such a process we state the first law in differential form as

$$dU = dQ - dW \qquad \text{(first law of thermodynamics, infinitesimal process)} \qquad (19.6)$$

For the systems we will discuss, the work dW is given by $dW = p\,dV$, so we can also state the first law as

$$dU = dQ - p\,dV \qquad (19.7)$$

Test Your Understanding of Section 19.4 Rank the following thermodynamic processes according to the change in internal energy in each process, from most positive to most negative. (i) As you do 250 J of work on a system, it transfers 250 J of heat to its surroundings; (ii) as you do 250 J of work on a system, it absorbs 250 J of heat from its surroundings; (iii) as a system does 250 J of work on you, it transfers 250 J of heat to its surroundings; (iv) as a system does 250 J of work on you, it absorbs 250 J of heat from its surroundings. ❚

19.5 Kinds of Thermodynamic Processes

In this section we describe four specific kinds of thermodynamic processes that occur often in practical situations. These can be summarized briefly as "no heat transfer" or *adiabatic,* "constant volume" or *isochoric,* "constant pressure" or *isobaric,* and "constant temperature" or *isothermal.* For some of these processes we can use a simplified form of the first law of thermodynamics.

Adiabatic Process

An **adiabatic process** (pronounced "ay-dee-ah-*bat*-ic") is defined as one with no heat transfer into or out of a system; $Q = 0$. We can prevent heat flow either by surrounding the system with thermally insulating material or by carrying out the process so quickly that there is not enough time for appreciable heat flow. From the first law we find that for every adiabatic process,

$$U_2 - U_1 = \Delta U = -W \qquad \text{(adiabatic process)} \qquad (19.8)$$

When a system expands adiabatically, W is positive (the system does work on its surroundings), so ΔU is negative and the internal energy decreases. When a system is *compressed* adiabatically, W is negative (work is done on the system by its surroundings) and U increases. In many (but not all) systems an increase of internal energy is accompanied by a rise in temperature, and a decrease in internal energy by a drop in temperature (Fig. 19.14).

The compression stroke in an internal-combustion engine is an approximately adiabatic process. The temperature rises as the air–fuel mixture in the cylinder is compressed. The expansion of the burned fuel during the power stroke is also an approximately adiabatic expansion with a drop in temperature. In Section 19.8 we'll consider adiabatic processes in an ideal gas.

Isochoric Process

An **isochoric process** (pronounced "eye-so-*kor*-ic") is a *constant-volume* process. When the volume of a thermodynamic system is constant, it does no work on its surroundings. Then $W = 0$ and

$$U_2 - U_1 = \Delta U = Q \qquad \text{(isochoric process)} \qquad (19.9)$$

In an isochoric process, all the energy added as heat remains in the system as an increase in internal energy. Heating a gas in a closed constant-volume container is an example of an isochoric process. The processes *ab* and *cd* in Example 19.4 are also examples of isochoric processes. (Note that there are types of work that do not involve a volume change. For example, we can do work on a fluid by stirring it. In some literature, "isochoric" is used to mean that no work of any kind is done.)

Isobaric Process

An **isobaric process** (pronounced "eye-so-*bear*-ic") is a *constant-pressure* process. In general, none of the three quantities ΔU, Q, and W is zero in an isobaric process, but calculating W is easy nonetheless. From Eq. (19.3),

$$W = p(V_2 - V_1) \qquad \text{(isobaric process)} \qquad (19.10)$$

Example 19.5 concerns an isobaric process, boiling water at constant pressure (Fig. 19.15).

Isothermal Process

An **isothermal process** is a *constant-temperature* process. For a process to be isothermal, any heat flow into or out of the system must occur slowly enough that thermal equilibrium is maintained. In general, none of the quantities ΔU, Q, or W is zero in an isothermal process.

In some special cases the internal energy of a system depends *only* on its temperature, not on its pressure or volume. The most familiar system having this special property is an ideal gas, as we'll discuss in the next section. For such systems, if the temperature is constant, the internal energy is also constant; $\Delta U = 0$ and $Q = W$. That is, any energy entering the system as heat Q must leave it again as work W done by the system. Example 19.1, involving an ideal gas, is an example of an isothermal process in which U is also constant. For most systems other than ideal gases, the internal energy depends on pressure as well as temperature, so U may vary even when T is constant.

19.14 When the cork is popped on a bottle of champagne, the pressurized gases inside the bottle expand rapidly and do work on the outside air ($W > 0$). There is no time for the gases to exchange heat with their surroundings, so the expansion is adiabatic ($Q = 0$). Hence the internal energy of the expanding gases decreases ($\Delta U = -W < 0$) and their temperature drops. This makes water vapor condense and form a miniature cloud.

19.15 Most cooking involves isobaric processes. That's because the air pressure above a saucepan or frying pan, or inside a microwave oven, remains essentially constant while the food is being heated.

19.16 Four different processes for 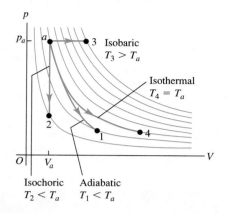 a constant amount of an ideal gas, all starting at state *a*. For the adiabatic process, $Q = 0$; for the isochoric process, $W = 0$; and for the isothermal process, $\Delta U = 0$. The temperature increases only during the isobaric expansion.

19.17 The partition is broken (or removed) to start the free expansion of gas into the vacuum region.

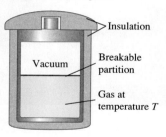

Figure 19.16 shows a *pV*-diagram for these four processes for a constant amount of an ideal gas. The path followed in an adiabatic process (*a* to 1) is called an **adiabat.** A vertical line (constant volume) is an **isochor,** a horizontal line (constant pressure) is an **isobar,** and a curve of constant temperature (shown as light blue lines in Fig. 19.16) is an **isotherm.**

Test Your Understanding of Section 19.5 Which of the processes in Fig. 19.7 are isochoric? Which are isobaric? Is it possible to tell if any of the processes are isothermal or adiabatic?

19.6 Internal Energy of an Ideal Gas

We now show that for an ideal gas, the internal energy *U* depends only on temperature, not on pressure or volume. Let's think again about the free-expansion experiment described in Section 19.3. A thermally insulated container with rigid walls is divided into two compartments by a partition (Fig. 19.17). One compartment has a quantity of an ideal gas and the other is evacuated.

When the partition is removed or broken, the gas expands to fill both parts of the container. The gas does no work on its surroundings because the walls of the container don't move, and there is no heat flow through the insulation. So both *Q* and *W* are zero and the internal energy *U* is constant. This is true of any substance, whether it is an ideal gas or not.

Does the *temperature* change during a free expansion? Suppose it *does* change, while the internal energy stays the same. In that case we have to conclude that the internal energy depends on both the temperature and the volume or on both the temperature and the pressure, but certainly not on the temperature alone. But if *T* is constant during a free expansion, for which we know that *U* is constant even though both *p* and *V* change, then we have to conclude that *U* depends only on *T*, not on *p* or *V*.

Many experiments have shown that when a low-density gas undergoes a free expansion, its temperature *does not* change. Such a gas is essentially an ideal gas. The conclusion is:

The internal energy of an ideal gas depends only on its temperature, not on its pressure or volume.

This property, in addition to the ideal-gas equation of state, is part of the ideal-gas model. Make sure you understand that *U* depends only on *T* for an ideal gas, for we will make frequent use of this fact.

For nonideal gases, some temperature change occurs during free expansions, even though the internal energy is constant. This shows that the internal energy cannot depend *only* on temperature; it must depend on pressure as well. From the microscopic viewpoint, in which internal energy *U* is the sum of the kinetic and potential energies for all the particles that make up the system, this is not surprising. Nonideal gases usually have attractive intermolecular forces, and when molecules move farther apart, the associated potential energies increase. If the total internal energy is constant, the kinetic energies must decrease. Temperature is directly related to molecular *kinetic* energy, and for such a gas a free expansion is usually accompanied by a *drop* in temperature.

Test Your Understanding of Section 19.6 Is the internal energy of a solid likely to be independent of its volume, as is the case for an ideal gas? Explain your reasoning. (*Hint:* See Fig. 18.20.)

19.7 Heat Capacities of an Ideal Gas

We defined specific heat and molar heat capacity in Section 17.5. We also remarked at the end of that section that the specific heat or molar heat capacity of a substance depends on the conditions under which the heat is added. It is usually easiest to measure the heat capacity of a gas in a closed container under constant-volume conditions. The corresponding heat capacity is the **molar heat capacity at constant volume,** denoted by C_V. Heat capacity measurements for solids and liquids are usually carried out in the atmosphere under constant atmospheric pressure, and we call the corresponding heat capacity the **molar heat capacity at constant pressure,** C_p. If neither p nor V is constant, we have an infinite number of possible heat capacities.

Let's consider C_V and C_p for an ideal gas. To measure C_V, we raise the temperature of an ideal gas in a rigid container with constant volume, neglecting its thermal expansion (Fig. 19.18a). To measure C_p, we let the gas expand just enough to keep the pressure constant as the temperature rises (Fig. 19.18b).

Why should these two molar heat capacities be different? The answer lies in the first law of thermodynamics. In a constant-volume temperature increase, the system does no work, and the change in internal energy ΔU equals the heat added Q. In a constant-pressure temperature increase, on the other hand, the volume *must* increase; otherwise, the pressure (given by the ideal-gas equation of state, $p = nRT/V$) could not remain constant. As the material expands, it does an amount of work W. According to the first law,

$$Q = \Delta U + W \qquad (19.11)$$

For a given temperature increase, the internal energy change ΔU of an ideal gas has the same value no matter what the process (remember that the internal energy of an ideal gas depends only on temperature, not on pressure or volume). Equation (19.11) then shows that the heat input for a constant-pressure process must be *greater* than that for a constant-volume process because additional energy must be supplied to account for the work W done during the expansion. So C_p is greater than C_V for an ideal gas. The pV-diagram in Fig. 19.19 shows this relationship. For air, C_p is 40% greater than C_V.

For a very few substances (one of which is water between 0°C and 4°C) the volume *decreases* during heating. In this case, W is negative, the heat input is *less* than in the constant-volume case, and C_p is *less* than C_V.

Relating C_p and C_V for an Ideal Gas

We can derive a simple relationship between C_p and C_V for an ideal gas. First consider the constant-*volume* process. We place n moles of an ideal gas at temperature T in a constant-volume container. We place it in thermal contact with a hotter body; an infinitesimal quantity of heat dQ flows into the gas, and its temperature increases by an infinitesimal amount dT. By the definition of C_V, the molar heat capacity at constant volume,

$$dQ = nC_V \, dT \qquad (19.12)$$

The pressure increases during this process, but the gas does no work ($dW = 0$) because the volume is constant. The first law in differential form, Eq. (19.6), is $dQ = dU + dW$. Since $dW = 0$, $dQ = dU$ and Eq. (19.12) can also be written as

$$dU = nC_V \, dT \qquad (19.13)$$

Now consider a constant-*pressure* process with the same temperature change dT. We place the same gas in a cylinder with a piston that we can allow to move just enough to maintain constant pressure, as shown in Fig. 19.18b. Again we bring the system into contact with a hotter body. As heat flows into

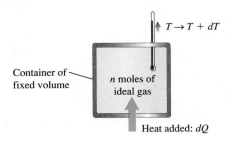

19.18 Measuring the molar heat capacity of an ideal gas **(a)** at constant volume and **(b)** at constant pressure.

(a) Constant volume: $dQ = nC_V \, dT$

(b) Constant pressure: $dQ = nC_p \, dT$

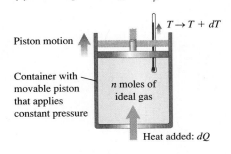

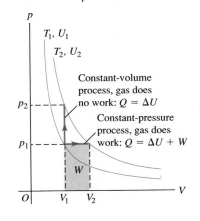

19.19 Raising the temperature of an ideal gas from T_1 to T_2 by a constant-volume or a constant-pressure process. For an ideal gas, U depends only on T, so ΔU is the same for both processes. But for the constant-pressure process, more heat Q must be added to both increase U and do work W. Hence $C_p > C_V$.

ActivPhysics 8.7: Heat Capacity
ActivPhysics 8.8: Isochoric Process
ActivPhysics 8.9: Isobaric Process
ActivPhysics 8.10: Isothermal Process

the gas, it expands at constant pressure and does work. By the definition of C_p, the molar heat capacity at constant pressure, the amount of heat dQ entering the gas is

$$dQ = nC_p \, dT \tag{19.14}$$

The work dW done by the gas in this constant-pressure process is

$$dW = p \, dV$$

We can also express dW in terms of the temperature change dT by using the ideal-gas equation of state, $pV = nRT$. Because p is constant, the change in V is proportional to the change in T:

$$dW = p \, dV = nR \, dT \tag{19.15}$$

Now we substitute Eqs. (19.14) and (19.15) into the first law, $dQ = dU + dW$. We obtain

$$nC_p \, dT = dU + nR \, dT \tag{19.16}$$

Now here comes the crux of the calculation. The internal energy change dU for the constant-pressure process is again given by Eq. (19.13), $dU = nC_V \, dT$, *even though now the volume is not constant.* Why is this so? Recall the discussion of Section 19.6; one of the special properties of an ideal gas is that its internal energy depends *only* on temperature. Thus the *change* in internal energy during any process must be determined only by the temperature change. If Eq. (19.13) is valid for an ideal gas for one particular kind of process, it must be valid for an ideal gas for *every* kind of process with the same dT. So we may replace dU in Eq. (19.16) by $nC_V \, dT$:

$$nC_p \, dT = nC_V \, dT + nR \, dT$$

When we divide each term by the common factor $n \, dT$, we get

$$C_p = C_V + R \qquad \text{(molar heat capacities of an ideal gas)} \tag{19.17}$$

As we predicted, the molar heat capacity of an ideal gas at constant pressure is *greater* than the molar heat capacity at constant volume; the difference is the gas constant R. (Of course, R must be expressed in the same units as C_p and C_V, such as J/mol · K.)

We have used the ideal-gas model to derive Eq. (19.17), but it turns out to be obeyed to within a few percent by many real gases at moderate pressures. Measured values of C_p and C_V are given in Table 19.1 for several real gases at low pressures; the difference in most cases is approximately $R = 8.314$ J/mol · K.

The table also shows that the molar heat capacity of a gas is related to its molecular structure, as we discussed in Section 18.4. In fact, the first two columns of Table 19.1 are the same as Table 18.1.

Table 19.1 Molar Heat Capacities of Gases at Low Pressure

Type of Gas	Gas	C_V (J/mol · K)	C_p (J/mol · K)	$C_p - C_V$ (J/mol · K)	$\gamma = C_p/C_V$
Monatomic	He	12.47	20.78	8.31	1.67
	Ar	12.47	20.78	8.31	1.67
Diatomic	H_2	20.42	28.74	8.32	1.41
	N_2	20.76	29.07	8.31	1.40
	O_2	20.85	29.17	8.31	1.40
	CO	20.85	29.16	8.31	1.40
Polyatomic	CO_2	28.46	36.94	8.48	1.30
	SO_2	31.39	40.37	8.98	1.29
	H_2S	25.95	34.60	8.65	1.33

The Ratio of Heat Capacities

The last column of Table 19.1 lists the values of the dimensionless **ratio of heat capacities,** C_p/C_V, denoted by γ (the Greek letter gamma):

$$\gamma = \frac{C_p}{C_V} \qquad \text{(ratio of heat capacities)} \qquad (19.18)$$

(This is sometimes called the "ratio of specific heats.") For gases, C_p is always greater than C_V and γ is always greater than unity. This quantity plays an important role in *adiabatic* processes for an ideal gas, which we will study in the next section.

We can use our kinetic-theory discussion of the molar heat capacity of an ideal gas (see Section 18.4) to predict values of γ. As an example, an ideal monatomic gas has $C_V = \frac{3}{2}R$. From Eq. (19.17),

$$C_p = C_V + R = \tfrac{3}{2}R + R = \tfrac{5}{2}R$$

so

$$\gamma = \frac{C_p}{C_V} = \frac{\frac{5}{2}R}{\frac{3}{2}R} = \tfrac{5}{3} = 1.67$$

As Table 19.1 shows, this agrees well with values of γ computed from measured heat capacities. For most diatomic gases near room temperature, $C_V = \frac{5}{2}R$, $C_p = C_V + R = \frac{7}{2}R$, and

$$\gamma = \frac{C_p}{C_V} = \frac{\frac{7}{2}R}{\frac{5}{2}R} = \tfrac{7}{5} = 1.40$$

also in good agreement with measured values.

Here's a final reminder: For an ideal gas the internal energy change in *any* process is given by $\Delta U = nC_V\,\Delta T$, *whether the volume is constant or not.* This relationship, which comes in handy in the following example, holds for other substances *only* when the volume is constant.

Example 19.6 | Cooling your room

A typical dorm room or bedroom contains about 2500 moles of air. Find the change in the internal energy of this much air when it is cooled from 35.0°C to 26.0°C at a constant pressure of 1.00 atm. Treat the air as an ideal gas with $\gamma = 1.400$.

SOLUTION

IDENTIFY and SET UP: Our target variable is the change in the internal energy ΔU of an ideal gas in a constant-pressure process. We are given the number of moles, the temperature change, and the value of γ for air. We use Eq. (19.13), $\Delta U = nC_V\,\Delta T$, which gives the internal energy change for an ideal gas in *any* process, *whether the volume is constant or not.* [See the discussion following Eq. (19.16).] We use Eqs. (19.17) and (19.18) to find C_V.

EXECUTE: From Eqs. (19.17) and (19.18),

$$\gamma = \frac{C_p}{C_V} = \frac{C_V + R}{C_V} = 1 + \frac{R}{C_V}$$

$$C_V = \frac{R}{\gamma - 1} = \frac{8.314 \text{ J/mol} \cdot \text{K}}{1.400 - 1} = 20.79 \text{ J/mol} \cdot \text{K}$$

Then from Eq. (19.13),

$$\Delta U = nC_V\,\Delta T$$
$$= (2500 \text{ mol})(20.79 \text{ J/mol} \cdot \text{K})(26.0°\text{C} - 35.0°\text{C})$$
$$= -4.68 \times 10^5 \text{ J}$$

EVALUATE: To cool 2500 moles of air from 35.0°C to 26.0°C, a room air conditioner must extract this much internal energy from the air and transfer it to the air outside. In Chapter 20 we'll discuss how this is done.

Test Your Understanding of Section 19.7 You want to cool a storage cylinder containing 10 moles of compressed gas from 30°C to 20°C. For which kind of gas would this be easiest? (i) a monatomic gas; (ii) a diatomic gas; (iii) a polyatomic gas; (iv) it would be equally easy for all of these.

19.8 Adiabatic Processes for an Ideal Gas

An adiabatic process, defined in Section 19.5, is a process in which no heat transfer takes place between a system and its surroundings. Zero heat transfer is an idealization, but a process is approximately adiabatic if the system is well insulated or if the process takes place so quickly that there is not enough time for appreciable heat flow to occur.

In an adiabatic process, $Q = 0$, so from the first law, $\Delta U = -W$. An adiabatic process for an ideal gas is shown in the pV-diagram of Fig. 19.20. As the gas expands from volume V_a to V_b, it does positive work, so its internal energy decreases and its temperature drops. If point a, representing the initial state, lies on an isotherm at temperature $T + dT$, then point b for the final state is on a different isotherm at a lower temperature T. For an ideal gas an adiabatic curve (adiabat) at any point is always *steeper* than the isotherm passing through the same point. For an adiabatic *compression* from V_b to V_a the situation is reversed and the temperature rises.

The air in the output hoses of air compressors used in gasoline stations, in paint-spraying equipment, and to fill scuba tanks is always warmer than the air entering the compressor; this is because the compression is rapid and hence approximately adiabatic. Adiabatic *cooling* occurs when you open a bottle of your favorite carbonated beverage. The gas just above the beverage surface expands rapidly in a nearly adiabatic process; the temperature of the gas drops so much that water vapor in the gas condenses, forming a miniature cloud (see Fig. 19.14).

CAUTION *"Heating" and "cooling" without heat* Keep in mind that when we talk about "adiabatic heating" and "adiabatic cooling," we really mean "raising the temperature" and "lowering the temperature," respectively. In an adiabatic process, the temperature change is due to work done by or on the system; there is *no* heat flow at all.

Adiabatic Ideal Gas: Relating V, T, and p

We can derive a relationship between volume and temperature changes for an infinitesimal adiabatic process in an ideal gas. Equation (19.13) gives the internal energy change dU for *any* process for an ideal gas, adiabatic or not, so we have $dU = nC_V\,dT$. Also, the work done by the gas during the process is given by $dW = p\,dV$. Then, since $dU = -dW$ for an adiabatic process, we have

$$nC_V\,dT = -p\,dV \qquad (19.19)$$

To obtain a relationship containing only the volume V and temperature T, we eliminate p using the ideal-gas equation in the form $p = nRT/V$. Substituting this into Eq. (19.19) and rearranging, we get

$$nC_V\,dT = -\frac{nRT}{V}\,dV$$

$$\frac{dT}{T} + \frac{R}{C_V}\frac{dV}{V} = 0$$

The coefficient R/C_V can be expressed in terms of $\gamma = C_p/C_V$. We have

$$\frac{R}{C_V} = \frac{C_p - C_V}{C_V} = \frac{C_p}{C_V} - 1 = \gamma - 1$$

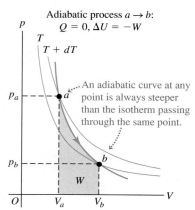

19.20 A pV-diagram of an adiabatic ($Q = 0$) process for an ideal gas. As the gas expands from V_a to V_b, it does positive work W on its environment, its internal energy decreases ($\Delta U = -W < 0$), and its temperature drops from $T + dT$ to T. (An adiabatic process is also shown in Fig. 19.16.)

Adiabatic process $a \rightarrow b$:
$Q = 0$, $\Delta U = -W$

An adiabatic curve at any point is always steeper than the isotherm passing through the same point.

Mastering**PHYSICS**

ActivPhysics 8.11: Adiabatic Process

$$\frac{dT}{T} + (\gamma - 1)\frac{dV}{V} = 0 \qquad (19.20)$$

Because γ is always greater than unity for a gas, $(\gamma - 1)$ is always positive. This means that in Eq. (19.20), dV and dT always have opposite signs. An adiabatic *expansion* of an ideal gas $(dV > 0)$ always occurs with a *drop* in temperature $(dT < 0)$, and an adiabatic *compression* $(dV < 0)$ always occurs with a *rise* in temperature $(dT > 0)$; this confirms our earlier prediction.

For finite changes in temperature and volume we integrate Eq. (19.20), obtaining

$$\ln T + (\gamma - 1)\ln V = \text{constant}$$

$$\ln T + \ln V^{\gamma-1} = \text{constant}$$

$$\ln (TV^{\gamma-1}) = \text{constant}$$

and finally,

$$TV^{\gamma-1} = \text{constant} \qquad (19.21)$$

Thus for an initial state (T_1, V_1) and a final state (T_2, V_2),

$$T_1 V_1^{\gamma-1} = T_2 V_2^{\gamma-1} \quad \text{(adiabatic process, ideal gas)} \qquad (19.22)$$

Because we have used the ideal-gas equation in our derivation of Eqs. (19.21) and (19.22), the T's must always be *absolute* (Kelvin) temperatures.

We can also convert Eq. (19.21) into a relationship between pressure and volume by eliminating T, using the ideal-gas equation in the form $T = pV/nR$. Substituting this into Eq. (19.21), we find

$$\frac{pV}{nR}V^{\gamma-1} = \text{constant}$$

or, because n and R are constant,

$$pV^\gamma = \text{constant} \qquad (19.23)$$

For an initial state (p_1, V_1) and a final state (p_2, V_2), Eq. (19.23) becomes

$$p_1 V_1^\gamma = p_2 V_2^\gamma \quad \text{(adiabatic process, ideal gas)} \qquad (19.24)$$

We can also calculate the *work* done by an ideal gas during an adiabatic process. We know that $Q = 0$ and $W = -\Delta U$ for *any* adiabatic process. For an ideal gas, $\Delta U = nC_V(T_2 - T_1)$. If the number of moles n and the initial and final temperatures T_1 and T_2 are known, we have simply

$$W = nC_V(T_1 - T_2) \quad \text{(adiabatic process, ideal gas)} \qquad (19.25)$$

We may also use $pV = nRT$ in this equation to obtain

$$W = \frac{C_V}{R}(p_1 V_1 - p_2 V_2) = \frac{1}{\gamma - 1}(p_1 V_1 - p_2 V_2) \quad \begin{array}{l}\text{(adiabatic process,} \\ \text{ideal gas)}\end{array} \qquad (19.26)$$

(We used the result $C_V = R/(\gamma - 1)$ from Example 19.6.) If the process is an expansion, the temperature drops, T_1 is greater than T_2, $p_1 V_1$ is greater than $p_2 V_2$, and the work is *positive,* as we should expect. If the process is a compression, the work is negative.

Throughout this analysis of adiabatic processes we have used the ideal-gas equation of state, which is valid only for *equilibrium* states. Strictly speaking, our results are valid only for a process that is fast enough to prevent appreciable heat exchange with the surroundings (so that $Q = 0$ and the process is adiabatic), yet slow enough that the system does not depart very much from thermal and mechanical equilibrium. Even when these conditions are not strictly satisfied, though, Eqs. (19.22), (19.24), and (19.26) give useful approximate results.

Example 19.7 **Adiabatic compression in a diesel engine**

The compression ratio of a diesel engine is 15.0 to 1; that is, air in a cylinder is compressed to $\frac{1}{(15.0)}$ of its initial volume (Fig. 19.21). (a) If the initial pressure is 1.01×10^5 Pa and the initial temperature is $27°C$ (300 K), find the final pressure and the temperature after adiabatic compression. (b) How much work does the gas do during the compression if the initial volume of the cylinder is $1.00 \text{ L} = 1.00 \times 10^{-3} \text{ m}^3$? Use the values $C_V = 20.8 \text{ J/mol} \cdot \text{K}$ and $\gamma = 1.400$ for air.

SOLUTION

IDENTIFY and SET UP: This problem involves the adiabatic compression of an ideal gas, so we can use the ideas of this section. In part (a) we are given the initial pressure and temperature $p_1 = 1.01 \times 10^5$ Pa and $T_1 = 300$ K; the ratio of initial and final volumes is $V_1/V_2 = 15.0$. We use Eq. (19.22) to find the final temperature T_2 and Eq. (19.24) to find the final pressure p_2. In part (b) our target variable is W, the work done *by* the gas during the adiabatic compression. We use Eq. (19.26) to calculate W.

19.21 Adiabatic compression of air in a cylinder of a diesel engine.

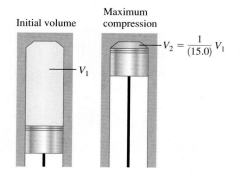

$$V_2 = \frac{1}{(15.0)} V_1$$

EXECUTE: (a) From Eqs. (19.22) and (19.24),

$$T_2 = T_1\left(\frac{V_1}{V_2}\right)^{\gamma-1} = (300 \text{ K})(15.0)^{0.40} = 886 \text{ K} = 613°C$$

$$p_2 = p_1\left(\frac{V_1}{V_2}\right)^{\gamma} = (1.01 \times 10^5 \text{ Pa})(15.0)^{1.40}$$

$$= 44.8 \times 10^5 \text{ Pa} = 44 \text{ atm}$$

(b) From Eq. (19.26), the work done is

$$W = \frac{1}{\gamma - 1}(p_1V_1 - p_2V_2)$$

Using $V_1/V_2 = 15.0$, this becomes

$$W = \frac{1}{1.400 - 1}\left[(1.01 \times 10^5 \text{ Pa})(1.00 \times 10^{-3} \text{ m}^3) - (44.8 \times 10^5 \text{ Pa})\left(\frac{1.00 \times 10^{-3} \text{ m}^3}{15.0}\right) \right]$$

$$= -494 \text{ J}$$

EVALUATE: If the compression had been isothermal, the final pressure would have been 15.0 atm. Because the temperature also increases during an adiabatic compression, the final pressure is much greater. When fuel is injected into the cylinders near the end of the compression stroke, the high temperature of the air attained during compression causes the fuel to ignite spontaneously without the need for spark plugs.

We can check our result in part (b) using Eq. (19.25). The number of moles of gas in the cylinder is

$$n = \frac{p_1V_1}{RT_1} = \frac{(1.01 \times 10^5 \text{ Pa})(1.00 \times 10^{-3} \text{ m}^3)}{(8.314 \text{ J/mol} \cdot \text{K})(300 \text{ K})} = 0.0405 \text{ mol}$$

Then Eq. (19.25) gives

$$W = nC_V(T_1 - T_2)$$

$$= (0.0405 \text{ mol})(20.8 \text{ J/mol} \cdot \text{K})(300 \text{ K} - 886 \text{ K})$$

$$= -494 \text{ J}$$

The work is negative because the gas is compressed.

Test Your Understanding of Section 19.8 You have four samples of ideal gas, each of which contains the same number of moles of gas and has the same initial temperature, volume, and pressure. You compress each sample to one-half of its initial volume. Rank the four samples in order from highest to lowest value of the final pressure. (i) a monatomic gas compressed isothermally; (ii) a monatomic gas compressed adiabatically; (iii) a diatomic gas compressed isothermally; (iv) a diatomic gas compressed adiabatically.

Heat and work in thermodynamic processes: A thermodynamic system has the potential to exchange energy with its surroundings by heat transfer or by mechanical work. When a system at pressure p changes volume from V_1 to V_2, it does an amount of work W given by the integral of p with respect to volume. If the pressure is constant, the work done is equal to p times the change in volume. A negative value of W means that work is done on the system. (See Example 19.1.)

In any thermodynamic process, the heat added to the system and the work done by the system depend not only on the initial and final states, but also on the path (the series of intermediate states through which the system passes).

$$W = \int_{V_1}^{V_2} p\, dV \qquad (19.2)$$

$$W = p(V_2 - V_1) \qquad (19.3)$$
(constant pressure only)

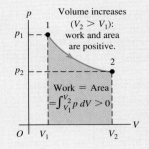

The first law of thermodynamics: The first law of thermodynamics states that when heat Q is added to a system while the system does work W, the internal energy U changes by an amount equal to $Q - W$. This law can also be expressed for an infinitesimal process. (See Examples 19.2, 19.3, and 19.5.)

The internal energy of any thermodynamic system depends only on its state. The change in internal energy in any process depends only on the initial and final states, not on the path. The internal energy of an isolated system is constant. (See Example 19.4.)

$$\Delta U = Q - W \qquad (19.4)$$

$$dU = dQ - dW \qquad (19.6)$$
(infinitesimal process)

Surroundings
(environment)

$Q = 150 \text{ J}$ $W = 100 \text{ J}$

System

$\Delta U = Q - W = + 50 \text{ J}$

Important kinds of thermodynamic processes:

- Adiabatic process: No heat transfer into or out of a system; $Q = 0$.
- Isochoric process: Constant volume; $W = 0$.
- Isobaric process: Constant pressure; $W = p(V_2 - V_1)$.
- Isothermal process: Constant temperature.

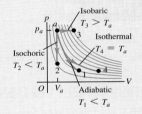

Thermodynamics of ideal gases: The internal energy of an ideal gas depends only on its temperature, not on its pressure or volume. For other substances the internal energy generally depends on both pressure and temperature.

The molar heat capacities C_V and C_p of an ideal gas differ by R, the ideal-gas constant. The dimensionless ratio of heat capacities, C_p/C_V, is denoted by γ. (See Example 19.6.)

$$C_p = C_V + R \qquad (19.17)$$

$$\gamma = \frac{C_p}{C_V} \qquad (19.18)$$

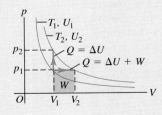

Adiabatic processes in ideal gases: For an adiabatic process for an ideal gas, the quantities $TV^{\gamma-1}$ and pV^{γ} are constant. The work done by an ideal gas during an adiabatic expansion can be expressed in terms of the initial and final values of temperature, or in terms of the initial and final values of pressure and volume. (See Example 19.7.)

$$W = nC_V(T_1 - T_2)$$

$$= \frac{C_V}{R}(p_1 V_1 - p_2 V_2) \qquad (19.25)$$

$$= \frac{1}{\gamma - 1}(p_1 V_1 - p_2 V_2) \qquad (19.26)$$

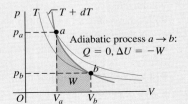

The van der Waals equation of state, an approximate representation of the behavior of gases at high pressure, is given by Eq. (18.7): $[p + (an^2/V^2)](V - nb) = nRT$, where a and b are constants having different values for different gases. (In the special case of $a = b = 0$, this is the ideal-gas equation.) (a) Calculate the work done by a gas with this equation of state in an isothermal expansion from V_1 to V_2. (b) For ethane gas (C_2H_6), $a = 0.554$ J·m^3/mol^2 and $b = 6.38 \times 10^{-5}$ m^3/mol. Calculate the work W done by 1.80 mol of ethane when it expands from 2.00×10^{-3} m^3 to 4.00×10^{-3} m^3 at a constant temperature of 300 K. Do the calculation using (i) the van der Waals equation of state and (ii) the ideal-gas equation of state. (c) For which equation of state is W larger? Why should this be so?

SOLUTION GUIDE

See MasteringPhysics® study area for a Video Tutor solution.

IDENTIFY and SET UP

1. Review the discussion of the van der Waals equation of state in Section 18.1. What is the significance of the quantities a and b?
2. Decide how to find the work done by an expanding gas whose pressure p does not depend on V in the same way as for an ideal gas. (*Hint:* See Section 19.2.)
3. How will you find the work done by an expanding ideal gas?

EXECUTE

4. Find the general expression for the work done by a van der Waals gas as it expands from volume V_1 to volume V_2. (*Hint:* If you set $a = b = 0$ in your result, it should reduce to the expression for the work done by an expanding ideal gas.)
5. Use your result from step 4 to solve part (b) for ethane treated as a van der Waals gas.
6. Use the formula you chose in step 3 to solve part (b) for ethane treated as an ideal gas.

EVALUATE

7. Is the difference between W for the two equations of state large enough to be significant?
8. Does the term with a in the van der Waals equation of state increase or decrease the amount of work done? What about the term with b? Which one is more important for the ethane in this problem?

Problems

For instructor-assigned homework, go to www.masteringphysics.com

•, ••, •••: Problems of increasing difficulty. **CP**: Cumulative problems incorporating material from earlier chapters. **CALC**: Problems requiring calculus. **BIO**: Biosciences problems.

DISCUSSION QUESTIONS

Q19.1 For the following processes, is the work done by the system (defined as the expanding or contracting gas) on the environment positive or negative? (a) expansion of the burned gasoline–air mixture in the cylinder of an automobile engine; (b) opening a bottle of champagne; (c) filling a scuba tank with compressed air; (d) partial crumpling of a sealed, empty water bottle, as you drive from the mountains down to sea level.

Q19.2 It is not correct to say that a body contains a certain amount of heat, yet a body can transfer heat to another body. How can a body give away something it does not have in the first place?

Q19.3 In which situation must you do more work: inflating a balloon at sea level or inflating the same balloon to the same volume at the summit of Mt. McKinley? Explain in terms of pressure and volume change.

Q19.4 If you are told the initial and final states of a system and the associated change in internal energy, can you determine whether the internal energy change was due to work or to heat transfer? Explain.

Q19.5 Discuss the application of the first law of thermodynamics to a mountaineer who eats food, gets warm and perspires a lot during a climb, and does a lot of mechanical work in raising herself to the summit. The mountaineer also gets warm during the descent. Is the source of this energy the same as the source during the ascent?

Q19.6 When ice melts at 0°C, its volume decreases. Is the internal energy change greater than, less than, or equal to the heat added? How can you tell?

Q19.7 You hold an inflated balloon over a hot-air vent in your house and watch it slowly expand. You then remove it and let it cool back to room temperature. During the expansion, which was larger: the heat added to the balloon or the work done by the air inside it? Explain. (Assume that air is an ideal gas.) Once the balloon has returned to room temperature, how does the net heat gained or lost by the air inside it compare to the net work done on or by the surrounding air?

Q19.8 You bake chocolate chip cookies and put them, still warm, in a container with a loose (not airtight) lid. What kind of process does the air inside the container undergo as the cookies gradually cool to room temperature (isothermal, isochoric, adiabatic, isobaric, or some combination)? Explain your answer.

Q19.9 Imagine a gas made up entirely of negatively charged electrons. Like charges repel, so the electrons exert repulsive forces on each other. Would you expect that the temperature of such a gas would rise, fall, or stay the same in a free expansion? Why?

Q19.10 There are a few materials that contract when their temperature is increased, such as water between 0°C and 4°C. Would you expect C_p for such materials to be greater or less than C_V? Explain?

Q19.11 When you blow on the back of your hand with your mouth wide open, your breath feels warm. But if you partially close your mouth to form an "o" and then blow on your hand, your breath feels cool. Why?

Q19.12 An ideal gas expands while the pressure is kept constant. During this process, does heat flow into the gas or out of the gas? Justify your answer.

Q19.13 A liquid is irregularly stirred in a well-insulated container and thereby undergoes a rise in temperature. Regard the liquid as the system. Has heat been transferred? How can you tell? Has work been done? How can you tell? Why is it important that the stirring is irregular? What is the sign of ΔU? How can you tell?

Q19.14 When you use a hand pump to inflate the tires of your bicycle, the pump gets warm after a while. Why? What happens to the temperature of the air in the pump as you compress it? Why does this happen? When you raise the pump handle to draw outside air into the pump, what happens to the temperature of the air taken in? Again, why does this happen?

Q19.15 In the carburetor of an aircraft or automobile engine, air flows through a relatively small aperture and then expands. In cool, foggy weather, ice sometimes forms in this aperture even though the outside air temperature is above freezing. Why?

Q19.16 On a sunny day, large "bubbles" of air form on the sun-warmed earth, gradually expand, and finally break free to rise through the atmosphere. Soaring birds and glider pilots are fond of using these "thermals" to gain altitude easily. This expansion is essentially an adiabatic process. Why?

Q19.17 The prevailing winds on the Hawaiian island of Kauai blow from the northeast. The winds cool as they go up the slope of Mt. Waialeale (elevation 1523 m), causing water vapor to condense and rain to fall. There is much more precipitation at the summit than at the base of the mountain. In fact, Mt. Waialeale is the rainiest spot on earth, averaging 11.7 m of rainfall a year. But what makes the winds cool?

Q19.18 Applying the same considerations as in Question Q19.17, explain why the island of Niihau, a few kilometers to the southwest of Kauai, is almost a desert and farms there need to be irrigated.

Q19.19 In a constant-volume process, $dU = nC_V\,dT$. But in a constant-pressure process, it is *not* true that $dU = nC_p\,dT$. Why not?

Q19.20 When a gas surrounded by air is compressed adiabatically, its temperature rises even though there is no heat input to the gas. Where does the energy come from to raise the temperature?

Q19.21 When a gas expands adiabatically, it does work on its surroundings. But if there is no heat input to the gas, where does the energy come from to do the work?

Q19.22 The gas used in separating the two uranium isotopes ^{235}U and ^{238}U has the formula UF_6. If you added heat at equal rates to a mole of UF_6 gas and a mole of H_2 gas, which one's temperature would you expect to rise faster? Explain.

EXERCISES

Section 19.2 Work Done During Volume Changes and Section 19.3 Paths Between Thermodynamic States

19.1 •• Two moles of an ideal gas are heated at constant pressure from $T = 27°C$ to $T = 107°C$. (a) Draw a pV-diagram for this process. (b) Calculate the work done by the gas.

19.2 • Six moles of an ideal gas are in a cylinder fitted at one end with a movable piston. The initial temperature of the gas is 27.0°C and the pressure is constant. As part of a machine design project, calculate the final temperature of the gas after it has done 2.40×10^3 J of work.

19.3 •• **CALC** Two moles of an ideal gas are compressed in a cylinder at a constant temperature of 65.0°C until the original pressure has tripled. (a) Sketch a pV-diagram for this process. (b) Calculate the amount of work done.

19.4 •• **BIO** **Work Done by the Lungs.** The graph in Fig. E19.4 shows a pV-diagram of the air in a human lung when a person is inhaling and then exhaling a deep breath. Such graphs, obtained in clinical practice, are normally somewhat curved, but we have modeled one as a set of straight lines of the same general shape. (*Important:* The

Figure **E19.4**

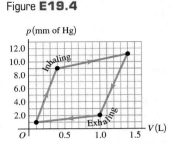

pressure shown is the *gauge* pressure, *not* the absolute pressure.) (a) How many joules of *net* work does this person's lung do during one complete breath? (b) The process illustrated here is somewhat different from those we have been studying, because the pressure change is due to changes in the amount of gas in the lung, not to temperature changes. (Think of your own breathing. Your lungs do not expand because they've gotten hot.) If the temperature of the air in the lung remains a reasonable 20°C, what is the maximum number of moles in this person's lung during a breath?

19.5 •• **CALC** During the time 0.305 mol of an ideal gas undergoes an isothermal compression at 22.0°C, 468 J of work is done on it by the surroundings. (a) If the final pressure is 1.76 atm, what was the initial pressure? (b) Sketch a pV-diagram for the process.

19.6 •• A gas undergoes two processes. In the first, the volume remains constant at 0.200 m³ and the pressure increases from 2.00×10^5 Pa to 5.00×10^5 Pa. The second process is a compression to a volume of 0.120 m³ at a constant pressure of 5.00×10^5 Pa. (a) In a pV-diagram, show both processes. (b) Find the total work done by the gas during both processes.

19.7 • **Work Done in a Cyclic Process.** (a) In Fig. 19.7a, consider the closed loop $1 \rightarrow 3 \rightarrow 2 \rightarrow 4 \rightarrow 1$. This is a *cyclic* process in which the initial and final states are the same. Find the total work done by the system in this cyclic process, and show that it is equal to the area enclosed by the loop. (b) How is the work done for the process in part (a) related to the work done if the loop is traversed in the opposite direction, $1 \rightarrow 4 \rightarrow 2 \rightarrow 3 \rightarrow 1$? Explain.

Section 19.4 Internal Energy and the First Law of Thermodynamics

19.8 •• Figure E19.8 shows a pV-diagram for an ideal gas in which its absolute temperature at b is one-fourth of its absolute temperature at a. (a) What volume does this gas occupy at point b? (b) How many joules of work was done by or on the gas in this process?

Figure **E19.8**

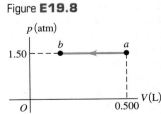

Was it done by or on the gas? (c) Did the internal energy of the gas increase or decrease from a to b? How do you know? (d) Did heat enter or leave the gas from a to b? How do you know?

19.9 • A gas in a cylinder expands from a volume of 0.110 m³ to 0.320 m³. Heat flows into the gas just rapidly enough to keep the pressure constant at 1.65×10^5 Pa during the expansion. The total heat added is 1.15×10^5 J. (a) Find the work done by the gas. (b) Find the change in internal energy of the gas. (c) Does it matter whether the gas is ideal? Why or why not?

19.10 •• Five moles of an ideal monatomic gas with an initial temperature of 127°C expand and, in the process, absorb 1200 J of heat and do 2100 J of work. What is the final temperature of the gas?

19.11 •• The process *abc* shown in the *pV*-diagram in Fig. E19.11 involves 0.0175 mole of an ideal gas. (a) What was the lowest temperature the gas reached in this process? Where did it occur? (b) How much work was done by or on the gas from *a* to *b*? From *b* to *c*? (c) If 215 J of heat was put into the gas during *abc*, how many of those joules went into internal energy?

Figure **E19.11**

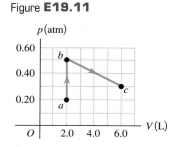

19.12 • A gas in a cylinder is held at a constant pressure of 1.80×10^5 Pa and is cooled and compressed from 1.70 m³ to 1.20 m³. The internal energy of the gas decreases by 1.40×10^5 J. (a) Find the work done by the gas. (b) Find the absolute value $|Q|$ of the heat flow into or out of the gas, and state the direction of the heat flow. (c) Does it matter whether the gas is ideal? Why or why not?

19.13 • **BIO** Doughnuts: Breakfast of Champions! A typical doughnut contains 2.0 g of protein, 17.0 g of carbohydrates, and 7.0 g of fat. The average food energy values of these substances are 4.0 kcal/g for protein and carbohydrates and 9.0 kcal/g for fat. (a) During heavy exercise, an average person uses energy at a rate of 510 kcal/h. How long would you have to exercise to "work off" one doughnut? (b) If the energy in the doughnut could somehow be converted into the kinetic energy of your body as a whole, how fast could you move after eating the doughnut? Take your mass to be 60 kg, and express your answer in m/s and in km/h.

19.14 • **Boiling Water at High Pressure.** When water is boiled at a pressure of 2.00 atm, the heat of vaporization is 2.20×10^6 J/kg and the boiling point is 120°C. At this pressure, 1.00 kg of water has a volume of 1.00×10^{-3} m³, and 1.00 kg of steam has a volume of 0.824 m³. (a) Compute the work done when 1.00 kg of steam is formed at this temperature. (b) Compute the increase in internal energy of the water.

19.15 • An ideal gas is taken from *a* to *b* on the *pV*-diagram shown in Fig. E19.15. During this process, 700 J of heat is added and the pressure doubles. (a) How much work is done by or on the gas? Explain. (b) How does the temperature of the gas at *a* compare to its temperature at *b*? Be specific. (c) How does the internal energy of the gas at *a* compare to the internal energy at *b*? Again, be specific and explain.

Figure **E19.15**

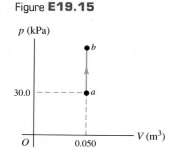

19.16 • A system is taken from state *a* to state *b* along the three paths shown in Fig. E19.16. (a) Along which path is the work done by the system the greatest? The least? (b) If $U_b > U_a$, along which path is

Figure **E19.16**

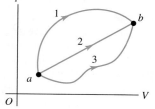

the absolute value $|Q|$ of the heat transfer the greatest? For this path, is heat absorbed or liberated by the system?

19.17 • A thermodynamic system undergoes a cyclic process as shown in Fig. E19.17. The cycle consists of two closed loops: I and II. (a) Over one complete cycle, does the system do positive or negative work? (b) In each of loops I and II, is the net work done by the system positive or negative? (c) Over one complete cycle, does heat flow into or out of the system? (d) In each of loops I and II, does heat flow into or out of the system?

Figure **E19.17**

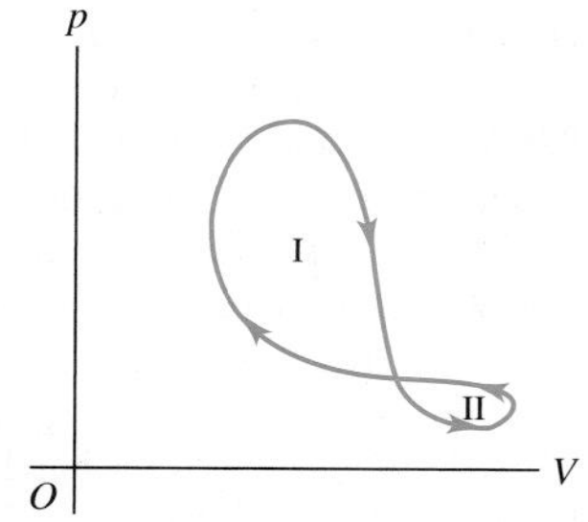

Section 19.5 Kinds of Thermodynamic Processes, Section 19.6 Internal Energy of an Ideal Gas, and Section 19.7 Heat Capacities of an Ideal Gas

19.18 • During an isothermal compression of an ideal gas, 335 J of heat must be removed from the gas to maintain constant temperature. How much work is done by the gas during the process?

19.19 • A cylinder contains 0.250 mol of carbon dioxide (CO_2) gas at a temperature of 27.0°C. The cylinder is provided with a frictionless piston, which maintains a constant pressure of 1.00 atm on the gas. The gas is heated until its temperature increases to 127.0°C. Assume that the CO_2 may be treated as an ideal gas. (a) Draw a *pV*-diagram for this process. (b) How much work is done by the gas in this process? (c) On what is this work done? (d) What is the change in internal energy of the gas? (e) How much heat was supplied to the gas? (f) How much work would have been done if the pressure had been 0.50 atm?

19.20 • A cylinder contains 0.0100 mol of helium at $T = 27.0°C$. (a) How much heat is needed to raise the temperature to 67.0°C while keeping the volume constant? Draw a *pV*-diagram for this process. (b) If instead the pressure of the helium is kept constant, how much heat is needed to raise the temperature from 27.0°C to 67.0°C? Draw a *pV*-diagram for this process. (c) What accounts for the difference between your answers to parts (a) and (b)? In which case is more heat required? What becomes of the additional heat? (d) If the gas is ideal, what is the change in its internal energy in part (a)? In part (b)? How do the two answers compare? Why?

19.21 • In an experiment to simulate conditions inside an automobile engine, 0.185 mol of air at a temperature of 780 K and a pressure of 3.00×10^6 Pa is contained in a cylinder of volume 40.0 cm³. Then 645 J of heat is transferred to the cylinder. (a) If the volume of the cylinder is constant while the heat is added, what is the final temperature of the air? Assume that the air is essentially nitrogen gas, and use the data in Table 19.1 even though the pressure is not low. Draw a *pV*-diagram for this process. (b) If instead the volume of the cylinder is allowed to increase while the pressure remains constant, find the final temperature of the air. Draw a *pV*-diagram for this process.

19.22 •• When a quantity of monatomic ideal gas expands at a constant pressure of 4.00×10^4 Pa, the volume of the gas increases from 2.00×10^{-3} m³ to 8.00×10^{-3} m³. What is the change in the internal energy of the gas?

19.23 • Heat Q flows into a monatomic ideal gas, and the volume increases while the pressure is kept constant. What fraction of the heat energy is used to do the expansion work of the gas?

19.24 • Three moles of an ideal monatomic gas expands at a constant pressure of 2.50 atm; the volume of the gas changes from 3.20×10^{-2} m^3 to 4.50×10^{-2} m^3. (a) Calculate the initial and final temperatures of the gas. (b) Calculate the amount of work the gas does in expanding. (c) Calculate the amount of heat added to the gas. (d) Calculate the change in internal energy of the gas.

19.25 • A cylinder with a movable piston contains 3.00 mol of N$_2$ gas (assumed to behave like an ideal gas). (a) The N$_2$ is heated at constant volume until 1557 J of heat have been added. Calculate the change in temperature. (b) Suppose the same amount of heat is added to the N$_2$, but this time the gas is allowed to expand while remaining at constant pressure. Calculate the temperature change. (c) In which case, (a) or (b), is the final internal energy of the N$_2$ higher? How do you know? What accounts for the difference between the two cases?

19.26 • Propane gas (C_3H_8) behaves like an ideal gas with $\gamma = 1.127$. Determine the molar heat capacity at constant volume and the molar heat capacity at constant pressure.

19.27 • **CALC** The temperature of 0.150 mol of an ideal gas is held constant at 77.0°C while its volume is reduced to 25.0% of its initial volume. The initial pressure of the gas is 1.25 atm. (a) Determine the work done by the gas. (b) What is the change in its internal energy? (c) Does the gas exchange heat with its surroundings? If so, how much? Does the gas absorb or liberate heat?

19.28 • An experimenter adds 970 J of heat to 1.75 mol of an ideal gas to heat it from 10.0°C to 25.0°C at constant pressure. The gas does +223 J of work during the expansion. (a) Calculate the change in internal energy of the gas. (b) Calculate γ for the gas.

Section 19.8 Adiabatic Processes for an Ideal Gas

19.29 • A monatomic ideal gas that is initially at a pressure of 1.50×10^5 Pa and has a volume of 0.0800 m^3 is compressed adiabatically to a volume of 0.0400 m^3. (a) What is the final pressure? (b) How much work is done by the gas? (c) What is the ratio of the final temperature of the gas to its initial temperature? Is the gas heated or cooled by this compression?

19.30 • In an adiabatic process for an ideal gas, the pressure decreases. In this process does the internal energy of the gas increase or decrease? Explain your reasoning.

19.31 •• Two moles of carbon monoxide (CO) start at a pressure of 1.2 atm and a volume of 30 liters. The gas is then compressed adiabatically to $\frac{1}{3}$ this volume. Assume that the gas may be treated as ideal. What is the change in the internal energy of the gas? Does the internal energy increase or decrease? Does the temperature of the gas increase or decrease during this process? Explain.

19.32 • The engine of a Ferrari F355 F1 sports car takes in air at 20.0°C and 1.00 atm and compresses it adiabatically to 0.0900 times the original volume. The air may be treated as an ideal gas with $\gamma = 1.40$. (a) Draw a pV-diagram for this process. (b) Find the final temperature and pressure.

19.33 • During an adiabatic expansion the temperature of 0.450 mol of argon (Ar) drops from 50.0°C to 10.0°C. The argon may be treated as an ideal gas. (a) Draw a pV-diagram for this process. (b) How much work does the gas do? (c) What is the change in internal energy of the gas?

19.34 •• A player bounces a basketball on the floor, compressing it to 80.0% of its original volume. The air (assume it is essentially N$_2$ gas) inside the ball is originally at a temperature of 20.0°C and a pressure of 2.00 atm. The ball's inside diameter is 23.9 cm. (a) What temperature does the air in the ball reach at its maximum compression? Assume the compression is adiabatic and treat the gas as ideal. (b) By how much does the internal energy of the air change between the ball's original state and its maximum compression?

19.35 •• On a warm summer day, a large mass of air (atmospheric pressure 1.01×10^5 Pa) is heated by the ground to a temperature of 26.0°C and then begins to rise through the cooler surrounding air. (This can be treated approximately as an adiabatic process; why?) Calculate the temperature of the air mass when it has risen to a level at which atmospheric pressure is only 0.850×10^5 Pa. Assume that air is an ideal gas, with $\gamma = 1.40$. (This rate of cooling for dry, rising air, corresponding to roughly 1°C per 100 m of altitude, is called the *dry adiabatic lapse rate*.)

19.36 • A cylinder contains 0.100 mol of an ideal monatomic gas. Initially the gas is at a pressure of 1.00×10^5 Pa and occupies a volume of 2.50×10^{-3} m^3. (a) Find the initial temperature of the gas in kelvins. (b) If the gas is allowed to expand to twice the initial volume, find the final temperature (in kelvins) and pressure of the gas if the expansion is (i) isothermal; (ii) isobaric; (iii) adiabatic.

PROBLEMS

19.37 •• One mole of ideal gas is slowly compressed to one-third of its original volume. In this compression, the work done on the gas has magnitude 600 J. For the gas, $C_p = 7R/2$. (a) If the process is isothermal, what is the heat flow Q for the gas? Does heat flow into or out of the gas? (b) If the process is isobaric, what is the change in internal energy of the gas? Does the internal energy increase or decrease?

19.38 •• **CALC** Figure P19.38 shows the pV-diagram for an isothermal expansion of 1.50 mol of an ideal gas, at a temperature of 15.0°C. (a) What is the change in internal energy of the gas? Explain. (b) Calculate the work done by (or on) the gas and the heat absorbed (or released) by the gas during the expansion.

Figure **P19.38**

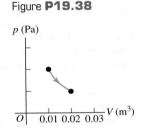

19.39 •• A quantity of air is taken from state a to state b along a path that is a straight line in the pV-diagram (Fig. P19.39). (a) In this process, does the temperature of the gas increase, decrease, or stay the same? Explain. (b) If $V_a = 0.0700$ m^3, $V_b = 0.1100$ m^3, $p_a = 1.00 \times 10^5$ Pa, and $p_b = 1.40 \times 10^5$ Pa, what is the work W done by the gas in this process? Assume that the gas may be treated as ideal.

Figure **P19.39**

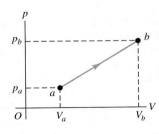

19.40 • One-half mole of an ideal gas is taken from state a to state c, as shown in Fig. P19.40. (a) Calculate the final temperature of the gas. (b) Calculate the work done on (or by) the gas as it moves from state a to state c. (c) Does heat leave the system or enter the system during this process? How much heat? Explain.

Figure **P19.40**

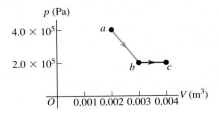

19.41 •• When a system is taken from state a to state b in Fig. P19.41 along the path acb, 90.0 J of heat flows into the system and 60.0 J of work is done by the system. (a) How much heat flows into the system along path adb if the work done by the system is 15.0 J? (b) When the system is returned from b to a along the curved path, the absolute value of the work done by the system is 35.0 J. Does the system absorb or liberate heat? How much heat? (c) If $U_a = 0$ and $U_d = 8.0$ J, find the heat absorbed in the processes ad and db.

Figure **P19.41**

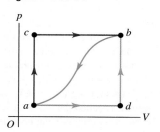

19.42 • A thermodynamic system is taken from state a to state c in Fig. P19.42 along either path abc or path adc. Along path abc, the work W done by the system is 450 J. Along path adc, W is 120 J. The internal energies of each of the four states shown in the figure are $U_a = 150$ J, $U_b = 240$ J, $U_c = 680$ J, and $U_d = 330$ J. Calculate the heat flow Q for each of the four processes ab, bc, ad, and dc. In each process, does the system absorb or liberate heat?

Figure **P19.42**

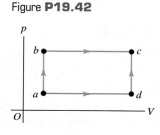

19.43 • A volume of air (assumed to be an ideal gas) is first cooled without changing its volume and then expanded without changing its pressure, as shown by the path abc in Fig. P19.43. (a) How does the final temperature of the gas compare with its initial temperature? (b) How much heat does the air exchange with its surroundings during the process abc? Does the air absorb heat or release heat during this process? Explain. (c) If the air instead expands from state a to state c by the straight-line path shown, how much heat does it exchange with its surroundings?

Figure **P19.43**

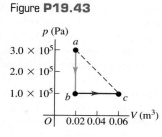

19.44 • Three moles of argon gas (assumed to be an ideal gas) originally at a pressure of 1.50×10^4 Pa and a volume of 0.0280 m³ are first heated and expanded at constant pressure to a volume of 0.0435 m³, then heated at constant volume until the pressure reaches 3.50×10^4 Pa, then cooled and compressed at constant pressure until the volume is again 0.0280 m³, and finally cooled at constant volume until the pressure drops to its original value of 1.50×10^4 Pa. (a) Draw the pV-diagram for this cycle. (b) Calculate the total work done by (or on) the gas during the cycle. (c) Calculate the net heat exchanged with the surroundings. Does the gas gain or lose heat overall?

19.45 •• Two moles of an ideal monatomic gas go through the cycle abc. For the complete cycle, 800 J of heat flows out of the gas. Process ab is at constant pressure, and process bc is at constant volume. States a and b have temperatures $T_a = 200$ K and $T_b = 300$ K. (a) Sketch the pV-diagram for the cycle. (b) What is the work W for the process ca?

19.46 •• Three moles of an ideal gas are taken around the cycle acb shown in Fig. P19.46. For this gas, $C_p = 29.1$ J/mol·K. Process ac is at constant pressure, process ba is at constant volume, and process cb is adiabatic. The temperatures of the gas in states a, c, and b are $T_a = 300$ K, $T_c = 492$ K, and $T_b = 600$ K. Calculate the total work W for the cycle.

Figure **P19.46**

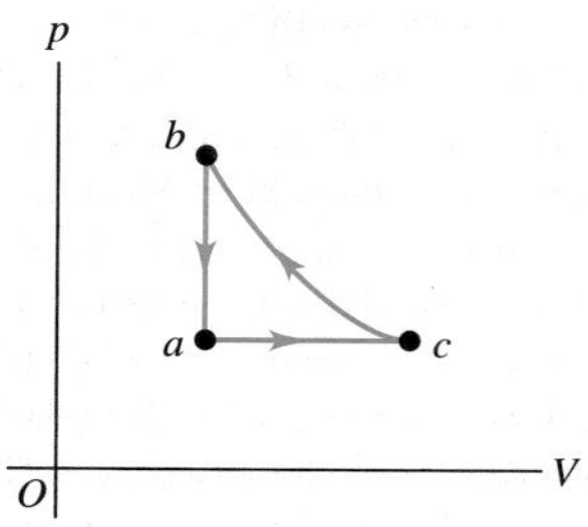

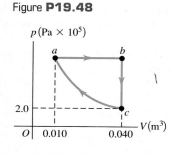

Figure **P19.47**

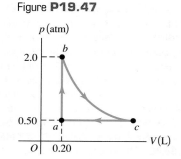

19.47 •• Figure P19.47 shows a pV-diagram for 0.0040 mole of ideal H_2 gas. The temperature of the gas does not change during segment bc. (a) What volume does this gas occupy at point c? (b) Find the temperature of the gas at points a, b, and c. (c) How much heat went into or out of the gas during segments ab, ca, and bc? Indicate whether the heat has gone into or out of the gas. (d) Find the change in the internal energy of this hydrogen during segments ab, bc, and ca. Indicate whether the internal energy increased or decreased during each of these segments.

19.48 •• The graph in Fig. P19.48 shows a pV-diagram for 3.25 moles of ideal helium (He) gas. Part ca of this process is isothermal. (a) Find the pressure of the He at point a. (b) Find the temperature of the He at points a, b, and c. (c) How much heat entered or left the He during segments ab, bc, and ca? In each segment, did the heat enter or leave? (d) By how much did the internal energy of the He change from a to b, from b to c, and from c to a? Indicate whether this energy increased or decreased.

Figure **P19.48**

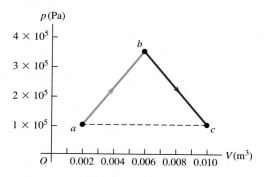

19.49 •• (a) One-third of a mole of He gas is taken along the path abc shown as the solid line in Fig. P19.49. Assume that the gas may be treated as ideal. How much heat is transferred into or out of the gas? (b) If the gas instead went from state a to state c along the horizontal dashed line in Fig. P19.49, how much heat would be transferred into or out of the gas? (c) How does Q in part (b) compare with Q in part (a)? Explain.

Figure **P19.49**

19.50 •• Two moles of helium are initially at a temperature of 27.0°C and occupy a volume of 0.0300 m³. The helium first expands at constant pressure until its volume has doubled. Then it expands adiabatically until the temperature returns to its initial value. Assume that the helium can be treated as an ideal gas. (a) Draw a diagram of the process in the pV-plane. (b) What is the total heat supplied to the helium in the process? (c) What is the total change in internal energy of the helium? (d) What is the total work done by the helium? (e) What is the final volume of the helium?

19.51 ••• Starting with 2.50 mol of N_2 gas (assumed to be ideal) in a cylinder at 1.00 atm and 20.0°C, a chemist first heats the gas at constant volume, adding 1.52×10^4 J of heat, then continues heating and allows the gas to expand at constant pressure to twice its original volume. (a) Calculate the final temperature of the gas. (b) Calculate the amount of work done by the gas. (c) Calculate the amount of heat added to the gas while it was expanding. (d) Calculate the change in internal energy of the gas for the whole process.

19.52 •• Nitrogen gas in an expandable container is cooled from 50.0°C to 10.0°C with the pressure held constant at 3.00×10^5 Pa. The total heat liberated by the gas is 2.50×10^4 J. Assume that the gas may be treated as ideal. (a) Find the number of moles of gas. (b) Find the change in internal energy of the gas. (c) Find the work done by the gas. (d) How much heat would be liberated by the gas for the same temperature change if the volume were constant?

19.53 • In a certain process, 2.15×10^5 J of heat is liberated by a system, and at the same time the system contracts under a constant external pressure of 9.50×10^5 Pa. The internal energy of the system is the same at the beginning and end of the process. Find the change in volume of the system. (The system is *not* an ideal gas.)

19.54 • **CALC** A cylinder with a frictionless, movable piston like that shown in Fig. 19.5 contains a quantity of helium gas. Initially the gas is at a pressure of 1.00×10^5 Pa, has a temperature of 300 K, and occupies a volume of 1.50 L. The gas then undergoes two processes. In the first, the gas is heated and the piston is allowed to move to keep the temperature equal to 300 K. This continues until the pressure reaches 2.50×10^4 Pa. In the second process, the gas is compressed at constant pressure until it returns to its original volume of 1.50 L. Assume that the gas may be treated as ideal. (a) In a pV-diagram, show both processes. (b) Find the volume of the gas at the end of the first process, and find the pressure and temperature at the end of the second process. (c) Find the total work done by the gas during both processes. (d) What would you have to do to the gas to return it to its original pressure and temperature?

19.55 •• **CP A Thermodynamic Process in a Liquid.** A chemical engineer is studying the properties of liquid methanol (CH_3OH). She uses a steel cylinder with a cross-sectional area of 0.0200 m^2 and containing 1.20×10^{-2} m^3 of methanol. The cylinder is equipped with a tightly fitting piston that supports a load of 3.00×10^4 N. The temperature of the system is increased from 20.0°C to 50.0°C. For methanol, the coefficient of volume expansion is 1.20×10^{-3} K^{-1}, the density is 791 kg/m^3, and the specific heat at constant pressure is $c_p = 2.51 \times 10^3$ J/kg·K. You can ignore the expansion of the steel cylinder. Find (a) the increase in volume of the methanol; (b) the mechanical work done by the methanol against the 3.00×10^4 N force; (c) the amount of heat added to the methanol; (d) the change in internal energy of the methanol. (e) Based on your results, explain whether there is any substantial difference between the specific heats c_p (at constant pressure) and c_V (at constant volume) for methanol under these conditions.

19.56 • **CP A Thermodynamic Process in a Solid.** A cube of copper 2.00 cm on a side is suspended by a string. (The physical properties of copper are given in Tables 14.1, 17.2, and 17.3.) The cube is heated with a burner from 20.0°C to 90.0°C. The air surrounding the cube is at atmospheric pressure (1.01×10^5 Pa). Find (a) the increase in volume of the cube; (b) the mechanical work done by the cube to expand against the pressure of the surrounding air; (c) the amount of heat added to the cube; (d) the change in internal energy of the cube. (e) Based on your results, explain whether there is any substantial difference between the specific heats c_p (at constant pressure) and c_V (at constant volume) for copper under these conditions.

19.57 • **BIO A Thermodynamic Process in an Insect.** The African bombardier beetle (*Stenaptinus insignis*) can emit a jet of defensive spray from the movable tip of its abdomen (Fig. P19.57). The beetle's body has reservoirs of two different chemicals; when the beetle is disturbed, these chemicals are combined in a reaction chamber, producing a compound that is warmed from 20°C to 100°C by the heat of reaction. The high pressure produced allows the compound to be sprayed out at speeds up to 19 m/s (68 km/h), scaring away predators of all kinds. (The beetle shown in the figure is 2 cm long.) Calculate the heat of reaction of the two chemicals (in J/kg). Assume that the specific heat of the two chemicals and the spray is the same as that of water, 4.19×10^3 J/kg·K, and that the initial temperature of the chemicals is 20°C.

Figure **P19.57**

19.58 ••• **High-Altitude Research.** A large research balloon containing 2.00×10^3 m^3 of helium gas at 1.00 atm and a temperature of 15.0°C rises rapidly from ground level to an altitude at which the atmospheric pressure is only 0.900 atm (Fig. P19.58). Assume the helium behaves like an ideal gas and the balloon's ascent is too rapid to permit much heat exchange with the surrounding air. (a) Calculate the volume of the gas at the higher altitude. (b) Calculate the temperature of the gas at the higher altitude. (c) What is the change in internal energy of the helium as the balloon rises to the higher altitude?

Figure **P19.58**

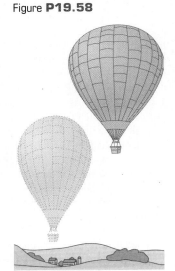

19.59 ••• **Chinook.** During certain seasons strong winds called chinooks blow from the west across the eastern slopes of the Rockies and downhill into Denver and nearby areas. Although the mountains are cool, the wind in Denver is very hot; within a few minutes after the chinook wind arrives, the temperature can climb 20 C° ("chinook" is a Native American word meaning "snow eater"). Similar winds occur in the Alps (called foehns) and in southern California (called Santa Anas). (a) Explain why the temperature of the chinook wind rises as it descends the slopes. Why is it important that the wind be fast moving? (b) Suppose a strong wind is blowing toward Denver (elevation 1630 m) from Grays Peak (80 km west of Denver, at an elevation of 4350 m), where the air pressure is

5.60×10^4 Pa and the air temperature is $-15.0°C$. The temperature and pressure in Denver before the wind arrives are $2.0°C$ and 8.12×10^4 Pa. By how many Celsius degrees will the temperature in Denver rise when the chinook arrives?

19.60 •• A certain ideal gas has molar heat capacity at constant volume C_V. A sample of this gas initially occupies a volume V_0 at pressure p_0 and absolute temperature T_0. The gas expands isobarically to a volume $2V_0$ and then expands further adiabatically to a final volume $4V_0$. (a) Draw a pV-diagram for this sequence of processes. (b) Compute the total work done by the gas for this sequence of processes. (c) Find the final temperature of the gas. (d) Find the absolute value $|Q|$ of the total heat flow into or out of the gas for this sequence of processes, and state the direction of heat flow.

19.61 ••• An air pump has a cylinder 0.250 m long with a movable piston. The pump is used to compress air from the atmosphere (at absolute pressure 1.01×10^5 Pa) into a very large tank at 4.20×10^5 Pa gauge pressure. (For air, $C_V = 20.8$ J/mol·K.) (a) The piston begins the compression stroke at the open end of the cylinder. How far down the length of the cylinder has the piston moved when air first begins to flow from the cylinder into the tank? Assume that the compression is adiabatic. (b) If the air is taken into the pump at $27.0°C$, what is the temperature of the compressed air? (c) How much work does the pump do in putting 20.0 mol of air into the tank?

19.62 •• **Engine Turbochargers and Intercoolers.** The power output of an automobile engine is directly proportional to the mass of air that can be forced into the volume of the engine's cylinders to react chemically with gasoline. Many cars have a *turbocharger*, which compresses the air before it enters the engine, giving a greater mass of air per volume. This rapid, essentially adiabatic compression also heats the air. To compress it further, the air then passes through an *intercooler* in which the air exchanges heat with its surroundings at essentially constant pressure. The air is then drawn into the cylinders. In a typical installation, air is taken into the turbocharger at atmospheric pressure (1.01×10^5 Pa), density $\rho = 1.23$ kg/m^3, and temperature $15.0°C$. It is compressed adiabatically to 1.45×10^5 Pa. In the intercooler, the air is cooled to the original temperature of $15.0°C$ at a constant pressure of 1.45×10^5 Pa. (a) Draw a pV-diagram for this sequence of processes. (b) If the volume of one of the engine's cylinders is 575 cm^3, what mass of air exiting from the intercooler will fill the cylinder at 1.45×10^5 Pa? Compared to the power output of an engine that takes in air at 1.01×10^5 Pa at $15.0°C$, what percentage increase in power is obtained by using the turbocharger and intercooler? (c) If the intercooler is not used, what mass of air exiting from the turbocharger will fill the cylinder at 1.45×10^5 Pa? Compared to the power output of an engine that takes in air at 1.01×10^5 Pa at $15.0°C$, what percentage increase in power is obtained by using the turbocharger alone?

19.63 • A monatomic ideal gas expands slowly to twice its original volume, doing 300 J of work in the process. Find the heat added to the gas and the change in internal energy of the gas if the process is (a) isothermal; (b) adiabatic; (c) isobaric.

19.64 •• **CALC** A cylinder with a piston contains 0.250 mol of oxygen at 2.40×10^5 Pa and 355 K. The oxygen may be treated as an ideal gas. The gas first expands isobarically to twice its original volume. It is then compressed isothermally back to its original volume, and finally it is cooled isochorically to its original pressure. (a) Show the series of processes on a pV-diagram. (b) Compute the temperature during the isothermal compression. (c) Compute the maximum pressure. (d) Compute the total work done by the piston on the gas during the series of processes.

19.65 • Use the conditions and processes of Problem 19.64 to compute (a) the work done by the gas, the heat added to it, and its internal energy change during the initial expansion; (b) the work done, the heat added, and the internal energy change during the final cooling; (c) the internal energy change during the isothermal compression.

19.66 •• **CALC** A cylinder with a piston contains 0.150 mol of nitrogen at 1.80×10^5 Pa and 300 K. The nitrogen may be treated as an ideal gas. The gas is first compressed isobarically to half its original volume. It then expands adiabatically back to its original volume, and finally it is heated isochorically to its original pressure. (a) Show the series of processes in a pV-diagram. (b) Compute the temperatures at the beginning and end of the adiabatic expansion. (c) Compute the minimum pressure.

19.67 • Use the conditions and processes of Problem 19.66 to compute (a) the work done by the gas, the heat added to it, and its internal energy change during the initial compression; (b) the work done by the gas, the heat added to it, and its internal energy change during the adiabatic expansion; (c) the work done, the heat added, and the internal energy change during the final heating.

19.68 • **Comparing Thermodynamic Processes.** In a cylinder, 1.20 mol of an ideal monatomic gas, initially at 3.60×10^5 Pa and 300 K, expands until its volume triples. Compute the work done by the gas if the expansion is (a) isothermal; (b) adiabatic; (c) isobaric. (d) Show each process in a pV-diagram. In which case is the absolute value of the work done by the gas greatest? Least? (e) In which case is the absolute value of the heat transfer greatest? Least? (f) In which case is the absolute value of the change in internal energy of the gas greatest? Least?

CHALLENGE PROBLEMS

19.69 ••• **CP Oscillations of a Piston.** A vertical cylinder of radius r contains a quantity of ideal gas and is fitted with a piston with mass m that is free to move (Fig. P19.69). The piston and the walls of the cylinder are frictionless, and the entire cylinder is placed in a constant-temperature bath. The outside air pressure is p_0. In equilibrium, the piston sits at a height h above the bottom of the cylinder. (a) Find the absolute pressure of the gas trapped below the piston when in equilibrium. (b) The piston is pulled up by a small distance and released. Find the net force acting on the piston when its base is a distance $h + y$ above the bottom of the cylinder, where y is much less than h. (c) After the piston is displaced from equilibrium and released, it oscillates up and down. Find the frequency of these small oscillations. If the displacement is not small, are the oscillations simple harmonic? How can you tell?

Figure **P19.69**

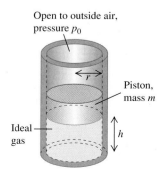

Open to outside air, pressure p_0

Piston, mass m

Ideal gas

h

Answers

Chapter Opening Question ?

No. The work done by a gas as its volume changes from V_1 to V_2 is equal to the integral $\int p\,dV$ between those two volume limits. If the volume of the gas contracts, the final volume V_2 is less than the initial volume V_1 and the gas does negative work. Propelling the locomotive requires that the gas do positive work, so the gas doesn't contribute to propulsion while contracting.

Test Your Understanding Questions

19.1 Answers: negative, positive, positive Heat flows out of the coffee, so $Q_{coffee} < 0$; heat flows into the aluminum cup, so $Q_{aluminum} > 0$. In mechanics, we would say that negative work is done *on* the block, since the surface exerts a force on the block that opposes the block's motion. But in thermodynamics we use the opposite convention and say that $W > 0$, which means that positive work is done *by* the block on the surface.

19.2 Answer: (ii) The work done in an expansion is represented by the area under the curve of pressure p versus volume V. In an isothermal expansion the pressure decreases as the volume increases, so the pV-diagram looks like Fig. 19.6a and the work done equals the shaded area under the blue curve from point 1 to point 2. If, however, the expansion is at constant pressure, the curve of p versus V would be the same as the dashed horizontal line at pressure p_2 in Fig. 19.6a. The area under this dashed line is smaller than the area under the blue curve for an isothermal expansion, so less work is done in the constant-pressure expansion than in the isothermal expansion.

19.3 Answer: (i) and (iv) (tie), (ii) and (iii) (tie) The accompanying figure shows the pV-diagrams for each of the four processes. The trapezoidal area under the curve, and hence the absolute value of the work, is the same in all four cases. In cases (i) and (iv) the volume increases, so the system does positive work as it expands against its surroundings. In cases (ii) and (iii) the volume decreases, so the system does negative work (shown by cross-hatching) as the surroundings push inward on it.

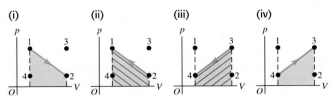

19.4 Answer: (ii), (i) and (iv) (tie), (iii) In the expression $\Delta U = Q - W$, Q is the heat *added* to the system and W is the work done *by* the system. If heat is transferred from the system to its surroundings, Q is negative; if work is done on the system, W is negative. Hence we have (i) $Q = -250$ J, $W = -250$ J, $\Delta U = -250$ J $- (-250$ J$) = 0$; (ii) $Q = 250$ J, $W = -250$ J, $\Delta U = 250$ J $- (-250$ J$) = 500$ J; (iii) $Q = -250$ J, $W = 250$ J, $\Delta U = -250$ J $- 250$ J $= -500$ J; and (iv) $Q = 250$ J, $W = 250$ J, $\Delta U = 250$ J $- 250$ J $= 0$.

19.5 Answers: $1 \to 4$ and $3 \to 2$ are isochoric; $1 \to 3$ and $4 \to 2$ are isobaric; no In a pV-diagram like those shown in Fig. 19.7, isochoric processes are represented by vertical lines (lines of constant volume) and isobaric processes are represented by horizontal lines (lines of constant pressure). The process $1 \to 2$ in Fig. 19.7 is shown as a curved line, which superficially resembles the adiabatic and isothermal processes for an ideal gas in Fig. 19.16. Without more information we can't tell whether process $1 \to 2$ is isothermal, adiabatic, or neither.

19.6 Answer: no Using the model of a solid in Fig. 18.20, we can see that the internal energy of a solid *does* depend on its volume. Compressing the solid means compressing the "springs" between the atoms, thereby increasing their stored potential energy and hence the internal energy of the solid.

19.7 Answer: (i) For a given number of moles n and a given temperature change ΔT, the amount of heat that must be transferred out of a fixed volume of air is $Q = nC_V\Delta T$. Hence the amount of heat transfer required is least for the gas with the smallest value of C_V. From Table 19.1, C_V is smallest for monatomic gases.

19.8 Answer: (ii), (iv), (i) and (iii) (tie) Samples (i) and (iii) are compressed isothermally, so pV = constant. The volume of each sample decreases to one-half of its initial value, so the final pressure is twice the initial pressure. Samples (ii) and (iv) are compressed adiabatically, so pV^γ = constant and the pressure increases by a factor of 2^γ. Sample (ii) is a monatomic gas for which $\gamma = \frac{5}{3}$, so its final pressure is $2^{5/3} = 3.17$ times greater than the initial pressure. Sample (iv) is a diatomic gas for which $\gamma = \frac{7}{5}$, so its final pressure is greater than the initial pressure by a factor of $2^{7/5} = 2.64$.

Bridging Problem

Answers: (a) $W = nRT \ln\left[\dfrac{V_2 - nb}{V_1 - nb}\right] + an^2\left[\dfrac{1}{V_2} - \dfrac{1}{V_1}\right]$

(b) (i) $W = 2.80 \times 10^3$ J, (ii) $W = 3.11 \times 10^3$ J

(c) Ideal gas, for which there is no attraction between molecules

20 THE SECOND LAW OF THERMODYNAMICS

? The second law of thermodynamics tells us that heat naturally flows from a hot body (such as molten lava, shown here flowing into the ocean in Hawaii) to a cold one (such as ocean water, which is heated to make steam). Is it *ever* possible for heat to flow from a cold body to a hot one?

Many thermodynamic processes proceed naturally in one direction but not the opposite. For example, heat by itself always flows from a hot body to a cooler body, never the reverse. Heat flow from a cool body to a hot body would not violate the first law of thermodynamics; energy would be conserved. But it doesn't happen in nature. Why not? As another example, note that it is easy to convert mechanical energy completely into heat; this happens every time we use a car's brakes to stop it. In the reverse direction, there are plenty of devices that convert heat *partially* into mechanical energy. (An automobile engine is an example.) But no one has ever managed to build a machine that converts heat *completely* into mechanical energy. Again, why not?

The answer to both of these questions has to do with the *directions* of thermodynamic processes and is called the *second law of thermodynamics*. This law places fundamental limitations on the efficiency of an engine or a power plant. It also places limitations on the minimum energy input needed to operate a refrigerator. So the second law is directly relevant for many important practical problems.

We can also state the second law in terms of the concept of *entropy*, a quantitative measure of the degree of disorder or randomness of a system. The idea of entropy helps explain why ink mixed with water never spontaneously unmixes and why we never observe a host of other seemingly possible processes.

20.1 Directions of Thermodynamic Processes

Thermodynamic processes that occur in nature are all **irreversible processes.** These are processes that proceed spontaneously in one direction but not the other (Fig. 20.1a). The flow of heat from a hot body to a cooler body is irreversible, as is the free expansion of a gas discussed in Sections 19.3 and 19.6. Sliding a book across a table converts mechanical energy into heat by friction;

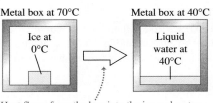

(a) A block of ice melts *irreversibly* when we place it in a hot (70°C) metal box.

Heat flows from the box into the ice and water, never the reverse.

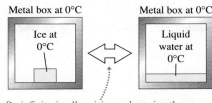

(b) A block of ice at 0°C can be melted *reversibly* if we put it in a 0°C metal box.

By infinitesimally raising or lowering the temperature of the box, we can make heat flow into the ice to melt it or make heat flow out of the water to refreeze it.

20.1 Reversible and irreversible processes.

this process is irreversible, for no one has ever observed the reverse process (in which a book initially at rest on the table would spontaneously start moving and the table and book would cool down). Our main topic for this chapter is the *second law of thermodynamics,* which determines the preferred direction for such processes.

Despite this preferred direction for every natural process, we can think of a class of idealized processes that *would* be reversible. A system that undergoes such an idealized **reversible process** is always very close to being in thermodynamic equilibrium within itself and with its surroundings. Any change of state that takes place can then be reversed by making only an infinitesimal change in the conditions of the system. For example, we can reverse heat flow between two bodies whose temperatures differ only infinitesimally by making only a very small change in one temperature or the other (Fig. 20.1b).

Reversible processes are thus **equilibrium processes,** with the system always in thermodynamic equilibrium. Of course, if a system were *truly* in thermodynamic equilibrium, no change of state would take place. Heat would not flow into or out of a system with truly uniform temperature throughout, and a system that is truly in mechanical equilibrium would not expand and do work against its surroundings. A reversible process is an idealization that can never be precisely attained in the real world. But by making the temperature gradients and the pressure differences in the substance very small, we can keep the system very close to equilibrium states and make the process nearly reversible.

By contrast, heat flow with finite temperature difference, free expansion of a gas, and conversion of work to heat by friction are all *irreversible* processes; no small change in conditions could make any of them go the other way. They are also all *nonequilibrium* processes, in that the system is not in thermodynamic equilibrium at any point until the end of the process.

Disorder and Thermodynamic Processes

There is a relationship between the direction of a process and the *disorder* or *randomness* of the resulting state. For example, imagine a thousand names written on file cards and arranged in alphabetical order. Throw the alphabetized stack of cards into the air, and they will likely come down in a random, disordered state. In the free expansion of a gas discussed in Sections 19.3 and 19.6, the air is more disordered after it has expanded into the entire box than when it was confined in one side, just as your clothes are more disordered when scattered all over your floor than when confined to your closet.

Similarly, macroscopic kinetic energy is energy associated with organized, coordinated motions of many molecules, but heat transfer involves changes in energy of random, disordered molecular motion. Therefore conversion of mechanical energy into heat involves an increase of randomness or disorder.

In the following sections we will introduce the second law of thermodynamics by considering two broad classes of devices: *heat engines,* which are partly successful in converting heat into work, and *refrigerators,* which are partly successful in transporting heat from cooler to hotter bodies.

Test Your Understanding of Section 20.1 Your left and right hands are normally at the same temperature, just like the metal box and ice in Fig. 20.1b. Is rubbing your hands together to warm them (i) a reversible process or (ii) an irreversible process? ❙

20.2 Heat Engines

The essence of our technological society is the ability to use sources of energy other than muscle power. Sometimes, mechanical energy is directly available; water power and wind power are examples. But most of our energy comes from the burning of fossil fuels (coal, oil, and gas) and from nuclear reactions. They supply energy that is transferred as *heat.* This is directly useful for heating buildings, for cooking, and for chemical processing, but to operate a machine or propel a vehicle, we need *mechanical* energy.

Thus it's important to know how to take heat from a source and convert as much of it as possible into mechanical energy or work. This is what happens in gasoline engines in automobiles, jet engines in airplanes, steam turbines in electric power plants, and many other systems. Closely related processes occur in the animal kingdom; food energy is "burned" (that is, carbohydrates combine with oxygen to yield water, carbon dioxide, and energy) and partly converted to mechanical energy as an animal's muscles do work on its surroundings.

Any device that transforms heat partly into work or mechanical energy is called a **heat engine** (Fig. 20.2). Usually, a quantity of matter inside the engine undergoes inflow and outflow of heat, expansion and compression, and sometimes change of phase. We call this matter the **working substance** of the engine. In internal-combustion engines, such as those used in automobiles, the working substance is a mixture of air and fuel; in a steam turbine it is water.

The simplest kind of engine to analyze is one in which the working substance undergoes a **cyclic process,** a sequence of processes that eventually leaves the substance in the same state in which it started. In a steam turbine the water is recycled and used over and over. Internal-combustion engines do not use the same air over and over, but we can still analyze them in terms of cyclic processes that approximate their actual operation.

20.2 All motorized vehicles other than purely electric vehicles use heat engines for propulsion. (Hybrid vehicles use their internal-combustion engine to help charge the batteries for the electric motor.)

Hot and Cold Reservoirs

All heat engines *absorb* heat from a source at a relatively high temperature, perform some mechanical work, and *discard* or *reject* some heat at a lower temperature. As far as the engine is concerned, the discarded heat is wasted. In internal-combustion engines the waste heat is that discarded in the hot exhaust gases and the cooling system; in a steam turbine it is the heat that must flow out of the used steam to condense and recycle the water.

When a system is carried through a cyclic process, its initial and final internal energies are equal. For any cyclic process, the first law of thermodynamics requires that

$$U_2 - U_1 = 0 = Q - W \qquad \text{so} \qquad Q = W$$

That is, the net heat flowing into the engine in a cyclic process equals the net work done by the engine.

When we analyze heat engines, it helps to think of two bodies with which the working substance of the engine can interact. One of these, called the *hot reservoir,* represents the heat source; it can give the working substance large amounts of heat at a constant temperature T_H without appreciably changing its own temperature. The other body, called the *cold reservoir,* can absorb large amounts of discarded heat from the engine at a constant lower temperature T_C. In a steam-turbine system the flames and hot gases in the boiler are the hot reservoir, and the cold water and air used to condense and cool the used steam are the cold reservoir.

We denote the quantities of heat transferred from the hot and cold reservoirs as Q_H and Q_C, respectively. A quantity of heat Q is positive when heat is transferred *into* the working substance and is negative when heat leaves the working substance. Thus in a heat engine, Q_H is positive but Q_C is negative, representing heat *leaving* the working substance. This sign convention is consistent with the rules we stated in Section 19.1; we will continue to use those rules here. For clarity, we'll often state the relationships in terms of the absolute values of the Q's and W's because absolute values are always positive.

Energy-Flow Diagrams and Efficiency

We can represent the energy transformations in a heat engine by the *energy-flow diagram* of Fig. 20.3. The engine itself is represented by the circle. The amount of heat Q_H supplied to the engine by the hot reservoir is proportional to the width of the incoming "pipeline" at the top of the diagram. The width of the outgoing pipeline at the bottom is proportional to the magnitude $|Q_C|$ of the heat rejected in the exhaust. The branch line to the right represents the portion of the heat supplied that the engine converts to mechanical work, W.

When an engine repeats the same cycle over and over, Q_H and Q_C represent the quantities of heat absorbed and rejected by the engine *during one cycle*; Q_H is positive, and Q_C is negative. The *net* heat Q absorbed per cycle is

$$Q = Q_H + Q_C = |Q_H| - |Q_C| \tag{20.1}$$

The useful output of the engine is the net work W done by the working substance. From the first law,

$$W = Q = Q_H + Q_C = |Q_H| - |Q_C| \tag{20.2}$$

Ideally, we would like to convert *all* the heat Q_H into work; in that case we would have $Q_H = W$ and $Q_C = 0$. Experience shows that this is impossible; there is always some heat wasted, and Q_C *is never zero.* We define the **thermal efficiency** of an engine, denoted by e, as the quotient

$$e = \frac{W}{Q_H} \tag{20.3}$$

The thermal efficiency e represents the fraction of Q_H that *is* converted to work. To put it another way, e is what you get divided by what you pay for. This is always less than unity, an all-too-familiar experience! In terms of the flow diagram of Fig. 20.3, the most efficient engine is one for which the branch pipeline representing the work output is as wide as possible and the exhaust pipeline representing the heat thrown away is as narrow as possible.

When we substitute the two expressions for W given by Eq. (20.2) into Eq. (20.3), we get the following equivalent expressions for e:

$$e = \frac{W}{Q_H} = 1 + \frac{Q_C}{Q_H} = 1 - \left| \frac{Q_C}{Q_H} \right| \quad \text{(thermal efficiency of an engine)} \tag{20.4}$$

Note that e is a quotient of two energy quantities and thus is a pure number, without units. Of course, we must always express W, Q_H, and Q_C in the same units.

20.3 Schematic energy-flow diagram for a heat engine.

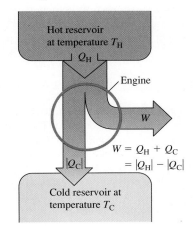

Hot reservoir at temperature T_H

Q_H

Engine

W

$W = Q_H + Q_C$
$= |Q_H| - |Q_C|$

$|Q_C|$

Cold reservoir at temperature T_C

Application **Biological Efficiency**
Although a biological organism is not a heat engine, the concept of efficiency still applies: Here e is the ratio of the work done to the energy that was used to do that work. To exercise on a stationary bike, your body must first convert the chemical-bond energy in glucose to chemical-bond energy in ATP (adenosine triphosphate), then convert energy from ATP into motion of your leg muscles, and finally convert muscular motion into motion of the pedals. The overall efficiency of this entire process is only about 25%. The remaining 75% of the energy liberated from glucose goes into heating your body.

Problem-Solving Strategy 20.1 **Heat Engines**

Problems involving heat engines are, fundamentally, problems in the first law of thermodynamics. You should review Problem-Solving Strategy 19.1 (Section 19.4).

IDENTIFY *the relevant concepts:* A heat engine is any device that converts heat partially to work, as shown schematically in Fig. 20.3. We will see in Section 20.4 that a refrigerator is essentially a heat engine running in reverse, so many of the same concepts apply.

SET UP *the problem* as suggested in Problem-Solving Strategy 19.1. Use Eq. (20.4) if the thermal efficiency of the engine is relevant. Sketch an energy-flow diagram like Fig. 20.3.

EXECUTE *the solution* as follows:
1. Be careful with the sign conventions for W and the various Q's. W is positive when the system expands and does work; W is

negative when the system is compressed and work is done on it. Each Q is positive if it represents heat entering the system and is negative if it represents heat leaving the system. When you know that a quantity is negative, such as Q_C in the above discussion, it sometimes helps to write it as $Q_C = -|Q_C|$.
2. Power is work per unit time ($P = W/t$), and rate of heat transfer (heat current) H is heat transfer per unit time ($H = Q/t$). In problems involving these concepts it helps to ask, "What is W or Q in one second (or one hour)?"
3. Keeping steps 1 and 2 in mind, solve for the target variables.

EVALUATE *your answer:* Use the first law of thermodynamics to check your results. Pay particular attention to algebraic signs.

Example 20.1 **Analyzing a heat engine**

A gasoline truck engine takes in 10,000 J of heat and delivers 2000 J of mechanical work per cycle. The heat is obtained by burning gasoline with heat of combustion $L_c = 5.0 \times 10^4$ J/g. (a) What is the thermal efficiency of this engine? (b) How much heat is discarded in each cycle? (c) If the engine goes through 25 cycles per second, what is its power output in watts? In horsepower? (d) How much gasoline is burned in each cycle? (e) How much gasoline is burned per second? Per hour?

SOLUTION

IDENTIFY and SET UP: This problem concerns a heat engine, so we can use the ideas of this section. Figure 20.4 is our energy-flow diagram for one cycle. In each cycle the engine does $W = 2000$ J of work and takes in heat $Q_H = 10,000$ J. We use Eq. (20.4), in the form $e = W/Q_H$, to find the thermal efficiency. We use Eq. (20.2) to find the amount of heat Q_C rejected per cycle. The heat of combustion tells us how much gasoline must be burned per cycle and hence per unit time. The power output is the time rate at which the work W is done.

20.4 Our sketch for this problem.

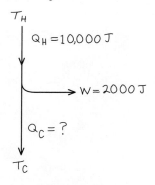

EXECUTE: (a) From Eq. (20.4), the thermal efficiency is

$$e = \frac{W}{Q_H} = \frac{2000 \text{ J}}{10,000 \text{ J}} = 0.20 = 20\%$$

(b) From Eq. (20.2), $W = Q_H + Q_C$, so

$$Q_C = W - Q_H = 2000 \text{ J} - 10,000 \text{ J} = -8000 \text{ J}$$

That is, 8000 J of heat leaves the engine during each cycle.

(c) The power P equals the work per cycle multiplied by the number of cycles per second:

$$P = (2000 \text{ J/cycle})(25 \text{ cycles/s}) = 50,000 \text{ W} = 50 \text{ kW}$$

$$= (50,000 \text{ W})\frac{1 \text{ hp}}{746 \text{ W}} = 67 \text{ hp}$$

(d) Let m be the mass of gasoline burned during each cycle. Then $Q_H = mL_c$ and

$$m = \frac{Q_H}{L_c} = \frac{10,000 \text{ J}}{5.0 \times 10^4 \text{ J/g}} = 0.20 \text{ g}$$

(e) The mass of gasoline burned per second equals the mass per cycle multiplied by the number of cycles per second:

$$(0.20 \text{ g/cycle})(25 \text{ cycles/s}) = 5.0 \text{ g/s}$$

The mass burned per hour is

$$(5.0 \text{ g/s})\frac{3600 \text{ s}}{1 \text{ h}} = 18,000 \text{ g/h} = 18 \text{ kg/h}$$

EVALUATE: An efficiency of 20% is fairly typical for cars and trucks if W includes only the work delivered to the wheels. We can check the mass burned per hour by expressing it in miles per gallon ("mileage"). The density of gasoline is about 0.70 g/cm^3, so this is about 25,700 cm^3, 25.7 L, or 6.8 gallons of gasoline per hour. If the truck is traveling at 55 mi/h (88 km/h), this represents fuel consumption of 8.1 miles/gallon (3.4 km/L). This is a fairly typical mileage for large trucks.

Test Your Understanding of Section 20.2 Rank the following heat engines in order from highest to lowest thermal efficiency. (i) an engine that in one cycle absorbs 5000 J of heat and rejects 4500 J of heat; (ii) an engine that in one cycle absorbs 25,000 J of heat and does 2000 J of work; (iii) an engine that in one cycle does 400 J of work and rejects 2800 J of heat.

20.3 Internal-Combustion Engines

The gasoline engine, used in automobiles and many other types of machinery, is a familiar example of a heat engine. Let's look at its thermal efficiency. Figure 20.5 shows the operation of one type of gasoline engine. First a mixture of air and gasoline vapor flows into a cylinder through an open intake valve while the piston descends, increasing the volume of the cylinder from a minimum of V (when the piston is all the way up) to a maximum of rV (when it is all the way down). The quantity r is called the **compression ratio;** for present-day automobile engines its value is typically 8 to 10. At the end of this *intake stroke,* the intake valve closes and the mixture is compressed, approximately adiabatically, to volume V during the *compression stroke.* The mixture is then ignited by the spark plug, and the heated gas expands, approximately adiabatically, back to volume rV, pushing on the piston and doing work; this is the *power stroke.* Finally, the exhaust valve opens, and the combustion products are pushed out (during the *exhaust stroke*), leaving the cylinder ready for the next intake stroke.

The Otto Cycle

Figure 20.6 is a pV-diagram for an idealized model of the thermodynamic processes in a gasoline engine. This model is called the **Otto cycle.** At point a the gasoline–air mixture has entered the cylinder. The mixture is compressed adiabatically to point b and is then ignited. Heat Q_H is added to the system by the burning gasoline along line bc, and the power stroke is the adiabatic expansion to d. The gas is cooled to the temperature of the outside air along line da; during this process, heat $|Q_C|$ is rejected. This gas leaves the engine as exhaust and does not enter the engine again. But since an equivalent amount of gasoline and air enters, we may consider the process to be cyclic.

20.5 Cycle of a four-stroke internal-combustion engine.

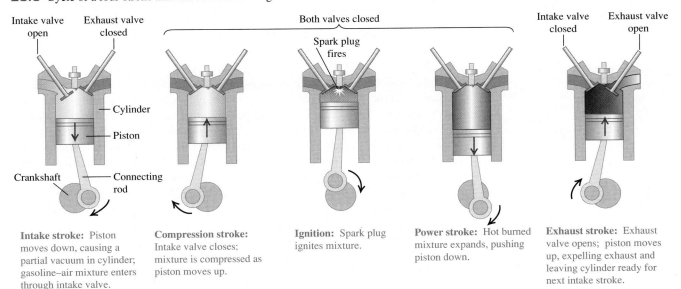

Intake stroke: Piston moves down, causing a partial vacuum in cylinder; gasoline–air mixture enters through intake valve.

Compression stroke: Intake valve closes; mixture is compressed as piston moves up.

Ignition: Spark plug ignites mixture.

Power stroke: Hot burned mixture expands, pushing piston down.

Exhaust stroke: Exhaust valve opens; piston moves up, expelling exhaust and leaving cylinder ready for next intake stroke.

20.6 The pV-diagram for the Otto cycle, an idealized model of the thermodynamic processes in a gasoline engine.

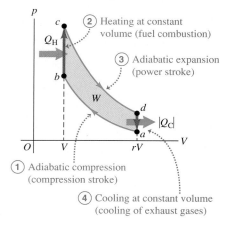

Otto cycle

② Heating at constant volume (fuel combustion)

③ Adiabatic expansion (power stroke)

① Adiabatic compression (compression stroke)

④ Cooling at constant volume (cooling of exhaust gases)

We can calculate the efficiency of this idealized cycle. Processes bc and da are constant-volume, so the heats Q_H and Q_C are related simply to the temperatures:

$$Q_H = nC_V(T_c - T_b) > 0$$
$$Q_C = nC_V(T_a - T_d) < 0$$

The thermal efficiency is given by Eq. (20.4). Inserting the above expressions and cancelling out the common factor nC_V, we find

$$e = \frac{Q_H + Q_C}{Q_H} = \frac{T_c - T_b + T_a - T_d}{T_c - T_b} \qquad (20.5)$$

To simplify this further, we use the temperature–volume relationship for adiabatic processes for an ideal gas, Eq. (19.22). For the two adiabatic processes ab and cd,

$$T_a(rV)^{\gamma-1} = T_b V^{\gamma-1} \qquad \text{and} \qquad T_d(rV)^{\gamma-1} = T_c V^{\gamma-1}$$

We divide each of these equations by the common factor $V^{\gamma-1}$ and substitute the resulting expressions for T_b and T_c back into Eq. (20.5). The result is

$$e = \frac{T_d r^{\gamma-1} - T_a r^{\gamma-1} + T_a - T_d}{T_d r^{\gamma-1} - T_a r^{\gamma-1}} = \frac{(T_d - T_a)(r^{\gamma-1} - 1)}{(T_d - T_a)r^{\gamma-1}}$$

Dividing out the common factor $(T_d - T_a)$, we get

$$e = 1 - \frac{1}{r^{\gamma-1}} \qquad \text{(thermal efficiency in Otto cycle)} \qquad (20.6)$$

The thermal efficiency given by Eq. (20.6) is always less than unity, even for this idealized model. With $r = 8$ and $\gamma = 1.4$ (the value for air) the theoretical efficiency is $e = 0.56$, or 56%. The efficiency can be increased by increasing r. However, this also increases the temperature at the end of the adiabatic compression of the air–fuel mixture. If the temperature is too high, the mixture explodes spontaneously during compression instead of burning evenly after the spark plug ignites it. This is called *pre-ignition* or *detonation;* it causes a knocking sound and can damage the engine. The octane rating of a gasoline is a measure of its antiknock qualities. The maximum practical compression ratio for high-octane, or "premium," gasoline is about 10 to 13.

The Otto cycle is a highly idealized model. It assumes that the mixture behaves as an ideal gas; it neglects friction, turbulence, loss of heat to cylinder walls, and many other effects that reduce the efficiency of an engine. Efficiencies of real gasoline engines are typically around 35%.

The Diesel Cycle

The Diesel engine is similar in operation to the gasoline engine. The most important difference is that there is no fuel in the cylinder at the beginning of the compression stroke. A little before the beginning of the power stroke, the injectors start to inject fuel directly into the cylinder, just fast enough to keep the pressure approximately constant during the first part of the power stroke. Because of the high temperature developed during the adiabatic compression, the fuel ignites spontaneously as it is injected; no spark plugs are needed.

Figure 20.7 shows the idealized **Diesel cycle.** Starting at point a, air is compressed adiabatically to point b, heated at constant pressure to point c, expanded adiabatically to point d, and cooled at constant volume to point a. Because there is no fuel in the cylinder during most of the compression stroke, pre-ignition cannot occur, and the compression ratio r can be much higher than for a gasoline engine. This improves efficiency and ensures reliable ignition when the fuel is injected (because of the high temperature reached during the

20.7 The pV-diagram for the idealized Diesel cycle.

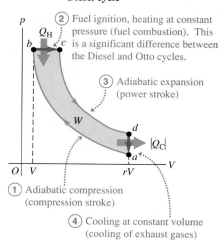

Diesel cycle

② Fuel ignition, heating at constant pressure (fuel combustion). This is a significant difference between the Diesel and Otto cycles.

③ Adiabatic expansion (power stroke)

① Adiabatic compression (compression stroke)

④ Cooling at constant volume (cooling of exhaust gases)

adiabatic compression). Values of r of 15 to 20 are typical; with these values and $\gamma = 1.4$, the theoretical efficiency of the idealized Diesel cycle is about 0.65 to 0.70. As with the Otto cycle, the efficiency of any actual engine is substantially less than this. While Diesel engines are very efficient, they must be built to much tighter tolerances than gasoline engines and the fuel-injection system requires careful maintenance.

Test Your Understanding of Section 20.3 For an Otto-cycle engine with cylinders of a fixed size and a fixed compression ratio, which of the following aspects of the pV-diagram in Fig. 20.6 would change if you doubled the amount of fuel burned per cycle? (There may be more than one correct answer.) (i) the vertical distance between points b and c; (ii) the vertical distance between points a and d; (iii) the horizontal distance between points b and a.

20.4 Refrigerators

We can think of a **refrigerator** as a heat engine operating in reverse. A heat engine takes heat from a hot place and gives off heat to a colder place. A refrigerator does the opposite; it takes heat from a cold place (the inside of the refrigerator) and gives it off to a warmer place (usually the air in the room where the refrigerator is located). A heat engine has a net *output* of mechanical work; the refrigerator requires a net *input* of mechanical work. Using the sign conventions from Section 20.2, for a refrigerator Q_C is positive but both W and Q_H are negative; hence $|W| = -W$ and $|Q_H| = -Q_H$.

Figure 20.8 shows an energy-flow diagram for a refrigerator. From the first law for a cyclic process,

$$Q_H + Q_C - W = 0 \quad \text{or} \quad -Q_H = Q_C - W$$

or, because both Q_H and W are negative,

$$|Q_H| = Q_C + |W| \tag{20.7}$$

Thus, as the diagram shows, the heat $|Q_H|$ leaving the working substance and given to the hot reservoir is always *greater* than the heat Q_C taken from the cold reservoir. Note that the absolute-value relationship

$$|Q_H| = |Q_C| + |W| \tag{20.8}$$

is valid for both heat engines and refrigerators.

From an economic point of view, the best refrigeration cycle is one that removes the greatest amount of heat $|Q_C|$ from the inside of the refrigerator for the least expenditure of mechanical work, $|W|$. The relevant ratio is therefore $|Q_C|/|W|$; the larger this ratio, the better the refrigerator. We call this ratio the **coefficient of performance,** denoted by K. From Eq. (20.8), $|W| = |Q_H| - |Q_C|$, so

$$K = \frac{|Q_C|}{|W|} = \frac{|Q_C|}{|Q_H| - |Q_C|} \quad \begin{array}{l}\text{(coefficient of performance} \\ \text{of a refrigerator)}\end{array} \tag{20.9}$$

As always, we measure Q_H, Q_C, and W all in the same energy units; K is then a dimensionless number.

Practical Refrigerators

The principles of the common refrigeration cycle are shown schematically in Fig. 20.9a. The fluid "circuit" contains a refrigerant fluid (the working substance). The left side of the circuit (including the cooling coils inside the refrigerator) is at low temperature and low pressure; the right side (including the condenser coils outside the refrigerator) is at high temperature and high pressure. Ordinarily, both sides contain liquid and vapor in phase equilibrium.

20.8 Schematic energy-flow diagram of a refrigerator.

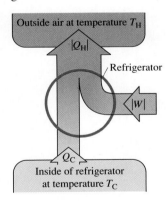

Outside air at temperature T_H

$|Q_H|$

Refrigerator

$|W|$

Q_C

Inside of refrigerator at temperature T_C

20.9 (a) Principle of the mechanical refrigeration cycle. (b) How the key elements are arranged in a practical refrigerator.

(a)   (b)

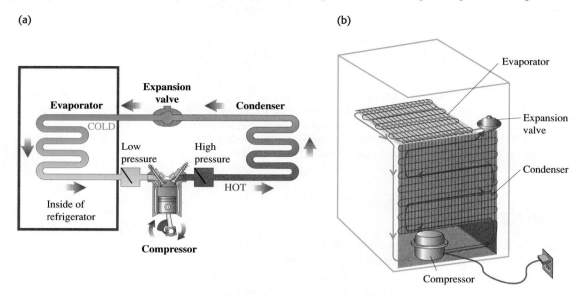

The compressor takes in fluid, compresses it adiabatically, and delivers it to the condenser coil at high pressure. The fluid temperature is then higher than that of the air surrounding the condenser, so the refrigerant gives off heat $|Q_H|$ and partially condenses to liquid. The fluid then expands adiabatically into the evaporator at a rate controlled by the expansion valve. As the fluid expands, it cools considerably, enough that the fluid in the evaporator coil is colder than its surroundings. It absorbs heat $|Q_C|$ from its surroundings, cooling them and partially vaporizing. The fluid then enters the compressor to begin another cycle. The compressor, usually driven by an electric motor (Fig. 20.9b), requires energy input and does work $|W|$ *on* the working substance during each cycle.

An air conditioner operates on exactly the same principle. In this case the refrigerator box becomes a room or an entire building. The evaporator coils are inside, the condenser is outside, and fans circulate air through these (Fig. 20.10). In large installations the condenser coils are often cooled by water. For air conditioners the quantities of greatest practical importance are the *rate* of heat removal (the heat current H from the region being cooled) and the *power* input $P = W/t$

20.10 An air conditioner works on the same principle as a refrigerator.

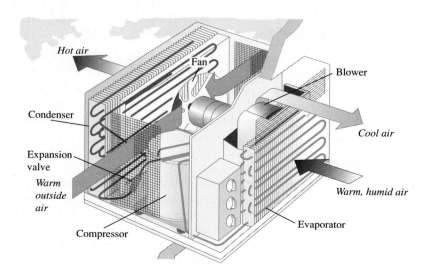

to the compressor. If heat $|Q_C|$ is removed in time t, then $H = |Q_C|/t$. Then we can express the coefficient of performance as

$$K = \frac{|Q_C|}{|W|} = \frac{Ht}{Pt} = \frac{H}{P}$$

Typical room air conditioners have heat removal rates H of 5000 to 10,000 Btu/h, or about 1500–3000 W, and require electric power input of about 600 to 1200 W. Typical coefficients of performance are about 3; the actual values depend on the inside and outside temperatures.

A variation on this theme is the **heat pump,** used to heat buildings by cooling the outside air. It functions like a refrigerator turned inside out. The evaporator coils are outside, where they take heat from cold air, and the condenser coils are inside, where they give off heat to the warmer air. With proper design, the heat $|Q_H|$ delivered to the inside per cycle can be considerably greater than the work $|W|$ required to get it there.

Work is *always* needed to transfer heat from a colder to a hotter body. Heat flows spontaneously from hotter to colder, and to reverse this flow requires the addition of work from the outside. Experience shows that it is impossible to make a refrigerator that transports heat from a colder body to a hotter body without the addition of work. If no work were needed, the coefficient of performance would be infinite. We call such a device a *workless refrigerator;* it is a mythical beast, like the unicorn and the free lunch.

Test Your Understanding of Section 20.4 Can you cool your house by leaving the refrigerator door open? ❙

20.5 The Second Law of Thermodynamics

Experimental evidence suggests strongly that it is *impossible* to build a heat engine that converts heat completely to work—that is, an engine with 100% thermal efficiency. This impossibility is the basis of one statement of the **second law of thermodynamics,** as follows:

> **It is impossible for any system to undergo a process in which it absorbs heat from a reservoir at a single temperature and converts the heat completely into mechanical work, with the system ending in the same state in which it began.**

We will call this the "engine" statement of the second law. (It is also known to physicists as the *Kelvin–Planck statement* of this law.)

The basis of the second law of thermodynamics is the difference between the nature of internal energy and that of macroscopic mechanical energy. In a moving body the molecules have random motion, but superimposed on this is a coordinated motion of every molecule in the direction of the body's velocity. The kinetic energy associated with this *coordinated* macroscopic motion is what we call the kinetic energy of the moving body. The kinetic and potential energies associated with the *random* motion constitute the internal energy.

When a body sliding on a surface comes to rest as a result of friction, the organized motion of the body is converted to random motion of molecules in the body and in the surface. Since we cannot control the motions of individual molecules, we cannot convert this random motion completely back to organized motion. We can convert *part* of it, and this is what a heat engine does.

If the second law were *not* true, we could power an automobile or run a power plant by cooling the surrounding air. Neither of these impossibilities violates the *first* law of thermodynamics. The second law, therefore, is not a deduction from the first but stands by itself as a separate law of nature. The first law denies the possibility of creating or destroying energy; the second law limits the *availability* of energy and the ways in which it can be used and converted.

Restating the Second Law

Our analysis of refrigerators in Section 20.4 forms the basis for an alternative statement of the second law of thermodynamics. Heat flows spontaneously from hotter to colder bodies, never the reverse. A refrigerator does take heat from a colder to a hotter body, but its operation requires an input of mechanical energy or work. Generalizing this observation, we state:

> **It is impossible for any process to have as its sole result the transfer of heat from a cooler to a hotter body.**

We'll call this the "refrigerator" statement of the second law. (It is also known as the *Clausius statement.*) It may not seem to be very closely related to the "engine" statement. In fact, though, the two statements are completely equivalent. For example, if we could build a workless refrigerator, violating the second or "refrigerator" statement of the second law, we could use it in conjunction with a heat engine, pumping the heat rejected by the engine back to the hot reservoir to be reused. This composite machine (Fig. 20.11a) would violate the "engine" statement of the second law because its net effect would be to take a net quantity of heat $Q_H - |Q_C|$ from the hot reservoir and convert it completely to work W.

Alternatively, if we could make an engine with 100% thermal efficiency, in violation of the first statement, we could run it using heat from the hot reservoir and use the work output to drive a refrigerator that pumps heat from the cold reservoir to the hot (Fig. 20.11b). This composite device would violate the "refrigerator" statement because its net effect would be to take heat Q_C from the

20.11 Energy-flow diagrams showing that the two forms of the second law are equivalent.

(a) The "engine" statement of the second law of thermodynamics

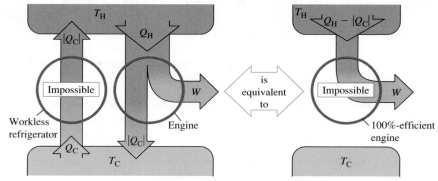

If a workless refrigerator were possible, it could be used in conjunction with an ordinary heat engine to form a 100%-efficient engine, converting heat $Q_H - |Q_C|$ completely to work.

(b) The "refrigerator" statement of the second law of thermodynamics

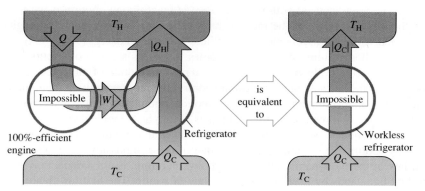

If a 100%-efficient engine were possible, it could be used in conjunction with an ordinary refrigerator to form a workless refrigerator, transferring heat Q_C from the cold to the hot reservoir with no input of work.

cold reservoir and deliver it to the hot reservoir without requiring any input of work. Thus any device that violates one form of the second law can be used to make a device that violates the other form. If violations of the first form are impossible, so are violations of the second!

The conversion of work to heat and the heat flow from hot to cold across a finite temperature gradient are *irreversible* processes. The "engine" and "refrigerator" statements of the second law state that these processes can be only partially reversed. We could cite other examples. Gases naturally flow from a region of high pressure to a region of low pressure; gases and miscible liquids left by themselves always tend to mix, not to unmix. The second law of thermodynamics is an expression of the inherent one-way aspect of these and many other irreversible processes. Energy conversion is an essential aspect of all plant and animal life and of human technology, so the second law of thermodynamics is of fundamental importance.

Test Your Understanding of Section 20.5   Would a 100%-efficient engine (Fig. 20.11a) violate the *first* law of thermodynamics? What about a workless refrigerator (Fig. 20.11b)? ❚

20.6 The Carnot Cycle

According to the second law, no heat engine can have 100% efficiency. How great an efficiency *can* an engine have, given two heat reservoirs at temperatures T_H and T_C? This question was answered in 1824 by the French engineer Sadi Carnot (1796–1832), who developed a hypothetical, idealized heat engine that has the maximum possible efficiency consistent with the second law. The cycle of this engine is called the **Carnot cycle.**

To understand the rationale of the Carnot cycle, we return to *reversibility* and its relationship to directions of thermodynamic processes. Conversion of work to heat is an irreversible process; the purpose of a heat engine is a *partial* reversal of this process, the conversion of heat to work with as great an efficiency as possible. For maximum heat-engine efficiency, therefore, *we must avoid all irreversible processes* (Fig. 20.12).

Heat flow through a finite temperature drop is an irreversible process. Therefore, during heat transfer in the Carnot cycle there must be *no* finite temperature difference. When the engine takes heat from the hot reservoir at temperature T_H, the working substance of the engine must also be at T_H; otherwise, irreversible heat flow would occur. Similarly, when the engine discards heat to the cold reservoir at T_C, the engine itself must be at T_C. That is, every process that involves heat transfer must be *isothermal* at either T_H or T_C.

Conversely, in any process in which the temperature of the working substance of the engine is intermediate between T_H and T_C, there must be *no* heat transfer between the engine and either reservoir because such heat transfer could not be reversible. Therefore any process in which the temperature T of the working substance changes must be *adiabatic.*

The bottom line is that every process in our idealized cycle must be either isothermal or adiabatic. In addition, thermal and mechanical equilibrium must be maintained at all times so that each process is completely reversible.

Steps of the Carnot Cycle

The Carnot cycle consists of two reversible isothermal and two reversible adiabatic processes. Figure 20.13 shows a Carnot cycle using as its working substance an ideal gas in a cylinder with a piston. It consists of the following steps:

1. The gas expands isothermally at temperature T_H, absorbing heat Q_H (*ab*).
2. It expands adiabatically until its temperature drops to T_C (*bc*).
3. It is compressed isothermally at T_C, rejecting heat $|Q_C|$ (*cd*).
4. It is compressed adiabatically back to its initial state at temperature T_H (*da*).

20.12 The temperature of the firebox of a steam engine is much higher than the temperature of water in the boiler, so heat flows irreversibly from firebox to water. Carnot's quest to understand the efficiency of steam engines led him to the idea that an ideal engine would involve only *reversible* processes.

Mastering**PHYSICS**

ActivPhysics 8.14: Carnot Cycle

20.13 The Carnot cycle for an ideal gas. The light blue lines in the *pV*-diagram are isotherms (curves of constant temperature) and the dark blue lines are adiabats (curves of zero heat flow).

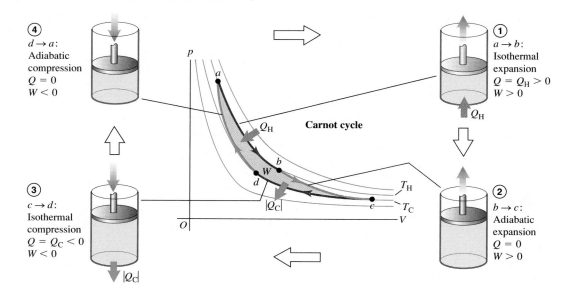

We can calculate the thermal efficiency *e* of a Carnot engine in the special case shown in Fig. 20.13 in which the working substance is an *ideal gas*. To carry out this calculation, we will first find the ratio Q_C/Q_H of the quantities of heat transferred in the two isothermal processes and then use Eq. (20.4) to find *e*.

For an ideal gas the internal energy *U* depends only on temperature and is thus constant in any isothermal process. For the isothermal expansion *ab*, $\Delta U_{ab} = 0$ and Q_H is equal to the work W_{ab} done by the gas during its isothermal expansion at temperature T_H. We calculated this work in Example 19.1 (Section 19.2); using that result, we have

$$Q_H = W_{ab} = nRT_H \ln \frac{V_b}{V_a} \tag{20.10}$$

Similarly,

$$Q_C = W_{cd} = nRT_C \ln \frac{V_d}{V_c} = -nRT_C \ln \frac{V_c}{V_d} \tag{20.11}$$

Because V_d is less than V_c, Q_C is negative ($Q_C = -|Q_C|$); heat flows out of the gas during the isothermal compression at temperature T_C.

The ratio of the two quantities of heat is thus

$$\frac{Q_C}{Q_H} = -\left(\frac{T_C}{T_H}\right) \frac{\ln (V_c/V_d)}{\ln (V_b/V_a)} \tag{20.12}$$

This can be simplified further by use of the temperature–volume relationship for an adiabatic process, Eq. (19.22). We find for the two adiabatic processes:

$$T_H V_b^{\gamma-1} = T_C V_c^{\gamma-1} \quad \text{and} \quad T_H V_a^{\gamma-1} = T_C V_d^{\gamma-1}$$

Dividing the first of these by the second, we find

$$\frac{V_b^{\gamma-1}}{V_a^{\gamma-1}} = \frac{V_c^{\gamma-1}}{V_d^{\gamma-1}} \quad \text{and} \quad \frac{V_b}{V_a} = \frac{V_c}{V_d}$$

Thus the two logarithms in Eq. (20.12) are equal, and that equation reduces to

$$\frac{Q_C}{Q_H} = -\frac{T_C}{T_H} \quad \text{or} \quad \frac{|Q_C|}{|Q_H|} = \frac{T_C}{T_H} \qquad \text{(heat transfer in a Carnot engine)} \quad \text{(20.13)}$$

The ratio of the heat rejected at T_C to the heat absorbed at T_H is just equal to the ratio T_C/T_H. Then from Eq. (20.4) the efficiency of the Carnot engine is

$$e_{\text{Carnot}} = 1 - \frac{T_C}{T_H} = \frac{T_H - T_C}{T_H} \qquad \text{(efficiency of a Carnot engine)} \quad \text{(20.14)}$$

This simple result says that the efficiency of a Carnot engine depends only on the temperatures of the two heat reservoirs. The efficiency is large when the temperature *difference* is large, and it is very small when the temperatures are nearly equal. The efficiency can never be exactly unity unless $T_C = 0$; we'll see later that this, too, is impossible.

CAUTION **Use Kelvin temperature in Carnot calculations** In all calculations involving the Carnot cycle, you must make sure that you use *absolute* (Kelvin) temperatures only. That's because Eqs. (20.10) through (20.14) come from the ideal-gas equation $pV = nRT$, in which T is absolute temperature. ▌

Example 20.2   **Analyzing a Carnot engine I**

A Carnot engine takes 2000 J of heat from a reservoir at 500 K, does some work, and discards some heat to a reservoir at 350 K. How much work does it do, how much heat is discarded, and what is its efficiency?

SOLUTION

IDENTIFY and SET UP: This problem involves a Carnot engine, so we can use the ideas of this section and those of Section 20.2 (which apply to heat engines of all kinds). Figure 20.14 shows the energy-flow diagram. We have $Q_H = 2000$ J, $T_H = 500$ K, and $T_C = 350$ K. We use Eq. (20.13) to find Q_C, and then use the first law of thermodynamics as given by Eq. (20.2) to find W. We find the efficiency e from T_C and T_H using Eq. (20.14).

EXECUTE: From Eq. (20.13),

$$Q_C = -Q_H \frac{T_C}{T_H} = -(2000 \text{ J}) \frac{350 \text{ K}}{500 \text{ K}} = -1400 \text{ J}$$

Then from Eq. (20.2), the work done is

$$W = Q_H + Q_C = 2000 \text{ J} + (-1400 \text{ J}) = 600 \text{ J}$$

From Eq. (20.14), the thermal efficiency is

$$e = 1 - \frac{T_C}{T_H} = 1 - \frac{350 \text{ K}}{500 \text{ K}} = 0.30 = 30\%$$

EVALUATE: The negative sign of Q_C is correct: It shows that 1400 J of heat flows *out* of the engine and into the cold reservoir. We can check our result for e by using the basic definition of thermal efficiency, Eq. (20.3):

$$e = \frac{W}{Q_H} = \frac{600 \text{ J}}{2000 \text{ J}} = 0.30 = 30\%$$

20.14 Our sketch for this problem.

$T_H = 500$ K
$Q_H = 2000$ J
$W = ?$
$e = ?$
$Q_C = ?$
$T_C = 350$ K

Example 20.3   **Analyzing a Carnot engine II**

Suppose 0.200 mol of an ideal diatomic gas ($\gamma = 1.40$) undergoes a Carnot cycle between 227°C and 27°C, starting at $p_a = 10.0 \times 10^5$ Pa at point a in the pV-diagram of Fig. 20.13. The volume doubles during the isothermal expansion step $a \to b$. (a) Find the pressure and volume at points a, b, c, and d. (b) Find Q, W, and ΔU for each step and for the entire cycle. (c) Find the efficiency directly from the results of part (b), and compare with the value calculated from Eq. (20.14).

Continued

SOLUTION

IDENTIFY and SET UP: This problem involves the properties of the Carnot cycle and those of an ideal gas. We are given the number of moles n and the pressure and temperature at point a (which is at the higher of the two reservoir temperatures); we can find the volume at a using the ideal-gas equation $pV = nRT$. We then find the pressure and volume at points b, c, and d from the known doubling of volume in step $a \rightarrow b$, from equations given in this section, and from $pV = nRT$. In each step we use Eqs. (20.10) and (20.11) to find the heat flow and work done and Eq. (19.13) to find the internal energy change.

EXECUTE: (a) With $T_H = (227 + 273.15)$ K $= 500$ K and $T_C = (27 + 273.15)$ K $= 300$ K, $pV = nRT$ yields

$$V_a = \frac{nRT_H}{p_a} = \frac{(0.200 \text{ mol})(8.314 \text{ J/mol} \cdot \text{K})(500 \text{ K})}{10.0 \times 10^5 \text{ Pa}}$$

$$= 8.31 \times 10^{-4} \text{ m}^3$$

The volume doubles during the isothermal expansion $a \rightarrow b$:

$$V_b = 2V_a = 2(8.31 \times 10^{-4} \text{ m}^3) = 16.6 \times 10^{-4} \text{ m}^3$$

Because the expansion $a \rightarrow b$ is isothermal, $p_a V_a = p_b V_b$, so

$$p_b = \frac{p_a V_a}{V_b} = 5.00 \times 10^5 \text{ Pa}$$

For the adiabatic expansion $b \rightarrow c$, we use the equation $T_H V_b^{\gamma-1} = T_C V_c^{\gamma-1}$ that follows Eq. (20.12) as well as the ideal-gas equation:

$$V_c = V_b \left(\frac{T_H}{T_C} \right)^{1/(\gamma-1)} = (16.6 \times 10^{-4} \text{ m}^3) \left(\frac{500 \text{ K}}{300 \text{ K}} \right)^{2.5}$$

$$= 59.6 \times 10^{-4} \text{ m}^3$$

$$p_c = \frac{nRT_C}{V_c} = \frac{(0.200 \text{ mol})(8.314 \text{ J/mol} \cdot \text{K})(300 \text{ K})}{59.6 \times 10^{-4} \text{ m}^3}$$

$$= 0.837 \times 10^5 \text{ Pa}$$

For the adiabatic compression $d \rightarrow a$ we have $T_C V_d^{\gamma-1} = T_H V_a^{\gamma-1}$ and so

$$V_d = V_a \left(\frac{T_H}{T_C} \right)^{1/(\gamma-1)} = (8.31 \times 10^{-4} \text{ m}^3) \left(\frac{500 \text{ K}}{300 \text{ K}} \right)^{2.5}$$

$$= 29.8 \times 10^{-4} \text{ m}^3$$

$$p_d = \frac{nRT_C}{V_d} = \frac{(0.200 \text{ mol})(8.314 \text{ J/mol} \cdot \text{K})(300 \text{ K})}{29.8 \times 10^{-4} \text{ m}^3}$$

$$= 1.67 \times 10^5 \text{ Pa}$$

(b) For the isothermal expansion $a \rightarrow b$, $\Delta U_{ab} = 0$. From Eq. (20.10),

$$W_{ab} = Q_H = nRT_H \ln \frac{V_b}{V_a}$$

$$= (0.200 \text{ mol})(8.314 \text{ J/mol} \cdot \text{K})(500 \text{ K})(\ln 2) = 576 \text{ J}$$

For the adiabatic expansion $b \rightarrow c$, $Q_{bc} = 0$. From the first law of thermodynamics, $\Delta U_{bc} = Q_{bc} - W_{bc} = -W_{bc}$; the work W_{bc} done by the gas in this adiabatic expansion equals the negative of the change in internal energy of the gas. From Eq. (19.13) we have $\Delta U = nC_V \Delta T$, where $\Delta T = T_C - T_H$. Using $C_V = 20.8$ J/mol $\cdot$ K for an ideal diatomic gas, we find

$$W_{bc} = -\Delta U_{bc} = -nC_V(T_C - T_H) = nC_V(T_H - T_C)$$

$$= (0.200 \text{ mol})(20.8 \text{ J/mol} \cdot \text{K})(500 \text{ K} - 300 \text{ K}) = 832 \text{ J}$$

For the isothermal compression $c \rightarrow d$, $\Delta U_{cd} = 0$; Eq. (20.11) gives

$$W_{cd} = Q_C = nRT_C \ln \frac{V_d}{V_c}$$

$$= (0.200 \text{ mol})(8.314 \text{ J/mol} \cdot \text{K})(300 \text{ K}) \left(\ln \frac{29.8 \times 10^{-4} \text{ m}^3}{59.6 \times 10^{-4} \text{ m}^3} \right)$$

$$= -346 \text{ J}$$

For the adiabatic compression $d \rightarrow a$, $Q_{da} = 0$ and

$$W_{da} = -\Delta U_{da} = -nC_V(T_H - T_C) = nC_V(T_C - T_H)$$

$$= (0.200 \text{ mol})(20.8 \text{ J/mol} \cdot \text{K})(300 \text{ K} - 500 \text{ K}) = -832 \text{ J}$$

We can tabulate these results as follows:

Process	Q	W	ΔU
$a \rightarrow b$	576 J	576 J	0
$b \rightarrow c$	0	832 J	-832 J
$c \rightarrow d$	-346 J	-346 J	0
$d \rightarrow a$	0	-832 J	832 J
Total	230 J	230 J	0

(c) From the above table, $Q_H = 576$ J and the total work is 230 J. Thus

$$e = \frac{W}{Q_H} = \frac{230 \text{ J}}{576 \text{ J}} = 0.40 = 40\%$$

We can compare this to the result from Eq. (20.14),

$$e = \frac{T_H - T_C}{T_H} = \frac{500 \text{ K} - 300 \text{ K}}{500 \text{ K}} = 0.40 = 40\%$$

EVALUATE: The table in part (b) shows that for the entire cycle $Q = W$ and $\Delta U = 0$, just as we would expect: In a complete cycle, the *net* heat input is used to do work, and there is zero net change in the internal energy of the system. Note also that the quantities of work in the two adiabatic processes are negatives of each other. Can you show from the analysis leading to Eq. (20.13) that this must *always* be the case in a Carnot cycle?

The Carnot Refrigerator

Because each step in the Carnot cycle is reversible, the *entire cycle* may be reversed, converting the engine into a refrigerator. The coefficient of performance of the Carnot refrigerator is obtained by combining the general definition of K, Eq. (20.9), with Eq. (20.13) for the Carnot cycle. We first rewrite Eq. (20.9) as

$$K = \frac{|Q_C|}{|Q_H| - |Q_C|} = \frac{|Q_C|/|Q_H|}{1 - |Q_C|/|Q_H|}$$

Then we substitute Eq. (20.13), $|Q_C|/|Q_H| = T_C/T_H$, into this expression. The result is

$$K_{Carnot} = \frac{T_C}{T_H - T_C} \qquad \text{(coefficient of performance of a Carnot refrigerator)} \qquad (20.15)$$

When the temperature difference $T_H - T_C$ is small, K is much larger than unity; in this case a lot of heat can be "pumped" from the lower to the higher temperature with only a little expenditure of work. But the greater the temperature difference, the smaller the value of K and the more work is required to transfer a given quantity of heat.

Example 20.4   **Analyzing a Carnot refrigerator**

If the cycle described in Example 20.3 is run backward as a refrigerator, what is its coefficient of performance?

SOLUTION

IDENTIFY and SET UP: This problem uses the ideas of Section 20.3 (for refrigerators in general) and the above discussion of Carnot refrigerators. Equation (20.9) gives the coefficient of performance K of *any* refrigerator in terms of the heat Q_C extracted from the cold reservoir per cycle and the work W that must be done per cycle.

EXECUTE: In Example 20.3 we found that in one cycle the Carnot engine rejects heat $Q_C = -346$ J to the cold reservoir and does work $W = 230$ J. When run in reverse as a refrigerator, the system

extracts heat $Q_C = +346$ J from the cold reservoir while requiring a work input of $W = -230$ J. From Eq. (20.9),

$$K = \frac{|Q_C|}{|W|} = \frac{346 \text{ J}}{230 \text{ J}} = 1.50$$

Because this is a Carnot cycle, we can also use Eq. (20.15):

$$K = \frac{T_C}{T_H - T_C} = \frac{300 \text{ K}}{500 \text{ K} - 300 \text{ K}} = 1.50$$

EVALUATE: Equations (20.14) and (20.15) show that e and K for a Carnot cycle depend only on T_H and T_C, and we don't need to calculate Q and W. For cycles containing irreversible processes, however, these two equations are not valid, and more detailed calculations are necessary.

The Carnot Cycle and the Second Law

We can prove that **no engine can be more efficient than a Carnot engine operating between the same two temperatures.** The key to the proof is the above observation that since each step in the Carnot cycle is reversible, the *entire cycle* may be reversed. Run backward, the engine becomes a refrigerator. Suppose we have an engine that is more efficient than a Carnot engine (Fig. 20.15). Let the Carnot engine, run backward as a refrigerator by negative work $-|W|$, take

20.15 Proving that the Carnot engine has the highest possible efficiency. A "superefficient" engine (more efficient than a Carnot engine) combined with a Carnot refrigerator could convert heat completely into work with no net heat transfer to the cold reservoir. This would violate the second law of thermodynamics.

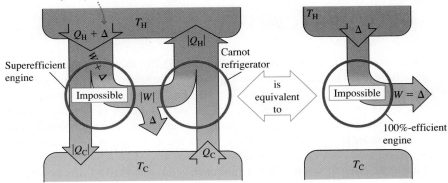

If a superefficient engine were possible, it could be used in conjunction with a Carnot refrigerator to convert the heat Δ completely to work, with no net transfer to the cold reservoir.

in heat Q_C from the cold reservoir and expel heat $|Q_H|$ to the hot reservoir. The superefficient engine expels heat $|Q_C|$, but to do this, it takes in a greater amount of heat $Q_H + \Delta$. Its work output is then $W + \Delta$, and the net effect of the two machines together is to take a quantity of heat Δ and convert it completely into work. This violates the engine statement of the second law. We could construct a similar argument that a superefficient engine could be used to violate the refrigerator statement of the second law. Note that we don't have to assume that the superefficient engine is reversible. In a similar way we can show that *no refrigerator can have a greater coefficient of performance than a Carnot refrigerator operating between the same two temperatures.*

Thus the statement that no engine can be more efficient than a Carnot engine is yet another equivalent statement of the second law of thermodynamics. It also follows directly that **all Carnot engines operating between the same two temperatures have the same efficiency, irrespective of the nature of the working substance.** Although we derived Eq. (20.14) for a Carnot engine using an ideal gas as its working substance, it is in fact valid for *any* Carnot engine, no matter what its working substance.

Equation (20.14), the expression for the efficiency of a Carnot engine, sets an upper limit to the efficiency of a real engine such as a steam turbine. To maximize this upper limit and the actual efficiency of the real engine, the designer must make the intake temperature T_H as high as possible and the exhaust temperature T_C as low as possible (Fig. 20.16).

The exhaust temperature cannot be lower than the lowest temperature available for cooling the exhaust. For a steam turbine at an electric power plant, T_C may be the temperature of river or lake water; then we want the boiler temperature T_H to be as high as possible. The vapor pressures of all liquids increase rapidly with temperature, so we are limited by the mechanical strength of the boiler. At 500°C the vapor pressure of water is about 240×10^5 Pa (235 atm); this is about the maximum practical pressure in large present-day steam boilers.

The Kelvin Temperature Scale

In Chapter 17 we expressed the need for a temperature scale that doesn't depend on the properties of any particular material. We can now use the Carnot cycle to define such a scale. The thermal efficiency of a Carnot engine operating between two heat reservoirs at temperatures T_H and T_C is independent of the nature of the working substance and depends only on the temperatures. From Eq. (20.4), this thermal efficiency is

$$e = \frac{Q_H + Q_C}{Q_H} = 1 + \frac{Q_C}{Q_H}$$

Therefore the ratio Q_C/Q_H is the same for *all* Carnot engines operating between two given temperatures T_H and T_C.

Kelvin proposed that we *define* the ratio of the temperatures, T_C/T_H, to be equal to the magnitude of the ratio Q_C/Q_H of the quantities of heat absorbed and rejected:

$$\frac{T_C}{T_H} = \frac{|Q_C|}{|Q_H|} = -\frac{Q_C}{Q_H} \qquad \text{(definition of Kelvin temperature)} \quad (20.16)$$

Equation (20.16) looks identical to Eq. (20.13), but there is a subtle and crucial difference. The temperatures in Eq. (20.13) are based on an ideal-gas thermometer, as defined in Section 17.3, while Eq. (20.16) *defines* a temperature scale based on the Carnot cycle and the second law of thermodynamics and is independent of the behavior of any particular substance. Thus the **Kelvin temperature scale** is truly *absolute*. To complete the definition of the Kelvin scale, we assign, as in Section 17.3, the arbitrary value of 273.16 K to the temperature of the triple point of water. When a substance is taken around a Carnot cycle, the

20.16 To maximize efficiency, the temperatures inside a jet engine are made as high as possible. Exotic ceramic materials are used that can withstand temperatures in excess of 1000°C without melting or becoming soft.

ratio of the heats absorbed and rejected, $|Q_H|/|Q_C|$, is equal to the ratio of the temperatures of the reservoirs *as expressed on the gas-thermometer scale* defined in Section 17.3. Since the triple point of water is chosen to be 273.16 K in both scales, it follows that *the Kelvin and ideal-gas scales are identical.*

The zero point on the Kelvin scale is called **absolute zero.** At absolute zero a system has its *minimum* possible total internal energy (kinetic plus potential). Because of quantum effects, however, it is *not* true that at $T = 0$, all molecular motion ceases. There are theoretical reasons for believing that absolute zero cannot be attained experimentally, although temperatures below 10^{-7} K have been achieved. The more closely we approach absolute zero, the more difficult it is to get closer. One statement of the *third law of thermodynamics* is that it is impossible to reach absolute zero in a finite number of thermodynamic steps.

Test Your Understanding of Section 20.6 An inventor looking for financial support comes to you with an idea for a gasoline engine that runs on a novel type of thermodynamic cycle. His design is made entirely of copper and is air-cooled. He claims that the engine will be 85% efficient. Should you invest in this marvelous new engine? (*Hint:* See Table 17.4.)   ❙

20.7 Entropy

The second law of thermodynamics, as we have stated it, is not an equation or a quantitative relationship but rather a statement of *impossibility.* However, the second law *can* be stated as a quantitative relationship with the concept of *entropy,* the subject of this section.

We have talked about several processes that proceed naturally in the direction of increasing disorder. Irreversible heat flow increases disorder because the molecules are initially sorted into hotter and cooler regions; this sorting is lost when the system comes to thermal equilibrium. Adding heat to a body increases its disorder because it increases average molecular speeds and therefore the randomness of molecular motion. Free expansion of a gas increases its disorder because the molecules have greater randomness of position after the expansion than before. Figure 20.17 shows another process in which disorder increases.

20.17 When firecrackers explode, disorder increases: The neatly packaged chemicals within each firecracker are dispersed in all directions, and the stored chemical energy is converted to random kinetic energy of the fragments.

Entropy and Disorder

Entropy provides a *quantitative* measure of disorder. To introduce this concept, let's consider an infinitesimal isothermal expansion of an ideal gas. We add heat dQ and let the gas expand just enough to keep the temperature constant. Because the internal energy of an ideal gas depends only on its temperature, the internal energy is also constant; thus from the first law, the work dW done by the gas is equal to the heat dQ added. That is,

$$dQ = dW = p\,dV = \frac{nRT}{V}\,dV \qquad \text{so} \qquad \frac{dV}{V} = \frac{dQ}{nRT}$$

The gas is more disordered after the expansion than before: The molecules are moving in a larger volume and have more randomness of position. Thus the fractional volume change dV/V is a measure of the increase in disorder, and the above equation shows that it is proportional to the quantity dQ/T. We introduce the symbol S for the entropy of the system, and we define the infinitesimal entropy change dS during an infinitesimal reversible process at absolute temperature T as

$$dS = \frac{dQ}{T} \qquad \text{(infinitesimal reversible process)} \qquad (20.17)$$

If a total amount of heat Q is added during a reversible isothermal process at absolute temperature T, the total entropy change $\Delta S = S_2 - S_1$ is given by

$$\Delta S = S_2 - S_1 = \frac{Q}{T} \qquad \text{(reversible isothermal process)} \qquad (20.18)$$

Entropy has units of energy divided by temperature; the SI unit of entropy is 1 J/K.

We can see how the quotient Q/T is related to the increase in disorder. Higher temperature means greater randomness of motion. If the substance is initially cold, with little molecular motion, adding heat Q causes a substantial fractional increase in molecular motion and randomness. But if the substance is already hot, the same quantity of heat adds relatively little to the greater molecular motion already present. So Q/T is an appropriate characterization of the increase in randomness or disorder when heat flows into a system.

Example 20.5 Entropy change in melting

What is the change of entropy of 1 kg of ice that is melted reversibly at 0°C and converted to water at 0°C? The heat of fusion of water is $L_f = 3.34 \times 10^5$ J/kg.

SOLUTION

IDENTIFY and SET UP: The melting occurs at a constant temperature $T = 0°C = 273$ K, so this is an *isothermal* reversible process. We can calculate the added heat Q required to melt the ice, then calculate the entropy change ΔS using Eq. (20.18).

EXECUTE: The heat needed to melt the ice is $Q = mL_f = 3.34 \times 10^5$ J. Then from Eq. (20.18),

$$\Delta S = S_2 - S_1 = \frac{Q}{T} = \frac{3.34 \times 10^5 \text{ J}}{273 \text{ K}} = 1.22 \times 10^3 \text{ J/K}$$

EVALUATE: This entropy increase corresponds to the increase in disorder when the water molecules go from the highly ordered state of a crystalline solid to the much more disordered state of a liquid. In *any* isothermal reversible process, the entropy change equals the heat transferred divided by the absolute temperature. When we refreeze the water, Q has the opposite sign, and the entropy change is $\Delta S = -1.22 \times 10^3$ J/K. The water molecules rearrange themselves into a crystal to form ice, so disorder and entropy both decrease.

Entropy in Reversible Processes

We can generalize the definition of entropy change to include *any* reversible process leading from one state to another, whether it is isothermal or not. We represent the process as a series of infinitesimal reversible steps. During a typical step, an infinitesimal quantity of heat dQ is added to the system at absolute temperature T. Then we sum (integrate) the quotients dQ/T for the entire process; that is,

$$\Delta S = \int_1^2 \frac{dQ}{T} \qquad \text{(entropy change in a reversible process)} \quad (20.19)$$

The limits 1 and 2 refer to the initial and final states.

Because entropy is a measure of the disorder of a system in any specific state, it must depend only on the current state of the system, not on its past history. (We will verify this later.) When a system proceeds from an initial state with entropy S_1 to a final state with entropy S_2, the change in entropy $\Delta S = S_2 - S_1$ defined by Eq. (20.19) does not depend on the path leading from the initial to the final state but is the same for *all possible* processes leading from state 1 to state 2. Thus the entropy of a system must also have a definite value for any given state of the system. *Internal energy,* introduced in Chapter 19, also has this property, although entropy and internal energy are very different quantities.

Since entropy is a function only of the state of a system, we can also compute entropy changes in *irreversible* (nonequilibrium) processes for which Eqs. (20.17) and (20.19) are not applicable. We simply invent a path connecting the given initial and final states that *does* consist entirely of reversible equilibrium processes and compute the total entropy change for that path. It is not the actual path, but the entropy change must be the same as for the actual path.

As with internal energy, the above discussion does not tell us how to calculate entropy itself, but only the change in entropy in any given process. Just as with internal energy, we may arbitrarily assign a value to the entropy of a system in a specified reference state and then calculate the entropy of any other state with reference to this.

Example 20.6 | **Entropy change in a temperature change**

One kilogram of water at 0°C is heated to 100°C. Compute its change in entropy. Assume that the specific heat of water is constant at 4190 J/kg·K over this temperature range.

SOLUTION

IDENTIFY and SET UP: The entropy change of the water depends only on the initial and final states of the system, no matter whether the process is reversible or irreversible. We can imagine a reversible process in which the water temperature is increased in a sequence of infinitesimal steps dT. We can use Eq. (20.19) to integrate over all these steps and calculate the entropy change for such a reversible process. (Heating the water on a stove whose cooking surface is maintained at 100°C would be an irreversible process. The entropy change would be the same, however.)

EXECUTE: From Eq. (17.14) the heat required to carry out each infinitesimal step is $dQ = mc\,dT$. Substituting this into Eq. (20.19) and integrating, we find

$$\Delta S = S_2 - S_1 = \int_1^2 \frac{dQ}{T} = \int_{T_1}^{T_2} mc\frac{dT}{T} = mc\,\ln\frac{T_2}{T_1}$$

$$= (1.00\ \text{kg})(4190\ \text{J/kg}\cdot\text{K})\left(\ln\frac{373\ \text{K}}{273\ \text{K}}\right)$$

$$= 1.31 \times 10^3\ \text{J/K}$$

EVALUATE: The entropy change is positive, as it must be for a process in which the system absorbs heat. Our assumption about the specific heat is a pretty good one, since c for water increases by only 1% between 0°C and 100°C.

CAUTION **When $\Delta S = Q/T$ can (and cannot) be used** In solving this problem you might be tempted to avoid doing an integral by using the simpler expression in Eq. (20.18), $\Delta S = Q/T$. This would be incorrect, however, because Eq. (20.18) is applicable only to *isothermal* processes, and the initial and final temperatures in our example are *not* the same. The *only* correct way to find the entropy change in a process with different initial and final temperatures is to use Eq. (20.19). ▮

Conceptual Example 20.7 | **Entropy change in a reversible adiabatic process**

A gas expands adiabatically and reversibly. What is its change in entropy?

SOLUTION

In an adiabatic process, no heat enters or leaves the system. Hence $dQ = 0$ and there is *no* change in entropy in this reversible

process: $\Delta S = 0$. Every *reversible* adiabatic process is a constant-entropy process. (That's why such processes are also called *isentropic* processes.) The increase in disorder resulting from the gas occupying a greater volume is exactly balanced by the decrease in disorder associated with the lowered temperature and reduced molecular speeds.

Example 20.8 | **Entropy change in a free expansion**

A partition divides a thermally insulated box into two compartments, each of volume V (Fig. 20.18). Initially, one compartment contains n moles of an ideal gas at temperature T, and the other compartment is evacuated. We break the partition and the gas expands, filling both compartments. What is the entropy change in this free-expansion process?

SOLUTION

IDENTIFY and SET UP: For this process, $Q = 0$, $W = 0$, $\Delta U = 0$, and therefore (because the system is an ideal gas) $\Delta T = 0$. We might think that the entropy change is zero because there is no heat exchange. But Eq. (20.19) can be used to calculate entropy changes for *reversible* processes only; this free expansion is *not* reversible, and there *is* an entropy change. As we mentioned at the beginning of this section, entropy increases in a free expansion because the positions of the molecules are more random than before the expansion. To calculate ΔS, we recall that the entropy change depends only on the initial and final states. We can devise a

20.18 (a, b) Free expansion of an insulated ideal gas. (c) The free-expansion process doesn't pass through equilibrium states from a to b. However, the entropy change $S_b - S_a$ can be calculated by using the isothermal path shown or *any* reversible path from a to b.

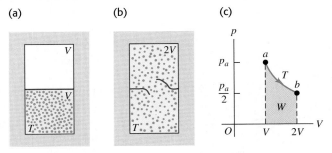

reversible process having the same endpoints as this free expansion, and in general we can then use Eq. (20.19) to calculate its entropy change, which will be the same as for the free expansion.

Continued

An appropriate reversible process is an *isothermal* expansion from V to $2V$ at temperature T, which allows us to use the simpler Eq. (20.18) to calculate ΔS. The gas does work W during this expansion, so an equal amount of heat Q must be supplied to keep the internal energy constant.

EXECUTE: We saw in Example 19.1 that the work done by n moles of ideal gas in an isothermal expansion from V_1 to V_2 is $W = nRT \ln(V_2/V_1)$. With $V_1 = V$ and $V_2 = 2V$, we have

$$Q = W = nRT \ln\frac{2V}{V} = nRT \ln 2$$

From Eq. (20.18), the entropy change is

$$\Delta S = \frac{Q}{T} = nR \ln 2$$

EVALUATE: For 1 mole, $\Delta S = (1 \text{ mol})(8.314 \text{ J/mol} \cdot \text{K})(\ln 2) = 5.76 \text{ J/K}$. The entropy change is positive, as we predicted. The factor $(\ln 2)$ in our answer is a result of the volume having increased by a factor of 2, from V to $2V$. Can you show that if the volume increases in a free expansion from V to xV, where x is an arbitrary number, the entropy change is $\Delta S = nR \ln x$?

Example 20.9   Entropy and the Carnot cycle

For the Carnot engine in Example 20.2 (Section 20.6), what is the total entropy change during one cycle?

SOLUTION

IDENTIFY and SET UP: All four steps in the Carnot cycle (see Fig. 20.13) are reversible, so we can use our expressions for the entropy change ΔS in a reversible process. We find ΔS for each step and add them to get ΔS for the complete cycle.

EXECUTE: There is no entropy change during the adiabatic expansion $b \rightarrow c$ or the adiabatic compression $d \rightarrow a$. During the isothermal expansion $a \rightarrow b$ at $T_H = 500$ K, the engine takes in 2000 J of heat, and from Eq. (20.18),

$$\Delta S_H = \frac{Q_H}{T_H} = \frac{2000 \text{ J}}{500 \text{ K}} = 4.0 \text{ J/K}$$

During the isothermal compression $c \rightarrow d$ at $T_C = 350$ K, the engine gives off 1400 J of heat, and

$$\Delta S_C = \frac{Q_C}{T_C} = \frac{-1400 \text{ J}}{350 \text{ K}} = -4.0 \text{ J/K}$$

The total entropy change in the engine during one cycle is $\Delta S_{tot} = \Delta S_H + \Delta S_C = 4.0 \text{ J/K} + (-4.0 \text{ J/K}) = 0$.

EVALUATE: The result $\Delta S_{total} = 0$ tells us that when the Carnot engine completes a cycle, it has the same entropy as it did at the beginning of the cycle. We'll explore this result in the next subsection.

What is the total entropy change of the engine's *environment* during this cycle? During the reversible isothermal expansion $a \rightarrow b$, the hot (500 K) reservoir gives off 2000 J of heat, so its entropy change is $(-2000 \text{ J})/(500 \text{ K}) = -4.0 \text{ J/K}$. During the reversible isothermal compression $c \rightarrow d$, the cold (350 K) reservoir absorbs 1400 J of heat, so its entropy change is $(+1400 \text{ J})/(350 \text{ K}) = +4.0 \text{ J/K}$. Thus the hot and cold reservoirs each have an entropy change, but the sum of these changes—that is, the total entropy change of the system's environment—is zero.

These results apply to the special case of the Carnot cycle, for which *all* of the processes are reversible. In this case we find that the total entropy change of the system and the environment together is zero. We will see that if the cycle includes irreversible processes (as is the case for the Otto and Diesel cycles of Section 20.3), the total entropy change of the system and the environment *cannot* be zero, but rather must be positive.

Entropy in Cyclic Processes

Example 20.9 showed that the total entropy change for a cycle of a particular Carnot engine, which uses an ideal gas as its working substance, is zero. This result follows directly from Eq. (20.13), which we can rewrite as

$$\frac{Q_H}{T_H} + \frac{Q_C}{T_C} = 0 \qquad (20.20)$$

The quotient Q_H/T_H equals ΔS_H, the entropy change of the engine that occurs at $T = T_H$. Likewise, Q_C/T_C equals ΔS_C, the (negative) entropy change of the engine that occurs at $T = T_C$. Hence Eq. (20.20) says that $\Delta S_H + \Delta S_C = 0$; that is, there is zero net entropy change in one cycle.

What about Carnot engines that use a different working substance? According to the second law, *any* Carnot engine operating between given temperatures T_H and T_C has the same efficiency $e = 1 - T_C/T_H$ [Eq. (20.14)]. Combining this expression for e with Eq. (20.4), $e = 1 + Q_C/Q_H$, just reproduces Eq. (20.20). So Eq. (20.20) is valid for any Carnot engine working between these temperatures, whether its working substance is an ideal gas or not. We conclude that *the total entropy change in one cycle of any Carnot engine is zero.*

20.19 (a) A reversible cyclic process for an ideal gas is shown as a red closed path on a pV-diagram. Several ideal-gas isotherms are shown in blue. (b) We can approximate the path in (a) by a series of long, thin Carnot cycles; one of these is highlighted in gold. The total entropy change is zero for each Carnot cycle and for the actual cyclic process. (c) The entropy change between points a and b is independent of the path.

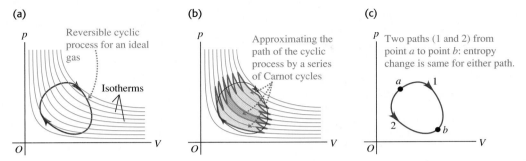

This result can be generalized to show that the total entropy change during *any* reversible cyclic process is zero. A reversible cyclic process appears on a pV-diagram as a closed path (Fig. 20.19a). We can approximate such a path as closely as we like by a sequence of isothermal and adiabatic processes forming parts of many long, thin Carnot cycles (Fig. 20.19b). The total entropy change for the full cycle is the sum of the entropy changes for each small Carnot cycle, each of which is zero. So **the total entropy change during *any* reversible cycle is zero:**

$$\int \frac{dQ}{T} = 0 \qquad \text{(reversible cyclic process)} \qquad (20.21)$$

It follows that when a system undergoes a reversible process leading from any state a to any other state b, *the entropy change of the system is independent of the path* (Fig. 20.19c). If the entropy change for path 1 were different from the change for path 2, the system could be taken along path 1 and then backward along path 2 to the starting point, with a nonzero net change in entropy. This would violate the conclusion that the total entropy change in such a cyclic process must be zero. Because the entropy change in such processes is independent of path, we conclude that in any given state, the system has a definite value of entropy that depends only on the state, not on the processes that led to that state.

Entropy in Irreversible Processes

In an idealized, reversible process involving only equilibrium states, the total entropy change of the system and its surroundings is zero. But all *irreversible* processes involve an increase in entropy. Unlike energy, *entropy is not a conserved quantity.* The entropy of an isolated system *can* change, but as we shall see, it can never decrease. The free expansion of a gas, described in Example 20.8, is an irreversible process in an isolated system in which there is an entropy increase.

Application **Entropy Changes in a Living Organism**
When a kitten or other growing animal eats, it takes ordered chemical energy from the food and uses it to make new cells that are even more highly ordered. This process alone lowers entropy. But most of the energy in the food is either excreted in the animal's feces or used to generate heat, processes that lead to a large increase in entropy. So while the entropy of the animal alone decreases, the *total* entropy of animal plus food *increases*.

Example 20.10 | **Entropy change in an irreversible process**

Suppose 1.00 kg of water at 100°C is placed in thermal contact with 1.00 kg of water at 0°C. What is the total change in entropy? Assume that the specific heat of water is constant at 4190 J/kg·K over this temperature range.

SOLUTION

IDENTIFY and SET UP: This process involves irreversible heat flow because of the temperature differences. There are equal masses of 0°C water and 100°C water, so the final temperature is the average of these two temperatures: 50°C = 323 K. Although the processes are irreversible, we can calculate the entropy changes for the (initially) hot water and the (initially) cold water by assuming that the process occurs reversibly. As in Example 20.6, we must use Eq. (20.19) to calculate ΔS for each substance because the temperatures are not constant.

Continued

EXECUTE: The entropy changes of the hot water (subscript H) and the cold water (subscript C) are

$$\Delta S_H = mc \int_{T_1}^{T_2} \frac{dT}{T} = (1.00 \text{ kg})(4190 \text{ J/kg} \cdot \text{K}) \int_{373 \text{ K}}^{323 \text{ K}} \frac{dT}{T}$$

$$= (4190 \text{ J/K})\left(\ln \frac{323 \text{ K}}{373 \text{ K}}\right) = -603 \text{ J/K}$$

$$\Delta S_C = (4190 \text{ J/K})\left(\ln \frac{323 \text{ K}}{273 \text{ K}}\right) = +705 \text{ J/K}$$

The *total* entropy change of the system is

$$\Delta S_{\text{tot}} = \Delta S_H + \Delta S_C = (-603 \text{ J/K}) + 705 \text{ J/K} = +102 \text{ J/K}$$

EVALUATE: An irreversible heat flow in an isolated system is accompanied by an increase in entropy. We could reach the same end state by mixing the hot and cold water, which is also an irreversible process; the total entropy change, which depends only on the initial and final states of the system, would again be 102 J/K.

Note that the entropy of the system increases *continuously* as the two quantities of water come to equilibrium. For example, the first 4190 J of heat transferred cools the hot water to 99°C and warms the cold water to 1°C. The net change in entropy for this step is approximately

$$\Delta S = \frac{-4190 \text{ J}}{373 \text{ K}} + \frac{4190 \text{ J}}{273 \text{ K}} = +4.1 \text{ J/K}$$

Can you show in a similar way that the net entropy change is positive for *any* one-degree temperature change leading to the equilibrium condition?

Entropy and the Second Law

The results of Example 20.10 about the flow of heat from a higher to a lower temperature are characteristic of *all* natural (that is, irreversible) processes. When we include the entropy changes of all the systems taking part in the process, the increases in entropy are always greater than the decreases. In the special case of a *reversible* process, the increases and decreases are equal. Hence we can state the general principle: **When all systems taking part in a process are included, the entropy either remains constant or increases.** In other words: **No process is possible in which the total entropy decreases, when all systems taking part in the process are included.** This is an alternative statement of the second law of thermodynamics in terms of entropy. Thus it is equivalent to the "engine" and "refrigerator" statements discussed earlier. Figure 20.20 shows a specific example of this general principle.

The increase of entropy in every natural, irreversible process measures the increase of disorder or randomness in the universe associated with that process. Consider again the example of mixing hot and cold water (Example 20.10). We *might* have used the hot and cold water as the high- and low-temperature reservoirs of a heat engine. While removing heat from the hot water and giving heat to the cold water, we could have obtained some mechanical work. But once the hot and cold water have been mixed and have come to a uniform temperature, this opportunity to convert heat to mechanical work is lost irretrievably. The lukewarm water will never *unmix* itself and separate into hotter and colder portions. No decrease in *energy* occurs when the hot and cold water are mixed. What has been lost is the *opportunity* to convert part of the heat from the hot water into mechanical work. Hence when entropy increases, energy becomes less *available*, and the universe becomes more random or "run down."

20.20 The mixing of colored ink and water starts from a state of relative order (low entropy) in which each fluid is separate and distinct from the other. The final state after mixing is more disordered (has greater entropy). Spontaneous unmixing of the ink and water, a process in which there would be a net decrease in entropy, is never observed.

Test Your Understanding of Section 20.7   Suppose 2.00 kg of water at 50°C spontaneously changes temperature, so that half of the water cools to 0°C while the other half spontaneously warms to 100°C. (All of the water remains liquid, so it doesn't freeze or boil.) What would be the entropy change of the water? Is this process possible? (*Hint:* See Example 20.10.)

20.8 Microscopic Interpretation of Entropy

We described in Section 19.4 how the internal energy of a system could be calculated, at least in principle, by adding up all the kinetic energies of its constituent particles and all the potential energies of interaction among the particles. This is called a *microscopic* calculation of the internal energy. We can also make a microscopic calculation of the entropy S of a system. Unlike energy, however, entropy is not something that belongs to each individual particle or pair of particles in the system. Rather, entropy is a measure of the disorder of the system as a whole. To see how to calculate entropy microscopically, we first have to introduce the idea of *macroscopic* and *microscopic states.*

Suppose you toss N identical coins on the floor, and half of them show heads and half show tails. This is a description of the large-scale or **macroscopic state** of the system of N coins. A description of the **microscopic state** of the system includes information about each individual coin: Coin 1 was heads, coin 2 was tails, coin 3 was tails, and so on. There can be many microscopic states that correspond to the same macroscopic description. For instance, with $N = 4$ coins there are six possible states in which half are heads and half are tails (Fig. 20.21). The number of microscopic states grows rapidly with increasing N; for $N = 100$ there are $2^{100} = 1.27 \times 10^{30}$ microscopic states, of which 1.01×10^{29} are half heads and half tails.

The least probable outcomes of the coin toss are the states that are either all heads or all tails. It is certainly possible that you could throw 100 heads in a row, but don't bet on it; the probability of doing this is only 1 in 1.27×10^{30}. The most probable outcome of tossing N coins is that half are heads and half are tails. The reason is that this *macroscopic* state has the greatest number of corresponding *microscopic* states, as Fig. 20.21 shows.

To make the connection to the concept of entropy, note that N coins that are all heads constitute a completely ordered macroscopic state; the description "all heads" completely specifies the state of each one of the N coins. The same is true if the coins are all tails. But the macroscopic description "half heads, half tails" by itself tells you very little about the state (heads or tails) of each individual coin. We say that the system is *disordered* because we know so little about its microscopic state. Compared to the state "all heads" or "all tails," the state "half heads, half tails" has a much greater number of possible microscopic states, much greater disorder, and hence much greater entropy (which is a quantitative measure of disorder).

Now instead of N coins, consider a mole of an ideal gas containing Avogadro's number of molecules. The macroscopic state of this gas is given by its pressure p, volume V, and temperature T; a description of the microscopic state involves stating the position and velocity for each molecule in the gas. At a given pressure, volume, and temperature, the gas may be in any one of an astronomically large number of microscopic states, depending on the positions and velocities of its 6.02×10^{23} molecules. If the gas undergoes a free expansion into a greater volume, the range of possible positions increases, as does the number of possible microscopic states. The system becomes more disordered, and the entropy increases as calculated in Example 20.8 (Section 20.7).

We can draw the following general conclusion: **For any system, the most probable macroscopic state is the one with the greatest number of corresponding microscopic states, which is also the macroscopic state with the greatest disorder and the greatest entropy.**

20.21 All possible microscopic states of four coins. There can be several possible microscopic states for each macroscopic state.

Macroscopic state	Corresponding microscopic states
Four heads	
Three heads, one tails	
Two heads, two tails	
One heads, three tails	
Four tails	

Calculating Entropy: Microscopic States

Let w represent the number of possible microscopic states for a given macroscopic state. (For the four coins shown in Fig. 20.21 the state of four heads has $w = 1$, the state of three heads and one tails has $w = 4$, and so on.) Then the entropy S of a macroscopic state can be shown to be given by

$$S = k \ln w \quad \text{(microscopic expression for entropy)} \quad (20.22)$$

where $k = R/N_A$ is the Boltzmann constant (gas constant per molecule) introduced in Section 18.3. As Eq. (20.22) shows, increasing the number of possible microscopic states w increases the entropy S.

What matters in a thermodynamic process is not the absolute entropy S but the *difference* in entropy between the initial and final states. Hence an equally valid and useful definition would be $S = k \ln w + C$, where C is a constant, since C cancels in any calculation of an entropy difference between two states. But it's convenient to set this constant equal to zero and use Eq. (20.22). With this choice, since the smallest possible value of w is unity, the smallest possible value of S for any system is $k \ln 1 = 0$. Entropy can *never* be negative.

In practice, calculating w is a difficult task, so Eq. (20.22) is typically used only to calculate the absolute entropy S of certain special systems. But we can use this relationship to calculate *differences* in entropy between one state and another. Consider a system that undergoes a thermodynamic process that takes it from macroscopic state 1, for which there are w_1 possible microscopic states, to macroscopic state 2, with w_2 associated microscopic states. The change in entropy in this process is

$$\Delta S = S_2 - S_1 = k \ln w_2 - k \ln w_1 = k \ln \frac{w_2}{w_1} \quad (20.23)$$

The *difference* in entropy between the two macroscopic states depends on the *ratio* of the numbers of possible microscopic states.

As the following example shows, using Eq. (20.23) to calculate a change in entropy from one macroscopic state to another gives the same results as considering a reversible process connecting those two states and using Eq. (20.19).

Example 20.11 **A microscopic calculation of entropy change**

Use Eq. (20.23) to calculate the entropy change in the free expansion of n moles of gas at temperature T described in Example 20.8 (Fig. 20.22).

SOLUTION

IDENTIFY and SET UP: We are asked to calculate the entropy change using the number of microstates in the initial and final macroscopic states (Figs. 20.22a and b). When the partition is broken, no work is done, so the velocities of the molecules are unaffected. But each molecule now has twice as much volume in which it can move and hence has twice the number of possible positions. This is all we need to calculate the entropy change using Eq. (20.23).

EXECUTE: Let w_1 be the number of microscopic states of the system as a whole when the gas occupies volume V (Fig. 20.22a). The number of molecules is $N = nN_A$, and each of these N molecules has twice as many possible states after the partition is broken. Hence the number w_2 of microscopic states when the gas occupies volume $2V$ (Fig. 20.22b) is greater by a factor of 2^N; that is, $w_2 = 2^N w_1$. The change in entropy in this process is

$$\Delta S = k \ln \frac{w_2}{w_1} = k \ln \frac{2^N w_1}{w_1} = k \ln 2^N = Nk \ln 2$$

Since $N = nN_A$ and $k = R/N_A$, this becomes

$$\Delta S = (nN_A)(R/N_A) \ln 2 = nR \ln 2$$

EVALUATE: We found the same result as in Example 20.8, but without any reference to the thermodynamic path taken.

20.22 In a free expansion of N molecules in which the volume doubles, the number of possible microscopic states increases by 2^N.

(a) Gas occupies volume V; number of microstates = w_1.

(b) Gas occupies volume $2V$; number of microstates = $w_2 = 2^N w_1$.

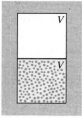
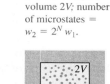
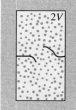

Microscopic States and the Second Law

The relationship between entropy and the number of microscopic states gives us new insight into the entropy statement of the second law of thermodynamics: that the entropy of a closed system can never decrease. From Eq. (20.22) this means that a closed system can never spontaneously undergo a process that decreases the number of possible microscopic states.

An example of such a forbidden process would be if all of the air in your room spontaneously moved to one half of the room, leaving a vacuum in the other half. Such a "free compression" would be the reverse of the free expansion of Examples 20.8 and 20.11. This would decrease the number of possible microscopic states by a factor of 2^N. Strictly speaking, this process is not impossible! The probability of finding a given molecule in one half of the room is $\frac{1}{2}$, so the probability of finding all of the molecules in one half of the room at once is $\left(\frac{1}{2}\right)^N$. (This is exactly the same as the probability of having a tossed coin come up heads N times in a row.) This probability is *not* zero. But lest you worry about suddenly finding yourself gasping for breath in the evacuated half of your room, consider that a typical room might hold 1000 moles of air, and so $N = 1000N_A = 6.02 \times 10^{26}$ molecules. The probability of all the molecules being in the same half of the room is therefore $\left(\frac{1}{2}\right)^{6.02 \times 10^{26}}$. Expressed as a decimal, this number has more than 10^{26} zeros to the right of the decimal point!

Because the probability of such a "free compression" taking place is so vanishingly small, it has almost certainly never occurred anywhere in the universe since the beginning of time. We conclude that for all practical purposes the second law of thermodynamics is never violated.

Test Your Understanding of Section 20.8 A quantity of N molecules of an ideal gas initially occupies volume V. The gas then expands to volume $2V$. The number of microscopic states of the gas increases in this expansion. Under which of the following circumstances will this number increase the most? (i) if the expansion is reversible and isothermal; (ii) if the expansion is reversible and adiabatic; (iii) the number will change by the same amount for both circumstances.

Reversible and irreversible processes: A reversible process is one whose direction can be reversed by an infinitesimal change in the conditions of the process, and in which the system is always in or very close to thermal equilibrium. All other thermodynamic processes are irreversible.

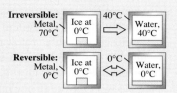

Heat engines: A heat engine takes heat Q_H from a source, converts part of it to work W, and discards the remainder $|Q_C|$ at a lower temperature. The thermal efficiency e of a heat engine measures how much of the absorbed heat is converted to work. (See Example 20.1.)

$$e = \frac{W}{Q_H} = 1 + \frac{Q_C}{Q_H} = 1 - \left|\frac{Q_C}{Q_H}\right| \quad (20.4)$$

The Otto cycle: A gasoline engine operating on the Otto cycle has a theoretical maximum thermal efficiency e that depends on the compression ratio r and the ratio of heat capacities γ of the working substance.

$$e = 1 - \frac{1}{r^{\gamma-1}} \quad (20.6)$$

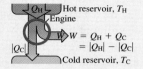

Refrigerators: A refrigerator takes heat Q_C from a colder place, has a work input $|W|$, and discards heat $|Q_H|$ at a warmer place. The effectiveness of the refrigerator is given by its coefficient of performance K.

$$K = \frac{|Q_C|}{|W|} = \frac{|Q_C|}{|Q_H| - |Q_C|} \quad (20.9)$$

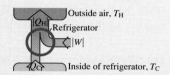

The second law of thermodynamics: The second law of thermodynamics describes the directionality of natural thermodynamic processes. It can be stated in several equivalent forms. The *engine* statement is that no cyclic process can convert heat completely into work. The *refrigerator* statement is that no cyclic process can transfer heat from a colder place to a hotter place with no input of mechanical work.

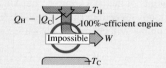

The Carnot cycle: The Carnot cycle operates between two heat reservoirs at temperatures T_H and T_C and uses only reversible processes. Its thermal efficiency depends only on T_H and T_C. An additional equivalent statement of the second law is that no engine operating between the same two temperatures can be more efficient than a Carnot engine. (See Examples 20.2 and 20.3.)

A Carnot engine run backward is a Carnot refrigerator. Its coefficient of performance depends only on T_H and T_C. Another form of the second law states that no refrigerator operating between the same two temperatures can have a larger coefficient of performance than a Carnot refrigerator. (See Example 20.4.)

$$e_{Carnot} = 1 - \frac{T_C}{T_H} = \frac{T_H - T_C}{T_H} \quad (20.14)$$

$$K_{Carnot} = \frac{T_C}{T_H - T_C} \quad (20.15)$$

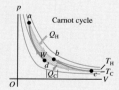

Entropy: Entropy is a quantitative measure of the disorder of a system. The entropy change in any reversible process depends on the amount of heat flow and the absolute temperature T. Entropy depends only on the state of the system, and the change in entropy between given initial and final states is the same for all processes leading from one state to the other. This fact can be used to find the entropy change in an irreversible process. (See Examples 20.5–20.10.)

$$\Delta S = \int_1^2 \frac{dQ}{T} \quad (20.19)$$
(reversible process)

An important statement of the second law of thermo-dynamics is that the entropy of an isolated system may increase but can never decrease. When a system interacts with its surroundings, the total entropy change of system and surroundings can never decrease. When the interaction involves only reversible processes, the total entropy is constant and $\Delta S = 0$; when there is any irreversible process, the total entropy increases and $\Delta S > 0$.

Entropy and microscopic states: When a system is in a particular macroscopic state, the particles that make up the system may be in any of w possible microscopic states. The greater the number w, the greater the entropy. (See Example 20.11.)

$$S = k \ln w \qquad (20.22)$$

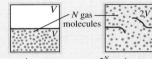

w microstates $2^N w$ microstates

BRIDGING PROBLEM   **Entropy Changes: Cold Ice in Hot Water**

An insulated container of negligible mass holds 0.600 kg of water at 45.0°C. You put a 0.0500-kg ice cube at −15.0°C in the water. (a) Calculate the final temperature of the water once the ice has melted. (b) Calculate the change in entropy of the system.

SOLUTION GUIDE

See MasteringPhysics® study area for a Video Tutor solution.

IDENTIFY and SET UP

1. Make a list of the known and unknown quantities, and identify the target variables.
2. How will you find the final temperature of the ice–water mixture? How will you decide whether or not all the ice melts?
3. Once you find the final temperature of the mixture, how will you determine the changes in entropy of (i) the ice initially at −15.0°C and (ii) the water initially at 45.0°C?

EXECUTE

4. Use the methods of Chapter 17 to calculate the final temperature T. (*Hint:* First assume that all of the ice melts, then write

an equation which says that the heat that flows into the ice equals the heat that flows out of the water. If your assumption is correct, the final temperature that you calculate will be greater than 0°C. If your assumption is incorrect, the final temperature will be 0°C or less, which means that some ice remains. You'll then need to redo the calculation to account for this.)

5. Use your result from step 4 to calculate the entropy changes of the ice and the water. (*Hint:* You must include the heat flow associated with temperature changes, as in Example 20.6, as well as the heat flow associated with the change of phase.)
6. Find the total change in entropy of the system.

EVALUATE

7. Do the signs of the entropy changes make sense? Why or why not?

Problems   For instructor-assigned homework, go to www.masteringphysics.com

•, ••, •••: Problems of increasing difficulty. **CP**: Cumulative problems incorporating material from earlier chapters. **CALC**: Problems requiring calculus. **BIO**: Biosciences problems.

DISCUSSION QUESTIONS

Q20.1 A pot is half-filled with water, and a lid is placed on it, forming a tight seal so that no water vapor can escape. The pot is heated on a stove, forming water vapor inside the pot. The heat is then turned off and the water vapor condenses back to liquid. Is this cycle reversible or irreversible? Why?

Q20.2 Give two examples of reversible processes and two examples of irreversible processes in purely mechanical systems, such as blocks sliding on planes, springs, pulleys, and strings. Explain what makes each process reversible or irreversible.

Q20.3 What irreversible processes occur in a gasoline engine? Why are they irreversible?

Q20.4 Suppose you try to cool the kitchen of your house by leaving the refrigerator door open. What happens? Why? Would the result be the same if you left open a picnic cooler full of ice? Explain the reason for any differences.

Q20.5 A member of the U.S. Congress proposed a scheme to produce energy as follows. Water molecules (H_2O) are to be broken apart to produce hydrogen and oxygen. The hydrogen is then burned (that is, combined with oxygen), releasing energy in the process. The only product of this combustion is water, so there is no pollution. In light of the second law of thermodynamics, what do you think of this energy-producing scheme?

Q20.6 Is it a violation of the second law of thermodynamics to convert mechanical energy completely into heat? To convert heat completely into work? Explain your answers.

Q20.7 Imagine a special air filter placed in a window of a house. The tiny holes in the filter allow only air molecules moving faster than a certain speed to exit the house, and allow only air molecules moving slower than that speed to enter the house from outside. Explain why such an air filter would cool the house, and why the second law of thermodynamics makes building such a filter an impossible task.

Q20.8 An electric motor has its shaft coupled to that of an electric generator. The motor drives the generator, and some current from the generator is used to run the motor. The excess current is used to light a home. What is wrong with this scheme?

Q20.9 When a wet cloth is hung up in a hot wind in the desert, it is cooled by evaporation to a temperature that may be 20 C° or so below that of the air. Discuss this process in light of the second law of thermodynamics.

Q20.10 Compare the pV-diagram for the Otto cycle in Fig. 20.6 with the diagram for the Carnot heat engine in Fig. 20.13. Explain some of the important differences between the two cycles.

Q20.11 If no real engine can be as efficient as a Carnot engine operating between the same two temperatures, what is the point of developing and using Eq. (20.14)?

Q20.12 The efficiency of heat engines is high when the temperature difference between the hot and cold reservoirs is large. Refrigerators, on the other hand, work better when the temperature difference is small. Thinking of the mechanical refrigeration cycle shown in Fig. 20.9, explain in physical terms why it takes less work to remove heat from the working substance if the two reservoirs (the inside of the refrigerator and the outside air) are at nearly the same temperature, than if the outside air is much warmer than the interior of the refrigerator.

Q20.13 What would be the efficiency of a Carnot engine operating with $T_H = T_C$? What would be the efficiency if $T_C = 0$ K and T_H were any temperature above 0 K? Interpret your answers.

Q20.14 Real heat engines, like the gasoline engine in a car, always have some friction between their moving parts, although lubricants keep the friction to a minimum. Would a heat engine with completely frictionless parts be 100% efficient? Why or why not? Does the answer depend on whether or not the engine runs on the Carnot cycle? Again, why or why not?

Q20.15 Does a refrigerator full of food consume more power if the room temperature is 20°C than if it is 15°C? Or is the power consumption the same? Explain your reasoning.

Q20.16 In Example 20.4, a Carnot refrigerator requires a work input of only 230 J to extract 346 J of heat from the cold reservoir. Doesn't this discrepancy imply a violation of the law of conservation of energy? Explain why or why not.

Q20.17 Explain why each of the following processes is an example of increasing disorder or randomness: mixing hot and cold water; free expansion of a gas; irreversible heat flow; developing heat by mechanical friction. Are entropy increases involved in all of these? Why or why not?

Q20.18 The free expansion of a gas is an adiabatic process and so no heat is transferred. No work is done, so the internal energy does not change. Thus, $Q/T = 0$, yet the disorder of the system and thus its entropy have increased after the expansion. Why does Eq. (20.19) not apply to this situation?

Q20.19 Are the earth and sun in thermal equilibrium? Are there entropy changes associated with the transmission of energy from the sun to the earth? Does radiation differ from other modes of heat transfer with respect to entropy changes? Explain your reasoning.

Q20.20 Discuss the entropy changes involved in the preparation and consumption of a hot fudge sundae.

Q20.21 If you run a movie film backward, it is as if the direction of time were reversed. In the time-reversed movie, would you see processes that violate conservation of energy? Conservation of linear momentum? Would you see processes that violate the second law of thermodynamics? In each case, if law-breaking processes could occur, give some examples.

Q20.22 **BIO** Some critics of biological evolution claim that it violates the second law of thermodynamics, since evolution involves simple life forms developing into more complex and more highly ordered organisms. Explain why this is not a valid argument against evolution.

Q20.23 **BIO** A growing plant creates a highly complex and organized structure out of simple materials such as air, water, and trace minerals. Does this violate the second law of thermodynamics? Why or why not? What is the plant's ultimate source of energy? Explain your reasoning.

EXERCISES

Section 20.2 Heat Engines

20.1 • A diesel engine performs 2200 J of mechanical work and discards 4300 J of heat each cycle. (a) How much heat must be supplied to the engine in each cycle? (b) What is the thermal efficiency of the engine?

20.2 • An aircraft engine takes in 9000 J of heat and discards 6400 J each cycle. (a) What is the mechanical work output of the engine during one cycle? (b) What is the thermal efficiency of the engine?

20.3 • **A Gasoline Engine.** A gasoline engine takes in 1.61×10^4 J of heat and delivers 3700 J of work per cycle. The heat is obtained by burning gasoline with a heat of combustion of 4.60×10^4 J/g. (a) What is the thermal efficiency? (b) How much heat is discarded in each cycle? (c) What mass of fuel is burned in each cycle? (d) If the engine goes through 60.0 cycles per second, what is its power output in kilowatts? In horsepower?

20.4 • A gasoline engine has a power output of 180 kW (about 241 hp). Its thermal efficiency is 28.0%. (a) How much heat must be supplied to the engine per second? (b) How much heat is discarded by the engine per second?

20.5 •• The pV-diagram in Fig. E20.5 shows a cycle of a heat engine that uses 0.250 mole of an ideal gas having $\gamma = 1.40$. The curved part ab

Figure **E20.5**

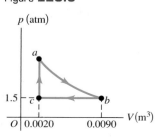

of the cycle is adiabatic. (a) Find the pressure of the gas at point *a*. (b) How much heat enters this gas per cycle, and where does it happen? (c) How much heat leaves this gas in a cycle, and where does it occur? (d) How much work does this engine do in a cycle? (e) What is the thermal efficiency of the engine?

Section 20.3 Internal-Combustion Engines

20.6 • (a) Calculate the theoretical efficiency for an Otto-cycle engine with $\gamma = 1.40$ and $r = 9.50$. (b) If this engine takes in 10,000 J of heat from burning its fuel, how much heat does it discard to the outside air?

20.7 •• The Otto-cycle engine in a Mercedes-Benz SLK230 has a compression ratio of 8.8. (a) What is the ideal efficiency of the engine? Use $\gamma = 1.40$. (b) The engine in a Dodge Viper GT2 has a slightly higher compression ratio of 9.6. How much increase in the ideal efficiency results from this increase in the compression ratio?

Section 20.4 Refrigerators

20.8 • The coefficient of performance $K = H/P$ is a dimensionless quantity. Its value is independent of the units used for H and P, as long as the same units, such as watts, are used for both quantities. However, it is common practice to express H in Btu/h and P in watts. When these mixed units are used, the ratio H/P is called the energy efficiency rating (EER). If a room air conditioner has a coefficient of performance $K = 3.0$, what is its EER?

20.9 • A refrigerator has a coefficient of performance of 2.10. In each cycle it absorbs 3.40×10^4 J of heat from the cold reservoir. (a) How much mechanical energy is required each cycle to operate the refrigerator? (b) During each cycle, how much heat is discarded to the high-temperature reservoir?

20.10 • A room air conditioner has a coefficient of performance of 2.9 on a hot day and uses 850 W of electrical power. (a) How many joules of heat does the air conditioner remove from the room in one minute? (b) How many joules of heat does the air conditioner deliver to the hot outside air in one minute? (c) Explain why your answers to parts (a) and (b) are not the same.

20.11 •• A refrigerator has a coefficient of performance of 2.25, runs on an input of 95 W of electrical power, and keeps its inside compartment at 5°C. If you put a dozen 1.0-L plastic bottles of water at 31°C into this refrigerator, how long will it take for them to be cooled down to 5°C? (Ignore any heat that leaves the plastic.)

20.12 •• A freezer has a coefficient of performance of 2.40. The freezer is to convert 1.80 kg of water at 25.0°C to 1.80 kg of ice at −5.0°C in one hour. (a) What amount of heat must be removed from the water at 25.0°C to convert it to ice at −5.0°C? (b) How much electrical energy is consumed by the freezer during this hour? (c) How much wasted heat is delivered to the room in which the freezer sits?

Section 20.6 The Carnot Cycle

20.13 • A Carnot engine whose high-temperature reservoir is at 620 K takes in 550 J of heat at this temperature in each cycle and gives up 335 J to the low-temperature reservoir. (a) How much mechanical work does the engine perform during each cycle? (b) What is the temperature of the low-temperature reservoir? (c) What is the thermal efficiency of the cycle?

20.14 • A Carnot engine is operated between two heat reservoirs at temperatures of 520 K and 300 K. (a) If the engine receives 6.45 kJ of heat energy from the reservoir at 520 K in each cycle, how many joules per cycle does it discard to the reservoir at 300 K? (b) How much mechanical work is performed by the engine during each cycle? (c) What is the thermal efficiency of the engine?

20.15 • A Carnot engine has an efficiency of 59% and performs 2.5×10^4 J of work in each cycle. (a) How much heat does the engine extract from its heat source in each cycle? (b) Suppose the engine exhausts heat at room temperature (20.0°C). What is the temperature of its heat source?

20.16 •• An ice-making machine operates in a Carnot cycle. It takes heat from water at 0.0°C and rejects heat to a room at 24.0°C. Suppose that 85.0 kg of water at 0.0°C are converted to ice at 0.0°C. (a) How much heat is discharged into the room? (b) How much energy must be supplied to the device?

20.17 • A Carnot refrigerator is operated between two heat reservoirs at temperatures of 320 K and 270 K. (a) If in each cycle the refrigerator receives 415 J of heat energy from the reservoir at 270 K, how many joules of heat energy does it deliver to the reservoir at 320 K? (b) If the refrigerator completes 165 cycles each minute, what power input is required to operate it? (c) What is the coefficient of performance of the refrigerator?

20.18 •• A certain brand of freezer is advertised to use 730 kW · h of energy per year. (a) Assuming the freezer operates for 5 hours each day, how much power does it require while operating? (b) If the freezer keeps its interior at a temperature of −5.0°C in a 20.0°C room, what is its theoretical maximum performance coefficient? (c) What is the theoretical maximum amount of ice this freezer could make in an hour, starting with water at 20.0°C?

20.19 •• A Carnot heat engine has a thermal efficiency of 0.600, and the temperature of its hot reservoir is 800 K. If 3000 J of heat is rejected to the cold reservoir in one cycle, what is the work output of the engine during one cycle?

20.20 •• A Carnot heat engine uses a hot reservoir consisting of a large amount of boiling water and a cold reservoir consisting of a large tub of ice and water. In 5 minutes of operation, the heat rejected by the engine melts 0.0400 kg of ice. During this time, how much work W is performed by the engine?

20.21 •• You design an engine that takes in 1.50×10^4 J of heat at 650 K in each cycle and rejects heat at a temperature of 350 K. The engine completes 240 cycles in 1 minute. What is the theoretical maximum power output of your engine, in horsepower?

Section 20.7 Entropy

20.22 • A 4.50-kg block of ice at 0.00°C falls into the ocean and melts. The average temperature of the ocean is 3.50°C, including all the deep water. By how much does the melting of this ice change the entropy of the world? Does it make it larger or smaller? (*Hint:* Do you think that the ocean will change temperature appreciably as the ice melts?)

20.23 • A sophomore with nothing better to do adds heat to 0.350 kg of ice at 0.0°C until it is all melted. (a) What is the change in entropy of the water? (b) The source of heat is a very massive body at a temperature of 25.0°C. What is the change in entropy of this body? (c) What is the total change in entropy of the water and the heat source?

20.24 • **CALC** You decide to take a nice hot bath but discover that your thoughtless roommate has used up most of the hot water. You fill the tub with 270 kg of 30.0°C water and attempt to warm it further by pouring in 5.00 kg of boiling water from the stove. (a) Is this a reversible or an irreversible process? Use physical reasoning to explain. (b) Calculate the final temperature of the bath water. (c) Calculate the net change in entropy of the system (bath water + boiling water), assuming no heat exchange with the air or the tub itself.

20.25 •• A 15.0-kg block of ice at 0.0°C melts to liquid water at 0.0°C inside a large room that has a temperature of 20.0°C. Treat the ice and the room as an isolated system, and assume that the room is large enough for its temperature change to be ignored. (a) Is the melting of the ice reversible or irreversible? Explain, using simple physical reasoning without resorting to any equations. (b) Calculate the net entropy change of the system during this process. Explain whether or not this result is consistent with your answer to part (a).

20.26 •• **CALC** You make tea with 0.250 kg of 85.0°C water and let it cool to room temperature (20.0°C) before drinking it. (a) Calculate the entropy change of the water while it cools. (b) The cooling process is essentially isothermal for the air in your kitchen. Calculate the change in entropy of the air while the tea cools, assuming that all the heat lost by the water goes into the air. What is the total entropy change of the system tea + air?

20.27 • Three moles of an ideal gas undergo a reversible isothermal compression at 20.0°C. During this compression, 1850 J of work is done on the gas. What is the change of entropy of the gas?

20.28 •• What is the change in entropy of 0.130 kg of helium gas at the normal boiling point of helium when it all condenses isothermally to 1.00 L of liquid helium? (*Hint:* See Table 17.4 in Section 17.6.)

20.29 • (a) Calculate the change in entropy when 1.00 kg of water at 100°C is vaporized and converted to steam at 100°C (see Table 17.4). (b) Compare your answer to the change in entropy when 1.00 kg of ice is melted at 0°C, calculated in Example 20.5 (Section 20.7). Is the change in entropy greater for melting or for vaporization? Interpret your answer using the idea that entropy is a measure of the randomness of a system.

20.30 • (a) Calculate the change in entropy when 1.00 mol of water (molecular mass 18.0 g/mol) at 100°C evaporates to form water vapor at 100°C. (b) Repeat the calculation of part (a) for 1.00 mol of liquid nitrogen, 1.00 mol of silver, and 1.00 mol of mercury when each is vaporized at its normal boiling point. (See Table 17.4 for the heats of vaporization, and Appendix D for the molar masses. Note that the nitrogen molecule is N_2.) (c) Your results in parts (a) and (b) should be in relatively close agreement. (This is called the *rule of Drepez and Trouton.*) Explain why this should be so, using the idea that entropy is a measure of the randomness of a system.

20.31 •• A 10.0-L gas tank containing 3.20 moles of ideal He gas at 20.0°C is placed inside a completely evacuated, insulated bell jar of volume 35.0 L. A small hole in the tank allows the He to leak out into the jar until the gas reaches a final equilibrium state with no more leakage. (a) What is the change in entropy of this system due to the leaking of the gas? (b) Is the process reversible or irreversible? How do you know?

Section 20.8 Microscopic Interpretation of Entropy

20.32 • A box is separated by a partition into two parts of equal volume. The left side of the box contains 500 molecules of nitrogen gas; the right side contains 100 molecules of oxygen gas. The two gases are at the same temperature. The partition is punctured, and equilibrium is eventually attained. Assume that the volume of the box is large enough for each gas to undergo a free expansion and not change temperature. (a) On average, how many molecules of each type will there be in either half of the box? (b) What is the change in entropy of the system when the partition is punctured? (c) What is the probability that the molecules will be found in the same distribution as they were before the partition was punctured—that is, 500 nitrogen molecules in the left half and 100 oxygen molecules in the right half?

20.33 • **CALC** Two moles of an ideal gas occupy a volume V. The gas expands isothermally and reversibly to a volume $3V$. (a) Is the velocity distribution changed by the isothermal expansion? Explain. (b) Use Eq. (20.23) to calculate the change in entropy of the gas. (c) Use Eq. (20.18) to calculate the change in entropy of the gas. Compare this result to that obtained in part (b).

20.34 • **CALC** A lonely party balloon with a volume of 2.40 L and containing 0.100 mol of air is left behind to drift in the temporarily uninhabited and depressurized International Space Station. Sunlight coming through a porthole heats and explodes the balloon, causing the air in it to undergo a free expansion into the empty station, whose total volume is 425 m³. Calculate the entropy change of the air during the expansion.

PROBLEMS

20.35 •• **CP** An ideal Carnot engine operates between 500°C and 100°C with a heat input of 250 J per cycle. (a) How much heat is delivered to the cold reservoir in each cycle? (b) What minimum number of cycles is necessary for the engine to lift a 500-kg rock through a height of 100 m?

20.36 • You are designing a Carnot engine that has 2 mol of CO_2 as its working substance; the gas may be treated as ideal. The gas is to have a maximum temperature of 527°C and a maximum pressure of 5.00 atm. With a heat input of 400 J per cycle, you want 300 J of useful work. (a) Find the temperature of the cold reservoir. (b) For how many cycles must this engine run to melt completely a 10.0-kg block of ice originally at 0.0°C, using only the heat rejected by the engine?

20.37 •• **CP** A certain heat engine operating on a Carnot cycle absorbs 150 J of heat per cycle at its hot reservoir at 135°C and has a thermal efficiency of 22.0%. (a) How much work does this engine do per cycle? (b) How much heat does the engine waste each cycle? (c) What is the temperature of the cold reservoir? (d) By how much does the engine change the entropy of the world each cycle? (e) What mass of water could this engine pump per cycle from a well 35.0 m deep?

20.38 •• **BIO** **Entropy of Metabolism.** An average sleeping person metabolizes at a rate of about 80 W by digesting food or burning fat. Typically, 20% of this energy goes into bodily functions, such as cell repair, pumping blood, and other uses of mechanical energy, while the rest goes to heat. Most people get rid of all this excess heat by transferring it (by conduction and the flow of blood) to the surface of the body, where it is radiated away. The normal internal temperature of the body (where the metabolism takes place) is 37°C, and the skin is typically 7 C° cooler. By how much does the person's entropy change per second due to this heat transfer?

20.39 •• **BIO** **Entropy Change from Digesting Fat.** Digesting fat produces 9.3 food calories per gram of fat, and typically 80% of this energy goes to heat when metabolized. (One food calorie is 1000 calories and therefore equals 4186 J.) The body then moves all this heat to the surface by a combination of thermal conductivity and motion of the blood. The internal temperature of the body (where digestion occurs) is normally 37°C, and the surface is usually about 30°C. By how much do the digestion and metabolism of a 2.50-g pat of butter change your body's entropy? Does it increase or decrease?

20.40 • A heat engine takes 0.350 mol of a diatomic ideal gas around the cycle shown in the *pV*-diagram of Fig. P20.40. Process 1 → 2 is at constant volume, process 2 → 3 is adiabatic, and process 3 → 1 is at a constant pressure of 1.00 atm. The value of γ for this gas is 1.40. (a) Find the pressure and volume at points 1, 2, and 3. (b) Calculate Q, W, and ΔU for each of the three processes. (c) Find the net work done by the gas in the cycle. (d) Find the net heat flow into the engine in one cycle. (e) What is the thermal efficiency of the engine? How does this compare to the efficiency of a Carnot-cycle engine operating between the same minimum and maximum temperatures T_1 and T_2?

Figure **P20.40**

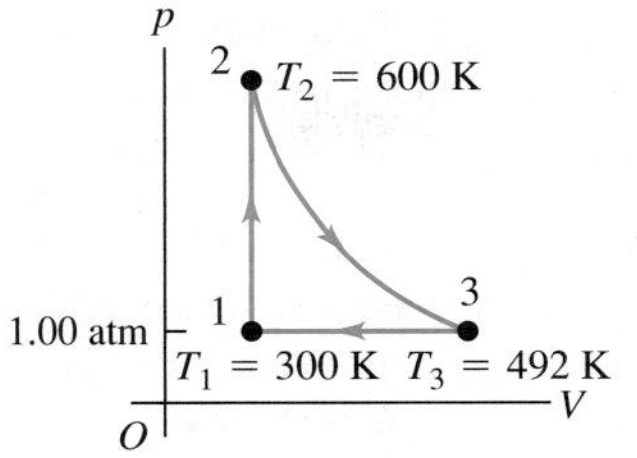

p
2 ● $T_2 = 600$ K

1.00 atm ┤ 1 ● 3
$T_1 = 300$ K $T_3 = 492$ K
O ───────────────── V

20.41 •• **CALC** You build a heat engine that takes 1.00 mol of an ideal diatomic gas through the cycle shown in Fig. P20.41. (a) Show that segment ab is an isothermal compression. (b) During which segment(s) of the cycle is heat absorbed by the gas? During which segment(s) is heat rejected? How do you know? (c) Calculate the temperature at points a, b, and c. (d) Calculate the net heat exchanged with the surroundings and the net work done by the engine in one cycle. (e) Calculate the thermal efficiency of the engine.

Figure **P20.41**

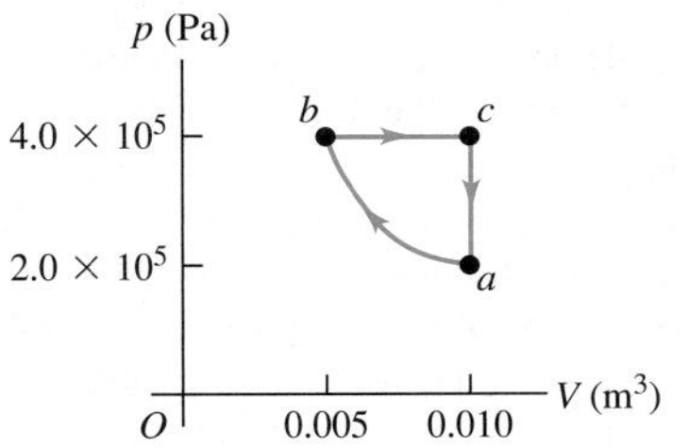

p (Pa)

4.0×10^5 ┤ b ●──→● c

2.0×10^5 ┤   ● a

O ───────┬──────┬──── V (m³)
      0.005   0.010

20.42 • **Heat Pump.** A heat pump is a heat engine run in reverse. In winter it pumps heat from the cold air outside into the warmer air inside the building, maintaining the building at a comfortable temperature. In summer it pumps heat from the cooler air inside the building to the warmer air outside, acting as an air conditioner. (a) If the outside temperature in winter is $-5.0°C$ and the inside temperature is $17.0°C$, how many joules of heat will the heat pump deliver to the inside for each joule of electrical energy used to run the unit, assuming an ideal Carnot cycle? (b) Suppose you have the option of using electrical resistance heating rather than a heat pump. How much electrical energy would you need in order to deliver the same amount of heat to the inside of the house as in part (a)? Consider a Carnot heat pump delivering heat to the inside of a house to maintain it at $68°F$. Show that the heat pump delivers less heat for each joule of electrical energy used to operate the unit as the outside temperature decreases. Notice that this behavior is opposite to the dependence of the efficiency of a Carnot heat engine on the difference in the reservoir temperatures. Explain why this is so.

20.43 • **CALC** A heat engine operates using the cycle shown in Fig. P20.43. The working substance is 2.00 mol of helium gas, which reaches a maximum temperature of $327°C$. Assume the helium can be treated as an ideal gas. Process bc is isothermal. The pressure in states a and c is 1.00×10^5 Pa, and the pressure in state b is 3.00×10^5 Pa. (a) How much heat enters the gas and how much

Figure **P20.43**

p
       ● b

a ●────←────● c
O ─────────────── V

leaves the gas each cycle? (b) How much work does the engine do each cycle, and what is its efficiency? (c) Compare this engine's efficiency with the maximum possible efficiency attainable with the hot and cold reservoirs used by this cycle.

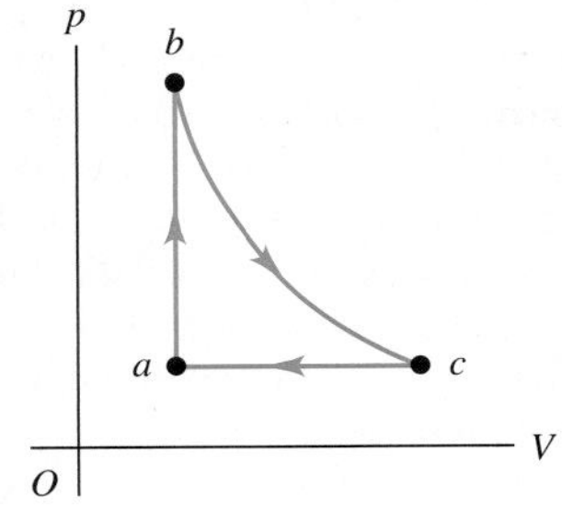

20.44 • **CP** As a budding mechanical engineer, you are called upon to design a Carnot engine that has 2.00 mol of a monatomic ideal gas as its working substance and operates from a high-temperature reservoir at $500°C$. The engine is to lift a 15.0-kg weight 2.00 m per cycle, using 500 J of heat input. The gas in the engine chamber can have a minimum volume of 5.00 L during the cycle. (a) Draw a pV-diagram for this cycle. Show in your diagram where heat enters and leaves the gas. (b) What must be the temperature of the cold reservoir? (c) What is the thermal efficiency of the engine? (d) How much heat energy does this engine waste per cycle? (e) What is the maximum pressure that the gas chamber will have to withstand?

20.45 ••• An experimental power plant at the Natural Energy Laboratory of Hawaii generates electricity from the temperature gradient of the ocean. The surface and deep-water temperatures are $27°C$ and $6°C$, respectively. (a) What is the maximum theoretical efficiency of this power plant? (b) If the power plant is to produce 210 kW of power, at what rate must heat be extracted from the warm water? At what rate must heat be absorbed by the cold water? Assume the maximum theoretical efficiency. (c) The cold water that enters the plant leaves it at a temperature of $10°C$. What must be the flow rate of cold water through the system? Give your answer in kg/h and in L/h.

20.46 •• What is the thermal efficiency of an engine that operates by taking n moles of diatomic ideal gas through the cycle $1 \rightarrow 2 \rightarrow 3 \rightarrow 4 \rightarrow 1$ shown in Fig. P20.46?

Figure **P20.46**

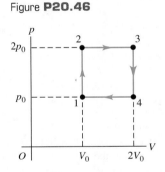

20.47 • **CALC** A cylinder contains oxygen at a pressure of 2.00 atm. The volume is 4.00 L, and the temperature is 300 K. Assume that the oxygen may be treated as an ideal gas. The oxygen is carried through the following processes:

(i) Heated at constant pressure from the initial state (state 1) to state 2, which has $T = 450$ K.
(ii) Cooled at constant volume to 250 K (state 3).
(iii) Compressed at constant temperature to a volume of 4.00 L (state 4).
(iv) Heated at constant volume to 300 K, which takes the system back to state 1.

(a) Show these four processes in a pV-diagram, giving the numerical values of p and V in each of the four states. (b) Calculate Q and W for each of the four processes. (c) Calculate the net work done by the oxygen in the complete cycle. (d) What is the efficiency of this device as a heat engine? How does this compare to the efficiency of a Carnot-cycle engine operating between the same minimum and maximum temperatures of 250 K and 450 K?

20.48 •• **CP BIO Human Entropy.** A person who has skin of surface area 1.85 m² and temperature $30.0°C$ is resting in an insulated room where the ambient air temperature is $20.0°C$. In this state, a person gets rid of excess heat by radiation. By how much does the person change the entropy of the air in this room each second?

(Recall that the room radiates back into the person and that the emissivity of the skin is 1.00.)

20.49 •• **CP BIO A Human Engine.** You decide to use your body as a Carnot heat engine. The operating gas is in a tube with one end in your mouth (where the temperature is 37.0°C) and the other end at the surface of your skin, at 30.0°C. (a) What is the maximum efficiency of such a heat engine? Would it be a very useful engine? (b) Suppose you want to use this human engine to lift a 2.50-kg box from the floor to a tabletop 1.20 m above the floor. How much must you increase the gravitational potential energy, and how much heat input is needed to accomplish this? (c) If your favorite candy bar has 350 food calories (1 food calorie = 4186 J) and 80% of the food energy goes into heat, how many of these candy bars must you eat to lift the box in this way?

20.50 •• **CP Entropy Change Due to the Sun.** Our sun radiates from a surface at 5800 K (with an emissivity of 1.0) into the near-vacuum of space, which is at a temperature of 3 K. (a) By how much does our sun change the entropy of the universe every second? (Consult Appendix F.) (b) Is the process reversible or irreversible? Is your answer to part (a) consistent with this conclusion? Explain.

20.51 • A monatomic ideal gas is taken around the cycle shown in Fig. P20.51 in the direction shown in the figure. The path for process $c \rightarrow a$ is a straight line in the pV-diagram. (a) Calculate Q, W, and ΔU for each process $a \rightarrow b$, $b \rightarrow c$, and $c \rightarrow a$. (b) What are Q, W, and ΔU for one complete cycle? (c) What is the efficiency of the cycle?

Figure **P20.51**

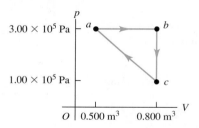

20.52 •• **CALC A Stirling-Cycle Engine.** The *Stirling cycle* is similar to the Otto cycle, except that the compression and expansion of the gas are done at constant temperature, not adiabatically as in the Otto cycle. The Stirling cycle is used in *external* combustion engines (in fact, burning fuel is not necessary; *any* way of producing a temperature difference will do—solar, geothermal, ocean temperature gradient, etc.), which means that the gas inside the cylinder is not used in the combustion process. Heat is supplied by burning fuel steadily outside the cylinder, instead of explosively inside the

Figure **P20.52**

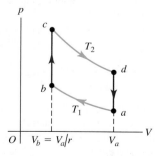

cylinder as in the Otto cycle. For this reason Stirling-cycle engines are quieter than Otto-cycle engines, since there are no intake and exhaust valves (a major source of engine noise). While small Stirling engines are used for a variety of purposes, Stirling engines for automobiles have not been successful because they are larger, heavier, and more expensive than conventional automobile engines. In the cycle, the working fluid goes through the following sequence of steps (Fig. P20.52):

(i) Compressed isothermally at temperature T_1 from the initial state a to state b, with a compression ratio r.
(ii) Heated at constant volume to state c at temperature T_2.
(iii) Expanded isothermally at T_2 to state d.
(iv) Cooled at constant volume back to the initial state a.

Assume that the working fluid is n moles of an ideal gas (for which C_V is independent of temperature). (a) Calculate Q, W, and ΔU for each of the processes $a \rightarrow b$, $b \rightarrow c$, $c \rightarrow d$, and $d \rightarrow a$. (b) In the Stirling cycle, the heat transfers in the processes $b \rightarrow c$ and $d \rightarrow a$ do not involve external heat sources but rather use *regeneration*: The same substance that transfers heat to the gas inside the cylinder in the process $b \rightarrow c$ also absorbs heat back from the gas in the process $d \rightarrow a$. Hence the heat transfers $Q_{b \rightarrow c}$ and $Q_{d \rightarrow a}$ do not play a role in determining the efficiency of the engine. Explain this last statement by comparing the expressions for $Q_{b \rightarrow c}$ and $Q_{d \rightarrow a}$ calculated in part (a). (c) Calculate the efficiency of a Stirling-cycle engine in terms of the temperatures T_1 and T_2. How does this compare to the efficiency of a Carnot-cycle engine operating between these same two temperatures? (Historically, the Stirling cycle was devised before the Carnot cycle.) Does this result violate the second law of thermodynamics? Explain. Unfortunately, actual Stirling-cycle engines cannot achieve this efficiency due to problems with the heat-transfer processes and pressure losses in the engine.

20.53 •• A Carnot engine operates between two heat reservoirs at temperatures T_H and T_C. An inventor proposes to increase the efficiency by running one engine between T_H and an intermediate temperature T' and a second engine between T' and T_C, using as input the heat expelled by the first engine. Compute the efficiency of this composite system, and compare it to that of the original engine.

20.54 ••• A typical coal-fired power plant generates 1000 MW of usable power at an overall thermal efficiency of 40%. (a) What is the rate of heat input to the plant? (b) The plant burns anthracite coal, which has a heat of combustion of 2.65×10^7 J/kg. How much coal does the plant use per day, if it operates continuously? (c) At what rate is heat ejected into the cool reservoir, which is the nearby river? (d) The river's temperature is 18.0°C before it reaches the power plant and 18.5°C after it has received the plant's waste heat. Calculate the river's flow rate, in cubic meters per second. (e) By how much does the river's entropy increase each second?

20.55 • **Automotive Thermodynamics.** A Volkswagen Passat has a six-cylinder Otto-cycle engine with compression ratio $r = 10.6$. The diameter of each cylinder, called the *bore* of the engine, is 82.5 mm. The distance that the piston moves during the compression in Fig. 20.5, called the *stroke* of the engine, is 86.4 mm. The initial pressure of the air–fuel mixture (at point a in Fig. 20.6) is 8.50×10^4 Pa, and the initial temperature is 300 K (the same as the outside air). Assume that 200 J of heat is added to each cylinder in each cycle by the burning gasoline, and that the gas has $C_V = 20.5$ J/mol·K and $\gamma = 1.40$. (a) Calculate the total work done in one cycle in each cylinder of the engine, and the heat released when the gas is cooled to the temperature of the outside air. (b) Calculate the volume of the air–fuel mixture at point a in the cycle. (c) Calculate the pressure, volume, and temperature of the gas at points b, c, and d in the cycle. In a pV-diagram, show the numerical values of p, V, and T for each of the four states. (d) Compare the efficiency of this engine with the efficiency of a Carnot-cycle engine operating between the same maximum and minimum temperatures.

20.56 • An air conditioner operates on 800 W of power and has a performance coefficient of 2.80 with a room temperature of 21.0°C and an outside temperature of 35.0°C. (a) Calculate the rate of heat removal for this unit. (b) Calculate the rate at which heat is discharged to the outside air. (c) Calculate the total entropy change in the room if the air conditioner runs for 1 hour. Calculate the total entropy change in the outside air for the same time period. (d) What is the net change in entropy for the system (room + outside air)?

20.57 •• **CALC** **Unavailable Energy.** The discussion of entropy and the second law that follows Example 20.10 (Section 20.7) says that the increase in entropy in an irreversible process is associated with energy becoming less available. Consider a Carnot cycle that uses a low-temperature reservoir with Kelvin temperature T_c. This is a true reservoir—that is, large enough not to change temperature when it accepts heat from the engine. Let the engine accept heat from an object of temperature T', where $T' > T_c$. The object is of finite size, so it cools as heat is extracted from it. The engine continues to operate until $T' = T_c$. (a) Show that the total magnitude of heat rejected to the low-temperature reservoir is $T_c|\Delta S_h|$, where ΔS_h is the change in entropy of the high-temperature reservoir. (b) Apply the result of part (a) to 1.00 kg of water initially at a temperature of 373 K as the heat source for the engine and $T_c = 273$ K. How much total mechanical work can be performed by the engine until it stops? (c) Repeat part (b) for 2.00 kg of water at 323 K. (d) Compare the amount of work that can be obtained from the energy in the water of Example 20.10 before and after it is mixed. Discuss whether your result shows that energy has become less available.

20.58 ••• **CP** The maximum power that can be extracted by a wind turbine from an air stream is approximately

$$P = kd^2v^3$$

where d is the blade diameter, v is the wind speed, and the constant $k = 0.5 \ \text{W} \cdot \text{s}^3/\text{m}^5$. (a) Explain the dependence of P on d and on v by considering a cylinder of air that passes over the turbine blades in time t (Fig. P20.58). This cylinder has diameter d, length $L = vt$, and density ρ. (b) The Mod-5B wind turbine at Kahaku on the Hawaiian island of Oahu has a blade diameter of 97 m (slightly longer than a football field) and sits atop a 58-m tower. It can produce 3.2 MW of electric power. Assuming 25% efficiency, what wind speed is required to produce this amount of power? Give your answer in m/s and in km/h. (c) Commercial wind turbines are commonly located in or downwind of mountain passes. Why?

Figure **P20.58**

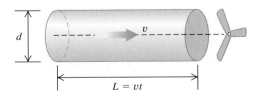

20.59 •• **CALC** (a) For the Otto cycle shown in Fig. 20.6, calculate the changes in entropy of the gas in each of the constant-volume processes $b \to c$ and $d \to a$ in terms of the temperatures T_a, T_b, T_c, and T_d and the number of moles n and the heat capacity C_V of the

gas. (b) What is the total entropy change in the engine during one cycle? (*Hint:* Use the relationships between T_a and T_b and between T_d and T_c.) (c) The processes $b \to c$ and $d \to a$ occur irreversibly in a real Otto engine. Explain how can this be reconciled with your result in part (b).

20.60 •• **CALC** **A TS-Diagram.** (a) Graph a Carnot cycle, plotting Kelvin temperature vertically and entropy horizontally. This is called a temperature–entropy diagram, or TS-diagram. (b) Show that the area under any curve representing a reversible path in a temperature–entropy diagram represents the heat absorbed by the system. (c) Derive from your diagram the expression for the thermal efficiency of a Carnot cycle. (d) Draw a temperature–entropy diagram for the Stirling cycle described in Problem 20.52. Use this diagram to relate the efficiencies of the Carnot and Stirling cycles.

20.61 • A physics student immerses one end of a copper rod in boiling water at 100°C and the other end in an ice–water mixture at 0°C. The sides of the rod are insulated. After steady-state conditions have been achieved in the rod, 0.120 kg of ice melts in a certain time interval. For this time interval, find (a) the entropy change of the boiling water; (b) the entropy change of the ice–water mixture; (c) the entropy change of the copper rod; (d) the total entropy change of the entire system.

20.62 •• **CALC** To heat 1 cup of water (250 cm³) to make coffee, you place an electric heating element in the cup. As the water temperature increases from 20°C to 78°C, the temperature of the heating element remains at a constant 120°C. Calculate the change in entropy of (a) the water; (b) the heating element; (c) the system of water and heating element. (Make the same assumption about the specific heat of water as in Example 20.10 in Section 20.7, and ignore the heat that flows into the ceramic coffee cup itself.) (d) Is this process reversible or irreversible? Explain.

20.63 •• **CALC** An object of mass m_1, specific heat c_1, and temperature T_1 is placed in contact with a second object of mass m_2, specific heat c_2, and temperature $T_2 > T_1$. As a result, the temperature of the first object increases to T and the temperature of the second object decreases to T'. (a) Show that the entropy increase of the system is

$$\Delta S = m_1 c_1 \ln\frac{T}{T_1} + m_2 c_2 \ln\frac{T'}{T_2}$$

and show that energy conservation requires that

$$m_1 c_1 (T - T_1) = m_2 c_2 (T_2 - T')$$

(b) Show that the entropy change ΔS, considered as a function of T, is a *maximum* if $T = T'$, which is just the condition of thermodynamic equilibrium. (c) Discuss the result of part (b) in terms of the idea of entropy as a measure of disorder.

CHALLENGE PROBLEM

20.64 ••• Consider a Diesel cycle that starts (at point a in Fig. 20.7) with air at temperature T_a. The air may be treated as an ideal gas. (a) If the temperature at point c is T_c, derive an expression for the efficiency of the cycle in terms of the compression ratio r. (b) What is the efficiency if $T_a = 300$ K, $T_c = 950$ K, $\gamma = 1.40$, and $r = 21.0$?

Answers

Chapter Opening Question ?

Yes. That's what a refrigerator does: It makes heat flow from the cold interior of the refrigerator to the warm outside. The second law of thermodynamics says that heat cannot *spontaneously* flow from a cold body to a hot one. A refrigerator has a motor that does work on the system to *force* the heat to flow in that way.

Test Your Understanding Questions

20.1 Answer: (ii) Like sliding a book across a table, rubbing your hands together uses friction to convert mechanical energy into heat. The (impossible) reverse process would involve your hands spontaneously getting colder, with the released energy forcing your hands to move rhythmically back and forth!

20.2 Answer: (iii), (i), (ii) From Eq. (20.4) the efficiency is $e = W/Q_H$, and from Eq. (20.2) $W = Q_H + Q_C = |Q_H| - |Q_C|$. For engine (i) $Q_H = 5000$ J and $Q_C = -4500$ J, so $W = 5000$ J $+ (-4500$ J$) = 500$ J and $e = (500$ J$)/(5000$ H$) = 0.100$. For engine (ii) $Q_H = 25{,}000$ J and $W = 2000$ J, so $e = (2000$ J$)/(25{,}000$ J$) = 0.080$. For engine (iii) $W = 400$ J and $Q_C = -2800$ J, so $Q_H = W - Q_C = 400$ J $- (-2800$ J$) = 3200$ J and $e = (400$ J$)/(3200$ J$) = 0.125$.

20.3 Answers: (i), (ii) Doubling the amount of fuel burned per cycle means that Q_H is doubled, so the resulting pressure increase from *b* to *c* in Fig. 20.6 is greater. The compression ratio and hence the efficiency remain the same, so $|Q_C|$ (the amount of heat rejected to the environment) must increase by the same factor as Q_H. Hence the pressure drop from *d* to *a* in Fig. 20.6 is also greater. The volume *V* and the compression ratio *r* don't change, so the horizontal dimensions of the *pV*-diagram don't change.

20.4 Answer: no A refrigerator uses an input of work to transfer heat from one system (the refrigerator's interior) to another system (its exterior, which includes the house in which the refrigerator is installed). If the door is open, these two systems are really the *same* system and will eventually come to the same temperature. By the first law of thermodynamics, all of the work input to the refrigerator motor will be converted into heat and the temperature in your house will actually *increase*. To cool the house you need a system that will transfer heat from it to the outside world, such as an air conditioner or heat pump.

20.5 Answers: no, no Both the 100%-efficient engine of Fig. 20.11a and the workless refrigerator of Fig. 20.11b return to the same state at the end of a cycle as at the beginning, so the net change in internal energy of each system is zero ($\Delta U = 0$). For the 100%-efficient engine, the net heat flow into the engine equals the net work done, so $Q = W$, $Q - W = 0$, and the first law ($\Delta U = Q - W$) is obeyed. For the workless refrigerator, no net work is done (so $W = 0$) and as much heat flows into it as out (so $Q = 0$), so again $Q - W = 0$ and $\Delta U = Q - W$ in accordance with the first law. It is the *second* law of thermodynamics that tells us that both the 100%-efficient engine and the workless refrigerator are impossible.

20.6 Answer: no The efficiency can be no better than that of a Carnot engine running between the same two temperature limits, $e_{\text{Carnot}} = 1 - (T_C/T_H)$ [Eq. (20.14)]. The temperature T_C of the cold reservoir for this air-cooled engine is about 300 K (ambient temperature), and the temperature T_H of the hot reservoir cannot exceed the melting point of copper, 1356 K (see Table 17.4). Hence the maximum possible Carnot efficiency is $e = 1 - (300$ K$)/(1356$ K$) = 0.78$, or 78%. The temperature of any real engine would be less than this, so it would be impossible for the inventor's engine to attain 85% efficiency. You should invest your money elsewhere.

20.7 Answers: -102 J/K, no The process described is exactly the opposite of the process used in Example 20.10. The result violates the second law of thermodynamics, which states that the entropy of an isolated system cannot decrease.

20.8 Answer: (i) For case (i), we saw in Example 20.8 (Section 20.7) that for an ideal gas, the entropy change in a free expansion is the same as in an isothermal expansion. From Eq. (20.23), this implies that the ratio of the number of microscopic states after and before the expansion, w_2/w_1, is also the same for these two cases. From Example 20.11, $w_2/w_1 = 2^N$, so the number of microscopic states increases by a factor 2^N. For case (ii), in a reversible expansion the entropy change is $\Delta S = \int dQ/T = 0$; if the expansion is adiabatic there is no heat flow, so $\Delta S = 0$. From Eq. (20.23), $w_2/w_1 = 1$ and there is *no* change in the number of microscopic states. The difference is that in an adiabatic expansion the temperature drops and the molecules move more slowly, so they have fewer microscopic states available to them than in an isothermal expansion.

Bridging Problem

Answers: (a) 34.83°C **(b)** +12.1 J/K

APPENDIX A

THE INTERNATIONAL SYSTEM OF UNITS

The Système International d'Unités, abbreviated SI, is the system developed by the General Conference on Weights and Measures and adopted by nearly all the industrial nations of the world. The following material is adapted from the National Institute of Standards and Technology (**http://physics.nist.gov/cuu**).

Quantity	Name of unit	Symbol	
SI base units			
length	meter	m	
mass	kilogram	kg	
time	second	s	
electric current	ampere	A	
thermodynamic temperature	kelvin	K	
amount of substance	mole	mol	
luminous intensity	candela	cd	
SI derived units			**Equivalent units**
area	square meter	m^2	
volume	cubic meter	m^3	
frequency	hertz	Hz	s^{-1}
mass density (density)	kilogram per cubic meter	kg/m^3	
speed, velocity	meter per second	m/s	
angular velocity	radian per second	rad/s	
acceleration	meter per second squared	m/s^2	
angular acceleration	radian per second squared	rad/s^2	
force	newton	N	$kg \cdot m/s^2$
pressure (mechanical stress)	pascal	Pa	N/m^2
kinematic viscosity	square meter per second	m^2/s	
dynamic viscosity	newton-second per square meter	$N \cdot s/m^2$	
work, energy, quantity of heat	joule	J	$N \cdot m$
power	watt	W	J/s
quantity of electricity	coulomb	C	$A \cdot s$
potential difference, electromotive force	volt	V	J/C, W/A
electric field strength	volt per meter	V/m	N/C
electric resistance	ohm	Ω	V/A
capacitance	farad	F	$A \cdot s/V$
magnetic flux	weber	Wb	$V \cdot s$
inductance	henry	H	$V \cdot s/A$
magnetic flux density	tesla	T	Wb/m^2
magnetic field strength	ampere per meter	A/m	
magnetomotive force	ampere	A	
luminous flux	lumen	lm	$cd \cdot sr$
luminance	candela per square meter	cd/m^2	
illuminance	lux	lx	lm/m^2
wave number	1 per meter	m^{-1}	
entropy	joule per kelvin	J/K	
specific heat capacity	joule per kilogram-kelvin	$J/kg \cdot K$	
thermal conductivity	watt per meter-kelvin	$W/m \cdot K$	

Quantity	Name of unit	Symbol	Equivalent units
radiant intensity	watt per steradian	W/sr	
activity (of a radioactive source)	becquerel	Bq	s^{-1}
radiation dose	gray	Gy	J/kg
radiation dose equivalent	sievert	Sv	J/kg
SI supplementary units			
plane angle	radian	rad	
solid angle	steradian	sr	

Definitions of SI Units

meter (m)   The *meter* is the length equal to the distance traveled by light, in vacuum, in a time of 1/299,792,458 second.

kilogram (kg)   The *kilogram* is the unit of mass; it is equal to the mass of the international prototype of the kilogram. (The international prototype of the kilogram is a particular cylinder of platinum-iridium alloy that is preserved in a vault at Sévres, France, by the International Bureau of Weights and Measures.)

second (s)   The *second* is the duration of 9,192,631,770 periods of the radiation corresponding to the transition between the two hyperfine levels of the ground state of the cesium-133 atom.

ampere (A)   The *ampere* is that constant current that, if maintained in two straight parallel conductors of infinite length, of negligible circular cross section, and placed 1 meter apart in vacuum, would produce between these conductors a force equal to 2×10^{-7} newton per meter of length.

kelvin (K)   The *kelvin*, unit of thermodynamic temperature, is the fraction 1/273.16 of the thermodynamic temperature of the triple point of water.

ohm (Ω)   The *ohm* is the electric resistance between two points of a conductor when a constant difference of potential of 1 volt, applied between these two points, produces in this conductor a current of 1 ampere, this conductor not being the source of any electromotive force.

coulomb (C)   The *coulomb* is the quantity of electricity transported in 1 second by a current of 1 ampere.

candela (cd)   The *candela* is the luminous intensity, in a given direction, of a source that emits monochromatic radiation of frequency 540×10^{12} hertz and that has a radiant intensity in that direction of 1/683 watt per steradian.

mole (mol)   The *mole* is the amount of substance of a system that contains as many elementary entities as there are carbon atoms in 0.012 kg of carbon 12. The elementary entities must be specified and may be atoms, molecules, ions, electrons, other particles, or specified groups of such particles.

newton (N)   The *newton* is that force that gives to a mass of 1 kilogram an acceleration of 1 meter per second per second.

joule (J)   The *joule* is the work done when the point of application of a constant force of 1 newton is displaced a distance of 1 meter in the direction of the force.

watt (W)   The *watt* is the power that gives rise to the production of energy at the rate of 1 joule per second.

volt (V)   The *volt* is the difference of electric potential between two points of a conducting wire carrying a constant current of 1 ampere, when the power dissipated between these points is equal to 1 watt.

weber (Wb)   The *weber* is the magnetic flux that, linking a circuit of one turn, produces in it an electromotive force of 1 volt as it is reduced to zero at a uniform rate in 1 second.

lumen (lm)   The *lumen* is the luminous flux emitted in a solid angle of 1 steradian by a uniform point source having an intensity of 1 candela.

farad (F)   The *farad* is the capacitance of a capacitor between the plates of which there appears a difference of potential of 1 volt when it is charged by a quantity of electricity equal to 1 coulomb.

henry (H)   The *henry* is the inductance of a closed circuit in which an electromotive force of 1 volt is produced when the electric current in the circuit varies uniformly at a rate of 1 ampere per second.

radian (rad)   The *radian* is the plane angle between two radii of a circle that cut off on the circumference an arc equal in length to the radius.

steradian (sr)   The *steradian* is the solid angle that, having its vertex in the center of a sphere, cuts off an area of the surface of the sphere equal to that of a square with sides of length equal to the radius of the sphere.

SI Prefixes   To form the names of multiples and submultiples of SI units, apply the prefixes listed in Appendix F.

APPENDIX B

USEFUL MATHEMATICAL RELATIONS

Algebra

$$a^{-x} = \frac{1}{a^x} \qquad a^{(x+y)} = a^x a^y \qquad a^{(x-y)} = \frac{a^x}{a^y}$$

Logarithms: If $\log a = x$, then $a = 10^x$. $\log a + \log b = \log(ab)$ $\log a - \log b = \log(a/b)$ $\log(a^n) = n\log a$

If $\ln a = x$, then $a = e^x$. $\ln a + \ln b = \ln(ab)$ $\ln a - \ln b = \ln(a/b)$ $\ln(a^n) = n\ln a$

Quadratic formula: If $ax^2 + bx + c = 0$, $x = \dfrac{-b \pm \sqrt{b^2 - 4ac}}{2a}$.

Binomial Theorem

$$(a + b)^n = a^n + na^{n-1}b + \frac{n(n-1)a^{n-2}b^2}{2!} + \frac{n(n-1)(n-2)a^{n-3}b^3}{3!} + \cdots$$

Trigonometry

In the right triangle ABC, $x^2 + y^2 = r^2$.

Definitions of the trigonometric functions:
$\sin \alpha = y/r \qquad \cos \alpha = x/r \qquad \tan \alpha = y/x$

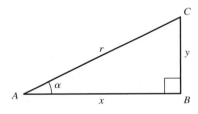

Identities: $\sin^2\alpha + \cos^2\alpha = 1$ $\tan\alpha = \dfrac{\sin\alpha}{\cos\alpha}$

$\sin 2\alpha = 2\sin\alpha\cos\alpha$ $\cos 2\alpha = \cos^2\alpha - \sin^2\alpha = 2\cos^2\alpha - 1$
$$= 1 - 2\sin^2\alpha$$

$\sin\tfrac{1}{2}\alpha = \sqrt{\dfrac{1 - \cos\alpha}{2}}$ $\cos\tfrac{1}{2}\alpha = \sqrt{\dfrac{1 + \cos\alpha}{2}}$

$\sin(-\alpha) = -\sin\alpha$ $\sin(\alpha \pm \beta) = \sin\alpha\cos\beta \pm \cos\alpha\sin\beta$
$\cos(-\alpha) = \cos\alpha$ $\cos(\alpha \pm \beta) = \cos\alpha\cos\beta \mp \sin\alpha\sin\beta$
$\sin(\alpha \pm \pi/2) = \pm\cos\alpha$ $\sin\alpha + \sin\beta = 2\sin\tfrac{1}{2}(\alpha + \beta)\cos\tfrac{1}{2}(\alpha - \beta)$
$\cos(\alpha \pm \pi/2) = \mp\sin\alpha$ $\cos\alpha + \cos\beta = 2\cos\tfrac{1}{2}(\alpha + \beta)\cos\tfrac{1}{2}(\alpha - \beta)$

For *any* triangle $A'B'C'$ (not necessarily a right triangle) with sides a, b, and c and angles α, β, and γ:

Law of sines: $\dfrac{\sin\alpha}{a} = \dfrac{\sin\beta}{b} = \dfrac{\sin\gamma}{c}$

Law of cosines: $c^2 = a^2 + b^2 - 2ab\cos\gamma$

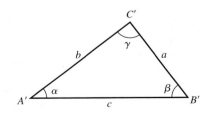

Geometry

Circumference of circle of radius r: $C = 2\pi r$ Surface area of sphere of radius r: $A = 4\pi r^2$
Area of circle of radius r: $A = \pi r^2$ Volume of cylinder of radius r and height h: $V = \pi r^2 h$
Volume of sphere of radius r: $V = 4\pi r^3/3$

Calculus

Derivatives:

$$\frac{d}{dx}x^n = nx^{n-1}$$

$$\frac{d}{dx}\ln ax = \frac{1}{x}$$

$$\frac{d}{dx}e^{ax} = ae^{ax}$$

$$\frac{d}{dx}\sin ax = a\cos ax$$

$$\frac{d}{dx}\cos ax = -a\sin ax$$

Integrals:

$$\int x^n\,dx = \frac{x^{n+1}}{n+1} \quad (n \neq -1)$$

$$\int \frac{dx}{x} = \ln x$$

$$\int e^{ax}\,dx = \frac{1}{a}e^{ax}$$

$$\int \sin ax\,dx = -\frac{1}{a}\cos ax$$

$$\int \cos ax\,dx = \frac{1}{a}\sin ax$$

$$\int \frac{dx}{\sqrt{a^2-x^2}} = \arcsin\frac{x}{a}$$

$$\int \frac{dx}{\sqrt{x^2+a^2}} = \ln\left(x + \sqrt{x^2+a^2}\right)$$

$$\int \frac{dx}{x^2+a^2} = \frac{1}{a}\arctan\frac{x}{a}$$

$$\int \frac{dx}{(x^2+a^2)^{3/2}} = \frac{1}{a^2}\frac{x}{\sqrt{x^2+a^2}}$$

$$\int \frac{x\,dx}{(x^2+a^2)^{3/2}} = -\frac{1}{\sqrt{x^2+a^2}}$$

Power series (convergent for range of x shown):

$$(1+x)^n = 1 + nx + \frac{n(n-1)x^2}{2!} + \frac{n(n-1)(n-2)}{3!}x^3 + \cdots \quad (|x| < 1)$$

$$\tan x = x + \frac{x^3}{3} + \frac{2x^2}{15} + \frac{17x^7}{315} + \cdots \quad (|x| < \pi/2)$$

$$e^x = 1 + x + \frac{x^2}{2!} + \frac{x^3}{3!} + \cdots \quad (\text{all } x)$$

$$\sin x = x - \frac{x^3}{3!} + \frac{x^5}{5!} - \frac{x^7}{7!} + \cdots \quad (\text{all } x)$$

$$\ln(1+x) = x - \frac{x^2}{2} + \frac{x^3}{3} - \frac{x^4}{4} + \cdots \quad (|x| < 1)$$

$$\cos x = 1 - \frac{x^2}{2!} + \frac{x^4}{4!} - \frac{x^6}{6!} + \cdots \quad (\text{all } x)$$

APPENDIX C

THE GREEK ALPHABET

Name	Capital	Lowercase	Name	Capital	Lowercase	Name	Capital	Lowercase
Alpha	A	α	Iota	I	ι	Rho	P	ρ
Beta	B	β	Kappa	K	κ	Sigma	Σ	σ
Gamma	Γ	γ	Lambda	Λ	λ	Tau	T	τ
Delta	Δ	δ	Mu	M	μ	Upsilon	Υ	υ
Epsilon	E	ϵ	Nu	N	ν	Phi	Φ	ϕ
Zeta	Z	ζ	Xi	Ξ	ξ	Chi	X	χ
Eta	H	η	Omicron	O	o	Psi	Ψ	ψ
Theta	Θ	θ	Pi	Π	π	Omega	Ω	ω

PERIODIC TABLE OF THE ELEMENTS

Group 1 2 3 4 5 6 7 8 9 10 11 12 13 14 15 16 17 18
Period

Period	1	2	3	4	5	6	7	8	9	10	11	12	13	14	15	16	17	18
1	1 **H** 1.008																	2 **He** 4.003
2	3 **Li** 6.941	4 **Be** 9.012											5 **B** 10.811	6 **C** 12.011	7 **N** 14.007	8 **O** 15.999	9 **F** 18.998	10 **Ne** 20.180
3	11 **Na** 22.990	12 **Mg** 24.305											13 **Al** 26.982	14 **Si** 28.086	15 **P** 30.974	16 **S** 32.065	17 **Cl** 35.453	18 **Ar** 39.948
4	19 **K** 39.098	20 **Ca** 40.078	21 **Sc** 44.956	22 **Ti** 47.867	23 **V** 50.942	24 **Cr** 51.996	25 **Mn** 54.938	26 **Fe** 55.845	27 **Co** 58.933	28 **Ni** 58.693	29 **Cu** 63.546	30 **Zn** 65.409	31 **Ga** 69.723	32 **Ge** 72.64	33 **As** 74.922	34 **Se** 78.96	35 **Br** 79.904	36 **Kr** 83.798
5	37 **Rb** 85.468	38 **Sr** 87.62	39 **Y** 88.906	40 **Zr** 91.224	41 **Nb** 92.906	42 **Mo** 95.94	43 **Tc** (98)	44 **Ru** 101.07	45 **Rh** 102.906	46 **Pd** 106.42	47 **Ag** 107.868	48 **Cd** 112.411	49 **In** 114.818	50 **Sn** 118.710	51 **Sb** 121.760	52 **Te** 127.60	53 **I** 126.904	54 **Xe** 131.293
6	55 **Cs** 132.905	56 **Ba** 137.327	71 **Lu** 174.967	72 **Hf** 178.49	73 **Ta** 180.948	74 **W** 183.84	75 **Re** 186.207	76 **Os** 190.23	77 **Ir** 192.217	78 **Pt** 195.078	79 **Au** 196.967	80 **Hg** 200.59	81 **Tl** 204.383	82 **Pb** 207.2	83 **Bi** 208.980	84 **Po** (209)	85 **At** (210)	86 **Rn** (222)
7	87 **Fr** (223)	88 **Ra** (226)	103 **Lr** (262)	104 **Rf** (261)	105 **Db** (262)	106 **Sg** (266)	107 **Bh** (264)	108 **Hs** (269)	109 **Mt** (268)	110 **Ds** (271)	111 **Rg** (272)	112 **Uub** (285)	113 **Uut** (284)	114 **Uuq** (289)	115 **Uup** (288)	116 **Uuh** (292)	117 **Uus** (294)	118 **Uuo**

Lanthanoids	57 **La** 138.905	58 **Ce** 140.116	59 **Pr** 140.908	60 **Nd** 144.24	61 **Pm** (145)	62 **Sm** 150.36	63 **Eu** 151.964	64 **Gd** 157.25	65 **Tb** 158.925	66 **Dy** 162.500	67 **Ho** 164.930	68 **Er** 167.259	69 **Tm** 168.934	70 **Yb** 173.04
Actinoids	89 **Ac** (227)	90 **Th** (232)	91 **Pa** (231)	92 **U** (238)	93 **Np** (237)	94 **Pu** (244)	95 **Am** (243)	96 **Cm** (247)	97 **Bk** (247)	98 **Cf** (251)	99 **Es** (252)	100 **Fm** (257)	101 **Md** (258)	102 **No** (259)

For each element the average atomic mass of the mixture of isotopes occurring in nature is shown. For elements having no stable isotope, the approximate atomic mass of the longest-lived isotope is shown in parentheses. For elements that have been predicted but not yet confirmed, no atomic mass is given. All atomic masses are expressed in atomic mass units ($1\ u = 1.660538782(83) \times 10^{-27}\ kg$), equivalent to grams per mole (g/mol).

APPENDIX E

UNIT CONVERSION FACTORS

Length
1 m = 100 cm = 1000 mm = $10^6 \, \mu$m = 10^9 nm
1 km = 1000 m = 0.6214 mi
1 m = 3.281 ft = 39.37 in.
1 cm = 0.3937 in.
1 in. = 2.540 cm
1 ft = 30.48 cm
1 yd = 91.44 cm
1 mi = 5280 ft = 1.609 km
1 Å = 10^{-10} m = 10^{-8} cm = 10^{-1} nm
1 nautical mile = 6080 ft
1 light year = 9.461 × 10^{15} m

Area
1 cm^2 = 0.155 in.^2
1 m^2 = 10^4 cm^2 = 10.76 ft^2
1 in.^2 = 6.452 cm^2
1 ft^2 = 144 in.^2 = 0.0929 m^2

Volume
1 liter = 1000 cm^3 = 10^{-3} m^3 = 0.03531 ft^3 = 61.02 in.^3
1 ft^3 = 0.02832 m^3 = 28.32 liters = 7.477 gallons
1 gallon = 3.788 liters

Time
1 min = 60 s
1 h = 3600 s
1 d = 86,400 s
1 y = 365.24 d = 3.156 × 10^7 s

Angle
1 rad = 57.30° = 180°/π
1° = 0.01745 rad = π/180 rad
1 revolution = 360° = 2π rad
1 rev/min (rpm) = 0.1047 rad/s

Speed
1 m/s = 3.281 ft/s
1 ft/s = 0.3048 m/s
1 mi/min = 60 mi/h = 88 ft/s
1 km/h = 0.2778 m/s = 0.6214 mi/h
1 mi/h = 1.466 ft/s = 0.4470 m/s = 1.609 km/h
1 furlong/fortnight = 1.662 × 10^{-4} m/s

Acceleration
1 m/s^2 = 100 cm/s^2 = 3.281 ft/s^2
1 cm/s^2 = 0.01 m/s^2 = 0.03281 ft/s^2
1 ft/s^2 = 0.3048 m/s^2 = 30.48 cm/s^2
1 mi/h·s = 1.467 ft/s^2

Mass
1 kg = 10^3 g = 0.0685 slug
1 g = 6.85 × 10^{-5} slug
1 slug = 14.59 kg
1 u = 1.661 × 10^{-27} kg
1 kg has a weight of 2.205 lb when g = 9.80 m/s^2

Force
1 N = 10^5 dyn = 0.2248 lb
1 lb = 4.448 N = 4.448 × 10^5 dyn

Pressure
1 Pa = 1 N/m^2 = 1.450 × 10^{-4} lb/in.^2 = 0.209 lb/ft^2
1 bar = 10^5 Pa
1 lb/in.^2 = 6895 Pa
1 lb/ft^2 = 47.88 Pa
1 atm = 1.013 × 10^5 Pa = 1.013 bar
     = 14.7 lb/in.^2 = 2117 lb/ft^2
1 mm Hg = 1 torr = 133.3 Pa

Energy
1 J = 10^7 ergs = 0.239 cal
1 cal = 4.186 J (based on 15° calorie)
1 ft·lb = 1.356 J
1 Btu = 1055 J = 252 cal = 778 ft·lb
1 eV = 1.602 × 10^{-19} J
1 kWh = 3.600 × 10^6 J

Mass–Energy Equivalence
1 kg ↔ 8.988 × 10^{16} J
1 u ↔ 931.5 MeV
1 eV ↔ 1.074 × 10^{-9} u

Power
1 W = 1 J/s
1 hp = 746 W = 550 ft·lb/s
1 Btu/h = 0.293 W

APPENDIX F

NUMERICAL CONSTANTS

Fundamental Physical Constants*

Name	Symbol	Value
Speed of light in vacuum	c	2.99792458×10^8 m/s
Magnitude of charge of electron	e	$1.602176487(40) \times 10^{-19}$ C
Gravitational constant	G	$6.67428(67) \times 10^{-11}$ N $\cdot$ m^2/kg^2
Planck's constant	h	$6.62606896(33) \times 10^{-34}$ J $\cdot$ s
Boltzmann constant	k	$1.3806504(24) \times 10^{-23}$ J/K
Avogadro's number	N_A	$6.02214179(30) \times 10^{23}$ molecules/mol
Gas constant	R	$8.314472(15)$ J/mol $\cdot$ K
Mass of electron	m_e	$9.10938215(45) \times 10^{-31}$ kg
Mass of proton	m_p	$1.672621637(83) \times 10^{-27}$ kg
Mass of neutron	m_n	$1.674927211(84) \times 10^{-27}$ kg
Permeability of free space	μ_0	$4\pi \times 10^{-7}$ Wb/A $\cdot$ m
Permittivity of free space	$\epsilon_0 = 1/\mu_0 c^2$	$8.854187817\ldots \times 10^{-12}$ C^2/N $\cdot$ m^2
	$1/4\pi\epsilon_0$	$8.987551787\ldots \times 10^9$ N $\cdot$ m^2/C^2

Other Useful Constants*

Mechanical equivalent of heat		4.186 J/cal ($15°$ calorie)
Standard atmospheric pressure	1 atm	1.01325×10^5 Pa
Absolute zero	0 K	$-273.15°$C
Electron volt	1 eV	$1.602176487(40) \times 10^{-19}$ J
Atomic mass unit	1 u	$1.660538782(83) \times 10^{-27}$ kg
Electron rest energy	$m_e c^2$	$0.510998910(13)$ MeV
Volume of ideal gas ($0°$C and 1 atm)		$22.413996(39)$ liter/mol
Acceleration due to gravity (standard)	g	9.80665 m/s^2

*Source: National Institute of Standards and Technology (**http://physics.nist.gov/cuu**). Numbers in parentheses show the uncertainty in the final digits of the main number; for example, the number 1.6454(21) means 1.6454 $\pm$ 0.0021. Values shown without uncertainties are exact.

Astronomical Data[†]

Body	Mass (kg)	Radius (m)	Orbit radius (m)	Orbit period
Sun	1.99×10^{30}	6.96×10^{8}	—	—
Moon	7.35×10^{22}	1.74×10^{6}	3.84×10^{8}	27.3 d
Mercury	3.30×10^{23}	2.44×10^{6}	5.79×10^{10}	88.0 d
Venus	4.87×10^{24}	6.05×10^{6}	1.08×10^{11}	224.7 d
Earth	5.97×10^{24}	6.38×10^{6}	1.50×10^{11}	365.3 d
Mars	6.42×10^{23}	3.40×10^{6}	2.28×10^{11}	687.0 d
Jupiter	1.90×10^{27}	6.91×10^{7}	7.78×10^{11}	11.86 y
Saturn	5.68×10^{26}	6.03×10^{7}	1.43×10^{12}	29.45 y
Uranus	8.68×10^{25}	2.56×10^{7}	2.87×10^{12}	84.02 y
Neptune	1.02×10^{26}	2.48×10^{7}	4.50×10^{12}	164.8 y
Pluto[‡]	1.31×10^{22}	1.15×10^{6}	5.91×10^{12}	247.9 y

[†]Source: NASA Jet Propulsion Laboratory Solar System Dynamics Group (**http://ssd.jpl.nasa.gov**), and P. Kenneth Seidelmann, ed., ***Explanatory Supplement to the Astronomical Almanac*** (University Science Books, Mill Valley, CA, 1992), pp. 704–706. For each body, "radius" is its radius at its equator and "orbit radius" is its average distance from the sun or (for the moon) from the earth.

[‡]In August 2006, the International Astronomical Union reclassified Pluto and other small objects that orbit the sun as "dwarf planets."

Prefixes for Powers of 10

Power of ten	Prefix	Abbreviation	Pronunciation
10^{-24}	yocto-	y	*yoc*-toe
10^{-21}	zepto-	z	*zep*-toe
10^{-18}	atto-	a	*at*-toe
10^{-15}	femto-	f	*fem*-toe
10^{-12}	pico-	p	*pee*-koe
10^{-9}	nano-	n	*nan*-oe
10^{-6}	micro-	μ	*my*-crow
10^{-3}	milli-	m	*mil*-i
10^{-2}	centi-	c	*cen*-ti
10^{3}	kilo-	k	*kil*-oe
10^{6}	mega-	M	*meg*-a
10^{9}	giga-	G	*jig*-a or *gig*-a
10^{12}	tera-	T	*ter*-a
10^{15}	peta-	P	*pet*-a
10^{18}	exa-	E	*ex*-a
10^{21}	zetta-	Z	*zet*-a
10^{24}	yotta-	Y	*yot*-a

Examples:

1 femtometer = 1 fm = 10^{-15} m

1 picosecond = 1 ps = 10^{-12} s

1 nanocoulomb = 1 nC = 10^{-9} C

1 microkelvin = 1μK = 10^{-6} K

1 millivolt = 1 mV = 10^{-3} V

1 kilopascal = 1 kPa = 10^{3} Pa

1 megawatt = 1 MW = 10^{6} W

1 gigahertz = 1 GHz = 10^{9} Hz

ANSWERS TO ODD-NUMBERED PROBLEMS

Chapter 1

1.1 a) 1.61 km   b) 3.28×10^3 ft
1.3 1.02 ns
1.5 5.36 L
1.7 31.7 y
1.9 a) 23.4 km/L   b) 1.4 tanks
1.11 9.0 cm
1.13 a) $1.1 \times 10^{-3}\%$ b) no
1.15 0.45%
1.17 a) no   b) no   c) no   d) no   e) no
1.19 $\approx 10^6$
1.21 $\approx 4 \times 10^8$
1.23 $\approx \$70$ million
1.25 $\approx 10^4$
1.27 7.8 km, 38° north of east
1.29 144 m, 41° south of west
1.31 $A_x = 0, A_y = -8.00$ m, $B_x = 7.50$ m, $B_y = 13.0$ m, $C_x = -10.9$ m, $C_y = -5.07$ m, $D_x = -7.99$ m, $D_y = 6.02$ m
1.33 a) -8.12 m   b) 15.3 m
1.35 a) 9.01 m, 33.8°   b) 9.01 m, 33.7°
c) 22.3 m, 250°   d) 22.3 m, 70.3°
1.37 3.39 km, 31.1° north of west
1.39 a) 2.48 cm, 18.4°   b) 4.09 cm, 83.7°
c) 4.09 cm, 264°
1.41 $\vec{A} = -(8.00 \text{ m})\hat{\jmath}$,
$\vec{B} = (7.50 \text{ m})\hat{\imath} + (13.0 \text{ m})\hat{\jmath}$,
$\vec{C} = (-10.9 \text{ m})\hat{\imath} + (-5.07 \text{ m})\hat{\jmath}$,
$\vec{D} = (-7.99)\hat{\imath} + (6.02 \text{ m})\hat{\jmath}$
1.43 a) $\vec{A} = (1.23 \text{ m})\hat{\imath} + (3.38 \text{ m})\hat{\jmath}$,
$\vec{B} = (-2.08 \text{ m})\hat{\imath} + (-1.20 \text{ m})\hat{\jmath}$
b) $\vec{C} = (12.0 \text{ m})\hat{\imath} + (14.9 \text{ m})\hat{\jmath}$
c) 19.2 m, 51.2°
1.45 a) -104 m²   b) -148 m²   c) 40.6 m²
1.47 a) 165°   b) 28°   c) 90°
1.49 a) -63.9 m²$\hat{k}$ b) 63.9 m²$\hat{k}$
1.51 a) -6.62 m²   b) 5.55 m²$\hat{k}$
1.53 a) $A = 5.38, B = 4.36$
b) $-5.00\hat{\imath} + 2.00\hat{\jmath} + 7.00\hat{k}$ c) 8.83, yes
1.55 a) 1.64×10^4 km   b) $2.57r_E$
1.57 a) 2200 g   b) 2.1 m
1.59 a) (2.8 ± 0.3) cm³   b) 170 ± 20
1.61 $\approx 6 \times 10^{27}$
1.63 $\$9 \times 10^{14}, \$3 \times 10^6$ per person
1.65 196 N, 392 N, 57.7° east of north; 360 N, 720 N, 57.7° east of south
1.67 b) $A_x = 3.03$ cm, $A_y = 8.10$ cm   c) 8.65 cm, 69.5°
1.69 144 m, 41° south of west
1.71 954 N, 16.8° above the forward direction
1.73 3.30 N
1.75 a) 45.5 N   b) 139°
1.77 a) (87, 258)   b) 136, 25° below straight left
1.79 160 N, 13° below horizontal
1.81 911 m, 8.9° west of south
1.83 29.6 m, 18.6° east of south
1.85 26.2 m, 34.2° east of south
1.87 124°
1.89 170 m²
1.91 a) 54.7°   b) 35.3°
1.93 28.0 m
1.95 $C_x = 8.0, C_y = 6.1$
1.97 b) 72.2
1.99 38.5 yd, 24.6° to the right of downfield
1.101 a) 76.2 ly   b) 129°

Chapter 2

2.1 25.0 m
2.3 1 hr 10 min
2.5 a) 0.312 m/s   b) 1.56 m/s
2.7 a) 12.0 m/s
b) (i) 0   (ii) 15.0 m/s   (iii) 12.0 m/s
c) 13.3 m/s
2.9 a) 2.33 m/s, 2.33 m/s
b) 2.33 m/s, 0.33 m/s

2.11 6.7 m/s, 6.7 m/s, 0, -40.0 m/s, -40.0 m/s, -40.0 m/s, 0
2.13 a) no
b) (i) 12.8 m/s²   (ii) 3.50 m/s²
(iii) 0.718 m/s²
2.15 a) 2.00 cm/s, 50.0 cm, -0.125 cm/s²
b) 16.0 s   c) 32.0 s
d) 6.20 s, 1.23 cm/s; 25.8 s, -1.23 cm/s; 36.4 s, -2.55 cm/s
2.17 a) 0.500 m/s²   b) 0, 1.00 m/s²
2.19 a) 5.0 m/s   b) 1.43 m/s²
2.21 a) 675 m/s²   b) 0.0667 s
2.23 1.70 m
2.25 38 cm
2.27 a) 3.1×10^6 m/s² $= 3.2g$
b) 1.6 ms   c) no
2.29 a) (i) 5.59 m/s²   (ii) 7.74 m/s²
b) (i) 179 m   (ii) 1.28×10^4 m
2.31 a) 0, 6.3 m/s², -11.2 m/s²
b) 100 m, 230 m, 320 m
2.33 a) 20.5 m/s² upward, 3.8 m/s² upward, 53.0 m/s² upward
b) 722 km
2.35 a) 2.94 m/s   b) 0.600 s
2.37 1.67 s
2.39 a) 33.5 m/s   b) 15.8 m/s
2.41 a) $t = \sqrt{2d/g}$ b) 0.190 s
2.43 a) 646 m   b) 16.4 s, 112 m/s
2.45 a) 249 m/s²   b) 25.4
c) 101 m   d) no (if a is constant)
2.47 0.0868 m/s²
2.49 a) 3.3 s   b) $9H$
2.51 a) 467 m   b) 110 m/s
2.53 a) $v_x = (0.75 \text{ m/s}^3)t^2 - (0.040 \text{ m/s}^4)t^3$, $x = (0.25 \text{ m/s}^3)t^3 - (0.010 \text{ m/s}^4)t^4$
b) 39.1 m/s
2.55 a) 10.0 m
b) (i) 8.33 m/s   (ii) 9.09 m/s   (iii) 9.52 m/s
2.57 b) 0.627 s, 1.59 s
c) negative at 0.627 s, positive at 1.59 s
d) 1.11 s   e) 2.45 m   f) 2.00 s, 0
2.59 250 km
2.61 a) 197 m/s   b) 169 m/s
2.63 a) 82 km/h   b) 31 km/h
2.65 a) 3.5 m/s²   b) 0   c) 1.5 m/s²
2.67 a) 92.0 m   b) 92.0 m
2.69 50.0 m
2.71 4.6 m/s²
2.73 a) 6.17 s   b) 24.8 m
c) auto: 21.0 m/s, truck: 13.0 m/s
2.75 a) 7.85 m/s   b) 5.00 cm/s
2.77 a) 15.9 s   b) 393 m   c) 29.5 m/s
2.79 a) -4.00 m/s   b) 12.0 m/s
2.81 a) $2.64H$ b) $2.64T$
2.83 a) no
b) yes, 14.4 m/s, not physically attainable
2.85 a) 6.69 m/s   b) 4.49 m   c) 1.42 s
2.87 a) 7.7 m/s   b) 0.78 s   c) 0.59 s   d) 1.3 m
2.89 a) 380 m   b) 184 m
2.91 a) 20.5 m/s   b) yes
2.93 a) 945 m   b) 393 m
2.95 a) car A b) 2.27 s, 5.73 s
c) 1.00 s, 4.33 s   d) 2.67 s
2.97 a) 9.55 s, 47.8 m
b) 1.62 m/s   d) 8.38 m/s
e) no
f) 3.69 m/s, 21.7 s, 80.0 m
2.99 a) 8.18 m/s   b) (i) 0.411 m   (ii) 1.15 km
c) 9.80 m/s   d) 4.90 m/s

Chapter 3

3.1 a) 1.4 m/s, -1.3 m/s
b) 1.9 m/s, 317°
3.3 a) 7.1 cm/s, 45°
b) 5.0 cm/s, 90°; 7.1 cm/s, 45°;
11 cm/s, 27°

3.5 b) -8.67 m/s², -2.33 m/s²
c) 8.98 m/s², 195°
3.7 b) $\vec{v} = \alpha\hat{\imath} - 2\beta t\hat{\jmath}, \vec{a} = -2\beta\hat{\jmath}$
c) 5.4 m/s, 297°; 2.4 m/s², 270°
d) speeding up and turning right
3.9 a) 0.600 m   b) 0.385 m
c) $v_x = 1.10$ m/s, $v_y = -3.43$ m/s, 3.60 m/s, 72.2° below the horizontal
3.11 3.32 m
3.13 a) 30.6 m/s   b) 36.3 m/s
3.15 1.28 m/s²
3.17 a) 0.683 s, 2.99 s
b) 24.0 m/s, 11.3 m/s; 24.0 m/s, -11.3 m/s
c) 30.0 m/s, 36.9° below the horizontal
3.19 a) 1.5 m   b) -0.89 m/s
3.21 a) 13.6 m/s   b) 34.6 m/s   c) 103 m
3.23 a) 296 m   b) 176 m   c) 198 m
d) (i) $v_x = 15.0$ m/s, $v_y = -58.8$ m/s
(ii) $v_x = 15.0$ m/s, $v_y = -78.8$ m/s
3.25 a) 0.034 m/s² = $0.0034g$ b) 1.4 h
3.27 140 m/s = 310 mph
3.29 a) 3.50 m/s² upward
b) 3.50 m/s² downward
c) 12.6 s
3.31 a) 14 s   b) 70 s
3.33 0.36 m/s, 52.5° south of west
3.35 a) 4.7 m/s, 25° south of east
b) 190 s   c) 380 m
3.37 b) -7.1 m/s, -42 m/s
c) 43 m/s, 9.6° west of south
3.39 a) 24° west of south   b) 5.5 h
3.41 a) $A = 0, B = 2.00$ m/s², $C = 50.0$ m, $D = 0.500$ m/s³
b) $\vec{v} = 0, \vec{a} = (4.00 \text{ m/s}^2)\hat{\imath}$
c) $v_x = 40.0$ m/s, $v_y = 150$ m/s, 155 m/s
d) $\vec{r} = (200 \text{ m})\hat{\imath} + (550 \text{ m})\hat{\jmath}$
3.43 $2b/3c$
3.45 4.41 s
3.47 a) 123 m
b) 280 m
3.49 22 m/s
3.51 31 m/s
3.53 274 m
3.55 795 m
3.57 33.7 m
3.59 a) 42.8 m/s   b) 42.0 m
3.61 a) $\sqrt{2gh}$ b) 30.0°   c) $6.93h$
3.63 a) 1.50 m/s
b) 4.66 m
3.65 a) 6.91 m   c) no
3.67 a) 17.8 m/s
b) in the river, 28.4 m horizontally from his launch point
3.69 a) 81.6 m   b) in the cart
c) 245 m   d) 53.1°
3.71 a) 49.5 m/s   b) 50 m
3.73 a) 2000 m   b) 2180 m
3.75 $\pm 25.4°$
3.77 61.2 km/h, 140 km/h
3.79 b) $v_x = R\omega(1 - \cos\omega t)$,
$v_y = R\omega\sin\omega t; a_x = R\omega^2\sin\omega t$,
$a_y = R\omega^2\cos\omega t$) $t = 0, 2\pi/\omega, 4\pi/\omega, \dots$;
$x = 0, 2\pi R, 4\pi R, \dots$;
$y = 0; a = R\omega^2$ in the $+y$-direction   d) no
3.81 a) 44.7 km/h, 26.6° west of south
b) 10.5° north of west
3.83 7.39 m/s, 12.4° north of east
3.85 a) 0.659 s   b) (i) 9.09 m/s   (ii) 6.46 m/s
c) 3.00 m, 2.13 m
3.87 a) 49.3°, 17.5° for level ground   b) $-17.0°$
3.89 a) 1.5 km/h   b) 3.5 km/h

Chapter 4

4.1 a) 0°   b) 90°   c) 180°
4.3 3.15 N
4.5 494 N, 31.8°

4.7   46.7 N, opposite to the motion of the skater
4.9   16.0 kg
4.11 a) 3.12 m, 3.12 m/s
      b) 21.9 m, 6.24 m/s
4.13 a) 45.0 N, between 2.0 s and 4.0 s
      b) between 2.0 s and 4.0 s
      c) 0 s, 6.0 s
4.15 a) $A = 100$ N, $B = 12.5$ N/s^2
      b) (i) 21.6 N, 2.70 m/s^2 (ii) 134 N, 16.8 m/s^2
      c) 26.6 m/s^2
4.17 2940 N
4.19 a) 4.49 kg   b) 4.49 kg, 8.13 N
4.21 825 N, blocks
4.23 20 N
4.25 7.4×10^{-23} m/s^2
4.27 b) yes
4.29 a) yes   b) no
4.31 b) 142 N
4.33 2.03 s
4.35 1840 N, 135°
4.37 a) 17 N, 90° clockwise from the $+x$-axis
      b) 840 N
4.39 a) 4.85 m/s
      b) 16.2 m/s^2 upward
      c) 1470 N upward (on him), 2360 N downward
      (on ground)
4.41 a) 153 N
4.43 a) 2.50 m/s^2 b) 10.0 N
      c) to the right, $F > T$
      d) 25.0 N
4.45 a) 4.4 m   b) 300 m/s
      c) (i) 2.7×10^4 N   (ii) 9.0×10^3 N
4.47 a) $T > mg$
      b) 79.6 N
4.49 b) 0.049 N   c) 410mg
4.51 a) 7.79 m/s
      b) 50.6 m/s^2 upward
      c) 4530 N upward, 6.16mg
4.53 a) w b) 0   c) $w/2$
4.55 b) 1395 N
4.57 a) 4.34 kg
      b) 5.30 kg
4.59 $F_x(t) = -6mBt$
4.61 7.78 m

Chapter 5

5.1   a) 25.0 N   b) 50.0 N
5.3   a) 990 N, 735 N
      b) 926 N
5.5   48°
5.7   a) $T_A = 0.732w$, $T_B = 0.897w$, $T_C = w$
      b) $T_A = 2.73w$, $T_B = 3.35w$, $T_C = w$
5.9   a) 337 N   b) 343 N
5.11 a) 1.10×10^8 N   b) 5w
      c) 8.4 s
5.13 a) 4610 m/s^2 = 470g
      b) 9.70×10^5 N = 471w
      c) 0.0187 s
5.15 b) 2.96 m/s^2 c) 191 N
5.17 b) 2.50 m/s^2 c) 1.37 kg
      d) 0.75mg
5.19 a) 0.832 m/s^2 b) 17.3 s
5.21 a) 3.4 m/s   (c) 2.2w
5.23 a) 14.0 m
      b) 18.0 m/s
5.25 50°
5.27 a) 22 N   b) 3.1 m
5.29 a) 0.710, 0.472   b) 258 N
      c) (i) 51.8 N   (ii) 4.97 m/s^2
5.31 a) 57.1 N
      b) 146 N up the ramp
5.33 a) 54.0 m   b) 16.3 m/s
5.35 a) $\mu_k(m_A + m_B)g$ b) $\mu_k m_A g$
5.37 a) 0.218 m/s
      b) 11.7 N
5.39 a) $\dfrac{\mu_k mg}{\cos\theta - \mu_k \sin\theta}$
      b) 1/tanθ
5.41 a) 0.44 kg/m   b) 42 m/s
5.43 a) 3.61 m/s   b) bottom   c) 3.33 m/s
5.45 a) 21.0°, no   b) 11,800 N; 23,600 N
5.47 1410 N, 8370 N

5.49 a) 1.5 rev/min
      b) 0.92 rev/min
5.51 a) 38.3 m/s = 138 km/h
      b) 3580 N
5.53 2.42 m/s
5.55 a) 1.73 m/s^2
      c) 0.0115 N upward
      d) 0.0098 N
5.57 a) rope making 60° angle   b) 6400 N
5.59 a) 470 N   b) 163 N
5.61 762 N
5.63 a) (i) -3.80 m/s   (ii) 24.6 m/s
      b) 4.36 m   c) 2.45 s
5.65 a) 11.4 N   b) 2.57 kg
5.67 10.4 kg
5.69 0.0259 (low pressure), 0.00505 (high pressure)
5.71 a) $m_1(\sin\alpha + \mu_k \cos\alpha)$
      b) $m_1(\sin\alpha - \mu_k \cos\alpha)$
      c) $m_1(\sin\alpha - \mu_s \cos\alpha) \le m_2 \le$
      $m_1(\sin\alpha + \mu_s \cos\alpha)$
5.73 a) 1.80 N   b) 2.52 N
5.75 a) 1.3×10^{-4} N = 62mg
      b) 2.9×10^{-4} N, at $t = 1.2$ ms
      c) 1.2 m/s
5.77 920 N
5.79 a) 11.5 m/s   b) 7.54 m/s
5.81 0.40
5.83 a) $g\dfrac{m_B + m_{\text{rope}}(d/L)}{m_A + m_B + m_{\text{rope}}}$, increase
      b) 0.63 m
      c) will not move for any value of d
5.85 a) 88.0 N northward
      b) 78 N southward
5.87 a) 294 N, 152 N, 152 N   b) 40.0 N
5.89 3.0 N
5.91 a) 12.9 kg
      b) $T_{AB} = 47.2$ N, $T_{BC} = 101$ N
5.93 $a_1 = \dfrac{2m_2 g}{4m_1 + m_2}$, $a_2 = \dfrac{m_2 g}{4m_1 + m_2}$
5.95 1.46 m above the floor
5.97 g/μ_s
5.99 b) 0.452
5.101 0.34
5.103 b) 8.8 N (c) 31.0 N
      d) 1.54 m/s^2
5.105 a) moves up
      b) remains constant
      c) remains constant
      d) slows down at the same rate as the monkey
5.107 a) 6.00 m/s^2
      b) 3.80 m/s^2
      c) 7.36 m/s
      d) 8.18 m/s
      e) 7.78 m, 6.29 m/s, 1.38 m/s^2
      f) 3.14 s
5.109 a) 0.015, 0.36 N · s^2/m^2
      b) 29 m/s
      c) $v/v_t = \sqrt{\sin\beta - (0.015)\cos\beta}$
5.111 a) $v_y(t) = v_0 e^{-kt/m} + v_t(1 - e^{-kt/m})$
5.113 a) 120 N
      b) 3.79 m/s
5.115 b) 0.28   c) no
5.117 a) right   b) 120 m
5.119 a) 81.1°   b) no
      c) The bead rides at the bottom of the hoop.
5.121 a) $F = \dfrac{\mu_k w}{\cos\theta + \mu_k \sin\theta}$
      b) $\theta = \tan^{-1}(\mu_k)$, 14.0°
5.123 $F = (M + m)g \tan\alpha$
5.125 a) $g\dfrac{-4m_1 m_2 + m_2 m_3 + m_1 m_3}{4m_1 m_2 + m_2 m_3 + m_1 m_3}$
      b) $a_B = -a_3$
      d) $g\dfrac{4m_1 m_2 - 3m_2 m_3 + m_1 m_3}{4m_1 m_2 + m_2 m_3 + m_1 m_3}$
      $g\dfrac{4m_1 m_2 - 3m_1 m_3 + m_2 m_3}{4m_1 m_2 + m_2 m_3 + m_1 m_3}$
      e) $g\dfrac{4m_1 m_2 m_3}{4m_1 m_2 + m_2 m_3 + m_1 m_3}$

f) $g\dfrac{8m_1 m_2 m_3}{4m_1 m_2 + m_2 m_3 + m_1 m_3}$
g) All the accelerations are zero, $T_A = m_2 g$,
$T_C = 2m_2 g$.
5.127 $\cos^2\beta$

Chapter 6

6.1   a) 3.60 J   b) -0.900 J
      c) 0   d) 0
      e) 2.70 J
6.3   a) 74 N   b) 333 J
      c) -330 J   d) 0, 0
      e) 0
6.5   a) -1750 J   b) no
6.7   a) (i) 9.00 J   (ii) -9.00 J
      b) (i) 0   (ii) 9.00 J   (iii) -9.00 J   (iv) 0
      c) zero for each block
6.9   a) (i) 0   (ii) 0
      b) (i) 0   (ii) -25.1 J
6.11 a) 324 J   b) -324 J
      c) 0
6.13 a) 36,000 J   b) 4
6.15 a) 1.0×10^{16} J   b) 2.4
6.17 a) 7.50 N   b) (i) 9.00 J   (ii) 5.40 J
      c) 14.4 J, same   d) 2.97 m/s
6.19 a) 43.2 m/s   b) 101 m/s
      c) 5.80 m   d) 3.53 m/s
      e) 7.35 m
6.21 $\sqrt{2gh(1 + \mu_k/\tan\alpha)}$
6.23 32.0 N
6.25 a) 4.48 m/s   b) 3.61 m/s
6.27 a) 4.96 m/s   b) 1.43 m/s^2, 4.96 m/s
6.29 a) $\dfrac{v_0^2}{2\mu_k g}$ b) (i) $^1/_2$ (ii) 4   (iii) 2
6.31 a) 40.0 N/m   b) 0.456 N
6.33 b) 14.4 cm, 13.6 cm, 12.8 cm
6.35 a) 2.83 m/s   b) 3.46 m/s
6.37 8.5 cm
6.39 a) 1.76   b) 0.666 m/s
6.41 a) 4.0 J   b) 0
      c) -1.0 J   d) 3.0 J
      e) -1.0 J
6.43 a) 2.83 m/s   b) 2.40 m/s
6.45 a) 0.0565 m   b) 0.57 J, no
6.47 8.17 m/s
6.49 a) 360,000 J   b) 100 m/s
6.51 $3.9 \times 10^{13}P$
6.53 745 W ≈ 1 hp
6.55 a) 84.6/min   b) 22.7/min
6.57 29.6 kW
6.59 0.20 W
6.61 877 J
6.63 a) 532 J   b) -315 J
      c) 0   d) -202 J
      e) 15 J   f) 1.2 m/s
6.65 a) 987 J   b) 3.02 s
6.67 a) 2.59×10^{12} J   b) 4800 J
6.69 a) 1.8 m/s = 4.0 mph
      b) 180 m/s^2 ≈ 18g, 900 N
6.71 a) $k\left(\dfrac{1}{x_2} - \dfrac{1}{x_1}\right)$, negative
      b) $k\left(\dfrac{1}{x_1} - \dfrac{1}{x_2}\right)$, positive
      c) same magnitude but opposite signs because
      the net work is zero
6.73 a) 5.11 m   b) 0.304
      c) 10.3 m
6.75 a) 0.11 N   b) 7.1 N   c) 0.33 J
6.77 a) 2.56 m/s   b) 3.52 N   c) 13.1 J
6.79 6.3×10^4 N/m
6.81 1.1 m
6.83 a) 1.02×10^4 N/m, 8.16 m
6.85 a) 0.600 m   b) 1.50 m/s
6.87 0.786
6.89 1.3 m
6.91 a) 1.10×10^5 J   b) 1.30×10^5 J
      c) 3.99 kW
6.93 3.6 h
6.95 a) 1.26×10^5 J   b) 1.46 W
6.97 a) 2.4 MW   b) 61 MW   c) 6.0 MW
6.99 a) 513 W   b) 354 W   c) 52.1 W

6.101 a) 358 N b) 47.1 hp c) 4.06 hp
d) 2.03%

6.103 a) $\dfrac{Mv^2}{6}$ b) 6.1 m/s
c) 3.9 m/s d) 0.40 J, 0.60 J

Chapter 7

7.1 a) 6.6×10^5 J b) -7.7×10^5 J
7.3 a) 820 N b) (i) 0 (ii) 740 J
7.5 a) 24.0 m/s b) 24.0 m/s c) part (b)
7.7 a) 2.0 m/s b) 9.8×10^{-7} J, 2.0 J/kg
c) 200 m, 63 m/s d) 5.9 J/kg
7.9 a) (i) 0 (ii) 0.98 J b) 2.8 m/s
c) only gravity is constant d) 5.1 N
7.11 −5400 J
7.13 a) 880 J b) −157 J c) 470 J d) 253 J
e) 3.16 m/s², 7.11 m/s, 253 J
7.15 a) 80.0 J b) 5.0 J
7.17 a) (i) $4U_0$ (ii) $U_0/4$
b) (i) $x_0\sqrt{2}$ (ii) $x_0/\sqrt{2}$
7.19 a) 6.32 cm b) 12 cm
7.21 ±0.092 m
7.23 a) 3.03 m/s, as it leaves the spring
b) 95.9 m/s², when the spring has its maximum compression
7.25 a) 4.46×10^5 N/m b) 0.128 m
7.27 a) −308 J b) −616 J
c) nonconservative
7.29 a) −3.6 J b) −3.6 J
c) −7.2 J d) nonconservative
7.31 a) −59 J b) −42 J
c) −59 J d) nonconservative
7.33 a) 8.41 m/s b) 638 J
7.35 2.46 N, +x-direction
7.37 130 m/s², 132° counterclockwise from the +x-axis
7.39 a) $F(r) = (12a/r^{13}) - (6b/r^7)$
b) $(2a/b)^{1/6}$, stable c) $b^2/4a$
d) $a = 6.67 \times 10^{-138}$ J · m¹², $b = 6.41 \times 10^{-78}$ J · m⁶
7.41 a) zero, 637 N b) 2.99 m/s
7.43 0.41
7.45 a) 16.0 m/s b) 11,500 N
7.47 a) 20.0 m along the rough bottom
b) −78.4 J
7.49 a) 22.2 m/s b) 16.4 m c) no
7.51 0.602 m
7.53 15.5 m/s
7.55 4.4 m/s
7.57 a) no b) yes, $150
7.59 a) 7.00 m/s b) 8.82 N
7.61 a) $mg(1 - h/d)$ b) 441 N
c) $\sqrt{2gh(1 - y/d)}$
7.63 48.2°
7.65 a) 0.392 b) −0.83 J
7.67 a) $U(x) = \frac{1}{2}\alpha x^2 + \frac{1}{3}\beta x^3$ b) 7.85 m/s
7.69 7.01 m/s
7.71 a) $\dfrac{m(g + a)^2}{2gh}$ b) $\dfrac{2gh}{g + a}$
7.73 a) 0.480 m/s b) 0.566 m/s
7.75 a) 3.87 m/s b) 0.10 m
7.77 0.456 N
7.79 a) 4.4×10^{12} J
b) 2.7×10^3 m³, 9.0×10^{-4} m
7.81 119 J
7.83 a) −50.6 J b) −67.5 J c) nonconservative
7.85 b) 0, 3.38 J, 0, 0; 3.38 J c) nonconservative
7.87 b) $v(x) = \sqrt{\dfrac{2\alpha}{mx_0^2}\left[\dfrac{x_0}{x} - \left(\dfrac{x_0}{x}\right)^2\right]}$
c) $x = 2x_0$, $v = \sqrt{\dfrac{\alpha}{2mx_0^2}}$ d) 0
e) $v(x) = \sqrt{\dfrac{2\alpha}{mx_0^2}\left[\dfrac{x_0}{x} - \left(\dfrac{x_0}{x}\right)^2 - \dfrac{2}{9}\right]}$
f) first case: x_0, ∞; second case: $3x_0/2$, $3x_0$

Chapter 8

8.1 a) 1.20×10^5 kg · m/s
b) (i) 60.0 m/s (ii) 26.8 m/s
8.3 b) 0.526, baseball c) 0.641, woman
8.5 a) 22.5 kg · m/s, to the left
b) 838 J
8.7 562 N, not significant
8.9 a) 10.8 m/s, to the right
b) 0.750 m/s, to the left
8.11 a) 500 N/s² b) 5810 N · s
c) 2.70 m/s
8.13 a) 2.50 N · s, in the direction of the force
b) (i) 6.25 m/s, to the right (ii) 3.75 m/s, to the right
8.15 0.593 kg · m/s
8.17 0.87 kg · m/s, in the same direction as the bullet is traveling
8.19 a) 6.79 m/s b) 55.2 J
8.21 a) 0.790 m/s b) −0.0023 J
8.23 1.53 m/s for both
8.25 a) 0.0559 m/s b) 0.0313 m/s
8.27 a) 7.20 m/s, 38.0° from Rebecca's original direction b) −680 J
8.29 a) 3.56 m/s
8.31 a) 29.3 m/s, 20.7 m/s b) 19.6%
8.33 a) 0.846 m/s b) 2.10 J
8.35 a) -1.4×10^{-6} km/h, no
b) -6.7×10^{-8} km/h, no
8.37 5.9 m/s, 58° north of east
8.39 a) Both cars have the same magnitude momentum change, but the lighter car has a greater velocity change.
b) $2.50\Delta v$ c) occupants of small car
8.41 19.5 m/s, 21.9 m/s
8.43 a) 2.93 cm b) 866 J c) 1.73 J
8.45 186 N
8.47 a) 3.33 J, 0.333 m/s b) 1.33 m/s, 0.667 m/s
8.49 a) $v_1/3$ b) $K_1/9$ c) 10
8.51 (0.0444 m, 0.0556 m)
8.53 2520 km
8.55 0.700 m to the right and 0.700 m upward
8.57 0.47 m/s
8.59 $F_x = -(1.50$ N/s$)t$, $F_y = 0.25$ N, $F_z = 0$
8.61 a) 0.053 kg b) 5.19 m/s
8.63 a) 0.442 b) 800 m/s c) 530 m/s
8.65 45.2
8.67 a) 0.474 kg · m/s, upward
b) 237 N, upward
8.69 a) −1.14 N · s, 0.330 N · s
b) 0.04 m/s, 1.8 m/s
8.71 2.40 m/s, 3.12 m/s
8.73 a) 1.75 m/s, 0.260 m/s b) −0.092 J
8.75 3.65×10^5 m/s
8.77 0.946 m
8.79 1.8 m
8.81 12 m/s, 21 m/s
8.83 a) 2.60 m/s b) 325 m/s
8.85 a) 5.3 m/s b) 5.7 m
8.87 53.7°
8.89 102 N
8.91 a) 0.125 b) 248 J c) 0.441 J
8.93 a) $M = m$ c) zero
8.95 a) 9.35 m/s b) 3.29 m/s
8.97 a) 3.56 m/s b) 5.22 m/s c) 4.66 m/s
8.99 13.6 m/s, 6.34 m/s, 65.0°
8.101 0.0544%
8.103 1.61×10^{-22} kg · m/s, to the left
8.105 1.33 m
8.107 0.400 m/s
8.109 a) 71.6 m/s, 14.3 m/s b) 347 m
8.111 a) yes b) decreases by 4800 J
8.113 a) $1.37v_{ex}$ b) $1.18v_{ex}$
c) $2.38v_{ex}$ d) 2.94 km/s
8.115 b) $2L/3$

Chapter 9

9.1 a) 34.4° b) 6.27 cm c) 1.05 m
9.3 a) rad/s, rad/s³ b) (i) 0 (ii) 15.0 rad/s²
c) 9.50 rad
9.5 a) $\omega_z = \gamma + 3\beta t^2$ b) 0.400 rad/s
c) 1.30 rad/s, 0.700 rad/s

9.7 a) $\pi/4$ rad, 2.00 rad/s, −0.139 rad/s³ b) 0
c) 19.5 rad, 9.36 rad/s
9.9 a) 2.25 rad/s b) 4.69 rad
9.11 a) 24.0 s b) 68.8 rev
9.13 10.5 rad/s
9.15 a) 300 rpm b) 75.0 s, 312 rev
9.17 9.00 rev
9.19 a) 1.99×10^{-7} rad/s b) 7.27×10^{-5} rad/s
c) 2.98×10^4 m/s d) 464 m/s
e) 0.0337 m/s², 0
9.21 a) 15.1 m/s² b) 15.1 m/s²
9.23 a) 0.180 m/s², 0, 0.180 m/s²
b) 0.180 m/s², 0.377 m/s², 0.418 m/s²
c) 0.180 m/s², 0.754 m/s², 0.775 m/s²
9.25 0.107 m, no
9.27 a) 0.831 m/s b) 109 m/s²
9.29 a) 2.29 b) 1.51 c) 15.7 m/s, 108g
9.31 a) (i) 0.469 kg · m² (ii) 0.117 kg · m²
(iii) 0
b) (i) 0.0433 kg · m² (ii) 0.0722 kg · m²
c) (i) 0.0288 kg · m² (ii) 0.0144 kg · m²
9.33 a) 2.33 kg · m² b) 7.33 kg · m²
c) 0 d) 1.25 kg · m²
9.35 0.193 kg · m²
9.37 8.52 kg · m²
9.39 5.61 m/s
9.41 a) 3.15×10^{23} J b) 158 y, no
9.43 0.600 kg · m²
9.45 7.35×10^4 J
9.47 a) 0.673 m b) 45.5%
9.49 46.5 kg
9.51 a) f^5 b) 6.37×10^8 J
9.53 an axis that is parallel to a diameter and is $0.516R$ from the center
9.55 $\dfrac{1}{3}M(a^2 + b^2)$
9.57 a) $ML^2/12$ b) $ML^2/12$
9.59 $^1/_2 MR^2$
9.61 a) 14.2 rad/s b) 59.6 rad
9.63 9.41 m
9.65 a) 0.600 m/s³ b) $\alpha = (2.40$ rad/s³$)t$
c) 3.54 s d) 17.7 rad
9.67 a) 0.0333 rad/s² b) 0.200 rad/s
c) 2.40 m/s² e) 3.12 m/s², 3.87 kN
f) 50.2°
9.69 a) 1.70 m/s b) 94.2 rad/s
9.71 2.99 cm
9.73 b) 1.50 m/s² d) 0.208 kg · m²
9.75 a) 7.36 m b) 327 m/s²
9.77 a) 2.14×10^{29} J b) 2.66×10^{33} J
9.79 a) $Mb^2/6$ b) 182 J
9.81 a) −0.882 J
b) 5.42 rad/s
c) 5.42 m/s
d) 5.42 m/s compared to 4.43 m/s
9.83 $\sqrt{\dfrac{2gd(m_B - \mu_k m_A)}{m_A + m_B + I/R^2}}$
9.85 $\sqrt{g(1 - \cos\beta)/R}$
9.87 a) 2.25×10^{-3} kg · m²
b) 3.40 m/s c) 4.95 m/s
9.89 13.9 m
9.91 a) $\dfrac{247}{512}MR^2$ b) $\dfrac{383}{512}MR^2$
9.93 a) 1.05 rad/s b) 5.0 J c) 78.5 J d) 6.4%
9.95 $\dfrac{1}{4}M(R_1^2 + R_2^2)$
9.97 a) $\dfrac{3}{5}MR^2$ b) larger
9.99 a) 55.3 kg b) 0.804 kg · m²
9.101 a) $s(\theta) = r_0\theta + \dfrac{\beta}{2}\theta^2$
b) $\theta(t) = \dfrac{1}{\beta}(\sqrt{r_0^2 + 2\beta vt} - r_0)$
c) $\omega_z(t) = \dfrac{v}{\sqrt{r_0^2 + 2\beta vt}}$,
$\alpha_z(t) = -\dfrac{\beta v^2}{(r_0^2 + 2\beta vt)^{3/2}}$, no
d) 25.0 mm, 0.247 μm/rad, 2.13×10^4 rev

Chapter 10

10.1 a) 40.0 N·m, out of the page
b) 34.6 N·m, out of the page
c) 20.0 N·m, out of the page
d) 17.3 N·m, into the page
e) 0 f) 0
10.3 2.50 N·m, out of the page
10.5 b) $-\hat{k}$ c) $(-1.05\ \text{N}\cdot\text{m})\hat{k}$
10.7 a) 8.7 N·m counterclockwise, 0,
5.0 N·m clockwise, 10.0 N·m clockwise
b) 6.3 N·m clockwise
10.9 13.1 N·m
10.11 a) 14.8 rad/s^2 b) 1.52 s
10.13 a) 7.5 N, 18.2 N b) 0.016 kg·m^2
10.15 0.255 kg·m^2
10.17 a) 32.6 N, 35.4 N b) 2.72 m/s^2
c) 32.6 N, 55.0 N
10.19 a) 1.80 m/s b) 7.13 J
c) (i) 3.60 m/s to the right   (ii) 0
(iii) 2.55 m/s at 45° below the horizontal
d) (i) 1.80 m/s to the right   (ii) 1.80 m/s to the
left   (iii) 1.80 m/s downward
10.21 a) 1/3 b) 2/7 c) 2/5 d) 5/13
10.23 a) 0.613 b) no c) no slipping
10.25 11.7 m
10.27 a) 3.76 m b) 8.58 m/s
10.29 a) 67.9 rad/s b) 8.35 J
10.31 a) 0.309 rad/s b) 100 J c) 6.67 W
10.33 a) 0.377 N·m b) 157 rad
c) 59.2 J d) 59.2 J
10.35 a) 358 N·m b) 1790 N c) 83.8 m/s
10.37 a) 115 kg·m^2/s into the page
b) 125 kg·m^2/s^2 out of the page
10.39 4.71×10^{-6} kg·m^2/s
10.41 4600 rad/s
10.43 1.14 rev/s
10.45 a) 1.38 rad/s b) 1080 J, 495 J
10.47 a) 0.120 rad/s b) 3.20×10^{-4} J
c) work done by bug
10.49 a) 5.88 rad/s
10.51 a) 1.71 rad/s
10.53 a) 1.62 N b) 1800 rev/min
10.55 a) halved b) doubled
c) halved d) doubled e) unchanged
10.57 a) 67.6 N b) 62.9 N c) 3.27 s
10.59 0.483
10.61 7.47 N
10.63 a) 16.3 rad/s^2 b) decreases c) 5.70 rad/s
10.65 a) FR b) FR c) $\sqrt{4F/MR}$
d) $2F/M$ e) $4F/M$
10.67 0.730 m/s^2, 6.08 rad/s^2, 36.3 N, 21.1 N
10.69 a) 293 N b) 16.2 rad/s^2
10.71 a) 2.88 m/s^2 b) 6.13 m/s^2
10.73 270 N

10.75 $a = \dfrac{2g}{2 + (R/b)^2}$, $\alpha = \dfrac{2g}{2b + R^2/b}$,

$T = \dfrac{2mg}{2(b/R)^2 + 1}$

10.77 a) 1.41 s, 70.5 m/s b) t larger, v smaller

10.79 $\dfrac{3}{5}H_0$

10.81 29.0 m/s
10.83 a) 26.0 m/s b) unchanged
10.85 a) $\sqrt{20hy/7}$ b) no
c) rolling friction d) $\sqrt{8hy/3}$
10.87 $g/3$
10.89 1.87 m

10.91 a) $\dfrac{6}{19}v/L$ b) 3/19

10.93 a) 5.46 rad/s b) 3.17 cm c) 1010 m/s
10.95 a) 2.00 rad/s b) 6.58 rad/s
10.97 0.30 rad/s clockwise
10.99 0.710 m

10.101 a) $a = \mu_k g$, $\alpha = \dfrac{2\mu_k g}{R}$

b) $\dfrac{R^2\omega_0^2}{18\mu_k g}$ c) $-\dfrac{MR^2\omega_0^2}{6}$

10.103 a) $mv_1^2 r_1^2/r^3$

b) $\dfrac{mv_1^2}{2}r_1^2\left(\dfrac{1}{r_2^2} - \dfrac{1}{r_1^2}\right)$

c) same

Chapter 11

11.1 29.8 cm
11.3 1.35 m
11.5 5.45 kN
11.7 a) 1000 N, 0.800 m from the end where the
600-N force is applied
b) 800 N, 0.75 m from the end where the
600-N force is applied
11.9 a) 550 N
b) 0.614 m from A
11.11 a) 1920 N b) 1140 N
11.13 a) $T = 2.60w$; $3.28w$, 37.6°
b) $T = 4.10w$; $5.39w$, 48.8°
11.15 272 N on each hand, 130 N on each foot
11.17 246 N, 0.34 m from the front feet
11.19 270 N, 303 N, 40°
11.21 a) 0.800 m b) clockwise
c) 0.800 m, clockwise
11.23 a) 208 N
11.25 1.4 mm
11.27 2.0×10^{11} Pa
11.29 a) 3.1×10^{-3}(upper),
2.0×10^{-3} (lower)
b) 1.6 mm (upper), 1.0 mm (lower)
11.31 a) 150 atm b) 1.5 km, no
11.33 8.6°
11.35 4.8×10^9 Pa, 2.1×10^{-10} Pa^{-1}
11.37 b) 6.6×10^5 N c) 1.8 mm
11.39 3.41×10^7 Pa
11.41 10.2 m/s^2
11.43 20.0 kg
11.45 a) 525 N b) 222 N, 328 N c) 1.48
11.47 tail: 600 N down, wing: 7300 N up
11.49 a) 140 N b) 6 cm to the right
11.51 a) 379 N b) 141 N
11.53 160 N to the right, 213 N upward
11.55 49.9 cm
11.57 a) 370 N b) when he starts to raise
his leg c) no
11.59 a) $V = mg + w$, $H = T = \left(w + \dfrac{mg}{4}\right)\cot\theta$
b) 926 N c) 6.00°
11.61 4900 N
11.63 b) 2000 N = 2.72mg c) 4.4 mm
11.65 a) 4.90 m b) 60 N
11.67 a) 175 N at each hand, 200 N at each foot
b) 91 N at each hand and at each foot
11.69 a) 1150 N b) 1940 N
c) 918 N d) 0.473
11.71 590 N (person above), 1370 N (person below);
person above

11.73 a) $\dfrac{T_{\max} hD}{L\sqrt{h^2 + D^2}}$

b) positive, $\dfrac{T_{\max} h}{L\sqrt{h^2 + D^2}}\left(1 - \dfrac{D^2}{h^2 + D^2}\right)$

11.75 a) 7140 N, tall walls b) 7900 N
11.77 a) 268 N b) 232 N
c) 366 N
11.79 a) 0.424 N (A), 1.47 N (B), 0.424 N (C)
b) 0.848 N
11.81 a) 27° to tip, 31° to slip, tips first
b) 27° to tip, 22° to slip, slips first
11.83 a) 80 N (A), 870 N (B) b) 1.92 m
11.85 a) $T = 3700$ N, 2000 N upward
11.87 a) 0.36 mm b) 0.045 mm
c) 0.33 mm
11.89 a) 0.54 cm b) 0.42 cm
11.91 a) 0.70 m from A b) 0.60 m from A
11.93 a) 1.63 m
b) brass: 2.00×10^8 Pa, nickel: 4.00×10^8 Pa
c) brass: 2.22×10^{-3}, nickel: 1.90×10^{-3}
11.95 0.0542 L
11.97 a) 600 N b) 13.5 kN

11.97 c) $F = \dfrac{\mu_s w}{\sin\theta - \mu_s \cos\theta}$ (to slide),

$F = \dfrac{w}{(1/9)\cos\theta + 2\sin\theta}$ (to tip), 66°

11.99 $h^2/L + L/2$; L if $h > L/\sqrt{2}$
11.101 a) 0.66 mm b) 0.022 J c) 8.35×10^{-3} J
d) -3.04×10^{-2} J e) 3.04×10^{-2} J

Chapter 12

12.1 41.8 N, no
12.3 7020 kg/m^3, yes
12.5 1.6
12.7 61.6 N
12.9 a) 1.86×10^6 Pa b) 184 m
12.11 0.581 m
12.13 a) 1.90×10^4 Pa
b) causes additional force on the walls of the
blood vessels
12.15 2.8 m
12.17 6.0×10^4 Pa
12.19 2.27×10^5 N
12.21 a) 636 Pa b) (i) 1170 Pa   (ii) 1170 Pa
12.23 10.9
12.25 0.107 m
12.27 6.43×10^{-4} m^3, 2.78×10^3 kg/m^3
12.29 a) $\rho < \rho_{\text{fluid}}$

c) above: $1 - \dfrac{\rho}{\rho_{\text{fluid}}}$, submerged: $\dfrac{\rho}{\rho_{\text{fluid}}}$

d) 32%
12.31 a) 116 Pa b) 921 Pa c) 0.822 kg, 822 kg/m^3
12.33 1910 kg/m^3
12.35 9.6 m/s
12.37 a) 17.0 m/s b) 0.317 m
12.39 0.956 m
12.41 28.4 m/s
12.43 1.47×10^5 Pa
12.45 2.03×10^4 Pa
12.47 2.25×10^5 Pa
12.49 $1.19D$

12.51 a) $(p_0 - p)\pi\dfrac{D^2}{4}$ b) 776 N

12.53 a) 5.9×10^5 N b) 1.8×10^5 N
12.55 c) independent of surface area
12.57 0.964 cm, rises
12.59 a) 1470 Pa b) 13.9 cm
12.61 9.8×10^6 kg, yes
12.63 a) 0.30 b) 0.70
12.65 3.50×10^{-4} m^3, 3.95 kg
12.67 a) 8.27×10^3 m^3 b) 83.8 kN
12.69 2.05 m
12.71 a) $H/2$ b) H
12.73 0.116 kg
12.75 33.4 N
12.77 b) 12.2 N c) 11.8 N
12.79 b) 2.52×10^{-4} m^3, 0.124
12.81 5.57×10^{-4} m

12.83 a) $1 - \dfrac{\rho_B}{\rho_L}$ b) $\left(\dfrac{\rho_L - \rho_B}{\rho_L - \rho_w}\right)L$ c) 4.60 cm

12.85 a) al/g b) $\omega^2 l^2/2g$
12.89 a) $2\sqrt{h(H-h)}$ b) h
12.91 a) 0.200 m^3/s b) 6.97×10^4 Pa
12.93 $3h_1$

12.95 a) $r = \dfrac{r_0\sqrt{v_0}}{(v_0^2 + 2gy)^{1/4}}$ b) 1.10 m

12.97 a) 80.4 N

Chapter 13

13.1 2.18
13.3 a) 1.2×10^{-11} m/s^2 b) 15 days
c) increase
13.5 2.1×10^{-9} m/s^2, downward
13.7 a) 2.4×10^{-3} N
b) $F_{\text{moon}}/F_{\text{earth}} = 3.5 \times 10^{-6}$
13.9 a) 0.634 m from $3m$
b) (i) unstable   (ii) stable

13.11 1.38×10^7 m
13.13 a) 0.37 m/s² b) 1700 kg/m³
13.15 610 N, 735 N (on earth)
13.17 a) 5020 m/s b) 60,600 m/s
13.19 a) 7460 m/s b) 1.68 h
13.21 6200 m/s
13.23 a) 4.7 m/s = 11 mph, easy to achieve
         b) 2.23 h
13.25 a) 82,700 m/s b) 14.5 days
13.27 b) Pluto: 4.45×10^{12} m,
         Neptune: 4.55×10^{12} m c) 248 y
13.29 2.3×10^{30} kg $= 1.2 M_S$
13.31 a) (i) 5.31×10^{-9} N (ii) 2.67×10^{-9} N
13.33 a) $-\dfrac{GmM}{\sqrt{x^2 + a^2}}$ b) $-GmM/x$
         c) $\dfrac{GmMx}{(x^2 + a^2)^{3/2}}$, attractive d) GmM/x^2
         e) $U = -GMm/a$, $F_x = 0$
13.35 a) 53 N b) 52 N
13.37 a) 4.3×10^{37} kg $= 2.1 \times 10^7 M_S$
         b) no c) 6.32×10^{10} m, yes
13.39 a) 4.64×10^{11} m
         b) 6.26×10^{36} kg $= 3.15 \times 10^6 M_S$
         c) 9.28×10^9 m
13.41 9.16×10^{13} N
13.43 a) 9.67×10^{-12} N, at 45° above the +x-axis
         b) 3.02×10^{-5} m/s
13.45 a) 1.62 m/s² b) 0.69 N c) 4.2 N
13.47 2.9×10^{15} kg, 0.0077 m/s² b) 6.2 m/s
13.49 b) (i) 1.49×10^{-5} m/s, 7.46×10^{-6} m/s
         (ii) 2.24×10^{-5} m/s c) 26.4 m
13.51 a) 3.59×10^7 m
13.53 177 m/s
13.55 a) 1.39×10^7 m b) 3.59×10^7 m
13.57 $(0.01)R_E = 6.4 \times 10^4$ m
13.59 1.83×10^{27} kg
13.61 0.28%
13.63 6060 km/h
13.65 $v_2 = \sqrt{\dfrac{2Gm_E h}{R_E(R_E + h)}}$
13.67 a) 13,700 m/s b) 13,300 m/s
         c) 13,200 m/s
13.69 a) (i) 2.84 y (ii) 6.11 y
         b) 4.90×10^{11} m c) 4.22×10^{11} m
13.71 a) $GM^2/4R^2$
         b) $v = \sqrt{GM/4R}$, $T = 4\pi\sqrt{R^3/GM}$
         c) $GM^2/4R$
13.73 6.8×10^4 m/s
13.75 a) 12,700 kg/m³ (at $r = 0$), 3150 kg/m³
         (at $r = R$)
13.77 a) 7910 s b) 1.53 c) 5510 m/s (apogee),
         8430 m/s (perigee) d) 2410 m/s (perigee),
         3250 m/s (apogee); perigee
13.79 5.36×10^9 J
13.81 9.36 m/s²
13.83 $GmMx/(a^2 + x^2)^{3/2}$
13.85 a) $U(r) = \dfrac{Gm_E m}{2R_E^3} r^2$ b) 7.90×10^3 m/s
13.87 a) against the direction of motion in both cases
         b) 259 days c) 44.1°
13.89 $\dfrac{2GMm}{a^2}\left(1 - \dfrac{x}{\sqrt{a^2 + x^2}}\right)$

Chapter 14

14.1 a) 2.15 ms, 2930 rad/s
        b) 2.00×10^4 Hz, 1.26×10^5 rad/s
        c) 4.3×10^{14} Hz $\le f \le 7.5 \times 10^{14}$ Hz;
           1.3×10^{-15} s $\le T \le 2.3 \times 10^{-15}$ s
        d) 2.0×10^{-7} s, 3.1×10^7 rad/s
14.3 5530 rad/s, 1.14 ms

14.5 0.0500 s
14.7 a) 0.167 s b) 37.7 rad/s c) 0.0844 kg
14.9 a) 0.150 s b) 0.0750 s
14.11 a) 0.98 m b) $\pi/2$ rad
         c) $x = (-0.98$ m$)\sin\left[(12.2 \text{ rad/s})t\right]$
14.13 a) -2.71 m/s²
         b) $x = (1.46$ cm$)\cos\left[(15.7 \text{ rad/s})t\right.$
            $+ 0.715$ rad$]$,
         $v_x = (-22.9$ cm/s$)\sin\left[(15.7 \text{ rad/s})t\right.$
            $+ 0.715$ rad$]$,
         $a_x = (-359$ cm/s²$)\cos\left[(15.7 \text{ rad/s})t\right.$
            $+ 0.715$ rad$]$
14.15 120 kg
14.17 a) 0.253 kg b) 1.21 cm
         c) 3.03 N
14.19 a) 1.51 s b) 26.0 N/m
         c) 30.8 cm/s d) 1.92 N
         e) -0.0125 m, 30.4 cm/s, 0.216 m/s²
         f) 0.324 N
14.21 a) $x = (0.0030$ m$)\cos\left[(2760 \text{ rad/s})t\right]$
         b) 8.3 m/s, 2.3×10^4 m/s²
         c) $da_x/dt = (6.3 \times 10^7 \text{ m/s}^3)$
            $\times \sin\left[(2760 \text{ rad/s})t\right]$, 6.3×10^7 m/s³
14.23 127 m/s²
14.25 a) 1.48 m/s b) 2.96×10^{-5} J
14.27 a) 1.20 m/s b) 1.11 m/s
         c) 36 m/s² d) 13.5 m/s² e) 0.36 J
14.29 $3M$, $^3/_4$
14.31 0.240 m
14.33 $A/\sqrt{2}$
14.35 a) 0.0778 m b) 1.28 Hz c) 0.624 m/s
14.37 a) 4.06 cm b) 1.21 m/s c) 29.8 rad/s
14.39 b) 23.9 cm, 1.45 Hz
14.41 a) 2.7×10^{-8} kg·m²
         b) 4.3×10^{-6} N·m/rad
14.43 0.0512 kg·m²
14.45 a) 0.25 s b) 0.25 s
14.47 0.407 swings per second
14.49 10.7 m/s²
14.51 a) 2.84 s b) 2.89 s
         c) 2.89 s, -2%
14.53 0.129 kg·m²
14.55 A: $2\pi\sqrt{\dfrac{L}{g}}$, B: $\dfrac{2\sqrt{2}}{3}\left(2\pi\sqrt{\dfrac{L}{g}}\right)$, pendulum A
14.57 A: $2\pi\sqrt{\dfrac{L}{g}}$, B: $\sqrt{\dfrac{11}{10}}\left(2\pi\sqrt{\dfrac{L}{g}}\right)$, pendulum B
14.59 a) 0.393 Hz b) 1.73 kg/s
14.61 a) A b) $-Ab/2m$
         c) $A\left(\dfrac{b^2}{2m^2} - \dfrac{k}{m}\right)$; negative if $b < \sqrt{2km}$,
         zero if $b = \sqrt{2km}$, positive if $b > \sqrt{2km}$
14.63 a) kg/s
         c) (i) $5.0\dfrac{F_{\max}}{k}$ (ii) $2.5\dfrac{F_{\max}}{k}$
14.65 0.353 m
14.67 a) 1.11×10^4 m/s² b) 5.00×10^3 N
         c) 23.6 m/s, 125 J d) 37.5 kW
         e) 1.21×10^4 N, 36.7 m/s, 302 J, 141 kW
14.69 a) none of them change
         b) $^1/_4$ as great c) $^1/_2$ as great
         d) $1/\sqrt{5}$ as great
         e) potential energy is the same, kinetic energy
            is 1/5 as great
14.71 a) 24.4 cm b) 0.221 s c) 1.19 m/s
14.73 a) 0.373 Hz, 0.426 m, 2.68 s b) 1.34 s
14.75 2.00 m
14.77 a) 0.107 m b) 2.42 s
14.79 $(0.921)\left(\dfrac{1}{2\pi}\sqrt{\dfrac{g}{L}}\right)$
14.81 a) 1.49 s b) -2.12×10^{-4} s per s, shorter
         c) 0.795 s

14.83 a) 0.150 m/s b) 0.112 m/s² downward
         c) 0.700 s d) 4.38 m
14.85 a) 2.6 m/s b) 0.21 m c) 0.49 s
14.87 9.08×10^{24} kg
14.89 1.17 s
14.91 0.505 s
14.93 c) -7.57×10^{-19} J e) 8.39×10^{12} Hz
14.95 0.705 Hz, 14.5°
14.97 $2\pi\sqrt{\dfrac{M}{3k}}$
14.99 $\dfrac{1}{4\pi}\sqrt{\dfrac{6g}{\sqrt{2}L}}$
14.101 a) $k_1 + k_2$ b) $k_1 + k_2$
         c) $\dfrac{k_1 k_2}{k_1 + k_2}$ d) $\sqrt{2}$
14.103 a) $Mv^2/6$ c) $\omega = \sqrt{\dfrac{3k}{M}}$, $M' = M/3$

Chapter 15

15.1 a) 0.439 m, 1.28 ms
        b) 0.219 m
15.3 220 m/s = 800 km/h
15.5 a) 1.7 cm to 17 m
        b) 4.3×10^{14} Hz to 7.5×10^{14} Hz
        c) 1.5 cm d) 6.4 cm
15.7 a) 25.0 Hz, 0.0400 s, 19.6 rad/m
        b) $y(x, t) = (0.0700 \text{ m})\cos\left[(19.6 \text{ m}^{-1})x + (157 \text{ rad/s})t\right]$ c) 4.95 cm
        d) 0.0050 s
15.9 a) yes b) yes c) no
        d) $v_y = \omega A\cos(kx + \omega t)$,
        $a_y = -\omega^2 A\sin(kx + \omega t)$
15.11 a) 4 mm b) 0.040 s c) 0.14 m, 3.5 m/s
         d) 0.24 m, 6.0 m/s e) no
15.13 b) +x-direction
15.15 a) 16.3 m/s b) 0.136 m
         c) both increase by a factor of $\sqrt{2}$
15.17 0.337 kg
15.19 a) 18.6 N b) 29.1 m/s
15.21 a) 10.0 m/s b) 0.250 m
         c) $y(x, t) = (3.00 \text{ cm})\cos\left[\pi(8.00 \text{ rad/m})x - (80.0\pi \text{ rad/s})t\right]$ d) 1890 m/s² e) yes
15.23 4.51 mm
15.25 a) 95 km b) 0.25 µW/m²
         c) 110 kW
15.27 a) 0.050 W/m² b) 22 kJ
15.29 9.48×10^{27} W
15.37 a) $(1.33 \text{ m})n$, $n = 0, 1, 2, \ldots$
         b) $(1.33 \text{ m})(n + ^1/_2)$, $n = 0, 1, 2, \ldots$
15.41 a) 96.0 m/s b) 461 N
         c) 1.13 m/s, 426 m/s²
15.43 b) 2.80 cm c) 277 cm
         d) 185 cm, 7.96 Hz, 0.126 s, 1470 cm/s
         e) 280 cm/s
         f) $y(x, t) = (5.60 \text{ cm}) \times$
            $\sin\left[(0.0906 \text{ rad/cm})x\right]\sin\left[(133 \text{ rad/s})t\right]$
15.45 a) $y(x, t) = (4.60 \text{ mm}) \times$
            $\sin\left[(6.98 \text{ rad/m})x\right]\sin\left[(742 \text{ rad/s})t\right]$
         b) 3rd harmonic c) 39.4 Hz
15.47 a) 45.0 cm b) no
15.49 a) 311 m/s b) 246 Hz
         c) 245 Hz, 1.40 m
15.51 a) 20.0 Hz, 126 rad/s, 3.49 rad/m
         b) $y(x, t) = (2.50 \times 10^{-3} \text{ m}) \times$
            $\cos\left[(3.49 \text{ rad/m})x - (126 \text{ rad/s})t\right]$
         c) $y(0, t) = (2.50 \times 10^{-3} \text{ m}) \times$
            $\cos\left[(126 \text{ rad/s})t\right]$
         d) $y(1.35 \text{ m}, t) = (2.50 \times 10^{-3} \text{ m}) \times$
            $\cos\left[(126 \text{ rad/s})t - 3\pi/2 \text{ rad}\right]$
         e) 0.315 m/s
         f) -2.50×10^{-3} m, 0
15.53 a) $\dfrac{7L}{2}\sqrt{\dfrac{\mu_1}{F}}$ b) no

15.55 a) $\dfrac{2\pi A}{\lambda}\sqrt{\dfrac{FL}{M}}$ b) increase F by a factor of 4

15.57 a) $\dfrac{4\pi^2 F\Delta x}{\lambda^2}$

15.59 32.4 Hz

15.61 1.83 m

15.63 330 Hz (copper), 447 Hz (aluminum)

15.65 c) C/B

15.67 b) ω must be decreased by a factor of $1/\sqrt{2}$, k must be decreased by a factor of $1/\sqrt{8}$

15.69 a) 7.07 cm b) 0.400 kW

15.71 d) $P(x, t) = -Fk\omega A^2 \sin^2(kx + \omega t)$

15.73 (0.800 Hz)n, $n = 1, 2, 3, \ldots$

15.75 c) $2A$, $2A\omega$, $2A\omega^2$

15.77 233 N

15.79 a) $0, L$ b) $0, L/2, L$ d) no

15.81 1780 kg/m^3

15.83 a) $r = 0.640$ mm, $L = 0.40$ m
b) 380 Hz

15.85 b) $u_k = \dfrac{1}{2}\mu\omega^2 A^2 \sin^2(kx - \omega t)$

e) $u_p = \dfrac{1}{2}Fk^2 A^2 \sin^2(kx - \omega t)$

Chapter 16

16.1 a) 0.344 m b) 1.2×10^{-5} m
c) 6.9 m, 50 Hz

16.3 a) 7.78 Pa b) 77.8 Pa c) 778 Pa

16.5 a) 90 m b) 102 kHz c) 1.4 cm
d) 4.4 mm to 8.8 mm e) 6.2 MHz

16.7 90.8 m

16.9 81.4°C

16.11 0.208 s

16.13 a) 5.5×10^{-15} J b) 0.074 mm/s

16.15 a) 9.44×10^{-11} m, 0.434 m
b) 5.66×10^{-9} m, 0.100 m

16.17 a) 1.95 Pa b) 4.58×10^{-3} W/m^2
c) 96.6 dB

16.19 a) 4.4×10^{-12} W/m^2 b) 6.4 dB
c) 5.8×10^{-11} m

16.21 14.0 dB

16.23 a) 2.0×10^{-7} W/m^2 b) 6.0 m c) 290 m

16.25 a) *fundamental:* displacement node at 0.60 m, pressure nodes at 0 and 1.20 m; *first overtone:* displacement nodes at 0.30 m and 0.90 m, pressure nodes at 0, 0.60 m, 1.20 m; *second overtone:* displacement nodes at 0.20 m, 0.60 m, 1.00 m, pressure nodes at 0, 0.40 m, 0.80 m, 1.20 m
b) *fundamental:* displacement node at 0, pressure node at 1.20 m; *first overtone:* displacement nodes at 0 and 0.80 m, pressure nodes at 0.40 m and 1.20 m; *second overtone:* displacement nodes at 0, 0.48 m, 0.96 m, pressure nodes at 0.24 m, 0.72 m, 1.20 m

16.27 506 Hz, 1517 Hz, 2529 Hz

16.29 a) 767 Hz b) no

16.31 a) 614 Hz b) 1230 Hz

16.33 a) 172 Hz b) 86 Hz

16.35 0.125 m

16.37 destructive

16.39 a) 433 Hz b) loosen

16.41 1.3 Hz

16.43 780 m/s

16.45 a) 375 Hz b) 371 Hz c) 4 Hz

16.47 a) 0.25 m/s b) 0.91 m

16.49 19.8 m/s

16.51 a) 1910 Hz b) 0.188 m

16.53 0.0950c, toward us

16.55 a) 36.0° b) 2.23 s

16.57 b) 0.68%

16.59 a) 1.00 b) 8.00
c) 4.73×10^{-8} m = 47.3 nm

16.61 b) $3f_0$

16.63 flute harmonic $3N$ resonates with string harmonic $4N$, $N = 1, 3, 5, \ldots$

16.65 a) stopped b) 7th and 9th c) 0.439 m

16.67 a) 375 m/s b) 1.39 c) 0.8 cm

16.69 1.27

16.71 a) 548 Hz b) 652 Hz

16.73 a) 2186 Hz, 0.157 m b) 2920 Hz, 0.118 m
c) 734 Hz

16.75 a) 0.0674 m b) 147 Hz

16.77 b) 2.0 m/s

16.79 a) 1.2×10^6 m/s
b) 3.6×10^{16} m = 3.8 ly
c) 5200 ly, about 4100 BCE

16.81 a) $f_0\left(\dfrac{2v_w}{v - v_w}\right)$ b) $f_0\left(\dfrac{2v_w}{v + v_w}\right)$

16.83 a) 9.69 cm/s, 667 m/s^2

Chapter 17

17.1 a) $-81.0°$F b) 134.1°F c) 88.0°F

17.3 a) 27.2 C° b) -55.6 C°

17.5 a) -18.0 F° b) -10.0 C°

17.7 0.964 atm

17.9 a) -282°C b) 47,600 Pa, no

17.11 0.39 m

17.13 Death Valley: 1.9014 cm, Greenland: 1.8964 cm

17.15 0.26 mm

17.17 49.4°C

17.19 1.7×10^{-5} (C°)$^{-1}$

17.21 a) 1.431 cm^2 b) 1.436 cm^2

17.23 a) 3.2×10^{-5} (C°)$^{-1}$ b) 2.6×10^9 Pa

17.25 a) 5.0 mm b) -8.4×10^7 Pa

17.27 5.79×10^5 J

17.29 240 J/kg·K

17.31 23 min

17.33 a) -1.54 kJ b) 0.0121 C°

17.35 45.2 C°

17.37 0.0613 C°

17.39 a) 215 J/kg·K b) water c) too small

17.41 27.5°C

17.43 a) 5.9 C° b) yes

17.45 150°C

17.47 7.6 min

17.49 36.4 kJ, 8.70 kcal, 34.5 Btu

17.51 357 m/s

17.53 3.45 L

17.55 5.05×10^{15} kg

17.57 0.0674 kg

17.59 2.10 kg

17.61 190 g

17.63 a) 222 K/m b) 10.7 W c) 73.3°C

17.65 a) -5.8°C b) 11 W/m^2

17.67 4.0×10^{-3} W/m·C°

17.69 105.5°C

17.71 a) 21 kW b) 6.4 kW

17.73 2.1 cm^2

17.75 a) 1.61×10^{11} m b) 5.43×10^6 m

17.77 a) 35.1°M b) 39.6 C°

17.79 53.3°C

17.81 35.0°C

17.83 23.0 cm, 7.0 cm

17.85 b) 1.9×10^8 Pa

17.87 a) 99.4 N c) -4.2 Hz, falls

17.89 a) 87°C b) -80°C

17.91 20.2°C

17.93 a) 54.3

17.95 a) 83.6 J b) 1.86 J/mol·K
c) 5.60 J/mol·K

17.97 a) 2.70×10^7 J b) 6.89 C° c) 19.3 C°

17.99 2.5 cm

17.101 a) 86.1°C b) no ice, no steam, 0.130 kg liquid water

17.103 a) 100°C b) 0.0214 kg steam, 0.219 kg liquid water

17.105 1.743 kg

17.107 a) 93.9 W b) 1.35

17.109 2.9

17.111 c) 170 h d) 1.5×10^{10} s $\approx$ 500 y, no

17.113 0.106 W/m·K

17.115 5.82 g

17.117 a) 1.04 kW b) 87.1 W c) 1.13 kW
d) 28 g e) 1.1 bottles

17.119 a) 69.6°C

17.121 1.76 C°

17.123 b) 0°C d) 3140 C°/m e) 121 W f) zero

g) 1.1×10^{-4} m^2/s h) -11 C°/s
i) 9.17 s j) decrease k) -7.71 C°/s

17.125 a) 103°C b) 27 W

17.127 a) (i) 280 W (ii) 0.248 W (iii) 2.10 kW
(iv) 116 W; radiation from the sun
b) 3.72 L/h c) 1.4 L/h

Chapter 18

18.1 a) 0.122 mol b) 14,700 Pa, 0.145 atm

18.3 0.100 atm

18.5 a) 0.0136 kg/m^3 (Mars), 67.6 kg/m^3 (Venus), 5.39 kg/m^3 (Titan)

18.7 503°C

18.9 16.8 kPa

18.11 0.159 L

18.13 0.0508V

18.15 a) 70.2°C b) yes

18.17 850 m

18.19 a) 6.95×10^{-16} kg b) 2.32×10^{-13} kg/m^3

18.21 22.8 kPa

18.23 a) \$8720 b) 3.88 cm

18.25 a) 8.2×10^{-17} atm b) no

18.27 55.6 mol, 3.35×10^{25} molecules

18.29 a) 9.00×10^{-5} m^3 b) 3.1×10^{-10} m

18.31 b) 1.004

18.33 (d) must be true, the others could be true

18.35 a) 1.93×10^6 m/s, no b) 7.3×10^{10} K

18.37 a) 6.21×10^{-21} J b) 2.34×10^5 m^2/s^2
c) 484 m/s d) 2.57×10^{-23} kg·m/s
e) 1.24×10^{-19} N f) 1.24×10^{-17} Pa
g) 8.17×10^{21} molecules
h) 2.45×10^{22} molecules

18.39 3800°C

18.41 a) 2600 J b) 1560 J

18.43 a) 741 J/kg·K, $c_w = 5.65c_{N_2}$
b) 5.65 kg c) 4.85 m^3

18.45 a) 923 J/kg·K
b) The value calculated is too large by about 1.4%.

18.47 a) 337 m/s b) 380 m/s c) 412 m/s

18.49 a) 610 Pa b) 22.12 MPa

18.51 no, no

18.53 a) 11.8 kPa b) 0.566 L

18.55 272°C

18.57 0.213 kg

18.59 a) -179°C b) 1.2×10^{26} molecules/m^3
c) The atmosphere of Titan is 4.8 times denser than that of the earth.

18.61 1.92 atm

18.63 a) 30.7 cylinders b) 8420 N c) 7800 N

18.65 a) 26.2 m/s b) 16.1 m/s, 5.44 m/s
c) 1.74 m

18.67 $\approx 5 \times 10^{27}$ atoms

18.69 a) A b) B c) 4250°C d) B

18.71 a) 4.65×10^{-26} kg b) 6.11×10^{-21} J
c) 2.04×10^{24} molecules d) 12.5 kJ

18.73 b) r_2 c) $r_1 = \dfrac{R_0}{2^{1/6}}$, $r_2 = R_0$, $2^{-1/6}$ d) U_0

18.75 a) 517 m/s b) 298 m/s

18.77 b) 1.40×10^5 K (N), 1.01×10^4 K (H)
c) 6370 K (N), 459 K (H)

18.79 a) 1.24×10^{-14} kg
b) 4.16×10^{11} molecules
c) 2.95 μm, no

18.81 a) $2R = 16.6$ J/mol·K b) less

18.83 CO$_2$: 20.79 J/mol·K, 27%; SO$_2$: 24.94 J/mol·K, 21%; H$_2$S: 24.94 J/mol·K, 3.9%

18.85 $3kT/m$

18.87 b) 0.0421N c) $2.94 \times 10^{-21}N$
d) 0.0297N, $2.08 \times 10^{-21}N$
e) 0.0595N, $4.15 \times 10^{-21}N$

18.89 42.6%

18.91 a) 4.5×10^{11} m
b) 703 m/s, 6.4×10^8 s ($\approx$20 y)
c) 1.4×10^{-14} Pa d) 650 m/s, evaporate
f) 2×10^5 K, >3 times the temperature of the sun, no

Chapter 19

19.1   b) 1330 J
19.3   b) -6180 J
19.5   a) 0.942 atm
19.7   a) $(p_1 - p_2)(V_2 - V_1)$ b) Negative of work done in reverse direction
19.9   a) 34.7 kJ   b) 80.4 kJ   c) no
19.11   a) 278 K   b) 0, 162 J   c) 53 J
19.13   a) 16.4 min   b) 139 m/s $= 501$ km/h
19.15   a) 0   b) $T_b = 2T_a$ c) $U_b = U_a + 700$ J
19.17   a) positive   b) $W_{\mathrm{I}} > 0$, $W_{\mathrm{II}} < 0$
        c) into the system   d) into the system for loop I, out of the system for loop II
19.19   b) 208 J   c) on the piston   d) 712 J
        e) 920 J   f) 208 J
19.21   a) 948 K   b) 900 K
19.23   2/5
19.25   a) 25.0 K   b) 17.9 K   c) higher for (a)
19.27   a) -605 J   b) 0   c) liberates 605 J
19.29   a) 476 kPa   b) -10.6 kJ   c) 1.59, heated
19.31   5.05 kJ, internal energy and temperature both increase
19.33   b) 224 J   c) -224 J
19.35   11.6°C
19.37   a) 600 J out of the gas
        b) -1500 J, decreases
19.39   a) increases   b) 4800 J
19.41   a) 45.0 J   b) liberates 65.0 J
        c) 23.0 J, 22.0 J
19.43   a) the same   b) absorbs 4.0 kJ   c) 8.0 kJ
19.45   b) -2460 J
19.47   a) 0.80 L   b) 305 K, 1220 K, 1220 K
        c) *ab:* 76 J, into the gas;
           ca: -107 J, out of the gas
           bc: 56 J, into the gas
        d) *ab:* 76 J, increase
           bc: 0, no change
           ca: -76 J, decrease
19.49   a) 3.00 kJ, into the gas
        b) 2.00 kJ, into the gas   c) $Q_a > Q_b$

19.51   a) 899°C   b) 12.2 kJ
        c) 42.6 kJ   d) 45.6 kJ
19.53 -0.226 m^3
19.55   a) 4.32×10^{-4} m^3 b) 648 J   c) 715 kJ
        d) 715 kJ   e) no substantial difference
19.57 3.4×10^5 J/kg
19.59   b) 11.9 C°
19.61   a) 0.173 m   b) 207°C   c) 74.7 kJ
19.63   a) $Q = 300$ J, $\Delta U = 0$
        b) $Q = 0$, $\Delta U = -300$ J
        c) $Q = 750$ J, $\Delta U = 450$ J
19.65   a) $W = 738$ J, $Q = 2590$ J, $\Delta U = 1850$ J
        b) $W = 0$, $Q = -1850$ J, $\Delta U = -1850$ J
        c) $\Delta U = 0$
19.67   a) $W = -187$ J, $Q = -654$ J, $\Delta U = -467$ J
        b) $W = 113$ J, $Q = 0$, $\Delta U = -113$ J
        c) $W = 0$, $Q = 580$ J, $\Delta U = 580$ J
19.69   a) $p_0 + \dfrac{mg}{\pi r^2}$

        b) $-\left(\dfrac{y}{h}\right)(p_0 \pi r^2 + mg)$

        c) $\dfrac{1}{2\pi}\sqrt{\dfrac{g}{h}\left(1 + \dfrac{p_0 \pi r^2}{mg}\right)}$, no

Chapter 20

20.1   a) 6500 J   b) 34%
20.3   a) 23%   b) 12,400 J
        c) 0.350 g   d) 222 kW $= 298$ hp
20.5   a) 12.3 atm   b) 5470 J (*ca*)   c) 3723 J (*bc*)
        d) 1747 J   e) 31.9%
20.7   a) 58%   b) 1.4%
20.9   a) 16.2 kJ   b) 50.2 kJ
20.11   1.7 h
20.13   a) 215 J   b) 378 K   c) 39.0%
20.15   a) 42.4 kJ   b) 441°C
20.17   a) 492 J   b) 212 W   c) 5.4
20.19   4.5 kJ

20.21   37.1 hp
20.23   a) 429 J/K   b) -393 J/k   c) 36 J/K
20.25   a) irreversible   b) 1250 J/K
20.27 -6.31 J/K
20.29   a) 6.05 kJ/K
        b) about five times greater for vaporization
20.31   a) 33.3 J/K   b) irreversible
20.33   a) no   b) 18.3 J/K   c) 18.3 J/K
20.35   a) 121 J   b) 3800 cycles
20.37   a) 33 J   b) 117 J   c) 45°C   d) 0   e) 96.2 g
20.39 -5.8 J/K, decrease
20.41   b) absorbed during *bc*, rejected during *ab* and *ca*   c) $T_a = T_b = 241$ K, $T_c = 481$ K
        d) 610 J, 610 J   e) 8.7%
20.43   a) 21.0 kJ (enters), 16.6 kJ (leaves)
        b) 4.4 kJ, 21%   c) 67%
20.45   a) 7.0%   b) 3.0 MW, 2.8 MW
        c) 6×10^5 kg/h $= 6 \times 10^5$ L/h
20.47   a) 2.00 atm, 4.00 L; 2.00 atm, 6.00 L; 1.11 atm, 6.00 L; 1.67 atm, 4.00 L
        b) $1 \to 2$: 1422 J, 405 J; $2 \to 3$: -1355 J, 0;
           $3 \to 4$: -274 J, -274 J; $4 \to 1$: 339 J, 0
        c) 131 J   d) 7.44%, 44.4%
20.49   a) 2.26%   b) 29.4 J (gravitational), 1.30 kJ
        c) 1.11×10^{-3} candy bars
20.51   a) *ab*: 225 kJ, 90 kJ, 135 kJ;
           bc: -240 kJ, 0, -240 kJ;
           ca: 45 kJ, -60 kJ, 105 kJ
        b) 30 kJ, 30 kJ, 0   c) 11.1%
20.53 $1 - T_{\mathrm{C}}/T_{\mathrm{H}}$
20.55   a) 122 J, -78 J   b) 5.10×10^{-4} m^3
        c) at *b*: 2.32 MPa, 4.81×10^{-5} m^3, 771 K;
           at *c*: 4.01 MPa, 4.81×10^{-5} m^3, 1332 K;
           at *d*: 0.147 MPa, 5.10×10^{-4} m^3, 518 K
        d) 61.1%, 77.5%
20.57   b) 357 kJ, 62 kJ   c) 385 kJ, 34 kJ
20.59   a) $nC_{\mathrm{V}} \ln (T_c/T_b)$, $nC_{\mathrm{V}} \ln (T_a/T_d)$ b) zero
20.61   a) -107 J/K   b) 147 J/K   c) 0
        d) 39.4 J/K

PHOTO CREDITS

About the Author Hugh D. Young; Roger Freedman

Chapter 1 Opener: Ralph Wetmore/Getty Images; 1.1a: Shutterstock; 1.1b: CERN/European Organization for Nuclear Research; 1.4: National Institute of Standards and Technology (NIST); 1.5a: R. Williams (STScI), the HDF-S Team, and NASA; 1.5b: SOHO (ESA & NASA); 1.5c: Courtesy of NASA/JPL/Caltech; 1.5d: Photodisc Green/Getty Images; 1.5e: PhotoDisc/Getty Images; 1.5f: Purdue University. Veeco Instruments, Inc.; 1.5g: SPL/Photo Researchers, Inc.; 1.6: Pearson Education; 1.7: Liaison/Getty Images; App. 1.1: Scott Bauer - USDA Agricultural Research Service

Chapter 2 Opener: Terje Rakke/Getty Images; 2.4: Bob Thomas/iStockphoto; 2.5: Shutterstock; App.2.1: NASA; 2.22: Richard Megna/Fundamental Photographs; 2.26: Corbis; 2.27: iStockphoto

Chapter 3 Opener: Feng Li/Getty Images; App. 3.1: Luca Lozzi/Getty Images; 3.8: PhotoAlto/Getty Images; 3.16: Richard Megna/Fundamental Photographs; 3.19a: Richard Megna/Fundamental Photographs; 3.19b: Fotolia; App. 3.2: David Wall/Alamy; 3.31: Hart Matthews/Reuters

Chapter 4 Opener: Brian Cleary/Getty Images; App. 4.1: Fotolia; 4.12: Wayne Eastep/Getty Images; 4.17: Shutterstock; App.4.2: Stan Kujawa/Alamy; 4.20: Shutterstock; 4.29: Shutterstock; 4.30a: John McDonough/Time Inc. Magazines/Sports Illustrated; 4.30b: John McDonough/Getty Images; 4.30c: Mark M. Lawrence/Corbis

Chapter 5 Opener: U.S. Air Force photo by Tech. Sgt. Jeremy Lock/Released; 5.11: NASA; 5.16: Dave Sandford/Getty Images; App. 5.1: Radim Spitzer/iStockphoto; App. 5.2: David Scharf /Photo Researchers, Inc.; 5.26b: Joggie Botma/iStockphoto; 5.35: AOPA/Aircraft Owners and Pilots Association; 5.38a: NASA; 5.38b: Helen Hansma, University of California, Santa Barbara; 5.38c: Nicholas Roemmelt/iStockphoto; 5.38d: David Malin, Anglo-Australian Observatory

Chapter 6 Opener: Flickr RM/Getty Images; 6.1: Allison Michael Orenstein/Getty Images; App. 6.1: Steve Gschmeissner/Photo Researchers, Inc.; 6.13: Bojan Fatur/iStockphoto; App. 6.2: Steve Gschmeissner/Photo Researchers, Inc.; 6.26: Hulton Archive/Getty Images; 6.27a: Keystone/Getty Images; 6.27b: Adrian Pingstone; 6.28: Shutterstock

Chapter 7 Opener: Geoff de Feu/Getty Images; 7.1: Shutterstock; App. 7.1: Yuji Sakai/Getty Images; 7.3: Robert F. Bukaty/AP; 7.5: Joggie Botma/iStockphoto; 7.12: Jupiterimages/Brand X/Corbis; App. 7.2: iStockphoto; 7.15: William Loy/iStockphoto; 7.21: Bill Grove/iStockphoto; App. 7.3: Matt Naylor/iStockphoto; App. 7.4: Peter Menzel/Photo Researchers, Inc.

Chapter 8 Opener: Flickr RM/Getty Images; 8.2: David Woods/Corbis; App. 8.1: Shutterstock; 8.4: Jim Cummins/Getty Images; 8.6: Andrew Davidhazy; 8.16: Hulton Archive/Getty Images; 8.21: David Leah/Getty Images; 8.29: Richard Megna/Fundamental Photographs; 8.31b: Shutterstock; App. 8.2: Dr. James B. Wood; 8.33: NASA

Chapter 9 Opener: Todd Klassy; 9.3: Mike Powell/Getty Images; App. 9.1: Hybrid Medical Animation/Photo Researchers, Inc.; App. 9.2a: Shutterstock; App. 9.2b: Jakob Leitner/iStockphoto; 9.18: David J. Phillip/AP; 9.21: NASA/Johnson Space Center; P9.98: NASA

Chapter 10 Opener: Ilia Shalamaev/Photolibrary; 10.7: Corbis; App. 10.1 (right): Shutterstock; App. 10.1 (left): David Lentink/SPL/Photo Researchers, Inc.; 10.14: James Warren/iStockphoto; 10.17: Loic Bernard/iStockphoto; 10.22: Images Source/Getty Images; 10.28: Gerard Lacz/Natural History Photographic Agency

Chapter 11 Opener: iStockphoto; 11.3: Jeremy Woodhouse/Getty Images; 11.12a: Shutterstock; 11.12b: Jonathan Blair/Corbis; 11.12c: Photodisc Green/Getty Images; App. 11.2: Dante Fenolio/Photo Researchers, Inc.

Chapter 12 Opener: Digital Vision/AGE Fotostock; App. 12.1: Sean Locke/iStockphoto; 12.6: Cenco Physics; 12.9: Shutterstock; 12.14: Fotolia; 12.19: Pearson Education; 12.20: Cordelia Molloy/Photo Researchers, Inc.; 12.27: Shutterstock; App. 12.2: Sean Locke/iStockphoto; 12.29a: Photodisc Green/Getty Images; 12.29b: Colin Barker/Getty Images; 12.30f: The Harold E. Edgerton 1992 Trust, Palm Press, Inc.

Chapter 13 Opener: NASA/JPL/Space Science Institute; 13.3: NASA/JPL/Caltech; 13.6: NASA; App. 13.1: NASA; 13.7: George Hall/Corbis; 13.13: NASA and Hubble Space Telescope (STScI); 13.16: NASA; 13.17: NASA; App. 13.2: NASA; 13.20b: NASA; 13.26: NASA; 13.27: NASA; 13.28: Prof. Andrea Ghez/UCLA - W. M. Keck Telescopes

Chapter 14 Opener: Tim McCaig/iStockphoto; App. 14.1: Nigel Cattlin/Photo Researchers, Inc.; 14.7: American Diagnostic Corporation; 14.21: Frank Herholdt/Getty Images; 14.25: Christopher Griffin/Alamy; App. 14.2: Shutterstock

Chapter 15 Opener: REUTERS/Nicky Loh; App. 15.1: MARCO POLO COLLECTION/Alamy; 15.2: REUTERS/Charles Platiau; 15.5: David Parker/Photo Researchers, Inc.; 15.12: iStockphoto; App. 15.2: iStockphoto; 15.18: Reproduced from *PSSC Physics*, 2nd ed. (1965), D.C. Heath & Company with Educational Development Center, Inc., Newton Massachusets; 15.23: Richard Megna/Fundamental Photographs; 15.25: iStockphoto; 15.27: National Optical Astronomy Observatories

Chapter 16 Opener: iStockphoto; App. 16.1: Steve Thorne/Getty Images; 16.5a: Lisa Pines/Getty Images; 16.5c: David Young-Wolff/PhotoEdit Inc.; 16.6: Dorling Kindersley; 16.9: Kretztechnik/Photo Researchers, Inc.; 16.10: Eastcott-Momatiuk/The Image Works; 16.15: Dorothy Y Riess/The Image Works; App. 16.2: iStockphoto; 16.20: Martin Bough/Fundamental Photographs; 16.25: Shutterstock; 16.28: Mark Reinstein/The Image Works; 16.35c: NASA/Langley Research Center; 16.36: NASA/Robert A. Hoover, Dryden Flight Research Center

Chapter 17 Opener: Shutterstock; 17.4: Exergen Corporation; App.17.1: Shutterstock; 17.5: Sargent-Welch/VWR International, Courtesy of Cenco Physics; 17.11: NASA/Jim Ross, Dryden Flight Research Center; 17.13: Eric Schrader - Pearson Science; 17.16: Roger Freedman; 17.18: Paul Seheult/Corbis; 17.19: Adam Hart-Davis/Photo Researchers, Inc.; 17.20: Richard Megna/Fundamental Photographs; 17.22: Shutterstock; App. 17.2a: iStockphoto; App. 17.2b: Paul Nicklen/National Geographic Creative/Getty Image; 17.27: Hugh D. Young; 17.28: Dr. Arthur Tucker/Photo Researchers, Inc.

Chapter 18 Opener: iStockphoto; 18.2: John Powell/The Image Works; 18.10: ThermoMicroscopes/Park Scientific Instruments; 18.13: Stone/Getty Images; 18.14: ARCTIC IMAGES/Alamy; 18.16: David Grossman/The Image Works; App. 18.2: Ray Coleman/Photo Researchers, Inc.; 18.25: PhotoDisc/Getty Images; E18.25: David Malin, Royal Observatory, Edinburgh and Anglo-Australian Telescope Board; P18.91: David Malin, Royal Observatory, Edinburgh and Anglo-Australian Telescope Board

Chapter 19 Opener: Richard A. Cooke III/Getty Images; 19.1: John P. Surey; 19.2a: PhotoDisc/StockTrek/Getty Images; 19.2b: iStockphoto; App. 19.1: iStockphoto; 19.10: Shutterstock; 19.14: Tom Branch/Photo Researchers, Inc.; 19.15: Shutterstock; P19.57: Photolibrary

Chapter 20 Opener: G. Brad Lewis/Getty Images; 20.2: Shutterstock; App. 20.1: Shutterstock; 20.12: Bill Bachman/Photo Researchers, Inc.; 20.16: U.S. Air Force photo by Staff Sgt. Robert Zoellner; 20.17: Erich Schrempp/Photo Researchers, Inc.; App. 20.2: iStockphoto; 20.20: Eric Schrader - Pearson Science

INDEX

For users of the three-volume edition: pages 1–686 are in Volume 1; pages 687–1260 are in Volume 2; and pages 1223–1522 are in Volume 3. Pages 1261–1522 are not in the Standard Edition.

Note: Page numbers followed by f indicate figures; those followed by t indicate tables.

A

Abdus, Salam, 1500
Absolute conservation laws, 1495
Absolute pressure, 377–378
Absolute temperature, 517
Absolute temperature scale, 556, 668–669
Absolute zero, 556, 669
Absorption lines, 1203
Absorption spectra
  line, 1293, 1297–1300, 1310–1314
  X-ray, 1396
AC source, 1022. *See also* Alternating-current circuits
Acceleration
  angular, 282–285, 284t, 311–314
  around curve, 74, 88
  average, 42–43. *See also* Average acceleration
  calculating by integration, 55–57
  centripetal, 86–87, 154
  centripetal component of, 286–287
  changing, 55–57
  circular motion and, 85–87
  constant, 46–52. *See also* Constant acceleration
  definition of, 43
  fluid resistance and, 152–154
  inertial frame of reference and, 110–112
  instantaneous, 43–44. *See also* Instantaneous acceleration
  linear, 282, 284t
  mass and, 113, 114, 118–120
  net force and, 112–118
  Newton's first law and, 108–112
  Newton's second law and, 112–117, 140–146
  of particle in wave, 480–482
  projectile motion and, 77–80, 87
  of rocket, 262–264
  of rolling sphere, 319–320
  signs for, 45, 46
  in simple harmonic motion, 444, 448
  tangential component of, 286
  units of, 117
  vs. velocity, 42
  on v_x-t graph, 44–46
  weight and, 118–119
  of yo-yo, 319
Acceleration due to gravity, 700
  apparent weight and, 143, 422
  definition of, 52
  at different latitudes and elevations, 422t
  in free fall, 52–55, 118, 143
  magnitude of, 405
  mass vs. weight and, 118–120
  variation with location and, 119
  vs. gravitation, 403
  weightlessness and, 143
Acceleration vectors, 35, 72–77, 283
  average, 73–75
  instantaneous, 73–75
  parallel and perpendicular components of, 75–77
Accelerators, 1485–1488. *See also* Particle accelerators
Acceptor level, 1425
Acceptors, 1425
Accretion disk, 425
Accuracy, 8
  vs. precision, 9
Acrobats, in unstable equilibrium, 228
Action-reaction pairs, 120–123
  gravitational forces as, 403
Activation energy, 610
Activity, in radioactive decay, 1456
Addition

significant figures in, 9
of vectors, 12–18
Adiabatic process, 634–635
  Carnot cycle and, 663
  for ideal gas, 640–642
Aging, relativity and, 1233
Air
  dielectric strength of, 768, 805
  as insulator, 571
  ionization of, 768–771
Air conditioners, 660–661
Air drag, 152–154
Air pressure, 375–376
Air resistance, projectile motion and, 77, 79–80
Airplanes
  banked curves and, 158
  noise control for, 531, 532
  sonic boom from, 538
  wing lift and, 388–389
  wing resonance in, 460
Airy disk, 1209
Airy, George, 1209
Alkali metals, 1390
  Bohr atomic model for, 1306
Alkaline earth elements, 1390
Alkaline earth metals, 1390
Alpha decay, 1450–1452
Alpha particles, 1294–1295
  emission of, 1450–1451
  tunneling and, 1349–1350
Alternating current, 822, 850, 1021
  applications of, 868
  dangers of, 1040
  lagging, 1036
  measurement of, 1022–1024
  rectified, 1022–1023
  rectified average, 1023
  root-mean-square value of, 1023–1024
Alternating-current circuits
  capacitors in, 1027–1028, 1029–1030, 1029t
  complex numbers in, 1049–1050
  impedance of, 1031–1033
  inductors in, 1025–1027, 1029, 1029t
  L-R-C series, 1030–1034
  phase angle and, 1026, 1031–1032
  phasors and, 1022
  power in, 1034–1037
  resistance and reactance in, 1024–1030
  resistors in, 1025, 1029, 1029t
  resonance in, 1037–1039
  tailoring, 1038–1039
  transformers and, 1040–1042
Alternators, 963–964, 1021–1022
Ammeters, 831, 860–861
  voltmeters and, 862–863, 1024
Amorphous solids, 1412
Ampere, 695, 820, 931–932
Ampère, André, 885
Ampère's law, 935–941. *See also* Maxwell's equations
  applications of, 938–941
  displacement current and, 975–977
  electromagnetic waves and, 1052, 1057, 1063
  general statement of, 937–938
  generalization of, 975–976
Amplitude
  displacement, 510, 518–519
  of electromagnetic waves, 1061
  of oscillation, 438
  of pendulum, 454–455
  pressure, 511–512, 519–521
  of sound waves, 510–513
Analyzers, 1095–1096
Anderson, Carl D., 1482, 1485
Angle(s)

notation for, 71
  polarizing, 1097
  radians and, 279, 287
Angle of deviation, 1111
Angle of incidence, critical, 1089
Angle of reflection, 1084
Angular acceleration, 282–285
  angular velocity and, 282
  calculation of, 283
  constant, 283–285, 284t
  torque and, 311–314
  as vector, 283
  vs. linear acceleration, 284t
Angular displacement, 279
  torque and, 320–322
Angular frequency, 438–439
  of electromagnetic waves, 1061
  natural, 459–460
  of particle waves, 1331
  period and, 438–439
  in simple harmonic motion, 441–442
  vs. frequency, 442
Angular magnification, vs. lateral magnification, 1147
  of microscope, 1147, 1148–1149
Angular momentum
  axis of symmetry and, 323–324
  of the body, 324
  conservation of, 325–328
  definition of, 322
  of electrons, 942
  of gyroscope, 328–330
  nuclear, 1442
  orbital, 1373–1374, 1384, 1387
  precession and, 328–330
  rate of change of, 323, 324
  rotation and, 322–328
  spin, 1384–1385, 1387, 1442
  torque and, 323, 324
  total, 1387, 1442
  as vector, 324, 328
Angular simple harmonic motion, 451
Angular size, 1146
Angular speed, 280
  instantaneous, 286
  precession, 329
  rate of change of, 286
Angular velocity, 279–282
  angular acceleration and, 282
  average, 279
  calculation of, 281
  instantaneous, 280
  rate of change of, 286
  as vector, 281–282
  vs. linear velocity, 280
Angular vs. linear kinematics, 285–288
Anomalous magnetic moment, 1443
Antimatter, 1516
Antineutrinos, 1492
Antineutrons, 1492
Antinodal curves, 1166
Antinodal planes, 1070
Antinodes, 492
  displacement, 523
  pressure, 523
Antiparallel vectors, 11, 12
Antiparticles, 1483
Antiprotons, 1491–1492
Antiquarks, 1496, 1499
Aphelion, 415
Apparent weight, 142
  acceleration due to gravity and, 143, 422
  Earth's rotation and, 421–423
  magnitude of, 422
Appliances, power distribution systems in, 868–872